8 BJT Small-Signal Analysis Common-emitter: $A_v = -R_C/r_e$, $Z_i = R_B \| \beta r_e$, $Z_o = R_C$, $A_i \simeq \beta$; voltage-divider: $R' = R_1 \| R_2$, $A_v = -R_C/r_e$, $Z_i = R' \| \beta r_e$, $Z_o = R_C$; emitter-bias: $Z_b = \beta(r_e + R_E) \simeq \beta R_E$, $A_v = -\beta R_C/Z_b = -R_C/(r_e + R_E) \simeq -R_C/R_E$; emitter-follower: $Z_b \simeq \beta(r_e + R_E)$, $A_v \simeq 1$, $Z_o \simeq r_e$; common-base: $A_v \simeq R_C/r_e$, $Z_i = R_E \| r_e$, $Z_o = R_C$; collector feedback: $A_v = -R_C/r_e$, $Z_i = \beta r_e \| R_F/|A_v|$, $Z_o \simeq R_C \| R_F$; collector dc feedback: $A_v = -(R_{F_2} \| R_C)/r_e$, $Z_i = R_{F_1} \| \beta r_e$, $Z_o = R_C \| R_{F_2}$; hybrid parameters: $A_i = h_f/(1 + h_o R_L)$, $A_v = -h_f R_L/[h_i + (h_i h_o - h_f h_r)R_L]$, $Z_i = h_i - h_f h_r R_L/(1 + h_o R_L)$, $Z_o = 1/[h_o - (h_f h_r/(h_i + R_s))]$

9 FET Small-Signal Analysis $g_m = g_{mo}(1 - V_{GS}/V_P)$, $g_{mo} = 2I_{DSS}/|V_P|$; basic configuration: $A_v = -g_m R_D$; unbypassed source resistance: $A_v = -g_m R_D/(1 + g_m R_S)$; source follower: $A_v = g_m R_S/(1 + g_m R_S)$; common gate: $A_v = g_m(R_D \| r_d)$

10 Systems Approach—Effects of R_S and R_L BJT: $A_v = R_L A_{v_{NL}}/(R_L + R_o)$, $A_i = -A_v Z_i/R_L$, $V_i = R_i V_s/(R_i + R_s)$; fixed-bias: $A_v = -(R_C \| R_L)/r_e$, $A_{v_s} = Z_i A_v/(Z_i + R_s)$, $Z_i = \beta r_e$, $Z_o = R_C$; voltage-divider: $A_v = -(R_C \| R_L)/r_e$, $A_{v_s} = Z_i A_v/(Z_i + R_s)$, $Z_i \simeq R_1 \| R_2 \| \beta r_e$, $Z_o = R_C$; emitter-bias: $A_v = -(R_C \| R_L)/R_E$, $A_{v_s} = Z_i A_v/(Z_i + R_s)$, $Z_i \simeq R_B \| \beta R_E$, $Z_o = R_C$; collector-feedback: $A_v = -(R_C \| R_L)/r_e$, $A_{v_s} = Z_i A_v/(Z_i + R_s)$, $Z_i = \beta r_e \| R_F/|A_v|$, $Z_o \simeq R_C \| R_F$; emitter-follower: $R_E' = R_E \| R_L$, $A_v = R_E'/(R_E' + r_e)$, $A_{v_s} = R_E'/(R_E' + R_s/\beta + r_e)$, $Z_i = R_B \| \beta(r_e + R_E')$, $Z_o = R_E \|(R_s/\beta + r_e)$; common-base: $A_v \simeq (R_C \| R_L)/r_e$, $A_i \simeq 1$, $Z_i \simeq r_e$, $Z_o = R_C$; FET: bypassed R_S: $A_v = -g_m(R_D \| R_L)$, $Z_i = R_G$, $Z_o = R_D$; unbypassed R_S: $A_v = -g_m(R_D \| R_L)/(1 + g_m R_S)$, $Z_i = R_G$, $Z_o = R_D$; source-follower: $A_v = g_m(R_S \| R_L)/[1 + g_m(R_S \| R_L)]$, $Z_i = R_G$, $Z_o = R_S \| r_d \| 1/g_m$; common gate: $A_v = g_m(R_D \| R_L)$, $Z_i = R_S \| 1/g_m$, $Z_o = R_O$; cascaded: $A_{v_T} = A_{v_1} \cdot A_{v_2} \cdot A_{v_3} \cdots A_{v_n}$, $A_{i_T} = \pm A_{v_T} Z_{i_1}/R_L$

11 BJT and JFET Frequency Response $\log_e a = 2.3 \log_{10} a$, $\log_{10} 1 = 0$, $\log_{10} a/b = \log_{10} a - \log_{10} b$, $\log_{10} 1/b = -\log_{10} b$, $\log_{10} ab = \log_{10} a + \log_{10} b$, $G_{dB} = 10 \log_{10} P_2/P_1$, $G_{dBm} = 10 \log_{10} P_2/1\,\text{mW}|_{600\Omega}$, $G_{dB} = 20 \log_{10} V_2/V_1$, $G_v = G_{v_1} + G_{v_2} + G_{v_3} + \cdots + G_{v_n}$, $P_{o_{HPF}} = 0.5 P_{o_{mid}}$, BW $= f_1 - f_2$; low frequency (BJT): $f_{L_s} = 1/2\pi(R_s + R_i)C_s$, $f_{L_C} = 1/2\pi(R_o + R_L)C_C$, $f_{L_E} = 1/2\pi R_e C_E$, $R_e = R_E \|(R_s'/\beta + r_e)$, $R_s' = R_s \| R_1 \| R_2$, FET: $f_{L_G} = 1/2\pi(R_{sig} + R_i)C_G$, $f_{L_C} = 1/2\pi R_o C_C$, $f_{L_S} = 1/2\pi R_{eq} C_S$, $R_{eq} = R_S \| 1/g_m(r_d \simeq \infty\,\Omega)$; Miller effect: $C_{M_i} = (1 - A_v)C_f$, $C_{M_o} = (1 - 1/A_v)C_f$; high frequency (BJT): $f_{H_i} = 1/2\pi R_{Th_1} C_i$, $R_{Th_1} = R_s \| R_1 \| R_2 \| R_i$, $C_i = C_{W_i} + C_{be} + C_{M_i}$, $f_{H_o} = 1/2\pi R_{Th_2} C_o$, $R_{Th_2} = R_C \| R_L \| r_o$, $C_o = C_{W_o} + C_{ce} + C_{M_o}$, $f_\beta \simeq 1/2\pi \beta_{mid} r_e(C_{be} + C_{bc})$, $f_T = \beta_{mid} f_\beta$; FET: $f_{H_i} = 1/2\pi R_{Th_1} C_i$, $R_{Th_1} = R_{sig} \| R_G$, $C_i = C_{W_i} + C_{gs} + C_{M_i}$, $f_{H_o} = 1/2\pi R_{Th_2} C_o$, $R_{Th_2} = R_D \| R_L \| r_d$, $C_o = C_{W_o} + C_{ds} + C_{M_o}$; multistage: $f_1' = f_1/\sqrt{2^{1/n} - 1}$, $f_2' = (\sqrt{2^{1/n} - 1})f_2$; square-wave testing: $f_{H_i} = 0.35/t_r$, % tilt $= [(V - V')/V] \times 100\%$, $f_{L_o} = (P/\pi)f_s$, $P = (V - V')/V$

12 Compound Configurations Differential voltage gain: $A_v = \beta R_C/2r_i$; common-mode voltage gain: $\beta R_C/[r_i + 2(\beta + 1)R_E]$

13 Operational Amplifiers CMRR $= A_d/A_c$; CMRR(log) $= 20 \log_{10}(A_d/A_c)$; constant-gain multiplier: $V_o/V_1 = -R_f/R_1$; noninverting amplifier: $V_o/V_1 = 1 + R_f/R_1$; unity follower: $V_o = V_1$; summing amplifier: $V_o = -[(R_f/R_1)V_1 + (R_f/R_2)V_2 + (R_f/R_3)V_3]$; integrator: $v_o(t) = -(1/R_1 C_1)\int v_1 dt$

14 Op-Amp Applications Constant-gain multiplier: $A = -R_f/R_1$; noninverting: $A = 1 + R_f/R_1$; voltage summing: $V_o = -[(R_f/R_1)V_1 + (R_f/R_2)V_2 + (R_f/R_3)V_3]$; high-pass active filter: $f_{oL} = 1/2\pi R_1 C_1$; low-pass active filter: $f_{oH} = 1/2\pi R_1 C_1$

15 Power Amplifiers

Power in: $P_i = V_{CC}I_{CQ}$

power out: $P_o = V_{CE}I_C = I_C^2 R_C = V_{CE}^2/R_C$ rms

$\qquad = V_{CE}I_C/2 = (I_C^2/2)R_C = V_{CE}^2/(2R_C)$ peak

$\qquad = V_{CE}I_C/8 = (I_C^2/8)R_C = V_{CE}^2/(8R_C)$ peak-to-peak

efficiency: $\%\eta = (P_o/P_i) \times 100\%$

maximum efficiency: Class A, series-fed $= 25\%$

$\qquad\qquad\qquad$ Class A, transformer-coupled $= 50\%$

$\qquad\qquad\qquad$ Class B, push-pull $= 78.5\%$

transformer relations: $V_2/V_1 = N_2/N_1 = I_1/I_2$, $R_2 = (N_2/N_1)^2 R_1$; power output: $P_o = [(V_{CE_{max}} - V_{CE_{min}})(I_{C_{max}} - I_{C_{min}})]/8$; class B power amplifier: $P_i = V_{CC}[(2/\pi)I_{peak}]$; $P_o = V_L^2(peak)/(2R_L)$; $\%\,\eta = (\pi/4)[V_L(peak)/V_{CC}] \times 100\%$; $P_Q = P_{2Q}/2 = (P_i - P_o)/2$; maximum $P_o = V_{CC}^2/2R_L$; maximum $P_i = 2V_{CC}^2/\pi R_L$; maximum $P_{2_Q} = 2V_{CC}^2/\pi^2 R_L$; % total harmonic distortion (% THD) $= \sqrt{D_2^2 + D_3^2 + D_4^2 + \cdots} \times 100\%$; heat-sink: $T_j = P_D\theta_{JA} + T_A$, $\theta_{JA} = 40°C/W$ (free air); $P_D = (T_J - T_A)/(\Theta_{JC} + \Theta_{CS} + \Theta_{SA})$

16 Linear-Digital ICs

Ladder network: $V_o = [(D_0 \times 2^0 + D_1 \times 2^1 + D_2 \times 2^2 + \cdots + D_n \times 2^n)/2^n]V_{ref}$; 555 oscillator: $f = 1.44(R_A + 2R_B)C$; 555 monostable: $T_{high} = 1.1R_AC$; VCO: $f_o = (2/R_1C_1)[(V^+ - V_C)/V^+]$; phase-locked loop (PLL): $f_o = 0.3/R_1C_1$, $f_L = \pm 8f_o/V$, $f_C = \pm(1/2\pi)\sqrt{2\pi f_L/(3.6 \times 10^3)C_2}$

17 Feedback and Oscillator Circuits

$A_f = A/(1 + \beta A)$; series feedback; $Z_{if} = Z_i(1 + \beta A)$; shunt feedback: $Z_{if} = Z_i/(1 + \beta A)$; voltage feedback: $Z_{of} = Z_o/(1 + \beta A)$; current feedback: $Z_{of} = Z_o(1 + \beta A)$; gain stability: $dA_f/A_f = 1/(|1 + \beta A|)(dA/A)$; oscillator; $\beta A = 1$; phase shift: $f = 1/2\pi RC\sqrt{6}$, $\beta = 1/29$, $A > 29$; FET phase shift: $|A| = g_m R_L$, $R_L = R_D r_d/(R_D + r_d)$; transistor phase shift: $f = (1/2\pi RC)[1/\sqrt{6 + 4(R_C/R)}]$, $h_{fe} > 23 + 29(R_C/R) + 4(R/R_C)$; Wien bridge: $R_3/R_4 = R_1/R_2 + C_2/C_1$, $f_o = 1/2\pi\sqrt{R_1C_1R_2C_2}$; tuned: $f_o = 1/2\pi\sqrt{LC_{eq}}$, $C_{eq} = C_1C_2/(C_1 + C_2)$, Hartley: $L_{eq} = L_1 + L_2 + 2M$, $f_o = 1/2\pi\sqrt{L_{eq}C}$

18 Power Supplies (Voltage Regulators)

Filters: $r = V_r(rms)/V_{dc} \times 100\%$, V.R. $= (V_{NL} - V_{FL})/V_{FL} \times 100\%$, $V_{dc} = V_m - V_r\,(p\text{-}p)/2$, $V_r(rms) = V_r(p\text{-}p)/2\sqrt{3}$, $V_r(rms) \simeq (I_{dc}/4\sqrt{3})(V_{dc}/V_m)$; full-wave, light load $V_r(rms) = 2.4I_{dc}/C$, $V_{dc} = V_m - 4.17I_{dc}/C$, $r = (2.4I_{dc}CV_{dc}) \times 100\% = 2.4/R_LC \times 100\%$, $I_{peak} = T/T_1 \times I_{dc}$; RC filter: $V'_{dc} = R_L V_{dc}/(R + R_L)$, $X_C = 2.653/C$ (half-wave), $X_C = 1.326/C$ (Full-wave), $V'_r(rms) = (X_C/\sqrt{R^2 + X_C^2})$; regulators: $IR = (I_{NL} - I_{FL})/I_{FL} \times 100\%$, $V_L = V_Z(1 + R_1/R_2)$, $V_o = V_{ref}(1 + R_2/R_1) + I_{adj}R_2$

19 Other Two-Terminal Devices

Varactor diode: $C_T = C(0)/(1 + |V_r/V_T|)^n$, $T_{C_c} = (\Delta C/C_o(T_1 - T_0)) \times 100\%$; photodiode: $W = hf$, $\lambda = v/f$, $1\,lm = 1.496 \times 10^{-10}$ W

20 pnpn and Other Devices

UJT: $R_{BB} = (R_{B_1} + R_{B_2})|_{I_E=0}$, $V_{R_{B_1}} = \eta V_{BB}|_{I_E=0}$, $\eta = R_{B_1}/(R_{B_1} + R_{B_2})|_{I_E=0}$, $V_P = \eta V_{BB} + V_D$; phototransistor: $I_C \simeq h_{fe}I_\lambda$; PUT: $\eta = R_{B_1}/(R_{B_1} + R_{B_2})$, $V_P = \eta V_{BB} + V_D$

EIGHTH EDITION

ELECTRONIC DEVICES AND CIRCUIT THEORY

ROBERT L. BOYLESTAD
LOUIS NASHELSKY

Upper Saddle River, New Jersey
Columbus, Ohio

Library of Congress Cataloging-in-Publication Data

Boylestad, Robert L.
    Electronic devices and circuit theory / Robert Boylestad, Louis
Nashelsky.—8th ed.
        p.   cm.
        ISBN 0-13-028483-1 (alk. paper)
        1. Electronic circuits.   2. Electronic apparatus and appliances.
I. Nashelsky, Louis. II. Title.
TK7867 .B66   2002
621.3815—dc21   2001021973

Editor in Chief: Stephen Helba
Product Manager: Scott J. Sambucci
Development Editor: Kate Linsner
Production Manager: Pat Tonneman
Production Editor: Rex Davidson
Design Coordinator: Karrie Converse-Jones
Cover Art: Painting by Sigmund Årseth, Artist and Teacher, Valdres, Norway

This book was set in Times Roman by **TECH**BOOKS and was printed and bound by Courier
Kendallville, Inc. The cover was printed by Phoenix Color Corp.

Pearson Education Ltd., *London*
Pearson Education Australia Pty. Limited, *Sydney*
Pearson Education Singapore Pte. Ltd.
Pearson Education North Asia Ltd., *Hong Kong*
Pearson Education Canada, Ltd., *Toronto*
Pearson Educación de Mexico, S.A. de C.V.
Pearson Education—Japan, *Tokyo*
Pearson Education Malaysia Pte. Ltd.
Pearson Education, *Upper Saddle River, New Jersey*

10 9 8 7 6 5 4 3
ISBN: 0–13–028483–1

Dedicated to
ELSE MARIE; ALISON, MARK, KELCY, and MORGAN; ERIC, RACHEL,
and SAMANTHA; STACEY, JONATHAN, and BRITT; JOHANNA
and to
KATRIN; KIRA, TOMMY, JUSTIN, and TYLER; LARREN, PATTY,
BRENDAN, and OWEN

Preface

In this edition we have written additional practical examples and summaries at the end of each chapter, and have expanded coverage of computer software. The chapter on IC construction was deleted and replaced with a well-written description of the process that first appeared in *Smithsonian Magazine*. It has some stunning photographs and content that is excellent for the new students of this rapidly changing field.

Over the years we have learned that improved readability can be obtained through the general appearance of the text, so we are committed to the format you find in this and recent editions of the text. We hope you agree that it makes the text material appear "friendlier" to the broad range of students using the text. As in the past, we continue to be committed to the strong pedagogical sense of the text, accuracy, completeness, and a broad range of ancillary materials that support the educational process.

PEDAGOGY

Reviewers and current users appear to be quite satisfied with the manner in which the content lends itself to a typical course syllabus. The improved pedagogy of the last two editions seems to support the instructor's lecture and helps students build the foundation necessary for future studies. The number of examples continues to grow, and isolated boldface statements continue to identify important concepts and conclusions. Color continues to be employed in a manner that helps define important regions of characteristics, or identifies important regions or parameters of a network. Icons at the top of the page, developed for each chapter of the text, facilitate referencing a particular area of text as quickly as possible. Problems, which have been developed for each section of the text, progress from the simple to the more complex. The title of each section is repeated in the problem section to identify the problems associated with a particular subject matter.

SYSTEMS APPROACH

There is no question that the growing development of packaged systems requires that the student become aware at the earliest opportunity of a "systems approach" to the design and analysis of electronic systems. Isolated no-load networks are first discussed in Chapters 8 and 9 to introduce the important parameters of any package and develop the important equations for the configuration. The impact of a source or load impedance on the individual package is then defined in Chapter 10 on a general basis before examining specific networks. Finally, the impact of tying the individual packages together is examined in the same chapter to establish some understanding of the systems approach. The later chapters on op-amps and IC units further develop the concepts introduced in these early chapters.

ACCURACY

The goal of any educational publication is to be absolutely free of errors. There is nothing more distressing to a student than to find that he or she has suffered for hours over a simple printing error. In fact, after all the hours that go into preparing a manuscript and checking every word, number, or letter there is nothing more distressing to an author than to find that errors have crept into the publication. Based on past history and the effort put into this publication, we believe you will find the highest level of accuracy obtainable for a publication of this kind.

SUMMARIES

In response to current users, summaries are added at the end of each chapter, reviewing the salient concepts and conclusions. To emphasize specific words and phrases, boldface lettering is used in much the same manner as a student would use a highlighting marker. The list of equations appearing with each summary was limited to those an instructor realistically hopes the student will bring away from the course.

PRACTICAL EXAMPLES

While the text now has over 80 practical examples, over 40 were added to this edition and they appear in their own sections. They provide an understanding of the design process that is normally not available at this level. Practical considerations associated with using the electronic devices introduced in this text are discussed as experienced by professionals in the field. The level of coverage is well beyond the surface description of the operation of a particular product. Networks are reduced for clarity and equations are developed to explain why specific response levels are obtained. An effort was made to give some idea of the range of application for each device introduced. Too often the student believes that each electronic device serves a particular purpose, and that's it. In general, the authors are pleased with the results of this demanding effort and invite your comments and suggestions so that the content can be improved upon in the future.

TRANSISTOR MODELING

BJT transistor modeling is an area that can be approached in a variety of ways. Some institutions employ the r_e model exclusively, while others lean toward the hybrid approach or to a combination of the two. This edition will emphasize the r_e model with sufficient coverage of the hybrid model to permit a comparison between the results obtained with each approach. An entire chapter (Chapter 7) has been devoted to the introduction of the models to ensure a clear, correct understanding of each and the relationships that exist between the two.

EQUATION DEVELOPMENT

For years the development of the equations for small-signal BJT and JFET networks avoided the impact of the output parameter r_o. In addition, results were often provided with no idea how they were obtained. Further, approximate equations were provided with no idea what conditions had to be satisfied to permit use of the equations. For these reasons, and probably others, the details of each derivation are provided in this text. The effect of r_o was separated for each development to first permit a less complex development. The effect of r_o was then demonstrated and the conditions under which the effect of r_o can be ignored were introduced. In most cases, the derivations are unique to any publication of this type. They were the result of extensive hours searching for the best path for the analysis. However, the result is a complete development of each equation that we hope will remove any doubt as to their validity.

COMPUTER SOFTWARE

In recent editions, both PSpice and Electronics Workbench examples were included. For this edition Mathcad was added to demonstrate the versatility of the package for an area such as electronics. Not only can it be used to quickly solve simultaneous equations, but also long series of calculations can be placed in storage for retrieval when a particular configuration is encountered. Numerous examples appear throughout the text, and we believe the student and instructor will find them quite interesting. The detailed coverage of PSpice was expanded slightly, but there is a larger expansion of the coverage of Electronics Workbench due to its growing popularity. For all the software packages there is no requirement that the student become versed in their use to proceed through the text. Although sufficient detail is provided for each application to permit a student to apply each to a variety of configurations, there is no requirement that the packages actually be used.

TROUBLESHOOTING

Troubleshooting is undoubtedly one of the most difficult subjects to discuss and develop in an introductory text. A student is just becoming familiar with the characteristics and operation of a device and now is asked to find an answer to an unexpected result. It is an art that has to develop with experience and exposure. The content of this text is essentially a review of situations that frequently occur in the laboratory environment. Some general hints as to how to isolate a problem are introduced along with a list of typical causes.

ANCILLARIES

The range of ancillary material is quite extensive, including a laboratory manual to which new experiments have been added. There is also an instructor's resource manual, which contains solutions to the in-text problems and the laboratory experiments as well as a test item file. PowerPoint® transparencies and a Prentice Hall Test Manager are also available.

    The CD-ROM included with every copy of the book contains Electronics Workbench Version 5 and Multisim circuit files and CircuitMaker Student Version Software and circuit files. Circuits appearing on the CD-ROM are designated in the text by a special icon next to the selected illustration.

Additional support for the student can be found at www.prenhall.com/boylestad in the form of an online student study guide. CourseCompass and Blackboard complete the supplements package.

USE OF THE TEXT

In general the text is divided into two main components: the dc analysis and the ac or frequency response. For some schools the dc section is sufficient for a one-semester introductory sequence, while for others the entire text may be covered in one semester by picking and choosing specific topics. In any event, the text is one that "builds" from the earlier chapters. Superfluous material is relegated to the later chapters to avoid excessive content on a particular subject early in the development stage. For each device the text covers a majority of the important configurations and applications—the text is very complete! By choosing specific examples and applications the instructor can reduce the content of a course without losing the progressive building characteristics of the text. Then again, if an instructor feels that a specific area is particularly important, the detail is provided for a more extensive review.

Robert L. Boylestad

Louis Nashelsky

Acknowledgments

Our sincerest appreciation must be extended to the instructors who have used the text and sent in comments, corrections, and suggestions. We also want to thank Rex Davidson, Production Editor at Prentice Hall, for keeping together the many detailed aspects of production, and Maggie Diehl for the copyediting. Our sincerest thanks to Scott Sambucci, Product Manager, and Kate Linsner, Development Editor, at Prentice Hall for their editorial support of the eighth edition of this text.

For the new Appendix A, "Making the Chips that Run the World," we thank Jake Page (author) and Kay Chernush (photographer) for their article from *Smithsonian Magazine*.

For the cover art, we thank Sigmund Årseth.

We wish to thank those individuals who have shared their suggestions and evaluations of this text throughout its many editions. The comments from these individuals have enabled us to present *Electronic Devices and Circuit Theory* in this eighth edition:

Ernest Lee Abbott	Napa College
Phillip D. Anderson	Muskegon Community College
Al Anthony	EG&G VACTEC Inc.
A. Duane Bailey	Southern Alberta Institute of Technology
Joe Baker	University of Southern California
Jerrold Barrosse	Pennsylvania State University
Ambrose Barry	University of North Carolina
Arthur Birch	Hartford State Technical College
Scott Bisland	SEMATECH
Edward Bloch	The Perkin-Elmer Corporation
Gary C. Bocksch	Charles S. Mott Community College
Jeffrey Bowe	Bunker Hill Community College
Alfred D. Buerosse	Waukesha County Technical College
Lila Caggiano	MicroSim Corporation
Mauro J. Caputi	Hofstra University
Robert Casiano	International Rectifier Corporation
Nathan Chao	Queensborough Community College, CUNY
Alan H. Czarapata	Montgomery College
Mohammad Dabbas	ITT Technical Institute
John Darlington	Humber College
Lucius B. Day	Metropolitan State College
Mike Durren	Lake Michigan College

Dr. Stephen Evanson	Bradford University
George Fredericks	Northeast State Technical Community College
F. D. Fuller	Humber College
Phil Golden	DeVry Institute of Technology
Joseph Grabinski	Hartford State Technical College
Thomas K. Grady	Western Washington University
Mohamad S. Haj-Mohamadi	North Carolina A & T State University
William Hill	ITT Technical Institute
Albert L. Ickstadt	San Diego Mesa College
Jeng-Nan Juang	Mercer University
Karen Karger	Tektronix Inc.
Kenneth E. Kent	DeKalb Technical Institute
Donald E. King	ITT Technical Institute
Charles Lewis	APPLIED MATERIALS, INC.
Donna Liverman	Texas Instruments Inc.
William Mack	Harrisburg Area Community College
Robert Martin	Northern Virginia Community College
George T. Mason	Indiana Vocational Technical College
William Maxwell	Nashville State Technical Institute
Abraham Michelen	Hudson Valley Community College
John MacDougall	University of Western Ontario
Donald E. McMillan	Southwest State University
Thomas E. Newman	L. H. Bates Vocational-Technical Institute
Byron Paul	Bismarck State College
Dr. Robert Payne	University of Glamorgan
Dr. Robert A. Powell	Oakland Community College
E. F. Rockafellow	Southern-Alberta Institute of Technology
Saeed A. Shaikh	Miami-Dade Community College
Dr. Noel Shammas	School of Engineering
Ken Simpson	Stark State College of Technology
Jerry Sitbon	Queensborough Community College
Eric Sung	Computronics Technology Inc.
Donald P. Szymanski	Owens Technical College
Parker M. Tabor	Greenville Technical College
Peter Tampas	Michigan Technological University
Chuck Tinney	University of Utah
Katherine L. Usik	Mohawk College of Applied Art & Technology
Domingo Uy	Hampton University
Richard J. Walters	DeVry Institute of Technology
Larry J. Wheeler	PSE&G Nuclear
Julian Wilson	Southern College of Technology
Syd R. Wilson	Motorola Inc.
Jean Younes	ITT Technical Institute
Charles E. Yunghans	Western Washington University
Ulrich E. Zeisler	Salt Lake Community College

We thank the following individuals for assisting in the review process for this eighth edition:

Joseph Booker	DeVry Institute of Technology
Charles F. Bunting	Old Dominion University
Mauro J. Caputi	Hofstra University
Kevin Ford	Alvin Community College
David Krispinsky	Rochester Institute of Technology
William Mack	Harrisburg Area Community College
John Sherrick	Rochester Institute of Technology

Contents

3   BIPOLAR JUNCTION TRANSISTORS   131

4   DC BIASING—BJTs   163

5 FIELD-EFFECT TRANSISTORS 245

6 FET BIASING 289

7 BJT TRANSISTOR MODELING 355

11 BJT AND JFET FREQUENCY RESPONSE   569

12 COMPOUND CONFIGURATIONS   627

13 OPERATIONAL AMPLIFIERS   675

14 OP-AMP APPLICATIONS

715

15 POWER AMPLIFIERS

747

16 LINEAR-DIGITAL ICs

791

17 FEEDBACK AND OSCILLATOR CIRCUITS

821

18 POWER SUPPLIES (VOLTAGE REGULATORS)   859

19 OTHER TWO-TERMINAL DEVICES   889

20 *pnpn* AND OTHER DEVICES   923

Semiconductor Diodes

1.1 INTRODUCTION

It is now over 50 years since the first transistor was introduced on December 23, 1947. For those of us who experienced the change from glass envelope tubes to the solid-state era, it still seems like a few short years ago. The first edition of this text contained heavy coverage of tubes, with succeeding editions involving the important decision of how much coverage should be dedicated to tubes and how much to semiconductor devices. It no longer seems valid to mention tubes at all or to compare the advantages of one over the other—we are firmly in the solid-state era.

The miniaturization that has resulted leaves us to wonder about its limits. Complete systems now appear on wafers thousands of times smaller than the single element of earlier networks. Integrated circuits (ICs) now have over 10 million transistors in an area no larger than a thumbnail.* New designs and systems surface weekly. The engineer becomes more and more limited in his or her knowledge of the broad range of advances—it is difficult enough simply to stay abreast of the changes in one area of research or development. We have also reached a point at which the primary purpose of the container is simply to provide some means of handling the device or system and to provide a mechanism for attachment to the remainder of the network. Miniaturization appears to be limited by three factors (each of which will be addressed in this text): the quality of the semiconductor material itself, the network design technique, and the limits of the manufacturing and processing equipment.

1.2 IDEAL DIODE

The first electronic device to be introduced is called the *diode*. It is the simplest of semiconductor devices but plays a very vital role in electronic systems, having characteristics that closely match those of a simple switch. It will appear in a range of applications, extending from the simple to the very complex. In addition to the details of its construction and characteristics, the very important data and graphs to be found on specification sheets will also be covered to ensure an understanding of the terminology employed and to demonstrate the wealth of information typically available from manufacturers.

The term *ideal* will be used frequently in this text as new devices are introduced. It refers to any device or system that has ideal characteristics—perfect in every way. It provides a basis for comparison, and it reveals where improvements can still be made. The *ideal diode* is a *two-terminal* device having the symbol and characteristics shown in Figs. 1.1a and b, respectively.

---

*When time permits, read Appendix A, "Making the Chips That Run the World."

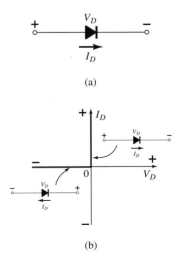

Figure 1.1 Ideal diode: (a) symbol; (b) characteristics.

Ideally, a diode will conduct current in the direction defined by the arrow in the symbol and act like an open circuit to any attempt to establish current in the opposite direction. In essence:

The characteristics of an ideal diode are those of a switch that can conduct current in only one direction.

In the description of the elements to follow, it is critical that the various *letter symbols, voltage polarities*, and *current directions* be defined. If the polarity of the applied voltage is consistent with that shown in Fig. 1.1a, the portion of the characteristics to be considered in Fig. 1.1b is to the right of the vertical axis. If a reverse voltage is applied, the characteristics to the left are pertinent. If the current through the diode has the direction indicated in Fig. 1.1a, the portion of the characteristics to be considered is above the horizontal axis, while a reversal in direction would require the use of the characteristics below the axis. For the majority of the device characteristics that appear in this book, the *ordinate* (or "y" axis) will be the *current* axis, while the *abscissa* (or "x" axis) will be the *voltage* axis.

One of the important parameters for the diode is the resistance at the point or region of operation. If we consider the conduction region defined by the direction of I_D and polarity of V_D in Fig. 1.1a (upper-right quadrant of Fig. 1.1b), we will find that the value of the forward resistance, R_F, as defined by Ohm's law is

$$R_F = \frac{V_F}{I_F} = \frac{0 \text{ V}}{2, 3, \text{ mA}, \ldots, \text{ or any positive value}} = \mathbf{0\ \Omega} \quad \text{(short circuit)}$$

where V_F is the forward voltage across the diode and I_F is the forward current through the diode.

The ideal diode, therefore, is a short circuit for the region of conduction.

Consider the region of negatively applied potential (third quadrant) of Fig. 1.1b,

$$R_R = \frac{V_R}{I_R} = \frac{-5, -20, \text{ or any reverse-bias potential}}{0 \text{ mA}} = \mathbf{\infty\ \Omega} \quad \text{(open-circuit)}$$

where V_R is reverse voltage across the diode and I_R is reverse current in the diode.

The ideal diode, therefore, is an open circuit in the region of nonconduction.

In review, the conditions depicted in Fig. 1.2 are applicable.

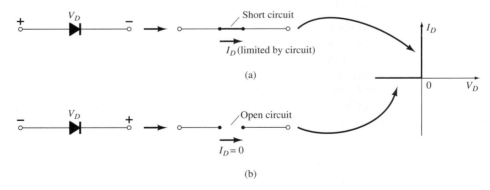

Figure 1.2 (a) Conduction and (b) nonconduction states of the ideal diode as determined by the applied bias.

In general, it is relatively simple to determine whether a diode is in the region of conduction or nonconduction simply by noting the direction of the current I_D established by an applied voltage. For conventional flow (opposite to that of electron flow), if the resultant diode current has the same direction as the arrowhead of the diode symbol, the diode is operating in the conducting region as depicted in Fig. 1.3a. If

Chapter 1 Semiconductor Diodes

the resulting current has the opposite direction, as shown in Fig. 1.3b, the open-circuit equivalent is appropriate.

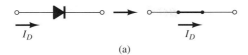

(a)

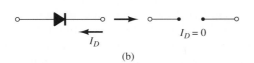

(b)

Figure 1.3 (a) Conduction and (b) nonconduction states of the ideal diode as determined by the direction of conventional current established by the network.

As indicated earlier, the primary purpose of this section is to introduce the characteristics of an ideal device for comparison with the characteristics of the commercial variety. As we progress through the next few sections, keep the following questions in mind:

How close will the forward or "on" resistance of a practical diode compare with the desired 0-Ω level?

Is the reverse-bias resistance sufficiently large to permit an open-circuit approximation?

1.3 SEMICONDUCTOR MATERIALS

The label *semiconductor* itself provides a hint as to its characteristics. The prefix *semi-* is normally applied to a range of levels midway between two limits.

The term conductor is applied to any material that will support a generous flow of charge when a voltage source of limited magnitude is applied across its terminals.

An insulator is a material that offers a very low level of conductivity under pressure from an applied voltage source.

A semiconductor, therefore, is a material that has a conductivity level somewhere between the extremes of an insulator and a conductor.

Inversely related to the conductivity of a material is its resistance to the flow of charge, or current. That is, the higher the conductivity level, the lower the resistance level. In tables, the term *resistivity* (ρ, Greek letter rho) is often used when comparing the resistance levels of materials. In metric units, the resistivity of a material is measured in Ω-cm or Ω-m. The units of Ω-cm are derived from the substitution of the units for each quantity of Fig. 1.4 into the following equation (derived from the basic resistance equation $R = \rho l/A$):

$$\rho = \frac{RA}{l} = \frac{(\Omega)(\text{cm}^2)}{\text{cm}} \Rightarrow \Omega\text{-cm} \qquad (1.1)$$

In fact, if the area of Fig. 1.4 is 1 cm^2 and the length 1 cm, the magnitude of the resistance of the cube of Fig. 1.4 is equal to the magnitude of the resistivity of the material as demonstrated below:

$$|R| = \rho \frac{l}{A} = \rho \frac{(1\text{ cm})}{(1\text{ cm}^2)} = |\rho|\text{ohms}$$

This fact will be helpful to remember as we compare resistivity levels in the discussions to follow.

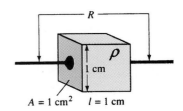

Figure 1.4 Defining the metric units of resistivity.

TABLE 1.1 Typical Resistivity Values		
Conductor	*Semiconductor*	*Insulator*
$\rho \cong 10^{-6}$ Ω-cm (copper)	$\rho \cong 50$ Ω-cm (germanium) $\rho \cong 50 \times 10^3$ Ω-cm (silicon)	$\rho \cong 10^{12}$ Ω-cm (mica)

In Table 1.1, typical resistivity values are provided for three broad categories of materials. Although you may be familiar with the electrical properties of copper and mica from your past studies, the characteristics of the semiconductor materials of germanium (Ge) and silicon (Si) may be relatively new. As you will find in the chapters to follow, they are certainly not the only two semiconductor materials. They are, however, the two materials that have received the broadest range of interest in the development of semiconductor devices. In recent years the shift has been steadily toward silicon and away from germanium, but germanium is still in modest production.

Note in Table 1.1 the extreme range between the conductor and insulating materials for the 1-cm length (1-cm² area) of the material. Eighteen places separate the placement of the decimal point for one number from the other. Ge and Si have received the attention they have for a number of reasons. One very important consideration is the fact that they can be manufactured to a very high purity level. In fact, recent advances have reduced impurity levels in the pure material to 1 part in 10 billion (1:10,000,000,000). One might ask if these low impurity levels are really necessary. They certainly are if you consider that the addition of one part impurity (of the proper type) per million in a wafer of silicon material can change that material from a relatively poor conductor to a good conductor of electricity. We are obviously dealing with a whole new spectrum of comparison levels when we deal with the semiconductor medium. The ability to change the characteristics of the material significantly through this process, known as "doping," is yet another reason why Ge and Si have received such wide attention. Further reasons include the fact that their characteristics can be altered significantly through the application of heat or light—an important consideration in the development of heat- and light-sensitive devices.

Some of the unique qualities of Ge and Si noted above are due to their atomic structure. The atoms of both materials form a very definite pattern that is periodic in nature (i.e., continually repeats itself). One complete pattern is called a *crystal* and the periodic arrangement of the atoms a *lattice*. For Ge and Si the crystal has the three-dimensional diamond structure of Fig. 1.5. Any material composed solely of repeating crystal structures of the same kind is called a *single-crystal* structure. For semiconductor materials of practical application in the electronics field, this single-crystal feature exists, and, in addition, the periodicity of the structure does not change significantly with the addition of impurities in the doping process.

Let us now examine the structure of the atom itself and note how it might affect the electrical characteristics of the material. As you are aware, the atom is composed of three basic particles: the *electron*, the *proton*, and the *neutron*. In the atomic lattice, the neutrons and protons form the *nucleus*, while the electrons revolve around the nucleus in a fixed *orbit*. The Bohr models of the two most commonly used semiconductors, *germanium* and *silicon*, are shown in Fig. 1.6.

As indicated by Fig. 1.6a, the germanium atom has 32 orbiting electrons, while silicon has 14 orbiting electrons. In each case, there are 4 electrons in the outermost (*valence*) shell. The potential (*ionization potential*) required to remove any one of these 4 valence electrons is lower than that required for any other electron in the structure. In a pure germanium or silicon crystal these 4 valence electrons are bonded to 4 adjoining atoms, as shown in Fig. 1.7 for silicon. Both Ge and Si are referred to as *tetravalent atoms* because they each have four valence electrons.

A bonding of atoms, strengthened by the sharing of electrons, is called covalent bonding.

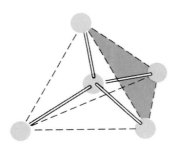

Figure 1.5 Ge and Si single-crystal structure.

Chapter 1 Semiconductor Diodes

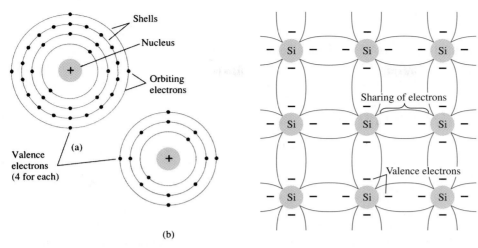

Figure 1.6 Atomic structure: (a) germanium; (b) silicon.

Figure 1.7 Covalent bonding of the silicon atom.

Although the covalent bond will result in a stronger bond between the valence electrons and their parent atom, it is still possible for the valence electrons to absorb sufficient kinetic energy from natural causes to break the covalent bond and assume the "free" state. The term *free* reveals that their motion is quite sensitive to applied electric fields such as established by voltage sources or any difference in potential. These natural causes include effects such as light energy in the form of photons and thermal energy from the surrounding medium. At room temperature there are approximately 1.5×10^{10} free carriers in a cubic centimeter of intrinsic silicon material.

Intrinsic materials are those semiconductors that have been carefully refined to reduce the impurities to a very low level—essentially as pure as can be made available through modern technology.

The free electrons in the material due only to natural causes are referred to as *intrinsic carriers*. At the same temperature, intrinsic germanium material will have approximately 2.5×10^{13} free carriers per cubic centimeter. The ratio of the number of carriers in germanium to that of silicon is greater than 10^3 and would indicate that germanium is a better conductor at room temperature. This may be true, but both are still considered poor conductors in the intrinsic state. Note in Table 1.1 that the resistivity also differs by a ratio of about 1000:1. with silicon having the larger value. This should be the case, of course, since resistivity and conductivity are inversely related.

An increase in temperature of a semiconductor can result in a substantial increase in the number of free electrons in the material.

As the temperature rises from absolute zero (0 K), an increasing number of valence electrons absorb sufficient thermal energy to break the covalent bond and contribute to the number of free carriers as described above. This increased number of carriers will increase the conductivity index and result in a lower resistance level.

Semiconductor materials such as Ge and Si that show a reduction in resistance with increase in temperature are said to have a negative temperature coefficient.

You will probably recall that the resistance of most conductors will increase with temperature. This is due to the fact that the numbers of carriers in a conductor will not increase significantly with temperature, but their vibration pattern about a relatively fixed location will make it increasingly difficult for electrons to pass through. An increase in temperature therefore results in an increased resistance level and a *positive temperature coefficient.*

1.4 ENERGY LEVELS

In the isolated atomic structure there are discrete (individual) energy levels associated with each orbiting electron, as shown in Fig. 1.8a. Each material will, in fact, have its own set of permissible energy levels for the electrons in its atomic structure.

The more distant the electron from the nucleus, the higher the energy state, and any electron that has left its parent atom has a higher energy state than any electron in the atomic structure.

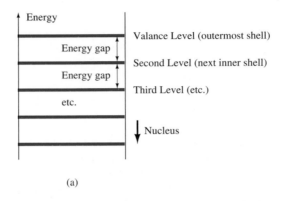

(a)

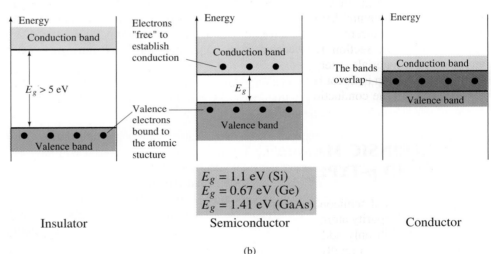

$E_g = 1.1$ eV (Si)
$E_g = 0.67$ eV (Ge)
$E_g = 1.41$ eV (GaAs)

Insulator   Semiconductor   Conductor

(b)

Figure 1.8 Energy levels: (a) discrete levels in isolated atomic structures; (b) conduction and valence bands of an insulator, semiconductor, and conductor.

Between the discrete energy levels are gaps in which no electrons in the isolated atomic structure can appear. As the atoms of a material are brought closer together to form the crystal lattice structure, there is an interaction between atoms that will result in the electrons in a particular orbit of one atom having slightly different energy levels from electrons in the same orbit of an adjoining atom. The net result is an expansion of the discrete levels of possible energy states for the valence electrons to that of bands as shown in Fig. 1.8b. Note that there are boundary levels and maximum energy states in which any electron in the atomic lattice can find itself, and there remains a *forbidden region* between the valence band and the ionization level. Recall that ionization is the mechanism whereby an electron can absorb sufficient energy to break away from the atomic structure and enter the conduction band. You will note that the energy associated with each electron is measured in *electron volts* (eV). The unit of measure is appropriate, since

$$\boxed{W = QV} \qquad \text{eV} \qquad\qquad (1.2)$$

as derived from the defining equation for voltage $V = W/Q$. The charge Q is the charge associated with a single electron.

Substituting the charge of an electron and a potential difference of 1 volt into Eq. (1.2) will result in an energy level referred to as one *electron volt*. Since energy is also measured in joules and the charge of one electron $= 1.6 \times 10^{-19}$ coulomb,

$$W = QV = (1.6 \times 10^{-19} \text{ C})(1 \text{ V})$$

and
$$\boxed{1 \text{ eV} = 1.6 \times 10^{-19} \text{ J}} \qquad\qquad (1.3)$$

At 0 K or absolute zero ($-273.15°C$), all the valence electrons of semiconductor materials find themselves locked in their outermost shell of the atom with energy levels associated with the valence band of Fig. 1.8b. However, at room temperature (300 K, 25°C) a large number of valence electrons have acquired sufficient energy to leave the valence band, cross the energy gap defined by E_g in Fig. 1.8b and enter the conduction band. For silicon E_g is 1.1 eV, for germanium 0.67 eV, and for gallium arsenide 1.41 eV. The obviously lower E_g for germanium accounts for the increased number of carriers in that material as compared to silicon at room temperature. Note for the insulator that the energy gap is typically 5 eV or more, which severely limits the number of electrons that can enter the conduction band at room temperature. The conductor has electrons in the conduction band even at 0 K. Quite obviously, therefore, at room temperature there are more than enough free carriers to sustain a heavy flow of charge, or current.

We will find in Section 1.5 that if certain impurities are added to the intrinsic semiconductor materials, energy states in the forbidden bands will occur which will cause a net reduction in E_g for both semiconductor materials—consequently, increased carrier density in the conduction band at room temperature!

1.5 EXTRINSIC MATERIALS— *n*- AND *p*-TYPE

The characteristics of semiconductor materials can be altered significantly by the addition of certain impurity atoms into the relatively pure semiconductor material. These impurities, although only added to perhaps 1 part in 10 million, can alter the band structure sufficiently to totally change the electrical properties of the material.

A semiconductor material that has been subjected to the doping process is called an extrinsic material.

There are two extrinsic materials of immeasurable importance to semiconductor device fabrication: *n*-type and *p*-type. Each will be described in some detail in the following paragraphs.

n-Type Material

Both the *n*- and *p*-type materials are formed by adding a predetermined number of impurity atoms into a germanium or silicon base. The *n*-type is created by introducing those impurity elements that have *five* valence electrons (*pentavalent*), such as *antimony, arsenic*, and *phosphorus*. The effect of such impurity elements is indicated in Fig. 1.9 (using antimony as the impurity in a silicon base). Note that the four covalent bonds are still present. There is, however, an additional fifth electron due to

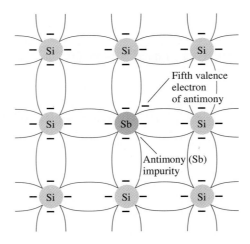

Figure 1.9 Antimony impurity in *n*-type material.

the impurity atom, which is *unassociated* with any particular covalent bond. This remaining electron, loosely bound to its parent (antimony) atom, is relatively free to move within the newly formed *n*-type material. Since the inserted impurity atom has donated a relatively "free" electron to the structure:

Diffused impurities with five valence electrons are called donor atoms.

It is important to realize that even though a large number of "free" carriers have been established in the *n*-type material, it is still electrically *neutral* since ideally the number of positively charged protons in the nuclei is still equal to the number of "free" and orbiting negatively charged electrons in the structure.

The effect of this doping process on the relative conductivity can best be described through the use of the energy-band diagram of Fig. 1.10. Note that a discrete energy level (called the *donor level*) appears in the forbidden band with an E_g significantly less than that of the intrinsic material. Those "free" electrons due to the added impurity sit at this energy level and have less difficulty absorbing a sufficient measure of thermal energy to move into the conduction band at room temperature. The result is that at room temperature, there are a large number of carriers (electrons) in the conduction level and the conductivity of the material increases significantly. At room temperature in an intrinsic Si material there is about one free electron for every 10^{12} atoms (1 to 10^9 for Ge). If our dosage level were 1 in 10 million (10^7), the ratio ($10^{12}/10^7 = 10^5$) would indicate that the carrier concentration has increased by a ratio of $100,000 : 1$.

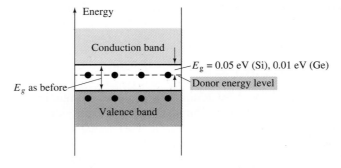

Figure 1.10 Effect of donor impurities on the energy band structure.

p-Type Material

The *p*-type material is formed by doping a pure germanium or silicon crystal with impurity atoms having *three* valence electrons. The elements most frequently used for this purpose are *boron, gallium,* and *indium.* The effect of one of these elements, boron, on a base of silicon is indicated in Fig. 1.11.

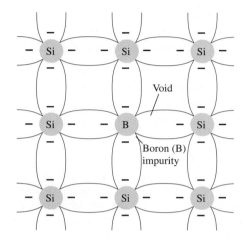

Figure 1.11 Boron impurity in *p*-type material.

Note that there is now an insufficient number of electrons to complete the covalent bonds of the newly formed lattice. The resulting vacancy is called a *hole* and is represented by a small circle or positive sign due to the absence of a negative charge. Since the resulting vacancy will readily *accept* a "free" electron:

The diffused impurities with three valence electrons are called acceptor atoms.

The resulting *p*-type material is electrically neutral, for the same reasons described for the *n*-type material.

Electron versus Hole Flow

The effect of the hole on conduction is shown in Fig. 1.12. If a valence electron acquires sufficient kinetic energy to break its covalent bond and fills the void created by a hole, then a vacancy, or hole, will be created in the covalent bond that released the electron. There is, therefore, a transfer of holes to the left and electrons to the right, as shown in Fig. 1.12. The direction to be used in this text is that of *conventional flow*, which is indicated by the direction of hole flow.

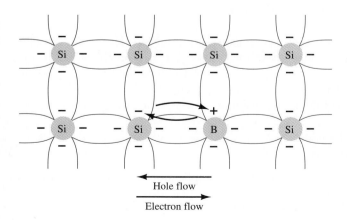

Figure 1.12 Electron versus hole flow.

Majority and Minority Carriers

In the intrinsic state, the number of free electrons in Ge or Si is due only to those few electrons in the valence band that have acquired sufficient energy from thermal or light sources to break the covalent bond or to the few impurities that could not be removed. The vacancies left behind in the covalent bonding structure represent our very limited supply of holes. In an *n*-type material, the number of holes has not changed significantly from this intrinsic level. The net result, therefore, is that the number of electrons far outweighs the number of holes. For this reason:

In an n-type material (Fig. 1.13a) the electron is called the majority carrier and the hole the minority carrier.

For the *p*-type material the number of holes far outweighs the number of electrons, as shown in Fig. 1.13b. Therefore:

In a p-type material the hole is the majority carrier and the electron is the minority carrier.

When the fifth electron of a donor atom leaves the parent atom, the atom remaining acquires a net positive charge: hence the positive sign in the donor-ion representation. For similar reasons, the negative sign appears in the acceptor ion.

The *n*- and *p*-type materials represent the basic building blocks of semiconductor devices. We will find in the next section that the "joining" of a single *n*-type material with a *p*-type material will result in a semiconductor element of considerable importance in electronic systems.

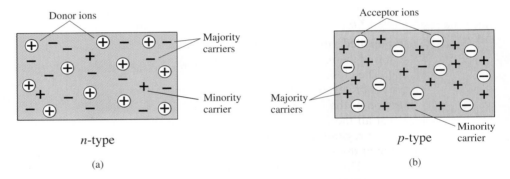

Figure 1.13   (a) *n*-type material; (b) *p*-type material.

1.6 SEMICONDUCTOR DIODE

In Section 1.5 both the *n*- and *p*-type materials were introduced. The semiconductor diode is formed by simply bringing these materials together (constructed from the same base—Ge or Si), as shown in Fig. 1.14, using techniques to be described in Chapter 19. At the instant the two materials are "joined" the electrons and holes in the region of the junction will combine, resulting in a lack of carriers in the region near the junction.

This region of uncovered positive and negative ions is called the depletion region due to the depletion of carriers in this region.

Since the diode is a two-terminal device, the application of a voltage across its terminals leaves three possibilities: *no bias* ($V_D = 0$ V), *forward bias* ($V_D > 0$ V), and *reverse bias* ($V_D < 0$ V). Each is a condition that will result in a response that the user must clearly understand if the device is to be applied effectively.

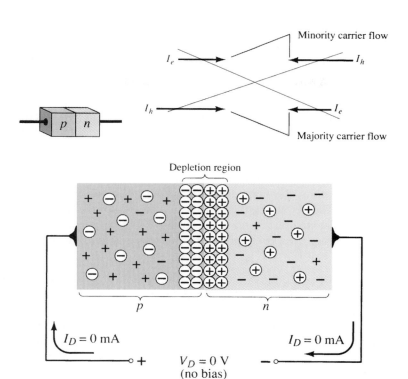

Figure 1.14 *p-n* junction with no external bias.

No Applied Bias ($V_D = 0$ V)

Under no-bias (no applied voltage) conditions, any minority carriers (holes) in the *n*-type material that find themselves within the depletion region will pass directly into the *p*-type material. The closer the minority carrier is to the junction, the greater the attraction for the layer of negative ions and the less the opposition of the positive ions in the depletion region of the *n*-type material. For the purposes of future discussions we shall assume that all the minority carriers of the *n*-type material that find themselves in the depletion region due to their random motion will pass directly into the *p*-type material. Similar discussion can be applied to the minority carriers (electrons) of the *p*-type material. This carrier flow has been indicated in Fig. 1.14 for the minority carriers of each material.

The majority carriers (electrons) of the *n*-type material must overcome the attractive forces of the layer of positive ions in the *n*-type material and the shield of negative ions in the *p*-type material to migrate into the area beyond the depletion region of the *p*-type material. However, the number of majority carriers is so large in the *n*-type material that there will invariably be a small number of majority carriers with sufficient kinetic energy to pass through the depletion region into the *p*-type material. Again, the same type of discussion can be applied to the majority carriers (holes) of the *p*-type material. The resulting flow due to the majority carriers is also shown in Fig. 1.14.

A close examination of Fig. 1.14 will reveal that the relative magnitudes of the flow vectors are such that the net flow in either direction is zero. This cancellation of vectors has been indicated by crossed lines. The length of the vector representing hole flow has been drawn longer than that for electron flow to demonstrate that the magnitude of each need not be the same for cancellation and that the doping levels for each material may result in an unequal carrier flow of holes and electrons. In summary, therefore:

In the absence of an applied bias voltage, the net flow of charge in any one direction for a semiconductor diode is zero.

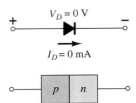

Figure 1.15 No-bias conditions for a semiconductor diode.

The symbol for a diode is repeated in Fig. 1.15 with the associated *n*- and *p*-type regions. Note that the arrow is associated with the *p*-type component and the bar with the *n*-type region. As indicated, for $V_D = 0$ V, the current in any direction is 0 mA.

Reverse-Bias Condition ($V_D < 0$ V)

If an external potential of *V* volts is applied across the *p-n* junction such that the positive terminal is connected to the *n*-type material and the negative terminal is connected to the *p*-type material as shown in Fig. 1.16, the number of uncovered positive ions in the depletion region of the *n*-type material will increase due to the large number of "free" electrons drawn to the positive potential of the applied voltage. For similar reasons, the number of uncovered negative ions will increase in the *p*-type material. The net effect, therefore, is a widening of the depletion region. This widening of the depletion region will establish too great a barrier for the majority carriers to overcome, effectively reducing the majority carrier flow to zero as shown in Fig. 1.16.

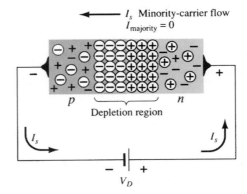

Figure 1.16 Reverse-biased *p-n* junction.

The number of minority carriers, however, that find themselves entering the depletion region will not change, resulting in minority-carrier flow vectors of the same magnitude indicated in Fig. 1.14 with no applied voltage.

> *The current that exists under reverse-bias conditions is called the reverse saturation current and is represented by I_s.*

The reverse saturation current is seldom more than a few microamperes except for high-power devices. In fact, in recent years its level is typically in the nanoampere range for silicon devices and in the low-microampere range for germanium. The term *saturation* comes from the fact that it reaches its maximum level quickly and does not change significantly with increase in the reverse-bias potential, as shown on the diode characteristics of Fig. 1.19 for $V_D < 0$ V. The reverse-biased conditions are depicted in Fig. 1.17 for the diode symbol and *p-n* junction. Note, in particular, that the direction of I_s is against the arrow of the symbol. Note also that the <u>n</u>egative potential is connected to the *p*-type material and the <u>p</u>ositive potential to the *n*-type material—the difference in underlined letters for each region revealing a reverse-bias condition.

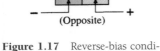

Figure 1.17 Reverse-bias conditions for a semiconductor diode.

Forward-Bias Condition ($V_D > 0$ V)

A *forward-bias* or "on" condition is established by applying the positive potential to the *p*-type material and the negative potential to the *n*-type material as shown in Fig. 1.18. For future reference, therefore:

> *A semiconductor diode is forward-biased when the association* p*-type and positive and* n*-type and negative has been established.*

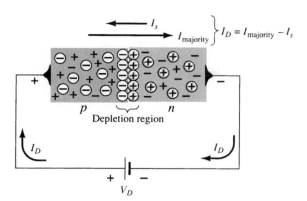

Figure 1.18 Forward-biased
p-n junction.

The application of a forward-bias potential V_D will "pressure" electrons in the *n*-type material and holes in the *p*-type material to recombine with the ions near the boundary and reduce the width of the depletion region as shown in Fig. 1.18. The resulting minority-carrier flow of electrons from the *p*-type material to the *n*-type material (and of holes from the *n*-type material to the *p*-type material) has not changed in magnitude (since the conduction level is controlled primarily by the limited number of impurities in the material), but the reduction in the width of the depletion region has resulted in a heavy majority flow across the junction. An electron of the *n*-type material now "sees" a reduced barrier at the junction due to the reduced depletion region and a strong attraction for the positive potential applied to the *p*-type material. As the applied bias increases in magnitude the depletion region will continue to decrease in width until a flood of electrons can pass through the junction, resulting in an exponential rise in current as shown in the forward-bias region of the characteristics of Fig. 1.19. Note that the vertical scale of Fig. 1.19 is measured in milliamperes (although some semiconductor diodes will have a vertical scale measured in amperes) and the horizontal scale in the forward-bias region has a maximum of 1 V. Typically, therefore, the voltage across a forward-biased diode will be less than 1 V. Note also, how quickly the current rises beyond the knee of the curve.

It can be demonstrated through the use of solid-state physics that the general characteristics of a semiconductor diode can be defined by the following equation for the forward- and reverse-bias regions:

$$I_D = I_s(e^{kV_D/T_K} - 1) \tag{1.4}$$

where I_s = reverse saturation current
k = 11,600/η with η = 1 for Ge and η = 2 for Si for relatively low levels of diode current (at or below the knee of the curve) and η = 1 for Ge and Si for higher levels of diode current (in the rapidly increasing section of the curve)
$T_K = T_C + 273°$

A plot of Eq. (1.4) is provided in Fig. 1.19. If we expand Eq. (1.4) into the following form, the contributing component for each region of Fig. 1.19 can easily be described:

$$I_D = I_s e^{kV_D/T_K} - I_s$$

For positive values of V_D the first term of the equation above will grow very quickly and overpower the effect of the second term. The result is that for positive values of V_D, I_D will be positive and grow as the function $y = e^x$ appearing in Fig. 1.20. At $V_D = 0$ V, Eq. (1.4) becomes $I_D = I_s(e^0 - 1) = I_s(1 - 1) = 0$ mA as appearing in Fig. 1.19. For negative values of V_D the first term will quickly drop off below I_s, resulting in $I_D = -I_s$, which is simply the horizontal line of Fig. 1.19. The break in the characteristics at $V_D = 0$ V is simply due to the dramatic change in scale from mA to μA.

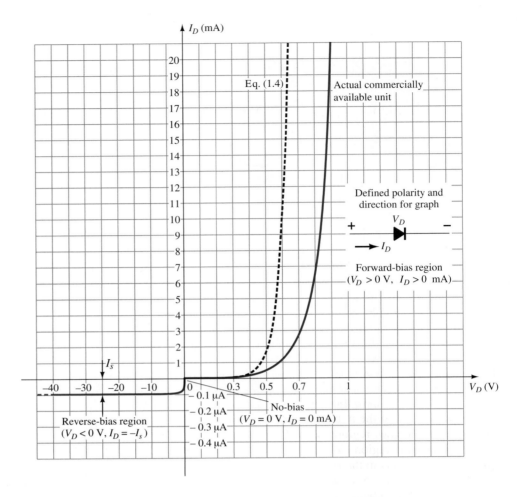

Figure 1.19 Silicon semiconductor diode characteristics.

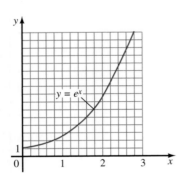

Figure 1.20 Plot of e^x.

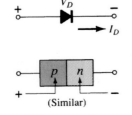

Figure 1.21 Forward-bias conditions for a semiconductor diode.

Note in Fig. 1.19 that the commercially available unit has characteristics that are shifted to the right by a few tenths of a volt. This is due to the internal "body" resistance and external "contact" resistance of a diode. Each contributes to an additional voltage at the same current level as determined by Ohm's law ($V = IR$). In time, as production methods improve, this difference will decrease and the actual characteristics approach those of Eq. (1.4).

It is important to note the change in scale for the vertical and horizontal axes. For positive values of I_D the scale is in milliamperes and the current scale below the axis is in microamperes (or possibly nanoamperes). For V_D the scale for positive values is in tenths of volts and for negative values the scale is in tens of volts.

Initially, Eq. (1.4) does appear somewhat complex and one may develop an unwarranted fear that it will be applied for all the diode applications to follow. Fortunately, however, a number of approximations will be made in a later section that will negate the need to apply Eq. (1.4) and provide a solution with a minimum of mathematical difficulty.

Before leaving the subject of the forward-bias state the conditions for conduction (the "on" state) are repeated in Fig. 1.21 with the required biasing polarities and the resulting direction of majority-carrier flow. Note in particular how the direction of conduction matches the arrow in the symbol (as revealed for the ideal diode).

Zener Region

Even though the scale of Fig. 1.19 is in tens of volts in the negative region, there is a point where the application of too negative a voltage will result in a sharp change

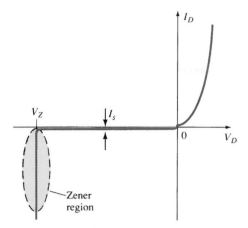

Figure 1.22 Zener region.

in the characteristics, as shown in Fig. 1.22. The current increases at a very rapid rate in a direction opposite to that of the positive voltage region. The reverse-bias potential that results in this dramatic change in characteristics is called the *Zener potential* and is given the symbol V_Z.

As the voltage across the diode increases in the reverse-bias region, the velocity of the minority carriers responsible for the reverse saturation current I_s will also increase. Eventually, their velocity and associated kinetic energy ($W_K = \frac{1}{2}mv^2$) will be sufficient to release additional carriers through collisions with otherwise stable atomic structures. That is, an *ionization* process will result whereby valence electrons absorb sufficient energy to leave the parent atom. These additional carriers can then aid the ionization process to the point where a high *avalanche* current is established and the *avalanche breakdown* region determined.

The avalanche region (V_Z) can be brought closer to the vertical axis by increasing the doping levels in the *p*- and *n*-type materials. However, as V_Z decreases to very low levels, such as -5 V, another mechanism, called *Zener breakdown,* will contribute to the sharp change in the characteristic. It occurs because there is a strong electric field in the region of the junction that can disrupt the bonding forces within the atom and "generate" carriers. Although the Zener breakdown mechanism is a significant contributor only at lower levels of V_Z, this sharp change in the characteristic at any level is called the *Zener region* and diodes employing this unique portion of the characteristic of a *p-n* junction are called *Zener diodes*. They are described in detail in Section 1.15.

The Zener region of the semiconductor diode described must be avoided if the response of a system is not to be completely altered by the sharp change in characteristics in this reverse-voltage region.

The maximum reverse-bias potential that can be applied before entering the Zener region is called the peak inverse voltage (referred to simply as the PIV rating) or the peak reverse voltage (denoted by PRV rating).

If an application requires a PIV rating greater than that of a single unit, a number of diodes of the same characteristics can be connected in series. Diodes are also connected in parallel to increase the current-carrying capacity.

Silicon versus Germanium

Silicon diodes have, in general, higher PIV and current rating and wider temperature ranges than germanium diodes. PIV ratings for silicon can be in the neighborhood of 1000 V, whereas the maximum value for germanium is closer to 400 V. Silicon can be used for applications in which the temperature may rise to about 200°C (400°F), whereas germanium has a much lower maximum rating (100°C). The disadvantage

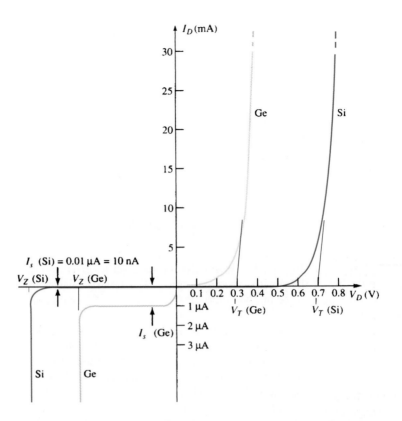

Figure 1.23 Comparison of Si and Ge semiconductor diodes.

of silicon, however, as compared to germanium, as indicated in Fig. 1.23, is the higher forward-bias voltage required to reach the region of upward swing. It is typically of the order of magnitude of 0.7 V for *commercially* available silicon diodes and 0.3 V for germanium diodes when rounded off to the nearest tenths. The increased offset for silicon is due primarily to the factor η in Eq. (1.4). This factor plays a part in determining the shape of the curve only at very low current levels. Once the curve starts its vertical rise, the factor η drops to 1 (the continuous value for germanium). This is evidenced by the similarities in the curves once the offset potential is reached. The potential at which this rise occurs is commonly referred to as the *offset, threshold,* or *firing potential*. Frequently, the first letter of a term that describes a particular quantity is used in the notation for that quantity. However, to ensure a minimum of confusion with other terms, such as output voltage (V_o) and forward voltage (V_F), the notation V_T has been adopted for this book, from the word "threshold."

In review:

$$
\boxed{
\begin{aligned}
V_T &\cong 0.7\,\text{V (Si)} \\
V_T &\cong 0.3\,\text{V (Ge)}
\end{aligned}
} \quad \text{V}
$$

Obviously, the closer the upward swing is to the vertical axis, the more "ideal" the device. However, the other characteristics of silicon as compared to germanium still make it the choice in the majority of commercially available units.

Temperature Effects

Temperature can have a marked effect on the characteristics of a silicon semiconductor diode as witnessed by a typical silicon diode in Fig. 1.24. It has been found experimentally that:

> *The reverse saturation current I_s will just about double in magnitude for every 10°C increase in temperature.*

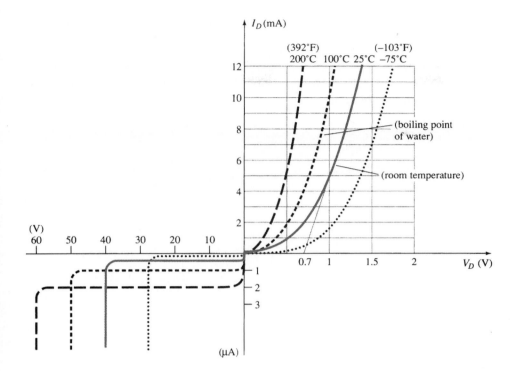

Figure 1.24 Variation in diode characteristics with temperature change.

It is not uncommon for a germanium diode with an I_s in the order of 1 or 2 μA at 25°C to have a leakage current of 100 μA = 0.1 mA at a temperature of 100°C. Current levels of this magnitude in the reverse-bias region would certainly question our desired open-circuit condition in the reverse-bias region. Typical values of I_s for silicon are much lower than that of germanium for similar power and current levels as shown in Fig. 1.23. The result is that even at high temperatures the levels of I_s for silicon diodes do not reach the same high levels obtained for germanium—a very important reason that silicon devices enjoy a significantly higher level of development and utilization in design. Fundamentally, the open-circuit equivalent in the reverse-bias region is better realized at any temperature with silicon than with germanium.

The increasing levels of I_s with temperature account for the lower levels of threshold voltage, as shown in Fig. 1.24. Simply increase the level of I_s in Eq. (1.4) and note the earlier rise in diode current. Of course, the level of T_K also will be increasing in the same equation, but the increasing level of I_s will overpower the smaller percent change in T_K. As the temperature increases the forward characteristics are actually becoming more "ideal," but we will find when we review the specifications sheets that temperatures beyond the normal operating range can have a very detrimental effect on the diode's maximum power and current levels. In the reverse-bias region the breakdown voltage is increasing with temperature, but note the undesirable increase in reverse saturation current.

1.7 MATHCAD

Throughout the text a mathematical software package called **Mathcad**® will be used to introduce the student to the variety of operations this popular package can perform and the advantages associated with its use. There is no need to obtain a copy of the software program unless you feel inclined to learn and use it after this brief introduction. In general, however, the coverage is only at the very introductory level to simply introduce the scope and power of the package. All the exercises appearing at the end of each chapter can be done without resorting to Mathcad.

The usefulness of Mathcad extends well beyond that of a hand-held scientific calculator. Mathcad can plot graphs, perform matrix algebra, permit the addition of text to any calculation, communicate with other data sources such as Excel® and MATLAB® or the Internet, store data, store information, etc.—the list is quite extensive and impressive. The more you learn about the package, the more uses you will find for it on a daily basis.

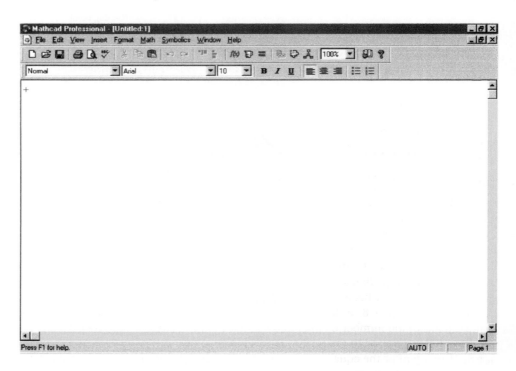

Figure 1.25 Basic Mathcad screen.

Once the package is installed, all operations begin with the basic screen of Fig. 1.25; labels have been added to this screen to identify the components of the display. In general, all the mathematical operations are performed in specific sequence such as shown in Fig. 1.26, that is, from left to right and then from top to bottom. For example, if line 2 is to operate on a variable, the value of the variable must be defined to the left on the same line or on line 1. Note that Mathcad is very sensitive to this order of things. For example, if you define a series of quantities on the same line but place one a little bit higher than the others, it will not be recognized by the other variables if it happens to be part of their definition. In other words, when writing on the same line, be absolutely sure that you stay on that same line for each new entry. Fortunately, Mathcad is well equipped to tell you when something is wrong. When you first use the program, you will get tired of seeing things in red, indicating that something was not entered or defined correctly. But, in time, as with any learning process, you will become quite comfortable with the software.

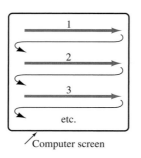

Figure 1.26 Defining the order of mathematical operations for Mathcad.

$$20 - 2 \cdot \frac{8}{6} = 17.333$$

Figure 1.27 Basic mathematical operation.

To perform basic arithmetic operations, simply click on the screen **at any point** to establish a cross hair on the display (the location of the first entry). If you decide you don't like the location, simply move the arrow to another location, and a simple click will move the cross hair. Then type in the mathematical operation $20 - 2 \times 8/6$ as shown in Fig. 1.27. The instant the equal sign is typed, the result will appear as shown in Fig. 1.27. The equal sign can come from the keyboard or the menu bar at the top of the screen. In fact, by going to **View-Tool bars-Calculator,** you can enter the entire expression and get the result using the mouse as your finger on a typical calculator. All the other mathematical operations such as powers, square roots, sine, tangent, etc., found in a typical scientific calculator are also available.

$$n := 2 \qquad TC := 27 \qquad VD := 0.6 \qquad IS := 50 \cdot 10^{-9}$$

$$k := \frac{11600}{n} \qquad TK := TC + 273 \qquad x := \frac{k}{TK} \cdot VD$$

$$x = 11.6$$

$$ID := IS \cdot \left(e^x - 1 \right)$$

$$ID = 5.455 \bullet 10^{-3}$$

Figure 1.28 Determining the diode current I_D using Eq. (1.4).

For practice using variables, let us calculate the current of a diode using Eq. (1.4). For equations with variables, the letter or symbol applied to the variable is first typed as shown in Fig. 1.28 followed by a colon sign. When the colon sign is entered, an equal sign will also appear as shown in the same figure. The value of the variable for the first series of calculations can then be entered. Next enter the remaining variables on the same line, and continue by calculating additional variables on the second line that are a function of those on the first line. Note that **x** requires that **k, TK,** and **VD** first be defined on the previous line or to the left on the same line. On the next line the value of **x** can be found by simply typing **x** followed by an equal sign. The correct response of 11.6 will appear immediately. Now Eq. (1.4) must be entered. As you enter each quantity, a bracket will appear around the quantity defining the quantity to be entered. In time, it becomes a friendly asset. Multiplication is entered with the asterisk above the letter 8 on the keyboard, and exponents are entered using the exponent sign above the number 6. Once the equation is correctly entered, **ID** can be written on the next line (or to the right of the equation), and the result of 5.455 mA will appear directly after the equal sign is selected. The result is that for a voltage of 0.6 V the current for this diode is 5.455 mA.

The beauty of Mathcad can now be effectively demonstrated by simply changing the voltage **VD** to 0.5 V. The instant the value is changed, the new level of **x** and **ID** will appear as shown in Fig. 1.29. A reduction in **VD** has obviously reduced the diode current to 0.789 mA. There is no need to enter the entire sequence of calculations again or to calculate all the quantities over again with a calculator. The results appear immediately.

$$\mathbf{n} := 2 \qquad TC := 27 \qquad VD := 0.5 \qquad IS := 50 \cdot 10^{-9}$$

$$k := \frac{11600}{\mathbf{n}} \qquad TK := TC + 273 \qquad x := \frac{k}{TK} \cdot VD$$

$$x = 9.667$$

$$ID := IS \cdot \left(e^x - 1 \right)$$

$$ID = 7.891 \bullet 10^{-4}$$

Figure 1.29 Demonstrating the effect of changing a parameter of Eq. (1.4).

Additional examples using Mathcad will appear throughout the text, but keep in mind that it is not necessary to become proficient in its use to grasp the material of this text—our purpose is simply to introduce the available software.

1.8 RESISTANCE LEVELS

As the operating point of a diode moves from one region to another the resistance of the diode will also change due to the nonlinear shape of the characteristic curve. It will be demonstrated in the next few paragraphs that the type of applied voltage or signal will define the resistance level of interest. Three different levels will be introduced in this section that will appear again as we examine other devices. It is therefore paramount that their determination be clearly understood.

DC or Static Resistance

The application of a dc voltage to a circuit containing a semiconductor diode will result in an operating point on the characteristic curve that will not change with time. The resistance of the diode at the operating point can be found simply by finding the corresponding levels of V_D and I_D as shown in Fig. 1.30 and applying the following equation:

$$R_D = \frac{V_D}{I_D}$$

(1.5)

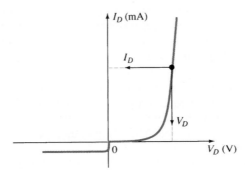

Figure 1.30 Determining the dc resistance of a diode at a particular operating point.

The dc resistance levels at the knee and below will be greater than the resistance levels obtained for the vertical rise section of the characteristics. The resistance levels in the reverse-bias region will naturally be quite high. Since ohmmeters typically employ a relatively constant-current source, the resistance determined will be at a preset current level (typically, a few milliamperes).

In general, therefore, the lower the current through a diode the higher the dc resistance level.

EXAMPLE 1.1

Determine the dc resistance levels for the diode of Fig. 1.31 at
(a) $I_D = 2$ mA
(b) $I_D = 20$ mA
(c) $V_D = -10$ V

Solution

(a) At $I_D = 2$ mA, $V_D = 0.5$ V (from the curve) and

$$R_D = \frac{V_D}{I_D} = \frac{0.5 \text{ V}}{2 \text{ mA}} = \mathbf{250\ \Omega}$$

(b) At $I_D = 20$ mA, $V_D = 0.8$ V (from the curve) and

Chapter 1   Semiconductor Diodes

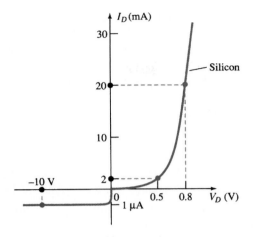

Figure 1.31 Example 1.1.

$$R_D = \frac{V_D}{I_D} = \frac{0.8 \text{ V}}{20 \text{ mA}} = \textbf{40 } \boldsymbol{\Omega}$$

(c) At $V_D = -10$ V, $I_D = -I_s = -1\,\mu\text{A}$ (from the curve) and

$$R_D = \frac{V_D}{I_D} = \frac{10 \text{ V}}{1\,\mu\text{A}} = \textbf{10 M}\boldsymbol{\Omega}$$

clearly supporting some of the earlier comments regarding the dc resistance levels of a diode.

AC or Dynamic Resistance

It is obvious from Eq. 1.5 and Example 1.1 that the dc resistance of a diode is independent of the shape of the characteristic in the region surrounding the point of interest. If a sinusoidal rather than dc input is applied, the situation will change completely. The varying input will move the instantaneous operating point up and down a region of the characteristics and thus defines a specific change in current and voltage as shown in Fig. 1.32. With no applied varying signal, the point of operation would be the *Q*-point appearing on Fig. 1.32 determined by the applied dc levels. The designation *Q-point* is derived from the word *quiescent,* which means "still or unvarying."

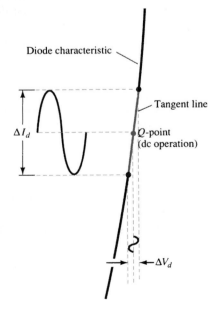

Figure 1.32 Defining the dynamic or ac resistance.

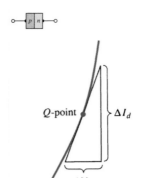

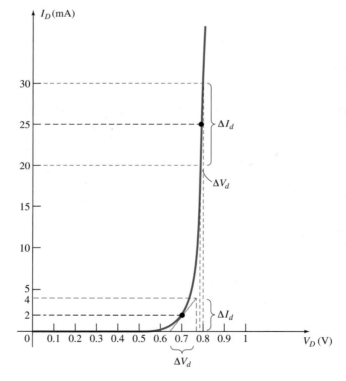

A straight line drawn tangent to the curve through the Q-point as shown in Fig. 1.33 will define a particular change in voltage and current that can be used to determine the *ac* or *dynamic* resistance for this region of the diode characteristics. An effort should be made to keep the change in voltage and current as small as possible and equidistant to either side of the Q-point. In equation form,

$$r_d = \frac{\Delta V_d}{\Delta I_d}$$

where Δ signifies a finite change in the quantity.   (1.6)

Figure 1.33 Determining the ac resistance at a Q-point.

The steeper the slope, the less the value of ΔV_d for the same change in ΔI_d and the less the resistance. The ac resistance in the vertical-rise region of the characteristic is therefore quite small, while the ac resistance is much higher at low current levels.

In general, therefore, the lower the Q-point of operation (smaller current or lower voltage) the higher the ac resistance.

EXAMPLE 1.2

For the characteristics of Fig. 1.34:
(a) Determine the ac resistance at $I_D = 2$ mA.
(b) Determine the ac resistance at $I_D = 25$ mA.
(c) Compare the results of parts (a) and (b) to the dc resistances at each current level.

Figure 1.34 Example 1.2.

Solution

(a) For $I_D = 2$ mA; the tangent line at $I_D = 2$ mA was drawn as shown in the figure and a swing of 2 mA above and below the specified diode current was chosen. At $I_D = 4$ mA, $V_D = 0.76$ V, and at $I_D = 0$ mA, $V_D = 0.65$ V. The resulting changes in current and voltage are

$$\Delta I_d = 4 \text{ mA} - 0 \text{ mA} = 4 \text{ mA}$$

and

$$\Delta V_d = 0.76 \text{ V} - 0.65 \text{ V} = 0.11 \text{ V}$$

and the ac resistance:

$$r_d = \frac{\Delta V_d}{\Delta I_d} = \frac{0.11 \text{ V}}{4 \text{ mA}} = \textbf{27.5 } \boldsymbol{\Omega}$$

(b) For I_D = 25 mA, the tangent line at I_D = 25 mA was drawn as shown in the figure and a swing of 5 mA above and below the specified diode current was chosen. At I_D = 30 mA, V_D = 0.8 V, and at I_D = 20 mA, V_D = 0.78 V. The resulting changes in current and voltage are

$$\Delta I_d = 30 \text{ mA} - 20 \text{ mA} = 10 \text{ mA}$$

and

$$\Delta V_d = 0.8 \text{ V} - 0.78 \text{ V} = 0.02 \text{ V}$$

and the ac resistance is

$$r_d = \frac{\Delta V_d}{\Delta I_d} = \frac{0.02 \text{ V}}{10 \text{ mA}} = \textbf{2 } \boldsymbol{\Omega}$$

(c) For I_D = 2 mA, V_D = 0.7 V and

$$R_D = \frac{V_D}{I_D} = \frac{0.7 \text{ V}}{2 \text{ mA}} = \textbf{350 } \boldsymbol{\Omega}$$

which far exceeds the r_d of 27.5 Ω.
   For I_D = 25 mA, V_D = 0.79 V and

$$R_D = \frac{V_D}{I_D} = \frac{0.79 \text{ V}}{25 \text{ mA}} = \textbf{31.62 } \boldsymbol{\Omega}$$

which far exceeds the r_d of 2 Ω.

We have found the dynamic resistance graphically, but there is a basic definition in differential calculus which states:

The derivative of a function at a point is equal to the slope of the tangent line drawn at that point.

Equation (1.6), as defined by Fig. 1.33, is, therefore, essentially finding the derivative of the function at the Q-point of operation. If we find the derivative of the general equation (1.4) for the semiconductor diode with respect to the applied forward bias and then invert the result, we will have an equation for the dynamic or ac resistance in that region. That is, taking the derivative of Eq. (1.4) with respect to the applied bias will result in

$$\frac{d}{dV_D}(I_D) = \frac{d}{dV}[I_s(e^{kV_D/T_K} - 1)]$$

and

$$\frac{dI_D}{dV_D} = \frac{k}{T_K}(I_D + I_s)$$

following a few basic maneuvers of differential calculus. In general, $I_D \gg I_s$ in the vertical slope section of the characteristics and

$$\frac{dI_D}{dV_D} \cong \frac{k}{T_K}I_D$$

Substituting $\eta = 1$ for Ge and Si in the vertical-rise section of the characteristics, we obtain

$$k = \frac{11,600}{\eta} = \frac{11,600}{1} = 11,600$$

and at room temperature,

$$T_K = T_C + 273° = 25° + 273° = 298°$$

so that

$$\frac{k}{T_K} = \frac{11,600}{298} \cong 38.93$$

and

$$\frac{dI_D}{dV_D} = 38.93 I_D$$

Flipping the result to define a resistance ratio $(R = V/I)$ gives us

$$\frac{dV_D}{dI_D} \cong \frac{0.026}{I_D}$$

or

$$\boxed{r_d = \frac{26 \text{ mV}}{I_D}}_{\text{Ge,Si}} \qquad (1.7)$$

The significance of Eq. (1.7) must be clearly understood. It implies that the dynamic resistance can be found simply by substituting the quiescent value of the diode current into the equation. There is no need to have the characteristics available or to worry about sketching tangent lines as defined by Eq. (1.6). It is important to keep in mind, however, that Eq. (1.7) is accurate only for values of I_D in the vertical-rise section of the curve. For lesser values of I_D, $\eta = 2$ (silicon) and the value of r_d obtained must be multiplied by a factor of 2. For small values of I_D below the knee of the curve, Eq. (1.7) becomes inappropriate.

All the resistance levels determined thus far have been defined by the p-n junction and do not include the resistance of the semiconductor material itself (called *body* resistance) and the resistance introduced by the connection between the semiconductor material and the external metallic conductor (called *contact* resistance). These additional resistance levels can be included in Eq. (1.7) by adding resistance denoted by r_B as appearing in Eq. (1.8). The resistance r'_d, therefore, includes the dynamic resistance defined by Eq. 1.7 and the resistance r_B just introduced.

$$\boxed{r'_d = \frac{26 \text{ mV}}{I_D} + r_B} \qquad \text{ohms} \qquad (1.8)$$

The factor r_B can range from typically 0.1 Ω for high-power devices to 2 Ω for some low-power, general-purpose diodes. For Example 1.2 the ac resistance at 25 mA was calculated to be 2 Ω. Using Eq. (1.7), we have

$$r_d = \frac{26 \text{ mV}}{I_D} = \frac{26 \text{ mV}}{25 \text{ mA}} = \mathbf{1.04 \ \Omega}$$

The difference of about 1 Ω could be treated as the contribution of r_B.

For Example 1.2 the ac resistance at 2 mA was calculated to be 27.5 Ω. Using Eq. (1.7) but multiplying by a factor of 2 for this region (in the knee of the curve $\eta = 2$),

$$r_d = 2\left(\frac{26 \text{ mV}}{I_D}\right) = 2\left(\frac{26 \text{ mV}}{2 \text{ mA}}\right) = 2(13 \ \Omega) = \mathbf{26 \ \Omega}$$

The difference of 1.5 Ω could be treated as the contribution due to r_B.

In reality, determining r_d to a high degree of accuracy from a characteristic curve using Eq. (1.6) is a difficult process at best and the results have to be treated with a grain of salt. At low levels of diode current the factor r_B is normally small enough compared to r_d to permit ignoring its impact on the ac diode resistance. At high levels of current the level of r_B may approach that of r_d, but since there will frequently be other resistive elements of a much larger magnitude in series with the diode we will assume in this book that the ac resistance is determined solely by r_d and the impact of r_B will be ignored unless otherwise noted. Technological improvements of recent years suggest that the level of r_B will continue to decrease in magnitude and eventually become a factor that can certainly be ignored in comparison to r_d.

The discussion above has centered solely on the forward-bias region. In the reverse-bias region we will assume that the change in current along the I_s line is nil from 0 V to the Zener region and the resulting ac resistance using Eq. (1.6) is sufficiently high to permit the open-circuit approximation.

Average AC Resistance

If the input signal is sufficiently large to produce a broad swing such as indicated in Fig. 1.35, the resistance associated with the device for this region is called the *average ac resistance*. The average ac resistance is, by definition, the resistance determined by a straight line drawn between the two intersections established by the maximum and minimum values of input voltage. In equation form (note Fig. 1.35),

$$r_{av} = \left.\frac{\Delta V_d}{\Delta I_d}\right|_{\text{pt. to pt.}} \tag{1.9}$$

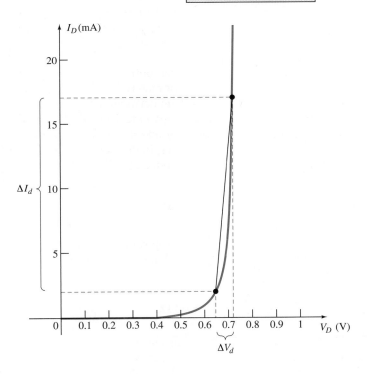

Figure 1.35 Determining the average ac resistance between indicated limits.

For the situation indicated by Fig. 1.35,

$$\Delta I_d = 17 \text{ mA} - 2 \text{ mA} = 15 \text{ mA}$$

and

$$\Delta V_d = 0.725 \text{ V} - 0.65 \text{ V} = 0.075 \text{ V}$$

with

$$r_{av} = \frac{\Delta V_d}{\Delta I_d} = \frac{0.075 \text{ V}}{15 \text{ mA}} = \mathbf{5 \ \Omega}$$

If the ac resistance (r_d) were determined at $I_D = 2$ mA its value would be more than 5 Ω, and if determined at 17 mA it would be less. In between the ac resistance would make the transition from the high value at 2 mA to the lower value at 17 mA.

Equation (1.9) has defined a value that is considered the average of the ac values from 2 to 17 mA. The fact that one resistance level can be used for such a wide range of the characteristics will prove quite useful in the definition of equivalent circuits for a diode in a later section.

As with the dc and ac resistance levels, the lower the level of currents used to determine the average resistance the higher the resistance level.

Summary Table

Table 1.2 was developed to reinforce the important conclusions of the last few pages and to emphasize the differences among the various resistance levels. As indicated earlier, the content of this section is the foundation for a number of resistance calculations to be performed in later sections and chapters.

TABLE 1.2 Resistance Levels

Type	Equation	Special Characteristics	Graphical Determination
DC or static	$R_D = \dfrac{V_D}{I_D}$	Defined as a *point* on the characteristics	
AC or dynamic	$r_d = \dfrac{\Delta V_d}{\Delta I_d} = \dfrac{26\text{ mV}}{I_D}$	Defined by a tangent line at the *Q*-point	
Average ac	$r_{\text{av}} = \dfrac{\Delta V_d}{\Delta I_d}\bigg|_{\text{pt. to pt.}}$	Defined by a straight line between limits of operation	

1.9 DIODE EQUIVALENT CIRCUITS

An equivalent circuit is a combination of elements properly chosen to best represent the actual terminal characteristics of a device, system, or such in a particular operating region.

In other words, once the equivalent circuit is defined, the device symbol can be removed from a schematic and the equivalent circuit inserted in its place without severely affecting the actual behavior of the system. The result is often a network that can be solved using traditional circuit analysis techniques.

Piecewise-Linear Equivalent Circuit

One technique for obtaining an equivalent circuit for a diode is to approximate the characteristics of the device by straight-line segments, as shown in Fig. 1.31. The resulting equivalent circuit is naturally called the *piecewise-linear equivalent circuit*. It should be obvious from Fig. 1.36 that the straight-line segments do not result in an exact duplication of the actual characteristics, especially in the knee region. However, the resulting segments are sufficiently close to the actual curve to establish an equivalent circuit that will provide an excellent first approximation to the actual behavior of the device. For the sloping section of the equivalence the average ac resistance as introduced in Section 1.7 is the resistance level appearing in the equivalent circuit of Fig. 1.37 next to the actual device. In essence, it defines the resistance level of the device when it is in the "on" state. The ideal diode is included to establish that there is only one direction of conduction through the device, and a reverse-bias condition will result in the open-circuit state for the device. Since a silicon semiconductor diode does not reach the conduction state until V_D reaches 0.7 V with a forward bias (as shown in Fig. 1.36), a battery V_T opposing the conduction direction must appear in the equivalent circuit as shown in Fig. 1.37. The battery simply specifies that the voltage across the device must be greater than the threshold battery voltage before conduction through the device in the direction dictated by the ideal diode can be established. When conduction is established the resistance of the diode will be the specified value of r_{av}.

Keep in mind, however, that V_T in the equivalent circuit is not an independent voltage source. If a voltmeter is placed across an isolated diode on the top of a lab bench, a reading of 0.7 V will not be obtained. The battery simply represents the horizontal offset of the characteristics that must be exceeded to establish conduction.

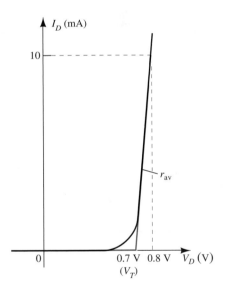

Figure 1.36 Defining the piecewise-linear equivalent circuit using straight-line segments to approximate the characteristic curve.

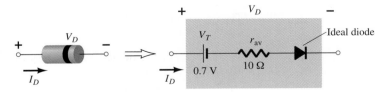

Figure 1.37 Components of the piecewise-linear equivalent circuit.

The approximate level of r_{av} can usually be determined from a specified operating point on the specification sheet (to be discussed in Section 1.10). For instance, for a silicon semiconductor diode, if $I_F = 10$ mA (a forward conduction current for the diode) at $V_D = 0.8$ V, we know for silicon that a shift of 0.7 V is required before the characteristics rise and

$$r_{av} = \frac{\Delta V_d}{\Delta I_d}\bigg|_{\text{pt. to pt.}} = \frac{0.8 \text{ V} - 0.7 \text{ V}}{10 \text{ mA} - 0 \text{ mA}} = \frac{0.1 \text{ V}}{10 \text{ mA}} = \mathbf{10 \ \Omega}$$

as obtained for Fig. 1.36.

Simplified Equivalent Circuit

For most applications, the resistance r_{av} is sufficiently small to be ignored in comparison to the other elements of the network. The removal of r_{av} from the equivalent circuit is the same as implying that the characteristics of the diode appear as shown in Fig. 1.38. Indeed, this approximation is frequently employed in semiconductor circuit analysis as demonstrated in Chapter 2. The reduced equivalent circuit appears in the same figure. It states that a forward-biased silicon diode in an electronic system under dc conditions has a drop of 0.7 V across it in the conduction state at any level of diode current (within rated values, of course).

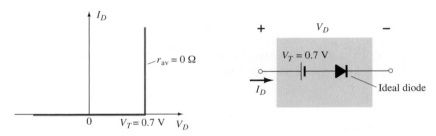

Figure 1.38   Simplified equivalent circuit for the silicon semiconductor diode.

Ideal Equivalent Circuit

Now that r_{av} has been removed from the equivalent circuit let us take it a step further and establish that a 0.7-V level can often be ignored in comparison to the applied voltage level. In this case the equivalent circuit will be reduced to that of an ideal diode as shown in Fig. 1.39 with its characteristics. In Chapter 2 we will see that this approximation is often made without a serious loss in accuracy.

In industry a popular substitution for the phrase "diode equivalent circuit" is diode *model*—a model by definition being a representation of an existing device, object, system, and so on. In fact, this substitute terminology will be used almost exclusively in the chapters to follow.

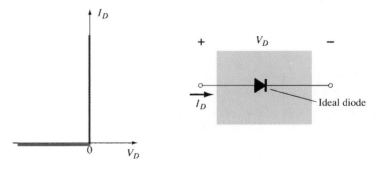

Figure 1.39   Ideal diode and its characteristics.

Chapter 1   Semiconductor Diodes

Summary Table

For clarity, the diode models employed for the range of circuit parameters and applications are provided in Table 1.3 with their piecewise-linear characteristics. Each will be investigated in greater detail in Chapter 2. There are always exceptions to the general rule, but it is fairly safe to say that the simplified equivalent model will be employed most frequently in the analysis of electronic systems while the ideal diode is frequently applied in the analysis of power supply systems where larger voltages are encountered.

TABLE 1.3 Diode Equivalent Circuits (Models)

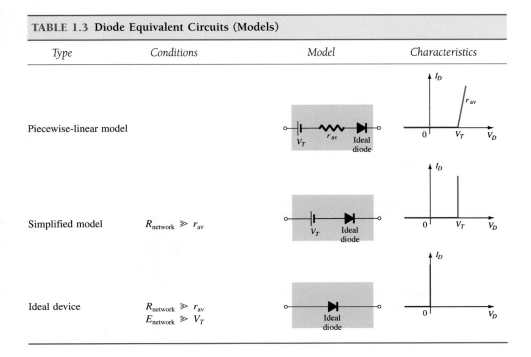

Type	Conditions	Model	Characteristics
Piecewise-linear model			
Simplified model	$R_{network} \gg r_{av}$		
Ideal device	$R_{network} \gg r_{av}$ $E_{network} \gg V_T$		

1.10 DIODE SPECIFICATION SHEETS

Data on specific semiconductor devices are normally provided by the manufacturer in one of two forms. Most frequently, it is a very brief description limited to perhaps one page. Otherwise, it is a thorough examination of the characteristics using graphs, artwork, tables, and so on. In either case, there are specific pieces of data that must be included for proper utilization of the device. They include:

1. The forward voltage V_F (at a specified current and temperature)
2. The maximum forward current I_F (at a specified temperature)
3. The reverse saturation current I_R (at a specified voltage and temperature)
4. The reverse-voltage rating [PIV or PRV or V(BR), where BR comes from the term "breakdown" (at a specified temperature)]
5. The maximum power dissipation level at a particular temperature
6. Capacitance levels (as defined in Section 1.11)
7. Reverse recovery time t_{rr} (as defined in Section 1.12)
8. Operating temperature range

Depending on the type of diode being considered, additional data may also be provided, such as frequency range, noise level, switching time, thermal resistance levels, and peak repetitive values. For the application in mind, the significance of the data will usually be self-apparent. If the maximum power or dissipation rating is also provided, it is understood to be equal to the following product:

$$P_{D\,\mathrm{max}} = V_D I_D \tag{1.10}$$

where I_D and V_D are the diode current and voltage at a particular point of operation.

If we apply the simplified model for a particular application (a common occurrence), we can substitute $V_D = V_T = 0.7$ V for a silicon diode in Eq. (1.10) and determine the resulting power dissipation for comparison against the maximum power rating. That is,

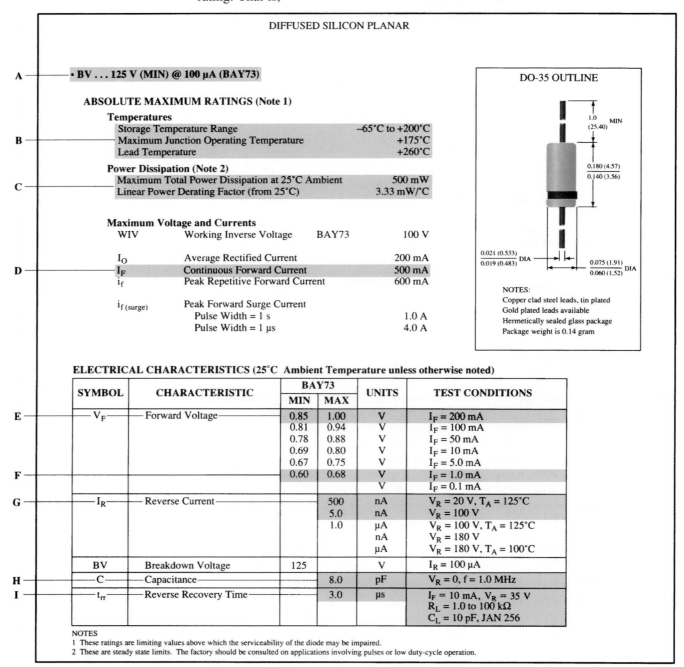

Figure 1.40 Electrical characteristics of a high-voltage, low-leakage diode.

$$P_{\text{dissipated}} \cong (0.7 \text{ V})I_D \qquad (1.11)$$

An exact copy of the data provided for a high-voltage/low-leakage diode appears in Figs. 1.40 and 1.41. This example would represent the expanded list of data and characteristics. The term *rectifier* is applied to a diode when it is frequently used in a *rectification* process to be described in Chapter 2.

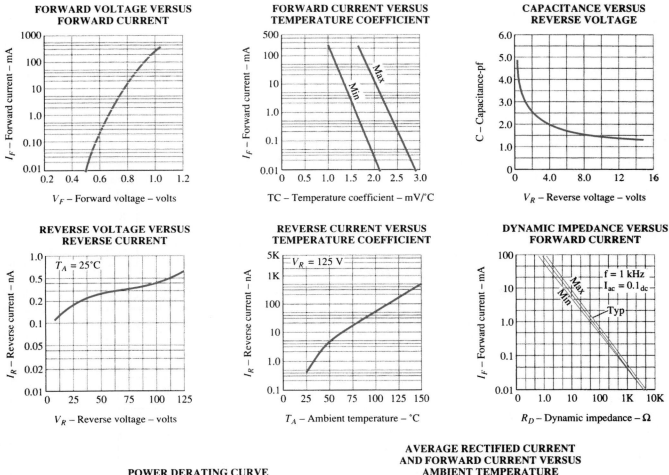

TYPICAL ELECTRICAL CHARACTERISTIC CURVES
at 25°C ambient temperature unless otherwise noted

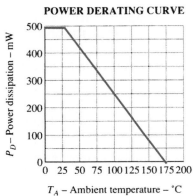

Figure 1.41 Terminal characteristics of a high-voltage diode.

Specific areas of the specification sheet have been highlighted in blue with a letter identification corresponding with the following description:

A: The *minimum* reverse-bias voltage (PIVs) for a diode at a specified reverse saturation current.

B: Temperature characteristics as indicated. Note the use of the Celsius scale and the wide range of utilization [recall that $32°F = 0°C = $ freezing (H_2O) and $212°F = 100°C = $ boiling (H_2O)].

C: Maximum power dissipation level $P_D = V_D I_D = 500$ mW. The maximum power rating decreases at a rate of 3.33 mW per degree increase in temperature above room temperature (25°C), as clearly indicated by the *power derating curve* of Fig. 1.41.

D: Maximum continuous forward current $I_{F_{max}} = 500$ mA (note I_F versus temperature in Fig. 1.41).

E: Range of values of V_F at $I_F = 200$ mA. Note that it exceeds $V_T = 0.7$ V for both devices.

F: Range of values of V_F at $I_F = 1.0$ mA. Note in this case how the upper limits surround 0.7 V.

G: At $V_R = 20$ V and a typical operating temperature $I_R = 500$ nA $= 0.5$ μA, while at a higher reverse voltage I_R drops to 5 nA $= 0.005$ μA.

H: The capacitance level between terminals is about 8 pF for the diode at $V_R = V_D = 0$ V (no-bias) and an applied frequency of 1 MHz.

I: The reverse recovery time is 3 μs for the list of operating conditions.

A number of the curves of Fig. 1.41 employ a log scale. A brief investigation of Section 11.2 should help with the reading of the graphs. Note in the top left figure how V_F increased from about 0.5 V to over 1 V as I_F increased from 10 μA to over 100 mA. In the figure below we find that the reverse saturation current does change slightly with increasing levels of V_R but remains at less than 1 nA at room temperature up to $V_R = 125$ V. As noted in the adjoining figure, however, note how quickly the reverse saturation current increases with increase in temperature (as forecasted earlier).

In the top right figure note how the capacitance decreases with increase in reverse bias voltage, and in the figure below note that the ac resistance (r_d) is only about 1 Ω at 100 mA and increases to 100 Ω at currents less than 1 mA (as expected from the discussion of earlier sections).

The average rectified current, peak repetitive forward current, and peak forward surge current as they appear on the specification sheet are defined as follows:

1. *Average rectified current.* A half-wave-rectified signal (described in Section 2.8) has an average value defined by $I_{av} = 0.318 I_{peak}$. The average current rating is lower than the continuous or peak repetitive forward currents because a half-wave current waveform will have instantaneous values much higher than the average value.

2. *Peak repetitive forward current.* This is the maximum instantaneous value of repetitive forward current. Note that since it is at this level for a brief period of time, its level can be higher than the continuous level.

3. *Peak forward surge current.* On occasion during turn-on, malfunctions, and so on, there will be very high currents through the device for very brief intervals of time (that are not repetitive). This rating defines the maximum value and the time interval for such surges in current level.

The more one is exposed to specification sheets, the "friendlier" they will become, especially when the impact of each parameter is clearly understood for the application under investigation.

1.11 TRANSITION AND DIFFUSION CAPACITANCE

Electronic devices are inherently sensitive to very high frequencies. Most shunt capacitive effects can be ignored at lower frequencies because the reactance $X_C = 1/2\pi fC$ is very large (open-circuit equivalent). This, however, cannot be ignored at very high frequencies. X_C will become sufficiently small due to the high value of f to introduce a low-reactance "shorting" path. In the p-n semiconductor diode, there are two capacitive effects to be considered. Both types of capacitance are present in the forward- and reverse-bias regions, but one so outweighs the other in each region that we consider the effects of only one in each region.

In the reverse-bias region we have the transition- or depletion-region capacitance (C_T), while in the forward-bias region we have the diffusion (C_D) or storage capacitance.

Recall that the basic equation for the capacitance of a parallel-plate capacitor is defined by $C = \epsilon A/d$, where ϵ is the permittivity of the dielectric (insulator) between the plates of area A separated by a distance d. In the reverse-bias region there is a depletion region (free of carriers) that behaves essentially like an insulator between the layers of opposite charge. Since the depletion width (d) will increase with increased reverse-bias potential, the resulting transition capacitance will decrease, as shown in Fig. 1.42. The fact that the capacitance is dependent on the applied reverse-bias potential has application in a number of electronic systems. In fact, in Chapter 19 a diode will be introduced whose operation is wholly dependent on this phenomenon.

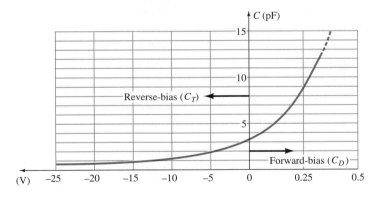

Figure 1.42 Transition and diffusion capacitance versus applied bias for a silicon diode.

Although the effect described above will also be present in the forward-bias region, it is overshadowed by a capacitance effect directly dependent on the rate at which charge is injected into the regions just outside the depletion region. The result is that increased levels of current will result in increased levels of diffusion capacitance. However, increased levels of current result in reduced level of associated resistance (to be demonstrated shortly), and the resulting time constant ($\tau = RC$), which is very important in high-speed applications, does not become excessive.

The capacitive effects described above are represented by a capacitor in parallel with the ideal diode, as shown in Fig. 1.43. For low- or mid-frequency applications (except in the power area), however, the capacitor is normally not included in the diode symbol.

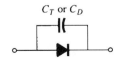

Figure 1.43 Including the effect of the transition or diffusion capacitance on the semiconductor diode.

1.12 REVERSE RECOVERY TIME

There are certain pieces of data that are normally provided on diode specification sheets provided by manufacturers. One such quantity that has not been considered yet is the reverse recovery time, denoted by t_{rr}. In the forward-bias state it was shown earlier that there are a large number of electrons from the n-type material progressing through the p-type material and a large number of holes in the n-type—a requirement for conduction. The electrons in the p-type and holes progressing through the n-type material establish a large number of minority carriers in each material. If the applied voltage should be reversed to establish a reverse-bias situation, we would ideally like to see the diode change instantaneously from the conduction state to the nonconduction state. However, because of the large number of minority carriers in each material, the diode current will simply reverse as shown in Fig. 1.44. and stay at this measurable level for the period of time t_s (storage time) required for the minority carriers to return to their majority-carrier state in the opposite material. In essence, the diode will remain in the short-circuit state with a current $I_{reverse}$ determined by the network parameters. Eventually, when this storage phase has passed, the current will reduce in level to that associated with the nonconduction state. This second period of time is denoted by t_t (transition interval). The reverse recovery time is the sum of these two intervals: $t_{rr} = t_s + t_t$. Naturally, it is an important consideration in high-speed switching applications. Most commercially available switching diodes have a t_{rr} in the range of a few nanoseconds to 1 μs. Units are available, however, with a t_{rr} of only a few hundred picoseconds (10^{-12}).

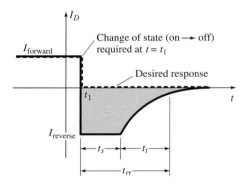

Figure 1.44   Defining the reverse recovery time.

1.13 SEMICONDUCTOR DIODE NOTATION

The notation most frequently used for semiconductor diodes is provided in Fig. 1.45. For most diodes any marking such as a dot or band, as shown in Fig. 1.45, appears at the cathode end. The terminology anode and cathode is a carryover from vacuum-tube notation. The anode refers to the higher or positive potential, and the cathode refers to the lower or negative terminal. This combination of bias levels will result in a forward-bias or "on" condition for the diode. A number of commercially available semiconductor diodes appear in Fig. 1.46. Some details of the actual construction of devices such as those appearing in Fig. 1.46 are provided in Chapters 12 and 19.

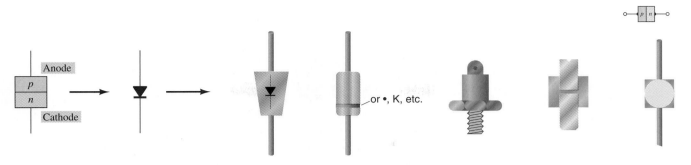

Figure 1.45 Semiconductor diode notation.

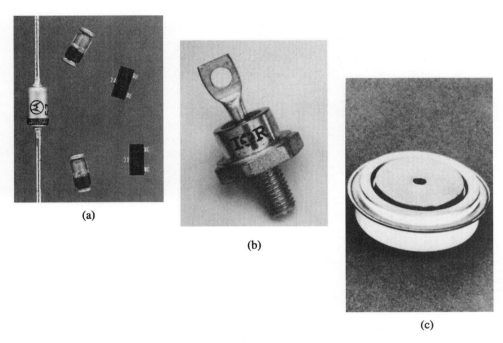

(a)

(b)

(c)

Figure 1.46 Various types of junction diodes. [(a) Courtesy of Motorola Inc.; and (b) and (c) Courtesy International Rectifier Corporation.]

1.14 DIODE TESTING

The condition of a semiconductor diode can be determined quickly using (1) a digital display meter (DDM) with a *diode checking function,* (2) the *ohmmeter section* of a multimeter, or (3) a *curve tracer.*

Diode Checking Function

A digital display meter with a diode checking capability appears in Fig. 1.47. Note the small diode symbol as the bottom option of the rotating dial. When set in this position and hooked up as shown in Fig. 1.48a, the diode should be in the "on" state and the display will provide an indication of the forward-bias voltage such as 0.67 V (for Si). The meter has an internal constant current source (about 2 mA) that will define the voltage level as indicated in Fig. 1.48b. An OL indication with the hookup of Fig. 1.48a reveals an open (defective) diode. If the leads are reversed, an OL indication should result due to the expected open-circuit equivalence for the diode. In general, therefore, an OL indication in both directions is an indication of an open or defective diode.

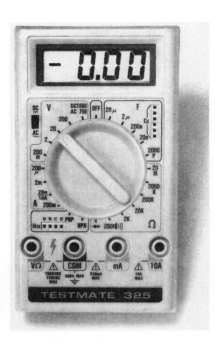

Figure 1.47 Digital display meter with diode checking capability. (Courtesy Computronics Technology, Inc.)

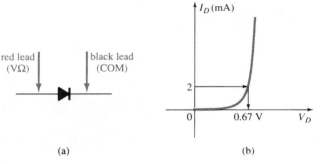

(a)

(b)

Figure 1.48 Checking a diode in the forward-bias state.

Ohmmeter Testing

In Section 1.8 we found that the forward-bias resistance of a semiconductor diode is quite low compared to the reverse-bias level. Therefore, if we measure the resistance of a diode using the connections indicated in Fig. 1.49a, we can expect a relatively low level. The resulting ohmmeter indication will be a function of the current established through the diode by the internal battery (often 1.5 V) of the ohmmeter circuit. The higher the current, the less the resistance level. For the reverse-bias situation the reading should be quite high, requiring a high resistance scale on the meter, as indicated in Fig. 1.49b. A high resistance reading in both directions obviously indicates an open (defective device) condition, while a very low resistance reading in both directions will probably indicate a shorted device.

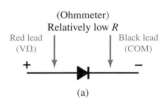

(a)

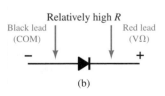

(b)

Figure 1.49 Checking a diode with an ohmmeter.

Curve Tracer

The curve tracer of Fig. 1.50 can display the characteristics of a host of devices, including the semiconductor diode. By properly connecting the diode to the test panel at the bottom center of the unit and adjusting the controls, the display of Fig. 1.51 can be obtained. Note that the vertical scaling is 1 mA/div, resulting in the levels indicated. For the horizontal axis the scaling is 100 mV/div, resulting in the voltage levels indicated. For a 2-mA level as defined for a DDM, the resulting voltage would be about 625 mV = 0.625 V. Although the instrument initially appears quite complex, the instruction manual and a few moments of exposure will reveal that the desired results can usually be obtained without an excessive amount of effort and time.

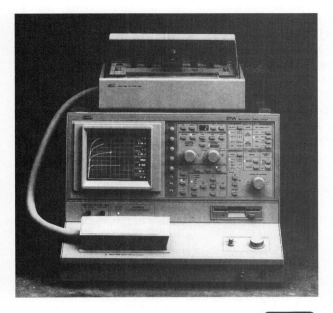

Figure 1.50 Curve tracer. (Courtesy of Tektronix, Inc.)

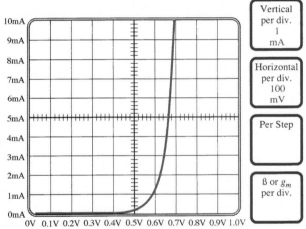

Figure 1.51 Curve tracer response to IN4007 silicon diode.

The same instrument will appear on more than one occasion in the chapters to follow as we investigate the characteristics of the variety of devices.

1.15 ZENER DIODES

The Zener region of Fig. 1.52 was discussed in some detail in Section 1.6. The characteristic drops in an almost vertical manner at a reverse-bias potential denoted V_Z. The fact that the curve drops down and away from the horizontal axis rather than up and away for the positive V_D region reveals that the current in the Zener region has a direction opposite to that of a forward-biased diode.

This region of unique characteristics is employed in the design of *Zener diodes,* which have the graphic symbol appearing in Fig. 1.53a. Both the semiconductor diode and zener diode are presented side by side in Fig. 1.53 to ensure that the direction of conduction of each is clearly understood together with the required polarity of the applied voltage. For the semiconductor diode the "on" state will support a current in the direction of the arrow in the symbol. For the Zener diode the direction of conduction is opposite to that of the arrow in the symbol as pointed out in the introduction to this section. Note also that the polarity of V_D and V_Z are the same as would be obtained if each were a resistive element.

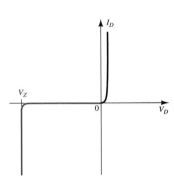

Figure 1.52 Reviewing the Zener region.

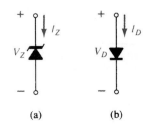

Figure 1.53 Conduction direction: (a) Zener diode; (b) semiconductor diode.

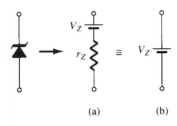

Figure 1.54 Zener equivalent circuit: (a) complete; (b) approximate.

The location of the Zener region can be controlled by varying the doping levels. An increase in doping, producing an increase in the number of added impurities, will decrease the Zener potential. Zener diodes are available having Zener potentials of 1.8 to 200 V with power ratings from $\frac{1}{4}$ to 50 W. Because of its higher temperature and current capability, silicon is usually preferred in the manufacture of Zener diodes.

The complete equivalent circuit of the Zener diode in the Zener region includes a small dynamic resistance and dc battery equal to the Zener potential, as shown in Fig. 1.54. For all applications to follow, however, we shall assume as a first approximation that the external resistors are much larger in magnitude than the Zener-equivalent resistor and that the equivalent circuit is simply the one indicated in Fig. 1.54b.

A larger drawing of the Zener region is provided in Fig. 1.55 to permit a description of the Zener nameplate data appearing in Table 1.4 for a 10-V, 500-mW, 20% diode. The term *nominal* associated with V_Z indicates that it is a typical average value. Since this is a 20% diode, the Zener potential can be expected to vary as 10 V $\pm$ 20% or from 8 to 12 V in its range of application. Also available are 10% and 5% diodes with the same specifications. The test current I_{ZT} is the current defined by the $\frac{1}{4}$ power level, and Z_{ZT} is the dynamic impedance at this current level. The maximum knee impedance occurs at the knee current of I_{ZK}. The reverse saturation current is provided at a particular potential level, and I_{ZM} is the maximum current for the 20% unit.

The temperature coefficient reflects the percent change in V_Z with temperature. It is defined by the equation

$$T_C = \frac{\Delta V_Z}{V_Z(T_1 - T_0)} \times 100\% \qquad \%/°C \qquad (1.12)$$

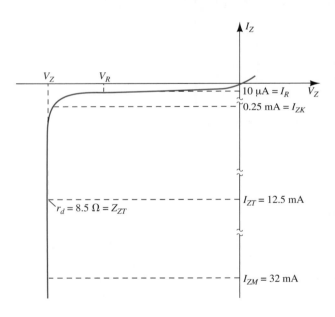

Figure 1.55 Zener test characteristics.

TABLE 1.4 Electrical Characteristics (25°C Ambient Temperature Unless Otherwise Noted)

Zener Voltage Nominal, V_Z (V)	Test Current, I_{ZT} (mA)	Max Dynamic Impedance, Z_{ZT} at I_{ZT} (Ω)	Maximum Knee Impedance, Z_{ZK} at I_{ZK} (Ω) (mA)	Maximum Reverse Current, I_R at V_R (μA)	Test Voltage, V_R (V)	Maximum Regulator Current, I_{ZM} (mA)	Typical Temperature Coefficient (%/°C)
10	12.5	8.5	700 0.25	10	7.2	32	+0.072

where ΔV_Z is the resulting change in Zener potential due to the temperature variation. Note in Fig. 1.56a that the temperature coefficient can be positive, negative, or even zero for different Zener levels. A positive value would reflect an increase in V_Z with an increase in temperature, while a negative value would result in a decrease in value with increase in temperature. The 24-V, 6.8-V, and 3.6-V levels refer to three Zener diodes having these nominal values within the same family of Zeners. The curve for the 10-V Zener would naturally lie between the curves of the 6.8-V and 24-V devices. Returning to Eq. (1.12), T_0 is the temperature at which V_Z is provided (normally room temperature—25°C), and T_1 is the new level. Example 1.3 will demonstrate the use of Eq. (1.12).

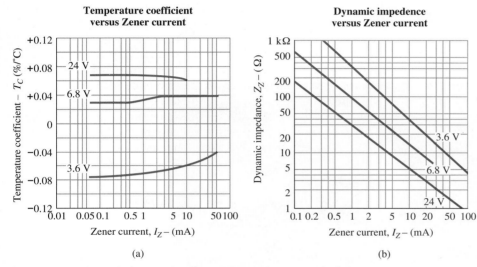

Figure 1.56 Electrical characteristics for a 10-V, 500-mW Zener diode.

Determine the nominal voltage for the Zener diode of Table 1.4 at a temperature of 100°C.

EXAMPLE 1.3

Solution

From Eq. 1.12,

$$\Delta V_Z = \frac{T_C V_Z}{100}(T_1 - T_0)$$

Substitution values from Table 1.4 yield

$$\Delta V_Z = \frac{(0.072)(10 \text{ V})}{100}(100°C - 25°C)$$

$$= (0.0072)(75)$$

$$= 0.54 \text{ V}$$

and because of the positive temperature coefficient, the new Zener potential, defined by V_Z', is

$$V_Z' = V_Z + 0.54 \text{ V}$$

$$= \mathbf{10.54 \text{ V}}$$

The variation in dynamic impedance (fundamentally, its series resistance) with current appears in Fig. 1.56b. Again, the 10-V Zener appears between the 6.8-V and

24-V Zeners. Note that the heavier the current (or the farther up the vertical rise you are in Fig. 1.52), the less the resistance value. Also note that as you drop below the knee of the curve, the resistance increases to significant levels.

The terminal identification and the casing for a variety of Zener diodes appear in Fig. 1.57. Figure 1.58 is an actual photograph of a variety of Zener devices. Note that their appearance is very similar to the semiconductor diode. A few areas of application for the Zener diode will be examined in Chapter 2.

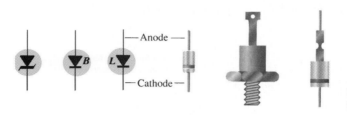

Figure 1.57 Zener terminal identification and symbols.

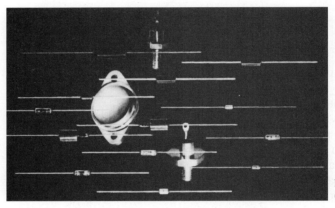

Figure 1.58 Zener diodes. (Courtesy Siemens Corporation.)

1.16 LIGHT-EMITTING DIODES

The increasing use of digital displays in calculators, watches, and all forms of instrumentation has contributed to the current extensive interest in structures that will emit light when properly biased. The two types in common use today to perform this function are the *light-emitting diode* (LED) and the *liquid-crystal display* (LCD). Since the LED falls within the family of *p-n* junction devices and will appear in some of the networks in the next few chapters, it will be introduced in this chapter. The LCD display is described in Chapter 19.

As the name implies, the light-emitting diode (LED) is a diode that will give off visible light when it is energized. In any forward-biased *p-n* junction there is, within the structure and primarily close to the junction, a recombination of holes and electrons. This recombination requires that the energy possessed by the unbound free electron be transferred to another state. In all semiconductor *p-n* junctions some of this energy will be given off as heat and some in the form of photons. In silicon and germanium the greater percentage is given up in the form of heat and the emitted light is insignificant. In other materials, such as gallium arsenide phosphide (GaAsP) or gallium phosphide (GaP), the number of photons of light energy emitted is sufficient to create a very visible light source.

The process of giving off light by applying an electrical source of energy is called electroluminescence.

As shown in Fig. 1.59 with its graphic symbol, the conducting surface connected to the *p*-material is much smaller, to permit the emergence of the maximum number of photons of light energy. Note in the figure that the recombination of the injected

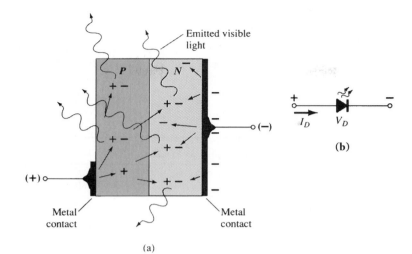

Figure 1.59 (a) Process of electroluminescence in the LED; (b) graphic symbol.

carriers due to the forward-biased junction results in emitted light at the site of recombination. There may, of course, be some absorption of the packages of photon energy in the structure itself, but a very large percentage are able to leave, as shown in the figure.

The appearance and characteristics of a subminiature high-efficiency solid-state lamp manufactured by Hewlett-Packard appears in Fig. 1.60. Note in Fig. 1.60b that the peak forward current is 60 mA, with 20 mA the typical average forward current. The test conditions listed in Fig. 1.60c, however, are for a forward current of 10 mA. The level of V_D under forward-bias conditions is listed as V_F and extends from 2.2 to 3 V. In other words, one can expect a typical operating current of about 10 mA at 2.5 V for good light emission.

Two quantities yet undefined appear under the heading Electrical/Optical Characteristics at $T_A = 25°C$. They are the *axial luminous intensity* (I_V) and the *luminous efficacy* (η_v). Light intensity is measured in *candela*. One candela emits a light flux of 4π lumens and establishes an illumination of 1 footcandle on a 1-ft^2 area 1 ft from the light source. Even though this description may not provide a clear understanding of the candela as a unit of measure, its level can certainly be compared between similar devices. The term *efficacy* is, by definition, a measure of the ability of a device to produce a desired effect. For the LED this is the ratio of the number of lumens generated per applied watt of electrical energy. The relative efficiency is defined by the luminous intensity per unit current, as shown in Fig. 1.60g. The relative intensity of each color versus wavelength appears in Fig. 1.60d.

Since the LED is a *p-n* junction device, it will have a forward-biased characteristic (Fig. 1.60e) similar to the diode response curves. Note the almost linear increase in relative luminous intensity with forward current (Fig. 1.60f). Figure 1.60h reveals that the longer the pulse duration at a particular frequency, the lower the permitted peak current (after you pass the break value of t_p). Figure 1.60i simply reveals that the intensity is greater at 0° (or head on) and the least at 90° (when you view the device from the side).

LED displays are available today in many different sizes and shapes. The light-emitting region is available in lengths from 0.1 to 1 in. Numbers can be created by segments such as shown in Fig. 1.61. By applying a forward bias to the proper *p*-type material segment, any number from 0 to 9 can be displayed.

There are also two-lead LED lamps that contain two LEDs, so that a reversal in biasing will change the color from green to red, or vice versa. LEDs are presently available in red, green, yellow, orange, and white, and white with blue soon to be commercially available. In general, LEDs operate at voltage levels from 1.7 to 3.3 V, which makes them completely compatible with solid-state circuits. They have a fast response time (nanoseconds) and offer good contrast ratios for visibility. The power requirement is typically from 10 to 150 mW with a lifetime of 100,000+ hours. Their semiconductor construction adds a significant ruggedness factor.

(a)

Absolute Maximum Ratings at $T_A = 25°C$

Parameter	High Eff. Red 4160	Units
Power dissipation	120	mW
Average forward current	20[1]	mA
Peak forward current	60	mA
Operating and storage temperature range	−55°C to 100°C	
Lead soldering temperature [1.6 mm (0.063 in.) from body]	230°C for 3 seconds	

[1] Derate from 50°C at 0.2mV/°C

(b)

Electrical/Optical Characteristics at $T_A = 25°C$

Symbol	Description	High Eff. Red 4160 Min.	Typ.	Max.	Units	Test Conditions
						$I_F = 10$ mA
I_V	Axial luminous intensity	1.0	3.0		mcd	
$2\theta_{1/2}$	Included angle between half luminous intensity points		80		deg.	Note 1
λ_{peak}	Peak wavelength		635		nm	Measurement at peak
λ_d	Dominant wavelength		628		nm	Note 2
τ_s	Speed of response		90		ns	
C	Capacitance		11		pF	$V_F = 0$; $f = 1$Mhz
θ_{JC}	Thermal resistance		120		°C/W	Junction to cathode lead at 0.79 mm (.031 in) from body
V_F	Forward voltage		2.2	3.0	V	$I_F = 10$ mA
BV_R	Reverse breakdown voltage	5.0			V	$I_R = 100$ μA
η_v	Luminous efficacy		147		lm/W	Note 3

NOTES:
1. $\theta_{1/2}$ is the off-axis angle at which the luminous intensity is half the axial luminous intensity.
2. The dominant wavelength, λ_d, is derived from the CIE chromaticity diagram and represents the single wavelength that defines the color of the device.
3. Radiant intensity, I_e, in watts/steradian, may be found from the equation $I_e = I_v/\eta_v$, where I_v is the luminous intensity in candelas and η_v is the luminous efficacy in lumens/watt.

(c)

Figure 1.60 Hewlett-Packard subminiature high-efficiency red solid-state lamp: (a) appearance; (b) absolute maximum ratings; (c) electrical/optical characteristics; (d) relative intensity versus wavelength; (e) forward current versus forward voltage; (f) relative luminous intensity versus forward current; (g) relative efficiency versus peak current; (h) maximum peak current versus pulse duration; (i) relative luminous intensity versus angular displacement. (Courtesy Hewlett-Packard Corporation.)

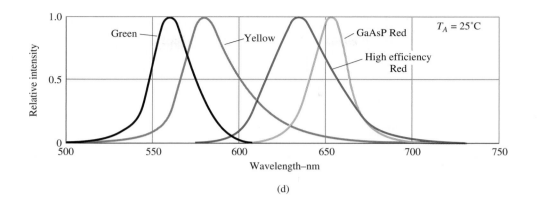

(d)

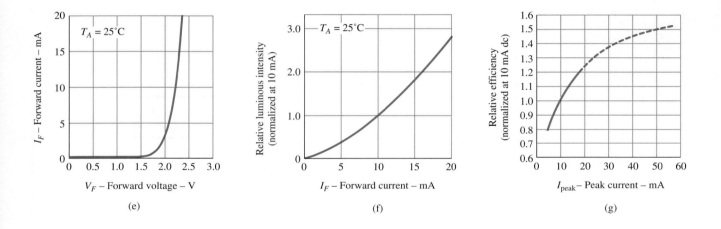

(e)   (f)   (g)

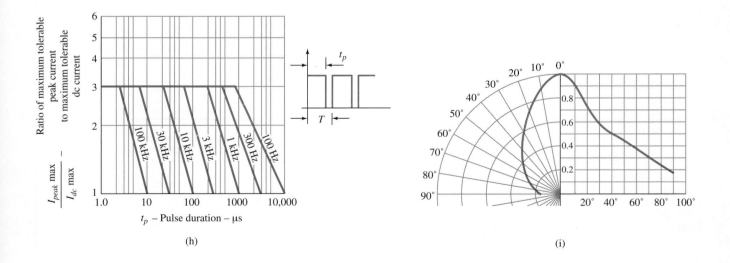

(h)   (i)

Figure 1.60 Continued.

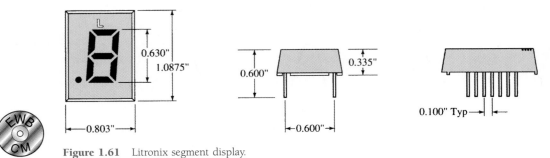

Figure 1.61 Litronix segment display.

PLANAR AIR–ISOLATED MONOLITHIC DIODE ARRAY

- C . . . 5.0 pF (MAX)
- ΔV_F . . . 15 mv (MAX) @ 10 mA

ABSOLUTE MAXIMUM RATINGS (Note 1)

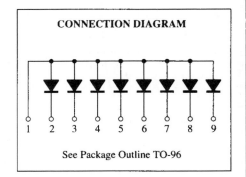

Temperatures

Storage Temperature Range	−55°C to +200°C
Maximum Junction Operating Temperature	+150°C
Lead Temperature	+260°C

Power Dissipation (Note 2)

Maximum Dissipation per Junction at 25°C Ambient	400 mW
per Package at 25°C Ambient	600 mW
Linear Derating Factor (from 25°C) Junction	3.2 mW/°C
Package	4.8 mW/°C

Maximum Voltage and Currents

WIV	Working Inverse Voltage	55 V
I_F	Continuous Forward Current	350 mA
$i_{f\,(surge)}$	Peak Forward Surge Current	
	Pulse Width = 1.0 s	1.0 A
	Pulse Width = 1.0 µs	2.0 A

ELECTRICAL CHARACTERISTICS (25°C Ambient Temperature unless otherwise noted)

SYMBOL	CHARACTERISTIC	MIN	MAX	UNITS	TEST CONDITIONS
B_V	Breakdown Voltage	60		V	$I_R = 10$ µA
V_F	Forward Voltage (Note 3)		1.5	V	$I_F = 500$ mA
			1.1	V	$I_F = 200$ mA
			1.0	V	$I_F = 100$ mA
I_R	Reverse Current		100	nA	$V_R = 40$ V
	Reverse Current ($T_A = 150$°C)		100	µA	$V_R = 40$ V
C	Capacitance		5.0	pF	$V_R = 0$, f = 1 MHz
V_{FM}	Peak Forward Voltage		4.0	V	$I_f = 500$ mA, $t_r < 10$ ns
t_{fr}	Forward Recovery Time		40	ns	$I_f = 500$ mA, $t_r < 10$ ns
t_{rr}	Reverse Recovery Time		10	ns	$I_f = I_r = 10 - 200$ mA $R_L = 100\,\Omega$, Rec. to 0.1 I_r
			50	ns	$I_f = 500$ mA, $I_r = 50$ mA $R_L = 100\,\Omega$, Rec. to 5 mA
ΔV_F	Forward Voltage Match		15	mV	$I_F = 10$ mA

NOTES
1 These ratings are limiting values above which life or satisfactory performance may be impaired.
2 These are steady state limits. The factory should be consulted on applications involving pulsed or low duty cycle operation.
3 V_F is measured using an 8 ms pulse.

Figure 1.62 Monolithic diode array.

1.17 DIODE ARRAYS—INTEGRATED CIRCUITS

The unique characteristics of integrated circuits are introduced in Appendix A, "Making the Chips That Run the World." When time permits, the content deserves a careful reading to establish a broad understanding of the manufacturing process. You will find that the integrated circuit is not a unique device with characteristics totally different from those we examine in these introductory chapters. It is simply a packaging technique that permits a significant reduction in the size of electronic systems. In other words, internal to the integrated circuit are systems and discrete devices that were available long before the integrated circuit as we know it today became a reality.

One possible array appears in Fig. 1.62 (see page 44). Note that eight diodes are internal to the diode array. That is, in the container shown in Fig. 1.63 there are diodes set in a single silicon wafer that have all the anodes connected to pin 1 and the cathodes of each to pins 2 through 9. Note in the same figure that pin 1 can be determined as being to the left of the small projection in the case if we look from the bottom toward the case. The other numbers then follow in sequence. If only one diode is to be used, then only pins 1 and 2 (or any number from 3 to 9) would be used. The remaining diodes would be left hanging and not affect the network to which pins 1 and 2 are connected.

Another diode array appears in Fig. 1.64. In this case the package is different but the numbering sequence appears in the outline. Pin 1 is the pin directly above the small indentation as you look down on the device.

Jack St. Clair Kilby, inventor of the integrated circuit and co-inventor of the electronic hand-held calculator. (Courtesy of Texas Instruments, Inc.)

Born: Jefferson City, Missouri, 1923 M.S. University of Wisconsin Director of Engineering and Technology, Components Group. Texas Instruments Fellow of the IEEE
Holds more than 60 U.S. patents

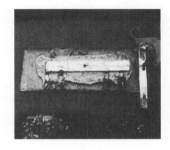

The first integrated circuit, a phase-shift oscillator, invented by Jack S. Kilby in 1958. (Courtesy of Texas Instruments, Inc.)

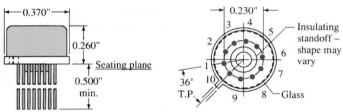

Figure 1.63 Package outline TO-96 for a diode array. All dimensions are in inches.

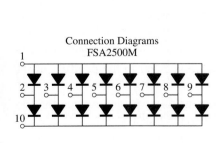

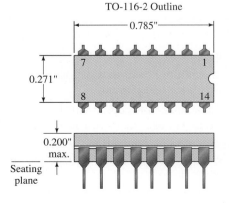

Figure 1.64 Monolithic diode array. All dimensions are in inches.

1.18 SUMMARY

Important Conclusions and Concepts

1. The characteristics of an ideal diode are a close match with those of a **simple switch** except for the important fact that an ideal diode can **conduct in only one direction.**

2. The ideal diode is a **short** in the region of conduction and an **open circuit** in the region of nonconduction.

3. A semiconductor is a material that has a conductivity level somewhere **between** that of a good conductor and that of an insulator.

4. A bonding of atoms, strengthened by the **sharing of electrons** between neighboring atoms, is called covalent bonding.

5. Increasing temperatures can cause a **significant increase** in the number of free electrons in a semiconductor material.

6. Most semiconductor materials used in the electronics industry have **negative temperature coefficients;** that is, the resistance drops with an increase in temperature.

7. Intrinsic materials are those semiconductors that have very **low level of impurities,** whereas extrinsic materials are semiconductors that have been **exposed to a doping process.**

8. An *n*-type material is formed by adding **donor** atoms that have **five** valence electrons to establish a high level of relatively free electrons. In an *n*-type material, the **electron is the majority carrier,** and the hole the minority carrier.

9. A *p*-type material is formed by adding **acceptor** atoms with **three** valence electrons that establish a high level of holes in the material. In a *p*-type material, the hole is the majority carrier, and the electron the minority carrier.

10. The region near the junction of a diode that has very few carriers is called the **depletion** region.

11. In the **absence** of any externally applied bias, the diode current is zero.

12. In the forward-bias region the diode current will **increase exponentially** with increase in voltage across the diode.

13. In the reverse-bias region the diode current is the **very small reverse saturation current** until Zener breakdown is reached and current will flow in the opposite direction through the diode.

14. The reverse saturation current I_s will just about **double** in magnitude for every 10-fold increase in temperature.

15. The dc resistance of a diode is determined by the **ratio** of the diode voltage and current at the point of interest and is **not sensitive** to the shape of the curve. The dc resistance **decreases** with increase in diode current or voltage.

16. The ac resistance of a diode is sensitive to the shape of the curve in the region of interest and decreases for higher levels of diode current or voltage.

17. The threshold voltage is about **0.7 V** for silicon diodes and **0.3 V** for germanium diodes.

18. The maximum power dissipation level of a diode is equal to the **product** of the diode voltage and current.

19. The capacitance of a diode will **increase exponentially** with increase in the forward-bias voltage. Its lowest levels are in the reverse-bias region.

20. The direction of conduction for a Zener diode is **opposite** to that of the arrow in the symbol, and the Zener voltage has a polarity opposite to that of a forward-biased diode.

21. Light-emitting diodes (LEDs) emit light under **forward-bias conditions** but require 2 V to 4 V for good emission.

Equations

$$I_D = I_s(e^{kV_D/T_K} - 1) \qquad k = \frac{11{,}600}{\eta} \qquad T_K = T_C + 273°$$

$$V_T \cong 0.7 \text{ V (Si)}$$

$$V_T \cong 0.3 \text{ V (Ge)}$$

$$R_D = \frac{V_D}{I_D}$$

$$r_d = \frac{\Delta V_d}{\Delta I_d} = \frac{26 \text{ mV}}{I_D}$$

$$r_{av} = \frac{\Delta V_d}{\Delta I_d} \bigg|_{\text{pt. to pt.}}$$

$$P_{D_{max}} = V_D I_D$$

1.19 COMPUTER ANALYSIS

The computer has now become such an integral part of the electronics industry that the capabilities of this working "tool" must be introduced at the earliest possible opportunity. For those students with no prior computer experience there is a common initial fear of this seemingly complicated powerful system. With this in mind the computer analysis of this book was designed to make the computer system more "friendly" by revealing the relative ease with which it can be applied to perform some very helpful and special tasks in a minimum amount of time with a high degree of accuracy. The content was written with the assumption that the reader has no prior computer experience or exposure to the terminology to be applied. There is also no suggestion that the content of this book is sufficient to permit a complete understanding of the "hows" and "whys" that will surface. The purpose here is solely to introduce some of the terminology, discuss a few of its capabilities, reveal the possibilities available, touch on some of its limitations, and demonstrate its versatility with a number of carefully chosen examples.

In general, the computer analysis of electronic systems can take one of two approaches: using a *language* such as C++, Pascal, FORTRAN, or QBASIC; or utilizing a *software package* such as PSpice, Electronics Workbench (EWB), MicroCap II, Breadboard, or Circuit Master, to name a few. A language, through its symbolic notation, forms a bridge between the user and the computer that permits a dialogue between the two for establishing the operations to be performed.

In earlier editions of this text, the chosen language was BASIC, primarily because it uses a number of familiar words and phrases from the English language that in themselves reveal the operation to be performed. When a language is employed to analyze a system, a *program* is developed that sequentially defines the operations to be performed—in much the same order in which we perform the same analysis in longhand. As with the longhand approach, one wrong step and the result obtained can be completely meaningless. Programs typically develop with time and application as more efficient paths toward a solution become obvious. Once established in its "best" form it can be cataloged for future use. The important advantage of the language approach is that a program can be tailored to meet all the special needs of the user. It permits innovative "moves" by the user that can result in printouts of data in an informative and interesting manner.

Figure 1.65 Mathcad 200 package.

Figure 1.66 PSpice Design package. (Courtesy of the OrCAD-MicroSim Corporation.)

Figure 1.67 Electronics Workbench, Version 6.2.

The alternative approach referred to above utilizes a software package to perform the desired investigation. A software package is a program written and tested over a period of time designed to perform a particular type of analysis or synthesis in an efficient manner with a high level of accuracy. The package itself cannot be altered by the user, and its application is limited to the operations built into the system. A user must adjust his or her desire for output information to the range of possibilities offered by the package. In addition, the user must input information exactly as requested by the package or the data may be misinterpreted.

The software packages available today have become so extensive in their coverage and range of operations that extensive exposure is now required to become truly proficient in their use. In fact, an associate with the broadest exposure to a particular software package is always an important source of information for those just starting out. The help that such an associate can initially provide is often invaluable in the time and effort it can save. But always keep in mind that at one time that local expert also had to pick his/her way through the provided manuals and sources of help to get a task done. Becoming proficient in the use of any software package is simply the end result of many hours of exposure, with the ability to ask questions and seek help when needed.

In this text, three software packages will be used extensively. However, the coverage is very introductory in nature, so the guidance provided by this text and the software manuals should be more than enough to enable readers to clearly understand the examples and work through the exercises. In Section 1.7, Mathcad was introduced to provide an awareness of the type of available mathematical assistance that extends well beyond the capability of the typical scientific calculator. Although the Mathcad 2000 package appearing in Fig. 1.65 is used in this text, the level of coverage is such that all the operations can be accomplished with older versions of Mathcad. For the electronic networks to be explored in this text, two software packages were employed: PSpice* and Electronics Workbench. A photograph of the Version 8.0 package for PSpice appears in Fig. 1.66 in the CD-ROM format (also available in 3.5" diskettes). A more sophisticated version, referred to simply as SPICE, has widespread application in industry. The package for Version 6.2 of Electronics Workbench appears in Fig. 1.67. Again, the coverage of this text is such that older versions can also be used to complete the exercises. For all the software packages, an effort was made to provide sufficient detail in the text to take the reader through each step in the analysis process. If questions do arise, first consult with your instructor and the software manuals, and as a last resort use the help-line provided with each package.

PSpice Windows

When using PSpice Windows, the network is first drawn on the screen followed by an analysis dictated by the needs of the user. This text will be using **Version 8.0,** though the differences between this and earlier Windows versions are so few and relatively minor for this level of application that one should not be concerned if using an earlier edition. The first step, of course, is to install PSpice into the hard-disk memory of your computer following the directions provided by MicroSim. Next, the **Schematics** screen must be obtained using a control mechanism such as **Windows 95.** Once established, the elements for the network must be obtained and placed on the screen to build the network. In this text, the procedure for each element will be described following the discussion of the characteristics and analysis of each device.

Since we have just finished covering the diode in detail, the procedure for finding the diodes stored in the library will be introduced along with the method for placing them on the screen. The next chapter will introduce the procedure for analyzing

---

*PSpice is a registered trademark of the OrCAD-MicroSim Corporation.

a complete network with diodes using PSpice. There are several ways to proceed, but the most direct path is to click on the picture symbol with the binoculars on the top right of the schematics screen. As you bring the marker close to the box using the mouse, a message **Get New Part** will be displayed. Left click on the symbol and a **Part Browser Basic** dialog box will appear. By choosing **Libraries, a Library Browser** dialog box will appear and the **EVAL.slb** library should be chosen. When selected, all available parts in this library will appear in the **Part** listing. Next, scroll the **Part** list and choose the **D1N4148** diode. The result is that the **Part Name** will appear above and the **Description** will indicate it is a diode. Once set, click **OK** and the **Part Browser Basic** dialog box will reappear with the full review of the chosen element. To place the device on the screen and close the dialog box, simply click on the **Place & Close** option. The result is that the diode will appear on the screen and can be put in place with a left click of the mouse. Once located, two labels will appear—one indicating how any diodes have been placed (**D1, D2, D3,** and so on) and the other the name of the chosen diode (**D1N4148**). The same diode can be placed in other places on the same screen by simply moving the pointer and left clicking the mouse. The process can be ended by a single right click of the mouse. Any of the diodes can be removed by simply clicking on them to make them red and pressing the **Delete** key. If preferred, the **Edit** choice of the menu bar at the top of the screen also can be chosen, followed by using the **Delete** command.

Another path for obtaining an element is to choose **Draw** on the menu bar, followed by **Get New Part.** Once chosen, the **Part Browser Basic** dialog box will appear as before and the same procedure can be followed. Now that we know the D1N4148 diode exists, it can be obtained directly once the **Part Browser Basic** dialog box appears. Simply type D1N4148 in the **Part Name** box, followed by **Place & Close,** and the **diode** will appear on the screen.

If a diode has to be moved, simply left click on it once, until it turns red. Then, click on it again and hold the clicker down on the mouse. At the same time, move the diode to any location you prefer and, when set, lift up on the clicker. Remember that anything in red can be operated on. To remove the red status, simply remove the pointer from the element and click it once. The diode will turn green and blue, indicating that its location and associated information is set in memory. For all the above and for the chapters to follow, if you happen to have a monochromatic (black-and-white) screen, you will simply have to remember whether the device is in the active state.

If the label or parameters of the diode are to be changed, simply click on the element once (to make it red) and choose **Edit,** followed by **Model.** An **Edit Model** dialog box will appear with a choice of changing the **model reference** (D1N4148), the **text** associated with each parameter, or the **parameters** that define the characteristics of the diode.

As mentioned above, additional comments regarding use of the diode will be made in the chapters to follow. For the moment, we are at least aware of how to find and place an element on the screen. If time permits, review the other elements available within the various libraries to prepare yourself for the work to follow.

Electronics Workbench (EWB)

Fortunately, there are a number of similarities between PSpice and Electronics Workbench (EWB). Of course, there are also an extensive number of differences, but the point is that once you become proficient in the use of one software package, the other will be much easier to learn.

Once the Multisim icon is chosen, the screen of Fig. 1.68 will appear. At first exposure, the menu bars seem quite extensive. In fact, simply familiarizing yourself with the range of options already available can take some time. Keep in mind, however, that for each item on a menu bar there is probably a subset that can be chosen, so the

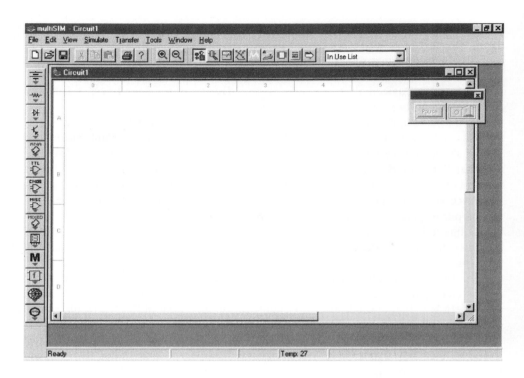

Figure 1.68 Basic Multisim (Electronics Workbench) screen.

list of options is quite extensive. First note at the top of the screen that the menu bar has been broken up into five distinct sections. The **system toolbar,** which includes the first four sections starting from the left, should be somewhat familiar from other software packages such as Microsoft Word. The remaining set of key pads (nine in all), called the **Multisim design bar,** is specifically designed for Electronics Workbench. The first key pad of the design bar has four different elements displayed to indicate that it is the source of components for any design. When you first enter EWB,

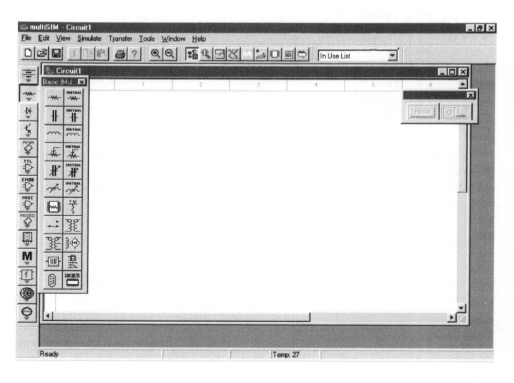

Figure 1.69 Electronics Workbench component family toolbar.

the key will probably be depressed to provide the menu bar of components on the left edge of the screen. Note the variety of components available, with the diode appearing as the third element down. The component menu bar can be removed or inserted through the use of the **component** key pad in the design bar. To add to the list of components, simply bring the cursor to the resistive component of the parts bin. The result is an expanded list of components as shown in Fig. 1.69.

The key pad to the right of the component pad resembles a capacitor. Its purpose is to permit **modification** of the components. The next key pad permits the selection of a variety of **instruments** to insert in the network when constructed. In fact, there are eleven instruments in all, extending from a simple multimeter to an oscilloscope.

The next key with its crosshatch of lines and components is the **simulation** pad for controlling the analysis being performed. Its options include **RUN/STOP** or **PAUSE/RESUME.** The next key controls the type of **analysis** to be performed, extending from DC to Noise Figure Analysis. The remaining keys will be left for the reader's investigation since they will not be needed for the intended coverage of this text.

The next logical step, which is to actually build a simple circuit and perform an analysis, will be left for the next chapter when dc biasing is applied to diode networks.

§ 1.2 Ideal Diode

1. Describe in your own words the meaning of the word *ideal* as applied to a device or system.

2. Describe in your own words the characteristics of the *ideal* diode and how they determine the on and off states of the device. That is, describe why the short-circuit and open-circuit equivalents are appropriate.

3. What is the one important difference between the characteristics of a simple switch and those of an ideal diode?

§ 1.3 Semiconductor Materials

4. In your own words, define *semiconductor, resistivity, bulk resistance,* and *ohmic contact resistance*.

5. (a) Using Table 1.1, determine the resistance of a silicon sample having an area of 1 cm^2 and a length of 3 cm.
   (b) Repeat part (a) if the length is 1 cm and the area 4 cm^2.
   (c) Repeat part (a) if the length is 8 cm and the area 0.5 cm^2.
   (d) Repeat part (a) for copper and compare the results.

6. Sketch the atomic structure of copper and discuss why it is a good conductor and how its structure is different from germanium and silicon.

7. In your own words, define an intrinsic material, a negative temperature coefficient, and covalent bonding.

8. Consult your reference library and list three materials that have a negative temperature coefficient and three that have a positive temperature coefficient.

§ 1.4 Energy Levels

9. How much energy in joules is required to move a charge of 6 C through a difference in potential of 3 V?

10. If 48 eV of energy is required to move a charge through a potential difference of 12 V, determine the charge involved.

11. Consult your reference library and determine the level of E_g for GaP and ZnS, two semiconductor materials of practical value. In addition, determine the written name for each material.

§ 1.5 Extrinsic Materials—*n*- and *p*-Type

12. Describe the difference between *n*-type and *p*-type semiconductor materials.

13. Describe the difference between donor and acceptor impurities.

14. Describe the difference between majority and minority carriers.

15. Sketch the atomic structure of silicon and insert an impurity of arsenic as demonstrated for silicon in Fig. 1.9.

16. Repeat Problem 15 but insert an impurity of indium.

17. Consult your reference library and find another explanation of hole versus electron flow. Using both descriptions, describe in your own words the process of hole conduction.

§ 1.6 Semiconductor Diode

18. Describe in your own words the conditions established by forward- and reverse-bias conditions on a *p-n* junction diode and how the resulting current is affected.

19. Describe how you will remember the forward- and reverse-bias states of the *p-n* junction diode. That is, how you will remember which potential (positive or negative) is applied to which terminal?

20. Using Eq. (1.4), determine the diode current at 20°C for a silicon diode with $I_s = 50$ nA and an applied forward bias of 0.6 V.

21. Repeat Problem 20 for $T = 100$°C (boiling point of water). Assume that I_s has increased to 5.0 μA.

22. (a) Using Eq. (1.4), determine the diode current at 20°C for a silicon diode with $I_s = 0.1$ μA at a reverse-bias potential of -10 V.
    (b) Is the result expected? Why?

23. (a) Plot the function $y = e^x$ for x from 0 to 5.
    (b) What is the value of $y = e^x$ at $x = 0$?
    (c) Based on the results of part (b), why is the factor -1 important in Eq. (1.4)?

24. In the reverse-bias region the saturation current of a silicon diode is about 0.1 μA ($T = 20$°C). Determine its approximate value if the temperature is increased 40°C.

25. Compare the characteristics of a silicon and a germanium diode and determine which you would prefer to use for most practical applications. Give some details. Refer to a manufacturer's listing and compare the characteristics of a germanium and a silicon diode of similar maximum ratings.

26. Determine the forward voltage drop across the diode whose characteristics appear in Fig. 1.24 at temperatures of -75°C, 25°C, 100°C, and 200°C and a current of 10 mA. For each temperature, determine the level of saturation current. Compare the extremes of each and comment on the ratio of the two.

§ 1.8 Resistance Levels

27. Determine the static or dc resistance of the commercially available diode of Fig. 1.19 at a forward current of 2 mA.

28. Repeat Problem 26 at a forward current of 15 mA and compare results.

29. Determine the static or dc resistance of the commercially available diode of Fig. 1.19 at a reverse voltage of -10 V. How does it compare to the value determined at a reverse voltage of -30 V?

30. (a) Determine the dynamic (ac) resistance of the diode of Fig. 1.34 at a forward current of 10 mA using Eq. (1.6).
    (b) Determine the dynamic (ac) resistance of the diode of Fig. 1.34 at a forward current of 10 mA using Eq. (1.7).
    (c) Compare solutions of parts (a) and (b).

31. Calculate the dc and ac resistance for the diode of Fig. 1.34 at a forward current of 10 mA and compare their magnitudes.

32. Using Eq. (1.6), determine the ac resistance at a current of 1 mA and 15 mA for the diode of Fig. 1.34. Compare the solutions and develop a general conclusion regarding the ac resistance and increasing levels of diode current.

33. Using Eq. (1.7), determine the ac resistance at a current of 1 mA and 15 mA for the diode of Fig. 1.19. Modify the equation as necessary for low levels of diode current. Compare to the solutions obtained in Problem 32.

34. Determine the average ac resistance for the diode of Fig. 1.19 for the region between 0.6 and 0.9 V.

35. Determine the ac resistance for the diode of Fig. 1.19 at 0.75 V and compare to the average ac resistance obtained in Problem 34.

§ 1.9 Diode Equivalent Circuits

36. Find the piecewise-linear equivalent circuit for the diode of Fig. 1.19. Use a straight line segment that intersects the horizontal axis at 0.7 V and best approximates the curve for the region greater than 0.7 V.

37. Repeat Problem 36 for the diode of Fig. 1.34.

§ 1.10 Diode Specification Sheets

* **38.** Plot I_F versus V_F using linear scales for the diode of Fig. 1.41. Note that the provided graph employs a log scale for the vertical axis (log scales are covered in sections 11.2 and 11.3).

39. Comment on the change in capacitance level with increase in reverse-bias potential for the diode of Fig. 1.41.

40. Does the reverse saturation current of the diode of Fig. 1.41 change significantly in magnitude for reverse-bias potentials in the range -25 to -100 V?

* **41.** For the diode of Fig. 1.41 determine the level of I_R at room temperature (25°C) and the boiling point of water (100°C). Is the change significant? Does the level just about double for every 10°C increase in temperature?

42. For the diode of Fig. 1.41, determine the maximum ac (dynamic) resistance at a forward current of 0.1, 1.5, and 20 mA. Compare levels and comment on whether the results support conclusions derived in earlier sections of this chapter.

43. Using the characteristics of Fig. 1.41, determine the maximum power dissipation levels for the diode at room temperature (25°C) and 100°C. Assuming that V_F remains fixed at 0.7 V, how has the maximum level of I_F changed between the two temperature levels?

44. Using the characteristics of Fig. 1.41, determine the temperature at which the diode current will be 50% of its value at room temperature (25°C).

§ 1.11 Transition and Diffusion Capacitance

* **45.** (a) Referring to Fig. 1.42, determine the transition capacitance at reverse-bias potentials of -25 and -10 V. What is the ratio of the change in capacitance to the change in voltage?
  (b) Repeat part (a) for reverse-bias potentials of -10 and -1 V. Determine the ratio of the change in capacitance to the change in voltage.
  (c) How do the ratios determined in parts (a) and (b) compare? What does it tell you about which range may have more areas of practical application?

46. Referring to Fig. 1.42, determine the diffusion capacitance at 0 and 0.25 V.

47. Describe in your own words how diffusion and transition capacitances differ.

48. Determine the reactance offered by a diode described by the characteristics of Fig. 1.42 at a forward potential of 0.2 V and a reverse potential of -20 V if the applied frequency is 6 MHz.

§ 1.12 Reverse Recovery Time

49. Sketch the waveform for i of the network of Fig. 1.70 if $t_t = 2t_s$ and the total reverse recovery time is 9 ns.

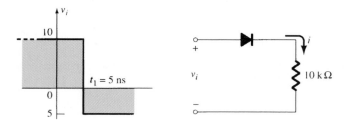

Figure 1.70 Problem 49

§ 1.15 Zener Diodes

50. The following characteristics are specified for a particular Zener diode: $V_Z = 29$ V, $V_R = 16.8$ V, $I_{ZT} = 10$ mA, $I_R = 20$ μA, and $I_{ZM} = 40$ mA. Sketch the characteristic curve in the manner displayed in Fig. 1.55.

* **51.** At what temperature will the 10-V Zener diode of Fig. 1.55 have a nominal voltage of 10.75 V? (*Hint*: Note the data in Table 1.4.)

52. Determine the temperature coefficient of a 5-V Zener diode (rated 25°C value) if the nominal voltage drops to 4.8 V at a temperature of 100°C.

53. Using the curves of Fig. 1.56a, what level of temperature coefficient would you expect for a 20-V diode? Repeat for a 5-V diode. Assume a linear scale between nominal voltage levels and a current level of 0.1 mA.

54. Determine the dynamic impedance for the 24-V diode at $I_Z = 10$ mA for Fig. 1.56b. Note that it is a log scale.

* **55.** Compare the levels of dynamic impedance for the 24-V diode of Fig. 1.56b at current levels of 0.2, 1, and 10 mA. How do the results relate to the shape of the characteristics in this region?

§ 1.16 Light-Emitting Diodes

56. Referring to Fig. 1.60e, what would appear to be an appropriate value of V_T for this device? How does it compare to the value of V_T for silicon and germanium?

57. Using the information provided in Fig. 1.60, determine the forward voltage across the diode if the relative luminous intensity is 1.5.

* **58.** (a) What is the percent increase in relative efficiency of the device of Fig. 1.60 if the peak current is increased from 5 to 10 mA?
  (b) Repeat part (a) for 30 to 35 mA (the same increase in current).
  (c) Compare the percent increase from parts (a) and (b). At what point on the curve would you say there is little gained by further increasing the peak current?

* **59.** (a) Referring to Fig. 1.60h, determine the maximum tolerable peak current if the period of the pulse duration is 1 ms, the frequency is 300 Hz, and the maximum tolerable dc current is 20 mA.
  (b) Repeat part (a) for a frequency of 100 Hz.

60. (a) If the luminous intensity at 0° angular displacement is 3.0 mcd for the device of Fig. 1.60, at what angle will it be 0.75 mcd?
  (b) At what angle does the loss of luminous intensity drop below the 50% level?

* **61.** Sketch the current derating curve for the average forward current of the high-efficiency red LED of Fig. 1.60 as determined by temperature. (Note the absolute maximum ratings.)

---

*Please Note: Asterisks indicate more difficult problems.

Diode Applications

2.1 INTRODUCTION

The construction, characteristics, and models of semiconductor diodes were introduced in Chapter 1. The primary goal of this chapter is to develop a working knowledge of the diode in a variety of configurations using models appropriate for the area of application. By chapter's end, the fundamental behavior pattern of diodes in dc and ac networks should be clearly understood. The concepts learned in this chapter will have significant carryover in the chapters to follow. For instance, diodes are frequently employed in the description of the basic construction of transistors and in the analysis of transistor networks in the dc and ac domains.

The content of this chapter will reveal an interesting and very positive side of the study of a field such as electronic devices and systems—once the basic behavior of a device is understood, its function and response in an infinite variety of configurations can be determined. The range of applications is endless, yet the characteristics and models remain the same. The analysis will proceed from one that employs the actual diode characteristic to one that utilizes the approximate models almost exclusively. It is important that the role and response of various elements of an electronic system be understood without continually having to resort to lengthy mathematical procedures. This is usually accomplished through the approximation process, which can develop into an art itself. Although the results obtained using the actual characteristics may be slightly different from those obtained using a series of approximations, keep in mind that the characteristics obtained from a specification sheet may in themselves be slightly different from the device in actual use. In other words, the characteristics of a 1N4001 semiconductor diode may vary from one element to the next in the same lot. The variation may be slight, but it will often be sufficient to validate the approximations employed in the analysis. Also consider the other elements of the network: Is the resistor labeled 100 Ω exactly 100 Ω? Is the applied voltage exactly 10 V or perhaps 10.08 V? All these tolerances contribute to the general belief that a response determined through an appropriate set of approximations can often be "as accurate" as one that employs the full characteristics. In this book the emphasis is toward developing a working knowledge of a device through the use of appropriate approximations, thereby avoiding an unnecessary level of mathematical complexity. Sufficient detail will normally be provided, however, to permit a detailed mathematical analysis if desired.

2.2 LOAD-LINE ANALYSIS

The applied load will normally have an important impact on the point or region of operation of a device. If the analysis is performed in a graphical manner, a line can be drawn on the characteristics of the device that represents the applied load. The intersection of the load line with the characteristics will determine the point of operation of the system. Such an analysis is, for obvious reasons, called *load-line analysis*. Although the majority of the diode networks analyzed in this chapter do not employ the load-line approach, the technique is one used quite frequently in subsequent chapters, and this introduction offers the simplest application of the method. It also permits a validation of the approximate technique described throughout the remainder of this chapter.

Consider the network of Fig. 2.1a employing a diode having the characteristics of Fig. 2.1b. Note in Fig. 2.1a that the "pressure" established by the battery is to establish a current through the series circuit in the clockwise direction. The fact that this current and the defined direction of conduction of the diode are a "match" reveals that the diode is in the "on" state and conduction has been established. The resulting polarity across the diode will be as shown and the first quadrant (V_D and I_D positive) of Fig. 2.1b will be the region of interest—the forward-bias region.

Applying Kirchhoff's voltage law to the series circuit of Fig. 2.1a will result in

$$E - V_D - V_R = 0$$

or

$$\boxed{E = V_D + I_D R} \tag{2.1}$$

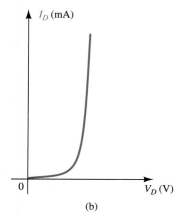

Figure 2.1 Series diode configuration: (a) circuit; (b) characteristics.

The two variables of Eq. (2.1) (V_D and I_D) are the same as the diode axis variables of Fig. 2.1b. This similarity permits a plotting of Eq. (2.1) on the same characteristics of Fig. 2.1b.

The intersections of the load line on the characteristics can easily be determined if one simply employs the fact that anywhere on the horizontal axis $I_D = 0$ A and anywhere on the vertical axis $V_D = 0$ V.

If we *set* $V_D = 0$ V in Eq. (2.1) and solve for I_D, we have the magnitude of I_D *on* the vertical axis. Therefore, with $V_D = 0$ V, Eq. (2.1) becomes

$$E = V_D + I_D R$$

$$= 0\text{ V} + I_D R$$

and

$$\boxed{I_D = \frac{E}{R}\Big|_{V_D = 0\text{ V}}} \tag{2.2}$$

as shown in Fig. 2.2. If we *set* $I_D = 0$ A in Eq. (2.1) and solve for V_D, we have the magnitude of V_D *on* the horizontal axis. Therefore, with $I_D = 0$ A, Eq. (2.1) becomes

$$E = V_D + I_D R$$

$$= V_D + (0\text{ A})R$$

and

$$\boxed{V_D = E\big|_{I_D = 0\text{ A}}} \tag{2.3}$$

as shown in Fig. 2.2. A straight line drawn between the two points will define the load line as depicted in Fig. 2.2. Change the level of R (the load) and the intersection on the vertical axis will change. The result will be a change in the slope of the load line and a different point of intersection between the load line and the device characteristics.

We now have a load line defined by the network and a characteristic curve defined by the device. The point of intersection between the two is the point of operation for

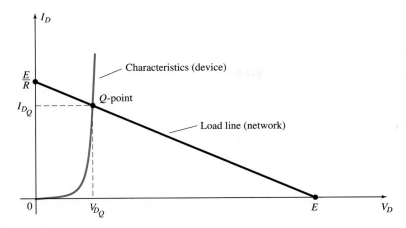

Figure 2.2 Drawing the load line and finding the point of operation.

this circuit. By simply drawing a line down to the horizontal axis the diode voltage V_{D_Q} can be determined, whereas a horizontal line from the point of intersection to the vertical axis will provide the level of I_{D_Q}. The current I_D is actually the current through the entire series configuration of Fig. 2.1a. The point of operation is usually called the *quiescent point* (abbreviated "*Q*-pt.") to reflect its "still, unmoving" qualities as defined by a dc network.

The solution obtained at the intersection of the two curves is the same that would be obtained by a simultaneous mathematical solution of Eqs. (2.1) and (1.4) $[I_D = I_s(e^{kV_D/T_K} - 1)]$ as demonstrated later in this section in a Mathcad example. Since the curve for a diode has nonlinear characteristics the mathematics involved would require the use of nonlinear techniques that are beyond the needs and scope of this book. The load-line analysis described above provides a solution with a minimum of effort and a "pictorial" description of why the levels of solution for V_{D_Q} and I_{D_Q} were obtained. The next two examples will demonstrate the techniques introduced above and reveal the relative ease with which the load line can be drawn using Eqs. (2.2) and (2.3).

For the series diode configuration of Fig. 2.3a employing the diode characteristics of Fig. 2.3b determine:
(a) V_{D_Q} and I_{D_Q}.
(b) V_R.

EXAMPLE 2.1

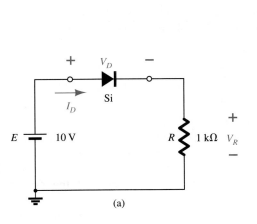

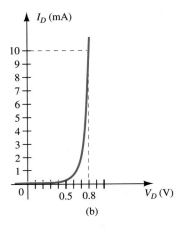

Figure 2.3 (a) Circuit; (b) characteristics.

Solution

(a) Eq. (2.2): $\quad I_D = \dfrac{E}{R}\bigg|_{V_D=0\text{ V}} = \dfrac{10\text{ V}}{1\text{ k}\Omega} = 10\text{ mA}$

Eq. (2.3): $\quad V_D = E|_{I_D=0\text{ A}} = 10\text{ V}$

The resulting load line appears in Fig. 2.4. The intersection between the load line and the characteristic curve defines the Q-point as

$$V_{D_Q} \cong \mathbf{0.78\ V}$$

$$I_{D_Q} \cong \mathbf{9.25\ mA}$$

The level of V_D is certainly an estimate, and the accuracy of I_D is limited by the chosen scale. A higher degree of accuracy would require a plot that would be much larger and perhaps unwieldy.

(b) $V_R = I_R R = I_{D_Q} R = (9.25\text{ mA})(1\text{ k}\Omega) = \mathbf{9.25\ V}$

or $\ V_R = E - V_D = 10\text{ V} - 0.78\text{ V} = \mathbf{9.22\ V}$

The difference in results is due to the accuracy with which the graph can be read. Ideally, the results obtained either way should be the same.

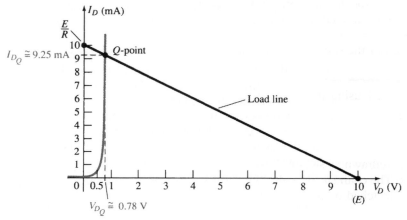

Figure 2.4   Solution to Example 2.1.

EXAMPLE 2.2

Repeat the analysis of Example 2.1 with $R = 2\text{ k}\Omega$.

Solution

(a) Eq. (2.2): $\quad I_D = \dfrac{E}{R}\bigg|_{V_D=0\text{ V}} = \dfrac{10\text{ V}}{2\text{ k}\Omega} = 5\text{ mA}$

Eq. (2.3): $\quad V_D = E|_{I_D=0\text{ A}} = 10\text{ V}$

The resulting load line appears in Fig. 2.5. Note the reduced slope and levels of diode current for increasing loads. The resulting Q-point is defined by

$$V_{D_Q} \cong \mathbf{0.7\ V}$$

$$I_{D_Q} \cong \mathbf{4.6\ mA}$$

(b) $V_R = I_R R = I_{D_Q} R = (4.6\text{ mA})(2\text{ k}\Omega) = \mathbf{9.2\ V}$

with $V_R = E - V_D = 10\text{ V} - 0.7\text{ V} = \mathbf{9.3\ V}$

The difference in levels is again due to the accuracy with which the graph can be read. Certainly, however, the results provide an expected magnitude for the voltage V_R.

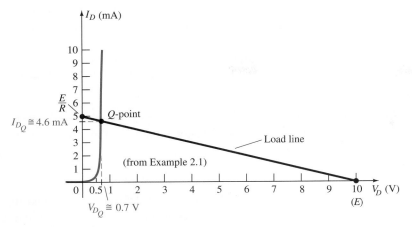

Figure 2.5 Solution to Example 2.2.

As noted in the examples above, the load line is determined solely by the applied network while the characteristics are defined by the chosen device. If we turn to our approximate model for the diode and do not disturb the network, the load line will be exactly the same as obtained in the examples above. In fact, the next two examples repeat the analysis of Examples 2.1 and 2.2 using the approximate model to permit a comparison of the results.

EXAMPLE 2.3

Repeat Example 2.1 using the approximate equivalent model for the silicon semiconductor diode.

Solution

The load line is redrawn as shown in Fig. 2.6 with the same intersections as defined in Example 2.1. The characteristics of the approximate equivalent circuit for the diode have also been sketched on the same graph. The resulting Q-point:

$$V_{D_Q} = \textbf{0.7 V}$$

$$I_{D_Q} = \textbf{9.25 mA}$$

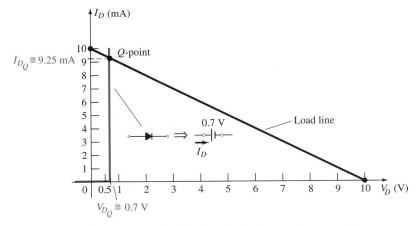

Figure 2.6 Solution to Example 2.1 using the diode approximate model.

The results obtained in Example 2.3 are quite interesting. The level of I_{D_Q} is exactly the same as obtained in Example 2.1 using a characteristic curve that is a great deal easier to draw than that appearing in Fig. 2.4. The level of $V_D = 0.7$ V versus 0.78 V from Example 2.1 is of a different magnitude to the hundredths place, but they are certainly in the same neighborhood if we compare their magnitudes to the magnitudes of the other voltages of the network.

<table>
<tr><td>EXAMPLE 2.4</td><td>

Repeat Example 2.2 using the approximate equivalent model for the silicon semi-conductor diode.

Solution

The load line is redrawn as shown in Fig. 2.7 with the same intersections defined in Example 2.2. The characteristics of the approximate equivalent circuit for the diode have also been sketched on the same graph. The resulting Q-point:

$$V_{D_Q} = \textbf{0.7 V}$$

$$I_{D_Q} = \textbf{4.6 mA}$$

</td></tr>
</table>

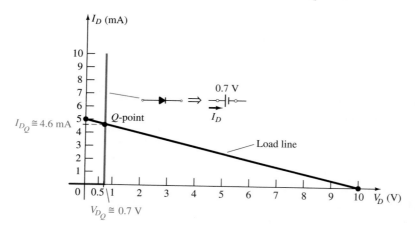

Figure 2.7 Solution to Example 2.2 using the diode approximate model.

In Example 2.4 the results obtained for both V_{D_Q} and I_{D_Q} are the same as those obtained using the full characteristics in Example 2.2. The examples above have demonstrated that the current and voltage levels obtained using the approximate model have been very close to those obtained using the full characteristics. It suggests, as will be applied in the sections to follow, that the use of appropriate approximations can result in solutions that are very close to the actual response with a reduced level of concern about properly reproducing the characteristics and choosing a large-enough scale. In the next example we go a step further and substitute the ideal model. The results will reveal the conditions that must be satisfied to apply the ideal equivalent properly.

EXAMPLE 2.5

Repeat Example 2.1 using the ideal diode model.

Solution

As shown in Fig. 2.8 the load line continues to be the same, but the ideal characteristics now intersect the load line on the vertical axis. The Q-point is therefore defined by

$$V_{D_Q} = \textbf{0 V}$$

$$I_{D_Q} = \textbf{10 mA}$$

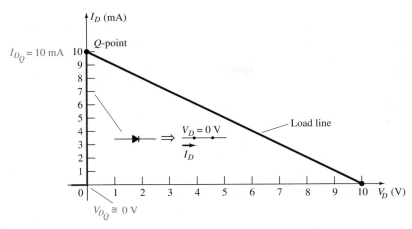

Figure 2.8 Solution to Example 2.1 using the ideal diode model.

The results are sufficiently different from the solutions of Example 2.1 to cause some concern about their accuracy. Certainly, they do provide some indication of the level of voltage and current to be expected relative to the other voltage levels of the network, but the additional effort of simply including the 0.7-V offset suggests that the approach of Example 2.3 is more appropriate.

Use of the ideal diode model therefore should be reserved for those occasions when the role of a diode is more important than voltage levels that differ by tenths of a volt and in those situations where the applied voltages are considerably larger than the threshold voltage V_T. In the next few sections the approximate model will be employed exclusively since the voltage levels obtained will be sensitive to variations that approach V_T. In later sections the ideal model will be employed more frequently since the applied voltages will frequently be quite a bit larger than V_T and the authors want to ensure that the role of the diode is correctly and clearly understood.

Mathcad

Mathcad will now be used to find the solution of the two simultaneous equations defined by the diode and network of Fig. 2.9.

The diode's characteristics are defined by

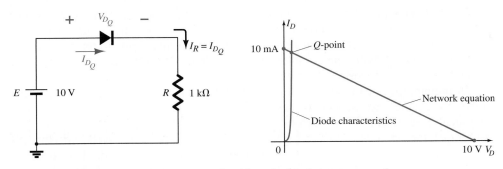

Figure. 2.9 Finding the operating point defined by a diode's characteristics and the network.

$$I_D = I_s(e^{kV_D/T_K} - 1) = 50 \text{ mA}(e^{19.33V_D} - 1)$$

Applying Kirchhoff's voltage law around the closed loop, we have

$$E - V_D - V_R = 0 \Rightarrow E - V_D = I_R R \Rightarrow E - V_D = I_D R$$

and solving for the diode current will result in

$$I_D = \frac{E - V_D}{R} = \frac{E}{R} - \frac{V_D}{R}$$

$$I_D = \frac{10 \text{ V}}{1 \text{ k}} - \frac{V_D}{1 \text{ k}} = 10 \text{ mA} - 110^3 V_D$$

Since we now have two equations and two unknowns (I_D and V_D), we can solve for each unknown using Mathcad as follows:

When using Mathcad to solve simultaneous equations, you must **guess** a value for each quantity to give the computer some direction in its **iterative process.** In other words, the computer will test solutions and work its way toward the actual solution by responding to the results obtained.

For our situation the initial guesses for **ID** and **VD** were 9 mA and 0.7 V as shown on the top of Fig. 2.10. Then, following the word **Given** (required), the two equations are entered using the equal sign obtained from **Ctrl =**. Next, type **Find(ID, VD)** to tell the computer what needs to be determined. Once the equal sign is entered, the results will appear as shown in Fig. 2.10. As indicated in Fig. 2.10 and in Fig. 2.9, the results are $I_D = 9.372$ mA and $V_D = 0.628$ V.

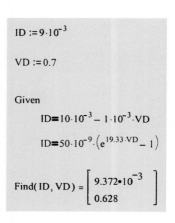

ID := $9 \cdot 10^{-3}$

VD := 0.7

Given

ID = $10 \cdot 10^{-3} - 1 \cdot 10^{-3} \cdot$ VD

ID = $50 \cdot 10^{-9} \cdot \left(e^{19.33 \cdot \text{VD}} - 1 \right)$

Find(ID, VD) = $\begin{bmatrix} 9.372 \cdot 10^{-3} \\ 0.628 \end{bmatrix}$

Figure 2.10 Defining the procedure for using Mathcad to find the solution for two simultaneous equations.

2.3 DIODE APPROXIMATIONS

In Section 2.2 we revealed that the results obtained using the approximate piecewise-linear equivalent model were quite close, if not equal, to the response obtained using the full characteristics. In fact, if one considers all the variations possible due to tolerances, temperature, and so on, one could certainly consider one solution to be "as accurate" as the other. Since the use of the approximate model normally results in a reduced expenditure of time and effort to obtain the desired results, it is the approach that will be employed in this book unless otherwise specified. Recall the following:

The primary purpose of this book is to develop a general knowledge of the behavior, capabilities, and possible areas of application of a device in a manner that will minimize the need for extensive mathematical developments.

The complete piecewise-linear equivalent model introduced in Chapter 1 was not employed in the load-line analysis because r_{av} is typically much less than the other series elements of the network. If r_{av} should be close in magnitude to the other series elements of the network, the complete equivalent model can be applied in much the same manner as described in Section 2.2.

In preparation for the analysis to follow, Table 2.1 was developed to review the important characteristics, models, and conditions of application for the approximate and ideal diode models. Although the silicon diode is used almost exclusively due to its temperature characteristics, the germanium diode is still employed and is therefore included in Table 2.1. As with the silicon diode, a germanium diode is approximated by an open-circuit equivalent for voltages less than V_T. It will enter the "on" state when $V_D \geq V_T = 0.3$ V.

TABLE 2.1 Approximate and Ideal Semiconductor Diode Models

Silicon

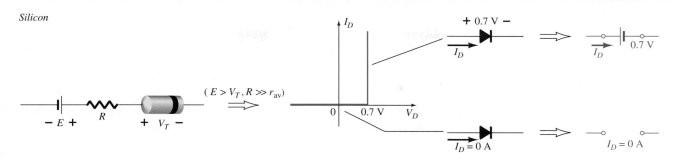

Germanium

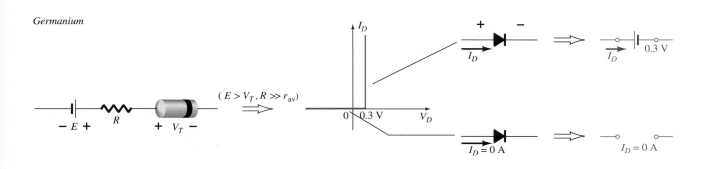

Ideal model (Si or Ge)

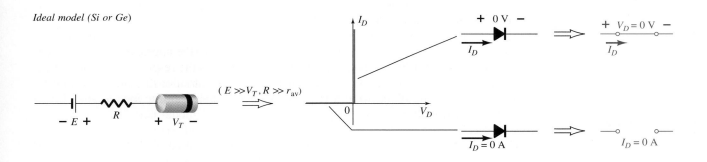

Keep in mind that the 0.7 and 0.3 V in the equivalent circuits are not *independent* sources of energy but are there simply to remind us that there is a "price to pay" to turn on a diode. An isolated diode on a laboratory table will not indicate 0.7 or 0.3 V if a voltmeter is placed across its terminals. The supplies specify the voltage drop across each when the device is "on" and specify that the diode voltage must be at least the indicated level before conduction can be established.

In the next few sections we demonstrate the impact of the models of Table 2.1 on the analysis of diode configurations. For those situations where the approximate equivalent circuit will be employed, the diode symbol will appear as shown in Fig. 2.11a for the silicon and germanium diodes. If conditions are such that the ideal diode model can be employed, the diode symbol will appear as shown in Fig. 2.11b.

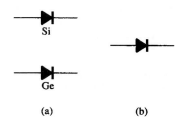

Figure 2.11 (a) Approximate model notation; (b) ideal diode notation.

2.4 SERIES DIODE CONFIGURATIONS WITH DC INPUTS

In this section the approximate model is utilized to investigate a number of series diode configurations with dc inputs. The content will establish a foundation in diode analysis that will carry over into the sections and chapters to follow. The procedure described can, in fact, be applied to networks with any number of diodes in a variety of configurations.

For each configuration the state of each diode must first be determined. Which diodes are "on" and which are "off"? Once determined, the appropriate equivalent as defined in Section 2.3 can be substituted and the remaining parameters of the network determined.

In general, a diode is in the "on" state if the current established by the applied sources is such that its direction matches that of the arrow in the diode symbol, and $V_D \geq 0.7$ V for silicon and $V_D \geq 0.3$ V for germanium.

For each configuration, *mentally* replace the diodes with resistive elements and note the resulting current direction as established by the applied voltages ("pressure"). If the resulting direction is a "match" with the arrow in the diode symbol, conduction through the diode will occur and the device is in the "on" state. The description above is, of course, contingent on the supply having a voltage greater than the "turn-on" voltage (V_T) of each diode.

If a diode is in the "on" state, one can either place a 0.7-V drop across the element, or the network can be redrawn with the V_T equivalent circuit as defined in Table 2.1. In time the preference will probably simply be to include the 0.7-V drop across each "on" diode and draw a line through each diode in the "off" or open state. Initially, however, the substitution method will be utilized to ensure that the proper voltage and current levels are determined.

The series circuit of Fig. 2.12 described in some detail in Section 2.2 will be used to demonstrate the approach described in the paragraphs above. The state of the diode is first determined by mentally replacing the diode with a resistive element as shown in Fig. 2.13. The resulting direction of I is a match with the arrow in the diode symbol, and since $E > V_T$ the diode is in the "on" state. The network is then redrawn as shown in Fig. 2.14 with the appropriate equivalent model for the forward-biased silicon diode. Note for future reference that the polarity of V_D is the same as would result if in fact the diode were a resistive element. The resulting voltage and current levels are the following:

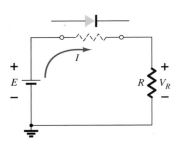

Figure 2.12 Series diode configuration.

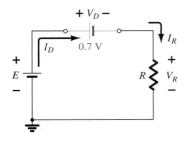

Figure 2.13 Determining the state of the diode of Fig. 2.12.

$$\boxed{V_D = V_T} \tag{2.4}$$

$$\boxed{V_R = E - V_T} \tag{2.5}$$

$$\boxed{I_D = I_R = \frac{V_R}{R}} \tag{2.6}$$

Figure 2.14 Substituting the equivalent model for the "on" diode of Fig. 2.12

In Fig. 2.15 the diode of Fig. 2.12 has been reversed. Mentally replacing the diode with a resistive element as shown in Fig. 2.16 will reveal that the resulting current direction does not match the arrow in the diode symbol. The diode is in the "off" state, resulting in the equivalent circuit of Fig. 2.17. Due to the open circuit, the diode current is 0 A and the voltage across the resistor R is the following:

$$V_R = I_R R = I_D R = (0 \text{ A})R = \mathbf{0 \text{ V}}$$

The fact that $V_R = 0$ V will establish E volts across the open circuit as defined by

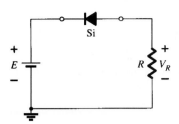

Figure 2.15 Reversing the diode of Fig. 2.12.

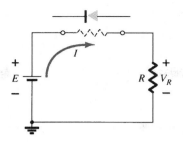

Figure 2.16 Determining the state of the diode of Fig. 2.15.

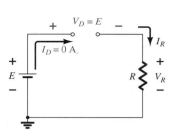

Figure 2.17 Substituting the equivalent model for the "off" diode of Figure 2.15.

Kirchhoff's voltage law. Always keep in mind that under any circumstances—dc, ac instantaneous values, pulses, and so on—Kirchhoff's voltage law must be satisfied!

For the series diode configuration of Fig. 2.18, determine V_D, V_R, and I_D.

EXAMPLE 2.6

Solution

Since the applied voltage establishes a current in the clockwise direction to match the arrow of the symbol and the diode is in the "on" state,

$$V_D = \mathbf{0.7\ V}$$

$$V_R = E - V_D = 8\ V - 0.7\ V = \mathbf{7.3\ V}$$

$$I_D = I_R = \frac{V_R}{R} = \frac{7.3\ V}{2.2\ k\Omega} \cong \mathbf{3.32\ mA}$$

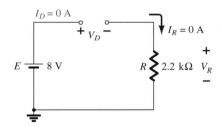

Figure 2.18 Circuit for Example 2.6.

Repeat Example 2.6 with the diode reversed.

EXAMPLE 2.7

Solution

Removing the diode, we find that the direction of I is opposite to the arrow in the diode symbol and the diode equivalent is the open circuit no matter which model is employed. The result is the network of Fig. 2.19, where $I_D = \mathbf{0\ A}$ due to the open circuit. Since $V_R = I_R R$, $V_R = (0)R = 0\ V$. Applying Kirchhoff's voltage law around the closed loop yields

$$E - V_D - V_R = 0$$

and

$$V_D = E - V_R = E - 0 = E = \mathbf{8\ V}$$

Figure 2.19 Determining the unknown quantities for Example 2.7.

In particular, note in Example 2.7 the high voltage across the diode even though it is an "off" state. The current is zero, but the voltage is significant. For review purposes, keep the following in mind for the analysis to follow:

1. An open circuit can have any voltage across its terminals, but the current is always 0 A.

2. A short circuit has a 0-V drop across its terminals, but the current is limited only by the surrounding network.

In the next example the notation of Fig. 2.20 will be employed for the applied voltage. It is a common industry notation and one with which the reader should become very familiar. Such notation and other defined voltage levels are treated further in Chapter 4.

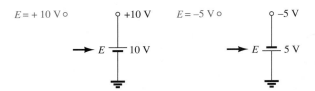

Figure 2.20   Source notation.

EXAMPLE 2.8

For the series diode configuration of Fig. 2.21, determine V_D, V_R, and I_D.

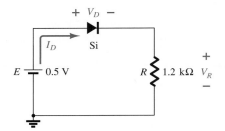

Figure 2.21   Series diode circuit for Example 2.8.

Solution

Although the "pressure" establishes a current with the same direction as the arrow symbol, the level of applied voltage is insufficient to turn the silicon diode "on." The point of operation on the characteristics is shown in Fig. 2.22, establishing the open-circuit equivalent as the appropriate approximation. The resulting voltage and current levels are therefore the following:

$$I_D = \mathbf{0\ A}$$

$$V_R = I_R R = I_D R = (0\ A)1.2\ k\Omega = \mathbf{0\ V}$$

and

$$V_D = E = \mathbf{0.5\ V}$$

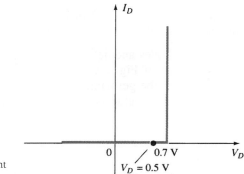

Figure 2.22   Operating point with $E = 0.5$ V.

EXAMPLE 2.9

Determine V_o and I_D for the series circuit of Fig. 2.23.

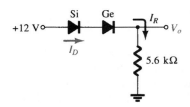

Figure 2.23   Circuit for Example 2.9.

Solution

An attack similar to that applied in Example 2.6 will reveal that the resulting current has the same direction as the arrowheads of the symbols of both diodes, and the network of Fig. 2.24 results because $E = 12$ V $> (0.7$ V $+ 0.3$ V$) = 1$ V. Note the redrawn supply of 12 V and the polarity of V_o across the 5.6-kΩ resistor. The resulting voltage

$$V_o = E - V_{T_1} - V_{T_2} = 12 \text{ V} - 0.7 \text{ V} - 0.3 \text{ V} = \mathbf{11 \text{ V}}$$

and

$$I_D = I_R = \frac{V_R}{R} = \frac{V_o}{R} = \frac{11 \text{ V}}{5.6 \text{ k}\Omega} \cong \mathbf{1.96 \text{ mA}}$$

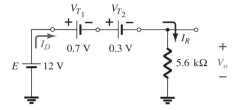

Figure 2.24 Determining the unknown quantities for Example 2.9.

Determine I_D, V_{D_2}, and V_o for the circuit of Fig. 2.25.   *EXAMPLE 2.10*

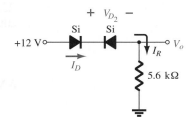

Figure 2.25 Circuit for Example 2.10.

Solution

Removing the diodes and determining the direction of the resulting current I will result in the circuit of Fig. 2.26. There is a match in current direction for the silicon diode but not for the germanium diode. The combination of a short circuit in series with an open circuit always results in an open circuit and $I_D = \mathbf{0 \text{ A}}$, as shown in Fig. 2.27.

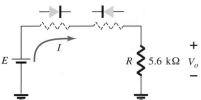

Figure 2.26 Determining the state of the diodes of Figure 2.25.

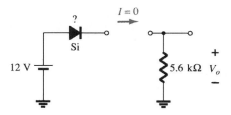

Figure 2.27 Substituting the equivalent state for the open diode.

The question remains as to what to substitute for the silicon diode. For the analysis to follow in this and succeeding chapters, simply recall for the actual practical diode that when $I_D = 0$ A, $V_D = 0$ V (and vice versa), as described for the no-bias situation in Chapter 1. The conditions described by $I_D = 0$ A and $V_{D_1} = 0$ V are indicated in Fig. 2.28.

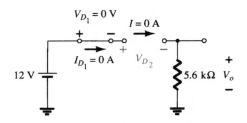

Figure 2.28 Determining the unknown quantities for the circuit of Example 2.10.

$$V_o = I_R R = I_D R = (0 \text{ A})R = 0 \text{ V}$$

and
$$V_{D_2} = V_{\text{open circuit}} = E = \mathbf{12 \text{ V}}$$

Applying Kirchhoff's voltage law in a clockwise direction gives us

$$E - V_{D_1} - V_{D_2} - V_o = 0$$

and
$$V_{D_2} = E - V_{D_1} - V_o = 12 \text{ V} - 0 - 0$$

$$= \mathbf{12 \text{ V}}$$

with
$$V_o = \mathbf{0 \text{ V}}$$

EXAMPLE 2.11

Determine I, V_1, V_2, and V_o for the series dc configuration of Fig. 2.29.

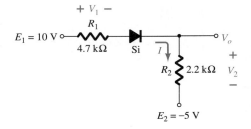

Figure 2.29 Circuit for Example 2.11.

Solution

The sources are drawn and the current direction indicated as shown in Fig. 2.30. The diode is in the "on" state and the notation appearing in Fig. 2.31 is included to indicate this state. Note that the "on" state is noted simply by the additional $V_D = 0.7$ V on the figure. This eliminates the need to redraw the network and avoids any confusion that may result from the appearance of another source. As indicated in the introduction to this section, this is probably the path and notation that one will take when a level

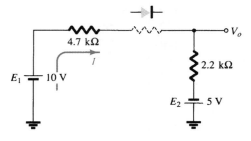

Figure 2.30 Determining the state of the diode for the network of Fig. 2.29.

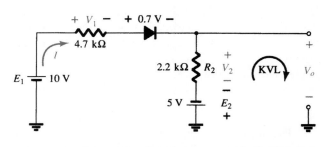

Figure 2.31 Determining the unknown quantities for the network of Fig. 2.29.

of confidence has been established in the analysis of diode configurations. In time the entire analysis will be performed simply by referring to the original network. Recall that a reverse-biased diode can simply be indicated by a line through the device.

The resulting current through the circuit is,

$$I = \frac{E_1 + E_2 - V_D}{R_1 + R_2} = \frac{10\ V + 5\ V - 0.7\ V}{4.7\ k\Omega + 2.2\ k\Omega} = \frac{14.3\ V}{6.9\ k\Omega}$$

$$\cong \mathbf{2.072\ mA}$$

and the voltages are

$$V_1 = IR_1 = (2.072\ mA)(4.7\ k\Omega) = \mathbf{9.74\ V}$$

$$V_2 = IR_2 = (2.072\ mA)(2.2\ k\Omega) = \mathbf{4.56\ V}$$

Applying Kirchhoff's voltage law to the output section in the clockwise direction will result in

$$-E_2 + V_2 - V_o = 0$$

and

$$V_o = V_2 - E_2 = 4.56\ V - 5\ V = \mathbf{-0.44\ V}$$

The minus sign indicates that V_o has a polarity opposite to that appearing in Fig. 2.29.

2.5 PARALLEL AND SERIES–PARALLEL CONFIGURATIONS

The methods applied in Section 2.4 can be extended to the analysis of parallel and series–parallel configurations. For each area of application, simply match the sequential series of steps applied to series diode configurations.

Determine V_o, I_1, I_{D_1}, and I_{D_2} for the parallel diode configuration of Fig. 2.32.

EXAMPLE 2.12

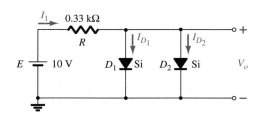

Figure 2.32 Network for Example 2.12.

Solution

For the applied voltage the "pressure" of the source is to establish a current through each diode in the same direction as shown in Fig. 2.33. Since the resulting current direction matches that of the arrow in each diode symbol and the applied voltage is greater than 0.7 V, both diodes are in the "on" state. The voltage across parallel elements is always the same and

$$V_o = \mathbf{0.7\ V}$$

The current

$$I_1 = \frac{V_R}{R} = \frac{E - V_D}{R} = \frac{10\ V - 0.7\ V}{0.33\ k\Omega} = \mathbf{28.18\ mA}$$

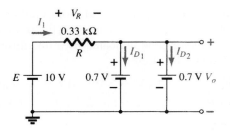

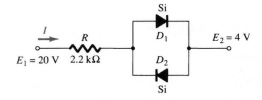

Figure 2.33 Determining the unknown quantities for the network of Example 2.12.

Assuming diodes of similar characteristics, we have

$$I_{D_1} = I_{D_2} = \frac{I_1}{2} = \frac{28.18 \text{ mA}}{2} = \textbf{14.09 mA}$$

Example 2.12 demonstrated one reason for placing diodes in parallel. If the current rating of the diodes of Fig. 2.32 is only 20 mA, a current of 28.18 mA would damage the device if it appeared alone in Fig. 2.32. By placing two in parallel, the current is limited to a safe value of 14.09 mA with the same terminal voltage.

EXAMPLE 2.13

Determine the current *I* for the network of Fig. 2.34.

Figure 2.34 Network for Example 2.13.

Solution

Redrawing the network as shown in Fig. 2.35 reveals that the resulting current direction is such as to turn on diode D_1 and turn off diode D_2. The resulting current *I* is then

$$I = \frac{E_1 - E_2 - V_D}{R} = \frac{20 \text{ V} - 4 \text{ V} - 0.7 \text{ V}}{2.2 \text{ k}\Omega} \cong \textbf{6.95 mA}$$

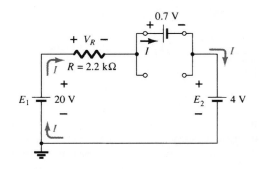

Figure 2.35 Determining the unknown quantities for the network of Example 2.13.

EXAMPLE 2.14

Determine the voltage V_o for the network of Fig. 2.36.

Solution

Initially, it would appear that the applied voltage will turn both diodes "on." However, if both were "on," the 0.7-V drop across the silicon diode would not match the

0.3 V across the germanium diode as required by the fact that the voltage across parallel elements must be the same. The resulting action can be explained simply by realizing that when the supply is turned on it will increase from 0 to 12 V over a period of time—although probably measurable in milliseconds. At the instant during the rise that 0.3 V is established across the germanium diode it will turn "on" and maintain a level of 0.3 V. The silicon diode will never have the opportunity to capture its required 0.7 V and therefore remains in its open-circuit state as shown in Fig. 2.37. The result:

$$V_o = 12 \text{ V} - 0.3 \text{ V} = \textbf{11.7 V}$$

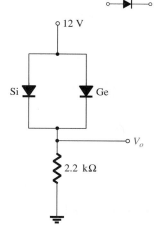

Figure 2.36 Network for Example 2.14.

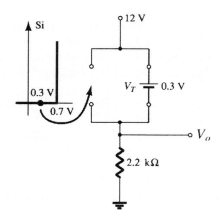

Figure 2.37 Determining V_o for the network of Fig. 2.36.

Determine the currents I_1, I_2, and I_{D_2} for the network of Fig. 2.38.

EXAMPLE 2.15

Solution

The applied voltage (pressure) is such as to turn both diodes on, as noted by the resulting current directions in the network of Fig. 2.39. Note the use of the abbreviated notation for "on" diodes and that the solution is obtained through an application of techniques applied to dc series–parallel networks.

$$I_1 = \frac{V_{T_2}}{R_1} = \frac{0.7 \text{ V}}{3.3 \text{ k}\Omega} = \textbf{0.212 mA}$$

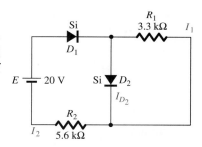

Figure 2.38 Network for Example 2.15.

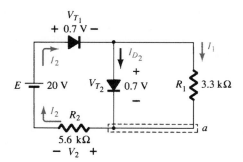

Figure 2.39 Determining the unknown quantities for Example 2.15.

Applying Kirchhoff's voltage law around the indicated loop in the clockwise direction yields

$$-V_2 + E - V_{T_1} - V_{T_2} = 0$$

and $V_2 = E - V_{T_1} - V_{T_2} = 20 \text{ V} - 0.7 \text{ V} - 0.7 \text{ V} = 18.6 \text{ V}$

with

$$I_2 = \frac{V_2}{R_2} = \frac{18.6 \text{ V}}{5.6 \text{ k}\Omega} = \textbf{3.32 mA}$$

At the bottom node (a),

$$I_{D_2} + I_1 = I_2$$

and

$$I_{D_2} = I_2 - I_1 = 3.32 \text{ mA} - 0.212 \text{ mA} = \textbf{3.108 mA}$$

2.6 AND/OR GATES

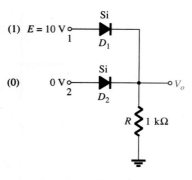

(1) $E = 10$ V $\quad$ Si $\quad D_1$

(0) $\quad$ 0 V $\quad$ Si $\quad D_2 \quad \circ V_o$

$R \lessgtr 1$ kΩ

Figure 2.40 Positive logic OR gate.

The tools of analysis are now at our disposal, and the opportunity to investigate a computer configuration is one that will demonstrate the range of applications of this relatively simple device. Our analysis will be limited to determining the voltage levels and will not include a detailed discussion of Boolean algebra or positive and negative logic.

The network to be analyzed in Example 2.16 is an OR gate for positive logic. That is, the 10-V level of Fig. 2.40 is assigned a "1" for Boolean algebra while the 0-V input is assigned a "0." An OR gate is such that the output voltage level will be a 1 if either *or* both inputs is a 1. The output is a 0 if both inputs are at the 0 level.

The analysis of AND/OR gates is made measurably easier by using the approximate equivalent for a diode rather than the ideal because we can stipulate that the voltage across the diode must be 0.7 V positive for the silicon diode (0.3 V for Ge) to switch to the "on" state.

In general, the best approach is simply to establish a "gut" feeling for the state of the diodes by noting the direction and the "pressure" established by the applied potentials. The analysis will then verify or negate your initial assumptions.

EXAMPLE 2.16

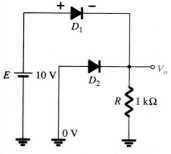

$+ \blacktriangleright -$
D_1

$E \overset{+}{=} 10$ V $\quad D_2 \quad \circ V_o$

$R \lessgtr 1$ kΩ

0 V

Figure 2.41 Redrawn network of Fig. 2.40.

Determine V_o for the network of Fig. 2.40.

Solution

First note that there is only one applied potential; 10 V at terminal 1. Terminal 2 with a 0-V input is essentially at ground potential, as shown in the redrawn network of Fig. 2.41. Figure 2.41 "suggests" that D_1 is probably in the "on" state due to the applied 10 V while D_2 with its "positive" side at 0 V is probably "off." Assuming these states will result in the configuration of Fig. 2.42.

The next step is simply to check that there is no contradiction to our assumptions. That is, note that the polarity across D_1 is such as to turn it on and the polarity across D_2 is such as to turn it off. For D_1 the "on" state establishes V_o at $V_o = E - V_D = 10 \text{ V} - 0.7 \text{ V} = \textbf{9.3 V.}$ With 9.3 V at the cathode $(-)$ side of D_2 and 0 V at the an-

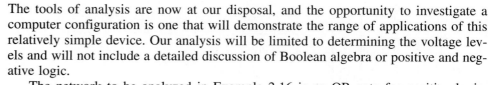

V_D
$+ \; \vert\vert \; -$
0.7 V

I

$E \overset{+}{=} 10$ V $\qquad V_o = E - V_D = V_R = IR$

$R \lessgtr 1$ kΩ

I

Figure 2.42 Assumed diode states for Fig. 2.40.

ode $(+)$ side, D_2 is definitely in the "off" state. The current direction and the resulting continuous path for conduction further confirm our assumption that D_1 is conducting. Our assumptions seem confirmed by the resulting voltages and current, and our initial analysis can be assumed to be correct. The output voltage level is not 10 V as defined for an input of 1, but the 9.3 V is sufficiently large to be considered a 1 level. The output is therefore at a 1 level with only one input, which suggests that the gate is an OR gate. An analysis of the same network with two 10-V inputs will result in both diodes being in the "on" state and an output of 9.3 V. A 0-V input at both inputs will not provide the 0.7 V required to turn the diodes on, and the output will be a 0 due to the 0-V output level. For the network of Fig. 2.42 the current level is determined by

$$I = \frac{E - V_D}{R} = \frac{10 \text{ V} - 0.7 \text{ V}}{1 \text{ k}\Omega} = \textbf{9.3 mA}$$

Determine the output level for the positive logic AND gate of Fig. 2.43.

EXAMPLE 2.17

Solution

Note in this case that an independent source appears in the grounded leg of the network. For reasons soon to become obvious it is chosen at the same level as the input logic level. The network is redrawn in Fig. 2.44 with our initial assumptions regarding the state of the diodes. With 10 V at the cathode side of D_1 it is assumed that D_1 is in the "off" state even though there is a 10-V source connected to the anode of D_1 through the resistor. However, recall that we mentioned in the introduction to this section that the use of the approximate model will be an aid to the analysis. For D_1, where will the 0.7 V come from if the input and source voltages are at the same level and creating opposing "pressures"? D_2 is assumed to be in the "on" state due to the low voltage at the cathode side and the availability of the 10-V source through the 1-kΩ resistor.

For the network of Fig. 2.44 the voltage at V_o is 0.7 V due to the forward-biased diode D_2. With 0.7 V at the anode of D_1 and 10 V at the cathode, D_1 is definitely in the "off" state. The current I will have the direction indicated in Fig. 2.44 and a magnitude equal to

$$I = \frac{E - V_D}{R} = \frac{10 \text{ V} - 0.7 \text{ V}}{1 \text{ k}\Omega} = \textbf{9.3 mA}$$

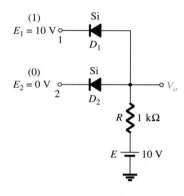

Figure 2.43 Positive logic AND gate.

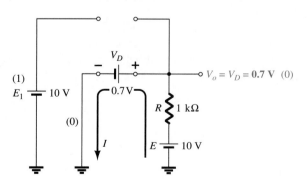

Figure 2.44 Substituting the assumed states for the diodes of Fig. 2.43.

The state of the diodes is therefore confirmed and our earlier analysis was correct. Although not 0 V as earlier defined for the 0 level, the output voltage is sufficiently small to be considered a 0 level. For the AND gate, therefore, a single input will result in a 0-level output. The remaining states of the diodes for the possibilities of two inputs and no inputs will be examined in the problems at the end of the chapter.

2.7 SINUSOIDAL INPUTS; HALF-WAVE RECTIFICATION

The diode analysis will now be expanded to include time-varying functions such as the sinusoidal waveform and the square wave. There is no question that the degree of difficulty will increase, but once a few fundamental maneuvers are understood, the analysis will be fairly direct and follow a common thread.

The simplest of networks to examine with a time-varying signal appears in Fig. 2.45. For the moment we will use the ideal model (note the absence of the Si or Ge label to denote ideal diode) to ensure that the approach is not clouded by additional mathematical complexity.

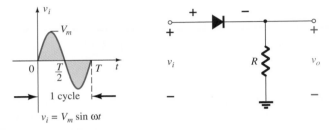

Figure 2.45 Half-wave rectifier.

Over one full cycle, defined by the period T of Fig. 2.45, the average value (the algebraic sum of the areas above and below the axis) is zero. The circuit of Fig. 2.45, called a *half-wave rectifier,* will generate a waveform v_o that will have an average value of particular use in the ac-to-dc conversion process. When employed in the rectification process, a diode is typically referred to as a *rectifier*. Its power and current ratings are typically much higher than those of diodes employed in other applications, such as computers and communication systems.

During the interval $t = 0 \rightarrow T/2$ in Fig. 2.45 the polarity of the applied voltage v_i is such as to establish "pressure" in the direction indicated and turn on the diode with the polarity appearing above the diode. Substituting the short-circuit equivalence for the ideal diode will result in the equivalent circuit of Fig. 2.46, where it is fairly obvious that the output signal is an exact replica of the applied signal. The two terminals defining the output voltage are connected directly to the applied signal via the short-circuit equivalence of the diode.

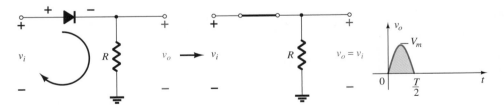

Figure 2.46 Conduction region $(0 \rightarrow T/2)$.

For the period $T/2 \rightarrow T$, the polarity of the input v_i is as shown in Fig. 2.47 and the resulting polarity across the ideal diode produces an "off" state with an open-circuit equivalent. The result is the absence of a path for charge to flow and $v_o = iR = (0)R = 0$ V for the period $T/2 \rightarrow T$. The input v_i and the output v_o were sketched together in Fig. 2.48 for comparison purposes. The output signal v_o now has a net positive area above the axis over a full period and an average value determined by

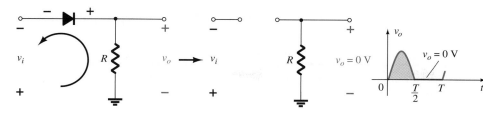

Figure 2.47 Nonconduction region ($T/2 \rightarrow T$).

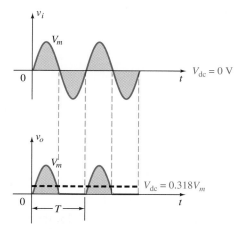

Figure 2.48 Half-wave rectified signal.

$$\boxed{V_{dc} = 0.318V_m} \quad \text{half-wave} \qquad (2.7)$$

The process of removing one-half the input signal to establish a dc level is aptly called *half-wave rectification*.

The effect of using a silicon diode with $V_T = 0.7$ V is demonstrated in Fig. 2.49 for the forward-bias region. The applied signal must now be at least 0.7 V before the diode can turn "on." For levels of v_i less than 0.7 V, the diode is still in an open-circuit state and $v_o = 0$ V as shown in the same figure. When conducting, the difference between v_o and v_i is a fixed level of $V_T = 0.7$ V and $v_o = v_i - V_T$, as shown in the figure. The net effect is a reduction in area above the axis, which naturally reduces the resulting dc voltage level. For situations where $V_m \gg V_T$, Eq. 2.8 can be applied to determine the average value with a relatively high level of accuracy.

$$\boxed{V_{dc} \cong 0.318(V_m - V_T)} \qquad (2.8)$$

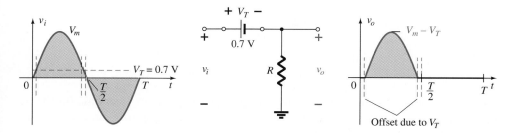

Figure 2.49 Effect of V_T on half-wave rectified signal.

In fact, if V_m is sufficiently greater than V_T, Eq. 2.7 is often applied as a first approximation for V_{dc}.

EXAMPLE 2.18

(a) Sketch the output v_o and determine the dc level of the output for the network of Fig. 2.50.
(b) Repeat part (a) if the ideal diode is replaced by a silicon diode.
(c) Repeat parts (a) and (b) if V_m is increased to 200 V and compare solutions using Eqs. (2.7) and (2.8).

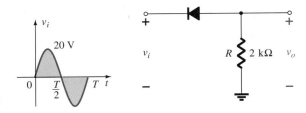

Figure 2.50 Network for Example 2.18.

Solution

(a) In this situation the diode will conduct during the negative part of the input as shown in Fig. 2.51, and v_o will appear as shown in the same figure. For the full period, the dc level is

$$V_{dc} = -0.318V_m = -0.318(20 \text{ V}) = \mathbf{-6.36 \text{ V}}$$

The negative sign indicates that the polarity of the output is opposite to the defined polarity of Fig. 2.50.

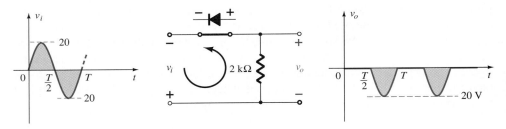

Figure 2.51 Resulting v_o for the circuit of Example 2.18.

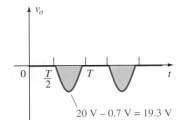

Figure 2.52 Effect of V_T on output of Fig. 2.51.

(b) Using a silicon diode, the output has the appearance of Fig. 2.52 and

$$V_{dc} \cong -0.318(V_m - 0.7 \text{ V}) = -0.318(19.3 \text{ V}) \cong \mathbf{-6.14 \text{ V}}$$

The resulting drop in dc level is 0.22 V or about 3.5%.

(c) Eq. (2.7): $V_{dc} = -0.318V_m = -0.318(200 \text{ V}) = \mathbf{-63.6 \text{ V}}$

Eq. (2.8): $V_{dc} = -0.318(V_m - V_T) = -0.318(200 \text{ V} - 0.7 \text{ V})$

$$= -(0.318)(199.3 \text{ V}) = \mathbf{-63.38 \text{ V}}$$

which is a difference that can certainly be ignored for most applications. For part c the offset and drop in amplitude due to V_T would not be discernible on a typical oscilloscope if the full pattern is displayed.

PIV (PRV)

The peak inverse voltage (PIV) [or PRV (peak reverse voltage)] rating of the diode is of primary importance in the design of rectification systems. Recall that it is the voltage rating that must not be exceeded in the reverse-bias region or the diode will enter the Zener avalanche region. The required PIV rating for the half-wave rectifier can be determined from Fig. 2.53, which displays the reverse-biased diode of Fig. 2.45 with maximum applied voltage. Applying Kirchhoff's voltage law, it is fairly obvious that the PIV rating of the diode must equal or exceed the peak value of the applied voltage. Therefore,

$$\boxed{\text{PIV rating} \geq V_m} \quad \text{half-wave rectifier} \qquad (2.9)$$

$V\,(\text{PIV})$

$I = 0$

V_m

R

$V_o = IR = (0)R = 0\text{ V}$

Figure 2.53 Determining the required PIV rating for the half-wave rectifier.

2.8 FULL-WAVE RECTIFICATION

Bridge Network

The dc level obtained from a sinusoidal input can be improved 100% using a process called *full-wave rectification*. The most familiar network for performing such a function appears in Fig. 2.54 with its four diodes in a *bridge* configuration. During the period $t = 0$ to $T/2$ the polarity of the input is as shown in Fig. 2.55. The resulting polarities across the ideal diodes are also shown in Fig. 2.55 to reveal that D_2 and D_3 are conducting while D_1 and D_4 are in the "off" state. The net result is the configuration of Fig. 2.56, with its indicated current and polarity across R. Since the diodes are ideal the load voltage is $v_o = v_i$, as shown in the same figure.

For the negative region of the input the conducting diodes are D_1 and D_4, resulting in the configuration of Fig. 2.57. The important result is that the polarity across

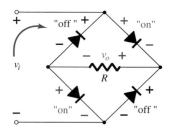

Figure 2.55 Network of Fig. 2.54 for the period $0 \rightarrow T/2$ of the input voltage v_i.

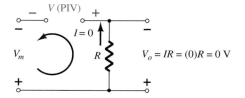

Figure 2.54 Full-wave bridge rectifier.

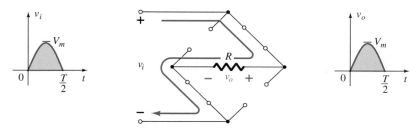

Figure 2.56 Conduction path for the positive region of v_i.

the load resistor R is the same as in Fig. 2.55, establishing a second positive pulse, as shown in Fig. 2.57. Over one full cycle the input and output voltages will appear as shown in Fig. 2.58.

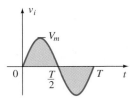 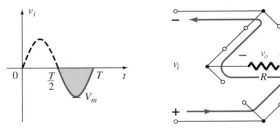 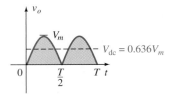

Figure 2.57 Conduction path for the negative region of v_i.

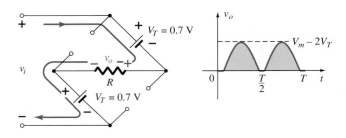

Figure 2.58 Input and output waveforms for a full-wave rectifier.

Since the area above the axis for one full cycle is now twice that obtained for a half-wave system, the dc level has also been doubled and

$$V_{dc} = 2(\text{Eq. 2.7}) = 2(0.318V_m)$$

or

$$\boxed{V_{dc} = 0.636V_m} \quad \text{full-wave} \qquad (2.10)$$

If silicon rather than ideal diodes are employed as shown in Fig. 2.59, an application of Kirchhoff's voltage law around the conduction path would result in

$$v_i - V_T - v_o - V_T = 0$$

and

$$v_o = v_i - 2V_T$$

The peak value of the output voltage v_o is therefore

$$V_{o_{max}} = V_m - 2V_T$$

For situations where $V_m \gg 2V_T$, Eq. (2.11) can be applied for the average value with a relatively high level of accuracy.

$$\boxed{V_{dc} \cong 0.636(V_m - 2V_T)} \qquad (2.11)$$

Then again, if V_m is sufficiently greater than $2V_T$, then Eq. (2.10) is often applied as a first approximation for V_{dc}.

Figure 2.59 Determining $V_{o_{max}}$ for silicon diodes in the bridge configuration.

PIV

The required PIV of each diode (ideal) can be determined from Fig. 2.60 obtained at the peak of the positive region of the input signal. For the indicated loop the maximum voltage across R is V_m and the PIV rating is defined by

$$\boxed{\text{PIV} \geqq V_m} \quad \text{full-wave bridge rectifier} \qquad (2.12)$$

Figure 2.60 Determining the required PIV for the bridge configuration.

Center-Tapped Transformer

A second popular full-wave rectifier appears in Fig. 2.61 with only two diodes but requiring a center-tapped (CT) transformer to establish the input signal across each section of the secondary of the transformer. During the positive portion of v_i applied to the primary of the transformer, the network will appear as shown in Fig. 2.62. D_1 assumes the short-circuit equivalent and D_2 the open-circuit equivalent, as determined by the secondary voltages and the resulting current directions. The output voltage appears as shown in Fig. 2.62.

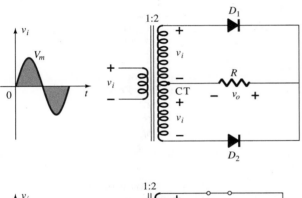

Figure 2.61 Center-tapped transformer full-wave rectifier.

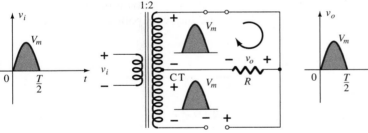

Figure 2.62 Network conditions for the positive region of v_i.

During the negative portion of the input the network appears as shown in Fig. 2.63, reversing the roles of the diodes but maintaining the same polarity for the voltage across the load resistor R. The net effect is the same output as that appearing in Fig. 2.58 with the same dc levels.

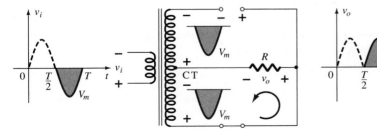

Figure 2.63 Network conditions for the negative region of v_i.

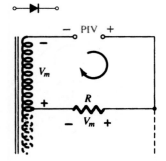

Figure 2.64 Determining the PIV level for the diodes of the CT transformer full-wave rectifier.

PIV

The network of Fig. 2.64 will help us determine the net PIV for each diode for this full-wave rectifier. Inserting the maximum voltage for the secondary voltage and V_m as established by the adjoining loop will result in

$$\text{PIV} = V_{\text{secondary}} + V_R$$
$$= V_m + V_m$$

and

$$\boxed{\text{PIV} \geqq 2V_m} \quad \text{CT transformer, full-wave rectifier} \tag{2.13}$$

EXAMPLE 2.19

Determine the output waveform for the network of Fig. 2.65 and calculate the output dc level and the required PIV of each diode.

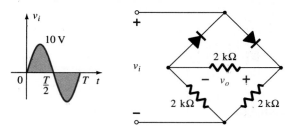

Figure 2.65 Bridge network for Example 2.19.

Solution

The network will appear as shown in Fig. 2.66 for the positive region of the input voltage. Redrawing the network will result in the configuration of Fig. 2.67, where $v_o = \frac{1}{2}v_i$ or $V_{o_{\max}} = \frac{1}{2}V_{i_{\max}} = \frac{1}{2}(10\text{ V}) = 5\text{ V}$, as shown in Fig. 2.67. For the negative part of the input the roles of the diodes will be interchanged and v_o will appear as shown in Fig. 2.68.

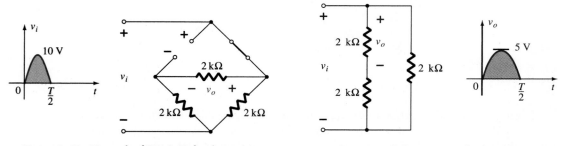

Figure 2.66 Network of Fig. 2.65 for the positive region of v_i.

Figure 2.67 Redrawn network of Fig. 2.66.

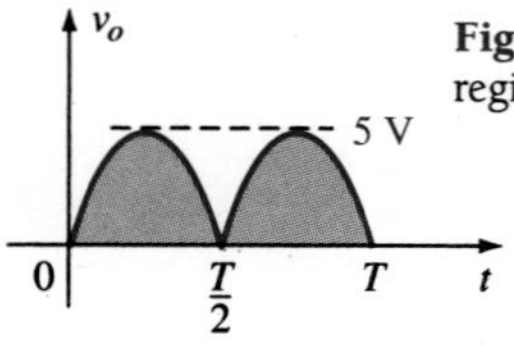

Figure 2.68 Resulting output for Example 2.19.

The effect of removing two diodes from the bridge configuration was therefore to reduce the available dc level to the following:

$$V_{\text{dc}} = 0.636(5\text{ V}) = \textbf{3.18 V}$$

or that available from a half-wave rectifier with the same input. However, the PIV as determined from Fig. 2.60 is equal to the maximum voltage across R, which is 5 V or half of that required for a half-wave rectifier with the same input.

2.9 CLIPPERS

There are a variety of diode networks called *clippers* that have the ability to "clip" off a portion of the input signal without distorting the remaining part of the alternating waveform. The half-wave rectifier of Section 2.7 is an example of the simplest form of diode clipper—one resistor and diode. Depending on the orientation of the diode, the positive or negative region of the input signal is "clipped" off.

There are two general categories of clippers: *series* and *parallel*. The series configuration is defined as one where the diode is in series with the load, while the parallel variety has the diode in a branch parallel to the load.

Series

The response of the series configuration of Fig. 2.69a to a variety of alternating waveforms is provided in Fig. 2.69b. Although first introduced as a half-wave rectifier (for sinusoidal waveforms), there are no boundaries on the type of signals that can be applied to a clipper. The addition of a dc supply such as shown in Fig. 2.70 can have a pronounced effect on the output of a clipper. Our initial discussion will be limited to ideal diodes, with the effect of V_T reserved for a concluding example.

There is no general procedure for analyzing networks such as the type in Fig. 2.70, but there are a few thoughts to keep in mind as you work toward a solution.

1. *Make a mental sketch of the response of the network based on the direction of the diode and the applied voltage levels.*

For the network of Fig. 2.70, the direction of the diode suggests that the signal v_i must be positive to turn it on. The dc supply further requires that the voltage v_i be greater than V volts to turn the diode on. The negative region of the input signal is "pressuring" the diode into the "off" state, supported further by the dc supply. In general, therefore, we can be quite sure that the diode is an open circuit ("off" state) for the negative region of the input signal.

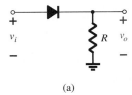

(a)

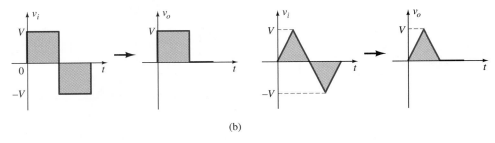

(b)

Figure 2.69   Series clipper.

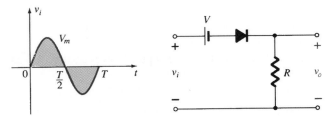

Figure 2.70   Series clipper with a dc supply.

2. *Determine the applied voltage (transition voltage) that will cause a change in state for the diode.*

For the ideal diode the transition between states will occur at the point on the characteristics where $v_d = 0$ V and $i_d = 0$ A. Applying the condition $i_d = 0$ at $v_d = 0$ to the network of Fig. 2.70 will result in the configuration of Fig. 2.71, where it is recognized that the level of v_i that will cause a transition in state is

$$\boxed{v_i = V} \tag{2.14}$$

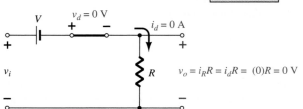

Figure 2.71 Determining the transition level for the circuit of Fig. 2.70.

$$v_o = i_R R = i_d R = (0)R = 0 \text{ V}$$

For an input voltage greater than V volts the diode is in the short-circuit state, while for input voltages less than V volts it is in the open-circuit or "off" state.

3. *Be continually aware of the defined terminals and polarity of v_o.*

When the diode is in the short-circuit state, such as shown in Fig. 2.72, the output voltage v_o can be determined by applying Kirchhoff's voltage law in the clockwise direction:

$$v_i - V - v_o = 0 \text{ (CW direction)}$$

and

$$\boxed{v_o = v_i - V} \tag{2.15}$$

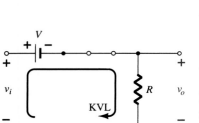

Figure 2.72 Determining v_o.

4. *It can be helpful to sketch the input signal above the output and determine the output at instantaneous values of the input.*

It is then possible that the output voltage can be sketched from the resulting data points of v_o as demonstrated in Fig. 2.73. Keep in mind that at an instantaneous value of v_i the input can be treated as a dc supply of that value and the corresponding dc value (the instantaneous value) of the output determined. For instance, at $v_i = V_m$ for the network of Fig. 2.70, the network to be analyzed appears in Fig. 2.74. For $V_m > V$ the diode is in the short-circuit state and $v_o = V_m - V$, as shown in Fig. 2.73.

At $v_i = V$ the diodes change state; at $v_i = -V_m$, $v_o = 0$ V; and the complete curve for v_o can be sketched as shown in Fig. 2.75.

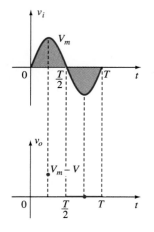

Figure 2.73 Determining levels of v_o.

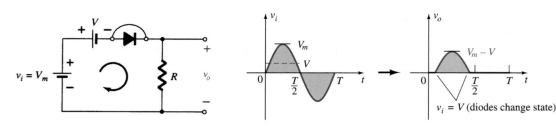

Figure 2.74 Determining v_o when $v_i = V_m$.   **Figure 2.75** Sketching v_o.

EXAMPLE 2.20

Determine the output waveform for the network of Fig. 2.76.

Solution

Past experience suggests that the diode will be in the "on" state for the positive region of v_i—especially when we note the aiding effect of $V = 5$ V. The network will

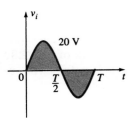

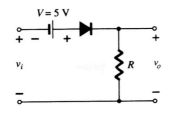

Figure 2.76 Series clipper for Example 2.20.

then appear as shown in Fig. 2.77 and $v_o = v_i + 5$ V. Substituting $i_d = 0$ at $v_d = 0$ for the transition levels, we obtain the network of Fig. 2.78 and $v_i = -5$ V.

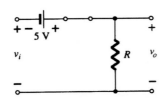

Figure 2.77 v_o with diode in the "on" state.

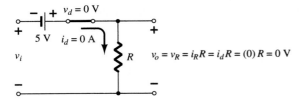

Figure 2.78 Determining the transition level for the clipper of Fig. 2.76.

For v_i more negative than -5 V the diode will enter its open-circuit state, while for voltage more positive than -5 V the diode is in the short-circuit state. The input and output voltages appear in Fig. 2.79.

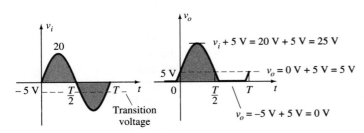

Figure 2.79 Sketching v_o for Example 2.20.

The analysis of clipper networks with square-wave inputs is actually easier to analyze than with sinusoidal inputs because only two levels have to be considered. In other words, the network can be analyzed as if it had two dc level inputs with the resulting output v_o plotted in the proper time frame.

Repeat Example 2.20 for the square-wave input of Fig. 2.80.

EXAMPLE 2.21

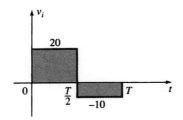

Figure 2.80 Applied signal for Example 2.21.

Solution

For $v_i = 20$ V $(0 \rightarrow T/2)$ the network of Fig. 2.81 will result. The diode is in the short-circuit state and $v_o = 20$ V $+ 5$ V $= 25$ V. For $v_i = -10$ V the network of Fig. 2.82 will result, placing the diode in the "off" state and $v_o = i_R R = (0)R = 0$ V. The resulting output voltage appears in Fig. 2.83.

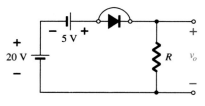

Figure 2.81 v_o at $v_i = +20$ V.

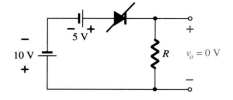

Figure 2.82 v_o at $v_i = -10$ V.

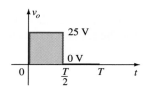

Figure 2.83 Sketching v_o for Example 2.21.

Note in Example 2.21 that the clipper not only clipped off 5 V from the total swing but raised the dc level of the signal by 5 V.

Parallel

The network of Fig. 2.84 is the simplest of parallel diode configurations with the output for the same inputs of Fig. 2.69. The analysis of parallel configurations is very similar to that applied to series configurations, as demonstrated in the next example.

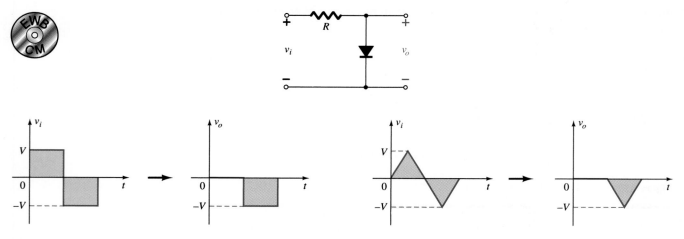

Figure 2.84 Response to a parallel clipper.

EXAMPLE 2.22 Determine v_o for the network of Fig. 2.85.

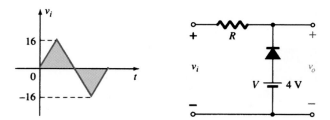

Figure 2.85 Example 2.22.

Chapter 2 Diode Applications

Solution

The polarity of the dc supply and the direction of the diode strongly suggest that the diode will be in the "on" state for the negative region of the input signal. For this region the network will appear as shown in Fig. 2.86, where the defined terminals for v_o require that $v_o = V = 4$ V.

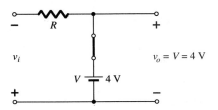

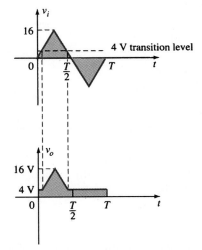

Figure 2.86 v_o for the negative region of v_i.

The transition state can be determined from Fig. 2.87, where the condition $i_d = 0$ A at $v_d = 0$ V has been imposed. The result is v_i (transition) $= V = 4$ V.

Since the dc supply is obviously "pressuring" the diode to stay in the short-circuit state, the input voltage must be greater than 4 V for the diode to be in the "off" state. Any input voltage less than 4 V will result in a short-circuited diode.

For the open-circuit state the network will appear as shown in Fig. 2.88, where $v_o = v_i$. Completing the sketch of v_o results in the waveform of Fig. 2.89.

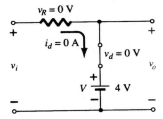

Figure 2.87 Determining the transition level for Example 2.22.

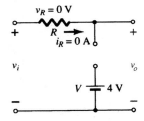

Figure 2.88 Determining v_o for the open state of the diode.

Figure 2.89 Sketching v_o for Example 2.22.

To examine the effects of V_T on the output voltage, the next example will specify a silicon diode rather than an ideal diode equivalent.

EXAMPLE 2.23   Repeat Example 2.22 using a silicon diode with $V_T = 0.7$ V.

Solution

The transition voltage can first be determined by applying the condition $i_d = 0$ A at $v_d = V_D = 0.7$ V and obtaining the network of Fig. 2.90. Applying Kirchhoff's voltage law around the output loop in the clockwise direction, we find that

$$v_i + V_T - V = 0$$

and

$$v_i = V - V_T = 4 \text{ V} - 0.7 \text{ V} = \mathbf{3.3 \text{ V}}$$

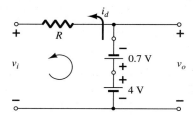

Figure 2.90 Determining the transition level for the network of Fig. 2.85.

For input voltages greater than 3.3 V, the diode will be an open circuit and $v_o = v_i$. For input voltages of less than 3.3 V, the diode will be in the "on" state and the network of Fig. 2.91 results, where

$$v_o = 4 \text{ V} - 0.7 \text{ V} = \mathbf{3.3 \text{ V}}$$

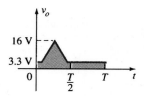

Figure 2.91 Determining v_o for the diode of Fig. 2.85 in the "on" state.

The resulting output waveform appears in Fig. 2.92. Note that the only effect of V_T was to drop the transition level to 3.3 from 4 V.

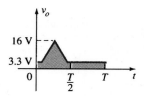

Figure 2.92 Sketching v_o for Example 2.23.

There is no question that including the effects of V_T will complicate the analysis somewhat, but once the analysis is understood with the ideal diode, the procedure, including the effects of V_T, will not be that difficult.

Summary

A variety of series and parallel clippers with the resulting output for the sinusoidal input are provided in Fig. 2.93. In particular, note the response of the last configuration, with its ability to clip off a positive and a negative section as determined by the magnitude of the dc supplies.

Simple Series Clippers (Ideal Diodes)

POSITIVE

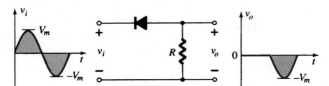

NEGATIVE

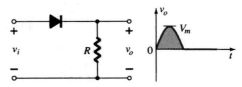

Biased Series Clippers (Ideal Diodes)

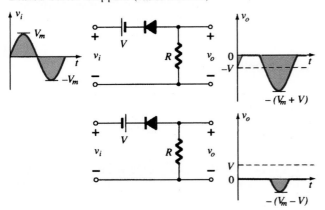

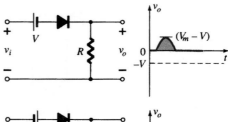

Simple Parallel Clippers (Ideal Diodes)

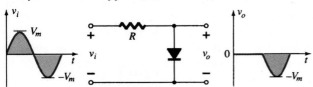

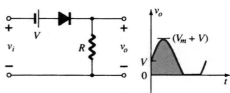

Biased Parallel Clippers (Ideal Diodes)

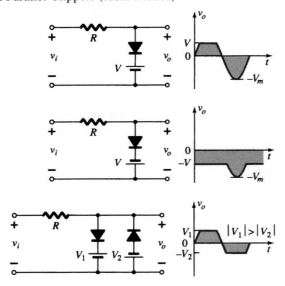

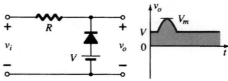

Figure 2.93 Clipping circuits.

2.10 CLAMPERS

The *clamping* network is one that will "clamp" a signal to a different dc level. The network must have a capacitor, a diode, and a resistive element, but it can also empoly an independent dc supply to introduce an additional shift. The magnitude of R and C must be chosen such that the time constant $\tau = RC$ is large enough to ensure that the voltage across the capacitor does not discharge significantly during the interval the diode is nonconducting. Throughout the analysis we will assume that for all practical purposes the capacitor will fully charge or discharge in five time constants.

The network of Fig. 2.94 will clamp the input signal to the zero level (for ideal diodes). The resistor R can be the load resistor or a parallel combination of the load resistor and a resistor designed to provide the desired level of R.

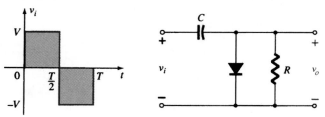

Figure 2.94 Clamper.

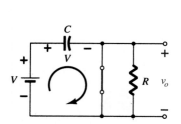

Figure 2.95 Diode "on" and the capacitor charging to V volts.

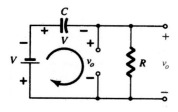

Figure 2.96 Determining v_o with the diode "off."

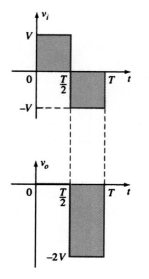

Figure 2.97 Sketching v_o for the network of Fig. 2.94.

During the interval $0 \rightarrow T/2$ the network will appear as shown in Fig. 2.95, with the diode in the "on" state effectively "shorting out" the effect of the resistor R. The resulting RC time constant is so small (R determined by the inherent resistance of the network) that the capacitor will charge to V volts very quickly. During this interval the output voltage is directly across the short circuit and $v_o = 0$ V.

When the input switches to the $-V$ state, the network will appear as shown in Fig. 2.96, with the open-circuit equivalent for the diode determined by the applied signal and stored voltage across the capacitor—both "pressuring" current through the diode from cathode to anode. Now that R is back in the network the time constant determined by the RC product is sufficiently large to establish a discharge period 5τ much greater than the period $T/2 \rightarrow T$, and it can be assumed on an approximate basis that the capacitor holds onto all its charge and, therefore, voltage (since $V = Q/C$) during this period.

Since v_o is in parallel with the diode and resistor, it can also be drawn in the alternative position shown in Fig. 2.96. Applying Kirchhoff's voltage law around the input loop will result in

$$-V - V - v_o = 0$$

and

$$v_o = -2V$$

The negative sign resulting from the fact that the polarity of $2V$ is opposite to the polarity defined for v_o. The resulting output waveform appears in Fig. 2.97 with the input signal. The output signal is clamped to 0 V for the interval 0 to $T/2$ but maintains the same total swing ($2V$) as the input.

For a clamping network:

The total swing of the output is equal to the total swing of the input signal.

This fact is an excellent checking tool for the result obtained.

In general, the following steps may be helpful when analyzing clamping networks:

1. *Start the analysis of clamping networks by considering that part of the input signal that will forward bias the diode.*

The statement above may require skipping an interval of the input signal (as demonstrated in an example to follow), but the analysis will not be extended by an unnecessary measure of investigation.

2. *During the period that the diode is in the "on" state, assume that the capacitor will charge up instantaneously to a voltage level determined by the network.*

3. *Assume that during the period when the diode is in the "off" state the capacitor will hold on to its established voltage level.*

4. *Throughout the analysis maintain a continual awareness of the location and reference polarity for v_o to ensure that the proper levels for v_o are obtained.*

5. *Keep in mind the general rule that the total swing of the total output must match the swing of the input signal.*

Determine v_o for the network of Fig. 2.98 for the input indicated.

EXAMPLE 2.24

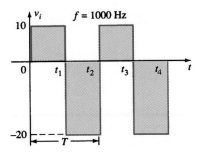

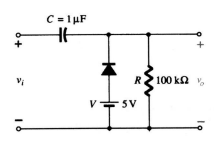

Figure 2.98 Applied signal and network for Example 2.24.

Solution

Note that the frequency is 1000 Hz, resulting in a period of 1 ms and an interval of 0.5 ms between levels. The analysis will begin with the period $t_1 \rightarrow t_2$ of the input signal since the diode is in its short-circuit state as recommended by comment 1. For this interval the network will appear as shown in Fig. 2.99. The output is across R, but it is also directly across the 5-V battery if you follow the direct connection between the defined terminals for v_o and the battery terminals. The result is $v_o = 5$ V for this interval. Applying Kirchhoff's voltage law around the input loop will result in

$$-20 \text{ V} + V_C - 5 \text{ V} = 0$$

and
$$V_C = 25 \text{ V}$$

The capacitor will therefore charge up to 25 V, as stated in comment 2. In this case the resistor R is not shorted out by the diode but a Thévenin equivalent circuit of that portion of the network which includes the battery and the resistor will result in $R_{\text{Th}} = 0 \ \Omega$ with $E_{\text{Th}} = V = 5$ V. For the period $t_2 \rightarrow t_3$ the network will appear as shown in Fig. 2.100.

The open-circuit equivalent for the diode will remove the 5-V battery from having any effect on v_o, and applying Kirchhoff's voltage law around the outside loop of the network will result in

$$+10 \text{ V} + 25 \text{ V} - v_o = 0$$

and
$$v_o = 35 \text{ V}$$

The time constant of the discharging network of Fig. 2.100 is determined by the product RC and has the magnitude

$$\tau = RC = (100 \text{ k}\Omega)(0.1 \ \mu\text{F}) = 0.01 \text{ s} = 10 \text{ ms}$$

The total discharge time is therefore $5\tau = 5(10 \text{ ms}) = 50 \text{ ms}$.

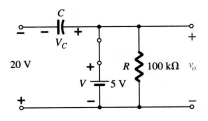

Figure 2.99 Determining v_o and V_C with the diode in the "on" state.

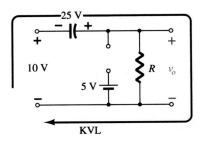

Figure 2.100 Determining v_o with the diode in the "off" state.

Since the interval $t_2 \rightarrow t_3$ will only last for 0.5 ms, it is certainly a good approximation that the capacitor will hold its voltage during the discharge period between pulses of the input signal. The resulting output appears in Fig. 2.101 with the input signal. Note that the output swing of 30 V matches the input swing as noted in step 5.

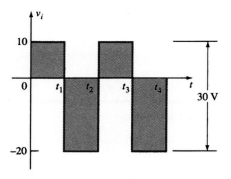

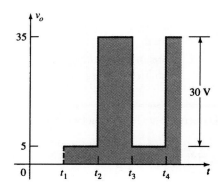

Figure 2.101 v_i and v_o for the clamper of Fig. 2.98.

EXAMPLE 2.25

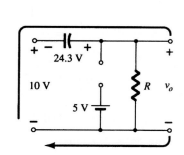

Figure 2.102 Determining v_o and V_C with the diode in the "on" state.

Repeat Example 2.24 using a silicon diode with $V_T = 0.7$ V.

Solution

For the short-circuit state the network now takes on the appearance of Fig. 2.102 and v_o can be determined by Kirchhoff's voltage law in the output section.

$$+5\text{ V} - 0.7\text{ V} - v_o = 0$$

and

$$v_o = 5\text{V} - 0.7\text{ V} = 4.3\text{ V}$$

For the input section Kirchhoff's voltage law will result in

$$-20\text{ V} + V_C + 0.7\text{ V} - 5\text{ V} = 0$$

and

$$V_C = 25\text{ V} - 0.7\text{ V} = 24.3\text{ V}$$

For the period $t_2 \rightarrow t_3$ the network will now appear as in Fig. 2.103, with the only change being the voltage across the capacitor. Applying Kirchhoff's voltage law yields

$$+10\text{ V} + 24.3\text{ V} - v_o = 0$$

and

$$v_o = 34.3\text{ V}$$

The resulting output appears in Fig. 2.104, verifying the statement that the input and output swings are the same.

Figure 2.103 Determining v_o with the diode in the open state.

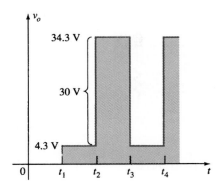

Figure 2.104 Sketching v_o for the clamper of Fig. 2.98 with a silicon diode.

A number of clamping circuits and their effect on the input signal are shown in Fig. 2.105. Although all the waveforms appearing in Fig. 2.105 are square waves, clamping networks work equally well for sinusoidal signals. In fact, one approach to the analysis of clamping networks with sinusoidal inputs is to replace the sinusoidal signal by a square wave of the same peak values. The resulting output will then form an envelope for the sinusoidal response as shown in Fig. 2.106 for a network appearing in the bottom right of Fig. 2.105.

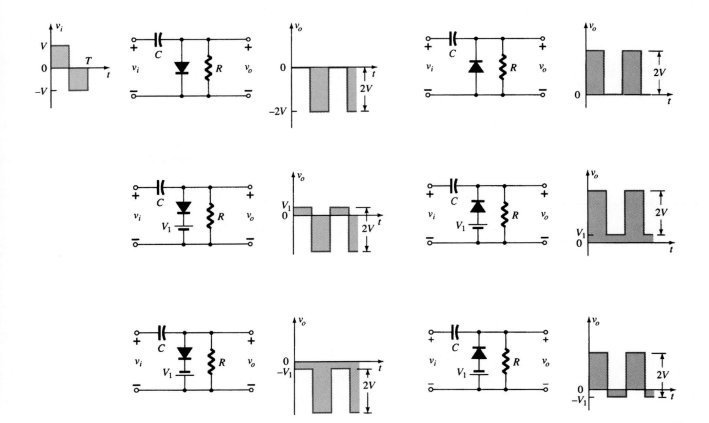

Figure 2.105 Clamping circuits with ideal diodes ($5\tau = 5RC \gg T/2$)

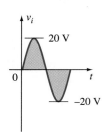

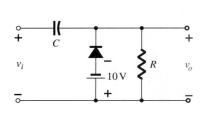

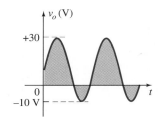

Figure 2.106 Clamping network with a sinusoidal input.

2.11 ZENER DIODES

The analysis of networks employing Zener diodes is quite similar to that applied to the analysis of semiconductor diodes in previous sections. First the state of the diode must be determined followed by a substitution of the appropriate model and a determination of the other unknown quantities of the network. Unless otherwise specified, the Zener model to be employed for the "on" state will be as shown in Fig. 2.107a. For the "off" state as defined by a voltage less than V_Z but greater than 0 V with the polarity indicated in Fig. 2.107b, the Zener equivalent is the open circuit that appears in the same figure.

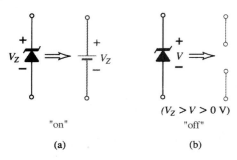

"on"

"off"

$(V_Z > V > 0 \text{ V})$

(a)

(b)

Figure 2.107 Zener diode equivalents for the (a) "on" and (b) "off" states.

V_i and R Fixed

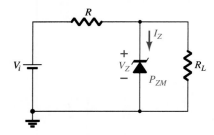

Figure 2.108 Basic Zener regulator.

The simplest of Zener diode networks appears in Fig. 2.108. The applied dc voltage is fixed, as is the load resistor. The analysis can fundamentally be broken down into two steps.

1. *Determine the state of the Zener diode by removing it from the network and calculating the voltage across the resulting open circuit.*

Applying step 1 to the network of Fig. 2.108 will result in the network of Fig. 2.109, where an application of the voltage divider rule will result in

$$V = V_L = \frac{R_L V_i}{R + R_L} \qquad (2.16)$$

If $V \geq V_Z$, the Zener diode is "on" and the equivalent model of Fig. 2.107a can be substituted. If $V < V_Z$, the diode is "off" and the open-circuit equivalence of Fig. 2.107b is substituted.

2. *Substitute the appropriate equivalent circuit and solve for the desired unknowns.*

For the network of Fig. 2.108, the "on" state will result in the equivalent network of Fig. 2.110. Since voltages across parallel elements must be the same, we find that

$$V_L = V_Z \qquad (2.17)$$

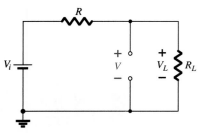

Figure 2.109 Determining the state of the Zener diode.

The Zener diode current must be determined by an application of Kirchhoff's current law. That is,

$$I_R = I_Z + I_L$$

and

$$\boxed{I_Z = I_R - I_L} \tag{2.18}$$

where

$$I_L = \frac{V_L}{R_L} \quad and \quad I_R = \frac{V_R}{R} = \frac{V_i - V_L}{R}$$

The power dissipated by the Zener diode is determined by

$$\boxed{P_Z = V_Z I_Z} \tag{2.19}$$

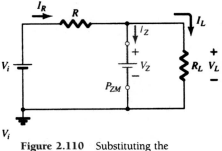

Figure 2.110 Substituting the Zener equivalent for the "on" situation.

which must be less than the P_{ZM} specified for the device.

Before continuing it is particularly important to realize that the first step was employed only to determine the *state of the Zener diode*. If the Zener diode is in the "on" state, the voltage across the diode is not V volts. When the system is turned on, the Zener diode will turn "on" as soon as the voltage across the Zener diode is V_Z volts. It will then "lock in" at this level and never reach the higher level of V volts.

Zener diodes are most frequently used in *regulator* networks or as a *reference* voltage. Figure 2.108 is a simple regulator designed to maintain a fixed voltage across the load R_L. For values of applied voltage greater than required to turn the Zener diode "on," the voltage across the load will be maintained at V_Z volts. If the Zener diode is employed as a reference voltage, it will provide a level for comparison against other voltages.

(a) For the Zener diode network of Fig. 2.111, determine V_L, V_R, I_Z, and P_Z.
(b) Repeat part (a) with $R_L = 3\ \text{k}\Omega$.

EXAMPLE 2.26

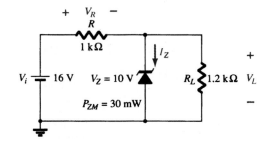

Figure 2.111 Zener diode regulator for Example 2.26.

Solution

(a) Following the suggested procedure the network is redrawn as shown in Fig. 2.112.

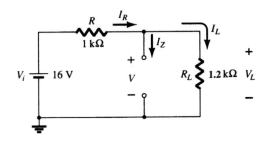

Figure 2.112 Determining V for the regulator of Fig. 2.111.

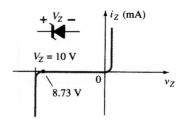

Figure 2.113 Resulting operating point for the network of Fig. 2.111.

Applying Eq. (2.16) gives

$$V = \frac{R_L V_i}{R + R_L} = \frac{1.2 \text{ k}\Omega(16 \text{ V})}{1 \text{ k}\Omega + 1.2 \text{ k}\Omega} = 8.73 \text{ V}$$

Since $V = 8.73$ V is less than $V_Z = 10$ V, the diode is in the "off" state as shown on the characteristics of Fig. 2.113. Substituting the open-circuit equivalent will result in the same network as in Fig. 2.112, where we find that

$$V_L = V = \textbf{8.73 V}$$

$$V_R = V_i - V_L = 16 \text{ V} - 8.73 \text{ V} = \textbf{7.27 V}$$

$$I_Z = \textbf{0 A}$$

and

$$P_Z = V_Z I_Z = V_Z(0 \text{ A}) = \textbf{0 W}$$

(b) Applying Eq. (2.16) will now result in

$$V = \frac{R_L V_i}{R + R_L} = \frac{3 \text{ k}\Omega(16 \text{ V})}{1 \text{ k}\Omega + 3 \text{ k}\Omega} = 12 \text{ V}$$

Since $V = 12$ V is greater than $V_Z = 10$ V, the diode is in the "on" state and the network of Fig. 2.114 will result. Applying Eq. (2.17) yields

$$V_L = V_Z = \textbf{10 V}$$

and

$$V_R = V_i - V_L = 16 \text{ V} - 10 \text{ V} = \textbf{6 V}$$

with

$$I_L = \frac{V_L}{R_L} = \frac{10 \text{ V}}{3 \text{ k}\Omega} = 3.33 \text{ mA}$$

and

$$I_R = \frac{V_R}{R} = \frac{6 \text{ V}}{1 \text{ k}\Omega} = 6 \text{ mA}$$

so that

$$I_Z = I_R - I_L \;[\text{Eq. (2.18)}]$$

$$= 6 \text{ mA} - 3.33 \text{ mA}$$

$$= \textbf{2.67 mA}$$

The power dissipated,

$$P_Z = V_Z I_Z = (10 \text{ V})(2.67 \text{ mA}) = \textbf{26.7 mW}$$

which is less than the specified $P_{ZM} = 30$ mW.

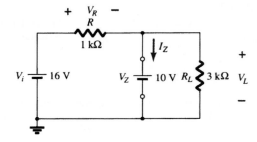

Figure 2.114 Network of Fig. 2.111 in the "on" state.

Fixed V_i, Variable R_L

Due to the offset voltage V_Z, there is a specific range of resistor values (and therefore load current) which will ensure that the Zener is in the "on" state. Too small a load resistance R_L will result in a voltage V_L across the load resistor less than V_Z, and the Zener device will be in the "off" state.

To determine the minimum load resistance of Fig. 2.108 that will turn the Zener diode on, simply calculate the value of R_L that will result in a load voltage $V_L = V_Z$. That is,

$$V_L = V_Z = \frac{R_L V_i}{R_L + R}$$

Solving for R_L, we have

$$R_{L_{min}} = \frac{R V_Z}{V_i - V_Z} \tag{2.20}$$

Any load resistance value greater than the R_L obtained from Eq. (2.20) will ensure that the Zener diode is in the "on" state and the diode can be replaced by its V_Z source equivalent.

The condition defined by Eq. (2.20) establishes the minimum R_L but in turn specifies the maximum I_L as

$$I_{L_{max}} = \frac{V_L}{R_L} = \frac{V_Z}{R_{L_{min}}} \tag{2.21}$$

Once the diode is in the "on" state, the voltage across R remains fixed at

$$V_R = V_i - V_Z \tag{2.22}$$

and I_R remains fixed at

$$I_R = \frac{V_R}{R} \tag{2.23}$$

The Zener current

$$I_Z = I_R - I_L \tag{2.24}$$

resulting in a minimum I_Z when I_L is a maximum and a maximum I_Z when I_L is a minimum value since I_R is constant.

Since I_Z is limited to I_{ZM} as provided on the data sheet, it does affect the range of R_L and therefore I_L. Substituting I_{ZM} for I_Z establishes the minimum I_L as

$$I_{L_{min}} = I_R - I_{ZM} \tag{2.25}$$

and the maximum load resistance as

$$R_{L_{max}} = \frac{V_Z}{I_{L_{min}}} \tag{2.26}$$

EXAMPLE 2.27

(a) For the network of Fig. 2.115, determine the range of R_L and I_L that will result in V_{RL} being maintained at 10 V.
(b) Determine the maximum wattage rating of the diode.

Solution

(a) To determine the value of R_L that will turn the Zener diode on, apply Eq. (2.20):

$$R_{L_{min}} = \frac{R V_Z}{V_i - V_Z} = \frac{(1 \text{ k}\Omega)(10 \text{ V})}{50 \text{ V} - 10 \text{ V}} = \frac{10 \text{ k}\Omega}{40} = \textbf{250 } \boldsymbol{\Omega}$$

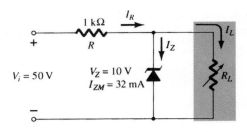

Figure 2.115 Voltage regulator for Example 2.27.

The voltage across the resistor R is then determined by Eq. (2.22):

$$V_R = V_i - V_Z = 50 \text{ V} - 10 \text{ V} = \textbf{40 V}$$

and Eq. (2.23) provides the magnitude of I_R:

$$I_R = \frac{V_R}{R} = \frac{40 \text{ V}}{1 \text{ k}\Omega} = \textbf{40 mA}$$

The minimum level of I_L is then determined by Eq. (2.25):

$$I_{L_{\min}} = I_R - I_{ZM} = 40 \text{ mA} - 32 \text{ mA} = \textbf{8 mA}$$

with Eq. (2.26) determining the maximum value of R_L:

$$R_{L_{\max}} = \frac{V_Z}{I_{L_{\min}}} = \frac{10 \text{ V}}{8 \text{ mA}} = \textbf{1.25 k}\Omega$$

A plot of V_L versus R_L appears in Fig. 2.116a and for V_L versus I_L in Fig. 2.116b.

(b) $P_{\max} = V_Z I_{ZM}$
$$= (10 \text{ V})(32 \text{ mA}) = \textbf{320 mW}$$

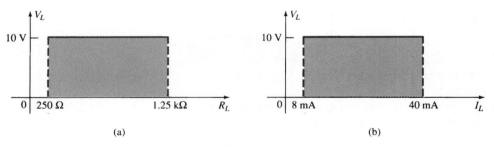

(a)   (b)

Figure 2.116 V_L versus R_L and I_L for the regulator of Fig. 2.115.

Fixed R_L, Variable V_i

For fixed values of R_L in Fig. 2.108, the voltage V_i must be sufficiently large to turn the Zener diode on. The minimum turn-on voltage $V_i = V_{i_{\min}}$ is determined by

$$V_L = V_Z = \frac{R_L V_i}{R_L + R}$$

and

$$\boxed{V_{i_{\min}} = \frac{(R_L + R)V_Z}{R_L}} \tag{2.27}$$

The maximum value of V_i is limited by the maximum Zener current I_{ZM}. Since $I_{ZM} = I_R - I_L$,

$$\boxed{I_{R_{\max}} = I_{ZM} + I_L} \tag{2.28}$$

Since I_L is fixed at V_Z/R_L and I_{ZM} is the maximum value of I_Z, the maximum V_i is defined by

$$V_{i_{max}} = V_{R_{max}} + V_Z$$

$$\boxed{V_{i_{max}} = I_{R_{max}}R + V_Z} \qquad (2.29)$$

Determine the range of values of V_i that will maintain the Zener diode of Fig. 2.117 in the "on" state.

EXAMPLE 2.28

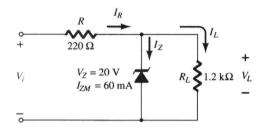

Figure 2.117 Regulator for Example 2.28.

Solution

Eq. (2.27): $\quad V_{i_{min}} = \dfrac{(R_L + R)V_Z}{R_L} = \dfrac{(1200\ \Omega + 220\ \Omega)(20\ V)}{1200\ \Omega} = \mathbf{23.67\ V}$

$$I_L = \frac{V_L}{R_L} = \frac{V_Z}{R_L} = \frac{20\ V}{1.2\ k\Omega} = 16.67\ mA$$

Eq. (2.28): $\quad I_{R_{max}} = I_{ZM} + I_L = 60\ mA + 16.67\ mA$

$$= 76.67\ mA$$

Eq. (2.29): $\quad V_{i_{max}} = I_{R_{max}}R + V_Z$

$$= (76.67\ mA)(0.22\ k\Omega) + 20\ V$$

$$= 16.87\ V + 20\ V$$

$$= \mathbf{36.87\ V}$$

A plot of V_L versus V_i is provided in Fig. 2.118.

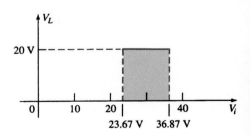

Figure 2.118 V_L versus V_i for the regulator of Fig. 2.117.

The results of Example 2.28 reveal that for the network of Fig. 2.117 with a fixed R_L, the output voltage will remain fixed at 20 V for a range of input voltage that extends from 23.67 to 36.87 V.

In fact, the input could appear as shown in Fig. 2.119 and the output would remain constant at 20 V, as shown in Fig. 2.118. The waveform appearing in Fig. 2.119 is obtained by *filtering* a half-wave- or full-wave-rectified output—a process described in detail in a later chapter. The net effect, however, is to establish a steady dc voltage (for a defined range of V_i) such as that shown in Fig. 2.118 from a sinusoidal source with 0 average value.

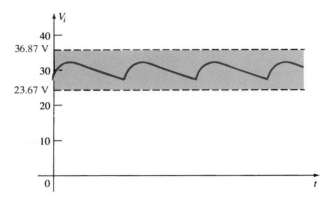

Figure 2.119 Waveform generated by a filtered rectified signal.

2.12 VOLTAGE-MULTIPLIER CIRCUITS

Voltage-multiplier circuits are employed to maintain a relatively low transformer peak voltage while stepping up the peak output voltage to two, three, four, or more times the peak rectified voltage.

Voltage Doubler

The network of Figure 2.120 is a half-wave voltage doubler. During the positive voltage half-cycle across the transformer, secondary diode D_1 conducts (and diode D_2 is cut off), charging capacitor C_1 up to the peak rectified voltage (V_m). Diode D_1 is ideally a short during this half-cycle, and the input voltage charges capacitor C_1 to V_m with the polarity shown in Fig. 2.121a. During the negative half-cycle of the secondary voltage, diode D_1 is cut off and diode D_2 conducts charging capacitor C_2. Since diode D_2 acts as a short during the negative half-cycle (and diode D_1 is open), we can sum the voltages around the outside loop (see Fig. 2.121b):

$$-V_m - V_{C_1} + V_{C_2} = 0$$

$$-V_m - V_m + V_{C_2} = 0$$

from which

$$V_{C_2} = 2V_m$$

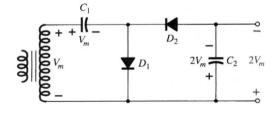

Figure 2.120 Half-wave voltage doubler.

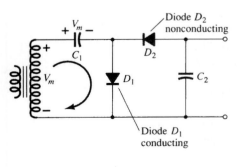

(a)

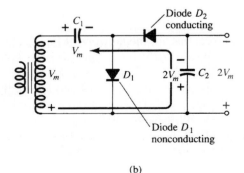

(b)

Figure 2.121 Double operation, showing each half-cycle of operation: (a) positive half-cycle; (b) negative half cycle.

On the next positive half-cycle, diode D_2 is nonconducting and capacitor C_2 will discharge through the load. If no load is connected across capacitor C_2, both capacitors stay charged—C_1 to V_m and C_2 to $2V_m$. If, as would be expected, there is a load connected to the output of the voltage doubler, the voltage across capacitor C_2 drops during the positive half-cycle (at the input) and the capacitor is recharged up to $2V_m$ during the negative half-cycle. The output waveform across capacitor C_2 is that of a half-wave signal filtered by a capacitor filter. The peak inverse voltage across each diode is $2V_m$.

Another doubler circuit is the full-wave doubler of Fig. 2.122. During the positive half-cycle of transformer secondary voltage (see Fig. 2.123a) diode D_1 conducts charging capacitor C_1 to a peak voltage V_m. Diode D_2 is nonconducting at this time.

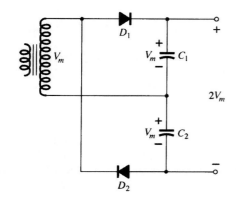

Figure 2.122 Full-wave voltage doubler.

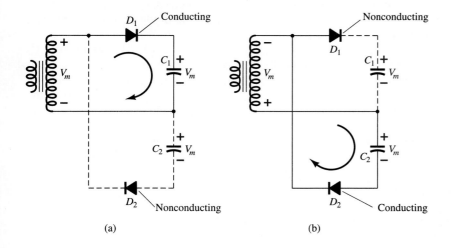

(a) (b)

Figure 2.123 Alternate half-cycles of operation for full-wave voltage doubler.

During the negative half-cycle (see Fig. 2.123b) diode D_2 conducts charging capacitor C_2 while diode D_1 is nonconducting. If no load current is drawn from the circuit, the voltage across capacitors C_1 and C_2 is $2V_m$. If load current is drawn from the circuit, the voltage across capacitors C_1 and C_2 is the same as that across a capacitor fed by a full-wave rectifier circuit. One difference is that the effective capacitance is that of C_1 and C_2 in series, which is less than the capacitance of either C_1 or C_2 alone. The lower capacitor value will provide poorer filtering action than the single-capacitor filter circuit.

The peak inverse voltage across each diode is $2V_m$, as it is for the filter capacitor circuit. In summary, the half-wave or full-wave voltage-doubler circuits provide twice the peak voltage of the transformer secondary while requiring no center-tapped transformer and only $2V_m$ PIV rating for the diodes.

Voltage Tripler and Quadrupler

Figure 2.124 shows an extension of the half-wave voltage doubler, which develops three and four times the peak input voltage. It should be obvious from the pattern of the circuit connection how additional diodes and capacitors may be connected so that the output voltage may also be five, six, seven, and so on, times the basic peak voltage (V_m).

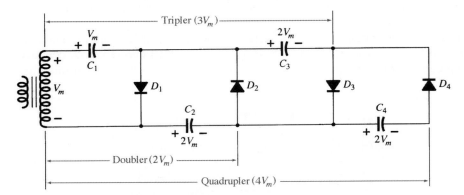

Figure 2.124   Voltage tripler and quadrupler.

In operation capacitor C_1 charges through diode D_1 to a peak voltage, V_m, during the positive half-cycle of the transformer secondary voltage. Capacitor C_2 charges to twice the peak voltage $2V_m$ developed by the sum of the voltages across capacitor C_1 and the transformer, during the negative half-cycle of the transformer secondary voltage.

During the positive half-cycle, diode D_3 conducts and the voltage across capacitor C_2 charges capacitor C_3 to the same $2V_m$ peak voltage. On the negative half-cycle, diodes D_2 and D_4 conduct with capacitor C_3, charging C_4 to $2V_m$.

The voltage across capacitor C_2 is $2V_m$, across C_1 and C_3 it is $3V_m$, and across C_2 and C_4 it is $4V_m$. If additional sections of diode and capacitor are used, each capacitor will be charged to $2V_m$. Measuring from the top of the transformer winding (Fig. 2.124) will provide odd multiples of V_m at the output, whereas measuring the output voltage from the bottom of the transformer will provide even multiples of the peak voltage, V_m.

The transformer rating is only V_m, maximum, and each diode in the circuit must be rated at $2V_m$ PIV. If the load is small and the capacitors have little leakage, extremely high dc voltages may be developed by this type of circuit, using many sections to step up the dc voltage.

2.13   PRACTICAL APPLICATIONS

The range of practical applications for diodes is so broad that it would be virtually impossible to consider all the options in one section. However, to develop some feeling for the use of the device in everyday networks, a number of the more common areas of application are introduced below. In particular, note that the use of diodes extends well beyond the important switching characteristic that was introduced earlier in this chapter.

Rectification

Battery chargers are a common household piece of equipment used to charge everything from small flashlight batteries to heavy-duty, marine, lead-acid batteries. Since all are plugged into a 120-V ac outlet such as found in the home, the basic construction of each is quite similar. In every charging system a *transformer* must be included to cut the ac voltage to a level appropriate for the dc level to be established. A *diode* (also called *rectifier*) arrangement must be included to convert the ac voltage which varies with time to a fixed dc level such as described in this chapter. Some dc chargers will also include a *regulator* to provide an improved dc level (one that varies less with time or load). Since the car battery charger is one of the most common, it will be described in the next few paragraphs.

The outside appearance and the internal construction of a Sears 6/2 AMP Manual Battery Charger are provided in Fig. 2.125. Note in Fig. 2.125b that the transformer (as in most chargers) takes up most of the internal space. The additional air space and the holes in the casing are there to ensure an outlet for the heat that will develop due to the resulting current levels.

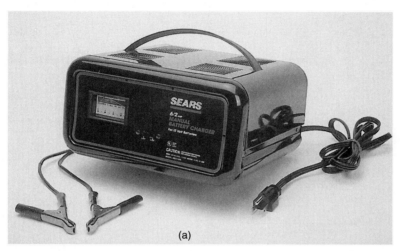

(a)

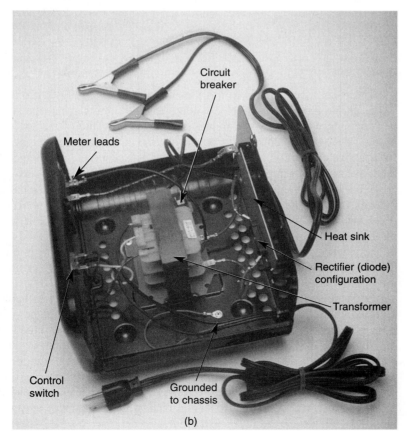

(b)

Figure 2.125 Battery charger: (a) external appearance; (b) internal construction.

The schematic of Fig. 2.126 includes all the basic components of the charger. Note first that the 120 V from the outlet are applied directly across the primary of the transformer. The charging rate of 6 A or 2 A is determined by the switch, which simply controls how many windings of the primary will be in the circuit for the chosen charging rate. If the battery is charging at the 2-A level, the full primary will be in the circuit and the ratio of the turns in the primary to the turns in the secondary will be a maximum. If it is charging at the 6-A level, fewer turns of the primary are in the circuit, and the ratio drops. When you study transformers, you will find that the voltage at the primary and secondary is directly related to the *turns ratio*. If the ratio from primary to secondary drops, then the voltage drops also. The reverse effect occurs if the turns on the secondary exceed those on the primary.

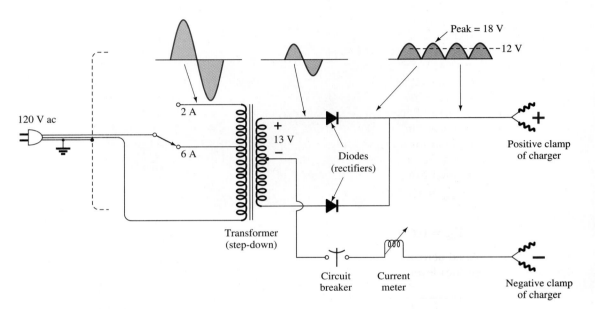

Figure 2.126 Electrical schematic for the battery charger of Fig. 2.125.

The general appearance of the waveforms appears in Fig. 2.126 for the 6-A charging level. Note that so far, the ac voltage has the same wave shape across the primary and secondary. The only difference is in the peak value of the waveforms. Now the diodes take over and convert the ac waveform which has zero average value (the waveform above equals the waveform below) to one that has an average value (all above the axis) as shown in the same figure. For the moment simply recognize that diodes are semiconductor electronic devices that permit only conventional current to flow through them in the direction indicated by the arrow in the symbol. Even though the waveform resulting from the diode action has a pulsing appearance with a peak value of about 18 V, it will charge the 12-V battery whenever its voltage is greater than that of the battery, as shown by the shaded area. Below the 12-V level the battery cannot discharge back into the charging network because the diodes permit current flow in only one direction.

In particular, note in Fig. 2.125b the large plate that carries the current from the rectifier (diode) configuration to the positive terminal of the battery. Its primary purpose is to provide a *heat sink* (a place for the heat to be distributed to the surrounding air) for the diode configuration. Otherwise the diodes would eventually melt down and self-destruct due to the resulting current levels. Each component of Fig. 2.126 has been carefully labeled in Fig. 2.125b for reference.

When current is first applied to a battery at the 6-A charge rate, the current demand as indicated by the meter on the face of the instrument may rise to 7 A or almost 8 A. However, the level of current will decrease as the battery charges until it drops to a level of 2 A or 3 A. For units such as this that do not have an automatic shutoff, it is important to disconnect the charger when the current drops to the fully charged level; otherwise, the battery will become overcharged and may be damaged. A battery that is at its 50% level can take as long as 10 hours to charge, so don't expect it to be a 10-minute operation. In addition, if a battery is in very bad shape with a lower than normal voltage, the initial charging current may be too high for the design. To protect against such situations, the circuit breaker will open and stop the charging process. Because of the high current levels, it is important that the directions provided with the charger be carefully read and applied.

In an effort to compare the theoretical world with the real world, a load (in the form of a headlight) was applied to the charger to permit a viewing of the actual output waveform. It is important to note and remember that **a diode with zero current through it will not display its rectifying capabilities.** In other words, the output from the charger of Fig. 2.125 will not be a rectified signal unless a load is applied to the system to draw current through the diode. Recall from the diode characteristics that when $I_D = 0$ A, $V_D = 0$ V.

By applying the headlamp as a load, however, sufficient current is drawn through the diode for it to behave like a switch and convert the ac waveform to a pulsating one as shown in Fig. 2.127 for the 6-A setting. First note that the waveform is slightly distorted by the nonlinear characteristics of the transformer and the nonlinear characteristics of the diode at low currents. The waveform, however, is certainly close to what is expected when we compare it to the theoretical patterns of Fig. 2.125. The peak value is determined from the vertical sensitivity as

$$V_{peak} = (3.3 \text{ divisions})(5 \text{ V/division}) = 16.5 \text{ V}$$

with a dc level of

$$V_{dc} = 0.636 V_{peak} = 0.636(16.5 \text{ V}) = 10.49 \text{ V}$$

A dc meter connected across the load registered 10.41 V, which is very close to the theoretical average (dc) level of 10.49 V.

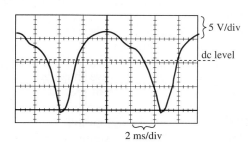

Figure 2.127 Pulsating response of the charger of Fig. 2.126 to the application of a headlamp as a load.

One may wonder how a charger having a dc level of 10.49 V can charge a 12-V battery to a typical level of 14 V. It is simply a matter of realizing that (as shown in Fig. 2.127) for a good deal of each pulse, the voltage across the battery will be greater than 12 V and the battery will be charging—a process referred to as **trickle charging.** In other words, charging does not occur during the entire cycle, but only when the charging voltage is more than the voltage of the battery.

Protective Configurations

Diodes are used in a variety of ways to protect elements and systems from excessive voltages or currents, polarity reversals, arcing, and shorting, to name a few. In Fig. 2.128a, the switch on a simple *R-L* circuit has been closed, and the current will rise to a level determined by the applied voltage and series resistor *R* as shown on the plot. Problems arise when the switch is quickly opened as in Fig. 2.128b to essentially tell the circuit that the current must drop to zero almost instantaneously. You will remember from your basic circuits courses, however, that the inductor will not

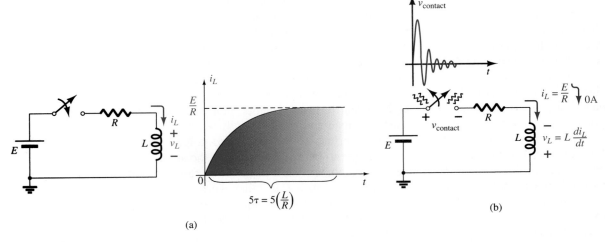

Figure 2.128 (a) Transient phase of a simple *R-L* circuit; (b) arcing that results across a switch when opened in series with an *R-L* circuit.

permit an instantaneous change in current through the coil. A conflict results which will establish arcing across the contacts of the switch as the coil tries to find a path for discharge. Recall also that the voltage across an inductor is directly related to the rate of change in current through the coil ($v_L = L \, di_L/dt$). When the switch is opened, it is trying to dictate that the current change almost instantaneously, causing a very high voltage to develop across the coil that will then appear across the contacts to establish this arcing current. Levels in the thousands of volts will develop across the contacts which will shortly, if not immediately, damage the contacts and, thereby, the switch. The effect is referred to as an "inductive kick." Note also that the polarity of the voltage across the coil during the "build-up" phase is opposite to that during the "release" phase. This is due to the fact that the current must maintain the same direction before and after the switch is opened. During the "build-up" phase, the coil appears as a load, whereas during the release phase, it has the characteristics of a source. In general, therefore, always keep in mind that

> *trying to change the current through an inductive element too quickly may result in an inductive kick that could damage surrounding elements or the system itself.*

In Fig. 2.129a the simple network above may be controlling the action of a relay. When the switch is closed, the coil will be energized, and steady-state current levels will be established. However, when the switch is opened to deenergize the network, we have the problem introduced above because the electromagnet controlling the relay action will appear as a coil to the energizing network. One of the cheapest but

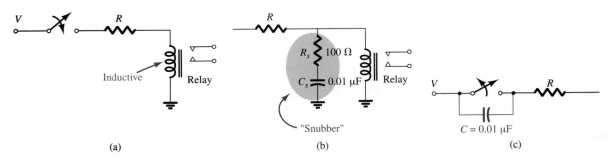

Figure 2.129 (a) Inductive characteristics of a relay; (b) snubber protection for the configuration of Fig. 2.129a; (c) capacitive protection for a switch.

most effective ways to protect the switching system is to place a capacitor (called a "snubber") across the terminals of the coil. When the switch is opened, the capacitor will initially appear as a short to the coil and will provide a current path that will bypass the dc supply and switch. The capacitor has the characteristics of a short (very low resistance) because of the high-frequency characteristics of the surge voltage as shown in Fig. 2.128b. Recall that the reactance of a capacitor is determined by $X_C = 1/2\pi fC$, so the higher the frequency, the less the resistance. Normally, because of the high surge voltages and relatively low cost, ceramic capacitors of about 0.01 μF are used. You don't want to use large capacitors because the voltage across the capacitor will build up too slowly and will essentially slow down the performance of the system. The resistor of 100 Ω in series with the capacitor is introduced solely to limit the surge current that will result when a change in state is called for. Often, the resistor does not appear because of the internal resistance of the coil as established by many turns of fine wire. On occasion, you may find the capacitor across the switch as shown in Fig. 2.129c. In this case, the shorting characteristics of the capacitor at high frequencies will bypass the contacts with the switch and extend its life. Recall that the voltage across a capacitor cannot change instantaneously.

In general, therefore, **capacitors in parallel with inductive elements or across switches are often there to act as protective elements, not as typical network capacitive elements.**

Finally, the diode is often used as a protective device for situations such as above. In Fig. 2.130, a diode has been placed in parallel with the inductive element of the relay configuration. When the switch is opened or the voltage source quickly disengaged, the polarity of the voltage across the coil is such as to turn the diode on and conduct in the direction indicated. The inductor now has a conduction path through the diode rather than through the supply and switch, thereby saving both. Since the current established through the coil must now switch directly to the diode, the diode must be able to carry **the same level of current** that was passing through the coil before the switch was opened. The rate at which the current collapses will be controlled by the resistance of the coil and the diode. It can be reduced by placing an additional resistor in series with the diode. The advantage of the diode configuration over that of the snubber is that the diode reaction and behavior are not frequency dependent. However, the protection offered by the diode will not work if the applied voltage is an alternating one such as ac or a square wave since the diode will conduct for one of the applied polarities. For such alternating systems, the "snubber" arrangement would be the best option.

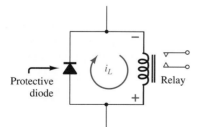

Figure 2.130 Diode protection for an *R-L* circuit.

In the next chapter we will find that the base-to-emitter junction of a transistor is forward-biased. That is, the voltage V_{BE} of Fig. 2.131a will be about 0.7 V positive. To prevent a situation where the emitter terminal would be made more positive than the base terminal by a voltage that could damage the transistor, the diode shown in Fig. 2.131a is added. The diode will prevent the reverse-bias voltage V_{EB} from exceeding 0.7 V. On occasion, you may also find a diode in series with the collector terminal of a transistor as shown in Fig. 2.131b. Normal transistor action requires that

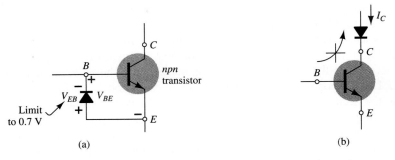

Figure 2.131 (a) Diode protection to limit the emitter-to-base voltage of a transistor; (b) diode protection to prevent a reversal in collector current.

the collector be more positive than the base or emitter terminal to establish a collector current in the direction shown. However, if a situation arises where the emitter or base terminal is at a higher potential than the collector terminal, the diode will prevent conduction in the opposite direction. In general, therefore,

> *diodes are often used to prevent the voltage between two points from exceeding 0.7 V or to prevent conduction in a particular direction.*

As shown in Fig. 2.132, diodes are often used at the input terminals of systems such as op-amps (Chapter 14) to limit the swing of the applied voltage. For the 400-mV level

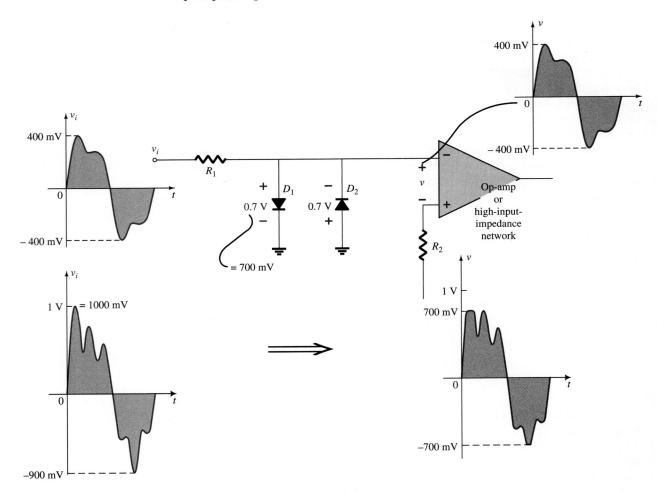

Figure 2.132 Diode control of the input swing to an op-amp or a high-input-impedance network.

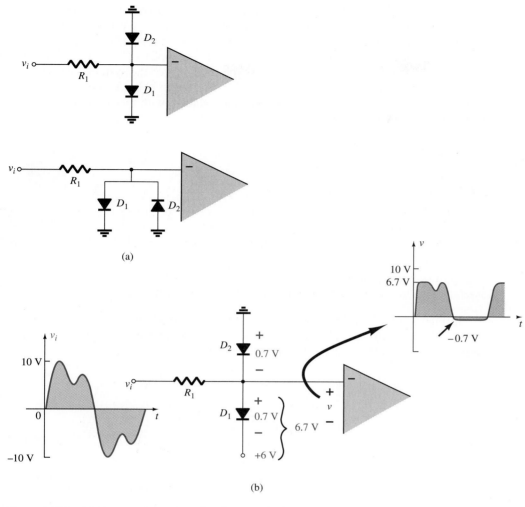

Figure 2.133 (a) Alternate appearances for the network of Fig. 2.132; (b) establishing random levels of control with separate dc supplies.

the signal will pass undisturbed to the input terminals of the op-amp. However, if the voltage jumps to a level of 1 V, the top and bottom peaks will be clipped off before appearing at the input terminals of the op-amp. Any clipped-off voltage will appear across the series resistor R_1.

The controlling diodes of Fig. 2.132 may also be drawn as shown in Fig. 2.133 to control the signal appearing at the input terminals of the op-amp. In this example, the diodes are acting more like shaping elements than as limiters as in Fig. 2.132. However, the point is that

the placement of elements may change, but their function may still be the same. Do not expect every network to appear exactly as you studied it for the first time.

In general, therefore, don't always assume that diodes are used simply as switches. There are a wide variety of uses for the diode as protective and limiting devices.

Polarity Insurance

There are numerous systems that are very sensitive to the polarity of the applied voltage. For instance, in Fig. 2.134a, assume for the moment that there is a very expensive piece of equipment that would be damaged by an incorrectly applied bias. In Fig. 2.134b the correct applied bias is shown on the left. As a result, the diode is reverse-biased, but the system works just fine—the diode has no effect. However, if the wrong polarity is applied as shown in Fig. 2.134c, the diode will conduct and ensure that no more than 0.7 V will appear across the terminals of the system—protecting it from excessive voltages of the wrong polarity. For either polarity, the difference between the applied voltage and the load or diode voltage will appear across the series source or network resistance.

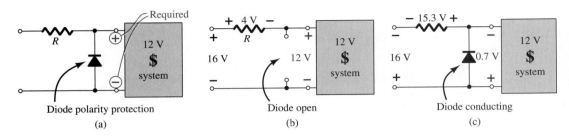

Figure 2.134 (a) Polarity protection for an expensive, sensitive piece of equipment; (b) correctly applied polarity; (c) application of the wrong polarity.

In Fig. 2.135 a sensitive measuring movement cannot withstand voltages greater than 1 V of the wrong polarity. With this simple design the sensitive movement is protected from voltages of the wrong polarity of more than 0.7 V.

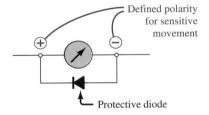

Figure 2.135 Protection for a sensitive meter movement.

Controlled Battery-Powered Backup

In numerous situations a system should have a backup power source to ensure that the system will still be operational in case of a loss of power. This is especially true of security systems and lighting systems that must turn on during a power failure. It is also important when a system such as a computer or radio is disconnected from its ac-to-dc power conversion source to a portable mode for traveling. In Fig. 2.136 the 12-V car radio operating off the 12-V dc power source has a 9-V battery backup system in a small compartment in the back of the radio, ready to take over the role of saving the clock mode and the channels stored in memory when the radio is removed from the car. With the full 12 V available from the car, D_1 is conducting, and the voltage at the radio is about 11.3 V. D_2 is reverse-biased (an open circuit), and the reserve 9-V battery inside the radio is disengaged. However, when the radio is removed from the car, D_1 will no longer be conducting because the 12-V source is no longer avail-

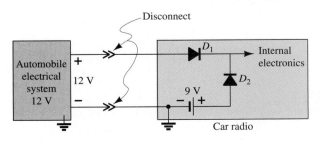

Figure 2.136 Backup system designed to prevent the loss of memory in a car radio when the radio is removed from a car.

able to forward-bias the diode. However, D_2 will be forward-biased by the 9-V battery, and the radio will continue to receive about 8.3 V to maintain the memory that has been set for components such as the clock and the channel selections.

Polarity Detector

Through the use of LEDs of different colors, the simple network of Fig. 2.137 can be used to check the polarity at any point in a dc network. When the polarity is as indicated for the applied 6 V the top terminal is positive, D_1 will conduct along with LED1, and a green light will result. Both D_2 and LED2 will be back-biased for the above polarity. However, if the polarity at the input is reversed, D_2 and LED2 will conduct, and a red light will appear, defining the top lead as the lead at the negative potential. It would appear that the network would work without diodes D_1 and D_2. However, in general, LEDs do not like to be reverse-biased because of sensitivity built-in during the doping process. Diodes D_1 and D_2 offer a series open-circuit condition that provides some protection to the LEDs. In the forward-bias state, the additional diodes D_1 and D_2 reduce the voltage across the LEDs to more common operating levels.

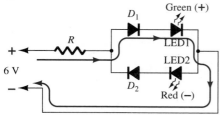

Figure 2.137 Polarity dector using diodes and LEDs.

Offering Longer Life and Durability

Some of the primary concerns of using electric light bulbs in exit signs are their limited lifetime (requiring frequent replacement); their sensitivity to heat, fire, etc.; their durability factor when catastrophic accidents occur; and their high voltage and power requirements. For this reason LEDs are often used to provide the longer life span, higher durability levels, and lower-demand voltage and power levels (especially when the reserve dc battery system has to take over).

In Fig. 2.138 a control network determines when the EXIT light should be on. When it is on, all the LEDs in series will be on, and the EXIT sign will be fully lit. Obviously, if one of the LEDs should burn out and open up, the entire section will turn off. However, this situation can be improved by simply placing parallel LEDs between every two points. Lose one, and you will still have the other parallel path. Parallel diodes will, of course, reduce the current through each LED, but two at a lower level of current can have a luminescence similar to one at twice the current.

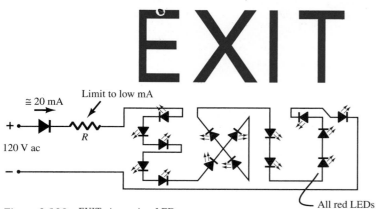

Figure 2.138 EXIT sign using LEDs.

Even though the applied voltage is ac, which means that the diodes will turn on and off as the 60-Hz voltage swings positive and negative, the persistence of the LEDs will provide a steady light for the sign.

Setting Voltage Reference Levels

Diodes and Zeners can be used to set reference levels as shown in Fig. 2.139. The network, through the use of two diodes and one Zener diode, is providing three different voltage levels.

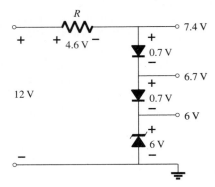

Figure 2.139 Providing different reference levels using diodes.

Establishing a Voltage Level Insensitive to the Load Current

As an example that clearly demonstrates the difference between a resistor and a diode in a voltage-divider network, consider the situation of Fig. 2.140a, where a load requires about 6 V to operate properly but a 9-V battery is all that is available. For the moment let us assume that operating conditions are such that the load has an internal resistance of 1 kΩ. Using the voltage-divider rule, we can easily determine that the series resistor should be 470 Ω (commercially available value) as shown in Fig. 2.140b. The result is a voltage across the load of 6.1 V, an acceptable situation for most 6-V loads. However, if the operating conditions of the load change and the load now has an internal resistance of only 600 Ω, the load voltage will drop to about 4.9 V, and the system will not operate correctly. This sensitivity to the load resistance can be eliminated by connecting four diodes in series with the load as shown in Fig.

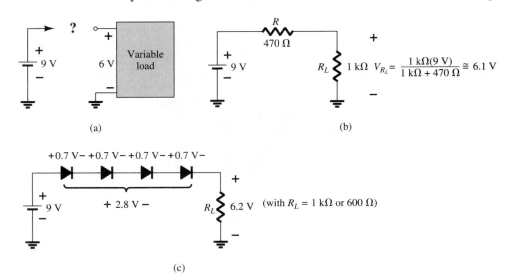

Figure 2.140 (a) How to drive a 6-V load with a 9-V supply? (b) using a fixed resistor value; (c) using a series combination of diodes.

2.140c. When all four diodes conduct, the load voltage will be about 6.2 V, irrespective of the load impedance (within device limits, of course)—the sensitivity to the changing load characteristics has been removed.

AC Regulator and Square-Wave Generator

Two back-to-back Zeners can also be used as an ac regulator as shown in Fig. 2.141a. For the sinusoidal signal v_i the circuit will appear as shown in Fig. 2.141b at the instant $v_i = 10$ V. The region of operation for each diode is indicated in the adjoining figure. Note that Z_1 is in a low-impedance region, while the impedance of Z_2 is quite large, corresponding with the open-circuit representation. The result is that $v_o = v_i$ when $v_i = 10$ V. The input and output will continue to duplicate each other until v_i reaches 20 V. Z_2 will then "turn on" (as a Zener diode), while Z_1 will be in a region of conduction with a resistance level sufficiently small compared to the series 5-kΩ resistor to be considered a short circuit. The resulting output for the full range of v_i is provided in Fig. 2.141a. Note that the waveform is not purely sinusoidal, but its rms value is lower than that associated with a full 22-V peak signal. The network is effectively limiting the rms value of the available voltage. The network of Fig. 2.141a can be extended to that of a simple square-wave generator (due to the clipping action) if the signal v_i is increased to perhaps a 50-V peak with 10-V Zeners as shown in Fig. 2.142 with the resulting output waveform.

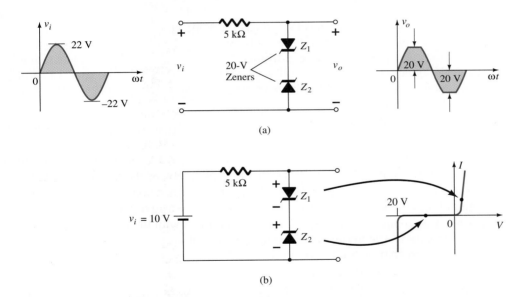

Figure 2.141 Sinusoidal ac regulation: (a) 40-V peak-to-peak sinusoidal ac regulator; (b) circuit operation at $v_i = 10$ V.

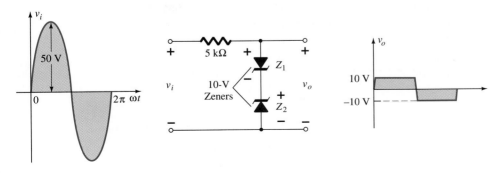

Figure 2.142 Simple square-wave generator.

2.14 SUMMARY

Important Conclusions and Concepts

1. The characteristics of a device are **unaltered** by the network in which it is employed. The network simply determines the point of operation of the device.

2. The operating point of a network is determined by the **intersection** of the network equation and an equation defining the characteristics of the device.

3. For most applications, the characteristics of a diode can be defined simply by the **threshold voltage in the forward-bias region** and an open circuit for applied voltages less than the threshold value.

4. To determine the state of a diode, simply **think of it initially as a resistor,** and find the polarity of the voltage across it and the direction of conventional current through it. If the voltage across it has a forward-bias polarity and the **current has a direction that matches the arrow in the symbol,** the diode is conducting.

5. To determine the state of diodes used in a logic gate, first make an **educated guess** about the state of the diodes, and then **test your assumptions.** If your estimate is incorrect, refine your guess and try again until the analysis verifies the conclusions.

6. Rectification is a process whereby an applied waveform of **zero average value** is changed to one that **has a dc level.** For applied signals more than a few volts, the ideal diode approximations can normally be applied.

7. It is very important that the PIV rating of a diode be checked when choosing a diode for a particular application. Simply determine the **maximum voltage** across the diode under **reverse-bias conditions,** and compare it to the nameplate rating. For the typical half-wave and full-wave bridge rectifiers, it is the peak value of the applied signal. For the CT transformer full-wave rectifier, it is twice the peak value (which can get quite high).

8. Clippers are networks that **"clip"** away part of the applied signal either to create a specific type of signal or to limit the voltage that can be applied to a network.

9. Clampers are networks that **"clamp"** the input signal to a different dc level. In any event, the peak-to-peak swing of the applied signal will remain the same.

10. Zener diodes are diodes that make effective use of the **Zener breakdown potential** of an ordinary p-n junction characteristic to provide a device of wide importance and application. For Zener conduction, the direction of conventional flow is **opposite to the arrow in the symbol.** The polarity under conduction is also **opposite to that of the conventional diode.**

11. To determine the state of a Zener diode in a dc network, simply remove the Zener from the network, and determine the **open-circuit voltage** between the two points where the Zener diode was originally connected. If it is **more than the Zener potential** and has the correct polarity, the Zener diode is in the "on" state.

12. A half-wave or full-wave voltage doubler employs two capacitors; a tripler, three capacitors; and a quadrupler, four capacitors. In fact, for each, the number of diodes equals the number of capacitors.

Equations

Approximate:

$$\text{Silicon:} \quad V_T = 0.7 \text{ V}; \quad I_D \text{ is determined by network.}$$

$$\text{Germanium:} \quad V_T = 0.3 \text{ V}; \quad I_D \text{ is determined by network.}$$

Ideal:

$$V_T = 0 \text{ V}; \quad I_D \text{ is determined by network.}$$

For conduction:

$$V_D \geq V_T$$

Half-wave rectifier:

$$V_{dc} = 0.318V_m$$

Full-wave rectifier:

$$V_{dc} = 0.636V_m$$

2.15   COMPUTER ANALYSIS

PSpice Windows

SERIES DIODE CONFIGURATION

PSpice Windows will now be applied to the network of Fig. 2.29 to permit a comparison with the hand-calculated solution. As briefly described in Chapter 1, the application of PSpice Windows requires that the network first be constructed on the schematics screen. The next few paragraphs will examine the basics of setting up the network on the screen, assuming no prior experience with the process. It might be helpful to reference the completed network of Fig. 2.143 as we progress through the discussion.

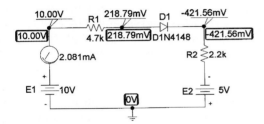

Figure 2.143 PSpice Windows analysis of a series diode configuration.

In general, it is easier to draw the network if the grid is on the screen and the stipulation is made that all elements be on the grid. This will ensure that all the connections are made between the elements. The screen can be set up by first choosing **Options** at the heading of the schematics screen, followed by **Display Options.** The **Display Options** dialog box will permit you to make all the choices necessary regarding the type of display desired. For our purposes, we will choose **Grid On, Stay on Grid,** and **Grid Spacing** of 0.1 in.

R

The resistor **R** will be the first to be positioned. By clicking on the **Get New Part** icon (the icon in the top right area with the binoculars) followed by **Libraries,** we can choose the **Analog.slb** library of basic elements. We can then scroll the **Part** list until we find **R.** Clicking on **R** followed by **OK** will result in the **Part Browser Basic** dialog box reflecting our choice of a resistive element. Choosing the **Place & Close** option will place the resistive element on the screen and close the dialog box. The resistor will appear horizontal, which is perfect for the R_1 of Fig. 2.29 (note Fig. 2.143). Move the resistor to a logical location, and click the left button of the mouse—the resistor R_1 is in place. Note that it snaps to the grid structure. The resistor R_2 must now be placed to the right of R_1. By simply moving the mouse to the right, the second resistor will appear, and R_2 can be placed in the proper location with a subsequent click of the mouse. Since the network only has two resistors, the depositing of resistors can be ended by a right click of the mouse. The resistor R_2 can be rotated by pressing the keys **Ctrl** and **R** simultaneously or by choosing **Edit** on the menu bar, followed by **Rotate.**

The result of the above is two resistors with the right labels but the worng values. To change a value, double click on the value of the screen (first **R1**). **A Set Attribute Value** dialog box will appear. Type in the correct value, and send the value to the screen with **OK.** The 4.7 kΩ will appear within a box that can be moved by simply clicking on the small box and, while holding the clicker down, moving the 4.7 kΩ to the desired location. Release the clicker, and the 4.7 kΩ label will remain where placed. Once located, an additional click anywhere on the screen will remove the boxes and end the process. If you want to move the 4.7 kΩ in the future, simply click once on the value and the boxes will reappear. Repeat the above for the value of the resistor R_2.

To remove (clip) an element, simply click on it (to establish the red or active color), and then click the **scissors** icon or use the sequence **Edit-Delete.**

E

The voltage sources are set by going to the **source.slb** library of **Library Browser** and choosing **VDC.** Clicking **OK** results in the source symbol appearing on the schematic. This symbol can be placed as required. After clicking it in the appropriate place, a **V1** label will appear. To change the label to **E1** simply click the **V1** twice and an **Edit Reference Designator** dialog box will appear. Change the label to **E1** and click **OK,** and then **E1** will appear on the screen within a box. The box can be moved in the same manner as the labels for resistors. When you have the correct position, simply click the mouse once more and place E_1 as desired.

To set the value of E_1, click the value twice and the **Set Attribute Value** will appear. Set the value to 10 V and click **OK.** The new value will appear on the schematic. The value can also be set by clicking the battery symbol itself twice, after which a dialog box will appear labeled **E1 PartName:VDC.** By choosing **DC = 0 V, DC** and **Value** will appear in the designated areas at the top of the dialog box. Using the mouse, bring the marker to the **Value** box and change it to 10 V. Then click **Save Attr.** to be sure and save the new value, and an **OK** will result in E_1 being changed to 10 V. E_1 can now be set, but be sure to turn it 180° with the appropriate operations.

DIODE

The diode is found in the **EVAL.slb** library of the **Library Browser** dialog box. Choosing the **D1N4148** diode followed by an **OK** and **Close & Place** will place the diode symbol on the screen. Move the diode to the correct position, click it in place with a left click, and end the operation with a right click of the mouse. The labels **D1** and **D1N4148** will appear near the diode. Clicking on either label will provide the boxes that permit movement of the labels.

Let us now take a look at the diode specs by clicking the diode symbol once, followed by the **Edit-Model-Edit Instance Model** sequence. For the moment, we will leave the parameters listed. In particular, note that I_s = 2.682nA and the terminal capacitance (important when the applied frequency becomes a factor) is 4pF.

IPROBE

One or more currents of a network can be displayed by inserting an **IPROBE** in the desired path. **IPROBE** is found in the **SPECIAL.slb** library and appears as a meter face on the screen. **IPROBE** will respond with a positive answer if the current (conventional) enters the symbol at the end with the arc representing the scale. Since we are looking for a positive answer in this investigation. **IPROBE** should be installed as shown in Fig. 2.143. When the symbol first appears, it is 180° out of phase with the desired current. Therefore, it is necessary to use the **Ctrl-R** sequence twice to rotate the symbol before finalizing its position. As with the elements described above, once it is in place a single click will place the meter and a right click will complete the insertion process.

LINE

The elements now need to be connected by choosing the icon with the thin line and pencil or by the sequence **Draw-Wire.** A pencil will appear that can draw the desired connections in the following manner: Move the pencil to the beginning of the line, and click the left side of the mouse. The pencil is now ready to draw. Draw the desired line (connection), and click the left side again when the connection is complete. The line will appear in red, waiting for another random click of the mouse or the insertion of another line. It will then turn geen to indicate it is in memory. For additional lines, simply repeat the procedure. When done, simply click the right side of the mouse.

EGND

The system must have a ground to serve as a reference point for the nodal voltages. Earth ground **(EGND)** is part of the **PORT.slb** library and can be placed in the same manner as the elements described above.

VIEWPOINT

Nodal voltages can be displayed on the diagram after the simulation using **VIEWPOINTS,** which is found in the **SPECIAL.slb** library. Simply place the arrow of the **VIEWPOINT** symbol where you desire the voltage with respect to ground. A **VIEWPOINT** can be placed at every node of the network if necessary, although only three are placed in Fig. 2.143. The network is now complete, as shown in Fig. 2.143.

ANALYSIS

The network is now ready to be analyzed. To expedite the process, click on **Analysis** and choose **Probe Setup.** By selecting **Do not auto-run Probe** you save intermediary steps that are inappropriate for this analysis; it is an option that will be discussed later in this chapter. After **OK,** go to **Analysis** and choose **Simulation.** If the network was installed properly, a **PSpiceAD** dialog box will appear and reveal that the bias (dc) points have been calculated. If we now exit the box by clicking on the small **x** in the top right corner, you will obtain the results appearing in Fig. 2.143. Note that the program has automatically provided four dc voltages of the network (in addition to the **VIEWPOINT** voltages). This occurred because an option under analysis was enabled. For future analysis we will want control over what is displayed so follow the path through **Analysis-Display Results on Schematic** and slide over to the adjoining **Enable** box. Clicking the **Enable** box will remove the check, and the dc voltages will not automatically appear. They will only appear where **VIEWPOINTS** have been inserted. A more direct path toward controlling the appearance of the dc voltages is to use the icon on the menu bar with the large captial **V.** By clicking it on and off, you can control whether the dc levels of the network will appear. The icon with the large capital **I** will permit all the dc currents of the network to be shown if desired. For practice, click it on and off and note the effect on the schematic. If you want to remove selected dc voltages on the schematic, simply click the nodal voltage of interest, then click the icon with the smaller capital **V** in the same grouping. Clicking it once will remove the selected dc voltage. The same can be done for selected currents with the remaining icon of the group. For the future, it should be noted that an analysis can also be initiated by simply clicking the **Simulation** icon having the yellow background and the two waveforms (square wave and sinusoidal).

Note also that the results are not an exact match with those obtained in Example 2.11. The **VIEWPOINT** voltage at the far right is -421.56 rather than the -454.2 mV obtained in Example 2.11. In addition, the current is 2.081 rather than the 2.066 mA obtained in the same example. Further, the voltage across the diode is 281.79 mV $+$ 421.56 mV $= 0.64$ V rather than the 0.7 V assumed for all silicon diodes. This all results from our using a real diode with a long list of variables defining its operation.

However, it is important to remember that the analysis of Example 2.11 was an approximate one and, therefore, it is expected that the results are only close to the actual response. On the other hand, the results obtained for the nodal voltage and current are quite close. If taken to the tenths place, the currents (2.1 mA) are an exact match.

The results obtained in Fig. 2.143 can be improved (in the sense that they will be a closer match to the hand-written solution) by clicking on the diode (to make it red) and using the sequence **Edit-Model-Edit Instance Model (Text)** to obtain the **Model Editor** dialog box. Choose **Is = 3.5E-15A** (a value determined by trial and error), and delete all the other parameters for the device. Then, follow with **OK-Simulate** icon to obtain the results of Fig. 2.144. Note that the voltage across the diode now is 260.17 mV + 440.93 mV = 0.701 V, or almost exactly 0.7 V. The **VIEWPOINT** voltage is -440.93 V or, again, an almost perfect match with the hand-written solution of -0.44 V. In either case, the results obtained are very close to the expected values. One is more accurate as far as the actual device is concerned, while the other provides an almost exact match with the hand-written solution. One cannot expect a perfect match for every diode network by simply setting I_s to 3.5E-15A. As the current through the diode changes, the level of I_s must also change if an exact match with the handwritten solution is to be obtained. However, rather than worry about the current in each system, it is suggested that I_s = 3.5E-15A be used as the standard value if the PSpice solution is desired to be a close match with the hand-written solution. The results will not always be perfect, but in most cases they will be closer than if the parameters of the diode are left at their default values. For transistors in the chapters to follow, it will be set to 2E-15A to obtain a suitable match with the hand-written solution. Note also that the **Bias Current Display** was enabled to show that the current is indeed the same everywhere in the circuit.

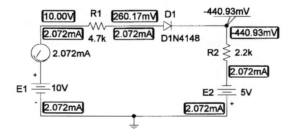

Figure 2.144 The circuit of Figure 2.143 reexamined with I_s set to 3.5E-15A.

The results can also be viewed in tabulated form by returning to **Analysis** and choosing **Examine Output.** The result is the long listing of Fig. 2.145. The **Schematics Netlist** describes the network in terms of numbered nodes. The 0 refers to ground level, with the 10 V source from node 0 to 5. The source **E2** is from 0 to node 3. The resistor **R2** is connected from node 3 to 4, and so on. Scrolling down the output file, we find the **Diode MODEL PARAMETERS** clearly showing that I_s is set at 3.5E-15A and is the only parameter listed. Next is the **SMALL SIGNAL BIAS SOLUTION** or dc solution with the voltages at the various nodes. In addition, the current through the sources of the network is shown. The negative sign reveals that it is reflecting the direction of electron flow (into the positive terminal). The total power dissipation of the elements is 31.1 mW. Finally, the **OPERATING POINT INFORMATION** reveals that the current through the diode is 2.07 mA and the voltage across the diode 0.701 V.

The analysis is now complete for the diode circuit of interest. We have not touched on all the alternative paths available through PSpice Windows, but sufficient coverage has been provided to examine any of the networks covered in this chapter with a dc source. For practice, the other examples should be examined using the Windows approach since the results are provided for comparison. The same can be said for the odd-numbered exercises at the end of this chapter.

```
****      CIRCUIT DESCRIPTION


*************************************************************************

* Schematics Netlist *

 R_R1        $N_0002 $N_0001  4.7k
 V_E2        0 $N_0003 5V
 R_R2        $N_0003 $N_0004  2.2k
 V_E1        $N_0005 0 10V
 D_D1        $N_0001 $N_0004 D1N4148-X2
 v_V3        $N_0005 $N_0002 0


****      Diode MODEL PARAMETERS

*************************************************************************

           D1N4148-X2
       IS   3.500000E-15


****      SMALL SIGNAL BIAS SOLUTION        TEMPERATURE =   27.000 DEG C

*************************************************************************


 NODE   VOLTAGE     NODE   VOLTAGE     NODE   VOLTAGE     NODE   VOLTAGE

($N_0001)   .2602           ($N_0002)   10.0000

($N_0003)   -5.0000          ($N_0004)    -.4409

($N_0005)   10.0000


     VOLTAGE SOURCE CURRENTS
     NAME           CURRENT

     V_E2       -2.072E-03
     V_E1       -2.072E-03
     v_V3        2.072E-03

   TOTAL POWER DISSIPATION   3.11E-02   WATTS

****      OPERATING POINT INFORMATION      TEMPERATURE =   27.000 DEG C

*************************************************************************


**** DIODES

     NAME        D_D1
     MODEL       D1N4148-X2
     ID          2.07E-03
     VD          7.01E-01
     REQ         1.25E+01
     CAP         0.00E+00
```

Figure 2.145 Output file for PSpice Windows analysis of the circuit of Figure 2.144.

DIODE CHARACTERISTICS

The characteristics of the D1N4148 diode used in the above analysis will now be obtained using a few maneuvers somewhat more sophisticated than those employed previously. First, the network in Fig. 2.146 is constructed using the procedures described above. Note, however, the **Vd** appearing above the diode **D1.** A point in the network (representing the voltage from anode to ground for the diode) has been identified as a particular voltage by double-clicking on the wire above the device and typing Vd in the **Set Attribute Value** as the **LABEL.** The resulting voltage V_d is, in this case, the voltage across the diode.

Next, **Analysis Setup** is chosen by either clicking on the Analysis Setup icon (at the top left edge of the schematic with the horizontal blue bar and the two small squares and rectangles) or by using the sequence **Analysis-Setup.** Within the **Analysis-Setup** dialog box the **DC Sweep** is enabled (the only one necessary for this exercise), followed by a single click of the **DC Sweep** rectangle. The **DC Sweep** dialog box will appear with various inquiries. In this case, we plan to sweep the source voltage from 0 to 10 V in 0.01-V increments, so the **Swept Var. Type** is Voltage Source, the **Sweep Type** will be linear, the **Name** E, and the **Start Value** 0 V, the **End Value** 10 V, and the **Increment** 0.01 V. Then, with an **OK** followed by a **Close** of the **Analysis Setup**

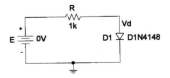

Figure 2.146 Network to obtain the characteristics of the D1N4148 diode.

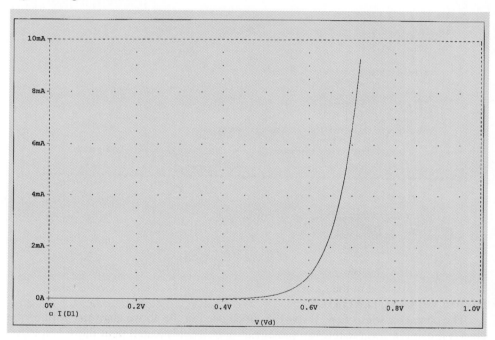

box, we are set to obtain the solution. The analysis to be performed will obtain a complete solution for the network for each value of E from 0 to 10 V in 0.01-V increments. In other words, the network will be analyzed 1000 times and the resulting data stored for the plot to be obtained. The analysis is performed by the sequence **Analysis-Run Probe,** followed by an immediate appearance of the **MicroSim Probe** graph showing only a horizontal axis of the source voltage E running from 0 to 10 V.

Since the plot we want is of I_D versus V_D, we have to change the horizontal (x-axis) to V_D. This is accomplished by selecting **Plot** and then **X-Axis Settings** to obtain the **X Axis Settings** dialog box. Next, we click **Axis Variable** and select **V(Vd)** from the listing. After **OK,** we return to the dialog box to set the horizontal scale. Choose **User Defined,** then enter 0 V to 1 V since this is the range of interest for Vd with a **Linear** scale. Click **OK** and you will find that the horizontal axis is now V(Vd) with a range of 0 to 1.0 V. The vertical axis must now be set to I_D by first choosing **Trace** (or the **Trace** icon, which is the red waveform with two sharp peaks and a set of axes) and then **Add** to obtain **Add Traces.** Choosing **I(D1)** and clicking **OK** will result in the plot of Fig. 2.147. In this case, the resulting plot extended from 0 to 10 mA. The range can be reduced or expanded by simply going to **Plot-Y-Axis Setting** and defining the range of interest.

Figure 2.147 Characteristics of the D1N4148 diode.

In the previous analysis, the voltage across the diode was 0.64 V, corresponding to a current of about 2 mA on the graph (recall the solution of 2.07 mA for the current). If the resulting current had been closer to 6.5 mA, the voltage across the diode would have been about 0.7 V and the PSpice solution closer to the hand-written approach. If I_s had been set to 3.5E-15A and all other parameters removed from the diode listing, the curve would have shifted to the right and an intersection of 0.7 V and 2.07 mA would have been obtained.

Electronics Workbench

The procedure for entering a circuit into EWB will now be described by checking the results of Example 2.15 which contained two diodes in a series-parallel configuration.

First the **Multisim** icon is chosen and the cursor placed on the **Sources** parts bin button. Left-click the mouse, and a list of sources will appear on which the dc voltage source symbol will appear. Once chosen, a dc source symbol will appear on the schematic with three rectangles. Simply moving the mouse will permit placing the source at any point on the screen. After choosing a position somewhere in the middle of the screen, do one additional left click, and the source will be set with a label such as **V1** and **12 V.** The default value of the source is 12 V, but the voltage can be changed by simply double-clicking on the voltage source symbol to obtain a **Battery** dialog box. Choose **Value** at the top, and change to 20 V for our example. Click **OK,** and **20V** will appear next to the battery on the schematic as shown in Fig. 2.148. If we call up the **Battery** dialog box again and choose **Label,** we can change the label from V1 to E to match our example. Finally, the labels can be placed anywhere around the battery symbol simply by clicking on the label or value of interest and then, while holding the left clicker down, moving it to the desired location. Resistors can now be set by selecting the **Basic** Parts Bin button and, when the toolbar appears, choosing the **Virtual** resistor so that the values can be set.

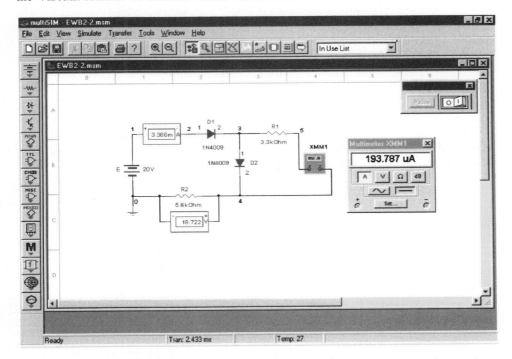

Figure 2.148 Verifying the results of Example 2.15 using Electronics Workbench.

Choosing the straight resistor option will result in a listing of manufactured resistors. Then proceed in the same manner as with the battery to place the resistor, and set the label and value. For diodes, choose the diode symbol, and under the **Component Browser** dialog box pick the IN4446 diode and place it as above. If you pick the **Virtual** diode, you will be limited to an ideal diode. Finally, return to the **Sources** parts bin button, and select the ground at the top of the toolbar and place in an appropriate location. Lastly, meters or indicators must be placed to obtain the quantities we want from the analysis.

Electronics Workbench is a very close simulation of the laboratory experience in that meters can be called up from the **Instruments** icon from the top toolbar. Once selected, eleven different meters will appear on a horizontal toolbar from which **multimeter** can be chosen to permit the choice of an ammeter, voltmeter, or ohmmeter. For this example, an ammeter (**A**) was selected along with the horizontal bar to indicate dc reading. Once **multimeter** is selected, two meters will appear on the schematic, one for the network and one for displaying the results. Another choice for obtaining voltage and current levels is to select **indicators** from the toolbar on the left of the screen. It is the fifth option from the bottom and looks like a number 8 LCD display. Once the options appear, choose **A** for ammeter, and the options are listed for how the meter should be placed. Indicators will appear as shown in Fig. 2.148 for the ammeter and voltmeter.

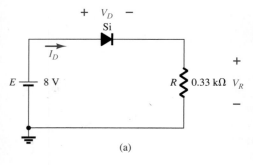

The components can then be connected using **Automatic Wiring.** When the cursor is placed at the end of the element followed by a left click of the mouse, the result is an **x** at the end of any component. Then move to the end of the other element, and left-click the mouse again—the wire will appear with the most direct route between the elements. During the finishing-off process, elements and wires will have to be moved. Simply click on the element or wire, and, while holding the clicker down, move it to the new location.

Finally, the software package must be told to perform the analysis, a process which can be initiated in a number of ways. One option is to select **Simulate** from the top toolbar followed by **Run/Stop.** Another is to select the fourth button from the left of the **Design** bar which looks like a number of branches of a network with diodes. Again the **Run/Stop** option will appear which will initiate the analysis. Once the option is chosen, the results will appear as shown in Fig. 2.148. When the analysis is complete, be sure to turn the process off by returning to the **Simulate** option and selecting **Run/Stop** again (remove the check).

The results obtained are a very close match with the solution obtained in Example 2.15. The differences result from the fact that the approximation of 0.7 V for the silicon diodes in the "on" state is not made here — these are actual diodes with terminal voltages that are sensitive to the current. You will notice in Fig. 2.148 that the node numbers are defined for the elements and network. The network has a total of five nodes due to the added instruments. Remove the instruments and the number drops to three. The node labels can be removed by simply going to **EDIT/User Preferences/Circuit** and removing the check from Show node names.

PROBLEMS

§ 2.2 Load-Line Analysis

1. (a) Using the characteristics of Fig. 2.149b, determine I_D, V_D, and V_R for the circuit of Fig. 2.149a.
  (b) Repeat part (a) using the approximate model for the diode and compare results.
  (c) Repeat part (a) using the ideal model for the diode and compare results.

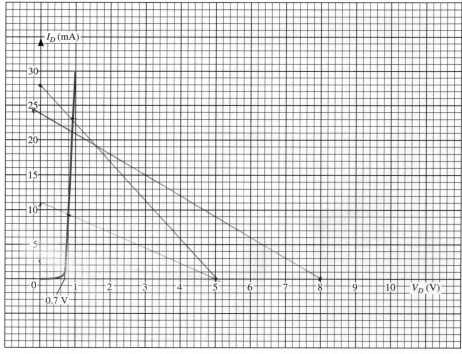

Figure 2.149 Problems 1, 2

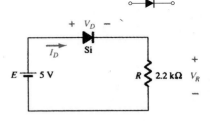

2. (a) Using the characteristics of Fig. 2.149b, determine I_D and V_D for the circuit of Fig. 2.150.
   (b) Repeat part (a) with $R = 0.47$ kΩ.
   (c) Repeat part (a) with $R = 0.18$ kΩ.
   (d) Is the level of V_D relatively close to 0.7 V in each case?

How do the resulting levels of I_D compare? Comment accordingly.

3. Determine the value of R for the circuit of Fig. 2.150 that will result in a diode current of 10 mA if $E = 7$ V. Use the characteristics of Fig. 2.149b for the diode.

Figure 2.150 Problems 2, 3

4. (a) Using the approximate characteristics for the Si diode, determine the level of V_D, I_D, and V_R for the circuit of Fig. 2.151.
   (b) Perform the same analysis as part (a) using the ideal model for the diode.
   (c) Do the results obtained in parts (a) and (b) suggest that the ideal model can provide a good approximation for the actual response under some conditions?

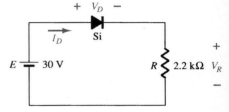

§ **2.4 Series Diode Configurations with DC Inputs**

5. Determine the current I for each of the configurations of Fig. 2.152 using the approximate equivalent model for the diode.

Figure 2.151 Problem 4

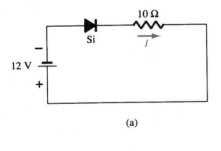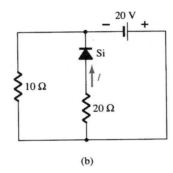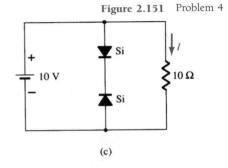

(a)　　　　　(b)　　　　　(c)

Figure 2.152 Problem 5

6. Determine V_o and I_D for the networks of Fig. 2.153.

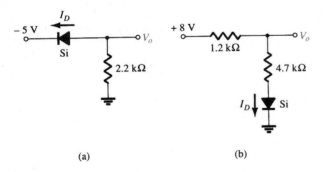

(a)　　　　　(b)

Figure 2.153 Problems 6, 49

*7. Determine the level of V_o for each network of Fig. 2.154.

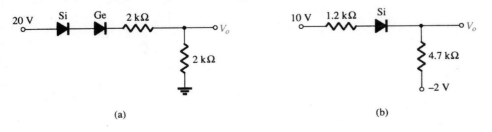

(a)　　　　　(b)

Figure 2.154 Problem 7

*8. Determine V_o and I_D for the networks of Fig. 2.155.

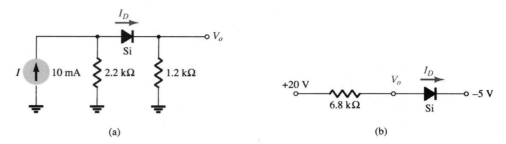

(a)

(b)

Figure 2.155   Problem 8

*9. Determine V_{o_1} and V_{o_2} for the networks of Fig. 2.156.

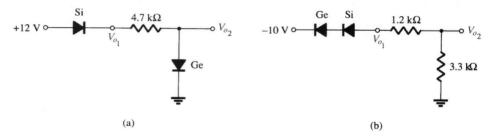

(a)

(b)

Figure 2.156   Problem 9

§ 2.5 Parallel and Series−Parallel Configurations

10. Determine V_o and I_D for the networks of Fig. 2.157.

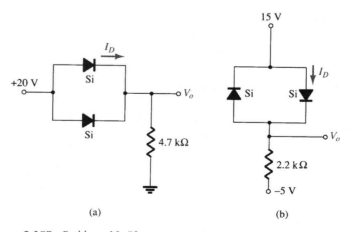

(a)

(b)

Figure 2.157   Problems 10, 50

*11. Determine V_o and I for the networks of Fig. 2.158.

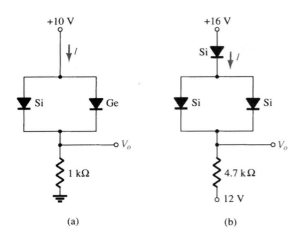

(a)　　　　　(b)

Figure 2.158　Problem 11

12. Determine V_{o_1}, V_{o_2}, and I for the network of Fig. 2.159.

*13. Determine V_o and I_D for the network of Fig. 2.160.

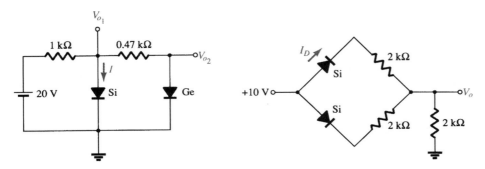

Figure 2.159　Problem 12

Figure 2.160　Problems 13, 51

§ 2.6 AND/OR Gates

14. Determine V_o for the network of Fig. 2.40 with 0 V on both inputs.

15. Determine V_o for the network of Fig. 2.40 with 10 V on both inputs.

16. Determine V_o for the network of Fig. 2.43 with 0 V on both inputs.

17. Determine V_o for the network of Fig. 2.43 with 10 V on both inputs.

18. Determine V_o for the negative logic OR gate of Fig. 2.161.

19. Determine V_o for the negative logic AND gate of Fig. 2.162.

20. Determine the level of V_o for the gate of Fig. 2.163.

21. Determine V_o for the configuration of Fig. 2.164.

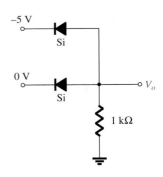

Figure 2.161　Problem 18

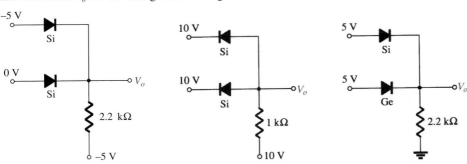

Figure 2.162　Problem 19　　　　**Figure 2.163**　Problem 20　　　　**Figure 2.164**　Problem 21

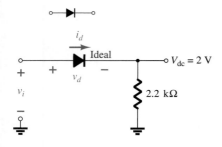

Figure 2.165   Problems 22, 23, 24

§ 2.7 Sinusoidal Inputs; Half-Wave Rectification

22. Assuming an ideal diode, sketch v_i, v_d, and i_d for the half-wave rectifier of Fig. 2.165. The input is a sinusoidal waveform with a frequency of 60 Hz

23. Repeat Problem 22 with a silicon diode ($V_T = 0.7$ V).

24. Repeat Problem 22 with a 6.8-kΩ load applied as shown in Fig. 2.166. Sketch v_L and i_L.

25. For the network of Fig. 2.167, sketch v_o and determine V_{dc}.

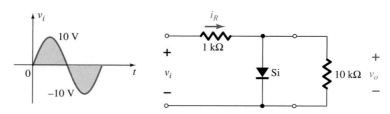

Figure 2.166   Problem 24   Figure 2.167   Problem 25

*26. For the network of Fig. 2.168, sketch v_o and i_R.

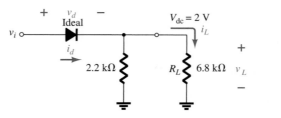

Figure 2.168   Problem 26

*27. (a) Given $P_{\text{max}} = 14$ mW for each diode of Fig. 2.169, determine the maximum current rating of each diode (using the approximate equivalent model).
   (b) Determine I_{max} for $V_{i_{\text{max}}} = 160$ V.
   (c) Determine the current through each diode at $V_{i_{\text{max}}}$ using the results of part (b).
   (d) If only one diode were present, determine the diode current and compare it to the maximum rating.

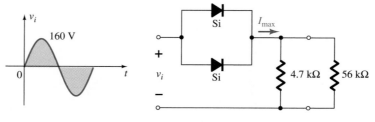

Figure 2.169   Problem 27

§ 2.8 Full-Wave Rectification

28. A full-wave bridge rectifier with a 120-V rms sinusoidal input has a load resistor of 1 kΩ.
    (a) If silicon diodes are employed, what is the dc voltage available at the load?
    (b) Determine the required PIV rating of each diode.
    (c) Find the maximum current through each diode during conduction.
    (d) What is the required power rating of each diode?

29. Determine v_o and the required PIV rating of each diode for the configuration of Fig. 2.170.

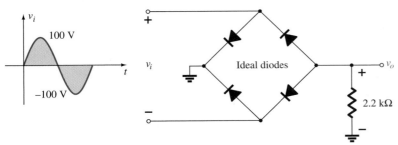

Figure 2.170   Problem 29

*30. Sketch v_o for the network of Fig. 2.171 and determine the dc voltage available.

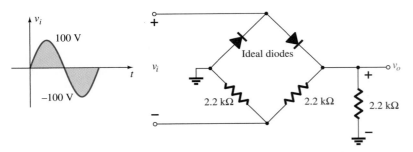

Figure 2.171   Problem 30

*31. Sketch v_o for the network of Fig. 2.172 and determine the dc voltage available.

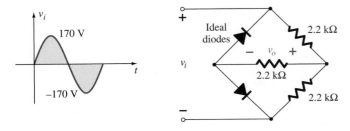

Figure 2.172   Problem 31

§ 2.9 Clippers

32. Determine v_o for each network of Fig. 2.173 for the input shown.

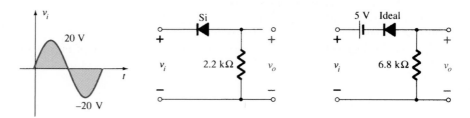

Figure 2.173   Problem 32

33. Determine v_o for each network of Fig. 2.174 for the input shown.

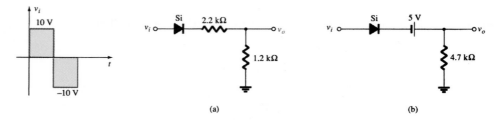

Figure 2.174   Problem 33

34. Determine v_o for each network of Fig. 2.175 for the input shown.

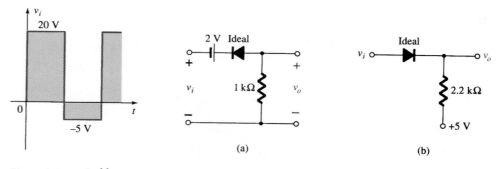

Figure 2.175   Problem 34

*35. Determine v_o for each network of Fig. 2.176 for the input shown.

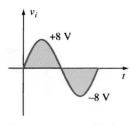

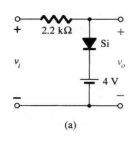

(a)

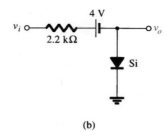

(b)

Figure 2.176   Problem 35

36. Sketch i_R and v_o for the network of Fig. 2.177 for the input shown.

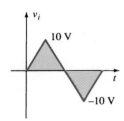

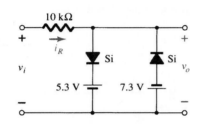

Figure 2.177   Problem 36

§ 2.10 Clampers

37. Sketch v_o for each network of Fig. 2.178 for the input shown.

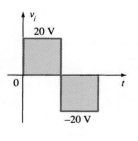

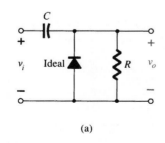

(a)

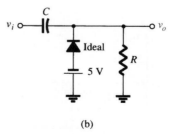

(b)

Figure 2.178   Problem 37

38. Sketch v_o for each network of Fig. 2.179 for the input shown. Would it be a good approximation to consider the diode to be ideal for both configurations? Why?

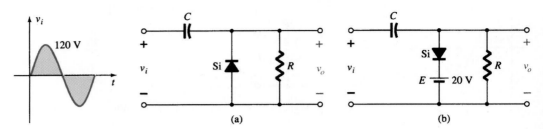

Figure 2.179 Problem 38

*   **39.** For the network of Fig. 2.180:
    (a) Calculate 5τ.
    (b) Compare 5τ to half the period of the applied signal.
    (c) Sketch v_o.

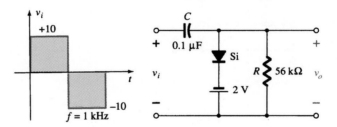

Figure 2.180 Problem 39

*   **40.** Design a clamper to perform the function indicated in Fig. 2.181.

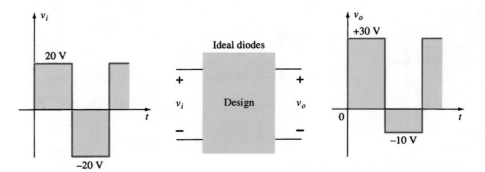

Figure 2.181 Problem 40

128   Chapter 2   Diode Applications

*41. Design a clamper to perform the function indicated in Fig. 2.182.

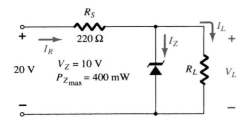
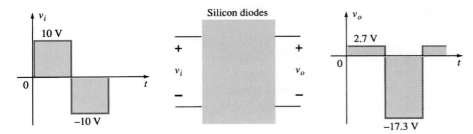

Figure 2.182   Problem 41

§ **2.11 Zener Diodes**

*42. (a) Determine V_L, I_L, I_Z, and I_R for the network Fig. 2.183 if $R_L = 180 \, \Omega$.
   (b) Repeat part (a) if $R_L = 470 \, \Omega$.
   (c) Determine the value of R_L that will establish maximum power conditions for the Zener diode.
   (d) Determine the minimum value of R_L to ensure that the Zener diode is in the "on" state.

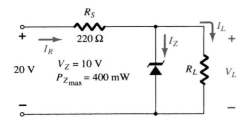

Figure 2.183   Problem 42

*43. (a) Design the network of Fig. 2.184 to maintain V_L at 12 V for a load variation (I_L) from 0 to 200 mA. That is, determine R_s and V_Z.
   (b) Determine $P_{Z_{max}}$ for the Zener diode of part (a).

*44. For the network of Fig. 2.185, determine the range of V_i that will maintain V_L at 8 V and not exceed the maximum power rating of the Zener diode.

45. Design a voltage regulator that will maintain an output voltage of 20 V across a 1-kΩ load with an input that will vary between 30 and 50 V. That is, determine the proper value of R_s and the maximum current I_{ZM}.

46. Sketch the output of the network of Fig. 2.142 if the input is a 50-V square wave. Repeat for a 5-V square wave.

§ **2.12 Voltage-Multiplier Circuits**

47. Determine the voltage available from the voltage doubler of Fig. 2.120 if the secondary voltage of the transformer is 120 V (rms).

48. Determine the required PIV ratings of the diodes of Fig. 2.120 in terms of the peak secondary voltage V_m.

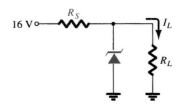

Figure 2.184   Problem 43

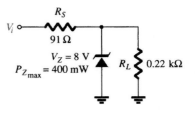

Figure 2.185   Problems 44, 52

§ 2.15 Computer Analysis

49. Perform an analysis of the network of Fig. 2.153 using PSpice Windows.

50. Perform an analysis of the network of Fig. 2.157 using PSpice Windows.

51. Perform an analysis of the network of Fig. 2.160 using PSpice Windows.

52. Perform a general analysis of the Zener network of Fig. 2.185 using PSpice Windows.

53. Repeat Problem 49 using Electronics Workbench.

54. Repeat Problem 50 using Electronics Workbench.

55. Repeat Problem 51 using Electronics Workbench.

56. Repeat Problem 52 using Electronics Workbench.

*Please Note: Asterisks indicate more difficult problems.

Bipolar Junction Transistors

CHAPTER

3

β

3.1 INTRODUCTION

During the period 1904–1947, the vacuum tube was undoubtedly the electronic device of interest and development. In 1904, the vacuum-tube diode was introduced by J. A. Fleming. Shortly thereafter, in 1906, Lee De Forest added a third element, called the *control grid,* to the vacuum diode, resulting in the first amplifier, the *triode*. In the following years, radio and television provided great stimulation to the tube industry. Production rose from about 1 million tubes in 1922 to about 100 million in 1937. In the early 1930s the four-element tetrode and five-element pentode gained prominence in the electron-tube industry. In the years to follow, the industry became one of primary importance and rapid advances were made in design, manufacturing techniques, high-power and high-frequency applications, and miniaturization.

On December 23, 1947, however, the electronics industry was to experience the advent of a completely new direction of interest and development. It was on the afternoon of this day that Walter H. Brattain and John Bardeen demonstrated the amplifying action of the first transistor at the Bell Telephone Laboratories. The original transistor (a point-contact transistor) is shown in Fig. 3.1. The advantages of this three-terminal solid-state device over the tube were immediately obvious: It was smaller

Co-inventors of the first transistor at Bell Laboratories: Dr. William Shockley (seated); Dr. John Bardeen (left); Dr. Walter H. Brattain. (Courtesy of AT&T Archives.)

Dr. Shockley Born: London, England, 1910 PhD Harvard, 1936

Dr. Bardeen Born: Madison, Wisconsin, 1908 PhD Princeton, 1936

Dr. Brattain Born: Amoy, China, 1902 PhD University of Minnesota, 1928

All shared the Nobel Prize in 1956 for this contribution.

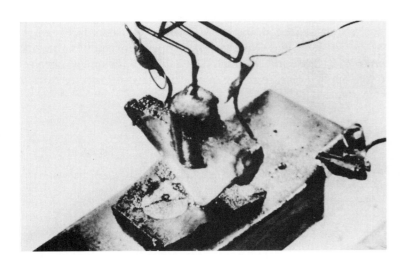

Figure 3.1 The first transistor. (Courtesy Bell Telephone Laboratories.)

131

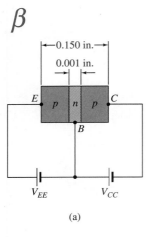

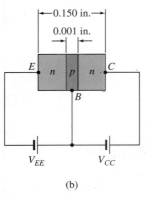

Figure 3.2 Types of transistors: (a) *pnp*; (b) *npn*.

and lightweight; had no heater requirement or heater loss; had rugged construction; and was more efficient since less power was absorbed by the device itself; it was instantly available for use, requiring no warm-up period; and lower operating voltages were possible. Note in the discussion above that this chapter is our first discussion of devices with three or more terminals. You will find that all amplifiers (devices that increase the voltage, current, or power level) will have at least three terminals with one controlling the flow between two other terminals.

3.2 TRANSISTOR CONSTRUCTION

The transistor is a three-layer semiconductor device consisting of either two *n*- and one *p*-type layers of material or two *p*- and one *n*-type layers of material. The former is called an *npn transistor*, while the latter is called a *pnp transistor*. Both are shown in Fig. 3.2 with the proper dc biasing. We will find in Chapter 4 that the dc biasing is necessary to establish the proper region of operation for ac amplification. The emitter layer is heavily doped, the base lightly doped, and the collector only lightly doped. The outer layers have widths much greater than the sandwiched *p*- or *n*-type material. For the transistors shown in Fig. 3.2 the ratio of the total width to that of the center layer is $0.150/0.001 = 150:1$. The doping of the sandwiched layer is also considerably less than that of the outer layers (typically, 10:1 or less). This lower doping level decreases the conductivity (increases the resistance) of this material by limiting the number of "free" carriers.

For the biasing shown in Fig. 3.2 the terminals have been indicated by the capital letters *E* for *emitter*, *C* for *collector*, and *B* for *base*. An appreciation for this choice of notation will develop when we discuss the basic operation of the transistor. The abbreviation BJT, from *bipolar junction transistor*, is often applied to this three-terminal device. The term *bipolar* reflects the fact that holes *and* electrons participate in the injection process into the oppositely polarized material. If only one carrier is employed (electron or hole), it is considered a *unipolar* device. The Schottky diode of Chapter 19 is such a device.

3.3 TRANSISTOR OPERATION

The basic operation of the transistor will now be described using the *pnp* transistor of Fig. 3.2a. The operation of the *npn* transistor is exactly the same if the roles played by the electron and hole are interchanged. In Fig. 3.3 the *pnp* transistor has been redrawn without the base-to-collector bias. Note the similarities between this situation and that of the *forward-biased* diode in Chapter 1. The depletion region has been reduced in width due to the applied bias, resulting in a heavy flow of majority carriers from the *p*- to the *n*-type material.

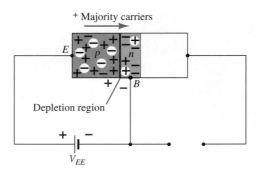

Figure 3.3 Forward-biased junction of a *pnp* transistor.

Let us now remove the base-to-emitter bias of the *pnp* transistor of Fig. 3.2a as shown in Fig. 3.4. Consider the similarities between this situation and that of the *reverse-biased* diode of Section 1.6. Recall that the flow of majority carriers is zero, resulting in only a minority-carrier flow, as indicated in Fig. 3.4. In summary, therefore:

> *One p-n junction of a transistor is reverse biased, while the other is forward biased.*

In Fig. 3.5 both biasing potentials have been applied to a *pnp* transistor, with the resulting majority- and minority-carrier flow indicated. Note in Fig. 3.5 the widths of the depletion regions, indicating clearly which junction is forward-biased and which is reverse-biased. As indicated in Fig. 3.5, a large number of majority carriers will diffuse across the forward-biased *p-n* junction into the *n*-type material. The question then is whether these carriers will contribute directly to the base current I_B or pass directly into the *p*-type material. Since the sandwiched *n*-type material is very thin and has a low conductivity, a very small number of these carriers will take this path of high resistance to the base terminal. The magnitude of the base current is typically on the order of microamperes as compared to milliamperes for the emitter and collector currents. The larger number of these majority carriers will diffuse across the reverse-biased junction into the *p*-type material connected to the collector terminal as indicated in Fig. 3.5. The reason for the relative ease with which the majority carriers can cross the reverse-biased junction is easily understood if we consider that for the reverse-biased diode the injected majority carriers will appear as minority carriers in the *n*-type material. In other words, there has been an *injection* of minority carriers into the *n*-type base region material. Combining this with the fact that all the minority carriers in the depletion region will cross the reverse-biased junction of a diode accounts for the flow indicated in Fig. 3.5.

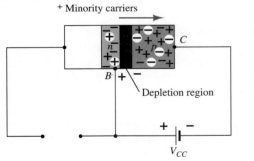

Figure 3.4 Reverse-biased junction of a *pnp* transistor.

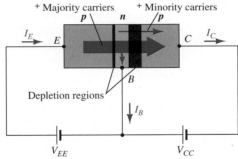

Figure 3.5 Majority and minority carrier flow of a *pnp* transistor.

Applying Kirchhoff's current law to the transistor of Fig. 3.5 as if it were a single node, we obtain

$$I_E = I_C + I_B \tag{3.1}$$

and find that the emitter current is the sum of the collector and base currents. The collector current, however, is comprised of two components—the majority and minority carriers as indicated in Fig. 3.5. The minority-current component is called the *leakage current* and is given the symbol I_{CO} (I_C current with emitter terminal *O*pen). The collector current, therefore, is determined in total by Eq. (3.2).

$$I_C = I_{C_{\text{majority}}} + I_{CO_{\text{minority}}} \tag{3.2}$$

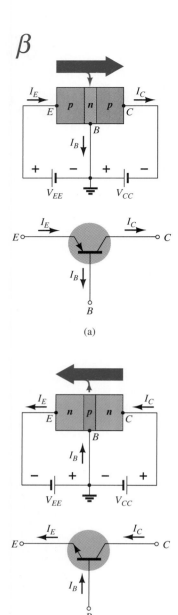

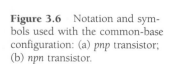

Figure 3.6 Notation and symbols used with the common-base configuration: (a) *pnp* transistor; (b) *npn* transistor.

For general-purpose transistors, I_C is measured in milliamperes, while I_{CO} is measured in microamperes or nanoamperes. I_{CO}, like I_s for a reverse-biased diode, is temperature sensitive and must be examined carefully when applications of wide temperature ranges are considered. It can severely affect the stability of a system at high temperature if not considered properly. Improvements in construction techniques have resulted in significantly lower levels of I_{CO}, to the point where its effect can often be ignored.

3.4 COMMON-BASE CONFIGURATION

The notation and symbols used in conjunction with the transistor in the majority of texts and manuals published today are indicated in Fig. 3.6 for the common-base configuration with *pnp* and *npn* transistors. The common-base terminology is derived from the fact that the base is common to both the input and output sides of the configuration. In addition, the base is usually the terminal closest to, or at, ground potential. Throughout this book all current directions will refer to conventional (hole) flow rather than electron flow. This choice was based primarily on the fact that the vast amount of literature available at educational and industrial institutions employs conventional flow and the arrows in all electronic symbols have a direction defined by this convention. Recall that the arrow in the diode symbol defined the direction of conduction for conventional current. For the transistor:

The arrow in the graphic symbol defines the direction of emitter current (conventional flow) through the device.

All the current directions appearing in Fig. 3.6 are the actual directions as defined by the choice of conventional flow. Note in each case that $I_E = I_C + I_B$. Note also that the applied biasing (voltage sources) are such as to establish current in the direction indicated for each branch. That is, compare the direction of I_E to the polarity of V_{EE} for each configuration and the direction of I_C to the polarity of V_{CC}.

To fully describe the behavior of a three-terminal device such as the common-base amplifiers of Fig. 3.6 requires two sets of characteristics—one for the *driving point* or *input* parameters and the other for the *output* side. The input set for the common-base amplifier as shown in Fig. 3.7 will relate an input current (I_E) to an input voltage (V_{BE}) for various levels of output voltage (V_{CB}).

The output set will relate an output current (I_C) to an output voltage (V_{CB}) for various levels of input current (I_E) as shown in Fig. 3.8. The output or *collector* set of characteristics has three basic regions of interest, as indicated in Fig. 3.8: the *active,*

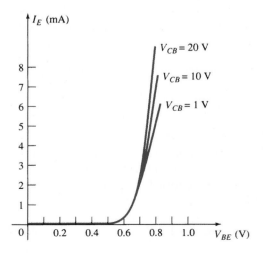

Figure 3.7 Input or driving point characteristics for a common-base silicon transistor amplifier.

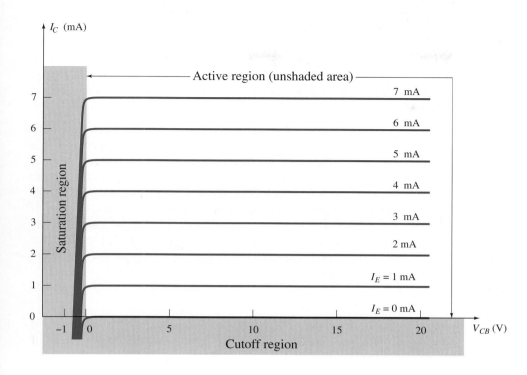

Figure 3.8 Output or collector characteristics for a common-base transistor amplifier.

cutoff, and *saturation* regions. The active region is the region normally employed for linear (undistorted) amplifiers. In particular:

In the active region the base-emitter junction is forward-biased, while the collector-base junction is reverse-biased.

The active region is defined by the biasing arrangements of Fig. 3.6. At the lower end of the active region the emitter current (I_E) is zero, the collector current is simply that due to the reverse saturation current I_{CO}, as indicated in Fig. 3.9. The current I_{CO} is so small (microamperes) in magnitude compared to the vertical scale of I_C (milliamperes) that it appears on virtually the same horizontal line as $I_C = 0$. The circuit conditions that exist when $I_E = 0$ for the common-base configuration are shown in Fig. 3.9. The notation most frequently used for I_{CO} on data and specification sheets is, as indicated in Fig. 3.9, I_{CBO}. Because of improved construction techniques, the level of I_{CBO} for general-purpose transistors (especially silicon) in the low- and mid-power ranges is usually so low that its effect can be ignored. However, for higher power units I_{CBO} will still appear in the microampere range. In addition, keep in mind that I_{CBO}, like I_s, for the diode (both reverse leakage currents) is temperature sensitive. At higher temperatures the effect of I_{CBO} may become an important factor since it increases so rapidly with temperature.

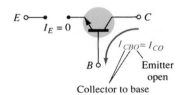

Figure 3.9 Reverse saturation current.

Note in Fig. 3.8 that as the emitter current increases above zero, the collector current increases to a magnitude essentially equal to that of the emitter current as determined by the basic transistor-current relations. Note also the almost negligible effect of V_{CB} on the collector current for the active region. The curves clearly indicate that *a first approximation to the relationship between I_E and I_C in the active region is given by*

$$\boxed{I_C \cong I_E} \tag{3.3}$$

As inferred by its name, the cutoff region is defined as that region where the collector current is 0 A, as revealed on Fig. 3.8. In addition:

In the cutoff region the base-emitter and collector-base junctions of a transistor are both reverse-biased.

β

The saturation region is defined as that region of the characteristics to the left of $V_{CB} = 0$ V. The horizontal scale in this region was expanded to clearly show the dramatic change in characteristics in this region. Note the exponential increase in collector current as the voltage V_{CB} increases toward 0 V.

In the saturation region the base-emitter and collector-base junctions are forward-biased.

The input characteristics of Fig. 3.7 reveal that for fixed values of collector voltage (V_{CB}), as the base-to-emitter voltage increases, the emitter current increases in a manner that closely resembles the diode characteristics. In fact, increasing levels of V_{CB} have such a small effect on the characteristics that as a first approximation the change due to changes in V_{CB} can be ignored and the characteristics drawn as shown in Fig. 3.10a. If we then apply the piecewise-linear approach, the characteristics of Fig. 3.10b will result. Taking it a step further and ignoring the slope of the curve and therefore the resistance associated with the forward-biased junction will result in the characteristics of Fig. 3.10c. For the analysis to follow in this book the equivalent model of Fig. 3.10c will be employed for all dc analysis of transistor networks. That is, once a transistor is in the "on" state, the base-to-emitter voltage will be assumed to be the following:

$$V_{BE} = 0.7 \text{ V} \qquad\qquad (3.4)$$

In other words, the effect of variations due to V_{CB} and the slope of the input characteristics will be ignored as we strive to analyze transistor networks in a manner that will provide a good approximation to the actual response without getting too involved with parameter variations of less importance.

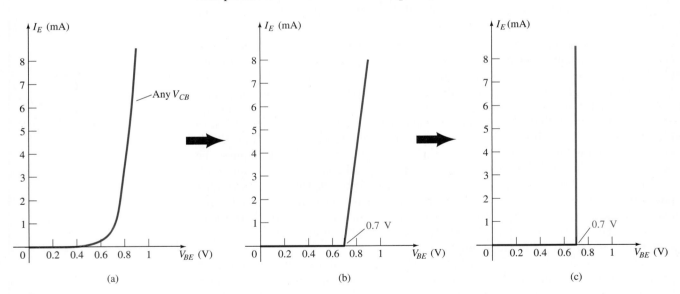

Figure 3.10 Developing the equivalent model to be employed for the base-to-emitter region of an amplifier in the dc mode.

It is important to fully appreciate the statement made by the characteristics of Fig. 3.10c. They specify that with the transistor in the "on" or active state the voltage from base to emitter will be 0.7 V at *any* level of emitter current as controlled by the external network. In fact, at the first encounter of any transistor configuration in the dc mode, one can now immediately specify that the voltage from base to emitter is 0.7 V if the device is in the active region—a very important conclusion for the dc analysis to follow.

(a) Using the characteristics of Fig. 3.8, determine the resulting collector current if $I_E = 3$ mA and $V_{CB} = 10$ V.
(b) Using the characteristics of Fig. 3.8, determine the resulting collector current if I_E remains at 3 mA but V_{CB} is reduced to 2 V.
(c) Using the characteristics of Figs. 3.7 and 3.8, determine V_{BE} if $I_C = 4$ mA and $V_{CB} = 20$ V.
(d) Repeat part (c) using the characteristics of Figs. 3.8 and 3.10c.

Solution

(a) The characteristics clearly indicate that $I_C \cong I_E = $ **3 mA.**
(b) The effect of changing V_{CB} is negligible and I_C continues to be **3 mA.**
(c) From Fig. 3.8, $I_E \cong I_C = 4$ mA. On Fig. 3.7 the resulting level of V_{BE} is about **0.74 V.**
(d) Again from Fig. 3.8, $I_E \cong I_C = 4$ mA. However, on Fig. 3.10c, V_{BE} is **0.7 V** for any level of emitter current.

Alpha (α)

In the dc mode the levels of I_C and I_E due to the majority carriers are related by a quantity called *alpha* and defined by the following equation:

$$\alpha_{dc} = \frac{I_C}{I_E} \tag{3.5}$$

where I_C and I_E are the levels of current at the point of operation. Even though the characteristics of Fig. 3.8 would suggest that $\alpha = 1$, for practical devices the level of alpha typically extends from 0.90 to 0.998, with most approaching the high end of the range. Since alpha is defined solely for the majority carriers, Eq. (3.2) becomes

$$I_C = \alpha I_E + I_{CBO} \tag{3.6}$$

For the characteristics of Fig. 3.8 when $I_E = 0$ mA, I_C is therefore equal to I_{CBO}, but as mentioned earlier, the level of I_{CBO} is usually so small that it is virtually undetectable on the graph of Fig. 3.8. In other words, when $I_E = 0$ mA on Fig. 3.8, I_C also appears to be 0 mA for the range of V_{CB} values.

For ac situations where the point of operation moves on the characteristic curve, an ac alpha is defined by

$$\alpha_{ac} = \frac{\Delta I_C}{\Delta I_E}\bigg|_{V_{CB} = \text{constant}} \tag{3.7}$$

The ac alpha is formally called the *common-base, short-circuit, amplification factor*, for reasons that will be more obvious when we examine transistor equivalent circuits in Chapter 7. For the moment, recognize that Eq. (3.7) specifies that a relatively small change in collector current is divided by the corresponding change in I_E with the collector-to-base voltage held constant. For most situations the magnitudes of α_{ac} and α_{dc} are quite close, permitting the use of the magnitude of one for the other. The use of an equation such as (3.7) will be demonstrated in Section 3.6.

Biasing

The proper biasing of the common-base configuration in the active region can be determined quickly using the approximation $I_C \cong I_E$ and assuming for the moment that

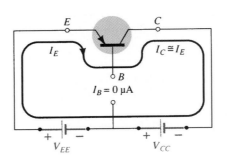

Figure 3.11 Establishing the proper biasing management for a common-base *pnp* transistor in the active region.

$I_B \cong 0 \ \mu A$. The result is the configuration of Fig. 3.11 for the *pnp* transistor. The arrow of the symbol defines the direction of conventional flow for $I_E \cong I_C$. The dc supplies are then inserted with a polarity that will support the resulting current direction. For the *npn* transistor the polarities will be reversed.

Some students feel that they can remember whether the arrow of the device symbol is pointing in or out by matching the letters of the transistor type with the appropriate letters of the phrases "pointing in" or "not pointing in." For instance, there is a match between the letters *npn* and the italic letters of *n*ot *p*ointing i*n* and the letters *pnp* with *p*ointing i*n*.

3.5 TRANSISTOR AMPLIFYING ACTION

Now that the relationship between I_C and I_E has been established in Section 3.4, the basic amplifying action of the transistor can be introduced on a surface level using the network of Fig. 3.12. The dc biasing does not appear in the figure since our interest will be limited to the ac response. For the common-base configuration the ac input resistance determined by the characteristics of Fig. 3.7 is quite small and typically varies from 10 to 100 Ω. The output resistance as determined by the curves of Fig. 3.8 is quite high (the more horizontal the curves the higher the resistance) and typically varies from 50 kΩ to 1 MΩ (100 kΩ for the transistor of Fig. 3.12). The difference in resistance is due to the forward-biased junction at the input (base to emitter) and the reverse-biased junction at the output (base to collector). Using a common value of 20 Ω for the input resistance, we find that

$$I_i = \frac{V_i}{R_i} = \frac{200 \ mV}{20 \ \Omega} = \textbf{10 mA}$$

If we assume for the moment that $\alpha_{ac} = 1 \ (I_c = I_e)$,

$$I_L = I_i = 10 \ mA$$

and
$$V_L = I_L R$$
$$= (10 \ mA)(5 \ k\Omega)$$
$$= \textbf{50 V}$$

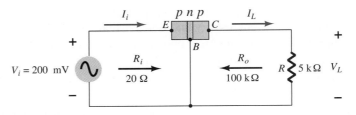

Figure 3.12 Basic voltage amplification action of the common-base configuration.

The voltage amplification is

$$A_v = \frac{V_L}{V_i} = \frac{50 \text{ V}}{200 \text{ mV}} = \textbf{250}$$

Typical values of voltage amplification for the common-base configuration vary from 50 to 300. The current amplification (I_C/I_E) is always less than 1 for the common-base configuration. This latter characteristic should be obvious since $I_C = \alpha I_E$ and α is always less than 1.

The basic amplifying action was produced by *transferring* a current I from a low- to a high-*resistance* circuit. The combination of the two terms in italics results in the label *transistor;* that is,

$$tran\text{sfer} + re\text{sistor} \rightarrow transistor$$

3.6 COMMON-EMITTER CONFIGURATION

The most frequently encountered transistor configuration appears in Fig. 3.13 for the *pnp* and *npn* transistors. It is called the *common-emitter configuration* since the emitter is common or reference to both the input and output terminals (in this case common to both the base and collector terminals). Two sets of characteristics are again necessary to describe fully the behavior of the common-emitter configuration: one for the *input* or *base-emitter* circuit and one for the *output* or *collector-emitter* circuit. Both are shown in Fig. 3.14.

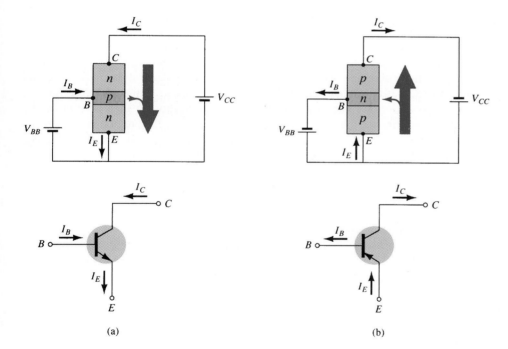

Figure 3.13 Notation and symbols used with the common-emitter configuration: (a) *npn* transistor; (b) *pnp* transistor.

The emitter, collector, and base currents are shown in their actual conventional current direction. Even though the transistor configuration has changed, the current relations developed earlier for the common-base configuration are still applicable. That is, $I_E = I_C + I_B$ and $I_C = \alpha I_E$.

For the common-emitter configuration the output characteristics are a plot of the output current (I_C) versus output voltage (V_{CE}) for a range of values of input current (I_B). The input characteristics are a plot of the input current (I_B) versus the input voltage (V_{BE}) for a range of values of output voltage (V_{CE}).

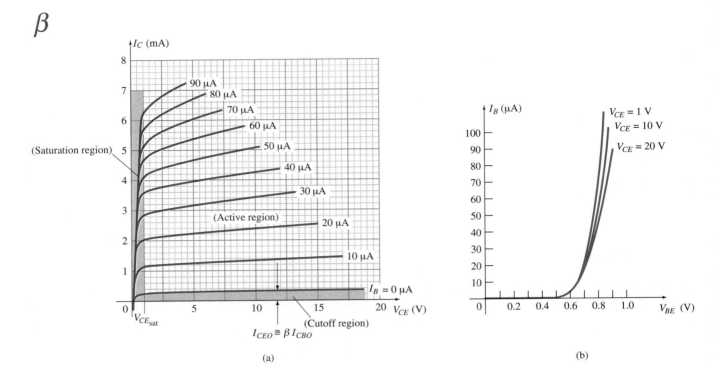

Figure 3.14 Characteristics of a silicon transistor in the common-emitter configuration: (a) collector characteristics; (b) base characteristics.

Note that on the characteristics of Fig. 3.14 the magnitude of I_B is in microamperes, compared to milliamperes of I_C. Consider also that the curves of I_B are not as horizontal as those obtained for I_E in the common-base configuration, indicating that the collector-to-emitter voltage will influence the magnitude of the collector current.

The active region for the common-emitter configuration is that portion of the upper-right quadrant that has the greatest linearity, that is, that region in which the curves for I_B are nearly straight and equally spaced. In Fig. 3.14a this region exists to the right of the vertical dashed line at $V_{CE_{sat}}$ and above the curve for I_B equal to zero. The region to the left of $V_{CE_{sat}}$ is called the saturation region.

In the active region of a common-emitter amplifier, the base-emitter junction is forward-biased, while the collector-base junction is reverse-biased.

You will recall that these were the same conditions that existed in the active region of the common-base configuration. The active region of the common-emitter configuration can be employed for voltage, current, or power amplification.

The cutoff region for the common-emitter configuration is not as well defined as for the common-base configuration. Note on the collector characteristics of Fig. 3.14 that I_C is not equal to zero when I_B is zero. For the common-base configuration, when the input current I_E was equal to zero, the collector current was equal only to the reverse saturation current I_{CO}, so that the curve $I_E = 0$ and the voltage axis were, for all practical purposes, one.

The reason for this difference in collector characteristics can be derived through the proper manipulation of Eqs. (3.3) and (3.6). That is,

$$\text{Eq. (3.6):}\quad I_C = \alpha I_E + I_{CBO}$$

Substitution gives $\qquad \text{Eq. (3.3):}\quad I_C = \alpha(I_C + I_B) + I_{CBO}$

Rearranging yields $\qquad\qquad I_C = \dfrac{\alpha I_B}{1 - \alpha} + \dfrac{I_{CBO}}{1 - \alpha}$ $\qquad\qquad$ (3.8)

If we consider the case discussed above, where $I_B = 0$ A, and substitute a typical value of α such as 0.996, the resulting collector current is the following:

$$I_C = \frac{\alpha(0\ A)}{1 - \alpha} + \frac{I_{CBO}}{1 - 0.996}$$

$$= \frac{I_{CBO}}{0.004} = 250 I_{CBO}$$

If I_{CBO} were 1 μA, the resulting collector current with $I_B = 0$ A would be $250(1\ \mu A) = 0.25$ mA, as reflected in the characteristics of Fig. 3.14.

For future reference, the collector current defined by the condition $I_B = 0\ \mu$A will be assigned the notation indicated by Eq. (3.9).

$$I_{CEO} = \frac{I_{CBO}}{1 - \alpha}\bigg|_{I_B = 0\,\mu A} \tag{3.9}$$

In Fig. 3.15 the conditions surrounding this newly defined current are demonstrated with its assigned reference direction.

For linear (least distortion) amplification purposes, cutoff for the common-emitter configuration will be defined by $I_C = I_{CEO}$.

In other words, the region below $I_B = 0\ \mu$A is to be avoided if an undistorted output signal is required.

When employed as a switch in the logic circuitry of a computer, a transistor will have two points of operation of interest: one in the cutoff and one in the saturation region. The cutoff condition should ideally be $I_C = 0$ mA for the chosen V_{CE} voltage. Since I_{CEO} is typically low in magnitude for silicon materials, *cutoff will exist for switching purposes when $I_B = 0\ \mu$A or $I_C = I_{CEO}$ for silicon transistors only. For germanium transistors, however, cutoff for switching purposes will be defined as those conditions that exist when $I_C = I_{CBO}$.* This condition can normally be obtained for germanium transistors by reverse-biasing the base-to-emitter junction a few tenths of a volt.

Recall for the common-base configuration that the input set of characteristics was approximated by a straight-line equivalent that resulted in $V_{BE} = 0.7$ V for any level of I_E greater than 0 mA. For the common-emitter configuration the same approach can be taken, resulting in the approximate equivalent of Fig. 3.16. The result supports our earlier conclusion that for a transistor in the "on" or active region the base-to-emitter voltage is 0.7 V. In this case the voltage is fixed for any level of base current.

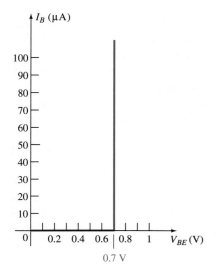

Figure 3.16 Piecewise-linear equivalent for the diode characteristics of Fig. 3.14b.

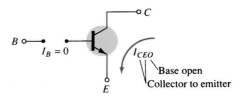

Figure 3.15 Circuit conditions related to I_{CEO}.

EXAMPLE 3.2

(a) Using the characteristics of Fig. 3.14, determine I_C at $I_B = 30 \, \mu A$ and $V_{CE} = 10$ V.
(b) Using the characteristics of Fig. 3.14, determine I_C at $V_{BE} = 0.7$ V and $V_{CE} = 15$ V.

Solution

(a) At the intersection of $I_B = 30 \, \mu A$ and $V_{CE} = 10$ V, $I_C = $ **3.4 mA.**
(b) Using Fig. 3.14b, $I_B = 20 \, \mu A$ at $V_{BE} = 0.7$ V. From Fig. 3.14a we find that $I_C = $ **2.5 mA** at the intersection of $I_B = 20 \, \mu A$ and $V_{CE} = 15$ V.

Beta (β)

In the dc mode the levels of I_C and I_B are related by a quantity called *beta* and defined by the following equation:

$$\beta_{dc} = \frac{I_C}{I_B} \tag{3.10}$$

where I_C and I_B are determined at a particular operating point on the characteristics. For practical devices the level of β typically ranges from about 50 to over 400, with most in the midrange. As for α, β certainly reveals the relative magnitude of one current to the other. For a device with a β of 200, the collector current is 200 times the magnitude of the base current.

On specification sheets β_{dc} is usually included as h_{FE} with the h derived from an ac *h*ybrid equivalent circuit to be introduced in Chapter 7. The subscripts *FE* are derived from *f*orward-current amplification and common-*e*mitter configuration, respectively.

For ac situations an ac beta has been defined as follows:

$$\beta_{ac} = \frac{\Delta I_C}{\Delta I_B}\bigg|_{V_{CE} = \text{constant}} \tag{3.11}$$

The formal name for β_{ac} is *common-emitter, forward-current, amplification factor*. Since the collector current is usually the output current for a common-emitter configuration and the base current the input current, the term *amplification* is included in the nomenclature above.

Equation (3.11) is similar in format to the equation for α_{ac} in Section 3.4. The procedure for obtaining α_{ac} from the characteristic curves was not described because of the difficulty of actually measuring changes of I_C and I_E on the characteristics. Equation (3.11), however, is one that can be described with some clarity, and in fact, the result can be used to find α_{ac} using an equation to be derived shortly.

On specification sheets β_{ac} is normally referred to as h_{fe}. Note that the only difference between the notation used for the dc beta, specifically, $\beta_{dc} = h_{FE}$, is the type of lettering for each subscript quantity. The lowercase letter h continues to refer to the hybrid equivalent circuit to be described in Chapter 7 and the *fe* to the *f*orward current gain in the common-*e*mitter configuration.

The use of Eq. (3.11) is best described by a numerical example using an actual set of characteristics such as appearing in Fig. 3.14a and repeated in Fig. 3.17. Let us determine β_{ac} for a region of the characteristics defined by an operating point of $I_B = 25 \, \mu A$ and $V_{CE} = 7.5$ V as indicated on Fig. 3.17. The restriction of $V_{CE} = $ constant requires that a vertical line be drawn through the operating point at $V_{CE} = 7.5$ V. At any location on this vertical line the voltage V_{CE} is 7.5 V, a constant. The change

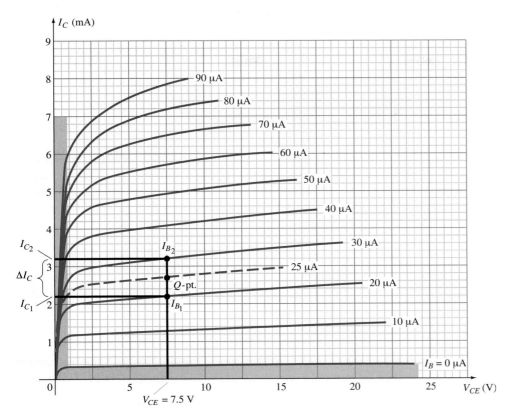

Figure 3.17 Determining β_{ac} and β_{dc} from the collector characteristics.

in $I_B (\Delta I_B)$ as appearing in Eq. (3.11) is then defined by choosing two points on either side of the Q-point along the vertical axis of about equal distances to either side of the Q-point. For this situation the $I_B = 20\ \mu$A and $30\ \mu$A curves meet the requirement without extending too far from the Q-point. They also define levels of I_B that are easily defined rather than have to interpolate the level of I_B between the curves. It should be mentioned that the best determination is usually made by keeping the chosen ΔI_B as small as possible. At the two intersections of I_B and the vertical axis, the two levels of I_C can be determined by drawing a horizontal line over to the vertical axis and reading the resulting values of I_C. The resulting β_{ac} for the region can then be determined by

$$\beta_{ac} = \frac{\Delta I_C}{\Delta I_B}\bigg|_{V_{CE}=\text{constant}} = \frac{I_{C_2} - I_{C_1}}{I_{B_2} - I_{B_1}}$$

$$= \frac{3.2\ \text{mA} - 2.2\ \text{mA}}{30\ \mu\text{A} - 20\ \mu\text{A}} = \frac{1\ \text{mA}}{10\ \mu\text{A}}$$

$$= \mathbf{100}$$

The solution above reveals that for an ac input at the base, the collector current will be about 100 times the magnitude of the base current.

If we determine the dc beta at the Q-point:

$$\beta_{dc} = \frac{I_C}{I_B} = \frac{2.7\ \text{mA}}{25\ \mu\text{A}} = \mathbf{108}$$

Although not exactly equal, the levels of β_{ac} and β_{dc} are usually reasonably close and are often used interchangeably. That is, if β_{ac} is known, it is assumed to be about the same magnitude as β_{dc}, and vice versa. Keep in mind that in the same lot, the value of β_{ac} will vary somewhat from one transistor to the next even though each transistor has the same number code. The variation may not be significant but for the majority of applications, it is certainly sufficient to validate the approximate approach above. Generally, the smaller the level of I_{CEO}, the closer the magnitude of the two betas. Since the trend is toward lower and lower levels of I_{CEO}, the validity of the foregoing approximation is further substantiated.

If the characteristics had the appearance of those appearing in Fig. 3.18, the level of β_{ac} would be the same in every region of the characteristics. Note that the step in I_B is fixed at 10 μA and the vertical spacing between curves is the same at every point in the characteristics—namely, 2 mA. Calculating the β_{ac} at the Q-point indicated will result in

$$\beta_{ac} = \frac{\Delta I_C}{\Delta I_B}\bigg|_{V_{CE} = \text{constant}} = \frac{9 \text{ mA} - 7 \text{ mA}}{45 \text{ } \mu\text{A} - 35 \text{ } \mu\text{A}} = \frac{2 \text{ mA}}{10 \text{ } \mu\text{A}} = \textbf{200}$$

Determining the dc beta at the same Q-point will result in

$$\beta_{dc} = \frac{I_C}{I_B} = \frac{8 \text{ mA}}{40 \text{ } \mu\text{A}} = \textbf{200}$$

revealing that if the characteristics have the appearance of Fig. 3.18, the magnitude of β_{ac} and β_{dc} *will be the same* at every point on the characteristics. In particular, note that $I_{CEO} = 0$ μA.

Although a true set of transistor characteristics will never have the exact appearance of Fig. 3.18, it does provide a set of characteristics for comparison with those obtained from a curve tracer (to be described shortly).

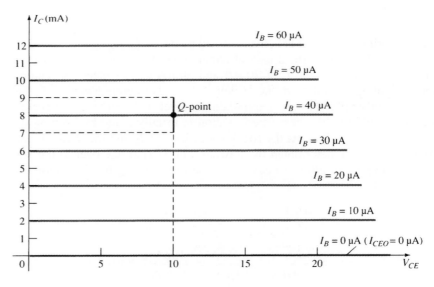

Figure 3.18 Characteristics in which β_{ac} is the same everywhere and $\beta_{ac} = \beta_{dc}$.

For the analysis to follow the subscript dc or ac will not be included with β to avoid cluttering the expressions with unnecessary labels. For dc situations it will simply be recognized as β_{dc} and for any ac analysis as β_{ac}. If a value of β is specified for a particular transistor configuration, it will normally be used for both the dc and ac calculations.

A relationship can be developed between β and α using the basic relationships introduced thus far. Using $\beta = I_C/I_B$ we have $I_B = I_C/\beta$, and from $\alpha = I_C/I_E$ we have $I_E = I_C/\alpha$. Substituting into

$$I_E = I_C + I_B$$

we have

$$\frac{I_C}{\alpha} = I_C + \frac{I_C}{\beta}$$

and dividing both sides of the equation by I_C will result in

$$\frac{1}{\alpha} = 1 + \frac{1}{\beta}$$

or

$$\beta = \alpha\beta + \alpha = (\beta + 1)\alpha$$

so that

$$\boxed{\alpha = \frac{\beta}{\beta + 1}} \qquad (3.12a)$$

or

$$\boxed{\beta = \frac{\alpha}{1 - \alpha}} \qquad (3.12b)$$

In addition, recall that

$$I_{CEO} = \frac{I_{CBO}}{1 - \alpha}$$

but using an equivalence of

$$\frac{1}{1 - \alpha} = \beta + 1$$

derived from the above, we find that

$$I_{CEO} = (\beta + 1)I_{CBO}$$

or

$$\boxed{I_{CEO} \cong \beta I_{CBO}} \qquad (3.13)$$

as indicated on Fig. 3.14a. Beta is a particularly important parameter because it provides a direct link between current levels of the input and output circuits for a common-emitter configuration. That is,

$$\boxed{I_C = \beta I_B} \qquad (3.14)$$

and since

$$I_E = I_C + I_B$$
$$= \beta I_B + I_B$$

we have

$$\boxed{I_E = (\beta + 1)I_B} \qquad (3.15)$$

Both of the equations above play a major role in the analysis in Chapter 4.

Biasing

The proper biasing of a common-emitter amplifier can be determined in a manner similar to that introduced for the common-base configuration. Let us assume that we are presented with an *npn* transistor such as shown in Fig. 3.19a and asked to apply the proper biasing to place the device in the active region.

The first step is to indicate the direction of I_E as established by the arrow in the transistor symbol as shown in Fig. 3.19b. Next, the other currents are introduced as

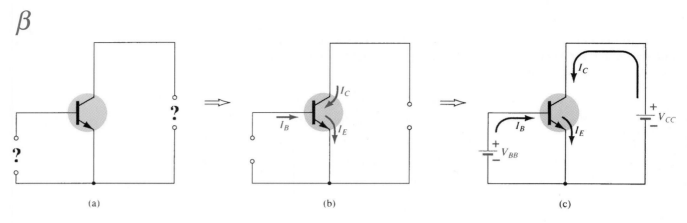

Figure 3.19 Determining the proper biasing arrangement for a common-emitter *npn* transistor configuration.

shown, keeping in mind the Kirchhoff's current law relationship: $I_C + I_B = I_E$. Finally, the supplies are introduced with polarities that will support the resulting directions of I_B and I_C as shown in Fig. 3.19c to complete the picture. The same approach can be applied to *pnp* transistors. If the transistor of Fig. 3.19 was a *pnp* transistor, all the currents and polarities of Fig. 3.19c would be reversed.

3.7 COMMON-COLLECTOR CONFIGURATION

The third and final transistor configuration is the *common-collector configuration,* shown in Fig. 3.20 with the proper current directions and voltage notation. The common-collector configuration is used primarily for impedance-matching purposes since it has a high input impedance and low output impedance, opposite to that of the common-base and common-emitter configurations.

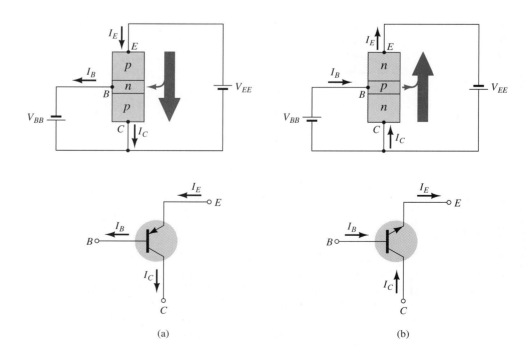

Figure 3.20 Notation and symbols used with the common-collector configuration: (a) *pnp* transistor; (b) *npn* transistor.

A common-collector circuit configuration is provided in Fig. 3.21 with the load resistor connected from emitter to ground. Note that the collector is tied to ground even though the transistor is connected in a manner similar to the common-emitter configuration. From a design viewpoint, there is no need for a set of common-collector characteristics to choose the parameters of the circuit of Fig. 3.21. It can be designed using the common-emitter characteristics of Section 3.6. For all practical purposes, the output characteristics of the common-collector configuration are the same as for the common-emitter configuration. For the common-collector configuration the output characteristics are a plot of I_E versus V_{CE} for a range of values of I_B. The input current, therefore, is the same for both the common-emitter and common-collector characteristics. The horizontal voltage axis for the common-collector configuration is obtained by simply changing the sign of the collector-to-emitter voltage of the common-emitter characteristics. Finally, there is an almost unnoticeable change in the vertical scale of I_C of the common-emitter characteristics if I_C is replaced by I_E for the common-collector characteristics (since $\alpha \cong 1$). For the input circuit of the common-collector configuration the common-emitter base characteristics are sufficient for obtaining the required information.

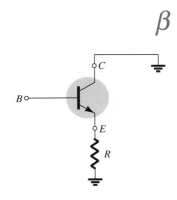

Figure 3.21 Common-collector configuration used for impedance-matching purposes.

3.8 LIMITS OF OPERATION

For each transistor there is a region of operation on the characteristics which will ensure that the maximum ratings are not being exceeded and the output signal exhibits minimum distortion. Such a region has been defined for the transistor characteristics of Fig. 3.22. All of the limits of operation are defined on a typical transistor specification sheet described in Section 3.9.

Some of the limits of operation are self-explanatory, such as maximum collector current (normally referred to on the specification sheet as *continuous* collector current) and maximum collector-to-emitter voltage (often abbreviated as V_{CEO} or $V_{(BR)CEO}$ on the specification sheet). For the transistor of Fig. 3.22, $I_{C_{max}}$ was specified as 50 mA and V_{CEO} as 20 V. The vertical line on the characteristics defined as $V_{CE_{sat}}$ specifies the

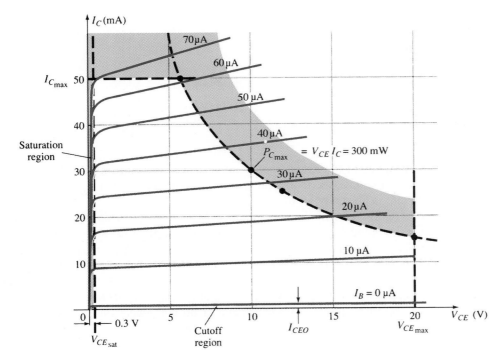

Figure 3.22 Defining the linear (undistorted) region of operation for a transistor.

minimum V_{CE} that can be applied without falling into the nonlinear region labeled the *saturation* region. The level of $V_{CE_{sat}}$ is typically in the neighborhood of the 0.3 V specified for this transistor.

The maximum dissipation level is defined by the following equation:

$$\boxed{P_{C_{max}} = V_{CE}I_C}$$

(3.16)

For the device of Fig. 3.22, the collector power dissipation was specified as 300 mW. The question then arises of how to plot the collector power dissipation curve specified by the fact that

$$P_{C_{max}} = V_{CE}I_C = 300 \text{ mW}$$

or

$$V_{CE}I_C = 300 \text{ mW}$$

At any point on the characteristics the product of V_{CE} and I_C must be equal to 300 mW. If we choose I_C to be the maximum value of 50 mA and substitute into the relationship above, we obtain

$$V_{CE}I_C = 300 \text{ mW}$$

$$V_{CE}(50 \text{ mA}) = 300 \text{ mW}$$

$$V_{CE} = \frac{300 \text{ mW}}{50 \text{ mA}} = \mathbf{6 \text{ V}}$$

As a result we find that if $I_C = 50$ mA, then $V_{CE} = 6$ V on the power dissipation curve as indicated in Fig. 3.22. If we now choose V_{CE} to be its maximum value of 20 V, the level of I_C is the following:

$$(20 \text{ V})I_C = 300 \text{ mW}$$

$$I_C = \frac{300 \text{ mW}}{20 \text{ V}} = \mathbf{15 \text{ mA}}$$

defining a second point on the power curve.

If we now choose a level of I_C in the midrange such as 25 mA, and solve for the resulting level of V_{CE}, we obtain

$$V_{CE}(25 \text{ mA}) = 300 \text{ mW}$$

and

$$V_{CE} = \frac{300 \text{ mW}}{25 \text{ mA}} = \mathbf{12 \text{ V}}$$

as also indicated on Fig. 3.22.

A rough estimate of the actual curve can usually be drawn using the three points defined above. Of course, the more points you have, the more accurate the curve, but a rough estimate is normally all that is required.

The *cutoff* region is defined as that region below $I_C = I_{CEO}$. This region must also be avoided if the output signal is to have minimum distortion. On some specification sheets only I_{CBO} is provided. One must then use the equation $I_{CEO} = \beta I_{CBO}$ to establish some idea of the cutoff level if the characteristic curves are unavailable. Operation in the resulting region of Fig. 3.22 will ensure minimum distortion of the output signal and current and voltage levels that will not damage the device.

If the characteristic curves are unavailable or do not appear on the specification sheet (as is often the case), one must simply be sure that I_C, V_{CE}, and their product $V_{CE}I_C$ fall into the range appearing in Eq. (3.17).

$$\boxed{\begin{aligned} I_{CEO} &\leqq I_C \leqq I_{C_{\max}} \\ V_{CE_{\text{sat}}} &\leqq V_{CE} \leqq V_{CE_{\max}} \\ V_{CE}I_C &\leqq P_{C_{\max}} \end{aligned}} \qquad\qquad (3.17)$$

For the common-base characteristics the maximum power curve is defined by the following product of output quantities:

$$\boxed{P_{C_{\max}} = V_{CB}I_C} \qquad\qquad (3.18)$$

3.9 TRANSISTOR SPECIFICATION SHEET

Since the specification sheet is the communication link between the manufacturer and user, it is particularly important that the information provided be recognized and correctly understood. Although all the parameters have not been introduced, a broad number will now be familiar. The remaining parameters will be introduced in the chapters that follow. Reference will then be made to this specification sheet to review the manner in which the parameter is presented.

The information provided as Fig. 3.23 is taken directly from the *Small-Signal Transistors, FETs, and Diodes* publication prepared by Motorola Inc. The 2N4123 is a general-purpose *npn* transistor with the casing and terminal identification appearing in the top-right corner of Fig. 3.23a. Most specification sheets are broken down into *maximum ratings, thermal characteristics, and electrical characteristics*. The electrical characteristics are further broken down into "on," "off," and small-signal characteristics. The "on" and "off" characteristics refer to dc limits, while the small-signal characteristics include the parameters of importance to ac operation.

Note in the maximum rating list that $V_{CE_{\max}} = V_{CEO} = 30$ V with $I_{C_{\max}} = 200$ mA. The maximum collector dissipation $P_{C_{\max}} = P_D = 625$ mW. The derating factor under the maximum rating specifies that the maximum rating must be decreased 5 mW for every 1° rise in temperature above 25°C. In the "off" characteristics I_{CBO} is specified as 50 nA and in the "on" characteristics $V_{CE_{\text{sat}}} = 0.3$ V. The level of h_{FE} has a range of 50 to 150 at $I_C = 2$ mA and $V_{CE} = 1$ V and a minimum value of 25 at a higher current of 50 mA at the same voltage.

The limits of operation have now been defined for the device and are repeated below in the format of Eq. (3.17) using $h_{FE} = 150$ (the upper limit) and $I_{CEO} \cong \beta I_{CBO} = (150)(50 \text{ nA}) = 7.5\ \mu\text{A}$. Certainly, for many applications the 7.5 μA = 0.0075 mA can be considered to be 0 mA on an approximate basis.

Limits of Operation

$$7.5\ \mu\text{A} \leqq I_C \leqq 200 \text{ mA}$$

$$0.3 \text{ V} \leqq V_{CE} \leqq 30 \text{ V}$$

$$V_{CE}I_C \leqq 650 \text{ mW}$$

In the small-signal characteristics the level of h_{fe} (β_{ac}) is provided along with a plot of how it varies with collector current in Fig. 3.23f. In Fig. 3.23j the effect of temperature and collector current on the level of h_{FE} (β_{ac}) is demonstrated. At room temperature (25°C), note that h_{FE} (β_{dc}) is a maximum value of 1 in the neighborhood of about 8 mA. As I_C increases beyond this level, h_{FE} drops off to one-half the value with I_C equal to 50 mA. It also drops to this level if I_C decreases to the low level of 0.15 mA. Since this is a *normalized* curve, if we have a transistor with $\beta_{\text{dc}} = h_{FE} = 50$ at room temperature, the maximum value at 8 mA is 50. At $I_C = 50$ mA it has dropped to 50/2 = 25. In other words, normalizing reveals that the actual level of h_{FE} at any level of I_C has been divided by the maximum value of h_{FE} at that temperature and $I_C = 8$ mA. Note also that the horizontal scale of Fig. 3.23j is a log scale. Log scales are examined in depth in Chapter 11. You may want to look back at the plots of this section when you find time to review the first few sections of Chapter 11.

MAXIMUM RATINGS

Rating	Symbol	2N4123	Unit
Collector-Emitter Voltage	V_{CEO}	30	Vdc
Collector-Base Voltage	V_{CBO}	40	Vdc
Emitter-Base Voltage	V_{EBO}	5.0	Vdc
Collector Current – Continuous	I_C	200	mAdc
Total Device Dissipation @ $T_A = 25°C$ Derate above 25°C	P_D	625 5.0	mW mW°C
Operating and Storage Junction Temperature Range	T_j, T_{stg}	−55 to +150	°C

THERMAL CHARACTERISTICS

Characteristic	Symbol	Max	Unit
Thermal Resistance, Junction to Case	$R_{\theta JC}$	83.3	°C W
Thermal Resistance, Junction to Ambient	$R_{\theta JA}$	200	°C W

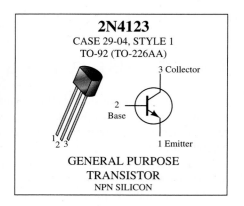

2N4123

CASE 29-04, STYLE 1
TO-92 (TO-226AA)

GENERAL PURPOSE
TRANSISTOR

NPN SILICON

ELECTRICAL CHARACTERISTICS ($T_A = 25°C$ unless otherwise noted)

Characteristic	Symbol	Min	Max	Unit
OFF CHARACTERISTICS				
Collector-Emitter Breakdown Voltage (1) ($I_C = 1.0$ mAdc, $I_E = 0$)	$V_{(BR)CEO}$	30		Vdc
Collector-Base Breakdown Voltage ($I_C = 10$ μAdc, $I_E = 0$)	$V_{(BR)CBO}$	40		Vdc
Emitter-Base Breakdown Voltage ($I_E = 10$ μAdc, $I_C = 0$)	$V_{(BR)EBO}$	5.0	–	Vdc
Collector Cutoff Current ($V_{CB} = 20$ Vdc, $I_E = 0$)	I_{CBO}	–	50	nAdc
Emitter Cutoff Current ($V_{BE} = 3.0$ Vdc, $I_C = 0$)	I_{EBO}	–	50	nAdc
ON CHARACTERISTICS				

DC Current Gain(1)
    ($I_C = 2.0$ mAdc, $V_{CE} = 1.0$ Vdc)
    ($I_C = 50$ mAdc, $V_{CE} = 1.0$ Vdc)
	h_{FE}	50
		25
			150
			–
				–

Collector-Emitter Saturation Voltage(1) ($I_C = 50$ mAdc, $I_B = 5.0$ mAdc)	$V_{CE(sat)}$	–	0.3	Vdc
Base-Emitter Saturation Voltage(1) ($I_C = 50$ mAdc, $I_B = 5.0$ mAdc)	$V_{BE(sat)}$	–	0.95	Vdc
SMALL-SIGNAL CHARACTERISTICS				
Current-Gain – Bandwidth Product ($I_C = 10$ mAdc, $V_{CE} = 20$ Vdc, f = 100 MHz)	f_T	250		MHz
Output Capacitance ($V_{CB} = 5.0$ Vdc, $I_E = 0$, f = 100 MHz)	C_{obo}	–	4.0	pF
Input Capacitance ($V_{BE} = 0.5$ Vdc, $I_C = 0$, f = 100 kHz)	C_{ibo}	–	8.0	pF
Collector-Base Capacitance ($I_E = 0$, $V_{CB} = 5.0$ V, f = 100 kHz)	C_{cb}	–	4.0	pF
Small-Signal Current Gain ($I_C = 2.0$ mAdc, $V_{CE} = 10$ Vdc, f = 1.0 kHz)	h_{fe}	50	200	–

Current Gain – High Frequency
    ($I_C = 10$ mAdc, $V_{CE} = 20$ Vdc, f = 100 MHz)
    ($I_C = 2.0$ mAdc, $V_{CE} = 10$ V, f = 1.0 kHz)
	h_{fe}	2.5
		50
			–
			200
				–

Noise Figure ($I_C = 100$ μAdc, $V_{CE} = 5.0$ Vdc, $R_S = 1.0$ k ohm, f = 1.0 kHz)	NF	–	6.0	dB

(1) Pulse Test: Pulse Width = 300 μs. Duty Cycle = 2.0%

(a)

Figure 3.23 Transistor specification sheet.

Before leaving this description of the characteristics, take note of the fact that the actual collector characteristics are not provided. In fact, most specification sheets as provided by the range of manufacturers fail to provide the full characteristics. It is expected that the data provided are sufficient to use the device effectively in the design process.

As noted in the introduction to this section, all the parameters of the specification sheet have not been defined in the preceding sections or chapters. However, the specification sheet provided in Fig. 3.23 will be referenced continually in the chapters to follow as parameters are introduced. The specification sheet can be a very valuable tool in the design or analysis mode, and every effort should be made to be aware of the importance of each parameter and how it may vary with changing levels of current, temperature, and so on.

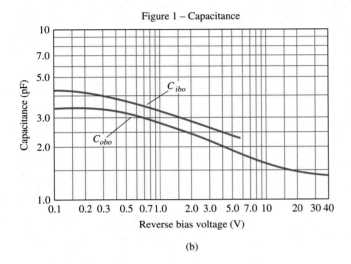

(b)

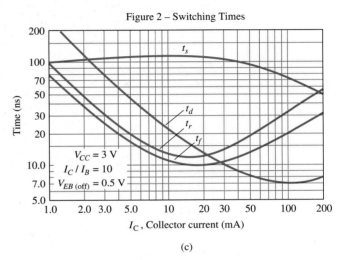

(c)

AUDIO SMALL SIGNAL CHARACTERISTICS

NOISE FIGURE

$(V_{CE} = 5 \text{ Vdc}, T_A = 25°C)$
Bandwidth = 1.0 Hz

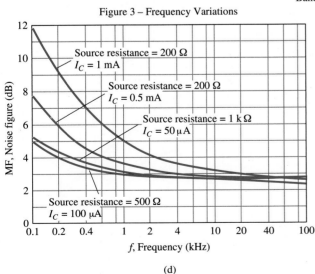

(d)

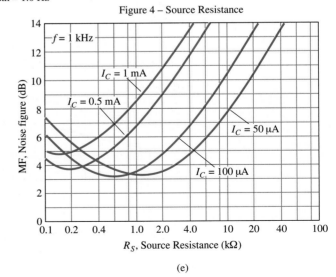

(e)

Figure 3.23 Continued.

h PARAMETERS
$V_{CE} = 10 \text{ V}, f = 1 \text{ kHz}, T_A = 25°\text{C}$

Figure 5 – Current Gain

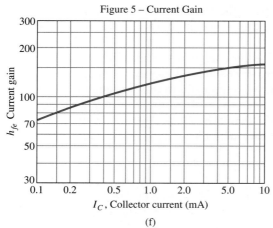

(f)

Figure 6 – Output Admittance

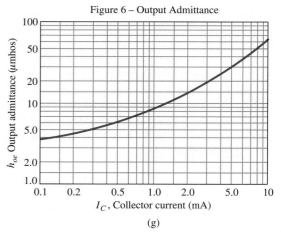

(g)

Figure 7 – Input Impedance

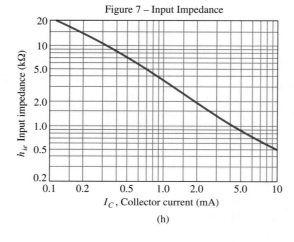

(h)

Figure 8 – Voltage Feedback Ratio

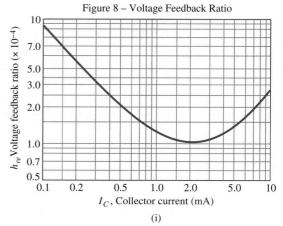

(i)

STATIC CHARACTERISTICS

Figure 9 – DC Current Gain

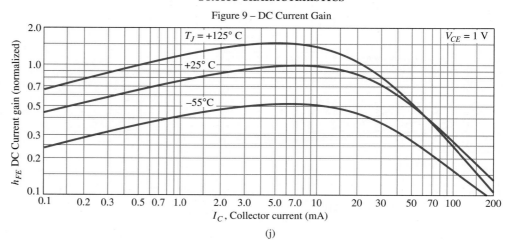

(j)

Figure 3.23 Continued.

3.10 TRANSISTOR TESTING

As with diodes, there are three routes one can take to check a transistor: *curve tracer,* *digital meter,* and *ohmmeter.*

Curve Tracer

The curve tracer of Fig. 1.50 will provide the display of Fig. 3.24 once all the controls have been properly set. The smaller displays to the right reveal the scaling to be applied to the characteristics. The vertical sensitivity is 2 mA/div, resulting in the scale shown to the left of the monitor's display. The horizontal sensitivity is 1 V/div, resulting in the scale shown below the characteristics. The step function reveals that the curves are separated by a difference of 10 μA, starting at 0 μA for the bottom curve. The last scale factor provided can be used to quickly determine the β_{ac} for any region of the characteristics. Simply multiply the displayed factor by the number of divisions between I_B curves in the region of interest. For instance, let us determine β_{ac} at a Q-point of $I_C = 7$ mA and $V_{CE} = 5$ V. In this region of the display, the distance between I_B curves is $\frac{9}{10}$ of a division, as indicated on Fig. 3.25. Using the factor specified, we find that

$$\beta_{ac} = \frac{9}{10} \, \text{div} \left(\frac{200}{\text{div}} \right) = \mathbf{180}$$

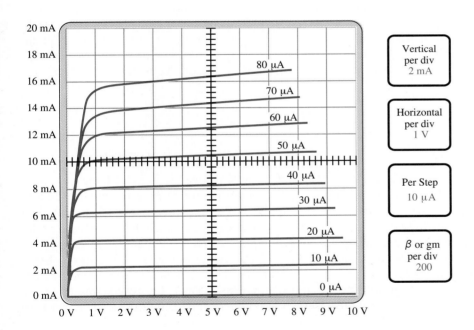

Figure 3.24 Curve tracer response to 2N3904 *npn* transistor.

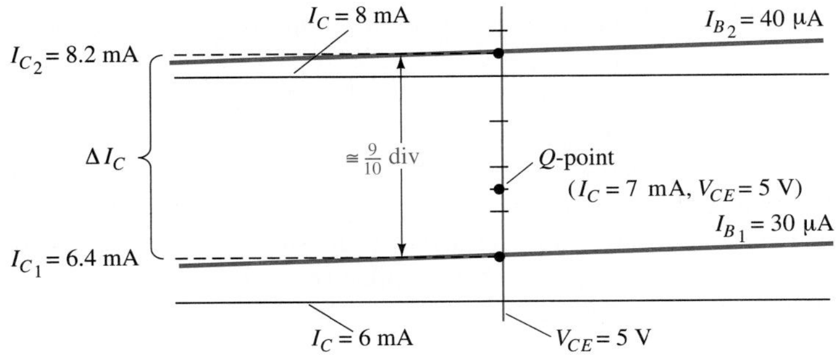

Figure 3.25 Determining β_{ac} for the transistor characteristics of Fig. 3.24 at $I_C = 7$ mA and $V_{CE} = 5$ V.

Figure 3.26 Transistor tester. (Courtesy Computronics Technology, Inc.)

Using Eq. (3.11) gives us

$$\beta_{ac} = \frac{\Delta I_C}{\Delta I_B}\bigg|_{V_{CE} = \text{constant}} = \frac{I_{C_2} - I_{C_1}}{I_{B_2} - I_{B_1}} = \frac{8.2 \text{ mA} - 6.4 \text{ mA}}{40 \text{ } \mu\text{A} - 30 \text{ } \mu\text{A}}$$

$$= \frac{1.8 \text{ mA}}{10 \text{ } \mu\text{A}} = \mathbf{180}$$

verifying the determination above.

Advanced Digital Meters

Advanced digital meters such as that shown in Fig. 3.26 are now available that can provide the level of h_{FE} using the lead sockets appearing at the bottom left of the dial. Note the choice of *pnp* or *npn* and the availability of two emitter connections to handle the sequence of leads as connected to the casing. The level of h_{FE} is determined at a collector current of 2 mA for the Testmate 175A, which is also provided on the digital display. Note that this versatile instrument can also check a diode. It can measure capacitance and frequency in addition to the normal functions of voltage, current, and resistance measurements.

In fact, in the diode testing mode it can be used to check the *p-n* junctions of a transistor. With the collector open the base-to-emitter junction should result in a low voltage of about 0.7 V with the red (positive) lead connected to the base and the black (negative) lead connected to the emitter. A reversal of the leads should result in an OL indication to represent the reverse-biased junction. Similarly, with the emitter open, the forward- and reverse-bias states of the base-to-collector junction can be checked.

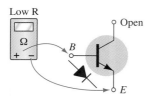

Figure 3.27 Checking the forward-biased base-to-emitter junction of an *npn* transistor.

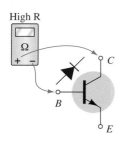

Figure 3.28 Checking the reverse-biased base-to-collector junction of an *npn* transistor.

Ohmmeter

An ohmmeter or the resistance scales of a DMM can be used to check the state of a transistor. Recall that for a transistor in the active region the base-to-emitter junction is forward-biased and the base-to-collector junction is reverse-biased. Essentially, therefore, the forward-biased junction should register a relatively low resistance while the reverse-biased junction shows a much higher resistance. For an *npn* transistor, the forward-biased junction (biased by the internal supply in the resistance mode) from base to emitter should be checked as shown in Fig. 3.27 and result in a reading that will typically fall in the range of 100 Ω to a few kilohms. The reverse-biased base-to-collector junction (again reverse-biased by the internal supply) should be checked as shown in Fig. 3.28 with a reading typically exceeding 100 kΩ. For a *pnp* transistor the leads are reversed for each junction. Obviously, a large or small resistance in both directions (reversing the leads) for either junction of an *npn* or *pnp* transistor indicates a faulty device.

If both junctions of a transistor result in the expected readings the type of transistor can also be determined by simply noting the polarity of the leads as applied to the base-emitter junction. If the positive (+) lead is connected to the base and the negative lead (−) to the emitter a low resistance reading would indicate an *npn* transistor. A high resistance reading would indicate a *pnp* transistor. Although an ohmmeter can also be used to determine the leads (base, collector, and emitter) of a transistor it is assumed that this determination can be made by simply looking at the orientation of the leads on the casing.

3.11 TRANSISTOR CASING AND TERMINAL IDENTIFICATION

After the transistor has been manufactured using one of the techniques described in Chapter 12, leads of, typically, gold, aluminum, or nickel are then attached and the entire structure is encapsulated in a container such as that shown in Fig. 3.29. Those with the heavy duty construction are high-power devices, while those with the small can (top hat) or plastic body are low- to medium-power devices.

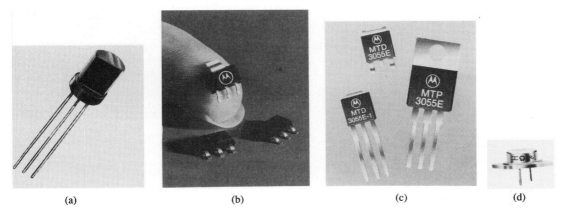

(a)   (b)   (c)   (d)

Figure 3.29 Various types of transistors: (a) Courtesy General Electric Company; (b) and (c) Courtesy of Motorola Inc.; (d) Courtesy International Rectifier Corporation.

Whenever possible, the transistor casing will have some marking to indicate which leads are connected to the emitter, collector, or base of a transistor. A few of the methods commonly used are indicated in Fig. 3.30.

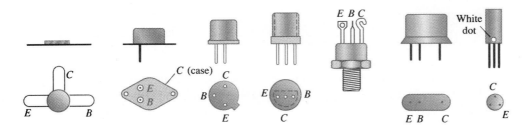

Figure 3.30 Transistor terminal identification.

The internal construction of a TO-92 package in the Fairchild line appears in Fig. 3.31. Note the very small size of the actual semiconductor device. There are gold bond wires, a copper frame, and an epoxy encapsulation.

Four (quad) individual *pnp* silicon transistors can be housed in the 14-pin plastic dual-in-line package appearing in Fig. 3.32a. The internal pin connections appear in Fig. 3.32b. As with the diode IC package, the indentation in the top surface reveals the number 1 and 14 pins.

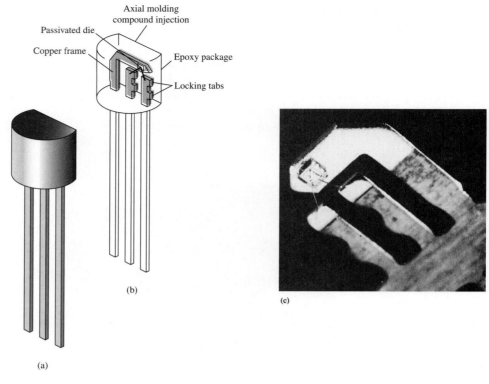

Figure 3.31 Internal construction of a Fairchild transistor in a TO-92 package. (Courtesy Fairchild Camera and Instrument Corporation.)

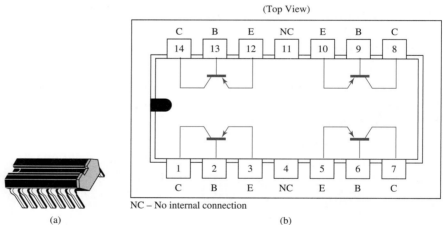

NC – No internal connection

Figure 3.32 Type Q2T2905 Texas Instruments quad *pnp* silicon transistors: (a) appearance; (b) pin connections. (Courtesy Texas Instruments Incorporated.)

3.12 SUMMARY

Important Conclusions and Concepts

1. Semiconductor devices have the following advantages over vacuum tubes: They are (1) of **smaller size**; (2) more **lightweight**; (3) more **rugged**; and (4) more **efficient**. In addition, they have (1) **no warm-up period**; (2) **no heater requirement**; and (3) **lower operating voltages**.

2. Transistors are **three-terminal devices** of three semiconductor layers having a base or center layer a great deal **thinner** than the other two layers. The outer two layers are either *n*- or *p*-type materials with the sandwiched layer the opposite type.

3. One *p-n* junction of a transistor is **forward-biased** while the other is **reverse-biased**.

4. The dc emitter current is always the **largest current** of a transistor whereas the base current is always the **smallest**. The emitter current is always the **sum** of the other two.

5. The collector current is made up of **two components**: the **majority component** and the **minority current** (also called the *leakage* current).

6. The arrow in the transistor symbol defines the direction of **conventional current flow for the emitter current** and thereby defines the direction for the other currents of the device.

7. A three-terminal device needs **two sets of characteristics** to completely define its characteristics.

8. In the active region of a transistor, the base-emitter junction is **forward-biased** while the collector-base junction is reverse-biased.

9. In the cutoff region the base-emitter and collector-base junctions of a transistor are **both reverse-biased**.

10. In the saturation region the base-emitter and collector-base junctions are **forward-biased**.

11. On an average basis, as a first approximation, the base-to-emitter voltage of an operating transistor can be assumed to be **0.7 V**.

12. The quantity alpha (α) relates the collector and emitter currents and is always close to **one**.

13. The impedance between terminals of a forward-biased junction is always relatively **small** while the impedance between terminals of a reverse-biased junction is usually **quite large**.

14. The arrow in the symbol of an *npn* transistor points out of the device (**not** pointing **in**), while the arrow points in to the center of the symbol for a *pnp* transistor (**pointing in**).

15. For linear amplification purposes, cutoff for the common-emitter configuration will be defined by $I_C = I_{CEO}$.

16. The quantity beta (β) provides an important relationship between the base and collector currents and is usually between **50 and 400**.

17. The dc beta is defined by a simple **ratio of dc currents at an operating point**, while the ac beta is **sensitive to the characteristics** in the region of interest. For most applications, however, the two are considered equivalent as a first approximation.

18. To ensure that a transistor is operating within its maximum power level rating, simply find the **product of the collector-to-emitter voltage and collector current**, and compare it to the rated value.

Equations

$$I_E = I_C + I_B$$

$$I_C = I_{C_{\text{majority}}} + I_{CO_{\text{minority}}}$$

$$V_{BE} = 0.7 \text{ V}$$

$$\alpha_{\text{dc}} = \frac{I_C}{I_E}$$

$$\alpha_{\text{ac}} = \frac{\Delta I_C}{\Delta I_E}\bigg|_{V_{CB} = \text{constant}}$$

$$I_{CEO} = \left.\frac{I_{CBO}}{1 - \alpha}\right|_{I_B = 0 \, \mu A}$$

$$\beta_{dc} = \frac{I_C}{I_B}$$

$$\beta_{ac} = \left.\frac{\Delta I_C}{\Delta I_B}\right|_{V_{CE} = \text{constant}}$$

$$\alpha = \frac{\beta}{\beta + 1}$$

$$I_C = \beta I_B$$

$$I_E = (\beta + 1)I_B$$

$$P_{C_{\max}} = V_{CE}I_C$$

3.13   COMPUTER ANALYSIS

PSpice Windows

Since the transistor characteristics were introduced in this chapter it seems appropriate that a procedure for obtaining those characteristics using PSpice Windows should be examined. The transistors are listed in the **EVAL.slb** library and start with the letter **Q.** The library includes two *npn* transistors and two *pnp* transistors. The fact that there are a series of curves defined by the levels of I_B will require that a sweep of I_B values (a *nested sweep*) occur within a sweep of collector-to-emitter voltages. This is unnecessary for the diode, however, since only one curve would result.

First, the network in Fig. 3.33 is established using the same procedure defined in Chapter 2. The voltage V_{CC} will establish our main sweep while the voltage V_{BB} will determine the nested sweep. For future reference, note the panel at the top right of the menu bar with the scroll control when building networks. This option allows you to retrieve elements that have been used in the past. For instance, if you placed a resistor a few elements ago, simply return to the scroll bar and scroll until the resistor **R** appears. Click the location once, and the resistor will appear on the screen.

Next, choose the **Analysis Setup** icon and enable the **DC Sweep.** Click on **DC Sweep,** and choose **Voltage Source** and **Linear.** Type in the **Name** V_{CC} with a **Start Value** of 0 V and an **End Value** of 10 V. Use an **Increment** of 0.01 V to ensure a continuous, detailed plot. Rather than click **OK,** this time we have to choose the **Nested Sweep** at the bottom left of the dialog box. When chosen, a **DC Nested Sweep** dialog box will appear and ask us to repeat the choices just made for the voltage V_{BB}. Again, Voltage Source and Linear are chosen, and the name is inserted as V_{BB}. The **Start Value** will now be 2.7 V to correspond with an initial current of 20 μA as determined by

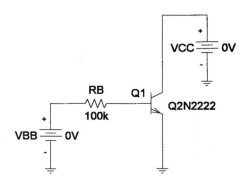

Figure 3.33 Network employed to obtain the collector characteristics of the Q2N2222 transistor.

$$I_B = \frac{V_{BB} - V_{BE}}{R_B} = \frac{2.7 \text{ V} - 0.7 \text{ V}}{100 \text{ k}\Omega} = 20 \ \mu\text{A}$$

The Increment will be 2V, corresponding with a change in base current of 20 μA between I_B levels. The final value will be 10.7 V, corresponding with a current of 100 μA. Before leaving the dialog box, be sure to enable the nested sweep. Then, choose **OK,** followed by a closing of the **Analysis Setup,** and we are ready for the analysis. This time we will automatically Run Probe after the analysis by choosing **Analysis-Probe Setup,** followed by selecting **Automatically run Probe after simulation.** After choosing **OK,** followed by a clicking of the **Simulation icon** (recall that it was the icon with the yellow background and two waveforms), the **OrCAD MicroSim Probe** screen will automatically appear. This time, since V_{CC} is the collector-to-emitter voltage, there is no need to label the voltage at the collector. In fact, since it appears as the horizontal axis of the Probe response, there is no need to touch the **X-Axis Settings** at all if we recognize that V_{CC} is the collector-to-emitter voltage. For the vertical axis, we turn to **Trace-Add** and obtain the **Add Traces** dialog box. Choosing **IC(Q1)** and **OK,** we obtain the transistor characteristics. Unfortunately, however, they extend from -10 to $+20$ mA on the vertical axis. This can be corrected by choosing **Plot** and then **Y-Axis Settings** to obtain the **Y-Axis Settings** dialog box. By choosing **User Defined,** the range can be set from 0 to 20 mA with a Linear scale. Choose **OK** again, and the characteristics of Fig. 3.34 result.

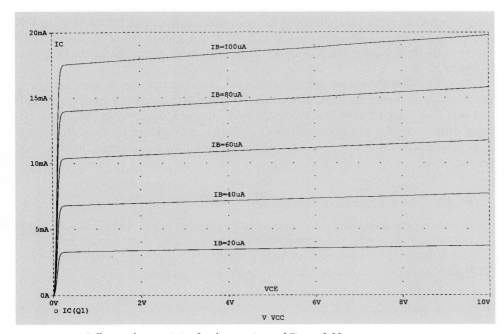

Figure 3.34 Collector characteristics for the transistor of Figure 3.33.

Using the **ABC** icon on the menu bar, the various levels of I_B can be inserted along with the axis labels V_{CE} and I_C. Simply click on the icon, and a dialog box appears asking for the text material. Enter the desired text, click **OK,** and it will appear on the screen. It can then be placed in the desired location.

If the ac beta is determined in the middle of the graph, you will find that its value is about 190—even though **Bf** in the list of specifications is 255.9. Again, like the diode, the other parameters of the element have a noticeable effect on the total operation. However, if we return to the diode specifications through **Edit-Model-Edit Instance Model (Text)** and remove all parameters of the device except Bf = 255.9

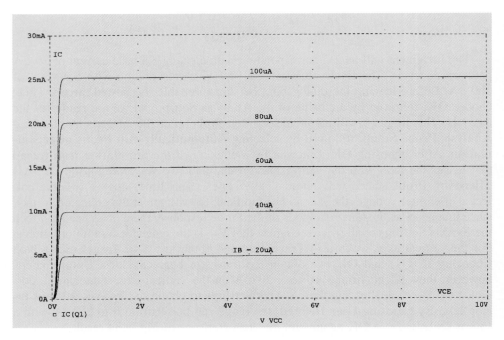

Figure 3.35 Ideal collector characteristics for the transistor of Figure 3.33.

(don't forget the close parentheses at the end of the listing) and follow with an **OK** and a **Simulation,** a new set of curves will result. An adjustment of the range of the *y*-axis to 0–30 mA using the **Y-Axis Settings** will result in the characteristic curves of Fig. 3.35.

Note first that the curves are all horizontal, meaning that the element is void of any resistive elements. In addition, the equal spacing of the curves throughout reveals that beta is the same everywhere (as specified by our new device characteristics). Using a difference of 5 mA between any two curves and dividing by the difference in I_B of 20 μA will result in a β of 250, which is essentially the same as that specified for the device.

PROBLEMS

§ 3.2 Transistor Construction

1. What names are applied to the two types of BJT transistors? Sketch the basic construction of each and label the various minority and majority carriers in each. Draw the graphic symbol next to each. Is any of this information altered by changing from a silicon to a germanium base?

2. What is the major difference between a bipolar and a unipolar device?

§ 3.3 Transistor Operation

3. How must the two transistor junctions be biased for proper transistor amplifier operation?

4. What is the source of the leakage current in a transistor?

5. Sketch a figure similar to Fig. 3.3 for the forward-biased junction of an *npn* transistor. Describe the resulting carrier motion.

6. Sketch a figure similar to Fig. 3.4 for the reverse-biased junction of an *npn* transistor. Describe the resulting carrier motion.

7. Sketch a figure similar to Fig. 3.5 for the majority- and minority-carrier flow of an *npn* transistor. Describe the resulting carrier motion.

8. Which of the transistor currents is always the largest? Which is always the smallest? Which two currents are relatively close in magnitude?

9. If the emitter current of a transistor is 8 mA and I_B is 1/100 of I_C, determine the levels of I_C and I_B.

§ 3.4 Common-Base Configuration

10. From memory, sketch the transistor symbol for a *pnp* and an *npn* transistor, and then insert the conventional flow direction for each current.

11. Using the characteristics of Fig. 3.7, determine V_{BE} at $I_E = 5$ mA for $V_{CB} = 1$, 10, and 20 V. Is it reasonable to assume on an approximate basis that V_{CB} has only a slight effect on the relationship between V_{BE} and I_E?

12. (a) Determine the average ac resistance for the characteristics of Fig. 3.10b.
    (b) For networks in which the magnitude of the resistive elements is typically in kilohms, is the approximation of Fig. 3.10c a valid one [based on the results of part (a)]?

13. (a) Using the characteristics of Fig. 3.8, determine the resulting collector current if $I_E = 4.5$ mA and $V_{CB} = 4$ V.
    (b) Repeat part (a) for $I_E = 4.5$ mA and $V_{CB} = 16$ V.
    (c) How have the changes in V_{CB} affected the resulting level of I_C?
    (d) On an approximate basis, how are I_E and I_C related based on the results above?

14. (a) Using the characteristics of Figs. 3.7 and 3.8, determine I_C if $V_{CB} = 10$ V and $V_{BE} = 800$ mV.
    (b) Determine V_{BE} if $I_C = 5$ mA and $V_{CB} = 10$ V.
    (c) Repeat part (b) using the characteristics of Fig. 3.10b.
    (d) Repeat part (b) using the characteristics of Fig. 3.10c.
    (e) Compare the solutions for V_{BE} for parts (b), (c), and (d). Can the difference be ignored if voltage levels greater than a few volts are typically encountered?

15. (a) Given an α_{dc} of 0.998, determine I_C if $I_E = 4$ mA.
    (b) Determine α_{dc} if $I_E = 2.8$ mA and $I_B = 20$ μA.
    (c) Find I_E if $I_B = 40$ μA and α_{dc} is 0.98.

16. From memory, and memory only, sketch the common-base BJT transistor configuration (for *npn* and *pnp*) and indicate the polarity of the applied bias and resulting current directions.

§ 3.5 Transistor Amplifying Action

17. Calculate the voltage gain ($A_v = V_L/V_i$) for the network of Fig. 3.12 if $V_i = 500$ mV and $R = 1$ kΩ. (The other circuit values remain the same.)

18. Calculate the voltage gain ($A_v = V_L/V_i$) for the network of Fig. 3.12 if the source has an internal resistance of 100 Ω in series with V_i.

§ 3.6 Common-Emitter Configuration

19. Define I_{CBO} and I_{CEO}. How are they different? How are they related? Are they typically close in magnitude?

20. Using the characteristics of Fig. 3.14:
    (a) Find the value of I_C corresponding to $V_{BE} = +750$ mV and $V_{CE} = +5$ V.
    (b) Find the value of V_{CE} and V_{BE} corresponding to $I_C = 3$ mA and $I_B = 30$ μA.

*21. (a) For the common-emitter characteristics of Fig. 3.14, find the dc beta at an operating point of $V_{CE} = +8$ V and $I_C = 2$ mA.
    (b) Find the value of α corresponding to this operating point.
    (c) At $V_{CE} = +8$ V, find the corresponding value of I_{CEO}.
    (d) Calculate the approximate value of I_{CBO} using the dc beta value obtained in part (a).

*22. (a) Using the characteristics of Fig. 3.14a, determine I_{CEO} at $V_{CE} = 10$ V.
    (b) Determine β_{dc} at $I_B = 10$ μA and $V_{CE} = 10$ V.
    (c) Using the β_{dc} determined in part (b), calculate I_{CBO}.

23. (a) Using the characteristics of Fig. 3.14a, determine β_{dc} at $I_B = 80$ μA and $V_{CE} = 5$ V.
    (b) Repeat part (a) at $I_B = 5$ μA and $V_{CE} = 15$ V.
    (c) Repeat part (a) at $I_B = 30$ μA and $V_{CE} = 10$ V.
    (d) Reviewing the results of parts (a) through (c), does the value of β_{dc} change from point to point on the characteristics? Where were the higher values found? Can you develop any general conclusions about the value of β_{dc} on a set of characteristics such as those provided in Fig. 3.14a?

*24. (a) Using the characteristics of Fig. 3.14a, determine β_{ac} at $I_B = 80\ \mu A$ and $V_{CE} = 5$ V.
   (b) Repeat part (a) at $I_B = 5\ \mu A$ and $V_{CE} = 15$ V.
   (c) Repeat part (a) at $I_B = 30\ \mu A$ and $V_{CE} = 10$ V.
   (d) Reviewing the results of parts (a) through (c), does the value of β_{ac} change from point to point on the characteristics? Where are the high values located? Can you develop any general conclusions about the value of β_{ac} on a set of collector characteristics?
   (e) The chosen points in this exercise are the same as those employed in Problem 23. If Problem 23 was performed, compare the levels of β_{dc} and β_{ac} for each point and comment on the trend in magnitude for each quantity.

25. Using the characteristics of Fig. 3.14a, determine β_{dc} at $I_B = 25\ \mu A$ and $V_{CE} = 10$ V. Then calculate α_{dc} and the resulting level of I_E. (Use the level of I_C determined by $I_C = \beta_{dc}I_B$.)

26. (a) Given that $\alpha_{dc} = 0.987$, determine the corresponding value of β_{dc}.
   (b) Given $\beta_{dc} = 120$, determine the corresponding value of α.
   (c) Given that $\beta_{dc} = 180$ and $I_C = 2.0$ mA, find I_E and I_B.

27. From memory, and memory only, sketch the common-emitter configuration (for *npn* and *pnp*) and insert the proper biasing arrangement with the resulting current directions for I_B, I_C, and I_E.

§ 3.7 Common-Collector Configuration

28. An input voltage of 2 V rms (measured from base to ground) is applied to the circuit of Fig. 3.21. Assuming that the emitter voltage follows the base voltage exactly and that V_{be} (rms) = 0.1 V, calculate the circuit voltage amplification ($A_v = V_o/V_i$) and emitter current for $R_E = 1$ kΩ.

29. For a transistor having the characteristics of Fig. 3.14, sketch the input and output characteristics of the common-collector configuration.

§ 3.8 Limits of Operation

30. Determine the region of operation for a transistor having the characteristics of Fig. 3.14 if $I_{C_{max}} = 7$ mA, $V_{CE_{max}} = 17$ V, and $P_{C_{max}} = 40$ mW.

31. Determine the region of operation for a transistor having the characteristics of Fig. 3.8 if $I_{C_{max}} = 6$ mA, $V_{CB_{max}} = 15$ V, and $P_{C_{max}} = 30$ mW.

§ 3.9 Transistor Specification Sheet

32. Referring to Fig. 3.23, determine the temperature range for the device in degrees Fahrenheit.

33. Using the information provided in Fig. 3.23 regarding $P_{D_{max}}$, $V_{CE_{max}}$, $I_{C_{max}}$, and $V_{CE_{sat}}$, sketch the boundaries of operation for the device.

34. Based on the data of Fig. 3.23, what is the expected value of I_{CEO} using the average value of β_{dc}?

35. How does the range of h_{FE} (Fig. 3.23(j), normalized from $h_{FE} = 100$) compare with the range of h_{fe} (Fig. 3.23(f)) for the range of I_C from 0.1 to 10 mA?

36. Using the characteristics of Fig. 3.23b, determine whether the input capacitance in the common-base configuration increases or decreases with increasing levels of reverse-bias potential. Can you explain why?

*37. Using the characteristics of Fig. 3.23f, determine how much the level of h_{fe} has changed from its value at 1 mA to its value at 10 mA. Note that the vertical scale is a log scale that may require reference to Section 11.2. Is the change one that should be considered in a design situation?

*38. Using the characteristics of Fig. 3.23j, determine the level of β_{dc} at $I_C = 10$ mA at the three levels of temperature appearing in the figure. Is the change significant for the specified temperature range? Is it an element to be concerned about in the design process?

§ 3.10 Transistor Testing

39. (a) Using the characteristics of Fig. 3.24, determine β_{ac} at $I_C = 14$ mA and $V_{CE} = 3$ V.
   (b) Determine β_{dc} at $I_C = 1$ mA and $V_{CE} = 8$ V.
   (c) Determine β_{ac} at $I_C = 14$ mA and $V_{CE} = 3$ V.
   (d) Determine β_{dc} at $I_C = 1$ mA and $V_{CE} = 8$ V.
   (e) How does the level of β_{ac} and β_{dc} compare in each region?
   (f) Is the approximation $\beta_{dc} \cong \beta_{ac}$ a valid one for this set of characteristics?

---

*Please Note: Asterisks indicate more difficult problems.

DC Biasing—BJTs

4.1 INTRODUCTION

The analysis or design of a transistor amplifier requires a knowledge of both the dc and ac response of the system. Too often it is assumed that the transistor is a magical device that can raise the level of the applied ac input without the assistance of an external energy source. In actuality, the improved output ac power level is the result of a transfer of energy from the applied dc supplies. The analysis or design of any electronic amplifier therefore has two components: the dc portion and the ac portion. Fortunately, the superposition theorem is applicable and the investigation of the dc conditions can be totally separated from the ac response. However, one must keep in mind that during the design or synthesis stage the choice of parameters for the required dc levels will affect the ac response, and vice versa.

The dc level of operation of a transistor is controlled by a number of factors, including the range of possible operating points on the device characteristics. In Section 4.2 we specify the range for the BJT amplifier. Once the desired dc current and voltage levels have been defined, a network must be constructed that will establish the desired operating point—a number of these networks are analyzed in this chapter. Each design will also determine the stability of the system, that is, how sensitive the system is to temperature variations—another topic to be investigated in a later section of this chapter.

Although a number of networks are analyzed in this chapter, there is an underlying similarity between the analysis of each configuration due to the recurring use of the following important basic relationships for a transistor:

$$V_{BE} = 0.7 \text{ V} \qquad (4.1)$$

$$I_E = (\beta + 1)I_B \cong I_C \qquad (4.2)$$

$$I_C = \beta I_B \qquad (4.3)$$

In fact, once the analysis of the first few networks is clearly understood, the path toward the solution of the networks to follow will begin to become quite apparent. In most instances the base current I_B is the first quantity to be determined. Once I_B is known, the relationships of Eqs. (4.1) through (4.3) can be applied to find the remaining quantities of interest. The similarities in analysis will be immediately obvious as we progress through the chapter. The equations for I_B are so similar for a number of configurations that one equation can be derived from another simply by dropping or adding a term or two. The primary function of this chapter is to develop a level of familiarity with the BJT transistor that would permit a dc analysis of any system that might employ the BJT amplifier.

4.2 OPERATING POINT

The term *biasing* appearing in the title of this chapter is an all-inclusive term for the application of dc voltages to establish a fixed level of current and voltage. For transistor amplifiers the resulting dc current and voltage establish an *operating point* on the characteristics that define the region that will be employed for amplification of the applied signal. Since the operating point is a fixed point on the characteristics, it is also called the *quiescent point* (abbreviated Q-point). By definition, *quiescent* means quiet, still, inactive. Figure 4.1 shows a general output device characteristic with four operating points indicated. The biasing circuit can be designed to set the device operation at any of these points or others within the *active region*. The maximum ratings are indicated on the characteristics of Fig. 4.1 by a horizontal line for the maximum collector current $I_{C_{max}}$ and a vertical line at the maximum collector-to-emitter voltage $V_{CE_{max}}$. The maximum power constraint is defined by the curve $P_{C_{max}}$ in the same figure. At the lower end of the scales are the *cutoff region*, defined by $I_B \leq 0 \ \mu A$, and the *saturation region*, defined by $V_{CE} \leq V_{CE_{sat}}$.

The BJT device could be biased to operate outside these maximum limits, but the result of such operation would be either a considerable shortening of the lifetime of the device or destruction of the device. Confining ourselves to the *active* region, one can select many different operating areas or points. The chosen Q-point often depends on the intended use of the circuit. Still, we can consider some differences among the various points shown in Fig. 4.1 to present some basic ideas about the operating point and, thereby, the bias circuit.

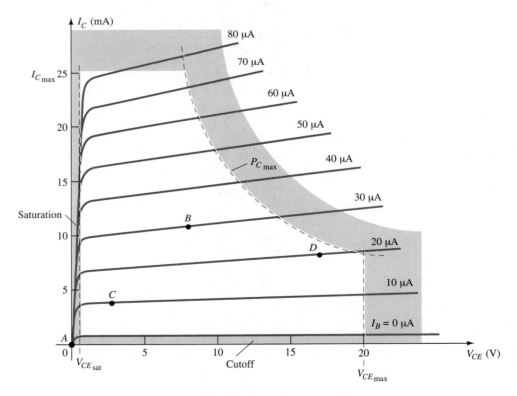

Figure 4.1 Various operating points within the limits of operation of a transistor.

If no bias were used, the device would initially be completely off, resulting in a Q-point at A—namely, zero current through the device (and zero voltage across it). Since it is necessary to bias a device so that it can respond to the entire range of an input signal, point A would not be suitable. For point B, if a signal is applied to the circuit, the device will vary in current and voltage from operating point, allowing the device to react to (and possibly amplify) both the positive and negative excursions of the input signal. If the input signal is properly chosen, the voltage and current of the device will vary but not enough to drive the device into *cutoff* or *saturation*. Point C would allow some positive and negative variation of the output signal, but the peak-to-peak value would be limited by the proximity of $V_{CE} = 0$ V and $I_C = 0$ mA. Operating at point C also raises some concern about the nonlinearities introduced by the fact that the spacing between I_B curves is rapidly changing in this region. In general, it is preferable to operate where the gain of the device is fairly constant (or linear) to ensure that the amplification over the entire swing of input signal is the same. Point B is a region of more linear spacing and therefore more linear operation, as shown in Fig. 4.1. Point D sets the device operating point near the maximum voltage and power level. The output voltage swing in the positive direction is thus limited if the maximum voltage is not to be exceeded. Point B therefore seems the best operating point in terms of linear gain and largest possible voltage and current swing. This is usually the desired condition for small-signal amplifiers (Chapter 8) but not the case necessarily for power amplifiers, which will be considered in Chapter 15. In this discussion, we will be concentrating primarily on biasing the transistor for *small-signal* amplification operation.

One other very important biasing factor must be considered. Having selected and biased the BJT at a desired operating point, the effect of temperature must also be taken into account. Temperature causes the device parameters such as the transistor current gain (β_{ac}) and the transistor leakage current (I_{CEO}) to change. Higher temperatures result in increased leakage currents in the device, thereby changing the operating condition set by the biasing network. The result is that the network design must also provide a degree of *temperature stability* so that temperature changes result in minimum changes in the operating point. This maintenance of the operating point can be specified by a *stability factor*, S, which indicates the degree of change in operating point due to a temperature variation. A highly stable circuit is desirable, and the stability of a few basic bias circuits will be compared.

For the BJT to be biased in its linear or active operating region the following must be true:

1. The base–emitter junction *must* be forward-biased (p-region voltage more *positive*), with a resulting forward-bias voltage of about 0.6 to 0.7 V.

2. The base–collector junction *must* be reverse-biased (n-region more *positive*), with the reverse-bias voltage being any value within the maximum limits of the device.

[Note that for forward bias the voltage across the *p-n* junction is *p*-positive, while for reverse bias it is opposite (reverse) with *n*-positive. This emphasis on the initial letter should provide a means of helping memorize the necessary voltage polarity.]

Operation in the cutoff, saturation, and linear regions of the BJT characteristic are provided as follows:

1. *Linear-region operation:*
   Base–emitter junction forward biased
   Base–collector junction reverse biased

2. *Cutoff-region operation:*
   Base–emitter junction reverse biased
   Base–collector junction reverse biased

3. *Saturation-region operation:*
   Base–emitter junction forward biased
   Base–collector junction forward biased

4.3 FIXED-BIAS CIRCUIT

The fixed-bias circuit of Fig. 4.2 provides a relatively straightforward and simple introduction to transistor dc bias analysis. Even though the network employs an *npn* transistor, the equations and calculations apply equally well to a *pnp* transistor configuration merely by changing all current directions and voltage polarities. The current directions of Fig. 4.2 are the actual current directions, and the voltages are defined by the standard double-subscript notation. For the dc analysis the network can be isolated from the indicated ac levels by replacing the capacitors with an open-circuit equivalent. In addition, the dc supply V_{CC} can be separated into two supplies (for analysis purposes only) as shown in Fig. 4.3 to permit a separation of input and output circuits. It also reduces the linkage between the two to the base current I_B. The separation is certainly valid, as we note in Fig. 4.3 that V_{CC} is connected directly to R_B and R_C just as in Fig. 4.2.

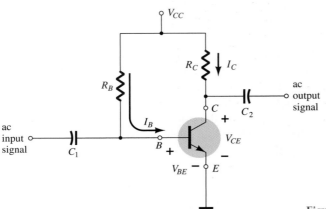

Figure 4.2   Fixed-bias circuit.

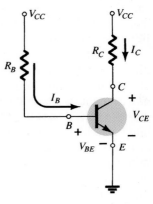

Figure 4.3   dc equivalent of Fig. 4.2.

Forward Bias of Base–Emitter

Consider first the base–emitter circuit loop of Fig. 4.4. Writing Kirchhoff's voltage equation in the clockwise direction for the loop, we obtain

$$+V_{CC} - I_B R_B - V_{BE} = 0$$

Note the polarity of the voltage drop across R_B as established by the indicated direction of I_B. Solving the equation for the current I_B will result in the following:

$$\boxed{I_B = \frac{V_{CC} - V_{BE}}{R_B}} \tag{4.4}$$

Equation (4.4) is certainly not a difficult one to remember if one simply keeps in mind that the base current is the current through R_B and by Ohm's law that current is the voltage across R_B divided by the resistance R_B. The voltage across R_B is the applied voltage V_{CC} at one end less the drop across the base-to-emitter junction (V_{BE}). In addition, since the supply voltage V_{CC} and the base–emitter voltage V_{BE} are constants, the selection of a base resistor, R_B, sets the level of base current for the operating point.

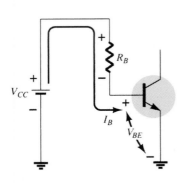

Figure 4.4   Base–emitter loop.

Collector–Emitter Loop

The collector–emitter section of the network appears in Fig. 4.5 with the indicated direction of current I_C and the resulting polarity across R_C. The magnitude of the collector current is related directly to I_B through

$$I_C = \beta I_B \qquad (4.5)$$

It is interesting to note that since the base current is controlled by the level of R_B and I_C is related to I_B by a constant β, the magnitude of I_C is not a function of the resistance R_C. Change R_C to any level and it will not affect the level of I_B or I_C as long as we remain in the active region of the device. However, as we shall see, the level of R_C will determine the magnitude of V_{CE}, which is an important parameter.

Applying Kirchhoff's voltage law in the clockwise direction around the indicated closed loop of Fig. 4.5 will result in the following:

$$V_{CE} + I_C R_C - V_{CC} = 0$$

and

$$V_{CE} = V_{CC} - I_C R_C \qquad (4.6)$$

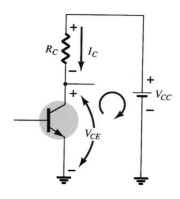

Figure 4.5 Collector–emitter loop.

which states in words that the voltage across the collector–emitter region of a transistor in the fixed-bias configuration is the supply voltage less the drop across R_C.

As a brief review of single- and double-subscript notation recall that

$$V_{CE} = V_C - V_E \qquad (4.7)$$

where V_{CE} is the voltage from collector to emitter and V_C and V_E are the voltages from collector and emitter to ground respectively. But *in this case*, since $V_E = 0$ V, we have

$$V_{CE} = V_C \qquad (4.8)$$

In addition, since

$$V_{BE} = V_B - V_E \qquad (4.9)$$

and $V_E = 0$ V, then

$$V_{BE} = V_B \qquad (4.10)$$

Keep in mind that voltage levels such as V_{CE} are determined by placing the red (positive) lead of the voltmeter at the collector terminal with the black (negative) lead at the emitter terminal as shown in Fig. 4.6. V_C is the voltage from collector to ground and is measured as shown in the same figure. In this case the two readings are identical, but in the networks to follow the two can be quite different. Clearly understanding the difference between the two measurements can prove to be quite important in the troubleshooting of transistor networks.

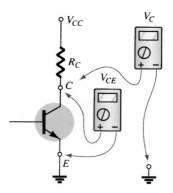

Figure 4.6 Measuring V_{CE} and V_C.

Determine the following for the fixed-bias configuration of Fig. 4.7.
(a) I_{B_Q} and I_{C_Q}.
(b) V_{CE_Q}.
(c) V_B and V_C.
(d) V_{BC}.

EXAMPLE 4.1

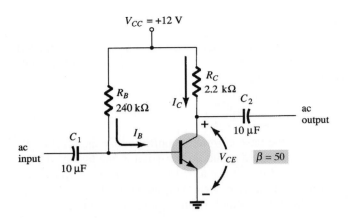

Figure 4.7 dc fixed-bias circuit for Example 4.1.

Solution

(a) Eq. (4.4): $I_{B_Q} = \dfrac{V_{CC} - V_{BE}}{R_B} = \dfrac{12\ V - 0.7\ V}{240\ k\Omega} = \mathbf{47.08\ \mu A}$

   Eq. (4.5): $I_{C_Q} = \beta I_{B_Q} = (50)(47.08\ \mu A) = \mathbf{2.35\ mA}$

(b) Eq. (4.6): $V_{CE_Q} = V_{CC} - I_C R_C$

   $= 12\ V - (2.35\ mA)(2.2\ k\Omega)$

   $= \mathbf{6.83\ V}$

(c) $V_B = V_{BE} = \mathbf{0.7\ V}$

   $V_C = V_{CE} = \mathbf{6.83\ V}$

(d) Using double-subscript notation yields

$$V_{BC} = V_B - V_C = 0.7\ V - 6.83\ V$$

$$= \mathbf{-6.13\ V}$$

with the negative sign revealing that the junction is reversed-biased, as it should be for linear amplification.

Transistor Saturation

The term *saturation* is applied to any system where levels have reached their maximum values. A saturated sponge is one that cannot hold another drop of liquid. For a transistor operating in the saturation region, the current is a maximum value *for the particular design.* Change the design and the corresponding saturation level may rise or drop. Of course, the highest saturation level is defined by the maximum collector current as provided by the specification sheet.

Saturation conditions are normally avoided because the base–collector junction is no longer reverse-biased and the output amplified signal will be distorted. An operating point in the saturation region is depicted in Fig. 4.8a. Note that it is in a region where the characteristic curves join and the collector-to-emitter voltage is at or below $V_{CE_{sat}}$. In addition, the collector current is relatively high on the characteristics.

If we approximate the curves of Fig. 4.8a by those appearing in Fig. 4.8b, a quick, direct method for determining the saturation level becomes apparent. In Fig. 4.8b, the current is relatively high and the voltage V_{CE} is assumed to be zero volts. Applying Ohm's law the resistance between collector and emitter terminals can be determined as follows:

$$R_{CE} = \frac{V_{CE}}{I_C} = \frac{0\ V}{I_{C_{sat}}} = 0\ \Omega$$

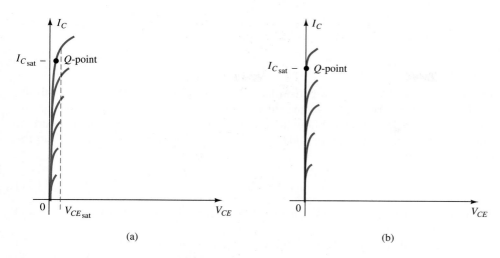

Figure 4.8 Saturation regions: (a) actual; (b) approximate.

Applying the results to the network schematic would result in the configuration of Fig. 4.9.

For the future, therefore, if there were an immediate need to know the approximate maximum collector current (saturation level) for a particular design, simply insert a short-circuit equivalent between collector and emitter of the transistor and calculate the resulting collector current. In short, set $V_{CE} = 0$ V. For the fixed-bias configuration of Fig. 4.10, the short circuit has been applied, causing the voltage across R_C to be the applied voltage V_{CC}. The resulting saturation current for the fixed-bias configuration is

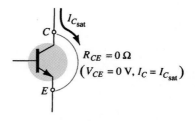

Figure 4.9 Determining $I_{C_{sat}}$.

$$I_{C_{sat}} = \frac{V_{CC}}{R_C} \qquad (4.11)$$

Figure 4.10 Determining $I_{C_{sat}}$ for the fixed-bias configuration.

Once $I_{C_{sat}}$ is known, we have some idea of the maximum possible collector current for the chosen design and the level to stay below if we expect linear amplification.

Determine the saturation level for the network of Fig. 4.7.

EXAMPLE 4.2

Solution

$$I_{C_{sat}} = \frac{V_{CC}}{R_C} = \frac{12 \text{ V}}{2.2 \text{ k}\Omega} = \textbf{5.45 mA}$$

The design of Example 4.1 resulted in $I_{C_Q} = 2.35$ mA, which is far from the saturation level and about one-half the maximum value for the design.

Load-Line Analysis

The analysis thus far has been performed using a level of β corresponding with the resulting Q-point. We will now investigate how the network parameters define the possible range of Q-points and how the actual Q-point is determined. The network of Fig. 4.11a establishes an output equation that relates the variables I_C and V_{CE} in the following manner:

$$V_{CE} = V_{CC} - I_C R_C \qquad (4.12)$$

The output characteristics of the transistor also relate the same two variables I_C and V_{CE} as shown in Fig. 4.11b.

In essence, therefore, we have a network equation and a set of characteristics that employ the same variables. The common solution of the two occurs where the constraints established by each are satisfied simultaneously. In other words, this is similar to finding the solution of two simultaneous equations: one established by the network and the other by the device characteristics.

The device characteristics of I_C versus V_{CE} are provided in Fig. 4.11b. We must now superimpose the straight line defined by Eq. (4.12) on the characteristics. The most direct method of plotting Eq. (4.12) on the output characteristics is to use the fact that a straight line is defined by two points. If we *choose* I_C to be 0 mA, we are specifying the horizontal axis as the line on which one point is located. By substituting $I_C = 0$ mA into Eq. (4.12), we find that

$$V_{CE} = V_{CC} - (0)R_C$$

and

$$V_{CE} = V_{CC}|_{I_C = 0\,\text{mA}} \qquad (4.13)$$

defining one point for the straight line as shown in Fig. 4.12.

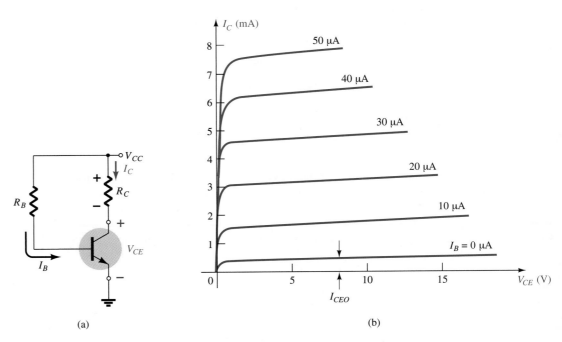

(a)

(b)

Figure 4.11 Load-line analysis: (a) the network; (b) the device characteristics.

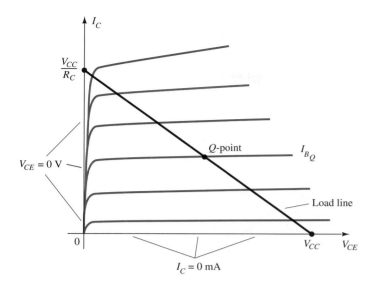

Figure 4.12 Fixed-bias load line.

If we now *choose* V_{CE} to be 0 V, which establishes the vertical axis as the line on which the second point will be defined, we find that I_C is determined by the following equation:

$$0 = V_{CC} - I_C R_C$$

and

$$I_C = \frac{V_{CC}}{R_C}\bigg|_{V_{CE} = 0 \text{ V}} \qquad (4.14)$$

as appearing on Fig. 4.12.

By joining the two points defined by Eqs. (4.13) and (4.14), the straight line established by Eq. (4.12) can be drawn. The resulting line on the graph of Fig. 4.12 is called the *load line* since it is defined by the load resistor R_C. By solving for the resulting level of I_B, the actual Q-point can be established as shown in Fig. 4.12.

If the level of I_B is changed by varying the value of R_B the Q-point moves up or down the load line as shown in Fig. 4.13. If V_{CC} is held fixed and R_C changed, the load line will shift as shown in Fig. 4.14. If I_B is held fixed, the Q-point will move as shown in the same figure. If R_C is fixed and V_{CC} varied, the load line shifts as shown in Fig. 4.15.

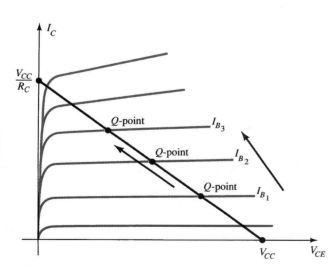

Figure 4.13 Movement of Q-point with increasing levels of I_B.

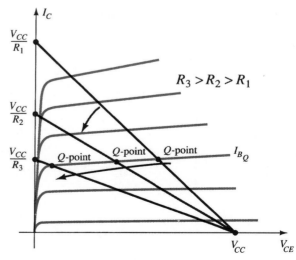

Figure 4.14 Effect of increasing levels of R_C on the load line and Q-point.

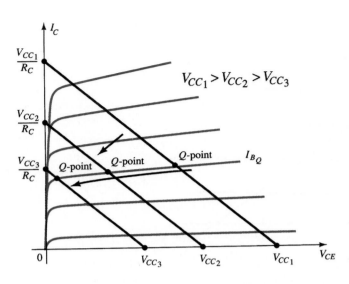

Figure 4.15 Effect of lower values of V_{CC} on the load line and Q-point.

EXAMPLE 4.3

Given the load line of Fig. 4.16 and the defined Q-point, determine the required values of V_{CC}, R_C, and R_B for a fixed-bias configuration.

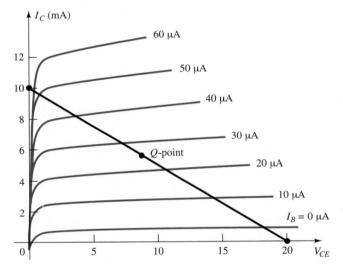

Figure 4.16 Example 4.3

Solution

From Fig. 4.16,

$$V_{CE} = V_{CC} = \textbf{20 V} \text{ at } I_C = 0 \text{ mA}$$

$$I_C = \frac{V_{CC}}{R_C} \text{ at } V_{CE} = 0 \text{ V}$$

and

$$R_C = \frac{V_{CC}}{I_C} = \frac{20 \text{ V}}{10 \text{ mA}} = \textbf{2 k}\boldsymbol{\Omega}$$

$$I_B = \frac{V_{CC} - V_{BE}}{R_B}$$

and

$$R_B = \frac{V_{CC} - V_{BE}}{I_B} = \frac{20 \text{ V} - 0.7 \text{ V}}{25 \text{ } \mu\text{A}} = \textbf{772 k}\boldsymbol{\Omega}$$

Chapter 4   DC Biasing—BJTs

4.4 EMITTER-STABILIZED BIAS CIRCUIT

The dc bias network of Fig. 4.17 contains an emitter resistor to improve the stability level over that of the fixed-bias configuration. The improved stability will be demonstrated through a numerical example later in the section. The analysis will be performed by first examining the base–emitter loop and then using the results to investigate the collector–emitter loop.

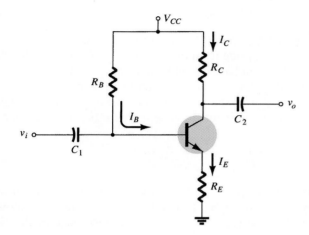

Figure 4.17 BJT bias circuit with emitter resistor.

Base–Emitter Loop

The base–emitter loop of the network of Fig. 4.17 can be redrawn as shown in Fig. 4.18. Writing Kirchhoff's voltage law around the indicated loop in the clockwise direction will result in the following equation:

$$+V_{CC} - I_B R_B - V_{BE} - I_E R_E = 0 \tag{4.15}$$

Recall from Chapter 3 that

$$I_E = (\beta + 1)I_B \tag{4.16}$$

Substituting for I_E in Eq. (4.15) will result in

$$V_{CC} - I_B R_B - V_{BE} - (\beta + 1)I_B R_E = 0$$

Grouping terms will then provide the following:

$$-I_B(R_B + (\beta + 1)R_E) + V_{CC} - V_{BE} = 0$$

Multiplying through by (-1) we have

$$I_B(R_B + (\beta + 1)R_E) - V_{CC} + V_{BE} = 0$$

with

$$I_B(R_B + (\beta + 1)R_E) = V_{CC} - V_{BE}$$

and solving for I_B gives

$$\boxed{I_B = \frac{V_{CC} - V_{BE}}{R_B + (\beta + 1)R_E}} \tag{4.17}$$

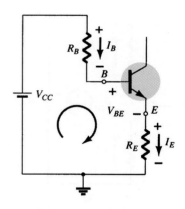

Figure 4.18 Base–emitter loop.

Note that the only difference between this equation for I_B and that obtained for the fixed-bias configuration is the term $(\beta + 1)R_E$.

There is an interesting result that can be derived from Eq. (4.17) if the equation is used to sketch a series network that would result in the same equation. Such is the

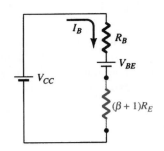

Figure 4.19 Network derived from Eq. (4.17).

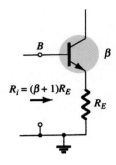

Figure 4.20 Reflected impedance level of R_E.

case for the network of Fig. 4.19. Solving for the current I_B will result in the same equation obtained above. Note that aside from the base-to-emitter voltage V_{BE}, the resistor R_E is *reflected* back to the input base circuit by a factor $(\beta + 1)$. In other words, the emitter resistor, which is part of the collector–emitter loop, "appears as" $(\beta + 1)R_E$ in the base–emitter loop. Since β is typically 50 or more, the emitter resistor appears to be a great deal larger in the base circuit. In general, therefore, for the configuration of Fig. 4.20,

$$R_i = (\beta + 1)R_E \qquad (4.18)$$

Equation (4.18) is one that will prove useful in the analysis to follow. In fact, it provides a fairly easy way to remember Eq. (4.17). Using Ohm's law, we know that the current through a system is the voltage divided by the resistance of the circuit. For the base–emitter circuit the net voltage is $V_{CC} - V_{BE}$. The resistance levels are R_B plus R_E reflected by $(\beta + 1)$. The result is Eq. (4.17).

Collector–Emitter Loop

The collector–emitter loop is redrawn in Fig. 4.21. Writing Kirchhoff's voltage law for the indicated loop in the clockwise direction will result in

$$+I_E R_E + V_{CE} + I_C R_C - V_{CC} = 0$$

Substituting $I_E \cong I_C$ and grouping terms gives

$$V_{CE} - V_{CC} + I_C(R_C + R_E) = 0$$

and
$$V_{CE} = V_{CC} - I_C(R_C + R_E) \qquad (4.19)$$

The single-subscript voltage V_E is the voltage from emitter to ground and is determined by

$$V_E = I_E R_E \qquad (4.20)$$

while the voltage from collector to ground can be determined from

$$V_{CE} = V_C - V_E$$

and
$$V_C = V_{CE} + V_E \qquad (4.21)$$

or
$$V_C = V_{CC} - I_C R_C \qquad (4.22)$$

The voltage at the base with respect to ground can be determined from

$$V_B = V_{CC} - I_B R_B \qquad (4.23)$$

or
$$V_B = V_{BE} + V_E \qquad (4.24)$$

Figure 4.21 Collector–emitter loop.

For the emitter bias network of Fig. 4.22, determine:

(a) I_B.

(b) I_C.

(c) V_{CE}.

(d) V_C.

(e) V_E.

(f) V_B.

(g) V_{BC}.

EXAMPLE 4.4

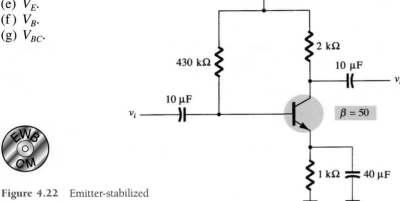

Figure 4.22 Emitter-stabilized bias circuit for Example 4.4.

Solution

(a) Eq. (4.17): $I_B = \dfrac{V_{CC} - V_{BE}}{R_B + (\beta + 1)R_E} = \dfrac{20\,V - 0.7\,V}{430\,k\Omega + (51)(1\,k\Omega)}$

$= \dfrac{19.3\,V}{481\,k\Omega} = \mathbf{40.1\,\mu A}$

(b) $I_C = \beta I_B$

$= (50)(40.1\,\mu A)$

$\cong \mathbf{2.01\,mA}$

(c) Eq. (4.19): $V_{CE} = V_{CC} - I_C(R_C + R_E)$

$= 20\,V - (2.01\,mA)(2\,k\Omega + 1\,k\Omega) = 20\,V - 6.03\,V$

$= \mathbf{13.97\,V}$

(d) $V_C = V_{CC} - I_C R_C$

$= 20\,V - (2.01\,mA)(2\,k\Omega) = 20\,V - 4.02\,V$

$= \mathbf{15.98\,V}$

(e) $V_E = V_C - V_{CE}$

$= 15.98\,V - 13.97\,V$

$= \mathbf{2.01\,V}$

or $V_E = I_E R_E \cong I_C R_E$

$= (2.01\,mA)(1\,k\Omega)$

$= \mathbf{2.01\,V}$

(f) $V_B = V_{BE} + V_E$

$= 0.7\,V + 2.01\,V$

$= \mathbf{2.71\,V}$

(g) $V_{BC} = V_B - V_C$

$= 2.71\,V - 15.98\,V$

$= \mathbf{-13.27\,V}$ (reverse-biased as required)

Improved Bias Stability

The addition of the emitter resistor to the dc bias of the BJT provides improved stability, that is, the dc bias currents and voltages remain closer to where they were set by the circuit when outside conditions, such as temperature, and transistor beta, change. While a mathematical analysis is provided in Section 4.12, some comparison of the improvement can be obtained as demonstrated by Example 4.5.

EXAMPLE 4.5

Prepare a table and compare the bias voltage and currents of the circuits of Fig. 4.7 and Fig. 4.22 for the given value of $\beta = 50$ and for a new value of $\beta = 100$. Compare the changes in I_C and V_{CE} for the same increase in β.

Solution

Using the results calculated in Example 4.1 and then repeating for a value of $\beta = 100$ yields the following:

β	$I_B\ (\mu A)$	$I_C\ (mA)$	$V_{CE}\ (V)$
50	47.08	2.35	6.83
100	47.08	4.71	1.64

The BJT collector current is seen to change by 100% due to the 100% change in the value of β. I_B is the same and V_{CE} decreased by 76%.

Using the results calculated in Example 4.4 and then repeating for a value of $\beta = 100$, we have the following:

β	$I_B\ (\mu A)$	$I_C\ (mA)$	$V_{CE}\ (V)$
50	40.1	2.01	13.97
100	36.3	3.63	9.11

Now the BJT collector current increases by about 81% due to the 100% increase in β. Notice that I_B decreased, helping maintain the value of I_C—or at least reducing the overall change in I_C due to the change in β. The change in V_{CE} has dropped to about 35%. The network of Fig. 4.22 is therefore more stable than that of Fig. 4.7 for the same change in β.

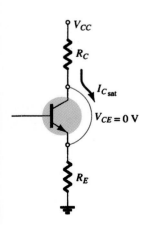

Figure 4.23 Determining $I_{C_{sat}}$ for the emitter-stabilized bias circuit.

Saturation Level

The collector saturation level or maximum collector current for an emitter-bias design can be determined using the same approach applied to the fixed-bias configuration: Apply a short circuit between the collector–emitter terminals as shown in Fig. 4.23 and calculate the resulting collector current. For Fig. 4.23:

$$I_{C_{sat}} = \frac{V_{CC}}{R_C + R_E} \tag{4.25}$$

The addition of the emitter resistor reduces the collector saturation level below that obtained with a fixed-bias configuration using the same collector resistor.

Determine the saturation current for the network of Example 4.4.

EXAMPLE 4.6

Solution

$$I_{C_{sat}} = \frac{V_{CC}}{R_C + R_E}$$

$$= \frac{20\text{ V}}{2\text{ k}\Omega + 1\text{ k}\Omega} = \frac{20\text{ V}}{3\text{ k}\Omega}$$

$$= \mathbf{6.67\text{ mA}}$$

which is about three times the level of I_{C_Q} for Example 4.4.

Load-Line Analysis

The load-line analysis of the emitter-bias network is only slightly different from that encountered for the fixed-bias configuration. The level of I_B as determined by Eq. (4.17) defines the level of I_B on the characteristics of Fig. 4.24 (denoted I_{B_Q}).

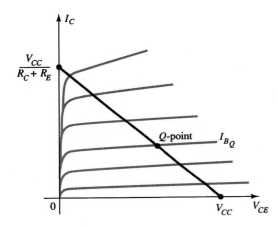

Figure 4.24 Load line for the emitter-bias configuration.

The collector–emitter loop equation that defines the load line is the following:

$$V_{CE} = V_{CC} - I_C(R_C + R_E)$$

Choosing $I_C = 0$ mA gives

$$\boxed{V_{CE} = V_{CC}|_{I_C = 0\text{ mA}}} \qquad (4.26)$$

as obtained for the fixed-bias configuration. Choosing $V_{CE} = 0$ V gives

$$\boxed{I_C = \frac{V_{CC}}{R_C + R_E}\bigg|_{V_{CE} = 0\text{ V}}} \qquad (4.27)$$

as shown in Fig. 4.24. Different levels of I_{B_Q} will, of course, move the Q-point up or down the load line.

4.5 VOLTAGE-DIVIDER BIAS

In the previous bias configurations the bias current I_{C_Q} and voltage V_{CE_Q} were a function of the current gain (β) of the transistor. However, since β is temperature sensitive, especially for silicon transistors, and the actual value of beta is usually not well defined, it would be desirable to develop a bias circuit that is less dependent, or in

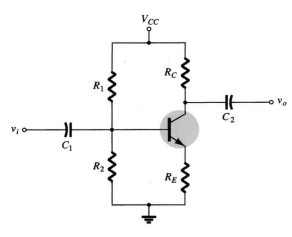

Figure 4.25 Voltage-divider bias configuration.

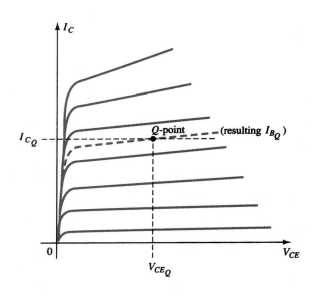

Figure 4.26 Defining the Q-point for the voltage-divider bias configuration.

fact, independent of the transistor beta. The voltage-divider bias configuration of Fig. 4.25 is such a network. If analyzed on an exact basis the sensitivity to changes in beta is quite small. If the circuit parameters are properly chosen, the resulting levels of I_{C_Q} and V_{CE_Q} can be almost totally independent of beta. Recall from previous discussions that a Q-point is defined by a fixed level of I_{C_Q} and V_{CE_Q} as shown in Fig. 4.26. The level of I_{B_Q} will change with the change in beta, but the operating point on the characteristics defined by I_{C_Q} and V_{CE_Q} can remain fixed if the proper circuit parameters are employed.

As noted above, there are two methods that can be applied to analyze the voltage-divider configuration. The reason for the choice of names for this configuration will become obvious in the analysis to follow. The first to be demonstrated is the *exact method* that can be applied to *any* voltage-divider configuration. The second is referred to as the *approximate method* and can be applied only if specific conditions are satisfied. The approximate approach permits a more direct analysis with a savings in time and energy. It is also particularly helpful in the design mode to be described in a later section. All in all, the approximate approach can be applied to the majority of situations and therefore should be examined with the same interest as the exact method.

Exact Analysis

The input side of the network of Fig. 4.25 can be redrawn as shown in Fig. 4.27 for the dc analysis. The Thévenin equivalent network for the network to the left of the base terminal can then be found in the following manner:

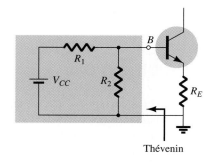

Thévenin

Figure 4.27 Redrawing the input side of the network of Fig. 4.25.

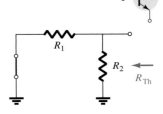

R_{Th}: The voltage source is replaced by a short-circuit equivalent as shown in Fig. 4.28.

$$R_{Th} = R_1 \| R_2 \qquad (4.28)$$

E_{Th}: The voltage source V_{CC} is returned to the network and the open-circuit Thévenin voltage of Fig. 4.29 determined as follows:

Applying the voltage-divider rule:

$$E_{Th} = V_{R_2} = \frac{R_2 V_{CC}}{R_1 + R_2} \qquad (4.29)$$

Figure 4.28 Determining R_{TH}.

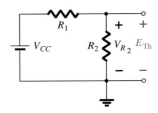

The Thévenin network is then redrawn as shown in Fig. 4.30, and I_{B_Q} can be determined by first applying Kirchhoff's voltage law in the clockwise direction for the loop indicated:

$$E_{Th} - I_B R_{Th} - V_{BE} - I_E R_E = 0$$

Substituting $I_E = (\beta + 1)I_B$ and solving for I_B yields

$$I_B = \frac{E_{Th} - V_{BE}}{R_{Th} + (\beta + 1)R_E} \qquad (4.30)$$

Figure 4.29 Determining E_{Th}.

Although Eq. (4.30) initially appears different from those developed earlier, note that the numerator is again a difference of two voltage levels and the denominator is the base resistance plus the emitter resistor reflected by $(\beta + 1)$—certainly very similar to Eq. (4.17).

Once I_B is known, the remaining quantities of the network can be found in the same manner as developed for the emitter-bias configuration. That is,

$$V_{CE} = V_{CC} - I_C(R_C + R_E) \qquad (4.31)$$

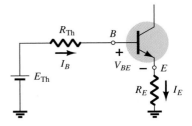

which is exactly the same as Eq. (4.19). The remaining equations for V_E, V_C, and V_B are also the same as obtained for the emitter-bias configuration.

Figure 4.30 Inserting the Thévenin equivalent circuit.

Determine the dc bias voltage V_{CE} and the current I_C for the voltage-divider configuration of Fig. 4.31.

EXAMPLE 4.7

Solution

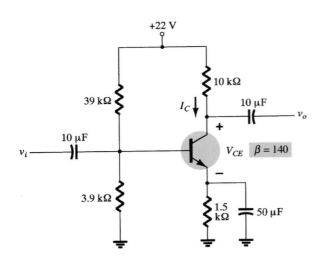

Figure 4.31 Beta-stabilized circuit for Example 4.7.

$$\text{Eq. (4.28):} \quad R_{Th} = R_1 \| R_2$$

$$= \frac{(39 \text{ k}\Omega)(3.9 \text{ k}\Omega)}{39 \text{ k}\Omega + 3.9 \text{ k}\Omega} = 3.55 \text{ k}\Omega$$

$$\text{Eq. (4.29):} \quad E_{Th} = \frac{R_2 V_{CC}}{R_1 + R_2}$$

$$= \frac{(3.9 \text{ k}\Omega)(22 \text{ V})}{39 \text{ k}\Omega + 3.9 \text{ k}\Omega} = 2 \text{ V}$$

$$\text{Eq. (4.30):} \quad I_B = \frac{E_{Th} - V_{BE}}{R_{Th} + (\beta + 1)R_E}$$

$$= \frac{2 \text{ V} - 0.7 \text{ V}}{3.55 \text{ k}\Omega + (141)(1.5 \text{ k}\Omega)} = \frac{1.3 \text{ V}}{3.55 \text{ k}\Omega + 211.5 \text{ k}\Omega}$$

$$= 6.05 \text{ }\mu\text{A}$$

$$I_C = \beta I_B$$

$$= (140)(6.05 \text{ }\mu\text{A})$$

$$= \mathbf{0.85 \text{ mA}}$$

$$\text{Eq. (4.31):} \quad V_{CE} = V_{CC} - I_C(R_C + R_E)$$

$$= 22 \text{ V} - (0.85 \text{ mA})(10 \text{ k}\Omega + 1.5 \text{ k}\Omega)$$

$$= 22 \text{ V} - 9.78 \text{ V}$$

$$= \mathbf{12.22 \text{ V}}$$

Approximate Analysis

The input section of the voltage-divider configuration can be represented by the network of Fig. 4.32. The resistance R_i is the equivalent resistance between base and ground for the transistor with an emitter resistor R_E. Recall from Section 4.4 [Eq. (4.18)] that the reflected resistance between base and emitter is defined by $R_i = (\beta + 1)R_E$. If R_i is much larger than the resistance R_2, the current I_B will be much smaller than I_2 (current always seeks the path of least resistance) and I_2 will be approximately equal to I_1. If we accept the approximation that I_B is essentially zero

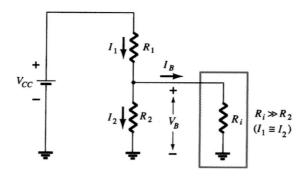

Figure 4.32 Partial-bias circuit for calculating the approximate base voltage V_B.

amperes compared to I_1 or I_2, then $I_1 = I_2$ and R_1 and R_2 can be considered series elements. The voltage across R_2, which is actually the base voltage, can be determined using the voltage-divider rule (hence the name for the configuration). That is,

$$V_B = \frac{R_2 V_{CC}}{R_1 + R_2} \qquad (4.32)$$

Since $R_i = (\beta + 1)R_E \cong \beta R_E$ the condition that will define whether the approximate approach can be applied will be the following:

$$\beta R_E \geq 10R_2 \qquad (4.33)$$

In other words, if β times the value of R_E is at least 10 times the value of R_2, the approximate approach can be applied with a high degree of accuracy.

Once V_B is determined, the level of V_E can be calculated from

$$V_E = V_B - V_{BE} \qquad (4.34)$$

and the emitter current can be determined from

$$I_E = \frac{V_E}{R_E} \qquad (4.35)$$

and

$$I_{C_Q} \cong I_E \qquad (4.36)$$

The collector-to-emitter voltage is determined by

$$V_{CE} = V_{CC} - I_C R_C - I_E R_E$$

but since $I_E \cong I_C$,

$$V_{CE_Q} = V_{CC} - I_C(R_C + R_E) \qquad (4.37)$$

Note in the sequence of calculations from Eq. (4.33) through Eq. (4.37) that β does not appear and I_B was not calculated. The Q-point (as determined by I_{C_Q} and V_{CE_Q}) is therefore independent of the value of β.

Repeat the analysis of Fig. 4.31 using the approximate technique, and compare solutions for I_{C_Q} and V_{CE_Q}.

EXAMPLE 4.8

Solution

Testing:

$$\beta R_E \geq 10R_2$$

$$(140)(1.5 \text{ k}\Omega) \geq 10(3.9 \text{ k}\Omega)$$

$$210 \text{ k}\Omega \geq 39 \text{ k}\Omega \ (\textit{satisfied})$$

$$\text{Eq. (4.32):} \quad V_B = \frac{R_2 V_{CC}}{R_1 + R_2}$$

$$= \frac{(3.9 \text{ k}\Omega)(22 \text{ V})}{39 \text{ k}\Omega + 3.9 \text{ k}\Omega}$$

$$= 2 \text{ V}$$

Note that the level of V_B is the same as E_{Th} determined in Example 4.7. Essentially, therefore, the primary difference between the exact and approximate techniques is the effect of R_{Th} in the exact analysis that separates E_{Th} and V_B.

$$\text{Eq. (4.34):} \quad V_E = V_B - V_{BE}$$

$$= 2\,\text{V} - 0.7\,\text{V}$$

$$= 1.3\,\text{V}$$

$$I_{CQ} \cong I_E = \frac{V_E}{R_E} = \frac{1.3\,\text{V}}{1.5\,\text{k}\Omega} = \mathbf{0.867\,mA}$$

compared to 0.85 mA with the exact analysis. Finally,

$$V_{CE_Q} = V_{CC} - I_C(R_C + R_E)$$

$$= 22\,\text{V} - (0.867\,\text{mA})(10\,\text{kV} + 1.5\,\text{k}\Omega)$$

$$= 22\,\text{V} - 9.97\,\text{V}$$

$$= \mathbf{12.03\,V}$$

versus 12.22 V obtained in Example 4.7.

The results for I_{C_Q} and V_{CE_Q} are certainly close, and considering the actual variation in parameter values one can certainly be considered as accurate as the other. The larger the level of R_i compared to R_2, the closer the approximate to the exact solution. Example 4.10 will compare solutions at a level well below the condition established by Eq. (4.33).

EXAMPLE 4.9

Repeat the exact analysis of Example 4.7 if β is reduced to 70, and compare solutions for I_{C_Q} and V_{CE_Q}.

Solution

This example is not a comparison of exact versus approximate methods but a testing of how much the Q-point will move if the level of β is cut in half. R_{Th} and E_{Th} are the same:

$$R_{Th} = 3.55\,\text{k}\Omega, \qquad E_{Th} = 2\,\text{V}$$

$$I_B = \frac{E_{Th} - V_{BE}}{R_{Th} + (\beta + 1)R_E}$$

$$= \frac{2\,\text{V} - 0.7\,\text{V}}{3.55\,\text{k}\Omega + (71)(1.5\,\text{k}\Omega)} = \frac{1.3\,\text{V}}{3.55\,\text{k}\Omega + 106.5\,\text{k}\Omega}$$

$$= 11.81\,\mu\text{A}$$

$$I_{C_Q} = \beta I_B$$

$$= (70)(11.81\,\mu\text{A})$$

$$= 0.83\,\text{mA}$$

$$V_{CE_Q} = V_{CC} - I_C(R_C + R_E)$$

$$= 22\,\text{V} - (0.83\,\text{mA})(10\,\text{k}\Omega + 1.5\,\text{k}\Omega)$$

$$= \mathbf{12.46\,V}$$

Tabulating the results, we have:

β	I_{C_Q} (mA)	V_{CE_Q} (V)
140	0.85	12.22
70	0.83	12.46

The results clearly show the relative insensitivity of the circuit to the change in β. Even though β is drastically cut in half, from 140 to 70, the levels of I_{C_Q} and V_{CE_Q} are essentially the same.

Determine the levels of I_{C_Q} and V_{CE_Q} for the voltage-divider configuration of Fig. 4.33 using the exact and approximate techniques and compare solutions. In this case, the conditions of Eq. (4.33) will not be satisfied but the results will reveal the difference in solution if the criterion of Eq. (4.33) is ignored.

EXAMPLE 4.10

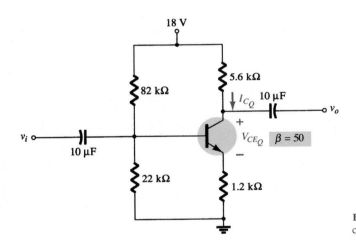

Figure 4.33 Voltage-divider configuration for Example 4.10.

Solution

Exact Analysis

Eq. (4.33): $\quad \beta R_E \geq 10R_2$

$\qquad (50)(1.2\text{ k}\Omega) \geq 10(22\text{ k}\Omega)$

$\qquad\quad 60\text{ k}\Omega \ngeq 220\text{ k}\Omega \ (\textit{not satisfied})$

$$R_{\text{Th}} = R_1 \| R_2 = 82\text{ k}\Omega \| 22\text{ k}\Omega = 17.35\text{ k}\Omega$$

$$E_{\text{Th}} = \frac{R_2 V_{CC}}{R_1 + R_2} = \frac{22\text{ k}\Omega(18\text{ V})}{82\text{ k}\Omega + 22\text{ k}\Omega} = 3.81\text{ V}$$

$$I_B = \frac{E_{\text{Th}} - V_{BE}}{R_{\text{Th}} + (\beta + 1)R_E} = \frac{3.81\text{ V} - 0.7\text{ V}}{17.35\text{ k}\Omega + (51)(1.2\text{ k}\Omega)} = \frac{3.11\text{ V}}{78.55\text{ k}\Omega} = 39.6\text{ }\mu\text{A}$$

4.5 Voltage-Divider Bias   **183**

$$I_{C_Q} = \beta I_B = (50)(39.6 \, \mu A) = \textbf{1.98 mA}$$

$$V_{CE_Q} = V_{CC} - I_C(R_C + R_E)$$

$$= 18 \, V - (1.98 \, mA)(5.6 \, k\Omega + 1.2 \, k\Omega)$$

$$= \textbf{4.54 V}$$

Approximate Analysis

$$V_B = E_{Th} = 3.81 \, V$$

$$V_E = V_B - V_{BE} = 3.81 \, V - 0.7 \, V = 3.11 \, V$$

$$I_{C_Q} \cong I_E = \frac{V_E}{R_E} = \frac{3.11 \, V}{1.2 \, k\Omega} = \textbf{2.59 mA}$$

$$V_{CE_Q} = V_{CC} - I_C(R_C + R_E)$$

$$= 18 \, V - (2.59 \, mA)(5.6 \, k\Omega + 1.2 \, k\Omega)$$

$$= \textbf{3.88 V}$$

Tabulating the results, we have:

	I_{C_Q} (mA)	V_{CE_Q} (V)
Exact	1.98	4.54
Approximate	2.59	3.88

The results reveal the difference between exact and approximate solutions. I_{C_Q} is about 30% greater with the approximate solution, while V_{CE_Q} is about 10% less. The results are notably different in magnitude, but even though βR_E is only about three times larger than R_2, the results are still relatively close to each other. For the future, however, our analysis will be dictated by Eq. (4.33) to ensure a close similarity between exact and approximate solutions.

Transistor Saturation

The output collector–emitter circuit for the voltage-divider configuration has the same appearance as the emitter-biased circuit analyzed in Section 4.4. The resulting equation for the saturation current (when V_{CE} is set to zero volts on the schematic) is therefore the same as obtained for the emitter-biased configuration. That is,

$$\boxed{I_{C_{sat}} = I_{C_{max}} = \frac{V_{CC}}{R_C + R_E}} \qquad (4.38)$$

Load-Line Analysis

The similarities with the output circuit of the emitter-biased configuration result in the same intersections for the load line of the voltage-divider configuration. The load line will therefore have the same appearance as that of Fig. 4.24, with

$$I_C = \left.\frac{V_{CC}}{R_C + R_E}\right|_{V_{CE}=0\,V} \qquad\qquad (4.39)$$

and

$$V_{CE} = V_{CC}\big|_{I_C=0\,mA} \qquad\qquad (4.40)$$

The level of I_B is of course determined by a different equation for the voltage-divider bias and the emitter-bias configurations.

Mathcad

The power and usefulness of Mathcad can now be demonstrated for the network of Example 4.7. When using Mathcad, there is no need to worry about whether the exact or the approximate method should be applied to the voltage-divider bias network—Mathcad will always provide the most accurate results possible for the given data.

As shown in Fig. 4.34, all the parameters (variables) of the network are first entered, with no need to include the unit of measure. Although the listing will appear as shown in Fig. 4.34, in storage (internal hard disk or floppy disk) the parameters can easily be changed at any time with an immediate change in results. All the equations are then introduced in an order that permits using the results of one calculation to calculate the next quantity of interest. That is, the equations must be entered from left to right and down the display. In this example, **IB** is first determined because it will be used to find **IC** on the next line.

Using Mathcad, the results obtained are an exact match for **IB** and **IC** and only slightly different for **VCE** because the level of **IC** carried a higher order of accuracy in the Mathcad solution. The wonderful advantage of having this sequence of calculations in storage is that it can be called up for any voltage-divider network, and the desired results can be obtained quickly and accurately simply by changing the magnitude of specific variables.

$$R1 := 39 \cdot 10^3 \qquad R2 := 3.9 \cdot 10^3 \qquad RC := 10 \cdot 10^3 \qquad RE := 1.5 \cdot 10^3$$

$$VCC := 22 \qquad beta := 140 \qquad VBE := 0.7$$

$$RTh := R1 \cdot \frac{R2}{(R1+R2)} \qquad ETh := R2 \cdot \frac{VCC}{(R1+R2)}$$

$$IB := \frac{ETh - (VBE)}{(RTh + (beta + 1) \cdot RE)} \qquad IB = 6.045 \cdot 10^{-6}$$

$$IC := beta \cdot IB \qquad IC = 8.463 \cdot 10^{-4}$$

$$VCE := VCC - IC \cdot (RC + RE)$$

$$VCE = 12.267$$

Figure 4.34 Verifying the results of Example 4.7 using Mathcad.

4.6 DC BIAS WITH VOLTAGE FEEDBACK

An improved level of stability can also be obtained by introducing a feedback path from collector to base as shown in Fig. 4.35. Although the Q-point is not totally independent of beta (even under approximate conditions), the sensitivity to changes in beta or temperature variations is normally less than encountered for the fixed-bias or emitter-biased configurations. The analysis will again be performed by first analyzing the base–emitter loop with the results applied to the collector–emitter loop.

Base–Emitter Loop

Figure 4.36 shows the base–emitter loop for the voltage feedback configuration. Writing Kirchhoff's voltage law around the indicated loop in the clockwise direction will result in

$$V_{CC} - I'_C R_C - I_B R_B - V_{BE} - I_E R_E = 0$$

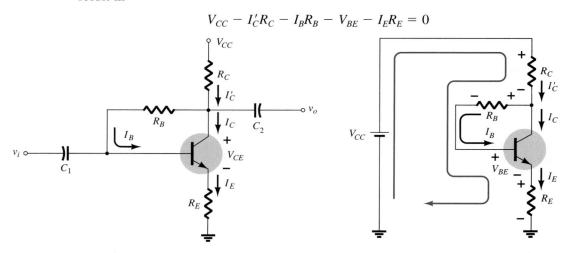

Figure 4.35 dc bias circuit with voltage feedback.

Figure 4.36 Base–emitter loop for the network of Fig. 4.35.

It is important to note that the current through R_C is not I_C but I'_C (where $I'_C = I_C + I_B$). However, the level of I_C and I'_C far exceeds the usual level of I_B and the approximation $I'_C \cong I_C$ is normally employed. Substituting $I'_C \cong I_C = \beta I_B$ and $I_E \cong I_C$ will result in

$$V_{CC} - \beta I_B R_C - I_B R_B - V_{BE} - \beta I_B R_E = 0$$

Gathering terms, we have

$$V_{CC} - V_{BE} - \beta I_B (R_C + R_E) - I_B R_B = 0$$

and solving for I_B yields

$$\boxed{I_B = \frac{V_{CC} - V_{BE}}{R_B + \beta(R_C + R_E)}} \qquad (4.41)$$

The result is quite interesting in that the format is very similar to equations for I_B obtained for earlier configurations. The numerator is again the difference of available voltage levels, while the denominator is the base resistance plus the collector and emitter resistors reflected by beta. In general, therefore, the feedback path results in a reflection of the resistance R_C back to the input circuit, much like the reflection of R_E.

In general, the equation for I_B has had the following format:

$$I_B = \frac{V'}{R_B + \beta R'}$$

with the absence of R' for the fixed-bias configuration, $R' = R_E$ for the emitter-bias setup (with $(\beta + 1) \cong \beta$), and $R' = R_C + R_E$ for the collector-feedback arrangement. The voltage V' is the difference between two voltage levels.

Since $I_C = \beta I_B$,

$$I_{C_Q} = \frac{\beta V'}{R_B + \beta R'}$$

In general, the larger $\beta R'$ is compared to R_B, the less the sensitivity of I_{C_Q} to variations in beta. Obviously, if $\beta R' \gg R_B$ and $R_B + \beta R' \cong \beta R'$, then

$$I_{C_Q} = \frac{\beta V'}{R_B + \beta R'} \cong \frac{\beta V'}{\beta R'} = \frac{V'}{R'}$$

and I_{C_Q} is independent of the value of beta. Since R' is typically larger for the voltage-feedback configuration than for the emitter-bias configuration, the sensitivity to variations in beta is less. Of course, R' is zero ohms for the fixed-bias configuration and is therefore quite sensitive to variations in beta.

Collector–Emitter Loop

The collector–emitter loop for the network of Fig. 4.35 is provided in Fig. 4.37. Applying Kirchhoff's voltage law around the indicated loop in the clockwise direction will result in

$$I_E R_E + V_{CE} + I_C' R_C - V_{CC} = 0$$

Since $I_C' \cong I_C$ and $I_E \cong I_C$, we have

$$I_C(R_C + R_E) + V_{CE} - V_{CC} = 0$$

and

$$\boxed{V_{CE} = V_{CC} - I_C(R_C + R_E)} \qquad (4.42)$$

which is exactly as obtained for the emitter-bias and voltage-divider bias configurations.

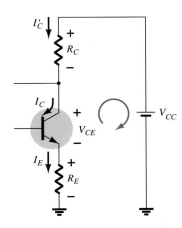

Figure 4.37 Collector–emitter loop for the network of Fig. 4.35.

Determine the quiescent levels of I_{C_Q} and V_{CE_Q} for the network of Fig. 4.38.

EXAMPLE 4.11

Solution

Eq. (4.41):
$$I_B = \frac{V_{CC} - V_{BE}}{R_B + \beta(R_C + R_E)}$$

$$= \frac{10\ V - 0.7\ V}{250\ k\Omega + (90)(4.7\ k\Omega + 1.2\ k\Omega)}$$

$$= \frac{9.3\ V}{250\ k\Omega + 531\ k\Omega} = \frac{9.3\ V}{781\ k\Omega}$$

$$= 11.91\ \mu A$$

$$I_{C_Q} = \beta I_B = (90)(11.91\ \mu A)$$

$$= \textbf{1.07 mA}$$

$$V_{CE_Q} = V_{CC} - I_C(R_C + R_E)$$

$$= 10\ V - (1.07\ mA)(4.7\ k\Omega + 1.2\ k\Omega)$$

$$= 10\ V - 6.31\ V$$

$$= \textbf{3.69 V}$$

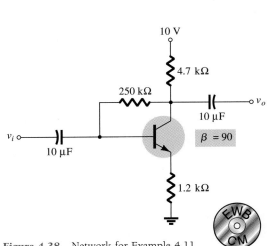

Figure 4.38 Network for Example 4.11.

EXAMPLE 4.12

Repeat Example 4.11 using a beta of 135 (50% more than Example 4.11).

Solution

It is important to note in the solution for I_B in Example 4.11 that the second term in the denominator of the equation is larger than the first. Recall in a recent discussion that the larger this second term is compared to the first, the less the sensitivity to changes in beta. In this example the level of beta is increased by 50%, which will increase the magnitude of this second term even more compared to the first. It is more important to note in these examples, however, that once the second term is relatively large compared to the first, the sensitivity to changes in beta is significantly less.

Solving for I_B gives

$$I_B = \frac{V_{CC} - V_{BE}}{R_B + \beta(R_C + R_E)}$$

$$= \frac{10\text{ V} - 0.7\text{ V}}{250\text{ k}\Omega + (135)(4.7\text{ k}\Omega + 1.2\text{ k}\Omega)}$$

$$= \frac{9.3\text{ V}}{250\text{ k}\Omega + 796.5\text{ k}\Omega} = \frac{9.3\text{ V}}{1046.5\text{ k}\Omega}$$

$$= 8.89\ \mu\text{A}$$

and

$$I_{C_Q} = \beta I_B$$

$$= (135)(8.89\ \mu\text{A})$$

$$= \mathbf{1.2\ mA}$$

and

$$V_{CE_Q} = V_{CC} - I_C(R_C + R_E)$$

$$= 10\text{ V} - (1.2\text{ mA})(4.7\text{ k}\Omega + 1.2\text{ k}\Omega)$$

$$= 10\text{ V} - 7.08\text{ V}$$

$$= \mathbf{2.92\ V}$$

Even though the level of β increased 50%, the level of I_{C_Q} only increased 12.1% while the level of V_{CE_Q} decreased about 20.9%. If the network were a fixed-bias design, a 50% increase in β would have resulted in a 50% increase in I_{C_Q} and a dramatic change in the location of the Q-point.

EXAMPLE 4.13

Determine the dc level of I_B and V_C for the network of Fig. 4.39.

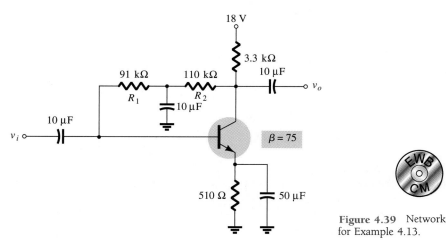

Figure 4.39 Network for Example 4.13.

Solution

In this case, the base resistance for the dc analysis is composed of two resistors with a capacitor connected from their junction to ground. For the dc mode, the capacitor assumes the open-circuit equivalence and $R_B = R_1 + R_2$.

Solving for I_B gives

$$I_B = \frac{V_{CC} - V_{BE}}{R_B + \beta(R_C + R_E)}$$

$$= \frac{18\text{ V} - 0.7\text{ V}}{(91\text{ k}\Omega + 110\text{ k}\Omega) + (75)(3.3\text{ k}\Omega + 0.51\text{ k}\Omega)}$$

$$= \frac{17.3\text{ V}}{201\text{ k}\Omega + 285.75\text{ k}\Omega} = \frac{17.3\text{ V}}{486.75\text{ k}\Omega}$$

$$= \textbf{35.5 } \boldsymbol{\mu}\textbf{A}$$

$$I_C = \beta I_B$$

$$= (75)(35.5\ \mu\text{A})$$

$$= 2.66\text{ mA}$$

$$V_C = V_{CC} - I'_C R_C \cong V_{CC} - I_C R_C$$

$$= 18\text{ V} - (2.66\text{ mA})(3.3\text{ k}\Omega)$$

$$= 18\text{ V} - 8.78\text{ V}$$

$$= \textbf{9.22 V}$$

Saturation Conditions

Using the approximation $I'_C = I_C$, the equation for the saturation current is the same as obtained for the voltage-divider and emitter-bias configurations. That is,

$$\boxed{I_{C_{\text{sat}}} = I_{C_{\text{max}}} = \frac{V_{CC}}{R_C + R_E}}$$

(4.43)

Load-Line Analysis

Continuing with the approximation $I'_C = I_C$ will result in the same load line defined for the voltage-divider and emitter-biased configurations. The level of I_{B_Q} will be defined by the chosen bias configuration.

4.7 MISCELLANEOUS BIAS CONFIGURATIONS

There are a number of BJT bias configurations that do not match the basic mold of those analyzed in the previous sections. In fact, there are variations in design that would require many more pages than is possible in a book of this type. However, the primary purpose here is to emphasize those characteristics of the device that permit a dc analysis of the configuration and to establish a general procedure toward the desired solution. For each configuration discussed thus far, the first step has been the derivation of an expression for the base current. Once the base current is known, the collector current and voltage levels of the output circuit can be determined quite directly. This is not to imply that all solutions will take this path, but it does suggest a possible route to follow if a new configuration is encountered.

The first example is simply one where the emitter resistor has been dropped from the voltage-feedback configuration of Fig. 4.35. The analysis is quite similar but does require dropping R_E from the applied equation.

EXAMPLE 4.14

For the network of Fig. 4.40:
(a) Determine I_{C_Q} and V_{CE_Q}.
(b) Find V_B, V_C, V_E, and V_{BC}.

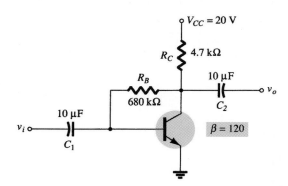

Figure 4.40 Collector feedback with $R_E = 0\ \Omega$.

Solution

(a) The absence of R_E reduces the reflection of resistive levels to simply that of R_C and the equation for I_B reduces to

$$I_B = \frac{V_{CC} - V_{BE}}{R_B + \beta R_C}$$

$$= \frac{20\ \text{V} - 0.7\ \text{V}}{680\ \text{k}\Omega + (120)(4.7\ \text{k}\Omega)} = \frac{19.3\ \text{V}}{1.244\ \text{M}\Omega}$$

$$= 15.51\ \mu\text{A}$$

$$I_{C_Q} = \beta I_B = (120)(15.51\ \mu\text{A})$$

$$= \textbf{1.86 mA}$$

$$V_{CE_Q} = V_{CC} - I_C R_C$$

$$= 20\ \text{V} - (1.86\ \text{mA})(4.7\ \text{k}\Omega)$$

$$= \textbf{11.26 V}$$

(b) $$V_B = V_{BE} = \textbf{0.7 V}$$

$$V_C = V_{CE} = \textbf{11.26 V}$$

$$V_E = \textbf{0 V}$$

$$V_{BC} = V_B - V_C = 0.7\ \text{V} - 11.26\ \text{V}$$

$$= \boldsymbol{-10.56\ \text{V}}$$

In the next example, the applied voltage is connected to the emitter leg and R_C is connected directly to ground. Initially, it appears somewhat unorthodox and quite different from those encountered thus far, but one application of Kirchhoff's voltage law to the base circuit will result in the desired base current.

Determine V_C and V_B for the network of Fig. 4.41.

EXAMPLE 4.15

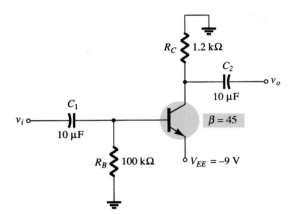

Figure 4.41 Example 4.15

Solution

Applying Kirchhoff's voltage law in the clockwise direction for the base–emitter loop will result in

$$-I_B R_B - V_{BE} + V_{EE} = 0$$

and

$$I_B = \frac{V_{EE} - V_{BE}}{R_B}$$

Substitution yields

$$I_B = \frac{9 \text{ V} - 0.7 \text{ V}}{100 \text{ k}\Omega}$$

$$= \frac{8.3 \text{ V}}{100 \text{ k}\Omega}$$

$$= 83 \text{ }\mu\text{A}$$

$$I_C = \beta I_B$$

$$= (45)(83 \text{ }\mu\text{A})$$

$$= 3.735 \text{ mA}$$

$$V_C = -I_C R_C$$

$$= -(3.735 \text{ mA})(1.2 \text{ k}\Omega)$$

$$= -\textbf{4.48 V}$$

$$V_B = -I_B R_B$$

$$= -(83 \text{ }\mu\text{A})(100 \text{ k}\Omega)$$

$$= -\textbf{8.3 V}$$

The next example employs a network referred to as an *emitter-follower* configuration. When the same network is analyzed on an ac basis, we will find that the output and input signals are in phase (one following the other) and the output voltage is slightly less than the applied signal. For the dc analysis the collector is grounded and the applied voltage is in the emitter leg.

EXAMPLE 4.16   Determine V_{CE_Q} and I_E for the network of Fig. 4.42.

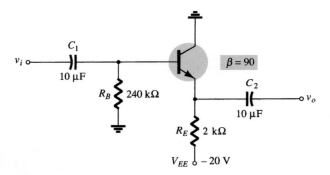

Figure 4.42   Common-collector (emitter-follower) configuration.

Solution

Applying Kirchhoff's voltage law to the input circuit will result in

$$-I_B R_B - V_{BE} - I_E R_E + V_{EE} = 0$$

but

$$I_E = (\beta + 1)I_B$$

and

$$V_{EE} - V_{BE} - (\beta + 1)I_B R_E - I_B R_B = 0$$

with

$$I_B = \frac{V_{EE} - V_{BE}}{R_B + (\beta + 1)R_E}$$

Substituting values yields

$$I_B = \frac{20 \text{ V} - 0.7 \text{ V}}{240 \text{ k}\Omega + (91)(2 \text{ k}\Omega)}$$

$$= \frac{19.3 \text{ V}}{240 \text{ k}\Omega + 182 \text{ k}\Omega} = \frac{19.3 \text{ V}}{422 \text{ k}\Omega}$$

$$= 45.73 \text{ }\mu\text{A}$$

$$I_C = \beta I_B$$

$$= (90)(45.73 \text{ }\mu\text{A})$$

$$= 4.12 \text{ mA}$$

Applying Kirchhoff's voltage law to the output circuit, we have

$$-V_{EE} + I_E R_E + V_{CE} = 0$$

but

$$I_E = (\beta + 1)I_B$$

and

$$V_{CE_Q} = V_{EE} - (\beta + 1)I_B R_E$$

$$= 20 \text{ V} - (91)(45.73 \text{ }\mu\text{A})(2 \text{ k}\Omega)$$

$$= \textbf{11.68 V}$$

$$I_E = \textbf{4.16 mA}$$

All of the examples thus far have employed a common-emitter or common-collector configuration. In the next example we investigate the common-base configuration. In this situation the input circuit will be employed to determine I_E rather than I_B. The collector current is then available to perform an analysis of the output circuit.

Determine the voltage V_{CB} and the current I_B for the common-base configuration of Fig. 4.43.

EXAMPLE 4.17

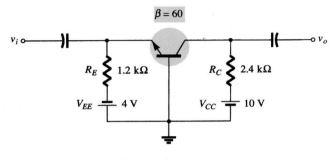

Figure 4.43 Common-base configuration.

Solution

Applying Kirchhoff's voltage law to the input circuit yields

$$-V_{EE} + I_E R_E + V_{BE} = 0$$

and

$$I_E = \frac{V_{EE} - V_{BE}}{R_E}$$

Substituting values, we obtain

$$I_E = \frac{4 \text{ V} - 0.7 \text{ V}}{1.2 \text{ k}\Omega} = 2.75 \text{ mA}$$

Applying Kirchhoff's voltage law to the output circuit gives

$$-V_{CB} + I_C R_C - V_{CC} = 0$$

and

$$V_{CB} = V_{CC} - I_C R_C \text{ with } I_C \cong I_E$$

$$= 10 \text{ V} - (2.75 \text{ mA})(2.4 \text{ k}\Omega)$$

$$= \mathbf{3.4 \text{ V}}$$

$$I_B = \frac{I_C}{\beta}$$

$$= \frac{2.75 \text{ mA}}{60}$$

$$= \mathbf{45.8 \text{ } \mu A}$$

Example 4.18 employs a split supply and will require the application of Thévenin's theorem to determine the desired unknowns.

EXAMPLE 4.18

Determine V_C and V_B for the network of Fig. 4.44.

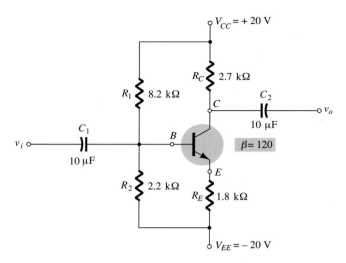

Figure 4.44   Example 4.18

Solution

The Thévenin resistance and voltage are determined for the network to the left of the base terminal as shown in Figs. 4.45 and 4.46.

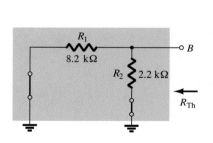

Figure 4.45   Determining R_{Th}.

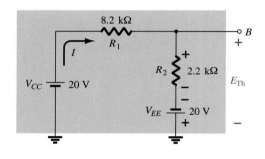

Figure 4.46   Determining E_{Th}.

R_{Th}:

$$R_{Th} = 8.2 \text{ k}\Omega \| 2.2 \text{ k}\Omega = 1.73 \text{ k}\Omega$$

E_{Th}:

$$I = \frac{V_{CC} + V_{EE}}{R_1 + R_2} = \frac{20 \text{ V} + 20 \text{ V}}{8.2 \text{ k}\Omega + 2.2 \text{ k}\Omega} = \frac{40 \text{ V}}{10.4 \text{ k}\Omega}$$

$$= 3.85 \text{ mA}$$

$$E_{Th} = IR_2 - V_{EE}$$

$$= (3.85 \text{ mA})(2.2 \text{ k}\Omega) - 20 \text{ V}$$

$$= -11.53 \text{ V}$$

The network can then be redrawn as shown in Fig. 4.47, where the application of Kirchhoff's voltage law will result in

$$-E_{Th} - I_B R_{Th} - V_{BE} - I_E R_E + V_{EE} = 0$$

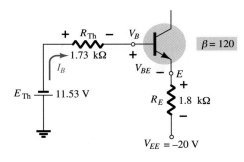

Figure 4.47 Substituting the Thévenin equivalent circuit.

Substituting $I_E = (\beta + 1)I_B$ gives

$$V_{EE} - E_{Th} - V_{BE} - (\beta + 1)I_B R_E - I_B R_{Th} = 0$$

and
$$I_B = \frac{V_{EE} - E_{Th} - V_{BE}}{R_{Th} + (\beta + 1)R_E}$$

$$= \frac{20\ V - 11.53\ V - 0.7\ V}{1.73\ k\Omega + (121)(1.8\ k\Omega)}$$

$$= \frac{7.77\ V}{219.53\ k\Omega}$$

$$= 35.39\ \mu A$$

$$I_C = \beta I_B$$

$$= (120)(35.39\ \mu A)$$

$$= 4.25\ mA$$

$$V_C = V_{CC} - I_C R_C$$

$$= 20\ V - (4.25\ mA)(2.7\ k\Omega)$$

$$= \mathbf{8.53\ V}$$

$$V_B = -E_{Th} - I_B R_{Th}$$

$$= -(11.53\ V) - (35.39\ \mu A)(1.73\ k\Omega)$$

$$= \mathbf{-11.59\ V}$$

4.8 DESIGN OPERATIONS

Discussions thus far have focused on the analysis of existing networks. All the elements are in place and it is simply a matter of solving for the current and voltage levels of the configuration. The design process is one where a current and/or voltage may be specified and the elements required to establish the designated levels must be determined. This synthesis process requires a clear understanding of the characteristics of the device, the basic equations for the network, and a firm understanding of the basic laws of circuit analysis, such as Ohm's law, Kirchhoff's voltage law, and so on. In most situations the thinking process is challenged to a higher degree in the design process than in the analysis sequence. The path toward a solution is less defined and in fact may require a number of basic assumptions that do not have to be made when simply analyzing a network.

The design sequence is obviously sensitive to the components that are already specified and the elements to be determined. If the transistor and supplies are specified, the design process will simply determine the required resistors for a particular design. Once the theoretical values of the resistors are determined, the nearest standard commercial values are normally chosen and any variations due to not using the exact resistance values are accepted as part of the design. This is certainly a valid approximation considering the tolerances normally associated with resistive elements and the transistor parameters.

If resistive values are to be determined, one of the most powerful equations is simply Ohm's law in the following form:

$$R_{unknown} = \frac{V_R}{I_R}$$ (4.44)

In a particular design the voltage across a resistor can often be determined from specified levels. If additional specifications define the current level, Eq. (4.44) can then be used to calculate the required resistance level. The first few examples will demonstrate how particular elements can be determined from specified levels. A complete design procedure will then be introduced for two popular configurations.

EXAMPLE 4.19

Given the device characteristics of Fig. 4.48a, determine V_{CC}, R_B, and R_C for the fixed-bias configuration of Fig. 4.48b.

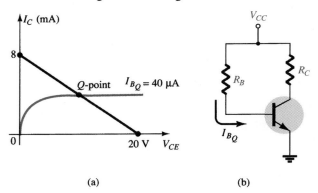

(a) (b) Figure 4.48 Example 4.19

Solution

From the load line

$$V_{CC} = \textbf{20 V}$$

$$I_C = \frac{V_{CC}}{R_C}\Big|_{V_{CE}=0\text{ V}}$$

and

$$R_C = \frac{V_{CC}}{I_C} = \frac{20\text{ V}}{8\text{ mA}} = \textbf{2.5 k}\boldsymbol{\Omega}$$

$$I_B = \frac{V_{CC} - V_{BE}}{R_B}$$

with

$$R_B = \frac{V_{CC} - V_{BE}}{I_B}$$

$$= \frac{20\text{ V} - 0.7\text{ V}}{40\ \mu\text{A}} = \frac{19.3\text{ V}}{40\ \mu\text{A}}$$

$$= \textbf{482.5 k}\boldsymbol{\Omega}$$

Standard resistor values:

$$R_C = 2.4 \text{ k}\Omega$$

$$R_B = 470 \text{ k}\Omega$$

Using standard resistor values gives

$$I_B = 41.1 \ \mu\text{A}$$

which is well within 5% of the value specified.

Given that $I_{C_Q} = 2$ mA and $V_{CE_Q} = 10$ V, determine R_1 and R_C for the network of Fig. 4.49.

EXAMPLE 4.20

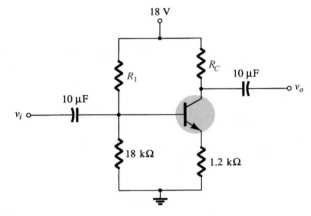

Figure 4.49 Example 4.20

Solution

$$V_E = I_E R_E \cong I_C R_E$$

$$= (2 \text{ mA})(1.2 \text{ k}\Omega) = 2.4 \text{ V}$$

$$V_B = V_{BE} + V_E = 0.7 \text{ V} + 2.4 \text{ V} = 3.1 \text{ V}$$

$$V_B = \frac{R_2 V_{CC}}{R_1 + R_2} = 3.1 \text{ V}$$

and

$$\frac{(18 \text{ k}\Omega)(18 \text{ V})}{R_1 + 18 \text{ k}\Omega} = 3.1 \text{ V}$$

$$324 \text{ k}\Omega = 3.1 R_1 + 55.8 \text{ k}\Omega$$

$$3.1 R_1 = 268.2 \text{ k}\Omega$$

$$R_1 = \frac{268.2 \text{ k}\Omega}{3.1} = \mathbf{86.52 \text{ k}\Omega}$$

Eq. (4.44): $$R_C = \frac{V_{R_C}}{I_C} = \frac{V_{CC} - V_C}{I_C}$$

with

$$V_C = V_{CE} + V_E = 10 \text{ V} + 2.4 \text{ V} = 12.4 \text{ V}$$

and

$$R_C = \frac{18 \text{ V} - 12.4 \text{ V}}{2 \text{ mA}}$$

$$= \mathbf{2.8 \text{ k}\Omega}$$

The nearest standard commercial values to R_1 are 82 and 91 kΩ. However, using the series combination of standard values of 82 kΩ and 4.7 kΩ = 86.7 kΩ would result in a value very close to the design level.

EXAMPLE 4.21

The emitter-bias configuration of Fig. 4.50 has the following specifications: $I_{C_Q} = \frac{1}{2}I_{\text{sat}}$, $I_{C_{\text{sat}}} = 8$ mA, $V_C = 18$ V, and $\beta = 110$. Determine R_C, R_E, and R_B.

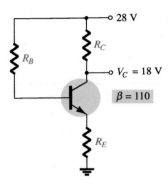

Figure 4.50 Example 4.21

Solution

$$I_{C_Q} = \tfrac{1}{2}I_{C_{\text{sat}}} = 4 \text{ mA}$$

$$R_C = \frac{V_{R_C}}{I_{C_Q}} = \frac{V_{CC} - V_C}{I_{C_Q}}$$

$$= \frac{28 \text{ V} - 18 \text{ V}}{4 \text{ mA}} = \mathbf{2.5 \text{ k}\Omega}$$

$$I_{C_{\text{sat}}} = \frac{V_{CC}}{R_C + R_E}$$

and

$$R_C + R_E = \frac{V_{CC}}{I_{C_{\text{sat}}}} = \frac{28 \text{ V}}{8 \text{ mA}} = 3.5 \text{ k}\Omega$$

$$R_E = 3.5 \text{ k}\Omega - R_C$$

$$= 3.5 \text{ k}\Omega - 2.5 \text{ k}\Omega$$

$$= \mathbf{1 \text{ k}\Omega}$$

$$I_{B_Q} = \frac{I_{C_Q}}{\beta} = \frac{4 \text{ mA}}{110} = 36.36 \text{ } \mu\text{A}$$

$$I_{B_Q} = \frac{V_{CC} - V_{BE}}{R_B + (\beta + 1)R_E}$$

and

$$R_B + (\beta + 1)R_E = \frac{V_{CC} - V_{BE}}{I_{B_Q}}$$

with

$$R_B = \frac{V_{CC} - V_{BE}}{I_{B_Q}} - (\beta + 1)R_E$$

$$= \frac{28 \text{ V} - 0.7 \text{ V}}{36.36 \text{ } \mu\text{A}} - (111)(1 \text{ k}\Omega)$$

$$= \frac{27.3 \text{ V}}{36.36 \text{ } \mu\text{A}} - 111 \text{ k}\Omega$$

$$= \mathbf{639.8 \text{ k}\Omega}$$

For standard values:

$$R_C = 2.4 \text{ k}\Omega$$

$$R_E = 1 \text{ k}\Omega$$

$$R_B = 620 \text{ k}\Omega$$

The discussion to follow will introduce one technique for designing an entire circuit to operate at a specified bias point. Often the manufacturer's specification (spec) sheets provide information on a suggested operating point (or operating region) for a particular transistor. In addition, other system components connected to the given amplifier stage may also define the current swing, voltage swing, value of common supply voltage, and so on, for the design.

In actual practice, many other factors may have to be considered that may affect the selection of the desired operating point. For the moment we shall concentrate, however, on determining the component values to obtain a specified operating point. The discussion will be limited to the emitter-bias and voltage-divider bias configurations, although the same procedure can be applied to a variety of other transistor circuits.

Design of a Bias Circuit with an Emitter Feedback Resistor

Consider first the design of the dc bias components of an amplifier circuit having emitter-resistor bias stabilization as shown in Fig. 4.51. The supply voltage and operating point were selected from the manufacturer's information on the transistor used in the amplifier.

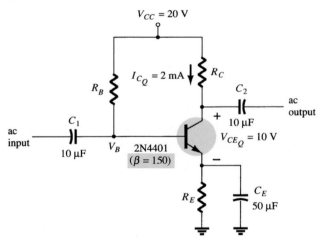

Figure 4.51 Emitter-stabilized bias circuit for design consideration.

The selection of collector and emitter resistors cannot proceed directly from the information just specified. The equation that relates the voltages around the collector–emitter loop has two unknown quantities present—the resistors R_C and R_E. At this point some engineering judgment must be made, such as the level of the emitter voltage compared to the applied supply voltage. Recall that the need for including a resistor from emitter to ground was to provide a means of dc bias stabilization so that the change of collector current due to leakage currents in the transistor and the transistor beta would not cause a large shift in the operating point. The emitter resistor cannot be unreasonably large because the voltage across it limits the range of voltage swing of the voltage from collector to emitter (to be noted when the ac response is discussed). The examples examined in this chapter reveal that the voltage from emitter to ground is typically around one-fourth to one-tenth of the supply voltage. Selecting the conservative case of one-tenth will permit calculating the emitter resistor R_E and the resistor R_C in a manner similar to the examples just completed. In the next example we perform a complete design of the network of Fig. 4.51 using the criteria just introduced for the emitter voltage.

EXAMPLE 4.22

Determine the resistor values for the network of Fig. 4.51 for the indicated operating point and supply voltage.

Solution

$$V_E = \tfrac{1}{10}V_{CC} = \tfrac{1}{10}(20 \text{ V}) = 2 \text{ V}$$

$$R_E = \frac{V_E}{I_E} \cong \frac{V_E}{I_C} = \frac{2 \text{ V}}{2 \text{ mA}} = \mathbf{1 \text{ k}\Omega}$$

$$R_C = \frac{V_{R_C}}{I_C} = \frac{V_{CC} - V_{CE} - V_E}{I_C} = \frac{20 \text{ V} - 10 \text{ V} - 2 \text{ V}}{2 \text{ mA}} = \frac{8 \text{ V}}{2 \text{ mA}}$$

$$= \mathbf{4 \text{ k}\Omega}$$

$$I_B = \frac{I_C}{\beta} = \frac{2 \text{ mA}}{150} = 13.33 \text{ }\mu\text{A}$$

$$R_B = \frac{V_{R_B}}{I_B} = \frac{V_{CC} - V_{BE} - V_E}{I_B} = \frac{20 \text{ V} - 0.7 \text{ V} - 2 \text{ V}}{13.33 \text{ }\mu\text{A}}$$

$$\cong \mathbf{1.3 \text{ M}\Omega}$$

Design of a Current-Gain-Stabilized (Beta-Independent) Circuit

The circuit of Fig. 4.52 provides stabilization both for leakage and current gain (beta) changes. The four resistor values shown must be obtained for the specified operating point. Engineering judgment in selecting a value of emitter voltage, V_E, as in the previous design consideration, leads to a direct straightforward solution for all the resistor values. The design steps are all demonstrated in the next example.

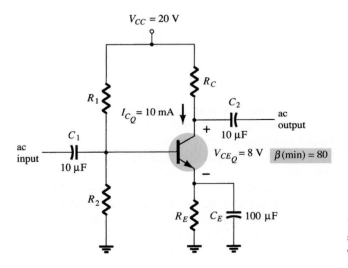

Figure 4.52 Current-gain-stabilized circuit for design considerations.

Determine the levels of R_C, R_E, R_1, and R_2 for the network of Fig. 4.52 for the operating point indicated.

EXAMPLE 4.23

Solution

$$V_E = \tfrac{1}{10}V_{CC} = \tfrac{1}{10}(20\text{ V}) = 2\text{ V}$$

$$R_E = \frac{V_E}{I_E} \cong \frac{V_E}{I_C} = \frac{2\text{ V}}{10\text{ mA}} = \mathbf{200\ \Omega}$$

$$R_C = \frac{V_{R_C}}{I_C} = \frac{V_{CC} - V_{CE} - V_E}{I_C} = \frac{20\text{ V} - 8\text{ V} - 2\text{ V}}{10\text{ mA}} = \frac{10\text{ V}}{10\text{ mA}}$$

$$= \mathbf{1\ k\Omega}$$

$$V_B = V_{BE} + V_E = 0.7\text{ V} + 2\text{ V} = 2.7\text{ V}$$

The equations for the calculation of the base resistors R_1 and R_2 will require a little thought. Using the value of base voltage calculated above and the value of the supply voltage will provide one equation—but there are two unknowns, R_1 and R_2. An additional equation can be obtained from an understanding of the operation of these two resistors in providing the necessary base voltage. For the circuit to operate efficiently, it is assumed that the current through R_1 and R_2 should be approximately equal to and much larger than the base current (at least 10:1). This fact and the voltage-divider equation for the base voltage provide the two relationships necessary to determine the base resistors. That is,

$$R_2 \leq \tfrac{1}{10}\beta R_E$$

and

$$V_B = \frac{R_2}{R_1 + R_2}V_{CC}$$

Substitution yields

$$R_2 \leq \tfrac{1}{10}(80)(0.2\text{ k}\Omega)$$

$$= \mathbf{1.6\ k\Omega}$$

$$V_B = 2.7\text{ V} = \frac{(1.6\text{ k}\Omega)(20\text{ V})}{R_1 + 1.6\text{ k}\Omega}$$

and

$$2.7R_1 + 4.32\text{ k}\Omega = 32\text{ k}\Omega$$

$$2.7R_1 = 27.68\text{ k}\Omega$$

$$R_1 = \mathbf{10.25\ k\Omega} \quad (\text{use }10\text{ k}\Omega)$$

4.9 TRANSISTOR SWITCHING NETWORKS

The application of transistors is not limited solely to the amplification of signals. Through proper design it can be used as a switch for computer and control applications. The network of Fig. 4.53a can be employed as an *inverter* in computer logic circuitry. Note that the output voltage V_C is opposite to that applied to the base or input terminal. In addition, note the absence of a dc supply connected to the base circuit. The only dc source is connected to the collector or output side and for computer applications is typically equal to the magnitude of the "high" side of the applied signal—in this case 5 V.

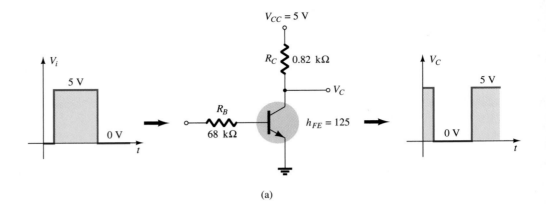

(a)

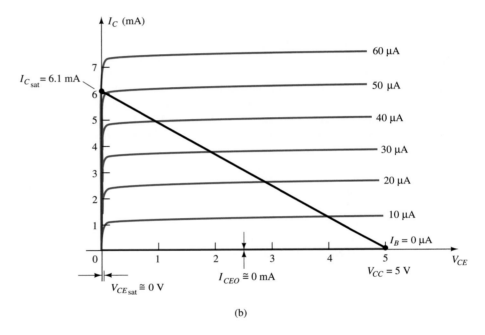

(b)

Figure 4.53 Transistor inverter.

Proper design for the inversion process requires that the operating point switch from cutoff to saturation along the load line depicted in Fig. 4.53b. For our purposes we will assume that $I_C = I_{CEO} = 0$ mA when $I_B = 0$ μA (an excellent approximation in light of improving construction techniques), as shown in Fig. 4.53b. In addition, we will assume that $V_{CE} = V_{CE_{sat}} = 0$ V rather than the typical 0.1- to 0.3-V level.

When $V_i = 5$ V, the transistor will be "on" and the design must ensure that the network is heavily saturated by a level of I_B greater than that associated with the I_B curve appearing near the saturation level. In Fig. 4.53b, this requires that $I_B > 50$ μA. The saturation level for the collector current for the circuit of Fig. 4.53a is defined by

$$I_{C_{sat}} = \frac{V_{CC}}{R_C} \tag{4.45}$$

The level of I_B in the active region just before saturation results can be approximated by the following equation:

$$I_{B_{max}} \cong \frac{I_{C_{sat}}}{\beta_{dc}}$$

For the saturation level we must therefore ensure that the following condition is satisfied:

$$\boxed{I_B > \frac{I_{C_{sat}}}{\beta_{dc}}} \tag{4.46}$$

For the network of Fig. 4.53b, when $V_i = 5$ V, the resulting level of I_B is the following:

$$I_B = \frac{V_i - 0.7 \text{ V}}{R_B} = \frac{5 \text{ V} - 0.7 \text{ V}}{68 \text{ k}\Omega} = 63 \text{ } \mu\text{A}$$

and

$$I_{C_{sat}} = \frac{V_{CC}}{R_C} = \frac{5 \text{ V}}{0.82 \text{ k}\Omega} \cong 6.1 \text{ mA}$$

Testing Eq. (4.46) gives

$$I_B = 63 \text{ } \mu\text{A} > \frac{I_{C_{sat}}}{\beta_{dc}} = \frac{6.1 \text{ mA}}{125} = 48.8 \text{ } \mu\text{A}$$

which is satisfied. Certainly, any level of I_B greater than 60 μA will pass through a Q-point on the load line that is very close to the vertical axis.

For $V_i = 0$ V, $I_B = 0$ μA, and since we are assuming that $I_C = I_{CEO} = 0$ mA, the voltage drop across R_C as determined by $V_{R_C} = I_C R_C = 0$ V, resulting in $V_C = +5$ V for the response indicated in Fig. 4.53a.

In addition to its contribution to computer logic, the transistor can also be employed as a switch using the same extremities of the load line. At saturation, the current I_C is quite high and the voltage V_{CE} very low. The result is a resistance level between the two terminals determined by

$$R_{sat} = \frac{V_{CE_{sat}}}{I_{C_{sat}}}$$

and depicted in Fig. 4.54.

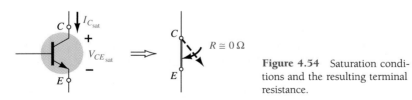

Figure 4.54 Saturation conditions and the resulting terminal resistance.

Using a typical average value of $V_{CE_{sat}}$ such as 0.15 V gives

$$R_{sat} = \frac{V_{CE_{sat}}}{I_{C_{sat}}} = \frac{0.15 \text{ V}}{6.1 \text{ mA}} = 24.6 \text{ } \Omega$$

which is a relatively low value and $\cong 0$ Ω when placed in series with resistors in the kilohm range.

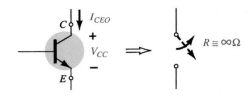

Figure 4.55 Cutoff conditions and the resulting terminal resistance.

For $V_i = 0$ V, as shown in Fig. 4.55, the cutoff condition will result in a resistance level of the following magnitude:

$$R_{\text{cutoff}} = \frac{V_{CC}}{I_{CEO}} = \frac{5 \text{ V}}{0 \text{ mA}} = \infty \text{ } \Omega$$

resulting in the open-circuit equivalence. For a typical value of $I_{CEO} = 10 \text{ } \mu\text{A}$, the magnitude of the cutoff resistance is

$$R_{\text{cutoff}} = \frac{V_{CC}}{I_{CEO}} = \frac{5 \text{ V}}{10 \text{ } \mu\text{A}} = 500 \text{ k}\Omega$$

which certainly approaches an open-circuit equivalence for many situations.

EXAMPLE 4.24

Determine R_B and R_C for the transistor inverter of Fig. 4.56 if $I_{C_{\text{sat}}} = 10$ mA.

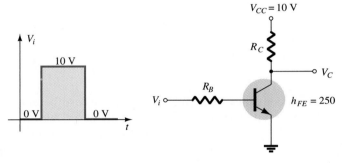

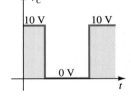

Figure 4.56 Inverter for Example 4.24.

Solution

At saturation:

$$I_{C_{\text{sat}}} = \frac{V_{CC}}{R_C}$$

and

$$10 \text{ mA} = \frac{10 \text{ V}}{R_C}$$

so that

$$R_C = \frac{10 \text{ V}}{10 \text{ mA}} = 1 \text{ k}\Omega$$

At saturation:

$$I_B \cong \frac{I_{C_{\text{sat}}}}{\beta_{\text{dc}}} = \frac{10 \text{ mA}}{250} = 40 \text{ } \mu\text{A}$$

Choosing $I_B = 60 \text{ } \mu\text{A}$ to ensure saturation and using

$$I_B = \frac{V_i - 0.7 \text{ V}}{R_B}$$

we obtain $\qquad R_B = \dfrac{V_i - 0.7\ \text{V}}{I_B} = \dfrac{10\ \text{V} - 0.7\ \text{V}}{60\ \mu\text{A}} = 155\ \text{k}\Omega$

Choose $R_B = 150\ \text{k}\Omega$, which is a standard value. Then

$$I_B = \frac{V_i - 0.7\ \text{V}}{R_B} = \frac{10\ \text{V} - 0.7\ \text{V}}{150\ \text{k}\Omega} = 62\ \mu\text{A}$$

and $\qquad I_B = 62\ \mu\text{A} > \dfrac{I_{C_{\text{sat}}}}{\beta_{\text{dc}}} = 40\ \mu\text{A}$

Therefore, use $R_B = \mathbf{150\ k\Omega}$ and $R_C = \mathbf{1\ k\Omega}$.

There are transistors that are referred to as *switching transistors* due to the speed with which they can switch from one voltage level to the other. In Fig. 3.23c the periods of time defined as t_s, t_d, t_r, and t_f are provided versus collector current. Their impact on the speed of response of the collector output is defined by the collector current response of Fig. 4.57. The total time required for the transistor to switch from the "off" to the "on" state is designated as t_{on} and defined by

$$\boxed{t_{\text{on}} = t_r + t_d} \tag{4.47}$$

with t_d the delay time between the changing state of the input and the beginning of a response at the output. The time element t_r is the rise time from 10% to 90% of the final value.

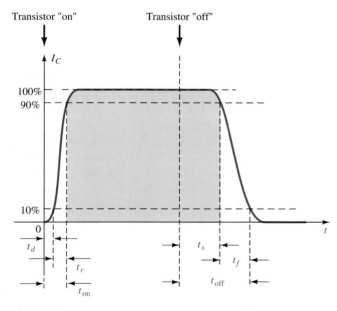

Figure 4.57 Defining the time intervals of a pulse waveform.

The total time required for a transistor to switch from the "on" to the "off" state is referred to as t_{off} and is defined by

$$\boxed{t_{\text{off}} = t_s + t_f} \tag{4.48}$$

where t_s is the storage time and t_f the fall time from 90% to 10% of the initial value.

For the general-purpose transistor of Fig. 3.23c at $I_C = 10$ mA, we find that

$$t_s = 120 \text{ ns}$$

$$t_d = 25 \text{ ns}$$

$$t_r = 13 \text{ ns}$$

and $$t_f = 12 \text{ ns}$$

so that $$t_{on} = t_r + t_d = 13 \text{ ns} + 25 \text{ ns} = \textbf{38 ns}$$

and $$t_{off} = t_s + t_f = 120 \text{ ns} + 12 \text{ ns} = \textbf{132 ns}$$

Comparing the values above with the following parameters of a BSV52L switching transistor reveals one of the reasons for choosing a switching transistor when the need arises.

$$t_{on} = \textbf{12 ns} \quad \text{and} \quad t_{off} = \textbf{18 ns}$$

4.10 TROUBLESHOOTING TECHNIQUES

The art of troubleshooting is such a broad topic that a full range of possibilities and techniques cannot be covered in a few sections of a book. However, the practitioner should be aware of a few basic maneuvers and measurements that can isolate the problem area and possibly identify a solution.

Quite obviously, the first step in being able to troubleshoot a network is to fully understand the behavior of the network and to have some idea of the expected voltage and current levels. For the transistor in the active region, the most important measurable dc level is the base-to-emitter voltage.

For an "on" transistor, the voltage V_{BE} should be in the neighborhood of 0.7 V.

The proper connections for measuring V_{BE} appear in Fig. 4.58. Note that the positive (red) lead is connected to the base terminal for an *npn* transistor and the negative (black) lead to the emitter terminal. Any reading totally different from the expected level of about 0.7 V, such as 0, 4, or 12 V, or negative in value would be suspect and the device or network connections should be checked. For a *pnp* transistor, the same connections can be used but a negative reading should be expected.

A voltage level of equal importance is the collector-to-emitter voltage. Recall from the general characteristics of a BJT that levels of V_{CE} in the neighborhood of 0.3 V suggest a saturated device—a condition that should not exist unless being employed in a switching mode. However:

For the typical transistor amplifier in the active region, V_{CE} is usually about 25% to 75% of V_{CC}.

For $V_{CC} = 20$ V, a reading of V_{CE} of 1 to 2 V or 18 to 20 V as measured in Fig. 4.59 is certainly an uncommon result, and unless knowingly designed for this response the design and operation should be investigated. If $V_{CE} = 20$ V (with $V_{CC} = 20$ V) at least two possibilities exist—either the device (BJT) is damaged and has the characteristics of an open circuit between collector and emitter terminals or a connection in the collector–emitter or base–emitter circuit loop is open as shown in Fig. 4.60, establishing I_C at 0 mA and $V_{R_C} = 0$ V. In Fig. 4.60, the black lead of the volt-

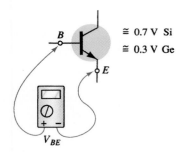

Figure 4.58 Checking the dc level of V_{BE}.

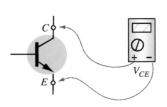

0.3 V = saturation
0 V = short-circuit state
    or poor connection
Normally a few volts
    or more

Figure 4.59 Checking the dc level of V_{CE}.

meter is connected to the common ground of the supply and the red lead to the bottom terminal of the resistor. The absence of a collector current and a resulting drop across R_C will result in a reading of 20 V. If the meter is connected to the collector terminal of the BJT, the reading will be 0 V since V_{CC} is blocked from the active device by the open circuit. One of the most common errors in the laboratory experience is the use of the wrong resistance value for a given design. Imagine the impact of using a 680-Ω resistor for R_B rather than the design value of 680 kΩ. For $V_{CC} = 20$ V and a fixed-bias configuration, the resulting base current would be

$$I_B = \frac{20 \text{ V} - 0.7 \text{ V}}{680 \text{ }\Omega} = 28.4 \text{ mA}$$

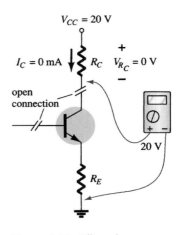

Figure 4.60 Effect of a poor connection or damaged device.

rather than the desired 28.4 μA—a significant difference!

A base current of 28.4 mA would certainly place the design in a saturation region and possibly damage the device. Since actual resistor values are often different from the nominal color-code value (recall the common tolerance levels for resistive elements), it is time well spent to measure a resistor before inserting it in the network. The result is actual values closer to theoretical levels and some insurance that the correct resistance value is being employed.

There are times when frustration will develop. You have checked the device on a curve tracer or other BJT testing instrumentation and it looks good. All resistor levels seem correct, the connections appear solid, and the proper supply voltage has been applied—what next? Now the troubleshooter must strive to attain a higher level of sophistication. Could it be that the internal connection between the wire and the end connection of a lead is faulty? How often has simply touching a lead at the proper point created a "make or break" situation between connections? Perhaps the supply was turned on and set at the proper voltage but the current-limiting knob was left in the zero position, preventing the proper level of current as demanded by the network design. Obviously, the more sophisticated the system, the broader the range of possibilities. In any case, one of the most effective methods of checking the operation of a network is to check various voltage levels with respect to ground by hooking up the black (negative) lead of a voltmeter to ground and "touching" the important terminals with the red (positive) lead. In Fig. 4.61, if the red lead is connected directly to V_{CC}, it should read V_{CC} volts since the network has one common ground for the supply and network parameters. At V_C the reading should be less, as determined by the drop across R_C, and V_E should be less than V_C by the collector–emitter voltage V_{CE}. The failure of any of these points to register what would appear to be a reasonable level may be sufficient in itself to define the faulty connection or element. If V_{R_C} and V_{R_E} are reasonable values but V_{CE} is 0 V, the possibility exists that the BJT is damaged and displays a short-circuit equivalence between collector and emitter terminals. As noted earlier, if V_{CE} registers a level of about 0.3 V as defined by $V_{CE} = V_C - V_E$ (the difference of the two levels as measured above), the network may be in saturation with a device that may or may not be defective.

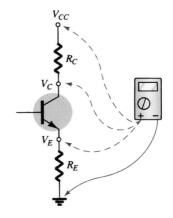

Figure 4.61 Checking voltage levels with respect to ground.

It should be somewhat obvious from the discussion above that the voltmeter section of the VOM or DMM is quite important in the troubleshooting process. Current levels are usually calculated from the voltage levels across resistors rather than "breaking" the network to insert the milliammeter section of a multimeter. On large schematics, specific voltage levels are provided with respect to ground for easy checking and identification of possible problem areas. Of course, for the networks covered in this chapter, one must simply be aware of typical levels within the system as defined by the applied potential and general operation of the network.

All in all, the troubleshooting process is a true test of your clear understanding of the proper behavior of a network and the ability to isolate problem areas using a few basic measurements with the appropriate instruments. Experience is the key, and that will come only with continued exposure to practical circuits.

EXAMPLE 4.25

Based on the readings provided in Fig. 4.62, determine whether the network is operating properly and, if not, the probable cause.

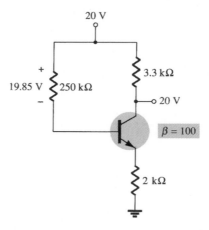

Figure 4.62 Network for Example 4.25.

Solution

The 20 V at the collector immediately reveals that $I_C = 0$ mA, due to an open circuit or a nonoperating transistor. The level of $V_{R_B} = 19.85$ V also reveals that the transistor is "off" since the difference of $V_{CC} - V_{R_B} = 0.15$ V is less than that required to turn "on" the transistor and provide some voltage for V_E. In fact, if we assume a short circuit condition from base to emitter, we obtain the following current through R_B:

$$I_{R_B} = \frac{V_{CC}}{R_B + R_E} = \frac{20 \text{ V}}{252 \text{ k}\Omega} = 79.4 \text{ } \mu A$$

which matches that obtained from

$$I_{R_B} = \frac{V_{R_B}}{R_B} = \frac{19.85 \text{ V}}{250 \text{ k}\Omega} = 79.4 \text{ } \mu A$$

If the network were operating properly, the base current should be

$$I_B = \frac{V_{CC} - V_{BE}}{R_B + (\beta + 1)R_E} = \frac{20 \text{ V} - 0.7 \text{ V}}{250 \text{ k}\Omega + (101)(2 \text{ k}\Omega)} = \frac{19.3 \text{ V}}{452 \text{ k}\Omega} = 42.7 \text{ } \mu A$$

The result, therefore, is that the transistor is in a damaged state, with a short-circuit condition between base and emitter.

EXAMPLE 4.26

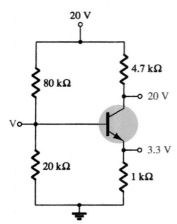

Figure 4.63 Network for Example 4.26.

Based on the readings appearing in Fig. 4.63, determine whether the transistor is "on" and the network is operating properly.

Solution

Based on the resistor values of R_1 and R_2 and the magnitude of V_{CC}, the voltage $V_B = 4$ V seems appropriate (and in fact it is). The 3.3 V at the emitter results in a 0.7-V drop across the base-to-emitter junction of the transistor, suggesting an "on" transistor. However, the 20 V at the collector reveals that $I_C = 0$ mA, although the connection to the supply must be "solid" or the 20 V would not appear at the collector of the device. Two possibilities exist—there can be a poor connection between R_C and the collector terminal of the transistor or the transistor has an open base-to-collector junction. First, check the continuity at the collector junction using an ohmmeter, and if okay, the transistor should be checked using one of the methods described in Chapter 3.

Chapter 4   DC Biasing—BJTs

4.11 PNP TRANSISTORS

The analysis thus far has been limited totally to *npn* transistors to ensure that the initial analysis of the basic configurations was as clear as possible and uncomplicated by switching between types of transistors. Fortunately, the analysis of *pnp* transistors follows the same pattern established for *npn* transistors. The level of I_B is first determined, followed by the application of the appropriate transistor relationships to determine the list of unknown quantities. In fact, the only difference between the resulting equations for a network in which an *npn* transistor has been replaced by a *pnp* transistor is the sign associated with particular quantities.

As noted in Fig. 4.64, the double-subscript notation continues as normally defined. The current directions, however, have been reversed to reflect the actual conduction directions. Using the defined polarities of Fig. 4.64, both V_{BE} and V_{CE} will be negative quantities.

Applying Kirchhoff's voltage law to the base–emitter loop will result in the following equation for the network of Fig. 4.64:

$$-I_E R_E + V_{BE} - I_B R_B + V_{CC} = 0$$

Substituting $I_E = (\beta + 1)I_B$ and solving for I_B yields

$$I_B = \frac{V_{CC} + V_{BE}}{R_B + (\beta + 1)R_E} \qquad (4.49)$$

The resulting equation is the same as Eq. (4.17) except for the sign for V_{BE}. However, in this case $V_{BE} = -0.7$ V and the substitution of values will result in the same sign for each term of Eq. (4.49) as Eq. (4.17). Keep in mind that the direction of I_B is now defined opposite of that for a *pnp* transistor as shown in Fig. 4.64.

For V_{CE} Kirchhoff's voltage law is applied to the collector–emitter loop, resulting in the following equation:

$$-I_E R_E + V_{CE} - I_C R_C + V_{CC} = 0$$

Substituting $I_E \cong I_C$ gives

$$V_{CE} = -V_{CC} + I_C(R_C + R_E) \qquad (4.50)$$

The resulting equation has the same format as Eq. (4.19), but the sign in front of each term on the right of the equal sign has changed. Since V_{CC} will be larger than the magnitude of the succeeding term, the voltage V_{CE} will have a negative sign, as noted in an earlier paragraph.

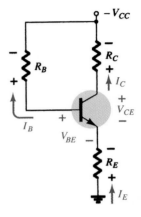

Figure 4.64 *pnp* transistor in an emitter-stabilized configuration.

Determine V_{CE} for the voltage-divider bias configuration of Fig. 4.65.

EXAMPLE 4.27

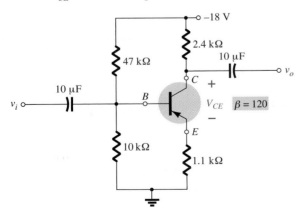

Figure 4.65 *pnp* transistor in a voltage-divider bias configuration.

Solution

Testing the condition

$$\beta R_E \geq 10R_2$$

results in

$$(120)(1.1 \text{ k}\Omega) \geq 10(10 \text{ k}\Omega)$$

$$132 \text{ k}\Omega \geq 100 \text{ k}\Omega \quad (satisfied)$$

Solving for V_B, we have

$$V_B = \frac{R_2 V_{CC}}{R_1 + R_2} = \frac{(10 \text{ k}\Omega)(-18 \text{ V})}{47 \text{ k}\Omega + 10 \text{ k}\Omega} = -3.16 \text{ V}$$

Note the similarity in format of the equation with the resulting negative voltage for V_B.

Applying Kirchhoff's voltage law around the base–emitter loop yields

$$+V_B - V_{BE} - V_E = 0$$

and

$$V_E = V_B - V_{BE}$$

Substituting values, we obtain

$$V_E = -3.16 \text{ V} - (-0.7 \text{ V})$$

$$= -3.16 \text{ V} + 0.7 \text{ V}$$

$$= -2.46 \text{ V}$$

Note in the equation above that the standard single- and double-subscript notation is employed. For an *npn* transistor the equation $V_E = V_B - V_{BE}$ would be exactly the same. The only difference surfaces when the values are substituted.

The current

$$I_E = \frac{V_E}{R_E} = \frac{2.46 \text{ V}}{1.1 \text{ k}\Omega} = 2.24 \text{ mA}$$

For the collector–emitter loop:

$$-I_E R_E + V_{CE} - I_C R_C + V_{CC} = 0$$

Substituting $I_E \cong I_C$ and gathering terms, we have

$$V_{CE} = -V_{CC} + I_C(R_C + R_E)$$

Substituting values gives

$$V_{CE} = -18 \text{ V} + (2.24 \text{ mA})(2.4 \text{ k}\Omega + 1.1 \text{ k}\Omega)$$

$$= -18 \text{ V} + 7.84 \text{ V}$$

$$= \mathbf{-10.16 \text{ V}}$$

4.12 BIAS STABILIZATION

The stability of a system is a measure of the sensitivity of a network to variations in its parameters. In any amplifier employing a transistor the collector current I_C is sensitive to each of the following parameters:

> β: *increases with increase in temperature*
>
> $|V_{BE}|$: *decreases about 7.5 mV per degree Celsius (°C) increase in temperature*
>
> I_{CO} *(reverse saturation current): doubles in value for every 10°C increase in temperature*

Any or all of these factors can cause the bias point to drift from the designed point of operation. Table 4.1 reveals how the level of I_{CO} and V_{BE} changed with increase in temperature for a particular transistor. At room temperature (about 25°C) $I_{CO} = 0.1$ nA, while at 100°C (boiling point of water) I_{CO} is about 200 times larger at 20 nA. For the same temperature variation, β increased from 50 to 80 and V_{BE} dropped from 0.65 to 0.48 V. Recall that I_B is quite sensitive to the level of V_{BE}, especially for levels beyond the threshold value.

TABLE 4.1	**Variation of Silicon Transistor Parameters with Temperature**		
$T\,(°C)$	$I_{co}\,(nA)$	β	$V_{BE}\,(V)$
−65	0.2×10^{-3}	20	0.85
25	0.1	50	0.65
100	20	80	0.48
175	3.3×10^3	120	0.3

The effect of changes in leakage current (I_{CO}) and current gain (β) on the dc bias point is demonstrated by the common-emitter collector characteristics of Fig. 4.66a and b. Figure 4.66 shows how the transistor collector characteristics change from a temperature of 25°C to a temperature of 100°C. Note that the significant increase in leakage current not only causes the curves to rise but also causes an increase in beta, as revealed by the larger spacing between curves.

An operating point may be specified by drawing the circuit dc load line on the graph of the collector characteristic and noting the intersection of the load line and the dc base current set by the input circuit. An arbitrary point is marked in Fig. 4.66a at $I_B = 30\ \mu A$. Since the fixed-bias circuit provides a base current whose value depends approximately on the supply voltage and base resistor, neither of which is af-

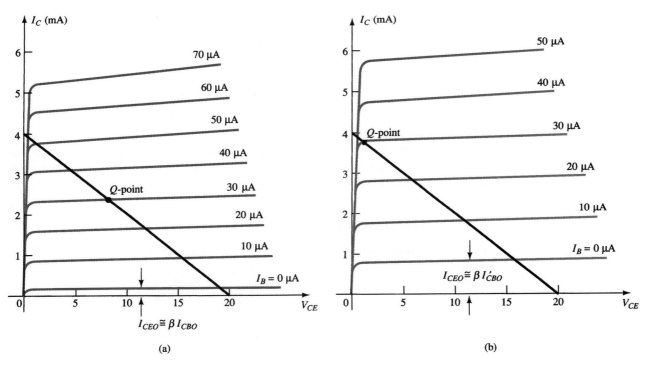

Figure 4.66 Shift in dc bias point (Q-point) due to change in temperature:
(a) 25°C; (b) 100°C.

fected by temperature or the change in leakage current or beta, the same base current magnitude will exist at high temperatures as indicated on the graph of Fig. 4.66b. As the figure shows, this will result in the dc bias point's shifting to a higher collector current and a lower collector–emitter voltage operating point. In the extreme, the transistor could be driven into saturation. In any case, the new operating point may not be at all satisfactory, and considerable distortion may result because of the bias-point shift. A better bias circuit is one that will stabilize or maintain the dc bias initially set, so that the amplifier can be used in a changing-temperature environment.

Stability Factors, $S(I_{CO})$, $S(V_{BE})$, and $S(\beta)$

A stability factor, S, is defined for each of the parameters affecting bias stability as listed below:

$$S(I_{CO}) = \frac{\Delta I_C}{\Delta I_{CO}} \tag{4.51}$$

$$S(V_{BE}) = \frac{\Delta I_C}{\Delta V_{BE}} \tag{4.52}$$

$$S(\beta) = \frac{\Delta I_C}{\Delta \beta} \tag{4.53}$$

In each case, the delta symbol (Δ) signifies change in that quantity. The numerator of each equation is the change in collector current as established by the change in the quantity in the denominator. For a particular configuration, if a change in I_{CO} fails to produce a significant change in I_C, the stability factor defined by $S(I_{CO}) = \Delta I_C/\Delta I_{CO}$ will be quite small. In other words:

Networks that are quite stable and relatively insensitive to temperature variations have low stability factors.

In some ways it would seem more appropriate to consider the quantities defined by Eqs. (4.51–4.53) to be sensitivity factors because:

The higher the stability factor, the more sensitive the network to variations in that parameter.

The study of stability factors requires the knowledge of differential calculus. Our purpose here, however, is to review the results of the mathematical analysis and to form an overall assessment of the stability factors for a few of the most popular bias configurations. A great deal of literature is available on this subject, and if time permits, you are encouraged to read more on the subject.

$S(I_{CO})$: Emitter-Bias Configuration

For the emitter-bias configuration, an analysis of the network will result in

$$S(I_{CO}) = (\beta + 1)\frac{1 + R_B/R_E}{(\beta + 1) + R_B/R_E} \tag{4.54}$$

For $R_B/R_E \gg (\beta + 1)$, Eq. (4.54) will reduce to the following:

$$S(I_{CO}) = \beta + 1 \tag{4.55}$$

as shown on the graph of $S(I_{CO})$ versus R_B/R_E in Fig. 4.67.

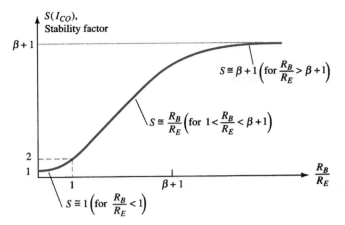

Figure 4.67 Variation of stability factor $S(I_{CO})$ with the resistor ratio R_B/R_E for the emitter-bias configuration.

For $R_B/R_E \ll 1$, Eq. (4.54) will approach the following level (as shown in Fig. 4.67):

$$S(I_{CO}) = (\beta + 1)\frac{1}{(\beta + 1)} = \rightarrow 1 \qquad (4.56)$$

revealing that the stability factor will approach its lowest level as R_E becomes sufficiently large. Keep in mind, however, that good bias control normally requires that R_B be greater than R_E. The result therefore is a situation where the best stability levels are associated with poor design criteria. Obviously, a trade-off must occur that will satisfy both the stability and bias specifications. It is interesting to note in Fig. 4.67 that the lowest value of $S(I_{CO})$ is 1, revealing that I_C will always increase at a rate equal to or greater than I_{CO}.

For the range where R_B/R_E ranges between 1 and $(\beta + 1)$, the stability factor will be determined by

$$S(I_{CO}) \cong \frac{R_B}{R_E} \qquad (4.57)$$

as shown in Fig. 4.67. The results reveal that the emitter-bias configuration is quite stable when the ratio R_B/R_E is as small as possible and the least stable when the same ratio approaches $(\beta + 1)$.

EXAMPLE 4.28

Calculate the stability factor and the change in I_C from 25°C to 100°C for the transistor defined by Table 4.1 for the following emitter-bias arrangements.
(a) $R_B/R_E = 250$ ($R_B = 250R_E$).
(b) $R_B/R_E = 10$ ($R_B = 10R_E$).
(c) $R_B/R_E = 0.01$ ($R_E = 100R_B$).

Solution

(a) $S(I_{CO}) = (\beta + 1)\dfrac{1 + R_B/R_E}{(\beta + 1) + R_B/R_E}$

$\qquad = 51\left(\dfrac{1 + 250}{51 + 250}\right) = 51\left(\dfrac{251}{301}\right)$

$\qquad \cong \mathbf{42.53}$

which begins to approach the level defined by $\beta + 1 = 51$.

$\quad \Delta I_C = [S(I_{CO})](\Delta I_{CO}) = (42.53)(19.9 \text{ nA})$

$\qquad \cong \mathbf{0.85 \ \mu A}$

(b) $S(I_{CO}) = (\beta + 1)\dfrac{1 + R_B/R_E}{1 + \beta + R_B/R_E}$

$\qquad = 51\left(\dfrac{1 + 10}{51 + 10}\right) = 51\left(\dfrac{11}{61}\right)$

$\qquad \cong \mathbf{9.2}$

$\Delta I_C = [S(I_{CO})](\Delta I_{CO}) = (9.2)(19.9 \text{ nA})$

$\qquad \cong \mathbf{0.18 \ \mu A}$

(c) $S(I_{CO}) = (\beta + 1)\dfrac{1 + R_B/R_E}{1 + \beta + R_B/R_E}$

$\qquad = 51\left(\dfrac{1 + 0.01}{51 + 0.01}\right) = 51\left(\dfrac{1.01}{51.01}\right)$

$\qquad \cong \mathbf{1.01}$

which is certainly very close to the level of 1 forecast if $R_B/R_E \ll 1$.

$\Delta I_C = [S(I_{CO})](\Delta I_{CO}) = 1.01(19.9 \text{ nA})$

$\qquad = \mathbf{20.1 \ nA}$

Example 4.28 reveals how lower and lower levels of I_{CO} for the modern-day BJT transistor have improved the stability level of the basic bias configurations. Even though the change in I_C is considerably different in a circuit having ideal stability ($S = 1$) from one having a stability factor of 42.53, the change in I_C is not that significant. For example, the amount of change in I_C from a dc bias current set at, say, 2 mA, would be from 2 to 2.085 mA in the worst case, which is obviously small enough to be ignored for most applications. Some power transistors exhibit larger leakage currents, but for most amplifier circuits the lower levels of I_{CO} have had a very positive impact on the stability question.

Fixed-Bias Configuration

For the fixed-bias configuration, if we multiply the top and bottom of Eq. (4.54) by R_E and then plug in $R_E = 0 \ \Omega$, the following equation will result:

$$\boxed{S(I_{CO}) = \beta + 1} \tag{4.58}$$

Note that the resulting equation matches the maximum value for the emitter-bias configuration. The result is a configuration with a poor stability factor and a high sensitivity to variations in I_{CO}.

Voltage-Divider Bias Configuration

Recall from Section 4.5 the development of the Thévenin equivalent network appearing in Fig. 4.68, for the voltage-divider bias configuration. For the network of Fig. 4.68, the equation for $S(I_{CO})$ is the following:

$$\boxed{S(I_{CO}) = (\beta + 1)\dfrac{1 + R_{Th}/R_E}{(\beta + 1) + R_{Th}/R_E}} \tag{4.59}$$

Note the similarities with Eq. (4.54), where it was determined that $S(I_{CO})$ had its lowest level and the network had its greatest stability when $R_E > R_B$. For Eq. (4.59), the corresponding condition is $R_E > R_{Th}$ or R_{Th}/R_E should be as small as possible.

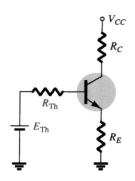

Figure 4.68 Equivalent circuit for the voltage-divider configuration.

Chapter 4   DC Biasing—BJTs

For the voltage-divider bias configuration, R_{Th} can be much less than the corresponding R_B of the emitter-bias configuration and still have an effective design.

Feedback-Bias Configuration ($R_E = 0\ \Omega$)

In this case,

$$S(I_{CO}) = (\beta + 1)\frac{1 + R_B/R_C}{(\beta + 1) + R_B/R_C} \tag{4.60}$$

Since the equation is similar in format to that obtained for the emitter-bias and voltage-divider bias configurations, the same conclusions regarding the ratio R_B/R_C can be applied here also.

Physical Impact

Equations of the type developed above often fail to provide a physical sense for why the networks perform as they do. We are now aware of the relative levels of stability and how the choice of parameters can affect the sensitivity of the network, but without the equations it may be difficult for us to explain in words why one network is more stable than another. The next few paragraphs attempt to fill this void through the use of some of the very basic relationships associated with each configuration.

For the fixed-bias configuration of Fig. 4.69a, the equation for the base current is the following:

$$I_B = \frac{V_{CC} - V_{BE}}{R_B}$$

with the collector current determined by

$$I_C = \beta I_B + (\beta + 1)I_{CO} \tag{4.61}$$

If I_C as defined by Eq. (4.61) should increase due to an increase in I_{CO}, there is nothing in the equation for I_B that would attempt to offset this undesirable increase in current level (assuming V_{BE} remains constant). In other words, the level of I_C would continue to rise with temperature, with I_B maintaining a fairly constant value—a very unstable situation.

For the emitter-bias configuration of Fig. 4.69b, however, an increase in I_C due to an increase in I_{CO} will cause the voltage $V_E = I_E R_E \cong I_C R_E$ to increase. The result is a drop in the level of I_B as determined by the following equation:

$$I_B\downarrow = \frac{V_{CC} - V_{BE} - V_E\uparrow}{R_B} \tag{4.62}$$

A drop in I_B will have the effect of reducing the level of I_C through transistor action and thereby offset the tendency of I_C to increase due to an increase in tempera-

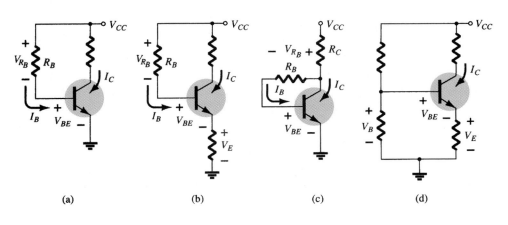

(a)   (b)   (c)   (d)

Figure 4.69 Review of biasing managements and the stability factor $S(I_{CO})$.

ture. In total, therefore, the configuration is such that there is a reaction to an increase in I_C that will tend to oppose the change in bias conditions.

The feedback configuration of Fig. 4.69c operates in much the same way as the emitter-bias configuration when it comes to levels of stability. If I_C should increase due to an increase in temperature, the level of V_{R_C} will increase in the following equation:

$$I_B \downarrow = \frac{V_{CC} - V_{BE} - V_{R_C}\uparrow}{R_B} \tag{4.63}$$

and the level of I_B will decrease. The result is a stabilizing effect as described for the emitter-bias configuration. One must be aware that the action described above does not happen in a step-by-step sequence. Rather, it is a simultaneous action to maintain the established bias conditions. In other words, the very instant I_C begins to rise the network will sense the change and the balancing effect described above will take place.

The most stable of the configurations is the voltage-divider bias network of Fig. 4.69d. If the condition $\beta R_E \gg 10R_2$ is satisfied, the voltage V_B will remain fairly constant for changing levels of I_C. The base-to-emitter voltage of the configuration is determined by $V_{BE} = V_B - V_E$. If I_C should increase, V_E will increase as described above, and for a constant V_B the voltage V_{BE} will drop. A drop in V_{BE} will establish a lower level of I_B, which will try to offset the increased level of I_C.

$S(V_{BE})$

The stability factor defined by

$$S(V_{BE}) = \frac{\Delta I_C}{\Delta V_{BE}}$$

will result in the following equation for the emitter-bias configuration:

$$S(V_{BE}) = \frac{-\beta}{R_B + (\beta + 1)R_E} \tag{4.64}$$

Substituting $R_E = 0\ \Omega$ as occurs for the fixed-bias configuration will result in

$$S(V_{BE}) = -\frac{\beta}{R_B} \tag{4.65}$$

Equation (4.64) can be written in the following form:

$$S(V_{BE}) = \frac{-\beta/R_E}{R_B/R_E + (\beta + 1)} \tag{4.66}$$

Substituting the condition $(\beta + 1) \gg R_B/R_E$ will result in the following equation for $S(V_{BE})$:

$$S(V_{BE}) \cong \frac{-\beta/R_E}{\beta + 1} \cong \frac{-\beta/R_E}{\beta} = -\frac{1}{R_E} \tag{4.67}$$

revealing that the larger the resistance R_E, the lower the stability factor and the more stable the system.

EXAMPLE 4.29   Determine the stability factor $S(V_{BE})$ and the change in I_C from 25°C to 100°C for the transistor defined by Table 4.1 for the following bias arrangements.
(a) Fixed-bias with $R_B = 240\ \text{k}\Omega$ and $\beta = 100$.
(b) Emitter-bias with $R_B = 240\ \text{k}\Omega$, $R_E = 1\ \text{k}\Omega$, and $\beta = 100$.
(c) Emitter-bias with $R_B = 47\ \text{k}\Omega$, $R_E = 4.7\ \text{k}\Omega$, and $\beta = 100$.

Solution

(a) Eq. (4.65): $S(V_{BE}) = -\dfrac{\beta}{R_B}$

$$= -\frac{100}{240 \text{ k}\Omega}$$

$$= -0.417 \times 10^{-3}$$

and $\Delta I_C = [S(V_{BE})](\Delta V_{BE})$

$$= (-0.417 \times 10^{-3})(0.48 \text{ V} - 0.65 \text{ V})$$

$$= (-0.417 \times 10^{-3})(-0.17 \text{ V})$$

$$= \textbf{70.9 } \boldsymbol{\mu}\textbf{A}$$

(b) In this case, $(\beta + 1) = 101$ and $R_B/R_E = 240$. The condition $(\beta + 1) \gg R_B/R_E$ is not satisfied, negating the use of Eq. (4.67) and requiring the use of Eq. (4.64).

Eq. (4.64): $S(V_{BE}) = \dfrac{-\beta}{R_B + (\beta + 1)R_E}$

$$= \frac{-100}{240 \text{ k}\Omega + (101)1 \text{ k}\Omega} = -\frac{100}{341 \text{ k}\Omega}$$

$$= -0.293 \times 10^{-3}$$

which is about 30% less than the fixed-bias value due to the additional $(\beta + 1)R_E$ term in the denominator of the $S(V_{BE})$ equation.

$$\Delta I_C = [S(V_{BE})](\Delta V_{BE})$$

$$= (-0.293 \times 10^{-3})(-0.17 \text{ V})$$

$$\cong \textbf{50 } \boldsymbol{\mu}\textbf{A}$$

(c) In this case,

$$(\beta + 1) = 101 \gg \frac{R_B}{R_E} = \frac{47 \text{ k}\Omega}{4.7 \text{ k}\Omega} = 10 \text{ (satisfied)}$$

Eq. (4.67): $S(V_{BE}) = -\dfrac{1}{R_E}$

$$= -\frac{1}{4.7 \text{ k}\Omega}$$

$$= -0.212 \times 10^{-3}$$

and $\Delta I_C = [S(V_{BE})](\Delta V_{BE})$

$$= (-0.212 \times 10^{-3})(-0.17 \text{ V})$$

$$= \textbf{36.04 } \boldsymbol{\mu}\textbf{A}$$

In Example 4.29, the increase of 70.9 μA will have some impact on the level of I_{C_Q}. For a situation where $I_{C_Q} = 2$ mA, the resulting collector current will increase to

$$I_{C_Q} = 2 \text{ mA} + 70.9 \text{ }\mu\text{A}$$

$$= 2.0709 \text{ mA}$$

a 3.5% increase.

For the voltage-divider configuration, the level of R_B will be changed to R_{Th} in Eq. (4.64) (as defined by Fig. 4.68). In Example 4.29, the use of $R_B = 47\ \text{k}\Omega$ is a questionable design. However, R_{Th} for the voltage-divider configuration can be this level or lower and still maintain good design characteristics. The resulting equation for $S(V_{BE})$ for the feedback network will be similar to that of Eq. (4.64) with R_E replaced by R_C.

$S(\beta)$

The last stability factor to be investigated is that of $S(\beta)$. The mathematical development is more complex than that encountered for $S(I_{CO})$ and $S(V_{BE})$, as suggested by the following equation for the emitter-bias configuration:

$$S(\beta) = \frac{\Delta I_C}{\Delta \beta} = \frac{I_{C_1}(1 + R_B/R_E)}{\beta_1(1 + \beta_2 + R_B/R_E)} \qquad (4.68)$$

The notation I_{C_1} and β_1 is used to define their values under one set of network conditions, while the notation β_2 is used to define the new value of beta as established by such causes as temperature change, variation in β for the same transistor, or a change in transistors.

EXAMPLE 4.30

Determine I_{C_Q} at a temperature of 100°C if $I_{C_Q} = 2\ \text{mA}$ at 25°C. Use the transistor described by Table 4.1, where $\beta_1 = 50$ and $\beta_2 = 80$, and a resistance ratio R_B/R_E of 20.

Solution

Eq. (4.68):
$$S(\beta) = \frac{I_{C_1}(1 + R_B/R_E)}{\beta_1(1 + \beta_2 + R_B/R_E)}$$

$$= \frac{(2 \times 10^{-3})(1 + 20)}{(50)(1 + 80 + 20)} = \frac{42 \times 10^{-3}}{5050}$$

$$= \mathbf{8.32 \times 10^{-6}}$$

and
$$\Delta I_C = [S(\beta)][\Delta \beta]$$

$$= (8.32 \times 10^{-6})(30)$$

$$\cong \mathbf{0.25\ mA}$$

In conclusion therefore the collector current changed from 2 mA at room temperature to 2.25 mA at 100°C, representing a change of 12.5%.

The fixed-bias configuration is defined by $S(\beta) = I_{C_1}/\beta_1$ and R_B of Eq. (4.68) can be replaced by R_{Th} for the voltage-divider configuration.

For the collector feedback configuration with $R_E = 0\ \Omega$,

$$S(\beta) = \frac{I_{C_1}(R_B + R_C)}{\beta_1(R_B + R_C(1 + \beta_2))} \qquad (4.69)$$

Summary

Now that the three stability factors of importance have been introduced, the total effect on the collector current can be determined using the following equation:

$$\Delta I_C = S(I_{CO})\Delta I_{CO} + S(V_{BE})\Delta V_{BE} + S(\beta)\Delta \beta \qquad (4.70)$$

The equation may initially appear quite complex, but take note that each component is simply a stability factor for the configuration multiplied by the resulting change in a parameter between the temperature limits of interest. In addition, the ΔI_C to be determined is simply the change in I_C from the level at room temperature.

For instance, if we examine the fixed-bias configuration, Eq. (4.70) becomes the following:

$$\Delta I_C = (\beta + 1)\Delta I_{CO} - \frac{\beta}{R_B}\Delta V_{BE} + \frac{I_{C_1}}{\beta_1}\Delta\beta \tag{4.71}$$

after substituting the stability factors as derived in this section. Let us now use Table 4.1 to find the change in collector current for a temperature change from 25°C (room temperature) to 100°C (the boiling point of water). For this range the table reveals that

$$\Delta I_{CO} = 20 \text{ nA} - 0.1 \text{ nA} = 19.9 \text{ nA}$$

$$\Delta V_{BE} = 0.48 \text{ V} - 0.65 \text{ V} = -0.17 \text{ V} \quad \text{(note the sign)}$$

and $\qquad \Delta\beta = 80 - 50 = 30$

Starting with a collector current of 2 mA with an R_B of 240 kΩ, the resulting change in I_C due to an increase in temperature of 75°C is the following:

$$\Delta I_C = (50 + 1)(19.9 \text{ nA}) - \frac{50}{240 \text{ k}\Omega}(-0.17 \text{ V}) + \frac{2 \text{ mA}}{50}(30)$$

$$= 1.01 \ \mu\text{A} + 35.42 \ \mu\text{A} + 1200 \ \mu\text{A}$$

$$= 1.236 \text{ mA}$$

which is a significant change due primarily to the change in β. The collector current has increased from 2 to 3.236 mA—but this was expected in the sense that we recognize from the content of this section that the fixed-bias configuration is the least stable.

If the more stable voltage-divider configuration were employed with a ratio $R_{Th}/R_E = 2$ and $R_E = 4.7$ kΩ, then

$$S(I_{CO}) = 2.89, \qquad S(V_{BE}) = -0.2 \times 10^{-3}, \qquad S(\beta) = 1.445 \times 10^{-6}$$

and $\qquad \Delta I_C = (2.89)(19.9 \text{ nA}) - 0.2 \times 10^{-3}(-0.17 \text{ V}) + 1.445 \times 10^{-6}(30)$

$$= 57.51 \text{ nA} + 34 \ \mu\text{A} + 43.4 \ \mu\text{A}$$

$$= 0.077 \text{ mA}$$

The resulting collector current is 2.077 mA, or essentially 2.1 mA, compared to the 2.0 mA at 25°C. The network is obviously a great deal more stable than the fixed-bias configuration, as mentioned in earlier discussions. In this case, $S(\beta)$ did not override the other two factors and the effects of $S(V_{BE})$ and $S(I_{CO})$ were equally important. In fact, at higher temperatures, the effects of $S(I_{CO})$ and $S(V_{BE})$ will be greater than $S(\beta)$ for the device of Table 4.1. For temperatures below 25°C, I_C will decrease with increasingly negative temperature levels.

The effect of $S(I_{CO})$ in the design process is becoming a lesser concern because of improved manufacturing techniques that continue to lower the level of $I_{CO} = I_{CBO}$. It should also be mentioned that for a particular transistor the variation in levels of I_{CBO} and V_{BE} from one transistor to another in a lot is almost negligible compared to the variation in beta. In addition, the results of the analysis above support the fact that for a good stabilized design:

The ratio R_B/R_E or R_{Th}/R_E should be as small as possible with due consideration to all aspects of the design, including the ac response.

Although the analysis above may have been clouded by some of the complex equations for some of the sensitivities, the purpose here was to develop a higher level of awareness of the factors that go into a good design and to be more intimate with the transistor parameters and their impact on the network's performance. The analysis of the earlier sections was for idealized situations with nonvarying parameter values. We are now more aware of how the dc response of the design can vary with the parameter variations of a transistor.

4.13 PRACTICAL APPLICATIONS

As with the diodes in Chapter 2, it would be virtually impossible to provide even a surface treatment of the broad areas of application of BJTs. However, a few applications were chosen here to demonstrate how different facets of the characteristics of BJTs are used to perform various functions.

Relay Driver

This application is in some ways a continuation of the discussion introduced for diodes about how the effects of inductive kick can be minimized through proper design. In Fig. 4.70a, a transistor is used to establish the current necessary to energize the relay in the collector circuit. With no input at the base of the transistor, the base current, collector current, and coil current are essentially 0 A, and the relay sits in the unenergized state (normally open, NO). However, when a positive pulse is applied to the base, the transistor turns on, establishing sufficient current through the coil of the electromagnet to close the relay. Problems can now develop when the signal is removed from the base to turn off the transistor and deenergize the relay. Ideally, the current through the coil and the transistor will quickly drop to zero, the arm of the relay will be released, and the relay will simply remain dormant until the next on signal. However, we know from our basic circuit courses that the current through a coil cannot change instantaneously, and, in fact, the more quickly it changes, the greater the induced voltage across the coil as defined by $v_L = L(di_L/dt)$. In this case, the rapidly changing current through the coil will develop a large voltage across the coil with the polarity shown in Fig. 4.70a that will appear directly across the output of the transistor. The chances are likely that its magnitude will exceed the maximum ratings of the transistor, and the semiconductor device will be permanently damaged. The voltage across the coil will not remain at its highest switching level but will oscillate as shown until its level drops to zero as the system settles down.

This destructive action can be subdued by placing a diode across the coil as shown in Fig. 4.70b. During the on state of the transistor, the diode is back-biased; it sits like an open circuit and doesn't affect a thing. However, when the transistor turns off, the voltage across the coil will reverse and will forward-bias the diode, placing the

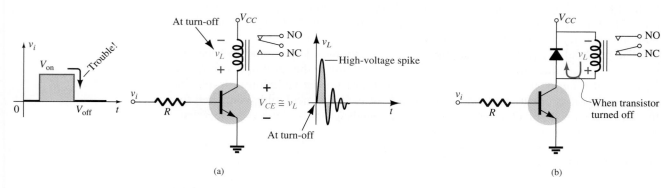

Figure 4.70 Relay driver: (a) absence of protective device; (b) with a diode across the relay coil.

diode in its on state. The current through the inductor established during the on state of the transistor can then continue to flow through the diode, eliminating the severe change in current level. Because the inductive current is switched to the diode almost instantaneously after the off state is established, the diode must have a current rating to match the current through the inductor and the transistor when in the on state. Eventually, because of the resistive elements in the loop, including the resistance of the coil windings and the diode, the high-frequency (quickly oscillating) variation in voltage level across the coil will decay to zero, and the system will settle down.

Transistor Switch

In Fig. 4.71a, a transistor is being used as a switch to control the on and off states of the light bulb in the collector branch of the network. When the switch is in the on position, we have a fixed-bias situation where the base-to-emitter voltage is at its 0.7-V level, and the base current is controlled by the resistor R_1 and the input impedance of the transistor. The current through the bulb will then be beta times the base current, and the bulb will light up. A problem can develop, however, if the bulb has not been on for a while. When a light bulb is first turned on, its resistance is quite low, even though the resistance will increase rapidly the longer the bulb is on. This can cause a momentary high level of collector current that could damage the bulb and the transistor over time. In Fig. 4.71b, for instance, the load line for the same network with a cold and a hot resistance for the bulb is included. Note that even though the base current is set by the base circuit, the intersection with the load line results in a higher current for the cold light bulb. Any concern about the turn-on level can easily be corrected by inserting an additional small resistor in series with the light bulb, as shown in Fig. 4.71c, just to ensure a limit on the initial surge in current when the bulb is first turned on.

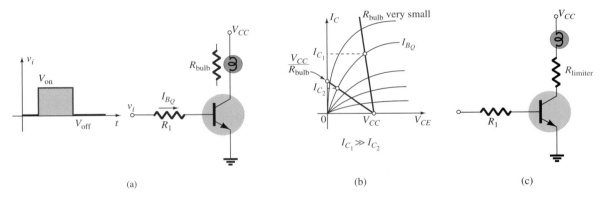

Figure 4.71 Using the transistor as a switch to control the on-off states of a bulb: (a) network; (b) effect of low bulb resistance on collector current; (c) limiting resistor.

Constant-Current Source (CCS)

If we assume that the characteristics of a transistor appear as shown in Fig. 4.72a (constant beta throughout), an excellent current source can be created using the simple transistor configuration shown in Fig. 4.72b because no matter what the load resistance, the collector or load current will remain the same as shown in Fig. 4.72c. The base current is fixed; and no matter where the load line is, the collector current remains the same. In other words, the collector current is independent of the load in the collector circuit—a perfect current source. However, because the actual characteristics are more like those in Fig. 4.71b, where beta will vary from point to point, and even though the base current may be fixed by the configuration, the beta will vary

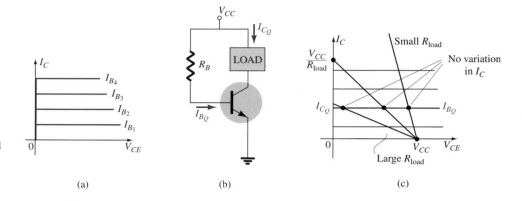

Figure 4.72 Building a constant-current source assuming ideal BJT characteristics: (a) ideal characteristics; (b) network; (c) demonstrating why I_C remains constant.

(a) (b) (c)

from point to point with the load intersection, and $I_C = I_L$ will vary—not characteristic of a good current source. Recall, however, that the voltage-divider configuration resulted in a low level of sensitivity to beta, so perhaps if that biasing arrangement is used, the current source equivalent is closer to reality. In fact, that is the case. If a biasing arrangement such as shown in Fig. 4.73 is employed, the sensitivity to changes in operating point due to varying loads is much less, and the collector current will remain fairly constant for changes in load resistance in the collector branch. In fact, the emitter voltage is determined by

$$V_E = V_B - 0.7 \text{ V}$$

with the collector or load current determined by

$$I_C \cong I_E = \frac{V_E}{R_E} = \frac{V_B - 0.7 \text{ V}}{R_E}$$

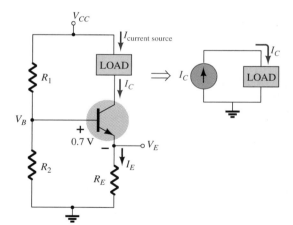

Figure 4.73 Network establishing a fairly constant current source due to its reduced sensitivity to changes in beta.

Using Fig. 4.73, the improved stability can be described by examining the case where I_C may be trying to rise for any number of reasons. The result is that $I_E = I_C$ will also rise and the voltage $V_{R_E} = I_E R_E$ will increase. However, if we assume V_B to be fixed (a good assumption since its level is determined by two fixed resistors and a voltage source), the base-to-emitter voltage $V_{BE} = V_B - V_{R_E}$ will drop. A drop in V_{BE} will cause I_B and therefore I_C ($= \beta I_B$) to drop. The result is a situation where any tendency for I_C to increase will be met with a network reaction that will work against the change to stabilize the system.

Alarm System with a CCS

An alarm system with a constant-current source of the type just introduced appears in Fig. 4.74. Since $\beta R_E = (100)(1 \text{ k}\Omega) = 100 \text{ k}\Omega$ is much greater than R_1, we can use the approximate approach and find the voltage V_{R_1} as follows:

$$V_{R_1} = \frac{2 \text{ k}\Omega(16 \text{ V})}{2 \text{ k}\Omega + 4.7 \text{ k}\Omega} = 4.78 \text{ V}$$

and then the voltage across R_E:

$$V_{R_E} = V_{R_1} - 0.7 \text{ V} = 4.78 \text{ V} - 0.7 \text{ V} = 4.08 \text{ V}$$

and finally the emitter and collector current:

$$I_E = \frac{V_{R_E}}{R_E} = \frac{4.08 \text{ V}}{1 \text{ k}\Omega} = 4.08 \text{ mA} \cong 4 \text{ mA} = I_C$$

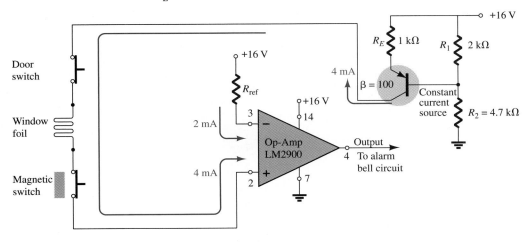

Figure 4.74 An alarm system with a constant-current source and an op-amp comparator.

Since the collector current is the current through the circuit, the 4-mA current will remain fairly constant for slight variations in network loading. Note that the current passes through a series of sensor elements and finally into an op-amp designed to compare the 4-mA level with the set level of 2 mA. (Although the op-amp may be a new device to you, it will be discussed in detail in Chapter 13—you will not need to know the details of its behavior for this application.)

The LM2900 operational amplifier of Fig. 4.74 is one of four found in the dual-in-line integrated circuit package appearing in Fig. 4.75a. Pins 2, 3, 4, 7, and 14 were used for the design of Fig. 4.74. For the sake of interest only, note in Fig. 4.75b the number of elements required to establish the desired terminal characteristics for the op-amp—as mentioned earlier, the details of its internal operation are left for another time. The 2 mA at terminal 3 of the op-amp is a *reference* current established by the 16-V source and R_{ref} at the negative side of the op-amp input. The 2-mA current level is required as a level against which the 4-mA current of the network is to be compared. As long as the 4-mA current on the positive input to the op-amp remains constant, the op-amp will provide a "high" output voltage that exceeds 13.5 V, with a typical level of 14.2 V (according to the specification sheets for the op-amp). However, if the sensor current drops from 4 mA to a level below 2 mA, the op-amp will respond with a "low" output voltage that is typically about 0.1 V. The output of the op-amp will then signal the alarm circuit about the disturbance. Note from the above that it is not necessary for the sensor current to drop to all the way down to 0 mA to signal the alarm circuit. Only a variation around the reference level that appears unusual is required—a good alarm feature.

Dual-in-line package

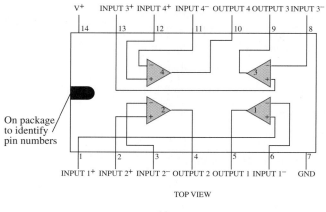

(a)

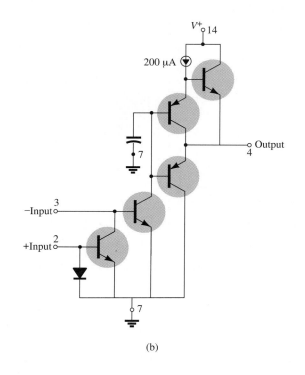

(b)

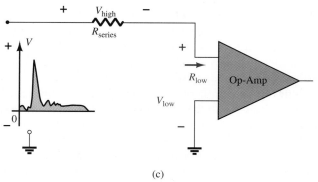

(c)

Figure 4.75 LM2900 operational amplifier: (a) dual-in-line package (DIP); (b) components; (c) impact of low-input impedance.

One very important characteristic of this particular op-amp is the low-input impedance as shown in Fig. 4.75c. This feature is important because you don't want alarm circuits reacting to every voltage spike or turbulence that comes down the line because of some external switching action or outside forces such as lightning. In Fig. 4.75c, for instance, if a high-voltage spike should appear at the input to the series configuration, most of the voltage will appear across the series resistor rather than the op-amp—thus preventing a false output and an activation of the alarm.

Logic Gates

By now it is probably a surprise to the reader that transistors in the dc mode are used for so many applications. For most students who have some prior awareness of transistors, the initial assumption is that a transistor is used only as an ac amplifier. In fact, most electronic components have a variety of applications in both the dc and the ac mode.

In this application, full use is made of the fact that the collector-to-emitter impedance of a transistor is quite low near or at saturation and large near or at cutoff. For instance, the load line defines *saturation* as the point where the current is quite high and the collector-to-emitter voltage quite low as shown in Fig. 4.76. The resulting resistance, defined by $R_{sat} = \dfrac{V_{CE_{sat}(low)}}{I_{C_{sat}(high)}}$, is quite low and is often approximated as a short circuit. At *cutoff*, the current is relatively low and the voltage near its maximum value as shown in Fig. 4.76, resulting in a very high impedance between the collector and emitter terminal which is often approximated by an open circuit.

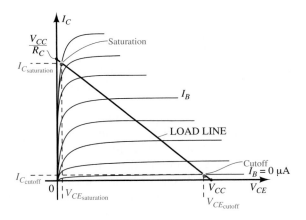

Figure 4.76 Points of operation for a BJT logic gate.

The above impedance levels established by on and off transistors make it relatively easy to understand the operation of the logic gates of Fig. 4.77. Because there are two inputs to each gate, there are four possible combinations of voltages at the input to the transistors. A 1 or on state is defined by a high voltage at the base terminal to turn the transistor on. A 0 or off state is defined by 0 V at the base, ensuring that transistor is off. If both A and B of the OR gate of Fig. 4.77a have a low or 0-V input, both transistors are off (cutoff), and the impedance between the collector and emitter of each transistor can be approximated by an open circuit. Mentally replacing both transistors by open circuits between the collector and emitter will remove any connection between the applied bias of 5 V and the output. The result is zero current through each transistor and through the 3.3-kΩ resistor. The output voltage is therefore 0 V or "low"—a 0 state. On the other hand, if transistor Q_1 is on and Q_2 is off due to a positive voltage at the base of Q_1 and 0 V at the base of Q_2, then the short-circuit equivalent between the collector and emitter for transistor Q_1 can be applied, and the voltage at the output is 5 V or "high"—a 1 state. Finally, if both transistors are turned on by a positive voltage applied to the base of each, they will

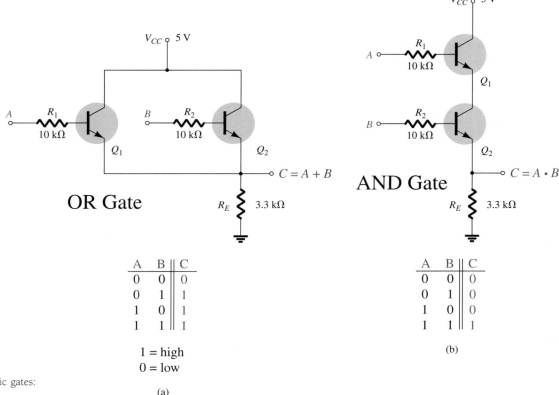

A	B	C
0	0	0
0	1	1
1	0	1
1	1	1

1 = high
0 = low

(a)

A	B	C
0	0	0
0	1	0
1	0	0
1	1	1

(b)

Figure 4.77 BJT logic gates: (a) OR; (b) AND.

both ensure that the output voltage is 5 V or "high"—a 1 state. The operation of the OR gate is properly defined: an output if either input terminal has applied turn-on voltage or if both are in the on state. A 0 state exists only if both do not have a 1 state at the input terminals.

The AND gate of Fig. 4.77b requires that the output be high only if both inputs have a turn-on voltage applied. If both are in the on state, a short-circuit equivalent can be used for the connection between the collector and emitter of each transistor, providing a direct path from the applied 5-V source to the output—thereby establishing a high or 1 state at the output terminal. If one or both transistors are off due to 0 V at the input terminal, an open circuit is placed in series with the path from the 5-V supply voltage to the output, and the output voltage is 0 V or an off state.

Current Mirror

The current mirror is a dc network in which the current through the load is the mirror image of another current of the same network. If the controlling current of the network is changed, the current through the load will change also.

A common current mirror constructed of two back-to-back *npn* transistors is shown in Fig. 4.78. The load current is the collector current of Q_2, and the controlling current is the collector current of Q_1. Note, in particular, that the collector of Q_1 is connected directly to the base of the same transistor, establishing the same potential at each point. The result is that $V_{C_1} = V_{B_1} = V_{B_2} = 0.7$ V for the on transistor. The controlling element is resistor R. If you change its value, you change the controlling current as determined by $I_R = I_{C_1} = (10\ \text{V} - 0.7\ \text{V})/R$ (ignoring the drop-off in I_{C_1} due to I_B as shown in Fig. 4.78).

Once the resistance is set, the collector current of Q_2 will immediately change to the new level. The operation of the mirror network is totally dependent on the fact that Q_1 and Q_2 are matched transistors, that is, transistors with very similar charac-

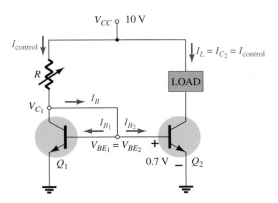

I_{control}

V_{CC} 10 V

R

V_{C_1}

I_B

I_{B_1} I_{B_2}

$V_{BE_1} = V_{BE_2}$ +

Q_1 0.7 V   − Q_2

$I_L = I_{C_2} = I_{\text{control}}$

LOAD

Figure 4.78   Current mirror using back-to-back BJTs.

teristics (ideally the same). In other words, a base current of 10 μA in either one will result in the same collector current for each; the base-to-emitter voltage of each in the on state will be the same; and so on.

The operation of the configuration is best defined by first setting the control current to the desired level, say, I_{R_1}. This will define the level of I_{C_1} and of I_{B_1} from $I_{C_1}/\beta_1 = I_{R_1}/\beta_1$ and will establish the level of V_{BE_1} as shown in Fig. 4.79. Since they are matched transistors, $V_{BE_1} = V_{BE_2}$, and the resulting level of I_{B_2} will be same as I_{B_1}. The result is the same collector (load current) defined by $I_L = I_{C_2} = \beta_2 I_{B_2}$ since both betas are the same. In general, therefore, $I_L = I_{C_2} = I_{C_1} = I_R$ for matched transistors.

The network also has a measure of built-in control that will try to ensure that any variation in load current will be corrected by the configuration itself. For instance, if I_L should try to increase for whatever reason, the base current of Q_2 will also increase due to the relationship $I_{B_2} = I_{C_2}/\beta_2 = I_L/\beta_2$. Returning to Fig. 4.79, we find that an increase in I_{B_2} will cause voltage V_{BE_2} to increase also. Since the base of Q_2 is connected directly to the collector of Q_1, the voltage V_{CE_1} will increase also. This action causes the voltage across the control resistor R to decrease, causing I_R to drop. But if I_R drops, the base current I_B will drop, causing both I_{B_1} and I_{B_2} to drop also. A drop in I_{B_2} will cause the collector current and therefore the load current to drop also. The result, therefore, is a sensitivity to unwanted changes that the network will make every effort to correct.

The entire sequence of events just described can be presented on a single line as shown below. Note that at one end the load current is trying to increase, and at the end of the sequence the load current is forced to return to its original level.

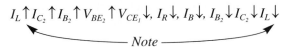

$$I_L \uparrow I_{C_2} \uparrow I_{B_2} \uparrow V_{BE_2} \uparrow V_{CE_1} \downarrow, \; I_R \downarrow, \; I_B \downarrow, \; I_{B_2} \downarrow I_{C_2} \downarrow I_L \downarrow$$

Note

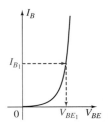

I_B

I_{B_1}

0 V_{BE_1} V_{BE}

Figure 4.79   Base characteristics for transistor Q_1 (and Q_2).

Voltage Level Indicator

The last application to be introduced in this section, the voltage level indicator, includes three of the elements introduced thus far in the text: the transistor, the Zener diode, and the LED. The voltage level indicator is a relatively simple network using a green LED to indicate when the source voltage is close to its monitoring level of 9 V. In Fig. 4.80 the potentiometer is set to establish 5.4 V at the point indicated in Fig. 4.80. The result is sufficient voltage to turn on both the 4.7-V Zener and the transistor and establish a collector current through the LED sufficient in magnitude to turn on the green LED.

Once the potentiometer is set, the LED will emit its green light as long as the supply voltage is near 9 V. However, if the terminal voltage of the 9-V battery should decrease, the voltage set up by the voltage-divider network may drop to 5 V from 5.4 V. At 5 V there is insufficient voltage to turn on both the Zener and the transistor, and

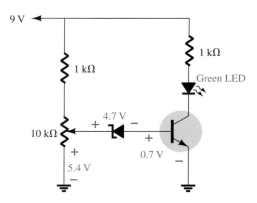

Figure 4.80 Voltage level indicator.

the transistor will be in the off state. The LED will immediately turn off, revealing that the supply voltage has dropped below 9 V or that the power source has been disconnected.

4.14 SUMMARY

Important Conclusions and Concepts

1. No matter what type of configuration a transistor is used in, the basic relationships between the currents are **always the same**, and the base-to-emitter voltage is the **threshold value** if the transistor is in the on state.

2. The operating point defines where the transistor will operate on its characteristic curves under **dc conditions**. For linear (minimum distortion) amplification, the dc operating point should not be too close to the maximum power, voltage, or current rating and should avoid the regions of saturation and cutoff.

3. For most configurations the dc analysis begins with a determination of the **base current**.

4. For the dc analysis of a transistor network, all capacitors are replaced by an **open-circuit equivalent**.

5. The fixed-bias configuration is the simplest of transistor biasing arrangements, but it is also quite unstable due its **sensitivity to beta** at the operating point.

6. Determining the saturation (maximum) collector current for any configuration can usually be done quite easily if an **imaginary short circuit** is superimposed between the collector and emitter terminals of the transistor. The resulting current through the short is then the saturation current.

7. The equation for the load line of a transistor network can be found by applying **Kirchhoff's voltage law** to the output or collector network. The *Q*-point is then determined by finding the **intersection** between the base current and the load line drawn on the device characteristics.

8. The emitter-stabilized biasing arrangement is less sensitive to changes in beta— providing more stability for the network. Keep in mind, however, that any resistance in the emitter leg is "seen" at the base of the transistor as a much **larger resistor**, a fact that will reduce the base current of the configuration.

9. The voltage-divider bias configuration is probably the most common of all the configurations. Its popularity is due primarily to its **low sensitivity** to changes in beta from one transistor to another of the same lot (with the same transistor label). The exact analysis can be applied to any configuration, but the approximate can be applied only if the reflected emitter resistance as seen at the base **is much larger** than the lower resistor of the voltage-divider bias arrangement connected to the base of the transistor.

10. When analyzing the dc bias with a voltage feedback configuration, be sure to remember that **both** the emitter resistor and the collector resistor are reflected back to the base circuit by beta. The least sensitivity to beta is obtained when the reflected resistance is much larger than the feedback resistor between the base and collector.

11. For the common-base configuration the **emitter current is normally determined first** due to the presence of the base-to-emitter junction in the same loop. Then the fact that the emitter and collector current are essentially of the same magnitude is employed.

12. A clear understanding of the procedure employed to analyze a dc transistor network will usually permit a design of the same configuration with a minimum of difficulty and confusion. Simply start with those relationships that **minimize the number of unknowns**, and then proceed to make some decisions about the unknown elements of the network.

13. In a switching configuration, a transistor quickly moves between **saturation and cutoff, or vice versa**. Essentially, the impedance between collector and emitter can be approximated as a short circuit for saturation and an open circuit for cutoff.

14. When checking the operation of a dc transistor network, first check that the base-to-emitter voltage is very close to **0.7 V** and that the collector-to-emitter voltage is between **25% and 75% of the applied voltage** V_{CC}.

15. The analysis of *pnp* configurations is exactly the same as that applied to *npn* transistors with the exception that current directions will **reverse** and voltages will have the **opposite** polarities.

16. Beta is very sensitive to **temperature**, and V_{BE} **decreases** about 7.5 mV (0.0075 V) for each 1° increase in temperature on a Celsius scale. The reverse saturation current typically **doubles** for every 10° increase in Celsius temperature.

17. Keep in mind that networks that are the **most stable** and least sensitive to temperature changes have the **smallest stability factors**.

Equations

$$V_{BE} = 0.7 \text{ V}$$

$$I_E = (\beta + 1)I_B \cong I_C$$

$$I_C = \beta I_B$$

Fixed bias:

$$I_B = \frac{V_{CC} - V_{BE}}{R_B}$$

$$I_C = \beta I_B$$

Emitter stabilized:

$$I_B = \frac{V_{CC} - V_{BE}}{R_B + (\beta + 1)R_E}$$

$$R_i = (\beta + 1)R_E$$

Voltage-divider bias:

$$\text{Exact:} \quad R_{\text{Th}} = R_1 \| R_2$$

$$E_{\text{Th}} = V_{R_2} = \frac{R_2 V_{CC}}{R_1 + R_2}$$

$$I_B = \frac{E_{\text{Th}} - V_{BE}}{R_{\text{Th}} + (\beta + 1)R_E}$$

Approximate: Test $\beta R_E \geq 10R_2$

$$V_B = \frac{R_2 V_{CC}}{R_1 + R_2}$$

$$V_E = V_B - V_{BE}$$

$$I_E = \frac{V_E}{R_E} \cong I_C$$

DC bias with voltage feedback:

$$I_B = \frac{V_{CC} - V_{BE}}{R_B + \beta(R_C + R_E)}$$

$$I_C' \cong I_C \cong I_E$$

Common base:

$$I_E = \frac{V_{EE} - V_{BE}}{R_E}$$

$$I_C \cong I_E$$

Transistor switching networks:

$$I_{C_{\text{sat}}} = \frac{V_{CC}}{R_C}$$

$$I_B > \frac{I_{C_{\text{sat}}}}{\beta_{\text{dc}}}$$

$$R_{\text{sat}} = \frac{V_{CE_{\text{sat}}}}{I_{C_{\text{sat}}}}$$

$$t_{\text{on}} = t_r + t_d$$

$$t_{\text{off}} = t_s + t_f$$

Stability factors:

$$S(I_{CO}) = \frac{\Delta I_C}{\Delta I_{CO}}$$

$$S(V_{BE}) = \frac{\Delta I_C}{\Delta V_{BE}}$$

$$S(\beta) = \frac{\Delta I_C}{\Delta \beta}$$

$S(I_{CO})$:

Fixed bias: $S(I_{CO}) = \beta + 1$

Emitter bias: $S(I_{CO}) = (\beta + 1)\dfrac{1 + R_B/R_E*}{(\beta + 1) + R_B/R_E}$

*Voltage-divider bias: Change R_B to R_{Th} in above equation.

*Feedback bias: Change R_E to R_C in above equation.

$S(V_{BE})$:

$$\text{Fixed bias:} \quad S(V_{BE}) = -\frac{\beta}{R_B}$$

$$\text{Emitter bias:} \quad S(V_{BE}) = \frac{-\beta^\dagger}{R_B + (\beta + 1)R_E}$$

†Voltage-divider bias: Change R_B to R_{Th} in above equation.

†Feedback bias: Change R_E to R_C in above equation.

$S(\beta)$:

$$\text{Fixed bias:} \quad S(\beta) = \frac{I_{C_1}}{\beta_1}$$

$$\text{Emitter bias:} \quad S(\beta) = \frac{I_{C_1}(1 + R_B/R_E)^\ddagger}{\beta_1(1 + \beta_2 + R_B/R_E)}$$

‡Voltage-divider bias: Change R_B to R_{Th} in above equation.

‡Feedback bias: Change R_E to R_C in above equation.

4.15 COMPUTER ANALYSIS

PSpice Windows

VOLTAGE-DIVIDER CONFIGURATION

The results of Example 4.7 will now be verified using PSpice Windows. Using methods described in previous chapters, the network of Fig. 4.81 can be constructed. Recall that the transistor can be found in the **EVAL.slb** library, the dc source under **SOURCE.slb,** and the resistor under **ANALOG.slb.** The capacitor will also appear in the **ANALOG.slb** library. Three **VIEWPOINTS** appear in Fig. 4.81 as obtained from the **SPECIAL.slb** library. The collector current will be sensed by the **IPROBE** option, also appearing in the **SPECIAL.slb** library. Recall that a positive result is obtained for **IPROBE** if the direction of conventional current enters that side of the symbol with the internal curve representing the scale of the meter. We will want to set the value of beta for the transistor to match that of the example. This is accomplished by clicking on the transistor symbol (to obtain the red outline) followed by **Edit-Model-Edit Instance Model (text)** to obtain the **Model Editor.** Then **Bf** is changed to 140 to match the value of Example 4.7. Click **OK,** and the network is set up for the analysis.

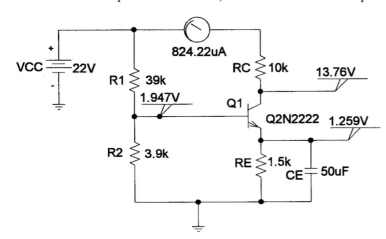

Figure 4.81 Applying PSpice Windows to the voltage-divider configuration of Example 4.7.

In this case, since we are only interested in the dc response, the **Probe Setup** under **Analysis** should enable **Do not auto-run Probe.** It will save us from having to deal with the Probe response before viewing the output file or screen. The sequence **Analysis-Simulate** will result in the dc levels appearing in Fig. 4.81, which closely match those of Example 4.7. The collector-to-emitter voltage is 13.76 V $-$ 1.259 V $=$ 12.5 V, versus 12.22 V of Example 4.7, and the collector current is 0.824 mA, versus 0.85 mA. Any differences are due to the fact that we are using an actual transistor with a host of parameters not considered in our analysis. Recall the difference in beta from the specification value and the value obtained from the plot of the previous chapter.

Since the voltage-divider network is one that is to have a low sensitivity to changes in beta, let us return to the transistor and replace the beta of 140 with the default value of 225.9 and examine the results. The analysis will result in the dc levels appearing in Fig. 4.82, which are very close to those of Fig. 4.81.

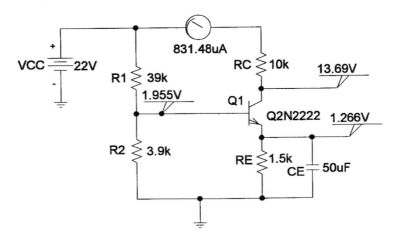

Figure 4.82 Response obtained after changing β from 140 to 255.9 for the network of Figure 4.81.

The collector-to-emitter voltage is 13.69 V $-$ 1.266 V $=$ 12.42 V, which is very close to that obtained with a much lower beta. The collector current is actually closer to the hand-calculated level, 0.832 mA versus 0.85 mA. There is no question, therefore, that the voltage-divider configuration demonstrates a low sensitivity to changes in beta. Recall, however, that the fixed-bias configuration was very sensitive to changes in beta, and let us proceed with the same type of analysis for the fixed-bias configuration and compare notes.

FIXED-BIAS CONFIGURATION

The fixed-bias configuration of Fig. 4.83 is from Example 4.1 to permit a comparison of results. Beta was set to 50 using the procedure described above. In this case, we will use a **VIEWPOINT** to read the collector-to-emitter voltage and enable the display of bias currents (using the icon with the large capital **I**). In addition, we will inhibit the display of some bias currents using the icon with the smaller capital **I** and the diode symbol. The final touch is to move some of the currents displayed to clean up the presentation.

A PSpice analysis of the network will result in the levels appearing in Fig. 4.83. These are a close match with the hand-written solution, with the collector voltage at 6.998 V versus 6.83 V, the collector current at 2.274 mA versus 2.35 mA, and the base current at 47.23 μA versus 47.08 μA.

Let us now test the sensitivity to changes in beta by changing to the default value of 255.9. The results appear in Fig. 4.84. Note the dramatic drop in V_C to 0.113 V compared to 6.83 V and the significant rise in I_D to 5.4 mA versus the solution of 2.35 mA. The fixed-bias configuration is obviously very beta-sensitive.

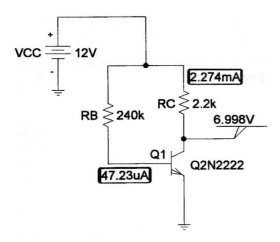

Figure 4.83 Fixed-bias configuration with a β of 50.

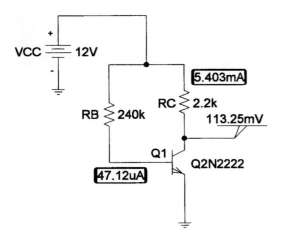

Figure 4.84 Network of Figure 4.83 with a β of 255.9.

Electronics Workbench

Electronics Workbench will now be applied to the fixed-bias network of Example 4.4 to provide an opportunity to review the transistor options internal to the software package and to compare results with the handwritten solution.

A full range of transistors is available by simply selecting the transistor option from the **Component Tool Bar.** The result is a **Component Browser** dialog box within which **Select Component** can be chosen. There is a long list of transistors under the **Component List** from which the 2N2712 transistor was chosen. Once the transistor has been selected, **Model Data** will appear with all the important parameters of the device. Click **OK,** and the transistor can be placed at any location on the screen as shown on Fig. 4.85. Double-click the device, and a **BJT NPN** dialog box will appear that will permit changing some of the parameters for our application. First choose **Edit,** change **Bf** to 50 and **Is** to 1nA, and select **Change Part Model.** Click **OK,** and the transistor on the screen will now have an asterisk on its label to indicate that a change in parameters has been made. The new value of beta can be displayed on the screen by the **Place Text** option under **Edit.** A left-click will set the location for the text to be entered. Once **BF=50** is typed, a double-click will establish the box around the entry. Give an additional left-click, and the label has been entered. To remove, simply reestablish the box with a left-click on the label, and a right-click will provide a list of options from which **Cut** can be chosen. The **Place Text** option can also be obtained at any time by a right-click of the mouse, after which the procedure is as described above.

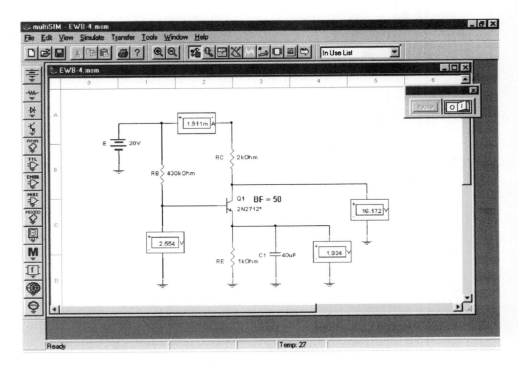

Figure 4.85 Verifying the results of Example 4.4 using Electronics Workbench.

Because it is always important to know the status of the **Simulation** control, it is recommended that the **Simulate Switch** be placed on the screen at all times. Keep in mind that once the **RUN** option is selected under **Simulation,** changes in the network cannot be made until the **Simulation** is stopped. The switch clearly shows the status of the simulation with a **0** and a **1.** To display the switch, simply select **Show Simulate Switch** under the **View** option.

Once the transistor is in place, the remaining elements can be placed as shown in Fig. 4.85 using the procedure described in detail in the earlier chapters. The one ammeter to measure the collector current is obtained through **Indicators-Ammeter H-OK,** while all the voltmeters are obtained through **Indicators-Voltmeters V-OK.**

Finally, the simulation switch is turned on by selecting the **1,** and the results of Fig. 4.85 will be obtained. In particular, note that the base-to-emitter voltage is 2.554 V − 1.934 V = 0.62 V rather than the 0.7 V assumed due to the relatively low level of current for this transistor. However, all the readings are very close to the reading obtained assuming a very simple model for the transistor, providing some validation of the approximations introduced in this chapter.

PROBLEMS

§ 4.3 Fixed-Bias Circuit

1. For the fixed-bias configuration of Fig. 4.86, determine:

   (a) I_{B_Q}.
   (b) I_{C_Q}.
   (c) V_{CE_Q}.
   (d) V_C.
   (e) V_B.
   (f) V_E.

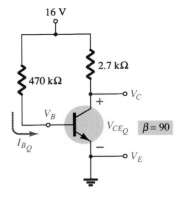

Figure 4.86 Problems 1, 4, 11, 47, 51, 52, 53

2. Given the information appearing in Fig. 4.87, determine:
   (a) I_C.
   (b) R_C.
   (c) R_B.
   (d) V_{CE}.

3. Given the information appearing in Fig. 4.88, determine:
   (a) I_C.
   (b) V_{CC}.
   (c) β.
   (d) R_B.

4. Find the saturation current $(I_{C_{sat}})$ for the fixed-bias configuration of Fig. 4.86.

*5. Given the BJT transistor characteristics of Fig. 4.89:
   (a) Draw a load line on the characteristics determined by $E = 21$ V and $R_C = 3$ kΩ for a fixed-bias configuration.
   (b) Choose an operating point midway between cutoff and saturation. Determine the value of R_B to establish the resulting operating point.
   (c) What are the resulting values of I_{C_Q} and V_{CE_Q}?
   (d) What is the value of β at the operating point?
   (e) What is the value of α defined by the operating point?
   (f) What is the saturation current $(I_{C_{sat}})$ for the design?
   (g) Sketch the resulting fixed-bias configuration.
   (h) What is the dc power dissipated by the device at the operating point?
   (i) What is the power supplied by V_{CC}?
   (j) Determine the power dissipated by the resistive elements by taking the difference between the results of parts (h) and (i).

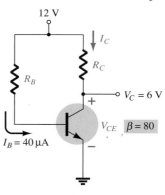

Figure 4.87   Problem 2

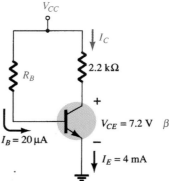

Figure 4.88   Problem 3

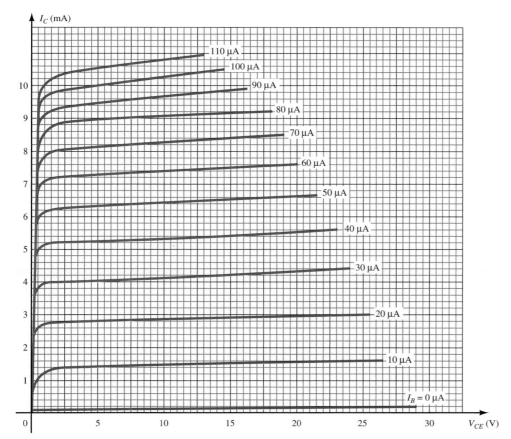

Figure 4.89   Problems 5, 10, 19, 35, 36

§ 4.4 Emitter-Stabilized Bias Circuit

6. For the emitter-stabilized bias circuit of Fig. 4.90, determine:
   (a) I_{B_Q}.
   (b) I_{C_Q}.
   (c) V_{CE_Q}.
   (d) V_C.
   (e) V_B.
   (f) V_E.

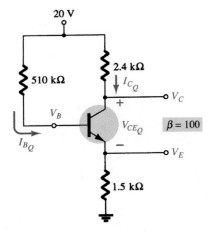

Figure 4.90 Problems 6, 9, 11, 48, 51, 54

7. Given the information provided in Fig. 4.91, determine:
   (a) R_C.
   (b) R_E.
   (c) R_B.
   (d) V_{CE}.
   (e) V_B.

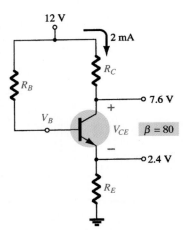

Figure 4.91 Problem 7

8. Given the information provided in Fig. 4.92, determine:
   (a) β.
   (b) V_{CC}.
   (c) R_B.

9. Determine the saturation current $(I_{C_{sat}})$ for the network of Fig. 4.90.

*10. Using the characteristics of Fig. 4.89, determine the following for an emitter-bias configuration if a Q-point is defined at $I_{C_Q} = 4$ mA and $V_{CE_Q} = 10$ V.
   (a) R_C if $V_{CC} = 24$ V and $R_E = 1.2$ kΩ.
   (b) β at the operating point.
   (c) R_B.
   (d) Power dissipated by the transistor.
   (e) Power dissipated by the resistor R_C.

*11. (a) Determine I_C and V_{CE} for the network of Fig. 4.86.
   (b) Change β to 135 and determine the new value of I_C and V_{CE} for the network of Fig. 4.86.
   (c) Determine the magnitude of the percent change in I_C and V_{CE} using the following equations:

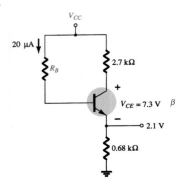

Figure 4.92 Problem 8

$$\%\Delta I_C = \left| \frac{I_{C_{(part\,b)}} - I_{C_{(part\,a)}}}{I_{C_{(part\,a)}}} \right| \times 100\%, \qquad \%\Delta V_{CE} = \left| \frac{V_{CE_{(part\,b)}} - V_{CE_{(part\,a)}}}{V_{CE_{(part\,a)}}} \right| \times 100\%$$

(d) Determine I_C and V_{CE} for the network of Fig. 4.90.

(e) Change β to 150 and determine the new value of I_C and V_{CE} for the network of Fig. 4.90.

(f) Determine the magnitude of the percent change in I_C and V_{CE} using the following equations:

$$\%\Delta I_C = \left| \frac{I_{C_{(part\,c)}} - I_{C_{(part\,d)}}}{I_{C_{(part\,d)}}} \right| \times 100\%, \quad \%\Delta V_{CE} = \left| \frac{V_{CE_{(part\,c)}} - V_{CE_{(part\,d)}}}{V_{CE_{(part\,d)}}} \right| \times 100\%$$

(g) In each of the above, the magnitude of β was increased 50%. Compare the percent change in I_C and V_{CE} for each configuration, and comment on which seems to be less sensitive to changes in β.

§ 4.5 Voltage-Divider Bias

12. For the voltage-divider bias configuration of Fig. 4.93, determine:
   (a) I_{B_Q}.
   (b) I_{C_Q}.
   (c) V_{CE_Q}.
   (d) V_C.
   (e) V_E.
   (f) V_B.

13. Given the information provided in Fig. 4.94, determine:
   (a) I_C.
   (b) V_E.
   (c) V_B.
   (d) R_1.

14. Given the information appearing in Fig. 4.95, determine:
   (a) I_C.
   (b) V_E.
   (c) V_{CC}.
   (d) V_{CE}.
   (e) V_B.
   (f) R_1.

15. Determine the saturation current $(I_{C_{sat}})$ for the network of Fig. 4.93.

*16. Determine the following for the voltage-divider configuration of Fig. 4.96 using the approximate approach if the condition established by Eq. (4.33) is satisfied.
   (a) I_C.
   (b) V_{CE}.
   (c) I_B.
   (d) V_E.
   (e) V_B.

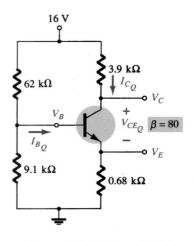

Figure 4.93 Problems 12, 15, 18, 20, 24, 49, 51, 52, 55

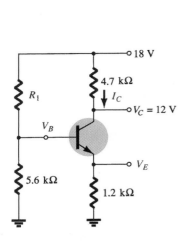

Figure 4.94 Problem 13

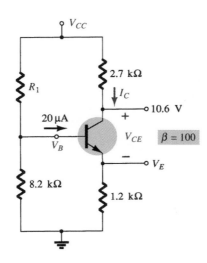

Figure 4.95 Problem 14

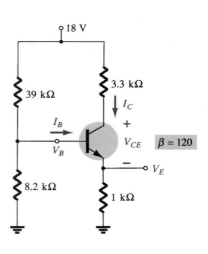

Figure 4.96 Problems 16, 17, 21

Problems   **237**

*17. Repeat Problem 16 using the exact (Thévenin) approach and compare solutions. Based on the results, is the approximate approach a valid analysis technique if Eq. (4.33) is satisfied?

18. (a) Determine I_{C_Q}, V_{CE_Q}, and I_{B_Q} for the network of Problem 12 (Fig. 4.93) using the approximate approach even though the condition established by Eq. (4.33) is not satisfied.
    (b) Determine I_{C_Q}, V_{CE_Q}, and I_{B_Q} using the exact approach.
    (c) Compare solutions and comment on whether the difference is sufficiently large to require standing by Eq. (4.33) when determining which approach to employ.

*19. (a) Using the characteristics of Fig. 4.89, determine R_C and R_E for a voltage-divider network having a Q-point of $I_{C_Q} = 5$ mA and $V_{CE_Q} = 8$ V. Use $V_{CC} = 24$ V and $R_C = 3R_E$.
    (b) Find V_E.
    (c) Determine V_B.
    (d) Find R_2 if $R_1 = 24$ kΩ assuming that $\beta R_E > 10R_2$.
    (e) Calculate β at the Q-point.
    (f) Test Eq. (4.33), and note whether the assumption of part (d) is correct.

*20. (a) Determine I_C and V_{CE} for the network of Fig. 4.93.
    (b) Change β to 120 (50% increase), and determine the new values of I_C and V_{CE} for the network of Fig. 4.93.
    (c) Determine the magnitude of the percent change in I_C and V_{CE} using the following equations:

$$\%\Delta I_C = \left| \frac{I_{C_{(\text{part b})}} - I_{C_{(\text{part a})}}}{I_{C_{(\text{part a})}}} \right| \times 100\%, \qquad \%\Delta V_{CE} = \left| \frac{V_{CE_{(\text{part b})}} - V_{CE_{(\text{part a})}}}{V_{CE_{(\text{part a})}}} \right| \times 100\%$$

    (d) Compare the solution to part (c) with the solutions obtained for parts (c) and (f) of Problem 11. If not performed, note the solutions provided in Appendix E.
    (e) Based on the results of part (d), which configuration is least sensitive to variations in β?

*21. (a) Repeat parts (a) through (e) of Problem 20 for the network of Fig. 4.96. Change β to 180 in part (b).
    (b) What general conclusions can be made about networks in which the condition $\beta R_E > 10R_2$ is satisfied and the quantities I_C and V_{CE} are to be determined in response to a change in β?

§ 4.6 DC Bias with Voltage Feedback

22. For the collector feedback configuration of Fig. 4.97, determine:
    (a) I_B.
    (b) I_C.
    (c) V_C.

23. For the voltage feedback network of Fig. 4.98, determine:
    (a) I_C.
    (b) V_C.
    (c) V_E.
    (d) V_{CE}.

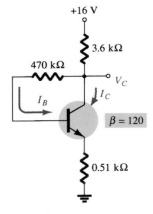

Figure 4.97 Problems 22, 56

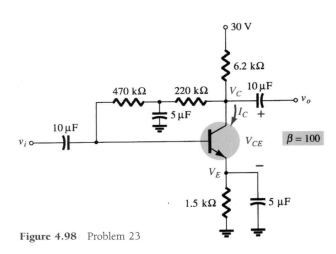

Figure 4.98 Problem 23

*24. (a) Determine the level of I_C and V_{CE} for the network of Fig. 4.99.
   (b) Change β to 135 (50% increase), and calculate the new levels of I_C and V_{CE}.
   (c) Determine the magnitude of the percent change in I_C and V_{CE} using the following equations:

$$\%\Delta I_C = \left| \frac{I_{C_{(part\,b)}} - I_{C_{(part\,a)}}}{I_{C_{(part\,a)}}} \right| \times 100\%, \qquad \%\Delta V_{CE} = \left| \frac{V_{CE_{(part\,b)}} - V_{CE_{(part\,a)}}}{V_{CE_{(part\,a)}}} \right| \times 100\%$$

   (d) Compare the results of part (c) with those of Problems 11(c), 11(f), and 20(c). How does the collector-feedback network stack up against the other configurations in sensitivity to changes in β?

25. Determine the range of possible values for V_C for the network of Fig. 4.100 using the 1-MΩ potentiometer.

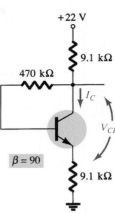

Figure 4.99 Problem 24

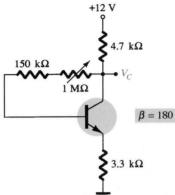

Figure 4.100 Problem 25

*26. Given $V_B = 4$ V for the network of Fig. 4.101, determine:
   (a) V_E.
   (b) I_C.
   (c) V_C.
   (d) V_{CE}.
   (e) I_B.
   (f) β.

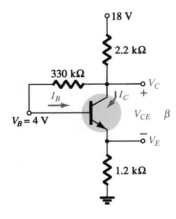

Figure 4.101 Problem 26

§ **4.7 Miscellaneous Bias Configurations**

27. Given $V_C = 8$ V for the network of Fig. 4.102, determine:
   (a) I_B.
   (b) I_C.
   (c) β.
   (d) V_{CE}.

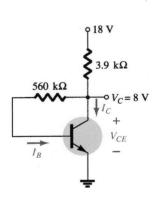

Figure 4.102 Problem 27

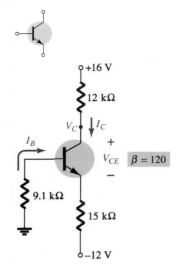

Figure 4.103 Problem 28

*28. For the network of Fig. 4.103, determine:
    (a) I_B.
    (b) I_C.
    (c) V_{CE}.
    (d) V_C.

*29. For the network of Fig. 4.104, determine:
    (a) I_B.
    (b) I_C.
    (c) V_E.
    (d) V_{CE}.

*30. Determine the level of V_E and I_E for the network of Fig. 4.105.

*31. For the network of Fig. 4.106, determine:
    (a) I_E.
    (b) V_C.
    (c) V_{CE}.

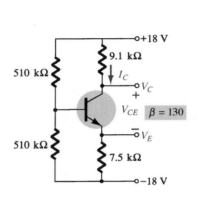

Figure 4.104 Problem 29

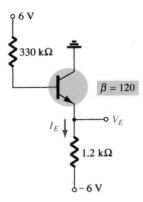

Figure 4.105 Problem 30

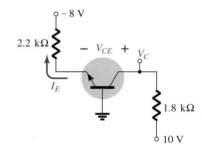

Figure 4.106 Problem 31

§ 4.8 Design Operations

32. Determine R_C and R_B for a fixed-bias configuration if $V_{CC} = 12$ V, $\beta = 80$, and $I_{C_Q} = 2.5$ mA with $V_{CE_Q} = 6$ V. Use standard values.

33. Design an emitter-stabilized network at $I_{C_Q} = \frac{1}{2}I_{C_{sat}}$ and $V_{CE_Q} = \frac{1}{2}V_{CC}$. Use $V_{CC} = 20$ V, $I_{C_{sat}} = 10$ mA, $\beta = 120$, and $R_C = 4R_E$. Use standard values.

34. Design a voltage-divider bias network using a supply of 24 V, a transistor with a beta of 110, and an operating point of $I_{C_Q} = 4$ mA and $V_{CE_Q} = 8$ V. Choose $V_E = \frac{1}{8}V_{CC}$. Use standard values.

*35. Using the characteristics of Fig. 4.89, design a voltage-divider configuration to have a saturation level of 10 mA and a Q-point one-half the distance between cutoff and saturation. The available supply is 28 V, and V_E is to be one-fifth of V_{CC}. The condition established by Eq. (4.33) should also be met to provide a high stability factor. Use standard values.

§ 4.9 Transistor Switching Networks

*36. Using the characteristics of Fig. 4.89, determine the appearance of the output waveform for the network of Fig. 4.107. Include the effects of $V_{CE_{sat}}$, and determine I_B, $I_{B_{max}}$, and $I_{C_{sat}}$ when $V_i = 10$ V. Determine the collector-to-emitter resistance at saturation and cutoff.

*37. Design the transistor inverter of Fig. 4.108 to operate with a saturation current of 8 mA using a transistor with a beta of 100. Use a level of I_B equal to 120% of $I_{B_{max}}$ and standard resistor values.

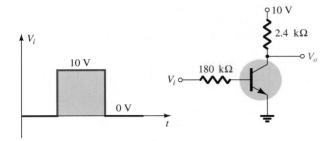

Figure 4.107   Problem 36

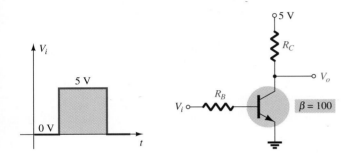

Figure 4.108   Problem 37

38. (a) Using the characteristics of Fig. 3.23c, determine t_{on} and t_{off} at a current of 2 mA. Note the use of log scales and the possible need to refer to Section 11.2.
   (b) Repeat part (a) at a current of 10 mA. How have t_{on} and t_{off} changed with increase in collector current?
   (c) For parts (a) and (b), sketch the pulse waveform of Fig. 4.57 and compare results.

§ 4.10 Troubleshooting Techniques

*39. The measurements of Fig. 4.109 all reveal that the network is not functioning correctly. List as many reasons as you can for the measurements obtained.

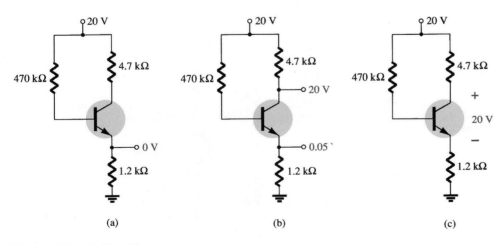

Figure 4.109   Problem 39

*40. The measurements appearing in Fig. 4.110 reveal that the networks are not operating properly. Be specific in describing why the levels obtained reflect a problem with the expected network behavior. In other words, the levels obtained reflect a very specific problem in each case.

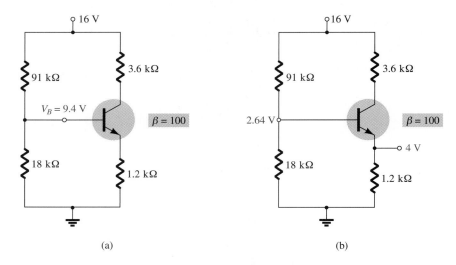

Figure 4.110   Problem 40

41. For the circuit of Fig. 4.111.
   (a) Does V_C increase or decrease if R_B is increased?
   (b) Does I_C increase or decrease if β is reduced?
   (c) What happens to the saturation current if β is increased?
   (d) Does the collector current increase or decrease if V_{CC} is reduced?
   (e) What happens to V_{CE} if the transistor is replaced by one with smaller β?

42. Answer the following questions about the circuit of Fig. 4.112.
   (a) What happens to the voltage V_C if the transistor is replaced by one having a larger value of β?
   (b) What happens to the voltage V_{CE} if the ground leg of resistor R_{B_2} opens (does not connect to ground)?
   (c) What happens to I_C if the supply voltage is low?
   (d) What voltage V_{CE} would occur if the transistor base–emitter junction fails by becoming open?
   (e) What voltage V_{CE} would result if the transistor base–emitter junction fails by becoming a short?

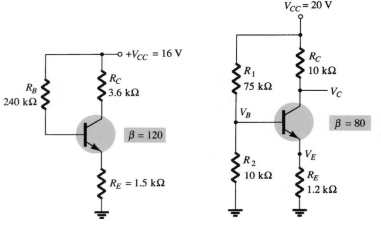

Figure 4.111   Problem 41   Figure 4.112   Problem 42

*43. Answer the following questions about the circuit of Fig. 4.113.
  (a) What happens to the voltage V_C if the resistor R_B is open?
  (b) What should happen to V_{CE} if β increases due to temperature?
  (c) How will V_E be affected when replacing the collector resistor with one whose resistance is at the lower end of the tolerance range?
  (d) If the transistor collector connection becomes open, what will happen to V_E?
  (e) What might cause V_{CE} to become nearly 18 V?

§ 4.11 *PNP* Transistors

44. Determine V_C, V_{CE}, and I_C for the network of Fig. 4.114.

45. Determine V_C and I_B for the network of Fig. 4.115.

46. Determine I_E and V_C for the network of Fig. 4.116.

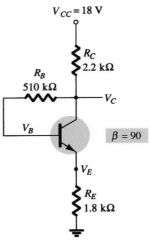

Figure 4.113 Problem 43

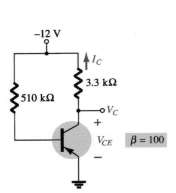

Figure 4.114 Problem 44

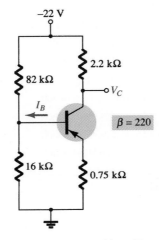

Figure 4.115 Problem 45

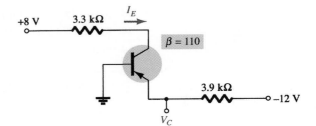

Figure 4.116 Problem 46

§ 4.12 Bias Stabilization

47. Determine the following for the network of Fig. 4.86.
  (a) $S(I_{CO})$.
  (b) $S(V_{BE})$.
  (c) $S(\beta)$ using T_1 as the temperature at which the parameter values are specified and $\beta(T_2)$ as 25% more than $\beta(T_1)$.
  (d) Determine the net change in I_C if a change in operating conditions results in I_{CO} increasing from 0.2 to 10 μA, V_{BE} drops from 0.7 to 0.5 V, and β increases 25%.

*48. For the network of Fig. 4.90, determine:
  (a) $S(I_{CO})$.
  (b) $S(V_{BE})$.
  (c) $S(\beta)$ using T_1 as the temperature at which the parameter values are specified and $\beta(T_2)$ as 25% more than $\beta(T_1)$.
  (d) Determine the net change in I_C if a change in operating conditions results in I_{CO} increasing from 0.2 to 10 μA, V_{BE} drops from 0.7 to 0.5 V, and β increases 25%.

*49. For the network of Fig. 4.93, determine:
  (a) $S(I_{CO})$.
  (b) $S(V_{BE})$.
  (c) $S(\beta)$ using T_1 as the temperature at which the parameter values are specified and $\beta(T_2)$ as 25% more than $\beta(T_1)$.
  (d) Determine the net change in I_C if a change in operating conditions results in I_{CO} increasing from 0.2 to 10 μA, V_{BE} drops from 0.7 to 0.5 V, and β increases 25%.

*50. For the network of Fig. 4.102, determine:
   (a) $S(I_{CO})$.
   (b) $S(V_{BE})$.
   (c) $S(\beta)$ using T_1 as the temperature at which the parameter values are specified and $\beta(T_2)$ as 25% more than $\beta(T_1)$.
   (d) Determine the net change in I_C if a change in operating conditions results in I_{CO} increasing from 0.2 to 10 μA, V_{BE} drops from 0.7 to 0.5 V, and β increases 25%.

*51. Compare the relative values of stability for Problems 47 through 50. The results for Exercises 47 and 49 can be found in Appendix E. Can any general conclusions be derived from the results?

*52. (a) Compare the levels of stability for the fixed-bias configuration of Problem 47.
   (b) Compare the levels of stability for the voltage-divider configuration of Problem 49.
   (c) Which factors of parts (a) and (b) seem to have the most influence on the stability of the system, or is there no general pattern to the results?

§ 4.15 Computer Analysis

53. Perform a PSpice analysis of the network of Fig. 4.86. That is, determine I_C, V_{CE}, and I_B.

54. Repeat Problem 53 for the network of Fig. 4.90.

55. Repeat Problem 53 for the network of Fig. 4.93.

56. Repeat Problem 53 for the network of Fig. 4.97.

57. Repeat Problem 53 using Electronics Workbench.

58. Repeat Problem 54 using Electronics Workbench.

59. Repeat Problem 55 using Electronics Workbench.

60. Repeat Problem 56 using Electronics Workbench.

*Please Note: Asterisks indicate more difficult problems.

5.1 INTRODUCTION

The field-effect transistor (FET) is a three-terminal device used for a variety of applications that match, to a large extent, those of the BJT transistor described in Chapters 3 and 4. Although there are important differences between the two types of devices, there are also many similarities that will be pointed out in the sections to follow.

The primary difference between the two types of transistors is the fact that the BJT transistor is a *current-controlled* device as depicted in Fig. 5.1a, while the JFET transistor is a *voltage-controlled* device as shown in Fig. 5.1b. In other words, the current I_C in Fig. 5.1a is a direct function of the level of I_B. For the FET the current I will be a function of the voltage V_{GS} applied to the input circuit as shown in Fig. 5.1b. In each case the current of the output circuit is being controlled by a parameter of the input circuit—in one case a current level and in the other an applied voltage.

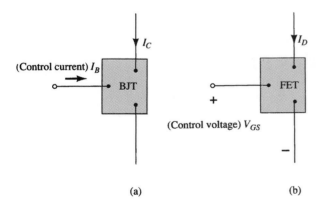

Figure 5.1 (a) Current-controlled and (b) voltage-controlled amplifiers.

Just as there are *npn* and *pnp* bipolar transistors, there are *n-channel* and *p-channel* field-effect transistors. However, it is important to keep in mind that the BJT transistor is a *bipolar* device—the prefix *bi-* revealing that the conduction level is a function of two charge carriers, electrons and holes. The FET is a *unipolar* device depending solely on either electron (*n*-channel) or hole (*p*-channel) conduction.

The term field-effect in the chosen name deserves some explanation. We are all familiar with the ability of a permanent magnet to draw metal filings to the magnet without the need for actual contact. The magnetic field of the permanent magnet has enveloped the filings and attracted them to the magnet through an effort on the part of the magnetic flux lines to be as short as possible. For the FET an *electric field* is established by the charges present that will control the conduction path of the output circuit without the need for direct contact between the controlling and controlled quantities.

There is a natural tendency when introducing a second device with a range of applications similar to one already introduced to compare some of the general characteristics of one versus the other. One of the most important characteristics of the FET is its *high input impedance*. At a level of 1 to several hundred megohms it far exceeds the typical input resistance levels of the BJT transistor configurations—a very important characteristic in the design of linear ac amplifier systems. On the other hand, the BJT transistor has a much higher sensitivity to changes in the applied signal. In other words, the variation in output current is typically a great deal more for BJTs than FETs for the same change in applied voltage. For this reason, typical ac voltage gains for BJT amplifiers are a great deal more than for FETs. In general, FETs are more temperature stable than BJTs, and FETs are usually smaller in construction than BJTs, making them particularly useful in *integrated-circuit (IC)* chips. The construction characteristics of some FETs, however, can make them more sensitive to handling than BJTs.

Two types of FETs will be introduced in this chapter: the *junction field-effect transistor* (JFET) and the *metal-oxide-semiconductor field-effect transistor (MOSFET)*. The MOSFET category is further broken down into depletion and enhancement types, which are both described. The MOSFET transistor has become one of the most important devices used in the design and construction of integrated circuits for digital computers. Its thermal stability and other general characteristics make it extremely popular in computer circuit design. However, as a discrete element in a typical top-hat container, it must be handled with care (to be discussed in a later section).

Once the FET construction and characteristics have been introduced, the biasing arrangements will be covered in Chapter 6. The analysis performed in Chapter 4 using BJT transistors will prove helpful in the derivation of the important equations and understanding the results obtained for FET circuits.

Drs. Ian Munro Ross (front) and G. C. Dacey jointly developed an experimental procedure for measuring the characteristics of a field-effect transistor in 1955. (Courtesy of AT&T Archives.)

Dr. Ross   Born: Southport, England
PhD Gonville and Caius College, Cambridge University President emeritus of AT&T Bell Labs Fellow—IEEE, Member of the National Science Board Chairman—National Advisory Committee on Semiconductors

Dr. Dacey   Born: Chicago, Illinois
PhD California Institute of Technology Director of Solid-State Electronics Research at Bell Labs Vice President, Research at Sandia Corporation Member IRE, Tau Beta Pi, Eta Kappa Nu

5.2   CONSTRUCTION AND CHARACTERISTICS OF JFETs

As indicated earlier, the JFET is a three-terminal device with one terminal capable of controlling the current between the other two. In our discussion of the BJT transistor the *npn* transistor was employed through the major part of the analysis and design sections, with a section devoted to the impact of using a *pnp* transistor. For the JFET transistor the *n*-channel device will appear as the prominent device, with paragraphs and sections devoted to the impact of using a *p*-channel JFET.

The basic construction of the *n*-channel JFET is shown in Fig. 5.2. Note that the major part of the structure is the *n*-type material that forms the channel between the embedded layers of *p*-type material. The top of the *n*-type channel is connected through an ohmic contact to a terminal referred to as the *drain (D)*, while the lower end of the same material is connected through an ohmic contact to a terminal referred to as the *source (S)*. The two *p*-type materials are connected together and to the *gate (G)* terminal. In essence, therefore, the drain and source are connected to the ends of the *n*-type channel and the gate to the two layers of *p*-type material. In the absence of any applied potentials the JFET has two *p-n* junctions under no-bias conditions. The result is a depletion region at each junction as shown in Fig. 5.2 that resembles the same region of a diode under no-bias conditions. Recall also that a depletion region is that region void of free carriers and therefore unable to support conduction through the region.

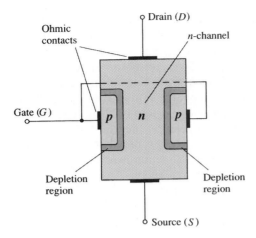

Figure 5.2 Junction field-effect transistor (JFET).

Analogies are seldom perfect and at times can be misleading, but the water analogy of Fig. 5.3 does provide a sense for the JFET control at the gate terminal and the appropriateness of the terminology applied to the terminals of the device. The source of water pressure can be likened to the applied voltage from drain to source that will establish a flow of water (electrons) from the spigot (source). The "gate," through an applied signal (potential), controls the flow of water (charge) to the "drain." The drain and source terminals are at opposite ends of the n-channel as introduced in Fig. 5.2 because the terminology is defined for electron flow.

Figure 5.3 Water analogy for the JFET control mechanism.

$V_{GS} = 0$ V, V_{DS} Some Positive Value

In Fig. 5.4, a positive voltage V_{DS} has been applied across the channel and the gate has been connected directly to the source to establish the condition $V_{GS} = 0$ V. The result is a gate and source terminal at the same potential and a depletion region in the low end of each p-material similar to the distribution of the no-bias conditions of Fig. 5.2. The instant the voltage V_{DD} ($= V_{DS}$) is applied, the electrons will be drawn to the drain terminal, establishing the conventional current I_D with the defined direction of Fig. 5.4. The path of charge flow clearly reveals that the drain and source currents are equivalent ($I_D = I_S$). Under the conditions appearing in Fig. 5.4, the flow of charge is relatively uninhibited and limited solely by the resistance of the n-channel between drain and source.

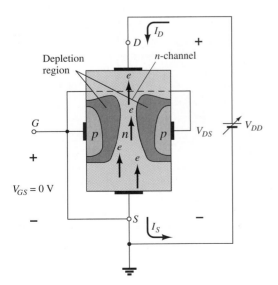

Figure 5.4 JFET in the $V_{GS} = 0$ V and $V_{DS} > 0$ V.

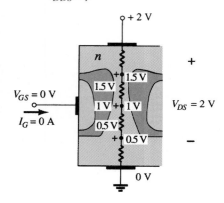

I_{DDS}/V_P

Figure 5.5 Varying reverse-bias potentials across the *p-n* junction of an *n*-channel JFET.

It is important to note that the depletion region is wider near the top of both *p*-type materials. The reason for the change in width of the region is best described through the help of Fig. 5.5. Assuming a uniform resistance in the *n*-channel, the resistance of the channel can be broken down to the divisions appearing in Fig. 5.5. The current I_D will establish the voltage levels through the channel as indicated on the same figure. The result is that the upper region of the *p*-type material will be reverse-biased by about 1.5 V, with the lower region only reverse-biased by 0.5 V. Recall from the discussion of the diode operation that the greater the applied reverse bias, the wider the depletion region—hence the distribution of the depletion region as shown in Fig. 5.5. The fact that the *p-n* junction is reverse-biased for the length of the channel results in a gate current of zero amperes as shown in the same figure. The fact that $I_G = 0$ A is an important characteristic of the JFET.

As the voltage V_{DS} is increased from 0 to a few volts, the current will increase as determined by Ohm's law and the plot of I_D versus V_{DS} will appear as shown in Fig. 5.6. The relative straightness of the plot reveals that for the region of low values of V_{DS}, the resistance is essentially constant. As V_{DS} increases and approaches a level referred to as V_P in Fig. 5.6, the depletion regions of Fig. 5.4 will widen, causing a noticeable reduction in the channel width. The reduced path of conduction causes the resistance to increase and the curve in the graph of Fig. 5.6 to occur. The more horizontal the curve, the higher the resistance, suggesting that the resistance is approaching "infinite" ohms in the horizontal region. If V_{DS} is increased to a level where it appears that the two depletion regions would "touch" as shown in Fig. 5.7, a condition referred to as *pinch-off* will result. The level of V_{DS} that establishes this condition is referred to as the *pinch-off voltage* and is denoted by V_P as shown in Fig. 5.6. In actuality, the term *pinch-off* is a misnomer in that it suggests the current I_D is pinched off and drops to 0 A. As shown in Fig. 5.6, however, this is hardly the case—I_D maintains a saturation level defined as I_{DSS} in Fig. 5.6. In reality a very small channel still exists, with a current of very high density. The fact that I_D does not drop off at pinch-off and maintains the saturation level indicated in Fig. 5.6 is verified by the following fact: The absence of a drain current would remove the possibility of different potential levels through the *n*-channel material to establish the varying levels of reverse bias along the *p-n* junction. The result would be a loss of the depletion region distribution that caused pinch-off in the first place.

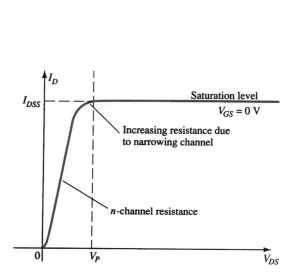

Figure 5.6 I_D versus V_{DS} for $V_{GS} = 0$ V.

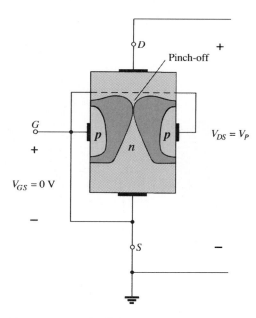

Figure 5.7 Pinch-off ($V_{GS} = 0$ V, $V_{DS} = V_P$).

As V_{DS} is increased beyond V_P, the region of close encounter between the two depletion regions will increase in length along the channel, but the level of I_D remains essentially the same. In essence, therefore, once $V_{DS} > V_P$ the JFET has the characteristics of a current source. As shown in Fig. 5.8, the current is fixed at $I_D = I_{DSS}$, but the voltage V_{DS} (for levels $> V_P$) is determined by the applied load.

The choice of notation I_{DSS} is derived from the fact that it is the *D*rain-to-*S*ource current with a *S*hort-circuit connection from gate to source. As we continue to investigate the characteristics of the device we will find that:

I_{DSS} is the maximum drain current for a JFET and is defined by the conditions $V_{GS} = 0\ V$ and $V_{DS} > |V_P|$.

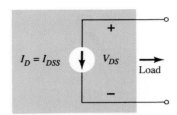

Figure 5.8 Current source equivalent for $V_{GS} = 0$ V, $V_{DS} > V_P$.

Note in Fig. 5.6 that $V_{GS} = 0$ V for the entire length of the curve. The next few paragraphs will describe how the characteristics of Fig. 5.6 are affected by changes in the level of V_{GS}.

$V_{GS} < 0$ V

The voltage from gate to source, denoted V_{GS}, is the controlling voltage of the JFET. Just as various curves for I_C versus V_{CE} were established for different levels of I_B for the BJT transistor, curves of I_D versus V_{DS} for various levels of V_{GS} can be developed for the JFET. For the *n*-channel device the controlling voltage V_{GS} is made more and more negative from its $V_{GS} = 0$ V level. In other words, the gate terminal will be set at lower and lower potential levels as compared to the source.

In Fig. 5.9 a negative voltage of -1 V has been applied between the gate and source terminals for a low level of V_{DS}. The effect of the applied negative-bias V_{GS} is to establish depletion regions similar to those obtained with $V_{GS} = 0$ V but at lower levels of V_{DS}. Therefore, the result of applying a negative bias to the gate is to reach the saturation level at a lower level of V_{DS} as shown in Fig. 5.10 for $V_{GS} = -1$ V. The resulting saturation level for I_D has been reduced and in fact will continue to decrease as V_{GS} is made more and more negative. Note also on Fig. 5.10 how the pinchoff voltage continues to drop in a parabolic manner as V_{GS} becomes more and more negative. Eventually, V_{GS} when $V_{GS} = -V_P$ will be sufficiently negative to establish a saturation level that is essentially 0 mA, and for all practical purposes the device has been "turned off." In summary:

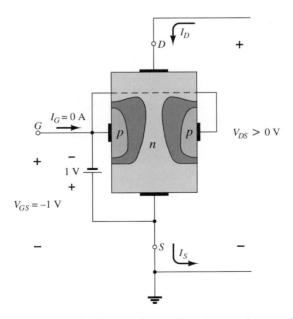

Figure 5.9 Application of a negative voltage to the gate of a JFET.

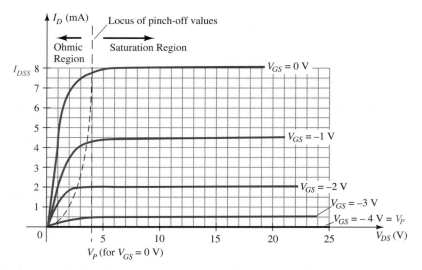

Figure 5.10 n-Channel JFET characteristics with $I_{DSS} = 8$ mA and $V_P = -4$ V.

The level of V_{GS} that results in $I_D = 0$ mA is defined by $V_{GS} = V_P$, with V_P being a negative voltage for n-channel devices and a positive voltage for p-channel JFETs.

On most specification sheets the pinch-off voltage is specified as $V_{GS(off)}$ rather than V_P. A specification sheet will be reviewed later in the chapter when the primary elements of concern have been introduced. The region to the right of the pinch-off locus of Fig. 5.10 is the region typically employed in linear amplifiers (amplifiers with minimum distortion of the applied signal) and is commonly referred to as the *constant-current, saturation,* or *linear amplification region.*

Voltage-Controlled Resistor

The region to the left of the pinch-off locus of Fig. 5.10 is referred to as the *ohmic* or *voltage-controlled resistance region.* In this region the JFET can actually be employed as a variable resistor (possibly for an automatic gain control system) whose resistance is controlled by the applied gate-to-source voltage. Note in Fig. 5.10 that the slope of each curve and therefore the resistance of the device between drain and source for $V_{DS} < V_P$ is a function of the applied voltage V_{GS}. As V_{GS} becomes more and more negative, the slope of each curve becomes more and more horizontal, corresponding with an increasing resistance level. The following equation will provide a good first approximation to the resistance level in terms of the applied voltage V_{GS}.

$$r_d = \frac{r_o}{(1 - V_{GS}/V_P)^2} \tag{5.1}$$

where r_o is the resistance with $V_{GS} = 0$ V and r_d the resistance at a particular level of V_{GS}.

For an n-channel JFET with r_o equal to 10 kΩ ($V_{GS} = 0$ V, $V_P = -6$ V), Eq. (5.1) will result in 40 kΩ at $V_{GS} = -3$ V.

p-Channel Devices

The p-channel JFET is constructed in exactly the same manner as the n-channel device of Fig. 5.2, but with a reversal of the p- and n-type materials as shown in Fig. 5.11.

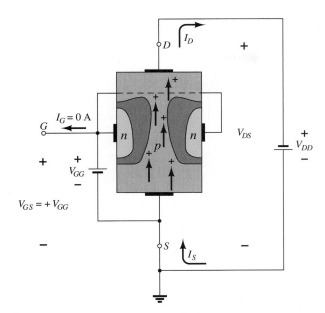

Figure 5.11 *p*-Channel JFET.

The defined current directions are reversed, as are the actual polarities for the voltages V_{GS} and V_{DS}. For the *p*-channel device, the channel will be constricted by increasing positive voltages from gate to source and the double-subscript notation for V_{DS} will result in negative voltages for V_{DS} on the characteristics of Fig. 5.12, which has an I_{DSS} of 6 mA and a pinch-off voltage of $V_{GS} = +6$ V. Do not let the minus signs for V_{DS} confuse you. They simply indicate that the source is at a higher potential than the drain.

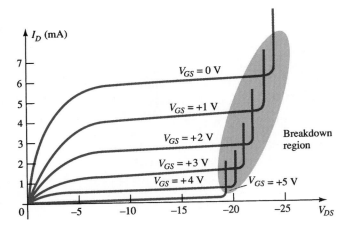

Figure 5.12 *p*-Channel JFET characteristics with $I_{DSS} = 6$ mA and $V_P = +6$ V.

Note at high levels of V_{DS} that the curves suddenly rise to levels that seem unbounded. The vertical rise is an indication that breakdown has occurred and the current through the channel (in the same direction as normally encountered) is now limited solely by the external circuit. Although not appearing in Fig. 5.10 for the *n*-channel device, they do occur for the *n*-channel device if sufficient voltage is applied. This region can be avoided if the level of $V_{DS_{max}}$ is noted on the specification sheet and the design is such that the actual level of V_{DS} is less than this value for *all* values of V_{GS}.

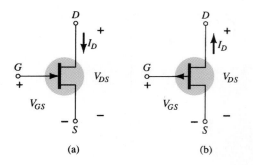

Figure 5.13 JFET symbols: (a) *n*-channel; (b) *p*-channel.

Symbols

The graphic symbols for the *n*-channel and *p*-channel JFETs are provided in Fig. 5.13. Note that the arrow is pointing in for the *n*-channel device of Fig. 5.13a to represent the direction in which I_G would flow if the *p-n* junction were forward-biased. For the *p*-channel device (Fig. 5.13b) the only difference in the symbol is the direction of the arrow.

Summary

A number of important parameters and relationships were introduced in this section. A few that will surface frequently in the analysis to follow in this chapter and the next for *n*-channel JFETs include the following:

The maximum current is defined as I_{DSS} and occurs when $V_{GS} = 0$ V and $V_{DS} \geq |V_P|$ as shown in Fig. 5.14a.

For gate-to-source voltages V_{GS} less than (more negative than) the pinch-off level, the drain current is 0 A ($I_D = 0$ A) as appearing in Fig. 5.14b.

For all levels of V_{GS} between 0 V and the pinch-off level, the current I_D will range between I_{DSS} and 0 A, respectively, as reviewed by Fig. 5.14c.

For p-channel JFETs a similar list can be developed.

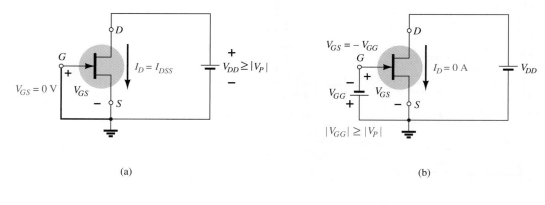

(a)   (b)

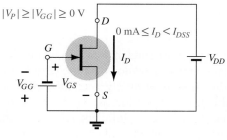

(c)

Figure 5.14 (a) $V_{GS} = 0$ V, $I_D = I_{DSS}$; (b) cutoff ($I_D = 0$ A) V_{GS} less than the pinch-off level; (c) I_D exists between 0 A and I_{DSS} for V_{GS} less than or equal to 0 V and greater than the pinch-off level.

5.3 TRANSFER CHARACTERISTICS

Derivation

For the BJT transistor the output current I_C and input controlling current I_B were related by beta, which was considered constant for the analysis to be performed. In equation form,

$$I_C = f(I_B) = \beta I_B \qquad \text{(5.2)}$$

control variable

constant

In Eq. (5.2) a linear relationship exists between I_C and I_B. Double the level of I_B and I_C will increase by a factor of two also.

Unfortunately, this linear relationship does not exist between the output and input quantities of a JFET. The relationship between I_D and V_{GS} is defined by *Shockley's equation:*

$$I_D = I_{DSS}\left(1 - \frac{V_{GS}}{V_P}\right)^2 \qquad \text{(5.3)}$$

control variable

constants

The squared term of the equation will result in a nonlinear relationship between I_D and V_{GS}, producing a curve that grows exponentially with decreasing magnitudes of V_{GS}.

For the dc analysis to be performed in Chapter 6, a graphical rather than mathematical approach will in general be more direct and easier to apply. The graphical approach, however, will require a plot of Eq. (5.3) to represent the device and a plot of the network equation relating the same variables. The solution is defined by the point of intersection of the two curves. It is important to keep in mind when applying the graphical approach that the device characteristics will be *unaffected* by the network in which the device is employed. The network equation may change along with the intersection between the two curves, but the transfer curve defined by Eq. (5.3) is unaffected. In general, therefore:

The transfer characteristics defined by Shockley's equation are unaffected by the network in which the device is employed.

The transfer curve can be obtained using Shockley's equation or from the output characteristics of Fig. 5.10. In Fig. 5.15 two graphs are provided, with the vertical

William Bradford Shockley (1910–1989), co-inventor of the first transistor and formulator of the "field-effect" theory employed in the development of the transistor and FET. (Courtesy of AT&T Archives.)

Born: London, England
PhD Harvard, 1936
Head, Transistor Physics Department–Bell Laboratories
President, Shockley Transistor Corp.
Poniatoff Professor of Engineering Science at Stanford University
Nobel Prize in physics in 1956 with Drs. Brattain and Bardeen

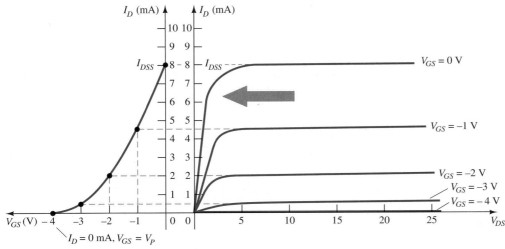

Figure 5.15 Obtaining the transfer curve from the drain characteristics.

scaling in milliamperes for each graph. One is a plot of I_D versus V_{DS}, while the other is I_D versus V_{GS}. Using the drain characteristics on the right of the "y" axis, a horizontal line can be drawn from the saturation region of the curve denoted $V_{GS} = 0$ V to the I_D axis. The resulting current level for both graphs is I_{DSS}. The point of intersection on the I_D versus V_{GS} curve will be as shown since the vertical axis is defined as $V_{GS} = 0$ V.

In review:

When $V_{GS} = 0$ V, $I_D = I_{DSS}$.

When $V_{GS} = V_P = -4$ V, the drain current is zero milliamperes, defining another point on the transfer curve. That is:

When $V_{GS} = V_P$, $I_D = 0$ mA.

Before continuing, it is important to realize that the drain characteristics relate one output (or drain) quantity to another output (or drain) quantity—both axes are defined by variables in the same region of the device characteristics. The transfer characteristics are a plot of an output (or drain) current versus an input-controlling quantity. There is therefore a direct "transfer" from input to output variables when employing the curve to the left of Fig. 5.15. If the relationship were linear, the plot of I_D versus V_{GS} would result in a straight line between I_{DSS} and V_P. However, a parabolic curve will result because the vertical spacing between steps of V_{GS} on the drain characteristics of Fig. 5.15 decreases noticeably as V_{GS} becomes more and more negative. Compare the spacing between $V_{GS} = 0$ V and $V_{GS} = -1$ V to that between $V_{GS} = -3$ V and pinch-off. The change in V_{GS} is the same, but the resulting change in I_D is quite different.

If a horizontal line is drawn from the $V_{GS} = -1$ V curve to the I_D axis and then extended to the other axis, another point on the transfer curve can be located. Note that $V_{GS} = -1$ V on the bottom axis of the transfer curve with $I_D = 4.5$ mA. Note in the definition of I_D at $V_{GS} = 0$ V and -1 V that the saturation levels of I_D are employed and the ohmic region ignored. Continuing with $V_{GS} = -2$ V and -3 V, the transfer curve can be completed. It is the transfer curve of I_D versus V_{GS} that will receive extended use in the analysis of Chapter 6 and not the drain characteristics of Fig. 5.15. The next few paragraphs will introduce a quick, efficient method of plotting I_D versus V_{GS} given only the levels of I_{DSS} and V_P and Shockley's equation.

Applying Shockley's Equation

The transfer curve of Fig. 5.15 can also be obtained directly from Shockley's equation (5.3) given simply the values of I_{DSS} and V_P. The levels of I_{DSS} and V_P define the limits of the curve on both axes and leave only the necessity of finding a few intermediate plot points. The validity of Eq. (5.3) as a source of the transfer curve of Fig. 5.15 is best demonstrated by examining a few specific levels of one variable and finding the resulting level of the other as follows:

Substituting $V_{GS} = 0$ V gives

$$\text{Eq. (5.3):} \quad I_D = I_{DSS}\left(1 - \frac{V_{GS}}{V_P}\right)^2$$

$$= I_{DSS}\left(1 - \frac{0}{V_P}\right)^2 = I_{DSS}(1 - 0)^2$$

and
$$\boxed{I_D = I_{DSS}\,|\,_{V_{GS}=0\,\text{V}}} \tag{5.4}$$

Substituting $V_{GS} = V_P$ yields

$$I_D = I_{DSS}\left(1 - \frac{V_P}{V_P}\right)^2$$

$$= I_{DSS}(1 - 1)^2 = I_{DSS}(0)$$

$$\boxed{I_D = 0\,A|_{V_{GS}=V_P}} \tag{5.5}$$

For the drain characteristics of Fig. 5.15, if we substitute $V_{GS} = -1\,V$,

$$I_D = I_{DSS}\left(1 - \frac{V_{GS}}{V_P}\right)^2$$

$$= 8\,mA\left(1 - \frac{-1\,V}{-4\,V}\right)^2 = 8\,mA\left(1 - \frac{1}{4}\right)^2 = 8\,mA(0.75)^2$$

$$= 8\,mA(0.5625)$$

$$= \mathbf{4.5\,mA}$$

as shown in Fig. 5.15. Note the care taken with the negative signs for V_{GS} and V_P in the calculations above. The loss of one sign would result in a totally erroneous result.

It should be obvious from the above that given I_{DSS} and V_P (as is normally provided on specification sheets) the level of I_D can be found for any level of V_{GS}. Conversely, by using basic algebra we can obtain [from Eq. (5.3)] an equation for the resulting level of V_{GS} for a given level of I_D. The derivation is quite straightforward and will result in

$$\boxed{V_{GS} = V_P\left(1 - \sqrt{\frac{I_D}{I_{DSS}}}\right)} \tag{5.6}$$

Let us test Eq. (5.6) by finding the level of V_{GS} that will result in a drain current of 4.5 mA for the device with the characteristics of Fig. 5.15.

$$V_{GS} = -4\,V\left(1 - \sqrt{\frac{4.5\,mA}{8\,mA}}\right)$$

$$= -4\,V(1 - \sqrt{0.5625}) = -4\,V(1 - 0.75)$$

$$= -4\,V(0.25)$$

$$= \mathbf{-1\,V}$$

as substituted in the above calculation and verified by Fig. 5.15.

Shorthand Method

Since the transfer curve must be plotted so frequently, it would be quite advantageous to have a shorthand method for plotting the curve in the quickest, most efficient manner while maintaining an acceptable degree of accuracy. The format of Eq. (5.3) is such that specific levels of V_{GS} will result in levels of I_D that can be memorized to provide the plot points needed to sketch the transfer curve. If we specify V_{GS} to be one-half the pinch-off value V_P, the resulting level of I_D will be the following, as determined by Shockley's equation:

$$I_D = I_{DSS}\left(1 - \frac{V_{GS}}{V_P}\right)^2$$

$$= I_{DSS}\left(\frac{1 - V_P/2}{V_P}\right)^2 = I_{DSS}\left(1 - \frac{1}{2}\right)^2 = I_{DSS}(0.5)^2$$

$$= I_{DSS}(0.25)$$

and
$$I_D = \frac{I_{DSS}}{4}\bigg|_{V_{GS}=V_P/2} \tag{5.7}$$

Now it is important to realize that Eq. (5.7) is not for a particular level of V_P. It is a general equation for any level of V_P as long as $V_{GS} = V_P/2$. The result specifies that the drain current will always be one-fourth of the saturation level I_{DSS} as long as the gate-to-source voltage is one-half the pinch-off value. Note the level of I_D for $V_{GS} = V_P/2 = -4\,\text{V}/2 = -2\,\text{V}$ in Fig. 5.15.

If we choose $I_D = I_{DSS}/2$ and substitute into Eq. (5.6), we find that

$$V_{GS} = V_P\left(1 - \sqrt{\frac{I_D}{I_{DSS}}}\right)$$
$$= V_P\left(1 - \sqrt{\frac{I_{DSS}/2}{I_{DSS}}}\right) = V_P(1 - \sqrt{0.5}) = V_P(0.293)$$

and
$$V_{GS} \cong 0.3V_P\big|_{I_D=I_{DSS}/2} \tag{5.8}$$

Additional points can be determined, but the transfer curve can be sketched to a satisfactory level of accuracy simply using the four plot points defined above and reviewed in Table 5.1. In fact, in the analysis of Chapter 6, a maximum of four plot points are used to sketch the transfer curves. On most occasions using just the plot point defined by $V_{GS} = V_P/2$ and the axis intersections at I_{DSS} and V_P will provide a curve accurate enough for most calculations.

TABLE 5.1 V_{GS} versus I_D Using Shockley's Equation

V_{GS}	I_D
0	I_{DSS}
$0.3\,V_P$	$I_{DSS}/2$
$0.5\,V_P$	$I_{DSS}/4$
V_P	$0\,mA$

EXAMPLE 5.1   Sketch the transfer curve defined by $I_{DSS} = 12\,\text{mA}$ and $V_P = -6\,\text{V}$.

Solution

Two plot points are defined by

$$I_{DSS} = \textbf{12 mA} \qquad \text{and} \qquad V_{GS} = \textbf{0 V}$$

and
$$I_D = \textbf{0 mA} \qquad \text{and} \qquad V_{GS} = V_P$$

At $V_{GS} = V_P/2 = -6\,\text{V}/2 = \textbf{-3 V}$ the drain current will be determined by $I_D = I_{DSS}/4 = 12\,\text{mA}/4 = \textbf{3 mA}$. At $I_D = I_{DSS}/2 = 12\,\text{mA}/2 = \textbf{6 mA}$ the gate-to-source voltage is determined by $V_{GS} \cong 0.3V_P = 0.3(-6\,\text{V}) = \textbf{-1.8 V}$. All four plot points are well defined on Fig. 5.16 with the complete transfer curve.

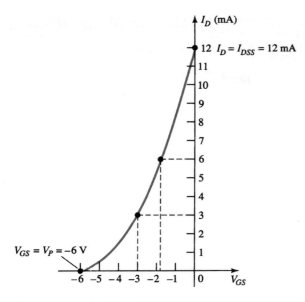

$V_{GS} = V_P = -6$ V

Figure 5.16 Transfer curve for Example 5.1.

For *p*-channel devices Shockley's equation (5.3) can still be applied exactly as it appears. In this case, both V_P and V_{GS} will be positive and the curve will be the mirror image of the transfer curve obtained with an *n*-channel and the same limiting values.

Sketch the transfer curve for a *p*-channel device with $I_{DSS} = 4$ mA and $V_P = 3$ V.

EXAMPLE 5.2

Solution

At $V_{GS} = V_P/2 = 3$ V$/2 = $ **1.5 V,** $I_D = I_{DSS}/4 = 4$ mA$/4 = $ **1 mA.** At $I_D = I_{DSS}/2 = 4$ mA$/2 = $ **2 mA,** $V_{GS} = 0.3V_P = 0.3(3$ V$) = $ **0.9 V.** Both plot points appear in Fig. 5.17 along with the points defined by I_{DSS} and V_P.

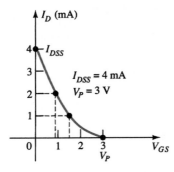

$I_{DSS} = 4$ mA
$V_P = 3$ V

Figure 5.17 Transfer curve for the *p*-channel device of Example 5.2.

Mathcad

Mathcad will now plot Shockley's equation using the **X-Y plot operator.** The plot operator can be selected using the **Graph = X-Y Plot** under the **Insert** option on the menu bar. Press the @ **key** or simply select the **X-Y Plot** button on the **Graph Palette.**

Once the plot is chosen, Mathcad will create a graph with six placeholders—three on each axis as shown in Fig. 5.18. To plot Shockley's equation, first select the place-holder in the middle of the horizontal axis, and enter the horizontal variable **VGS.** Then establish a range for **VGS** by first typing **VGS** followed by a colon and the

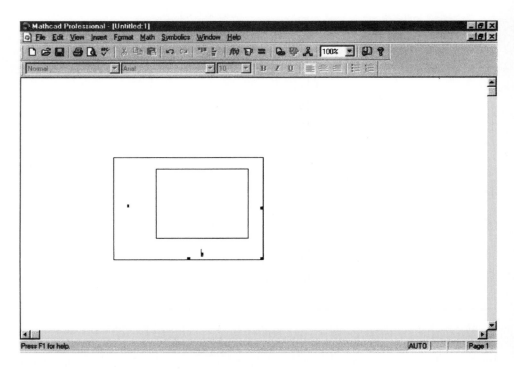

Figure 5.18 The resulting graph when the plotting routing using Mathcad is initiated.

range of values. The range of values is entered by first typing **0** (to represent **VGS = 0V**) followed by a comma and the next value to be substituted in the equation for **ID**. This range also defines the interval between data points for the plot. If the **−0.1** were omitted, Mathcad would have used −1 as the interval, and the plot would have appeared with straight-line segments between data points rather than the smooth curve normally associated with Schockley's equation. For this example the chosen interval is **− 0.1 V;** take careful note of the negative sign since **VGS** is getting more and more negative. Next the semicolon (**;**) key is chosen to tell the computer that a range is being defined. The computer response, however, is a double period, as shown in Fig. 5.19, followed by the last value of the range, the pinch-off voltage of **−4V.**

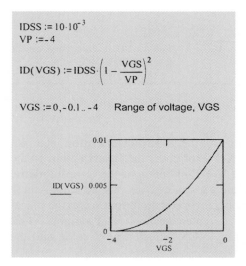

Figure 5.19 Plotting Shockley's equation using Mathcad.

To define the quantity to be graphed, select the placeholder at the middle of the vertical axis, and type **ID(VGS).** The range is also defined as shown in Fig. 5.19. Click anywhere outside the graph, and the plot of Fig. 5.19 will appear.

Since Shockley's equation is plotted so frequently in the dc analysis of JFET networks, it is very useful to have such a quick method to obtain the plot. Simply change the value of **IDSS** and/or **VP**, and the new plot will appear with the single click of the mouse.

5.4 SPECIFICATION SHEETS (JFETs)

Although the general content of specification sheets may vary from the absolute minimum to an extensive display of graphs and charts, there are a few fundamental parameters that will be provided by all manufacturers. A few of the most important are discussed in the following paragraphs. The specification sheet for the 2N5457 *n*-channel JFET as provided by Motorola is provided as Fig. 5.20.

2N5457

CASE 29-04, STYLE 5
TO-92 (TO-226AA)

JFETs
GENERAL PURPOSE
N-CHANNEL—DEPLETION

Refer to 2N4220 for graphs.

MAXIMUM RATINGS

Rating	Symbol	Value	Unit
Drain-Source Voltage	V_{DS}	25	Vdc
Drain-Gate Voltage	V_{DG}	25	Vdc
Reverse Gate-Source Voltage	V_{GSR}	−25	Vdc
Gate Current	I_G	10	mAdc
Total Device Dissipation @ T_A = 25°C Derate above 25°C	P_D	310 2.82	mW mW/°C
Junction Temperature Range	T_J	125	°C
Storage Channel Temperature Range	T_{stg}	−65 to +150	°C

ELECTRICAL CHARACTERISTICS (T_A = 25°C unless otherwise noted)

Characteristic		Symbol	Min	Typ	Max	Unit
OFF CHARACTERISTICS						
Gate-Source Breakdown Voltage (I_G = −10 µAdc, V_{DS} = 0)		$V_{(BR)GSS}$	−25	–	–	Vdc
Gate Reverse Current (V_{GS} = −15 Vdc, V_{DS} = 0) (V_{GS} = −15 Vdc, V_{DS} = 0, T_A = 100°C)		I_{GSS}	– –	– –	−1.0 −200	nAdc
Gate Source Cutoff Voltage (V_{DS} = 15 Vdc, I_D = 10 nAdc)	2N5457	$V_{GS(off)}$	−0.5	–	−6.0	Vdc
Gate Source Voltage (V_{DS} = 15 Vdc, I_D = 100 µAdc)	2N5457	V_{GS}	–	−2.5	–	Vdc
ON CHARACTERISTICS						
Zero-Gate-Voltage Drain Current* (V_{DS} = 15 Vdc, V_{GS} = 0)	2N5457	I_{DSS}	1.0	3.0	5.0	mAdc
SMALL-SIGNAL CHARACTERISTICS						
Forward Transfer Admittance Common Source* (V_{DS} = 15 Vdc, V_{GS} = 0, f = 1.0 kHz)	2N5457	$|y_{fs}|$	1000	–	5000	µmhos
Output Admittance Common Source* (V_{DS} = 15 Vdc, V_{GS} = 0, f = 1.0 kHz)		$|y_{os}|$	–	10	50	µmhos
Input Capacitance (V_{DS} = 15 Vdc, V_{GS} = 0, f = 1.0 MHz)		C_{iss}	–	4.5	7.0	pF
Reverse Transfer Capacitance (V_{DS} = 15 Vdc, V_{GS} = 0, f = 1.0 MHz)		C_{rss}	–	1.5	3.0	pF

*Pulse Test: Pulse Width ≤ 630 ms; Duty Cycle ≤ 10%

Figure 5.20 2N5457 Motorola *n*-channel JFET.

Maximum Ratings

The maximum rating list usually appears at the beginning of the specification sheet, with the maximum voltages between specific terminals, maximum current levels, and the maximum power dissipation level of the device. The specified maximum levels for V_{DS} and V_{DG} must not be exceeded at any point in the design operation of the device. The applied source V_{DD} can exceed these levels, but the actual level of voltage between these terminals must never exceed the level specified. Any good design will try to avoid these levels by a good margin of safety. The term *reverse* in V_{GSR} defines the maximum voltage with the source positive with respect to the gate (as normally biased for an *n*-channel device) before breakdown will occur. On some specification sheets it is referred to as BV_{DSS}—the *Breakdown Voltage* with the *Drain-Source Shorted* ($V_{DS} = 0$ V). Although normally designed to operate with $I_G = 0$ mA, if *forced* to accept a gate current it could withstand 10 mA before damage would occur. The total device dissipation at 25°C (room temperature) is the maximum power the device can dissipate under normal operating conditions and is defined by

$$P_D = V_{DS}I_D \tag{5.9}$$

Note the similarity in format with the maximum power dissipation equation for the BJT transistor.

The derating factor is discussed in detail in Chapter 3, but for the moment recognize that the 2.82 mW/°C rating reveals that the dissipation rating *decreases* by 2.82 mW for each *increase* in temperature of 1°C above 25°C.

Electrical Characteristics

The electrical characteristics include the level of V_P in the OFF CHARACTERISTICS and I_{DSS} in the ON CHARACTERISTICS. In this case $V_P = V_{GS(off)}$ has a range from -0.5 to -6.0 V and I_{DSS} from 1 to 5 mA. The fact that both will vary from device to device with the same nameplate identification must be considered in the design process. The other quantities are defined under conditions appearing in parentheses. The small-signal characteristics are discussed in Chapter 9.

Case Construction and Terminal Identification

This particular JFET has the appearance provided on the specification sheet of Fig. 5.20. The terminal identification is also provided directly under the figure. JFETs are also available in top-hat containers, as shown in Fig. 5.21 with its terminal identification.

2N2844

CASE 22-03, STYLE 12
TO-18 (TO-206AA)

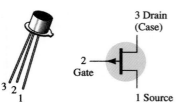

JFETs
GENERAL PURPOSE
P-CHANNEL

Figure 5.21 Top-hat container and terminal identification for a *p*-channel JFET.

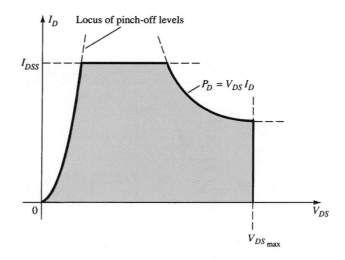

Figure 5.22 Normal operating region for linear amplifier design.

Operating Region

The specification sheet and the curve defined by the pinch-off levels at each level of V_{GS} define the region of operation for linear amplification on the drain characteristics as shown in Fig. 5.22. The ohmic region defines the minimum permissible values of V_{DS} at each level of V_{GS}, and $V_{DS_{max}}$ specifies the maximum value for this parameter. The saturation current I_{DSS} is the maximum drain current, and the maximum power dissipation level defines the curve drawn in the same manner as described for BJT transistors. The resulting shaded region is the normal operating region for amplifier design.

5.5 INSTRUMENTATION

Recall from Chapter 3 that hand-held instruments are available to measure the level of β_{dc} for the BJT transistor. Similar instrumentation is not available to measure the levels of I_{DSS} and V_P. However, the curve tracer introduced for the BJT transistor can also display the drain characteristics of the JFET transistor through a proper setting of the various controls. The vertical scale (in milliamperes) and the horizontal scale (in volts) have been set to provide a full display of the characteristics, as shown in Fig. 5.23. For the JFET of Fig. 5.23, each vertical division (in centimeters) reflects a 1-mA change in I_D while each horizontal division has a value of 1 V. The step voltage is 500 mV/step (0.5 V/step), revealing that the top curve is defined by $V_{GS} = 0$ V and the next curve down -0.5 V for the *n*-channel device. Using the same step voltage the next curve is -1 V, then -1.5 V, and finally -2 V. By drawing a line from the top curve over to the I_D axis, the level of I_{DSS} can be estimated to be about 9 mA. The level of V_P can be estimated by noting the V_{GS} value of the bottom curve and taking into account the shrinking distance between curves as V_{GS} becomes more and more negative. In this case, V_P is certainly more negative than -2 V and perhaps V_P is close to -2.5 V. However, keep in mind that the V_{GS} curves contract very quickly as they approach the cutoff condition, and perhaps $V_P = -3$ V is a better choice. It should

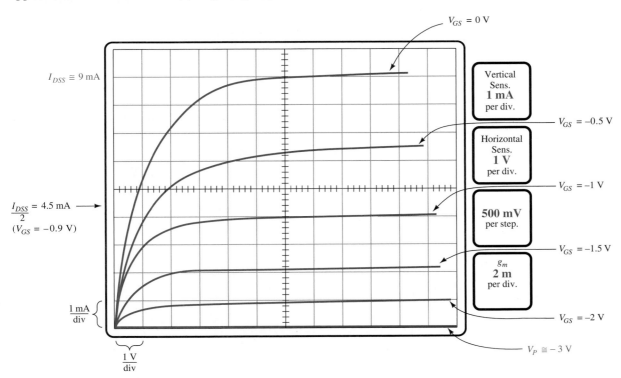

Figure 5.23 Drain characteristics for a 2N4416 JFET transistor as displayed on a curve tracer.

also be noted that the step control is set for a 5-step display, limiting the displayed curves to $V_{GS} = 0, -0.5, -1, -1.5$, and -2 V. If the step control had been increased to 10, the voltage per step could be reduced to $250 \text{ mV} = 0.25 \text{ V}$ and the curve for $V_{GS} = -2.25$ V would have been included as well as an additional curve between each step of Fig. 5.23. The $V_{GS} = -2.25$ V curve would reveal how quickly the curves are closing in on each other for the same step voltage. Fortunately, the level of V_P can be estimated to a reasonable degree of accuracy simply by applying a condition appearing in Table 5.1. That is, when $I_D = I_{DSS}/2$, then $V_{GS} = 0.3V_P$. For the characteristics of Fig. 5.23, $I_D = I_{DSS}/2 = 9 \text{ mA}/2 = 4.5 \text{ mA}$, and as visible from Fig. 5.23 the corresponding level of V_{GS} is about -0.9 V. Using this information we find that $V_P = V_{GS}/0.3 = -0.9 \text{ V}/0.3 = -3 \text{ V}$, which will be our choice for this device. Using this value we find that at $V_{GS} = -2$ V,

$$I_D = I_{DSS}\left(1 - \frac{V_{GS}}{V_P}\right)^2$$

$$= 9 \text{ mA}\left(1 - \frac{-2 \text{ V}}{-3 \text{ V}}\right)^2$$

$$\cong 1 \text{ mA}$$

as supported by Fig. 5.23.

At $V_{GS} = -2.5$ V, Shockley's equation will result in $I_D = 0.25 \text{ mA}$, with $V_P = -3$ V clearly revealing how quickly the curves contract near V_P. The importance of the parameter g_m and how it is determined from the characteristics of Fig. 5.23 are described in Chapter 8 when small-signal ac conditions are examined.

5.6 IMPORTANT RELATIONSHIPS

A number of important equations and operating characteristics have been introduced in the last few sections that are of particular importance for the analysis to follow for the dc and ac configurations. In an effort to isolate and emphasize their importance, they are repeated below next to a corresponding equation for the BJT transistor. The JFET equations are defined for the configuration of Fig. 5.24a, while the BJT equations relate to Fig. 5.24b.

JFET		**BJT**	
$I_D = I_{DSS}\left(1 - \dfrac{V_{GS}}{V_P}\right)^2$	$\Leftrightarrow$	$I_C = \beta I_B$	
$I_D = I_S$	$\Leftrightarrow$	$I_C \cong I_E$	(5.10)
$I_G \cong 0 \text{ A}$	$\Leftrightarrow$	$V_{BE} \cong 0.7 \text{ V}$	

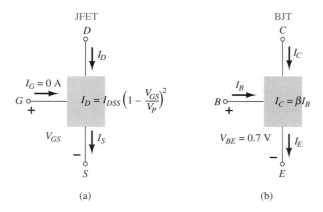

Figure 5.24 (a) JFET versus (b) BJT.

A clear understanding of the impact of each of the equations above is sufficient background to approach the most complex of dc configurations. Recall that $V_{BE} = 0.7$ V was often the key to initiating an analysis of a BJT configuration. Similarly, the condition $I_G = 0$ A is often the starting point for the analysis of a JFET configuration. For the BJT configuration, I_B is normally the first parameter to be determined. For the JFET, it is normally V_{GS}. The number of similarities between the analysis of BJT and JFET dc configurations will become quite apparent in Chapter 6.

5.7 DEPLETION-TYPE MOSFET

As noted in the chapter introduction, there are two types of FETs: JFETs and MOSFETs. MOSFETs are further broken down into *depletion type* and *enhancement type*. The terms *depletion* and *enhancement* define their basic mode of operation, while the label MOSFET stands for *metal-oxide-semiconductor-field-effect transistor*. Since there are differences in the characteristics and operation of each type of MOSFET, they are covered in separate sections. In this section we examine the depletion-type MOSFET, which happens to have characteristics similar to those of a JFET between cutoff and saturation at I_{DSS} but then has the added feature of characteristics that extend into the region of opposite polarity for V_{GS}.

Basic Construction

The basic construction of the *n*-channel depletion-type MOSFET is provided in Fig. 5.25. A slab of *p*-type material is formed from a silicon base and is referred to as the *substrate*. It is the foundation upon which the device will be constructed. In some cases the substrate is internally connected to the source terminal. However, many discrete devices provide an additional terminal labeled *SS*, resulting in a four-terminal device, such as that appearing in Fig. 5.25. The source and drain terminals are connected through metallic contacts to *n*-doped regions linked by an *n*-channel as shown in the figure. The gate is also connected to a metal contact surface but remains insulated from the *n*-channel by a very thin silicon dioxide (SiO$_2$) layer. SiO$_2$ is a particular type of insulator referred to as a *dielectric* that sets up opposing (as revealed by the prefix *di-*) electric fields within the dielectric when exposed to an externally applied field. The fact that the SiO$_2$ layer is an insulating layer reveals the following fact:

There is no direct electrical connection between the gate terminal and the channel of a MOSFET.

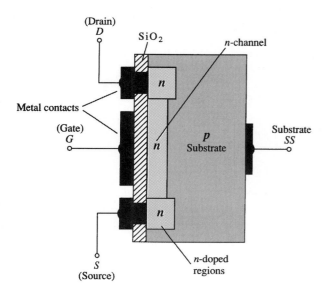

Figure 5.25 *n*-Channel depletion-type MOSFET.

In addition:

> It is the insulating layer of SiO_2 in the MOSFET construction that accounts for the very desirable high input impedance of the device.

In fact, the input resistance of a MOSFET is often that of the typical JFET, even though the input impedance of most JFETs is sufficiently high for most applications. The very high input impedance continues to fully support the fact that the gate current (I_G) is essentially zero amperes for dc-biased configurations.

The reason for the label metal-oxide-semiconductor FET is now fairly obvious: *metal* for the drain, source, and gate connections to the proper surface—in particular, the gate terminal and the control to be offered by the surface area of the contact, the *oxide* for the silicon dioxide insulating layer, and the *semiconductor* for the basic structure on which the *n*- and *p*-type regions are diffused. The insulating layer between the gate and channel has resulted in another name for the device: *insulated-gate FET* or *IGFET*, although this label is used less and less in current literature.

Basic Operation and Characteristics

In Fig. 5.26 the gate-to-source voltage is set to zero volts by the direct connection from one terminal to the other, and a voltage V_{DS} is applied across the drain-to-source terminals. The result is an attraction for the positive potential at the drain by the *free* electrons of the *n*-channel and a current similar to that established through the channel of the JFET. In fact, the resulting current with $V_{GS} = 0$ V continues to be labeled I_{DSS}, as shown in Fig. 5.27.

In Fig. 5.28, V_{GS} has been set at a negative voltage such as -1 V. The negative potential at the gate will tend to pressure electrons toward the *p*-type substrate (like charges repel) and attract holes from the *p*-type substrate (opposite charges attract) as shown in Fig. 5.28. Depending on the magnitude of the negative bias established by V_{GS}, a level of recombination between electrons and holes will occur that will reduce the number of free electrons in the *n*-channel available for conduction. The more negative the bias, the higher the rate of recombination. The resulting level of drain current is therefore reduced with increasing negative bias for V_{GS} as shown in Fig. 5.27 for $V_{GS} = -1$ V, -2 V, and so on, to the pinch-off level of -6 V. The resulting levels of drain current and the plotting of the transfer curve proceeds exactly as described for the JFET.

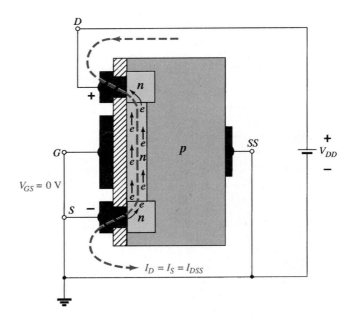

Figure 5.26 *n*-Channel depletion-type MOSFET with $V_{GS} = 0$ V and and applied voltage V_{DD}.

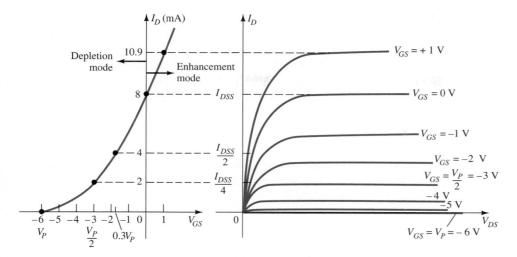

Figure 5.27 Drain and transfer characteristics for an *n*-channel depletion-type MOSFET.

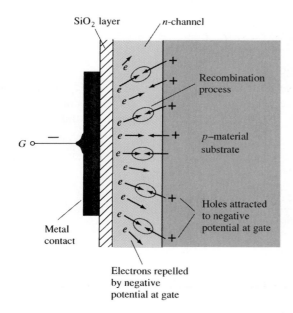

Figure 5.28 Reduction in free carriers in channel due to a negative potential at the gate terminal.

For positive values of V_{GS}, the positive gate will draw additional electrons (free carriers) from the *p*-type substrate due to the reverse leakage current and establish new carriers through the collisions resulting between accelerating particles. As the gate-to-source voltage continues to increase in the positive direction, Fig. 5.27 reveals that the drain current will increase at a rapid rate for the reasons listed above. The vertical spacing between the $V_{GS} = 0$ V and $V_{GS} = +1$ V curves of Fig. 5.27 is a clear indication of how much the current has increased for the 1-V change in V_{GS}. Due to the rapid rise, the user must be aware of the maximum drain current rating since it could be exceeded with a positive gate voltage. That is, for the device of Fig. 5.27, the application of a voltage $V_{GS} = +4$ V would result in a drain current of 22.2 mA, which could possibly exceed the maximum rating (current or power) for the device. As revealed above, the application of a positive gate-to-source voltage has "enhanced" the level of free carriers in the channel compared to that encountered with $V_{GS} = 0$ V.

5.7   Depletion-Type MOSFET   **265**

For this reason the region of positive gate voltages on the drain or transfer characteristics is often referred to as the *enhancement region,* with the region between cutoff and the saturation level of I_{DSS} referred to as the *depletion region.*

It is particularly interesting and helpful that Shockley's equation will continue to be applicable for the depletion-type MOSFET characteristics in both the depletion and enhancement regions. For both regions, it is simply necessary that the proper sign be included with V_{GS} in the equation and the sign be carefully monitored in the mathematical operations.

EXAMPLE 5.3

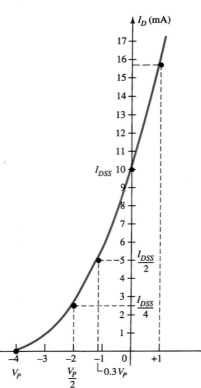

Figure 5.29 Transfer characteristics for an *n*-channel depletion-type MOSFET with $I_{DSS} = 10$ mA and $V_P = -4$ V.

Sketch the transfer characteristics for an *n*-channel depletion-type MOSFET with $I_{DSS} = 10$ mA and $V_P = -4$ V.

Solution

$$\text{At } V_{GS} = 0 \text{ V}, \qquad I_D = I_{DSS} = 10 \text{ mA}$$

$$V_{GS} = V_P = -4 \text{ V}, \qquad I_D = 0 \text{ mA}$$

$$V_{GS} = \frac{V_P}{2} = \frac{-4 \text{ V}}{2} = -2 \text{ V}, \qquad I_D = \frac{I_{DSS}}{4} = \frac{10 \text{ mA}}{4} = 2.5 \text{ mA}$$

and at $I_D = \dfrac{I_{DSS}}{2}, \qquad V_{GS} = 0.3V_P = 0.3(-4 \text{ V}) = -1.2 \text{ V}$

all of which appear in Fig. 5.29.

Before plotting the positive region of V_{GS}, keep in mind that I_D increases very rapidly with increasing positive values of V_{GS}. In other words, be conservative with the choice of values to be substituted into Shockley's equation. In this case, we will try $+1$ V as follows:

$$I_D = I_{DSS}\left(1 - \frac{V_{GS}}{V_P}\right)^2$$

$$= 10 \text{ mA}\left(1 - \frac{+1 \text{ V}}{-4 \text{ V}}\right)^2 = 10 \text{ mA}(1 + 0.25)^2 = 10 \text{ mA}(1.5625)$$

$$\cong 15.63 \text{ mA}$$

which is sufficiently high to finish the plot.

---

p-Channel Depletion-Type MOSFET

The construction of a *p*-channel depletion-type MOSFET is exactly the reverse of that appearing in Fig. 5.25. That is, there is now an *n*-type substrate and a *p*-type channel, as shown in Fig. 5.30a. The terminals remain as identified, but all the voltage polarities and the current directions are reversed, as shown in the same figure. The drain characteristics would appear exactly as in Fig. 5.27 but with V_{DS} having negative values, I_D having positive values as indicated (since the defined direction is now reversed), and V_{GS} having the opposite polarities as shown in Fig. 5.30c. The reversal in V_{GS} will result in a mirror image (about the I_D axis) for the transfer characteristics as shown in Fig. 5.30b. In other words, the drain current will increase from cutoff at $V_{GS} = V_P$ in the positive V_{GS} region to I_{DSS} and then continue to increase for increasingly negative values of V_{GS}. Shockley's equation is still applicable and requires simply placing the correct sign for both V_{GS} and V_P in the equation.

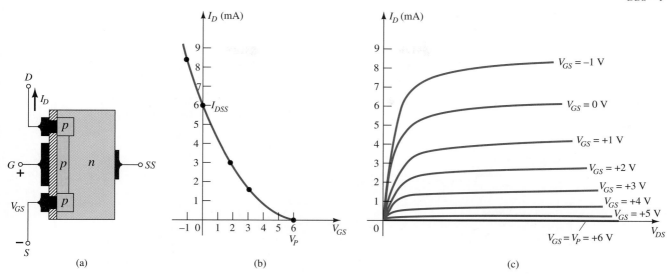

(a)   (b)   (c)

Figure 5.30 p-Channel depletion-type MOSFET with $I_{DSS} = 6$ mA and $V_P = +6$ V.

Symbols, Specification Sheets, and Case Construction

The graphic symbols for an n- and p-channel depletion-type MOSFET are provided in Fig. 5.31. Note how the symbols chosen try to reflect the actual construction of the device. The lack of a direct connection (due to the gate insulation) between the gate and channel is represented by a space between the gate and the other terminals of the symbol. The vertical line representing the channel is connected between the drain and source and is "supported" by the substrate. Two symbols are provided for each type of channel to reflect the fact that in some cases the substrate is externally available while in others it is not. For most of the analysis to follow in Chapter 6, the substrate and source will be connected and the lower symbols will be employed.

The device appearing in Fig. 5.32 has three terminals, with the terminal identification appearing in the same figure. The specification sheet for a depletion-type MOSFET is similar to that of a JFET. The levels of V_P and I_{DSS} are provided along with a list of maximum values and typical "on" and "off" characteristics. In addition, however, since I_D can extend beyond the I_{DSS} level, another point is normally provided that reflects a typical value of I_D for some positive voltage (for an n-channel device). For the unit of Fig. 5.32, I_D is specified as $I_{D(on)} = 9$ mA dc, with $V_{DS} = 10$ V and $V_{GS} = 3.5$ V.

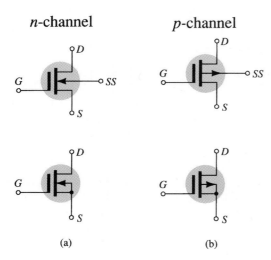

(a)   **(b)**

Figure 5.31   Graphic symbols for (a) n-channel depletion-type MOSFETs and (b) p-channel depletion-type MOSFETs.

2N3797

CASE 22-03, STYLE 2
TO-18 (TO-206AA)

3 Drain

Gate
2

3 2 1

1 Source

**MOSFETs
LOW POWER AUDIO**

N-CHANNEL – DEPLETION

MAXIMUM RATINGS

Rating	Symbol	Value	Unit
Drain–Source Voltage 2N3797	V_{DS}	20	Vdc
Gate–Source Voltage	V_{GS}	±10	Vdc
Drain Current	I_D	20	mAdc
Total Device Dissipation @ $T_A = 25°C$ Derate above 25°C	P_D	200 1.14	mW mW/°C
Junction Temperature Range	T_J	+175	°C
Storage Channel Temperature Range	T_{stg}	–65 to +200	°C

ELECTRICAL CHARACTERISTICS ($T_A = 25°C$ unless otherwise noted)

Characteristic	Symbol	Min	Typ	Max	Unit
OFF CHARACTERISTICS					
Drain Source Breakdown Voltage ($V_{GS} = -7.0$ V, $I_D = 5.0$ μA) 2N3797	$V_{(BR)DSX}$	20	25	–	Vdc
Gate Reverse Current (1) ($V_{GS} = -10$ V, $V_{DS} = 0$) ($V_{GS} = -10$ V, $V_{DS} = 0$, $T_A = 150°C$)	I_{GSS}	– –	– –	1.0 200	pAdc
Gate Source Cutoff Voltage ($I_D = 2.0$ μA, $V_{DS} = 10$ V) 2N3797	$V_{GS(off)}$	–	–5.0	–7.0	Vdc
Drain-Gate Reverse Current (1) ($V_{DG} = 10$ V, $I_S = 0$)	I_{DGO}	–	–	1.0	pAdc
ON CHARACTERISTICS					
Zero-Gate-Voltage Drain Current ($V_{DS} = 10$ V, $V_{GS} = 0$) 2N3797	I_{DSS}	2.0	2.9	6.0	mAdc
On-State Drain Current ($V_{DS} = 10$ V, $V_{GS} = +3.5$ V) 2N3797	$I_{D(on)}$	9.0	14	18	mAdc
SMALL-SIGNAL CHARACTERISTICS					
Forward Transfer Admittance ($V_{DS} = 10$ V, $V_{GS} = 0$, f = 1.0 kHz) 2N3797	$\|y_{fs}\|$	1500	2300	3000	μmhos
($V_{DS} = 10$ V, $V_{GS} = 0$, f = 1.0 MHz) 2N3797		1500	–	–	
Output Admittance ($I_{DS} = 10$ V, $V_{GS} = 0$, f = 1.0 kHz) 2N3797	$\|y_{os}\|$	–	27	60	μmhos
Input Capacitance ($V_{DS} = 10$ V, $V_{GS} = 0$, f = 1.0 MHz) 2N3797	C_{iss}	–	6.0	8.0	pF
Reverse Transfer Capacitance ($V_{DS} = 10$ V, $V_{GS} = 0$, f = 1.0 MHz)	C_{rss}	–	0.5	0.8	pF
FUNCTIONAL CHARACTERISTICS					
Noise Figure ($V_{DS} = 10$ V, $V_{GS} = 0$, f = 1.0 kHz, $R_S = 3$ megohms)	NF	–	3.8	–	dB

(1) This value of current includes both the FET leakage current as well as the leakage current associated with the test socket and fixture when measured under best attainable conditions.

Figure 5.32   2N3797 Motorola *n*-channel depletion-type MOSFET.

5.8 ENHANCEMENT-TYPE MOSFET

Although there are some similarities in construction and mode of operation between depletion-type and enhancement-type MOSFETs, the characteristics of the enhancement-type MOSFET are quite different from anything obtained thus far. The transfer curve is not defined by Shockley's equation, and the drain current is now cut off until the gate-to-source voltage reaches a specific magnitude. In particular, current control in an *n*-channel device is now effected by a positive gate-to-source voltage rather than the range of negative voltages encountered for *n*-channel JFETs and *n*-channel depletion-type MOSFETs.

Basic Construction

The basic construction of the *n*-channel enhancement-type MOSFET is provided in Fig. 5.33. A slab of *p*-type material is formed from a silicon base and is again referred to as the substrate. As with the depletion-type MOSFET, the substrate is sometimes internally connected to the source terminal, while in other cases a fourth lead is made available for external control of its potential level. The source and drain terminals are again connected through metallic contacts to *n*-doped regions, but note in Fig. 5.33 the absence of a channel between the two *n*-doped regions. This is the primary difference between the construction of depletion-type and enhancement-type MOSFETs—the absence of a channel as a constructed component of the device. The SiO₂ layer is still present to isolate the gate metallic platform from the region between the drain and source, but now it is simply separated from a section of the *p*-type material. In summary, therefore, the construction of an enhancement-type MOSFET is quite similar to that of the depletion-type MOSFET, except for the absence of a channel between the drain and source terminals.

Basic Operation and Characteristics

If V_{GS} is set at 0 V and a voltage applied between the drain and source of the device of Fig. 5.33, the absence of an *n*-channel (with its generous number of free carriers) will result in a current of effectively zero amperes—quite different from the depletion-type MOSFET and JFET where $I_D = I_{DSS}$. It is not sufficient to have a large accumulation of carriers (electrons) at the drain and source (due to the *n*-doped regions) if a path fails to exist between the two. With V_{DS} some positive voltage, V_{GS} at 0 V, and terminal *SS* directly connected to the source, there are in fact two reverse-biased *p-n* junctions between the *n*-doped regions and the *p*-substrate to oppose any significant flow between drain and source.

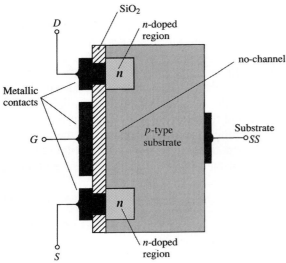

Figure 5.33 *n*-Channel enhancement-type MOSFET.

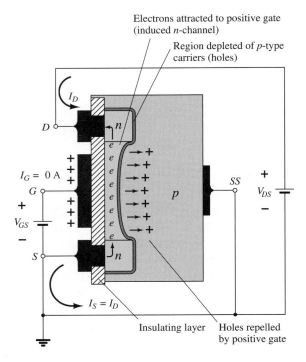

Figure 5.34 Channel formation in the *n*-channel enhancement-type MOSFET.

In Fig. 5.34 both V_{DS} and V_{GS} have been set at some positive voltage greater than 0 V, establishing the drain and gate at a positive potential with respect to the source. The positive potential at the gate will pressure the holes (since like charges repel) in the *p*-substrate along the edge of the SiO_2 layer to leave the area and enter deeper regions of the *p*-substrate, as shown in the figure. The result is a depletion region near the SiO_2 insulating layer void of holes. However, the electrons in the *p*-substrate (the minority carriers of the material) will be attracted to the positive gate and accumulate in the region near the surface of the SiO_2 layer. The SiO_2 layer and its insulating qualities will prevent the negative carriers from being absorbed at the gate terminal. As V_{GS} increases in magnitude, the concentration of electrons near the SiO_2 surface increases until eventually the induced *n*-type region can support a measurable flow between drain and source. The level of V_{GS} that results in the significant increase in drain current is called the *threshold voltage* and is given the symbol V_T. On specification sheets it is referred to as $V_{GS(Th)}$, although V_T is less unwieldy and will be used in the analysis to follow. Since the channel is nonexistent with $V_{GS} = 0$ V and "enhanced" by the application of a positive gate-to-source voltage, this type of MOSFET is called an *enhancement-type MOSFET*. Both depletion- and enhancement-type MOSFETs have enhancement-type regions, but the label was applied to the latter since it is its only mode of operation.

As V_{GS} is increased beyond the threshold level, the density of free carriers in the induced channel will increase, resulting in an increased level of drain current. However, if we hold V_{GS} constant and increase the level of V_{DS}, the drain current will eventually reach a saturation level as occurred for the JFET and depletion-type MOSFET. The leveling off of I_D is due to a pinching-off process depicted by the narrower channel at the drain end of the induced channel as shown in Fig. 5.35. Applying Kirchhoff's voltage law to the terminal voltages of the MOSFET of Fig. 5.35, we find that

$$V_{DG} = V_{DS} - V_{GS} \qquad (5.11)$$

If V_{GS} is held fixed at some value such as 8 V and V_{DS} is increased from 2 to 5 V, the voltage V_{DG} [by Eq. (5.11)] will drop from −6 to −3 V and the gate will become less and less positive with respect to the drain. This reduction in gate-to-drain

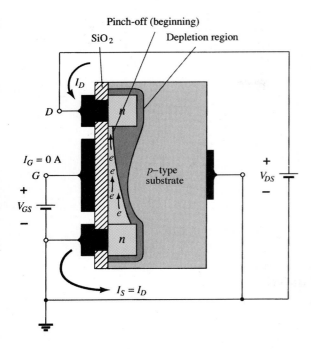

Figure 5.35 Change in channel and depletion region with increasing level of V_{DS} for a fixed value of V_{GS}.

voltage will in turn reduce the attractive forces for free carriers (electrons) in this region of the induced channel, causing a reduction in the effective channel width. Eventually, the channel will be reduced to the point of pinch-off and a saturation condition will be established as described earlier for the JFET and depletion-type MOSFET. In other words, any further increase in V_{DS} at the fixed value of V_{GS} will not affect the saturation level of I_D until breakdown conditions are encountered.

The drain characteristics of Fig. 5.36 reveal that for the device of Fig. 5.35 with $V_{GS} = 8$ V, saturation occurred at a level of $V_{DS} = 6$ V. In fact, the saturation level for V_{DS} is related to the level of applied V_{GS} by

$$V_{DS_{\text{sat}}} = V_{GS} - V_T \qquad (5.12)$$

Obviously, therefore, for a fixed value of V_T, then the higher the level of V_{GS}, the more the saturation level for V_{DS}, as shown in Fig. 5.35 by the locus of saturation levels.

For the characteristics of Fig. 5.35 the level of V_T is 2 V, as revealed by the fact that the drain current has dropped to 0 mA. In general, therefore:

For values of V_{GS} less than the threshold level, the drain current of an enhancement-type MOSFET is 0 mA.

Figure 5.36 clearly reveals that as the level of V_{GS} increased from V_T to 8 V, the resulting saturation level for I_D also increased from a level of 0 to 10 mA. In addition, it is quite noticeable that the spacing between the levels of V_{GS} increased as the magnitude of V_{GS} increased, resulting in ever-increasing increments in drain current.

For levels of $V_{GS} > V_T$, the drain current is related to the applied gate-to-source voltage by the following nonlinear relationship:

$$I_D = k(V_{GS} - V_T)^2 \qquad (5.13)$$

Again, it is the squared term that results in the nonlinear (curved) relationship between I_D and V_{GS}. The k term is a constant that is a function of the construction of the device. The value of k can be determined from the following equation [derived from Eq. (5.13)] where $I_{D(\text{on})}$ and $V_{GS(\text{on})}$ are the values for each at a particular point on the characteristics of the device.

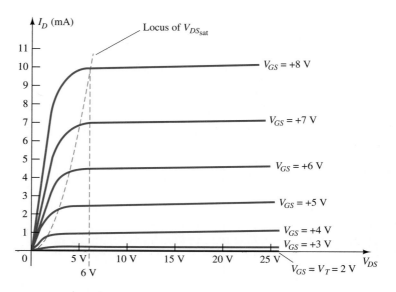

Figure 5.36 Drain characteristics of an n-channel enhancement-type MOSFET with $V_T = 2$ V and $k = 0.278 \times 10^{-3}$ A/V².

$$k = \frac{I_{D(\text{on})}}{(V_{GS(\text{on})} - V_T)^2} \tag{5.14}$$

Substituting $I_{D(\text{on})} = 10$ mA when $V_{GS(\text{on})} = 8$ V from the characteristics of Fig. 5.36 yields

$$k = \frac{10 \text{ mA}}{(8 \text{ V} - 2 \text{ V})^2} = \frac{10 \text{ mA}}{(6 \text{ V})^2} = \frac{10 \text{ mA}}{36 \text{ V}^2}$$

$$= \mathbf{0.278 \times 10^{-3} \text{ A/V}^2}$$

and a general equation for I_D for the characteristics of Fig. 5.36 results in:

$$I_D = 0.278 \times 10^{-3}(V_{GS} - 2 \text{ V})^2$$

Substituting $V_{GS} = 4$ V, we find that

$$I_D = 0.278 \times 10^{-3}(4 \text{ V} - 2 \text{ V})^2 = 0.278 \times 10^{-3}(2)^2$$

$$= 0.278 \times 10^{-3}(4) = \mathbf{1.11 \text{ mA}}$$

as verified by Fig. 5.36. At $V_{GS} = V_T$, the squared term is 0 and $I_D = 0$ mA.

For the dc analysis of enhancement-type MOSFETs to appear in Chapter 6, the transfer characteristics will again be the characteristics to be employed in the graphical solution. In Fig. 5.37 the drain and transfer characteristics have been set side by side to describe the transfer process from one to the other. Essentially, it proceeds as introduced earlier for the JFET and depletion-type MOSFETs. In this case, however, it must be remembered that the drain current is 0 mA for $V_{GS} \leq V_T$. At this point a measurable current will result for I_D and will increase as defined by Eq. (5.13). Note that in defining the points on the transfer characteristics from the drain characteristics, only the saturation levels are employed, thereby limiting the region of operation to levels of V_{DS} greater than the saturation levels as defined by Eq. (5.12).

The transfer curve of Fig. 5.37 is certainly quite different from those obtained earlier. For an n-channel (induced) device, it is now totally in the positive V_{GS} region and does not rise until $V_{GS} = V_T$. The question now surfaces as to how to plot the transfer characteristics given the levels of k and V_T as included below for a particular MOSFET:

$$I_D = 0.5 \times 10^{-3}(V_{GS} - 4 \text{ V})^2$$

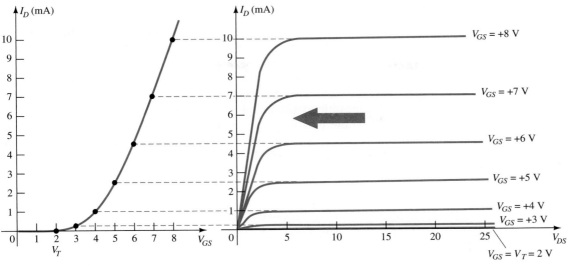

Figure 5.37 Sketching the transfer characteristics for an *n*-channel enhancement-type MOSFET from the drain characteristics.

First, a horizontal line is drawn at $I_D = 0$ mA from $V_{GS} = 0$ V to $V_{GS} = 4$ V as shown in Fig. 5.38a. Next, a level of V_{GS} greater than V_T such as 5 V is chosen and substituted into Eq. (5.13) to determine the resulting level of I_D as follows:

$$I_D = 0.5 \times 10^{-3}(V_{GS} - 4 \text{ V})^2$$
$$= 0.5 \times 10^{-3}(5 \text{ V} - 4 \text{ V})^2 = 0.5 \times 10^{-3}(1)^2$$
$$= \mathbf{0.5 \text{ mA}}$$

and a point on the plot is obtained as shown in Fig. 5.38b. Finally, additional levels of V_{GS} are chosen and the resulting levels of I_D obtained. In particular, at $V_{GS} = 6, 7$, and 8 V, the level of I_D is 2, 4.5, and 8 mA, respectively, as shown on the resulting plot of Fig. 5.38c.

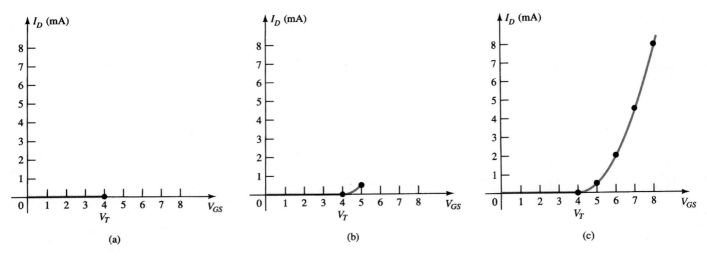

Figure 5.38 Plotting the transfer characteristics of an *n*-channel enhancement-type MOSFET with $k = 0.5 \times 10^{-3}$ A/V^2 and $V_T = 4$ V.

p-Channel Enhancement-Type MOSFETs

The construction of a p-channel enhancement-type MOSFET is exactly the reverse of that appearing in Fig. 5.33, as shown in Fig. 5.39a. That is, there is now an n-type substrate and p-doped regions under the drain and source connections. The terminals remain as identified, but all the voltage polarities and the current directions are reversed. The drain characteristics will appear as shown in Fig. 5.39c, with increasing levels of current resulting from increasingly negative values of V_{GS}. The transfer characteristics will be the mirror image (about the I_D axis) of the transfer curve of Fig. 5.37, with I_D increasing with increasingly negative values of V_{GS} beyond V_T, as shown in Fig. 5.39b. Equations (5.11) through (5.14) are equally applicable to p-channel devices.

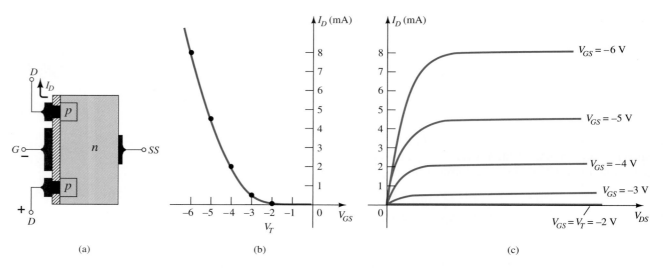

Figure 5.39   p-Channel enhancement-type MOSFET with $V_T = 2$ V and $k = 0.5 \times 10^{-3}$ A/V².

Symbols, Specification Sheets, and Case Construction

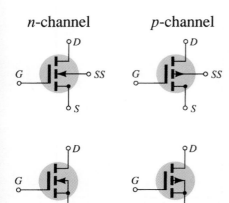

Figure 5.40   Symbols for (a) n-channel enhancement-type MOSFETs and (b) p-channel enhancement-type MOSFETs.

The graphic symbols for the n- and p-channel enhancement-type MOSFETs are provided as Fig. 5.40. Again note how the symbols try to reflect the actual construction of the device. The dashed line between drain and source was chosen to reflect the fact that a channel does not exist between the two under no-bias conditions. It is, in fact, the only difference between the symbols for the depletion-type and enhancement-type MOSFETs.

The specification sheet for a Motorola n-channel enhancement-type MOSFET is provided as Fig. 5.41. The case construction and terminal identification are provided next to the maximum ratings, which now include a maximum drain current of 30 mA dc. The specification sheet provides the level of I_{DSS} under "off" conditions, which is now simply 10 nA dc (at $V_{DS} = 10$ V and $V_{GS} = 0$ V) compared to the milliampere range for the JFET and depletion-type MOSFET. The threshold voltage is specified as $V_{GS(Th)}$ and has a range of 1 to 5 V dc, depending on the unit employed. Rather than provide a range of k in Eq. (5.13), a typical level of $I_{D(on)}$ (3 mA in this case) is specified at a particular level of $V_{GS(on)}$ (10 V for the specified I_D level). In other words, when $V_{GS} = 10$ V, $I_D = 3$ mA. The given levels of $V_{GS(Th)}, I_{D(on)}$ and $V_{GS(on)}$ permit a determination of k from Eq. (5.14) and a writing of the general equation for the transfer characteristics. The handling requirements of MOSFETs are reviewed in Section 5.9.

MAXIMUM RATINGS

Rating	Symbol	Value	Unit
Drain–Source Voltage	V_{DS}	25	Vdc
Drain–Gate Voltage	V_{DG}	30	Vdc
Gate–Source Voltage*	V_{GS}	30	Vdc
Drain Current	I_D	30	mAdc
Total Device Dissipation @ T_A = 25°C Derate above 25°C	P_D	300 1.7	mW mW/°C
Junction Temperature Range	T_J	175	°C
Storage Temperature Range	T_{stg}	–65 to +175	°C

* Transient potentials of ± 75 Volt will not cause gate-oxide failure.

2N4351
CASE 20-03, STYLE 2
TO-72 (TO-206AF)

**MOSFET
SWITCHING**
N-CHANNEL – ENHANCEMENT

ELECTRICAL CHARACTERISTICS (T_A = 25°C unless otherwise noted.)

Characteristic	Symbol	Min	Max	Unit
OFF CHARACTERISTICS				
Drain-Source Breakdown Voltage (I_D = 10 μA, V_{GS} = 0)	$V_{(BR)DSX}$	25	–	Vdc
Zero-Gate-Voltage Drain Current (V_{DS} = 10 V, V_{GS} = 0) T_A = 25°C T_A = 150°C	I_{DSS}	– –	10 10	nAdc μAdc
Gate Reverse Current (V_{GS} = ± 15 Vdc, V_{DS} = 0)	I_{GSS}	–	± 10	pAdc
ON CHARACTERISTICS				
Gate Threshold Voltage (V_{DS} = 10 V, I_D = 10 μA)	$V_{GS(Th)}$	1.0	5	Vdc
Drain-Source On-Voltage (I_D = 2.0 mA, V_{GS} = 10V)	$V_{DS(on)}$	–	1.0	V
On-State Drain Current (V_{GS} = 10 V, V_{DS} = 10 V)	$I_{D(on)}$	3.0	–	mAdc
SMALL-SIGNAL CHARACTERISTICS				
Forward Transfer Admittance (V_{DS} = 10 V, I_D = 2.0 mA, f = 1.0 kHz)	$\lvert y_{fs} \rvert$	1000	–	μmho
Input Capacitance (V_{DS} = 10 V, V_{GS} = 0, f = 140 kHz)	C_{iss}	–	5.0	pF
Reverse Transfer Capacitance (V_{DS} = 0, V_{GS} = 0, f = 140 kHz)	C_{rss}	–	1.3	pF
Drain-Substrate Capacitance ($V_{D(SUB)}$ = 10 V, f = 140 kHz)	$C_{d(sub)}$	–	5.0	pF
Drain-Source Resistance (V_{GS} = 10 V, I_D = 0, f = 1.0 kHz)	$r_{ds(on)}$	–	300	ohms
SWITCHING CHARACTERISTICS				
Turn-On Delay (Fig. 5)	t_{d1}	–	45	ns
Rise Time (Fig. 6)	t_r	–	65	ns
Turn-Off Delay (Fig. 7)	t_{d2}	–	60	ns
Fall Time (Fig. 8)	t_f	–	100	ns

I_D = 2.0 mAdc, V_{DS} = 10 Vdc, (V_{GS} = 10 Vdc) (See Figure 9; Times Circuit Determined)

Figure 5.41 2N4351 Motorola *n*-channel enhancement-type MOSFET.

Using the data provided on the specification sheet of Fig. 5.41 and an average thresh-old voltage of $V_{GS(Th)}$ = 3 V, determine:

(a) The resulting value of *k* for the MOSFET.

(b) The transfer characteristics.

EXAMPLE 5.4

Solution

(a) Eq. (5.14):

$$k = \frac{I_{D(on)}}{(V_{GS(on)} - V_{GS(Th)})^2}$$

$$= \frac{3 \text{ mA}}{(10 \text{ V} - 3 \text{ V})^2} = \frac{3 \text{ mA}}{(7 \text{ V})^2} = \frac{3 \times 10^{-3}}{49} \text{ A/V}^2$$

$$= \mathbf{0.061 \times 10^{-3} \text{ A/V}^2}$$

(b) Eq. (5.13):

$$I_D = k(V_{GS} - V_T)^2$$

$$= 0.061 \times 10^{-3}(V_{GS} - 3 \text{ V})^2$$

For $V_{GS} = 5$ V,

$$I_D = 0.061 \times 10^{-3}(5 \text{ V} - 3 \text{ V})^2 = 0.061 \times 10^{-3}(2)^2$$

$$= 0.061 \times 10^{-3}(4) = 0.244 \text{ mA}$$

For $V_{GS} = 8$, 10, 12, and 14 V, I_D will be 1.525, 3 (as defined), 4.94, and 7.38 mA, respectively. The transfer characteristics are sketched in Fig. 5.42.

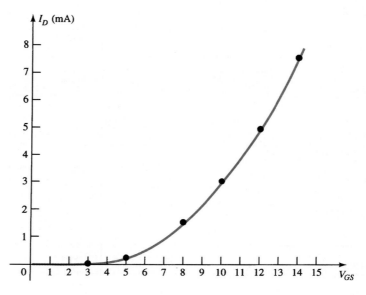

Figure 5.42 Solution to Example 5.4.

5.9 MOSFET HANDLING

The thin SiO_2 layer between the gate and channel of MOSFETs has the positive effect of providing a high-input-impedance characteristic for the device, but because of its extremely thin layer, it introduces a concern for its handling that was not present for the BJT or JFET transistors. There is often sufficient accumulation of static charge (that we pick up from our surroundings) to establish a potential difference across the thin layer that can break down the layer and establish conduction through it. It is therefore imperative that we leave the shorting (or conduction) shipping foil (or ring) connecting the leads of the device together until the device is to be inserted in the system. The shorting ring prevents the possibility of applying a potential across any two terminals of the device. With the ring the potential difference between any two terminals is maintained at 0 V. At the very least always touch ground to permit discharge of the accumulated static charge before handling the device, and always pick up the transistor by the casing.

There are often transients (sharp changes in voltage or current) in a network when elements are removed or inserted if the power is on. The transient levels can often be more than the device can handle, and therefore the power should always be off when network changes are made.

The maximum gate-to-source voltage is normally provided in the list of maximum ratings of the device. One method of ensuring that this voltage is not exceeded (perhaps by transient effects) for either polarity is to introduce two Zener diodes, as shown in Fig. 5.43. The Zeners are back to back to ensure protection for either polarity. If both are 30-V Zeners and a positive transient of 40 V appears, the lower Zener will "fire" at 30 V and the upper will turn on with a 0-V drop (ideally—for the positive "on" region of a semiconductor diode) across the other diode. The result is a maximum of 30 V for the gate-to-source voltage. One disadvantage introduced by the Zener protection is that the off resistance of a Zener diode is less than the input impedance established by the SiO₂ layer. The result is a reduction in input resistance, but even so it is still high enough for most applications. So many of the discrete devices now have the Zener protection that some of the concerns listed above are not as troublesome. However, it is still best to be somewhat cautious when handling discrete MOSFET devices.

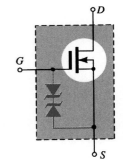

Figure 5.43 Zener-protected MOSFET.

5.10 VMOS

One of the disadvantages of the typical MOSFET is the reduced power-handling levels (typically, less than 1 W) compared to BJT transistors. This shortfall for a device with so many positive characteristics can be softened by changing the construction mode from one of a planar nature such as shown in Fig. 5.25 to one with a vertical structure as shown in Fig. 5.44. All the elements of the planar MOSFET are present in the vertical metal-oxide-silicon FET (VMOS)—the metallic surface connection to the terminals of the device—the SiO₂ layer between the gate and the p-type region between the drain and source for the growth of the induced n-channel (enhancement-mode operation). The term *vertical* is due primarily to the fact that the channel is now formed in the vertical direction rather than the horizontal direction for the planar device. However, the channel of Fig. 5.44 also has the appearance of a "V" cut in the semiconductor base, which often stands out as a characteristic for mental memorization of the name of the device. The construction of Fig. 5.44 is somewhat simplistic

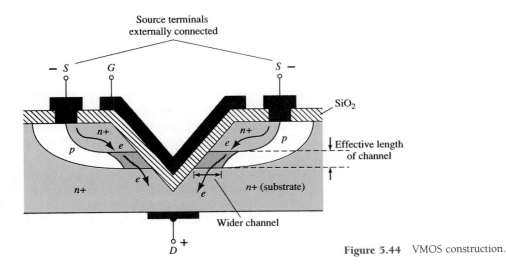

Figure 5.44 VMOS construction.

in nature, leaving out some of the transition levels of doping, but it does permit a description of the most important facets of its operation.

The application of a positive voltage to the drain and a negative voltage to the source with the gate at 0 V or some typical positive "on" level as shown in Fig. 5.44 will result in the induced n-channel in the narrow p-type region of the device. The length of the channel is now defined by the vertical height of the p-region, which can be made significantly less than that of a channel using planar construction. On a horizontal plane the length of the channel is limited to 1 to 2 μm (1 μm = 10^{-6} m). Diffusion layers (such as the p-region of Fig. 5.44) can be controlled to small fractions of a micrometer. Since decreasing channel lengths result in reduced resistance levels, the power dissipation level of the device (power lost in the form of heat) at operating current levels will be reduced. In addition, the contact area between the channel and the n^+ region is greatly increased by the vertical mode construction, contributing to a further decrease in the resistance level and an increased area for current between the doping layers. There is also the existence of two conduction paths between drain and source, as shown in Fig. 5.44, to further contribute to a higher current rating. The net result is a device with drain currents that can reach the ampere levels with power levels exceeding 10 W.

In general:

Compared with commercially available planar MOSFETs, VMOS FETs have reduced channel resistance levels and higher current and power ratings.

An additional important characteristic of the vertical construction is:

VMOS FETs have a positive temperature coefficient that will combat the possibility of thermal runaway.

If the temperature of a device should increase due to the surrounding medium or currents of the device, the resistance levels will increase, causing a reduction in drain current rather than an increase as encountered for a conventional device. Negative temperature coefficients result in decreased levels of resistance with increases in temperature that fuel the growing current levels and result in further temperature instability and thermal runaway.

Another positive characteristic of the VMOS configuration is:

The reduced charge storage levels result in faster switching times for VMOS construction compared to those for conventional planar construction.

In fact, VMOS devices typically have switching times less than one-half that encountered for the typical BJT transistor.

5.11 CMOS

A very effective logic circuit can be established by constructing a p-channel and an n-channel MOSFET on the same substrate as shown in Fig. 5.45. Note the induced p-channel on the left and the induced n-channel on the right for the p- and n-channel devices, respectively. The configuration is referred to as a *complementary MOSFET* arrangement (CMOS) that has extensive applications in computer logic design. The relatively high input impedance, fast switching speeds, and lower operating power levels of the CMOS configuration have resulted in a whole new discipline referred to as *CMOS logic design.*

One very effective use of the complementary arrangement is as an inverter, as shown in Fig. 5.46. As introduced for switching transistors, an inverter is a logic element that "inverts" the applied signal. That is, if the logic levels of operation are 0 V (0-state) and 5 V (1-state), an input level of 0 V will result in an output level of 5 V, and vice versa. Note in Fig. 5.46 that both gates are connected to the applied signal and both drain to the output V_o. The source of the p-channel MOSFET (Q_2) is con-

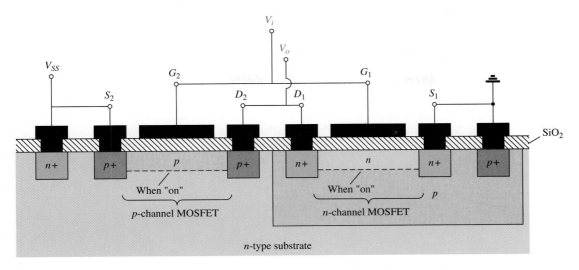

Figure 5.45 CMOS with the connections indicated in Fig. 5.46.

nected directly to the applied voltage V_{SS}, while the source of the *n*-channel MOS-FET (Q_1) is connected to ground. For the logic levels defined above, the application of 5 V at the input should result in approximately 0 V at the output. With 5 V at V_i (with respect to ground), $V_{GS_1} = V_i$ and Q_1 is "on," resulting in a relatively low resistance between drain and source as shown in Fig. 5.47. Since V_i and V_{SS} are at 5 V, $V_{GS_2} = 0$ V, which is less than the required V_T for the device, resulting in an "off" state. The resulting resistance level between drain and source is quite high for Q_2, as shown in Fig. 5.47. A simple application of the voltage-divider rule will reveal that V_o is very close to 0 V or the 0-state, establishing the desired inversion process. For an applied voltage V_i of 0 V (0-state), $V_{GS_1} = 0$ V and Q_1 will be off with $V_{SS_2} = -5$ V, turning on the *p*-channel MOSFET. The result is that Q_2 will present a small resistance level, Q_1 a high resistance, and $V_o = V_{SS} = 5$ V (the 1-state). Since the drain current that flows for either case is limited by the "off" transistor to the leakage value, the power dissipated by the device in either state is very low. Additional comment on the application of CMOS logic is presented in Chapter 16.

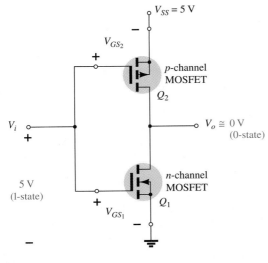

Figure 5.46 CMOS inverter.

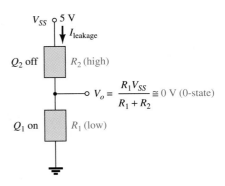

Figure 5.47 Relative resistance levels for $V_i = 5$ V (1-state).

5.12   SUMMARY TABLE

Since the transfer curves and some important characteristics vary from one type of FET to another, Table 5.2 was developed to clearly display the differences from one device to the next. A clear understanding of all the curves and parameters of the table will provide a sufficient background for the dc and ac analyses to follow in Chapters 6 and 8. Take a moment to ensure that each curve is recognizable and its derivation understood, and then establish a basis for comparison of the levels of the important parameters of R_i and C_i for each device.

TABLE 5.2 Field Effect Transistors

Type	-Symbol- Basic Relationships	Transfer Curve	Input Resistance and Capacitance
JFET (*n*-channel)	$I_G = 0\text{ A}, I_D = I_S$ $I_D = I_{DSS}\left(1 - \dfrac{V_{GS}}{V_P}\right)^2$		$R_i > 100\text{ M}\Omega$ $C_i: (1-10)\text{ pF}$
MOSFET depletion-type (*n*-channel)	$I_G = 0\text{ A}, I_D = I_S$ $I_D = I_{DSS}\left(1 - \dfrac{V_{GS}}{V_P}\right)^2$		$R_i > 10^{10}\ \Omega$ $C_i: (1-10)\text{ pF}$
MOSFET enhancement-type (*n*-channel)	$I_G = 0\text{ A}, I_D = I_S$ $I_D = k\,(V_{GS} - V_{GS\,(\text{Th})})^2$ $k = \dfrac{I_{D\,(\text{on})}}{(V_{GS(\text{on})} - V_{GS\,(\text{Th})})^2}$		$R_i > 10^{10}\ \Omega$ $C_i: (1-10)\text{ pF}$

5.13 SUMMARY

Important Conclusions and Concepts

1. A **current-controlled device** is one in which a current defines the operating conditions of the device, whereas a **voltage-controlled device** is one in which a particular voltage defines the operating conditions.

2. The JFET can actually be used as a **voltage-controlled resistor** because of a unique sensitivity of the drain-to-source impedance to the gate-to-source voltage.

3. The **maximum current** for any JFET is labeled I_{DSS} and occurs when $V_{GS} = 0$ V.

4. The **minimum current** for a JFET occurs at pinch-off defined by $V_{GS} = V_P$.

5. The relationship between the drain current and the gate-to-source voltage of a JFET is a **nonlinear one** defined by Shockley's equation. As the current level approaches I_{DSS}, the sensitivity of I_D to changes in V_{GS} increases significantly.

6. The transfer characteristics (I_D versus V_{GS}) are characteristics **of the device itself** and are not sensitive to the network in which the JFET is employed.

7. When $V_{GS} = V_P/2$, $I_D = I_{DSS}/4$; and at a point where $I_D = I_{DSS}/2$, $V_{GS} \cong 0.3$ V.

8. Maximum operating conditions are determined by the **product** of the drain-to-source voltage and the drain current.

9. MOSFETs are available in one of two types: **depletion and enhancement**.

10. The depletion-type MOSFET has the same transfer characteristics as a JFET for drain currents up to the I_{DSS} level. At this point the characteristics of a depletion-type MOSFET **continue to levels above I_{DSS}**, whereas those of the JFET will end.

11. The arrow in the symbol of n-channel JFETs or MOSFETs will **always point in to the center of the symbol**, whereas those of a p-channel device will always point out of the center of the symbol.

12. The transfer characteristics of an enhancement-type MOSFET are **not defined by Shockley's equation** but rather by a nonlinear equation controlled by the gate-to-source voltage, the threshold voltage, and a constant k defined by the device employed. The resulting plot of I_D versus V_{GS} is one that **rises exponentially with increasing values of V_{GS}**.

13. Always handle MOSFETs with **additional care** due to the static electricity that exists in places we might least suspect. Do not remove any shorting mechanism between the leads of the device until it is installed.

14. A CMOS (complementary MOSFET) device is one that employs a unique **combination of a p-channel and an n-channel MOSFET** with a single set of external leads. It has the advantages of a very high input impedance, fast switching speeds, and low operating power levels, all of which make it very useful in logic circuits.

Equations

JFET:

$$I_D = I_{DSS}\left(1 - \frac{V_{GS}}{V_P}\right)^2$$

$$I_D = I_{DSS}\big|_{V_{GS}=0\text{ V}}$$

$$I_D = 0\text{ mA}\big|_{V_{GS}=V_P}$$

$$I_D = \frac{I_{DSS}}{4}\bigg|_{V_{GS}=V_P/2}$$

$$V_{GS} \cong 0.3V_P|_{I_D = I_{DSS}/2}$$

$$V_{GS} = V_P\left(1 - \sqrt{\frac{I_D}{I_{DSS}}}\right)$$

$$P_D = V_{DS}I_D$$

$$r_d = \frac{r_o}{(1 - V_{GS}/V_P)^2}$$

MOSFET (enhancement):

$$I_D = k(V_{GS} - V_T)^2$$

$$k = \frac{I_{D(\text{on})}}{(V_{GS(\text{on})} - V_T)^2}$$

5.14 COMPUTER ANALYSIS

PSpice Windows

The characteristics of an n-channel JFET can be found in much the same manner as employed for the bipolar transistor. The series of curves for various levels of V will require a nested sweep under the main sweep for the drain-to-source voltage. The configuration required appears in Fig. 5.48. Note the absence of any resistors since the input impedance is assumed to be infinite, resulting in $I_G = 0$ A. Calling up the device specifications through **Edit-Model-Edit Instance Model (Text)** will result in a display having at the head of the listing a parameter **Beta**. For the junction-field-effect transistor **Beta** is defined by

$$\text{Beta} = \frac{I_{DSS}}{|V_P|^2} \tag{5.15}$$

The parameter **Vto** $= -3$ defines $V_{GS} = V_P = -3$ V as the pinch-off voltage—something to check when we obtain our characteristics. Choosing the **Setup Analysis** icon (recall that it has the horizontal blue line at the top), the **DC Sweep** is first enabled and then activated to produce the **DC Sweep** dialog box. Select **Voltage Source-Linear,** and insert the **Name:** VDD, the **Start Value** of 0 V, **End Value** of 10 V, and **Increment** of 0.01 V. Then, the **Nested Sweep** is chosen, and **Voltage** and **Linear** are chosen once more. Finally, the **Name:** VGG is entered, the **Start Value** of 0 V is chosen, the **End Value** of -5 V is entered, and the **Increment** is set at -1 V. Then, be sure to **Enable Nested Sweep** before clicking on **OK** and closing. With the **Automatically run Probe after Simulation** enabled, clicking on the analysis icon will result in the **OrCAD-MicroSim Probe** screen. There is no need to call up the **X-Axis Settings** because the horizontal axis has the correct range and the voltage V_{DD} is actually the drain-to-source voltage. By choosing the **Trace** icon, the **Add Traces** dialog box will appear. **ID(J1)** is chosen, followed by **OK**. The result is the set of characteristics appearing in Fig. 5.49. The remaining labels were added using the **ABC** icon.

Note that the pinch-off voltage is -3 V, as expected by the **Vto** parameter. The value of I_{DSS} is very close to 12 mA.

The transfer characteristics can be obtained by returning to the network configuration and choosing the **Analysis-Setup** icon. The **DC Sweep** is again enabled, and the **DC Sweep** is chosen. This time, since the result will only be one curve, a nested operation will not be performed. After choosing **Voltage Source** and **Linear,** the **Name** will be VGG, the **Start Value** -3 V (since we now know that $V_P = -3$ V), the **End**

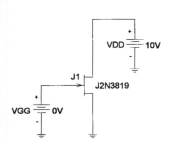

Figure 5.48 Network employed to obtain the characteristics of the n-channel J2N3819 JFET.

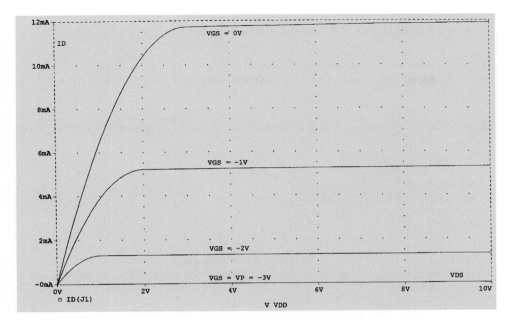

Figure 5.49 Drain characteristics for the *n*-channel J2N3819 JFET of Figure 5.48.

Value 0 V, and the **Increment** 0.01 V to get a good continuous plot. After an **OK** followed by a **Close,** the **Simulation** icon can be chosen. Once the **Probe** screen appears, choose **Plot-X-Axis Settings-Axis Variable** and choose **V(J1:g)** for the gate-to-source voltage. Choose **OK** and we're back to the **X-Axis Settings** dialog box to choose the **User Defined** range of −3 V to 0 V (which already appears because of our sweep settings). Choose **OK** again and the **Trace** ID(J1) can be chosen to result in the transfer characteristics of Fig. 5.50.

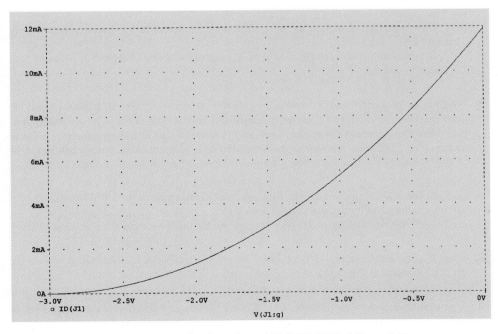

Figure 5.50 Transfer characteristics for the *n*-channel J2N3819 JFET of Figure 5.48.

PROBLEMS § 5.2 Construction and Characteristics of JFETs

1. (a) Draw the basic construction of a *p*-channel JFET.
   (b) Apply the proper biasing between drain and source and sketch the depletion region for $V_{GS} = 0$ V.

2. Using the characteristics of Fig. 5.10, determine I_D for the following levels of V_{GS} (with $V_{DS} > V_P$).
   (a) $V_{GS} = 0$ V.
   (b) $V_{GS} = -1$ V.
   (c) $V_{GS} = -1.5$ V.
   (d) $V_{GS} = -1.8$ V.
   (e) $V_{GS} = -4$ V.
   (f) $V_{GS} = -6$ V.

3. (a) Determine V_{DS} for $V_{GS} = 0$ V and $I_D = 6$ mA using the characteristics of Fig. 5.10.
   (b) Using the results of part (a), calculate the resistance of the JFET for the region $I_D = 0$ to 6 mA for $V_{GS} = 0$ V.
   (c) Determine V_{DS} for $V_{GS} = -1$ V and $I_D = 3$ mA.
   (d) Using the results of part (c), calculate the resistance of the JFET for the region $I_D = 0$ to 3 mA for $V_{GS} = -1$ V.
   (e) Determine V_{DS} for $V_{GS} = -2$ V and $I_D = 1.5$ mA.
   (f) Using the results of part (e), calculate the resistance of the JFET for the region $I_D = 0$ to 1.5 mA for $V_{GS} = -2$ V.
   (g) Defining the result of part (b) as r_o, determine the resistance for $V_{GS} = -1$ V using Eq. (5.1) and compare with the results of part (d).
   (h) Repeat part (g) for $V_{GS} = -2$ V using the same equation, and compare the results with part (f).
   (i) Based on the results of parts (g) and (h), does Eq. (5.1) appear to be a valid approximation?

4. Using the characteristics of Fig. 5.10:
   (a) Determine the difference in drain current (for $V_{DS} > V_P$) between $V_{GS} = 0$ V and $V_{GS} = -1$ V.
   (b) Repeat part (a) between $V_{GS} = -1$ and -2 V.
   (c) Repeat part (a) between $V_{GS} = -2$ and -3 V.
   (d) Repeat part (a) between $V_{GS} = -3$ and -4 V.
   (e) Is there a marked change in the difference in current levels as V_{GS} becomes increasingly negative?
   (f) Is the relationship between the change in V_{GS} and the resulting change in I_D linear or non-linear? Explain.

5. What are the major differences between the collector characteristics of a BJT transistor and the drain characteristics of a JFET transistor? Compare the units of each axis and the controlling variable. How does I_C react to increasing levels of I_B versus changes in I_D to increasingly negative values of V_{GS}? How does the spacing between steps of I_B compare to the spacing between steps of V_{GS}? Compare $V_{C_{sat}}$ to V_P in defining the nonlinear region at low levels of output voltage.

6. (a) Describe in your own words why I_G is effectively zero amperes for a JFET transistor.
   (b) Why is the input impedance to a JFET so high?
   (c) Why is the terminology *field effect* appropriate for this important three-terminal device?

7. Given $I_{DSS} = 12$ mA and $|V_P| = 6$ V, sketch a probable distribution of characteristic curves for the JFET (similar to Fig. 5.10).

8. In general, comment on the polarity of the various voltages and direction of the currents for an *n*-channel JFET versus a *p*-channel JFET.

§ 5.3 Transfer Characteristics

9. Given the characteristics of Fig. 5.51:
   (a) Sketch the transfer characteristics directly from the drain characteristics.
   (b) Using Fig. 5.51 to establish the values of I_{DSS} and V_P, sketch the transfer characteristics using Shockley's equation.
   (c) Compare the characteristics of parts (a) and (b). Are there any major differences?

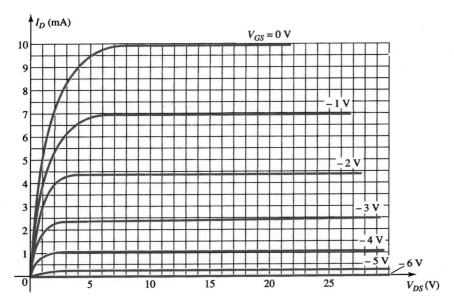

Figure 5.51   Problems 9, 17.

10. (a) Given $I_{DSS} = 12$ mA and $V_P = -4$ V, sketch the transfer characteristics for the JFET transistor.
    (b) Sketch the drain characteristics for the device of part (a).

11. Given $I_{DSS} = 9$ mA and $V_P = -3.5$ V, determine I_D when:
    (a) $V_{GS} = 0$ V.
    (b) $V_{GS} = -2$ V.
    (c) $V_{GS} = -3.5$ V.
    (d) $V_{GS} = -5$ V.

12. Given $I_{DSS} = 16$ mA and $V_P = -5$ V, sketch the transfer characteristics using the data points of Table 5.1. Determine the value of I_D at $V_{GS} = -3$ V from the curve, and compare it to the value determined using Shockley's equation. Repeat the above for $V_{GS} = -1$ V.

13. A p-channel JFET has device parameters of $I_{DSS} = 7.5$ mA and $V_P = 4$ V. Sketch the transfer characteristics.

14. Given $I_{DSS} = 6$ mA and $V_P = -4.5$ V:
    (a) Determine I_D at $V_{GS} = -2$ and -3.6 V.
    (b) Determine V_{GS} at $I_D = 3$ and 5.5 mA.

15. Given a Q-point of $I_{D_Q} = 3$ mA and $V_{GS} = -3$ V, determine I_{DSS} if $V_P = -6$ V.

§ 5.4 Specification Sheets (JFETs)

16. Define the region of operation for the 2N5457 JFET of Fig. 5.20 using the range of I_{DSS} and V_P provided. That is, sketch the transfer curve defined by the maximum I_{DSS} and V_P and the transfer curve for the minimum I_{DSS} and V_P. Then, shade in the resulting area between the two curves.

17. Define the region of operation for the JFET of Fig. 5.51 if $V_{DS_{max}} = 25$ V and $P_{D_{max}} = 120$ mW.

§ 5.5 Instrumentation

18. Using the characteristics of Fig. 5.23, determine I_D at $V_{GS} = -0.7$ V and $V_{DS} = 10$ V.

19. Referring to Fig. 5.23, is the locus of pinch-off values defined by the region of $V_{DS} < |V_P| = 3$ V?

20. Determine V_P for the characteristics of Fig. 5.23 using I_{DSS} and I_D at some value of V_{GS}. That is, simply substitute into Shockley's equation and solve for V_P. Compare the result to the assumed value of -3 V from the characteristics.

21. Using $I_{DSS} = 9$ mA and $V_P = -3$ V for the characteristics of Fig. 5.23, calculate I_D at $V_{GS} = -1$ V using Shockley's equation and compare to the level appearing in Fig. 5.23.

22. (a) Calculate the resistance associated with the JFET of Fig. 5.23 for $V_{GS} = 0$ V from $I_D = 0$ to 4 mA.
(b) Repeat part (a) for $V_{GS} = -0.5$ V from $I_D = 0$ to 3 mA.
(c) Assigning the label r_o to the result of part (a) and r_d to that of part (b), use Eq. (5.1) to determine r_d and compare to the result of part (b).

§ 5.7 Depletion-Type MOSFET

23. (a) Sketch the basic construction of a *p*-channel depletion-type MOSFET.
(b) Apply the proper drain-to-source voltage and sketch the flow of electrons for $V_{GS} = 0$ V.

24. In what ways is the construction of a depletion-type MOSFET similar to that of a JFET? In what ways is it different?

25. Explain in your own words why the application of a positive voltage to the gate of an *n*-channel depletion-type MOSFET will result in a drain current exceeding I_{DSS}.

26. Given a depletion-type MOSFET with $I_{DSS} = 6$ mA and $V_P = -3$ V, determine the drain current at $V_{GS} = -1, 0, 1$, and 2 V. Compare the difference in current levels between -1 and 0 V with the difference between 1 and 2 V. In the positive V_{GS} region, does the drain current increase at a significantly higher rate than for negative values? Does the I_D curve become more and more vertical with increasing positive values of V_{GS}? Is there a linear or a nonlinear relationship between I_D and V_{GS}? Explain.

27. Sketch the transfer and drain characteristics of an *n*-channel depletion-type MOSFET with $I_{DSS} = 12$ mA and $V_P = -8$ V for a range of $V_{GS} = -V_P$ to $V_{GS} = 1$ V.

28. Given $I_D = 14$ mA and $V_{GS} = 1$ V, determine V_P if $I_{DSS} = 9.5$ mA for a depletion-type MOSFET.

29. Given $I_D = 4$ mA at $V_{GS} = -2$ V, determine I_{DSS} if $V_P = -5$ V.

30. Using an average value of 2.9 mA for the I_{DSS} of the 2N3797 MOSFET of Fig. 5.32, determine the level of V_{GS} that will result in a maximum drain current of 20 mA if $V_P = -5$ V.

31. If the drain current for the 2N3797 MOSFET of Fig. 5.32 is 8 mA, what is the maximum permissible value of V_{DS} utilizing the maximum power rating?

§ 5.8 Enhancement-Type MOSFET

32. (a) What is the significant difference between the construction of an enhancement-type MOSFET and a depletion-type MOSFET?

(b) Sketch a p-channel enhancement-type MOSFET with the proper biasing applied ($V_{DS} > 0$ V, $V_{GS} > V_T$) and indicate the channel, the direction of electron flow, and the resulting depletion region.

(c) In your own words, briefly describe the basic operation of an enhancement-type MOSFET.

33. (a) Sketch the transfer and drain characteristics of an *n*-channel enhancement-type MOSFET if $V_T = 3.5$ V and $k = 0.4 \times 10^{-3}$ A/V^2.

(b) Repeat part (a) for the transfer characteristics if V_T is maintained at 3.5 V but k is increased by 100% to 0.8×10^{-3} A/V^2.

34. (a) Given $V_{GS(Th)} = 4$ V and $I_{D(on)} = 4$ mA at $V_{GS(on)} = 6$ V, determine k and write the general expression for I_D in the format of Eq. (5.13).

(b) Sketch the transfer characteristics for the device of part (a).

(c) Determine I_D for the device of part (a) at $V_{GS} = 2, 5$, and 10 V.

35. Given the transfer characteristics of Fig. 5.52, determine V_T and k and write the general equation for I_D.

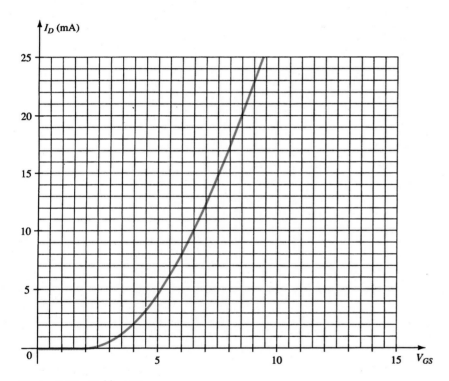

Figure 5.52   Problem 35

36. Given $k = 0.4 \times 10^{-3}$ A/V^2 and $I_{D(on)} = 3$ mA with $V_{GS(on)} = 4$ V, determine V_T.

37. The maximum drain current for the 2N4351 *n*-channel enhancement-type MOSFET is 30 mA. Determine V_{GS} at this current level if $k = 0.06 \times 10^{-3}$ A/V^2 and V_T is the maximum value.

38. Does the current of an enhancement-type MOSFET increase at about the same rate as a depletion-type MOSFET for the conduction region? Carefully review the general format of the equations, and if your mathematics background includes differential calculus, calculate dI_D/dV_{GS} and compare its magnitude.

39. Sketch the transfer characteristics of a *p*-channel enhancement-type MOSFET if $V_T = -5$ V and $k = 0.45 \times 10^{-3}$ A/V^2.

40. Sketch the curve of $I_D = 0.5 \times 10^{-3}(V_{GS}^2)$ and $I_D = 0.5 \times 10^{-3}(V_{GS} - 4)^2$ for V_{GS} from 0 to 10 V. Does $V_T = 4$ V have a significant impact on the level of I_D for this region?

§ 5.10 VMOS

41. (a) Describe in your own words why the VMOS FET can withstand a higher current and power rating than the standard construction technique.
   (b) Why do VMOS FETs have reduced channel resistance levels?
   (c) Why is a positive temperature coefficient desirable?

§ 5.11 CMOS

* **42.** (a) Describe in your own words the operation of the network of Fig. 5.46 with $V_i = 0$ V.
   (b) If the "on" MOSFET of Fig. 5.46 (with $V_i = 0$ V) has a drain current of 4 mA with $V_{DS} = 0.1$ V, what is the approximate resistance level of the device? If $I_D = 0.5 \mu$A for the "off" transistor, what is the approximate resistance of the device? Do the resulting resistance levels suggest that the desired output voltage level will result?

43. Research CMOS logic at your local or college library, and describe the range of applications and basic advantages of the approach.

---

*Please Note: Asterisks indicate more difficult problems.

FET Biasing

6.1 INTRODUCTION

In Chapter 5 we found that the biasing levels for a silicon transistor configuration can be obtained using the characteristic equations $V_{BE} = 0.7$ V, $I_C = \beta I_B$, and $I_C \cong I_E$. The linkage between input and output variables is provided by β, which is assumed to be fixed in magnitude for the analysis to be performed. The fact that beta is a constant establishes a *linear* relationship between I_C and I_B. Doubling the value of I_B will double the level of I_C, and so on.

For the field-effect transistor, the relationship between input and output quantities is *nonlinear* due to the squared term in Shockley's equation. Linear relationships result in straight lines when plotted on a graph of one variable versus the other, while nonlinear functions result in curves as obtained for the transfer characteristics of a JFET. The nonlinear relationship between I_D and V_{GS} can complicate the mathematical approach to the dc analysis of FET configurations. A graphical approach may limit solutions to tenths-place accuracy, but it is a quicker method for most FET amplifiers. Since the graphical approach is in general the most popular, the analysis of this chapter will have a graphical orientation rather than direct mathematical techniques.

Another distinct difference between the analysis of BJT and FET transistors is that the input controlling variable for a BJT transistor is a current level, while for the FET a voltage is the controlling variable. In both cases, however, the controlled variable on the output side is a current level that also defines the important voltage levels of the output circuit.

The general relationships that can be applied to the dc analysis of all FET amplifiers are

$$I_G \cong 0 \text{ A} \tag{6.1}$$

and

$$I_D = I_S \tag{6.2}$$

For JFETS and depletion-type MOSFETs, Shockley's equation is applied to relate the input and output quantities:

$$I_D = I_{DSS}\left(1 - \frac{V_{GS}}{V_P}\right)^2 \tag{6.3}$$

For enhancement-type MOSFETs, the following equation is applicable:

$$I_D = k(V_{GS} - V_T)^2 \qquad (6.4)$$

It is particularly important to realize that all of the equations above are for the *device only!* They do not change with each network configuration so long as the device is in the active region. The network simply defines the level of current and voltage associated with the operating point through its own set of equations. In reality, the dc solution of BJT and FET networks is the solution of simultaneous equations established by the device and network. The solution can be determined using a mathematical or graphical approach—a fact to be demonstrated by the first few networks to be analyzed. However, as noted earlier, the graphical approach is the most popular for FET networks and is employed in this book.

The first few sections of this chapter are limited to JFETs and the graphical approach to analysis. The depletion-type MOSFET will then be examined with its increased range of operating points, followed by the enhancement-type MOSFET. Finally, problems of a design nature are investigated to fully test the concepts and procedures introduced in the chapter.

6.2 FIXED-BIAS CONFIGURATION

The simplest of biasing arrangements for the *n*-channel JFET appears in Fig. 6.1. Referred to as the fixed-bias configuration, it is one of the few FET configurations that can be solved just as directly using either a mathematical or graphical approach. Both methods are included in this section to demonstrate the difference between the two philosophies and also to establish the fact that the same solution can be obtained using either method.

The configuration of Fig. 6.1 includes the ac levels V_i and V_o and the coupling capacitors (C_1 and C_2). Recall that the coupling capacitors are "open circuits" for the dc analysis and low impedances (essentially short circuits) for the ac analysis. The resistor R_G is present to ensure that V_i appears at the input to the FET amplifier for the ac analysis (Chapter 9). For the dc analysis,

$$I_G \cong 0 \text{ A}$$

and

$$V_{R_G} = I_G R_G = (0 \text{ A})R_G = 0 \text{ V}$$

The zero-volt drop across R_G permits replacing R_G by a short-circuit equivalent, as appearing in the network of Fig. 6.2 specifically redrawn for the dc analysis.

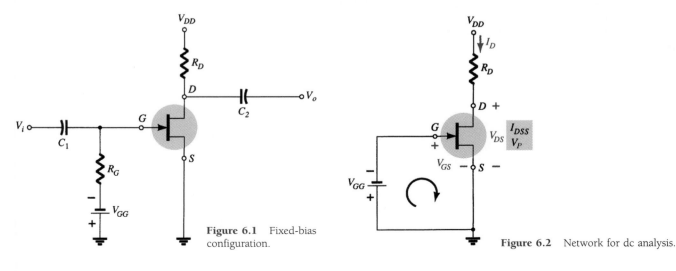

Figure 6.1 Fixed-bias configuration.

Figure 6.2 Network for dc analysis.

The fact that the negative terminal of the battery is connected directly to the defined positive potential of V_{GS} clearly reveals that the polarity of V_{GS} is directly opposite to that of V_{GG}. Applying Kirchhoff's voltage law in the clockwise direction of the indicated loop of Fig. 6.2 will result in

$$-V_{GG} - V_{GS} = 0$$

and

$$\boxed{V_{GS} = -V_{GG}} \tag{6.5}$$

Since V_{GG} is a fixed dc supply, the voltage V_{GS} is fixed in magnitude, resulting in the notation "fixed-bias configuration."

The resulting level of drain current I_D is now controlled by Shockley's equation:

$$I_D = I_{DSS}\left(1 - \frac{V_{GS}}{V_P}\right)^2$$

Since V_{GS} is a fixed quantity for this configuration, its magnitude and sign can simply be substituted into Shockley's equation and the resulting level of I_D calculated. This is one of the few instances in which a mathematical solution to a FET configuration is quite direct.

A graphical analysis would require a plot of Shockley's equation as shown in Fig. 6.3. Recall that choosing $V_{GS} = V_P/2$ will result in a drain current of $I_{DSS}/4$ when plotting the equation. For the analysis of this chapter, the three points defined by I_{DSS}, V_P, and the intersection just described will be sufficient for plotting the curve.

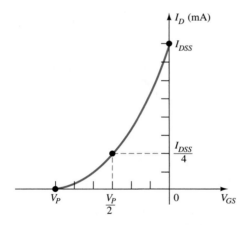

Figure 6.3 Plotting Shockley's equation.

In Fig. 6.4, the fixed level of V_{GS} has been superimposed as a vertical line at $V_{GS} = -V_{GG}$. At any point on the vertical line, the level of V_{GS} is $-V_{GG}$—the level of I_D must simply be determined on this vertical line. The point where the two curves

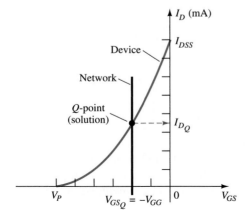

Figure 6.4 Finding the solution for the fixed-bias configuration.

intersect is the common solution to the configuration—commonly referred to as the *quiescent* or *operating point.* The subscript Q will be applied to drain current and gate-to-source voltage to identify their levels at the Q-point. Note in Fig. 6.4 that the quiescent level of I_D is determined by drawing a horizontal line from the Q-point to the vertical I_D axis as shown in Fig. 6.4. It is important to realize that once the network of Fig. 6.1 is constructed and operating, the dc levels of I_D and V_{GS} that will be measured by the meters of Fig. 6.5 are the quiescent values defined by Fig. 6.4.

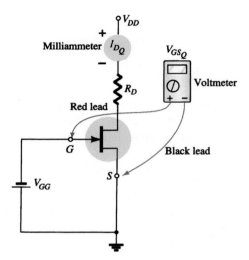

Figure 6.5 Measuring the quiescent values of I_D and V_{GS}.

The drain-to-source voltage of the output section can be determined by applying Kirchhoff's voltage law as follows:

$$+V_{DS} + I_D R_D - V_{DD} = 0$$

and

$$\boxed{V_{DS} = V_{DD} - I_D R_D} \qquad (6.6)$$

Recall that single-subscript voltages refer to the voltage at a point with respect to ground. For the configuration of Fig. 6.2,

$$\boxed{V_S = 0 \text{ V}} \qquad (6.7)$$

Using double-subscript notation:

$$V_{DS} = V_D - V_S$$

or

$$V_D = V_{DS} + V_S = V_{DS} + 0 \text{ V}$$

and

$$\boxed{V_D = V_{DS}} \qquad (6.8)$$

In addition,

$$V_{GS} = V_G - V_S$$

or

$$V_G = V_{GS} + V_S = V_{GS} + 0 \text{ V}$$

and

$$\boxed{V_G = V_{GS}} \qquad (6.9)$$

The fact that $V_D = V_{DS}$ and $V_G = V_{GS}$ is fairly obvious from the fact that $V_S = 0$ V, but the derivations above were included to emphasize the relationship that exists between double-subscript and single-subscript notation. Since the configuration requires two dc supplies, its use is limited and will not be included in the forthcoming list of the most common FET configurations.

Determine the following for the network of Fig. 6.6.

(a) V_{GS_Q}.
(b) I_{D_Q}.
(c) V_{DS}.
(d) V_D.
(e) V_G.
(f) V_S.

EXAMPLE 6.1

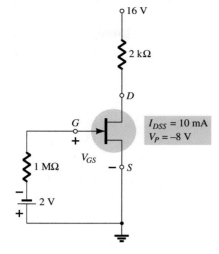

Figure 6.6   Example 6.1.

Solution

Mathematical Approach:

(a) $V_{GS_Q} = -V_{GG} = \mathbf{-2\ V}$

(b) $I_{D_Q} = I_{DSS}\left(1 - \dfrac{V_{GS}}{V_P}\right)^2 = 10\ \text{mA}\left(1 - \dfrac{-2\ \text{V}}{-8\ \text{V}}\right)^2$

$\qquad = 10\ \text{mA}(1 - 0.25)^2 = 10\ \text{mA}(0.75)^2 = 10\ \text{mA}(0.5625)$

$\qquad = \mathbf{5.625\ mA}$

(c) $V_{DS} = V_{DD} - I_D R_D = 16\ \text{V} - (5.625\ \text{mA})(2\ \text{k}\Omega)$

$\qquad = 16\ \text{V} - 11.25\ \text{V} = \mathbf{4.75\ V}$

(d) $V_D = V_{DS} = \mathbf{4.75\ V}$

(e) $V_G = V_{GS} = \mathbf{-2\ V}$

(f) $V_S = \mathbf{0\ V}$

Graphical Approach:

The resulting Shockley curve and the vertical line at $V_{GS} = -2$ V are provided in Fig. 6.7. It is certainly difficult to read beyond the second place without significantly

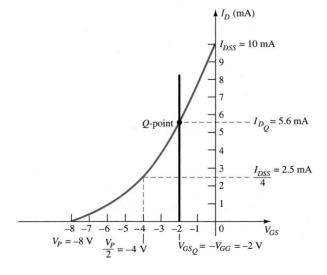

Figure 6.7   Graphical solution for the network of Fig. 6.6.

increasing the size of the figure, but a solution of 5.6 mA from the graph of Fig. 6.7 is quite acceptable. Therefore, for part (a),

$$V_{GS_Q} = -V_{GG} = \mathbf{-2\ V}$$

(b) $I_{D_Q} = \mathbf{5.6\ mA}$

(c) $V_{DS} = V_{DD} - I_D R_D = 16\ V - (5.6\ mA)(2\ k\Omega)$

$$= 16\ V - 11.2\ V = \mathbf{4.8\ V}$$

(d) $V_D = V_{DS} = \mathbf{4.8\ V}$

(e) $V_G = V_{GS} = \mathbf{-2\ V}$

(f) $V_S = \mathbf{0\ V}$

The results clearly confirm the fact that the mathematical and graphical approaches generate solutions that are quite close.

6.3 SELF-BIAS CONFIGURATION

The self-bias configuration eliminates the need for two dc supplies. The controlling gate-to-source voltage is now determined by the voltage across a resistor R_S introduced in the source leg of the configuration as shown in Fig. 6.8.

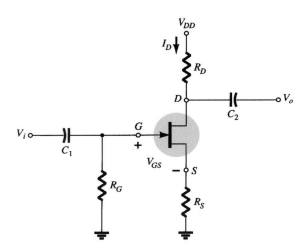

Figure 6.8 JFET self-bias configuration.

For the dc analysis, the capacitors can again be replaced by "open circuits" and the resistor R_G replaced by a short-circuit equivalent since $I_G = 0$ A. The result is the network of Fig. 6.9 for the important dc analysis.

The current through R_S is the source current I_S, but $I_S = I_D$ and

$$V_{R_S} = I_D R_S$$

For the indicated closed loop of Fig. 6.9, we find that

$$-V_{GS} - V_{R_S} = 0$$

and

$$V_{GS} = -V_{R_S}$$

or

$$\boxed{V_{GS} = -I_D R_S} \tag{6.10}$$

Note in this case that V_{GS} is a function of the output current I_D and not fixed in magnitude as occurred for the fixed-bias configuration.

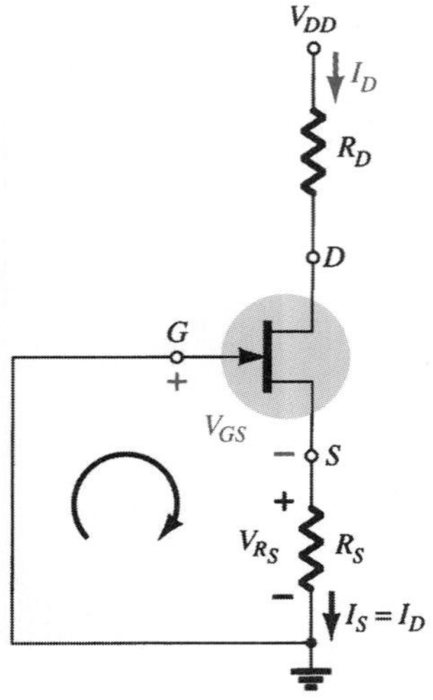

Figure 6.9 DC analysis of the self-bias configuration.

Chapter 6 FET Biasing

Equation (6.10) is defined by the network configuration, and Shockley's equation relates the input and output quantities of the device. Both equations relate the same two variables, permitting either a mathematical or graphical solution.

A mathematical solution could be obtained simply by substituting Eq. (6.10) into Shockley's equation as shown below:

$$I_D = I_{DSS}\left(1 - \frac{V_{GS}}{V_P}\right)^2$$

$$= I_{DSS}\left(1 - \frac{-I_D R_S}{V_P}\right)^2$$

or

$$I_D = I_{DSS}\left(1 + \frac{I_D R_S}{V_P}\right)^2$$

By performing the squaring process indicated and rearranging terms, an equation of the following form can be obtained:

$$I_D^2 + K_1 I_D + K_2 = 0$$

The quadratic equation can then be solved for the appropriate solution for I_D.

The sequence above defines the mathematical approach. The graphical approach requires that we first establish the device transfer characteristics as shown in Fig. 6.10. Since Eq. (6.10) defines a straight line on the same graph, let us now identify two points on the graph that are on the line and simply draw a straight line between the two points. The most obvious condition to apply is $I_D = 0$ A since it results in $V_{GS} = -I_D R_S = (0\ \text{A})R_S = 0$ V. For Eq. (6.10), therefore, one point on the straight line is defined by $I_D = 0$ A and $V_{GS} = 0$ V, as appearing on Fig. 6.10.

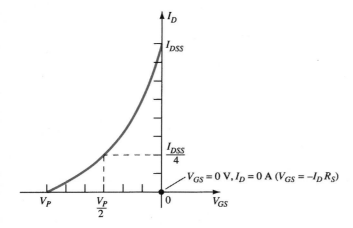

Figure 6.10 Defining a point on the self-bias line.

The second point for Eq. (6.10) requires that a level of V_{GS} or I_D be chosen and the corresponding level of the other quantity be determined using Eq. (6.10). The resulting levels of I_D and V_{GS} will then define another point on the straight line and permit an actual drawing of the straight line. Suppose, for example, that we choose a level of I_D equal to one-half the saturation level. That is,

$$I_D = \frac{I_{DSS}}{2}$$

then

$$V_{GS} = -I_D R_S = -\frac{I_{DSS} R_S}{2}$$

The result is a second point for the straight-line plot as shown in Fig. 6.11. The straight line as defined by Eq. (6.10) is then drawn and the quiescent point obtained at the

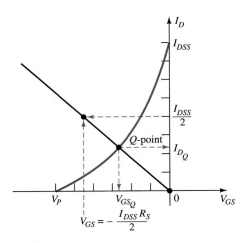

$$V_{GS} = -\frac{I_{DSS}R_S}{2}$$

Figure 6.11 Sketching the self-bias line.

intersection of the straight-line plot and the device characteristic curve. The quiescent values of I_D and V_{GS} can then be determined and used to find the other quantities of interest.

The level of V_{DS} can be determined by applying Kirchhoff's voltage law to the output circuit, with the result that

$$V_{R_S} + V_{DS} + V_{R_D} - V_{DD} = 0$$

and
$$V_{DS} = V_{DD} - V_{R_S} - V_{R_D} = V_{DD} - I_S R_S - I_D R_D$$

but
$$I_D = I_S$$

and
$$\boxed{V_{DS} = V_{DD} - I_D(R_S + R_D)} \tag{6.11}$$

In addition:

$$\boxed{V_S = I_D R_S} \tag{6.12}$$

$$\boxed{V_G = 0 \text{ V}} \tag{6.13}$$

and
$$\boxed{V_D = V_{DS} + V_S = V_{DD} - V_{R_D}} \tag{6.14}$$

EXAMPLE 6.2

Determine the following for the network of Fig. 6.12.
(a) V_{GS_Q}.
(b) I_{D_Q}.
(c) V_{DS}.
(d) V_S.
(e) V_G.
(f) V_D.

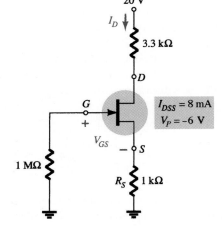

Figure 6.12 Example 6.2.

Solution

(a) The gate-to-source voltage is determined by

$$V_{GS} = -I_D R_S$$

Choosing $I_D = 4$ mA, we obtain

$$V_{GS} = -(4 \text{ mA})(1 \text{ k}\Omega) = -4 \text{ V}$$

The result is the plot of Fig. 6.13 as defined by the network.

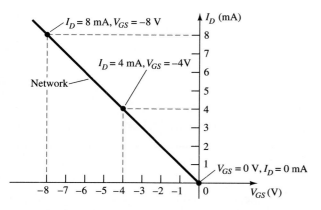

Figure 6.13 Sketching the self-bias line for the network of Fig. 6.12.

If we happen to choose $I_D = 8$ mA, the resulting value of V_{GS} would be -8 V, as shown on the same graph. In either case, the same straight line will result, clearly demonstrating that any appropriate value of I_D can be chosen as long as the corresponding value of V_{GS} as determined by Eq. (6.10) is employed. In addition, keep in mind that the value of V_{GS} could be chosen and the value of I_D calculated with the same resulting plot.

For Shockley's equation, if we choose $V_{GS} = V_P/2 = -3$ V, we find that $I_D = I_{DSS}/4 = 8$ mA$/4 = 2$ mA, and the plot of Fig. 6.14 will result, representing the characteristics of the device. The solution is obtained by superimposing the network characteristics defined by Fig. 6.13 on the device characteristics of Fig. 6.14 and finding the point of intersection of the two as indicated on Fig. 6.15. The resulting operating point results in a quiescent value of gate-to-source voltage of

$$V_{GS_Q} = \mathbf{-2.6 \text{ V}}$$

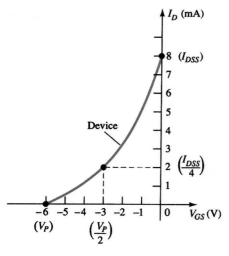

Figure 6.14 Sketching the device characteristics for the JFET of Fig. 6.12.

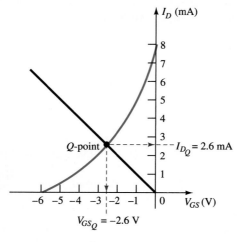

Figure 6.15 Determining the Q-point for the network of Fig. 6.12.

(b) At the quiescent point:

$$I_{D_Q} = \mathbf{2.6 \ mA}$$

(c) Eq. (6.11): $\quad V_{DS} = V_{DD} - I_D(R_S + R_D)$

$$= 20 \ V - (2.6 \ mA)(1 \ k\Omega + 3.3 \ k\Omega)$$

$$= 20 \ V - 11.18 \ V$$

$$= \mathbf{8.82 \ V}$$

(d) Eq. (6.12): $\quad V_S = I_D R_S$

$$= (2.6 \ mA)(1 \ k\Omega)$$

$$= \mathbf{2.6 \ V}$$

(e) Eq. (6.13): $\quad V_G = \mathbf{0 \ V}$

(f) Eq. (6.14): $\quad V_D = V_{DS} + V_S = 8.82 \ V + 2.6 \ V = \mathbf{11.42 \ V}$

or $\quad\quad\quad\quad V_D = V_{DD} - I_D R_D = 20 \ V - (2.6 \ mA)(3.3 \ k\Omega) = \mathbf{11.42 \ V}$

Mathcad

Mathcad will now be used to find the quiescent conditions for Example 6.2 using a process described in detail in Section 2.2. The two simultaneous equations that defined the Q-point for the network of Fig. 6.12 are

$$I_D = \frac{V_{GS}}{R_S} = \frac{V_{GS}}{1 \ k\Omega} = 1 \times 10^{-3} \ V_{GS}$$

and $\quad\quad I_D = I_{DSS}\left(1 - \frac{V_{GS}}{V_P}\right)^2 = 8 \times 10^{-3}\left(1 - \frac{V_{GS}}{-6}\right)^2$

After calling up Mathcad, we must first provide guesses for the two variables I_D and V_{GS}. The chosen values are 8 mA and −5 V, respectively. Each is entered by first entering the variable followed by **Shift :.** Next, the word **Given** must be entered, followed by the two simultaneous equations using the equal sign obtained from **Ctrl =.** Finally, the variables to be determined must be defined by **Find (ID, VGS)** as shown in Fig. 6.16. The results will appear once the equal sign is entered.

Mathcad has returned a value of −2.59 V for **VGS** which is very close to the calculated level of −2.6 V. In addition, the current of 2.59 mA is very close to the calculated level of 2.6 mA.

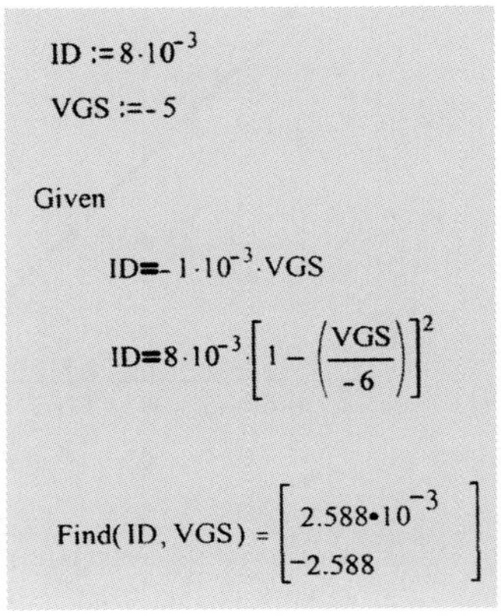

ID := 8·10⁻³

VGS := −5

Given

$$ID = -1 \cdot 10^{-3} \cdot VGS$$

$$ID = 8 \cdot 10^{-3} \cdot \left[1 - \left(\frac{VGS}{-6}\right)\right]^2$$

$$Find(ID, VGS) = \begin{bmatrix} 2.588 \cdot 10^{-3} \\ -2.588 \end{bmatrix}$$

Figure 6.16 Determining the quiescent point of operation for the network of Example 6.2.

Find the quiescent point for the network of Fig. 6.12 if:
(a) $R_S = 100\ \Omega$.
(b) $R_S = 10\ k\Omega$.

EXAMPLE 6.3

Solution

Note Fig. 6.17.

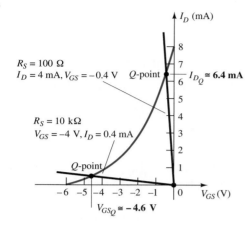

$R_S = 100\ \Omega$
$I_D = 4\ \text{mA}, V_{GS} = -0.4\ \text{V}$ Q-point $I_{D_Q} \cong 6.4\ \text{mA}$

$R_S = 10\ k\Omega$
$V_{GS} = -4\ \text{V}, I_D = 0.4\ \text{mA}$

Q-point

$V_{GS_Q} \cong -4.6\ \text{V}$

Figure 6.17   Example 6.3.

(a) With the I_D scale,

$$I_{D_Q} \cong \mathbf{6.4\ mA}$$

From Eq. (6.10),

$$V_{GS_Q} \cong \mathbf{-0.64\ V}$$

(b) With the V_{GS} scale,

$$V_{GS_Q} \cong \mathbf{-4.6\ V}$$

From Eq. (6.10),

$$I_{D_Q} \cong \mathbf{0.46\ mA}$$

In particular, note how lower levels of R_S bring the load line of the network closer to the I_D axis while increasing levels of R_S bring the load line closer to the V_{GS} axis.

Determine the following for the common-gate configuration of Fig. 6.18.
(a) V_{GS_Q}.
(b) I_{D_Q}.
(c) V_D.
(d) V_G.
(e) V_S.
(f) V_{DS}.

EXAMPLE 6.4

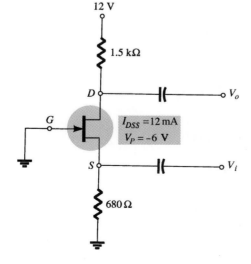

12 V

1.5 kΩ

D V_o

G $I_{DSS} = 12\ \text{mA}$
        $V_P = -6\ \text{V}$

S V_i

680 Ω

Figure 6.18   Example 6.4.

6.3   Self-Bias Configuration   299

Solution

The grounded gate terminal and the location of the input establish strong similarities with the common-base BJT amplifier. Although different in appearance from the basic structure of Fig. 6.8, the resulting dc network of Fig. 6.19 has the same basic structure as Fig. 6.9. The dc analysis can therefore proceed in the same manner as recent examples.

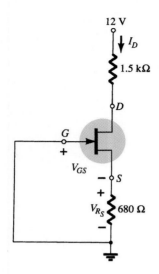

Figure 6.19 Sketching the dc equivalent of the network of Fig. 6.18.

(a) The transfer characteristics and load line appear in Fig. 6.20. In this case, the second point for the sketch of the load line was determined by choosing (arbitrarily) $I_D = 6$ mA and solving for V_{GS}. That is,

$$V_{GS} = -I_D R_S = -(6 \text{ mA})(680 \text{ } \Omega) = -4.08 \text{ V}$$

as shown in Fig. 6.20. The device transfer curve was sketched using

$$I_D = \frac{I_{DSS}}{4} = \frac{12 \text{ mA}}{4} = 3 \text{ mA}$$

and the associated value of V_{GS}:

$$V_{GS} = \frac{V_P}{2} = -\frac{6 \text{ V}}{2} = -3 \text{ V}$$

as shown on Fig. 6.20. Using the resulting quiescent point of Fig. 6.20 results in

$$V_{GS_Q} \cong -2.6 \text{ V}$$

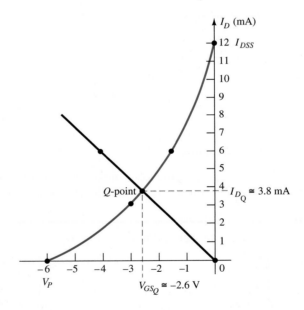

Figure 6.20 Determining the Q-point for the network of Fig. 6.18.

(b) From Fig. 6.20,

$$I_{D_Q} \cong 3.8 \text{ mA}$$

(c) $V_D = V_{DD} - I_D R_D$

$$= 12 \text{ V} - (3.8 \text{ mA})(1.5 \text{ k}\Omega) = 12 \text{ V} - 5.7 \text{ V}$$

$$= 6.3 \text{ V}$$

(d) $V_G = 0 \text{ V}$

(e) $V_S = I_D R_S = (3.8 \text{ mA})(680 \text{ } \Omega)$

$$= 2.58 \text{ V}$$

(f) $V_{DS} = V_D - V_S$

$= 6.3 \text{ V} - 2.58 \text{ V}$

$= \textbf{3.72 V}$

6.4 VOLTAGE-DIVIDER BIASING

The voltage-divider bias arrangement applied to BJT transistor amplifiers is also applied to FET amplifiers as demonstrated by Fig. 6.21. The basic construction is exactly the same, but the dc analysis of each is quite different. $I_G = 0$ A for FET amplifiers, but the magnitude of I_B for common-emitter BJT amplifiers can affect the dc levels of current and voltage in both the input and output circuits. Recall that I_B provided the link between input and output circuits for the BJT voltage-divider configuration while V_{GS} will do the same for the FET configuration.

The network of Fig. 6.21 is redrawn as shown in Fig. 6.22 for the dc analysis. Note that all the capacitors, including the bypass capacitor C_S, have been replaced by an "open-circuit" equivalent. In addition, the source V_{DD} was separated into two equivalent sources to permit a further separation of the input and output regions of the network. Since $I_G = 0$ A, Kirchhoff's current law requires that $I_{R_1} = I_{R_2}$ and the series equivalent circuit appearing to the left of the figure can be used to find the level of V_G. The voltage V_G, equal to the voltage across R_2, can be found using the voltage-divider rule as follows:

$$V_G = \frac{R_2 V_{DD}}{R_1 + R_2}$$ (6.15)

Applying Kirchhoff's voltage law in the clockwise direction to the indicated loop of Fig. 6.22 will result in

$$V_G - V_{GS} - V_{R_S} = 0$$

and

$$V_{GS} = V_G - V_{R_S}$$

Substituting $V_{R_S} = I_S R_S = I_D R_S$, we have

$$V_{GS} = V_G - I_D R_S$$ (6.16)

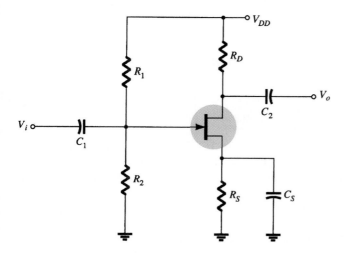

Figure 6.21 Voltage-divider bias arrangement.

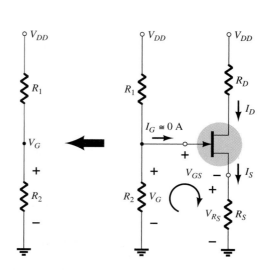

Figure 6.22 Redrawn network of Fig. 6.21 for dc analysis.

The result is an equation that continues to include the same two variables appearing in Shockley's equation: V_{GS} and I_D. The quantities V_G and R_S are fixed by the network construction. Equation (6.16) is still the equation for a straight line, but the origin is no longer a point in the plotting of the line. The procedure for plotting Eq. (6.16) is not a difficult one and will proceed as follows. Since any straight line requires two points to be defined, let us first use the fact that anywhere on the horizontal axis of Fig. 6.23 the current $I_D = 0$ mA. If we therefore select I_D to be 0 mA, we are in essence stating that we are somewhere on the horizontal axis. The exact location can be determined simply by substituting $I_D = 0$ mA into Eq. (6.16) and finding the resulting value of V_{GS} as follows:

$$V_{GS} = V_G - I_D R_S$$
$$= V_G - (0 \text{ mA})R_S$$

and
$$\boxed{V_{GS} = V_G|_{I_D = 0 \text{ mA}}} \qquad (6.17)$$

The result specifies that whenever we plot Eq. (6.16), if we choose $I_D = 0$ mA, the value of V_{GS} for the plot will be V_G volts. The point just determined appears in Fig. 6.23.

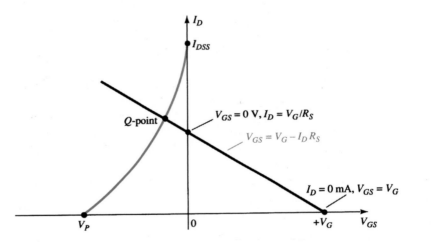

Figure 6.23   Sketching the network equation for the voltage-divider configuration.

For the other point, let us now employ the fact that at any point on the vertical axis $V_{GS} = 0$ V and solve for the resulting value of I_D:

$$V_{GS} = V_G - I_D R_S$$
$$0 \text{ V} = V_G - I_D R_S$$

and
$$\boxed{I_D = \frac{V_G}{R_S}\bigg|_{V_{GS} = 0 \text{ V}}} \qquad (6.18)$$

The result specifies that whenever we plot Eq. (6.16), if $V_{GS} = 0$ V, the level of I_D is determined by Eq. (6.18). This intersection also appears on Fig. 6.23.

The two points defined above permit the drawing of a straight line to represent Eq. (6.16). The intersection of the straight line with the transfer curve in the region to the left of the vertical axis will define the operating point and the corresponding levels of I_D and V_{GS}.

Since the intersection on the vertical axis is determined by $I_D = V_G/R_S$ and V_G is fixed by the input network, increasing values of R_S will reduce the level of the I_D intersection as shown in Fig. 6.24. It is fairly obvious from Fig. 6.24 that:

Increasing values of R_S result in lower quiescent values of I_D and more negative values of V_{GS}.

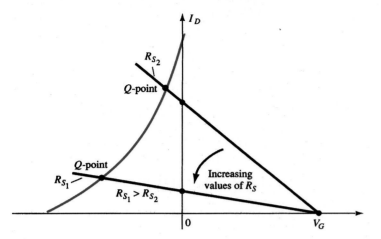

Figure 6.24 Effect of R_S on the resulting Q-point.

Once the quiescent values of I_{D_Q} and V_{GS_Q} are determined, the remaining network analysis can be performed in the usual manner. That is,

$$V_{DS} = V_{DD} - I_D(R_D + R_S)$$ (6.19)

$$V_D = V_{DD} - I_D R_D$$ (6.20)

$$V_S = I_D R_S$$ (6.21)

$$I_{R_1} = I_{R_2} = \frac{V_{DD}}{R_1 + R_2}$$ (6.22)

Determine the following for the network of Fig. 6.25.
(a) I_{D_Q} and V_{GS_Q}.
(b) V_D.
(c) V_S.
(d) V_{DS}.
(e) V_{DG}.

EXAMPLE 6.5

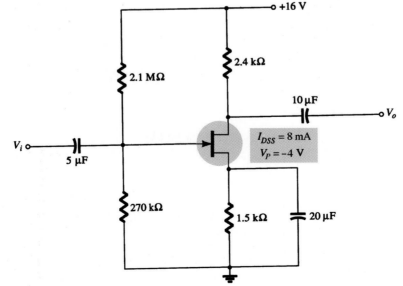

Figure 6.25 Example 6.5.

6.4 Voltage-Divider Biasing

303

Solution

(a) For the transfer characteristics, if $I_D = I_{DSS}/4 = 8 \text{ mA}/4 = 2 \text{ mA}$, then $V_{GS} = V_P/2 = -4 \text{ V}/2 = -2 \text{ V}$. The resulting curve representing Shockley's equation appears in Fig. 6.26. The network equation is defined by

$$V_G = \frac{R_2 V_{DD}}{R_1 + R_2}$$

$$= \frac{(270 \text{ k}\Omega)(16 \text{ V})}{2.1 \text{ M}\Omega + 0.27 \text{ M}\Omega}$$

$$= 1.82 \text{ V}$$

and
$$V_{GS} = V_G - I_D R_S$$

$$= 1.82 \text{ V} - I_D(1.5 \text{ k}\Omega)$$

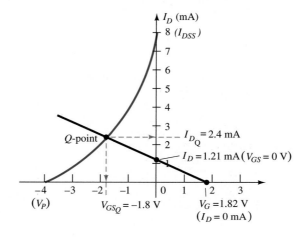

Figure 6.26 Determining the Q-point for the network of Fig. 6.25.

When $I_D = 0 \text{ mA}$:

$$V_{GS} = +1.82 \text{ V}$$

When $V_{GS} = 0 \text{ V}$:

$$I_D = \frac{1.82 \text{ V}}{1.5 \text{ k}\Omega} = 1.21 \text{ mA}$$

The resulting bias line appears on Fig. 6.26 with quiescent values of

$$I_{D_Q} = \textbf{2.4 mA}$$

and
$$V_{GS_Q} = \textbf{−1.8 V}$$

(b) $V_D = V_{DD} - I_D R_D$

$\qquad = 16 \text{ V} - (2.4 \text{ mA})(2.4 \text{ k}\Omega)$

$\qquad = \textbf{10.24 V}$

(c) $V_S = I_D R_S = (2.4 \text{ mA})(1.5 \text{ k}\Omega)$

$\qquad = \textbf{3.6 V}$

(d) $V_{DS} = V_{DD} - I_D(R_D + R_S)$

$\qquad = 16 \text{ V} - (2.4 \text{ mA})(2.4 \text{ k}\Omega + 1.5 \text{ k}\Omega)$

$\qquad = \textbf{6.64 V}$

or $V_{DS} = V_D - V_S = 10.24\,\text{V} - 3.6\,\text{V}$

$\qquad = \mathbf{6.64\ V}$

(e) Although seldom requested, the voltage V_{DG} can easily be determined using

$$V_{DG} = V_D - V_G$$

$$= 10.24\,\text{V} - 1.82\,\text{V}$$

$$= \mathbf{8.42\ V}$$

Although the basic construction of the network in the next example is quite different from the voltage-divider bias arrangement, the resulting equations require a solution very similar to that just described. Note that the network employs a supply at the drain and source.

Determine the following for the network of Fig. 6.27.
(a) I_{D_Q} and V_{GS_Q}.
(b) V_{DS}.
(c) V_D.
(d) V_S.

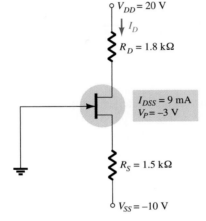

$V_{DD} = 20\,\text{V}$

I_D

$R_D = 1.8\,\text{k}\Omega$

$I_{DSS} = 9\,\text{mA}$
$V_P = -3\,\text{V}$

$R_S = 1.5\,\text{k}\Omega$

$V_{SS} = -10\,\text{V}$

Figure 6.27 Example 6.6.

Solution

(a) An equation for V_{GS} in terms of I_D is obtained by applying Kirchhoff's voltage law to the input section of the network as redrawn in Fig. 6.28.

$$-V_{GS} - I_S R_S + V_{SS} = 0$$

or

$$V_{GS} = V_{SS} - I_S R_S$$

but

$$I_S = I_D$$

and

$$\boxed{V_{GS} = V_{SS} - I_D R_S} \qquad (6.23)$$

The result is an equation very similar in format to Eq. (6.16) that can be superimposed on the transfer characteristics using the procedure described for Eq. (6.16). That is, for this example,

$$V_{GS} = 10\,\text{V} - I_D(1.5\,\text{k}\Omega)$$

For $I_D = 0\,\text{mA}$,

$$V_{GS} = V_{SS} = 10\,\text{V}$$

For $V_{GS} = 0\,\text{V}$,

$$0 = 10\,\text{V} - I_D(1.5\,\text{k}\Omega)$$

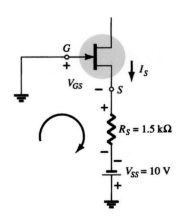

G

V_{GS}

I_S

S

$R_S = 1.5\,\text{k}\Omega$

$V_{SS} = 10\,\text{V}$

Figure 6.28 Determining the network equation for the configuration of Fig. 6.27.

and
$$I_D = \frac{10\ V}{1.5\ k\Omega} = 6.67\ mA$$

The resulting plot points are identified on Fig. 6.29.

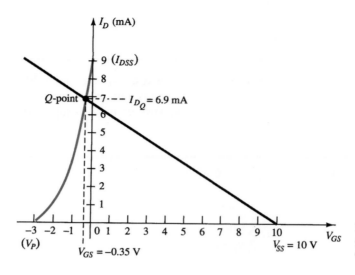

Figure 6.29 Determining the Q-point for the network of Fig. 6.27.

The transfer characteristics are sketched using the plot point established by $V_{GS} = V_P/2 = -3\ V/2 = -1.5\ V$ and $I_D = I_{DSS}/4 = 9\ mA/4 = 2.25\ mA$, as also appearing on Fig. 6.29. The resulting operating point establishes the following quiescent levels:

$$I_{D_Q} = \textbf{6.9 mA}$$

$$V_{GS_Q} = \textbf{-0.35 V}$$

(b) Applying Kirchhoff's voltage law to the output side of Fig. 6.27 will result in

$$-V_{SS} + I_S R_S + V_{DS} + I_D R_D - V_{DD} = 0$$

Substituting $I_S = I_D$ and rearranging gives

$$\boxed{V_{DS} = V_{DD} + V_{SS} - I_D(R_D + R_S)} \tag{6.24}$$

which for this example results in

$$V_{DS} = 20\ V + 10\ V - (6.9\ mA)(1.8\ k\Omega + 1.5\ k\Omega)$$

$$= 30\ V - 22.77\ V$$

$$= \textbf{7.23 V}$$

(c) $\quad V_D = V_{DD} - I_D R_D$

$$= 20\ V - (6.9\ mA)(1.8\ k\Omega) = 20\ V - 12.42\ V$$

$$= \textbf{7.58 V}$$

(d) $V_{DS} = V_D - V_S$

or $\quad V_S = V_D - V_{DS}$

$$= 7.58\ V - 7.23\ V$$

$$= \textbf{0.35 V}$$

6.5 DEPLETION-TYPE MOSFETs

The similarities in appearance between the transfer curves of JFETs and depletion-type MOSFETs permit a similar analysis of each in the dc domain. The primary difference between the two is the fact that depletion-type MOSFETs permit operating points with positive values of V_{GS} and levels of I_D that exceed I_{DSS}. In fact, for all the configurations discussed thus far, the analysis is the same if the JFET is replaced by a depletion-type MOSFET.

The only undefined part of the analysis is how to plot Shockley's equation for positive values of V_{GS}. How far into the region of positive values of V_{GS} and values of I_D greater than I_{DSS} does the transfer curve have to extend? For most situations, this required range will be fairly well defined by the MOSFET parameters and the resulting bias line of the network. A few examples will reveal the impact of the change in device on the resulting analysis.

For the n-channel depletion-type MOSFET of Fig. 6.30, determine: *EXAMPLE 6.7*
(a) I_{D_Q} and V_{GS_Q}.
(b) V_{DS}.

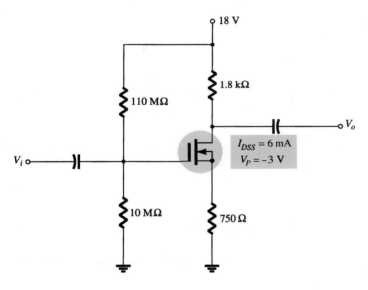

Figure 6.30 Example 6.7.

Solution

(a) For the transfer characteristics, a plot point is defined by $I_D = I_{DSS}/4 = 6\text{ mA}/4 = 1.5$ mA and $V_{GS} = V_P/2 = -3\text{ V}/2 = -1.5$ V. Considering the level of V_P and the fact that Shockley's equation defines a curve that rises more rapidly as V_{GS} becomes more positive, a plot point will be defined at $V_{GS} = +1$ V. Substituting into Shockley's equation yields

$$I_D = I_{DSS}\left(1 - \frac{V_{GS}}{V_P}\right)^2$$

$$= 6\text{ mA}\left(1 - \frac{+1\text{ V}}{-3\text{ V}}\right)^2 = 6\text{ mA}\left(1 + \frac{1}{3}\right)^2 = 6\text{ mA}(1.778)$$

$$= 10.67\text{ mA}$$

The resulting transfer curve appears in Fig. 6.31. Proceeding as described for JFETs, we have:

Eq. (6.15): $V_G = \dfrac{10 \text{ M}\Omega(18 \text{ V})}{10 \text{ M}\Omega + 110 \text{ M}\Omega} = 1.5 \text{ V}$

Eq. (6.16): $V_{GS} = V_G - I_D R_S = 1.5 \text{ V} - I_D(750 \text{ }\Omega)$

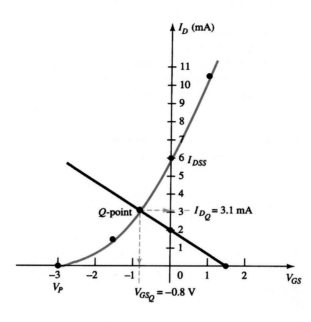

Figure 6.31 Determining the Q-point for the network of Fig. 6.30.

Setting $I_D = 0$ mA results in

$$V_{GS} = V_G = 1.5 \text{ V}$$

Setting $V_{GS} = 0$ V yields

$$I_D = \frac{V_G}{R_S} = \frac{1.5 \text{ V}}{750 \text{ }\Omega} = 2 \text{ mA}$$

The plot points and resulting bias line appear in Fig. 6.31. The resulting operating point:

$$I_{D_Q} = \textbf{3.1 mA}$$

$$V_{GS_Q} = \textbf{-0.8 V}$$

(b) Eq. (6.19): $V_{DS} = V_{DD} - I_D(R_D + R_S)$

$= 18 \text{ V} - (3.1 \text{ mA})(1.8 \text{ k}\Omega + 750 \text{ }\Omega)$

$\cong \textbf{10.1 V}$

EXAMPLE 6.8

Repeat Example 6.7 with $R_S = 150 \text{ }\Omega$.

Solution

(a) The plot points are the same for the transfer curve as shown in Fig. 6.32. For the bias line,

$$V_{GS} = V_G - I_D R_S = 1.5 \text{ V} - I_D(150 \text{ }\Omega)$$

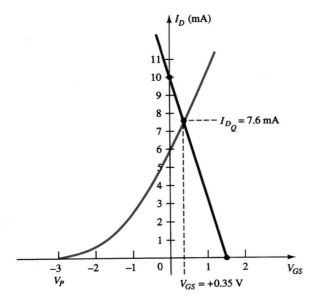

Figure 6.32 Example 6.8.

Setting $I_D = 0$ mA results in

$$V_{GS} = 1.5 \text{ V}$$

Setting $V_{GS} = 0$ V yields

$$I_D = \frac{V_G}{R_S} = \frac{1.5 \text{ V}}{150 \ \Omega} = 10 \text{ mA}$$

The bias line is included on Fig. 6.32. Note in this case that the quiescent point results in a drain current that exceeds I_{DSS}, with a positive value for V_{GS}. The result:

$$I_{D_Q} = \mathbf{7.6 \text{ mA}}$$

$$V_{GS_Q} = \mathbf{+0.35 \text{ V}}$$

(b) Eq. (6.19): $V_{DS} = V_{DD} - I_D(R_D + R_S)$

$$= 18 \text{ V} - (7.6 \text{ mA})(1.8 \text{ k}\Omega + 150 \ \Omega)$$

$$= \mathbf{3.18 \text{ V}}$$

Determine the following for the network of Fig. 6.33.
(a) I_{D_Q} and V_{GS_Q}.
(b) V_D.

EXAMPLE 6.9

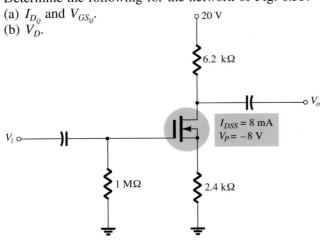

Figure 6.33 Example 6.9.

Solution

(a) The self-bias configuration results in

$$V_{GS} = -I_D R_S$$

as obtained for the JFET configuration, establishing the fact that V_{GS} must be less than zero volts. There is therefore no requirement to plot the transfer curve for positive values of V_{GS}, although it was done on this occasion to complete the transfer characteristics. A plot point for the transfer characteristics for $V_{GS} < 0$ V is

$$I_D = \frac{I_{DSS}}{4} = \frac{8 \text{ mA}}{4} = 2 \text{ mA}$$

and

$$V_{GS} = \frac{V_P}{2} = \frac{-8 \text{ V}}{2} = -4 \text{ V}$$

and for $V_{GS} > 0$ V, since $V_P = -8$ V, we will choose

$$V_{GS} = +2 \text{ V}$$

and

$$I_D = I_{DSS}\left(1 - \frac{V_{GS}}{V_P}\right)^2 = 8 \text{ mA}\left(1 - \frac{+2 \text{ V}}{-8 \text{ V}}\right)^2$$

$$= 12.5 \text{ mA}$$

The resulting transfer curve appears in Fig. 6.34. For the network bias line, at $V_{GS} = 0$ V, $I_D = 0$ mA. Choosing $V_{GS} = -6$ V gives

$$I_D = -\frac{V_{GS}}{R_S} = -\frac{-6 \text{ V}}{2.4 \text{ k}\Omega} = 2.5 \text{ mA}$$

The resulting Q-point:

$$I_{D_Q} = \mathbf{1.7 \text{ mA}}$$

$$V_{GS_Q} = \mathbf{-4.3 \text{ V}}$$

(b) $V_D = V_{DD} - I_D R_D$

$\qquad = 20 \text{ V} - (1.7 \text{ mA})(6.2 \text{ k}\Omega)$

$\qquad = \mathbf{9.46 \text{ V}}$

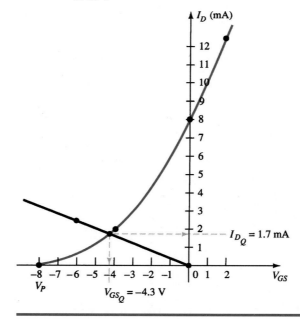

Figure 6.34 Determining the Q-point for the network of Fig. 6.33.

Chapter 6 FET Biasing

The example to follow employs a design that can also be applied to JFET transistors. At first impression it appears rather simplistic, but in fact it often causes some confusion when first analyzed due to the special point of operation.

Determine V_{DS} for the network of Fig. 6.35.

EXAMPLE 6.10

Solution

The direct connection between the gate and source terminals requires that

$$V_{GS} = 0 \text{ V}$$

Since V_{GS} is fixed at 0 V, the drain current must be I_{DSS} (by definition). In other words,

$$V_{GS_Q} = \mathbf{0 \text{ V}}$$

and

$$I_{D_Q} = \mathbf{10 \text{ mA}}$$

There is therefore no need to draw the transfer curve and

$$V_D = V_{DD} - I_D R_D = 20 \text{ V} - (10 \text{ mA})(1.5 \text{ k}\Omega)$$

$$= 20 \text{ V} - 15 \text{ V}$$

$$= \mathbf{5 \text{ V}}$$

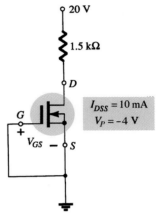

Figure 6.35 Example 6.10.

6.6 ENHANCEMENT-TYPE MOSFETs

The transfer characteristics of the enhancement-type MOSFET are quite different from those encountered for the JFET and depletion-type MOSFETs, resulting in a graphical solution quite different from the preceding sections. First and foremost, recall that for the n-channel enhancement-type MOSFET, the drain current is zero for levels of gate-to-source voltage less than the threshold level $V_{GS(Th)}$, as shown in Fig. 6.36. For levels of V_{GS} greater than $V_{GS(Th)}$, the drain current is defined by

$$\boxed{I_D = k(V_{GS} - V_{GS(Th)})^2} \tag{6.25}$$

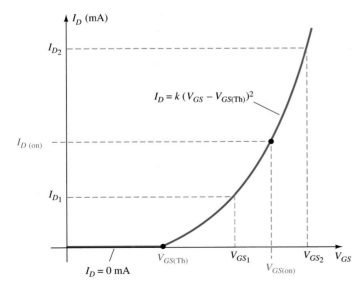

Figure 6.36 Transfer characteristics of an n-channel enhancement-type MOSFET.

Since specification sheets typically provide the threshold voltage and a level of drain current ($I_{D(\text{on})}$) and its corresponding level of $V_{GS(\text{on})}$, two points are defined immediately as shown in Fig. 6.36. To complete the curve, the constant k of Eq. (6.25) must be determined from the specification sheet data by substituting into Eq. (6.25) and solving for k as follows:

$$I_D = k(V_{GS} - V_{GS(\text{Th})})^2$$

$$I_{D(\text{on})} = k(V_{GS(\text{on})} - V_{GS(\text{Th})})^2$$

and

$$\boxed{k = \frac{I_{D(\text{on})}}{(V_{GS(\text{on})} - V_{GS(\text{Th})})^2}}$$

(6.26)

Once k is defined, other levels of I_D can be determined for chosen values of V_{GS}. Typically, a point between $V_{GS(\text{Th})}$ and $V_{GS(\text{on})}$ and one just greater than $V_{GS(\text{on})}$ will provide a sufficient number of points to plot Eq. (6.25) (note I_{D_1} and I_{D_2} on Fig. 6.36).

Feedback Biasing Arrangement

A popular biasing arrangement for enhancement-type MOSFETs is provided in Fig. 6.37. The resistor R_G brings a suitably large voltage to the gate to drive the MOSFET "on." Since $I_G = 0$ mA and $V_{R_G} = 0$ V, the dc equivalent network appears as shown in Fig. 6.38.

A direct connection now exists between drain and gate, resulting in

$$V_D = V_G$$

and

$$\boxed{V_{DS} = V_{GS}}$$

(6.27)

For the output circuit,

$$V_{DS} = V_{DD} - I_D R_D$$

which becomes the following after substituting Eq. (6.27):

$$\boxed{V_{GS} = V_{DD} - I_D R_D}$$

(6.28)

The result is an equation that relates the same two variables as Eq. (6.25), permitting the plot of each on the same set of axes.

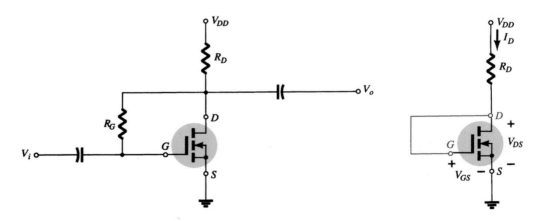

Figure 6.37 Feedback biasing arrangement.

Figure 6.38 DC equivalent of the network of Fig. 6.37.

Since Eq. (6.28) is that of a straight line, the same procedure described earlier can be employed to determine the two points that will define the plot on the graph. Substituting $I_D = 0$ mA into Eq. (6.28) gives

$$\boxed{V_{GS} = V_{DD}|_{I_D = 0\,\text{mA}}}$$ (6.29)

Substituting $V_{GS} = 0$ V into Eq. (6.28), we have

$$\boxed{I_D = \frac{V_{DD}}{R_D}\bigg|_{V_{GS} = 0\,\text{V}}}$$ (6.30)

The plots defined by Eqs. (6.25) and (6.28) appear in Fig. 6.39 with the resulting operating point.

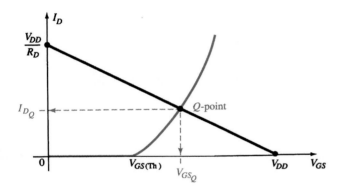

Figure 6.39 Determining the Q-point for the network of Fig. 6.37.

Determine I_{D_Q} and V_{DS_Q} for the enhancement-type MOSFET of Fig. 6.40.

EXAMPLE 6.11

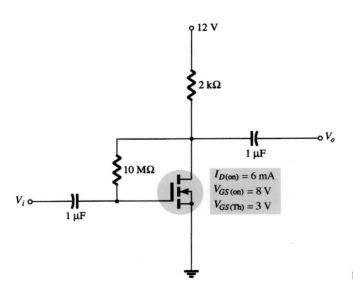

Figure 6.40 Example 6.11.

Solution

Plotting the Transfer Curve:

Two points are defined immediately as shown in Fig. 6.41. Solving for k:

$$\text{Eq. (6.26):} \quad k = \frac{I_{D(\text{on})}}{(V_{GS(\text{on})} - V_{GS(\text{Th})})^2}$$

$$= \frac{6\ \text{mA}}{(8\ \text{V} - 3\ \text{V})^2} = \frac{6 \times 10^{-3}}{25}\ \text{A/V}^2$$

$$= \mathbf{0.24 \times 10^{-3}\ A/V^2}$$

For $V_{GS} = 6\ \text{V}$ (between 3 and 8 V):

$$I_D = 0.24 \times 10^{-3}(6\ \text{V} - 3\ \text{V})^2 = 0.24 \times 10^{-3}(9)$$

$$= 2.16\ \text{mA}$$

as shown on Fig. 6.41. For $V_{GS} = 10\ \text{V}$ (slightly greater than $V_{GS(\text{Th})}$):

$$I_D = 0.24 \times 10^{-3}(10\ \text{V} - 3\ \text{V})^2 = 0.24 \times 10^{-3}(49)$$

$$= 11.76\ \text{mA}$$

as also appearing on Fig. 6.41. The four points are sufficient to plot the full curve for the range of interest as shown in Fig. 6.41.

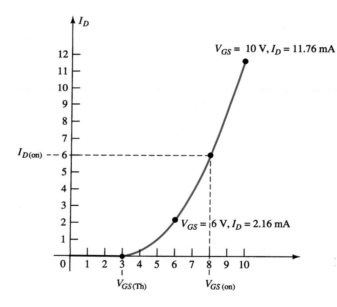

Figure 6.41 Plotting the transfer curve for the MOSFET of Fig. 6.40.

For the Network Bias Line:

$$V_{GS} = V_{DD} - I_D R_D$$

$$= 12\ \text{V} - I_D(2\ \text{k}\Omega)$$

$$\text{Eq. (6.29):} \quad V_{GS} = V_{DD} = 12\ \text{V}\big|_{I_D = 0\ \text{mA}}$$

$$\text{Eq. (6.30):} \quad I_D = \frac{V_{DD}}{R_D} = \frac{12\ \text{V}}{2\ \text{k}\Omega} = 6\ \text{mA}\big|_{V_{GS} = 0\ \text{V}}$$

The resulting bias line appears in Fig. 6.42.

At the operating point:

$$I_{D_Q} = \mathbf{2.75\ mA}$$

and
$$V_{GS_Q} = 6.4\ V$$

with
$$V_{DS_Q} = V_{GS_Q} = \mathbf{6.4\ V}$$

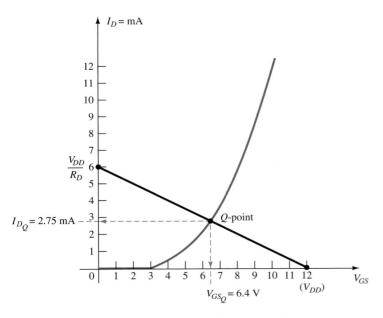

Figure 6.42   Determining the Q-point for the network of Fig. 6.40.

Voltage-Divider Biasing Arrangement

A second popular biasing arrangement for the enhancement-type MOSFET appears in Fig. 6.43. The fact that $I_G = 0$ mA results in the following equation for V_{GG} as derived from an application of the voltage-divider rule:

$$V_G = \frac{R_2 V_{DD}}{R_1 + R_2} \tag{6.31}$$

Applying Kirchhoff's voltage law around the indicated loop of Fig. 6.43 will result in

$$+V_G - V_{GS} - V_{R_S} = 0$$

and
$$V_{GS} = V_G - V_{R_S}$$

or
$$V_{GS} = V_G - I_D R_S \tag{6.32}$$

For the output section:

$$V_{R_S} + V_{DS} + V_{R_D} - V_{DD} = 0$$

and
$$V_{DS} = V_{DD} - V_{R_S} - V_{R_D}$$

or
$$V_{DS} = V_{DD} - I_D(R_S + R_D) \tag{6.33}$$

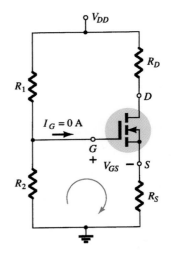

Figure 6.43   Voltage-divider biasing arrangement for an n-channel enhancement MOSFET.

6.6   **Enhancement-Type MOSFETs**   315

Since the characteristics are a plot of I_D versus V_{GS} and Eq. (6.32) relates the same two variables, the two curves can be plotted on the same graph and a solution determined at their intersection. Once I_{D_Q} and V_{GS_Q} are known, all the remaining quantities of the network such as V_{DS}, V_D, and V_S can be determined.

EXAMPLE 6.12   Determine I_{D_Q}, V_{GS_Q}, and V_{DS} for the network of Fig. 6.44.

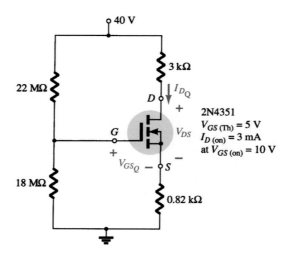

Figure 6.44 Example 6.12.

Solution

Network:

$$\text{Eq. (6.31):} \quad V_G = \frac{R_2 V_{DD}}{R_1 + R_2} = \frac{(18 \text{ M}\Omega)(40 \text{ V})}{22 \text{ M}\Omega + 18 \text{ M}\Omega} = 18 \text{ V}$$

$$\text{Eq. (6.32):} \quad V_{GS} = V_G - I_D R_S = 18 \text{ V} - I_D(0.82 \text{ k}\Omega)$$

When $I_D = 0$ mA,

$$V_{GS} = 18 \text{ V} - (0 \text{ mA})(0.82 \text{ k}\Omega) = 18 \text{ V}$$

as appearing on Fig. 6.45. When $V_{GS} = 0$ V,

$$V_{GS} = 18 \text{ V} - I_D(0.82 \text{ k}\Omega)$$

$$0 = 18 \text{ V} - I_D(0.82 \text{ k}\Omega)$$

$$I_D = \frac{18 \text{ V}}{0.82 \text{ k}\Omega} = 21.95 \text{ mA}$$

as appearing on Fig. 6.45.

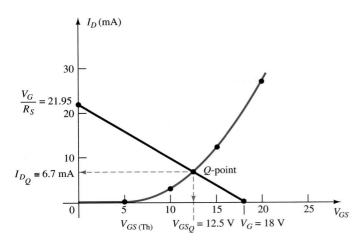

Figure 6.45 Determining the Q-point for the network of Example 6.12.

Device:

$$V_{GS(Th)} = 5 \text{ V}, \qquad I_{D(on)} = 3 \text{ mA with } V_{GS(on)} = 10 \text{ V}$$

$$\text{Eq. (6.26):} \quad k = \frac{I_{D(on)}}{(V_{GS(on)} - V_{GS(Th)})^2}$$

$$= \frac{3 \text{ mA}}{(10 \text{ V} - 5 \text{ V})^2} = 0.12 \times 10^{-3} \text{ A/V}^2$$

and

$$I_D = k(V_{GS} - V_{GS(Th)})^2$$
$$= 0.12 \times 10^{-3}(V_{GS} - 5)^2$$

which is plotted on the same graph (Fig. 6.45). From Fig. 6.45,

$$I_{D_Q} \cong \mathbf{6.7 \text{ mA}}$$

$$V_{GS_Q} = \mathbf{12.5 \text{ V}}$$

$$\text{Eq. (6.33):} \quad V_{DS} = V_{DD} - I_D(R_S + R_D)$$

$$= 40 \text{ V} - (6.7 \text{ mA})(0.82 \text{ k}\Omega + 3.0 \text{ k}\Omega)$$

$$= 40 \text{ V} - 25.6 \text{ V}$$

$$= \mathbf{14.4 \text{ V}}$$

6.7 SUMMARY TABLE

Now that the most popular biasing arrangements for the various FETs have been introduced, Table 6.1 reviews the basic results and demonstrates the similarity in approach for a number of configurations. It also reveals that the general analysis of dc configurations for FETs is not overly complex. Once the transfer characteristics are established, the network self-bias line can be drawn and the Q-point determined at the intersection of the device transfer characteristic and the network bias curve. The remaining analysis is simply an application of the basic laws of circuit analysis.

TABLE 6.1 FET Bias Configurations

Type	Configuration	Pertinent Equations	Graphical Solution
JFET Fixed-bias		$V_{GS_Q} = -V_{GG}$ $V_{DS} = V_{DD} - I_D R_S$	
JFET Self-bias		$V_{GS} = -I_D R_S$ $V_{DS} = V_{DD} - I_D(R_D + R_S)$	
JFET Voltage-divider bias		$V_G = \dfrac{R_2 V_{DD}}{R_1 + R_2}$ $V_{GS} = V_G - I_D R_S$ $V_{DS} = V_{DD} - I_D(R_D + R_S)$	
JFET Common-gate		$V_{GS} = V_{SS} - I_D R_S$ $V_{DS} = V_{DD} + V_{SS} - I_D(R_D + R_S)$	
JFET $(V_{GS_Q} = 0\text{ V})$		$V_{GS_Q} = 0\text{ V}$ $I_{D_Q} = I_{DSS}$	
JFET $(R_D = 0\ \Omega)$		$V_{GS} = -I_D R_S$ $V_D = V_{DD}$ $V_S = I_D R_S$ $V_{DS} = V_{DD} - I_S R_S$	
Depletion-type MOSFET Fixed-bias		$V_{GS_Q} = +V_{GG}$ $V_{DS} = V_{DD} - I_D R_S$	
Depletion-type MOSFET Voltage-divider bias		$V_G = \dfrac{R_2 V_{DD}}{R_1 + R_2}$ $V_{GS} = V_G - I_S R_S$ $V_{DS} = V_{DD} - I_D(R_D + R_S)$	
Enhancement type MOSFET Feedback configuration		$V_{GS} = V_{DS}$ $V_{GS} = V_{DD} - I_D R_D$	
Enhancement type MOSFET Voltage-divider bias		$V_G = \dfrac{R_2 V_{DD}}{R_1 + R_2}$ $V_{GS} = V_G - I_D R_S$	

6.8 COMBINATION NETWORKS

Now that the dc analysis of a variety of BJT and FET configurations is established, the opportunity to analyze networks with both types of devices presents itself. Fundamentally, the analysis simply requires that we *first* approach the device that will provide a terminal voltage or current level. The door is then usually open to calculate other quantities and concentrate on the remaining unknowns. These are usually particularly interesting problems due to the challenge of finding the opening and then using the results of the past few sections and Chapter 5 to find the important quantities for each device. The equations and relationships used are simply those we have now employed on more than one occasion—no need to develop any new methods of analysis.

Determine the levels of V_D and V_C for the network of Fig. 6.46.

EXAMPLE 6.13

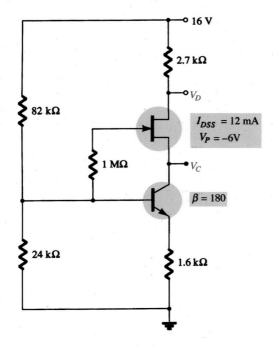

Figure 6.46   Example 6.13.

Solution

From past experience we now realize that V_{GS} is typically an important quantity to determine or write an equation for when analyzing JFET networks. Since V_{GS} is a level for which an immediate solution is not obvious, let us turn our attention to the transistor configuration. The voltage-divider configuration is one where the approximate technique can be applied ($\beta R_E = (180 \times 1.6 \text{ k}\Omega) = 288 \text{ k}\Omega > 10R_2 = 240 \text{ k}\Omega$), permitting a determination of V_B using the voltage-divider rule on the input circuit.
   For V_B:

$$V_B = \frac{24 \text{ k}\Omega(16 \text{ V})}{82 \text{ k}\Omega + 24 \text{ k}\Omega} = 3.62 \text{ V}$$

Using the fact that $V_{BE} = 0.7$ V results in

$$V_E = V_B - V_{BE} = 3.62 \text{ V} - 0.7 \text{ V}$$
$$= 2.92 \text{ V}$$

and
$$I_E = \frac{V_{RE}}{R_E} = \frac{V_E}{R_E} = \frac{2.92 \text{ V}}{1.6 \text{ k}\Omega} = 1.825 \text{ mA}$$

with
$$I_C \cong I_E = 1.825 \text{ mA}$$

Continuing, we find for this configuration that

$$I_D = I_S = I_C$$

and
$$V_D = 16 \text{ V} - I_D(2.7 \text{ k}\Omega)$$

$$= 16 \text{ V} - (1.825 \text{ mA})(2.7 \text{ k}\Omega) = 16 \text{ V} - 4.93 \text{ V}$$

$$= \mathbf{11.07 \text{ V}}$$

The question of how to determine V_C is not as obvious. Both V_{CE} and V_{DS} are unknown quantities preventing us from establishing a link between V_D and V_C or from V_E to V_D. A more careful examination of Fig. 6.46 reveals that V_C is linked to V_B by V_{GS} (assuming that $V_{R_G} = 0$ V). Since we know V_B if we can find V_{GS}, V_C can be determined from

$$V_C = V_B - V_{GS}$$

The question then arises as to how to find the level of V_{GS_Q} from the quiescent value of I_D. The two are related by Shockley's equation:

$$I_{D_Q} = I_{DSS}\left(1 - \frac{V_{GS_Q}}{V_P}\right)^2$$

and V_{GS_Q} could be found mathematically by solving for V_{GS_Q} and substituting numerical values. However, let us turn to the graphical approach and simply work in the reverse order employed in the preceding sections. The JFET transfer characteristics are first sketched as shown in Fig. 6.47. The level of I_{D_Q} is then established by a horizontal line as shown in the same figure. V_{GS_Q} is then determined by dropping a line down from the operating point to the horizontal axis, resulting in

$$V_{GS_Q} = \mathbf{-3.7 \text{ V}}$$

The level of V_C:

$$V_C = V_B - V_{GS_Q} = 3.62 \text{ V} - (-3.7 \text{ V})$$

$$= \mathbf{7.32 \text{ V}}$$

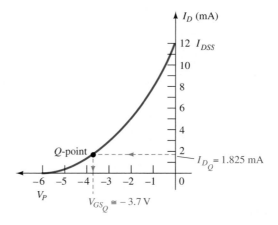

Figure 6.47 Determining the Q-point for the network of Fig. 6.46.

Determine V_D for the network of Fig. 6.48.

EXAMPLE 6.14

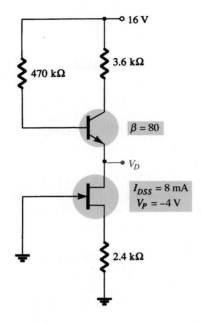

Figure 6.48 Example 6.14.

Solution

In this case, there is no obvious path to determine a voltage or current level for the transistor configuration. However, turning to the self-biased JFET, an equation for V_{GS} can be derived and the resulting quiescent point determined using graphical techniques. That is,

$$V_{GS} = -I_D R_S = -I_D(2.4 \text{ k}\Omega)$$

resulting in the self-bias line appearing in Fig. 6.49 that establishes a quiescent point at

$$V_{GS_Q} = -2.6 \text{ V}$$

$$I_{D_Q} = 1 \text{ mA}$$

For the transistor,

$$I_E \cong I_C = I_D = 1 \text{ mA}$$

and

$$I_B = \frac{I_C}{\beta} = \frac{1 \text{ mA}}{80} = 12.5 \ \mu\text{A}$$

$$V_B = 16 \text{ V} - I_B(470 \text{ k}\Omega)$$

$$= 16 \text{ V} - (12.5 \ \mu\text{A})(470 \text{ k}\Omega) = 16 \text{ V} - 5.875 \text{ V}$$

$$= 10.125 \text{ V}$$

and

$$V_E = V_D = V_B - V_{BE}$$

$$= 10.125 \text{ V} - 0.7 \text{ V}$$

$$= \mathbf{9.425 \text{ V}}$$

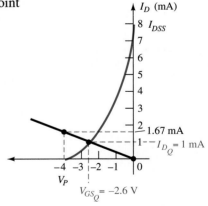

Figure 6.49 Determining the Q-point for the network of Fig. 6.48.

6.9 DESIGN

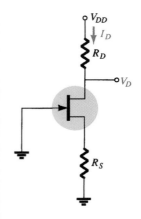

Figure 6.50 Self-bias configuration to be designed.

The design process is one that is not limited solely to dc conditions. The area of application, level of amplification desired, signal strength, and operating conditions are just a few of the conditions that enter into the total design process. However, we will first concentrate on establishing the chosen dc conditions.

For example, if the levels of V_D and I_D are specified for the network of Fig. 6.50, the level of V_{GS_Q} can be determined from a plot of the transfer curve and R_S can then be determined from $V_{GS} = -I_D R_S$. If V_{DD} is specified, the level of R_D can then be calculated from $R_D = (V_{DD} - V_D)/I_D$. Of course, the value of R_S and R_D may not be standard commercial values, requiring that the nearest commercial value be employed. However, with the tolerance (range of values) normally specified for the parameters of a network, the slight variation due to the choice of standard values will seldom cause a real concern in the design process.

The above is only one possibility for the design phase involving the network of Fig. 6.50. It is possible that only V_{DD} and R_D are specified together with the level of V_{DS}. The device to be employed may have to be specified along with the level of R_S. It appears logical that the device chosen should have a maximum V_{DS} greater than the specified value by a safe margin.

In general, it is good design practice for linear amplifiers to choose operating points that do not crowd the saturation level (I_{DSS}) or cutoff (V_P) regions. Levels of V_{GS_Q} close to $V_P/2$ or I_{D_Q} near $I_{DSS}/2$ are certainly reasonable starting points in the design. Of course, in every design procedure the maximum levels of I_D and V_{DS} as appearing on the specification sheet must not be considered as exceeded.

The examples to follow have a design or synthesis orientation in that specific levels are provided and network parameters such as R_D, R_S, V_{DD}, and so on, must be determined. In any case, the approach is in many ways the opposite of that described in previous sections. In some cases, it is just a matter of applying Ohm's law in its appropriate form. In particular, if resistive levels are requested, the result is often obtained simply by applying Ohm's law in the following form:

$$R_{\text{unknown}} = \frac{V_R}{I_R} \qquad (6.34)$$

where V_R and I_R are often parameters that can be found directly from the specified voltage and current levels.

EXAMPLE 6.15

For the network of Fig. 6.51, the levels of V_{D_Q} and I_{D_Q} are specified. Determine the required values of R_D and R_S. What are the closest standard commercial values?

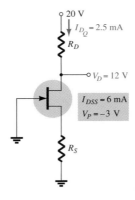

Figure 6.51 Example 6.15.

Chapter 6 FET Biasing

Solution

As defined by Eq. (6.34),

$$R_D = \frac{V_{R_D}}{I_{D_Q}} = \frac{V_{DD} - V_{D_Q}}{I_{D_Q}}$$

and

$$= \frac{20 \text{ V} - 12 \text{ V}}{2.5 \text{ mA}} = \frac{8 \text{ V}}{2.5 \text{ mA}} = \mathbf{3.2 \text{ k}\Omega}$$

Plotting the transfer curve in Fig. 6.52 and drawing a horizontal line at $I_{D_Q} = 2.5$ mA will result in $V_{GS_Q} = -1$ V, and applying $V_{GS} = -I_D R_S$ will establish the level of R_S:

$$R_S = \frac{-(V_{GS_Q})}{I_{D_Q}} = \frac{-(-1 \text{ V})}{2.5 \text{ mA}} = \mathbf{0.4 \text{ k}\Omega}$$

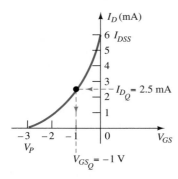

Figure 6.52 Determining V_{GS_Q} for the network of Fig. 6.51.

The nearest standard commercial values are

$$R_D = 3.2 \text{ k}\Omega \Rightarrow \mathbf{3.3 \text{ k}\Omega}$$

$$R_S = 0.4 \text{ k}\Omega \Rightarrow \mathbf{0.39 \text{ k}\Omega}$$

For the voltage-divider bias configuration of Fig. 6.53, if $V_D = 12$ V and $V_{GS_Q} = -2$ V, determine the value of R_S.

EXAMPLE 6.16

Solution

The level of V_G is determined as follows:

$$V_G = \frac{47 \text{ k}\Omega(16 \text{ V})}{47 \text{ k}\Omega + 91 \text{ k}\Omega} = 5.44 \text{ V}$$

with

$$I_D = \frac{V_{DD} - V_D}{R_D}$$

$$= \frac{16 \text{ V} - 12 \text{ V}}{1.8 \text{ k}\Omega} = 2.22 \text{ mA}$$

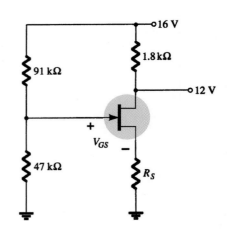

Figure 6.53 Example 6.16.

The equation for V_{GS} is then written and the known values substituted:

$$V_{GS} = V_G - I_D R_S$$

$$-2 \text{ V} = 5.44 \text{ V} - (2.22 \text{ mA})R_S$$

$$-7.44 \text{ V} = -(2.22 \text{ mA})R_S$$

and

$$R_S = \frac{7.44 \text{ V}}{2.22 \text{ mA}} = \mathbf{3.35 \text{ k}\Omega}$$

The nearest standard commercial value is 3.3 kΩ.

EXAMPLE 6.17

The levels of V_{DS} and I_D are specified as $V_{DS} = \frac{1}{2}V_{DD}$ and $I_D = I_{D(on)}$ for the network of Fig. 6.54. Determine the level of V_{DD} and R_D.

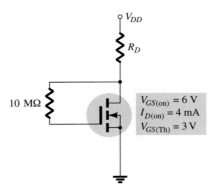

$$V_{GS(on)} = 6 \text{ V}$$
$$I_{D(on)} = 4 \text{ mA}$$
$$V_{GS(Th)} = 3 \text{ V}$$

Figure 6.54   Example 6.17.

Solution

Given $I_D = I_{D(on)} = 4$ mA and $V_{GS} = V_{GS(on)} = 6$ V, for this configuration,

$$V_{DS} = V_{GS} = \tfrac{1}{2}V_{DD}$$

and

$$6 \text{ V} = \tfrac{1}{2}V_{DD}$$

so that

$$V_{DD} = \mathbf{12 \text{ V}}$$

Applying Eq. (6.34) yields

$$R_D = \frac{V_{R_D}}{I_D} = \frac{V_{DD} - V_{DS}}{I_{D(on)}} = \frac{V_{DD} - \frac{1}{2}V_{DD}}{I_{D(on)}} = \frac{\frac{1}{2}V_{DD}}{I_{D(on)}}$$

and

$$R_D = \frac{6 \text{ V}}{4 \text{ mA}} = \mathbf{1.5 \text{ k}\Omega}$$

which is a standard commercial value.

6.10   TROUBLESHOOTING

How often has a network been carefully constructed only to find that when the power is applied, the response is totally unexpected and fails to match the theoretical calculations. What is the next step? Is it a bad connection? A misreading of the color code for a resistive element? An error in the construction process? The range of possibilities seems vast and often frustrating. The troubleshooting process first described in the analysis of BJT transistor configurations should narrow down the list of possibilities and isolate the problem area following a definite plan of attack. In general, the process begins with a rechecking of the network construction and the terminal connections. This is usually followed by the checking of voltage levels between specific terminals and ground or between terminals of the network. Seldom are current levels measured since such maneuvers require disturbing the network structure to insert the meter. Of course, once the voltage levels are obtained, current levels can be calculated using Ohm's law. In any case, some idea of the expected voltage or current level must be known for the measurement to have any importance. In total, therefore, the troubleshooting process can begin with some hope of success only if the basic operation of the network is understood along with some expected levels of voltage or current. For the *n*-channel JFET amplifier, it is clearly understood that the quiescent value of V_{GS_Q} is limited to 0 V or a negative voltage. For the network of Fig. 6.55, V_{GS_Q} is limited to negative values in the range 0 V to V_P. If a meter is hooked up as shown in Fig. 6.55, with the positive lead (normally red) to the gate and the negative lead (usually black) to the source, the resulting reading should have a negative sign and a magnitude of a few volts. Any other response should be considered suspicious and needs to be investigated.

The level of V_{DS} is typically between 25% and 75% of V_{DD}. A reading of 0 V for V_{DS} clearly indicates that either the output circuit has an "open" or the JFET is internally short-circuited between drain and source. If V_D is V_{DD} volts, there is obviously no drop across R_D due to the lack of current through R_D and the connections should be checked for continuity.

If the level of V_{DS} seems inappropriate, the continuity of the output circuit can easily be checked by grounding the negative lead of the voltmeter and measuring the voltage levels from V_{DD} to ground using the positive lead. If $V_D = V_{DD}$, the current through R_D may be zero, but there is continuity between V_D and V_{DD}. If $V_S = V_{DD}$, the device is not open between drain and source, but it is also not "on." The continuity through to V_S is confirmed, however. In this case, it is possible that there is a poor ground connection between R_S and ground that may not be obvious. The internal connection between the wire of your lead and the terminal connector may have separated. Other possibilities also exist, such as a shorted device from drain to source, but the troubleshooter will simply have to narrow down the possible causes for the malfunction.

The continuity of a network can also be checked simply by measuring the voltage across any resistor of the network (except for R_G in the JFET configuration). An indication of 0 V immediately reveals the lack of current through the element due to an open circuit in the network.

The most sensitive element in the BJT and JFET configurations is the amplifier itself. The application of excessive voltage during the construction or testing phase or the use of incorrect resistor values resulting in high current levels can destroy the device. If you question the condition of the amplifier, the best test for the FET is the curve tracer since it not only reveals whether the device is operable but also its range of current and voltage levels. Some testers may reveal that the device is still fundamentally sound but do not reveal whether its range of operation has been severely reduced.

The development of good troubleshooting techniques comes primarily from experience and a level of confidence in what to expect and why. There are, of course, times when the reasons for a strange response seem to disappear mysteriously when you check a network. In such cases, it is best not to breathe a sigh of relief and continue with the construction. The cause for such a sensitive "make or break" situation should be found and corrected, or it may reoccur at the most inopportune moment.

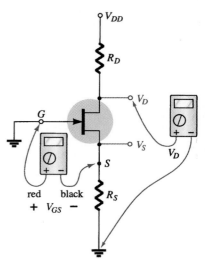

Figure 6.55 Checking the dc operation of the JFET self-bias configuration.

6.11 *P*-CHANNEL FETS

The analysis thus far has been limited solely to *n*-channel FETs. For *p*-channel FETs, a mirror image of the transfer curves is employed, and the defined current directions are reversed as shown in Fig. 6.56 for the various types of FETs.

Note for each configuration of Fig. 6.56 that each supply voltage is now a negative voltage drawing current in the indicated direction. In particular, note that the double-subscript notation for voltages continues as defined for the *n*-channel device: V_{GS}, V_{DS}, and so on. In this case, however, V_{GS} is positive (positive or negative for the depletion-type MOSFET) and V_{DS} negative.

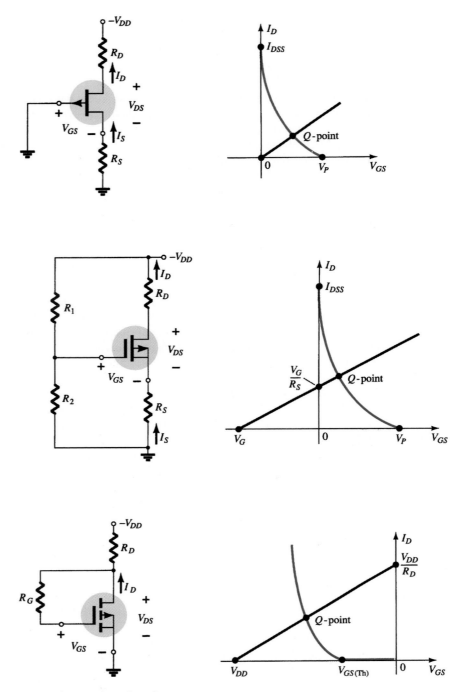

Figure 6.56 *p*-channel configurations.

Due to the similarities between the analysis of *n*-channel and *p*-channel devices, one can actually assume an *n*-channel device and reverse the supply voltage and perform the entire analysis. When the results are obtained, the magnitude of each quantity will be correct, although the current direction and voltage polarities will have to be reversed. However, the next example will demonstrate that with the experience gained through the analysis of *n*-channel devices, the analysis of *p*-channel devices is quite straightforward.

Determine I_{D_Q}, V_{GS_Q}, and V_{DS} for the *p*-channel JFET of Fig. 6.57.

EXAMPLE 6.18

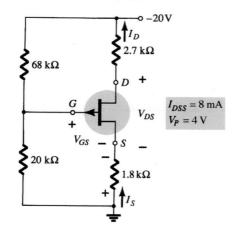

Figure 6.57 Example 6.18.

Solution

$$V_G = \frac{20 \text{ k}\Omega(-20 \text{ V})}{20 \text{ k}\Omega + 68 \text{ k}\Omega} = -4.55 \text{ V}$$

Applying Kirchhoff's voltage law gives

$$V_G - V_{GS} + I_D R_S = 0$$

and

$$V_{GS} = V_G + I_D R_S$$

Choosing $I_D = 0$ mA yields

$$V_{GS} = V_G = -4.55 \text{ V}$$

as appearing in Fig. 6.58.

Choosing $V_{GS} = 0$ V, we obtain

$$I_D = -\frac{V_G}{R_S} = -\frac{-4.55 \text{ V}}{1.8 \text{ k}\Omega} = 2.53 \text{ mA}$$

as also appearing in Fig. 6.58.

The resulting quiescent point from Fig. 6.58:

$$I_{D_Q} = \textbf{3.4 mA}$$

$$V_{GS_Q} = \textbf{1.4 V}$$

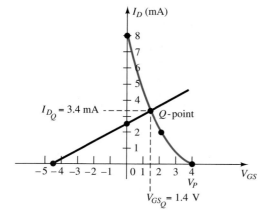

Figure 6.58 Determining the *Q*-point for the JFET configuration of Fig. 6.57.

For V_{DS}, Kirchhoff's voltage law will result in

$$-I_D R_S + V_{DS} - I_D R_D + V_{DD} = 0$$

and

$$V_{DS} = -V_{DD} + I_D(R_D + R_S)$$
$$= -20\ V + (3.4\ mA)(2.7\ k\Omega + 1.8\ k\Omega)$$
$$= -20\ V + 15.3\ V$$
$$= \mathbf{-4.7\ V}$$

6.12   UNIVERSAL JFET BIAS CURVE

Since the dc solution of a FET configuration requires drawing the transfer curve for each analysis, a universal curve was developed that can be used for any level of I_{DSS} and V_P. The universal curve for an n-channel JFET or depletion-type MOSFET (for negative values of V_{GS_Q}) is provided in Fig. 6.59. Note that the horizontal axis is not that of V_{GS} but of a normalized level defined by $V_{GS}/|V_P|$, the $|V_P|$ indicating that only the magnitude of V_P is to be employed, not its sign. For the vertical axis, the scale is also a normalized level of I_D/I_{DSS}. The result is that when $I_D = I_{DSS}$, the ratio is 1, and when $V_{GS} = V_P$, the ratio $V_{GS}/|V_P|$ is -1. Note also that the scale for I_D/I_{DSS} is on the left rather than on the right as encountered for I_D in past exercises. The additional two scales on the right need an introduction. The vertical scale labeled m can in itself be used to find the solution to fixed-bias configurations. The other scale, labeled M, is employed along with the m scale to find the solution to voltage-divider configura-

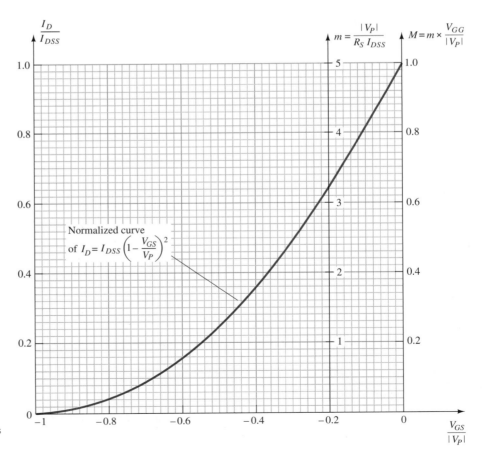

Figure 6.59   Universal JFET bias curve.

tions. The scaling for m and M come from a mathematical development involving the network equations and normalized scaling just introduced. The description to follow will not concentrate on why the m scale extends from 0 to 5 at $V_{GS}/|V_P| = -0.2$ and the M scale from 0 to 1 at $V_{GS}/|V_P| = 0$ but rather on how to use the resulting scales to obtain a solution for the configurations. The equations for m and M are the following, with V_G as defined by Eq. (6.15).

$$m = \frac{|V_P|}{I_{DSS}R_S} \qquad (6.35)$$

$$M = m \times \frac{V_G}{|V_P|} \qquad (6.36)$$

with
$$V_G = \frac{R_2 V_{DD}}{R_1 + R_2}$$

Keep in mind that the beauty of this approach is the elimination of the need to sketch the transfer curve for each analysis, that the superposition of the bias line is a great deal easier, and that the calculations are fewer. The use of the m and M axes is best described by examples employing the scales. Once the procedure is clearly understood, the analysis can be quite rapid, with a good measure of accuracy.

Determine the quiescent values of I_D and V_{GS} for the network of Fig. 6.60.

EXAMPLE 6.19

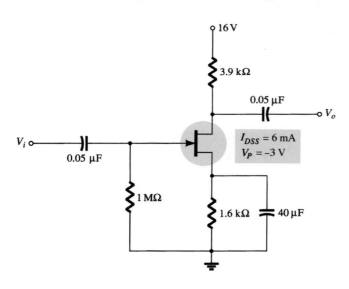

Figure 6.60 Example 6.19.

Solution

Calculating the value of m, we obtain

$$m = \frac{|V_P|}{I_{DSS}R_S} = \frac{|-3\text{ V}|}{(6\text{ mA})(1.6\text{ k}\Omega)} = 0.31$$

The self-bias line defined by R_S is plotted by drawing a straight line from the origin through a point defined by $m = 0.31$, as shown in Fig. 6.61.

The resulting Q-point:

$$\frac{I_D}{I_{DSS}} = 0.18 \qquad \text{and} \qquad \frac{V_{GS}}{|V_P|} = -0.575$$

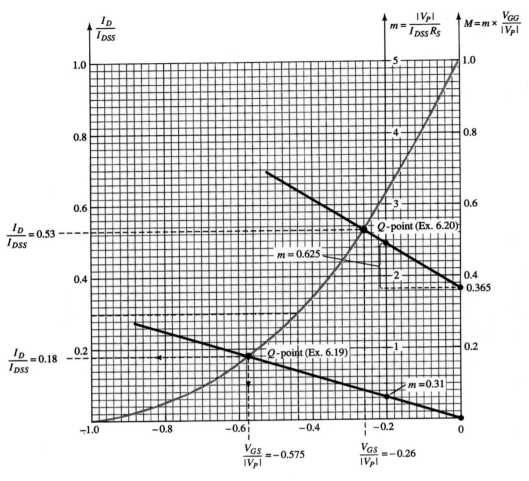

Figure 6.61 Universal curve for Examples 6.19 and 6.20.

The quiescent values of I_D and V_{GS} can then be determined as follows:

$$I_{D_Q} = 0.18 I_{DSS} = 0.18(6 \text{ mA}) = \mathbf{1.08 \text{ mA}}$$

and

$$V_{GS_Q} = -0.575|V_P| = -0.575(3 \text{ V}) = \mathbf{-1.73 \text{ V}}$$

EXAMPLE 6.20

Determine the quiescent values of I_D and V_{GS} for the network of Fig. 6.62.

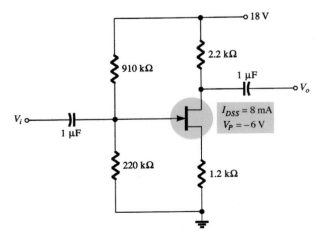

Figure 6.62 Example 6.20.

Solution

Calculating m gives

$$m = \frac{|V_P|}{I_{DSS}R_S} = \frac{|-6 \text{ V}|}{(8 \text{ mA})(1.2 \text{ k}\Omega)} = 0.625$$

Determining V_G yields

$$V_G = \frac{R_2 V_{DD}}{R_1 + R_2} = \frac{(220 \text{ k}\Omega)(18 \text{ V})}{910 \text{ k}\Omega + 220 \text{ k}\Omega} = 3.5 \text{ V}$$

Finding M, we have

$$M = m \times \frac{V_G}{|V_P|} = 0.625\left(\frac{3.5 \text{ V}}{6 \text{ V}}\right) = 0.365$$

Now that m and M are known, the bias line can be drawn on Fig. 6.61. In particular, note that even though the levels of I_{DSS} and V_P are different for the two networks, the same universal curve can be employed. First find M on the M axis as shown in Fig. 6.61. Then draw a horizontal line over to the m axis and, at the point of intersection, add the magnitude of m as shown in the figure. Using the resulting point on the m axis and the M intersection, draw the straight line to intersect with the transfer curve and define the Q-point:

That is,
$$\frac{I_D}{I_{DSS}} = 0.53 \quad \text{and} \quad \frac{V_{GS}}{|V_P|} = -0.26$$

and
$$I_{D_Q} = 0.53 I_{DSS} = 0.53(8 \text{ mA}) = \textbf{4.24 mA}$$

with
$$V_{GS_Q} = -0.26|V_P| = -0.26(6 \text{ V}) = \textbf{-1.56 V}$$

6.13 PRACTICAL APPLICATIONS

The applications described here take full advantage of the high input impedance of field-effect transistors, the isolation that exists between the gate and drain circuits, and the linear region of JFET characteristics that permit approximating the device by a resistive element between the drain and source terminals.

Voltage-Controlled Resistor (Noninverting Amplifier)

One of the most common applications of the JFET is as a variable resistor whose resistance value is controlled by the applied dc voltage at the gate terminal. In Fig. 6.63a, the linear region of a JFET transistor has been clearly indicated. Note that in this region the various curves all start at the origin and follow a fairly straight path as the drain-to-source voltage and drain current increase. Recall from your basic dc courses that **the plot of a fixed resistor is nothing more than a straight line with its origin at the intersection of the axes.**

In Fig. 6.63b, the linear region has been expanded to a maximum drain-to-source voltage of about 0.5 V. Note that even though the curves do have some curvature to them, they can easily be approximated by fairly straight lines, all having their origin at the intersection of the axes and a slope determined by the gate-to-source dc voltage. Recall from earlier discussions that **for an I-V plot where the current is the vertical axis and the voltage the horizontal axis, the steeper the slope, the less the resistance; and the more horizontal the curve, the greater the resistance.** The natural result is that a vertical line has 0 Ω resistance, and a horizontal line infinite resistance. At $V_{GS} = 0$ V, the slope is the steepest and the resistance the least. As the

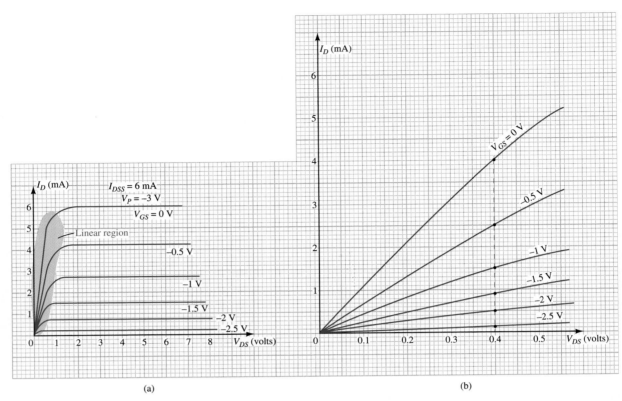

Figure 6.63 JFET characteristics: (a) defining the linear region; (b) expanding the linear region.

gate-to-source voltage becomes increasingly negative, the slope decreases until it is almost horizontal near the pinch-off voltage.

It is important to remember that this linear region is limited to levels of V_{DS} that are relatively small compared to the pinch-off voltage. In general, **the linear region of a JFET is defined by $V_{DS} \ll V_{DS_{max}}$ and $|V_{GS}| \ll |V_P|$.**

Using Ohm's law, let us calculate the resistance associated with each curve of Fig. 6.63b using the current that results at a drain-to-source voltage of 0.4 V.

$$V_{GS} = 0 \text{ V}: \qquad R_{DS} = \frac{V_{DS}}{I_{DS}} = \frac{0.4 \text{ V}}{4 \text{ mA}} = \mathbf{100 \ \Omega}$$

$$V_{GS} = -0.5 \text{ V}: \quad R_{DS} = \frac{V_{DS}}{I_{DS}} = \frac{0.4 \text{ V}}{2.5 \text{ mA}} = \mathbf{160 \ \Omega}$$

$$V_{GS} = -1 \text{ V}: \qquad R_{DS} = \frac{V_{DS}}{I_{DS}} = \frac{0.4 \text{ V}}{1.5 \text{ mA}} = \mathbf{267 \ \Omega}$$

$$V_{GS} = -1.5 \text{ V}: \quad R_{DS} = \frac{V_{DS}}{I_{DS}} = \frac{0.4 \text{ V}}{0.9 \text{ mA}} = \mathbf{444 \ \Omega}$$

$$V_{GS} = -2 \text{ V}: \qquad R_{DS} = \frac{V_{DS}}{I_{DS}} = \frac{0.4 \text{ V}}{0.5 \text{ mA}} = \mathbf{800 \ \Omega}$$

$$V_{GS} = -2.5 \text{ V}: \quad R_{DS} = \frac{V_{DS}}{I_{DS}} = \frac{0.4 \text{ V}}{0.12 \text{ mA}} = \mathbf{3.3 \ k\Omega}$$

In particular, note how **the drain-to-source resistance increases as the gate-to-source voltage approaches the pinch-off value.**

The results just obtained can be verified by Eq. (5.1) using the pinch-off voltage of -3 V and $R_o = 100$ Ω at $V_{GS} = 0$ V.

$$R_{DS} = \frac{R_o}{\left(1 - \dfrac{V_{GS}}{V_P}\right)^2} = \frac{100\ \Omega}{\left(1 - \dfrac{V_{GS}}{-3\ \text{V}}\right)^2}$$

$V_{GS} = -0.5$ V: $\quad R_{DS} = \dfrac{100\ \Omega}{\left(1 - \dfrac{-0.5\ \text{V}}{-3\ \text{V}}\right)^2} = \textbf{144}\ \boldsymbol{\Omega}$ (versus 160 Ω above)

$V_{GS} = -1$ V: $\quad R_{DS} = \dfrac{100\ \Omega}{\left(1 - \dfrac{-1\ \text{V}}{-3\ \text{V}}\right)^2} = \textbf{225}\ \boldsymbol{\Omega}$ (versus 267 Ω above)

$V_{GS} = -1.5$ V: $\quad R_{DS} = \dfrac{100\ \Omega}{\left(1 - \dfrac{-1.5\ \text{V}}{-3\ \text{V}}\right)^2} = \textbf{400}\ \boldsymbol{\Omega}$ (versus 444 Ω above)

$V_{GS} = -2$ V: $\quad R_{DS} = \dfrac{100\ \Omega}{\left(1 - \dfrac{-2\ \text{V}}{-3\ \text{V}}\right)^2} = \textbf{900}\ \boldsymbol{\Omega}$ (versus 800 Ω above)

$V_{GS} = -2.5$ V: $\quad R_{DS} = \dfrac{100\ \Omega}{\left(1 - \dfrac{-2.5\ \text{V}}{-3\ \text{V}}\right)^2} = \textbf{3.6 k}\boldsymbol{\Omega}$ (versus 3.3 kΩ above)

Although the results are not an exact match, for most applications Equation (5.1) provides an excellent approximation to the actual resistance level for R_{DS}.

Keep in mind that **the possible levels of V_{GS} between 0 V and pinch-off are infinite,** resulting in the full range of resistor values between 100 Ω and 3.3 kΩ. In general, therefore, the above discussion is summarized by Fig. 6.64a. For $V_{GS} = 0$ V, the equivalence of Fig. 6.64b would result; for $V_{GS} = -1.5$ V, the equivalence of Fig. 6.64c; and so on.

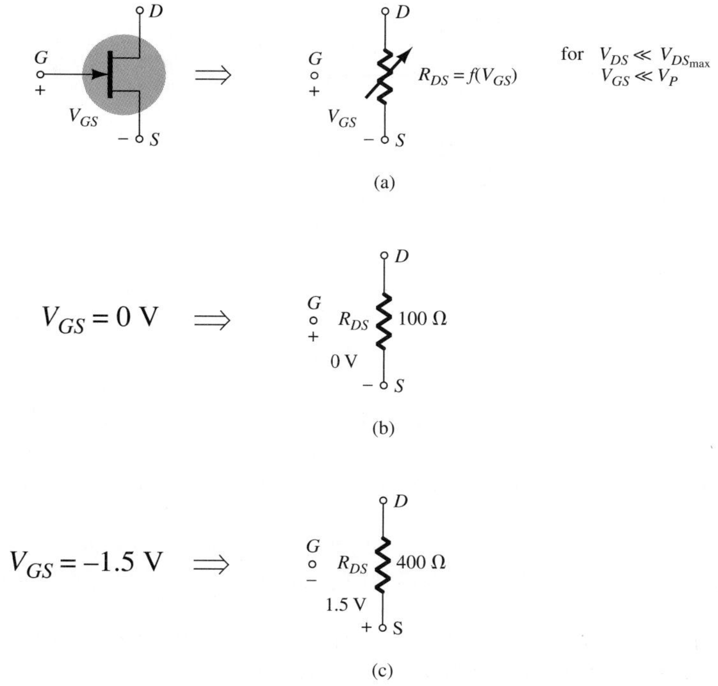

(a)

$V_{GS} = 0$ V $\implies$

(b)

$V_{GS} = -1.5$ V $\implies$

(c)

Figure 6.64 JFET voltage-controlled drain resistance: (a) general equivalence; (b) with $V_{GS} = 0$ **V**; (c) with $V_{GS} = -1.5$ **V**.

Let us now investigate the use of this voltage-controlled drain resistance in the noninverting amplifier of Fig. 6.65a—**noninverting, revealing that the input and output signals are in phase.** In Chapter 13, the op-amp of Fig. 6.65a will be discussed in detail, with the equation for the gain derived in Section 13.4.

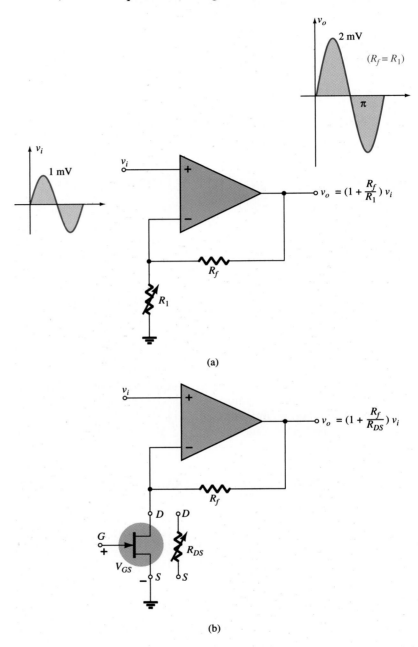

Figure 6.65 (a) Noninverting op-amp configuration; (b) using the voltage-controlled drain-to-source resistance of a JFET in the noninverting amplifier.

If $R_f = R_1$, the resulting gain is 2 as shown by the in-phase sinusoidal signals of Fig. 6.65a. In Fig. 6.65b, the variable resistor has been replaced by an *n*-channel JFET. If $R_f = 3.3$ kΩ and the transistor of Fig. 6.63 were employed, the gain could extend from $1 + 3.3$ kΩ/3.3 kΩ $= 2$ to $1 + 3.3$ kΩ/100 Ω $= 34$ for V_{GS} varying from -2.5 V to 0 V, respectively. In general, therefore, the gain of the amplifier can be set at any value between 2 and 34 by simply controlling the applied dc biasing voltage. The impact of this type of control can be extended to an infinite variety of applications. For instance, if the battery voltage of a radio should start to drop due to extended use, the dc level at the gate of the controlling JFET will drop, and the level of R_{DS} will decrease also. A drop in R_{DS} will result in an increase in gain for the same value of R_f, and the output volume of the radio can be maintained. A number of oscillators (networks designed to generate sinusoidal signals of specific frequencies) have a resistance factor in the equation for the frequency generated. If the frequency generated should start to drift, a feedback network can be designed that would change the dc level at the gate of a JFET and therefore its drain resistance. If that drain resistance is part of the resistance factor in the frequency equation, the frequency generated can be stabilized or maintained.

One of the most important factors that affect the stability of a system is temperature variation. As a system heats up, the usual tendency is for the gain to increase, which in turn will usually cause additional heating and may eventually result in a condition referred to as "thermal runaway." Through proper design, a thermistor can be introduced that will affect the biasing level of a voltage-controlled variable JFET resistor. As the resistance of the thermistor drops with increase in heat, the biasing control of the JFET could be such that the drain resistance would change in the amplifier design to reduce the gain—establishing a balancing effect.

Before leaving the subject of thermal problems, note that some design specifications (often military type) require that systems that are overly sensitive to temperature variations be placed in a "chamber" or "oven" to establish a constant heat level. For instance, a 1-W resistor may be placed in an enclosed area with an oscillator network simply to establish a constant ambient heat level in the region. The design would then center on this heat level that would be so high compared to the heat normally generated by the components that the variations in temperature levels of the elements could be ignored and a steady output frequency assured.

Other areas of application include any form of volume control, musical effects, meters, attenuators, filters, stability designs, etc. One general advantage of this type of stability is that it avoids the need for expensive regulators (Chapter 18) in the overall design, although it should be understood that the purpose of this type of control mechanism is to "fine-tune" rather than to provide the primary source of stability.

For the noninverting amplifier, **one of the most important advantages associated with using a JFET for control is the fact that it is dc rather than ac control.** For most systems, dc control not only results in a reduced chance of adding unwanted noise to the system but also lends itself well to remote control. For example, in Fig. 6.66a, a remote control panel is controlling the amplifier gain for the speaker by an ac line connected to the variable resistor. **The long line from the amplifier can easily pick up noise from the surrounding air as generated by fluorescent lights, local radio stations, operating equipment (even computers), motors, generators, etc.** The result may be a 2-mV signal on the line with a 1-mV noise level—a terrible signal-to-noise ratio that would only contribute to further deterioration of the signal coming in from the microphone due to the loop gain of the amplifier. In Fig. 6.66b, a dc line is controlling the gate voltage of the JFET and the variable resistance of the noninverting amplifier. Even though the dc line voltage on the line may be only -2 V, a ripple of 1 mV picked up by the long line will result in a very large signal-to-noise ratio that could essentially be ignored in the distortion process. In other words, the noise on the dc line would simply move the dc operating point slightly on the device characteristics and would have almost no effect on the resulting drain resistance—isolation between the noise on the line and the amplifier response would be almost ideal.

Even though Figures 6.66a and 6.66b have a relatively long control line, the control line may only be 6″ long as shown in the control panel of Fig. 6.66c, where all the elements of the amplifier are housed in the same container. Consider, however, **that just 1″ is enough to pick up RF noise** so dc control is a favorable characteristic for almost any system. Furthermore, since the control resistance in Fig. 6.66a is usually quite large (hundreds of kΩ) while the dc voltage control resistors of the dc system of Fig. 6.66b are usually quite small (a few kΩ), the volume control resistor for the ac system will absorb a great deal more ac noise than the dc design. This phenomenon is a result of the fact that **RF noise signals in the air have a very high internal resistance, and therefore the larger the pickup resistance, the greater the RF noise absorbed by the receiver.** Recall Thévenin's theorem which states that for maximum power transfer, the load resistance should equal the internal resistance of the source.

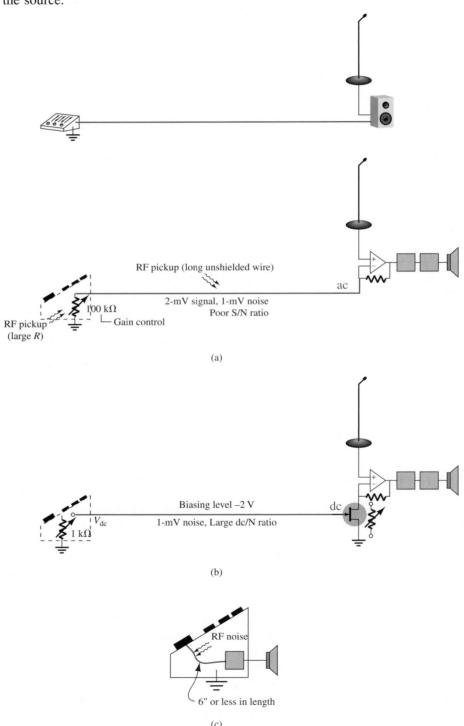

Figure 6.66 Demonstrating the benefits of dc control: system with (a) ac control; (b) dc control; (c) RF noise pickup.

As noted above, **dc control lends itself to computer and remote control systems** since they operate off specific fixed dc levels. For instance, when an infrared (*IR*) signal is sent out by a remote control to the receiver in a TV or VCR, the signal is passed through a decoder–counter sequence to define a particular dc voltage level on a staircase of voltage levels that can be fed into the gate of the JFET. If we're talking about a volume control, that gate voltage may control the drain resistance of a noninverting amplifier controlling the volume of the system.

JFET Voltmeter

The voltage-controlled resistor effect just described will now be put to good use in the JFET voltmeter of Fig. 6.67a. The drain resistance of the JFET provides one arm of a bridge network that when balanced will result in zero current through the sensitive movement appearing in the equivalent diagram of Fig. 6.67b. Because of the need to properly bias the JFET, the user must be particularly careful to hook up the leads

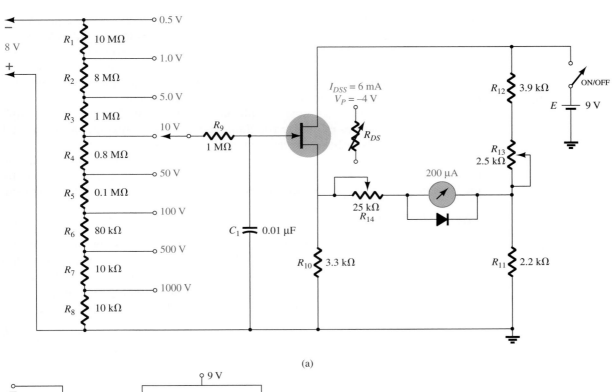

(a)

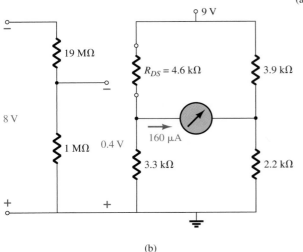

(b)

Figure 6.67 JFET voltmeter: (a) network; (b) reduced equivalent with an 8-V measurement. (Redrawn from International Rectifier Corporation)

as shown to the 8 V being measured. A more sophisticated design would have a polarity switch that reverses the polarity if the meter pins or the reading is erroneous. For the 8 V being measured, the 10-V scale was chosen, resulting in a voltage-divider configuration as shown in Fig. 6.67b that would result in -0.4 V from gate to source for the JFET. The resulting drain resistance of the JFET would then establish an unbalanced condition that would result in a current through the movement and an indication on the meter. Of course, for the reading to have any meaning, the meter would first have to be calibrated (movement set to zero under specific operating conditions), but this discussion is beyond the needs of this text.

For the special idealized situation of Fig. 6.67b with the balance resistors set on $0 \, \Omega$, a drain-to-source JFET resistance of about 4.6 kΩ would result in a meter current of about 160 μA, or 80% of full-scale reading (200 μA) as required for the 8 V being read on the 10-V scale. In addition, note that the gate-to-source voltage of -0.4 V is considerably less than the pinch-off voltage of -4 V and certainly much less than the maximum value of V_{DS} as required in the linear region of the JFET. The voltage-divider network of the input circuit will ensure that the gate-to-source voltage does not exceed the boundaries that permit the variable resistance equivalence. In Fig. 6.67a, the capacitor is included to remove any surges that may develop when hooked up to the dc voltage to be measured and to short to ground any erroneous pickup noise at the source. The diode is included to protect the movement from excessive voltages (greater than 0.7 V). The variable resistors are included to zero the meter and to calibrate the meter using a known voltage source.

Before leaving the meter, note that **all meter movements have an "air-damp" mechanism designed to minimize the damage from surge currents and external turbulence.** When you shake a meter, you will find that the movement does not follow the shaking motion directly, but seems to lag the motion in a slower, lethargic manner. The reason is that air is being pushed out of the "air-damping container" by the motion of the pointer, slowing down the response of the balanced mechanism.

Timer Network

The high isolation between gate and drain circuits permits the design of a relatively simple timer such as shown in Fig. 6.68. The switch is a normally open (NO) switch that when closed will short out the capacitor and cause its terminal voltage to quickly drop to 0 V. The switching network can handle the rapid discharge of voltage across the capacitor because the working voltages are relatively low and the discharge time

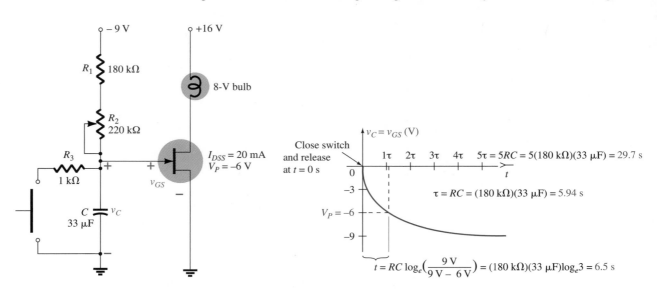

Figure 6.68 JFET timer network.

is extremely short. Some would say it is a poor design, but in the practical world it is frequently used and not looked upon as a "terrible crime."

When power is first applied, the capacitor will respond with its short-circuit equivalence since the **voltage across the capacitor cannot change instantaneously.** The result is that the gate-to-source voltage of the JFET will immediately be set to 0 V, the drain current I_D will equal I_{DSS}, and the bulb will turn on. However, with the switch in the normally open position, the capacitor will begin to charge to -9 V. **Because of the parallel high input impedance of the JFET, it has essentially no effect on the charging time constant of the capacitor.** Eventually, when the capacitor reaches the pinch-off level, the JFET and bulb will turn off. In general, therefore, when the system is first turned on, the bulb will light for a very short period of time and then turn off. It is now ready to perform its timing function.

When the switch is closed, it will short out the capacitor ($R_3 \ll R_1, R_2$) and will set the voltage at the gate to 0 V. The resulting drain current is I_{DSS}, and the bulb will burn brightly. When the switch is released, the capacitor will charge toward -9 V, and eventually when it reaches the pinch-off level, the JFET and bulb will turn off. The period during which the bulb is on will be determined by the time constant of the charging network determined by $\tau = (R_1 + R_2)C$ and the level of the pinch-off voltage. The more negative the pinch-off level, the longer the bulb will be on. Resistor R_1 is included to be sure that there is some resistance in the charging circuit when the power is turned on. Otherwise, a very heavy current could result that might damage the network. Resistor R_2 is a variable resistor, so the on time can be controlled. Resistor R_3 was added to limit the discharge current when the switch is closed. When the switch across the capacitor is closed, the discharge time of the capacitor will be only $5\tau = 5RC = 5(1\text{ k}\Omega)(33\ \mu\text{F}) = 165\ \mu\text{s} = 0.165\text{ ms} = 0.000165\text{ s}$. In summary, therefore, when the switch is pressed and released, the bulb will come on brightly, and then, as time goes on, it will become dimmer until it shuts off after a period of time determined by the network time constant.

One of the most obvious applications of such a timing system is in a hallway or travel corridor where you want light for a short period of time so that you can pass safely but then want the system to turn off on its own. When you enter or leave a car, you may want a light on for a short period of time but don't want to worry about turning it off. There are endless possibilities for a timing network such as just described. Just consider the variety of other electrical or electronic systems that you would like to turn on for specific periods of time, and the list of uses grow exponentially.

One might ask why a BJT would not be a good alternative to the JFET for the same application. First, the input resistance of the BJT may be only a few kilohms. That would affect not only the time constant of the charging network but also the maximum voltage to which the capacitor could charge. Just draw an equivalent network with the transistor replaced by a 1-kΩ resistor, and the above will immediately become clear. In addition, the control levels will have to be designed with a great deal more care since the BJT transistor turns on at about 0.7 V. The voltage swing from off to on is only 0.7 V rather than 4 V for the JFET configuration. One final note: You might have noticed the absence of a series resistor in the drain circuit for the situation when the bulb is first turned on and the resistance of the bulb is very low. The resulting current could be quite high until the bulb reaches its rated intensity. However, again, as described above for the switch across the capacitor, if the energy levels are small and the duration of stress minimal, such designs are often accepted. If there were any concern, adding a resistor of 0.1 to 1 Ω in series with the bulb would provide some security.

Fiber Optic Systems

The introduction of fiber optic technology has had a dramatic effect on the communications industry. **The information-carrying capacity of fiber optic cable is significantly**

greater than that provided by conventional methods with individual pairs of wire. In addition, **the cable size is reduced; the cable is less expensive; crosstalk due to electromagnetic effects between current-carrying conductors is eliminated; and noise pickup due to external disturbances such as lightning are eliminated.**

The entire fiber optic industry is based on the fact that information can be transmitted on a beam of light. Although the speed of light through free space is 3×10^8 meters per second or approximately 186,000 miles per second, its speed will be reduced by encounters with other media, causing reflection and refraction. When light information is passed through a fiber optic cable, it is expected to bounce off the walls of the cable. However, the angle at which the light is injected into the cable is critical, as well as the actual design of the cable. In Fig. 6.69 the basic elements of a fiber optic cable are defined. The glass or plastic core of the cable can be as small as 8 μm which is close to 0.1 the diameter of a human hair. The core is surrounded by an outer layer called the *cladding* that is also made of glass or plastic but has a different refractive index to ensure that the light in the core that hits the outer surface of the core is reflected back into the core. A protective coating is then added to protect the two layers from outside environmental effects.

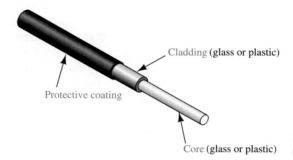

Cladding (glass or plastic)

Protective coating

Core (glass or plastic)

Figure 6.69 Basic elements of a fiber optic cable.

Most optical communication systems work in the infrared frequency range, which extends from 3×10^{11} to 5×10^{14} Hz. This spectrum is just below the visible light spectrum, which extends from 5×10^{14} to 7.7×10^{14} Hz. For most optical systems the frequency range of 1.87×10^{14} to 3.75×10^{14} Hz is used. Because of the very high frequencies, each carrier can be modulated by hundreds or thousands of voice channels simultaneously. In addition, **very-high-speed computer transmission is a possibility, although one must be sure that the electronic components of the modulators can also operate successfully at the same frequency.** For distances over 30 nautical miles, repeaters (a combination receiver, amplifier, and transmitter) must be used which require an additional electrical conductor in the cable that carries a current of about 1.5 A at 2500 V.

The basic components of an optical communication system are shown in Fig. 6.70. The input signal is applied to a light modulator whose sole purpose is to convert the input signal to one of corresponding levels of light intensity to be directed down the length of fiber optic cable. The information is then carried through the cable to the receiving station where a light demodulator will convert the varying light intensities back to voltage levels that will match those of the original signal.

An electronic equivalent for the transmission of computer TTL (transistor-transistor-logic) information is provided in Fig. 6.71a. With the Enable control in the on or 1-state, the TTL information at the input to the AND gate can pass through to the gate of the JFET configuration. The design is such that the discrete levels of voltage associated with the TTL logic will turn the JFET on and off (perhaps 0 V and -5 V, respectively, for a JFET with $V_P = -4$ V). The resulting change in current levels will result in two distinct levels of light intensity from the LED (Section 1.16) in the drain circuit. That emitted light will then be directed through the cable to the re-

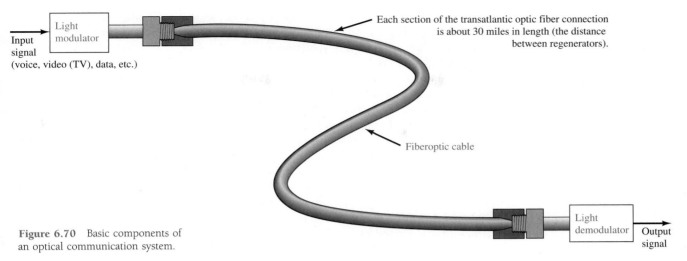

Figure 6.70 Basic components of an optical communication system.

ceiving station where a photodiode (Section 19.6) will react to the incident light and permit different levels of current to pass through as established by V and R. The current for photodiodes is a reverse current having the direction shown in Fig. 6.71a, but in the ac equivalent the photodiode and resistor R are in parallel as shown in Fig. 6.71b, establishing the desired signal with the polarity shown at the gate of the JFET. Capacitor C is simply an open circuit to dc to isolate the biasing arrangement for the photodiode from the JFET and a short circuit as shown for the signal v_s. The incoming signal will then be amplified and will appear at the drain terminal of the output JFET.

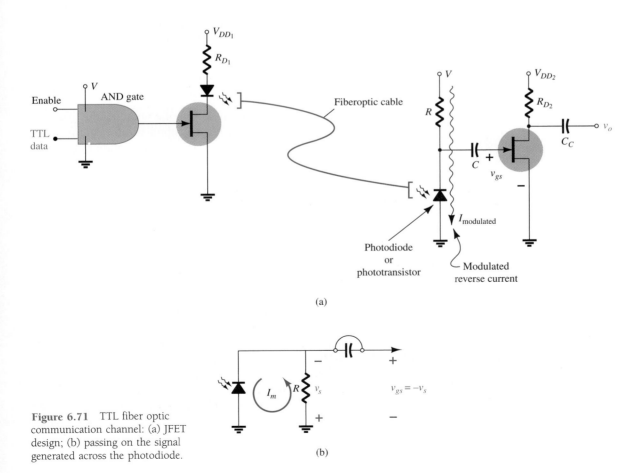

Figure 6.71 TTL fiber optic communication channel: (a) JFET design; (b) passing on the signal generated across the photodiode.

As mentioned above, all the elements of the design, including the JFETs, LED, photodiode, capacitors, etc., must be carefully chosen to ensure that they function properly at the high frequency of transmission. In fact, laser diodes are frequently used instead of LEDs in the modulator because they work at higher information rates and higher powers and have lower coupling and transmission losses. However, laser diodes are a great deal more expensive and more temperature-sensitive, and they typically have a shorter lifetime than LEDs. For the demodulator side, the photodiodes are either of the pin photodiode or the avalanche photodiode variety. The *pin* abbreviation comes from the *p-*intrinsic-*n* construction process, and the term *avalanche* from the rapidly growing ionization process that develops during operation.

In general, the JFET is excellent for this application because of its high isolation at the input side and its ability to quickly "snap" from one state to the other due to the TTL input. At the output side the isolation blocks any effect of the demodulator sensing circuit from affecting the ac response, and it provides some gain for the signal before it is passed on to the next stage.

MOSFET Relay Driver

The MOSFET relay driver to be described in this section is an excellent example of how the FETs can be used to **drive high-current/high-voltage networks without drawing current or power from the driving circuit. The high input impedance of FETs essentially isolates the two parts of the network without the need for optical or electromagnetic linkages.** The network to be described can be used for a variety of applications, but our application will be limited to an alarm system activated when someone or something passes the plane of the transmitted light.

The IR (infrared—not visible) LED of Fig. 6.72 is directing its light through a directional funnel to hit the face of a photoconductive cell (Section 19.7) of the controlling network. The photoconductive cell has a range of resistance from about 200 kΩ as its dark resistance level down to less than 1 kΩ at high illumination levels. Resistor R_1 is a variable resistance that can be used to set the threshold level of the depletion-type MOSFET. A medium-power MOSFET was employed because of the high level of drain current through the magnetizing coil. The diode is included as a protective device for reasons discussed in detail in Section 2.13.

When the system is on and the light consistently hitting the photoconductive cell, the resistance of the cell may drop to 10 kΩ. At this level an application of the voltage-divider rule will result in a voltage of about 0.54 V at the gate terminal (with the

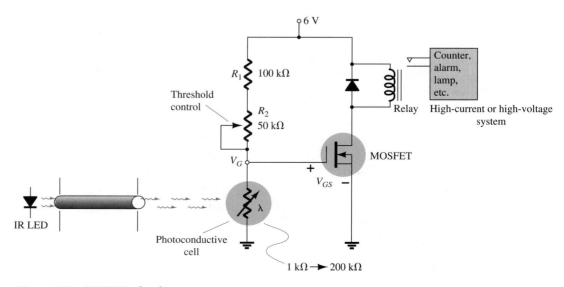

Figure 6.72 MOSFET relay driver.

Chapter 6   FET Biasing

50-kΩ potentiometer set to 0 kΩ). The MOSFET will be on but not at a drain current level that will cause the relay to change state. When someone passes by, the light source will be cut off, and the resistance of the cell may quickly (in a few ms) rise to 100 kΩ. The voltage at the gate will then rise to 3 V, heavily turning on the MOSFET and activating the relay and turning on the system under control. An alarm circuit would have its own control design which will ensure that it will not turn off when light returns to the photoconductive cell.

In essence, therefore, we have controlled a high-current network with a relatively small dc voltage level and a rather inexpensive design. The only obvious flaw in the design is the fact that the MOSFET will be on even when there is no intrusion. This can be remedied through the use of a more sophisticated design, but keep in mind that **MOSFETs are typically low-power-consumption devices,** so the power loss, even over time, is not that great.

6.14 SUMMARY

Important Conclusions and Concepts

1. A fixed-bias configuration has, as the label implies, a **fixed** dc voltage applied from gate to source to establish the operating point.

2. The **nonlinear** relationship between the gate-to-source voltage and the drain current of a JFET requires that a graphical or mathematical solution (involving the solution of two simultaneous equations) be used to determine the quiescent point of operation.

3. All voltages with a single subscript define a voltage from a specified point to **ground**.

4. The self-bias configuration is determined by an equation for V_{GS} that will *always* pass through the origin. Any other point determined by the biasing equation will establish a **straight** line to represent the biasing network.

5. For the voltage-divider biasing configuration, one can always assume that the gate current is 0 A to permit an **isolation** of the voltage-divider network from the output section. The resulting gate-to-ground voltage will always be **positive for an *n*-channel JFET** and **negative for a *p*-channel JFET. Increasing values of R_S** result in **lower quiescent values of I_D** and more **negative values of V_{GS} for an *n*-channel JFET**.

6. The method of analysis applied to depletion-type MOSFETs is the same as applied to JFETs, with the only difference being a possible operating point with an I_D level **above** the I_{DSS} value.

7. The characteristics and method of analysis applied to enhancement-type MOSFETs are **entirely different** from those of JFETs and depletion-type MOSFETs. For values of V_{GS} less than the threshold value, the drain current is 0 A.

8. When analyzing networks with a variety of devices, first work with the region of the network that will provide a **voltage or current level** using the basic relationships associated with those devices. Then use that level and the appropriate equations to find other voltage or current levels of the network in the surrounding region of the system.

9. The design process often requires finding a resistance level to establish the desired voltage or current level. With this in mind, remember that a resistance level is defined by the **voltage across the resistor divided by the current** through the resistor. In the design process, both of these quantities are often available for a particular resistive element.

10. The ability to troubleshoot a network requires a **clear, firm understanding** of the terminal behavior of each of the devices in the network. That knowledge will

provide an **estimate** of the working voltage levels of specific points of the network which can be checked with a voltmeter. The ohmmeter section of a multimeter is particularly helpful in ensuring that there is a **true connection** between all the elements of the network.

11. The analysis of *p*-channel FETs is the same as that applied to *n*-channel FETs except for the fact that all the voltages will have the **opposite polarity** and the currents the **opposite direction.**

Equations

JFETs/depletion-type MOSFETs:

$$\text{Fixed-bias configuration:} \quad V_{GS} = -V_{GG} = V_G$$

$$\text{Self-bias configuration:} \quad V_{GS} = -I_D R_S$$

$$\text{Voltage-divider biasing:} \quad V_G = \frac{R_2 V_{DD}}{R_1 + R_2}$$

$$V_{GS} = V_G - I_D R_S$$

Enhancement-type MOSFETs:

$$\text{Feedback biasing:} \quad V_{DS} = V_{GS}$$

$$V_{GS} = V_{DD} - I_D R_D$$

$$\text{Voltage-divider biasing:} \quad V_G = \frac{R_2 V_{DD}}{R_1 + R_2}$$

$$V_{GS} = V_G - I_D R_S$$

6.15 COMPUTER ANALYSIS

PSpice Windows

JFET VOLTAGE-DIVIDER CONFIGURATION

The results of Example 6.20 will now be verified using PSpice Windows. The network of Fig. 6.73 is constructed using computer methods described in the previous chapters. The J2N3819 JFET is obtained from the **EVAL.slb** library and, through **Edit-Model-Edit Instance Model (Text), Vto** is set to −6V and **Beta,** as defined by Beta = $I_{DSS}/|V_P|^2$ is set to 0.222 mA/V². After an **OK** followed by clicking the **Simulation** icon (the yellow background with the two waveforms) and clearing the **Message Viewer, PSpiceAD** screens will result in Fig. 6.73. The resulting drain current is 4.231 mA compared to the calculated level of 4.24 mA, and V_{GS} is 3.504 V − 5.077 V = −1.573 V versus the calculated value of −1.56 V—both excellent comparisons.

COMBINATION NETWORK

Next, the results of Example 6.13 with both a transistor and JFET will be verified. For the transistor, the **Model** must be altered to have a **Bf**(beta) of 180 to match the example, and for the JFET, **Vto** must be set to −6V and **Beta** to 0.333 mA/V². The results appearing in Fig. 6.74 are again an excellent comparison with the handwritten solution. V_D is 11.44 V compared to 11.07 V, V_C is 7.138 V compared to 7.32 V, and V_{GS} is −3.758 V compared to −3.7 V.

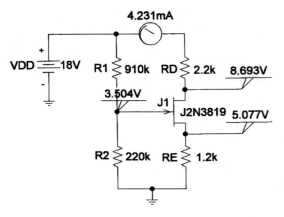

Figure 6.73 JFET voltage-divider configuration with PSpice Windows results for the dc levels.

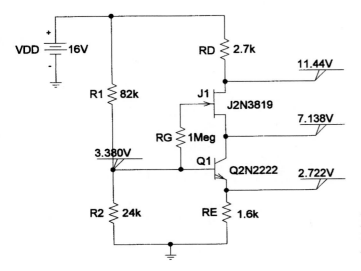

Figure 6.74 Verifying the hand-calculated solution of Example 6.13 using PSpice Windows.

ENHANCEMENT MOSFET

Next, the analysis procedure of Section 6.6 will be verified using the IRF150 enhancement-type n-channel MOSFET found in the **EVAL.slb** library. First, the device characteristics will be obtained by constructing the network of Fig. 6.75.

Clicking on the **Setup Analysis** icon (with the blue bar at the top in the left-hand corner of the screen), **DC Sweep** is chosen to obtain the **DC Sweep** dialog box. **Voltage Source** is chosen as the **Swept Var. Type,** and **Linear** is chosen for the **Sweep Type.** Since only one curve will be obtained, there is no need for a **Nested Sweep.** The voltage-drain voltage VDD will remain fixed at a value of 9 V (about three times the threshold value (**Vto**) of 2.831 V), while the gate-to-source voltage V_{GS}, which in

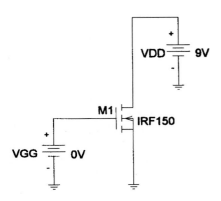

Figure 6.75 Network employed to obtain the characteristics of the IRF150 enhancement-type n-channel MOSFET.

this case is VGG, will be swept from 0 to 10 V. The **Name** therefore is VGG and the **Start Value** 0 V, the **End Value** 10 V, and the **Increment** 0.01 V. After an **OK** followed by a **Close** of the **Analysis Setup,** the analysis can be performed through the **Analysis** icon. If **Automatically run Probe after simulation** is chosen under the **Probe Setup Options** of **Analysis,** the **OrCAD-MicroSim Probe** screen will result, with the horizontal axis appearing with VGG as the variable and range from 0 to 10 V. Next, the **Add Traces** dialog box can be obtained by clicking the **Traces** icon (red pointed pattern on an axis) and the **ID(M1)** chosen to obtain the drain current versus the gate-to-source voltage. Click **OK,** and the characteristics will appear on the screen. To expand the scale of the resulting plot to 20 V, simply choose **Plot** followed by **X-Axis Settings** and set the **User Defined** range to 0 to 20 V. After another **OK,** the plot of Fig. 6.76 will result, revealing a rather high-current device. The labels **ID** and **VGS** were added using the **Text Label** icon with the letters A, B, and C. The hand-drawn load line will be described in the paragraph to follow.

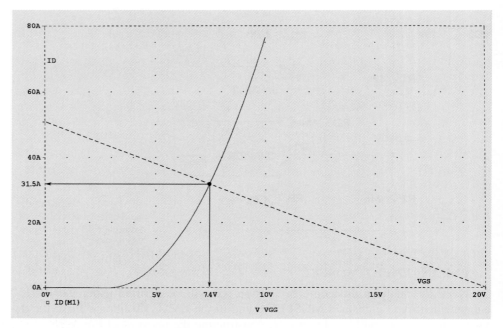

Figure 6.76 Characteristics of the IRF500 MOSFET of Figure 6.75 with a load line defined by the network of Figure 6.77.

The network of Fig. 6.77 was then established to provide a load line extending from I_D equal to 20 V/0.4 Ω = 50 A down to $V_{GS} = V_{GG}$ = 20 V as shown in Fig. 6.76. A simulation resulted in the levels shown, which match the solution of Fig. 6.76.

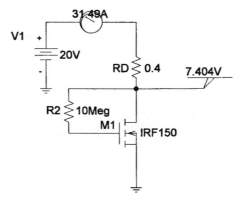

Figure 6.77 Feedback-biasing arrangement employing an IRF150 enhancement-type MOSFET.

Chapter 6   FET Biasing

Electronics Workbench

The results of Example 6.2 will now be verified using Electronics Workbench. The construction of the network of Fig. 6.78 is essentially the same as applied in the previous chapter for BJTs. The JFET is obtained by first going to the vertical component toolbar at the left of the screen and selecting **transistor** from the fourth box down. The result is the component family toolbar which has an *n*-channel JFET as the fourth selection up from the bottom. When it is selected, a **Component Browser** dialog box will appear from which the 2N3821 JFET can be selected. Once it is selected, all the characteristics of the device will appear under the **Model Data** heading. Click **OK,** and the JFET will appear on the screen to be placed like any other element. One left-click, and the JFET is set in place. To define the value of **Beta** as determined by

$$\text{Beta} = \frac{I_{DSS}}{|V_P|^2} = \frac{8 \text{ mA}}{|-6 \text{ V}|^2} = \frac{8 \text{ mA}}{36 \text{ V}^2} = 0.222 \text{ mA/V}^2$$

and to insert the pinch-off voltage, double-click on the JFET symbol to obtain the **JFET_N** dialog box. Select **Edit Model** and change the value of **Beta** and **Vto** to **0.222m** and **–6V,** respectively. Then select **Change Part Model,** followed by **OK,** and the parameters will be set. To label the JFET with the value of **IDSS** and **Vp** as shown in Fig. 6.78, simply use the **Edit-Place Text** sequence as introduced in earlier chapters. Finally, the **Indicator** option is selected on the vertical toolbar, and **VOLT-METER_V** is selected under the **Component Browser** dialog box. Click **OK,** and the indicators can be placed as shown in the figure.

    Simulate-Run/Stop, or selecting the **1** on the **Simulate Switch,** will result in the dc levels appearing in Fig. 6.78. Note that V_{GS} is an exact match with the hand-calculated solution of −2.6 V. The level of 11.364 V for the dc drain-to-ground voltage is very close to the hand-calculated solution of 11.42 V, providing a very satisfying demonstration of the power of Electronics Workbench.

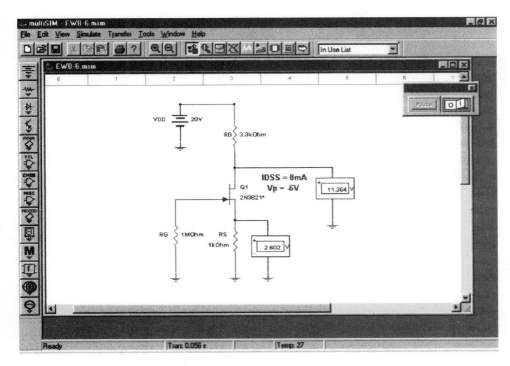

Figure 6.78 Verifying the results of Example 6.2 using Electronics Workbench.

PROBLEMS

§ 6.2 Fixed-Bias Configuration

1. For the fixed-bias configuration of Fig. 6.79:
   (a) Sketch the transfer characteristics of the device.
   (b) Superimpose the network equation on the same graph.
   (c) Determine I_{D_Q} and V_{DS_Q}.
   (d) Using Shockley's equation, solve for I_{D_Q} and then find V_{DS_Q}. Compare with the solutions of part (c).

2. For the fixed-bias configuration of Fig. 6.80, determine:
   (a) I_{D_Q} and V_{GS_Q} using a purely mathematical approach.
   (b) Repeat part (a) using a graphical approach and compare results.
   (c) Find V_{DS}, V_D, V_G, and V_S using the results of part (a).

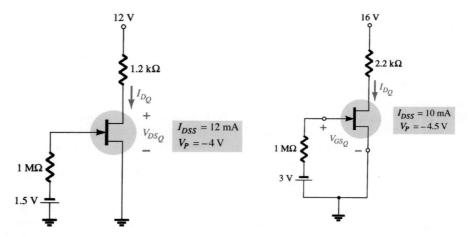

Figure 6.79 Problems 1, 35	Figure 6.80 Problem 2

3. Given the measured value of V_D in Fig. 6.81, determine:
   (a) I_D.
   (b) V_{DS}.
   (c) V_{GG}.

4. Determine V_D for the fixed-bias configuration of Fig. 6.82.

5. Determine V_D for the fixed-bias configuration of Fig. 6.83.

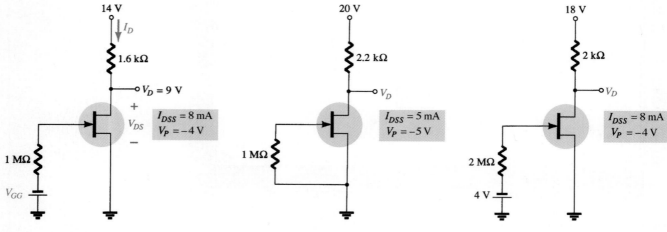

Figure 6.81 Problem 3

Figure 6.82 Problem 4

Figure 6.83 Problem 5

Chapter 6 FET Biasing

6. For the self-bias configuration of Fig. 6.84.
   (a) Sketch the transfer curve for the device.
   (b) Superimpose the network equation on the same graph.
   (c) Determine I_{D_Q} and V_{GS_Q}.
   (d) Calculate V_{DS}, V_D, V_G, and V_S.

*7. Determine I_{D_Q} for the network of Fig. 6.84 using a purely mathematical approach. That is, establish a quadratic equation for I_D and choose the solution compatible with the network characteristics. Compare to the solution obtained in Problem 6.

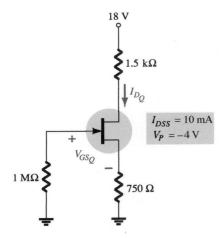

Figure 6.84   Problems 6, 7, 36

8. For the network of Fig. 6.85, determine:
   (a) V_{GS_Q} and I_{D_Q}.
   (b) V_{DS}, V_D, V_G, and V_S.

9. Given the measurement $V_S = 1.7$ V for the network of Fig. 6.86, determine:
   (a) I_{D_Q}.
   (b) V_{GS_Q}.
   (c) I_{DSS}.
   (d) V_D.
   (e) V_{DS}.

*10. For the network of Fig. 6.87, determine:
   (a) I_D.
   (b) V_{DS}.
   (c) V_D.
   (d) V_S.

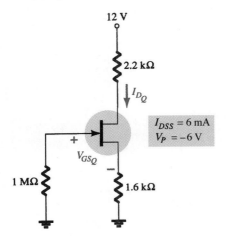

Figure 6.85   Problem 8

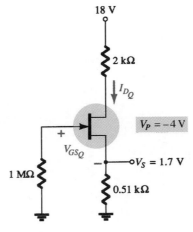

Figure 6.86   Problem 9

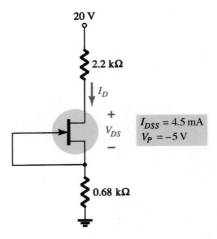

Figure 6.87   Problem 10

Problems   **349**

* **11.** Find V_S for the network of Fig. 6.88.

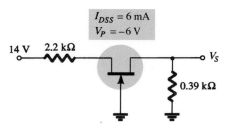

Figure 6.88   Problem 11

§   6.4 Voltage-Divider Biasing

12. For the network of Fig. 6.89, determine:
   (a) V_G.
   (b) I_{D_Q} and V_{GS_Q}.
   (c) V_D and V_S.
   (d) V_{DS_Q}.

13. (a) Repeat Problem 12 with $R_S = 0.51$ kΩ (about 50% of the value of 12). What is the effect of a smaller R_S on I_{D_Q} and V_{GS_Q}?
   (b) What is the minimum possible value of R_S for the network of Fig. 6.89?

14. For the network of Fig. 6.90, $V_D = 9$ V. Determine:
   (a) I_D.
   (b) V_S and V_{DS}.
   (c) V_G and V_{GS}.
   (d) V_P.

* **15.** For the network of Fig. 6.91, determine:
   (a) I_{D_Q} and V_{GS_Q}.
   (b) V_{DS} and V_S.

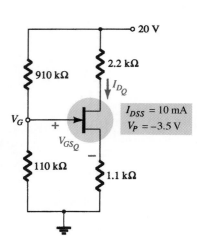

Figure 6.89   Problems 12, 13

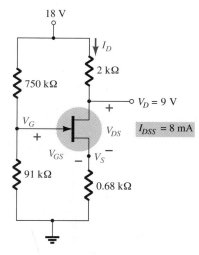

Figure 6.90   Problem 14

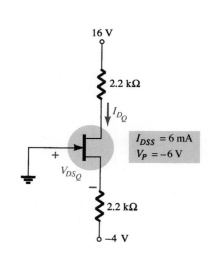

Figure 6.91   Problems 15, 37

Chapter 6   FET Biasing

16. Given $V_{DS} = 4$ V for the network of Fig. 6.92, determine:
  (a) I_D.
  (b) V_D and V_S.
  (c) V_{GS}.

§ 6.5 Depletion-Type MOSFETs

17. For the self-bias configuration of Fig. 6.93, determine:
  (a) I_{D_Q} and V_{GS_Q}.
  (b) V_{DS} and V_D.

18. For the network of Fig. 6.94, determine:
  (a) I_{D_Q} and V_{GS_Q}.
  (b) V_{DS} and V_S.

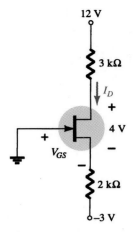

Figure 6.92 Problem 16

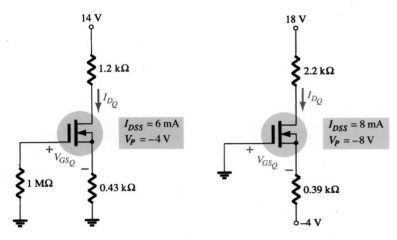

Figure 6.93 Problem 17

Figure 6.94 Problem 18

§ 6.6 Enhancement-Type MOSFETs

19. For the network of Fig. 6.95, determine:
  (a) I_{D_Q}.
  (b) V_{GS_Q} and V_{DS_Q}.
  (c) V_D and V_S.
  (d) V_{DS}.

20. For the voltage-divider configuration of Fig. 6.96, determine:
  (a) I_{D_Q} and V_{GS_Q}.
  (b) V_D and V_S.

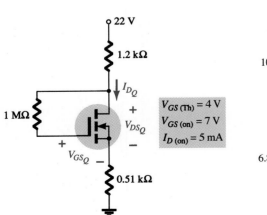

Figure 6.95 Problem 19

Figure 6.96 Problem 20

§ 6.8 Combination Networks

*21. For the network of Fig. 6.97, determine:
   (a) V_G.
   (b) V_{GS_Q} and I_{D_Q}.
   (c) I_E.
   (d) I_B.
   (e) V_D.
   (f) V_C.

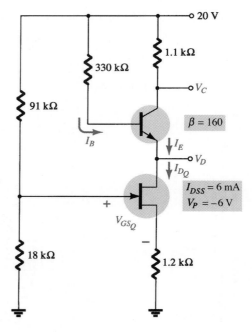

Figure 6.97   Problem 21

*22. For the combination network of Fig. 6.98, determine:
   (a) V_B and V_G.
   (b) V_E.
   (c) I_E, I_C, and I_D.
   (d) I_B.
   (e) V_C, V_S, and V_D.
   (f) V_{CE}.
   (g) V_{DS}.

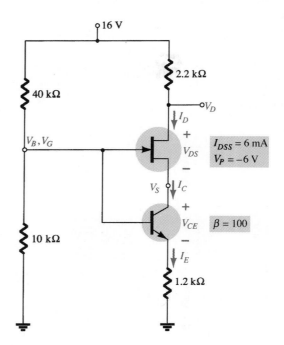

Figure 6.98   Problem 22

§ 6.9 Design

*23. Design a self-bias network using a JFET transistor with $I_{DSS} = 8$ mA and $V_P = -6$ V to have a Q-point at $I_{D_Q} = 4$ mA using a supply of 14 V. Assume that $R_D = 3R_S$ and use standard values.

*24. Design a voltage-divider bias network using a depletion-type MOSFET with $I_{DSS} = 10$ mA and $V_P = -4$ V to have a Q-point at $I_{D_Q} = 2.5$ mA using a supply of 24 V. In addition, set $V_G = 4$ V and use $R_D = 2.5R_S$ with $R_1 = 22$ MΩ. Use standard values.

25. Design a network such as appears in Fig. 6.40 using an enhancement-type MOSFET with $V_{GS(Th)} = 4$ V, $k = 0.5 \times 10^{-3}$ A/V^2 to have a Q-point of $I_{D_Q} = 6$ mA. Use a supply of 16 V and standard values.

§ 6.10 Troubleshooting

*26. What do the readings for each configuration of Fig. 6.99 suggest about the operation of the network?

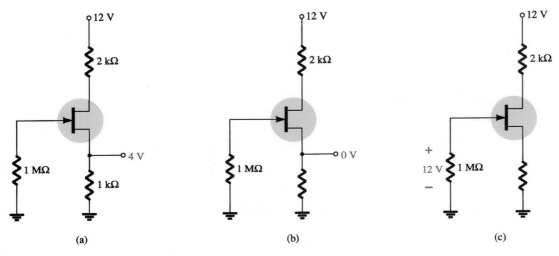

(a)   (b)   (c)

Figure 6.99   Problem 26

*27. Although the readings of Fig. 6.100 initially suggest that the network is behaving properly, determine a possible cause for the undesirable state of the network.

*28. The network of Fig. 6.101 is not operating properly. What is the specific cause for its failure?

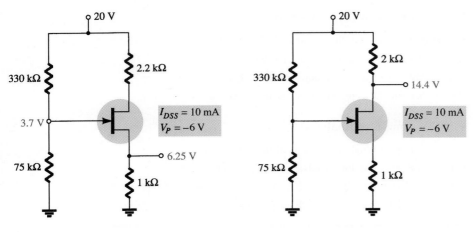

Figure 6.100   Problem 27   **Figure 6.101**   Problem 28

§ 6.11 *p*-Channel FETs

29. For the network of Fig. 6.102, determine:
    (a) I_{D_Q} and V_{GS_Q}.
    (b) V_{DS}.
    (c) V_D.

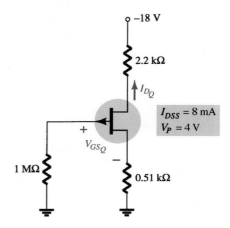

Figure 6.102 Problem 29

30. For the network of Fig. 6.103, determine:
    (a) I_{D_Q} and V_{GS_Q}.
    (b) V_{DS}.
    (c) V_D.

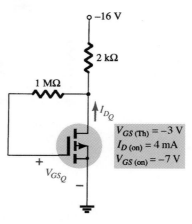

Figure 6.103 Problem 30

§ 6.12 Universal JFET Bias Curve

31. Repeat Problem 1 using the universal JFET bias curve.

32. Repeat Problem 6 using the universal JFET bias curve.

33. Repeat Problem 12 using the universal JFET bias curve.

34. Repeat Problem 15 using the universal JFET bias curve.

§ 6.15 Computer Analysis

35. Perform a PSpice Windows analysis of the network of Problem 1.

36. Perform a PSpice Windows analysis of the network of Problem 6.

37. Perform an Electronics Workbench analysis of the network of Problem 15.

38. Perform an Electronics Workbench analysis of the network of Problem 30.

*Please Note: Asterisks indicate more difficult problems.

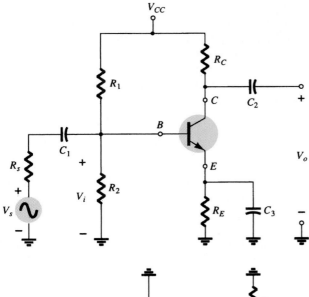

Figure 7.3 Transistor circuit under examination in this introductory discussion.

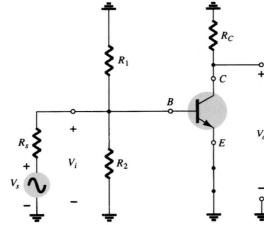

Figure 7.4 The network of Fig. 7.3 following removal of the dc supply and insertion of the short-circuit equivalent for the capacitors.

result in the "shorting out" of the dc biasing resistor R_E. Recall that capacitors assume an "open-circuit" equivalent under dc steady-state conditions, permitting an isolation between stages for the dc levels and quiescent conditions.

If we establish a common ground and rearrange the elements of Fig. 7.4, R_1 and R_2 will be in parallel and R_C will appear from collector to emitter as shown in Fig. 7.5. Since the components of the transistor equivalent circuit appearing in Fig. 7.5 employ familiar components such as resistors and independent controlled sources, analysis techniques such as superposition, Thévenin's theorem, and so on, can be applied to determine the desired quantities.

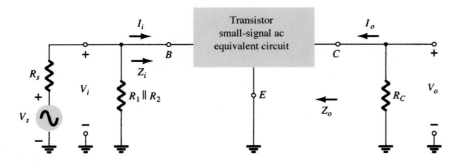

Figure 7.5 Circuit of Fig. 7.4 redrawn for small-signal ac analysis.

Let us further examine Fig. 7.5 and identify the important quantities to be determined for the system. Since we know that the transistor is an amplifying device, we would expect some indication of how the output voltage V_o is related to the input voltage V_i—the *voltage gain*. Note in Fig. 7.5 for this configuration that $I_i = I_b$ and $I_o = I_c$, which define the *current gain* $A_i = I_o/I_i$. The input impedance Z_i and output impedance Z_o will prove particularly important in the analysis to follow. A great deal more will be offered about these parameters in the sections to follow.

In summary, therefore, the ac equivalent of a network is obtained by:

1. Setting all dc sources to zero and replacing them by a short-circuit equivalent
2. Replacing all capacitors by a short-circuit equivalent
3. Removing all elements bypassed by the short-circuit equivalents introduced by steps 1 and 2
4. Redrawing the network in a more convenient and logical form

In the sections to follow, the r_e and hybrid equivalent circuits will be introduced to complete the ac analysis of the network of Fig. 7.5.

7.4 THE IMPORTANT PARAMETERS: Z_i, Z_o, A_v, A_i

Before investigating the equivalent circuits for BJTs in some detail, let us concentrate on those parameters of a two-port system that are of paramount importance from an analysis and design viewpoint. For the two-port (two pairs of terminals) system of Fig. 7.6, the input side (the side to which the signal is normally applied) is to the left and the output side (where the load is connected) is to the right. In fact, for most electrical and electronic systems, the general flow is usually from the left to the right. For both sets of terminals, the impedance between each pair of terminals under normal operating conditions is quite important.

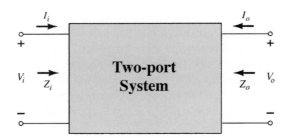

Figure 7.6 Two-port system.

Input Impedance, Z_i

For the input side, the input impedance Z_i is defined by Ohm's law as the following:

$$Z_i = \frac{V_i}{I_i} \tag{7.1}$$

If the input signal V_i is changed, the current I_i can be computed using the same level of input impedance. In other words:

For small-signal analysis, once the input impedance has been determined the same numerical value can be used for changing levels of applied signal.

In fact, we will find in the sections to follow that the input impedance of a transistor can be approximately determined by the dc biasing conditions—conditions that do not change simply because the magnitude of the applied ac signal has changed.

It is particularly noteworthy that for frequencies in the low to mid-range (typically ≤100 kHz):

The input impedance of a BJT transistor amplifier is purely resistive in nature and, depending on the manner in which the transistor is employed, can vary from a few ohms to megohms.

In addition:

An ohmmeter cannot be used to measure the small-signal ac input impedance since the ohmmeter operates in the dc mode.

Equation (7.1) is particularly useful in that it provides a method for measuring the input resistance in the ac domain. For instance, in Fig. 7.7 a sensing resistor has been added to the input side to permit a determination of I_i using Ohm's law. An oscilloscope or sensitive digital multimeter (DMM) can be used to measure the voltage V_s and V_i. Both voltages can be the peak-to-peak, peak, or rms values, as long as both levels use the same standard. The input impedance is then determined in the following manner:

$$I_i = \frac{V_s - V_i}{R_{sense}} \qquad (7.2)$$

and

$$Z_i = \frac{V_i}{I_i} \qquad (7.3)$$

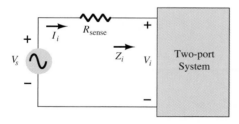

Figure 7.7 Determining Z_i.

The importance of the input impedance of a system can best be demonstrated by the network of Fig. 7.8. The signal source has an internal resistance of 600 Ω, and the system (possibly a transistor amplifier) has an input resistance of 1.2 kΩ. If the source were ideal ($R_s = 0\ \Omega$), the full 10 mV would be applied to the system, but

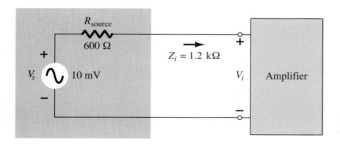

Figure 7.8 Demonstrating the impact of Z_i on an amplifier's response.

7.4   **The Important Parameters:** Z_i, Z_o, A_v, A_i

with a source impedance, the input voltage must be determined using the voltage-divider rule as follows:

$$V_i = \frac{Z_i V_s}{Z_i + R_{source}} = \frac{(1.2\ k\Omega)(10\ mV)}{1.2\ k\Omega + 0.6\ k\Omega} = 6.67\ mV$$

Thus, only 66.7% of the full-input signal is available at the input. If Z_i were only 600 Ω, then $V_i = \frac{1}{2}(10\ mV) = 5\ mV$ or 50% of the available signal. Of course, if $Z_i = 8.2\ k\Omega$, V_i will be 93.2% of the applied signal. The level of input impedance, therefore, can have a significant impact on the level of signal that reaches the system (or amplifier). In the sections and chapters to follow, it will be demonstrated that the ac input resistance is dependent on whether the transistor is in the common-base, common-emitter, or common-collector configuration and on the placement of the resistive elements.

EXAMPLE 7.1   For the system of Fig. 7.9, determine the level of input impedance.

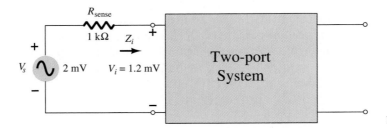

Figure 7.9   Example 7.1.

Solution

$$I_i = \frac{V_s - V_i}{R_{sense}} = \frac{2\ mV - 1.2\ mV}{1\ k\Omega} = \frac{0.8\ mV}{1\ k\Omega} = 0.8\ \mu A$$

and

$$Z_i = \frac{V_i}{I_i} = \frac{1.2\ mV}{0.8\ \mu A} = \mathbf{1.5\ k\Omega}$$

Output Impedance, Z_o

The output impedance is naturally defined at the output set of terminals, but the manner in which it is defined is quite different from that of the input impedance. That is:

The output impedance is determined at the output terminals looking back into the system with the applied signal set to zero.

In Fig. 7.10, for example, the applied signal has been set to zero volts. To determine Z_o, a signal, V_s, is applied to the output terminals and the level of V_o is measured with an oscilloscope or sensitive DMM. The output impedance is then determined in the following manner:

$$I_o = \frac{V - V_o}{R_{sense}} \tag{7.4}$$

and

$$Z_o = \frac{V_o}{I_o} \tag{7.5}$$

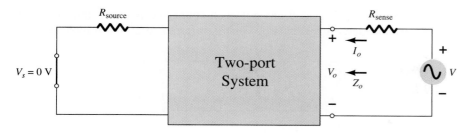

Figure 7.10 Determining Z_o.

In particular, for frequencies in the low to mid-range (typically ≤100 kHz):

The output impedance of a BJT transistor amplifier is resistive in nature and, depending on the configuration and the placement of the resistive elements, Z_o can vary from a few ohms to a level that can exceed 2 MΩ.

In addition:

An ohmmeter cannot be used to measure the small-signal ac output impedance since the ohmmeter operates in the dc mode.

For amplifier configurations where significant gain in current is desired, the level of Z_o should be as large as possible. As demonstrated by Fig. 7.11, if $Z_o \gg R_L$, the majority of the amplifier output current will pass on to the load. It will be demonstrated in the sections and chapters to follow that Z_o is frequently so large compared to R_L that it can be replaced by an open-circuit equivalent.

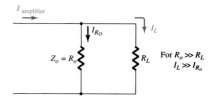

Figure 7.11 Effect of $Z_o = R_o$ on the load or output current I_L.

For the system of Fig. 7.12, determine the level of output impedance.

EXAMPLE 7.2

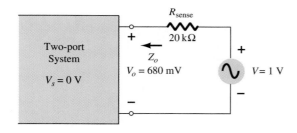

Figure 7.12 Example 7.2.

Solution

$$I_o = \frac{V - V_o}{R_{sense}} = \frac{1 \text{ V} - 680 \text{ mV}}{20 \text{ k}\Omega} = \frac{320 \text{ mV}}{20 \text{ k}\Omega} = 16 \ \mu\text{A}$$

and

$$Z_o = \frac{V_o}{I_o} = \frac{680 \text{ mV}}{16 \ \mu\text{A}} = \mathbf{42.5 \text{ k}\Omega}$$

Voltage Gain, A_v

One of the most important characteristics of an amplifier is the small-signal ac voltage gain as determined by

$$A_v = \frac{V_o}{V_i} \tag{7.6}$$

For the system of Fig. 7.13, a load has not been connected to the output terminals and the level of gain determined by Eq. (7.6) is referred to as the no-load voltage gain. That is,

$$A_{v_{NL}} = \frac{V_o}{V_i}\bigg|_{R_L = \infty \ \Omega \ (\text{open circuit})} \qquad (7.7)$$

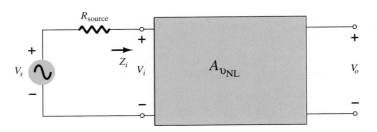

Figure 7.13   Determining the no-load voltage gain.

In Chapter 9 it will be demonstrated that:

For transistor amplifiers, the no-load voltage gain is greater than the loaded voltage gain.

For the system of Fig. 7.13 having a source resistance R_s, the level of V_i would first have to be determined using the voltage-divider rule before the gain V_o/V_s could be calculated. That is,

$$V_i = \frac{Z_i V_s}{Z_i + R_s}$$

with

$$\frac{V_i}{V_s} = \frac{Z_i}{Z_i + R_s}$$

and

$$A_{v_s} = \frac{V_o}{V_s} = \frac{V_i}{V_s} \cdot \frac{V_o}{V_i}$$

so that

$$A_{v_s} = \frac{V_o}{V_s} = \frac{Z_i}{Z_i + R_s} A_{v_{NL}} \qquad (7.8)$$

Experimentally, the voltage gain A_{v_s} or $A_{v_{NL}}$ can be determined simply by measuring the appropriate voltage levels with an oscilloscope or sensitive DMM and substituting into the appropriate equation.

Depending on the configuration, the magnitude of the voltage gain for a loaded single-stage transistor amplifier typically ranges from just less than 1 to a few hundred. A multistage (multiunit) system, however, can have a voltage gain in the thousands.

EXAMPLE 7.3

For the BJT amplifier of Fig. 7.14, determine:
(a) V_i.
(b) I_i.
(c) Z_i.
(d) A_{v_s}.

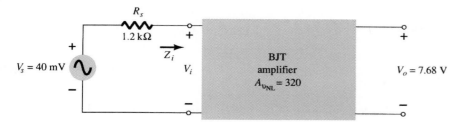

Figure 7.14 Example 7.3.

Solution

(a) $A_{v_{NL}} = \dfrac{V_o}{V_i}$ and $V_i = \dfrac{V_o}{A_{v_{NL}}} = \dfrac{7.68\text{ V}}{320} =$ **24 mV**

(b) $I_i = \dfrac{V_s - V_i}{R_s} = \dfrac{40\text{ mV} - 24\text{ mV}}{1.2\text{ k}\Omega} =$ **13.33 μA**

(c) $Z_i = \dfrac{V_i}{I_i} = \dfrac{24\text{ mV}}{13.33\ \mu\text{A}} =$ **1.8 kΩ**

(d) $A_{v_s} = \dfrac{Z_i}{Z_i + R_s} A_{v_{NL}}$

$= \dfrac{1.8\text{ k}\Omega}{1.8\text{ k}\Omega + 1.2\text{ k}\Omega}(320)$

$=$ **192**

Current Gain, A_i

The last numerical characteristic to be discussed is the current gain defined by

$$A_i = \frac{I_o}{I_i} \tag{7.9}$$

Although typically the recipient of less attention than the voltage gain, it is, however, an important quantity that can have significant impact on the overall effectiveness of a design. In general:

For BJT amplifiers, the current gain typically ranges from a level just less than 1 to a level that may exceed 100.

For the loaded situation of Fig. 7.15,

$$I_i = \frac{V_i}{Z_i} \quad \text{and} \quad I_o = -\frac{V_o}{R_L}$$

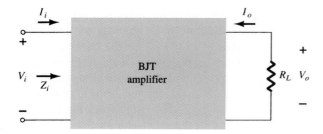

Figure 7.15 Determining the loaded current gain.

with
$$A_i = \frac{I_o}{I_i} = -\frac{V_o/R_L}{V_i/Z_i} = -\frac{V_o Z_i}{V_i R_L}$$

and
$$A_i = -A_v \frac{Z_i}{R_L} \tag{7.10}$$

Eq. (7.10) allows the determination of the current gain from the voltage gain and the impedance levels.

Phase Relationship

The phase relationship between input and output sinusoidal signals is important for a variety of practical reasons. Fortunately, however:

For the typical transistor amplifier at frequencies that permit ignoring the effects of the reactive elements, the input and output signals are either 180° out of phase or in phase.

The reason for the either–or situation will become quite clear in the chapters to follow.

Summary

The parameters of primary importance for an amplifier have now been introduced: the input impedance Z_i, the output impedance Z_o, the voltage gain A_v, the current gain A_i, and the resulting phase relationship. Other factors, such as the applied frequency at the low and high ends of the frequency spectrum, will affect some of these parameters, but this will be discussed in Chapter 11. In the sections and chapters to follow, all the parameters will be determined for a variety of transistor networks to permit a comparison of the strengths and weaknesses for each configuration.

7.5   THE r_e TRANSISTOR MODEL

The r_e model employs a diode and controlled current source to duplicate the behavior of a transistor in the region of interest. Recall that a current-controlled current source is one where the parameters of the current source are controlled by a current elsewhere in the network. In fact, in general:

BJT transistor amplifiers are referred to as current-controlled devices.

Common-Base Configuration

In Fig. 7.16a, a common-base *pnp* transistor has been inserted within the two-port structure employed in our discussion of the last few sections. In Fig. 7.16b, the r_e model for the transistor has been placed between the same four terminals. As noted in Section 7.3, the model (equivalent circuit) is chosen in such a way as to approximate the behavior of the device it is replacing in the operating region of interest. In other words, the results obtained with the model in place should be relatively close to those obtained with the actual transistor. You will recall from Chapter 3 that one junction of an operating transistor is forward-biased while the other is reverse-biased. The forward-biased junction will behave much like a diode (ignoring the effects of changing levels of V_{CE}) as verified by the curves of Fig. 3.7. For the base-to-emitter junction of the transistor of Fig. 7.16a, the diode equivalence of Fig. 7.16b between the same two terminals seems to be quite appropriate. For the output side, recall that the horizontal curves of Fig. 3.8 revealed that $I_c \cong I_e$ (as derived from $I_c = \alpha I_e$) for the range of values of V_{CE}. The current source of Fig. 7.16b establishes the fact that $I_c = \alpha I_e$, with the controlling current I_e appearing in the input side of the equivalent

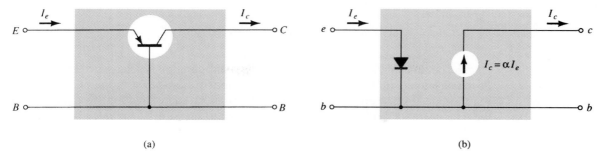

Figure 7.16 (a) Common-base BJT transistor; (b) r_e model for the configuration of Fig. 7.16a.

circuit as dictated by Fig. 7.16a. We have therefore established an equivalence at the input and output terminals with the current-controlled source, providing a link between the two—an initial review would suggest that the model of Fig. 7.16b is a valid model of the actual device.

Recall from Chapter 1 that the ac resistance of a diode can be determined by the equation $r_{ac} = 26$ mV/I_D, where I_D is the dc current through the diode at the Q (quiescent) point. This same equation can be used to find the ac resistance of the diode of Fig. 7.16b if we simply substitute the emitter current as follows:

$$r_e = \frac{26 \text{ mV}}{I_E} \qquad (7.11)$$

The subscript e of r_e was chosen to emphasize that it is the dc level of emitter current that determines the ac level of the resistance of the diode of Fig. 7.16b. Substituting the resulting value of r_e in Fig. 7.16b will result in the very useful model of Fig. 7.17.

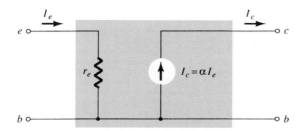

Figure 7.17 Common-base r_e equivalent circuit.

Due to the isolation that exists between input and output circuits of Fig. 7.17, it should be fairly obvious that the input impedance Z_i for the common-base configuration of a transistor is simply r_e. That is,

$$Z_i = r_e \Big|_{CB} \qquad (7.12)$$

For the common-base configuration, typical values of Z_i range from a few ohms to a maximum of about 50 Ω.

For the output impedance, if we set the signal to zero, then $I_e = 0$ A and $I_c = \alpha I_e = \alpha(0 \text{ A}) = 0$ A, resulting in an open-circuit equivalence at the output terminals. The result is that for the model of Fig. 7.17,

$$Z_o \cong \infty \ \Omega \Big|_{CB} \qquad (7.13)$$

7.5   The r_e Transistor Model   365

In actuality:

For the common-base configuration, typical values of Z_o are in the megohm range.

The output resistance of the common-base configuration is determined by the slope of the characteristic lines of the output characteristics as shown in Fig. 7.18. Assuming the lines to be perfectly horizontal (an excellent approximation) would result in the conclusion of Eq. (7.13). If care were taken to measure Z_o graphically or experimentally, levels typically in the range 1- to 2-MΩ would be obtained.

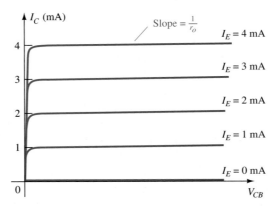

Figure 7.18 Defining Z_o.

In general, for the common-base configuration the input impedance is relatively small and the output impedance quite high.

The voltage gain will now be determined for the network of Fig. 7.19.

$$V_o = -I_o R_L = -(-I_c)R_L = \alpha I_e R_L$$

and

$$V_i = I_e Z_i = I_e r_e$$

so that

$$A_v = \frac{V_o}{V_i} = \frac{\alpha I_e R_L}{I_e r_e}$$

and

$$\boxed{A_v = \frac{\alpha R_L}{r_e} \cong \frac{R_L}{r_e}}_{CB} \qquad (7.14)$$

For the current gain,

$$A_i = \frac{I_o}{I_i} = \frac{-I_c}{I_e} = -\frac{\alpha I_e}{I_e}$$

and

$$\boxed{A_i = -\alpha \cong -1}_{CB} \qquad (7.15)$$

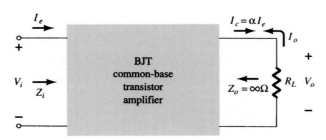

Figure 7.19 Defining $A_v = V_o/V_i$ for the common-base configuration.

The fact that the polarity of the voltage V_o as determined by the current I_c is the same as defined by Fig. 7.19 (i.e., the negative side is at ground potential) reveals that V_o and V_i are *in phase* for the common-base configuration. For an *npn* transistor in the common-base configuration, the equivalence would appear as shown in Fig. 7.20.

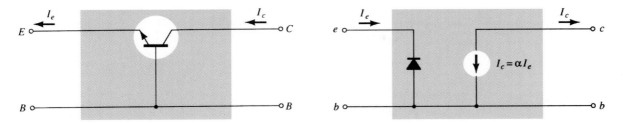

Figure 7.20 Approximate model for a common-base *npn* transistor configuration.

For a common-base configuration of Fig. 7.17 with $I_E = 4$ mA, $\alpha = 0.98$, and an ac signal of 2 mV applied between the base and emitter terminals:
(a) Determine the input impedance.
(b) Calculate the voltage gain if a load of 0.56 kΩ is connected to the output terminals.
(c) Find the output impedance and current gain.

EXAMPLE 7.4

Solution

(a) $r_e = \dfrac{26 \text{ mV}}{I_E} = \dfrac{26 \text{ mV}}{4 \text{ mA}} = \mathbf{6.5 \ \Omega}$

(b) $I_i = I_e = \dfrac{V_i}{Z_i} = \dfrac{2 \text{ mV}}{6.5 \ \Omega} = 307.69 \ \mu\text{A}$

$V_o = I_c R_L = \alpha I_e R_L = (0.98)(307.69 \ \mu\text{A})(0.56 \text{ k}\Omega)$

$\qquad = 168.86 \text{ mV}$

and $A_v = \dfrac{V_o}{V_i} = \dfrac{168.86 \text{ mV}}{2 \text{ mV}} = \mathbf{84.43}$

or from Eq. (7.14),

$$A_v = \frac{\alpha R_L}{r_e} = \frac{(0.98)(0.56 \text{ k}\Omega)}{6.5 \ \Omega} = \mathbf{84.43}$$

(c) $Z_o \cong \infty \ \Omega$

$A_i = \dfrac{I_o}{I_i} = -\alpha = \mathbf{-0.98} \qquad$ as defined by Eq. (7.15)

Common-Emitter Configuration

For the common-emitter configuration of Fig. 7.21a, the input terminals are the base and emitter terminals, but the output set is now the collector and emitter terminals. In addition, the emitter terminal is now common between the input and output ports of the amplifier. Substituting the r_e equivalent circuit for the *npn* transistor will result in the configuration of Fig. 7.21b. Note that the controlled-current source is still connected between the collector and base terminals and the diode between the base and

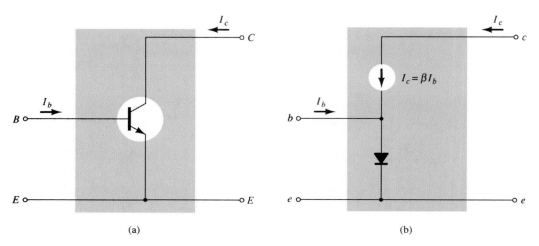

Figure 7.21 (a) Common-emitter BJT transistor; (b) approximate model for the configuration of Fig. 7.21a.

emitter terminals. In this configuration, the base current is the input current while the output current is still I_c. Recall from Chapter 3 that the base and collector currents are related by the following equation:

$$I_c = \beta I_b \tag{7.16}$$

The current through the diode is therefore determined by

$$I_e = I_c + I_b = \beta I_b + I_b$$

and

$$I_e = (\beta + 1)I_b \tag{7.17}$$

However, since the ac beta is typically much greater than 1, we will use the following approximation for the current analysis:

$$I_e \cong \beta I_b \tag{7.18}$$

The input impedance is determined by the following ratio:

$$Z_i = \frac{V_i}{I_i} = \frac{V_{be}}{I_b}$$

The voltage V_{be} is across the diode resistance as shown in Fig. 7.22. The level of r_e is still determined by the dc current I_E. Using Ohm's law gives

$$V_i = V_{be} = I_e r_e \cong \beta I_b r_e$$

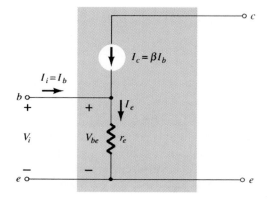

Figure 7.22 Determining Z_i using the approximate model.

Substituting yields

$$Z_i = \frac{V_{be}}{I_b} \cong \frac{\beta I_b r_e}{I_b}$$

and

$$\boxed{Z_i \cong \beta r_e}_{CE} \tag{7.19}$$

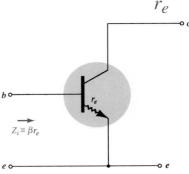

Figure 7.23 Impact of r_e on input impedance.

In essence, Eq. (7.19) states that the input impedance for a situation such as shown in Fig. 7.23 is beta times the value of r_e. In other words, a resistive element in the emitter leg is reflected into the input circuit by a multiplying factor β. For instance, if $r_e = 6.5\ \Omega$ as in Example 7.4 and $\beta = 160$ (quite typical), the input impedance has increased to a level of

$$Z_i \cong \beta r_e = (160)(6.5\ \Omega) = \mathbf{1.04\ k\Omega}$$

For the common-emitter configuration, typical values of Z_i defined by βr_e range from a few hundred ohms to the kilohm range, with maximums of about 6–7 kΩ.

For the output impedance, the characteristics of interest are the output set of Fig. 7.24. Note that the slope of the curves increases with increase in collector current. The steeper the slope, the less the level of output impedance (Z_o). The r_e model of Fig. 7.21 does not include an output impedance, but if available from a graphical analysis or from data sheets, it can be included as shown in Fig. 7.25.

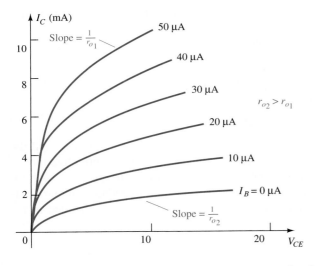

Figure 7.24 Defining r_o for the common-emitter configuration.

For the common-emitter configuration, typical values of Z_o are in the range of 40 to 50 kΩ.

For the model of Fig. 7.25, if the applied signal is set to zero, the current I_c is 0 A and the output impedance is

$$\boxed{Z_o = r_o}_{CE} \tag{7.20}$$

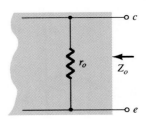

Figure 7.25 Including r_o in the transistor equivalent circuit.

Of course, if the contribution due to r_o is ignored as in the r_e model, the output impedance is defined by $Z_o = \infty\ \Omega$.

The voltage gain for the common-emitter configuration will now be determined for the configuration of Fig. 7.26 using the assumption that $Z_o = \infty\ \Omega$. The effect of including r_o will be considered in Chapter 8. For the defined direction of I_o and polarity of V_o,

$$V_o = -I_o R_L$$

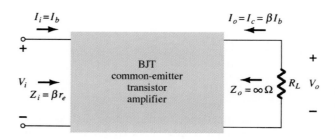

Figure 7.26 Determining the voltage and current gain for the common-emitter transistor amplifier.

The minus sign simply reflects the fact that the direction of I_o in Fig. 7.26 would establish a voltage V_o with the opposite polarity. Continuing gives

$$V_o = -I_oR_L = -I_cR_L = -\beta I_b R_L$$

and

$$V_i = I_i Z_i = I_b \beta r_e$$

so that

$$A_v = \frac{V_o}{V_i} = -\frac{\beta I_b R_L}{I_b \beta r_e}$$

and

$$\boxed{A_v = -\frac{R_L}{r_e}}_{\;CE, r_o = \infty\,\Omega} \qquad (7.21)$$

The resulting minus sign for the voltage gain reveals that the output and input voltages are $180°$ out of phase.

The current gain for the configuration of Fig. 7.26:

$$A_i = \frac{I_o}{I_i} = \frac{I_c}{I_b} = \frac{\beta I_b}{I_b}$$

and

$$\boxed{A_i = \beta}_{\;CE, r_o = \infty\,\Omega} \qquad (7.22)$$

Using the facts that the input impedance is βr_e, the collector current is βI_b, and the output impedance is r_o, the equivalent model of Fig. 7.27 can be an effective tool in the analysis to follow. For typical parameter values, the common-emitter configuration can be considered one that has a moderate level of input impedance, a high voltage and current gain, and an output impedance that may have to be included in the network analysis.

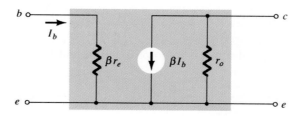

Figure 7.27 r_e model for the common-emitter transistor configuration.

EXAMPLE 7.5

Given $\beta = 120$ and $I_E = 3.2$ mA for a common-emitter configuration with $r_o = \infty\ \Omega$, determine:
(a) Z_i.
(b) A_v if a load of 2 kΩ is applied.
(c) A_i with the 2 kΩ load.

Solution

(a) $r_e = \dfrac{26\text{ mV}}{I_E} = \dfrac{26\text{ mV}}{3.2\text{ mV}} = 8.125\ \Omega$

and $Z_i = \beta r_e = (120)(8.125\ \Omega) = \mathbf{975\ \Omega}$

(b) Eq. (7.21): $A_v = -\dfrac{R_L}{r_e} = -\dfrac{2\text{ k}\Omega}{8.125\ \Omega} = \mathbf{-246.15}$

(c) $A_i = \dfrac{I_o}{I_i} = \beta = \mathbf{120}$

Common-Collector Configuration

For the common-collector configuration, the model defined for the common-emitter configuration of Fig. 7.21 is normally applied rather than defining a model for the common-collector configuration. In subsequent chapters, a number of common-collector configurations will be investigated and the impact of using the same model will become quite apparent.

7.6   THE HYBRID EQUIVALENT MODEL

It was pointed out in Section 7.5 that the r_e model for a transistor is sensitive to the dc level of operation of the amplifier. The result is an input resistance that will vary with the dc operating point. For the hybrid equivalent model to be described in this section, the parameters are defined at an operating point that may or may not reflect the actual operating conditions of the amplifier. This is due to the fact that specification sheets cannot provide parameters for an equivalent circuit at every possible operating point. They must choose operating conditions that they believe reflect the general characteristics of the device. The hybrid parameters as shown in Fig. 7.28 are drawn from the specification sheet for the 2N4400 transistor described in Chapter 3. The values are provided at a dc collector current of 1 mA and a collector-to-emitter voltage of 10 V. In addition, a range of values is provided for each parameter for guidance in the initial design or analysis of a system. One obvious advantage of the specification sheet listing is the immediate knowledge of typical levels for the parameters of the device as compared to other transistors.

The quantities h_{ie}, h_{re}, h_{fe}, and h_{oe} of Fig. 7.28 are called the hybrid parameters and are the components of a small-signal equivalent circuit to be described shortly. For years, the hybrid model with all its parameters was the chosen model for the educational and industrial communities. Presently, however, the r_e model is applied more frequently, but often with the h_{oe} parameter of the hybrid equivalent model to provide

			Min.	Max.	
Input impedance ($I_C = 1$ mA dc, $V_{CE} = 10$ V dc, $f = 1$ kHz) 2N4400	h_{ie}		0.5	7.5	kΩ
Voltage feedback ratio ($I_C = 1$ mA dc, $V_{CE} = 10$ V dc, $f = 1$ kHz)	h_{re}		0.1	8.0	$\times 10^{-4}$
Small-signal current gain ($I_C = 1$ mA dc, $V_{CE} = 10$ V dc, $f = 1$ kHz) 2N4400	h_{fe}		20	250	—
Output admittance ($I_C = 1$ mA dc, $V_{CE} = 10$ V dc, $f = 1$ kHz)	h_{oe}		1.0	30	1μS

Figure 7.28   Hybrid parameters for the 2N4400 transistor.

some measure for the output impedance. Since specification sheets do provide the hybrid parameters and the hybrid model continues to receive a good measure of attention, it is quite important that the hybrid model be covered in some detail in this book. Once developed, the similarities between the r_e and hybrid models will be quite apparent. In fact, once the components of one are defined for a particular operating point, the parameters of the other model are immediately available.

Our description of the hybrid equivalent model will begin with the general two-port system of Fig. 7.29. The following set of equations (7.23) is only one of a number of ways in which the four variables of Fig. 7.29 can be related. It is the most frequently employed in transistor circuit analysis, however, and therefore is discussed in detail in this chapter.

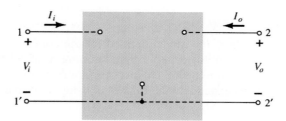

Figure 7.29 Two-port system.

$$V_i = h_{11}I_i + h_{12}V_o \tag{7.23a}$$

$$I_o = h_{21}I_i + h_{22}V_o \tag{7.23b}$$

The parameters relating the four variables are called *h-parameters* from the word "hybrid." The term *hybrid* was chosen because the mixture of variables (V and I) in each equation results in a "hybrid" set of units of measurement for the h-parameters. A clearer understanding of what the various h-parameters represent and how we can determine their magnitude can be developed by isolating each and examining the resulting relationship.

If we arbitrarily set $V_o = 0$ (short circuit the output terminals) and solve for h_{11} in Eq. (7.23a), the following will result:

$$h_{11} = \left. \frac{V_i}{I_i} \right|_{V_o = 0} \qquad \text{ohms} \tag{7.24}$$

The ratio indicates that the parameter h_{11} is an impedance parameter with the units of ohms. Since it is the ratio of the *input* voltage to the *input* current with the output terminals *shorted*, it is called the *short-circuit input-impedance parameter*. The subscript 11 of h_{11} defines the fact that the parameter is determined by a ratio of quantities measured at the input terminals.

If I_i is set equal to zero by opening the input leads, the following will result for h_{12}:

$$h_{12} = \left. \frac{V_i}{V_o} \right|_{I_i = 0} \qquad \text{unitless} \tag{7.25}$$

The parameter h_{12}, therefore, is the ratio of the input voltage to the output voltage with the input current equal to zero. It has no units since it is a ratio of voltage levels and is called the *open-circuit reverse transfer voltage ratio parameter*. The subscript 12 of h_{12} reveals that the parameter is a transfer quantity determined by a ratio of input to output measurements. The first integer of the subscript defines the measured quantity to appear in the numerator; the second integer defines the source of the quantity to appear in the denominator. The term *reverse* is included because the ratio is an input voltage over an output voltage rather than the reverse ratio typically of interest.

If in Eq. (7.23b) V_o is equal to zero by again shorting the output terminals, the following will result for h_{21}:

$$h_{21} = \left.\frac{I_o}{I_i}\right|_{V_o = 0} \qquad \text{unitless} \qquad (7.26)$$

Note that we now have the ratio of an output quantity to an input quantity. The term *forward* will now be used rather than *reverse* as indicated for h_{12}. The parameter h_{21} is the ratio of the output current to the input current with the output terminals shorted. This parameter, like h_{12}, has no units since it is the ratio of current levels. It is formally called the *short-circuit forward transfer current ratio parameter*. The subscript 21 again indicates that it is a transfer parameter with the output quantity in the numerator and the input quantity in the denominator.

The last parameter, h_{22}, can be found by again opening the input leads to set $I_1 = 0$ and solving for h_{22} in Eq. (7.23b):

$$h_{22} = \left.\frac{I_o}{V_o}\right|_{I_i = 0} \qquad \text{siemens} \qquad (7.27)$$

Since it is the ratio of the output current to the output voltage, it is the output conductance parameter and is measured in siemens (S). It is called the *open-circuit output admittance parameter*. The subscript 22 reveals that it is determined by a ratio of output quantities.

Since each term of Eq. (7.23a) has the unit volt, let us apply Kirchhoff's voltage law "in reverse" to find a circuit that "fits" the equation. Performing this operation will result in the circuit of Fig. 7.30. Since the parameter h_{11} has the unit ohm, it is represented by a resistor in Fig. 7.30. The quantity h_{12} is dimensionless and therefore simply appears as a multiplying factor of the "feedback" term in the input circuit.

Since each term of Eq. (7.23b) has the units of current, let us now apply Kirchhoff's current law "in reverse" to obtain the circuit of Fig. 7.31. Since h_{22} has the units of admittance, which for the transistor model is conductance, it is represented by the resistor symbol. Keep in mind, however, that the resistance in ohms of this resistor is equal to the reciprocal of conductance ($1/h_{22}$).

The complete "ac" equivalent circuit for the basic three-terminal linear device is indicated in Fig. 7.32 with a new set of subscripts for the *h*-parameters. The notation of Fig. 7.32 is of a more practical nature since it relates the *h*-parameters to the resulting ratio obtained in the last few paragraphs. The choice of letters is obvious from the following listing:

$$h_{11} \rightarrow \text{input resistance} \rightarrow h_i$$

$$h_{12} \rightarrow \text{reverse transfer voltage ratio} \rightarrow h_r$$

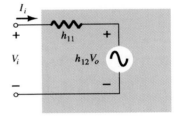

Figure 7.30 Hybrid input equivalent circuit.

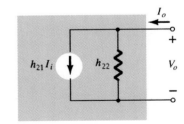

Figure 7.31 Hybrid output equivalent circuit.

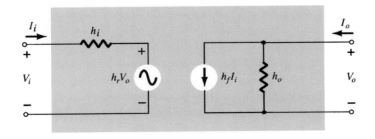

Figure 7.32 Complete hybrid equivalent circuit.

$$h_{21} \rightarrow \text{forward transfer current ratio} \rightarrow h_f$$

$$h_{22} \rightarrow \text{output conductance} \rightarrow h_o$$

The circuit of Fig. 7.32 is applicable to any linear three-terminal electronic device or system with no internal independent sources. For the transistor, therefore, even though it has three basic configurations, *they are all three-terminal configurations,* so that the resulting equivalent circuit will have the same format as shown in Fig. 7.32. In each case, the bottom of the input and output sections of the network of Fig. 7.32 can be connected as shown in Fig. 7.33 since the potential level is the same. Essentially, therefore, the transistor model is a three-terminal two-port system. The *h*-parameters, however, will change with each configuration. To distinguish which parameter has been used or which is available, a second subscript has been added to the *h*-parameter notation. For the common-base configuration, the lowercase letter *b* was added, while for the common-emitter and common-collector configurations, the letters *e* and *c* were added, respectively. The hybrid equivalent network for the common-emitter configuration appears with the standard notation in Fig. 7.33. Note that $I_i = I_b$, $I_o = I_c$, and through an application of Kirchhoff's current law, $I_e = I_b + I_c$. The input voltage is now V_{be}, with the output voltage V_{ce}. For the common-base configuration of Fig. 7.34, $I_i = I_e$, $I_o = I_c$ with $V_{eb} = V_i$ and $V_{cb} = V_o$. The networks of Figs. 7.33 and 7.34 are applicable for *pnp* or *npn* transistors.

The fact that both a Thévenin and Norton circuit appear in the circuit of Fig. 7.32 was further impetus for calling the resultant circuit a *hybrid* equivalent circuit. Two additional transistor equivalent circuits, not to be discussed in this text, called the

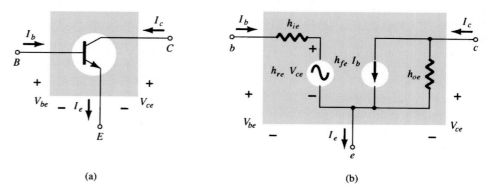

(a)   **(b)**

Figure 7.33 Common-emitter configuration: (a) graphical symbol; (b) hybrid equivalent circuit.

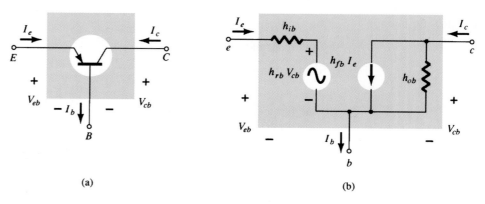

(a)   **(b)**

Figure 7.34 Common-base configuration: (a) graphical symbol; (b) hybrid equivalent circuit.

Chapter 7 BJT Transistor Modeling

z-parameter and y-parameter equivalent circuits, use either the voltage source or the current source, but not both, in the same equivalent circuit. In Section 7.7, the magnitudes of the various parameters will be found from the transistor characteristics in the region of operation resulting in the desired *small-signal equivalent network* for the transistor.

For the common-emitter and common-base configurations, the magnitude of h_r and h_o is often such that the results obtained for the important parameters such as Z_i, Z_o, A_v, and A_i are only slightly affected if they (h_r and h_o) are not included in the model.

Since h_r is normally a relatively small quantity, its removal is approximated by $h_r \cong 0$ and $h_r V_o = 0$, resulting in a short-circuit equivalent for the feedback element as shown in Fig. 7.35. The resistance determined by $1/h_o$ is often large enough to be ignored in comparison to a parallel load, permitting its replacement by an open-circuit equivalent for the CE and CB models, as shown in Fig. 7.35.

The resulting equivalent of Fig. 7.36 is quite similar to the general structure of the common-base and common-emitter equivalent circuits obtained with the r_e model. In fact, the hybrid equivalent and the r_e models for each configuration have been repeated in Fig. 7.37 for comparison. It should be reasonably clear from Fig. 7.37a that

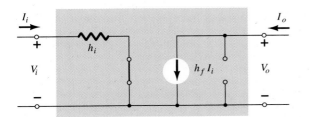

Figure 7.35 Effect of removing h_{re} and h_{oe} from the hybrid equivalent circuit.

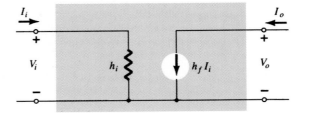

Figure 7.36 Approximate hybrid equivalent model.

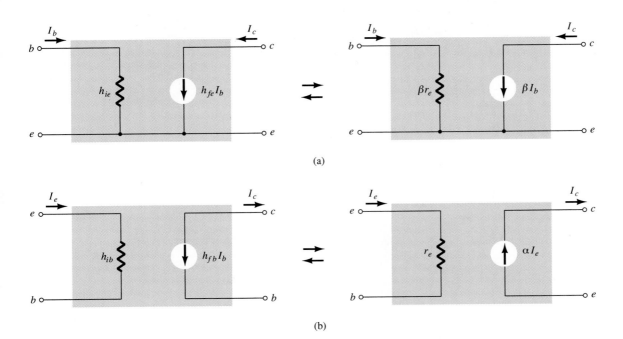

Figure 7.37 Hybrid versus r_e model: (a) common-emitter configuration; (b) common-base configuration.

$$h_{ie} = \beta r_e \qquad (7.28)$$

and

$$h_{fe} = \beta_{ac} \qquad (7.29)$$

From Fig. 7.37b,

$$h_{ib} = r_e \qquad (7.30)$$

and

$$h_{fb} = -\alpha \cong -1 \qquad (7.31)$$

In particular, note that the minus sign in Eq. (7.31) accounts for the fact that the current source of the standard hybrid equivalent circuit is pointing down rather than in the actual direction as shown in the r_e model of Fig. 7.37b.

EXAMPLE 7.6

Given $I_E = 2.5$ mA, $h_{fe} = 140$, $h_{oe} = 20$ μS (μmho), and $h_{ob} = 0.5$ μS, determine:
(a) The common-emitter hybrid equivalent circuit.
(b) The common-base r_e model.

Solution

(a) $r_e = \dfrac{26 \text{ mV}}{I_E} = \dfrac{26 \text{ mV}}{2.5 \text{ mA}} = 10.4 \ \Omega$

$h_{ie} = \beta r_e = (140)(10.4 \ \Omega) = \textbf{1.456 k}\boldsymbol{\Omega}$

$r_o = \dfrac{1}{h_{oe}} = \dfrac{1}{20 \ \mu\text{S}} = 50 \text{ k}\Omega$

Note Fig. 7.38.

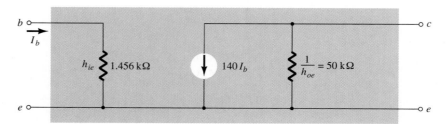

Figure 7.38 Common-emitter hybrid equivalent circuit for the parameters of Example 7.6.

(b) $r_e = \textbf{10.4} \ \boldsymbol{\Omega}$

$\alpha \cong 1, \qquad r_o = \dfrac{1}{h_{ob}} = \dfrac{1}{0.5 \ \mu\text{S}} = \textbf{2 M}\boldsymbol{\Omega}$

Note Fig. 7.39.

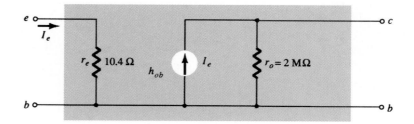

Figure 7.39 Common-base r_e model for the parameters of Example 7.6.

A series of equations relating the parameters of each configuration for the hybrid equivalent circuit is provided in Appendix B. In Section 7.8, we demonstrate that the hybrid parameter h_{fe} (β_{ac}) is the least sensitive of the hybrid parameters to a change in collector current. Assuming, therefore, that $h_{fe} = \beta$ is a constant for the range of interest, is a fairly good approximation. It is $h_{ie} = \beta r_e$ that will vary significantly with I_C and should be determined at operating levels, since it can have a real impact on the gain levels of a transistor amplifier.

7.7 GRAPHICAL DETERMINATION OF THE h-PARAMETERS

Using partial derivatives (calculus), it can be shown that the magnitude of the h-parameters for the small-signal transistor equivalent circuit in the region of operation for the common-emitter configuration can be found using the following equations:*

$$h_{ie} = \frac{\partial v_i}{\partial i_i} = \frac{\partial v_{be}}{\partial i_b} \cong \frac{\Delta v_{be}}{\Delta i_b}\bigg|_{V_{CE}=\text{constant}} \qquad \text{(ohms)} \qquad (7.32)$$

$$h_{re} = \frac{\partial v_i}{\partial v_o} = \frac{\partial v_{be}}{\partial v_{ce}} \cong \frac{\Delta v_{be}}{\Delta v_{ce}}\bigg|_{I_B=\text{constant}} \qquad \text{(unitless)} \qquad (7.33)$$

$$h_{fe} = \frac{\partial i_o}{\partial i_i} = \frac{\partial i_c}{\partial i_b} \cong \frac{\Delta i_c}{\Delta i_b}\bigg|_{V_{CE}=\text{constant}} \qquad \text{(unitless)} \qquad (7.34)$$

$$h_{oe} = \frac{\partial i_o}{\partial v_o} = \frac{\partial i_c}{\partial v_{ce}} \cong \frac{\Delta i_c}{\Delta v_{ce}}\bigg|_{I_B=\text{constant}} \qquad \text{(siemens)} \qquad (7.35)$$

In each case, the symbol Δ refers to a small change in that quantity around the quiescent point of operation. In other words, the h-parameters are determined in the region of operation for the applied signal so that the equivalent circuit will be the most accurate available. The constant values of V_{CE} and I_B in each case refer to a condition that must be met when the various parameters are determined from the characteristics of the transistor. For the common-base and common-collector configurations, the proper equation can be obtained by simply substituting the proper values of v_i, v_o, i_i, and i_o.

The parameters h_{ie} and h_{re} are determined from the input or base characteristics, while the parameters h_{fe} and h_{oe} are obtained from the output or collector characteristics. Since h_{fe} is usually the parameter of greatest interest, we shall discuss the operations involved with equations, such as Eqs. (7.32) through (7.35), for this parameter first. The first step in determining any of the four hybrid parameters is to find the quiescent point of operation as indicated in Fig. 7.40. In Eq. (7.34) the condition $V_{CE} = $ constant requires that the changes in base current and collector current be taken along a vertical straight line drawn through the Q-point representing a fixed collector-to-emitter voltage. Equation (7.34) then requires that a small change in collector current be divided by the corresponding change in base current. For the greatest accuracy, these changes should be made as small as possible.

---

*The partial derivative $\partial v_i/\partial i_i$ provides a measure of the instantaneous change in v_i due to an instantaneous change in i_i.

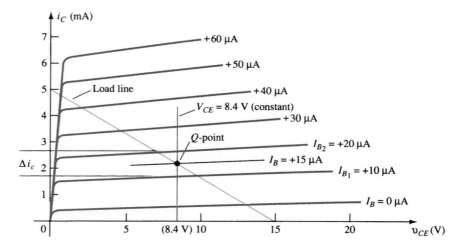

Figure 7.40 h_{fe} determination.

In Fig. 7.40, the change in i_b was chosen to extend from I_{B_1} to I_{B_2} along the perpendicular straight line at V_{CE}. The corresponding change in i_c is then found by drawing the horizontal lines from the intersections of I_{B_1} and I_{B_2} with $V_{CE} = $ constant to the vertical axis. All that remains is to substitute the resultant changes of i_b and i_c into Eq. (7.34). That is,

$$|h_{fe}| = \frac{\Delta i_c}{\Delta i_b}\bigg|_{V_{CE} = \text{constant}} = \frac{(2.7 - 1.7)\,\text{mA}}{(20 - 10)\,\mu\text{A}}\bigg|_{V_{CE} = 8.4\,\text{V}}$$

$$= \frac{10^{-3}}{10 \times 10^{-6}} = \mathbf{100}$$

In Fig. 7.41, a straight line is drawn tangent to the curve I_B through the Q-point to establish a line $I_B = $ constant as required by Eq. (7.35) for h_{oe}. A change in v_{CE} was then chosen and the corresponding change in i_C determined by drawing the horizontal lines to the vertical axis at the intersections on the $I_B = $ constant line. Substituting into Eq. (7.35), we get

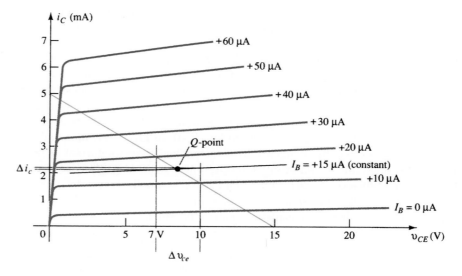

Figure 7.41 h_{oe} determination.

Chapter 7   BJT Transistor Modeling

$$|h_{oe}| = \frac{\Delta i_c}{\Delta v_{ce}}\bigg|_{I_B = \text{constant}} = \frac{(2.2 - 2.1)\,\text{mA}}{(10 - 7)\,\text{V}}\bigg|_{I_B = +15\,\mu\text{A}}$$

$$= \frac{0.1 \times 10^{-3}}{3} = 33\,\mu\text{A/V} = 33 \times 10^{-6}\,\text{S} = 33\,\mu\text{S}$$

To determine the parameters h_{ie} and h_{re} the Q-point must first be found on the input or base characteristics as indicated in Fig. 7.42. For h_{ie}, a line is drawn tangent to the curve $V_{CE} = 8.4$ V through the Q-point to establish a line $V_{CE} = $ constant as required by Eq. (7.32). A small change in v_{be} was then chosen, resulting in a corresponding change in i_b. Substituting into Eq. (7.32), we get

$$|h_{ie}| = \frac{\Delta v_{be}}{\Delta i_b}\bigg|_{V_{CE} = \text{constant}} = \frac{(733 - 718)\,\text{mV}}{(20 - 10)\,\mu\text{A}}\bigg|_{V_{CE} = 8.4\,\text{V}}$$

$$= \frac{15 \times 10^{-3}}{10 \times 10^{-6}} = 1.5\,\text{k}\Omega$$

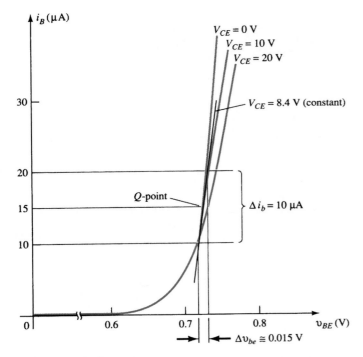

Figure 7.42 h_{ie} determination.

The last parameter, h_{re}, can be found by first drawing a horizontal line through the Q-point at $I_B = 15\,\mu$A. The natural choice then is to pick a change in v_{CE} and find the resulting change in v_{BE} as shown in Fig. 7.43.

Substituting into Eq. (7.33), we get

$$|h_{re}| = \frac{\Delta v_{be}}{\Delta v_{ce}}\bigg|_{I_B = \text{constant}} = \frac{(733 - 725)\,\text{mV}}{(20 - 0)\,\text{V}} = \frac{8 \times 10^{-3}}{20} = 4 \times 10^{-4}$$

For the transistor whose characteristics have appeared in Figs. 7.40 through 7.43, the resulting hybrid small-signal equivalent circuit is shown in Fig. 7.44.

r_e

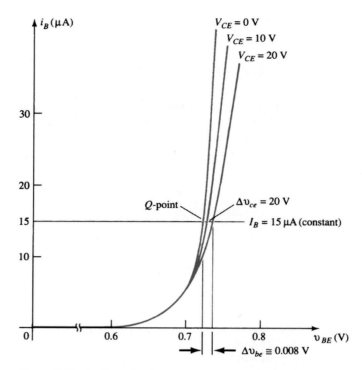

Figure 7.43 h_{re} determination.

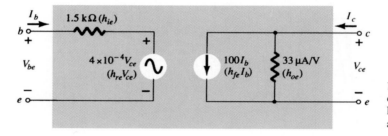

Figure 7.44 Complete hybrid equivalent circuit for a transistor having the characteristics that appear in Figs. 7.40 through 7.43.

As mentioned earlier, the hybrid parameters for the common-base and common-collector configurations can be found using the same basic equations with the proper variables and characteristics.

Table 7.1 lists typical parameter values in each of the three configurations for the broad range of transistors available today. The minus sign indicates that in Eq. (7.34) as one quantity increased in magnitude, within the change chosen, the other decreased in magnitude.

TABLE 7.1 Typical Parameter Values for the CE, CC, and CB Transistor Configurations

Parameter	CE	CC	CB
h_i	1 kΩ	1 kΩ	20 Ω
h_r	2.5×10^{-4}	$\cong 1$	3.0×10^{-4}
h_f	50	-50	-0.98
h_o	25 μA/V	25 μA/V	0.5 μA/V
$1/h_o$	40 kΩ	40 kΩ	2 MΩ

Note in retrospect (Section 3.5: Transistor Amplifying Action) that the input resistance of the common-base configuration is low, while the output resistance is high. Consider also that the short-circuit current gain is very close to 1. For the common-emitter and common-collector configurations, note that the input resistance is much higher than that of the common-base configuration and that the ratio of output to input resistance is about 40 : 1. Consider also for the common-emitter and common-base configurations that h_r is very small in magnitude. Transistors are available today with values of h_{fe} that vary from 20 to 600. For any transistor, the region of operation and conditions under which it is being used will have an effect on the various h-parameters. The effect of temperature and collector current and voltage on the h-parameters is discussed in Section 7.8.

7.8 VARIATIONS OF TRANSISTOR PARAMETERS

There are a large number of curves that can be drawn to show the variations of the h-parameters with temperature, frequency, voltage, and current. The most interesting and useful at this stage of the development include the h-parameter variations with junction temperature and collector voltage and current.

In Fig. 7.45, the effect of the collector current on the h-parameter has been indicated. Take careful note of the logarithmic scale on the vertical and horizontal axes. Logarithmic scales will be examined in Chapter 11. The parameters have all been normalized to unity so that the relative change in magnitude with collector current can easily be determined. On every set of curves, such as in Fig. 7.46, the operating point at which the parameters were found is always indicated. For this particular situation, the quiescent point is at the intersection of $V_{CE} = 5.0$ V and $I_C = 1.0$ mA. Since the frequency and temperature of operation will also affect the h-parameters, these quantities are also indicated on the curves. At 0.1 mA, h_{fe} is about 0.5 or 50% of its value at 1.0 mA, while at 3 mA, it is 1.5 or 150% of that value. In other words, if $h_{fe} = 50$ at $I_C = 1.0$ mA, h_{fe} has changed from a value of 0.5(50) = 25 to 1.5(50) = 75, with

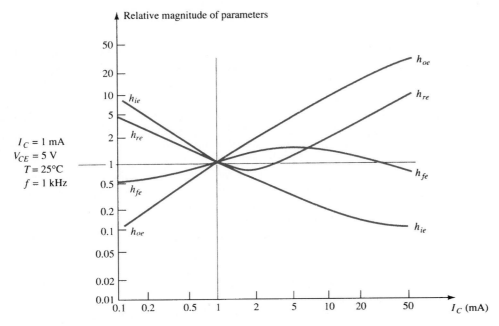

Figure 7.45 Hybrid parameter variations with collector current.

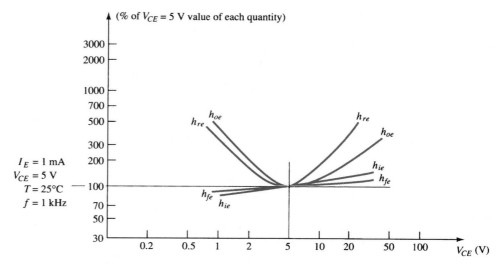

Figure 7.46 Hybrid parameter variations with collector-emitter potential.

a change of I_C from 0.1 to 3 mA. Consider, however, the point of operation at $I_C = 50$ mA. The magnitude of h_{re} is now approximately 11 times that at the defined Q-point, a magnitude that may not permit eliminating this parameter from the equivalent circuit. The parameter h_{oe} is approximately 35 times the normalized value. This increase in h_{oe} will decrease the magnitude of the output resistance of the transistor to a point where it may approach the magnitude of the load resistor. There would then be no justification in eliminating h_{oe} from the equivalent circuit on an approximate basis.

In Fig. 7.46, the variation in magnitude of the h-parameters on a normalized basis has been indicated with changes in collector voltage. This set of curves was normalized at the same operating point of the transistor discussed in Fig. 7.45 so that a comparison between the two sets of curves can be made. Note that h_{ie} and h_{fe} are relatively steady in magnitude while h_{oe} and h_{re} are much larger to the left and right of the chosen operating point. In other words, h_{oe} and h_{re} are much more sensitive to changes in collector voltage than are h_{ie} and h_{fe}.

It is interesting to note from Figs. 7.45 and 7.46 that the value of h_{fe} appears to change the least. Therefore, the specific value of current gain, whether h_{fe} or β, can, on an approximate and relative basis, be considered fairly constant for the range of collector current and voltage.

The value of $h_{ie} = \beta r_e$ does vary considerably with collector current as one might expect due to the sensitivity of r_e to emitter ($I_E \cong I_C$) current. It is therefore a quantity that should be determined as close to operating conditions as possible. For values below the specified V_{CE}, h_{re} is fairly constant, but it does increase measurably for higher values. It is indeed fortunate that for most applications the magnitude of h_{re} and h_{oe} are such that they can usually be ignored. They are quite sensitive to collector current and collector-to-emitter voltage.

In Fig. 7.47, the variation in h-parameters has been plotted for changes in junction temperature. The normalization value is taken to be room temperature: $T = 25°C$. The horizontal scale is a linear scale rather than a logarithmic scale as was employed for Figs. 7.45 and 7.46. In general, all the parameters increase in magnitude with temperature. The parameter least affected, however, is h_{oe}, while the input impedance h_{ie} changes at the greatest rate. The fact that h_{fe} will change from 50% of its normalized value at $-50°C$ to 150% of its normalized value at $+150°C$ indicates clearly that the operating temperature must be carefully considered in the design of transistor circuits.

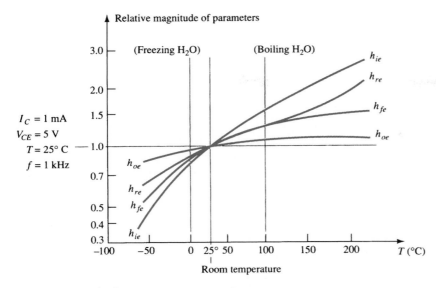

Figure 7.47 Hybrid parameter variations with temperature.

7.9 SUMMARY

Important Conclusions and Concepts

1. Amplification in the ac domain cannot be obtained **without the application of dc biasing level.**

2. For most applications the BJT amplifier can be considered linear, permitting the use of the **superposition theorem** to separate the dc and ac analyses and designs.

3. A **model** is the combination of circuit elements, carefully chosen, that best approximates the behavior of a BJT for a particular set of operating conditions.

4. When introducing the **ac model** for a BJT:

   a. All **dc sources are set to zero** and replaced by a short-circuit connection to ground.

   b. All **capacitors** are replaced by a **short-circuit equivalent.**

   c. All elements **in parallel with** an introduced short-circuit equivalent should be removed from the network.

   d. The network should be **redrawn** as often as possible.

5. The **input impedance** of an ac network **cannot be measured** with an ohmmeter.

6. The **output impedance** of an amplifier is measured with the **applied signal set to zero.** It cannot be measured with an ohmmeter.

7. For all transistor amplifiers, the no-load gain **is always greater** than the loaded gain.

8. The gain from source to load **is always reduced** by the internal resistance of the source.

9. The **current gain** of an amplifier is **very sensitive to** the input impedance of the amplifier and the applied load.

10. The r_e **model** for a transistor is **very sensitive** to the dc biasing network of the amplifier.

11. An **output impedance** for the r_e model **can be included** only if obtained from a data sheet or from a graphical measurement from the characteristic curves.

12. For the **common-base configuration,** the **input impedance** is generally **quite small** and the **output impedance quite large.** In addition, the **voltage gain** can be **quite large,** but the **current gain** is always very close to **1.**

13. For the **common-emitter configuration,** the **input impedance** generally is approximately a **few kilohms,** and the **output impedance** is relatively **large.** In addition, the common-emitter configuration can a have a **relatively high voltage and current gain.**

14. The **parameters** of a hybrid equivalent model for a transistor are provided for a **particular set of dc operating conditions.** However, four parameters are provided rather than the two that normally appear for the r_e model. For some applications the reverse transfer voltage ratio and the typical output impedance not normally found in the r_e model can be quite important.

15. The **amplification factor** (beta β, or h_{fe}) is the least sensitive to changes in **collector current,** whereas the **output impedance** parameter is the most sensitive. The output impedance is also quite sensitive to changes in V_{CE}, whereas the **amplification** factor is the **least sensitive.** However, the **output impedance** is the **least sensitive** to changes in **temperature,** whereas the amplification factor is somewhat sensitive.

Equations

$$r_e = \frac{26\text{ mV}}{I_E}$$

Common-base:

$$Z_i = r_e$$

$$A_v \cong \frac{R_L}{r_e}$$

$$A_i \cong -1$$

Common-emitter:

$$Z_i \cong \beta r_e$$

$$A_v = -\frac{R_L}{r_e}$$

$$A_i = \beta$$

Hybrid:

$$h_{ie} = \beta r_e$$

$$h_{fe} = \beta_{ac}$$

$$h_{ib} = r_e$$

$$h_{fb} = -\alpha \cong -1$$

§ 7.2 Amplification in the AC Domain

1. (a) What is the expected amplification of a BJT transistor amplifier if the dc supply is set to zero volts?
   (b) What will happen to the output ac signal if the dc level is insufficient? Sketch the effect on the waveform.
   (c) What is the conversion efficiency of an amplifier in which the effective value of the current through a 2.2-kΩ load is 5 mA and the drain on the 18-V dc supply is 3.8 mA?

2. Can you think of an analogy that would explain the importance of the dc level on the resulting ac gain?

§ 7.3 BJT Transistor Modeling

3. What is the reactance of a 10-μF capacitor at a frequency of 1 kHz? For networks in which the resistor levels are typically in the kilohm range, is it a good assumption to use the short-circuit equivalence for the conditions just described? How about at 100 kHz?

4. Given the common-base configuration of Fig. 7.48, sketch the ac equivalent using the notation for the transistor model appearing in Fig. 7.5.

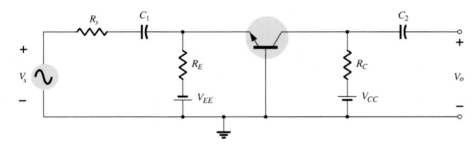

Figure 7.48 Problem 4

5. (a) Describe the differences between the r_e and hybrid equivalent models for a BJT transistor.
   (b) For each model, list the conditions under which it should be applied.

§ 7.4 The Important Parameters: Z_i, Z_o, A_v, A_i

6. (a) For the configuration of Fig. 7.7, determine Z_i if $V_s = 40$ mV, $R_{sense} = 0.5$ kΩ, and $I_i = 20$ μA.
   (b) Using the results of part (a), determine V_i if the applied source is changed to 12 mV with an internal resistance of 0.4 kΩ.

7. (a) For the network of Fig. 7.10, determine Z_o if $V = 600$ mV, $R_{sense} = 10$ kΩ, and $I_o = 10$ μA.
   (b) Using the Z_o obtained in part (a), determine I_L for the configuration of Fig. 7.11 if $R_L = 2.2$ kΩ and $I_{amplifier} = 6$ mA.

8. Given the BJT configuration of Fig. 7.49, determine:
   (a) V_i.
   (b) Z_i.
   (c) $A_{v_{NL}}$.
   (d) A_{v_s}.

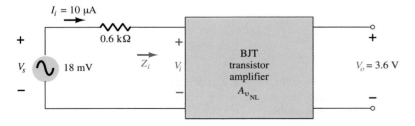

Figure 7.49 Problem 8

9. For the BJT amplifier of Fig. 7.50, determine:
   (a) I_i.
   (b) Z_i.
   (c) V_o.
   (d) I_o.
   (e) A_i using the results of parts (a) and (d).
   (f) A_i using Eq. (7.10).

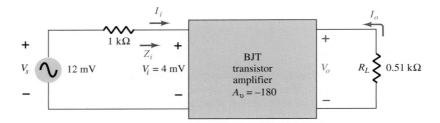

Figure 7.50 Problem 9

§ 7.5 The r_e Transistor Model

10. For the common-base configuration of Fig. 7.17, an ac signal of 10 mV is applied, resulting in an emitter current of 0.5 mA. If $\alpha = 0.980$, determine:
   (a) Z_i.
   (b) V_o if $R_L = 1.2$ kΩ.
   (c) $A_v = V_o/V_i$.
   (d) Z_o with $r_o = \infty$ Ω.
   (e) $A_i = I_o/I_i$.
   (f) I_b.

11. For the common-base configuration of Fig. 7.17, the emitter current is 3.2 mA and α is 0.99. Determine the following if the applied voltage is 48 mV and the load is 2.2 kΩ.
   (a) r_e.
   (b) Z_i.
   (c) I_c.
   (d) V_o.
   (e) A_v.
   (f) I_b.

12. Using the model of Fig. 7.27, determine the following for a common-emitter amplifier if $\beta = 80$, $I_E(\text{dc}) = 2$ mA, and $r_o = 40$ kΩ.
   (a) Z_i.
   (b) I_b.
   (c) $A_i = I_o/I_i = I_L/I_b$ if $R_L = 1.2$ kΩ.
   (d) A_v if $R_L = 1.2$ kΩ.

13. The input impedance to a common-emitter transistor amplifier is 1.2 kΩ with $\beta = 140$, $r_o = 50$ kΩ, and $R_L = 2.7$ kΩ. Determine:
   (a) r_e.
   (b) I_b if $V_i = 30$ mV.
   (c) I_c.
   (d) $A_i = I_o/I_i = I_L/I_b$.
   (e) $A_v = V_o/V_i$.

§ 7.6 The Hybrid Equivalent Model

14. Given $I_E(\text{dc}) = 1.2$ mA, $\beta = 120$, and $r_o = 40$ kΩ, sketch the:
   (a) Common-emitter hybrid equivalent model.
   (b) Common-emitter r_e equivalent model.
   (c) Common-base hybrid equivalent model.
   (d) Common-base r_e equivalent model.

15. Given $h_{ie} = 2.4$ kΩ, $h_{fe} = 100$, $h_{re} = 4 \times 10^{-4}$, and $h_{oe} = 25\,\mu S$, sketch the:
    (a) Common-emitter hybrid equivalent model.
    (b) Common-emitter r_e equivalent model.
    (c) Common-base hybrid equivalent model.
    (d) Common-base r_e equivalent model.

16. Redraw the common-emitter network of Fig. 7.3 for the ac response with the approximate hybrid equivalent model substituted between the appropriate terminals.

17. Redraw the network of Fig. 7.51 for the ac response with the r_e model inserted between the appropriate terminals. Include r_o.

18. Redraw the network of Fig. 7.52 for the ac response with the r_e model inserted between the appropriate terminals. Include r_o.

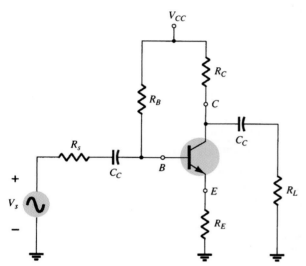

Figure 7.51 Problem 17

Figure 7.52 Problem 18

19. Given the typical values of $h_{ie} = 1$ kΩ, $h_{re} = 2 \times 10^{-4}$, and $A_v = -160$ for the input configuration of Fig. 7.53:
    (a) Determine V_o in terms of V_i.
    (b) Calculate I_b in terms of V_i.
    (c) Calculate I_b if $h_{re}V_o$ is ignored.
    (d) Determine the percent difference in I_b using the following equation:

    $$\% \text{ difference in } I_b = \frac{I_b(\text{without } h_{re}) - I_b(\text{with } h_{re})}{I_b(\text{without } h_{re})} \times 100\%$$

    (e) Is it a valid approach to ignore the effects of $h_{re}V_o$ for the typical values employed in this example?

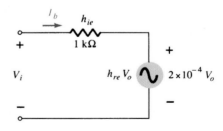

Figure 7.53 Problems 19, 21

20. Given the typical values of $R_L = 2.2$ kΩ and $h_{oe} = 20\,\mu S$, is it a good approximation to ignore the effects of $1/h_{oe}$ on the total load impedance? What is the percent difference in total loading on the transistor using the following equation?

    $$\% \text{ difference in total load} = \frac{R_L - R_L \| (1/h_{oe})}{R_L} \times 100\%$$

21. Repeat Problem 19 using the average values of the parameters of Fig. 7.28 with $A_v = -180$.

22. Repeat Problem 20 for $R_L = 3.3$ kΩ and the average value of h_{oe} in Fig. 7.28.

§ 7.7 Graphical Determination of the *h*-Parameters

23. (a) Using the characteristics of Fig. 7.40, determine h_{fe} at $I_C = 6$ mA and $V_{CE} = 5$ V.
    (b) Repeat part (a) at $I_C = 1$ mA and $V_{CE} = 15$ V.

24. (a) Using the characteristics of Fig. 7.41, determine h_{oe} at $I_C = 6$ mA and $V_{CE} = 5$ V.
    (b) Repeat part (a) at $I_C = 1$ mA and $V_{CE} = 15$ V.

25. (a) Using the characteristics of Fig. 7.42, determine h_{ie} at $I_B = 20$ μA and $V_{CE} = 20$ V.
    (b) Repeat part (a) at $I_B = 5$ μA and $V_{CE} = 10$ V.

26. (a) Using the characteristics of Fig. 7.43, determine h_{re} at $I_B = 20$ μA.
    (b) Repeat part (a) at $I_B = 30$ μA.

* 27. Using the characteristics of Figs. 7.40 and 7.42, determine the approximate CE hybrid equivalent model at $I_B = 25$ μA and $V_{CE} = 12.5$ V.

* 28. Determine the CE r_e model at $I_B = 25$ μA and $V_{CE} = 12.5$ V using the characteristics of Figs. 7.40 and 7.42.

* 29. Using the results of Fig. 7.44, sketch the r_e equivalent model for the transistor having the characteristics appearing in Figs. 7.40 through 7.43. Include r_o.

§ 7.8 Variations of Transistor Parameters

For Problems 30 through 34, use Figs. 7.45 through 7.47.

30. (a) Using Fig. 7.45, determine the magnitude of the percent change in h_{fe} for an I_C change from 0.2 mA to 1 mA using the equation

$$\% \text{ change} = \left| \frac{h_{fe}(0.2 \text{ mA}) - h_{fe}(1 \text{ mA})}{h_{fe}(0.2 \text{ mA})} \right| \times 100\%$$

    (b) Repeat part (a) for an I_C change from 1 mA to 5 mA.

31. Repeat Problem 30 for h_{ie} (same changes in I_C).

32. (a) If $h_{oe} = 20$ μS at $I_C = 1$ mA on Fig. 7.45, what is the approximate value of h_{oe} at $I_C = 0.2$ mA?
    (b) Determine its resistive value at 0.2 mA and compare to a resistive load of 6.8 kΩ. Is it a good approximation to ignore the effects of $1/h_{oe}$ in this case?

33. (a) If $h_{oe} = 20$ μS at $I_C = 1$ mA on Fig. 7.45, what is the approximate value of h_{oe} at $I_C = 10$ mA?
    (b) Determine its resistive value at 10 mA and compare to a resistive load of 6.8 kΩ. Is it a good approximation to ignore the effects of $1/h_{oe}$ in this case?

34. (a) If $h_{re} = 2 \times 10^{-4}$ at $I_C = 1$ mA on Fig. 7.45, determine the approximate value of h_{re} at 0.1 mA.
    (b) Using the value of h_{re} determined in part (a), can h_{re} be ignored as a good approximation if $A_v = 210$?

* 35. (a) Reviewing the characteristics of Fig. 7.45, which parameter changed the least for the full range of collector current?
    (b) Which parameter changed the most?
    (c) What are the maximum and minimum values of $1/h_{oe}$? Is the approximation $1/h_{oe} \| R_L \cong R_L$ more appropriate at high or low levels of collector current?
    (d) In which region of current spectrum is the approximation $h_{re}V_{ce} \cong 0$ the most appropriate?

36. (a) Reviewing the characteristics of Fig. 7.47, which parameter changed the most with increase in temperature?
    (b) Which changed the least?
    (c) What are the maximum and minimum values of h_{fe}? Is the change in magnitude significant? Was it expected?
    (d) How does r_e vary with increase in temperature? Simply calculate its level at three or four points and compare their magnitudes.
    (e) In which temperature range do the parameters change the least?

---

*Please Note: Asterisks indicate more difficult problems.

CHAPTER

8

BJT Small-Signal Analysis

8.1 INTRODUCTION

The transistor models introduced in Chapter 7 will now be used to perform a small-signal ac analysis of a number of standard transistor network configurations. The networks analyzed represent the majority of those appearing in practice today. Modifications of the standard configurations will be relatively easy to examine once the content of this chapter is reviewed and understood.

Since the r_e model is sensitive to the actual point of operation, it will be our primary model for the analysis to be performed. For each configuration, however, the effect of an output impedance is examined as provided by the h_{oe} parameter of the hybrid equivalent model. To demonstrate the similarities in analysis that exist between models, a section is devoted to the small-signal analysis of BJT networks using solely the hybrid equivalent model. The analysis of this chapter does not include a load resistance R_L or source resistance R_s. The effect of both parameters is reserved for a systems approach in Chapter 10.

The computer analysis section includes a brief description of the transistor model employed in the PSpice and Electronics Workbench software packages. It demonstrates the range and depth of the computer analysis systems available today and how relatively easy it is to enter a complex network and print out the desired results.

8.2 COMMON-EMITTER FIXED-BIAS CONFIGURATION

The first configuration to be analyzed in detail is the common-emitter *fixed-bias* network of Fig. 8.1. Note that the input signal V_i is applied to the base of the transistor while the output V_o is off the collector. In addition, recognize that the input current I_i is not the base current but the source current, while the output current I_o is the collector current. The small-signal ac analysis begins by removing the dc effects of V_{CC} and replacing the dc blocking capacitors C_1 and C_2 by short-circuit equivalents, resulting in the network of Fig. 8.2.

Note in Fig. 8.2 that the common ground of the dc supply and the transistor emitter terminal permits the relocation of R_B and R_C in parallel with the input and output sections of the transistor, respectively. In addition, note the placement of the important network parameters Z_i, Z_o, I_i, and I_o on the redrawn network. Substituting the r_e model for the common-emitter configuration of Fig. 8.2 will result in the network of Fig. 8.3.

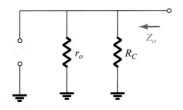

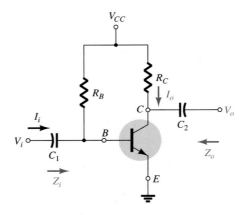

V_{CC}

R_C

I_o

R_B

C

V_o

C_2

I_i

V_i

C_1

B

Z_o

Z_i

E

Figure 8.1 Common-emitter fixed-bias configuration.

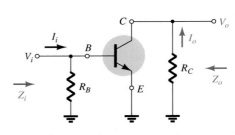

C

V_o

I_i

B

I_o

V_i

R_C

R_B

Z_o

Z_i

E

Figure 8.2 Network of Figure 8.1 following the removal of the effects of V_{CC}, C_1, and C_2.

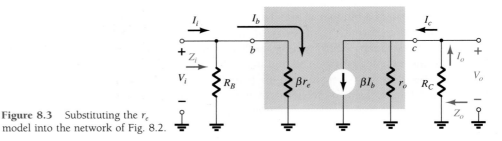

I_i I_b I_c

$+$ Z_i b c I_o $+$

V_i R_B βr_e βI_b r_o R_C V_o

$-$ Z_o $-$

Figure 8.3 Substituting the r_e model into the network of Fig. 8.2.

The next step is to determine β, r_e, and r_o. The magnitude of β is typically obtained from a specification sheet or by direct measurement using a curve tracer or transistor testing instrument. The value of r_e must be determined from a dc analysis of the system, and the magnitude of r_o is typically obtained from the specification sheet or characteristics. Assuming that β, r_e, and r_o have been determined will result in the following equations for the important two-port characteristics of the system.

Z_i: Figure 8.3 clearly reveals that

$$\boxed{Z_i = R_B \| \beta r_e} \qquad \text{ohms} \qquad (8.1)$$

For the majority of situations R_B is greater than βr_e by more than a factor of 10 (recall from the analysis of parallel elements that the total resistance of two parallel resistors is always less than the smallest and very close to the smallest if one is much larger than the other), permitting the following approximation:

$$\boxed{Z_i \cong \beta r_e} \bigg|_{R_B \geq 10\beta r_e} \qquad \text{ohms} \qquad (8.2)$$

Z_o: Recall that the output impedance of any system is defined as the impedance Z_o determined when $V_i = 0$. For Fig. 8.3, when $V_i = 0$, $I_i = I_b = 0$, resulting in an open-circuit equivalence for the current source. The result is the configuration of Fig. 8.4.

$$\boxed{Z_o = R_C \| r_o} \qquad \text{ohms} \qquad (8.3)$$

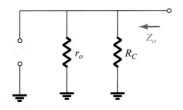

Z_o

r_o R_C

Figure 8.4 Determining Z_o for the network of Fig. 8.3.

If $r_o \geq 10R_C$, the approximation $R_C \| r_o \cong R_C$ is frequently applied and

$$\boxed{Z_o \cong R_C}_{\; r_o \geq 10R_C} \tag{8.4}$$

A_v: The resistors r_o and R_C are in parallel,

and

$$V_o = -\beta I_b (R_C \| r_o)$$

but

$$I_b = \frac{V_i}{\beta r_c}$$

so that

$$V_o = -\beta \left(\frac{V_i}{\beta r_e} \right)(R_C \| r_o)$$

and

$$\boxed{A_v = \frac{V_o}{V_i} = -\frac{(R_C \| r_o)}{r_e}} \tag{8.5}$$

If $r_o \geq 10R_C$,

$$\boxed{A_v = -\frac{R_C}{r_e}}_{\; r_o \geq 10R_C} \tag{8.6}$$

Note the explicit absence of β in Eqs. (8.5 and 8.6), although we recognize that β must be utilized to determine r_e.

A_i: The current gain is determined in the following manner: Applying the current-divider rule to the input and output circuits,

$$I_o = \frac{(r_o)(\beta I_b)}{r_o + R_C} \quad \text{and} \quad \frac{I_o}{I_b} = \frac{r_o \beta}{r_o + R_C}$$

with

$$I_b = \frac{(R_B)(I_i)}{R_B + \beta r_e} \quad \text{or} \quad \frac{I_b}{I_i} = \frac{R_B}{R_B + \beta r_e}$$

The result is

$$A_i = \frac{I_o}{I_i} = \left(\frac{I_o}{I_b} \right)\left(\frac{I_b}{I_i} \right) = \left(\frac{r_o \beta}{r_o + R_C} \right)\left(\frac{R_B}{R_B + \beta r_e} \right)$$

and

$$\boxed{A_i = \frac{I_o}{I_i} = \frac{\beta R_B r_o}{(r_o + R_C)(R_B + \beta r_e)}} \tag{8.7}$$

which is certainly an unwieldy, complex expression.

However, if $r_o \geq 10R_C$ and $R_B \geq 10\,\beta r_e$, which is often the case,

$$A_i = \frac{I_o}{I_i} \cong \frac{\beta R_B r_o}{(r_o)(R_B)}$$

and

$$\boxed{A_i \cong \beta}_{\; r_o \geq 10R_C,\, R_B \geq 10\beta r_e} \tag{8.8}$$

The complexity of Eq. (8.7) suggests that we may want to return to an equation such as Eq. (7.10), which utilizes A_o and Z_i. That is,

$$\boxed{A_i = -A_v \frac{Z_i}{R_C}} \tag{8.9}$$

Phase Relationship: The negative sign in the resulting equation for A_v reveals that a 180° phase shift occurs between the input and output signals, as shown in Fig. 8.5.

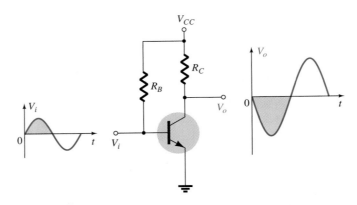

Figure 8.5 Demonstrating the 180° phase shift between input and output waveforms.

EXAMPLE 8.1

For the network of Fig. 8.6:
(a) Determine r_e.
(b) Find Z_i (with $r_o = \infty\ \Omega$).
(c) Calculate Z_o (with $r_o = \infty\ \Omega$).
(d) Determine A_v (with $r_o = \infty\ \Omega$).
(e) Find A_i (with $r_o = \infty\ \Omega$).
(f) Repeat parts (c) through (e) including $r_o = 50\ \text{k}\Omega$ in all calculations and compare results.

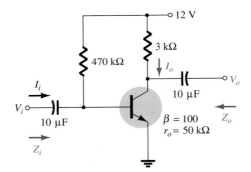

Figure 8.6 Example 8.1.

Solution

(a) DC analysis:

$$I_B = \frac{V_{CC} - V_{BE}}{R_B} = \frac{12\ \text{V} - 0.7\ \text{V}}{470\ \text{k}\Omega} = 24.04\ \mu\text{A}$$

$$I_E = (\beta + 1)I_B = (101)(24.04\ \mu\text{A}) = 2.428\ \text{mA}$$

$$r_e = \frac{26\ \text{mV}}{I_E} = \frac{26\ \text{mV}}{2.428\ \text{mA}} = \mathbf{10.71\ \Omega}$$

(b) $\beta r_e = (100)(10.71\ \Omega) = 1.071\ \text{k}\Omega$

$\quad Z_i = R_B\|\beta r_e = 470\ \text{k}\Omega\|1.071\ \text{k}\Omega = \mathbf{1.069\ k\Omega}$

(c) $Z_o = R_C = \mathbf{3\ k\Omega}$

(d) $A_v = -\dfrac{R_C}{r_e} = -\dfrac{3\ \text{k}\Omega}{10.71\ \Omega} = \mathbf{-280.11}$

(e) Since $R_B \geq 10\beta r_e (470\ \text{k}\Omega > 10.71\ \text{k}\Omega)$

$\quad A_i \cong \beta = \mathbf{100}$

(f) $Z_o = r_o \| R_C = 50 \text{ k}\Omega \| 3 \text{ k}\Omega = \mathbf{2.83 \text{ k}\Omega}$ vs. 3 kΩ

$$A_v = -\frac{r_o \| R_C}{r_e} = \frac{2.83 \text{ k}\Omega}{10.71 \Omega} = -264.24 \text{ vs. } -280.11$$

$$A_i = \frac{\beta R_B r_o}{(r_o + R_C)(R_B + \beta r_e)} = \frac{(100)(470 \text{ k}\Omega)(50 \text{ k}\Omega)}{(50 \text{ k}\Omega + 3 \text{ k}\Omega)(470 \text{ k}\Omega + 1.071 \text{ k}\Omega)}$$
$$= \mathbf{94.13} \text{ vs. } 100$$

As a check:
$$A_i = -A_v \frac{Z_i}{R_C} = \frac{-(-264.24)(1.069 \text{ k}\Omega)}{3 \text{ k}\Omega} = \mathbf{94.16}$$

which differs slightly only due to the accuracy carried through the calculations.

8.3 VOLTAGE-DIVIDER BIAS

The next configuration to be analyzed is the *voltage-divider* bias network of Fig. 8.7. Recall that the name of the configuration is a result of the voltage-divider bias at the input side to determine the dc level of V_B.

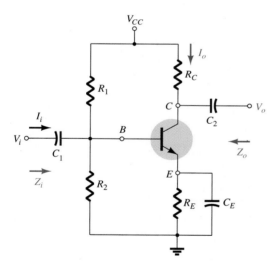

Figure 8.7 Voltage-divider bias configuration.

Substituting the r_e equivalent circuit will result in the network of Fig. 8.8. Note the absence of R_E due to the low-impedance shorting effect of the bypass capacitor, C_E. That is, at the frequency (or frequencies) of operation, the reactance of the capacitor is so small compared to R_E that it is treated as a short circuit across R_E. When

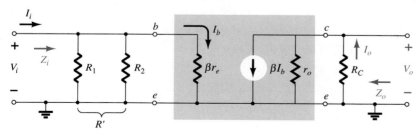

Figure 8.8 Substituting the r_e equivalent circuit into the ac equivalent network of Fig. 8.7.

V_{CC} is set to zero, it places one end of R_1 and R_C at ground potential as shown in Fig. 8.8. In addition, note that R_1 and R_2 remain part of the input circuit while R_C is part of the output circuit. The parallel combination of R_1 and R_2 is defined by

$$R' = R_1 \| R_2 = \frac{R_1 R_2}{R_1 + R_2} \tag{8.10}$$

Z_i: From Fig. 8.8,

$$Z_i = R' \| \beta r_e \tag{8.11}$$

Z_o: From Fig. 8.8 with V_i set to 0 V resulting in $I_b = 0\ \mu A$ and $\beta I_b = 0$ mA,

$$Z_o = R_C \| r_o \tag{8.12}$$

If $r_o \geq 10 R_C$,

$$Z_o \cong R_C \Big|_{r_o \geq 10 R_C} \tag{8.13}$$

A_v: Since R_C and r_o are in parallel,

$$V_o = -(\beta I_b)(R_C \| r_o)$$

and

$$I_b = \frac{V_i}{\beta r_e}$$

so that

$$V_o = -\beta \left(\frac{V_i}{\beta r_e} \right)(R_C \| r_o)$$

and

$$A_v = \frac{V_o}{V_i} = \frac{-R_C \| r_o}{r_e} \tag{8.14}$$

which you will note is an exact duplicate of the equation obtained for the fixed-bias configuration.

For $r_o \geq 10 R_C$,

$$A_v = \frac{V_o}{V_i} \cong -\frac{R_C}{r_e} \Big|_{r_o \geq 10 R_C} \tag{8.15}$$

A_i: Since the network of Fig. 8.8 is so similar to that of Fig. 8.3 except for the fact that $R' = R_1 \| R_2 = R_B$, the equation for the current gain will have the same format as Eq. (8.7). That is,

$$A_i = \frac{I_o}{I_i} = \frac{\beta R' r_o}{(r_o + R_C)(R' + \beta r_e)} \tag{8.16}$$

For $r_o \geq 10 R_C$,

$$A_i = \frac{I_o}{I_i} \cong \frac{\beta R' r_o}{r_o (R' + \beta r_e)}$$

and

$$A_i = \frac{I_o}{I_i} \cong \frac{\beta R'}{R' + \beta r_e} \Big|_{r_o \geq 10 R_C} \tag{8.17}$$

And if $R' \geq 10 \beta r_e$,

$$A_i = \frac{I_o}{I_i} = \frac{\beta R'}{R'}$$

and

$$A_i = \frac{I_o}{I_i} \cong \beta$$ (8.18)

$r_o \geq 10R_C,\ R' \geq 10\beta r_e$

As an option,

$$A_i = -A_v \frac{Z_i}{R_C}$$ (8.19)

Phase relationship: The negative sign of Eq. (8.14) reveals a 180° phase shift between V_o and V_i.

For the network of Fig. 8.9, determine:

(a) r_e.
(b) Z_i.
(c) $Z_o\ (r_o = \infty\ \Omega)$.
(d) $A_v\ (r_o = \infty\ \Omega)$.
(e) $A_i\ (r_o = \infty\ \Omega)$.
(f) The parameters of parts (b) through (e) if $r_o = 1/h_{oe} = 50\ \text{k}\Omega$ and compare results.

EXAMPLE 8.2

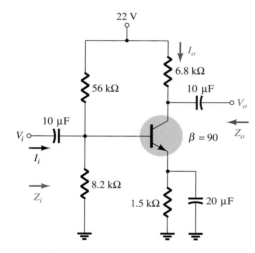

Figure 8.9 Example 8.2.

Solution

(a) DC: Testing $\beta R_E > 10R_2$

$$(90)(1.5\ \text{k}\Omega) > 10(8.2\ \text{k}\Omega)$$
$$135\ \text{k}\Omega > 82\ \text{k}\Omega\ (satisfied)$$

Using the approximate approach,

$$V_B = \frac{R_2}{R_1 + R_2}V_{CC} = \frac{(8.2\ \text{k}\Omega)(22\ \text{V})}{56\ \text{k}\Omega + 8.2\ \text{k}\Omega} = 2.81\ \text{V}$$

$$V_E = V_B - V_{BE} = 2.81\ \text{V} - 0.7\ \text{V} = 2.11\ \text{V}$$

$$I_E = \frac{V_E}{R_E} = \frac{2.11\ \text{V}}{1.5\ \text{k}\Omega} = 1.41\ \text{mA}$$

$$r_e = \frac{26\ \text{mV}}{I_E} = \frac{26\ \text{mV}}{1.41\ \text{mA}} = \mathbf{18.44\ \Omega}$$

(b) $R' = R_1 \| R_2 = (56 \text{ k}\Omega) \| (8.2 \text{ k}\Omega) = 7.15 \text{ k}\Omega$

$\quad Z_i = R' \| \beta r_e = 7.15 \text{ k}\Omega \| (90)(18.44 \ \Omega) = 7.15 \text{ k}\Omega \| 1.66 \text{ k}\Omega$

$\quad = \mathbf{1.35 \ k\Omega}$

(c) $Z_o = R_C = \mathbf{6.8 \ k\Omega}$

(d) $A_v = -\dfrac{R_C}{r_e} = -\dfrac{6.8 \text{ k}\Omega}{18.44 \ \Omega} = \mathbf{-368.76}$

(e) The condition $R' \geq 10\beta r_e$ $(7.15 \text{ k}\Omega \geq 10(1.66 \text{ k}\Omega) = 16.6 \text{ k}\Omega$ is *not* satisfied. Therefore,

$$A_i \cong \frac{\beta R'}{R' + \beta r_e} = \frac{(90)(7.15 \text{ k}\Omega)}{7.15 \text{ k}\Omega + 1.66 \text{ k}\Omega} = \mathbf{73.04}$$

(f) $Z_i = \mathbf{1.35 \ k\Omega}$

$\quad Z_o = R_C \| r_o = 6.8 \text{ k}\Omega \| 50 \text{ k}\Omega = \mathbf{5.98 \ k\Omega}$ vs. $6.8 \text{ k}\Omega$

$\quad A_v = -\dfrac{R_C \| r_o}{r_e} = -\dfrac{5.98 \text{ k}\Omega}{18.44 \ \Omega} = \mathbf{-324.3}$ vs. -368.76

The condition

$$r_o \geq 10R_C \ (50 \text{ k}\Omega \geq 10(6.8 \text{ k}\Omega) = 68 \text{ k}\Omega)$$

is *not* satisfied. Therefore,

$$A_i = \frac{\beta R' r_o}{(r_o + R_C)(R' + \beta r_e)} = \frac{(90)(7.15 \text{ k}\Omega)(50 \text{ k}\Omega)}{(50 \text{ k}\Omega + 6.8 \text{ k}\Omega)(7.15 \text{ k}\Omega + 1.66 \text{ k}\Omega)}$$

$$= \mathbf{64.3} \text{ vs. } 73.04$$

There was a measurable difference in the results for Z_o, A_v, and A_i because the condition $r_o \geq 10R_C$ was *not* satisfied.

8.4 CE EMITTER-BIAS CONFIGURATION

The networks examined in this section include an emitter resistor that may or may not be bypassed in the ac domain. We will first consider the unbypassed situation and then modify the resulting equations for the bypassed configuration.

Unbypassed

The most fundamental of unbypassed configurations appears in Fig. 8.10. The r_e equivalent model is substituted in Fig. 8.11, but note the absence of the resistance r_o. The effect of r_o is to make the analysis a great deal more complicated, and considering the fact that in most situations its effect can be ignored, it will not be included in the current analysis. However, the effect of r_o will be discussed later in this section.

Applying Kirchhoff's voltage law to the input side of Fig. 8.11 will result in

$$V_i = I_b \beta r_e + I_e R_E$$

or

$$V_i = I_b \beta r_e + (\beta + 1)I_b R_E$$

and the input impedance looking into the network to the right of R_B is

$$Z_b = \frac{V_i}{I_b} = \beta r_e + (\beta + 1)R_E$$

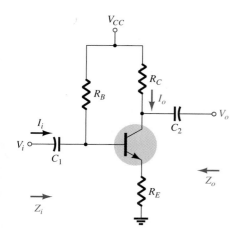

Figure 8.10   CE emitter-bias configuration.

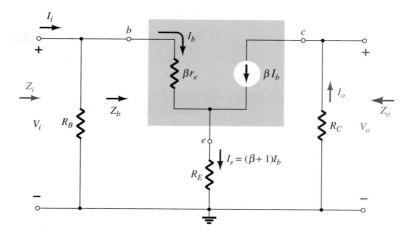

Figure 8.11   Substituting the r_e equivalent circuit into the ac equivalent network of Fig. 8.10.

The result as displayed in Fig. 8.12 reveals that the input impedance of a transistor with an unbypassed resistor R_E is determined by

$$Z_b = \beta r_e + (\beta + 1)R_E \qquad (8.20)$$

Since β is normally much greater than 1, the approximate equation is the following:

$$Z_b \cong \beta r_e + \beta R_E$$

and

$$Z_b \cong \beta(r_e + R_E) \qquad (8.21)$$

Since R_E is often much greater than r_e, Eq. (8.21) can be further reduced to

$$Z_b \cong \beta R_E \qquad (8.22)$$

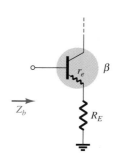

Figure 8.12   Defining the input impedance of a transistor with an unbypassed emitter resistor.

Z_i: Returning to Fig. 8.11, we have

$$Z_i = R_B \| Z_b \qquad (8.23)$$

Z_o: With V_i set to zero, $I_b = 0$ and βI_b can be replaced by an open-circuit equivalent. The result is

$$Z_o = R_C \qquad (8.24)$$

A_v:

$$I_b = \frac{V_i}{Z_b}$$

and

$$V_o = -I_o R_C = -\beta I_b R_C$$

$$= -\beta \left(\frac{V_i}{Z_b}\right) R_C$$

with

$$A_v = \frac{V_o}{V_i} = -\frac{\beta R_C}{Z_b} \qquad (8.25)$$

Substituting $Z_b = \beta(r_e + R_E)$ gives

$$A_v = \frac{V_o}{V_i} = -\frac{R_C}{r_e + R_E} \qquad (8.26)$$

and for the approximation $Z_b \cong \beta R_E$,

$$\boxed{A_v = \frac{V_o}{V_i} \cong -\frac{R_C}{R_E}}$$

(8.27)

Note again the absence of β from the equation for A_v.

A_i: The magnitude of R_B is often too close to Z_b to permit the approximation $I_b = I_i$. Applying the current-divider rule to the input circuit will result in

$$I_b = \frac{R_B I_i}{R_B + Z_b}$$

and

$$\frac{I_b}{I_i} = \frac{R_B}{R_B + Z_b}$$

In addition,

$$I_o = \beta I_b$$

and

$$\frac{I_o}{I_b} = \beta$$

so that

$$A_i = \frac{I_o}{I_i} = \frac{I_o}{I_b}\frac{I_b}{I_i}$$

$$= \beta \frac{R_B}{R_B + Z_b}$$

and

$$\boxed{A_i = \frac{I_o}{I_i} = \frac{\beta R_B}{R_B + Z_b}}$$

(8.28)

or

$$\boxed{A_i = -A_v\frac{Z_i}{R_C}}$$

(8.29)

Phase relationship: The negative sign in Eq. (8.25) again reveals a 180° phase shift between V_o and V_i.

Effect of r_o: The equations appearing below will clearly reveal the additional complexity resulting from including r_o in the analysis. Note in each case, however, that when certain conditions are met, the equations return to the form just derived. The derivation of each equation is beyond the needs of this text and is left as an exercise for the reader. Each equation can be derived through *careful* application of the basic laws of circuit analysis such as Kirchhoff's voltage and current laws, source conversions, Thévenin's theorem, and so on. The equations were included to remove the nagging question of the effect of r_o on the important parameters of a transistor configuration.

Z_i:

$$\boxed{Z_b = \beta r_e + \left[\frac{(\beta + 1) + R_C/r_o}{1 + (R_C + R_E)/r_o}\right]R_E}$$

(8.30)

Since the ratio R_C/r_o is always much less than $(\beta + 1)$,

$$Z_b \cong \beta r_e + \frac{(\beta + 1)R_E}{1 + (R_C + R_E)/r_o}$$

For $r_o \geq 10(R_C + R_E)$,

$$Z_b \cong \beta r_e + (\beta + 1)R_E$$

which compares directly with Eq. (8.20).

In other words, if $r_o \geq 10(R_C + R_E)$, all the equations derived earlier will result. Since $\beta + 1 \cong \beta$, the following equation is an excellent one for most applications:

$$Z_b \cong \beta(r_e + R_E) \qquad_{r_o \geq 10(R_C + R_E)} \qquad (8.31)$$

Z_o:

$$Z_o = R_C \| \left[r_o + \frac{\beta(r_o + r_e)}{1 + \dfrac{\beta r_e}{R_E}} \right] \qquad (8.32)$$

However, $r_o \gg r_e$, and

$$Z_o \cong R_C \| r_o \left[1 + \frac{\beta}{1 + \dfrac{\beta r_e}{R_E}} \right]$$

which can be written as

$$Z_o \cong R_C \| r_o \left[1 + \frac{1}{\dfrac{1}{\beta} + \dfrac{r_e}{R_E}} \right]$$

Typically $1/\beta$ and r_e/R_E are less than one with a sum usually less than one. The result is a multiplying factor for r_o greater than one. For $\beta = 100$, $r_e = 10 \ \Omega$, and $R_E = 1 \ k\Omega$:

$$\frac{1}{\dfrac{1}{\beta} + \dfrac{r_e}{R_E}} = \frac{1}{\dfrac{1}{100} + \dfrac{10 \ \Omega}{1000 \ \Omega}} = \frac{1}{0.02} = 50$$

and

$$Z_o = R_C \| 51 r_o$$

which is certainly simply R_C. Therefore,

$$Z_o = R_C \qquad_{\text{Any level of } r_o} \qquad (8.33)$$

which was obtained earlier.

A_v:

$$A_v = \frac{V_o}{V_i} = \frac{-\dfrac{\beta R_C}{Z_b}\left[1 + \dfrac{r_e}{r_o} \right] + \dfrac{R_C}{r_o}}{1 + \dfrac{R_C}{r_o}} \qquad (8.34)$$

The ratio $\dfrac{r_e}{r_o} \ll 1$

and

$$A_v = \frac{V_o}{V_i} \cong \frac{-\dfrac{\beta R_C}{Z_b} + \dfrac{R_C}{r_o}}{1 + \dfrac{R_C}{r_o}}$$

For $r_o \geq 10R_C$,

$$A_v = \frac{V_o}{V_i} \cong -\frac{\beta R_C}{Z_b} \qquad_{r_o \geq 10R_C} \qquad (8.35)$$

as obtained earlier.

A_i: The determination of A_i will be left to the equation

$$A_i = -A_v \frac{Z_i}{R_C}$$

(8.36)

using the above equations.

Bypassed

If R_E of Fig. 8.10 is bypassed by an emitter capacitor C_E, the complete r_e equivalent model can be substituted resulting in the same equivalent network as Fig. 8.3. Eqs. (8.1 through 8.9) are therefore applicable.

EXAMPLE 8.3

For the network of Fig. 8.13, without C_E (unbypassed), determine:
(a) r_e.
(b) Z_i.
(c) Z_o.
(d) A_v.
(e) A_i.

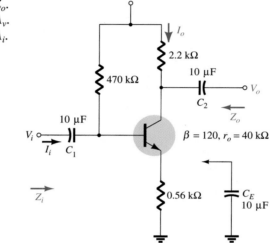

Figure 8.13 Example 8.3.

Solution

(a) DC: $I_B = \dfrac{V_{CC} - V_{BE}}{R_B + (\beta + 1)R_E} = \dfrac{20\text{ V} - 0.7\text{ V}}{470\text{ k}\Omega + (121)0.56\text{ k}\Omega} = 35.89\ \mu\text{A}$

$I_E = (\beta + 1)I_B = (121)(35.89\ \mu\text{A}) = 4.34\text{ mA}$

and $r_e = \dfrac{26\text{ mV}}{I_E} = \dfrac{26\text{ mV}}{4.34\text{ mA}} = \mathbf{5.99\ \Omega}$

(b) Testing the condition $r_o \geq 10(R_C + R_E)$,

$$40\text{ k}\Omega \geq 10(2.2\text{ k}\Omega + 0.56\text{ k}\Omega)$$
$$40\text{ k}\Omega \geq 10(2.76\text{ k}\Omega) = 27.6\text{ k}\Omega \ (\textit{satisfied})$$

Therefore,

$$Z_b \cong \beta(r_e + R_E) = 120(5.99\ \Omega + 560\ \Omega)$$
$$= 67.92\text{ k}\Omega$$

and $Z_i = R_B \| Z_b = 470\text{ k}\Omega \| 67.92\text{ k}\Omega$
$$= \mathbf{59.34\text{ k}\Omega}$$

(c) $Z_o = R_C = \mathbf{2.2\text{ k}\Omega}$

(d) $r_o \geq 10R_C$ is satisfied. Therefore,

$$A_v = \frac{V_o}{V_i} \cong -\frac{\beta R_C}{Z_b} = -\frac{(120)(2.2 \text{ k}\Omega)}{67.92 \text{ k}\Omega}$$

$$= -3.89$$

compared to -3.93 using Eq. (8.27): $A_v \cong -R_C/R_E$.

(e) $A_i = -A_v \dfrac{Z_i}{R_C} = -(-3.89)\left(\dfrac{59.34 \text{ k}\Omega}{2.2 \text{ k}\Omega}\right)$

$$= 104.92$$

compared to 104.85 using Eq. (8.28): $A_i \cong \beta R_B/(R_B + Z_b)$.

Mathcad

The lengthy equations resulting from the analysis of the CE emitter-bias configuration demonstrate the value of becoming proficient in the use of the Mathcad software package.

Priorities do not permit a detailed description of each step of the process, but a few general comments can certainly be made. First, all the parameters of the network that will appear in the equations must be defined as shown in Fig. 8.14. Next, the equations for each of the desired quantities are entered, being very careful to include parentheses in the proper places to ensure that the resulting equation is correct. Actually, more parentheses appear than necessary, but an effort was made to make the equations look as much like those in the text as possible. After each equation is de-

$\text{VCC} := 20 \qquad \text{VBE} := 0.7 \qquad \text{Bf} := 120$

$\text{RB} := 470 \cdot 10^3 \qquad \text{RC} := 2.2 \cdot 10^3 \qquad \text{RE} := 0.56 \cdot 10^3 \qquad \text{ro} := 40 \cdot 10^3$

$\text{IB} := \dfrac{(\text{VCC} - \text{VBE})}{[\text{RB} + (\text{Bf} + 1)\text{RE}]}$

$\text{IE} := (\text{Bf} + 1)\text{IB}$

$\text{IB} = 3.589 \times 10^{-5}$

$\text{IE} = 4.343 \times 10^{-3}$

$\text{re} := \dfrac{(26 \cdot 10^{-3})}{\text{IE}}$

$\text{re} = 5.987 \qquad \textbf{Ex.8.3 re = 5.99ohms}$

$\text{Zb} := \text{Bf} \cdot \text{re} + \left[\dfrac{\left[(\text{Bf} + 1) + \dfrac{\text{RC}}{\text{ro}} \right]}{\left[1 + \dfrac{(\text{RC} + \text{RE})}{\text{ro}} \right]} \right] \cdot \text{RE}$

$\text{Zi} := \dfrac{(\text{RB} \cdot \text{Zb})}{(\text{RB} + \text{Zb})}$

$\text{Zi} = 5.643 \times 10^4 \qquad \textbf{Ex.8.3 Zi = 59.34kilohms}$

$\text{Z} := \text{ro} + \dfrac{[\text{Bf} \cdot (\text{ro} + \text{re})]}{\left[1 + \dfrac{(\text{Bf} \cdot \text{re})}{\text{RE}} \right]}$

$\text{Zo} := \dfrac{(\text{RC} \cdot \text{Z})}{(\text{RC} + \text{Z})}$

$\text{Zo} = 2.198 \times 10^3 \qquad \textbf{Ex.8.3 Zo = 2.2kilohms}$

$\text{Av} := \dfrac{-\left[\dfrac{(\text{Bf} \cdot \text{RC})}{\text{Zb}} \cdot \left(1 + \dfrac{\text{re}}{\text{ro}} \right) + \dfrac{\text{RC}}{\text{ro}} \right]}{\left(1 + \dfrac{\text{RC}}{\text{ro}} \right)}$

$\text{Av} = -3.955 \qquad \textbf{Ex.8.3 Av = -3.89}$

$\text{Ai} := -\text{Av} \cdot \dfrac{\text{Zi}}{\text{RC}}$

$\text{Ai} = 101.439 \qquad \textbf{Ex.8.3 Ai = 104.92}$

Figure 8.14 Network parameters and equations.

fined, its value can be determined by simply entering the variable name again and pressing the equal sign. This is shown to the right of each equation at a level just below the defining equation. For the base current, for example, once **IB** is entered and the equal sign pressed, the base current of 35.89 μA will appear. Note as you progress down the page that as a variable is determined, it can be used in the equations to follow. In fact, it is a necessary sequence if the continuing line of equations is to have the specific numbers to deal with.

For each of the quantities calculated, a text message was added to permit a comparison with the results of Example 8.3. There is excellent correspondence between results when you consider that a number of approximations were used in Example 8.3. The largest difference occurred for the input impedance which has a lengthy equation for **Zb.** That difference is reflected in the current gain which has a greater difference than that obtained for the output impedance and voltage gain.

The real beauty of having entered all these equations properly is that the file can be saved and recalled at any time. As the parameters appearing on the first two lines are changed, all the quantities on the following lines will be recalculated—there is no need to reenter any of the equations, and this sequence even does the dc analysis before determining the ac response.

EXAMPLE 8.4

Repeat the analysis of Example 8.3 with C_E in place.

Solution

(a) The dc analysis is the same, and $r_e = 5.99\ \Omega$.
(b) R_E is "shorted out" by C_E for the ac analysis. Therefore,

$$Z_i = R_B \| Z_b = R_B \| \beta r_e = 470\ \text{k}\Omega \| (120)(5.99\ \Omega)$$
$$= 470\ \text{k}\Omega \| 718.8\ \Omega \cong \textbf{717.70}\ \Omega$$

(c) $Z_o = R_C = \textbf{2.2 k}\Omega$

(d) $A_v = -\dfrac{R_C}{r_e}$

$$= -\dfrac{2.2\ \text{k}\Omega}{5.99\ \Omega} = \textbf{-367.28} \quad \text{(a significant increase)}$$

(e) $A_i = \dfrac{\beta R_B}{R_B + Z_b} = \dfrac{(120)(470\ \text{k}\Omega)}{470\ \text{k}\Omega + 718.8\ \Omega}$

$$= \textbf{119.82}$$

EXAMPLE 8.5

For the network of Fig. 8.15, determine (using appropriate approximations):
(a) r_e.
(b) Z_i.
(c) Z_o.
(d) A_v.
(e) A_i.

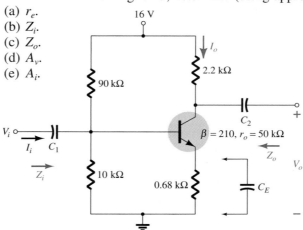

Figure 8.15   Example 8.5.

Solution

(a) Testing $\beta R_E > 10 R_2$

$$(210)(0.68 \text{ k}\Omega) > 10(10 \text{ k}\Omega)$$
$$142.8 \text{ k}\Omega > 100 \text{ k}\Omega \ (\textit{satisfied})$$

$$V_B = \frac{R_2}{R_1 + R_2} V_{CC} = \frac{10 \text{ k}\Omega}{90 \text{ k}\Omega + 10 \text{ k}\Omega} (16 \text{ V}) = 1.6 \text{ V}$$

$$V_E = V_B - V_{BE} = 1.6 \text{ V} - 0.7 \text{ V} = 0.9 \text{ V}$$

$$I_E = \frac{V_E}{R_E} = \frac{0.9 \text{ V}}{0.68 \text{ k}\Omega} = 1.324 \text{ mA}$$

$$r_e = \frac{26 \text{ mV}}{I_E} = \frac{26 \text{ mV}}{1.324 \text{ mA}} = \textbf{19.64 } \boldsymbol{\Omega}$$

(b) The ac equivalent circuit is provided in Fig. 8.16. The resulting configuration is now different from Fig. 8.11 only by the fact that now

$$R_B = R' = R_1 \| R_2 = 9 \text{ k}\Omega$$

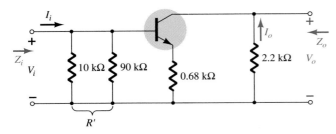

Figure 8.16   The ac equivalent circuit of Fig. 8.15.

The testing conditions of $r_o \geq 10(R_C + R_E)$ and $r_o \geq 10 R_C$ are both satisfied. Using the appropriate approximations yields

$$Z_b \cong \beta R_E = 142.8 \text{ k}\Omega$$
$$Z_i = R_B \| Z_b = 9 \text{ k}\Omega \| 142.8 \text{ k}\Omega$$
$$= \textbf{8.47 k}\boldsymbol{\Omega}$$

(c) $Z_o = R_C = \textbf{2.2 k}\boldsymbol{\Omega}$

(d) $A_v = -\dfrac{R_C}{R_E} = -\dfrac{2.2 \text{ k}\Omega}{0.68 \text{ k}\Omega} = \textbf{-3.24}$

(e) $A_i = -A_v \dfrac{Z_i}{R_C} = -(-3.24) \left(\dfrac{8.47 \text{ k}\Omega}{2.2 \text{ k}\Omega} \right)$

$$= \textbf{12.47}$$

Repeat Example 8.5 with C_E in place.

EXAMPLE 8.6

Solution

(a) The dc analysis is the same, and $r_e = \textbf{19.64 } \boldsymbol{\Omega}.$

(b) $Z_b = \beta r_e = (210)(19.64 \ \Omega) \cong 4.12 \text{ k}\Omega$

$Z_i = R_B \| Z_b = 9 \text{ k}\Omega \| 4.12 \text{ k}\Omega$

$= \textbf{2.83 k}\boldsymbol{\Omega}$

(c) $Z_o = R_C = \textbf{2.2 k}\boldsymbol{\Omega}$

(d) $A_v = -\dfrac{R_C}{r_e} = -\dfrac{2.2 \text{ k}\Omega}{19.64\Omega} = -112.02$ (a significant increase)

(e) $A_i = -A_v \dfrac{Z_i}{R_L} = -(-112.02)\left(\dfrac{2.83 \text{ k}\Omega}{2.2 \text{ k}\Omega}\right)$

$\qquad = \mathbf{144.1}$

Another variation of an emitter-bias configuration appears in Fig. 8.17. For the dc analysis, the emitter resistance is $R_{E_1} + R_{E_2}$, while for the ac analysis, the resistor R_E in the equations above is simply R_{E_1} with R_{E_2} bypassed by C_E.

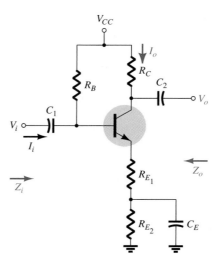

Figure 8.17 An emitter-bias configuration with a portion of the emitter-bias resistance bypassed in the ac domain.

8.5 EMITTER-FOLLOWER CONFIGURATION

When the output is taken from the emitter terminal of the transistor as shown in Fig. 8.18, the network is referred to as an *emitter-follower*. The output voltage is always slightly less than the input signal due to the drop from base to emitter, but the approximation $A_v \cong 1$ is usually a good one. Unlike the collector voltage, the emitter voltage is in phase with the signal V_i. That is, both V_o and V_i will attain their positive and negative peak values at the same time. The fact that V_o "follows" the magnitude of V_i with an in-phase relationship accounts for the terminology emitter-follower.

The most common emitter-follower configuration appears in Fig. 8.18. In fact, because the collector is grounded for ac analysis, it is actually a *common-collector* configuration. Other variations of Fig. 8.18 that draw the output off the emitter with $V_o \cong V_i$ will appear later in this section.

The emitter-follower configuration is frequently used for impedance-matching purposes. It presents a high impedance at the input and a low impedance at the output, which is the direct opposite of the standard fixed-bias configuration. The resulting effect is much the same as that obtained with a transformer, where a load is matched to the source impedance for maximum power transfer through the system.

Substituting the r_e equivalent circuit into the network of Fig. 8.18 will result in the network of Fig. 8.19. The effect of r_o will be examined later in the section.

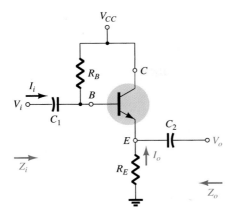

Figure 8.18 Emitter-follower configuration.

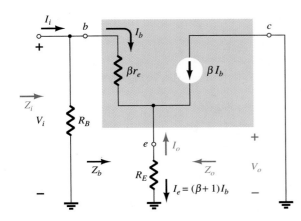

Figure 8.19 Substituting the r_e equivalent circuit into the ac equivalent network of Fig. 8.18.

Z_i: The input impedance is determined in the same manner as described in the preceding section:

$$Z_i = R_B \| Z_b \tag{8.37}$$

with

$$Z_b = \beta r_e + (\beta + 1)R_E \tag{8.38}$$

or

$$Z_b \cong \beta(r_e + R_E) \tag{8.39}$$

and

$$Z_b \cong \beta R_E \tag{8.40}$$

Z_o: The output impedance is best described by first writing the equation for the current I_b:

$$I_b = \frac{V_i}{Z_b}$$

and then multiplying by $(\beta + 1)$ to establish I_e. That is,

$$I_e = (\beta + 1)I_b = (\beta + 1)\frac{V_i}{Z_b}$$

Substituting for Z_b gives

$$I_e = \frac{(\beta + 1)V_i}{\beta r_e + (\beta + 1)R_E}$$

or
$$I_e = \frac{V_i}{[\beta r_e/(\beta + 1)] + R_E}$$

but
$$(\beta + 1) \cong \beta$$

and
$$\frac{\beta r_e}{\beta + 1} \cong \frac{\beta r_e}{\beta} = r_e$$

so that
$$I_e \cong \frac{V_i}{r_e + R_E} \qquad (8.41)$$

If we now construct the network defined by Eq. (8.41), the configuration of Fig. 8.20 will result.

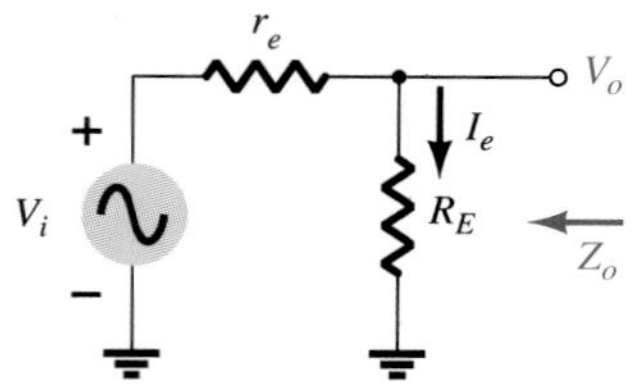

To determine Z_o, V_i is set to zero and
$$Z_o = R_E \| r_e \qquad (8.42)$$

Since R_E is typically much greater than r_e, the following approximation is often applied:
$$Z_o \cong r_e \qquad (8.43)$$

A_v: Figure 8.20 can be utilized to determine the voltage gain through an application of the voltage-divider rule:
$$V_o = \frac{R_E V_i}{R_E + r_e}$$

and
$$A_v = \frac{V_o}{V_i} = \frac{R_E}{R_E + r_e} \qquad (8.44)$$

Since R_E is usually much greater than r_e, $R_E + r_e \cong R_E$ and
$$A_v = \frac{V_o}{V_i} \cong 1 \qquad (8.45)$$

A_i: From Fig. 8.19,
$$I_b = \frac{R_B I_i}{R_B + Z_b}$$

or
$$\frac{I_b}{I_i} = \frac{R_B}{R_B + Z_b}$$

and
$$I_o = -I_e = -(\beta + 1)I_b$$

or
$$\frac{I_o}{I_b} = -(\beta + 1)$$

so that
$$A_i = \frac{I_o}{I_i} = \frac{I_o}{I_b}\frac{I_b}{I_i} = -(\beta + 1)\frac{R_B}{R_B + Z_b}$$

and since
$$(\beta + 1) \cong \beta,$$
$$A_i \cong -\frac{\beta R_B}{R_B + Z_b} \qquad (8.46)$$

or
$$A_i = -A_v \frac{Z_i}{R_E} \qquad (8.47)$$

Figure 8.20 Defining the output impedance for the emitter-follower configuration.

Phase relationship: As revealed by Eq. (8.44) and earlier discussions of this section, V_o and V_i are in phase for the emitter-follower configuration.

Effect of r_o:

Z_i:

$$Z_b = \beta r_e + \frac{(\beta + 1)R_E}{1 + \dfrac{R_E}{r_o}} \qquad (8.48)$$

If the condition $r_o \geq 10R_E$ is satisfied,

$$Z_b = \beta r_e + (\beta + 1)R_E$$

which matches earlier conclusions with

$$\boxed{Z_b \cong \beta(r_e + R_E)}\Big|_{r_o \geq 10R_E} \qquad (8.49)$$

Z_o:

$$Z_o = r_o\|R_E\|\frac{\beta r_e}{(\beta + 1)} \qquad (8.50)$$

Using $\beta + 1 \cong \beta$,

$$Z_o = r_o\|R_E\|r_e$$

and since $r_o \gg r_e$,

$$\boxed{Z_o \cong R_E\|r_e}\Big|_{\text{Any } r_o} \qquad (8.51)$$

A_v:

$$A_v = \frac{(\beta + 1)R_E/Z_b}{1 + \dfrac{R_E}{r_o}} \qquad (8.52)$$

If the condition $r_o \geq 10R_E$ is satisfied and we use the approximation $\beta + 1 \cong \beta$,

$$A_v \cong \frac{\beta R_E}{Z_b}$$

But

$$Z_b \cong \beta(r_e + R_E)$$

so that

$$A_v \cong \frac{\beta R_E}{\beta(r_e + R_E)}$$

and

$$\boxed{A_v \cong \frac{R_E}{r_e + R_E}}\Big|_{r_o \geq 10R_E} \qquad (8.53)$$

For the emitter-follower network of Fig. 8.21, determine:

(a) r_e.
(b) Z_i.
(c) Z_o.
(d) A_v.
(e) A_i.
(f) Repeat parts (b) through (e) with $r_o = 25 \text{ k}\Omega$ and compare results.

EXAMPLE 8.7

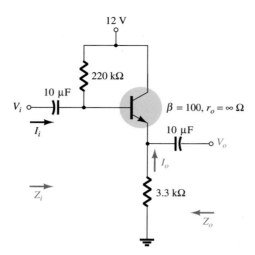

Figure 8.21 Example 8.7.

Solution

(a) $I_B = \dfrac{V_{CC} - V_{BE}}{R_B + (\beta + 1)R_E}$

$ = \dfrac{12\text{ V} - 0.7\text{ V}}{220\text{ k}\Omega + (101)3.3\text{ k}\Omega} = 20.42\ \mu\text{A}$

$I_E = (\beta + 1)I_B$

$ = (101)(20.42\ \mu\text{A}) = 2.062\text{ mA}$

$r_e = \dfrac{26\text{ mV}}{I_E} = \dfrac{26\text{ mV}}{2.062\text{ mA}} = \mathbf{12.61\ \Omega}$

(b) $Z_b = \beta r_e + (\beta + 1)R_E$

$ = (100)(12.61\ \Omega) + (101)(3.3\text{ k}\Omega)$

$ = 1.261\text{ k}\Omega + 333.3\text{ k}\Omega$

$ = 334.56\text{ k}\Omega \cong \beta R_E$

$Z_i = R_B \| Z_b = 220\text{ k}\Omega \| 334.56\text{ k}\Omega$

$ = \mathbf{132.72\ k\Omega}$

(c) $Z_o = R_E \| r_e = 3.3\text{ k}\Omega \| 12.61\ \Omega$

$ = \mathbf{12.56\ \Omega} \cong r_e$

(d) $A_v = \dfrac{V_o}{V_i} = \dfrac{R_E}{R_E + r_e} = \dfrac{3.3\text{ k}\Omega}{3.3\text{ k}\Omega + 12.61\ \Omega}$

$ = \mathbf{0.996} \cong \mathbf{1}$

(e) $A_i \cong -\dfrac{\beta R_B}{R_B + Z_b} = -\dfrac{(100)(220\text{ k}\Omega)}{220\text{ k}\Omega + 334.56\text{ k}\Omega} = \mathbf{-39.67}$

versus

$A_i = -A_v \dfrac{Z_i}{R_E} = -(0.996)\left(\dfrac{132.72\text{ k}\Omega}{3.3\text{ k}\Omega}\right) = \mathbf{-40.06}$

(f) Checking the condition $r_o \geq 10R_E$, we have

$$25\text{ k}\Omega \geq 10(3.3\text{ k}\Omega) = 33\text{ k}\Omega$$

which is *not* satisfied. Therefore,

$$Z_b = \beta r_e + \dfrac{(\beta + 1)R_E}{1 + \dfrac{R_E}{r_o}} = (100)(12.61\ \Omega) + \dfrac{(100 + 1)3.3\ \text{k}\Omega}{1 + \dfrac{3.3\text{ k}\Omega}{25\text{ k}\Omega}}$$

$$= 1.261 \text{ k}\Omega + 294.43 \text{ k}\Omega$$

$$= 295.7 \text{ k}\Omega$$

with $\quad Z_i = R_B \| Z_b = 220 \text{ k}\Omega \| 295.7 \text{ k}\Omega$

$$= \mathbf{126.15 \text{ k}\Omega} \text{ vs. } 132.72 \text{ k}\Omega \quad \text{obtained earlier}$$

$$Z_o = R_E \| r_e = \mathbf{12.56 \ \Omega} \quad \text{as obtained earlier}$$

$$A_v = \frac{(\beta + 1)R_E/Z_b}{\left[1 + \dfrac{R_E}{r_o}\right]} = \frac{(100 + 1)(3.3 \text{ k}\Omega)/295.7 \text{ k}\Omega}{\left[1 + \dfrac{3.3 \text{ k}\Omega}{25 \text{ k}\Omega}\right]}$$

$$= \mathbf{0.996} \cong 1$$

matching the earlier result.

In general, therefore, even though the condition $r_o \geq 10R_E$ was not satisfied, the results for Z_o and A_v are the same, with Z_i only slightly less. The results suggest that for most applications a good approximation for the actual results can be obtained by simply ignoring the effects of r_o for this configuration.

The network of Fig. 8.22 is a variation of the network of Fig. 8.18, which employs a voltage-divider input section to set the bias conditions. Equations (8.37) through (8.47) are changed only by replacing R_B by $R' = R_1 \| R_2$.

The network of Fig. 8.23 will also provide the input/output characteristics of an emitter-follower but includes a collector resistor R_C. In this case R_B is again replaced by the parallel combination of R_1 and R_2. The input impedance Z_i and output impedance Z_o are unaffected by R_C since it is not reflected into the base or emitter equivalent networks. In fact, the only effect of R_C will be to determine the Q-point of operation.

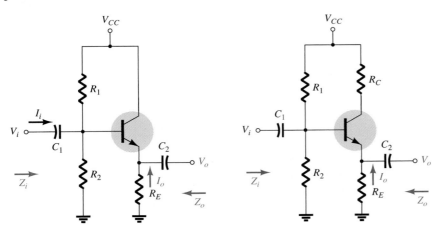

Figure 8.22 Emitter-follower configuration with a voltage-divider biasing arrangement.

Figure 8.23 Emitter-follower configuration with a collector resistor R_C.

8.6 COMMON-BASE CONFIGURATION

The common-base configuration is characterized as having a relatively low input and a high output impedance and a current gain less than 1. The voltage gain, however, can be quite large. The standard configuration appears in Fig. 8.24, with the common-base r_e equivalent model substituted in Fig. 8.25. The transistor output impedance r_o is not included for the common-base configuration because it is typically in the megohm range and can be ignored in parallel with the resistor R_C.

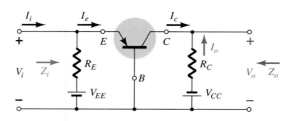

Figure 8.24 Common-base configuration.

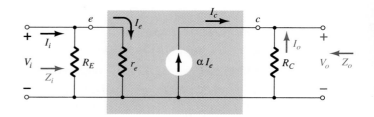

Figure 8.25 Substituting the r_e equivalent circuit into the ac equivalent network of Fig. 8.24.

Z_i:

$$Z_i = R_E \| r_e \tag{8.54}$$

Z_o:

$$Z_o = R_C \tag{8.55}$$

A_v:

$$V_o = -I_o R_C = -(-I_c)R_C = \alpha I_e R_C$$

with

$$I_e = \frac{V_i}{r_e}$$

so that

$$V_o = \alpha\left(\frac{V_i}{r_e}\right)R_C$$

and

$$A_v = \frac{V_o}{V_i} = \frac{\alpha R_C}{r_e} \cong \frac{R_C}{r_e} \tag{8.56}$$

A_i: Assuming that $R_E \gg r_e$ yields

$$I_e = I_i$$

and

$$I_o = -\alpha I_e = -\alpha I_i$$

with

$$A_i = \frac{I_o}{I_i} = -\alpha \cong -1 \tag{8.57}$$

Phase relationship: The fact that A_v is a positive number reveals that V_o and V_i are in phase for the common-base configuration.

Effect of r_o: For the common-base configuration, $r_o = 1/h_{ob}$ is typically in the megohm range and sufficiently larger than the parallel resistance R_C to permit the approximation $r_o \| R_C \cong R_C$.

EXAMPLE 8.8

For the network of Fig. 8.26, determine:
(a) r_e.
(b) Z_i.
(c) Z_o.
(d) A_v.
(e) A_i.

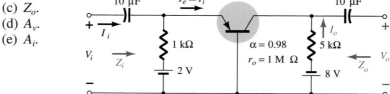

Figure 8.26 Example 8.8.

Solution

(a) $I_E = \dfrac{V_{EE} - V_{BE}}{R_E} = \dfrac{2\text{ V} - 0.7\text{ V}}{1\text{ k}\Omega} = \dfrac{1.3\text{ V}}{1\text{ k}\Omega} = 1.3\text{ mA}$

$r_e = \dfrac{26\text{ mV}}{I_E} = \dfrac{26\text{ mV}}{1.3\text{ mA}} = \mathbf{20\ \Omega}$

(b) $Z_i = R_E \| r_e = 1\text{ k}\Omega \| 20\ \Omega$

$\qquad = \mathbf{19.61\ \Omega} \cong r_e$

(c) $Z_o = R_C = \mathbf{5\text{ k}\Omega}$

(d) $A_v \cong \dfrac{R_C}{r_e} = \dfrac{5\text{ k}\Omega}{20\ \Omega} = \mathbf{250}$

(e) $A_i = \mathbf{-0.98} \cong -1$

8.7 COLLECTOR FEEDBACK CONFIGURATION

The collector feedback network of Fig. 8.27 employs a feedback path from collector to base to increase the stability of the system as discussed in Section 4.12. However, the simple maneuver of connecting a resistor from base to collector rather than base to dc supply has a significant impact on the level of difficulty encountered when analyzing the network.

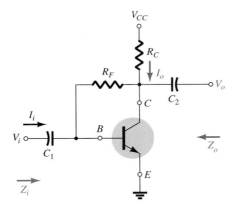

Figure 8.27 Collector feedback configuration.

Some of the steps to be performed below are the result of experience working with such configurations. It is not expected that a new student of the subject would choose the sequence of steps described below without taking a wrong step or two. Substituting the equivalent circuit and redrawing the network will result in the configuration of Fig. 8.28. The effects of a transistor output resistance r_o will be discussed later in the section.

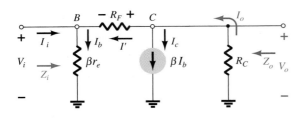

Figure 8.28 Substituting the r_e equivalent circuit into the ac equivalent network of Fig. 8.27.

Z_i:

$$I' = \frac{V_o - V_i}{R_F}$$

with

$$V_o = -I_o R_C$$

and

$$I_o = \beta I_b + I'$$

Since βI_b is normally much larger than I',

$$I_o \cong \beta I_b$$

and

$$V_o = -(\beta I_b)R_C = -\beta I_b R_C$$

but

$$I_b = \frac{V_i}{\beta r_e}$$

and

$$V_o = -\beta\left(\frac{V_i}{\beta r_e}\right)R_C = -\frac{R_C}{r_e}V_i$$

Therefore,

$$I' = \frac{V_o - V_i}{R_F} = \frac{V_o}{R_F} - \frac{V_i}{R_F} = -\frac{R_C V_i}{r_e R_F} - \frac{V_i}{R_F} = -\frac{1}{R_F}\left[1 + \frac{R_C}{r_e}\right]V_i$$

The result is

$$V_i = I_b \beta r_e = (I_i + I')\beta r_e = I_i \beta r_e + I'\beta r_e$$

$$V_i = I_i \beta r_e - \frac{1}{R_F}\left[1 + \frac{R_C}{r_e}\right]\beta r_e V_i$$

or

$$V_i\left[1 + \frac{\beta r_e}{R_F}\left[1 + \frac{R_C}{r_e}\right]\right] = I_i \beta r_e$$

and

$$Z_i = \frac{V_i}{I_i} = \frac{\beta r_e}{1 + \frac{\beta r_e}{R_F}\left[1 + \frac{R_C}{r_e}\right]}$$

but R_C is usually much greater than r_e and $1 + \frac{R_C}{r_e} \cong \frac{R_C}{r_e}$

so that

$$Z_i = \frac{\beta r_e}{1 + \frac{\beta R_C}{R_F}}$$

or

$$\boxed{Z_i = \frac{r_e}{\frac{1}{\beta} + \frac{R_C}{R_F}}}$$

(8.58)

Z_o: If we set V_i to zero as required to define Z_o, the network will appear as shown in Fig. 8.29. The effect of βr_e is removed and R_F appears in parallel with R_C and

$$\boxed{Z_o \cong R_C \| R_F}$$

(8.59)

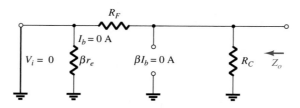

Figure 8.29 Defining Z_o for the collector feedback configuration.

A_v: At node C of Fig. 8.28,

$$I_o = \beta I_b + I'$$

For typical values, $\beta I_b >> I'$ and $I_o \cong \beta I_b$.

$$V_o = -I_o R_C = -(\beta I_b) R_C$$

Substituting $I_b = V_i / \beta r_e$ gives us

$$V_o = -\beta \frac{V_i}{\beta r_e} R_C$$

and

$$\boxed{A_v = \frac{V_o}{V_i} = -\frac{R_C}{r_e}} \qquad (8.60)$$

A_i: Applying Kirchhoff's voltage law around the outside network loop yields

$$V_i + V_{R_F} - V_o = 0$$

and

$$I_b \beta r_e + (I_b - I_i) R_F + I_o R_C = 0$$

Using $I_o \cong \beta I_b$, we have

$$I_b \beta r_e + I_b R_F - I_i R_F + \beta I_b R_C = 0$$

and

$$I_b(\beta r_e + R_F + \beta R_C) = I_i R_F$$

Substituting $I_b = I_o / \beta$ from $I_o \cong \beta I_b$ yields

$$\frac{I_o}{\beta}(\beta r_e + R_F + \beta R_C) = I_i R_F$$

and

$$I_o = \frac{\beta R_F I_i}{\beta r_e + R_F + \beta R_C}$$

Ignoring βr_e compared to R_F and βR_C gives us

$$\boxed{A_i = \frac{I_o}{I_i} = \frac{\beta R_F}{R_F + \beta R_C}} \qquad (8.61)$$

For $\beta R_C >> R_F$,

$$A_i = \frac{I_o}{I_i} = \frac{\beta R_F}{\beta R_C}$$

and

$$\boxed{A_i = \frac{I_o}{I_i} \cong \frac{R_F}{R_C}} \qquad (8.62)$$

Phase relationship: The negative sign of Eq. (8.60) reveals a 180° phase shift between V_o and V_i.

Effect of r_o:

Z_i: A complete analysis without applying approximations will result in

$$\boxed{Z_i = \frac{1 + \dfrac{R_C \| r_o}{R_F}}{\dfrac{1}{\beta r_e} + \dfrac{1}{R_F} + \dfrac{R_C \| r_o}{R_F r_e}}} \qquad (8.63)$$

Recognizing that $1/R_F \cong 0$ and applying the condition $r_o \geq 10R_C$,

$$Z_i = \frac{1 + \dfrac{R_C}{R_F}}{\dfrac{1}{\beta r_e} + \dfrac{R_C}{R_F r_e}}$$

but typically $R_C/R_F \ll 1$ and

$$Z_i = \frac{1}{\dfrac{1}{\beta r_e} + \dfrac{R_C}{R_F r_e}}$$

or

$$\boxed{Z_i = \frac{r_e}{\dfrac{1}{\beta} + \dfrac{R_C}{R_F}}}\Bigg|_{r_o \geq 10R_C} \tag{8.64}$$

as obtained earlier.

Z_o: Including r_o in parallel with R_C in Fig. 8.29 will result in

$$\boxed{Z_o = r_o \| R_C \| R_F} \tag{8.65}$$

For $r_o \geq 10R_C$,

$$\boxed{Z_o \cong R_C \| R_F}\Big|_{r_o \geq 10R_C} \tag{8.66}$$

as obtained earlier. For the common condition of $R_F \gg R_C$,

$$\boxed{Z_o \cong R_C}\Big|_{r_o \geq 10R_C,\ R_F \gg R_C} \tag{8.67}$$

A_v:

$$\boxed{A_v = -\frac{\left[\dfrac{1}{R_F} + \dfrac{1}{r_e}\right](r_o \| R_C)}{1 + \dfrac{r_o \| R_C}{R_F}}} \tag{8.68}$$

Since $R_F \gg r_e$,

$$A_v \cong -\frac{\dfrac{r_o \| R_C}{r_e}}{1 + \dfrac{r_o \| R_C}{R_F}}$$

For $r_o \geq 10R_C$,

$$\boxed{A_v \cong -\frac{\dfrac{R_C}{r_e}}{1 + \dfrac{R_C}{R_F}}}\Bigg|_{r_o \geq 10R_C} \tag{8.69}$$

and since R_C/R_F is typically much less than one,

$$\boxed{A_v \cong -\frac{R_C}{r_e}}\Big|_{r_o \geq 10R_C,\ R_F \gg R_C} \tag{8.70}$$

as obtained earlier.

For the network of Fig. 8.30, determine:

(a) r_e.

(b) Z_i.

(c) Z_o.

(d) A_v.

(e) A_i.

(f) Repeat parts (b) through (e) with $r_o = 20$ kΩ and compare results.

EXAMPLE 8.9

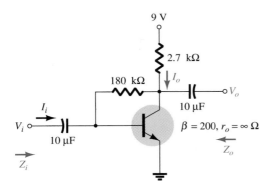

9 V

2.7 kΩ

180 kΩ

I_o

V_o

10 µF

I_i

V_i

10 µF

$β = 200, r_o = ∞$ Ω

Z_o

Z_i

Figure 8.30 Example 8.9.

Solution

(a) $I_B = \dfrac{V_{CC} - V_{BE}}{R_F + \beta R_C} = \dfrac{9 \text{ V} - 0.7 \text{ V}}{180 \text{ k}\Omega + (200)2.7 \text{ k}\Omega}$

$= 11.53 \text{ } \mu\text{A}$

$I_E = (\beta + 1)I_B = (201)(11.53 \text{ } \mu\text{A}) = 2.32 \text{ mA}$

$r_e = \dfrac{26 \text{ mV}}{I_E} = \dfrac{26 \text{ mV}}{2.32 \text{ mA}} = \mathbf{11.21 \text{ } \Omega}$

(b) $Z_i = \dfrac{r_e}{\dfrac{1}{\beta} + \dfrac{R_C}{R_F}} = \dfrac{11.21 \text{ } \Omega}{\dfrac{1}{200} + \dfrac{2.7 \text{ k}\Omega}{180 \text{ k}\Omega}} = \dfrac{11.21 \text{ } \Omega}{0.005 + 0.015}$

$= \dfrac{11.21 \text{ } \Omega}{0.02} = 50(11.21 \text{ } \Omega) = \mathbf{560.5 \text{ } \Omega}$

(c) $Z_o = R_C \| R_F = 2.7 \text{ k}\Omega \| 180 \text{ k}\Omega = \mathbf{2.66 \text{ k}\Omega}$

(d) $A_v = -\dfrac{R_C}{r_e} = -\dfrac{27 \text{ k}\Omega}{11.21 \text{ } \Omega} = \mathbf{-240.86}$

(e) $A_i = \dfrac{\beta R_F}{R_F + \beta R_C} = \dfrac{(200)(180 \text{ k}\Omega)}{180 \text{ k}\Omega + (200)(2.7 \text{ k}\Omega)}$

$= \mathbf{50}$

(f) Z_i: The condition $r_o \geq 10R_C$ is *not* satisfied. Therefore,

$Z_i = \dfrac{1 + \dfrac{R_C \| r_o}{R_F}}{\dfrac{1}{\beta r_e} + \dfrac{1}{R_F} + \dfrac{R_C \| r_o}{R_F r_e}} = \dfrac{1 + \dfrac{2.7 \text{ k}\Omega \| 20 \text{ k}\Omega}{180 \text{ k}\Omega}}{\dfrac{1}{(200)(11.21)} + \dfrac{1}{180 \text{ k}\Omega} + \dfrac{2.7 \text{ k}\Omega \| 20 \text{ k}\Omega}{(180 \text{ k}\Omega)(11.21 \text{ } \Omega)}}$

$= \dfrac{1 + \dfrac{2.38 \text{ k}\Omega}{180 \text{ k}\Omega}}{0.45 \times 10^{-3} + 0.006 \times 10^{-3} + 1.18 \times 10^{-3}} = \dfrac{1 + 0.013}{1.64 \times 10^{-3}}$

$= \mathbf{617.7 \text{ } \Omega}$ vs. 560.5 Ω above

Z_o:

$$Z_o = r_o \| R_C \| R_F = 20 \text{ k}\Omega \| 2.7 \text{ k}\Omega \| 180 \text{ k}\Omega$$
$$= \mathbf{2.35 \text{ k}\Omega} \text{ vs. } 2.66 \text{ k}\Omega \text{ above}$$

A_v:

$$A_v = \frac{-\left[\dfrac{1}{R_F} + \dfrac{1}{r_e}\right](r_o \| R_C)}{1 + \dfrac{r_o \| R_C}{R_F}} = \frac{-\left[\dfrac{1}{180 \text{ k}\Omega} + \dfrac{1}{11.21 \text{ }\Omega}\right](2.38 \text{ k}\Omega)}{1 + \dfrac{2.38 \text{ k}\Omega}{180 \text{ k}\Omega}}$$

$$= \frac{-[5.56 \times 10^{-6} - 8.92 \times 10^{-2}](2.38 \text{ k}\Omega)}{1 + 0.013}$$

$$= \mathbf{-209.56} \text{ vs. } -240.86 \text{ above}$$

A_i:

$$A_i = -A_v \frac{Z_i}{R_C}$$

$$= -(-209.56)\frac{617.7 \text{ }\Omega}{2.7 \text{ k}\Omega}$$

$$= \mathbf{47.94} \text{ vs. } 50 \text{ above}$$

For the configuration of Fig. 8.31, Eqs. (8.71) through (8.74) will determine the variables of interest. The derivations are left as an exercise at the end of the chapter.

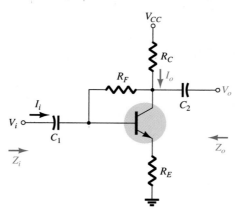

Figure 8.31 Collector feedback configuration with an emitter resistor R_E.

Z_i:

$$Z_i \cong \frac{R_E}{\left[\dfrac{1}{\beta} + \dfrac{(R_E + R_C)}{R_F}\right]} \tag{8.71}$$

Z_o:

$$Z_o \cong R_C \| R_F \tag{8.72}$$

A_v:

$$A_v \cong -\frac{R_C}{R_E} \tag{8.73}$$

A_i:

$$A_i \cong \frac{1}{\dfrac{1}{\beta} + \dfrac{(R_E + R_C)}{R_F}} \qquad (8.74)$$

8.8 COLLECTOR DC FEEDBACK CONFIGURATION

The network of Fig. 8.32 has a dc feedback resistor for increased stability, yet the capacitor C_3 will shift portions of the feedback resistance to the input and output sections of the network in the ac domain. The portion of R_F shifted to the input or output side will be determined by the desired ac input and output resistance levels.

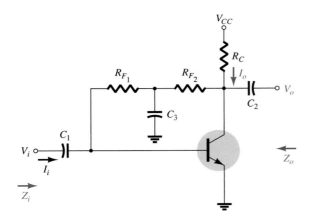

Figure 8.32 Collector dc feedback configuration.

At the frequency or frequencies of operation, the capacitor will assume a short-circuit equivalent to ground due to its low impedance level compared to the other elements of the network. The small-signal ac equivalent circuit will then appear as shown in Fig. 8.33.

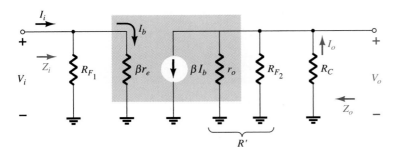

Figure 8.33 Substituting the r_e equivalent circuit into the ac equivalent network of Fig. 8.32.

Z_i:

$$\boxed{Z_i = R_{F_1} \| \beta r_e} \qquad (8.75)$$

Z_o:

$$\boxed{Z_o = R_C \| R_{F_2} \| r_o} \qquad (8.76)$$

For $r_o \geq 10R_C$,

$$\boxed{Z_o \cong R_C \| R_{F_2}}_{r_0 \geq 10R_C} \tag{8.77}$$

A_v:

$$R' = r_o \| R_{F_2} \| R_C$$

and

$$V_o = -\beta I_b R'$$

but

$$I_b = \frac{V_i}{\beta r_e}$$

and

$$V_o = -\beta \frac{V_i}{\beta r_e} R'$$

so that

$$\boxed{A_v = \frac{V_o}{V_i} = -\frac{r_o \| R_{F_2} \| R_C}{r_e}} \tag{8.78}$$

For $r_o \geq 10R_C$,

$$\boxed{A_v = \frac{V_o}{V_i} \cong -\frac{R_{F_2} \| R_C}{r_e}}_{r_o \geq 10R_C} \tag{8.79}$$

A_i: For the input side,

$$I_b = \frac{R_{F_1} I_i}{R_{F_1} + \beta r_e} \qquad \text{or} \qquad \frac{I_b}{I_i} = \frac{R_{F_1}}{R_{F_1} + \beta r_e}$$

and for the output side using $R' = r_o \| R_{F_2}$

$$I_o = \frac{R' \beta I_b}{R' + R_C} \qquad \text{or} \qquad \frac{I_o}{I_b} = \frac{R' \beta}{R' + R_C}$$

The current gain,

$$A_i = \frac{I_o}{I_i} = \frac{I_o}{I_b} \cdot \frac{I_b}{I_i}$$

$$= \frac{R' \beta}{R' + R_C} \cdot \frac{R_{F_1}}{R_{F_1} + \beta r_e}$$

and

$$\boxed{A_i = \frac{I_o}{I_i} = \frac{\beta R_{F_1} R'}{(R_{F_1} + \beta r_e)(R' + R_C)}}_{R' = r_o \| R_{F_2}} \tag{8.80}$$

Since R_{F_1} is usually much larger than βr_e, $R_{F_1} + \beta r_e \cong R_{F_1}$

and

$$A_i = \frac{I_o}{I_i} \cong \frac{\beta \cancel{R_{F_1}} (r_o \| R_{F_2})}{\cancel{R_{F_1}} (r_o \| R_{F_2} + R_C)}$$

so that

$$\boxed{A_i = \frac{I_o}{I_i} \cong \frac{\beta}{1 + \dfrac{R_C}{r_o \| R_{F_2}}}}_{R_{F_1} \geq 10\beta r_e} \tag{8.81}$$

or

$$\boxed{A_i = \frac{I_o}{I_i} = -A_v \frac{Z_i}{R_C}} \tag{8.82}$$

Phase relationship: The negative sign in Eq. (8.78) clearly reveals a 180° phase shift between input and output voltages.

For the network of Fig. 8.34, determine:

(a) r_e.
(b) Z_i.
(c) Z_o.
(d) A_v.
(e) A_i.

EXAMPLE 8.10

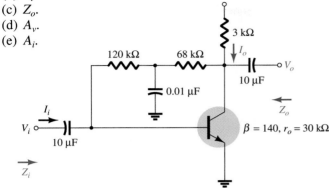

Figure 8.34 Example 8.10.

Solution

(a) DC: $\quad I_B = \dfrac{V_{CC} - V_{BE}}{R_F + \beta R_C}$

$\qquad = \dfrac{12\text{ V} - 0.7\text{ V}}{(120\text{ k}\Omega + 68\text{ k}\Omega) + (140)3\text{ k}\Omega}$

$\qquad = \dfrac{11.3\text{ V}}{608\text{ k}\Omega} = 18.6\ \mu\text{A}$

$\quad I_E = (\beta + 1)I_B = (141)(18.6\ \mu\text{A})$

$\qquad = 2.62\text{ mA}$

$\quad r_e = \dfrac{26\text{ mV}}{I_E} = \dfrac{26\text{ mV}}{2.62\text{ mA}} = \mathbf{9.92\ \Omega}$

(b) $\beta r_e = (140)(9.92\ \Omega) = 1.39\text{ k}\Omega$

The ac equivalent network appears in Fig. 8.35.

$\quad Z_i = R_{F_1}\|\beta r_e = 120\text{ k}\Omega\|1.39\text{ k}\Omega$

$\qquad \cong \mathbf{1.37\ k\Omega}$

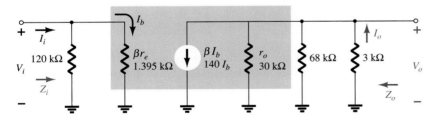

Figure 8.35 Substituting the r_e equivalent circuit into the ac equivalent network of Fig. 8.34.

(c) Testing the condition $r_o \geq 10R_C$, we find

$$30\text{ k}\Omega \geq 10(3\text{ k}\Omega) = 30\text{ k}\Omega$$

which is satisfied through the equals sign in the condition. Therefore,

$\quad Z_o \cong R_C\|R_{F_2} = 3\text{ k}\Omega\|68\text{ k}\Omega$

$\qquad = \mathbf{2.87\ k\Omega}$

(d) $r_o \geq 10 R_C$, therefore,

$$A_v \cong -\frac{R_{F_2} \| R_C}{r_e} = -\frac{68 \text{ k}\Omega \| 3 \text{ k}\Omega}{9.92 \text{ }\Omega}$$

$$\cong -\frac{2.87 \text{ k}\Omega}{9.92 \text{ }\Omega}$$

$$\cong -289.3$$

(e) Since the condition $R_{F_1} \gg \beta r_e$ is satisfied,

$$A_i \cong \frac{\beta}{1 + \dfrac{R_C}{r_o \| R_{F_2}}} = \frac{140}{1 + \dfrac{3 \text{ k}\Omega}{30 \text{ k}\Omega \| 68 \text{ k}\Omega}} = \frac{140}{1 + 0.14} = \frac{140}{1.14}$$

$$\cong 122.8$$

8.9 APPROXIMATE HYBRID EQUIVALENT CIRCUIT

The analysis using the approximate hybrid equivalent circuit of Fig. 8.36 for the common-emitter configuration and of Fig. 8.37 for the common-base configuration is very similar to that just performed using the r_e model. Although time and priorities do not permit a detailed analysis of all the configurations discussed thus far, a brief overview of some of the most important will be included in this section to demonstrate the similarities in approach and the resulting equations.

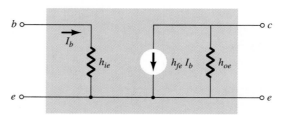

Figure 8.36 Approximate common-emitter hybrid equivalent circuit.

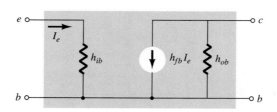

Figure 8.37 Approximate common-base hybrid equivalent circuit.

Since the various parameters of the hybrid model are specified by a data sheet or experimental analysis, the dc analysis associated with use of the r_e model is not an integral part of the use of the hybrid parameters. In other words, when the problem is presented, the parameters such as h_{ie}, h_{fe}, h_{ib}, and so on, are specified. Keep in mind, however, that the hybrid parameters and components of the r_e model are related by the following equations as discussed in detail in Chapter 7: $h_{ie} = \beta r_e$, $h_{fe} = \beta$, $h_{oe} = 1/r_o$, $h_{fb} = -\alpha$, and $h_{ib} = r_e$ (note Appendix B).

Fixed-Bias Configuration

For the fixed-bias configuration of Fig. 8.38, the small-signal ac equivalent network will appear as shown in Fig. 8.39 using the approximate common-emitter hybrid equivalent model. Compare the similarities in appearance with Fig. 8.3 and the r_e model analysis. The similarities suggest that the analysis will be quite similar, and the results of one can be directly related to the other.

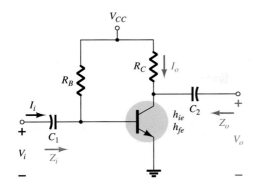

Figure 8.38 Fixed-bias configuration.

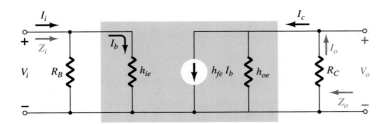

Figure 8.39 Substituting the approximate hybrid equivalent circuit into the ac equivalent network of Fig. 8.38.

Z_i: From Fig. 8.39,

$$Z_i = R_B \| h_{ie} \tag{8.83}$$

Z_o: From Fig. 8.39,

$$Z_o = R_C \| 1/h_{oe} \tag{8.84}$$

A_v: Using $R' = 1/h_{oe} \| R_C$,

$$V_o = -I_o R' = -I_C R'$$
$$= -h_{fe} I_b R'$$

and

$$I_b = \frac{V_i}{h_{ie}}$$

with

$$V_o = -h_{fe}\frac{V_i}{h_{ie}} R'$$

so that

$$A_v = \frac{V_o}{V_i} = -\frac{h_{fe}(R_C \| 1/h_{oe})}{h_{ie}} \tag{8.85}$$

A_i: Assuming that $R_B \gg h_{ie}$ and $1/h_{oe} \geq 10R_C$, then $I_b \cong I_i$ and $I_o = I_c = h_{fe}I_b = h_{fe}I_i$ with

$$A_i = \frac{I_o}{I_i} \cong h_{fe} \tag{8.86}$$

8.9 **Approximate Hybrid Equivalent Circuit**   421

EXAMPLE 8.11

For the network of Fig. 8.40, determine:
(a) Z_i.
(b) Z_o.
(c) A_v.
(d) A_i.

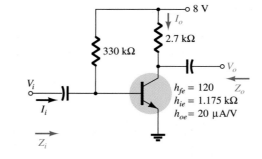

Figure 8.40 Example 8.11.

Solution

(a) $Z_i = R_B \| h_{ie} = 330 \text{ k}\Omega \| 1.175 \text{ k}\Omega$

   $\cong h_{ie} = \mathbf{1.171 \text{ k}\Omega}$

(b) $r_o = \dfrac{1}{h_{oe}} = \dfrac{1}{20 \text{ }\mu\text{A/V}} = 50 \text{ k}\Omega$

   $Z_o = \dfrac{1}{h_{oe}} \| R_C = 50 \text{ k}\Omega \| 2.7 \text{ k}\Omega = \mathbf{2.56 \text{ k}\Omega} \cong R_C$

(c) $A_v = -\dfrac{h_{fe}(R_C \| 1/h_{oe})}{h_{ie}} = -\dfrac{(120)(2.7 \text{ k}\Omega \| 50 \text{ k}\Omega)}{1.171 \text{ k}\Omega} = \mathbf{-262.34}$

(d) $A_i \cong h_{fe} = 120$

Voltage-Divider Configuration

For the voltage-divider bias configuration of Fig. 8.41, the resulting small-signal ac equivalent network will have the same appearance as Fig. 8.39, with R_B replaced by $R' = R_1 \| R_2$.

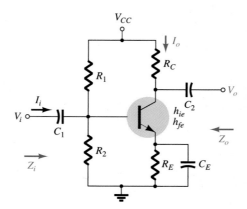

Figure 8.41 Voltage-divider bias configuration.

Z_i: From Fig. 8.39 with $R_B = R'$,

$$\boxed{Z_i = R' \| h_{ie}}$$ (8.87)

Z_o: From Fig. 8.39,

$$\boxed{Z_o \cong R_C}$$ (8.88)

A_v:

$$A_v = -\frac{h_{fe}(R_C \| 1/h_{oe})}{h_{ie}} \qquad (8.89)$$

A_i:

$$A_i = -\frac{h_{fe}R'}{R' + h_{ie}} \qquad (8.90)$$

Unbypassed Emitter-Bias Configuration

For the CE unbypassed emitter-bias configuration of Fig. 8.42, the small-signal ac model will be the same as Fig. 8.11, with βr_e replaced by h_{ie} and βI_b by $h_{fe}I_b$. The analysis will proceed in the same manner.

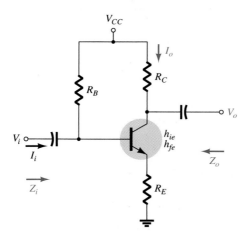

Figure 8.42 CE unbypassed emitter-bias configuration.

Z_i:

$$Z_b \cong h_{fe}R_E \qquad (8.91)$$

and

$$Z_i = R_B \| Z_b \qquad (8.92)$$

Z_o:

$$Z_o = R_C \qquad (8.93)$$

A_v:

$$A_v = -\frac{h_{fe}R_C}{Z_b} \cong -\frac{h_{fe}R_C}{h_{fe}R_E}$$

and

$$A_v \cong -\frac{R_C}{R_E} \qquad (8.94)$$

A_i:

$$A_i = \frac{h_{fe}R_B}{R_B + Z_b} \qquad (8.95)$$

or

$$A_i = -A_v \frac{Z_i}{R_C} \qquad (8.96)$$

Emitter-Follower Configuration

For the emitter-follower of Fig. 8.43, the small-signal ac model will match Fig. 8.19, with $\beta r_e = h_{ie}$ and $\beta = h_{fe}$. The resulting equations will therefore be quite similar.

Z_i:

$$Z_b \cong h_{fe} R_E \qquad (8.97)$$

$$Z_i = R_B \| Z_b \qquad (8.98)$$

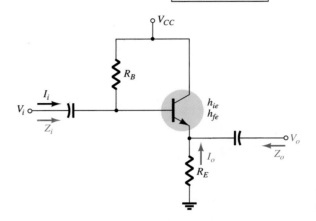

Figure 8.43 Emitter-follower configuration.

Z_o: For Z_o, the output network defined by the resulting equations will appear as shown in Fig. 8.44. Review the development of the equations in Section 8.5 and

$$Z_o = R_E \| \frac{h_{ie}}{1 + h_{fe}}$$

or since $1 + h_{fe} \cong h_{fe}$,

$$Z_o \cong R_E \| \frac{h_{ie}}{h_{fe}} \qquad (8.99)$$

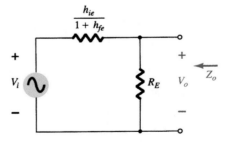

Figure 8.44 Defining Z_o for the emitter-follower configuration.

A_v: For the voltage gain, the voltage-divider rule can be applied to Fig. 8.44 as follows:

$$V_o = \frac{R_E(V_i)}{R_E + h_{ie}/(1 + h_{fe})}$$

but since $1 + h_{fe} \cong h_{fe}$,

$$A_v = \frac{V_o}{V_i} \cong \frac{R_E}{R_E + h_{ie}/h_{fe}} \qquad (8.100)$$

A_i:

$$A_i = \frac{h_{fe}R_B}{R_B + Z_b} \qquad (8.101)$$

or

$$A_i = -A_v\frac{Z_i}{R_E} \qquad (8.102)$$

Common-Base Configuration

The last configuration to be examined with the approximate hybrid equivalent circuit will be the common-base amplifier of Fig. 8.45. Substituting the approximate common-base hybrid equivalent model will result in the network of Fig. 8.46, which is very similar to Fig. 8.25.

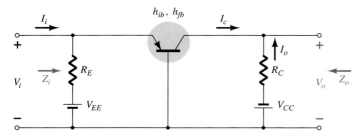

Figure 8.45 Common-base configuration.

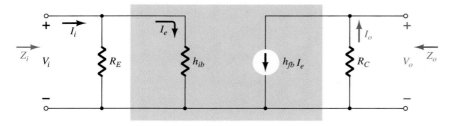

Figure 8.46 Substituting the approximate hybrid equivalent circuit into the ac equivalent network of Fig. 8.45.

From Fig. 8.46,

Z_i:

$$Z_i = R_E \| h_{ib} \qquad (8.103)$$

Z_o:

$$Z_o = R_C \qquad (8.104)$$

A_v:

$$V_o = -I_oR_C = -(h_{fb}I_e)R_C$$

with

$$I_e = \frac{V_i}{h_{ib}} \quad \text{and} \quad V_o = -h_{fb}\frac{V_i}{h_{ib}}R_C$$

so that

$$A_v = \frac{V_o}{V_i} = -\frac{h_{fb}R_C}{h_{ib}} \qquad (8.105)$$

A_i:

$$A_i = \frac{I_o}{I_i} = h_{fb} \cong -1 \qquad\qquad (8.106)$$

EXAMPLE 8.12

For the network of Fig. 8.47, determine:
(a) Z_i.
(b) Z_o.
(c) A_v.
(d) A_i.

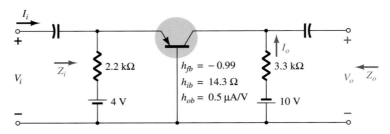

Figure 8.47   Example 8.12.

Solution

(a) $Z_i = R_E \| h_{ib} = 2.2 \text{ k}\Omega \| 14.3 \ \Omega = \mathbf{14.21 \ \Omega} \cong h_{ib}$

(b) $r_o = \dfrac{1}{h_{ob}} = \dfrac{1}{0.5 \ \mu\text{A/V}} = 2 \text{ M}\Omega$

$Z_o = \dfrac{1}{h_{ob}} \| R_C \cong R_C = \mathbf{3.3 \ k\Omega}$

(c) $A_v = -\dfrac{h_{fb} R_C}{h_{ib}} = -\dfrac{(-0.99)(3.3 \text{ k}\Omega)}{14.21} = \mathbf{229.91}$

(d) $A_i \cong h_{fb} = \mathbf{-1}$

The remaining configurations of Sections 8.1 through 8.8 that were not analyzed in this section are left as an exercise in the problem section of this chapter. It is assumed that the analysis above clearly reveals the similarities in approach using the r_e or approximate hybrid equivalent models, thereby removing any real difficulty with analyzing the remaining networks of the earlier sections.

8.10   COMPLETE HYBRID EQUIVALENT MODEL

The analysis of Section 8.9 was limited to the approximate hybrid equivalent circuit with some discussion about the output impedance. In this section, we employ the complete equivalent circuit to show the impact of h_r and define in more specific terms the impact of h_o. It is important to realize that since the hybrid equivalent model has the same appearance for the common-base, common-emitter, and common-collector configurations, the equations developed in this section can be applied to each configuration. It is only necessary to insert the parameters defined for each configuration. That is, for a common-base configuration, h_{fb}, h_{ib}, and so on, are employed, while for a common-emitter configuration, h_{fe}, h_{ie}, and so on, are utilized. Recall that

Appendix B permits a conversion from one set to the other if one set is provided and the other is required.

Consider the general configuration of Fig. 8.48 with the two-port parameters of particular interest. The complete hybrid equivalent model is then substituted in Fig. 8.49 using parameters that do not specify the type of configuration. In other words, the solutions will be in terms of h_i, h_r, h_f, and h_o. Unlike the analysis of previous sections of this chapter, the current gain A_i will be determined first since the equations developed will prove useful in the determination of the other parameters.

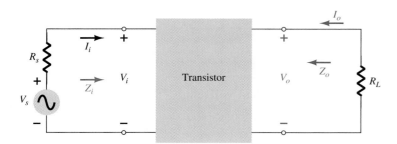

Figure 8.48 Two-port system.

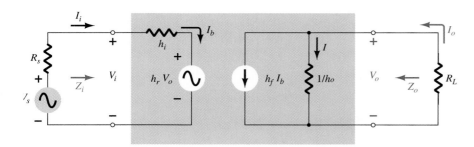

Figure 8.49 Substituting the complete hybrid equivalent circuit into the two-port system of Fig. 8.48.

Current Gain, $A_i = I_o/I_i$

Applying Kirchhoff's current law to the output circuit yields

$$I_o = h_f I_b + I = h_f I_i + \frac{V_o}{1/h_o} = h_f I_i + h_o V_o$$

Substituting $V_o = -I_o R_L$ gives us

$$I_o = h_f I_i - h_o R_L I_o$$

Rewriting the equation above, we have

$$I_o + h_o R_L I_o = h_f I_i$$

and

$$I_o(1 + h_o R_L) = h_f I_i$$

so that

$$A_i = \frac{I_o}{I_i} = \frac{h_f}{1 + h_o R_L} \qquad (8.107)$$

Note that the current gain will reduce to the familiar result of $A_i = h_f$ if the factor $h_o R_L$ is sufficiently small compared to 1.

Voltage Gain, $A_v = V_o/V_i$

Applying Kirchhoff's voltage law to the input circuit results in

$$V_i = I_i h_i + h_r V_o$$

Substituting $I_i = (1 + h_o R_L)I_o/h_f$ from Eq. (8.107) and $I_o = -V_o/R_L$ from above results in

$$V_i = \frac{-(1 + h_o R_L)h_i}{h_f R_L} V_o + h_r V_o$$

Solving for the ratio V_o/V_i yields

$$A_v = \frac{V_o}{V_i} = \frac{-h_f R_L}{h_i + (h_i h_o - h_f h_r)R_L} \qquad (8.108)$$

In this case, the familiar form of $A_v = -h_f R_L/h_i$ will return if the factor $(h_i h_o - h_f h_r)R_L$ is sufficiently small compared to h_i.

Input Impedance, $Z_i = V_i/I_i$

For the input circuit,

$$V_i = h_i I_i + h_r V_o$$

Substituting $\qquad V_o = -I_o R_L$

we have $\qquad V_i = h_i I_i - h_r R_L I_o$

Since $\qquad A_i = \dfrac{I_o}{I_i}$

$$I_o = A_i I_i$$

so that the equation above becomes

$$V_i = h_i I_i - h_r R_L A_i I_i$$

Solving for the ratio V_i/I_i, we obtain

$$Z_i = \frac{V_i}{I_i} = h_i - h_r R_L A_i$$

and substituting

$$A_i = \frac{h_f}{1 + h_o R_L}$$

yields $\qquad Z_i = \dfrac{V_i}{I_i} = h_i - \dfrac{h_f h_r R_L}{1 + h_o R_L} \qquad (8.109)$

The familiar form of $Z_i = h_i$ will be obtained if the second factor in the denominator $(h_o R_L)$ is sufficiently smaller than one.

Output Impedance, $Z_o = V_o/I_o$

The output impedance of an amplifier is defined to be the ratio of the output voltage to the output current with the signal V_s set to zero. For the input circuit with $V_s = 0$,

$$I_i = -\frac{h_r V_o}{R_s + h_i}$$

Substituting this relationship into the following equation obtained from the output circuit yields

$$I_o = h_f I_i + h_o V_o$$

$$= -\frac{h_f h_r V_o}{R_s + h_i} + h_o V_o$$

and

$$Z_o = \frac{V_o}{I_o} = \frac{1}{h_o - [h_f h_r/(h_i + R_s)]} \qquad (8.110)$$

In this case, the output impedance will reduce to the familiar form $Z_o = 1/h_o$ for the transistor when the second factor in the denominator is sufficiently smaller than the first.

For the network of Fig. 8.50, determine the following parameters using the complete hybrid equivalent model and compare to the results obtained using the approximate model.

(a) Z_i and Z_i'.
(b) A_v.
(c) $A_i = I_o/I_i$ and $A_i' = I_o/I_i'$.
(d) Z_o (within R_C) and Z_o' (including R_C).

EXAMPLE 8.13

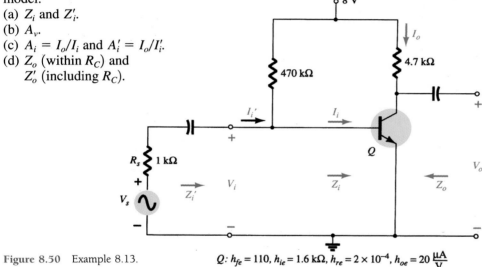

Figure 8.50 Example 8.13.

Q: $h_{fe} = 110$, $h_{ie} = 1.6$ kΩ, $h_{re} = 2 \times 10^{-4}$, $h_{oe} = 20 \frac{\mu A}{V}$

Solution

Now that the basic equations for each quantity have been derived, the order in which they are calculated is arbitrary. However, the input impedance is often a useful quantity to know and therefore will be calculated first. The complete common-emitter hybrid

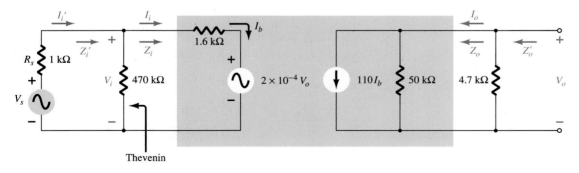

Figure 8.51 Substituting the complete hybrid equivalent circuit into the ac equivalent network of Fig. 8.50.

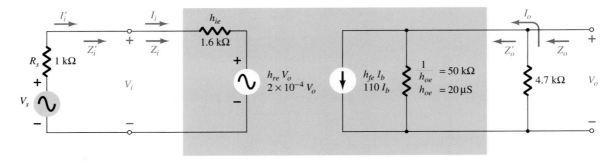

Figure 8.52 Replacing the input section of Fig. 8.51 with a Thévenin equivalent circuit.

equivalent circuit has been substituted and the network redrawn as shown in Fig. 8.51. A Thévenin equivalent circuit for the input section of Fig. 8.51 will result in the input equivalent of Fig. 8.52 since $E_{Th} \cong V_s$ and $R_{Th} \cong R_s = 1 \text{ k}\Omega$ (a result of $R_B = 470 \text{ k}\Omega$ being much greater than $R_s = 1 \text{ k}\Omega$). In this example, $R_L = R_C$ and I_o is defined as the current through R_C as in previous examples of this chapter. The output impedance Z_o as defined by Eq. (8.110) is for the output transistor terminals only. It does not include the effects of R_C. Z_o' is simply the parallel combination of Z_o and R_L. The resulting configuration of Fig. 8.52 is then an exact duplicate of the defining network of Fig. 8.49, and the equations derived above can be applied.

(a) Eq. (8.109): $\quad Z_i = \dfrac{V_i}{I_i} = h_{ie} - \dfrac{h_{fe}h_{re}R_L}{1 + h_{oe}R_L}$

$$= 1.6 \text{ k}\Omega - \dfrac{(110)(2 \times 10^{-4})(4.7 \text{ k}\Omega)}{1 + (20\ \mu S)(4.7 \text{ k}\Omega)}$$

$$= 1.6 \text{ k}\Omega - 94.52\ \Omega$$

$$= \mathbf{1.51 \text{ k}\Omega}$$

versus 1.6 kΩ using simply h_{ie}.

$$Z_i' = 470 \text{ k}\Omega \| Z_i \cong Z_i = \mathbf{1.51 \text{ k}\Omega}$$

(b) Eq. (8.108): $\quad A_v = \dfrac{V_o}{V_i} = \dfrac{-h_{fe}R_L}{h_{ie} + (h_{ie}h_{oe} - h_{fe}h_{re})R_L}$

$$= \dfrac{-(110)(4.7 \text{ k}\Omega)}{1.6 \text{ k}\Omega + [(1.6 \text{ k}\Omega)(20\ \mu S) - (110)(2 \times 10^{-4})]4.7 \text{ k}\Omega}$$

$$= \dfrac{-517 \times 10^3\ \Omega}{1.6 \text{ k}\Omega + (0.032 - 0.022)4.7 \text{ k}\Omega}$$

$$= \dfrac{-517 \times 10^3\ \Omega}{1.6 \text{ k}\Omega + 47\ \Omega}$$

$$= \mathbf{-313.9}$$

versus -323.125 using $A_v \cong -h_{fe}R_L/h_{ie}$.

(c) Eq. (8.107): $\quad A_i = \dfrac{I_o}{I_i} = \dfrac{h_{fe}}{1 + h_{oe}R_L} = \dfrac{110}{1 + (20\ \mu S)(4.7 \text{ k}\Omega)}$

$$= \dfrac{110}{1 + 0.094} = \mathbf{100.55}$$

versus 110 using simply h_{fe}. Since 470 k$\Omega \gg Z_i$, $I_i' \cong I_i$ and $A_i' \cong \mathbf{100.55}$ also.

(d) Eq. (8.110):

$$Z_o = \frac{V_o}{I_o} = \frac{1}{h_{oe} - [h_{fe}h_{re}/(h_{ie} + R_s)]}$$

$$= \frac{1}{20\ \mu S - [(110)(2 \times 10^{-4})/(1.6\ k\Omega + 1\ k\Omega)]}$$

$$= \frac{1}{20\ \mu S - 8.46\ \mu S}$$

$$= \frac{1}{11.54\ \mu S}$$

$$= \mathbf{86.66\ k\Omega}$$

which is greater than the value determined from $1/h_{oe} = 50\ k\Omega$.

$$Z_o' = R_C \| Z_o = 4.7\ k\Omega \| 86.66\ k\Omega = \mathbf{4.46\ k\Omega}$$

versus 4.7 kΩ using only R_C.

Note from the results above that the approximate solutions for A_v and Z_i were very close to those calculated with the complete equivalent model. In fact, even A_i was off by less than 10%. The higher value of Z_o only contributed to our earlier conclusion that Z_o is often so high that it can be ignored compared to the applied load. However, keep in mind that when there is a need to determine the impact of h_{re} and h_{oe}, the complete hybrid equivalent model must be used, as described above.

The specification sheet for a particular transistor typically provides the common-emitter parameters as noted in Fig. 7.28. The next example will employ the same transistor parameters appearing in Fig. 8.50 in a *pnp* common-base configuration to introduce the parameter conversion procedure and emphasize the fact that the hybrid equivalent model maintains the same layout.

For the common-base amplifier of Fig. 8.53, determine the following parameters using the complete hybrid equivalent model and compare the results to those obtained using the approximate model.
(a) Z_i and Z_i'.
(b) A_i and A_i'.
(c) A_v.
(d) Z_o and Z_o'.

EXAMPLE 8.14

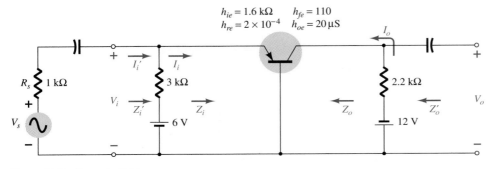

Figure 8.53 Example 8.14.

Solution

The common-base hybrid parameters are derived from the common-emitter parameters using the approximate equations of Appendix B:

$$h_{ib} \cong \frac{h_{ie}}{1 + h_{fe}} = \frac{1.6 \text{ k}\Omega}{1 + 110} = \textbf{14.41 } \boldsymbol{\Omega}$$

Note how closely the magnitude compares with the value determined from

$$h_{ib} = r_e = \frac{h_{ie}}{\beta} = \frac{1.6 \text{ k}\Omega}{110} = 14.55 \text{ }\Omega$$

$$h_{rb} \cong \frac{h_{ie}h_{oe}}{1 + h_{fe}} - h_{re} = \frac{(1.6 \text{ k}\Omega)(20 \text{ }\mu\text{S})}{1 + 110} - 2 \times 10^{-4}$$

$$= \textbf{0.883} \times \textbf{10}^{-4}$$

$$h_{fb} \cong \frac{-h_{fe}}{1 + h_{fe}} = \frac{-110}{1 + 110} = \textbf{-0.991}$$

$$h_{ob} \cong \frac{h_{oe}}{1 + h_{fe}} = \frac{20 \text{ }\mu\text{S}}{1 + 110} = \textbf{0.18 } \boldsymbol{\mu}\textbf{S}$$

Substituting the common-base hybrid equivalent circuit into the network of Fig. 8.53 will then result in the small-signal equivalent network of Fig. 8.54. The Thévenin network for the input circuit will result in $R_{\text{Th}} = 3 \text{ k}\Omega \| 1 \text{ k}\Omega = 0.75 \text{ k}\Omega$ for R_s in the equation for Z_o.

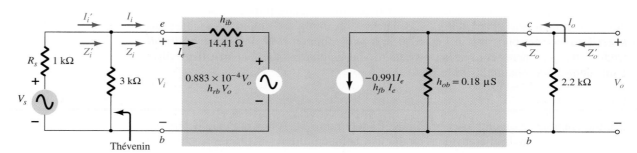

Figure 8.54 Small-signal equivalent for the network of Fig. 8.53.

(a) Eq. (8.109): $\quad Z_i = \dfrac{V_i}{I_i} = h_{ib} - \dfrac{h_{fb}h_{rb}R_L}{1 + h_{ob}R_L}$

$$= 14.41 \text{ }\Omega - \frac{(-0.991)(0.883 \times 10^{-4})(2.2 \text{ k}\Omega)}{1 + (0.18 \text{ }\mu\text{S})(2.2 \text{ k}\Omega)}$$

$$= 14.41 \text{ }\Omega + 0.19 \text{ }\Omega$$

$$= \textbf{14.60 } \boldsymbol{\Omega}$$

versus 14.41 Ω using $Z_i \cong h_{ib}$.

$$Z_i' = 3 \text{ k}\Omega \| Z_i \cong Z_i = \textbf{14.60 } \boldsymbol{\Omega}$$

(b) Eq. (8.107): $\quad A_i = \dfrac{I_o}{I_i} = \dfrac{h_{fb}}{1 + h_{ob}R_L}$

$$= \frac{-0.991}{1 + (0.18 \text{ }\mu\text{S})(2.2 \text{ k}\Omega)}$$

$$= \textbf{-0.991} = h_{fb}$$

Since $3 \text{ k}\Omega \gg Z_i$, $I_i' \cong I_i$ and $A_i' = I_o/I_i' \cong -1$ also.

(c) Eq. (8.108): $A_v = \dfrac{V_o}{V_i} = \dfrac{-h_{fb}R_L}{h_{ib} + (h_{ib}h_{ob} - h_{fb}h_{rb})R_L}$

$$= \dfrac{-(-0.991)(2.2 \text{ k}\Omega)}{14.41 \ \Omega + [(14.41 \ \Omega)(0.18 \ \mu\text{S}) - (-0.991)(0.883 \times 10^{-4})]2.2 \text{ k}\Omega}$$

$$= \mathbf{149.25}$$

versus 151.3 using $A_v \cong -h_{fb}R_L/h_{ib}$.

(d) Eq. (8.110): $Z_o = \dfrac{1}{h_{ob} - [h_{fb}h_{rb}/(h_{ib} + R_s)]}$

$$= \dfrac{1}{0.18 \ \mu\text{S} - [(-0.991)(0.883 \times 10^{-4})/(14.41 \ \Omega + 0.75 \text{ k}\Omega)]}$$

$$= \dfrac{1}{0.295 \ \mu\text{S}}$$

$$= \mathbf{3.39 \ M\Omega}$$

versus 5.56 MΩ using $Z_o \cong 1/h_{ob}$. For Z_o' as defined by Fig. 8.54:

$$Z_o' = R_C \| Z_o = 2.2 \text{ k}\Omega \| 3.39 \text{ M}\Omega = \mathbf{2.199 \ k\Omega}$$

versus 2.2 kΩ using $Z_o' \cong R_C$.

8.11   SUMMARY TABLE

Now that the most familiar configurations of the small-signal transistor amplifiers have been introduced, Table 8.1 is presented to review the general characteristics of each for immediate recall. It must be absolutely clear that the values listed are simply typical values to establish a basis for comparison. The levels obtained in an actual analysis will most likely be different, and certainly different from one configuration to another. Being able to repeat most of the information in the table is an important first step in developing a general familiarity with the subject matter. For instance, one should now be able to state with some assurance that the emitter-follower configuration typically has a high input impedance, low output impedance, and a voltage gain slightly less than 1. There should be no need to perform a variety of calculations to recall salient facts such as those above. For the future, it will permit the study of a network or system without becoming mathematically involved. The function of each component of a design will become increasingly familiar as general facts such as those above become part of your background.

One obvious advantage of being able to recall general facts like the above is an ability to check the results of a mathematical analysis. If the input impedance of a common-base configuration is in the kilohm range, there is good reason to recheck the analysis. However, on the other side of the coin, a result of 22 Ω suggests that the analysis may be correct.

8.12   TROUBLESHOOTING

Although the terminology *troubleshooting* suggests that the procedures to be described are designed simply to isolate a malfunction, it is important to realize that the same techniques can be applied to ensure that a system is operating properly. In any case, the testing, checking, or isolating procedures require an understanding of what to expect at various points in the network in both the dc and ac domains. In most cases, a network operating correctly in the dc mode will also behave properly in the ac domain. In addition, a network providing the expected ac response is most likely biased

TABLE 8.1 Relative Levels for the Important Parameters of the CE, CB, and CC Transistor Amplifiers

Configuration	Z_i	Z_o	A_v	A_i
Fixed-bias:	Medium (1 kΩ) $= \boxed{R_B \| \beta r_e}$ $\cong \boxed{\beta r_e}$ $(R_B \ge 10\beta r_e)$	Medium (2 kΩ) $= \boxed{R_C \| r_o}$ $\cong \boxed{R_C}$ $(r_o \ge 10R_C)$	High (-200) $= \boxed{-\dfrac{(R_C \| r_o)}{r_e}}$ $\cong \boxed{-\dfrac{R_C}{r_e}}$ $(r_o \ge 10R_C)$	High (100) $= \boxed{\dfrac{\beta R_B r_o}{(r_o + R_C)(R_B + \beta r_e)}}$ $\cong \boxed{\beta}$ $(r_o \ge 10R_C,$ $R_B \ge 10\beta r_e)$
Voltage-divider bias:	Medium (1 kΩ) $= \boxed{R_1 \| R_2 \| \beta r_e}$	Medium (2 kΩ) $= \boxed{R_C \| r_o}$ $\cong \boxed{R_C}$ $(r_o \ge 10R_C)$	High (-200) $= \boxed{-\dfrac{R_C \| r_o}{r_e}}$ $\cong \boxed{-\dfrac{R_C}{r_e}}$ $(r_o \ge 10R_C)$	High (50) $= \boxed{\dfrac{\beta(R_1\|R_2)r_o}{(r_o + R_C)(R_1\|R_2 + \beta r_e)}}$ $\cong \boxed{\dfrac{\beta\|(R_1\|R_2)}{R_1\|R_2 + \beta r_e}}$ $(r_o \ge 10R_C)$
Unbypassed emitter bias:	High (100 kΩ) $= \boxed{R_B \| Z_b}$ $Z_b \cong \beta(r_e + R_E)$ $\cong \boxed{R_B \| \beta R_E}$ $(R_E \gg r_e)$	Medium (2 kΩ) $= \boxed{R_C}$ (any level of r_o)	Low (-5) $= \boxed{-\dfrac{R_C}{r_e + R_E}}$ $\cong \boxed{-\dfrac{R_C}{R_E}}$ $(R_E \gg r_e)$	High (50) $\cong \boxed{-\dfrac{\beta R_B}{R_B + Z_b}}$
Emitter-follower:	High (100 kΩ) $= \boxed{R_B \| Z_b}$ $Z_b \cong \beta(r_e + R_E)$ $\cong \boxed{R_B \| \beta R_E}$ $(R_E \gg r_e)$	Low (20 Ω) $= \boxed{R_E \| r_e}$ $\cong \boxed{r_e}$ $(R_E \gg r_e)$	Low $(\cong 1)$ $= \boxed{\dfrac{R_E}{R_E + r_e}}$ $\cong \boxed{1}$	High (-50) $\cong \boxed{-\dfrac{\beta R_B}{R_B + Z_b}}$
Common-base:	Low (20 Ω) $= \boxed{R_E \| r_e}$ $\cong \boxed{r_e}$ $(R_E \gg r_e)$	Medium (2 kΩ) $= \boxed{R_C}$	High (200) $\cong \boxed{\dfrac{R_C}{r_e}}$	Low (-1) $\cong \boxed{-1}$
Collector feedback:	Medium (1 kΩ) $= \boxed{\dfrac{r_e}{\dfrac{1}{\beta} + \dfrac{R_C}{R_F}}}$ $(r_o \ge 10R_C)$	Medium (2 kΩ) $\cong \boxed{R_C \| R_F}$ $(r_o \ge 10R_C)$	High (-200) $\cong \boxed{-\dfrac{R_C}{r_e}}$ $(r_o \ge 10R_C)$ $(R_F \gg R_C)$	High (50) $= \boxed{\dfrac{\beta R_F}{R_F + \beta R_C}}$ $\cong \boxed{\dfrac{R_F}{R_C}}$

as planned. In a typical laboratory setting, both the dc and ac supplies are applied and the ac response at various points in the network is checked with an oscilloscope as shown in Fig. 8.55. Note that the black (gnd) lead of the oscilloscope is connected directly to ground and the red lead is moved from point to point in the network, providing the patterns appearing in Fig. 8.55. The vertical channels are set in the ac mode to remove any dc component associated with the voltage at a particular point. The small ac signal applied to the base is amplified to the level appearing from collector to ground. Note the difference in vertical scales for the two voltages. There is no ac response at the emitter terminal due to the short-circuit characteristics of the capacitor at the applied frequency. The fact that v_o is measured in volts and v_i in millivolts suggests a sizable gain for the amplifier. In general, the network appears to be operating properly. If desired, the dc mode of the multimeter could be used to check V_{BE} and the levels of V_B, V_{CE}, and V_E to review whether they lie in the expected range. Of course, the oscilloscope can also be used to compare dc levels simply by switching to the dc mode for each channel.

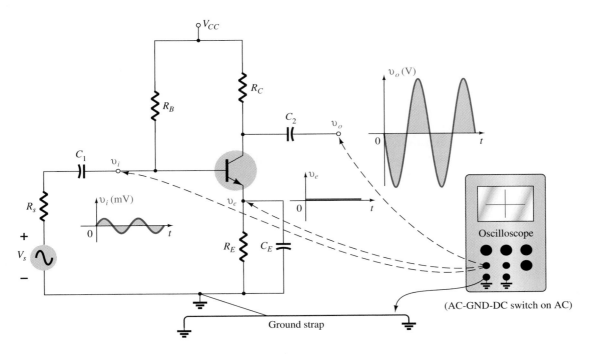

Figure 8.55 Using the oscilloscope to measure and display various voltages of a BJT amplifier.

Needless to say, a poor ac response can be due to a variety of reasons. In fact, there may be more than one problem area in the same system. Fortunately, however, with time and experience, the probability of malfunctions in some areas can be predicted and an experienced person can isolate problem areas fairly quickly.

In general, there is nothing mysterious about the general troubleshooting process. If you decide to follow the ac response, it is good procedure to start with the applied signal and progress through the system toward the load, checking critical points along the way. An unexpected response at some point suggests that the network is fine up to that area, thereby defining the region that must be investigated further. The waveform obtained on the oscilloscope will certainly help in defining the possible problems with the system.

If the response for the network of Fig. 8.55 is as appears in Fig. 8.56, the network has a malfunction that is probably in the emitter area. An ac response across the emitter is unexpected, and the gain of the system as revealed by v_o is much lower. Recall

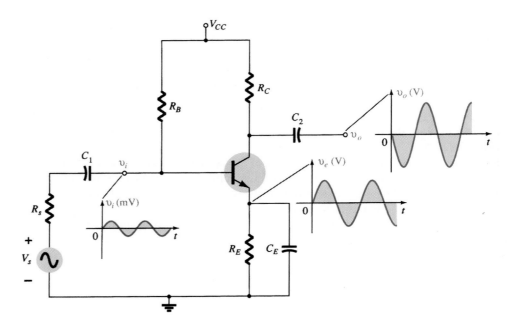

Figure 8.56 The waveforms resulting from a malfunction in the emitter area.

for this configuration that the gain is much greater if R_E is bypassed. The response obtained suggests that R_E is not bypassed by the capacitor and the terminal connections of the capacitor and the capacitor itself should be checked. In this case, a checking of the dc levels will probably not isolate the problem area since the capacitor has an "open-circuit" equivalent for dc. In general, a prior knowledge of what to expect, a familiarity with the instrumentation, and most important, experience are all factors that contribute to the development of an effective approach to the art of troubleshooting.

8.13 PRACTICAL APPLICATIONS

Audio Mixer

When two or more signals are to be combined into a single audio output, mixers such as appearing in Fig. 8.57 are employed. The potentiometers at the input are the volume controls for each channel, with potentiometer R_3 included to provide additional balance between the two signals. Resistors R_4 and R_5 are there to ensure that one channel does not load down the other, that is, to ensure that one signal does not appear as a load to the other, draw power, and affect the desired balance on the mixed signal.

The effect of resistors R_4 and R_5 is an important one that should be discussed in some detail. A dc analysis of the transistor configuration will result in $r_e = 11.71 \Omega$ which will establish an input impedance to the transistor of about 1.4 kΩ. The parallel combination of $R_6 \| Z_i$ is also approximately 1.4 kΩ. Setting both volume controls to their maximum value and the balance control R_3 to its midpoint will result in the equivalent network of Fig. 8.58a. The signal at v_1 is assumed to be a low-impedance microphone with an internal resistance of 1 kΩ. The signal at v_2 is assumed to be a guitar amplifier with a higher internal resistance of 10 kΩ. Since the 470-kΩ and 500-kΩ resistors are in parallel for the above conditions, they can be combined and replaced with a single resistor of about 242 kΩ. Each source will then have an equivalent such as shown in Fig. 8.58b for the microphone. Applying Thévenin's theorem will reveal that it is an excellent approximation to simply drop the 242 kΩ and assume that the equivalent network is as shown for each channel. The result is the equiv-

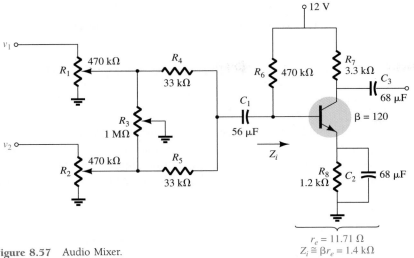

Figure 8.57   Audio Mixer.

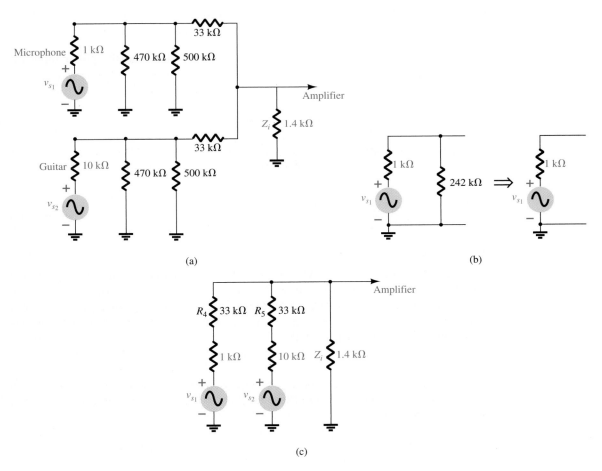

(a)

(b)

(c)

Figure 8.58   (a) Equivalent network with R_3 set at the midpoint and the volume controls on their maximum settings; (b) finding the Thévenin equivalent for channel 1; (c) substituting the Thévenin equivalent networks into Fig. 8.58a.

alent network of Fig. 8.58c for the input section of the mixer. Applying the superposition theorem will result in the following equation for the ac voltage at the base of the transistor:

$$v_b = \frac{(1.4 \text{ k}\Omega \| 43 \text{ k}\Omega)v_{s_1}}{34 \text{ k}\Omega + (1.4 \text{ k}\Omega \| 43 \text{ k}\Omega)} + \frac{(1.4 \text{ k}\Omega \| 34 \text{ k}\Omega)v_{s_2}}{43 \text{ k}\Omega + (1.4 \text{ k}\Omega \| 34 \text{ k}\Omega)}$$

$$= 38 \times 10^{-3}v_{s_1} + 30 \times 10^{-3}v_{s_2}$$

With $r_e = 11.71 \text{ }\Omega$, the gain of the amplifier will be $-R_C/r_e = 3.3 \text{ k}\Omega/11.71 \text{ }\Omega$ $= -281.8$, and the output voltage is

$$v_o = -10.7v_{s_1} - 8.45v_{s_2}$$

which provides a pretty good balance between the two signals even though they have a 10:1 ratio in internal impedance. In general, the system will respond quite well. However, if we now remove the 33-kΩ resistors from the diagram of Fig. 8.58c, the equivalent network of Fig. 8.59 will result and the following equation for v_b obtained using the superposition theorem:

$$v_b = \frac{(1.4 \text{ k}\Omega \| 10 \text{ k}\Omega)v_{s_1}}{1 \text{ k}\Omega + 1.4 \text{ k}\Omega \| 10 \text{ k}\Omega} + \frac{(1.4 \text{ k}\Omega \| 1 \text{ k}\Omega)v_{s_2}}{10 \text{ k}\Omega + (1.4 \text{ k}\Omega \| 1 \text{ k}\Omega)}$$

$$= 0.55v_{s_1} + 0.055v_{s_2}$$

Using the same gain as before, the output voltage is

$$v_o = 155v_{s_1} + 15.5v_{s_2} \cong 155v_{s_1}$$

revealing that the microphone will be quite loud and clear and the guitar input essentially lost.

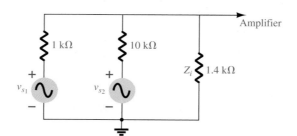

Figure. 8.59 Redrawing the network of Fig. 8.58c with the 33-kΩ resistors removed.

The importance of the 33-kΩ resistors is therefore defined. It makes each applied signal appear to have similar impedance levels so that there is good balance at the output. One might suggest that the larger resistor improves the balance. However, even though the balance at the base of the transistor may be better, the strength of the signal at the base of the transistor will be less, and the output level reduced accordingly. In other words, the choice of resistors R_4 and R_5 is a give-and-take situation between the input level at the base of the transistor and balance of the output signal.

To demonstrate that the capacitors are truly short-circuit equivalents in the audio range, substitute a very low audio frequency of 100 Hz into the reactance equation of a 56-μF capacitor:

$$X_C = \frac{1}{2\pi f C} = \frac{1}{2\pi(100 \text{ Hz})(56 \text{ }\mu\text{F})} = 28.42 \text{ }\Omega$$

A level of 28.42 Ω compared to any of the neighboring impedances is certainly small enough to be ignored. Higher frequencies will have even less effect.

A similar mixer will be discussed in the following JFET chapter. The major difference will be the fact that the input impedance of the JFET can be approximated by an open circuit rather than the rather low level input impedance of the BJT configuration. The result will be a higher signal level at the input to the JFET amplifier.

However, the gain of the FET is much less than that of the BJT transistor, resulting in output levels that are actually quite similar.

Preamplifier

The primary function of a **preamplifier** is as its name implies: **an amplifier used to pick up the signal from its primary source and then operate on it in preparation for its passage into the amplifier section.** Typically, a preamplifier will amplify the signal, control its volume, perhaps change its input impedance characteristics, and if necessary determine its route through the stages to follow, in total, a stage of any system with a multitude of functions.

A preamplifier such as shown in Fig. 8.60 is often used with dynamic microphones to bring the signal level up to levels that are suitable for further amplification or power amplifiers. Typically, dynamic microphones are low-impedance microphones since their internal resistance is determined primarily by the winding of the voice coil. The basic construction consists of a voice coil attached to a small diaphragm that is free to move within a permanent magnet. When one speaks into the microphone, the diaphragm will move accordingly and cause the voice coil to move in the same manner within the magnetic field. Through Faraday's law a voltage will be induced across the coil that will carry the audio signal.

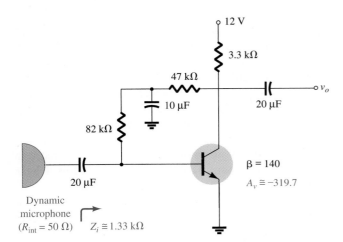

Figure 8.60 Preamplifier for a dynamic microphone.

Since it is a low-impedance microphone, the input impedance of the transistor amplifier does not have to be that high to pick up most of the signal. Since the internal impedance of a dynamic microphone may be as low as 20 Ω to 100 Ω, most of the signal would be picked up with an amplifier having an input impedance as low as 1 to 2 kΩ. This, in fact, is the case for the preamplifier of Fig. 8.60. For dc biasing conditions, the collector dc feedback configuration was chosen because of its high stablity characteristics.

In the ac domain, the 10-μF capacitor will assume a short-circuit state (on an approximate basis), placing the 82-kΩ resistor across the input impedance of the transistor and the 47 kΩ across the output of the transistor. A dc analysis of the transistor configuration will result in $r_e = 9.64$ Ω, resulting in an ac gain determined by

$$A_v = -\frac{(47 \text{ k}\Omega \| 3.3 \text{ k}\Omega)}{9.64 \ \Omega} = \mathbf{-319.7}$$

which is excellent for this application. Of course, the gain will drop when this pickup stage of the design is connected to the input of the amplifier section. That is, the input resistance of the next stage will appear in parallel with the 47-kΩ and 3.3-kΩ resistors and will drop the gain below the unloaded level of 319.7.

The input impedance of the preamplifier is determined by

$$Z_i = 82 \text{ k}\Omega \| \beta r_e = 82 \text{ k}\Omega \| (140)(9.64 \text{ }\Omega) = 82 \text{ k}\Omega \| 1.34 \text{ k}\Omega = \textbf{1.33 k}\boldsymbol{\Omega}$$

which is also fine for most low-impedance dynamic microphones. In fact, for a microphone with an internal impedance of 50 Ω, the signal at the base would be over 98% of that available. This discussion is important because if the impedance of the microphone is a great deal more, say, 1 kΩ, the preamplifier would have to be designed differently to ensure that the input impedance was at least 10 kΩ or more.

Random-Noise Generator

There is often a need for a random-noise generator to test the response of a speaker, microphone, filter, and, in fact, any system designed to work over a wide range of frequencies. A **random-noise generator** is just as its name implies: **a generator that generates signals of random amplitude and frequency.** The fact that these signals are usually totally unintelligible and unpredictable is the reason that they are simply referred to as *noise*. **Thermal noise** is noise generated due to thermal effects resulting from the interaction between free electrons and the vibrating ions of a material in conduction. The result is an uneven flow of electrons through the medium that will result in a varying potential across the medium. In most cases, these randomly generated signals are in the μV range, but with sufficient amplification they can wreak havoc on a system's response. This thermal noise is also called **Johnson noise** (named after the original researcher in the area) or **white noise** (because in optics, white light contains all frequencies). This type of noise has a fairly flat frequency response such as shown in Fig. 8.61a; that is, a plot of its power versus frequency from the very low to the very high end is fairly uniform. A second type of noise is called **shot noise,** a name derived from the fact that its noise sounds like a shower of lead shot hitting a solid surface or like heavy rain on a window. Its source is pockets of carriers passing through a medium at uneven rates. A third is **pink, flicker,** or **1/f noise** that is due to the variation in transit times for carriers crossing various junctions of semiconductor devices. It is called 1/f noise because its magnitude drops off with increase in frequency. **Its impact is usually the most dramatic for frequencies below 1 kHz** as shown in Fig. 8.61b.

The network of Fig. 8.62 is designed to generate both a white noise and a pink noise. Rather than a separate source for each, first white noise is developed (level across the entire frequency spectrum), and then a filter is applied to remove the mid- and high-frequency component, leaving only the low-frequency noise response. The filter is further designed to modify the flat response of the white noise in the low-frequency region (to create a 1/f drop-off) by having sections of the filter "drop in" as the frequency increases. The white noise is created by leaving the collector terminal of transistor Q_1 open and reverse-biasing the base-to-emitter junction. In

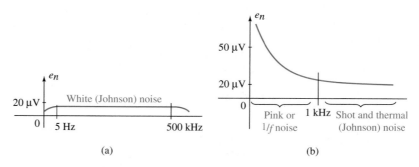

Figure 8.61 Typical noise frequency spectrums: (a) white or Johnson; (b) pink, thermal, and shot.

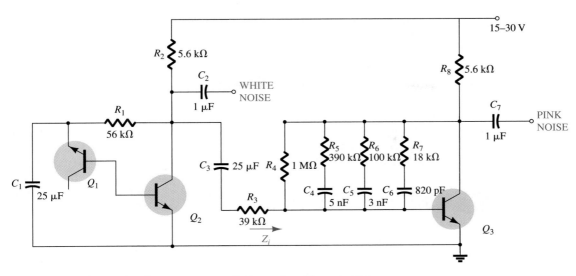

Figure 8.62 White-and pink-noise generator. (Redrawn from *Electronics Today International*)

essence, the transistor is being used as a diode biased in the Zener avalanche region. Biasing a transistor in this region creates a very unstable situation that is conductive to the generation of random white noise. The combination of the avalanche region with its rapidly changing charge levels, sensitivity of the current level to temperature, and quickly changing impedance levels all contribute to the level of noise voltage and current generated by the transistor. Germanium transistors are often used because the avalanche region is less defined and less stable than in silicon transistors. In addition, there are diodes and transistors designed specifically for random-noise generation.

There it is: The source of the noise is not some specially designed generator. It is simply due to the fact that current flow is not an ideal phenomenon but actually varies with time at a level that generates unwanted variations in the terminal voltage across elements. In fact, that variation in flow is so broad that it can generate frequencies that extend across a wide spectrum—a very interesting phenomenon.

The generated noise current of Q_1 will then be the base current for Q_2 which will be amplified to generate a white noise of perhaps 100 mV, which for this design would suggest an input noise voltage of about 170 μV. Capacitor C_1 will have a low impedance throughout the frequency range of interest to provide a "shorting effect" on any spurious signals in the air from contributing to the signal at the base of Q_1. The capacitor C_2 is there to isolate the dc biasing of the white-noise generator from the dc levels of the filter network to follow. The 39 kΩ and the input impedance of the next stage create the simple voltage-divider network of Fig. 8.63. If the 39 kΩ were not present, the parallel combination of R_2 and Z_i would load down the first stage and reduce the gain of Q_1 considerably. In the gain equation, R_2 and Z_i would appear in parallel (discussed in Chapter 11).

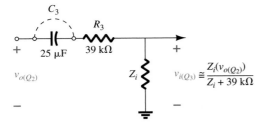

$$v_{i(Q_3)} \cong \frac{Z_i(v_{o(Q_2)})}{Z_i + 39 \text{ k}\Omega}$$

Figure 8.63 Input circuit for the second stage.

The filter network is actually part of the feedback loop from collector to base appearing in the collector feedback network of Section 8.7. To describe its behavior, let us first consider the extremes of the frequency spectrum. For very low frequencies all the capacitors can be approximated by an open circuit, and the only resistance from collector to base is the 1-MΩ resistor. Using a beta of 100, the gain of the section is about 280 and the input impedance about 1.28 kΩ. At a sufficiently high frequency all the capacitors could be replaced by short circuits, and the total resistance combination between collector and base would be reduced to about 14.5 kΩ which would result in a very high unloaded gain of about 731, more than twice just obtained with $R_F = 1$ MΩ. Since the $1/f$ filter is supposed to reduce the gain at high frequencies, it initially appears as though there is an error in design. However, when the input impedance is determined, it has dropped to about 19.33 Ω which is a 66:1 drop from the level obtained with $R_F = 1$ MΩ. This would have a significant impact on the input voltage appearing at the second stage when we consider the voltage-divider action of Fig. 8.63. In fact, when compared to the series 39-kΩ resistor, the signal at the second stage can be assumed to be negligible or at a level where even a gain in excess of 700 cannot raise it to a level of any consequence. In total, therefore, doubling the gain is totally lost due to the tremendous drop in Z_i, and the output at very high frequencies can be ignored entirely.

For the range of frequencies between the very low and the very high, the three capacitors of the filter will cause the gain to drop off with increase in frequency. First, capacitor C_4 will drop in and cause a reduction in gain (around 100 Hz). Then capacitor C_5 will be included and will place the three branches in parallel (around 500 Hz). Finally, capacitor C_6 will result in four parallel branches and the minimum feedback resistance (around 6 kHz).

The result is a network with an excellent random-noise signal for the full frequency spectrum (white) and the low-frequency spectrum (pink).

Sound-Modulated Light Source

The light from the 12-V bulb of Fig. 8.64 will vary at a frequency and an intensity sensitive to the applied signal. The applied signal may be the output of an acoustical amplifier, a musical instrument, or even a microphone. Of particular interest is the fact that the applied voltage is 12 V ac rather than the typical dc biasing supply. The immediate question, in the absence of a dc supply, is how the dc biasing levels for the transistor will be established. In actuality, the dc level is obtained through the use of diode D_1, which rectifies the ac signal, and capacitor C_2, which acts as a power supply filter to generate a dc level across the output branch of the transistor. The peak value of a 12-V rms supply is about 17 V, resulting in a dc level after the capacitive filtering in the neighborhood of 16 V. If the potentiometer is set so that R_1 is about

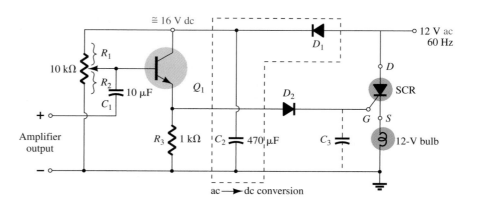

Figure 8.64 Sound-modulated light source. (Redrawn from *Electronics Today International*)

Chapter 8   BJT Small-Signal Analysis

320 Ω, the voltage from base to emitter of the transistor will be about 0.5 V, and the transistor will be in the off state. In this state the collector and emitter current are essentially 0 mA, and the voltage across resistor R_3 is approximately 0 V. The voltage at the junction of the collector terminal and the diode is therefore 0 V, resulting in D_2 being in the off state and 0 V at the gate terminal of the silicon-controlled rectifier (SCR). The SCR (see Section 20.3) is fundamentally a diode whose state is controlled by an applied voltage at the gate terminal. The absence of a voltage at the gate means that the SCR and bulb are "off."

If a signal is now applied to the gate terminal, the combination of the established biasing level and the applied signal can establish the required 0.7-V "turn-on" voltage, and the transistor will be turned on for periods of time dependent on the applied signal. When the transistor turns on, it will establish a collector current through resistor R_3 that will establish a voltage from collector to ground. If the voltage is more than the required 0.7 V for diode D_2, a voltage will appear at the gate of the SCR that may be sufficient to turn it "on" and establish conduction from the drain to source of the SCR. However, we must now examine one of the most interesting aspects of this design. Since the applied voltage across the SCR is ac, which will vary in magnitude with time as shown in Fig 8.65, the conduction strength of the SCR will vary with time also. As shown in the figure, if the SCR is turned "on" when the sinusoidal voltage is a maximum, the resulting current through the SCR will be a maximum also, and the bulb will be its brightest. If the SCR should turn on when the sinusoidal voltage is near its minimum, the bulb may turn on, but the lower current will result in considerably less illumination. The result is that the light bulb turns on in sync with when the input signal is peaking, but the strength of turn-on will be determined by where you are on the applied 12-V signal. One can imagine the interesting and varied responses of such a system. Each time you apply the same audio signal, the response will have a different character.

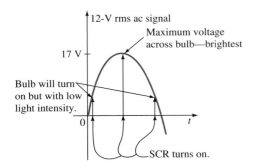

Figure 8.65 Demonstrating the effect of an ac voltage on the operation of the SCR of Fig. 8.64.

In the above action, the potentiometer was set below the turn-on voltage of the transistor. The potentiometer can also be adjusted so that the transistor is "just on," resulting in a low-level base current. The result is a low-level collector current and insufficient voltage to forward-bias diode D_2 and turn on the SCR at the gate. However, by setting the system up in this manner, the resultant light output will be more sensitive to lower-amplitude components of the applied signal. In the first case, the system acted more like a peak detector, whereas in the latter case it is sensitive to more components of the signal.

Diode D_2 was included to be sure that there is sufficient voltage to turn on both the diode and the SCR, in other words, to eliminate the possibility of noise or some other low-level unexpected voltage on the line turning the SCR on. Capacitor C_3 can be inserted to slow down the response by making the voltage charge across the capacitor before the gate will reach sufficient voltage to turn on the SCR.

8.14 SUMMARY

Important Conclusions and Concepts

1. The r_e **model** for a BJT in the ac domain is sensitive to the **actual dc operating conditions of the network.** This parameter is normally not provided on a specification sheet, although h_{ie} of the normally provided hybrid parameters is equal to βr_e, but only under specific operating conditions.

2. Most **specification sheets** for BJTs include a **list of hybrid parameters** to establish an ac model for the transistor. One must be aware, however, that they are provided for a particular set of dc operating conditions.

3. The **CE fixed-bias configuration** can have a **significant voltage gain** characteristic, although its **input impedance can be relatively low.** The approximate **current gain** is given by simply **beta,** and the **output impedance** is normally assumed to be R_C.

4. The **voltage-divider bias configuration** has a **higher stability** than the fixed-bias configuration, but it has about the **same voltage gain, current gain, and output impedance.** Due to the biasing resistors, its input impedance may be lower than that of the fixed-bias configuration.

5. The **CE emitter-bias configuration** with an unbypassed emitter resistor has a **larger input resistance** than the bypassed configuration, but it will have a **much smaller voltage gain** than the bypassed configuration. For the unbypassed or bypassed situation, the **output impedance** is normally assumed to be simply R_C.

6. The **emitter-follower configuration** will always have an **output voltage slightly less than the input signal.** However, the **input impedance** can be **very large,** making it very useful for situations where a high-input first stage is needed to "pick up" as much of the applied signal as possible. Its **output impedance** is **extremely low,** making it an excellent signal source for the second stage of a multistage amplifier.

7. The **common-base configuration** has a **very low input impedance,** but it can have a **significant voltage gain.** The **current gain** is just **less than 1,** and the **output impedance** is simply R_C.

8. The **collector feedback configuration** has an **input impedance** that is **sensitive to beta** and that can be quite low depending on the parameters of the configuration. However, the **voltage gain** can be **significant** and the **current gain** of **some magnitude** if the parameters are chosen properly. The **output impedance** is most often simply the collector resistance R_C.

9. The **collector dc feedback configuration** utilizes the dc feedback to **increase its stability** and the changing state of a capacitor from dc to ac to establish a **higher voltage gain** than obtained with a straight feedback connection. The **output impedance** is usually close to R_C and the **input impedance** relatively close to that obtained with the **basic common-emitter configuration.**

10. The **approximate hybrid equivalent network** is very **similar** in composition to that used with the r_e model. In fact, the **same methods** of analysis can be applied to both models. For the hybrid model the results will be in terms of the network parameters and the hybrid parameters, whereas for the r_e model they will be in terms of the network parameters and β, r_e, and r_o.

11. The **complete hybrid equivalent model** will **include** the effects of the **feedback parameter** h_{re} which typically **reduces** the **voltage gain** and the **input impedance** although it may **increase** the **output impedance.**

12. The **hybrid model** for common-emitter, common-base, and common-collector configurations **is the same.** The only difference will be the magnitude of the parameters of the equivalent network.

13. For BJT amplifiers that **fail to operate properly,** the first step should be to **check the dc levels** and be sure that they support the dc operation of the design.

14. Always keep in mind that **capacitors** are typically open circuits for the **dc analysis** and operation and essentially **short circuits** for the **ac response.**

Equations

CE fixed bias:

$$Z_i \cong \beta r_e$$
$$Z_o \cong R_C$$
$$A_v = -\frac{R_C}{r_e}$$
$$A_i = -A_v\frac{Z_i}{R_C} \cong \beta$$

Voltage-divider bias:

$$Z_i = R_1\|R_2\|\beta r_e$$
$$Z_o \cong R_C$$
$$A_v = -\frac{R_C}{r_e}$$
$$A_i = -A_v\frac{Z_i}{R_C} \cong \beta$$

CE emitter-bias:

$$Z_i \cong R_B\|\beta R_E$$
$$Z_o \cong R_C$$
$$A_v \cong -\frac{R_C}{R_E}$$
$$A_i \cong \frac{\beta R_B}{R_B + \beta R_E}$$

Emitter-follower:

$$Z_i \cong R_B\|\beta R_E$$
$$Z_o \cong r_e$$
$$A_v \cong 1$$
$$A_i = -A_v\frac{Z_i}{R_E}$$

Common-base:

$$Z_i \cong R_E\|r_e$$
$$Z_o \cong R_C$$
$$A_v \cong \frac{R_C}{r_e}$$
$$A_i \cong -1$$

Collector feedback:

$$Z_i \cong \frac{r_e}{\dfrac{1}{\beta} + \dfrac{R_C}{R_F}}$$

$$Z_o \cong R_C \| R_F$$

$$A_v = -\frac{R_C}{r_e}$$

$$A_i \cong \frac{R_F}{R_C}$$

Collector dc feedback:

$$Z_i \cong R_{F_1} \| \beta r_e$$
$$Z_o \cong R_C \| R_{F_2}$$
$$A_v = -\frac{R_{F_2} \| R_C}{r_e}$$
$$A_i = -A_v \frac{Z_i}{R_C}$$

8.15 COMPUTER ANALYSIS

PSpice Windows

VOLTAGE-DIVIDER CONFIGURATION USING THE SOFTWARE TRANSISTOR PARAMETERS

Now that the basic maneuvers for developing the network on the schematics grid have been introduced, the current description will concentrate on the variations introduced by the ac analysis.

Using schematics, the network of Fig. 8.9 (Example 8.2) is developed as shown in Fig. 8.66. Note the ac source of 1 mV and the printer symbol at the output terminal of the network.

The sinusoidal ac source is listed in the **SOURCE.slb** library as **VSIN.** Once placed on the diagram, double-clicking the symbol will result in the **PartName: VSIN** dialog box with a list of options. Each choice can be made by double-clicking the desired quantity, which will then appear in the **Name** and **Value** rectangles at the top of the box. The cursor appears in the **Value** box, and the desired value can be entered. After each entry, be sure to **Save Attr** to save the entered attribute. If done properly, the assigned value will appear in the listing.

For our analysis, the following choices will be made:

VAMPL = 1 mV (the peak value of the sinusoidal signal).

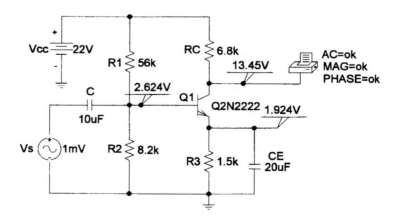

Figure 8.66 Using PSpice Windows to analyze the network of Figure 8.9 (Example 8.2).

FREQ $= 10$ kHz (the frequency of interest).

PHASE $= 0$ (no initial phase angle for the sinusoidal signal).

VOFF $= 0$ (no dc offset voltage for the sinusoidal signal).

AC $= 1$ mV.

If you want to display the value of the ac signal, simply click on **Change Display** after saving the attribute. For instance, if AC $= 1$ mV was just saved and **Change Display** was chosen, a **Change Attribute** dialog box would appear. Since AC is the name and 1 mV the value, choose **Value Only,** and only the 1 mV will be displayed after the sequence **OK-OK.**

The printer symbol on the collector of the transistor is listed as **VPRINT1** under the **SPECIAL.slb** library. When placed on the schematic, it dictates that the ac voltage at that point will be printed in the output file (*.out). Double-clicking on the printer symbol will result in a **PRINT1** dialog box, within which the following choices should be made:

AC $=$ ok.

MAG $=$ ok.

PHASE $=$ ok.

After each entry, be sure to **Save Attr** or the computer will remind you. The above choices can be listed next to the printer symbol on the schematic by simply clicking the **Change Display** option and choosing the **Display Value and Name** for each item.

The transistor is obtained through the sequence **Get New Part** icon-**Libraries-EVAL.slb-Q2N2222-OK-Place & Close.** Since we will want the parameters of the transistor to match those of the example as closely as possible, one must first click on the transistor to put it in the active mode (red) and then choose **Edit-Model-Edit Instance Model (Text).** Next, the beta (**Bf**) is set to 90 and **Is** is set to 2E-15A to result in a base-to-emitter voltage close to 0.7 V. This value of I is the result of numerous runs of the network to find that value of I_s that provided a level of V_{BE} closest to 0.7 V. For the remainder of this text, however, this chosen level of I_s will remain the same. In most cases, it provides the desired results.

VIEWPOINTs have been inserted to display the three dc voltages of interest. Since it has been used recently, **VIEWPOINT** can be found in the scroll listing at the top right of the menu bar rather than by returning to the library listing.

Choosing the **Setup Analysis** icon will result in the **Analysis Setup** dialog box, in which the **AC Sweep** must be chosen because of the applied ac source. Clicking on **AC Sweep** will result in an **AC Sweep and Noise Analysis** dialog box, in which **Linear** is chosen along with **Total Pts: 1**, **Start Freq:** 10kHz, and **End Freq:** 10kHz. The result will be an analysis at only one frequency. Our initial interest will simply be in the magnitude of the quantities of interest and not their shape or appearance. Therefore, we should turn to **Analysis-Probe Setup** and choose **Do not auto-run Probe** to save time getting to the desired results.

Clicking the **Analysis** icon will result in a **PSpiceAD** dialog box that will indicate the **AC Analysis** is finished. Note also the listing of the frequency applied at the bottom of the dialog box. Within this box, if we choose **File** followed by **Examine Output,** we will obtain a lengthy listing of input and output data on the analyzed network. Specific headings are duplicated in Fig. 8.67. Under **Schematics Netlist,** the nodes assigned to the network are revealed. Note that ground is always defined as the 0 node and the assumed node of higher potential listed first. The transistor is listed in the order Collector-Base-Emitter. Under **BJT MODEL PARAMETERS,** the defining parameters of the device are listed with the set values of $I_s = 2E-15A$ and $\beta = 90$. Under **SMALL-SIGNAL BIAS SOLUTION,** the dc levels at the various nodes are revealed, which compare directly with the **VIEWPOINT** values. In particular, note that V_{BE} is exactly 0.7 V.

```
****        CIRCUIT DESCRIPTION

**********************************************************
* Schematics Netlist *


V_V1          $N_0001 0 22
R_R1          $N_0001 $N_0002 56k
R_R4          $N_0002 0  8.2k
R_R2          $N_0001 $N_0003  6.8k
R_R3          $N_0004 0  1.5k
C_C1          $N_0004 0  20u
C_C2          $N_0005 $N_0002 10u
V_V2          $N_0005 0  AC 1m
+SIN 0 1m 10k 0 0 0

.PRINT        AC
+ VM([$N_0003])
+ VP([$N_0003])
Q_Q1          $N_0003 $N_0002 $N_0004 Q2N2222-X

**** RESUMING edc8a.cir ****


****        BJT MODEL PARAMETERS

**********************************************************



              Q2N2222-X
              NPN
        IS    2.000000E-15
        BF    90
        NF    1
        VAF   74.03
        IKF   .2847
        ISE   14.340000E-15
        NE    1.307
        BR    6.092
        NR    1
        RB    10
        RC    1
        CJE   22.010000E-12
        MJE   .377
        CJC   7.306000E-12
        MJC   .3416
        TF    411.100000E-12
        XTF   3
        VTF   1.7
        ITF   .6
        TR    46.910000E-09
        XTB   1.5


****        SMALL SIGNAL BIAS SOLUTION        TEMPERATURE =   27.000 DEG C

**********************************************************


NODE   VOLTAGE    NODE   VOLTAGE    NODE   VOLTAGE    NODE   VOLTAGE

($N_0001)   22.0000                  ($N_0002)   2.6239

($N_0003)   13.4530                  ($N_0004)   1.9244

($N_0005)    0.0000



       VOLTAGE SOURCE CURRENTS
       NAME         CURRENT

       V_V1         -1.603E-03
       V_V2          0.000E+00

       TOTAL POWER DISSIPATION   3.53E-02  WATTS
****        OPERATING POINT INFORMATION        TEMPERATURE =   27.000 DEG C

**********************************************************




**** BIPOLAR JUNCTION TRANSISTORS

NAME        Q_Q1
MODEL       Q2N2222-X
IB          2.60E-05
IC          1.26E-03
VBE         6.99E-01
VBC         -1.08E+01
VCE         1.15E+01
BETADC      4.83E+01
GM          4.84E-02
RPI         1.14E+03
RX          1.00E+01
RO          6.75E+04
CBE         5.78E-11
CBC         2.87E-12
CJS         0.00E+00
BETAAC      5.50E+01
CBX         0.00E+00
FT          1.27E+08

****        AC ANALYSIS        TEMPERATURE =   27.000 DEG C

**********************************************************


    FREQ        VM($N_0003) VP($N_0003)

  1.000E+04    2.961E-01  -1.780E+02
```

Figure 8.67 Output file for the network of Figure 8.66.

The next listing, **OPERATING POINT INFORMATION,** reveals that even though beta of the **BJT MODEL PARAMETERS** listing was set at 90, the operating conditions of the network resulted in a dc beta of 48.3 and an ac beta of 55. Fortunately, however, the voltage-divider configuration is less sensitive to changes in beta in the dc mode, and the dc results are excellent. However, the drop in ac beta had an effect on the resulting level of V_o: 296.1 mV versus the hand-written solution (with $r_o = 50$ kΩ) of 324.3 mV—a 9% difference. The results are certainly close, but probably not as close as one would like. A closer result (within 7%) could be obtained by setting all the parameters of the device except I_s and beta to zero. However, for the moment, the impact of the remaining parameters has been demonstrated, and the results will be accepted as sufficiently close to the hand-written levels. Later in this chapter, an ac model for the transistor will be introduced with results that will be an exact match with the hand-written solution. The phase angle is $-178°$ versus the ideal of $-180°$—a very close match.

A plot of the output waveform can be obtained using the **Probe** option. The sequence **Analysis-Probe Setup-Automatically run Probe after simulation-OK** will result in a **MicroSim Probe** screen when the **Analysis** icon is chosen. However, if we follow this procedure without setting the horizontal scale, we will simply end up with a plot point of 296 mV at a frequency of 10 kHz. The horizontal scale is set by the sequence **Analysis-Setup-Transient** with the **AC Sweep** disabled. Clicking the **Transient** option will result in a **Transient** dialog box, in which a number of choices have to be made based on the waveform to be viewed. The period of the applied signal of 10 kHz is 0.1 ms $= 100$ μs. The **Print Step** option refers to the time interval between printing or plotting the results of the transient analysis. For our example, we will choose 1 μs to provide 100 plot points per cycle. The **Final Time** is the last instant the network's response will be determined. Our choice is 500 μs or 0.5 ms to provide five full cycles of the waveform. The **No-Print Delay** was chosen as 0 since all the capacitors are essentially short circuits at 10 kHz. If we felt there was a transient phase between energizing the network and reaching a steady-state response, the **No-Print Delay** could be used to effectively eliminate this period of time. The last choice of **Step Ceiling** sets a maximum time period between response calculations for the system, which we will set at 1 μs. The time between calculations will be adjusted internally by the software package to ensure sufficient data at times when the response may change faster than usual. However, they will never be separated by a time period greater than that set by the **Step Ceiling.**

After **Simulation,** a **MicroSim Probe** screen will appear showing only the horizontal scale from 0 to 500 μs as specified in the **Transient** dialog box. To obtain a waveform, one can either choose **Trace** on the menu bar or the **Trace** icon (red pattern on a black axis). If the **Trace** on the menu bar is chosen, one must follow with **Add,** and the **Add Traces** dialog box will appear. Using the icon results in the dialog box immediately. Now one must choose the waveform to be displayed from the list of **Simulation Output Variables.** Since we want the collector-to-emitter voltage of the transistor, **V(Q1:c)**—an option under **Alias** names—will be chosen, followed by an **OK.** The result is the waveform of Fig. 8.68, with the waveform riding on the dc level of 13.45 V. The range of the vertical axis was automatically chosen by the computer. Five full cycles of the output waveform are displayed (as we expected) with 100 data points for each cycle. If you would like to see the data points (as shown in Fig. 8.68), simply turn to **Tools-Options-Probe Options** and choose **Mark Data Points.** Click **OK,** and the data points will appear. Using the scale of the graph, the peak-to-peak value of the curve is approximately 13.76 V $-$ 13.16 V $= 0.6$ V $=$ 600 mV, resulting in a peak value of 300 mV. Since a 1-mV signal was applied, the gain is 300, or very close to the values displayed above.

If a comparison is to be made between the input and output voltages on the same graph, the **Add Y-Axis** option under **Plot** can be chosen. After it is triggered, choose the **Add Trace** icon and select **V(Vs:+).** The result is that both waveforms will ap-

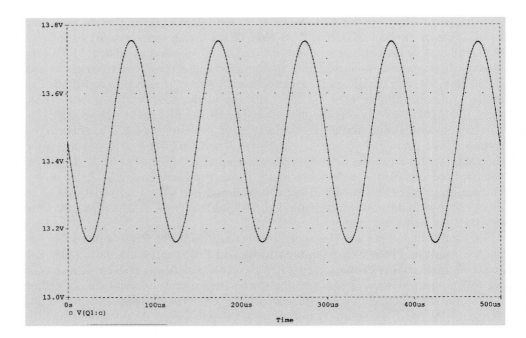

Figure 8.68 Voltage v_c for the network of Figure 8.66.

pear on the same screen, each with their own vertical scale. Labels can be added to the waveforms as shown in Fig. 8.69 using **Tools-Label-Text.** A **Text Label** dialog box will appear, in which the desired text can be entered. Click **OK,** and it can be placed with the mouse in any location on the graph. Lines can also be added with **Tools-Label-Line.** A pencil will appear, which can be used to draw the line with a left-click at the starting point and another click when the line is in place. Each plot can be printed with **File-Print-Copies-OK.**

If two separate graphs are preferred, we can choose the **Plot** option and select **Add Plot** after **V** has been displayed. Upon selection, another graph will appear, waiting for the next choice. The sequence **Trace-Add-V(Vs: +)** will then result in the graphs of Fig. 8.70. The labels **Vs** and **Vc** were added using the **Tools** option. If further operations are to be performed on either graph, the **SEL′ +** defines the active plot.

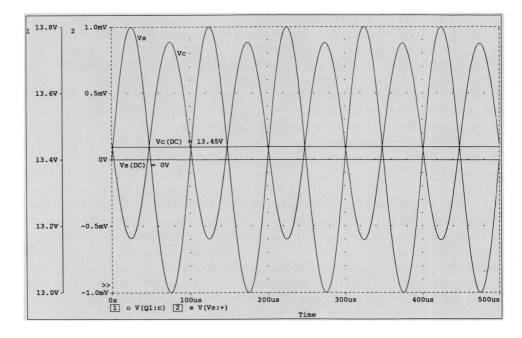

Figure 8.69 The voltages v_c and v_s for the network of Figure 8.66.

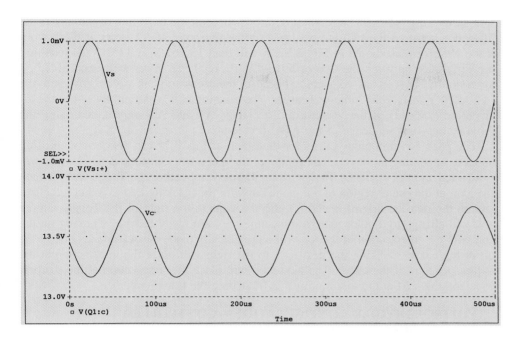

Figure 8.70 Two separate plots of v_c and v_s in Figure 8.66.

The last waveform of this section will demonstrate the use of the **Cursor** option that can be called up using the **Tools** menu choice or the **Cursor Point** icon (having a graph with an arrow drawn from the graph to the vertical axis). The sequence **Tools-Cursor-Display** will result in a line at the dc level of 13.453 V, as shown in the dialog box at the bottom right of the graph of Fig. 8.71. Left-clicking on the mouse once will result in a horizontal and vertical line intersecting at some point on the curve. By clicking on the vertical line and holding it down, the vertical line and corresponding horizontal line (on the graph) can be moved across the waveform. At each point, the vertical and horizontal intersections will appear in the dialog box. If moved to the first peak value, **A1** will be at 13.754 V and 74.825 μs. By right-clicking on the mouse, a second intersection, defined by **A2,** appears, which also has its location registered in the dialog box. These intersecting lines are moved by holding down the

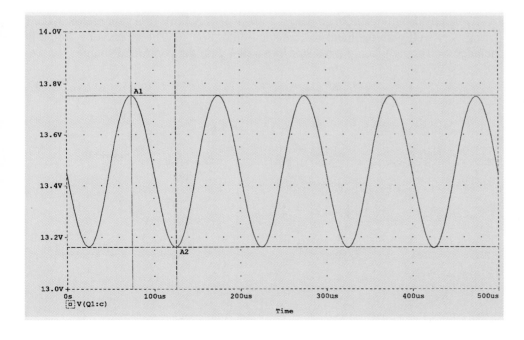

Figure 8.71 Demonstrating the use of cursors to read specific points on a plot.

right side of the mouse. The remaining information on the third line of the box is the difference between the two intersections on the two axis. If **A2** is set at the bottom of the waveform as shown in Fig. 8.71, it will read 13.162 V at 125.17 μs, resulting in a difference between the two of 591.999 mV, or 0.592 V vertically and 50.35 μs horizontally. This is as expected, because the peak-to-peak value matches the 2×0.296 V = 0.592 V obtained earlier. The time interval is essentially 1/4 of the total period (200 μs) of the waveform. The labels **A1** and **A2** were added using the **Tools-Label-Text** sequence or the **ABC** text icon.

The peak and minimum values for the graph of Fig. 8.71 can also be found using the icons appearing in the top right region of the menu bar. Once the desired waveform is obtained and the sequence **Tools-Cursor-Display** applied or the **Toggle cursor** icon (the icon in the center region of the menu bar with the black dashed axis and red curve passing through the origin) is chosen, the six icons to the right of the **Toggle cursor** icon will change to a color pattern indicating they are ready for use. Clicking on the icon with the intersection at the top will automatically place the **A1** intersection at the top of the curve. Clicking the next icon to the right will place the intersection at the bottom (trough) of the curve. The next icon will place the intersection at the steepest slope and the next at the minimum value (matching the trough value).

VOLTAGE-DIVIDER CONFIGURATION—CONTROLLED SOURCE SUBSTITUTION

The results obtained for any analysis using the transistors provided in the software package will always be different from those obtained with an equivalent model that only includes the effect of beta and r_e. This was demonstrated for the network of Fig. 8.66. If a solution is desired that is limited to the approximate model, then the transistor must be represented by a model such as appearing in Fig. 8.72.

For Example 8.2, β is 90, with $\beta r_e = 1.66$ kΩ. The current controlled current source **(CCCS)** is found in the **ANALOG.slb** library as **Part F.** When you click on **F,** the **Description** above will read **Current-controlled current source.** After **OK-Place & Close,** the graphical symbol for the **CCCS** will appear on the screen as shown in Fig. 8.73. Since βr_e does not appear within the basic structure of the **CCCS,** it must be added in series with the controlling current indicated by the arrow on the left. Once the source is in place, a double-click of the symbol will result in a **Part-Name: F** dialog box. The **GAIN** can then be called up and set at 90. Be sure to **Save Attr** and **Change Display** to display **Both name and value.** Nothing else need be set for the **CCCS.**

An **Analysis** will result in an output voltage of 368.3 mV or a gain of 368.3, comparing very well with the hand-written solution of 368.76. The effects of r_o could be included by simply placing a resistor in parallel with the controlled source.

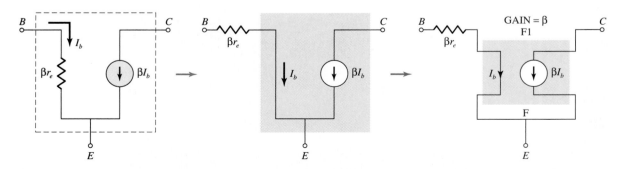

Figure 8.72 Using a controlled source to represent the transistor of Figure 8.66.

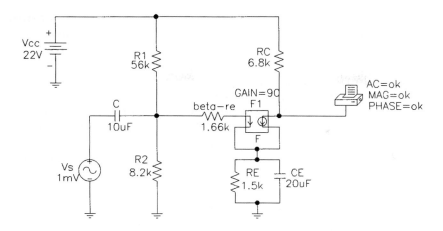

Figure 8.73 Substituting the controlled source of Figure 8.72 for the transistor of Figure 8.66.

Electronics Workbench

Since the collector feedback configuration generated the most complex equations for the various parameters of a BJT network, it seems appropriate that Electronics Workbench be used to verify the conclusions. The network of Example 8.9 will appear as shown in Fig. 8.74 using the **Virtual** transistor from the **Transistor** family toolbar. After the transistor is chosen, it should be placed in the general area of the screen where it will be used. Then double-click on the symbol, and the **BJT_NPN_VIR-TUAL** dialog box will appear in which **Edit Model** can be selected. The **Edit Model** dialog box will appear in which **BF** can be changed to 200 from the default value of 100. Then selecting **Change Part Model** will bring you back to the **BJT_NPN_VIR-TUAL** dialog box. Select **OK,** and you will return to the general schematic with the modified transistor parameter. The ac voltage source is obtained under the **Sources** family toolbar and placed in the desired location. Next a double-click of the source will result in the **AC Voltage** dialog box under which **Value** can be selected. For our purposes it will be set to 1 mV with the **Frequency** left on the default value of 1 kHz. Next **Label** is selected, and the **Reference ID** changed to **Vi.** The label **Bf = 200** is added through **Edit-Place Text.**

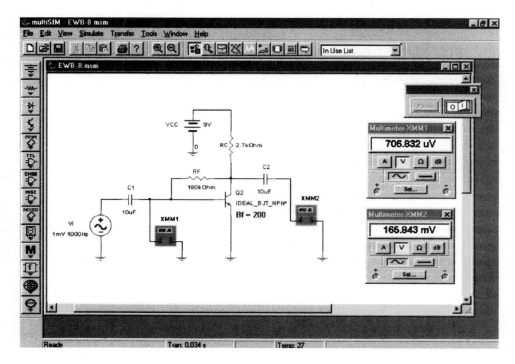

Figure 8.74 Network of Example 8.9 redrawn using EWB.

Finally, the meters are added as shown by using the **Simulate-Instruments-Multimeter** sequence or by simply selecting the meter on the **Systems** toolbar to obtain the **Instruments** toolbar. Since this is an ac analysis, the multimeters are set on the ac mode with **V** selected for the voltage levels desired. Turning the simulate switch "on" (the 1 position) will result in the meter readings of Fig. 8.74. The gain of 165.843 mV/706.832 μV = 234.6 is very close to the 240 obtained in Example 8.9 using the lengthy mathematical analysis. The multimeter **XMM1** reads 0.707 mV rather than the applied 1 mV because the meter reads effective values and Multisim defines its ac sources with the peak value. The difference in reading between the meter and the actual value of 707.1 μV is due to the drop across the capacitor at the frequency of 1 kHz.

PROBLEMS

§ 8.2 Common-Emitter Fixed-Bias Configuration

1. For the network of Fig. 8.75:
   (a) Determine Z_i and Z_o.
   (b) Find A_v and A_i.
   (c) Repeat part (a) with $r_o = 20$ kΩ.
   (d) Repeat part (b) with $r_o = 20$ kΩ.

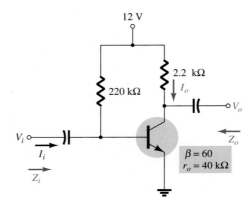

Figure 8.75   Problems 1 and 21

2. For the network of Fig. 8.76, determine V_{CC} for a voltage gain of $A_v = -200$.

∗ **3.** For the network of Fig. 8.77:
   (a) Calculate I_B, I_C, and r_e.
   (b) Determine Z_i and Z_o.
   (c) Calculate A_v and A_i.
   (d) Determine the effect of $r_o = 30$ kΩ on A_v and A_i.

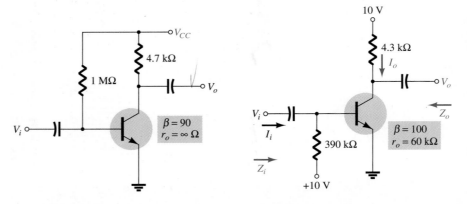

Figure 8.76   Problem 2   **Figure 8.77**   Problem 3

Chapter 8   BJT Small-Signal Analysis

§ 8.3 Voltage-Divider Bias

4. For the network of Fig. 8.78:
  (a) Determine r_e.
  (b) Calculate Z_i and Z_o.
  (c) Find A_v and A_i.
  (d) Repeat parts (b) and (c) with $r_o = 25$ kΩ.

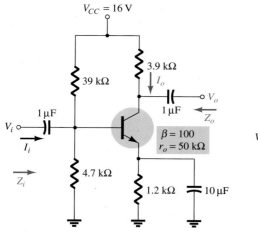

Figure 8.78 Problem 4

5. Determine V_{CC} for the network of Fig. 8.79 if $A_v = -160$ and $r_o = 100$ kΩ.

6. For the network of Fig. 8.80:
  (a) Determine r_e.
  (b) Calculate V_B and V_C.
  (c) Determine Z_i and $A_v = V_o/V_i$.

§ 8.4 CE Emitter-Bias Configuration

7. For the network of Fig. 8.81:
  (a) Determine r_e.
  (b) Find Z_i and Z_o.
  (c) Calculate A_v and A_i.
  (d) Repeat parts (b) and (c) with $r_o = 20$ kΩ.

Figure 8.79 Problem 5

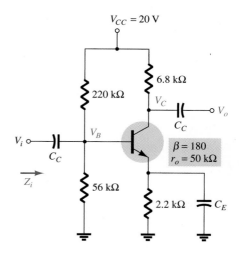

Figure 8.80 Problem 6

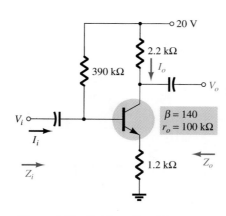

Figure 8.81 Problems 7 and 9

8. For the network of Fig. 8.82, determine R_E and R_B if $A_v = -10$ and $r_e = 3.8\ \Omega$. Assume that $Z_b = \beta R_E$.

9. Repeat Problem 7 with R_E bypassed. Compare results.

* **10.** For the network of Fig. 8.83:
  (a) Determine r_e.
  (b) Find Z_i and A_v.
  (c) Calculate A_i.

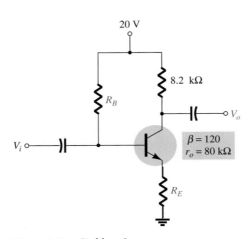

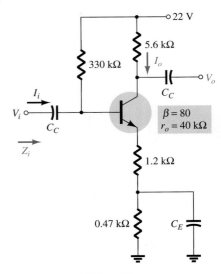

Figure 8.82 Problem 8   **Figure 8.83** Problem 10

§ 8.5 Emitter-Follower Configuration

11. For the network of Fig. 8.84:
  (a) Determine r_e and βr_e.
  (b) Find Z_i and Z_o.
  (c) Calculate A_v and A_i.

* **12.** For the network of Fig. 8.85:
  (a) Determine Z_i and Z_o.
  (b) Find A_v.
  (c) Calculate V_o if $V_i = 1$ mV.

* **13.** For the network of Fig. 8.86:
  (a) Calculate I_B and I_C.
  (b) Determine r_e.
  (c) Determine Z_i and Z_o.
  (d) Find A_v and A_i.

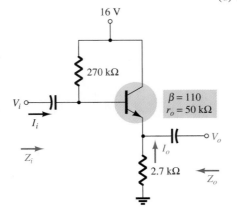

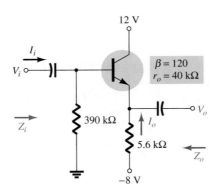

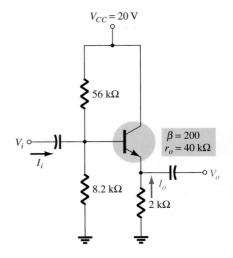

Figure 8.84 Problem 11   **Figure 8.85** Problem 12   **Figure 8.86** Problem 13

§ 8.6 Common-Base Configuration

14. For the common-base configuration of Fig. 8.87:
   (a) Determine r_e.
   (b) Find Z_i and Z_o.
   (c) Calculate A_v and A_i.

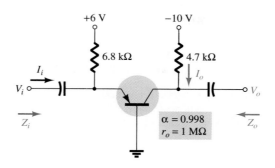

Figure 8.87 Problem 14 ·

* **15.** For the network of Fig. 8.88, determine A_v and A_i.

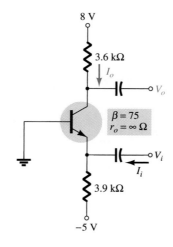

Figure 8.88 Problem 15

§ 8.7 Collector Feedback Configuration

16. For the collector FB configuration of Fig. 8.89:
   (a) Determine r_e.
   (b) Find Z_i and Z_o.
   (c) Calculate A_v and A_i.

* **17.** Given $r_e = 10\ \Omega$, $\beta = 200$, $A_v = -160$, and $A_i = 19$ for the network of Fig. 8.90, determine R_C, R_F, and V_{CC}.

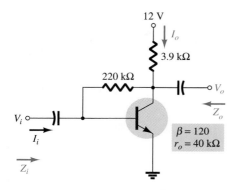

Figure 8.89 Problem 16

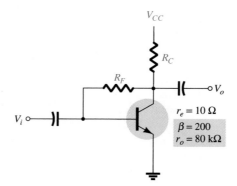

Figure 8.90 Problem 17

* **18.** For the network of Fig. 8.31:
    (a) Derive the approximate equation for A_v.
    (b) Derive the approximate equation for A_i.
    (c) Derive the approximate equations for Z_i and Z_o.
    (d) Given $R_C = 2.2$ kΩ, $R_F = 120$ kΩ, $R_E = 1.2$ kΩ, $\beta = 90$, and $V_{CC} = 10$ V, calculate the magnitudes of A_v, A_i, Z_i, and Z_o using the equations of parts (a) through (c).

§ 8.8 Collector DC Feedback Configuration

19. For the network of Fig. 8.91:
    (a) Determine Z_i and Z_o.
    (b) Find A_v and A_i.

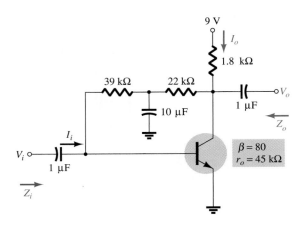

Figure 8.91   Problem 19

§ 8.9 Approximate Hybrid Equivalent Circuit

20. (a) Given $\beta = 120$, $r_e = 4.5$ Ω, and $r_o = 40$ kΩ, sketch the approximate hybrid equivalent circuit.
    (b) Given $h_{ie} = 1$ kΩ, $h_{re} = 2 \times 10^{-4}$, $h_{fe} = 90$, and $h_{oe} = 20$ μS, sketch the r_e model.

21. For the network of Problem 1:
    (a) Determine r_e.
    (b) Find h_{fe} and h_{ie}.
    (c) Find Z_i and Z_o using the hybrid parameters.
    (d) Calculate A_v and A_i using the hybrid parameters.
    (e) Determine Z_i and Z_o if $h_{oe} = 50$ μS.
    (f) Determine A_v and A_i if $h_{oe} = 50$ μS.
    (g) Compare the solutions above with those of Problem 1. (Note: The solutions are available in Appendix E if Problem 1 was not performed.)

22. For the network of Fig. 8.92:
    (a) Determine Z_i and Z_o.
    (b) Calculate A_v and A_i.
    (c) Determine r_e and compare βr_e to h_{ie}.

* **23.** For the common-base network of Fig. 8.93:
    (a) Determine Z_i and Z_o.
    (b) Calculate A_v and A_i.
    (c) Determine α, β, r_e, and r_o.

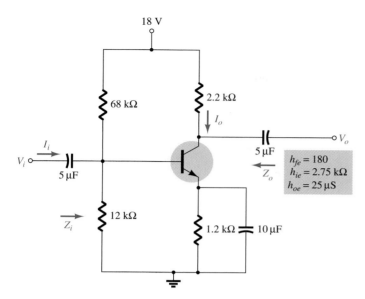

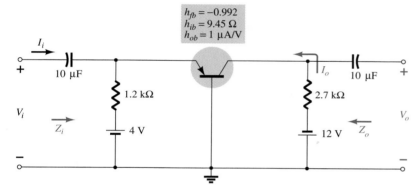

Figure 8.92 Problems 22 and 24

Figure 8.93 Problem 23

§ 8.10 Complete Hybrid Equivalent Model

* **24.** Repeat parts (a) and (b) of Problem 22 with $h_{re} = 2 \times 10^{-4}$ and compare results.

* **25.** For the network of Fig. 8.94, determine:
  (a) Z_i.
  (b) A_v.
  (c) $A_i = I_o/I_i$.
  (d) Z_o.

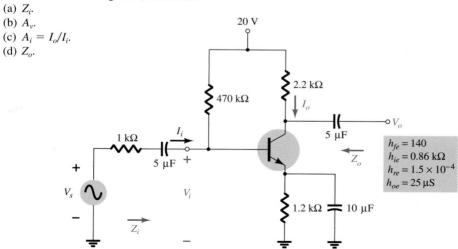

Figure 8.94 Problem 25

* **26.** For the common-base amplifier of Fig. 8.95, determine:
   (a) Z_i.
   (b) A_i.
   (c) A_v.
   (d) Z_o.

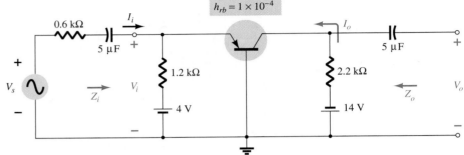

$h_{ib} = 9.45\ \Omega$
$h_{fb} = -0.997$
$h_{ob} = 0.5\ \mu A/V$
$h_{rb} = 1 \times 10^{-4}$

Figure 8.95   Problem 26

§ 8.12 Troubleshooting

* **27.** Given the network of Fig. 8.96:
   (a) Determine if the system is operating properly based on the voltage-divider bias levels and expected waveforms for v_o and v_E.
   (b) Determine the reason for the dc levels obtained and why the waveform for v_o was obtained.

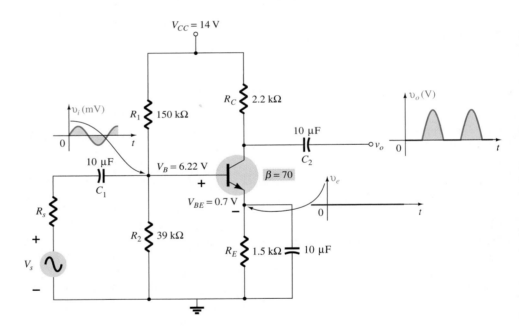

Figure 8.96   Problem 27

§ 8.15 Computer Analysis

28. Using PSpice Windows, determine the voltage gain for the network of Fig. 8.6. Use Probe to display the input and output waveforms.

29. Using PSpice Windows, determine the voltage gain for the network of Fig. 8.13. Use Probe to display the input and output waveforms.

30. Using PSpice Windows, determine the voltage gain for the network of Fig. 8.26. Use Probe to display the input and output waveforms.

31. Using Electronics Workbench, determine the voltage gain for the network of Fig. 8.9.

32. Using Electronics Workbench, determine the voltage gain for the network of Fig. 8.21.

*Please Note: Asterisks indicate more difficult problems.

FET Small-Signal Analysis

9.1 INTRODUCTION

Field-effect transistor amplifiers provide an excellent voltage gain with the added feature of a high input impedance. They are also considered low-power consumption configurations with good frequency range and minimal size and weight. Both JFET and depletion MOSFET devices can be used to design amplifiers having similar voltage gains. The depletion MOSFET circuit, however, has a much higher input impedance than a similar JFET configuration.

While a BJT device controls a large output (collector) current by means of a relatively small input (base) current, the FET device controls an output (drain) current by means of a small input (gate-voltage) voltage. In general, therefore, the BJT is a *current-controlled* device and the FET is a *voltage-controlled* device. In both cases, however, note that the output current is the controlled variable. Because of the high input characteristic of FETs, the ac equivalent model is somewhat simpler than that employed for BJTs. While the BJT had an amplification factor β (beta), the FET has a transconductance factor, g_m.

The FET can be used as a linear amplifier or as a digital device in logic circuits. In fact, the enhancement MOSFET is quite popular in digital circuitry, especially in CMOS circuits that require very low power consumption. FET devices are also widely used in high-frequency applications and in buffering (interfacing) applications. Table 9.1, located in Section 9.13, provides a summary of FET small-signal amplifier circuits and related formulas.

While the common-source configuration is the most popular providing an inverted, amplified signal, one also finds common-drain (source-follower) circuits providing unity gain with no inversion and common-gate circuits providing gain with no inversion. As with BJT amplifiers, the important circuit features described in this chapter include voltage gain, input impedance, and output impedance. Due to the very high input impedance, the input current is generally assumed to be 0 μA and the current gain is an undefined quantity. While the voltage gain of an FET amplifier is generally less than that obtained using a BJT amplifier, the FET amplifier provides a much higher input impedance than that of a BJT configuration. Output impedance values are comparable for both BJT and FET circuits.

FET ac amplifier networks can also be analyzed using computer software. Using PSpice or Electronics Workbench, one can perform a dc analysis to obtain the circuit bias conditions and an ac analysis to determine the small-signal voltage gain. Using PSpice transistor models, one can analyze the circuit using specific transistor models. On the other hand, one can develop a program using a language such as BASIC that can perform both the dc and ac analyses and provide the results in a very special format.

9.2 FET SMALL-SIGNAL MODEL

The ac analysis of an FET configuration requires that a small-signal ac model for the FET be developed. A major component of the ac model will reflect the fact that an ac voltage applied to the input gate-to-source terminals will control the level of current from drain to source.

The gate-to-source voltage controls the drain-to-source (channel) current of an FET.

Recall from Chapter 6 that a dc gate-to-source voltage controlled the level of dc drain current through a relationship known as Shockley's equation: $I_D = I_{DSS}(1 - V_{GS}/V_P)^2$. The *change* in collector current that will result from a *change* in gate-to-source voltage can be determined using the transconductance factor g_m in the following manner:

$$\Delta I_D = g_m \, \Delta V_{GS} \qquad (9.1)$$

The prefix *trans-* in the terminology applied to g_m reveals that it establishes a relationship between an output and input quantity. The root word *conductance* was chosen because g_m is determined by a voltage-to-current ratio similar to the ratio that defines the conductance of a resistor $G = 1/R = I/V$.

Solving for g_m in Eq. (9.1), we have:

$$g_m = \frac{\Delta I_D}{\Delta V_{GS}} \qquad (9.2)$$

Graphical Determination of g_m

If we now examine the transfer characteristics of Fig. 9.1, we find that g_m is actually the slope of the characteristics at the point of operation. That is,

$$g_m = m = \frac{\Delta y}{\Delta x} = \frac{\Delta I_D}{\Delta V_{GS}} \qquad (9.3)$$

Following the curvature of the transfer characteristics, it is reasonably clear that the slope and, therefore, g_m increase as we progress from V_P to I_{DSS}. Or, in other words, as V_{GS} approaches 0 V, the magnitude of g_m increases.

Equation (9.2) reveals that g_m can be determined at any Q-point on the transfer characteristics by simply choosing a finite increment in V_{GS} (or in I_D) about the Q-point and then finding the corresponding change in I_D (or V_{GS}, respectively). The resulting changes in each quantity are then substituted in Eq. (9.2) to determine g_m.

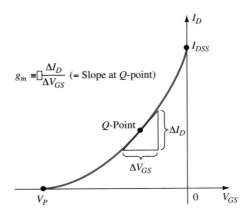

Figure 9.1 Definition of g_m using transfer characteristic.

Determine the magnitude of g_m for a JFET with $I_{DSS} = 8$ mA and $V_P = -4$ V at the following dc bias points:

EXAMPLE 9.1

(a) $V_{GS} = -0.5$ V.
(b) $V_{GS} = -1.5$ V.
(c) $V_{GS} = -2.5$ V.

Solution

The transfer characteristics are generated as Fig. 9.2 using the procedure defined in Chapter 6. Each operating point is then identified and a tangent line is drawn at each point to best reflect the slope of the transfer curve in this region. An appropriate increment is then chosen for V_{GS} to reflect a variation to either side of each Q-point. Equation (9.2) is then applied to determine g_m.

(a) $g_m = \dfrac{\Delta I_D}{\Delta V_{GS}} \cong \dfrac{2.1 \text{ mA}}{0.6 \text{ V}} = \textbf{3.5 mS}$

(b) $g_m = \dfrac{\Delta I_D}{\Delta V_{GS}} \cong \dfrac{1.8 \text{ mA}}{0.7 \text{ V}} \cong \textbf{2.57 mS}$

(c) $g_m = \dfrac{\Delta I_D}{\Delta V_{GS}} = \dfrac{1.5 \text{ mA}}{1.0 \text{ V}} = \textbf{1.5 mS}$

Note the decrease in g_m as V_{GS} approaches V_P.

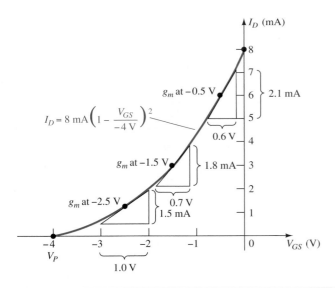

Figure 9.2 Calculating g_m at various bias points.

Mathematical Definition of g_m

The graphical procedure just described is limited by the accuracy of the transfer plot and the care with which the changes in each quantity can be determined. Naturally, the larger the graph the better the accuracy, but this can then become a cumbersome problem. An alternative approach to determining g_m employs the approach used to find the ac resistance of a diode in Chapter 1, where it was stated that:

> *The derivative of a function at a point is equal to the slope of the tangent line drawn at that point.*

If we therefore take the derivative of I_D with respect to V_{GS} (differential calculus) using Shockley's equation, an equation for g_m can be derived as follows:

$$g_m = \frac{\Delta I_D}{\Delta V_{GS}}\bigg|_{Q\text{-pt.}} = \frac{dI_D}{dV_{GS}}\bigg|_{Q\text{-pt.}} = \frac{d}{dV_{GS}}\left[I_{DSS}\left(1 - \frac{V_{GS}}{V_P}\right)^2\right]$$

$$= I_{DSS}\frac{d}{dV_{GS}}\left(1 - \frac{V_{GS}}{V_P}\right)^2 = 2I_{DSS}\left[1 - \frac{V_{GS}}{V_P}\right]\frac{d}{dV_{GS}}\left(1 - \frac{V_{GS}}{V_P}\right)$$

$$= 2I_{DSS}\left[1 - \frac{V_{GS}}{V_P}\right]\left[\frac{d}{dV_{GS}}(1) - \frac{1}{V_P}\frac{dV_{GS}}{dV_{GS}}\right] = 2I_{DSS}\left[1 - \frac{V_{GS}}{V_P}\right]\left[0 - \frac{1}{V_P}\right]$$

and
$$\boxed{g_m = \frac{2I_{DSS}}{|V_P|}\left[1 - \frac{V_{GS}}{V_P}\right]} \qquad (9.4)$$

where $|V_P|$ denotes magnitude only to ensure a positive value for g_m.

It was mentioned earlier that the slope of the transfer curve is a maximum at $V_{GS} = 0$ V. Plugging in $V_{GS} = 0$ V into Eq. (9.4) will result in the following equation for the maximum value of g_m for a JFET in which I_{DSS} and V_P have been specified:

$$g_m = \frac{2I_{DSS}}{|V_P|}\left[1 - \frac{0}{V_P}\right]$$

and
$$\boxed{g_{m0} = \frac{2I_{DSS}}{|V_P|}} \qquad (9.5)$$

where the added subscript 0 reminds us that it is the value of g_m when $V_{GS} = 0$ V. Equation (9.4) then becomes

$$\boxed{g_m = g_{m0}\left[1 - \frac{V_{GS}}{V_P}\right]} \qquad (9.6)$$

EXAMPLE 9.2

For the JFET having the transfer characteristics of Example 9.1:
(a) Find the maximum value of g_m.
(b) Find the value of g_m at each operating point of Example 9.1 using Eq. (9.6) and compare with the graphical results.

Solution

(a) $g_{m0} = \dfrac{2I_{DSS}}{|V_P|} = \dfrac{2(8 \text{ mA})}{4 \text{ V}} = \textbf{4 mS}$ \qquad (maximum possible value of g_m)

(b) At $V_{GS} = -0.5$ V,

$$g_m = g_{m0}\left[1 - \frac{V_{GS}}{V_P}\right] = 4 \text{ mS}\left[1 - \frac{-0.5 \text{ V}}{-4 \text{ V}}\right] = \textbf{3.5 mS} \qquad \text{(versus 3.5 mS graphically)}$$

At $V_{GS} = -1.5$ V,

$$g_m = g_{m0}\left[1 - \frac{V_{GS}}{V_P}\right] = 4 \text{ mS}\left[1 - \frac{-1.5 \text{ V}}{-4 \text{ V}}\right] = \textbf{2.5 mS} \qquad \text{(versus 2.57 mS graphically)}$$

At $V_{GS} = -2.5$ V,

$$g_m = g_{m0}\left[1 - \frac{V_{GS}}{V_P}\right] = 4 \text{ mS}\left[1 - \frac{-2.5 \text{ V}}{-4 \text{ V}}\right] = \textbf{1.5 mS} \qquad \text{(versus 1.5 mS graphically)}$$

The results of Example 9.2 are certainly sufficiently close to validate Eq. (9.4) through (9.6) for future use when g_m is required.

On specification sheets, g_m is provided as y_{fs} where y indicates it is part of an admittance equivalent circuit. The f signifies *forward* transfer parameter, and the s reveals that it is connected to the source terminal.

In equation form,

$$g_m = y_{fs}$$ (9.7)

For the JFET of Fig. 5.18, y_{fs} ranges from 1000 to 5000 μS or 1 to 5 mS.

Plotting g_m vs. V_{GS}

Since the factor $\left(1 - \dfrac{V_{GS}}{V_P}\right)$ of Eq. (9.6) is less than 1 for any value of V_{GS} other than

0 V, the magnitude of g_m will decrease as V_{GS} approaches V_P and the ratio $\dfrac{V_{GS}}{V_P}$

increases in magnitude. At $V_{GS} = V_P$, $g_m = g_{m0}(1 - 1) = 0$. Equation (9.6) defines a straight line with a minimum value of 0 and a maximum value of g_m as shown by the plot of Fig. 9.3.

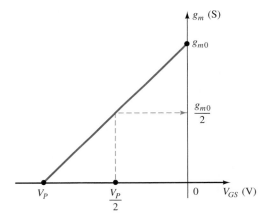

Figure 9.3 Plot of g_m vs. V_{GS}.

Figure 9.3 also reveals that when V_{GS} is one-half the pinch-off value, g_m will be one-half the maximum value.

Plot g_m vs. V_{GS} for the JFET of Examples 9.1 and 9.2.

EXAMPLE 9.3

Solution

Note Fig. 9.4.

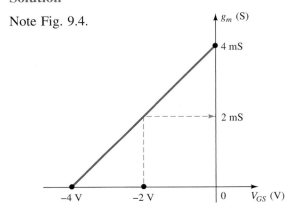

Figure 9.4 Plot of g_m vs. V_{GS} for a JFET with $I_{DSS} = 8$ mA and $V_P = -4$ V.

Impact of I_D on g_m

A mathematical relationship between g_m and the dc bias current I_D can be derived by noting that Shockley's equation can be written in the following form:

$$1 - \frac{V_{GS}}{V_P} = \sqrt{\frac{I_D}{I_{DSS}}} \qquad (9.8)$$

Substituting Eq. (9.8) into Eq. (9.6) will result in

$$g_m = g_{m0}\left(1 - \frac{V_{GS}}{V_P}\right) = g_{m0}\sqrt{\frac{I_D}{I_{DSS}}} \qquad (9.9)$$

Using Eq. (9.9) to determine g_m for a few specific values of I_D, the results are

(a) If $I_D = I_{DSS}$,

$$g_m = g_{m0}\sqrt{\frac{I_{DSS}}{I_{DSS}}} = \mathbf{g_{m0}}$$

(b) If $I_D = I_{DSS}/2$,

$$g_m = g_{m0}\sqrt{\frac{I_{DSS}/2}{I_{DSS}}} = \mathbf{0.707 g_{m0}}$$

(c) If $I_D = I_{DSS}/4$,

$$g_m = g_{m0}\sqrt{\frac{I_{DSS}/4}{I_{DSS}}} = \frac{g_{m0}}{2} = \mathbf{0.5 g_{m0}}$$

EXAMPLE 9.4

Plot g_m vs. I_D for the JFET of Examples 9.1 through 9.3.

Solution

See Fig. 9.5.

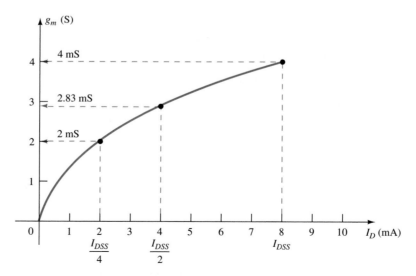

Figure 9.5 Plot of g_m vs. I_D for a JFET with $I_{DSS} = 8$ mA and $V_{GS} = -4$ V.

The plots of Examples 9.3 and 9.4 clearly reveal that the highest values of g_m are obtained when V_{GS} approaches 0 V and I_D its maximum value of I_{DSS}.

FET Input Impedance Z_i

The input impedance of all commercially available FETs is sufficiently large to assume that the input terminals approximate an open circuit. In equation form,

$$Z_i \, (\text{FET}) = \infty \ \Omega \qquad (9.10)$$

For a JFET a practical value of $10^9 \ \Omega$ (1000 MΩ) is typical, while a value of 10^{12} to $10^{15} \ \Omega$ is typical for MOSFETs.

FET Output Impedance Z_o

The output impedance of FETs is similar in magnitude to that of conventional BJTs. On FET specification sheets, the output impedance will typically appear as y_{os} with the units of μS. The parameter y_{os} is a component of an *admittance equivalent circuit,* with the subscript o signifying an *o*utput network parameter and s the terminal (source) to which it is attached in the model. For the JFET of Fig. 5.18, y_{os} has a range of 10 to 50 μS or 20 kΩ ($R = 1/G = 1/50 \ \mu S$) to 100 kΩ ($R = 1/G = 1/10 \ \mu S$).

In equation form,

$$Z_o \, (\text{FET}) = r_d = \frac{1}{y_{os}} \qquad (9.11)$$

The output impedance is defined on the characteristics of Fig. 9.6 as the slope of the horizontal characteristic curve at the point of operation. The more horizontal the curve, the greater the output impedance. If perfectly horizontal, the ideal situation is on hand with the output impedance being infinite (an open circuit)—an often applied approximation.

In equation form,

$$r_d = \left. \frac{\Delta V_{DS}}{\Delta I_D} \right|_{V_{GS} = \text{constant}} \qquad (9.12)$$

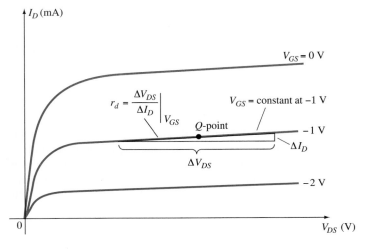

Figure 9.6 Definition of r_d using FET drain characteristics.

Note the requirement when applying Eq. (9.12) that the voltage V_{GS} remain constant when r_d is determined. This is accomplished by drawing a straight line approximating the V_{GS} line at the point of operation. A ΔV_{DS} or ΔI_D is then chosen and the other quantity measured off for use in the equation.

EXAMPLE 9.5

Determine the output impedance for the FET of Fig. 9.7 for $V_{GS} = 0$ V and $V_{GS} = -2$ V at $V_{DS} = 8$ V.

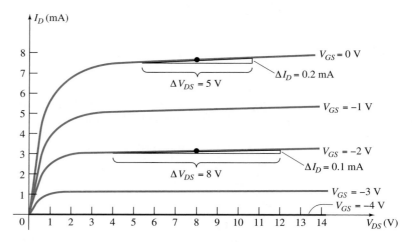

Figure 9.7 Drain characteristics used to calculate r_d in Example 9.5.

Solution

For $V_{GS} = 0$ V, a tangent line is drawn and ΔV_{DS} is chosen as 5 V, resulting in a ΔI_D of 0.2 mA. Substituting into Eq. (9.12),

$$r_d = \left.\frac{\Delta V_{DS}}{\Delta I_D}\right|_{V_{GS} = 0\,\text{V}} = \frac{5\,\text{V}}{0.2\,\text{mA}} = \mathbf{25\ k\Omega}$$

For $V_{GS} = -2$ V, a tangent line is drawn and ΔV_{DS} is chosen as 8 V, resulting in a ΔI_D of 0.1 mA. Substituting into Eq. (9.12),

$$r_d = \left.\frac{\Delta V_{DS}}{\Delta I_D}\right|_{V_{GS} = -2\,\text{V}} = \frac{8\,\text{V}}{0.1\,\text{mA}} = \mathbf{80\ k\Omega}$$

revealing that r_d does change from one operating region to another, with lower values typically occurring at lower levels of V_{GS} (closer to 0 V).

FET AC Equivalent Circuit

Now that the important parameters of an ac equivalent circuit have been introduced and discussed, a model for the FET transistor in the ac domain can be constructed. The control of I_d by V_{gs} is included as a current source $g_m V_{gs}$ connected from drain to source as shown in Fig. 9.8. The current source has its arrow pointing from drain to source to establish a 180° phase shift between output and input voltages as will occur in actual operation.

The input impedance is represented by the open circuit at the input terminals and the output impedance by the resistor r_d from drain to source. Note that the gate to source voltage is now represented by V_{gs} (lower-case subscripts) to distinguish it from

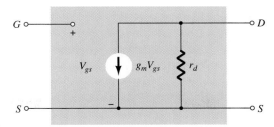

Figure 9.8 FET ac equivalent circuit.

dc levels. In addition, take note of the fact that the source is common to both input and output circuits while the gate and drain terminals are only in "touch" through the controlled current source $g_m V_{gs}$.

In situations where r_d is ignored (assumed sufficiently large to other elements of the network to be approximated by an open circuit), the equivalent circuit is simply a current source whose magnitude is controlled by the signal V_{gs} and parameter g_m—clearly a voltage-controlled device.

Given $y_{fs} = 3.8$ mS and $y_{os} = 20$ μS, sketch the FET ac equivalent model. *EXAMPLE 9.6*

Solution

$$g_m = y_{fs} = 3.8 \text{ mS} \quad \text{and} \quad r_d = \frac{1}{y_{os}} = \frac{1}{20 \text{ }\mu\text{S}} = 50 \text{ k}\Omega$$

resulting in the ac equivalent model of Fig. 9.9.

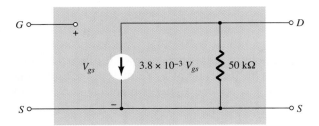

Figure 9.9 FET ac equivalent model for Example 9.6.

9.3 JFET FIXED-BIAS CONFIGURATION

Now that the FET equivalent circuit has been defined, a number of fundamental FET small-signal configurations will be investigated. The approach will parallel the ac analysis of BJT amplifiers with a determination of the important parameters of Z_i, Z_o, and A_v for each configuration.

The *fixed-bias* configuration of Fig. 9.10 includes the coupling capacitors C_1 and C_2 that isolate the dc biasing arrangement from the applied signal and load; they act as short-circuit equivalents for the ac analysis.

Once the level of g_m and r_d are determined from the dc biasing arrangement, specification sheet, or characteristics, the ac equivalent model can be substituted

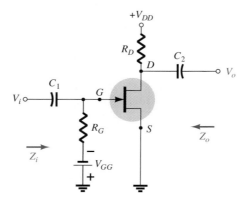

Figure 9.10   JFET fixed-bias configuration.

between the appropriate terminals as shown in Fig. 9.11. Note that both capacitors have the short-circuit equivalent because the reactance $X_C = 1/(2\pi fC)$ is sufficiently small compared to other impedance levels of the network, and the dc batteries V_{GG} and V_{DD} are set to zero volts by a short-circuit equivalent.

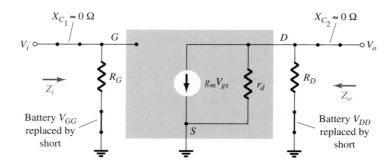

Figure 9.11   Substituting the JFET ac equivalent circuit unit into the network of Fig. 9.10.

The network of Fig. 9.11 is then carefully redrawn as shown in Fig. 9.12. Note the defined polarity of V_{gs}, which defines the direction of $g_m V_{gs}$. If V_{gs} is negative, the direction of the current source reverses. The applied signal is represented by V_i and the output signal across R_D by V_o.

   Z_i: Figure 9.12 clearly reveals that

$$\boxed{Z_i = R_G} \tag{9.13}$$

because of the open-circuit equivalence at the input terminals of the JFET.

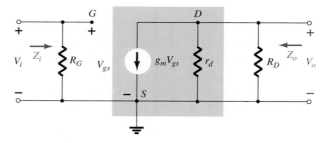

Figure 9.12   Redrawn network of Fig. 9.11.

Z_o: Setting $V_i = 0$ V as required by the definition of Z_o will establish V_{gs} as 0 V also. The result is $g_m V_{gs} = 0$ mA, and the current source can be replaced by an open-circuit equivalent as shown in Fig. 9.13. The output impedance is

$$Z_o = R_D \| r_d \qquad (9.14)$$

If the resistance r_d is sufficiently large (at least 10:1) compared to R_D, the approximation $r_d \| R_D \cong R_D$ can often be applied and

$$Z_o \cong R_D \qquad (9.15)$$
$$\scriptstyle r_d \geq 10R_D$$

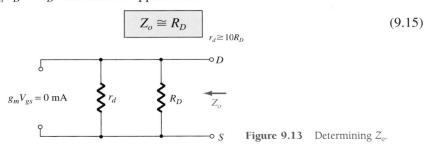

Figure 9.13   Determining Z_o.

A_v: Solving for V_o in Fig. 9.12, we find

$$V_o = -g_m V_{gs}(r_d \| R_D)$$

but

$$V_{gs} = V_i$$

and

$$V_o = -g_m V_i(r_d \| R_D)$$

so that

$$A_v = \frac{V_o}{V_i} = -g_m(r_d \| R_D) \qquad (9.16)$$

If $r_d \geq 10R_D$:

$$A_v = \frac{V_o}{V_i} = -g_m R_D \qquad (9.17)$$
$$\scriptstyle r_d \geq 10R_D$$

Phase Relationship: The negative sign in the resulting equation for A_v clearly reveals a phase shift of 180° between input and output voltages.

The fixed-bias configuration of Example 6.1 had an operating point defined by $V_{GS_Q} = -2$ V and $I_{D_Q} = 5.625$ mA, with $I_{DSS} = 10$ mA and $V_P = -8$ V. The network is redrawn as Fig. 9.14 with an applied signal V_i. The value of y_{os} is provided as 40 μS.
(a) Determine g_m.
(b) Find r_d.
(c) Determine Z_i.
(d) Calculate Z_o.
(e) Determine the voltage gain A_v.
(f) Determine A_v ignoring the effects of r_d.

EXAMPLE 9.7

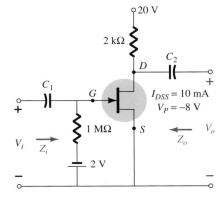

Figure 9.14   JFET configuration for Example 9.7.

Solution

(a) $g_{m0} = \dfrac{2I_{DSS}}{|V_P|} = \dfrac{2(10 \text{ mA})}{8 \text{ V}} = 2.5 \text{ mS}$

$g_m = g_{m0}\left(1 - \dfrac{V_{GS_Q}}{V_P}\right) = 2.5 \text{ mS}\left(1 - \dfrac{(-2 \text{ V})}{(-8 \text{ V})}\right) = \textbf{1.88 mS}$

(b) $r_d = \dfrac{1}{y_{os}} = \dfrac{1}{40 \ \mu\text{S}} = \textbf{25 k}\boldsymbol{\Omega}$

(c) $Z_i = R_G = \textbf{1 M}\boldsymbol{\Omega}$

(d) $Z_o = R_D \| r_d = 2 \text{ k}\Omega \| 25 \text{ k}\Omega = \textbf{1.85 k}\boldsymbol{\Omega}$

(e) $A_v = -g_m(R_D \| r_d) = -(1.88 \text{ mS})(1.85 \text{ k}\Omega)$

$\qquad = \textbf{-3.48}$

(f) $A_v = -g_m R_D = -(1.88 \text{ mS})(2 \text{ k}\Omega) = \textbf{-3.76}$

As demonstrated in part (f), a ratio of $25 \text{ k}\Omega : 2 \text{ k}\Omega = 12.5 : 1$ between r_d and R_D resulted in a difference of 8% in solution.

9.4 JFET SELF-BIAS CONFIGURATION

Bypassed R_S

The fixed-bias configuration has the distinct disadvantage of requiring two dc voltage sources. The *self-bias* configuration of Fig. 9.15 requires only one dc supply to establish the desired operating point.

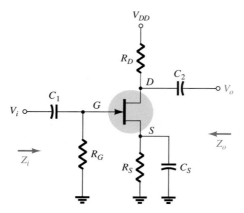

Figure 9.15 Self-bias JFET configuration.

The capacitor C_S across the source resistance assumes its short-circuit equivalence for dc, allowing R_S to define the operating point. Under ac conditions, the capacitor assumes the short-circuit state and "short circuits" the effects of R_S. If left in the ac, gain will be reduced as will be shown in the paragraphs to follow.

The JFET equivalent circuit is established in Fig. 9.16 and carefully redrawn in Fig. 9.17.

Since the resulting configuration is the same as appearing in Fig. 9.12, the resulting equations Z_i, Z_o, and A_v will be the same.

$\quad \textbf{\textit{Z}}_\textit{i}\textbf{:}$

$$\boxed{Z_i = R_G} \qquad\qquad (9.18)$$

Chapter 9 FET Small-Signal Analysis

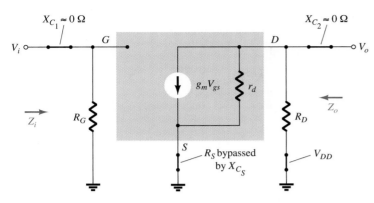

Figure 9.16 Network of Fig. 9.15 following the substitution of the JFET ac equivalent circuit.

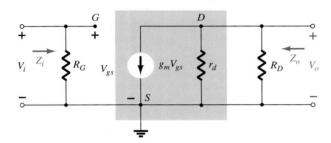

Figure 9.17 Redrawn network of Fig. 9.16.

Z_o:

$$Z_o = r_d \| R_D \qquad \text{(9.19)}$$

If $r_d \geq 10R_D$,

$$Z_o \cong R_D \Big|_{r_d \geq 10R_D} \qquad \text{(9.20)}$$

A_v:

$$A_v = -g_m(r_d \| R_D) \qquad \text{(9.21)}$$

If $r_d \geq 10R_D$,

$$A_v = -g_m R_D \Big|_{r_d \geq 10R_D} \qquad \text{(9.22)}$$

Phase relationship: The negative sign in the solutions for A_v again indicates a phase shift of $180°$ between V_i and V_o.

Unbypassed R_S

If C_S is removed from Fig 9.15, the resistor R_S will be part of the ac equivalent circuit as shown in Fig. 9.18. In this case, there is no obvious way to reduce the network to lower its level of complexity. In determining the levels of Z_i, Z_o, and A_v, one must

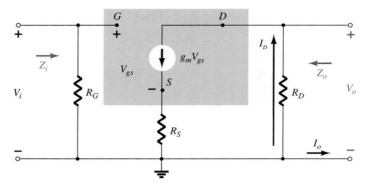

Figure 9.18 Self-bias JFET configuration including the effects of R_S with $r_d = \infty \, \Omega$.

simply be very careful with notation and defined polarities and direction. Initially, the resistance r_d will be left out of the analysis to form a basis for comparison.

Z_i: Due to the open-circuit condition between the gate and output network, the input remains the following:

$$\boxed{Z_i = R_G} \tag{9.23}$$

Z_o: The output impedance is defined by

$$Z_o = \frac{V_o}{I_o}\bigg|_{V_i=0}$$

Setting $V_i = 0$ V in Fig. 9.18 will result in the gate terminal being at ground potential (0 V). The voltage across R_G is then 0 V, and R_G has been effectively "shorted out" of the picture.

Applying Kirchhoff's current law will result in

$$I_o + I_D = g_m V_{gs}$$

with

$$V_{gs} = -(I_o + I_D)R_S$$

so that

$$I_o + I_D = -g_m(I_o + I_D)R_S = -g_m I_o R_S - g_m I_D R_S$$

or

$$I_o[1 + g_m R_S] = -I_D[1 + g_m R_S]$$

and $I_o = -I_D$ (the controlled current source $g_m V_{gs} = 0$ A
                                          for the applied conditions)

Since

$$V_o = -I_D R_D$$

then

$$V_o = -(-I_o)R_D = I_o R_D$$

and

$$\boxed{Z_o = \frac{V_o}{I_o} = R_D}\bigg|_{r_d=\infty \, \Omega} \tag{9.24}$$

If r_d is included in the network, the equivalent will appear as shown in Fig. 9.19.

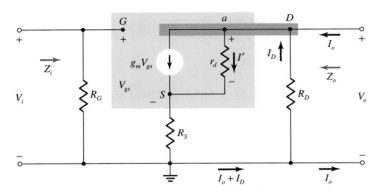

Figure 9.19 Including the effects of r_d in the self-bias JFET configuration.

Since
$$Z_o = \frac{V_o}{I_o}\bigg|_{V_i = 0 \text{ V}} = -\frac{I_D R_D}{I_o}$$

we should try to find an expression for I_o in terms of I_D.

Applying Kirchhoff's current law:
$$I_o = g_m V_{gs} + I_{r_d} - I_D$$

but
$$V_{r_d} = V_o + V_{gs}$$

and
$$I_o = g_m V_{gs} + \frac{V_o + V_{gs}}{r_d} - I_D$$

or
$$I_o = \left(g_m + \frac{1}{r_d}\right)V_{gs} - \frac{I_D R_D}{r_d} - I_D \text{ using } V_o = -I_D R_D$$

Now,
$$V_{gs} = -(I_D + I_o)R_S$$

so that
$$I_o = -\left(g_m + \frac{1}{r_d}\right)(I_D + I_o)R_S - \frac{I_D R_D}{r_d} - I_D$$

with the result that
$$I_o\left[1 + g_m R_S + \frac{R_S}{r_d}\right] = -I_D\left[1 + g_m R_S + \frac{R_S}{r_d} + \frac{R_D}{r_d}\right]$$

or
$$I_o = \frac{-I_D\left[1 + g_m R_S + \dfrac{R_S}{r_d} + \dfrac{R_D}{r_d}\right]}{1 + g_m R_S + \dfrac{R_S}{r_d}}$$

and
$$Z_o = \frac{V_o}{I_o} = \frac{-I_D R_D}{\dfrac{-I_D\left(1 + g_m R_S + \dfrac{R_S}{r_d} + \dfrac{R_D}{r_d}\right)}{1 + g_m R_S + \dfrac{R_S}{r_d}}}$$

and finally,
$$Z_o = \frac{\left[1 + g_m R_S + \dfrac{R_S}{r_d}\right]}{\left[1 + g_m R_S + \dfrac{R_S}{r_d} + \dfrac{R_D}{r_d}\right]}R_D \qquad (9.25a)$$

For $r_d \geq 10R_D$, $\left(1 + g_mR_S + \dfrac{R_S}{r_d}\right) \gg \dfrac{R_D}{r_d}$ and $1 + g_mR_S + \dfrac{R_S}{r_d} + \dfrac{R_D}{r_d}$

$\cong 1 + g_mR_S + \dfrac{R_S}{r_d}$ and

$$\boxed{Z_o = R_D}\Big|_{r_d \geq 10R_D} \tag{9.25b}$$

A_v: For the network of Fig. 9.19, an application of Kirchhoff's voltage law on the input circuit will result in

$$V_i - V_{gs} - V_{R_S} = 0$$
$$V_{gs} = V_i - I_DR_S$$

The voltage across r_d using Kirchhoff's voltage law is

$$V_o - V_{R_S}$$

and

$$I' = \frac{V_o - V_{R_S}}{r_d}$$

so that an application of Kirchhoff's current law will result in

$$I_D = g_mV_{gs} + \frac{V_o - V_{R_S}}{r_d}$$

Substituting for V_{gs} from above and substituting for V_o and V_{R_S} we have

$$I_D = g_m[V_i - I_DR_S] + \frac{(-I_DR_D) - (I_DR_S)}{r_d}$$

so that

$$I_D\left[1 + g_mR_S + \frac{R_D + R_S}{r_d}\right] = g_mV_i$$

or

$$I_D = \frac{g_mV_i}{1 + g_mR_S + \dfrac{R_D + R_S}{r_d}}$$

The output voltage is then

$$V_o = -I_DR_D = -\frac{g_mR_DV_i}{1 + g_mR_S + \dfrac{R_D + R_S}{r_d}}$$

and

$$\boxed{A_v = \frac{V_o}{V_i} = -\frac{g_mR_D}{1 + g_mR_S + \dfrac{R_D + R_S}{r_d}}} \tag{9.26}$$

Again, if $r_d \geq 10(R_D + R_S)$,

$$\boxed{A_v = \frac{V_o}{V_i} = -\frac{g_mR_D}{1 + g_mR_S}}\Big|_{r_d \geq 10(R_D + R_S)} \tag{9.27}$$

Phase Relationship: The negative sign in Eq. (9.26) again reveals that a 180° phase shift will exist between V_i and V_o.

The self-bias configuration of Example 6.2 has an operating point defined by $V_{GS_Q} = -2.6$ V and $I_{D_Q} = 2.6$ mA, with $I_{DSS} = 8$ mA and $V_P = -6$ V. The network is redrawn as Fig. 9.20 with an applied signal V_i. The value of y_{os} is given as 20 μS.
(a) Determine g_m.
(b) Find r_d.
(c) Find Z_i.
(d) Calculate Z_o with and without the effects of r_d. Compare the results.
(e) Calculate A_v with and without the effects of r_d. Compare the results.

EXAMPLE 9.8

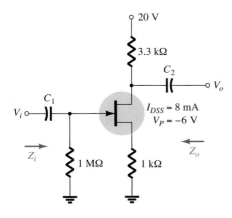

Figure 9.20 Network for Example 9.8.

Solution

(a) $g_{m0} = \dfrac{2I_{DSS}}{|V_P|} = \dfrac{2(8 \text{ mA})}{6 \text{ V}} = 2.67$ mS

$g_m = g_{m0}\left(1 - \dfrac{V_{GS_Q}}{V_P}\right) = 2.67 \text{ mS}\left(1 - \dfrac{(-2.6 \text{ V})}{(-6 \text{ V})}\right) = \mathbf{1.51 \text{ mS}}$

(b) $r_d = \dfrac{1}{y_{os}} = \dfrac{1}{20 \text{ }\mu\text{S}} = \mathbf{50 \text{ k}\Omega}$

(c) $Z_i = R_G = \mathbf{1 \text{ M}\Omega}$

(d) With r_d:

$$r_d = 50 \text{ k}\Omega > 10R_D = 33 \text{ k}\Omega$$

Therefore,

$$Z_o = R_D = \mathbf{3.3 \text{ k}\Omega}$$

If $r_d = \infty \text{ }\Omega$

$$Z_o = R_D = \mathbf{3.3 \text{ k}\Omega}$$

(e) With r_d:

$$A_v = \dfrac{-g_m R_D}{1 + g_m R_S + \dfrac{R_D + R_S}{r_d}} = \dfrac{-(1.51 \text{ mS})(3.3 \text{ k}\Omega)}{1 + (1.51 \text{ mS})(1 \text{ k}\Omega) + \dfrac{3.3 \text{ k}\Omega + 1 \text{ k}\Omega}{50 \text{ k}\Omega}}$$

$$= \mathbf{-1.92}$$

Without r_d:

$$A_v = \dfrac{-g_m R_D}{1 + g_m R_S} = \dfrac{-(1.51 \text{ mS})(3.3 \text{ k}\Omega)}{1 + (1.51 \text{ mS})(1 \text{ k}\Omega)} = \mathbf{-1.98}$$

As above, the effect of r_d was minimal because the condition $r_d \geq 10(R_D + R_S)$ was satisfied.

Note also that the typical gain of a JFET amplifier is less than that generally encountered for BJTs of similar configurations. Keep in mind, however, that Z_i is magnitudes greater than the typical Z_i of a BJT, which will have a very positive effect on the overall gain of a system.

Mathcad

The complexity of some of the equations for the unbypassed source resistance suggests that it might be a good opportunity to employ Mathcad. In fact, the results of Example 9.8 will be verified using equations that have been introduced in Chapter 6 and in this chapter. The analysis will proceed in a manner that permits a quick change in the parameter list so that any network of this configuration can be analyzed in short order. To this end, note that the first line of Fig. 9.21 is a list of all the elements of the network. In the future, the user simply has to change the parameter list, and Mathcad will quickly generate the new results without the user's having to enter all the equations again. The next line is a list of **Guess** values for the quantities to be calculated to help with the iteration process. As described in Chapter 6, **Given** must then be entered, followed by the equations that will generate the desired unknowns. Lastly, the **Find** statement in the format shown will direct the software to determine the quantities appearing within the brackets. As soon as the equal sign is selected, the results will appear in an order that matches the listing within the **Find** brackets. Note that the resulting drain current is 2.588 mA to match the 2.6 mA of Example 6.2, and V_{GS} is -2.588 V to match the -2.6 V of the same example. The resulting value of g_m is 1.517 mS to match the 1.51 mS of Example 9.8, and the overall gain is -1.923 which compares well with the calculated result of -1.98.

$$\text{IDSS} := 8 \cdot 10^{-3} \quad \text{VP} := -6 \quad \text{RS} := 1 \cdot 10^{3} \quad \text{RD} := 3.3 \cdot 10^{3} \quad \text{rd} := 50 \cdot 10^{3}$$

$$\text{ID} := 2 \cdot 10^{-3} \quad \text{VGS} := -2 \quad \text{GM} := 2 \cdot 10^{-3} \quad \text{AV} := -2$$

Given

$$\text{ID} = \frac{-\text{VGS}}{\text{RS}}$$

$$\text{ID} = \text{IDSS} \cdot \left[1 - \left(\frac{\text{VGS}}{\text{VP}} \right) \right]^{2}$$

$$\text{GM} = 2 \frac{\text{IDSS}}{|\text{VP}|} \cdot \left(1 - \frac{\text{VGS}}{\text{VP}} \right)$$

$$\text{AV} = -\text{GM} \cdot \frac{\text{RD}}{\left[1 + \text{GM} \cdot \text{RS} + \frac{(\text{RD} + \text{RS})}{\text{rd}} \right]}$$

$$\text{Find}(\text{ID}, \text{VGS}, \text{GM}, \text{AV}) = \begin{pmatrix} 2.588 \times 10^{-3} \\ -2.588 \\ 1.517 \times 10^{-3} \\ -1.923 \end{pmatrix}$$

Figure 9.21 Parameters and equations for Example 9.8 using Mathcad.

One of the most confusing elements of Mathcad can be which equal sign to use for which operation. All the parameters of the first two lines used the equal sign generated by **Shift:,** and all the boldface equal signs in the defining equations come from **Ctrl =.** Following the **Find** statement, the keyboard equal sign is employed.

There is no question that it takes some time to get used to entering complex equations such as appearing in Fig. 9.21, but be assured that the skills will develop quickly. As mentioned above, if any of the parameters such as **IDSS** and **VP** should change, all that has to be done is to retrieve the program from memory and change those two values; the new results for the four quantities will appear almost instantaneously—a real time saver.

9.5 JFET VOLTAGE-DIVIDER CONFIGURATION

The popular voltage-divider configuration for BJTs can also be applied to JFETs as demonstrated in Fig. 9.22.

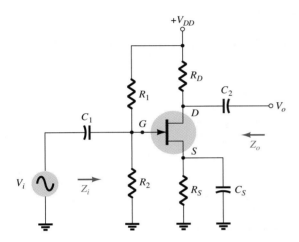

Figure 9.22 JFET voltage-divider configuration.

Substituting the ac equivalent model for the JFET will result in the configuration of Fig. 9.23. Replacing the dc supply V_{DD} by a short-circuit equivalent has grounded one end of R_1 and R_D. Since each network has a common ground, R_1 can be brought down in parallel with R_2 as shown in Fig. 9.24. R_D can also be brought down to ground but in the output circuit across r_d. The resulting ac equivalent network now has the basic format of some of the networks already analyzed.

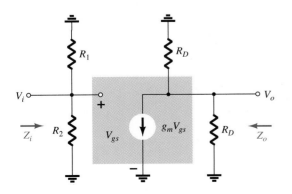

Figure 9.23 Network of Fig. 9.22 under ac conditions.

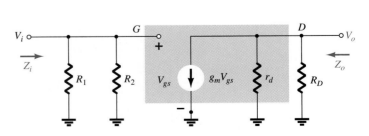

Figure 9.24 Redrawn network of Fig. 9.23.

Z_i: R_1 and R_2 are in parallel with the open-circuit equivalence of the JFET resulting in

$$Z_i = R_1 \| R_2 \tag{9.28}$$

Z_o: Setting $V_i = 0$ V will set V_{gs} and $g_m V_{gs}$ to zero and

$$Z_o = r_d \| R_D \tag{9.29}$$

For $r_d \geq 10 R_D$,

$$Z_o \cong R_D \Big|_{r_d \geq 10 R_D} \tag{9.30}$$

A_v:

$$V_{gs} = V_i$$

and

$$V_o = -g_m V_{gs}(r_d \| R_D)$$

so that

$$A_v = \frac{V_o}{V_i} = \frac{-g_m V_{gs}(r_d \| R_D)}{V_{gs}}$$

and

$$A_v = \frac{V_o}{V_i} = -g_m(r_d \| R_D) \tag{9.31}$$

If $r_d \geq 10 R_D$,

$$A_v = \frac{V_o}{V_i} \cong -g_m R_D \Big|_{r_d \geq 10 R_D} \tag{9.32}$$

Note that the equations for Z_o and A_v are the same as obtained for the fixed-bias and self-bias (with bypassed R_S) configurations. The only difference is the equation for Z_i, which is now sensitive to the parallel combination of R_1 and R_2.

9.6   JFET SOURCE-FOLLOWER (COMMON-DRAIN) CONFIGURATION

The JFET equivalent of the BJT emitter-follower configuration is the source-follower configuration of Fig. 9.25. Note that the output is taken off the source terminal and, when the dc supply is replaced by its short-circuit equivalent, the drain is grounded (hence, the terminology common-drain).

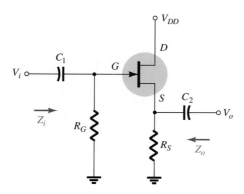

Figure 9.25   JFET source-follower configuration.

Substituting the JFET equivalent circuit will result in the configuration of Fig. 9.26. The controlled source and internal output impedance of the JFET are tied to ground at one end and R_S on the other, with V_o across R_S. Since $g_m V_{gs}$, r_d, and R_S are connected to the same terminal and ground, they can all be placed in parallel as shown in Fig. 9.27. The current source reversed direction but V_{gs} is still defined between the gate and source terminals.

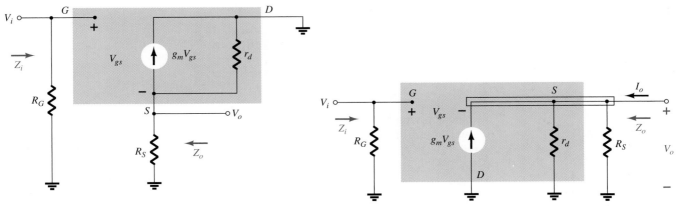

Figure 9.26 Network of Fig. 9.25 following the substitution of the JFET ac equivalent model.

Figure 9.27 Network of Fig. 9.26 redrawn.

Z_i: Figure 9.27 clearly reveals that Z_i is defined by

$$Z_i = R_G \qquad (9.33)$$

Z_o: Setting $V_i = 0$ V will result in the gate terminal being connected directly to ground as shown in Fig. 9.28.

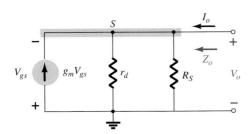

Figure 9.28 Determining Z_o for the network of Fig. 9.25.

The fact that V_{gs} and V_o are across the same parallel network results in $V_o = -V_{gs}$.
Applying Kirchhoff's current law at node S,

$$I_o + g_m V_{gs} = I_{r_d} + I_{R_S}$$

$$= \frac{V_o}{r_d} + \frac{V_o}{R_S}$$

The result is
$$I_o = V_o \left[\frac{1}{r_d} + \frac{1}{R_S} \right] - g_m V_{gs}$$

$$= V_o \left[\frac{1}{r_d} + \frac{1}{R_S} \right] - g_m [-V_o]$$

$$= V_o \left[\frac{1}{r_d} + \frac{1}{R_S} + g_m \right]$$

9.6 JFET Source-Follower (Common-Drain) Configuration

and $\quad Z_o = \dfrac{V_o}{I_o} = \dfrac{V_o}{V_o\left[\dfrac{1}{r_d} + \dfrac{1}{R_S} + g_m\right]} = \dfrac{1}{\dfrac{1}{r_d} + \dfrac{1}{R_S} + g_m} = \dfrac{1}{\dfrac{1}{r_d} + \dfrac{1}{R_S} + \dfrac{1}{1/g_m}}$

which has the same format as the total resistance of three parallel resistors. Therefore,

$$\boxed{Z_o = r_d \| R_S \| 1/g_m} \qquad (9.34)$$

For $r_d \geq 10R_S$,

$$\boxed{Z_o \cong R_S \| 1/g_m}\Big|_{r_d \geq 10R_S} \qquad (9.35)$$

A_v: The output voltage V_o is determined by

$$V_o = g_m V_{gs}(r_d \| R_S)$$

and applying Kirchhoff's voltage law around the perimeter of the network of Fig. 9.27 will result in

$$V_i = V_{gs} + V_o$$

and $\qquad\qquad\qquad\qquad V_{gs} = V_i - V_o$

so that $\qquad\qquad\qquad V_o = g_m(V_i - V_o)(r_d \| R_S)$

or $\qquad\qquad\qquad\quad V_o = g_m V_i(r_d \| R_S) - g_m V_o(r_d \| R_S)$

and $\qquad\qquad\qquad V_o[1 + g_m(r_d \| R_S)] = g_m V_i(r_d \| R_S)$

so that $\qquad\qquad \boxed{A_v = \dfrac{V_o}{V_i} = \dfrac{g_m(r_d \| R_S)}{1 + g_m(r_d \| R_S)}} \qquad (9.36)$

In the absence of r_d or if $r_d \geq 10R_S$,

$$\boxed{A_v = \dfrac{V_o}{V_i} \cong \dfrac{g_m R_S}{1 + g_m R_S}}\Big|_{r_d \geq 10R_S} \qquad (9.37)$$

Since the bottom of Eq. (9.36) is larger than the numerator by a factor of one, the gain can never be equal to or greater than one (as encountered for the emitter-follower BJT network).

Phase Relationship: Since A_v of Eq. (9.36) is a positive quantity, V_o and V_i are in phase for the JFET source-follower configuration.

EXAMPLE 9.9

A dc analysis of the source-follower network of Fig. 9.29 will result in $V_{GS_Q} = -2.86$ V and $I_{D_Q} = 4.56$ mA.
(a) Determine g_m.
(b) Find r_d.
(c) Determine Z_i.
(d) Calculate Z_o with and without r_d. Compare results.
(e) Determine A_v with and without r_d. Compare results.

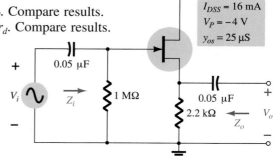

Figure 9.29 Network to be analyzed in Example 9.9.

Solution

(a) $g_{m0} = \dfrac{2I_{DSS}}{|V_P|} = \dfrac{2(16\text{ mA})}{4\text{ V}} = 8\text{ mS}$

$g_m = g_{m0}\left(1 - \dfrac{V_{GS_Q}}{V_P}\right) = 8\text{ mS}\left(1 - \dfrac{(-2.86\text{ V})}{(-4\text{ V})}\right) = \mathbf{2.28\text{ mS}}$

(b) $r_d = \dfrac{1}{y_{os}} = \dfrac{1}{25\ \mu\text{S}} = \mathbf{40\text{ k}\Omega}$

(c) $Z_i = R_G = \mathbf{1\text{ M}\Omega}$

(d) With r_d:

$$Z_o = r_d \| R_S \| 1/g_m = 40\text{ k}\Omega \| 2.2\text{ k}\Omega \| 1/2.28\text{ mS}$$

$$= 40\text{ k}\Omega \| 2.2\text{ k}\Omega \| 438.6\ \Omega$$

$$= \mathbf{362.52\ \Omega}$$

revealing that Z_o is often relatively small and determined primarily by $1/g_m$. Without r_d:

$$Z_o = R_S \| 1/g_m = 2.2\text{ k}\Omega \| 438.6\ \Omega = \mathbf{365.69\ \Omega}$$

revealing that r_d typically has little impact on Z_o.

(e) With r_d:

$$A_v = \dfrac{g_m(r_d \| R_S)}{1 + g_m(r_d \| R_S)} = \dfrac{(2.28\text{ mS})(40\text{ k}\Omega \| 2.2\text{ k}\Omega)}{1 + (2.28\text{ mS})(40\text{ k}\Omega \| 2.2\text{ k}\Omega)}$$

$$= \dfrac{(2.28\text{ mS})(2.09\text{ k}\Omega)}{1 + (2.28\text{ mS})(2.09\text{ k}\Omega)} = \dfrac{4.77}{1 + 4.77} = \mathbf{0.83}$$

which is less than 1 as predicted above.
Without r_d:

$$A_v = \dfrac{g_m R_S}{1 + g_m R_S} = \dfrac{(2.28\text{ mS})(2.2\text{ k}\Omega)}{1 + (2.28\text{ mS})(2.2\text{ k}\Omega)}$$

$$= \dfrac{5.02}{1 + 5.02} = \mathbf{0.83}$$

revealing that r_d usually has little impact on the gain of the configuration.

9.7 JFET COMMON-GATE CONFIGURATION

The last JFET configuration to be analyzed in detail is the common-gate configuration of Fig. 9.30, which parallels the common-base configuration employed with BJT transistors.

Substituting the JFET equivalent circuit will result in Fig. 9.31. Note the continuing requirement that the controlled source $g_m V_{gs}$ be connected from drain to source with r_d in parallel. The isolation between input and output circuits has obviously been lost since the gate terminal is now connected to the common ground of the network. In addition, the resistor connected between input terminals is no longer R_G but the resistor R_S connected from source to ground. Note also the location of the controlling voltage V_{gs} and the fact that it appears directly across the resistor R_S.

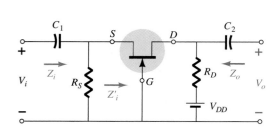

Figure 9.30 JFET common-gate configuration.

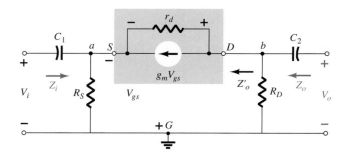

Figure 9.31 Network of Fig. 9.30 following substitution of JFET ac equivalent model.

$\mathbf{Z_i}$: The resistor R_S is directly across the terminals defining Z_i. Let us therefore find the impedance Z_i' of Fig. 9.30, which will simply be in parallel with R_S when Z_i is defined.

The network of interest is redrawn as Fig. 9.32. The voltage $V' = -V_{gs}$. Applying Kirchhoff's voltage law around the output perimeter of the network will result in

$$V' - V_{r_d} - V_{R_D} = 0$$

and

$$V_{r_d} = V' - V_{R_D} = V' - I'R_D$$

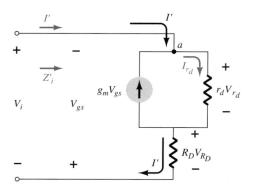

Figure 9.32 Determining Z_i' for the network of Fig. 9.30.

Applying Kirchhoff's current law at node a results in

$$I' + g_m V_{gs} = I_{r_d}$$

and

$$I' = I_{r_d} - g_m V_{gs} = \frac{(V' - I'R_D)}{r_d} - g_m V_{gs}$$

or

$$I' = \frac{V'}{r_d} - \frac{I'R_D}{r_d} - g_m[-V']$$

so that

$$I'\left[1 + \frac{R_D}{r_d}\right] = V'\left[\frac{1}{r_d} + g_m\right]$$

and

$$Z_i' = \frac{V'}{I'} = \frac{\left[1 + \dfrac{R_D}{r_d}\right]}{\left[g_m + \dfrac{1}{r_d}\right]}$$

(9.38)

or

$$Z_i' = \frac{V'}{I'} = \frac{r_d + R_D}{1 + g_m r_d}$$

and
$$Z_i = R_S \| Z_i'$$

results in

$$\boxed{Z_i = R_S \left\| \left[\frac{r_d + R_D}{1 + g_m r_d} \right] \right.} \qquad (9.39)$$

If $r_d \geq 10R_D$, Eq. (9.38) permits the following approximation since $R_D/r_d \ll 1$ and $1/r_d \ll g_m$:

$$Z_i' = \frac{\left[1 + \dfrac{R_D}{r_d} \right]}{\left[g_m + \dfrac{1}{r_d} \right]} \cong \frac{1}{g_m}$$

and

$$\boxed{Z_i \cong R_S \| 1/g_m}_{r_d \geq 10R_D} \qquad (9.40)$$

Z_o: Substituting $V_i = 0$ V in Fig. 9.31 will "short-out" the effects of R_S and set V_{gs} to 0 V. The result is $g_m V_{gs} = 0$, and r_d will be in parallel with R_D. Therefore,

$$\boxed{Z_o = R_D \| r_d} \qquad (9.41)$$

For $r_d \geq 10R_D$,

$$\boxed{Z_o \cong R_D}_{r_d \geq 10R_D} \qquad (9.42)$$

A_v: Figure 9.31 reveals that
$$V_i = -V_{gs}$$
and
$$V_o = I_D R_D$$
The voltage across r_d is
$$V_{r_d} = V_o - V_i$$
and
$$I_{r_d} = \frac{V_o - V_i}{r_d}$$

Applying Kirchhoff's current law at node b in Fig. 9.31 results in
$$I_{r_d} + I_D + g_m V_{gs} = 0$$
and
$$I_D = -I_{r_d} - g_m V_{gs}$$
$$= -\left[\frac{V_o - V_i}{r_d} \right] - g_m[-V_i]$$
$$I_D = \frac{V_i - V_o}{r_d} + g_m V_i$$

so that
$$V_o = I_D R_D = \left[\frac{V_i - V_o}{r_d} + g_m V_i \right] R_D$$
$$= \frac{V_i R_D}{r_d} - \frac{V_o R_D}{r_d} + g_m$$

and
$$V_o \left[1 + \frac{R_D}{r_d} \right] = V_i \left[\frac{R_D}{r_d} + g_m R_D \right]$$

9.7 JFET Common-Gate Configuration 485

with

$$A_v = \frac{V_o}{V_i} = \frac{\left[g_m R_D + \dfrac{R_D}{r_d} \right]}{\left[1 + \dfrac{R_D}{r_d} \right]} \qquad (9.43)$$

For $r_d \geq 10 R_D$, the factor R_D/r_d of Eq. (9.43) can be dropped as a good approximation and

$$\boxed{A_v = g_m R_D} \qquad\qquad (9.44)$$
$$\scriptstyle r_d \geq 10 R_D$$

Phase Relationship: The fact that A_v is a positive number will result in an *in-phase* relationship between V_o and V_i for the common-gate configuration.

EXAMPLE 9.10

Although the network of Fig. 9.33 may not initially appear to be of the common-gate variety, a close examination will reveal that it has all the characteristics of Fig. 9.30. If $V_{GS_Q} = -2.2$ V and $I_{D_Q} = 2.03$ mA:
(a) Determine g_m.
(b) Find r_d.
(c) Calculate Z_i with and without r_d. Compare results.
(d) Find Z_o with and without r_d. Compare results.
(e) Determine V_o with and without r_d. Compare results.

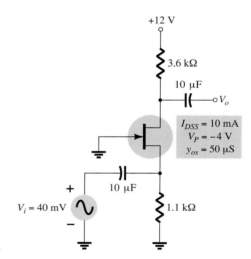

Figure 9.33 Network for Example 9.10.

Solution

(a) $g_{m0} = \dfrac{2 I_{DSS}}{|V_P|} = \dfrac{2(10 \text{ mA})}{4 \text{ V}} = 5 \text{ mS}$

$g_m = g_{m0}\left(1 - \dfrac{V_{GS_Q}}{V_P} \right) = 5 \text{ mS}\left(1 - \dfrac{(-2.2 \text{ V})}{(-4 \text{ V})} \right) = \mathbf{2.25 \text{ mS}}$

(b) $r_d = \dfrac{1}{y_{os}} = \dfrac{1}{50 \ \mu\text{S}} = \mathbf{20 \text{ k}\Omega}$

(c) With r_d:

$$Z_i = R_S \| \left[\frac{r_d + R_D}{1 + g_m r_d} \right] = 1.1 \text{ k}\Omega \| \left[\frac{20 \text{ k}\Omega + 3.6 \text{ k}\Omega}{1 + (2.25 \text{ ms})(20 \text{ k}\Omega)} \right]$$

$$= 1.1 \text{ k}\Omega \| 0.51 \text{ k}\Omega = \mathbf{0.35 \text{ k}\Omega}$$

Without r_d:

$$Z_i = R_S \| 1/g_m = 1.1 \text{ k}\Omega \| 1/2.25 \text{ ms} = 1.1 \text{ k}\Omega \| 0.44 \text{ k}\Omega$$
$$= \mathbf{0.31 \text{ k}\Omega}$$

Even though the condition,

$$r_d \geq 10R_D = > 20 \text{ k}\Omega \geq 10(3.6 \text{ k}\Omega) = > 20 \text{ k}\Omega \geq 36 \text{ k}\Omega$$

is *not* satisfied, both equations result in essentially the same level of impedance. In this case, $1/g_m$ was the predominant factor.

(d) With r_d:

$$Z_o = R_D \| r_d = 3.6 \text{ k}\Omega \| 20 \text{ k}\Omega = \mathbf{3.05 \text{ k}\Omega}$$

Without r_d:

$$Z_o = R_D = \mathbf{3.6 \text{ k}\Omega}$$

Again the condition $r_d \geq 10R_D$ is *not* satisfied, but both results are reasonably close. R_D is certainly the predominant factor in this example.

(e) With r_d:

$$A_v = \frac{\left[g_m R_D + \dfrac{R_D}{r_d} \right]}{\left[1 + \dfrac{R_D}{r_d} \right]} = \frac{\left[(2.25 \text{ mS})(3.6 \text{ k}\Omega) + \dfrac{3.6 \text{ k}\Omega}{20 \text{ k}\Omega} \right]}{\left[1 + \dfrac{3.6 \text{ k}\Omega}{20 \text{ k}\Omega} \right]}$$

$$= \frac{8.1 + 0.18}{1 + 0.18} = \mathbf{7.02}$$

and $\quad A_v = \dfrac{V_o}{V_i} = \blacktriangleright V_o = A_v V_i = (7.02)(40 \text{ mV}) = \mathbf{280.8 \text{ mV}}$

Without r_d:

$$A_v = g_m R_D = (2.25 \text{ mS})(3.6 \text{ k}\Omega) = \mathbf{8.1}$$

with $\quad V_o = A_v V_i = (8.1)(40 \text{ mV}) = \mathbf{324 \text{ mV}}$

In this case, the difference is a little more noticeable but not dramatically so.

Example 9.10 demonstrates that even though the condition $r_d \geq 10R_D$ was not satisfied, the results for the parameters given were not significantly different using the exact and approximate equations. In fact, in most cases, the approximate equations can be used to find a reasonable idea of particular levels with a reduced amount of effort.

9.8 DEPLETION-TYPE MOSFETs

The fact that Shockley's equation is also applicable to depletion-type MOSFETs results in the same equation for g_m. In fact, the ac equivalent model for D-MOSFETs is exactly the same as that employed for JFETs as shown in Fig. 9.34.

The only difference offered by D-MOSFETs is that V_{GS_Q} can be positive for *n*-channel devices and negative for *p*-channel units. The result is that g_m can be greater than g_{m0} as demonstrated by the example to follow. The range of r_d is very similar to that encountered for JFETs.

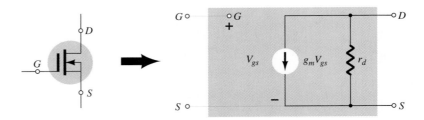

Figure 9.34 D-MOSFET ac equivalent model.

EXAMPLE 9.11

The network of Fig. 9.35 was analyzed as Example 6.8, resulting in $V_{GS_Q} = 0.35$ V and $I_{D_Q} = 7.6$ mA.

(a) Determine g_m and compare to g_{m0}.
(b) Find r_d.
(c) Sketch the ac equivalent network for Fig. 9.35.
(d) Find Z_i.
(e) Calculate Z_o.
(f) Find A_v.

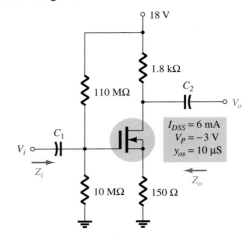

Figure 9.35 Network for Example 9.11.

Solution

(a) $g_{m0} = \dfrac{2I_{DSS}}{|V_P|} = \dfrac{2(6\text{ mA})}{3\text{ V}} = 4$ mS

$g_m = g_{m0}\left(1 - \dfrac{V_{GS_Q}}{V_P}\right) = 4\text{ mS}\left(1 - \dfrac{(+0.35\text{ V})}{(-3\text{ V})}\right) = 4\text{ mS}(1 + 0.117) = \textbf{4.47 mS}$

(b) $r_d = \dfrac{1}{y_{os}} = \dfrac{1}{10\ \mu\text{S}} = \textbf{100 k}\boldsymbol{\Omega}$

(c) See Fig. 9.36. Note the similarities with the network of Fig. 9.24. Equations (9.28) through (9.32) are therefore applicable.

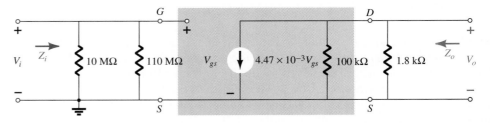

Figure 9.36 AC equivalent circuit for Fig. 9.35.

(d) Eq. (9.28): $Z_i = R_1 \| R_2 = 10 \text{ M}\Omega \| 110 \text{ M}\Omega = \mathbf{9.17 \text{ M}\Omega}$
(e) Eq. (9.29): $Z_o = r_d \| R_D = 100 \text{ k}\Omega \| 1.8 \text{ k}\Omega = \mathbf{1.77 \text{ k}\Omega} \cong R_D = \mathbf{1.8 \text{ k}\Omega}$
(f) $r_d \geq 10 R_D \rightarrow 100 \text{ k}\Omega \geq 18 \text{ k}\Omega$
    Eq. (9.32): $A_v = -g_m R_D = -(4.47 \text{ mS})(1.8 \text{ k}\Omega) = \mathbf{8.05}$

9.9 ENHANCEMENT-TYPE MOSFETS

The enhancement-type MOSFET can be either an *n*-channel (*n*MOS) or *p*-channel (*p*MOS) device, as shown in Fig. 9.37. The ac small-signal equivalent circuit of either device is shown in Fig. 9.37, revealing an open-circuit between gate and drain-source channel and a current source from drain to source having a magnitude dependent on the gate-to-source voltage. There is an output impedance from drain to source r_d, which is usually provided on specification sheets as an admittance y_{os}. The device transconductance, g_m, is provided on specification sheets as the forward transfer admittance, y_{fs}.

In our analysis of JFETs, an equation for g_m was derived from Shockley's equation.

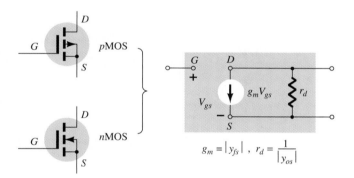

Figure 9.37 Enhancement MOSFET ac small-signal model.

For E-MOSFETs, the relationship between output current and controlling voltage is defined by

$$I_D = k(V_{GS} - V_{GS(\text{Th})})^2$$

Since g_m is still defined by

$$g_m = \frac{\Delta I_D}{\Delta V_{GS}}$$

we can take the derivative of the transfer equation to determine g_m as an operating point. That is,

$$g_m = \frac{dI_D}{dV_{GS}} = \frac{d}{dV_{GS}} k(V_{GS} - V_{GS(\text{Th})})^2 = k \frac{d}{dV_{GS}}(V_{GS} - V_{GS(\text{Th})})^2$$

$$= 2k(V_{GS} - V_{GS(\text{Th})}) \frac{d}{dV_{GS}}(V_{GS} - V_{GS(\text{Th})}) = 2k(V_{GS} - V_{GS(\text{Th})})(1 - 0)$$

and

$$g_m = 2k(V_{GS_Q} - V_{GS(\text{Th})}) \qquad (9.45)$$

Recall that the constant k can be determined from a given typical operating point on a specification sheet. In every other respect, the ac analysis is the same as that employed for JFETs or D-MOSFETs. Be aware, however, that the characteristics of an E-MOSFET are such that the biasing arrangements are somewhat limited.

9.10 E-MOSFET DRAIN-FEEDBACK CONFIGURATION

The E-MOSFET drain-feedback configuration appears in Fig. 9.38. Recall from dc calculations that R_G could be replaced by a short-circuit equivalent since $I_G = 0$ A and therefore $V_{R_G} = 0$ V. However, for ac situations it provides an important high impedance between V_o and V_i. Otherwise, the input and output terminals would be connected directly and $V_o = V_i$.

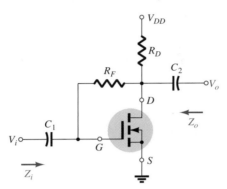

Figure 9.38 E-MOSFET drain-feedback configuration.

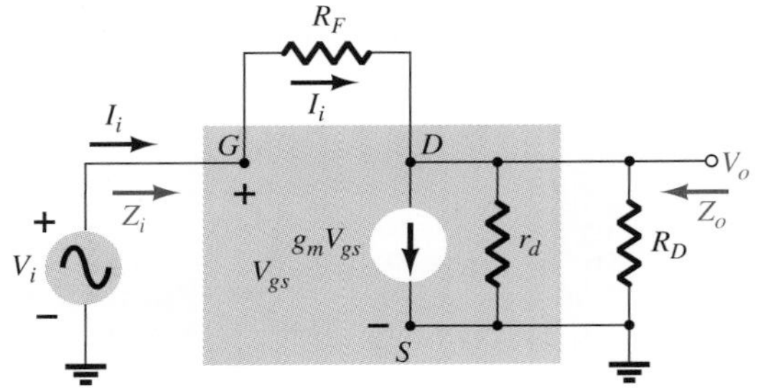

Figure 9.39 AC equivalent of the network of Fig. 9.38.

Substituting the ac equivalent model for the device will result in the network of Fig. 9.39. Note that R_F is not within the shaded area defining the equivalent model of the device but does provide a direct connection between input and output circuits.

Z_i: Applying Kirchhoff's current law to the output circuit (at node D in Fig. 9.39) results in

$$I_i = g_m V_{gs} + \frac{V_o}{r_d \| R_D}$$

and

$$V_{gs} = V_i$$

so that

$$I_i = g_m V_i + \frac{V_o}{r_d \| R_D}$$

or

$$I_i - g_m V_i = \frac{V_o}{r_d \| R_D}$$

Therefore,

$$V_o = (r_d \| R_D)(I_i - g_m V_i)$$

with

$$I_i = \frac{V_i - V_o}{R_F} = \frac{V_i - (r_d \| R_D)(I_i - g_m V_i)}{R_F}$$

and

$$I_i R_F = V_i - (r_d \| R_D)I_i + (r_d \| R_D)g_m V_i$$

so that

$$V_i[1 + g_m(r_d \| R_D)] = I_i[R_F + r_d \| R_D]$$

and finally,

$$\boxed{Z_i = \frac{V_i}{I_i} = \frac{R_F + r_d \| R_D}{1 + g_m(r_d \| R_D)}}$$ (9.46)

Typically, $R_F \gg r_d \| R_D$, so that

$$Z_i \cong \frac{R_F}{1 + g_m(r_d \| R_D)}$$

For $r_d \geq 10R_D$,

$$\boxed{Z_i \cong \frac{R_F}{1 + g_m R_D}} \quad {}_{R_F \gg r_d \| R_D, \, r_d \geq 10R_D} \qquad (9.47)$$

Z_o: Substituting $V_i = 0$ V will result in $V_{gs} = 0$ V and $g_m V_{gs} = 0$, with a short-circuit path from gate to ground as shown in Fig. 9.40. R_F, r_d, and R_D are then in parallel and

$$\boxed{Z_o = R_F \| r_d \| R_D} \qquad (9.48)$$

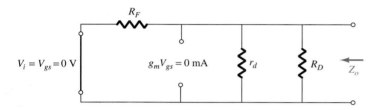

Figure 9.40   Determining Z_o for the network of Fig. 9.38.

Normally, R_F is so much larger than $r_d \| R_D$ that

$$Z_o \cong r_d \| R_D$$

and with $r_d \geq 10R_D$,

$$\boxed{Z_o \cong R_D} \quad {}_{R_F \gg r_d \| R_D, \, r_d \geq 10R_D} \qquad (9.49)$$

A_v: Applying Kirchhoff's current law at node D of Fig. 9.39 will result in

$$I_i = g_m V_{gs} + \frac{V_o}{r_d \| R_D}$$

but

$$V_{gs} = V_i \text{ and } I_i = \frac{V_i - V_o}{R_F}$$

so that

$$\frac{V_i - V_o}{R_F} = g_m V_i + \frac{V_o}{r_d \| R_D}$$

and

$$\frac{V_i}{R_F} - \frac{V_o}{R_F} = g_m V_i + \frac{V_o}{r_d \| R_D}$$

so that

$$V_o \left[\frac{1}{r_d \| R_D} + \frac{1}{R_F} \right] = V_i \left[\frac{1}{R_F} - g_m \right]$$

and

$$A_v = \frac{V_o}{V_i} = \frac{\left[\dfrac{1}{R_F} - g_m \right]}{\left[\dfrac{1}{r_d \| R_D} + \dfrac{1}{R_F} \right]}$$

but

$$\frac{1}{r_d\|R_D} + \frac{1}{R_F} = \frac{1}{R_F\|r_d\|R_D}$$

and

$$g_m \gg \frac{1}{R_F}$$

so that

$$\boxed{A_v = -g_m(R_F\|r_d\|R_D)} \tag{9.50}$$

Since R_F is usually $\gg r_d\|R_D$ and if $r_d \geq 10R_D$,

$$\boxed{A_v \cong -g_mR_D} \underset{R_F \gg r_d\|R_D,\ r_d \geq 10R_D}{} \tag{9.51}$$

Phase Relationship: The negative sign for A_v reveals that V_o and V_i are out of phase by 180°.

EXAMPLE 9.12

The E-MOSFET of Fig. 9.41 was analyzed in Example 6.11, with the result that $k = 0.24 \times 10^{-3}$ A/V², $V_{GS_Q} = 6.4$ V, and $I_{D_Q} = 2.75$ mA.
(a) Determine g_m.
(b) Find r_d.
(c) Calculate Z_i with and without r_d. Compare results.
(d) Find Z_o with and without r_d. Compare results.
(e) Find A_v with and without r_d. Compare results.

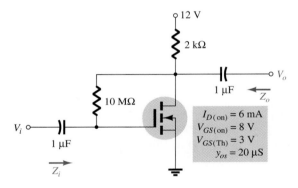

Figure 9.41 Drain-feedback amplifier from Example 6.11.

Solution

(a) $g_m = 2k(V_{GS_Q} - V_{GS(\text{Th})}) = 2(0.24 \times 10^{-3}\ \text{A/V}^2)(6.4\ \text{V} - 3\ \text{V})$
   $= \mathbf{1.63\ mS}$

(b) $r_d = \dfrac{1}{y_{os}} = \dfrac{1}{20\ \mu S} = \mathbf{50\ k\Omega}$

(c) With r_d:

$$Z_i = \frac{R_F + r_d\|R_D}{1 + g_m(r_d\|R_D)} = \frac{10\ \text{M}\Omega + 50\ \text{k}\Omega\|2\ \text{k}\Omega}{1 + (1.63\ \text{mS})(50\ \text{k}\Omega\|2\ \text{k}\Omega)}$$

$$= \frac{10\ \text{M}\Omega + 1.92\ \text{k}\Omega}{1 + 3.13} = \mathbf{2.42\ M\Omega}$$

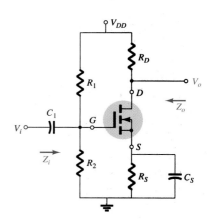

Without r_d:

$$Z_i \cong \frac{R_F}{1 + g_m R_D} = \frac{10 \text{ M}\Omega}{1 + (1.63 \text{ mS})(2 \text{ k}\Omega)} = \mathbf{2.53 \text{ M}\Omega}$$

revealing that since the condition $r_d \geq 10 R_D = 50 \text{ k}\Omega \geq 40 \text{ k}\Omega$ is satisfied, the results for Z_o with or without r_d will be quite close.

(d) With r_d:

$$Z_o = R_F \| r_d \| R_D = 10 \text{ M}\Omega \| 50 \text{ k}\Omega \| 2 \text{ k}\Omega = 49.75 \text{ k}\Omega \| 2 \text{ k}\Omega$$
$$= \mathbf{1.92 \text{ k}\Omega}$$

Without r_d:

$$Z_o \cong R_D = \mathbf{2 \text{ k}\Omega}$$

again providing very close results.

(e) With r_d:

$$A_v = -g_m(R_F \| r_d \| R_D)$$
$$= -(1.63 \text{ mS})(10 \text{ M}\Omega \| 50 \text{ k}\Omega \| 2 \text{ k}\Omega)$$
$$= -(1.63 \text{ mS})(1.92 \text{ k}\Omega)$$
$$= \mathbf{-3.21}$$

Without r_d:

$$A_v = -g_m R_D = -(1.63 \text{ mS})(2 \text{ k}\Omega)$$
$$= \mathbf{-3.26}$$

which is very close to the above result.

9.11 E-MOSFET VOLTAGE-DIVIDER CONFIGURATION

The last E-MOSFET configuration to be examined in detail is the voltage-divider network of Fig. 9.42. The format is exactly the same as appearing in a number of earlier discussions.

Substituting the ac equivalent network for the E-MOSFET will result in the configuration of Fig. 9.43, which is exactly the same as Fig. 9.24. The result is that Eqs. (9.28) through (9.32) are applicable as listed below for the E-MOSFET.

Figure 9.42 E-MOSFET voltage-divider configuration.

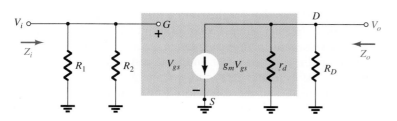

Figure 9.43 AC equivalent network for the configuration of Fig. 9.42.

Z_i:

$$\boxed{Z_i = R_1 \| R_2} \qquad (9.52)$$

Z_o:

$$\boxed{Z_o = r_d \| R_D} \qquad (9.53)$$

For $r_d \geq 10R_D$,

$$\boxed{Z_o \cong R_d}_{\;r_d \geq 10R_D} \qquad (9.54)$$

A_v:

$$\boxed{A_v = \frac{V_o}{V_i} = -g_m(r_d \| R_D)} \qquad (9.55)$$

and if $r_d \geq 10R_D$,

$$\boxed{A_v = \frac{V_o}{V_i} \cong -g_m R_D} \qquad (9.56)$$

9.12 DESIGNING FET AMPLIFIER NETWORKS

Design problems at this stage are limited to obtaining a desired dc bias condition or ac voltage gain. In most cases, the various equations developed are used "in reverse" to define the parameters necessary to obtain the desired gain, input impedance, or output impedance. To avoid unnecessary complexity during the initial stages of the design, the approximate equations are often employed because some variation will occur when calculated resistors are replaced by standard values. Once the initial design is completed, the results can be tested and refinements made using the complete equations.

Throughout the design procedure be aware that although superposition permits a separate analysis and design of the network from a dc and an ac viewpoint, a parameter chosen in the dc environment will often play an important role in the ac response. In particular, recall that the resistance R_G could be replaced by a short-circuit equivalent in the feedback configuration because $I_G \cong 0$ A for dc conditions, but for the ac analysis, it presents an important high impedance path between V_o and V_i. In addition, recall that g_m is larger for operating points closer to the I_D axis ($V_{GS} = 0$ V), requiring that R_S be relatively small. In the unbypassed R_S network, a small R_S will also contribute to a higher gain, but for the source-follower, the gain is reduced from its maximum value of 1. In total, simply keep in mind that network parameters can affect the dc and ac levels in different ways. Often a balance must be made between a particular operating point and its impact on the ac response.

In most situations, the available dc supply voltage is known, the FET to be employed has been determined, and the capacitors to be employed at the chosen frequency are defined. It is then necessary to determine the resistive elements necessary to establish the desired gain or impedance level. The next three examples will determine the required parameters for a specific gain.

Design the fixed-bias network of Fig. 9.44 to have an ac gain of 10. That is, determine the value of R_D.

EXAMPLE 9.13

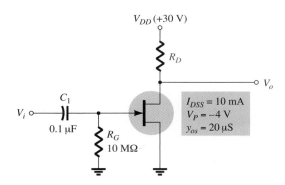

Figure 9.44 Circuit for desired voltage gain in Example 9.13.

Solution

Since $V_{GS_Q} = 0$ V, the level of g_m is g_{m0}. The gain is therefore determined by

$$A_v = -g_m(R_D \| r_d) = -g_{m0}(R_D \| r_d)$$

with

$$g_{m0} = \frac{2I_{DSS}}{|V_P|} = \frac{2(10 \text{ mA})}{4 \text{ V}} = 5 \text{ mS}$$

The result is

$$-10 = -5 \text{ mS}(R_D \| r_d)$$

and

$$R_D \| r_d = \frac{10}{5 \text{ mS}} = 2 \text{ k}\Omega$$

From the device specifications,

$$r_d = \frac{1}{y_{os}} = \frac{1}{20 \times 10^{-6} \text{ S}} = 50 \text{ k}\Omega$$

Substituting, we find

$$R_D \| r_d = R_D \| 50 \text{ k}\Omega = 2 \text{ k}\Omega$$

and

$$\frac{R_D(50 \text{ k}\Omega)}{R_D + 50 \text{ k}\Omega} = 2 \text{ k}\Omega$$

or

$$50R_D = 2(R_D + 50 \text{ k}\Omega) = 2R_D + 100 \text{ k}\Omega$$

with

$$48R_D = 100 \text{ k}\Omega$$

and

$$R_D = \frac{100 \text{ k}\Omega}{48} \cong 2.08 \text{ k}\Omega$$

The closest standard value is **2 kΩ** (Appendix D), which would be employed for this design.

The resulting level of V_{DS_Q} would then be determined as follows:

$$V_{DS_Q} = V_{DD} - I_{D_Q}R_D = 30 \text{ V} - (10 \text{ mA})(2 \text{ k}\Omega) = \mathbf{10 \text{ V}}$$

The levels of Z_i and Z_o are set by the levels of R_G and R_D, respectively. That is,

$$Z_i = R_G = \mathbf{10 \text{ M}\Omega}$$

$$Z_o = R_D \| r_d = 2 \text{ k}\Omega \| 50 \text{ k}\Omega = \mathbf{1.92 \text{ k}\Omega} \cong R_D = 2 \text{ k}\Omega$$

EXAMPLE 9.14

Choose the values of R_D and R_S for the network of Fig. 9.45 that will result in a gain of 8 using a relatively high level of g_m for this device defined at $V_{GS_Q} = \frac{1}{4}V_P$.

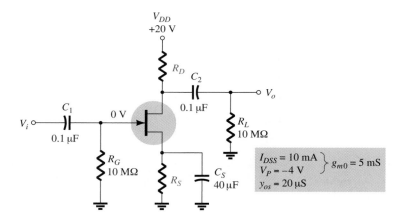

Figure 9.45 Network for desired voltage gain in Example 9.14.

Solution

The operating point is defined by

$$V_{GS_Q} = \frac{1}{4}V_P = \frac{1}{4}(-4\text{ V}) = -1\text{ V}$$

and

$$I_D = I_{DSS}\left(1 - \frac{V_{GS_Q}}{V_P}\right)^2 = 10\text{ mA}\left(1 - \frac{(-1\text{ V})}{(-4\text{ V})}\right)^2 = 5.625\text{ mA}$$

Determining g_m,

$$g_m = g_{m0}\left(1 - \frac{V_{GS_Q}}{V_P}\right)$$

$$= 5\text{ mS}\left(1 - \frac{(-1\text{ V})}{(-4\text{ V})}\right) = 3.75\text{ mS}$$

The magnitude of the ac voltage gain is determined by

$$|A_v| = g_m(R_D\|r_d)$$

Substituting known values will result in

$$8 = (3.75\text{ mS})(R_D\|r_d)$$

so that

$$R_D\|r_d = \frac{8}{3.75\text{ mS}} = 2.13\text{ k}\Omega$$

The level of r_d is defined by

$$r_d = \frac{1}{y_{os}} = \frac{1}{20\text{ }\mu\text{S}} = 50\text{ k}\Omega$$

and

$$R_D\|50\text{ k}\Omega = 2.13\text{ k}\Omega$$

with the result that

$$R_D = \mathbf{2.2\text{ k}\Omega}$$

which is a standard value.

The level of R_S is determined by the dc operating conditions as follows:

$$V_{GS_Q} = -I_D R_S$$

$$-1 \text{ V} = -(5.625 \text{ mA})R_S$$

and

$$R_S = \frac{1 \text{ V}}{5.625 \text{ mA}} = 177.8 \text{ }\Omega$$

The closest standard value is **180 Ω.** In this example, R_S does not appear in the ac design because of the shorting effect of C_S.

In the next example, R_S is unbypassed and the design becomes a bit more complicated.

Determine R_D and R_S for the network of Fig. 9.45 to establish a gain of 8 if the bypass capacitor C_S is removed.

EXAMPLE 9.15

Solution

V_{GS_Q} and I_{D_Q} are still -1 V and 5.625 mA, and since the equation $V_{GS} = -I_D R_S$ has not changed, R_S continues to equal the standard value of **180 Ω** obtained in Example 9.14.

The gain of an unbypassed self-bias configuration is

$$A_v = -\frac{g_m R_D}{1 + g_m R_S}$$

For the moment it is assumed that $r_d \geq 10(R_D + R_S)$. Using the full equation for A_v at this stage of the design would simply complicate the process unnecessarily.

Substituting (for the specified magnitude of 8 for the gain),

$$|8| = \left| \frac{-(3.75 \text{ mS})R_D}{1 + (3.75 \text{ mS})(180 \text{ }\Omega)} \right| = \frac{(3.75 \text{ mS})R_D}{1 + 0.675}$$

and

$$8(1 + 0.675) = (3.75 \text{ mS})R_D$$

so that

$$R_D = \frac{13.4}{3.75 \text{ mS}} = 3.573 \text{ k}\Omega$$

with the closest standard value at **3.6 kΩ.**

We can now test the condition:

$$r_d \geq 10(R_D + R_S)$$

$$50 \text{ k}\Omega \geq 10(3.6 \text{ k}\Omega + 0.18 \text{ k}\Omega) = 10(3.78 \text{ k}\Omega)$$

and

$$50 \text{ k}\Omega \geq 37.8 \text{ k}\Omega$$

which is satisfied—the solution stands!

9.13 SUMMARY TABLE

In an effort to provide a quick comparison between configurations and offer a listing that can be helpful for a variety of reasons, Table 9.1 was developed. The exact and approximate equation for each important parameter are provided with a typical range of values for each. Although all the possible configurations are not present, the majority of the most frequently encountered are included. In fact, any configuration not

TABLE 9.1 Z_i, Z_o, and A_v for various FET configurations

Configuration	Z_i	Z_o	$A_v = \dfrac{V_o}{V_i}$
Fixed-bias [JFET or D-MOSFET]	High (10 MΩ) $= \boxed{R_G}$	Medium (2 kΩ) $= \boxed{R_D \| r_d}$ $\cong \boxed{R_D}$ $_{(r_d \geq 10\,R_D)}$	Medium (-10) $= \boxed{-g_m(r_d \| R_D)}$ $\cong \boxed{-g_m R_D}$ $_{(r_d \geq 10\,R_D)}$
Self-bias bypassed R_S [JFET or D-MOSFET]	High (10 MΩ) $= \boxed{R_G}$	Medium (2 kΩ) $= \boxed{R_D \| r_d}$ $\cong \boxed{R_D}$ $_{(r_d \geq 10\,R_D)}$	Medium (-10) $= \boxed{-g_m(r_d \| R_D)}$ $\cong \boxed{-g_m R_D}$ $_{(r_d \geq 10\,R_D)}$
Self-bias unbypassed R_S [JFET or D-MOSFET]	High (10 MΩ) $= \boxed{R_G}$	$= \boxed{\dfrac{\left[1 + g_m R_S + \dfrac{R_S}{r_d}\right] R_D}{\left[1 + g_m R_S + \dfrac{R_S}{r_d} + \dfrac{R_D}{r_d}\right]}}$ $= \boxed{R_D}$ $_{r_d \geq 10\,R_D \text{ or } r_d = \infty\,\Omega}$	Low (-2) $= \boxed{\dfrac{g_m R_D}{1 + g_m R_S + \dfrac{R_D + R_S}{r_d}}}$ $\cong \boxed{-\dfrac{g_m R_D}{1 + g_m R_S}}$ $[r_d \geq 10\,(R_D + R_S)]$
Voltage-divider bias [JFET or D-MOSFET]	High (10 MΩ) $= \boxed{R_1 \| R_2}$	Medium (2 kΩ) $= \boxed{R_D \| r_d}$ $\cong \boxed{R_D}$ $_{(r_d \geq 10\,R_D)}$	Medium (-10) $= \boxed{-g_m(r_d \| R_D)}$ $\cong \boxed{-g_m R_D}$ $_{(r_d \geq 10\,R_D)}$

TABLE 9.1 (Continued)

Configuration	Z_i	Z_o	$A_v = \dfrac{V_o}{V_i}$
Source-follower [JFET or D-MOSFET]	High (10 MΩ) $= \boxed{R_G}$	Low (100 kΩ) $= \boxed{r_d\|R_S\|1/g_m}$ $\cong \boxed{R_S\|1/g_m}_{\,(r_d \ge 10\,R_S)}$	Low (<1) $= \boxed{\dfrac{g_m(r_d\|R_S)}{1 + g_m(r_d\|R_S)}}$ $\cong \boxed{\dfrac{g_mR_S}{1 + g_mR_S}}_{\,(r_d \ge 10\,R_S)}$
Common-gate [JFET or D-MOSFET]	Low (1 kΩ) $= \boxed{R_S\left[\dfrac{r_d + R_D}{1 + g_mr_d}\right]}$ $\cong \boxed{R_S\left\|\dfrac{1}{g_m}\right.}_{\,(r_d \ge 10\,R_D)}$	Medium (2 kΩ) $= \boxed{R_D\|r_d}$ $\cong \boxed{R_D}_{\,(r_d \ge 10\,R_D)}$	Medium ($+10$) $= \boxed{\dfrac{g_mR_D + \dfrac{R_D}{r_d}}{1 + \dfrac{R_D}{r_d}}}$ $\cong \boxed{g_mR_D}_{\,(r_d \ge 10\,R_D)}$
Drain-feedback bias E-MOSFET	Medium (1 MΩ) $\cong \boxed{\dfrac{R_F + r_d\|R_D}{1 + g_m(r_d\|R_D)}}$ $\cong \boxed{\dfrac{R_F}{1 + g_mR_D}}_{\,(r_d \ge 10\,R_D)}$	Medium (2 kΩ) $= \boxed{R_F\|r_d\|R_D}$ $\cong \boxed{R_D}_{\,(R_F,\; r_d \ge 10\,R_D)}$	Medium (-10) $= \boxed{-g_m(R_F\|r_d\|R_D)}$ $\cong \boxed{-g_mR_D}_{\,(R_F,\; r_d \ge 10\,R_D)}$
Voltage-divider bias E-MOSFET	Medium (1 MΩ) $= \boxed{R_1\|R_2}$	Medium (2 kΩ) $= \boxed{R_D\|r_d}$ $\cong \boxed{R_D}_{\,(R_d \ge 10\,R_D)}$	Medium (-10) $= \boxed{-g_m(r_d\|R_D)}$ $\cong \boxed{-g_mR_D}_{\,(r_d \ge 10\,R_D)}$

9.13 **Summary Table**   **499**

listed will probably be some variation of those appearing in the table, so at the very least, the listing will provide some insight as to what expected levels should be and which path will probably generate the desired equations. The format chosen was designed to permit a duplication of the entire table on the front and back of one $8\frac{1}{2}$ by 11 inch page.

9.14 TROUBLESHOOTING

As mentioned before, troubleshooting a circuit is a combination of knowing the theory and having experience using meters and an oscilloscope to check the operation of the circuit. A good troubleshooter has a "nose" for finding the trouble in a circuit—this ability to "see" what is happening being greatly developed through building, testing, and repairing many different circuits. For an FET small-signal amplifier, one could go about troubleshooting a circuit by performing a number of basic steps:

1. Look at the circuit board to see if any obvious problems can be seen: an area charred by excess heating of a component; a component that feels or seems too hot to touch; what appears to be a poor solder joint; any connection that appears to have come loose.
2. Use a dc meter: make some measurements as marked in a repair manual containing the circuit schematic diagram and a listing of test dc voltages.
3. Apply a test ac signal: measure the ac voltages starting at the input and working along toward the output.
4. If the problem is identified at a particular stage, the ac signal at various points should be checked using an oscilloscope to see the waveform, its polarity, amplitude, and frequency, as well as any unusual waveform "glitches" that may be present. In particular, observe that the signal is present for the full signal cycle.

Possible Symptoms and Actions

If there is no output ac voltage:

1. Check if the supply voltage is present.
2. Check if the output voltage at V_D is between 0 V and V_{DD}.
3. Check if there is any input ac signal at the gate terminal.
4. Check the ac voltage at each side of the coupling capacitor terminals.

  When building and testing an FET amplifier circuit in the laboratory:

1. Check the color code of resistor values to be sure that they are correct. Even better, measure the resistor value as components used repeatedly may get overheated when used incorrectly, causing the nominal value to change.
2. Check that all dc voltages are present at the component terminals. Be sure that all ground connections are made common.
3. Measure the ac input signal to be sure the expected value is provided to the circuit.

9.15 PRACTICAL APPLICATIONS

Three-Channel Audio Mixer

The basic components of a three-channel JFET audio mixer are shown in Fig. 9.46. The three input signals can come from different sources such as a microphone, a musical instrument, background sound generators, etc. All signals can be applied to the

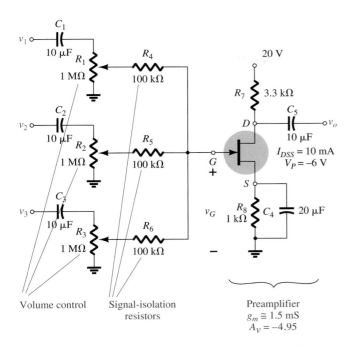

Volume control   Signal-isolation resistors   Preamplifier
$g_m \cong 1.5$ mS
$A_v = -4.95$

Figure 9.46 Basic components of a three-channel JFET audio mixer.

same gate terminal because the input impedance of the JFET is so high that it can be approximated by an open circuit. **In general, the input impedance is 1000 MΩ(10^9 Ω) or better for JFETs and 100 million MΩ (10^{14} Ω) or better for MOSFETs.** If BJTs were employed instead of JFETs, the lower input impedance would require a transistor amplifier for each channel or at least an emitter-follower as the first stage to provide a higher input impedance.

The 10 -μF capacitors are there to prevent any dc biasing levels on the input signal from appearing at the gate of the JFET, and the 1-MΩ potentiometers are the volume control for each channel. The need for the 100-kΩ resistors for each channel is less obvious. Their purpose is to ensure that one channel does not load down the other channels and severely reduce or distort the signal at the gate. For instance, in Fig. 9.47a, one channel has a high-impedance (10-kΩ) microphone, whereas another channel has a low-impedance (0.5-kΩ) guitar amplifier. Channel 3 is left open, and the 100-kΩ isolation resistors have been removed for the moment. Replacing the capacitors by their short-circuit equivalent for the frequency range of interest and ignoring the effects of the parallel 1-MΩ potentiometers (set at their maximum value) will result in the equivalent circuit of Fig. 9.47b at the gate of the JFET amplifier. Using the superposition theorem, the voltage at the gate of the JFET is determined by

$$v_G = \frac{0.5 \text{ k}\Omega(v_{\text{mic}})}{10.5 \text{ k}\Omega} + \frac{10 \text{ k}\Omega(v_{\text{guitar}})}{10.5 \text{ k}\Omega}$$

$$= 0.047v_{\text{mic}} + 0.95v_{\text{guitar}} \cong v_{\text{guitar}}$$

clearly showing that the guitar has swamped the signal of the microphone. The only response of the amplifier of Fig. 9.47 will be to the guitar. Now, with the 100-kΩ resistors in place, the situation of Fig. 9.47c results. Using the superposition theorem again, the equation for the voltage at the gate becomes the following:

$$v_G = \frac{101 \text{ k}\Omega(v_{\text{mic}})}{211 \text{ k}\Omega} + \frac{110 \text{ k}\Omega(v_{\text{guitar}})}{211 \text{ k}\Omega}$$

$$\cong 0.48v_{\text{mic}} + 0.52v_{\text{guitar}}$$

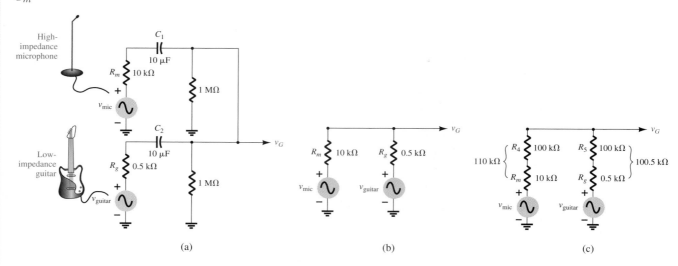

Figure 9.47 (a) Application of a high and low impedance source to the mixer of Fig. 9.46; (b) reduced equivalent without the 100-kΩ isolation resistors; (c) reduced equivalent with the 100-kΩ resistors.

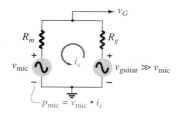

Figure 9.48 Demonstrating the fact that for parallel signals, the channel with the least internal impedance and most power controls the situation.

showing an even balance in the signals at the gate of the JFET. **In general, therefore, the 100-kΩ resistors compensate for any difference in signal impedance to ensure that one does not load down the other and develop a mixed level of signals at the amplifier. Technically, they are often called "signal isolation resistors."**

An interesting consequence of a situation such as described in Fig. 9.47b is depicted in Fig. 9.48 where a guitar of low impedance has a signal level of about 150 mV, whereas the microphone, having a larger internal impedance, has a signal strength of only 50 mV. As pointed out above, the majority of the signal at the "feed" point (v_G) is that of the guitar. The resulting direction of current and power flow is unquestionably from the guitar to the microphone. Further, **since the basic construction of a microphone and speaker is quite similar, the microphone may be forced to act like a speaker and broadcast the guitar signal.** New acoustical bands often face this problem as they learn the rudiments of good amplifier basics. **In general, for parallel signals, the channel with the least internal impedance controls the situation.**

For some it may come as quite a surprise that a microphone can actually behave like a speaker. However, the classical example of the use of one voice cone to act as a microphone and speaker is in the typical intercom system such as appearing in Fig. 9.49a. The 16-Ω, 0.5-W speaker of Fig. 9.49b is being used as both a microphone and a speaker, depending on the position of the activation switch. It is important to note, however, as in the microphone-guitar example above, that most speakers are designed to handle reasonable power levels; but most microphones are designed to simply accept the voice-activated input, and they cannot handle the power levels normally associated with speakers. Just compare the size of each in any audio system. In general, a situation such as described above, where the guitar signal is heard over the microphone, will ultimately damage the microphone. For an intercom system the speaker is designed to handle both types of excitation without difficulty.

In Fig. 9.46 the gain of the self-biased JFET is determined by $-g_m R_D$, which for this situation is

$$-g_m R_D = -(1.5 \text{ mS})(3.3 \text{ k}\Omega) = \mathbf{-4.95}$$

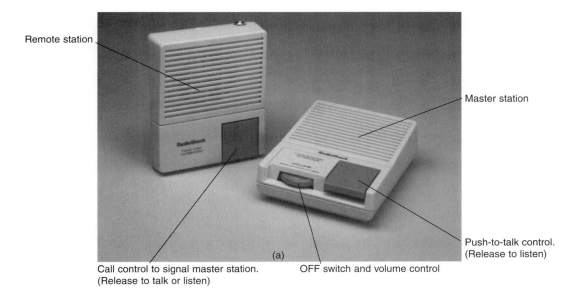

(a)

Remote station

Master station

Push-to-talk control.
(Release to listen)

Call control to signal master station.
(Release to talk or listen)

OFF switch and volume control

(b)

Cable connection   Remote station   Cable connection

Master station

16-Ω, 0.5-W
speaker (mic)

16-Ω, 0.5-W
speaker (mic)

9-V battery connection

Call control

Electrolytic capacitors

Integrated circuit

33-μF, 16-V
electrolytic
capacitor

Ceramic disc capacitors

Push-to-talk control

Potentiometric OFF
and volume control

Figure 9.49 Battery-powered
(9-V), two-station intercom: (a)
external appearance; (b) internal
construction.

Silent Switching

**Any electronic system that incorporates mechanical switching such as shown
in Fig. 9.50 is prone to developing noise on the line that will reduce the signal-
to-noise ratio.** When the switch of Fig. 9.50 is opened and closed, you will often get

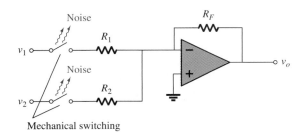

Noise

v_1 R_1 R_F v_o

Noise

v_2 R_2

Mechanical switching

Figure 9.50 Noise development
due to mechanical switching.

an annoying "pfft, pfft" sound as part of the output signal. In addition, the longer wires normally associated with mechanical switches will require that the switch be as close to the amplifier as possible to reduce the noise pickup on the line.

One effective method to essentially eliminate this source of noise is to use electronic switching such as shown in Fig. 9.51a for a two-channel mixing network.

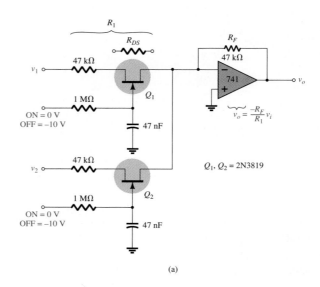

(a)

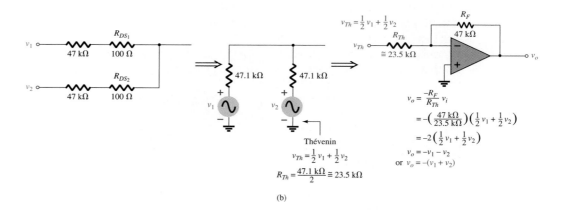

(b)

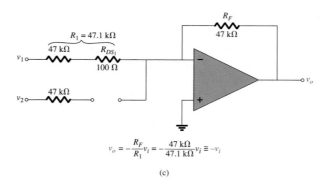

$$v_o = -\frac{R_F}{R_1}v_i = -\frac{47 \text{ k}\Omega}{47.1 \text{ k}\Omega}v_i \cong -v_i$$

(c)

Figure 9.51 Silent switching audio network: (a) JFET configuration; (b) with both signals present; (c) with one signal on. (Redrawn from *Electronics Today International*)

Recall from Chapter 6 that the drain to source of a JFET for low values of V_{DS} can be looked upon as a resistance whose value is determined by the applied gate-to-source voltage as described in detail in Section 6.13. In addition, recall that the resistance is the least at $V_{GS} = 0$ V and the highest near pinch-off. In Fig. 9.51a, the signals to be mixed are applied to the drain side of each JFET, and the dc control is connected directly to the gate terminal of each JFET. With 0 V at each control terminal, both JFETs are heavily "on," and the resistance from D_1 to S_1 and from D_2 to S_2 is relatively small, say, 100 Ω for this discussion. Although 100 Ω is not the 0 Ω assumed with an ideal switch, it is so small compared to the series 47-kΩ resistor that it can often be ignored. Both switches are therefore in the "on" position, and both input signals can make their way to the input of the inverting amplifier (to be introduced in Section 13.4) as shown in Fig. 9.51b. In particular, note that the chosen resistor values result in an output signal that is simply an inversion of the sum of the two signals. The amplifier stage to follow will then raise the summation to audio levels.

Both electronic switches can be put in the "off" state by applying a voltage that is more negative than the pinch-off level as indicated by the 10 V in Fig. 9.51a. The level of "off" resistance can approach 10,000 MΩ which certainly can be approximated by an open circuit for most applications. Since both channels are isolated, one can be "on" while the other is "off." The speed of operation of a JFET switch is controlled by the substrate (those due to the device construction) and stray capacitance levels and the low "on" resistance of the JFET. **Maximum speeds for JFETs are about 100 MHz, with 10 MHz being more typical.** However, this speed is critically reduced by the input resistance and capacitance of the design. In Fig. 9.51a, the 1-MΩ resistor and the 47-nF capacitors will have a time constant of $\tau = RC = 47$ ms $= 0.047$ s for the dc charging network that is controlling the voltage at the gate. If we assume two time constants to charge to the pinch-off level, the total time is 0.094 s, or a switching speed of $1/0.094$ s $\cong 10.6$ per second. Compared to the typical switching speed of the JFET at 10 million times in 1s, this number is extremely small. Keep in mind, however, that the application is the important consideration, and for a typical mixer, switching is not going to occur at speeds greater than 10.6 per second unless we have some radical input signals. One might ask why it is necessary to have the RC time constant at the gate at all. Why not let the applied dc level at the gate simply control the state of the JFET? In general, the RC time constant ensures that the control signal is not a spurious one generated by noise or "ringing" due to the sharply rising and falling applied pulses at the gate. By using a charging network, the dc level must be present for a period of time before the pinch-off level is reached. Any spike on the line will not be present long enough to charge the capacitor and switch the state of the JFET.

It is important to realize **that the JFET switch is a bilateral switch.** That is, signals in the "on" state can pass through the drain-source region in either direction. This, of course, is the way ordinary mechanical switches work, which makes it that much easier to replace mechanical switch designs with electronic switches. Remember that the diode is not a bilateral switch because it can conduct current at low voltages in only one direction.

It should be noted that **because the state of the JFETs can be controlled by a dc level, the design of Fig. 9.51a lends itself to remote and computer control** for the same reasons described in Chapter 6 when dc control was discussed.

The data sheet for a low-cost JFET analog switch is provided in Fig. 9.52. Note, in particular, that the pinch-off voltage is typically about -10 V at a drain-to-source voltage of 12 V. In addition, a current level of 10 nA is used to define the pinch-off level. The level of I_{DSS} is 15 mA, whereas the drain-to-source resistance is quite low at 150 Ω with $V_{GS} = 0$ V. The turn-on time is quite small at 10 ns $(t_d + t_r)$, whereas the turn-off time is 25 ns.

ON Semiconductor™

JFET Switching
N–Channel — Depletion

2N5555

CASE 29–11, STYLE 5
TO–92 (TO–226AA)

MAXIMUM RATINGS

Rating	Symbol	Value	Unit
Drain–Source Voltage	V_{DS}	25	Vdc
Drain–Gate Voltage	V_{DG}	25	Vdc
Gate–Source Voltage	V_{GS}	25	Vdc
Forward Gate Current	I_{GF}	10	mAdc
Total Device Dissipation @ T_C = 25°C Derate above 25°C	P_D	350 2.8	mW mW/°C
Junction Temperature Range	T_J	–65 to +150	°C
Storage Temperature Range	T_{stg}	–65 to +150	°C

1 DRAIN

3 GATE

2 SOURCE

ELECTRICAL CHARACTERISTICS (T_A = 25°C unless otherwise noted)

Characteristic	Symbol	Min	Max	Unit
OFF CHARACTERISTICS				
Gate–Source Breakdown Voltage (I_G = 10 µAdc, V_{DS} = 0)	$V_{(BR)GSS}$	25	—	Vdc
Gate Reverse Current (V_{GS} = 15 Vdc, V_{DS} = 0)	I_{GSS}	—	1.0	nAdc
Drain Cutoff Current (V_{DS} = 12 Vdc, V_{GS} = –10 V) (V_{DS} = 12 Vdc, V_{GS} = –10 V, T_A = 100°C)	$I_{D(off)}$	— —	10 2.0	nAdc µAdc
ON CHARACTERISTICS				
Zero–Gate–Voltage Drain Current[(1)] (V_{DS} = 15 Vdc, V_{GS} = 0)	I_{DSS}	15	—	mAdc
Gate–Source Forward Voltage ($I_{G(f)}$ = 1.0 mAdc, V_{DS} = 0)	$V_{GS(f)}$	—	1.0	Vdc
Drain–Source On–Voltage (I_D = 7.0 mAdc, V_{GS} = 0)	$V_{DS(on)}$	—	1.5	Vdc
Static Drain–Source On Resistance (I_D = 0.1 mAdc, V_{GS} = 0)	$r_{DS(on)}$	—	150	Ohms

1. Pulse Test: Pulse Width < 300 µs, Duty Cycle < 3.0%.

Characteristic	Symbol	Min	Max	Unit
SMALL–SIGNAL CHARACTERISTICS				
Small–Signal Drain–Source "ON" Resistance (V_{GS} = 0, I_D = 0, f = 1.0 kHz)	$r_{ds(on)}$	—	150	Ohms
Input Capacitance (V_{DS} = 15 Vdc, V_{GS} = 0, f = 1.0 MHz)	C_{iss}	—	5.0	pF
Reverse Transfer Capacitance (V_{DS} = 0, V_{GS} = 10 Vdc, f = 1.0 MHz)	C_{rss}	—	1.2	pF
SWITCHING CHARACTERISTICS				
Turn–On Delay Time	(V_{DD} = 10 Vdc, $I_{D(on)}$ = 7.0 mAdc,	$t_{d(on)}$	—	5.0	ns
Rise Time	$V_{GS(on)}$ = 0, $V_{GS(off)}$ = –10 Vdc) (See Figure 1)	t_r	—	5.0	ns
Turn–Off Delay Time	(V_{DD} = 10 Vdc, $I_{D(on)}$ = 7.0 mAdc,	$t_{d(off)}$	—	15	ns
Fall Time	$V_{GS(on)}$ = 0, $V_{GS(off)}$ = –10 Vdc) (See Figure 1)	t_f	—	10	ns

Figure 9.52   Specification sheet for a low-cost analog JFET current switch.
(Copyright of Semiconductor Components Industries, LLC. Used by permission.)

Phase-Shift Networks

Using the voltage-controlled drain-to-source resistance characteristic of a JFET, the phase angle of a signal can be controlled using the configurations of Fig. 9.53. The network of Fig. 9.53a is a phase-advance network that adds an angle to the applied signal, whereas the network of Fig. 9.53b is a phase-retard configuration that creates a negative phase shift.

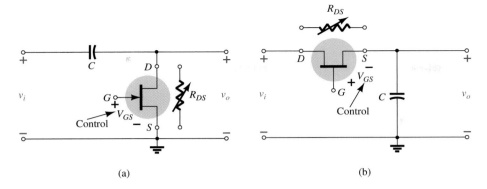

Figure 9.53 Phase-shift networks: (a) advance; (b) retard.

For example, let us consider the effect of R_{DS} on an input signal having a frequency such as 10 kHz if we apply it to the network of Fig. 9.53a. For discussion, let us assume that the drain-to-source resistance is 2 kΩ due to an applied gate-to-source voltage of -3 V. Drawing the equivalent network will result in the general configuration of Fig. 9.54. Solving for the output voltage will result in

$$\mathbf{V}_o = \frac{R_{DS}\angle 0°\, V_i \angle 0°}{R_{DS} - jX_C} = \frac{R_{DS}\, V_i \angle 0°}{\sqrt{R_{DS}^2 + X_C^2}\angle -\tan^{-1}\dfrac{X_C}{R_{DS}}}$$

$$= \frac{R_{DS} V_i}{\sqrt{R_{DS}^2 + X_C^2}}\angle \tan^{-1}\frac{X_C}{R_{DS}} = \left(\frac{R_{DS}}{\sqrt{R_{DS}^2 + X_C^2}}\right) V_i \angle \tan^{-1}\frac{X_C}{R_{DS}}$$

so that

$$\mathbf{V}_o = k_1 V_i \angle \theta_1$$

where

$$k_1 = \frac{R_{DS}}{\sqrt{R_{DS}^2 + X_C^2}} \quad \text{and} \quad \theta_1 = \tan^{-1}\frac{X_C}{R_{DS}} \tag{9.57}$$

Substituting the numerical values from above will result in

$$X_C = \frac{1}{2\pi f C} = \frac{1}{2\pi(10\text{ kHz})(0.01\ \mu\text{F})} = 1.592\text{ k}\Omega$$

and

$$k_1 = \frac{R_{DS}}{\sqrt{R_{DS}^2 + X_C^2}} = \frac{2\text{ k}\Omega}{\sqrt{(2\text{ k}\Omega)^2 + (1.592\text{ k}\Omega)^2}} = 0.782$$

with

$$\theta_1 = \tan^{-1}\frac{X_C}{R_{DS}} = \tan^{-1}\frac{1.592\text{ k}\Omega}{2\text{ k}\Omega} = \tan^{-1} 0.796 = 38.52°$$

so that

$$\mathbf{V}_o = 0.782 V_i \angle 38.52°$$

and an output signal that is 78.2% of its applied signal but with a phase shift of 38.52°.

In general, therefore, the network of Fig. 9.53a can introduce a positive phase shift extending from a few degrees (with X_C relatively small compared to R_{DS}) to almost 90° (with X_C relatively large compared to R_{DS}). Keep in mind, however, that for fixed values of R_{DS}, as the frequency increases, X_C will decrease and the phase shift will approach 0°. For decreasing frequencies and a fixed R_{DS}, the phase shift will approach 90°. It is also important to realize that for a fixed R_{DS}, an increasing level of X_C

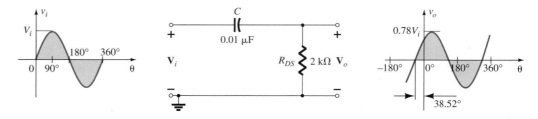

Figure 9.54 *R-C phase-advance network.*

results in diminishing magnitude for V_o. For such a network, a balance between gain and desired phase shift will have to be made.

For the network of Fig. 9.53b the concluding equations are

$$\mathbf{V}_o = k_2 V_i \angle \theta_2 \qquad (9.58)$$

where

$$k_2 = \frac{X_C}{\sqrt{R_{DS}^2 + X_C^2}} \quad \text{and} \quad \theta_2 = -\tan^{-1}\frac{R_{DS}}{X_C}$$

Motion-Detection System

The basic components of a passive infrared (PIR) motion-detection system are shown in Fig. 9.55. The heart of the system is **the pyroelectric detector that generates a voltage that varies with the amount of incident heat.** It filters out all but the infrared radiation from a particular area and focuses the energy onto a temperature-sensing element. Recall from Chapter 6 in the Practical Applications section that the infrared band is a nonvisible band just below the visible light spectrum. **Passive detectors do not emit a signal of any kind but simply respond to the energy flow of the environment.**

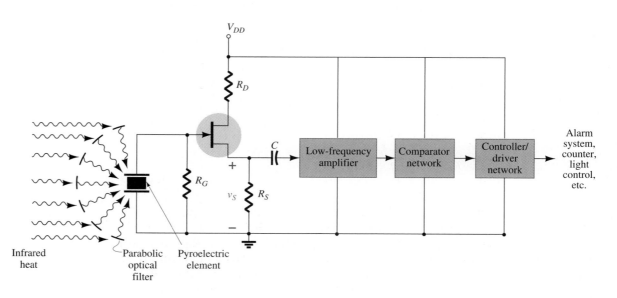

Figure 9.55 *Passive infrared (PIR) motion-detection system.*

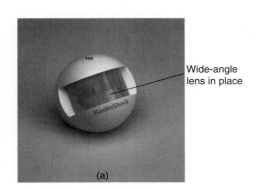

(a)

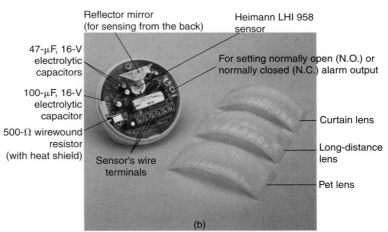

Reflector mirror
(for sensing from the back)

Heimann LHI 958 sensor

47-μF, 16-V electrolytic capacitors

For setting normally open (N.O.) or normally closed (N.C.) alarm output

100-μF, 16-V electrolytic capacitor

500-Ω wirewound resistor
(with heat shield)

Sensor's wire terminals

Curtain lens

Long-distance lens

Pet lens

(b)

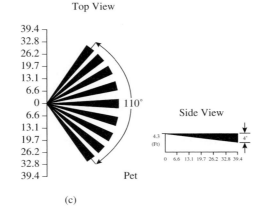

Lens No.	3
Type	Pet
Mounting Height	4.3 ft (1.3 m)
Zone's Coverage	110°

Top View

Side View

(c)

Figure 9.56 Commercially available PIR motion-detection unit: (a) external appearance; (b) internal construction; (c) pet option coverage.

An external and an internal view of a commercially available unit are provided in Fig. 9.56a and b. Four interchangeable lens are provided for different coverage areas. For our purposes the pet option was selected with the coverage indicated in Fig. 9.56c. Note the space under the ray for pet motion and the maximum distance of 39.4 ft. The unit is mounted at a height of 6.6 ft and operates at a dc voltage of 9 to 16 V, drawing a current of 25 mA at 12 V dc. In Fig. 9.56b, the Heimann LH1 958 pyroelectric detector is identified along with the deflector for side detection and the very prominent reed switch in the "can." The controlling ICs are on the other side of the printed circuit board.

To focus the incident ambient heat on the pyroelectric detector, the unit of Fig. 9.56 uses a parabolic deflector. As a person walks past a sensor, he will cut the various fields appearing in Fig. 9.56c, and the detector will sense the **rapid changes** in heat level. **The result is a changing dc level akin to a low-frequency ac signal of relatively high internal impedance appearing at the gate of the JFET.** One might then ask why turning a heating system on or turning on a lamp doesn't generate an alarm signal since heat will be generated. The answer is that both will generate a voltage at the detector that grows steadily with increasing heat level from the heating system or burning bulb. Remember that for the lamp, the detector is heat-sensitive and not light-sensitive. The resulting voltage is not oscillating between levels but simply climbing in level and will not set off the alarm—a varying ac voltage will not be generated by the pyroelectric detector!

Note in Fig. 9.55 that a JFET source-follower configuration was employed to ensure a very high input impedance to capture most of the pyroelectric signal. It is then passed through a low-frequency amplifier, followed by a peak detecting network and a comparator to determine whether the alarm should be set off. The dc voltage comparator is a network that "captures" the peak value of the generated ac voltage and compares it to a known dc voltage level. The output processor will determine whether the difference between the two levels is sufficient to tell the driver to energize the alarm.

9.16 SUMMARY

Important Conclusions and Concepts

1. **The transconductance parameter g_m** is determined by the **ratio of the change in drain current** associated with a particular **change in gate-to-source voltage** in the region of interest. The **steeper the slope** of the I_D-versus-V_{GS} curve, the **greater** the level of g_m. In addition, the **closer the point or region of interest to the saturation current I_{DSS}**, the **greater** the transconductance parameter.

2. On specification sheets, g_m is provided as y_{fs}.

3. When V_{GS} is **one-half the pinch-off value**, g_m is **one-half the maximum value**.

4. When I_D is **one-fourth the saturation level of I_{DSS}**, g_m **is one-half the value at saturation**.

5. The **output impedance** of FETs is **similar in magnitude** to that of **conventional BJTs.**

6. On specification sheets the **output impedance r_d is provided as $1/y_{os}$**. The **more horizontal** the characteristic curves on the drain characteristics, the **greater the output impedance.**

7. The **voltage gain** for the fixed-bias and self-bias JFET configurations (with a bypassed source capacitance) **is the same.**

8. The **ac analysis** of JFETs and depletion-type MOSFETs **is the same.**

9. The **ac equivalent network** for an enhancement-type MOSFET **is the same** as that employed for JFETs and depletion-type MOSFETs. The only difference is the equation for g_m.

10. The **magnitude of the gain** of FET networks is typically **between 2 and 20.** The **self-bias configuration** (without a bypass source capacitance) and the **source-follower** are **low-gain configurations.**

11. There is **no phase shift** between input and output for the **source-follower** and **common-gate configurations.** Most others have a 180° phase shift.

12. The **output impedance** for most FET configurations is **determined primarily by R_D.** For the **source-follower** configuration it is determined by R_S and g_m.

13. The **input impedance** for most FET configurations is **quite high.** However, it is **quite low** for the **common-gate configuration.**

14. Whenever **troubleshooting any electronic or mechanical system,** always check the **most obvious causes first.**

Equations

$$g_m = y_{fs} = \frac{\Delta I_D}{\Delta V_{GS}}$$

$$g_{m0} = \frac{2I_{DSS}}{|V_P|}$$

$$g_m = g_{m0} \left[1 - \frac{V_{GS}}{V_P} \right]$$

$$g_m = g_{m0} \sqrt{\frac{I_D}{I_{DSS}}}$$

$$r_d = \frac{1}{y_{os}} = \frac{\Delta V_{DS}}{\Delta I_D} \bigg|_{V_{GS} = \text{constant}}$$

JFETs and depletion-type MOSFETs ($r_d \geq 10R_D$, $r_d \geq 10R_S$):
   Fixed-bias:

$$Z_i = R_G$$

$$Z_o \cong R_D$$

$$A_v = -g_m R_D$$

Self-bias (bypassed R_S):

$$Z_i = R_G$$

$$Z_o \cong R_D$$

$$A_v = -g_m R_D$$

Self-bias (unbypassed R_S):

$$Z_i = R_G$$

$$Z_o = R_D$$

$$A_v \cong \frac{-g_m R_D}{1 + g_m R_S}$$

Voltage-divider bias:

$$Z_i = R_1 \| R_2$$

$$Z_o = R_D$$

$$A_v = -g_m R_D$$

Source-follower:

$$Z_i = R_G$$

$$Z_o = R_S \| 1/g_m$$

$$A_v = \frac{g_m R_S}{1 + g_m R_S}$$

Common-gate:

$$Z_i = R_S \| 1/g_m$$

$$Z_o \cong R_D$$

$$A_v = g_m R_D$$

Enhancement-type MOSFETs ($R_F \gg r_d \| R_D$; $r_d \geq 10R_D$):

$$g_m = 2k(V_{GS_Q} - V_{GS(\text{Th})})$$

Drain-feedback configuration:

$$Z_i = \frac{R_F}{1 + g_m R_D}$$

$$Z_o \cong R_D$$

$$A_v \cong -g_m R_D$$

Voltage-divider bias:

$$Z_i = R_1 \| R_2$$

$$Z_o \cong R_D$$

$$A_v = -g_m R_D$$

9.17   COMPUTER ANALYSIS

PSpice Windows

JFET FIXED-BIAS CONFIGURATION

The first JFET configuration to be analyzed using PSpice Windows is the fixed-bias configuration of Fig. 9.57, which has a JFET with $V_P = -4$ V and $I_{DSS} = 10$ mA. The 10-MΩ resistor was added to act as a path to ground for the capacitor but is essentially an open-circuit as a load. The **J2N3819 *n*-channel JFET** from the **EVAL.slb** library will be used, and the ac voltage will be determined at four different points for comparison and review.

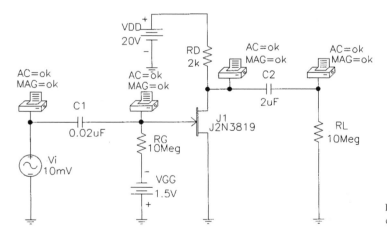

Figure 9.57   Fixed-bias JFET configuration with an ac source.

The constant **Beta** is determined by

$$\text{Beta} = \frac{I_{DSS}}{|V_P|^2} = \frac{10 \text{ mA}}{4^2} = 0.625 \text{ mA/V}^2$$

and inserted as a **Model Parameter** using the sequence **Edit-Model-Edit Instance Model (Text). Vto** must also be changed to -4 V. The remaining elements of the network are set as described for the transistor in Chapter 8.

An analysis of the network will result in the printout of Fig. 9.58. The **Schematics Netlist** reveals the nodes assigned to each parameter and defines the nodes for which the ac voltage is to be printed. In this case, note that **Vi** is set at 10 mV at a frequency of 10 kHz from node 2 to 0. In the list of **Junction FET MODEL PARAMETERS, VTO** is -4 V and **BETA** is 625E-6 as entered. The **SMALL-SIGNAL**

```
****        CIRCUIT DESCRIPTION

***********************************************************************

* Schematics Netlist *


V_VDD         $N_0001 0 20V
C_C1          $N_0002 $N_0003  0.02uF
R_RG          $N_0004 $N_0003  10Meg
R_RD          $N_0005 $N_0001  2k
V_VGG         0 $N_0004 1.5V

.PRINT        AC
+ VM([$N_0003])

.PRINT        AC
+ VM([$N_0002])

.PRINT        AC
+ VM([$N_0005])

.PRINT        AC
+ VM([$N_0006])
C_C2          $N_0005 $N_0006  2uF
V_Vi          $N_0002 0  AC 10mV
+SIN 0 10mV 10kHz 0 0 0
R_RL          0 $N_0006  10Meg
J_J1          $N_0005 $N_0003 0 J2N3819-X

****     Junction FET MODEL PARAMETERS

***********************************************************************


             J2N3819-X
             NJF
      VTO     -4
     BETA     625.000000E-06
   LAMBDA     2.250000E-03
       IS     33.570000E-15
      ISR     322.400000E-15
    ALPHA     311.700000E-06
       VK     243.6
       RD     1
       RS     1
      CGD     1.600000E-12
      CGS     2.414000E-12
        M     .3622
    VTOTC     -2.500000E-03
   BETATCE    -.5
       KF     9.882000E-18


****     SMALL SIGNAL BIAS SOLUTION      TEMPERATURE =   27.000 DEG C

***********************************************************************


NODE   VOLTAGE    NODE   VOLTAGE    NODE   VOLTAGE    NODE   VOLTAGE

($N_0001)   20.0000              ($N_0002)    0.0000

($N_0003)   -1.5000              ($N_0004)   -1.5000

($N_0005)   12.0020              ($N_0006)    0.0000



    VOLTAGE SOURCE CURRENTS
    NAME         CURRENT

    V_VDD        -3.999E-03
    V_VGG        -1.366E-12

****     OPERATING POINT INFORMATION     TEMPERATURE =   27.000 DEG C

***********************************************************************


**** JFETS

NAME        J_J1
MODEL       J2N3819-X
ID          4.00E-03
VGS         -1.50E+00
VDS         1.20E+01
GM          3.20E-03
GDS         8.76E-06
CGS         1.73E-12
CGD         6.07E-13 .
**** 10/08/97 11:23:59 ******** NT Evaluation PSpice (October 1996) *********

****     AC ANALYSIS                     TEMPERATURE =   27.000 DEG C

***********************************************************************

  FREQ        VM($N_0003)

  1.000E+04   9.997E-03


  FREQ        VM($N_0002)

  1.000E+04   1.000E-02


  FREQ        VM($N_0005)

  1.000E+04   6.275E-02


  FREQ        VM($N_0006)

  1.000E+04   6.275E-02
```

Figure 9.58 Output file for the network of Figure 9.57.

BIAS SOLUTION reveals that the voltage at both ends of R_G is -1.5 V, resulting in $V_{GS} = -1.5$ V. The voltage from drain to source (ground) is 12 V, leaving a drop of 8 V across R_D. The **AC ANALYSIS** at the end of the listing reveals that the voltage at the source (node 2) is 10 mV as set, but the voltage at the other end of the capacitor is 3 μV less due to the impedance of the capacitor at 10 kHz—certainly a drop to be ignored. The choice of 0.02 μF for this frequency was obviously a good one. The voltages before and after the capacitor on the output side are exactly the same (to three places), revealing that the larger the capacitor, the closer the characteristics to a short circuit. The output of 6.275E-2 = 62.75 mV reflects a gain of 6.275. The **OPERATING POINT INFORMATION** reveals that I_D is 4 mA and g_m is 3.2 mS. Calculating the value of g_m from:

$$g_m = \frac{2I_{DSS}}{|V_P|}\left(1 - \frac{V_{GS_Q}}{V_P}\right)$$

$$= \frac{2(10 \text{ mA})}{4 \text{ V}}\left(1 - \frac{(-1.5 \text{ V})}{(-4 \text{ V})}\right)$$

$$= 3.125 \text{ mS}$$

confirming our analysis.

JFET SELF-BIAS CONFIGURATION

The self-bias configuration of Fig. 9.59 will be analyzed using the **J2N3819 JFET** from the library and then using an approximate equivalent circuit. It will be interesting to see if there are any major differences in solution.

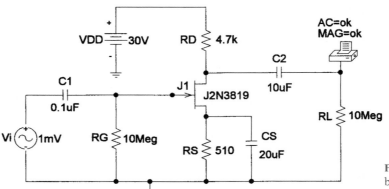

Figure 9.59 Self-bias configuration with an ac source.

Again, $V_P = -4$ V and $I_{DSS} = 10$ mA, resulting in a **Vto** of -4 and a **Beta** of 6.25E-4. The **Analysis** is run and the results of Fig. 9.60 obtained. The nodes are identified in the **Schematics Netlist** and the parameters in the **Junction FET MODEL PARAMETERS**. The **SMALL-SIGNAL BIAS SOLUTION** reveals that $V_{GS} = -1.7114$ V and $V_D = 14.228$ V—results that are very close to a hand-written solution of -1.68 V and 14.49 V. The **OPERATING POINT INFORMATION** reveals that I_D is 3.36 mA compared to a hand-calculated level of 3.3 mA and that g_m is 2.94 mS compared to a hand-calculated level of 2.90 mS. The **AC ANALYSIS** provides an output level of 13.3 mV at an angle of $-179.9°$, which compares well with a hand-calculated level of 13.63 mV at an angle of $-180°$. The results of JFETs are a lot closer than those obtained for transistors when we used the provided elements because of the special feature of having essentially infinite input impedance so that the gate current is zero ampere. Recall that for the transistor, V_{BE} is a function of the operating conditions.

```
****    CIRCUIT DESCRIPTION


***********************************************************************

* Schematics Netlist *


V_Vi          $N_0001 0   AC 1mV
+SIN 0V 1mV 10kHz 0 0 0
C_C1          $N_0001 $N_0002  0.1uF
R_RG          $N_0002 0  10Meg
V_VDD         $N_0003 0 30V
R_RD          $N_0003 $N_0004  4.7k
R_RS          $N_0005 0  510
C_CS          $N_0005 0  20uF
C_C2          $N_0004 $N_0006  10uF
R_RL          $N_0006 0  10Meg
J_J1          $N_0004 $N_0002 $N_0005 J2N3819-X

.PRINT        AC
+ VM([$N_0006])
+ VP([$N_0006])

****    Junction FET MODEL PARAMETERS


***********************************************************************

             J2N3819-X
              NJF
       VTO   -4
      BETA   625.000000E-06
    LAMBDA    2.250000E-03
        IS    33.570000E-15
       ISR   322.400000E-15
     ALPHA   311.700000E-06
        VK   243.6
        RD   1
        RS   1
       CGD    1.600000E-12
       CGS    2.414000E-12
         M    .3622
     VTOTC   -2.500000E-03
   BETATCE   -.5
        KF    9.882000E-18


****    SMALL SIGNAL BIAS SOLUTION      TEMPERATURE =   27.000 DEG C


***********************************************************************

NODE   VOLTAGE    NODE   VOLTAGE     NODE   VOLTAGE    NODE   VOLTAGE

($N_0001)    0.0000          ($N_0002) 13.95E-06

($N_0003)   30.0000          ($N_0004)   14.2280

($N_0005)    1.7114          ($N_0006)    0.0000


    VOLTAGE SOURCE CURRENTS
    NAME           CURRENT

    V_Vi          0.000E+00
    V_VDD        -3.356E-03

    TOTAL POWER DISSIPATION   1.01E-01  WATTS

****    OPERATING POINT INFORMATION     TEMPERATURE =   27.000 DEG C


***********************************************************************


**** JFETS

NAME       J_J1
MODEL      J2N3819-X
ID         3.36E-03
VGS       -1.71E+00
VDS        1.25E+01
GM         2.94E-03
GDS        7.34E-06
CGS        1.68E-12
CGD        5.97E-13 ·

****    AC ANALYSIS                     TEMPERATURE =   27.000 DEG C


***********************************************************************


FREQ         VM($N_0006) VP($N_0006)

 1.000E+04   1.330E-02  -1.799E+02
```

Figure 9.60 Output file for the network of Figure 9.59.

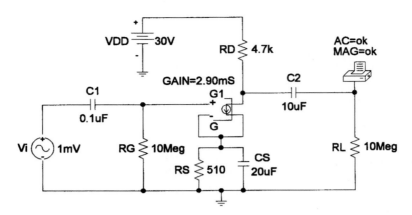

Figure 9.61 Network in Figure 9.59 following substitution of a VCCS for the JFET in the ac domain.

We will now investigate the self-bias configuration using the approximate model as done for the transistor and see if there is an improvement in the results (compared to the hand-calculated levels). In this case, we need the voltage controlled current source (**VCCS**) found in the **ANALOG.slb** library as **G.** When selected, the **Description** reads **Voltage-controlled current source.** When placed on the schematic, it will appear as shown in Fig. 9.61. The sensing voltage is between the plus and minus sign, while the controlled current is between the other two external terminals. Double-clicking on the schematic symbol will result in a **PartName: G** dialog box, in which the **GAIN**(g_m) can be set to the hand-calculated level of 2.90 mS.

The result of an analysis is a gain of 13.62—almost an exact match with the hand-written gain. This approach is certainly valid for an ac analysis, but if we examine the **SMALL-SIGNAL BIAS SOLUTION,** we find that the results are meaningless. Therefore, the equivalent appearing in Fig. 9.61 is only valid for the ac gain since the only parameter defined is the ac transconductance factor.

JFET VOLTAGE-DIVIDER CONFIGURATION

The last network to be analyzed in this PSpice Windows presentation is the voltage-divider configuration of Fig. 9.62. Note that the parameters chosen are different from those employed in earlier examples, with V_i at 24 mV and a frequency of 5 kHz. In addition, the dc levels are displayed and a plot of the output and input voltages will be obtained on the same screen.

After setting up the network, the source V_i must be set to the indicated parameters by double-clicking on the source and then sequentially double-clicking on each parameter and typing in the correct values. Each must be saved and then the display changed to print the magnitude of the ac voltage and the applied frequency. In this

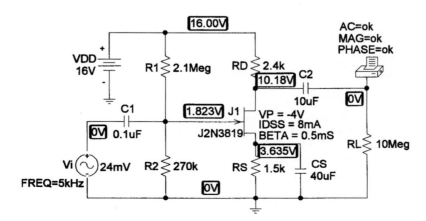

Figure 9.62 JFET voltage-divider configuration with an ac source.

example, the JFET parameters were printed on the screen using the **ABC** icon. **BETA** is of course calculated from $I_{DSS}/|V_P|^2$. Under **Analysis-Probe Setup,** the option **Do not auto-run Probe** was chosen, and under **Setup, AC Sweep** was chosen and the frequency of 5 kHz entered. Finally, since we want the dc levels to be displayed, the **Display Results on Schematic** option is chosen under **Analysis,** and **Enable Voltage Display** is enabled. The resulting dc levels of Fig. 9.62 reveal that V_{GS} is 1.823 V − 3.635 V = −1.812 V, comparing very well with the −1.8 V calculated in Example 6.5. V_D is 10.18 V compared to the calculated level of 10.24 V, and V_{DS} is 10.18 V − 3.635 V = 6.545 V compared to 6.64 V.

For the ac solution, we can choose **Examine Output** under **Analysis** and find under **OPERATING POINT INFORMATION** that g_m is 2.22 mS, comparing very well with the hand-calculated value of 2.2 mS, and under **AC ANALYSIS** that the output ac voltage is 125.8 mV, resulting in a gain of 125.8 mV/24 mV = 5.24. The hand-calculated level is $g_m R_D = (2.2 \text{ mS})(2.4 \text{ k}\Omega) = 5.28$. The ac waveform for the output can be obtained by first applying the sequence **Analysis-Probe Setup-Automatically run Probe after simulation.** Then, return to **Setup** under **Analysis,** and enable **Transient,** disable **AC Sweep,** and double-click **Transient** to obtain the **Transient** dialog box. For the frequency of 5 kHz, the period is 200 μs. A **Print Step** to 2 μs would then give us 100 plot points for each cycle. The **Final Time** will be 5 × 200 μs = 1 ms to show five cycles. The **No-Print Delay** will be 0s and the **Step Ceiling** 2 μs. Then, click the **Trace** icon, choose **V(J1:d),** and the output waveform of Fig. 9.63 will appear. Choose **Plot-Add Plot-Trace-Add-V(Vi:+),** and both waveforms will appear as shown. Shift **SEL** ≫ to the bottom waveform by simply bringing the pointer to the left of the lower waveform and left-clicking the mouse once. Click the **Toggle cursor** icon, and a horizontal line will appear at the dc level of the output voltage at 10.184 V. A left click of the mouse and an intersecting set of lines will appear. Choose the **Cursor Peak** icon, and the intersection will automatically go to the peak value of the waveform (**A1** in the dialog box). The difference appearing in the dialog box is 125.496 mV, comparing well with the printed value in the output file. The difference is simply due to the number of points chosen for the plot; an increased number of plot points would have brought the two levels closer together.

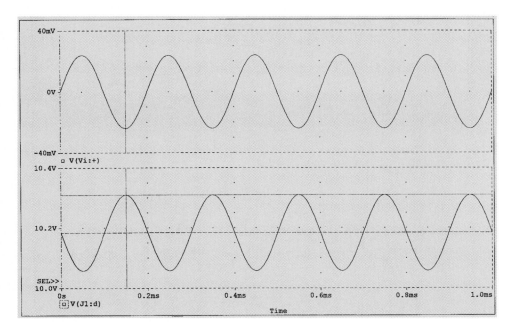

Figure 9.63 The ac drain and gate voltage for the voltage-divider JFET configuration of Figure 9.62.

Electronics Workbench

The ac gain for the JFET self-bias network of Fig. 9.64 will now be determined using Electronics Workbench. The entire procedure for setting up the network and obtaining the desired readings was described for BJT ac networks in Chapter 8. This particular network will appear again in Chapter 11 as Fig. 11.39 when we turn our attention to the frequency response of a loaded JFET amplifier. A detailed analysis is provided in Chapter 11, including determining the dc levels, the value for g_m, and the loaded gain. The drain current of Example 11.10 is 2 mA, resulting in a drain voltage of 10.6 V and a source voltage of 2 V which compare very well with the 10.497 V and 2.009 V of Fig. 9.64. When a load such as R_L is added to the network, it will appear in parallel with R_D of the network, changing the gain equation to $-g_m R_D \| R_L$. For Example 11.10, g_m is 2 mS, resulting in an overall gain of $-(2 \text{ mS})(2.2 \text{ k}\Omega \| 4.7 \text{ k}\Omega) = -2.997$. The meters of Fig. 9.64 provide effective values of the ac voltages, resulting in a gain of $-2.047 \text{ mV}/699.651 \ \mu\text{V} = -2.926$, which is nice match with the hand-calculated solution. Recall from the discussion of BJT ac networks that the voltage appearing with an ac voltage source is the peak value as indicated in Fig. 9.64. The reason the voltage at the gate is not $0.7071(1 \text{ mV}) = 707.1 \ \mu\text{V}$ is the drop across the internal impedance of the source (**Rsig**) and the isolation capacitor (**CG**).

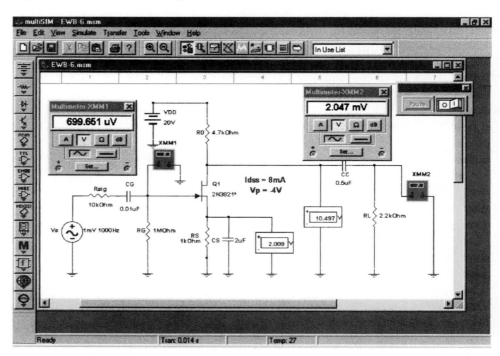

Figure 9.64 JFET self-bias network using EWB.

PROBLEMS

§ 9.2 FET Small-Signal Model

1. Calculate g_{m0} for a JFET having device parameters $I_{DSS} = 15$ mA and $V_P = -5$ V.

2. Determine the pinch-off voltage of a JFET with $g_{m0} = 10$ mS and $I_{DSS} = 12$ mA.

3. For a JFET having device parameters $g_{m0} = 5$ mS and $V_P = -3.5$ V, what is the device current at $V_{GS} = 0$ V?

4. Calculate the value of g_m for a JFET $(I_{DSS} = 12 \text{ mA}, V_P = -3 \text{ V})$ at a bias point of $V_{GS} = -1$ V.

5. For a JFET having $g_m = 6$ mS at $V_{GS_Q} = -1$ V, what is the value of I_{DSS} if $V_P = -2.5$ V?

6. A JFET $(I_{DSS} = 10 \text{ mA}, V_P = -5 \text{ V})$ is biased at $I_D = I_{DSS}/4$. What is the value of g_m at that bias point?

7. Determine the value of g_m for a JFET ($I_{DSS} = 8$ mA, $V_P = -5$ V) when biased at $V_{GS_Q} = V_P/4$.

8. A specification sheet provides the following data (at a listed drain-source current)

$$y_{fs} = 4.5 \text{ mS}, \qquad y_{os} = 25 \ \mu\text{S}$$

At the listed drain–source current, determine:
   (a) g_m.
   (b) r_d.

9. For a JFET having specified values of $y_{fs} = 4.5$ mS and $y_{os} = 25 \ \mu$S, determine the device output impedance, Z_o(FET), and device ideal voltage gain, A_v(FET).

10. If a JFET having a specified value of $r_d = 100$ kΩ has an ideal voltage gain of A_v(FET) $= -200$, what is the value of g_m?

11. Using the transfer characteristic of Fig. 9.65:
   (a) What is the value of g_{m0}?
   (b) Determine g_m at $V_{GS} = -1.5$ V graphically.
   (c) What is the value of g_m at $V_{GS_Q} = -1.5$ V using Eq. (9.6)? Compare with the solution to part (b).
   (d) Graphically determine g_m at $V_{GS} = -2.5$ V.
   (e) What is the value of g_m at $V_{GS_Q} = -2.5$ V using Eq. (9.6)? Compare with the solution to part (d).

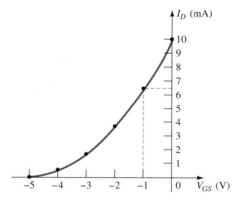

Figure 9.65 JFET transfer characteristic for Problem 11

12. Using the drain characteristic of Fig. 9.66:
   (a) What is the value of r_d for $V_{GS} = 0$ V?
   (b) What is the value of g_{m0} at $V_{DS} = 10$ V?

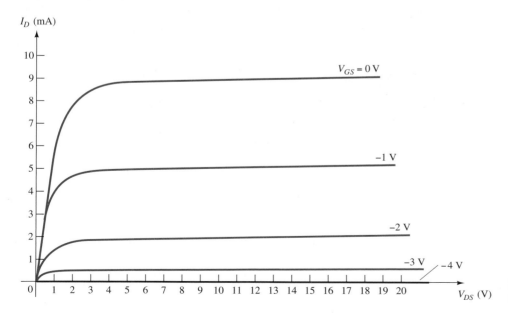

Figure 9.66 JFET drain characteristic for Problem 12

13. For a 2N4220 *n*-channel JFET (y_{fs}(minimum) $= 750 \, \mu\text{S}$, y_{os}(maximum) $= 10 \, \mu\text{S}$):
    (a) What is the value of g_m?
    (b) What is the value of r_d?

14. (a) Plot g_m vs. V_{GS} for an *n*-channel JFET with $I_{DSS} = 8 \, \text{mA}$ and $V_P = -6 \, \text{V}$.
    (b) Plot g_m vs. I_D for the same *n*-channel JFET as part (a).

15. Sketch the ac equivalent model for a JFET if $y_{fs} = 5.6 \, \text{mS}$ and $y_{os} = 15 \, \mu\text{S}$.

16. Sketch the ac equivalent model for a JFET if $I_{DSS} = 10 \, \text{mA}$, $V_P = -4 \, \text{V}$, $V_{GS_Q} = -2 \, \text{V}$, and $y_{os} = 25 \, \mu\text{S}$.

§ **9.3 JFET Fixed-Bias Configuration**

17. Determine Z_i, Z_o, and A_v for the network of Fig. 9.67 if $I_{DSS} = 10 \, \text{mA}$, $V_P = -4 \, \text{V}$, and $r_d = 40 \, \text{k}\Omega$.

18. Determine Z_i, Z_o, and A_v for the network of Fig. 9.67 if $I_{DSS} = 12 \, \text{mA}$, $V_P = -6 \, \text{V}$, and $y_{os} = 40 \, \mu\text{S}$.

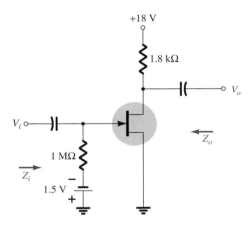

Figure 9.67 Fixed-bias amplifier for Problems 17 and 18

§ **9.4 JFET Self-Bias Configuration**

19. Determine Z_i, Z_o, and A_v for the network of Fig. 9.68 if $y_{fs} = 3000 \, \mu\text{S}$ and $y_{os} = 50 \, \mu\text{s}$.

20. Determine Z_i, Z_o, and A_v for the network of Fig. 9.69 if $I_{DSS} = 6 \, \text{mA}$, $V_P = -6 \, \text{V}$, and $y_{os} = 40 \, \mu\text{S}$.

21. Determine Z_i, Z_o, and A_v for the network of Fig. 9.68 if the 20-μF capacitor is removed and the parameters of the network are the same as in Problem 19. Compare results with those of Problem 19.

22. Repeat Problem 19 if y_{os} is $10 \, \mu\text{S}$. Compare the results to those of Problem 19.

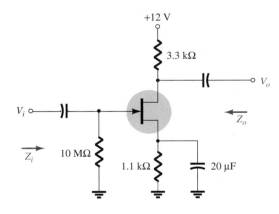

Figure 9.68 Problems 19, 21, 22 and 46

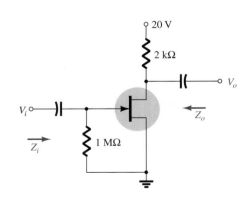

Figure 9.69 Self-bias configuration for Problems 20 and 47

§ 9.5 JFET Voltage-Divider Configuration

23. Determine Z_i, Z_o, and V_o for the network of Fig. 9.70 if $V_i = 20$ mV.

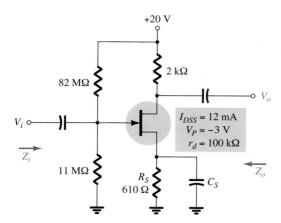

Figure 9.70 Problems 23–26 and 48

24. Determine Z_i, Z_o, and V_o for the network of Fig. 9.70 if $V_i = 20$ mV and the capacitor C_S is removed.

25. Repeat Problem 23 if $r_d = 20$ kΩ and compare results.

26. Repeat Problem 24 if $r_d = 20$ kΩ and compare results.

§ 9.6 JFET Source-Follower Configuration

27. Determine Z_i, Z_o, and A_v for the network of Fig. 9.71.

28. Repeat Problem 27 if $r_d = 20$ kΩ.

29. Determine Z_i, Z_o, and A_v for the network of Fig. 9.72.

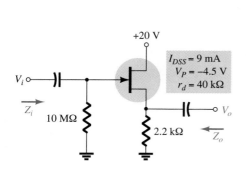

Figure 9.71 Problems 27 and 28

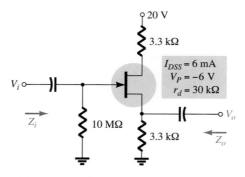

Figure 9.72 Problem 29

§ 9.7 JFET Common-Gate Configuration

30. Determine Z_i, Z_o, and V_o for the network of Fig. 9.73 if $V_i = 0.1$ mV.

31. Repeat Problem 30 if $r_d = 25$ kΩ.

32. Determine Z_i, Z_o, and A_v for the network of Fig. 9.74 if $r_d = 33$ kΩ.

g_m

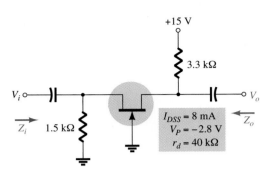

Figure 9.73   Problems 30, 31, and 49

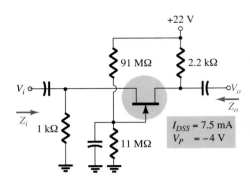

Figure 9.74   Problem 32

§ 9.8 Depletion-Type MOSFETs

33. Determine V_o for the network of Fig. 9.75 if $y_{os} = 20\ \mu S$.

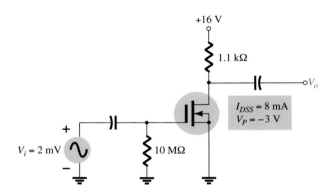

Figure 9.75   Problem 33

34. Determine Z_i, Z_o, and A_v for the network of Fig. 9.76 if $r_d = 60\ k\Omega$.

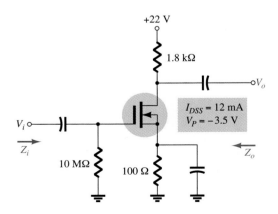

Figure 9.76   Problems 34, 35, and 50

35. Repeat Problem 34 if $r_d = 25$ kΩ.

36. Determine V_o for the network of Fig. 9.77 if $V_i = 4$ mV.

37. Determine Z_i, Z_o, and A_v for the network of Fig. 9.78.

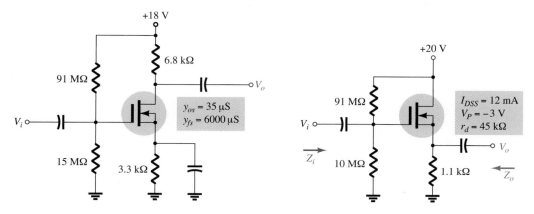

Figure 9.77   Problem 36

Figure 9.78   Problem 37

§ 9.10 E-MOSFET Drain-Feedback Configuration

38. Determine g_m for a MOSFET if $V_{GS(Th)} = 3$ V and it is biased at $V_{GS_Q} = 8$ V. Assume $k = 0.3 \times 10^{-3}$.

39. Determine Z_i, Z_o, and A_v for the amplifier of Fig. 9.79 if $k = 0.3 \times 10^{-3}$.

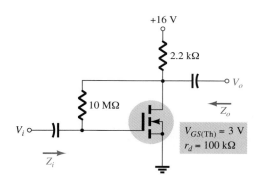

Figure 9.79   Problems 39, 40, and 51

40. Repeat Problem 39 if k drops to 0.2×10^{-3}. Compare results.

41. Determine V_o for the network of Fig. 9.80 if $V_i = 20$ mV.

42. Determine V_o for the network of Fig. 9.80 if $V_i = 4$ mV, $V_{GS(Th)} = 4$ V, and $I_{D(on)} = 4$ mA, with $V_{GS(on)} = 7$ V and $y_{os} = 20$ μS.

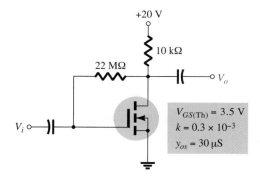

Figure 9.80   Problems 41 and 42

§ 9.11 E-MOSFET Voltage-Divider Configuration

43. Determine the output voltage for the network of Fig. 9.81 if $V_i = 0.8$ mV and $r_d = 40$ kΩ.

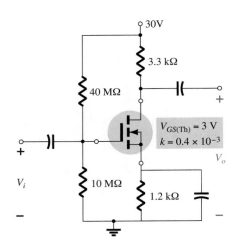

Figure 9.81 Problem 43

§ 9.12 Designing FET Amplifier Networks

44. Design the fixed-bias network of Fig. 9.82 to have a gain of 8.

45. Design the self-bias network of Fig. 9.83 to have a gain of 10. The device should be biased at $V_{GS_Q} = \frac{1}{3}V_P$.

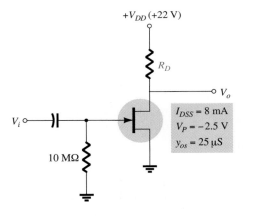

Figure 9.82 Problem 44

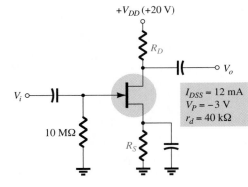

Figure 9.83 Problem 45

§ 9.17 Computer Analysis

46. Using PSpice Windows, determine the voltage gain for the network of Fig. 9.68.

47. Using Electronics Workbench, determine the voltage gain for the network of Fig. 9.69.

48. Using PSpice Windows, determine the voltage gain for the network of Fig. 9.70.

49. Using Electronics Workbench, determine the voltage gain for the network of Fig. 9.73.

50. Using PSpice Windows, determine the voltage gain for the network of Fig. 9.76.

51. Using PSpice Windows, determine the voltage gain for the network of Fig. 9.79.

CHAPTER 10

Systems Approach— Effects of R_s and R_L

R_s/R_L

10.1 INTRODUCTION

In recent years, the introduction of a wide variety of packaged networks and systems has generated an increasing interest in the systems approach to design and analysis. Fundamentally, this approach concentrates on the terminal characteristics of a package and treats each as a building block in the formation of the total package. The content of this chapter is a first step in developing some familiarity with this approach. The techniques introduced will be used in the remaining chapters and broadened as the need arises. The trend to packaged systems is quite understandable when you consider the enormous advances in the design and manufacturing of integrated circuits (ICs). The small IC packages contain stable, reliable, self-testing, sophisticated designs that would be quite bulky if built with discrete (individual) components. The systems approach is not a difficult one to apply once the basic definitions of the various parameters are correctly understood and the manner in which they are utilized is clearly demonstrated. In the next few sections, we develop the systems approach in a slow deliberate manner that will include numerous examples to make each salient point. If the content of this chapter is clearly and correctly understood, a first plateau in the understanding of system analysis will be accomplished.

10.2 TWO-PORT SYSTEMS

The description to follow can be applied to any two-port system—not only those containing BJTs and FETs—although the emphasis in this chapter is on these active devices. The emphasis in previous chapters on determining the two-port parameters for various configurations will be quite helpful in the analysis to follow. In fact, many of the results obtained in the last two chapters are utilized in the analysis to follow.

In Fig. 10.1, the important parameters of a two-port system have been identified. Note in particular the absence of a load and a source resistance. The impact of these important elements is considered in detail in a later section. For the moment recognize that the impedance levels and the gains of Fig. 10.1 are determined for no-load (absence of R_L) and no-source resistance (R_s) conditions.

If we take a "Thévenin look" at the output terminals we find with V_i set to zero that

$$\boxed{Z_{\text{Th}} = Z_o = R_o} \qquad (10.1)$$

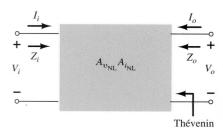

Figure 10.1 Two-port system.

E_{Th} is the open-circuit voltage between the output terminals identified as V_o. However,

$$A_{v_{\text{NL}}} = \frac{V_o}{V_i}$$

and

$$V_o = A_{v_{\text{NL}}} V_i$$

so that

$$\boxed{E_{\text{Th}} = A_{v_{\text{NL}}} V_i} \tag{10.2}$$

Note the use of the additional subscript notation "NL" to identify a no-load voltage gain.

Substituting the Thévenin equivalent circuit between the output terminals will result in the output configuration of Fig. 10.2. For the input circuit the parameters V_i and I_i are related by $Z_i = R_i$, permitting the use of R_i to represent the input circuit. Since our present interest is in BJT and FET amplifiers, both Z_o and Z_i can be represented by resistive elements.

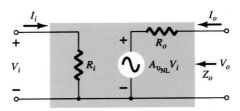

Figure 10.2 Substituting the internal elements for the two-port system of Fig. 10.1.

Before continuing let us check the results of Fig. 10.2 by finding Z_o and $A_{v_{\text{NL}}}$ in the usual manner. To find Z_o, V_i is set to zero, resulting in $A_{v_{\text{NL}}} V_i = 0$, permitting a short-circuit equivalent for the source. The result is an output impedance equal to R_o as originally defined. The absence of a load will result in $I_o = 0$, and the voltage drop across the impedance R_o will be 0 V. The open-circuit output voltage is therefore $A_{v_{\text{NL}}} V_i$, as it should be. Before looking at an example, take note of the fact that A_i does not appear in the two-port model of Fig. 10.2 and in fact is seldom part of the two-port system analysis of active devices. This is not to say that the quantity is seldom calculated, but it is most frequently calculated from the expression $A_i = -A_v(Z_i/R_L)$, where R_L is the defined load for the analysis of interest.

EXAMPLE 10.1

For the fixed-bias transistor network of Fig. 10.3 (Example 8.1), sketch the two-port equivalent of Fig. 10.2.

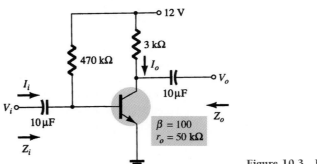

Figure 10.3 Example 10.1.

Solution

From Example 8.1,

$$Z_i = 1.069 \text{ k}\Omega$$

$$Z_o = 3 \text{ k}\Omega$$

$$A_{v_{\text{NL}}} = -280.11$$

Using the information above, the two-port equivalent of Fig. 10.4 can be drawn. Note in particular the negative sign associated with the controlled voltage source, revealing an opposite polarity for the controlled source than that indicated in the figure. It also reveals a 180° phase-shift between the input and output voltages.

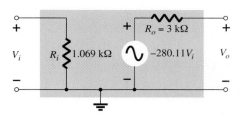

Figure 10.4 Two-port equivalent for the parameters specified in Example 10.1.

In Example 10.1, $R_C = 3 \text{ k}\Omega$ was included in defining the no-load voltage gain. Although this need not be the case (R_C could be defined as the load resistor in Chapter 8), the analysis of this chapter will assume that all biasing resistors are part of the no-load gain and that a loaded system requires an additional load R_L connected to the output terminals.

A second format for Fig. 10.2, particularly popular with op-amps (operational amplifiers), appears in Fig. 10.5. The only change is the general appearance of the model.

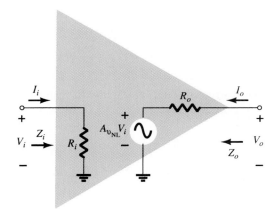

Figure 10.5 Operational amplifier (op-amp) notation.

10.3 EFFECT OF A LOAD IMPEDANCE (R_L)

In this section, the effect of an applied load is investigated using the two-port model of Fig. 10.2. The model can be applied to any current- or voltage-controlled amplifier. $A_{v_{\text{NL}}}$ is, as defined earlier, the gain of the system without an applied load. R_i and R_o are the input and output impedances of the amplifier as defined by the configuration. Ideally, all the parameters of the model are unaffected by changing loads or source resistances (as normally encountered for op-amps to be described in Chapter 14). However, for some transistor amplifier configurations, R_i can be quite sensitive to the applied load, while for others R_o can be sensitive to the source resistance. In any case, once $A_{v_{NL}}$, R_i, and R_o are defined for a particular configuration, the equations about to be derived can be employed.

Applying a load to the two-port system of Fig. 10.2 will result in the configuration of Fig. 10.6. Applying the voltage-divider rule to the output circuit will result in

$$V_o = \frac{R_L A_{v_{NL}} V_i}{R_L + R_o}$$

and

$$A_v = \frac{V_o}{V_i} = \frac{R_L}{R_L + R_o} A_{v_{NL}} \tag{10.3}$$

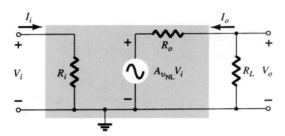

Figure 10.6 Applying a load to the two-port system of Fig. 10.2.

Since the ratio $R_L/(R_L + R_o)$ will always be less than 1:

The loaded voltage gain of an amplifier is always less than the no-load level.

Note also that the formula for the voltage gain does not include the input impedance or current gain.

Although the level of R_i may change with the configuration, the applied voltage and input current will always be related by

$$I_i = \frac{V_i}{Z_i} = \frac{V_i}{R_i} \tag{10.4}$$

Defining the output current as the current through the load will result in

$$I_o = -\frac{V_o}{R_L} \tag{10.5}$$

with the minus sign occurring due to the defined direction for I_o in Fig. 10.6.

The current gain is then determined by

$$A_i = \frac{I_o}{I_i} = \frac{-V_o/R_L}{V_i/Z_i} = -\frac{V_o}{V_i}\frac{Z_i}{R_L}$$

and

$$A_i = -A_v \frac{Z_i}{R_L} \tag{10.6}$$

for the unloaded situation. In general, therefore, the current gain can be obtained from the voltage gain and impedance parameters Z_i and R_L. The next example will demonstrate the usefulness and validity of Eqs. (10.3) through (10.6).

EXAMPLE 10.2

In Fig. 10.7, a load has been applied to the fixed-bias transistor amplifier of Example 10.1 (Fig. 10.3).
(a) Determine the voltage and current gain using the two-port systems approach defined by the model of Fig. 10.4.
(b) Determine the voltage and current gain using the r_e model and compare results.

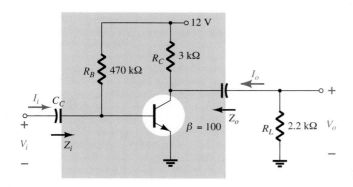

Figure 10.7 Example 10.2.

Solution

(a) Recall from Example 10.1 that

$$Z_i = 1.071 \text{ k}\Omega \qquad (\text{with } r_e = 10.71 \ \Omega \text{ and } \beta = 100)$$

$$Z_o = 3 \text{ k}\Omega$$

$$A_{v_{\text{NL}}} = -280.11$$

Applying Eq. (10.3) yields

$$A_v = \frac{R_L}{R_L + R_o} A_{v_{\text{NL}}}$$

$$= \frac{2.2 \text{ k}\Omega}{2.2 \text{ k}\Omega + 3 \text{ k}\Omega}(-280.11)$$

$$= (0.423)(-280.11)$$

$$= \mathbf{-118.5}$$

For the current gain,

$$A_i = -A_v \frac{Z_i}{R_L}$$

In this case, Z_i is unaffected by the applied load and

$$A_i = -(-118.5)\frac{1.071 \text{ k}\Omega}{2.2 \text{ k}\Omega} = \mathbf{57.69}$$

(b) Substituting the r_e model will result in the network of Fig. 10.8. Note in particular that the applied load is in parallel with the collector resistor R_C defining a net parallel resistance

$$R'_L = R_C \| R_L = 3 \text{ k}\Omega \| 2.2 \text{ k}\Omega = 1.269 \text{ k}\Omega$$

The output voltage

$$V_o = -\beta I_B R'_L$$

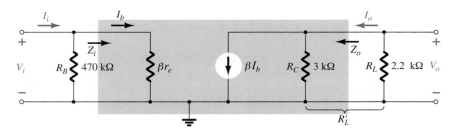

Figure 10.8 Substituting the r_e model in the ac equivalent network of Fig. 10.7.

with

$$I_b = \frac{V_i}{\beta r_e}$$

and

$$V_o = -\beta \frac{V_i}{\beta r_e} R_L'$$

so that

$$\boxed{A_v = \frac{V_o}{V_i} = -\frac{R_L'}{r_e} = -\frac{R_C \| R_L}{r_e}}$$

(10.7)

Substituting values gives

$$A_v = -\frac{1.269 \text{ k}\Omega}{10.71 \ \Omega} = \mathbf{-118.5}$$

as obtained above. For the current gain, by the current-divider rule,

$$I_b = \frac{(470 \text{ k}\Omega)I_i}{470 \text{ k}\Omega + 1.071 \text{ k}\Omega} = 0.9977 I_i \cong I_i$$

and

$$I_o = \frac{3 \text{ k}\Omega(\beta I_b)}{3 \text{ k}\Omega + 2.2 \text{ k}\Omega}$$

$$= 0.5769 \beta I_b$$

so that

$$A_i = \frac{I_o}{I_i} = \frac{0.5769 \beta I_b}{I_i} = \frac{0.5769 \beta I_i}{I_i}$$

$$= 0.5769(100) = \mathbf{57.69}$$

as obtained using Eq. (10.6).

Example 10.2 demonstrated two techniques to solve the same problem. Although any network can be solved using the r_e model approach, the advantage of the systems approach is that once the two-port parameters of a system are known, the effect of changing the load can be determined directly from Eq. (10.3). No need to go back to the ac equivalent model and analyze the entire network. The advantages of the systems approach are similar to those associated with applying Thévenin's theorem. They permit concentrating on the effects of the load without having to re-examine the entire network. Of course, if the network of Fig. 10.7 were presented for analysis without the unloaded parameters, it would be a toss-up as to which approach would yield the desired results in the most direct, efficient manner. However, keep in mind that the "package" approach is the developing trend. When you purchase a "system" the two-port parameters are provided, and as with any trend, the user must be aware of how to utilize the given data.

The AC Load Line

For a system such as appearing in Fig. 10.9a, the dc load line was drawn on the output characteristics as shown in Fig. 10.9b. The load resistance did not contribute to the dc load line since it was isolated from the biasing network by the coupling capacitor (C_C). For the ac analysis, the coupling capacitors are replaced by a short-circuit equivalence that will place the load and collector resistors in a parallel arrangement defined by

$$R'_L = R_C \| R_L$$

The effect on the load line is shown in Fig. 10.9b with the levels to determine the new axes intersections. Note of particular importance that the ac and dc load lines pass through the same Q-point—a condition that must be satisfied to ensure a common solution for the network under dc and/or ac conditions.

For the unloaded situation, the application of a relatively small sinusoidal signal to the base of the transistor could cause the base current to swing from a level of I_{B_2} to I_{B_4} as shown in Fig. 10.9b. The resulting output voltage v_{ce} would then have the swing appearing in the same figure. The application of the same signal for a loaded situation would result in the same swing in the I_B level, as shown in Fig. 10.9b. The result, however, of the steeper slope of the ac load line is a smaller output voltage

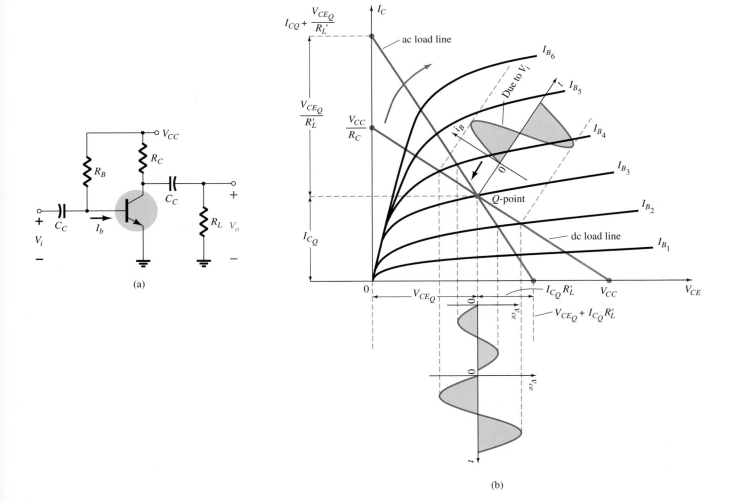

Figure 10.9 Demonstrating the differences between the dc and ac load lines.

swing (v_{ce}) and a drop in the gain of the system as demonstrated in the numerical analysis above. It should be obvious from the intersection of the ac load line on the vertical axis that the smaller the level of R'_L, the steeper the slope and the smaller the ac voltage gain. Since R'_L is smaller for reduced levels of R_L, it should be fairly clear that:

For a particular design, the smaller the level of R_L, the lower the level of ac voltage gain.

10.4 EFFECT OF THE SOURCE IMPEDANCE (R_s)

Our attention will now turn to the input side of the two-port system and the effect of an internal source resistance on the gain of an amplifier. In Fig. 10.10, a source with an internal resistance has been applied to the basic two-port system. The definitions of Z_i and $A_{v_{NL}}$ are such that:

The parameters Z_i and $A_{v_{NL}}$ of a two-port system are unaffected by the internal resistance of the applied source.

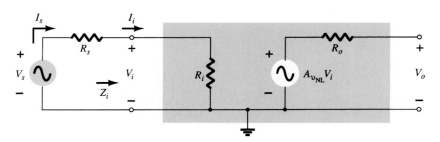

Figure 10.10 Including the effects of the source resistance R_s.

However:

The output impedance may be affected by the magnitude of R_s.

Recall Eq. (8.110) for the complete hybrid equivalent model. The fraction of the applied signal reaching the input terminals of the amplifier of Fig. 10.10 is determined by the voltage-divider rule. That is,

$$V_i = \frac{R_i V_s}{R_i + R_s} \tag{10.8}$$

Equation (10.8) clearly shows that the larger the magnitude of R_s, the less the voltage at the input terminals of the amplifier. In general, therefore:

For a particular amplifier, the larger the internal resistance of a signal source the less the overall gain of the system.

For the two-port system of Fig. 10.10,

$$V_o = A_{v_{NL}} V_i$$

and

$$V_i = \frac{R_i V_s}{R_i + R_s}$$

so that

$$V_o = A_{v_{NL}} \frac{R_i}{R_i + R_s} V_s$$

and

$$A_{v_s} = \frac{V_o}{V_s} = \frac{R_i}{R_i + R_s} A_{v_{NL}} \tag{10.9}$$

The result clearly supports the statement above regarding the reduction in gain with increase in R_s. Using Eq. (10.9), if $R_s = 0 \, \Omega$ (ideal voltage source), $A_{v_s} = A_{v_{NL}}$, which is the maximum possible value.

The source current:

$$I_s = I_i = \frac{V_s}{R_s + R_i} \qquad (10.10)$$

In Fig. 10.11, a source with an internal resistance has been applied to the fixed-bias transistor amplifier of Example 10.1 (Fig. 10.3).
(a) Determine the voltage gain $A_{v_s} = V_o/V_s$. What percent of the applied signal appears at the input terminals of the amplifier?
(b) Determine the voltage gain $A_{v_s} = V_o/V_s$ using the r_e model.

EXAMPLE 10.3

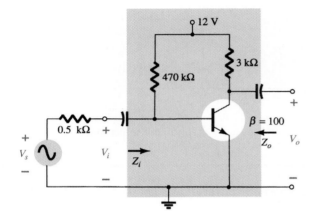

Figure 10.11 Example 10.3.

Solution

(a) The two-port equivalent for the network appears in Fig. 10.12.

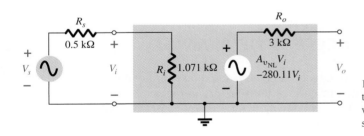

Figure 10.12 Substituting the two-port equivalent network for the fixed-bias transistor amplifier of Fig. 10.11.

$$\text{Eq. (10.9):} \quad A_{v_s} = \frac{V_o}{V_s} = \frac{R_i}{R_i + R_s} A_{v_{NL}} = \frac{1.071 \text{ k}\Omega}{1.071 \text{ k}\Omega + 0.5 \text{ k}\Omega}(-280.11)$$

$$= (0.6817)(-280.11)$$

$$= \mathbf{-190.96}$$

$$\text{Eq. (10.8):} \quad V_i = \frac{R_i V_s}{R_i + R_s} = \frac{(1.071 \text{ k}\Omega)V_s}{1.071 \text{ k}\Omega + 0.5 \text{ k}\Omega} = 0.6817 V_s$$

or **68.2%** of the available signal reached the amplifier and 31.8% was lost across the internal resistance of the source.

(b) Substituting the r_e model will result in the equivalent circuit of Fig. 10.13. Solving for V_o gives

$$V_o = -(100I_b)3 \text{ k}\Omega$$

with $$Z_i \cong \beta r_e \text{ and } I_b \cong I_i = \frac{V_s}{R_s + \beta r_e} = \frac{V_s}{1.571 \text{ k}\Omega}$$

and $$V_o = -100\left(\frac{V_s}{1.571 \text{ k}\Omega}\right)3 \text{ k}\Omega$$

so that $$A_{v_s} = \frac{V_o}{V_s} = -\frac{(100)(3 \text{ k}\Omega)}{1.57 \text{ k}\Omega}$$

$$= -190.96$$

as above.

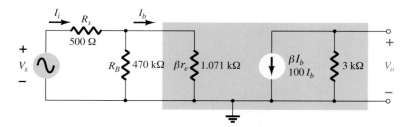

Figure 10.13   Substituting the r_e equivalent circuit for the fixed-bias transistor amplifier of Fig. 10.11.

Throughout the analysis above, note that R_s was not included in the definition of Z_i for the two-port system. Of course, the resistance "seen" by the source is now $R_s + Z_i$, but R_s remains a quantity associated only with the applied source.

Note again in Example 10.3 that the same results were obtained with the systems approach and using the r_e model. Certainly, if the two-port parameters are available, they should be applied. If not, the approach to the solution is simply a matter of preference.

10.5   COMBINED EFFECT OF R_s AND R_L

The effects of R_s and R_L have now been demonstrated on an individual basis. The next natural question is how the presence of both factors in the same network will affect the total gain. In Fig. 10.14, a source with an internal resistance R_s and a load R_L have been applied to a two-port system for which the parameters Z_i, $A_{v_{NL}}$, and Z_o have been specified. For the moment, let us assume that Z_i and Z_o are unaffected by R_L and R_s, respectively.

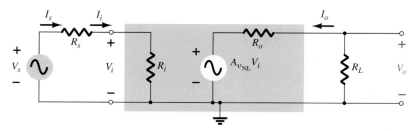

Figure 10.14   Considering the effects of R_s and R_L on the gain of an amplifier.

At the input side we find

$$\text{Eq. (10.8):} \quad V_i = \frac{R_i V_s}{R_i + R_s}$$

or

$$\boxed{\frac{V_i}{V_s} = \frac{R_i}{R_i + R_s}} \tag{10.11}$$

and at the output side,

$$V_o = \frac{R_L A_{v_{\text{NL}}} V_i}{R_L + R_o}$$

or

$$\boxed{A_v = \frac{V_o}{V_i} = \frac{R_L A_{v_{\text{NL}}}}{R_L + R_o}} \tag{10.12}$$

For the total gain $A_{v_s} = V_o/V_s$, the following mathematical steps can be performed:

$$A_{v_s} = \frac{V_o}{V_s} = \frac{V_o}{V_i} \frac{V_i}{V_s} \tag{10.13}$$

and substituting Eqs. (10.11) and (10.12) will result in

$$A_{v_s} = \frac{R_L A_{v_{\text{NL}}}}{R_L + R_o} \frac{R_i}{R_i + R_s}$$

and

$$\boxed{A_{v_s} = \frac{V_o}{V_s} = \frac{R_i}{R_i + R_s} \frac{R_L}{R_L + R_o} A_{v_{\text{NL}}}} \tag{10.14}$$

Since $I_i = V_i/R_i$, as before,

$$\boxed{A_i = -A_v \frac{R_i}{R_L}} \tag{10.15}$$

or using $I_s = V_s/(R_s + R_i)$,

$$\boxed{A_{i_s} = -A_{v_s} \frac{R_s + R_i}{R_L}} \tag{10.16}$$

However, $I_i = I_s$, so Eqs. (10.15) and (10.16) will generate the same result. Equation (10.14) clearly reveals that both the source and the load resistance will reduce the overall gain of the system. In fact:

The larger the source resistance and/or smaller the load resistance, the less the overall gain of an amplifier.

The two reduction factors of Eq. (10.14) form a product that has to be carefully considered in any design procedure. It is not sufficient to ensure that R_s is relatively small if the impact of the magnitude of R_L is ignored. For instance, in Eq. (10.14), if the first factor is 0.9 and the second factor is 0.2, the product of the two results in an overall reduction factor equal to $(0.9)(0.2) = 0.18$, which is close to the lower factor. The effect of the excellent 0.9 level was completely wiped out by the significantly lower second multiplier. If both were 0.9-level factors, the net result would be $(0.9)(0.9) = 0.81$, which is still quite high. Even if the first were 0.9 and the second 0.7, the net result of 0.63 would still be respectable. In general, therefore, for good overall gain the effect of both R_s and R_L must be evaluated individually and as a product.

EXAMPLE 10.4

For the single-stage amplifier of Fig. 10.15, with $R_L = 4.7$ kΩ and $R_s = 0.3$ kΩ, determine:

(a) A_{v_s}.

(b) $A_v = V_o/V_i$.

(c) A_i.

The two-port parameters for the fixed-bias configuration are $Z_i = 1.071$ kΩ, $Z_o = 3$ kΩ, and $A_{v_{NL}} = -280.11$.

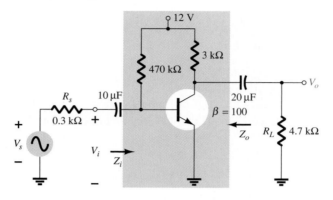

Figure 10.15   Example 10.4.

Solution

(a) Eq. (10.14): $A_{v_s} = \dfrac{V_o}{V_s} = \dfrac{R_i}{R_i + R_s} \dfrac{R_L}{R_L + R_o} A_{v_{NL}}$

$$= \left(\frac{1.071 \text{ k}\Omega}{1.071 \text{ k}\Omega + 0.3 \text{ k}\Omega} \right) \left(\frac{4.7 \text{ k}\Omega}{4.7 \text{ k}\Omega + 3 \text{ k}\Omega} \right)(-280.11)$$

$$= (0.7812)(0.6104)(-280.11)$$

$$= (0.4768)(-280.11)$$

$$= \mathbf{-133.57}$$

(b) $A_v = \dfrac{V_o}{V_i} = \dfrac{R_L A_{v_{NL}}}{R_L + R_o} = \dfrac{(4.7 \text{ k}\Omega)(-280.11)}{4.7 \text{ k}\Omega + 3 \text{ k}\Omega}$

$= (0.6104)(-280.11) = \mathbf{-170.98}$

(c) $A_i = -A_v \dfrac{R_i}{R_L} = -(-170.98)\left(\dfrac{1.071 \text{ k}\Omega}{4.7 \text{ k}\Omega} \right)$

$= \mathbf{38.96}$

or $A_{i_s} = -A_{v_s} \dfrac{R_s + R_i}{R_L} = -(-133.57)\left(\dfrac{0.3 \text{ k}\Omega + 1.071 \text{ k}\Omega}{4.7 \text{ k}\Omega} \right)$

$= \mathbf{38.96}$

as above.

10.6   BJT CE NETWORKDS

The fixed-bias configuration has been employed throughout the analysis of the early sections of this chapter to clearly show the effects of R_s and R_L. In this section, various CE configurations are examined with a load and a source resistance. A detailed analysis will not be performed for each configuration since they follow a very similar path to that demonstrated in the last few sections.

Fixed Bias

For the fixed-bias configuration examined in detail in recent sections, the system model with a load and source resistance will appear as shown in Fig. 10.16. In general,

$$V_o = \frac{R_L}{R_L + R_o} A_{v_{NL}} V_i$$

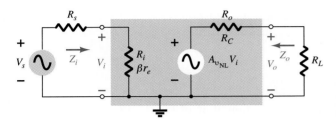

Figure 10.16 Fixed-bias configuration with R_s and R_L.

Substituting Eq. (8.6), $A_{v_{NL}} = -R_C/r_e$ and $R_o = R_C$,

$$V_o = -\frac{R_L(-R_C/r_e)V_i}{R_L + R_C}$$

and

$$A_v = \frac{V_o}{V_i} = -\frac{R_L R_C}{R_L + R_C}\frac{1}{r_e}$$

but

$$R_L \| R_C = \frac{R_L R_C}{R_L + R_C}$$

and

$$\boxed{A_v = -\frac{R_L \| R_C}{r_e}}$$ (10.17)

If the r_e model were substituted for the transistor in the fixed-bias configuration, the network of Fig. 10.17 would result, clearly revealing that R_C and R_L are in parallel.

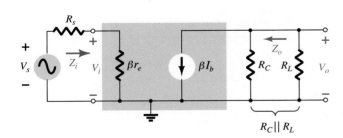

$$R_C \| R_L$$

Figure 10.17 Fixed-bias configuration with the substitution of the r_e model.

For the voltage gain A_{v_s} of Fig. 10.16,

$$V_i = \frac{Z_i V_s}{Z_i + R_s}$$

and

$$\frac{V_i}{V_s} = \frac{Z_i}{Z_i + R_s}$$

with

$$A_{v_s} = \frac{V_o}{V_s} = \frac{V_i}{V_s}\frac{V_o}{V_i}$$

so that

$$A_{v_s} = \frac{Z_i}{Z_i + R_s} A_v \qquad (10.18)$$

Since the load is connected to the collector terminal of the common-emitter configuration,

$$Z_i = \beta r_e \qquad (10.19)$$

and

$$Z_o = R_C \qquad (10.20)$$

as obtained earlier.

Voltage-Divider Bias

For the loaded voltage-divider bias configuration of Fig. 10.18, the load is again connected to the collector terminal and Z_i remains

$$Z_i \cong R' \| \beta r_e \qquad (R' = R_1 \| R_2) \qquad (10.21)$$

and for the system's output impedance

$$Z_o = R_C \qquad (10.22)$$

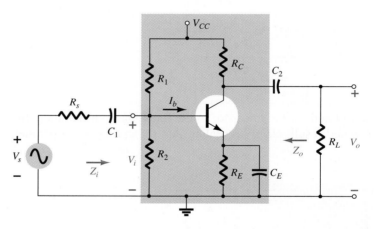

Figure 10.18 Voltage-divider bias configuration with R_s and R_L.

In the small-signal ac model, R_C and R_L will again be in parallel and

$$A_v = -\frac{R_C \| R_L}{r_e} \qquad (10.23)$$

with

$$A_{v_s} = \frac{Z_i}{Z_i + R_s} A_v \qquad (10.24)$$

Mathcad

Let us now use Mathcad to perform a complete analysis of the loaded voltage-divider configuration of Fig. 10.19. Once all the parameters are defined and the

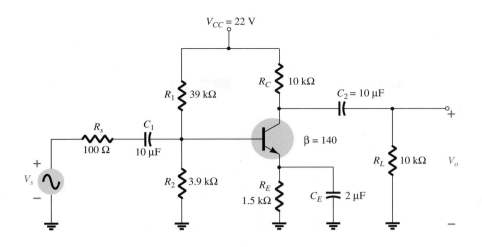

Figure 10.19 Loaded voltage-divider BJT configuration to be analyzed using Mathcad.

equations entered as shown in Fig. 10.20, a complete analysis (dc and ac) can be obtained for any combination of elements in a very short period of time (essentially instantaneously). Simply enter a new magnitude for any one parameter or any combination of parameters, and the new results will appear immediately. Compare this process to the time and effort required using a calculator if a single parameter such as R_1 is changed. The whole sequence of calculations would have to be repeated—the value of a software package such as Mathcad should now be fairly obvious.

The results appearing in Fig.10.20 reveal that Z_i is very close to 2 kΩ, Z_o is 10 kΩ, and A_v is about 164 with a load that matches the value of R_C. In Example 4.7 the levels of I_B and I_C are 6.05 μA and 0.85 mA, respectively, comparing very closely with the results appearing in Fig. 10.20. Note the small drop in gain due to R_s since its value is approximately 1/20 of the input impedance.

$R1 := 39 \cdot 10^3 \qquad R2 := 3.9 \cdot 10^3 \qquad RC := 10 \cdot 10^3 \qquad RE := 1.5 \cdot 10^3$

$VCC := 22 \qquad beta := 140 \qquad VBE := 0.7 \qquad RL := 10 \cdot 10^3 \qquad RS := 100$

$$RTh := \frac{R1 \cdot (R2)}{(R1 + R2)} \qquad ETh := \frac{R2 \cdot (VCC)}{(R1 + R2)}$$

$$IB := \frac{ETh - (VBE)}{(RTh + (beta + 1) \cdot RE)} \qquad IB = 6.045 \cdot 10^{-6}$$

$$IE := (beta + 1) \cdot IB \qquad IE = 8.524 \cdot 10^{-4}$$

$$re := \frac{26 \cdot (10^{-3})}{IE} \qquad re = 30.503$$

$$Zi := RTh \cdot beta \cdot \frac{re}{RTh + beta \cdot re} \qquad Zi = 1.937 \cdot 10^3$$

$$Zo := RC \qquad Zo = 1 \cdot 10^4$$

$$Av := -\frac{RC \cdot \frac{(RL)}{((RC + RL))}}{re} \qquad Av = -163.919$$

$$Avs := \left[\frac{Zi}{(Zi + RS)} \right] \cdot Av \qquad Avs = -155.872$$

Figure 10.20 Mathcad analysis of Fig. 10.19.

CE Unbypassed Emitter Bias

For the common-emitter unbypassed emitter-bias configuration of Fig. 10.21, Z_i remains independent of the applied load and

$$Z_i \cong R_B \| \beta R_E \qquad (10.25)$$

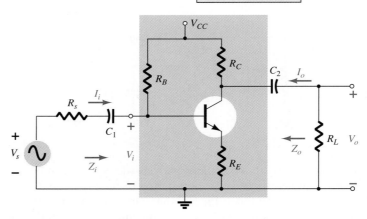

Figure 10.21 CE unbypassed emitter-bias configuration with R_s and R_L.

For the output impedance,

$$Z_o = R_C \qquad (10.26)$$

For the voltage gain, the resistance R_C will again drop down in parallel with R_L and

$$A_v = \frac{V_o}{V_i} = -\frac{R_C \| R_L}{R_E} \qquad (10.27)$$

with

$$A_{v_s} = \frac{V_o}{V_s} = \frac{Z_i}{Z_i + R_s} A_v \qquad (10.28)$$

and

$$A_i = \frac{I_o}{I_i} = -A_v \frac{Z_i}{R_L} \qquad (10.29)$$

but keep in mind that $I_i = I_s = V_s/(R_s + Z_i) = V_i/Z_i$.

Collector Feedback

To keep with our connection of the load to the collector terminal the next configuration to be examined is the collector feedback configuration of Fig. 10.22. In the small-signal ac model of the system, R_C and R_L will again drop down in parallel and

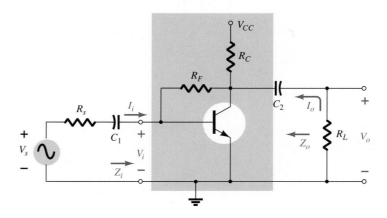

Figure 10.22 Collector feedback configuration with R_s and R_L.

$$A_v = -\frac{R_C\|R_L}{r_e} \tag{10.30}$$

with

$$A_{v_s} = \frac{Z_i}{Z_i + R_s}A_v \tag{10.31}$$

The output impedance

$$Z_o \cong R_C\|R_F \tag{10.32}$$

and

$$Z_i = \beta r_e\|\frac{R_F}{|A_v|} \tag{10.33}$$

The fact that A_v [Eq. (10.30)] is a function of R_L will alter the level of Z_i from the no-load value. Therefore, if the no-load model is available, the level of Z_i must be modified as demonstrated in the next example.

The collector feedback amplifier of Fig. 10.23 has the following no-load system parameters: $A_{v_{NL}} = -238.94$, $Z_o = R_C\|R_F = 2.66 \text{ k}\Omega$, and $Z_i = 0.553 \text{ k}\Omega$, with $r_e = 11.3 \Omega$, and $\beta = 200$. Using the systems approach, determine:

(a) A_v.

(b) A_{v_s}.

(c) A_i.

EXAMPLE 10.5

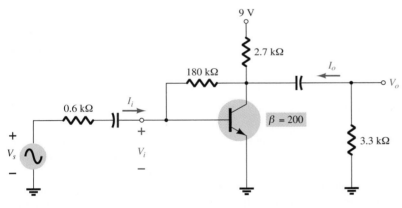

Figure 10.23 Example 10.5.

Solution

(a) For the two-port system:

$$A_v = -\frac{R_C\|R_L}{r_e} = -\frac{2.7 \text{ k}\Omega\|3.3 \text{ k}\Omega}{11.3 \Omega}$$

$$= -\frac{1.485 \text{ k}\Omega}{11.3 \Omega} = \mathbf{-131.42}$$

with

$$Z_i = \beta r_e\|\frac{R_F}{|A_v|} = (200)(11.3 \Omega)\|\frac{180 \text{ k}\Omega}{131.42}$$

$$= 2.26 \text{ k}\Omega\|1.37 \text{ k}\Omega$$

$$= 0.853 \text{ k}\Omega$$

The system approach will result in the configuration of Fig. 10.24 with the value of Z_i as controlled by R_L and the voltage gain. Now the two-port gain equation can be applied (slight difference in A_v due to approximation $\beta I_b \gg I_{R_F}$ in Section 8.7):

$$A_v = \frac{R_L A_{v_{NL}}}{R_L + R_o} = \frac{(3.3 \text{ k}\Omega)(-238.94)}{3.3 \text{ k}\Omega + 2.66 \text{ k}\Omega} = -132.3$$

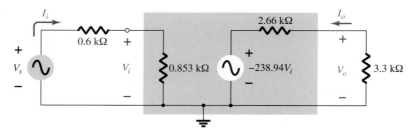

Figure 10.24 The ac equivalent circuit for the network of Fig. 10.23.

(b) $A_{v_s} = \dfrac{Z_i}{Z_i + R_s} A_v = \dfrac{0.853 \text{ k}\Omega}{0.853 \text{ k}\Omega + 0.6 \text{ k}\Omega}(-132.3)$

$\quad = -77.67$

(c) $A_i = -A_v \dfrac{Z_i}{R_L} = -(-132.3)\left(\dfrac{0.853 \text{ k}\Omega}{3.3 \text{ k}\Omega}\right) = \dfrac{(132.3)(0.853 \text{ k}\Omega)}{3.3 \text{ k}\Omega}$

$\quad = 34.2$

or $\quad A_i = -A_{v_s} \dfrac{Z_i + R_s}{R_L} = -(-77.67)\left(\dfrac{0.853 \text{ k}\Omega + 0.6 \text{ k}\Omega}{3.3 \text{ k}\Omega}\right)$

$\quad = 34.2$

10.7 BJT EMITTER-FOLLOWER NETWORKS

The input and output impedance parameters of the two-port model for the emitter-follower network are sensitive to the applied load and source resistance. For the emitter-follower configuration of Fig. 10.25, the small-signal ac model would appear as shown in Fig. 10.26. For the input section of Fig. 10.26, the resistance R_B is neglected because it is usually so much larger than the source resistance that a Thévenin equiv-

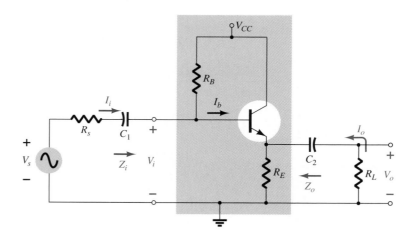

Figure 10.25 Emitter-follower configuration with R_s and R_L.

Chapter 10 Systems Approach—Effects of R_s and R_L

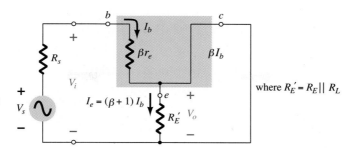

Figure 10.26 Emitter-follower configuration of Fig. 10.25 following the substitution of the r_e equivalent circuit.

where $R_E' = R_E \| R_L$

alent circuit for the configuration of Fig. 10.27 would result in simply R_s and V_s as shown in Fig. 10.26. Of course, if current levels are to be determined such as I_i in the original diagram, the effect of R_B must be included.

Applying Kirchhoff's voltage law to the input circuit of Fig. 10.26 will result in

$$V_s - I_b R_s - I_b \beta r_e - (\beta + 1)I_b R_E' = 0$$

and

$$V_s - I_b(R_s + \beta r_e + (\beta + 1)R_E') = 0$$

so that

$$I_b = \frac{V_s}{R_s + \beta r_e + (\beta + 1)R_E'}$$

Establishing I_e, we have

$$I_e = (\beta + 1)I_b = \frac{(\beta + 1)V_s}{R_s + \beta r_e + (\beta + 1)R_E'}$$

and

$$I_e = \frac{V_s}{[(R_s + \beta r_e)/(\beta + 1)] + R_E'}$$

Using $\beta + 1 \cong \beta$ yields

$$\boxed{I_e = \frac{V_s}{(R_s/\beta + r_e) + R_E'}} \qquad (10.34)$$

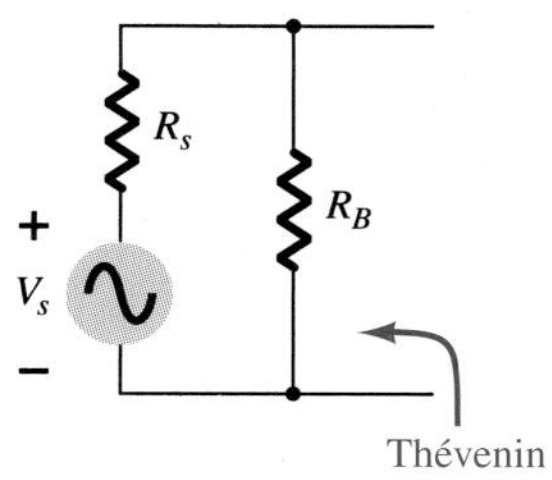

Drawing the network to "fit" Eq. (10.34) will result in the configuration of Fig. 10.28a. In Fig. 10.28b, R_E and the load resistance R_L have been separated to permit a definition of Z_o and I_o.

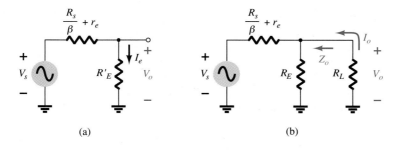

(a)   (b)

Figure 10.28 Networks resulting from the application of Kirchhoff's voltage law to the input circuit of Fig. 10.26.

The voltage gain can then be obtained directly from Fig. 10.28a using the voltage divider rule:

$$V_o = \frac{R_E' V_s}{R_E' + (R_s/\beta + r_e)}$$

or

$$A_{v_s} = \frac{V_o}{V_s} = \frac{R_E'}{R_E' + (R_s/\beta + r_e)}$$

Figure 10.27 Determining the Thévenin equivalent circuit for the input circuit of Fig. 10.25.

Thévenin

10.7   BJT Emitter-Follower Networks

543

and

$$A_{v_s} = \frac{V_o}{V_s} = \frac{R_E \| R_L}{R_E \| R_L + R_s/\beta + r_e} \qquad (10.35)$$

Setting $V_s = 0$ and solving for Z_o will result in

$$Z_o = R_E \| \left(\frac{R_s}{\beta} + r_e \right) \qquad (10.36)$$

For the input impedance,

$$Z_b = \beta(r_e + R_E')$$

and

$$Z_i = R_B \| Z_b$$

or

$$Z_i = R_B \| \beta(r_e + R_E \| R_L) \qquad (10.37)$$

For no-load conditions, the gain equation is

$$A_{v_{NL}} \cong \frac{R_E}{R_E + r_e}$$

while for loaded conditions,

$$A_v \cong \frac{V_o}{V_i} = \frac{R_E \| R_L}{R_E \| R_L + r_e} \qquad (10.38)$$

EXAMPLE 10.6

For the loaded emitter-follower configuration of Fig. 10.29 with a source resistance and the no-load two-port parameters of $Z_i = 155.83$ kΩ, $Z_o = 21.6$ Ω, and $A_{v_{NL}} = 0.993$ with $r_e = 21.74$ Ω and $\beta = 65$, determine:
(a) The new values of Z_i and Z_o as determined by the load and R_s, respectively.
(b) A_v using the systems approach.
(c) A_{v_s} using the systems approach.
(d) $A_i = I_o/I_i$.

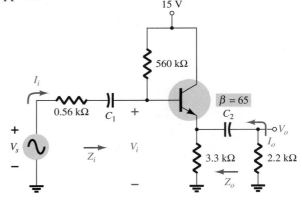

Figure 10.29 Example 10.6.

Solution

Eq. (10.37): $Z_i = R_B \| \beta(r_e + R_E \| R_L)$

$= 560 \text{ k}\Omega \| 65(21.74 \ \Omega + \underbrace{3.3 \text{ k}\Omega \| 2.2 \text{ k}\Omega})$

$= 560 \text{ k}\Omega \| 87.21 \text{ k}\Omega \qquad 1.32 \text{ k}\Omega$

$= \mathbf{75.46 \ k\Omega}$

versus 155.83 kΩ (no-load).

$$Z_o = R_E \left\| \left(\frac{R_s}{\beta} + r_e \right) \right.$$

$$= 3.3 \text{ k}\Omega \left\| \left(\frac{0.56 \text{ k}\Omega}{65} + 21.74 \text{ }\Omega \right) \right.$$

$$= 3.3 \text{ k}\Omega \| 30.36 \text{ }\Omega$$

$$= \mathbf{30.08 \text{ }\Omega}$$

versus 21.6 Ω (no R_s).

(b) Substituting the two-port equivalent network will result in the small-signal ac equivalent network of Fig. 10.30.

$$V_o = \frac{R_L A_{v_{NL}} V_i}{R_L + R_o} = \frac{(2.2 \text{ k}\Omega)(0.993) V_i}{2.2 \text{ k}\Omega + 30.08 \text{ }\Omega}$$

$$\cong 0.98 V_i$$

with $A_v = \dfrac{V_o}{V_i} \cong \mathbf{0.98}$

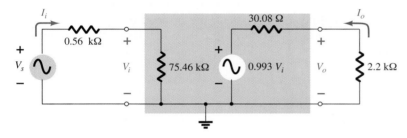

Figure 10.30 Small-signal ac equivalent circuit for the network of Fig. 10.29.

(c) $V_i = \dfrac{Z_i V_s}{Z_i + R_s} = \dfrac{(75.46 \text{ k}\Omega) V_s}{75.46 \text{ k}\Omega + 0.56 \text{ k}\Omega} = 0.993 V_s$

so that $\qquad A_{v_s} = \dfrac{V_o}{V_s} = \dfrac{V_o}{V_i} \dfrac{V_i}{V_s} = (0.98)(0.993) = \mathbf{0.973}$

(d) $A_i = \dfrac{I_o}{I_i} = -A_v \dfrac{Z_i}{R_L}$

$$= -(0.98) \left(\frac{75.46 \text{ k}\Omega}{2.2 \text{ k}\Omega} \right)$$

$$= \mathbf{-33.61}$$

10.8 BJT CB NETWORKS

A common-base amplifier with an applied load and source resistance appear in Fig. 10.31. The fact that the load is connected between the collector and base terminals isolates it from the input circuit, and Z_i remains essentially the same for no-load or loaded conditions. The isolation that exists between input and output circuits also maintains Z_o at a fixed level even though the level of R_s may change. The voltage gain is now determined by

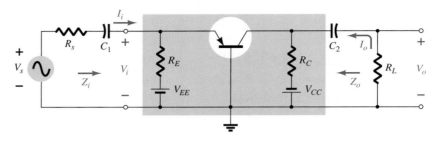

Figure 10.31 Common-base configuration with R_s and R_L.

$$A_v \cong \frac{R_C \| R_L}{r_e}$$

(10.39)

and the current gain:

$$A_i \cong -1$$

(10.40)

EXAMPLE 10.7

For the common-base amplifier of Fig. 10.32, the no-load two-port parameters are (using $\alpha \cong 1$) $Z_i \cong r_e = 20\ \Omega$, $A_{v_{NL}} = 250$, and $Z_o = 5\ \text{k}\Omega$. Using the two-port equivalent model, determine:

(a) A_v.
(b) A_{v_s}.
(c) A_i.

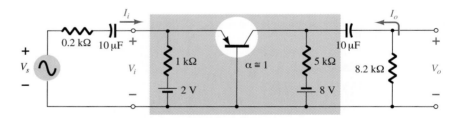

Figure 10.32 Example 10.7.

Solution

(a) The small-signal ac equivalent network appears in Fig. 10.33.

$$V_o = \frac{R_L A_{v_{NL}} V_i}{R_L + R_o} = \frac{(8.2\ \text{k}\Omega)(250)V_i}{8.2\ \text{k}\Omega + 5\ \text{k}\Omega} = 155.3 V_i$$

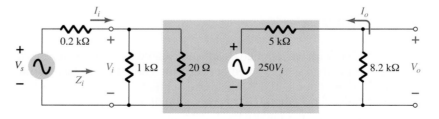

Figure 10.33 Small-signal ac equivalent circuit for the network of Fig. 10.32.

and
$$A_v = \frac{V_o}{V_i} = \mathbf{155.3}$$

or
$$A_v \cong \frac{R_C \| R_L}{r_e} = \frac{5 \text{ k}\Omega \| 8.2 \text{ k}\Omega}{20 \text{ }\Omega} = \frac{3.106 \text{ k}\Omega}{20 \text{ }\Omega}$$

$$= \mathbf{155.3}$$

(b) $A_{v_s} = \dfrac{V_o}{V_s} = \dfrac{V_i}{V_s}\dfrac{V_o}{V_i}$

$$= \frac{R_i}{R_i + R_s} A_v = \left(\frac{20 \text{ }\Omega}{20 \text{ }\Omega + 200 \text{ }\Omega}\right)(155.3)$$

$$= \mathbf{14.12}$$

Note the relatively low gain due to a source impedance much larger than the input impedance of the amplifier.

(c) $A_i = -A_v\dfrac{Z_i}{R_L} = -(155.3)\left(\dfrac{20 \text{ }\Omega}{8.2 \text{ k}\Omega}\right)$

$$= \mathbf{-0.379}$$

which is significantly less than 1 due to the division of output current between R_C and R_L.

10.9 FET NETWORKS

As noted in Chapter 9, the isolation that exists between gate and drain or source of an FET amplifier ensures that changes in R_L do not affect the level of Z_i and changes in R_{sig} do not affect R_o. In essence, therefore:

The no-load two-port model of Fig. 10.2 for an FET amplifier is unaffected by an applied load or source resistance.

Bypassed Source Resistance

For the FET amplifier of Fig. 10.34, the applied load will appear in parallel with R_D in the small-signal model, resulting in the following equation for the loaded gain:

$$\boxed{A_v = -g_m(R_D \| R_L)} \tag{10.41}$$

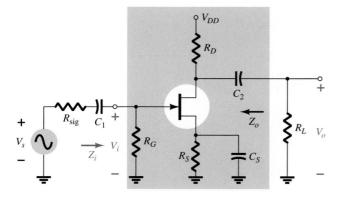

Figure 10.34 JFET amplifier with R_{sig} and R_L.

The impedance levels remain at

$$Z_i = R_G \qquad (10.42)$$

$$Z_o = R_D \qquad (10.43)$$

Unbypassed Source Resistance

For the FET amplifier of Fig. 10.35, the load will again appear in parallel with R_D and the loaded gain becomes

$$A_v = \frac{V_o}{V_i} = -\frac{g_m(R_D\|R_L)}{1 + g_m R_S} \qquad (10.44)$$

with

$$Z_i = R_G \qquad (10.45)$$

and

$$Z_o = R_D \qquad (10.46)$$

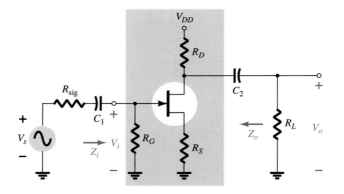

Figure 10.35 JFET amplifier with unbypassed R_S.

EXAMPLE 10.8

For the FET amplifier of Fig. 10.36, the no-load two-port parameters are $A_{v_{NL}} = -3.18$, $Z_i = R_1\|R_2 = 239$ kΩ, and $Z_o = 2.4$ kΩ, with $g_m = 2.2$ mS.
(a) Using the two-port parameters above, determine A_v and A_{v_s}.
(b) Using Eq. (10.44), calculate the loaded gain and compare to the result of part (a).

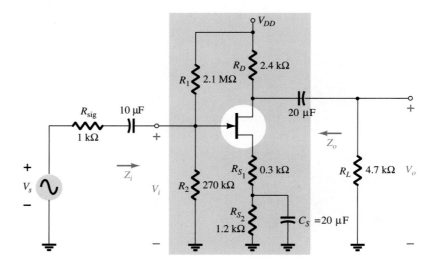

Figure 10.36 Example 10.8.

Solution

(a) The small-signal ac equivalent network appears in Fig. 10.37, and

$$A_v = \frac{V_o}{V_i} = \frac{R_L A_{v_{NL}}}{R_L + R_o} = \frac{(4.7 \text{ k}\Omega)(-3.18)}{4.7 \text{ k}\Omega + 2.4 \text{ k}\Omega}$$

$$= -2.105$$

$$A_{v_s} = \frac{V_o}{V_s} = \frac{V_i}{V_s} \frac{V_o}{V_i} = \frac{R_i}{R_i + R_{sig}} A_v$$

$$= \frac{(239 \text{ k}\Omega)(-2.105)}{239 \text{ k}\Omega + 1 \text{ k}\Omega}$$

$$= -2.096 \cong A_v$$

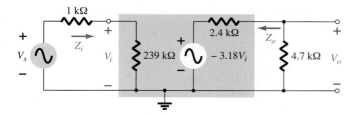

Figure 10.37 Small-signal ac equivalent circuit for the network of Fig. 10.36.

(b) Eq. (10.44): $\quad A_v = \dfrac{-g_m(R_D \| R_L)}{1 + g_m R_{S_1}}$

$$= \frac{-(2.2 \text{ mS})(2.4 \text{ k}\Omega \| 4.7 \text{ k}\Omega)}{1 + (2.2 \text{ mS})(0.3 \text{ k}\Omega)} = \frac{-3.498}{1.66}$$

$$= -2.105 \text{ as above}$$

Source Follower

For the source-follower configuration of Fig. 10.38, the level of Z_i is independent of the magnitude of R_L and determined by

$$\boxed{Z_i = R_G} \tag{10.47}$$

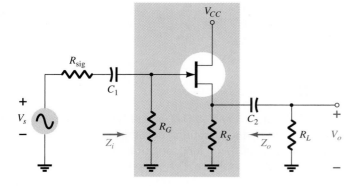

Figure 10.38 Source-follower configuration with R_{sig} and R_L.

The loaded voltage gain has the same format as the unloaded gain with R_S replaced by the parallel combination of R_S and R_L.

$$A_v = \frac{V_o}{V_i} = \frac{g_m(R_S\|R_L)}{1 + g_m(R_S\|R_L)} \qquad (10.48)$$

The level of output impedance is as determined in Chapter 9:

$$Z_o = R_S\|\frac{1}{g_m} \qquad (10.49)$$

revealing an insensitivity to the magnitude of the source resistance R_{sig}.

Common Gate

Even though the common-gate configuration of Fig. 10.39 is somewhat different from those described above with regard to the placement of R_L and R_{sig}, the input and output circuits remain isolated and

$$Z_i = \frac{R_S}{1 + g_m R_S} \qquad (10.50)$$

$$Z_o = R_D \qquad (10.51)$$

The loaded voltage gain is given by

$$A_v = g_m(R_D\|R_L) \qquad (10.52)$$

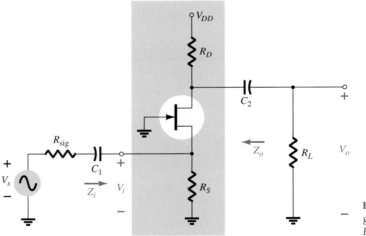

Figure 10.39 Common-gate configuration with R_{sig} and R_L.

10.10 SUMMARY TABLE

Now that the loaded and unloaded (Chapters 8 and 9) BJT and JFET amplifiers have been examined in some detail, a review of the equations developed is provided by Table 10.1. Although all the equations are for the loaded situation, the removal of R_L will result in the equations for the unloaded amplifier. The same is true for the effect of R_s (for BJTs) and R_{sig} (for JFETs) on Z_o. In each case, the phase relationship between the input and output voltages is also provided for quick reference. A review of the equations will reveal that the isolation provided by the JFET between the gate and channel by the SiO_2 layer results in a series of less complex equations than those encountered for the BJT configurations. The linkage provided by I_b between input and output circuits of the BJT transistor amplifier adds a touch of complexity to some of the equations.

TABLE 10.1 Summary of Transistor Configurations (A_v, Z_i, Z_o)

Configuration	$A_v = V_o/V_i$	Z_i	Z_o
	$\dfrac{-(R_L\|R_C)}{r_e}$	$R_B\|\beta r_e$	R_C
	$\dfrac{-h_{fe}}{h_{ie}}(R_L\|R_C)$	$R_B\|h_{ie}$	R_C
	Including r_o: $-\dfrac{(R_L\|R_C\|r_o)}{r_e}$	$R_B\|\beta r_e$	$R_C\|r_o$
	$\dfrac{-(R_L\|R_C)}{r_e}$	$R_1\|R_2\|\beta r_e$	R_C
	$\dfrac{-h_{fe}}{h_{ie}}(R_L\|R_C)$	$R_1\|R_2\|h_{ie}$	R_C
	Including r_o: $\dfrac{-(R_L\|R_C\|r_o)}{r_e}$	$R_1\|R_2\|\beta r_e$	$R_C\|r_o$
	$\cong 1$	$R'_E = R_L\|R_E$ $R_1\|R_2\|\beta(r_e + R'_E)$	$R'_s = R_s\|R_1\|R_2$ $R_E\|\left(\dfrac{R'_s}{\beta} + r_e\right)$
	$\cong 1$	$R_1\|R_2\|(h_{ie} + h_{fe}R'_E)$	$R_E\|\left(\dfrac{R'_s + h_{ie}}{h_{fe}}\right)$
	Including r_o: $\cong 1$	$R_1\|R_2\|\beta(r_e + R'_E)$	$R_E\|\left(\dfrac{R'_s}{\beta} + r_e\right)$
	$\cong \dfrac{-(R_L\|R_C)}{r_e}$	$R_E\|r_e$	R_C
	$\cong \dfrac{-h_{fb}}{h_{ib}}(R_L\|R_C)$	$R_E\|h_{ib}$	R_C
	Including r_o: $\cong \dfrac{-(R_L\|R_C\|r_o)}{r_e}$	$R_E\|r_e$	$R_C\|r_o$

TABLE 10.1 Summary of Transistor Configurations (A_v, Z_i, Z_o) (Continued)

Configuration	$A_v = V_o/V_i$	Z_i	Z_o
	$\dfrac{-(R_L\|R_C)}{R_E}$	$R_1\|R_2\|\beta(r_e + R_E)$	R_C
	$\dfrac{-(R_L\|R_C)}{R_E}$	$R_1\|R_2\|(h_{ie} + h_{fe}R_E)$	R_C
	Including r_o: $\dfrac{-(R_L\|R_C)}{R_E}$	$R_1\|R_2\|\beta(r_e + R_e)$	$\cong R_C$
	$\dfrac{-(R_L\|R_C)}{R_{E_1}}$	$R_B\|\beta(r_e + R_{E_1})$	R_C
	$\dfrac{-(R_L\|R_C)}{R_{E_1}}$	$R_B\|(h_{ie} + h_{fe}R_{E_1})$	R_C
	Including r_o: $\dfrac{-(R_L\|R_C)}{R_{E_1}}$	$R_B\|\beta(r_e + R_E)$	$\cong R_C$
	$\dfrac{-(R_L\|R_C)}{r_e}$	$\beta r_e\left\|\dfrac{R_F}{\|A_v\|}\right.$	R_C
	$\dfrac{-h_{fe}}{h_{ie}}(R_L\|R_C)$	$h_{ie}\left\|\dfrac{R_F}{\|A_v\|}\right.$	R_C
	Including r_o: $\dfrac{-(R_L\|R_C\|r_o)}{r_e}$	$\beta r_e\left\|\dfrac{R_F}{\|A_v\|}\right.$	$R_C\|R_F\|r_o$
	$\dfrac{-(R_L\|R_C)}{R_E}$	$\beta R_E\left\|\dfrac{R_F}{\|A_v\|}\right.$	$\cong R_C\|R_F$
	$\dfrac{-(R_L\|R_C)}{R_E}$	$h_{fe}R_E\left\|\dfrac{R_F}{\|A_v\|}\right.$	$\cong R_C\|R_F$
	Including r_o: $\cong \dfrac{-(R_L\|R_C)}{R_E}$	$\cong \beta R_E\left\|\dfrac{R_F}{\|A_v\|}\right.$	$\cong R_C\|R_F$

TABLE 10.1 (Continued)

Configuration	$A_v = V_o/V_i$	Z_i	Z_o
	$-g_m(R_D\|R_L)$ Including r_d: $-g_m(R_D\|R_L\|r_d)$	R_G R_G	R_D $R_D\|r_d$
	$\dfrac{-g_m(R_D\|R_L)}{1 + g_m R_S}$ Including r_d: $\dfrac{-g_m(R_D\|R_L)}{1 + g_m R_S + \dfrac{R_D + R_S}{r_d}}$	R_G R_G	$\dfrac{R_D}{1 + g_m R_S}$ $\cong \dfrac{R_D}{1 + g_m R_S}$
	$-g_m(R_D\|R_L)$ Including r_d: $-g_m(R_D\|R_L\|r_d)$	$R_1\|R_2$ $R_1\|R_2$	R_D $R_D\|r_{dp};$
	$\dfrac{g_m(R_S\|R_L)}{1 + g_m(R_S\|R_L)}$ Including r_d: $= \dfrac{g_m r_d(R_S\|R_L)}{r_d + R_D + g_m r_d(R_S\|R_L)}$	R_G R_G	$R_S\|1/g_m$ $\dfrac{R_S}{1 + \dfrac{g_m r_d R_S}{r_d + R_D}}$
	$g_m(R_D\|R_L)$ Including r_d: $\cong g_m(R_D\|R_L)$	$\dfrac{R_S}{1 + g_m R_S}$ $Z_i = \dfrac{R_S}{1 + \dfrac{g_m r_d R_S}{r_d + R_D\|R_L}}$	R_D $R_D\|r_d$

10.11 CASCADED SYSTEMS

The two-port systems approach is particularly useful for cascaded systems such as that appearing in Fig. 10.40, where $A_{v_1}, A_{v_2}, A_{v_3}$, and so on, are the voltage gains of each stage *under loaded conditions*. That is, A_{v_1} is determined with the input impedance to A_{v_2} acting as the load on A_{v_1}. For A_{v_2}, A_{v_1} will determine the signal strength and source impedance at the input to A_{v_2}. The total gain of the system is then determined by the product of the individual gains as follows:

$$A_{v_T} = A_{v_1} \cdot A_{v_2} \cdot A_{v_3} \cdots \cdots \tag{10.53}$$

and the total current gain by

$$A_{i_T} = -A_{v_T} \frac{Z_{i_1}}{R_L} \tag{10.54}$$

No matter how perfect the system design, the application of a load to a two-port system will affect the voltage gain. Therefore, there is no possibility of a situation where A_{v_1}, A_{v_2}, and so on, of Fig. 10.40 are simply the no-load values. The loading of each succeeding stage must be considered. The no-load parameters can be used to determine the loaded gains of Fig. 10.40, but Eq. (10.53) requires the loaded values.

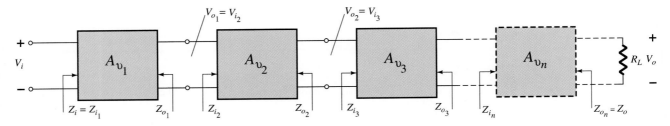

Figure 10.40 Cascaded system.

EXAMPLE 10.9

The two-stage system of Fig. 10.41 employed a transistor emitter-follower configuration prior to a common-base configuration to ensure that the maximum percent of the applied signal appears at the input terminals of the common-base amplifier. In Fig. 10.41, the no-load values are provided for each system, with the exception of Z_i and Z_o for the emitter-follower, which are the loaded values. For the configuration of Fig. 10.41, determine:

(a) The loaded gain for each stage.
(b) The total gain for the system, A_v and A_{v_s}.
(c) The total current gain for the system.
(d) The total gain for the system if the emitter-follower configuration were removed.

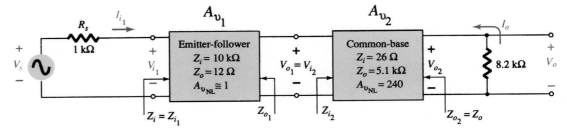

Figure 10.41 Example 10.9.

Solution

(a) For the emitter-follower configuration, the loaded gain is

$$V_{o_1} = \frac{Z_{i_2}A_{v_{NL}}V_{i_1}}{Z_{i_2} + Z_{o_1}} = \frac{(26\ \Omega)(1)V_{i_1}}{26\ \Omega + 12\ \Omega} = 0.684V_{i_1}$$

and $\qquad A_{v_1} = \dfrac{V_{o_1}}{V_{i_1}} = \mathbf{0.684}$

For the common-base configuration,

$$V_{o_2} = \frac{R_L A_{v_{NL}} V_{i_2}}{R_L + R_{o_2}} = \frac{(8.2\ \text{k}\Omega)(240)V_{i_2}}{8.2\ \text{k}\Omega + 5.1\ \text{k}\Omega} = 147.97V_{i_2}$$

and $\qquad A_{v_L} = \dfrac{V_{o_2}}{V_{i_2}} = \mathbf{147.97}$

(b) $A_{v_T} = A_{v_1}A_{v_2}$

$\qquad = (0.684)(147.97)$

$\qquad = \mathbf{101.20}$

$$A_{v_s} = \frac{Z_{i_1}}{Z_{i_1} + R_s}A_{v_T} = \frac{(10\ \text{k}\Omega)(101.20)}{10\ \text{k}\Omega + 1\ \text{k}\Omega}$$

$\qquad = \mathbf{92}$

(c) $A_{i_T} = -A_{v_T}\dfrac{Z_{i_1}}{R_L} = -(101.20)\left(\dfrac{10\ \text{k}\Omega}{8.2\ \text{k}\Omega}\right)$

$\qquad = \mathbf{-123.41}$

(d) $V_{i_{CB}} = \dfrac{Z_{i_{CB}}V_s}{Z_{i_{CB}} + R_s} = \dfrac{(26\ \Omega)V_s}{26\ \Omega + 1\ \text{k}\Omega} = 0.025V_s$

and $\qquad \dfrac{V_i}{V_s} = 0.025 \qquad$ with $\qquad \dfrac{V_o}{V_i} = 147.97 \qquad$ from above

and $\qquad A_{v_s} = \dfrac{V_i}{V_s}\dfrac{V_o}{V_i} = (0.025)(147.97) = \mathbf{3.7}$

In total, therefore, the gain is about 25 times greater with the emitter-follower configuration to draw the signal to the amplifier stages. Take note, however, that it was also important that the output impedance of the first stage was relatively close to the input impedance of the second stage or the signal would have been "lost" again by the voltage-divider action.

10.12 SUMMARY

Important Conclusions and Concepts

1. The quantities $A_{v_{NL}}$ and $A_{i_{NL}}$ are the gains of a two-port system with **no load applied.** Neither is affected by the applied load nor the internal resistance of the source.

2. The loaded voltage gain of an amplifier **is always less** than the no-load level.

3. One of the distinct advantages of the systems approach is that once the two-port parameters of a system are known, **the effect of changing the load** on the overall gain can quickly be determined. There is no need to reanalyze the entire network.

4. Elements that were isolated by capacitors for the dc analysis **will appear in the ac analysis** due to the short-circuit equivalent for the capacitive elements.

5. The **larger the internal resistance** of the applied source, the **less** the overall gain of the system.

6. The **common-base BJT amplifier** and the **common-gate JFET amplifier** have output waveforms **in phase** with the applied signal.

7. For all **common-emitter** and **common-source** systems, the **output signal** at the collector and drain, respectively, will be **180° out of phase** with the applied signal.

8. Outputs taken off the **emitter of a common-emitter BJT amplifier** or **source of a common-source JFET** amplifier are **in phase** with the applied signal.

9. The total gain of a cascaded system is determined by the **product of the gains of each stage.** The gain of each stage, however, must be determined **under loaded conditions.**

10. Since the total gain is the product of the individual gains of a cascaded system, the **weakest link** can have a major impact on the total gain.

Equations

Effect of load impedance:

$$A_v = \frac{V_o}{V_i} = \frac{R_L}{R_L + R_o} A_{v_{NL}}$$

$$A_i = \frac{I_o}{I_i} = -A_v \frac{Z_i}{R_L}$$

Effect of source impedance:

$$V_i = \frac{R_i V_s}{R_i + R_s}$$

$$A_{v_s} = \frac{V_o}{V_s} = \frac{R_i}{R_i + R_s} A_{v_{NL}}$$

$$I_s = \frac{V_s}{R_s + R_i}$$

Combined effect of load and source impedance:

$$A_v = \frac{V_o}{V_i} = \frac{R_L}{R_L + R_o} A_{v_{NL}}$$

$$A_{v_s} = \frac{V_o}{V_s} = \frac{R_i}{R_i + R_s} \cdot \frac{R_L}{R_L + R_o} A_{v_{NL}}$$

$$A_i = \frac{I_o}{I_i} = -A_v \frac{R_i}{R_L}$$

$$A_{i_s} = \frac{I_o}{I_s} = -A_{v_s} \frac{R_s + R_i}{R_L}$$

Consult Table 10.1 for an extensive list of equations.

10.13 COMPUTER ANALYSIS

PSpice Windows

LOADED VOLTAGE-DIVIDER BJT TRANSISTOR CONFIGURATION

The computer analysis of this section includes a PSpice Windows evaluation of the response of a loaded BJT and FET amplifier with a source resistance. The BJT network of Fig. 10.42 employs the same unloaded configuration examined in the PSpice analysis of Chapter 8, where the unloaded gain was 369 (Example 8.2, $r_e = 18.44\ \Omega$). For the transistor, all the parameters listed under **Model Editor** were removed except I_s and beta, which were set to 2E-15A and 90, respectively. In this way, the results will be as close to the hand-written solutions as possible without going to the controlled source equivalents. Note the placement of the **VPRINT1** option to pick up the voltage lost across the source resistance and to note if there is any drop in gain across the capacitor. The option **Do not auto-run Probe** was chosen, and under **Analysis Setup,** the **AC Sweep** was set at a fixed frequency of 10 kHz. In addition, **Display Results on Schematic** under **Analysis** was chosen and the **Voltage Display** enabled.

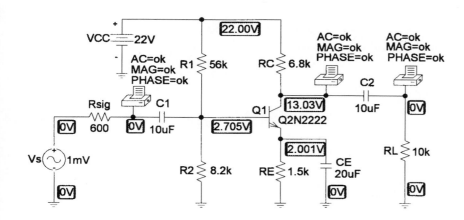

Figure 10.42 Loaded voltage-divider BJT transistor configuration.

An **Analysis** resulted in the dc levels appearing in Fig. 10.42. In particular, note the zero volt levels at the left side of C_1 and the right side of C_2. In addition, note that V_{BE} is essentially 0.7 V and the dc levels of each terminal of the transistor are very close to those calculated in Example 8.2 (using the approximate approach). Reviewing the output file following **Analysis-Examine Output** will result in the data listings of Fig. 10.43. The nodes are defined in the **Schematics Netlist,** and the **BJT MODEL PARAMETERS** reveal our choices for this run—although the last three are default values. The **SMALL-SIGNAL BIAS SOLUTION** simply confirms the levels printed on the schematic, and the **Operating Point Information** reveals that beta (dc and ac) is 90, that V_{BE} is 0.7 V, that I_C is 1.32 mA, and that I_B is 14.7 μA (in addition to a host of other levels). The **AC ANALYSIS** reveals that the voltage on the other side of R_{sig} is about 0.7 mV, resulting in a drop of about 0.3 mV (30% loss in signal voltage) of the applied signal across R_{sig}. The remaining two ac levels are the same, revealing that the capacitor is an effective short circuit for ac. The loaded gain from source to output is 144.9. The gain from the base of the transistor to the output is 144.9 mV/0.7 mV = 207. Both levels are certainly significantly less than the no-load level of 369. If we return to the network and change R_L to 10 MΩ, the output voltage will rise to 243.3 mV, resulting in a gain of 243.3 mV/0.7 mV = 347.57, which is quite close to the hand-calculated, approximate level of 369.

```
****        CIRCUIT DESCRIPTION


*********************************************************************************
* Schematics Netlist *


V_Vs          $N_0001 0   AC 1mV
+SIN 0V 1mV 10kHz 0 0 0
R_Rsig        $N_0001 $N_0002  600

.PRINT        AC
+ VM([$N_0002])
+ VP([$N_0002])
C_C1          $N_0002 $N_0003  10uF
V_VCC         $N_0004 0  22V
R_R1          $N_0004 $N_0003  56k
R_R2          $N_0003 0  8.2k
R_RC          $N_0004 $N_0005  6.8k
R_RE          $N_0006 0  1.5k
C_CE          $N_0006 0  20uF
R_RL          $N_0007 0  10k
C_C2          $N_0005 $N_0007  10uF

.PRINT        AC
+ VM([$N_0005])
+ VP([$N_0005])

.PRINT        AC
+ VM([$N_0007])
+ VP([$N_0007])
Q_Q1          $N_0005 $N_0003 $N_0006 Q2N2222-X


****        BJT MODEL PARAMETERS

*********************************************************************************




            Q2N2222-X
            NPN
      IS    2.000000E-15
      BF    90
      NF    1
      BR    1
      NR    1
****        SMALL SIGNAL BIAS SOLUTION       TEMPERATURE =   27.000 DEG C

*********************************************************************************


  NODE  VOLTAGE    NODE  VOLTAGE    NODE  VOLTAGE    NODE  VOLTAGE

($N_0001)   0.0000              ($N_0002)   0.0000

($N_0003)   2.7051              ($N_0004)  22.0000

($N_0005)  13.0280              ($N_0006)   2.0012

($N_0007)   0.0000

    VOLTAGE SOURCE CURRENTS
    NAME        CURRENT

    V_Vs        0.000E+00
    V_VCC      -1.664E-03

  TOTAL POWER DISSIPATION   3.66E-02  WATTS
****        OPERATING POINT INFORMATION      TEMPERATURE =   27.000 DEG C

*********************************************************************************



**** BIPOLAR JUNCTION TRANSISTORS

NAME        Q_Q1
MODEL       Q2N2222-X
IB          1.47E-05
IC          1.32E-03
VBE         7.04E-01
VBC        -1.03E+01
VCE         1.10E+01
BETADC      9.00E+01
GM          5.10E-02
RPI         1.76E+03
RX          0.00E+00
RO          1.00E+12
CBE         0.00E+00
CBC         0.00E+00
CJS         0.00E+00
BETAAC      9.00E+01
CBX         0.00E+00
FT          8.12E+17

    ****    AC ANALYSIS                  TEMPERATURE =   27.000 DEG C


*********************************************************************************


  FREQ        VM($N_0002) VP($N_0002)

  1.000E+04   7.025E-04  -5.801E-01

  FREQ        VM($N_0005) VP($N_0005)

  1.000E+04   1.449E-01  -1.782E+02

  FREQ        VM($N_0007) VP($N_0007)

  1.000E+04   1.449E-01  -1.782E+02
```

Figure 10.43 Output file for the network of Fig. 10.42.

For interest's sake, let us now calculate the loaded voltage gain and compare to the PSpice solution of 144.9.

$$r_e = 18.44 \ \Omega$$

and
$$Z_i \cong R_1\|R_2\|\beta r_e$$

$$= 56 \text{ k}\Omega\|8.2 \text{ k}\Omega\|(90)(18.44 \ \Omega)$$

$$\cong 1.35 \text{ k}\Omega$$

$$V_i = \frac{Z_i V_s}{Z_i + R_s} = \frac{(1.35 \text{ k}\Omega)V_s}{1.35 \text{ k}\Omega + 0.6 \text{ k}\Omega} = 0.69 V_s$$

and
$$\frac{V_i}{V_s} = 0.69$$

$$A_v = \frac{V_o}{V_i} = \frac{R_L A_{v_{\text{NL}}}}{R_L + R_o} = \frac{(10 \text{ k}\Omega)(-350.4)}{10 \text{ k}\Omega + 6.8 \text{ k}\Omega}$$

$$= -208.57$$

with
$$A_{v_s} = \frac{V_i}{V_s}\frac{V_o}{V_i} = (0.69)(-208.57)$$

$$\cong -144$$

which is an excellent comparison with the computer solution.

LOADED JFET SELF-BIAS TRANSISTOR CONFIGURATION

The network of Fig. 10.44 is a loaded version of the network examined in Chapter 9, which resulted in a no-load gain of 13.3. In the **Model Editor** dialog box, **Beta** was set to 0.625mA/V^2 and **Vto** $= -4$V. The remaining parameters were left alone to permit a close comparison with the Chapter 9 solution and because they have less effect on the response than for a BJT transistor.

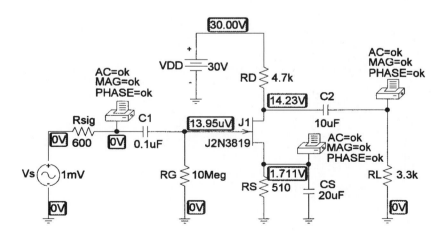

Figure 10.44 Loaded self-bias JFET transistor configuration.

Again, note the effectiveness of the capacitors to block the dc voltages. In addition, note the small voltage at the gate, indicating that the input impedance to the device is in reality not infinite (although for all practical purposes it is an excellent assumption.) Again, the frequency was set to 10 kHz and an Analysis called for without the **Probe** option. The sequence **Analysis-Examine Output** will result in the listing of Fig. 10.45. The **Schematics Netlist** provides a listing of assigned nodes, and the

```
****      CIRCUIT DESCRIPTION
*********************************************************************************

* Schematics Netlist *

V_Vs          $N_0001 0   AC 1mV
+SIN 0V 1mV 10kHz 0 0 0
R_Rsig        $N_0001 $N_0002  600

.PRINT        AC
+ VM([$N_0002])
+ VP([$N_0002])
C_C1          $N_0002 $N_0003  0.1uF
R_RG          $N_0003 0  10Meg
C_C2          $N_0004 $N_0005  10uF

.PRINT        AC
+ VM([$N_0006])
+ VP([$N_0006])

.PRINT        AC
+ VM([$N_0005])
+ VP([$N_0005])
V_VDD         $N_0007 0 30V
R_RD          $N_0007 $N_0004  4.7k
R_RS          $N_0006 0  510
C_CS          $N_0006 0  20uF
J_J1          $N_0004 $N_0003 $N_0006 J2N3819-X1
R_RL          $N_0005 0  3.3k

****      Junction FET MODEL PARAMETERS
*********************************************************************************
              J2N3819-X1
              NJF
       VTO    -4
      BETA    625.000000E-06
    LAMBDA    2.250000E-03
        IS    33.570000E-15
       ISR    322.400000E-15
     ALPHA    311.700000E-06
        VK    243.6
        RD    1
        RS    1
       CGD    1.600000E-12
       CGS    2.414000E-12
         M    .3622
     VTOTC    -2.500000E-03
    BETATCE   -.5
        KF    9.882000E-18

****      SMALL SIGNAL BIAS SOLUTION        TEMPERATURE =   27.000 DEG C
*********************************************************************************
NODE   VOLTAGE      NODE    VOLTAGE      NODE    VOLTAGE      NODE    VOLTAGE


($N_0001)   0.0000                   ($N_0002)   0.0000

($N_0003) 13.95E-06                  ($N_0004)   14.2280

($N_0005)   0.0000                   ($N_0006)   1.7114

($N_0007)   30.0000


    VOLTAGE SOURCE CURRENTS
    NAME         CURRENT

    V_Vs         0.000E+00
    V_VDD        -3.356E-03

    TOTAL POWER DISSIPATION   1.01E-01  WATTS
 ****      OPERATING POINT INFORMATION     TEMPERATURE =   27.000 DEG C
*********************************************************************************
**** JFETS


NAME          J_J1
MODEL         J2N3819-X1
ID            3.36E-03
VGS           -1.71E+00
VDS           1.25E+01
GM            2.94E-03
GDS           7.34E-06
CGS           1.68E-12
CGD           5.97E-13 ·
 ****      AC ANALYSIS                      TEMPERATURE =   27.000 DEG C
*********************************************************************************

  FREQ        VM($N_0002) VP($N_0002)


  1.000E+04   9.999E-04   -1.213E-02


  FREQ        VM($N_0006) VP($N_0006)


  1.000E+04   2.297E-06   -8.979E+01


  FREQ        VM($N_0005) VP($N_0005)


  1.000E+04   5.597E-03   -1.799E+02
```

Figure 10.45 Output file for the network of Fig. 10.44.

OPERATING POINT INFORMATION reveals that the drain current is 3.36 mA, that V_{GS} is -1.71 V, and that g_m is 2.94 mS. The **AC ANALYSIS** reveals that there is negligible drop across either capacitor at this frequency, and the short-circuit equivalency can be assumed. The output voltage is 5.597 mV resulting in a loaded gain of 5.597 compared to the unloaded gain of 13.3. Note also that the drop across R_{sig} is negligible due to the high input impedance of the device.

Using the value of g_m hand-calculated earlier, the equation for the loaded gain will result in a gain of 5.62 as shown below—an excellent comparison with the computer solution.

$$A_v = -g_m(R_D \| R_L)$$
$$= -(2.90 \text{ mS})(4.7 \text{ k}\Omega \| 3.3 \text{ k}\Omega)$$
$$= -5.62$$

The results obtained above have clearly substantiated the analysis and equations presented in this chapter for a loaded amplifier.

Electronics Workbench

The results obtained for Example 10.6 will now be verified using Electronics Workbench. The elements of the network were introduced as shown in Fig. 10.46 using procedures introduced in earlier chapters. The value of beta was set in the default listing for the *npn* transistor, and the value beta $= 65$ was introduced on the figure by using the text option.

The dc operation was checked with the introduction of a dc meter on the emitter leg. Remember that for the ac response, the ac meter will respond with the effective value of the ac signal. The result is 0.708 V at the input for the applied 1-V peak signal. The output of 0.691 V results in a gain of 691 mV/708 mV $= 0.975$ versus 0.973 in Example 10.6. The current gain is -315 μA/9.727 μA $= -32.38$ versus the -33.61 of Example 10.6.

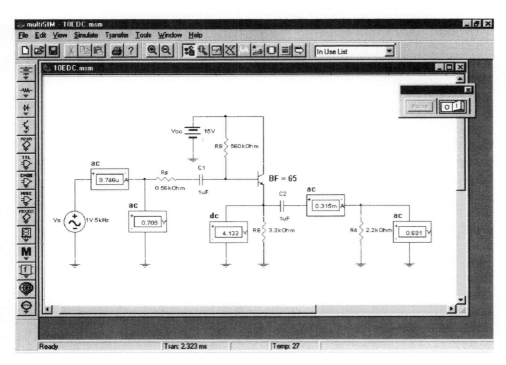

Figure 10.46 Circuit of Example 10.6 using EWB.

§ 10.3 Effect of a Load Impedance (R_L)

1. For the fixed-bias configuration of Fig. 10.47:
   (a) Determine $A_{v_{NL}}$, Z_i, and Z_o,
   (b) Sketch the two-port model of Fig. 10.2 with the parameters determined in part (a) in place.
   (c) Calculate the gain A_v using the model of part (b) and Eq. (10.3).
   (d) Determine the current gain using Eq. (10.6).
   (e) Determine A_v, Z_i, and Z_o using the r_e model and compare with the solutions above.

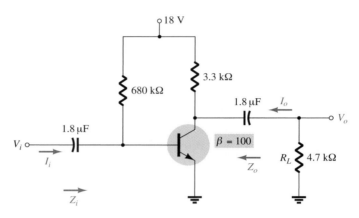

Figure 10.47   Problems 1, 2, and 3

* 2. (a) Draw the dc and ac load lines for the network of Fig. 10.47 on the characteristics of Fig. 10.48.
   (b) Determine the peak-to-peak value of I_c and V_{ce} from the graph if V_i has a peak value of 10 mV. Determine the voltage gain $A_v = V_o/V_i$ and compare with the solution obtained in Problem 1.

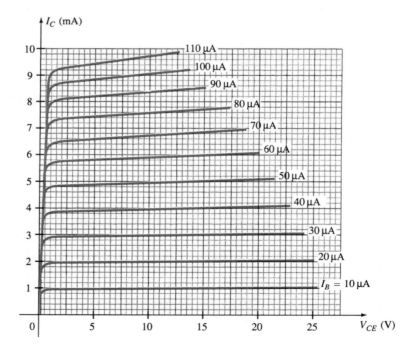

Figure 10.48   Problems 2 and 7

3. (a) Determine the voltage gain A_v for the network of Fig. 10.47 for $R_L = 4.7$, 2.2, and 0.5 kΩ. What is the effect of decreasing levels of R_L on the voltage gain?

   (b) How will Z_i, Z_o, and $A_{v_{NL}}$ change with decreasing values of R_L?

§ **10.4 Effect of a Source Impedance (R_s)**

* **4.** For the network of Fig. 10.49:

   (a) Determine $A_{v_{NL}}$, Z_i, and Z_o.

   (b) Sketch the two-port model of Fig. 10.2 with the parameters determined in part (a) in place.

   (c) Determine A_v using the results of part (b).

   (d) Determine A_{v_s}.

   (e) Determine A_{v_s} using the r_e model and compare the results to that obtained in part (d).

   (f) Change R_s to 1 kΩ and determine A_v. How does A_v change with the level of R_s?

   (g) Change R_s to 1 kΩ and determine A_{v_s}. How does A_{v_s} change with the level of R_s?

   (h) Change R_s to 1 kΩ and determine $A_{v_{NL}}$, Z_i, and Z_o. How do they change with change in R_s?

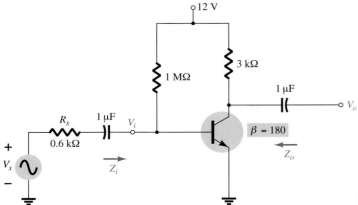

Figure 10.49 Problem 4

§ **10.5 Combined Effect of R_s and R_L**

* **5.** For the network of Fig. 10.50:

   (a) Determine $A_{v_{NL}}$, Z_i, and Z_o.

   (b) Sketch the two-port model of Fig. 10.2 with the parameters determined in part (a) in place.

   (c) Determine A_v and A_{v_s}.

   (d) Calculate A_i.

   (e) Change R_L to 5.6 kΩ and calculate A_{v_s}. What is the effect of increasing levels of R_L on the gain?

   (f) Change R_s to 0.5 kΩ (with R_L at 2.7 kΩ) and comment on the effect of reducing R_s on A_{v_s}.

   (g) Change R_L to 5.6 kΩ and R_s to 0.5 kΩ and determine the new levels of Z_i and Z_o. How are the impedance parameters affected by changing levels of R_L and R_s?

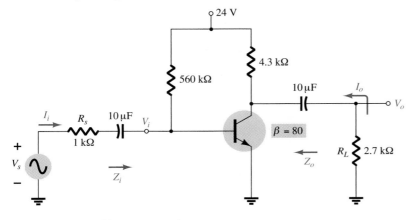

Figure 10.50 Problems 5, 17, and 21

§ 10.6 BJT CE Networks

6. For the voltage-divider configuration of Fig. 10.51:
   (a) Determine $A_{v_{NL}}$, Z_i, and Z_o.
   (b) Sketch the two-port model of Fig. 10.2 with the parameters determined in part (a) in place.
   (c) Calculate the gain A_v using the model of part (b).
   (d) Determine the current gain A_i.
   (e) Determine A_v, Z_i, and Z_o using the r_e model and compare solutions.

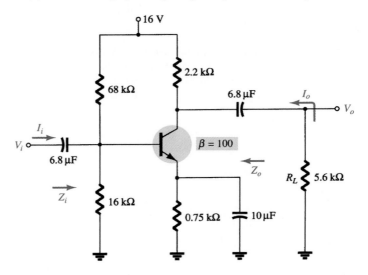

Figure 10.51   Problems 6, 7, and 8

*7. (a) Draw the dc and ac load lines for the network of Fig. 10.51 on the characteristics of Fig. 10.48.
   (b) Determine the peak-to-peak value of I_c and V_{ce} from the graph if V_i has a peak value of 10 mV. Determine the voltage gain $A_v = V_o/V_i$ and compare the solution with that obtained in Problem 6.

8. (a) Determine the voltage gain A_v for the network of Fig. 10.51 with $R_L = 4.7$, 2.2, and 0.5 kΩ. What is the effect of decreasing levels of R_L on the voltage gain?
   (b) How will Z_i, Z_o, and $A_{v_{NL}}$ change with decreasing levels of R_L?

9. For the emitter-stabilized network of Fig. 10.52:
   (a) Determine $A_{v_{NL}}$, Z_i, and Z_o.
   (b) Sketch the two-port model of Fig. 10.2 with the values determined in part (a).
   (c) Determine A_v and A_{v_s}.
   (d) Change R_s to 1 kΩ. What is the effect on $A_{v_{NL}}$, Z_i, and Z_o?
   (e) Change R_s to 1 kΩ and determine A_v and A_{v_s}. What is the effect of increasing levels of R_s on A_v and A_{v_s}?

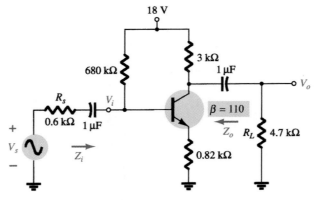

Figure 10.52   Problem 9

§ 10.7 BJT Emitter-Follower Networks

* **10.** For the network of Fig. 10.53:
  (a) Determine $A_{v_{NL}}$, Z_i, and Z_o.
  (b) Sketch the two-port model of Fig. 10.2 with the values determined in part (a).
  (c) Determine A_v and A_{v_s}.
  (d) Change R_s to 1 kΩ and determine A_v and A_{v_s}. What is the effect of increasing levels of R_s on the voltage gains?
  (e) Change R_s to 1 kΩ and determine $A_{v_{NL}}$, Z_i, and Z_o. What is the effect of increasing levels of R_s on the parameters?
  (f) Change R_L to 5.6 kΩ and determine A_v and A_{v_s}. What is the effect of increasing levels of R_L on the voltage gains? Maintain R_s at its original level of 0.6 kΩ.

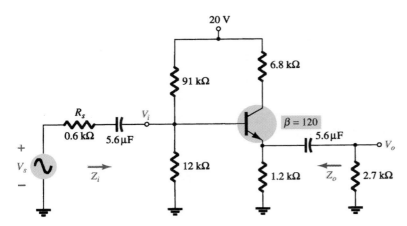

Figure 10.53 Problems 10 and 18

§ 10.8 BJT CB Networks

* **11.** For the common-base network of Fig. 10.54:
  (a) Determine Z_i, Z_o, and $A_{v_{NL}}$.
  (b) Sketch the two-port model of Fig. 10.2 with the parameters of part (a) in place.
  (c) Determine A_v and A_{v_s}.
  (d) Determine A_v and A_{v_s} using the r_e model and compare with the results of part (c).
  (e) Change R_s to 0.5 kΩ and R_L to 2.2 kΩ and calculate A_v and A_{v_s}. What is the effect of changing levels of R_s and R_L on the voltage gains?
  (f) Determine Z_o if R_s changed to 0.5 kΩ with all other parameters as appearing in Fig. 10.54. How is Z_o affected by changing levels of R_s?
  (g) Determine Z_i if R_L is reduced to 2.2 kΩ. What is the effect of changing levels of R_L on the input impedance?

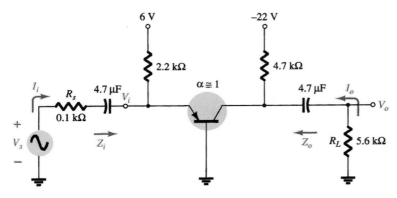

Figure 10.54 Problems 11, 19, and 22

§ 10.9 FET Networks

12. For the self-bias JFET network of Fig. 10.55:
   (a) Determine $A_{v_{NL}}$, Z_i, and Z_o.
   (b) Sketch the two-port model of Fig. 10.2 with the parameters determined in part (a) in place.
   (c) Determine A_v and A_{v_s}.
   (d) Change R_L to 6.8 kΩ and R_{sig} to 1 kΩ. and calculate the new levels of A_v and A_{v_s}. How are the voltage gains affected by changes in R_{sig} and R_L?
   (e) For the same changes as part (d), determine Z_i and Z_o. What was the impact on both impedances?

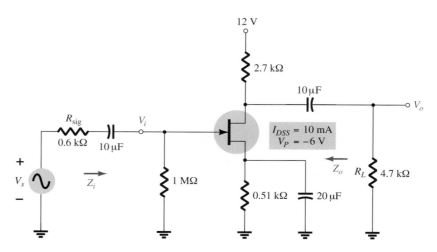

Figure 10.55 Problem 12 and 20

13. For the source-follower network of Fig. 10.56:
   (a) Determine $A_{v_{NL}}$, Z_i, and Z_o.
   (b) Sketch the two-port model of Fig. 10.2 with the parameters determined in part (a) in place.
   (c) Determine A_v and A_{v_s}.
   (d) Change R_L to 4.7 kΩ and calculate A_v and A_{v_s}. What was the effect of increasing levels of R_L on both voltage gains?
   (e) Change R_{sig} to 1 kΩ (with R_L at 2.2 kΩ) and calculate A_v and A_{v_s}. What was the effect of increasing levels of R_{sig} on both voltage gains?
   (f) Change R_L to 4.7 kΩ and R_{sig} to 1 kΩ and calculate Z_i and Z_o. What was the effect on both parameters?

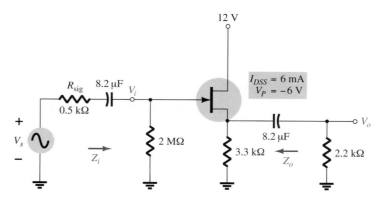

Figure 10.56 Problem 13

* **14.** For the common-gate configuration of Fig. 10.57:
  (a) Determine $A_{v_{NL}}$, Z_i, and Z_o.
  (b) Sketch the two-port model of Fig. 10.2 with the parameters determined in part (a) in place.
  (c) Determine A_v and A_{v_s}.
  (d) Change R_L to 2.2 kΩ and calculate A_v and A_{v_s}. What was the effect of changing R_L on the voltage gains?
  (e) Change R_{sig} to 0.5 kΩ (with R_L at 4.7 kΩ) and calculate A_v and A_{v_s}. What was the effect of changing R_{sig} on the voltage gains?
  (f) Change R_L to 2.2 kΩ and R_{sig} to 0.5 kΩ and calculate Z_i and Z_o. What was the effect on both parameters?

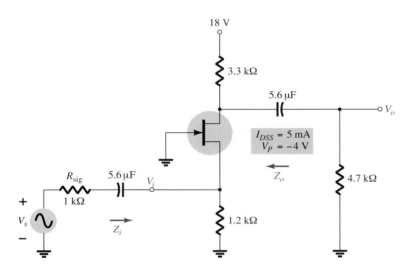

Figure 10.57 Problem 14

§ **10.11 Cascaded Systems**

* **15.** For the cascaded system of Fig. 10.58 with two identical stages, determine:
  (a) The loaded voltage gain of each stage.
  (b) The total gain of the system, A_v and A_{v_s}.
  (c) The loaded current gain of each stage.
  (d) The total current gain of the system.
  (e) How Z_i is affected by the second stage and R_L.
  (f) How Z_o is affected by the first stage and R_s.
  (g) The phase relationship between V_o and V_i.

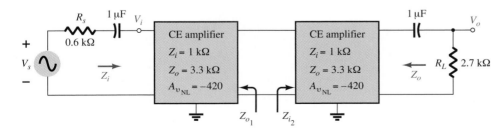

Figure 10.58 Problem 15

*16. For the cascaded system of Fig. 10.59, determine:
   (a) The loaded voltage gain of each stage.
   (b) The total gain of the system, A_v and A_{v_s}.
   (c) The loaded current gain of each stage.
   (d) The total current gain of the system.
   (e) How Z_i is affected by the second stage and R_L.
   (f) How Z_o is affected by the first stage and R_s.
   (g) The phase relationship between V_o and V_i.

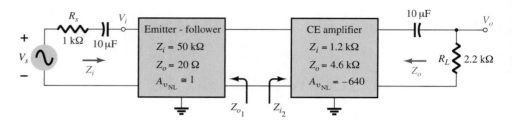

Figure 10.59   Problem 16

§ 10.13 Compter Analysis

17. Using PSpice Windows, determine the level of V_o for $V_s = 1$ mV for the network of Fig. 10.50. For the capacitive elements assume a frequency of 1 kHz.
18. Repeat Problem 17 for the network of Fig. 10.53 and compare the results with those of Problem 10.
19. Repeat Problem 17 for the network of Fig. 10.54 and compare with the results of Problem 11.
20. Repeat Problem 17 for the network of Fig. 10.55 and compare with the results of Problem 12.
21. Repeat Problem 17 using Electronics Workbench.
22. Repeat Problem 19 using Electronics Workbench.

*Please Note: Asterisks indicate more difficult problems.

BJT and JFET Frequency Response

f

11.1 INTRODUCTION

The analysis thus far has been limited to a particular frequency. For the amplifier, it was a frequency that normally permitted ignoring the effects of the capacitive elements, reducing the analysis to one that included only resistive elements and sources of the independent and controlled variety. We will now investigate the frequency effects introduced by the larger capacitive elements of the network at low frequencies and the smaller capacitive elements of the active device at the high frequencies. Since the analysis will extend through a wide frequency range, the logarithmic scale will be defined and used throughout the analysis. In addition, since industry typically uses a decibel scale on its frequency plots, the concept of the decibel is introduced in some detail. The similarities between the frequency response analyses of both BJTs and FETs permit a coverage of each in the same chapter.

11.2 LOGARITHMS

In this field, there is no escaping the need to become comfortable with the logarithmic function. The plotting of a variable between wide limits, comparing levels without unwieldy numbers, and identifying levels of particular importance in the design, review, and analysis procedures are all positive features of using the logarithmic function.

As a first step in clarifying the relationship between the variables of a logarithmic function, consider the following mathematical equations:

$$a = b^x, \qquad x = \log_b a \qquad (11.1)$$

The variables a, b, and x are the same in each equation. If a is determined by taking the base b to the x power, the same x will result if the log of a is taken to the base b. For instance, if $b = 10$ and $x = 2$,

$$a = b^x = (10)^2 = 100$$

but

$$x = \log_b a = \log_{10} 100 = 2$$

In other words, if you were asked to find the power of a number that would result in a particular level such as shown below:

$$10,000 = 10^x$$

f

the level of x could be determined using logarithms. That is,

$$x = \log_{10} 10{,}000 = 4$$

For the electrical/electronics industry and in fact for the vast majority of scientific research, the base in the logarithmic equation is limited to 10 and the number $e = 2.71828.\ldots$

Logarithms taken to the base 10 are referred to as *common logarithms,* while logarithms taken to the base e are referred to as *natural logarithms.* In summary:

$$\boxed{\text{Common logarithm:}\quad x = \log_{10} a} \qquad (11.2)$$

$$\boxed{\text{Natural logarithm:}\quad y = \log_e a} \qquad (11.3)$$

The two are related by

$$\boxed{\log_e a = 2.3 \log_{10} a} \qquad (11.4)$$

On today's scientific calculators, the common logarithm is typically denoted by the $\boxed{\text{log}}$ key and the natural logarithm by the $\boxed{\text{ln}}$ key.

EXAMPLE 11.1

Using the calculator, determine the logarithm of the following numbers to the base indicated.
(a) $\log_{10} 10^6$.
(b) $\log_e e^3$.
(c) $\log_{10} 10^{-2}$.
(d) $\log_e e^{-1}$.

Solution

(a) **6**   (b) **3**   (c) **−2**   (d) **−1**

The results in Example 11.1 clearly reveal that the logarithm of a number taken to a power is simply the power of the number if the number matches the base of the logarithm. In the next example, the base and the variable x are not related by an integer power of the base.

EXAMPLE 11.2

Using the calculator, determine the logarithm of the following numbers.
(a) $\log_{10} 64$.
(b) $\log_e 64$.
(c) $\log_{10} 1600$.
(d) $\log_{10} 8000$.

Solution

(a) **1.806**   (b) **4.159**   (c) **3.204**   (d) **3.903**

Note in parts (a) and (b) of Example 11.2 that the logarithms $\log_{10} a$ and $\log_e a$ are indeed related as defined by Eq. (11.4). In addition, note that the logarithm of a number does not increase in the same linear fashion as the number. That is, 8000 is 125 times larger than 64, but the logarithm of 8000 is only about 2.16 times larger than the magnitude of the logarithm of 64, revealing a very nonlinear relationship. In fact, Table 11.1 clearly shows how the logarithm of a number increases only as the exponent of the number. If the antilogarithm of a number is desired, the 10^x or e^x calculator functions are employed.

TABLE 11.1	
$\log_{10} 10^0$	$= 0$
$\log_{10} 10$	$= 1$
$\log_{10} 100$	$= 2$
$\log_{10} 1{,}000$	$= 3$
$\log_{10} 10{,}000$	$= 4$
$\log_{10} 100{,}000$	$= 5$
$\log_{10} 1{,}000{,}000$	$= 6$
$\log_{10} 10{,}000{,}000$	$= 7$
$\log_{10} 100{,}000{,}000$	$= 8$
and so on	

EXAMPLE 11.3

Using a calculator, determine the antilogarithm of the following expressions:
(a) $1.6 = \log_{10} a$.
(b) $0.04 = \log_e a$.

Solution

(a) $a = 10^{1.6}$
Calculator keys: 〔1〕〔.〕〔6〕〔2nd F〕〔10^x〕
and $a = \mathbf{39.81}$
(b) $a = e^{0.04}$
Calculator keys: 〔0〕〔.〕〔0〕〔4〕〔2nd F〕〔e^x〕
and $a = \mathbf{1.0408}$

Since the remaining analysis of this chapter employs the common logarithm, let us now review a few properties of logarithms using solely the common logarithm. In general, however, the same relationships hold true for logarithms to any base.

$$\boxed{\log_{10} 1 = 0} \tag{11.5}$$

As clearly revealed by Table 11.1, since $10^0 = 1$,

$$\boxed{\log_{10} \frac{a}{b} = \log_{10} a - \log_{10} b} \tag{11.6}$$

which for the special case of $a = 1$ becomes

$$\boxed{\log_{10} \frac{1}{b} = -\log_{10} b} \tag{11.7}$$

revealing that for any b greater than 1 the logarithm of a number less than 1 is always negative.

$$\boxed{\log_{10} ab = \log_{10} a + \log_{10} b} \tag{11.8}$$

In each case, the equations employing natural logarithms will have the same format.

f

EXAMPLE 11.4   Using a calculator, determine the logarithm of the following numbers:
(a) $\log_{10} 0.5$.

(b) $\log_{10} \dfrac{4000}{250}$.

(c) $\log_{10} (0.6 \times 30)$.

Solution

(a) **−0.3**

(b) $\log_{10} 4000 - \log_{10} 250 = 3.602 - 2.398 = \mathbf{1.204}$

$$\text{Check: } \log_{10} \frac{4000}{250} = \log_{10} 16 = \mathbf{1.204}$$

(c) $\log_{10} 0.6 + \log_{10} 30 = -0.2218 + 1.477 = \mathbf{1.255}$

$$\text{Check: } \log_{10} (0.6 \times 30) = \log_{10} 18 = \mathbf{1.255}$$

The use of log scales can significantly expand the range of variation of a particular variable on a graph. Most graph paper available is of the semilog or double-log (log-log) variety. The term *semi* (meaning one-half) indicates that only one of the two scales is a log scale, whereas double-log indicates that both scales are log scales. A semilog scale appears in Fig. 11.1. Note that the vertical scale is a linear scale with equal divisions. The spacing between the lines of the log plot is shown on the graph.

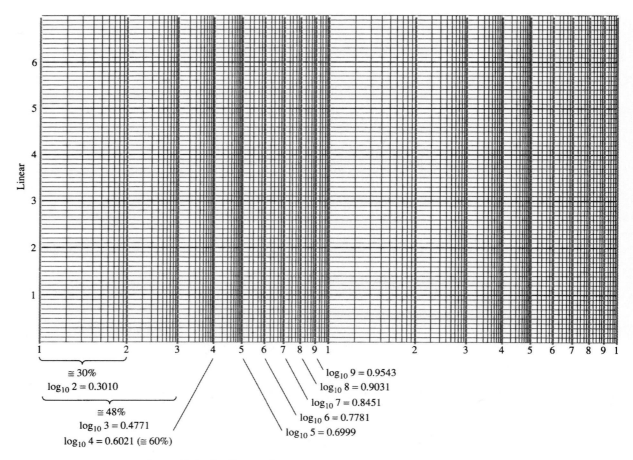

Figure 11.1 Semilog graph paper.

The log of 2 to the base 10 is approximately 0.3. The distance from 1 ($\log_{10} 1 = 0$) to 2 is therefore 30% of the span. The log of 3 to the base 10 is 0.4771 or almost 48% of the span (very close to one-half the distance between power of 10 increments on the log scale). Since $\log_{10} 5 \cong 0.7$, it is marked off at a point 70% of the distance. Note that between any two digits the same compression of the lines appears as you progress from the left to the right. It is important to note the resulting numerical value and the spacing, since plots will typically only have the tic marks indicated in Fig. 11.2 due to a lack of space. You must realize that the longer bars for this figure have the numerical values of 0.3, 3, and 30 associated with them, whereas the next shorter bars have values of 0.5, 5, and 50 and the shortest bars 0.7, 7, and 70.

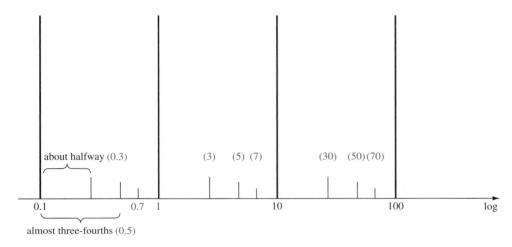

Figure 11.2   Identifying the numerical values of the tic marks on a log scale.

Be aware that plotting a function on a log scale can change the general appearance of the waveform as compared to a plot on a linear scale. A straight-line plot on a linear scale can develop a curve on a log scale, and a nonlinear plot on a linear scale can take on the appearance of a straight line on a log plot. The important point is that the results extracted at each level be correctly labeled by developing a familiarity with the spacing of Figs. 11.1 and 11.2. This is particularly true for some of the log-log plots that appear later in the book.

11.3   DECIBELS

The concept of the decibel (dB) and the associated calculations will become increasingly important in the remaining sections of this chapter. The background surrounding the term *decibel* has its origin in the established fact that power and audio levels are related on a logarithmic basis. That is, an increase in power level, say 4 to 16 W, does not result in an audio level increase by a factor of $16/4 = 4$. It will increase by a factor of 2 as derived from the power of 4 in the following manner: $(4)^2 = 16$. For a change of 4 to 64 W, the audio level will increase by a factor of 3 since $(4)^3 = 64$. In logarithmic form, the relationship can be written as $\log_4 64 = 3$.

The term *bel* was derived from the surname of Alexander Graham Bell. For standardization, the bel (B) was defined by the following equation to relate power levels P_1 and P_2:

$$G = \log_{10} \frac{P_2}{P_1} \qquad \text{bel} \qquad (11.9)$$

f

It was found, however, that the bel was too large a unit of measurement for practical purposes, so the decibel (dB) was defined such that 10 decibels = 1 bel. Therefore,

$$G_{dB} = 10 \log_{10} \frac{P_2}{P_1} \qquad \text{dB} \qquad (11.10)$$

The terminal rating of electronic communication equipment (amplifiers, microphones, etc.) is commonly rated in decibels. Equation (11.10) indicates clearly, however, that the decibel rating is a measure of the difference in magnitude between *two* power levels. For a specified terminal (output) power (P_2) there must be a reference power level (P_1). The reference level is generally accepted to be 1 mW, although on occasion, the 6-mW standard of earlier years is applied. The resistance to be associated with the 1-mW power level is 600 Ω, chosen because it is the characteristic impedance of audio transmission lines. When the 1-mW level is employed as the reference level, the decibel symbol frequently appears as dBm. In equation form,

$$G_{dBm} = 10 \log_{10} \frac{P_2}{1 \text{ mW}} \bigg|_{600 \Omega} \qquad \text{dBm} \qquad (11.11)$$

There exists a second equation for decibels that is applied frequently. It can be best described through the system of Fig. 11.3. For V_i equal to some value V_1, $P_1 = V_1^2/R_i$, where R_i is the input resistance of the system of Fig. 11.3. If V_i should be increased (or decreased) to some other level, V_2, then $P_2 = V_2^2/R_i$. If we substitute into Eq. (11.10) to determine the resulting difference in decibels between the power levels,

$$G_{dB} = 10 \log_{10} \frac{P_2}{P_1} = 10 \log_{10} \frac{V_2^2/R_i}{V_1^2/R_i} = 10 \log_{10} \left(\frac{V_2}{V_1}\right)^2$$

and

$$G_{dB} = 20 \log_{10} \frac{V_2}{V_1} \qquad \text{dB} \qquad (11.12)$$

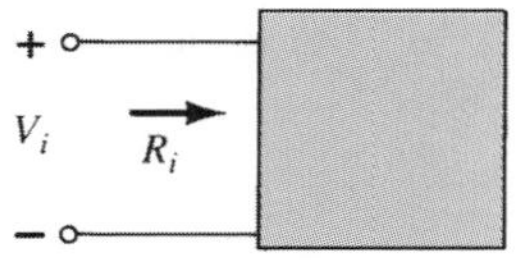

Figure 11.3 Configuration employed in the discussion of Eq. (11.12).

Frequently, the effect of different impedances ($R_1 \neq R_2$) is ignored and Eq. (11.12) applied simply to establish a basis of comparison between levels—voltage or current. For situations of this type, the decibel gain should more correctly be referred to as the voltage or current gain in decibels to differentiate it from the common usage of decibel as applied to power levels.

One of the advantages of the logarithmic relationship is the manner in which it can be applied to cascaded stages. For example, the magnitude of the overall voltage gain of a cascaded system is given by

$$|A_{v_T}| = |A_{v_1}||A_{v_2}||A_{v_3}| \cdots |A_{v_n}| \qquad (11.13)$$

Applying the proper logarithmic relationship results in

$$G_v = 20 \log_{10} |A_{v_T}| = 20 \log_{10} |A_{v_1}| + 20 \log_{10} |A_{v_2}|$$
$$+ 20 \log_{10} |A_{v_3}| + \cdots + 20 \log_{10} |A_{v_n}| \quad \text{(dB)} \quad (11.14)$$

In words, the equation states that the decibel gain of a cascaded system is simply the sum of the decibel gains of each stage, that is,

$$G_{dB_T} = G_{dB_1} + G_{dB_2} + G_{dB_3} + \cdots + G_{dB_n} \qquad \text{dB} \qquad (11.15)$$

In an effort to develop some association between dB levels and voltage gains, Table 11.2 was developed. First note that a gain of 2 results in a dB level of +6 dB while a drop to $\frac{1}{2}$ results in a -6-dB level. A change in V_o/V_i from 1 to 10, 10 to 100, or 100 to 1000 results in the same 20-dB change in level. When $V_o = V_i$, $V_o/V_i = 1$ and the dB level is 0. At a very high gain of 1000, the dB level is 60, while at the much higher gain of 10,000, the dB level is 80 dB, an increase of only 20 dB—a result of the logarithmic relationship. Table 11.2 clearly reveals that voltage gains of 50 dB or higher should immediately be recognized as being quite high.

TABLE 11.2	
Voltage Gain, V_o/V_i	dB Level
0.5	−6
0.707	−3
1	0
2	6
10	20
40	32
100	40
1000	60
10,000	80
etc.	

Find the magnitude gain corresponding to a decibel gain of 100.

EXAMPLE 11.5

Solution

By Eq. (11.10),

$$G_{dB} = 10 \log_{10} \frac{P_2}{P_1} = 100 \text{ dB} \rightarrow \log_{10} \frac{P_2}{P_1} = 10$$

so that

$$\frac{P_2}{P_1} = 10^{10} = 10,000,000,000$$

This example clearly demonstrates the range of decibel values to be expected from practical devices. Certainly, a future calculation giving a decibel result in the neighborhood of 100 should be questioned immediately.

The input power to a device is 10,000 W at a voltage of 1000 V. The output power is 500 W, while the output impedance is 20 Ω.
(a) Find the power gain in decibels.
(b) Find the voltage gain in decibels.
(c) Explain why parts (a) and (b) agree or disagree.

EXAMPLE 11.6

Solution

(a) $G_{dB} = 10 \log_{10} \dfrac{P_o}{P_i} = 10 \log_{10} \dfrac{500 \text{ W}}{10 \text{ kW}} = 10 \log_{10} \dfrac{1}{20} = -10 \log_{10} 20$

$= -10(1.301) = \mathbf{-13.01 \text{ dB}}$

(b) $G_v = 20 \log_{10} \dfrac{V_o}{V_i} = 20 \log_{10} \dfrac{\sqrt{PR}}{1000} = 20 \log_{10} \dfrac{\sqrt{(500 \text{ W})(20 \text{ }\Omega)}}{1000 \text{ V}}$

$= 20 \log_{10} \dfrac{100}{1000} = 20 \log_{10} \dfrac{1}{10} = -20 \log_{10} 10 = \mathbf{-20 \text{ dB}}$

(c) $R_i = \dfrac{V_i^2}{P_i} = \dfrac{(1 \text{ kV})^2}{10 \text{ kW}} = \dfrac{10^6}{10^4} = \mathbf{100 \text{ }\Omega} \neq R_o = \mathbf{20 \text{ }\Omega}$

EXAMPLE 11.7

An amplifier rated at 40-W output is connected to a 10-Ω speaker.
(a) Calculate the input power required for full power output if the power gain is 25 dB.
(b) Calculate the input voltage for rated output if the amplifier voltage gain is 40 dB.

Solution

(a) Eq. (11.10): $25 = 10 \log_{10} \dfrac{40 \text{ W}}{P_i} \Rightarrow P_i = \dfrac{40 \text{ W}}{\text{antilog } (2.5)} = \dfrac{40 \text{ W}}{3.16 \times 10^2}$

$$= \dfrac{40 \text{ W}}{316} \cong \textbf{126.5 mW}$$

(b) $G_v = 20 \log_{10} \dfrac{V_o}{V_i} \Rightarrow 40 = 20 \log_{10} \dfrac{V_o}{V_i}$

$\dfrac{V_o}{V_i} = \text{antilog } 2 = 100$

$V_o = \sqrt{PR} = \sqrt{(40 \text{ W})(10 \text{ V})} = 20 \text{ V}$

$V_i = \dfrac{V_o}{100} = \dfrac{20 \text{ V}}{100} = 0.2 \text{ V} = \textbf{200 mV}$

Mathcad

There are a number of ways to obtain the log of a number or an expression using Mathcad. The most direct is simply to type **log( ),** inserting the quantity of interest within the brackets and then selecting the equal sign. The result will appear immediately.

Another approach is to use the sequence **View-Toolbars-Calculator,** and the **Calculator** with all its options will appear on the screen. Selecting **log** will result in **log( )** with a request for the quantity in the brackets.

Lastly, the sequence **Insert-Function** will result in an **Insert Function** dialog box from which **Log and Exponential** can be chosen under **Function Category** and **Log** under **Function Name**.

For Example 11.6, part (a) will appear as shown in Fig. 11.4. In Example 11.7, the antilogarithm is required in part (b). Keeping in mind that if $x = \log_b a$, then $a = b^x$, we can simply insert $b = 10$ and $x = 2$ into the equation $a = 10^2 = 100$ as shown in Fig. 11.4. To obtain the power of a number using Mathcad, simply use the **Shift + ^** keys and insert the power followed by the equal sign.

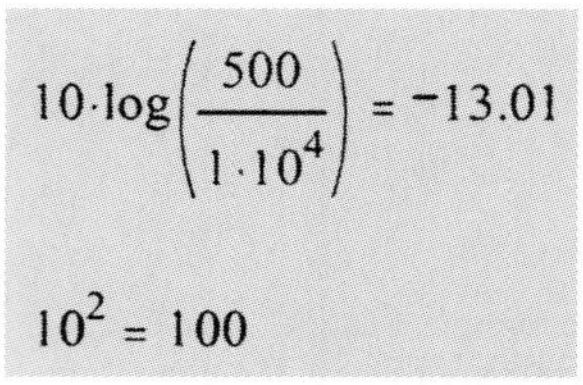

$$10 \cdot \log\left(\frac{500}{1 \cdot 10^4}\right) = {}^-13.01$$

$$10^2 = 100$$

Figure 11.4   Example 11.6, part (a), using Mathcad.

11.4   GENERAL FREQUENCY CONSIDERATIONS

The frequency of the applied signal can have a pronounced effect on the response of a single-stage or multistage network. The analysis thus far has been for the midfrequency spectrum. At low frequencies, we shall find that the coupling and by-pass capacitors can no longer be replaced by the short-circuit approximation because of the increase in reactance of these elements. The frequency-dependent parameters of the small-signal equivalent circuits and the stray capacitive elements associated with the active device and the network will limit the high-frequency response of the system. An increase in the number of stages of a cascaded system will also limit both the high- and low-frequency responses.

The magnitudes of the gain response curves of an *RC*-coupled, direct-coupled, and transformer-coupled amplifier system are provided in Fig. 11.5. Note that the horizontal scale is a logarithmic scale to permit a plot extending from the low- to the high-frequency regions. For each plot, a low-, high-, and mid-frequency region has been defined. In addition, the primary reasons for the drop in gain at low and high frequencies have also been indicated within the parentheses. For the *RC*-coupled amplifier, the drop at low frequencies is due to the increasing reactance of C_C, C_s, or C_E, while its upper frequency limit is determined by either the parasitic capacitive elements of the network and frequency dependence of the gain of the active device. An explanation of the drop in gain for the transformer-coupled system requires a basic understanding of "transformer action" and the transformer equivalent circuit. For the moment, let us say that it is simply due to the "shorting effect" (across the input

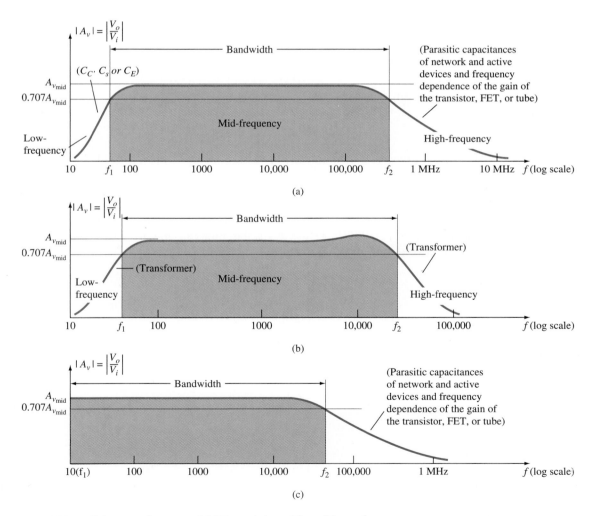

Figure 11.5 Gain versus frequency: (a) *RC*-coupled amplifiers; (b) transformer-coupled amplifiers; (c) direct-coupled amplifiers.

11.4 General Frequency Considerations

terminals of the transformer) of the magnetizing inductive reactance at low frequencies ($X_L = 2\pi f L$). The gain must obviously be zero at $f = 0$ since at this point there is no longer a changing flux established through the core to induce a secondary or output voltage. As indicated in Fig. 11.5, the high-frequency response is controlled primarily by the stray capacitance between the turns of the primary and secondary windings. For the direct-coupled amplifier, there are no coupling or bypass capacitors to cause a drop in gain at low frequencies. As the figure indicates, it is a flat response to the upper cutoff frequency, which is determined by either the parasitic capacitances of the circuit or the frequency dependence of the gain of the active device.

For each system of Fig. 11.5, there is a band of frequencies in which the magnitude of the gain is either equal or relatively close to the midband value. To fix the frequency boundaries of relatively high gain, $0.707A_{v_{mid}}$ was chosen to be the gain at the cutoff levels. The corresponding frequencies f_1 and f_2 are generally called the *corner, cutoff, band, break,* or *half-power frequencies.* The multiplier 0.707 was chosen because at this level the output power is half the midband power output, that is, at midfrequencies,

$$P_{o_{mid}} = \frac{|V_o^2|}{R_o} = \frac{|A_{v_{mid}}V_i|^2}{R_o}$$

and at the half-power frequencies,

$$P_{o_{HPF}} = \frac{|0.707\,A_{v_{mid}}V_i|^2}{R_o} = 0.5\frac{|A_{v_{mid}}V_i|^2}{R_o}$$

and

$$\boxed{P_{o_{HPF}} = 0.5\,P_{o_{mid}}} \tag{11.16}$$

The bandwidth (or passband) of each system is determined by f_1 and f_2, that is,

$$\boxed{\text{bandwidth (BW)} = f_2 - f_1} \tag{11.17}$$

For applications of a communications nature (audio, video), a decibel plot of the voltage gain versus frequency is more useful than that appearing in Fig. 11.5. Before obtaining the logarithmic plot, however, the curve is generally normalized as shown in Fig. 11.6. In this figure, the gain at each frequency is divided by the midband value. Obviously, the midband value is then 1 as indicated. At the half-power frequencies, the resulting level is $0.707 = 1/\sqrt{2}$. A decibel plot can now be obtained by applying Eq. (11.12) in the following manner:

$$\boxed{\left.\frac{A_v}{A_{v_{mid}}}\right|_{dB} = 20\log_{10}\frac{A_v}{A_{v_{mid}}}} \tag{11.18}$$

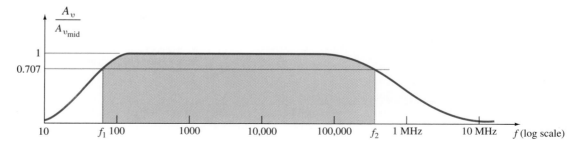

Figure 11.6 Normalized gain versus frequency plot.

At midband frequencies, $20 \log_{10} 1 = 0$, and at the cutoff frequencies, $20 \log_{10} 1/\sqrt{2} = -3$ dB. Both values are clearly indicated in the resulting decibel plot of Fig. 11.7. The smaller the fraction ratio, the more negative the decibel level.

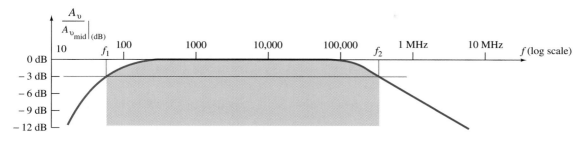

Figure 11.7   Decibel plot of the normalized gain versus frequency plot of Fig. 11.6.

For the greater part of the discussion to follow, a decibel plot will be made only for the low- and high-frequency regions. Keep Fig. 11.7 in mind, therefore, to permit a visualization of the broad system response.

It should be understood that most amplifiers introduce a 180° phase shift between input and output signals. This fact must now be expanded to indicate that this is the case only in the midband region. At low frequencies, there is a phase shift such that V_o lags V_i by an increased angle. At high frequencies, the phase shift will drop below 180°. Figure 11.8 is a standard phase plot for an *RC*-coupled amplifier.

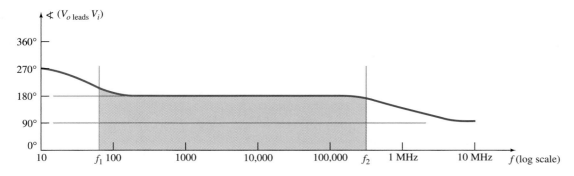

Figure 11.8   Phase plot for an *RC*-coupled amplifier system.

11.5   LOW-FREQUENCY ANALYSIS— BODE PLOT

In the low-frequency region of the single-stage BJT or FET amplifier, it is the *R-C* combinations formed by the network capacitors C_C, C_E, and C_s and the network resistive parameters that determine the cutoff frequencies. In fact, an *R-C* network similar to Fig. 11.9 can be established for each capacitive element and the frequency at which the output voltage drops to 0.707 of its maximum value determined. Once the cutoff frequencies due to each capacitor are determined, they can be compared to establish which will determine the low-cutoff frequency for the system.

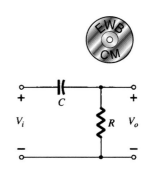

Figure 11.9   *R-C* combination that will define a low cutoff frequency.

Our analysis, therefore, will begin with the series *R-C* combination of Fig. 11.9 and the development of a procedure that will result in a plot of the frequency response with a minimum of time and effort. At very high frequencies,

$$X_C = \frac{1}{2\pi f C} \cong 0 \ \Omega$$

and the short-circuit equivalent can be substituted for the capacitor as shown in Fig. 11.10. The result is that $V_o \cong V_i$ at high frequencies. At $f = 0$ Hz,

$$X_C = \frac{1}{2\pi f C} = \frac{1}{2\pi(0)C} = \infty \ \Omega$$

and the open-circuit approximation can be applied as shown in Fig. 11.11, with the result that $V_o = 0$ V.

Between the two extremes, the ratio $A_v = V_o/V_i$ will vary as shown in Fig. 11.12. As the frequency increases, the capacitive reactance decreases and more of the input voltage appears across the output terminals.

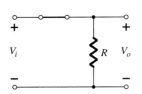

Figure 11.10   *R-C* circuit of Figure 11.9 at very high frequencies.

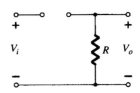

Figure 11.11   *R-C* circuit of Figure 11.9 at $f = 0$ Hz.

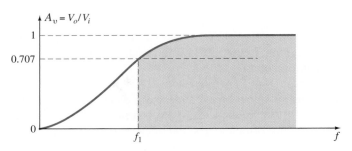

Figure 11.12   Low frequency response for the *R-C* circuit of Figure 11.9.

The output and input voltages are related by the voltage-divider rule in the following manner:

$$\mathbf{V}_o = \frac{\mathbf{R}\mathbf{V}_i}{\mathbf{R} + \mathbf{X}_C}$$

with the magnitude of V_o determined by

$$V_o = \frac{RV_i}{\sqrt{R^2 + X_C^2}}$$

For the special case where $X_C = R$,

$$V_o = \frac{RV_i}{\sqrt{R^2 X_C^2}} = \frac{RV_i}{\sqrt{R^2 + R^2}} = \frac{RV_i}{\sqrt{2R^2}} = \frac{RV_i}{\sqrt{2}R} = \frac{1}{\sqrt{2}}V_i$$

and

$$\boxed{|A_v| = \frac{V_o}{V_i} = \frac{1}{\sqrt{2}} = 0.707|_{X_C = R}}$$

(11.19)

the level of which is indicated on Fig. 11.12. In other words, at the frequency of which $X_C = R$, the output will be 70.7% of the input for the network of Fig. 11.9.

The frequency at which this occurs is determined from

$$X_C = \frac{1}{2\pi f_1 C} = R$$

and

$$\boxed{f_1 = \frac{1}{2\pi R C}}$$

(11.20)

In terms of logs,

$$G_v = 20 \log_{10} A_v = 20 \log_{10} \frac{1}{\sqrt{2}} = -3 \text{ dB}$$

while at $A_v = V_o/V_i = 1$ or $V_o = V_i$ (the maximum value),

$$G_v = 20 \log_{10} 1 = 20(0) = 0 \text{ dB}$$

In Fig. 11.7, we recognize that there is a 3-dB drop in gain from the midband level when $f = f_1$. In a moment, we will find that an RC network will determine the low-frequency cutoff frequency for a BJT transistor and f_1 will be determined by Eq. (11.20).

If the gain equation is written as

$$A_v = \frac{V_o}{V_i} = \frac{R}{R - jX_C} = \frac{1}{1 - j(X_C/R)} = \frac{1}{1 - j(1/\omega CR)} = \frac{1}{1 - j(1/2\pi fCR)}$$

and using the frequency defined above,

$$A_v = \frac{1}{1 - j(f_1/f)} \qquad (11.21)$$

In the magnitude and phase form,

$$A_v = \frac{V_o}{V_i} = \underbrace{\frac{1}{\sqrt{1 + (f_1/f)^2}}}_{\text{magnitude of } A_v} \underbrace{\angle \tan^{-1}(f_1/f)}_{\substack{\text{phase } \angle \text{ by which} \\ V_o \text{ leads } V_i}} \qquad (11.22)$$

For the magnitude when $f = f_1$,

$$|A_v| = \frac{1}{\sqrt{1 + (1)^2}} = \frac{1}{\sqrt{2}} = 0.707 \rightarrow -3 \text{ dB}$$

In the logarithmic form, the gain in dB is

$$A_{v(dB)} = 20 \log_{10} \frac{1}{\sqrt{1 + (f_1/f)^2}} \qquad (11.23)$$

$$A_{v(dB)} = -20 \log_{10} \left[1 + \left(\frac{f_1}{f} \right)^2 \right]^{1/2}$$

$$= -(\tfrac{1}{2})(20) \log_{10} \left[1 + \left(\frac{f_1}{f} \right)^2 \right]$$

$$= -10 \log_{10} \left[1 + \left(\frac{f_1}{f} \right)^2 \right]$$

For frequencies where $f \ll f_1$ or $(f_1/f)^2 \gg 1$, the equation above can be approximated by

$$A_{v(dB)} = -10 \log_{10} \left(\frac{f_1}{f} \right)^2$$

and finally,

$$A_{v(dB)} = -20 \log_{10} \frac{f_1}{f} \qquad (11.24)$$

$$\scriptstyle f \ll f_1$$

Ignoring the condition $f \ll f_1$ for a moment, a plot of Eq. (11.24) on a frequency log scale will yield a result of a very useful nature for future decibel plots.

$$\text{At } f = f_1: \frac{f_1}{f} = 1 \text{ and } -20 \log_{10} 1 = 0 \text{ dB}$$

$$\text{At } f = \tfrac{1}{2} f_1: \frac{f_1}{f} = 2 \text{ and } -20 \log_{10} 2 \cong -6 \text{ dB}$$

$$\text{At } f = \tfrac{1}{4} f_1: \frac{f_1}{f} = 4 \text{ and } -20 \log_{10} 4 \cong -12 \text{ dB}$$

$$\text{At } f = \tfrac{1}{10} f_1: \frac{f_1}{f} = 10 \text{ and } -20 \log_{10} 10 = -20 \text{ dB}$$

A plot of these points is indicated in Fig. 11.13 from $0.1f_1$ to f_1. Note that this results in a straight line when plotted against a log scale. In the same figure, a straight line is also drawn for the condition of 0 dB for $f \gg f_1$. As stated earlier, the straight-line segments (asymptotes) are only accurate for 0 dB when $f \gg f_1$ and the sloped line when $f_1 \gg f$. We know, however, that when $f = f_1$, there is a 3-dB drop from the mid-band level. Employing this information in association with the straight-line segments permits a fairly accurate plot of the frequency response as indicated in the same figure. The piecewise linear plot of the asymptotes and associated breakpoints is called a *Bode plot* of the magnitude versus frequency.

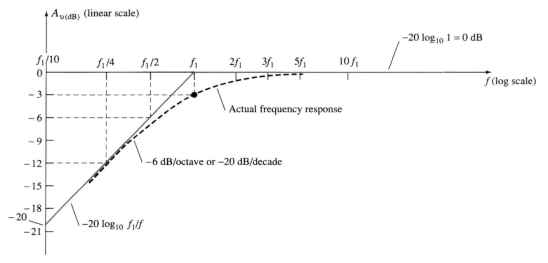

Figure 11.13 Bode plot for the low-frequency region.

The calculations above and the curve itself demonstrate clearly that:

A change in frequency by a factor of 2, equivalent to 1 octave, results in a 6-dB change in the ratio as noted by the change in gain from $f_1/2$ to f_1.

As noted by the change in gain from $f_1/2$ to f_1:

For a 10:1 change in frequency, equivalent to 1 decade, there is a 20-dB change in the ratio as demonstrated between the frequencies of $f_1/10$ and f_1.

In the future, therefore, a decibel plot can easily be obtained for a function having the format of Eq. (11.24). First, simply find f_1 from the circuit parameters and then sketch two asymptotes—one along the 0-dB line and the other drawn through f_1 sloped at 6 dB/octave or 20 dB/decade. Then, find the 3-dB point corresponding to f_1 and sketch the curve.

For the network of Fig. 11.14:
(a) Determine the break frequency.
(b) Sketch the asymptotes and locate the -3-dB point.
(c) Sketch the frequency response curve.

EXAMPLE 11.8

Solution

(a) $f_1 = \dfrac{1}{2\pi RC} = \dfrac{1}{(6.28)(5 \times 10^3\ \Omega)(0.1 \times 10^{-6}\ F)}$

$\cong$ **318.5 Hz**

(b) and (c). See Fig. 11.15.

Figure 11.14   Example 11.8.

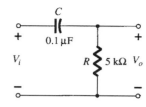

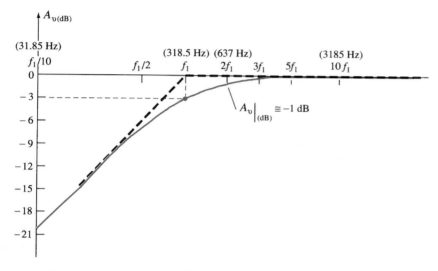

Figure 11.15   Frequency response for the *R-C* circuit of Figure 11.14.

Mathcad

Mathcad will now be used to obtain a plot of the dB gain for Example 11.8 using Eq. (11.23) without using the approximations introduced in the development of Eq. (11.24). In other words, the response obtained will be a point-to-point plot of the gain equation.

Using Mathcad, the first step is to set up a range for the horizontal variable, frequency (f). This is done by first typing **f** followed by **Shift:** to obtain the colon and equal sign appearing in Fig. 11.16. Next, in an attempt to match the curve of Fig. 11.15, the starting frequency was chosen as 10 Hz as also indicated on Fig. 11.16. Then a range was defined by selecting the semicolon key to obtain the two sequential dots that follow the 10. Finally, the upper limit of 10 kHz was selected with 1 followed by an ***** for multiplication and the 10 to the 4th power using the **^** key and the number **4.** The range of the variable f has now been defined for the equation to follow.

Using the capital letter **A** to represent amplification (gain), the variable against which the gain is to be determined **must** be defined by (f). Forgetting to add the (f) will result in a meaningless response. Next the equation must be entered, paying particular attention to the location of the **placeholders. A placeholder** can be moved using the left or right **directional** keys ($\leftarrow$ $\rightarrow$). In addition, you can move backward through an equation using the **backspace** key; but remember that as you backspace, the quantity to the left of the vertical component of the **placeholder** is lost—not so

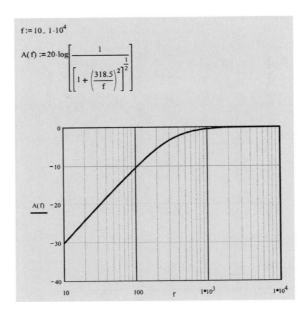

Figure 11.16 Plot of dB gain for Example 11.8 using Mathcad.

with the directional keys. The **spacebar** will also permit some backward motion through the equation.

Now we have to generate the desired plot. First position the cross hair in the area in which you want to generate the plot. Then type in **A(f)** and select **Insert-Graph-X-Y Plot** or **View-Toolbars-Graph** to obtain the **Graph palette** on which **X-Y Plot Shift+2** can be chosen. The result is a frame for the graph that has some solid squares at various points around the frame. Type **A(f)** at the location of the solid black square at the middle of the vertical line and **f** at the solid black square at the center of the horizontal line to define the variables to be plotted on each axis. Then simply click anywhere on the screen outside the defined area, and the plot will appear with a scaling chosen by Mathcad.

To change the horizontal axis to a log scale, first click anywhere on the graph to create the rectangular enclosure around the plot. Then use the sequence **Format-Graph-X-Y Plot** to obtain the **Setting Default Formats for X-Y Plots** dialog box. Choose **X-Y Axes** followed by **X-Axes-Log Scale** to set the log scale and **Grid Lines** to clearly show the log scale. After choosing **OK,** the plot of Fig. 11.16 will result.

Note that the results of Example 11.8 are verified by the intersection of the -3-dB level and $f = f_1 = 318.5$ Hz. In addition, note how closely the -20-dB level corresponds with $f = f_1/10 = 31.85$ Hz on Fig. 11.15. The results certainly verify the approximations applied to obtain a quick response with a minimum of mathematical difficulty.

The gain at any frequency can be determined from the frequency plot in the following manner:

$$A_{v(\text{dB})} = 20 \log_{10} \frac{V_o}{V_i}$$

but

$$\frac{A_{v(\text{dB})}}{20} = \log_{10} \frac{V_o}{V_i}$$

and

$$A_v = \frac{V_o}{V_i} = 10^{A_{v(\text{dB})}/20} \qquad (11.25)$$

For example, if $A_{v(dB)} = -3$ dB,

$$A_v = \frac{V_o}{V_i} = 10^{(-3/20)} = 10^{(-0.15)} \cong 0.707 \qquad \text{as expected}$$

The quantity $10^{-0.15}$ is determined using the 10^x function found on most scientific calculators.

From Fig. 11.15, $A_{v(dB)} \cong -1$ dB at $f = 2f_1 = 637$ Hz. The gain at this point is

$$A_v = \frac{V_o}{V_i} = 10^{A_{v(dB)}/20} = 10^{(-1/20)} = 10^{(-0.05)} = 0.891$$

and $$V_o = 0.891V_i$$

or V_o is 89.1% of V_i at $f = 637$ Hz.

The phase angle of θ is determined from

$$\theta = \tan^{-1}\frac{f_1}{f} \qquad\qquad (11.26)$$

from Eq. (11.22).

For frequencies $f \ll f_1$,

$$\theta = \tan^{-1}\frac{f_1}{f} \to 90°$$

For instance, if $f_1 = 100f$,

$$\theta = \tan^{-1}\frac{f_1}{f} = \tan^{-1}(100) = 89.4°$$

For $f = f_1$,

$$\theta = \tan^{-1}\frac{f_1}{f} = \tan^{-1}1 = 45°$$

For $f \gg f_1$,

$$\theta = \tan^{-1}\frac{f_1}{f} \to 0°$$

For instance, if $f = 100f_1$,

$$\theta = \tan^{-1}\frac{f_1}{f} = \tan^{-1}0.01 = 0.573°$$

A plot of $\theta = \tan^{-1}(f_1/f)$ is provided in Fig. 11.17. If we add the additional 180° phase shift introduced by an amplifier, the phase plot of Fig. 11.8 will be obtained.

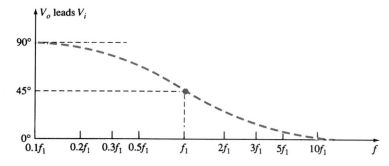

Figure 11.17 Phase response for the R-C circuit of Figure 11.9.

The magnitude and phase response for an *R-C* combination have now been established. In Section 11.6, each capacitor of importance in the low-frequency region will be redrawn in an *R-C* format and the cutoff frequency for each determined to establish the low-frequency response for the BJT amplifier.

11.6 LOW-FREQUENCY RESPONSE— BJT AMPLIFIER

The analysis of this section will employ the loaded voltage-divider BJT bias configuration, but the results can be applied to any BJT configuration. It will simply be necessary to find the appropriate equivalent resistance for the *R-C* combination. For the network of Fig. 11.18, the capacitors C_s, C_C, and C_E will determine the low-frequency response. We will now examine the impact of each independently in the order listed.

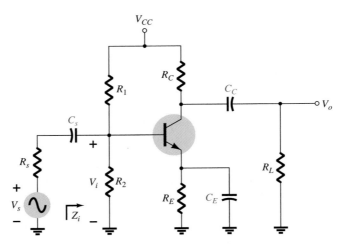

Figure 11.18 Loaded BJT amplifier with capacitors that affect the low-frequency response.

C_s

Since C_s is normally connected between the applied source and the active device, the general form of the *R-C* configuration is established by the network of Fig. 11.19. The total resistance is now $R_s + R_i$, and the cutoff frequency as established in Section 11.5 is

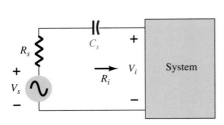

Figure 11.19 Determining the effect of C_s on the low frequency response.

$$f_{L_s} = \frac{1}{2\pi(R_s + R_i)C_s} \tag{11.27}$$

At mid or high frequencies, the reactance of the capacitor will be sufficiently small to permit a short-circuit approximation for the element. The voltage V_i will then be related to V_s by

$$V_i|_{mid} = \frac{R_i V_s}{R_i + R_s} \tag{11.28}$$

At f_{L_s}, the voltage V_i will be 70.7% of the value determined by Eq. (11.28), assuming that C_s is the only capacitive element controlling the low-frequency response.

For the network of Fig. 11.18, when we analyze the effects of C_s we must make the assumption that C_E and C_C are performing their designed function or the analysis becomes too unwieldy, that is, that the magnitude of the reactances of C_E and C_C permits employing a short-circuit equivalent in comparison to the magnitude of the other series impedances. Using this hypothesis, the ac equivalent network for the input section of Fig. 11.18 will appear as shown in Fig. 11.20.

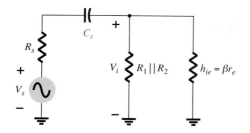

Figure 11.20 Localized ac equivalent for C_s.

The value of R_i for Eq. (11.27) is determined by

$$R_i = R_1 \| R_2 \| \beta r_e \qquad (11.29)$$

The voltage $\mathbf{V}_i$ applied to the input of the active device can be calculated using the voltage-divider rule:

$$\mathbf{V}_i = \frac{R_i \mathbf{V}_s}{R_s + R_i - jX_{C_s}} \qquad (11.30)$$

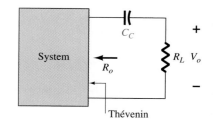

Figure 11.21 Determining the effect of C_C on the low-frequency response.

C_C

Since the coupling capacitor is normally connected between the output of the active device and the applied load, the R-C configuration that determines the low cutoff frequency due to C_C appears in Fig. 11.21. From Fig. 11.21, the total series resistance is now $R_o + R_L$ and the cutoff frequency due to C_C is determined by

$$f_{L_C} = \frac{1}{2\pi(R_o + R_L)C_C} \qquad (11.31)$$

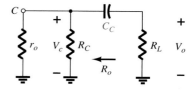

Figure 11.22 Localized ac equivalent for C_C with $V_i = 0$ V.

Ignoring the effects of C_s and C_E, the output voltage V_o will be 70.7% of its midband value at f_{L_C}. For the network of Fig. 11.18, the ac equivalent network for the output section with $V_i = 0$ V appears in Fig. 11.22. The resulting value for R_o in Eq. (11.31) is then simply

$$R_o = R_C \| r_o \qquad (11.32)$$

C_E

To determine f_{L_E}, the network "seen" by C_E must be determined as shown in Fig. 11.23. Once the level of R_e is established, the cutoff frequency due to C_E can be determined using the following equation:

$$f_{L_E} = \frac{1}{2\pi R_e C_E} \qquad (11.33)$$

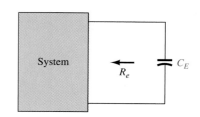

Figure 11.23 Determining the effect of C_E on the low-frequency response.

For the network of Fig. 11.18, the ac equivalent as "seen" by C_E appears in Fig. 11.24. The value of R_e is therefore determined by

$$R_e = R_E \left\| \left(\frac{R_s'}{\beta} + r_e \right) \right. \qquad (11.34)$$

where $R_s' = R_s \| R_1 \| R_2$.

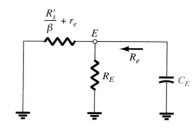

Figure 11.24 Localized ac equivalent of C_E.

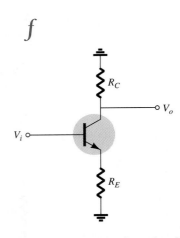

Figure 11.25 Network employed to describe the effect of C_E on the amplifier gain.

The effect of C_E on the gain is best described in a quantitative manner by recalling that the gain for the configuration of Fig. 11.25 is given by

$$A_v = \frac{-R_C}{r_e + R_E}$$

The maximum gain is obviously available where R_E is zero ohms. At low frequencies, with the bypass capacitor C_E in its "open-circuit" equivalent state, all of R_E appears in the gain equation above, resulting in the minimum gain. As the frequency increases, the reactance of the capacitor C_E will decrease, reducing the parallel impedance of R_E and C_E until the resistor R_E is effectively "shorted out" by C_E. The result is a maximum or midband gain determined by $A_v = -R_C/r_e$. At f_{L_E} the gain will be 3 dB below the midband value determined with R_E "shorted out."

Before continuing, keep in mind that C_s, C_C, and C_E will affect only the low-frequency response. At the midband frequency level, the short-circuit equivalents for the capacitors can be inserted. Although each will affect the gain $A_v = V_o/V_i$ in a similar frequency range, the highest low-frequency cutoff determined by C_s, C_C, or C_E will have the greatest impact since it will be the last encountered before the midband level. If the frequencies are relatively far apart, the highest cutoff frequency will essentially determine the lower cutoff frequency for the entire system. If there are two or more "high" cutoff frequencies, the effect will be to raise the lower cutoff frequency and reduce the resulting bandwidth of the system. In other words, there is an interaction between capacitive elements that can affect the resulting low cutoff frequency. However, if the cutoff frequencies established by each capacitor are sufficiently separated, the effect of one on the other can be ignored with a high degree of accuracy—a fact that will be demonstrated by the printouts to appear in the following example.

EXAMPLE 11.9

(a) Determine the lower cutoff frequency for the network of Fig. 11.18 using the following parameters:

$$C_s = 10\ \mu F, \qquad C_E = 20\ \mu F, \qquad C_C = 1\ \mu F$$
$$R_s = 1\ k\Omega, \qquad R_1 = 40\ k\Omega, \qquad R_2 = 10\ k\Omega, \qquad R_E = 2\ k\Omega, \qquad R_C = 4\ k\Omega,$$
$$R_L = 2.2\ k\Omega$$
$$\beta = 100, \qquad r_o = \infty\ \Omega, \qquad V_{CC} = 20\ V$$

(b) Sketch the frequency response using a Bode plot.

Solution

(a) Determining r_e for dc conditions:

$$\beta R_E = (100)(2\ k\Omega) = 200\ k\Omega \gg 10R_2 = 100\ k\Omega$$

The result is:

$$V_B \cong \frac{R_2 V_{CC}}{R_2 + R_1} = \frac{10\ k\Omega(20\ V)}{10\ k\Omega + 40\ k\Omega} = \frac{200\ V}{50} = 4\ V$$

with

$$I_E = \frac{V_E}{R_E} = \frac{4\ V - 0.7\ V}{2\ k\Omega} = \frac{3.3\ V}{2\ k\Omega} = 1.65\ mA$$

so that

$$r_e = \frac{26\ mV}{1.65\ mA} \cong \mathbf{15.76\ \Omega}$$

and

$$\beta r_e = 100(15.76\ \Omega) = 1576\ \Omega = \mathbf{1.576\ k\Omega}$$

Midband Gain:

$$A_v = \frac{V_o}{V_i} = \frac{-R_C \| R_L}{r_e} = -\frac{(4 \text{ k}\Omega)\|(2.2 \text{ k}\Omega)}{15.76 \ \Omega} \cong -90$$

The input impedance:

$$Z_i = R_i = R_1 \| R_2 \| \beta r_e$$
$$= 40 \text{ k}\Omega \| 10 \text{ k}\Omega \| 1.576 \text{ k}\Omega$$
$$\cong 1.32 \text{ k}\Omega$$

and from Fig. 11.26,

$$V_i = \frac{R_i V_s}{R_i + R_s}$$

or

$$\frac{V_i}{V_s} = \frac{R_i}{R_i + R_s} = \frac{1.32 \text{ k}\Omega}{1.32 \text{ k}\Omega + 1 \text{ k}\Omega} = 0.569$$

so that

$$A_{v_s} = \frac{V_o}{V_s} = \frac{V_o}{V_i}\frac{V_i}{V_s} = (-90)(0.569)$$
$$= \mathbf{-51.21}$$

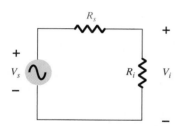

Figure 11.26 Determining the effect of R_s on the gain A_{v_s}.

C_s

$$R_i = R_1 \| R_2 \| \beta r_e = 40 \text{ k}\Omega \| 10 \text{ k}\Omega \| 1.576 \text{ k}\Omega \cong 1.32 \text{ k}\Omega$$

$$f_{L_S} = \frac{1}{2\pi(R_s + R_i)C_s} = \frac{1}{(6.28)(1 \text{ k}\Omega + 1.32 \text{ k}\Omega)(10 \ \mu\text{F})}$$

$$f_{L_S} \cong \mathbf{6.86 \ Hz}$$

The results just obtained will now be verified using PSpice Windows. The network with its various capacitors appears in Fig. 11.27. The **Model Editor** was used to set I_s to 2E-15A and beta to 100. The remaining parameters were removed from the listing to idealize the response to the degree possible. Under **Analysis Setup-AC Sweep,** the frequency was set to 10 kHz to establish a frequency in the midband region. A simulation of the network resulted in the dc levels of Fig. 11.27. Note that V_B is 3.9 V versus the calculated level of 4 V and that V_E is 3.2 V versus the calculated level of 3.3 V. It is very close when you consider that the approximate model was used. V_{BE} is very close to the 0.7 V at 0.71 V. The output file reveals that the ac voltage across the load at a frequency of 10 kHz is 49.67 mV, resulting in a gain of 49.67, which is very close to the calculated level of 51.21.

A plot of the gain versus frequency will now be obtained with only C_s as a determining factor. The other capacitors, C_C and C_E, will be set to very high values so they are essentially short circuits at any of the frequencies of interest. Setting C_C and C_E to 1 F will remove any effect they will have on the response on the low-frequency region. Here, one must be careful as the program does not recognize 1F as one Farad.

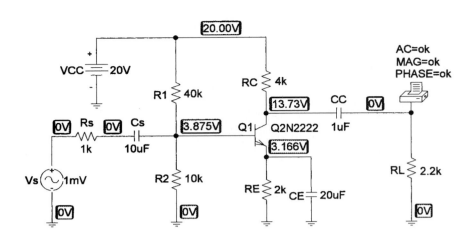

Figure 11.27   Network of Figure 11.18 with assigned values.

It must be entered as 1E6uF. Since the pattern desired is gain versus frequency, we must use the sequence **Analysis-Setup-Analysis Setup-Enable AC Sweep-AC Sweep** to obtain the **AC Sweep and Noise Analysis** dialog box. Since our interest will be in the low-frequency range, we will choose a range of 1 Hz (0 Hz is an invalid entry) to 100 Hz. If you want a frequency range starting close to 0 Hz, you would have to choose a frequency such as 0.001 Hz or something small enough not to be noticeable on the plot. The **Total Pts.:** will be set at 1000 for a good continuous plot, the **Start Freq.:** at 1 Hz, and the **End Freq.:** at 100 Hz. The **AC Sweep Type** will be left on **Linear.** A simulation followed by **Trace-Add-V(RL:1)** will result in the desired plot. However, the computer has selected a log scale for the horizontal axis that extends from 1 Hz to 1 kHz even though we requested a linear scale. If we choose **Plot-X-Axis Settings-Linear-OK,** we will get a linear plot to 120 Hz, but the curve of interest is all in the low end—the log axis obviously provided a better plot for our region of interest. Returning to **Plot-X-Axis Settings** and choosing **Log,** we return to the original plot. Our interest only lies in the region of 1 to 100 Hz, so the remaining frequencies to 1 kHz should be removed with **Plot-X-Axis Settings-User Defined-1Hz**

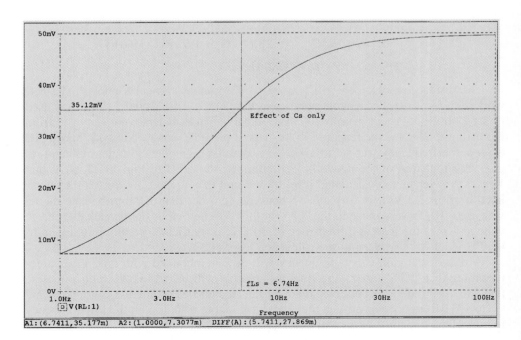

Figure 11.28   Low-frequency response due to C_S.

to 100Hz-OK. The vertical axis also goes to 60 mV, and we want to limit to 50 mV for this frequency range. This is accomplished with **Plot-Y-Axis Settings-User Defined-0V to 50mV-OK,** after which the pattern of Fig. 11.28 will be obtained.

Note how closely the curve approaches 50 mV in this range. The cutoff level is determined by 0.707(49.67 mV) = 35.12 mV, which can be found by clicking the **Toggle cursor** icon and moving the intersection up the graph until the 35.177-mV level is reached for **A1.** At this point, the frequency of the horizontal axis can be read as 6.74 Hz, comparing very well to the predicted value of 6.86 Hz. Note that **A2** remains at the lowest level of the plot, at 1 Hz.

C_C

$$f_{L_C} = \frac{1}{2\pi(R_C + R_L)C_C}$$

$$= \frac{1}{(6.28)(4 \text{ k}\Omega + 2.2 \text{ k}\Omega)(1 \text{ }\mu\text{F})}$$

$$\cong \textbf{25.68 Hz}$$

To investigate the effects of C_C on the lower cutoff frequency, both C_S and C_E must be set to 1 Farad as described above. Following the procedure outlined above will result in the plot of Fig. 11.29, with a cutoff frequency of 25.58 Hz, providing a close match with the calculated level of 25.68 Hz.

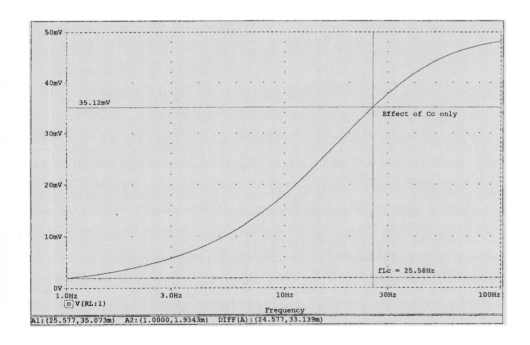

Figure 11.29 Low-frequency response due to C_C.

C_E

$$R'_s = R_s \| R_1 \| R_2 = 1 \text{ k}\Omega \| 40 \text{ k}\Omega \| 10 \text{ k}\Omega \cong 0.889 \text{ k}\Omega$$

$$R_e = R_E \left\| \left(\frac{R'_s}{\beta} + r_e \right) \right. = 2 \text{ k}\Omega \left\| \left(\frac{0.889 \text{ k}\Omega}{100} + 15.76 \text{ }\Omega \right) \right.$$

$$= 2 \text{ k}\Omega \| (8.89 \text{ }\Omega + 15.76 \text{ }\Omega) = 2 \text{ k}\Omega \| 24.65 \text{ }\Omega \cong 24.35 \text{ }\Omega$$

$$f_{L_E} = \frac{1}{2\pi R_e C_E} = \frac{1}{(6.28)(24.35 \text{ }\Omega)(20 \text{ }\mu\text{F})} = \frac{10^6}{3058.36} \cong \textbf{327 Hz}$$

The effect of C_E can be examined using PSpice Windows by setting both C_s and C_C to 1 Farad. In addition, since the frequency range is greater, the start frequency has to be changed to 10 Hz and the final frequency to 1 kHz. The result is the plot of Fig. 11.30, with a cutoff frequency of 321.17 Hz, providing a close match with the calculated value of 327 Hz.

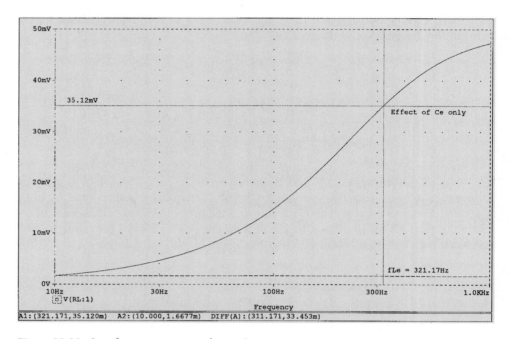

Figure 11.30 Low-frequency response due to C_E.

The fact that f_{L_E} is significantly higher than f_{L_S} or f_{L_C} suggests that it will be the predominant factor in determining the low-frequency response for the complete system. To test the accuracy of our hypothesis, the network is simulated with all the initial values of capacitance level to obtain the results of Fig. 11.31. Note the strong similarity with the waveform of Fig. 11.30, with the only visible difference being the higher gain at lower frequencies on Fig. 11.30. Without question, the plot supports the fact that the highest of the low cutoff frequencies will have the most impact on the low cutoff frequency for the system.

(b) It was mentioned earlier that dB plots are usually normalized by dividing the voltage gain A_v by the magnitude of the midband gain. For Fig. 11.18, the magnitude of the midband gain is 51.21, and naturally the ratio $|A_v/A_{v_{mid}}|$ will be 1 in the midband region. The result is a 0-dB asymptote in the midband region as shown in Fig. 11.32. Defining f_{L_E} as our lower cutoff frequency f_1, an asymptote at -6 dB/octave can be drawn as shown in Fig. 11.32 to form the Bode plot and our envelope for the actual response. At f_1, the actual curve is -3 dB down from the midband level as defined by the $0.707A_{v_{mid}}$ level, permitting a sketch of the actual frequency response curve as shown in Fig. 11.32. A -6-dB/octave asymptote was drawn at each frequency defined in the analysis above to demonstrate clearly that it is f_{L_E} for this network that will determine the -3-dB point. It is not until about -24 dB that f_{L_C} begins to affect the shape of the envelope. The magnitude plot shows that the slope of the resultant asymptote is the sum of the asymptotes having the same sloping direction in the same frequency interval. Note in Fig. 11.32 that the slope has dropped to -12 dB/octave for frequencies less than f_{L_C} and could drop to -18 dB/octave if the three defined cutoff frequencies of Fig. 11.32 were closer together.

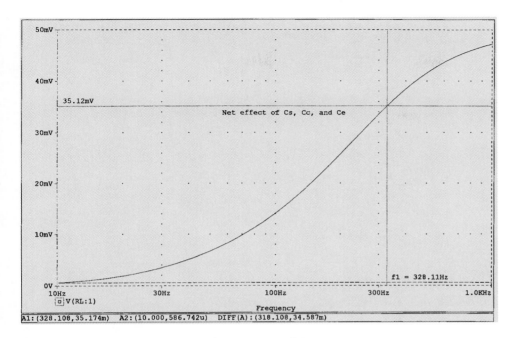

Figure 11.31 Low-frequency response due to C_S, C_E, and C_C.

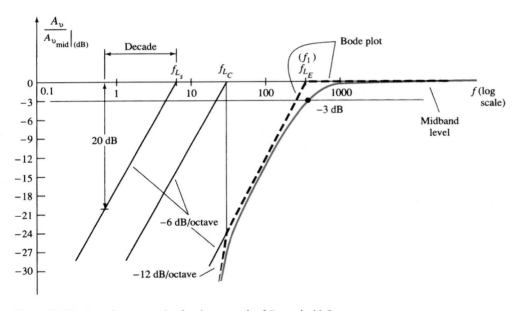

Figure 11.32 Low-frequency plot for the network of Example 11.9.

Using **PROBE,** a plot of $20 \log_{10}|A_v/A_{v_{\text{mid}}}| = A_v/A_{v_{\text{mid}}}|_{\text{dB}}$ can be obtained by recalling that if $V_s = 1$ mV, the magnitude of $|A_v/A_{v_{\text{mid}}}|$ is the same as $|V_o/A_{v_{\text{mid}}}|$ since V_o will have the same numerical value as A_v. The required **Trace Expression,** which is entered on the bottom of the **Add Traces** dialog box, appears on the horizontal axis of Fig. 11.33. The plot clearly reveals the change in slope of the asymptote at f_{L_C} and how the actual curve follows the envelope created by the Bode plot. In addition, note the 3-dB drop at f_1.

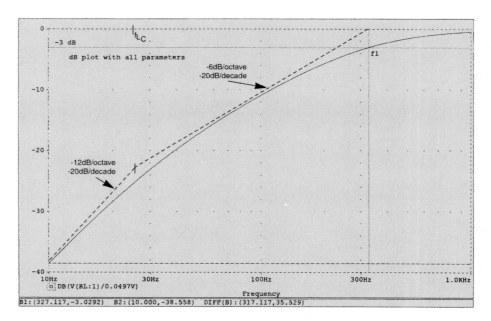

Figure 11.33 dB plot of the low-frequency response of the BJT amplifier of Fig. 11.27.

Keep in mind as we proceed to the next section that the analysis of this section is not limited to the network of Fig. 11.18. For any transistor configuration it is simply necessary to isolate each *R-C* combination formed by a capacitive element and determine the break frequencies. The resulting frequencies will then determine whether there is a strong interaction between capacitive elements in determining the overall response and which element will have the greatest impact on establishing the lower cutoff frequency. In fact, the analysis of the next section will parallel this section as we determine the low cutoff frequencies for the FET amplifier.

11.7 LOW-FREQUENCY RESPONSE— FET AMPLIFIER

The analysis of the FET amplifier in the low-frequency region will be quite similar to that of the BJT amplifier of Section 11.6. There are again three capacitors of primary concern as appearing in the network of Fig. 11.34: C_G, C_C, and C_S. Although Fig. 11.34 will be used to establish the fundamental equations, the procedure and conclusions can be applied to most FET configurations.

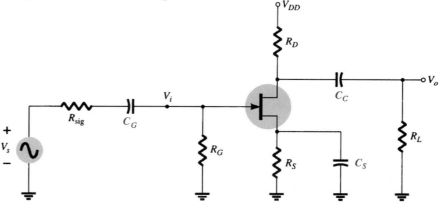

Figure 11.34 Capacitive elements that affect the low-frequency response of a JFET amplifier.

C_G

For the coupling capacitor between the source and the active device, the ac equivalent network will appear as shown in Fig. 11.35. The cutoff frequency determined by C_G will then be

$$f_{L_G} = \frac{1}{2\pi(R_{\text{sig}} + R_i)C_G} \qquad (11.35)$$

which is an exact match of Eq. (11.27). For the network of Fig. 11.34,

$$R_i = R_G \qquad (11.36)$$

Typically, $R_G \gg R_{\text{sig}}$, and the lower cutoff frequency will be determined primarily by R_G and C_G. The fact that R_G is so large permits a relatively low level of C_G while maintaining a low cutoff frequency level for f_{L_G}.

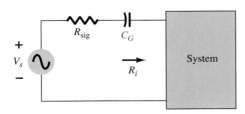

Figure 11.35 Determining the effect of C_G on the low-frequency response.

C_C

For the coupling capacitor between the active device and the load the network of Fig. 11.36 will result, which is also an exact match of Fig. 11.21. The resulting cutoff frequency is

$$f_{L_C} = \frac{1}{2\pi(R_o + R_L)C_C} \qquad (11.37)$$

For the network of Fig. 11.34,

$$R_o = R_D \| r_d \qquad (11.38)$$

C_S

For the source capacitor C_S, the resistance level of importance is defined by Fig. 11.37. The cutoff frequency will be defined by

$$f_{L_S} = \frac{1}{2\pi R_{\text{eq}}C_S} \qquad (11.39)$$

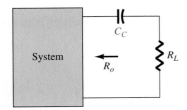

Figure 11.36 Determining the effect of C_C on the low-frequency response.

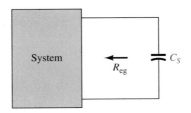

Figure 11.37 Determining the effect of C_S on the low-frequency response.

f

For Fig. 11.34, the resulting value of R_{eq} is

$$R_{eq} = \frac{R_S}{1 + R_S(1 + g_m r_d)/(r_d + R_D \| R_L)} \qquad (11.40)$$

which for $r_d \cong \infty \, \Omega$ becomes

$$R_{eq} = R_S \| \frac{1}{g_m} \qquad (11.41)$$

EXAMPLE 11.10

(a) Determine the lower cutoff frequency for the network of Fig. 11.34 using the following parameters:

$C_G = 0.01 \, \mu\text{F}, \qquad C_C = 0.5 \, \mu\text{F}, \qquad C_S = 2 \, \mu\text{F}$

$R_{\text{sig}} = 10 \, \text{k}\Omega, \quad R_G = 1 \, \text{M}\Omega, \quad R_D = 4.7 \, \text{k}\Omega, \quad R_S = 1 \, \text{k}\Omega, \quad R_L = 2.2 \, \text{k}\Omega$

$I_{DSS} = 8 \, \text{mA}, \qquad V_P = -4 \, \text{V} \qquad r_d = \infty \, \Omega, \qquad V_{DD} = 20 \, \text{V}$

(b) Sketch the frequency response using a Bode plot.

Solution

(a) DC Analysis: Plotting the transfer curve of $I_D = I_{DSS}(1 - V_{GS}/V_P)^2$ and superimposing the curve defined by $V_{GS} = -I_D R_S$ will result in an intersection at $V_{GS_Q} = -2 \, \text{V}$ and $I_{D_Q} = 2 \, \text{mA}$. In addition,

$$g_{m0} = \frac{2I_{DSS}}{|V_P|} = \frac{2(8 \, \text{mA})}{4 \, \text{V}} = 4 \, \text{mS}$$

$$g_m = g_{m0}\left(1 - \frac{V_{GS_Q}}{V_P}\right) = 4 \, \text{mS}\left(1 - \frac{-2 \, \text{V}}{-4 \, \text{V}}\right) = 2 \, \text{mS}$$

C_G

$$\text{Eq. (11.35):} \quad f_{L_G} = \frac{1}{2\pi(10 \, \text{k}\Omega + 1 \, \text{M}\Omega)(0.01 \, \mu\text{F})} \cong \mathbf{15.8 \, Hz}$$

C_C

$$\text{Eq. (11.37):} \quad f_{L_C} = \frac{1}{2\pi(4.7 \, \text{k}\Omega + 2.2 \, \text{k}\Omega)(0.5 \, \mu\text{F})} \cong \mathbf{46.13 \, Hz}$$

C_S

$$R_{eq} = R_S \| \frac{1}{g_m} = 1 \, \text{k}\Omega \| \frac{1}{2 \, \text{mS}} = 1 \, \text{k}\Omega \| 0.5 \, \text{k}\Omega = 333.33 \, \Omega$$

$$\text{Eq. (11.39):} \quad f_{L_S} = \frac{1}{2\pi(333.33 \, \Omega)(2 \, \mu\text{F})} = \mathbf{238.73 \, Hz}$$

Since f_{L_S} is the largest of the three cutoff frequencies, it defines the low cutoff frequency for the network of Fig. 11.34.

(b) The midband gain of the system is determined by

$$A_{v_{\text{mid}}} = \frac{V_o}{V_i} = -g_m(R_D \| R_L) = -(2 \, \text{mS})(4.7 \, \text{k}\Omega \| 2.2 \, \text{k}\Omega)$$

$$= -(2 \, \text{mS})(1.499 \, \text{k}\Omega)$$

$$\cong \mathbf{-3}$$

Using the midband gain to normalize the response for the network of Fig. 11.34 will result in the frequency plot of Fig. 11.38.

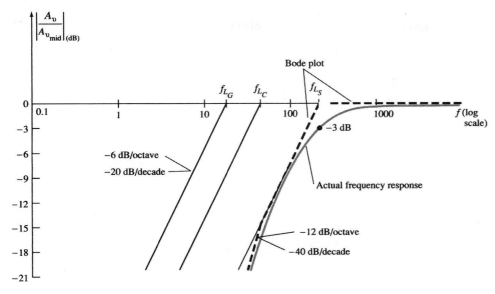

Figure 11.38   Low-frequency response for the JFET configuration of Example 11.10.

Using PSpice Windows, the network will appear as shown in Fig. 11.39, with the JFET parameters **Beta** set at 0.5 mA/V^2 and **Vto** at -4 V (all others set to zero) and the frequency of interest at a midband value of 10 kHz. The resulting dc levels confirm that V_{GS} is -2 V and place V_D at 10.60 V, which should be right in the middle of the linear active region since $V_{GS} = \frac{1}{2}V_D$ and $V_{DS} = \frac{1}{2}V_{DD}$. The 0-V levels clearly reveal that the capacitors have isolated the transistor for the dc biasing. The ac response results in an ac level of 2.993 mV across the load for a gain of 2.993, which is essentially equal to the calculated gain of 3.

Returning to **Analysis** and choosing **Automatically run Probe after simulation** followed by **Setup-AC Sweep-Decade-Pts/Decade** = 1000, **Start Freq.:** 10 Hz, and **End Freq.:** 10 kHz will setup **Simulation-Trace-Add-Trace Expression: DB (V(RL:1)/2.993mV)-OK,** which will result in the plot of Fig. 11.40, with a low cutoff frequency of 227.5 Hz primarily determined by the source capacitance.

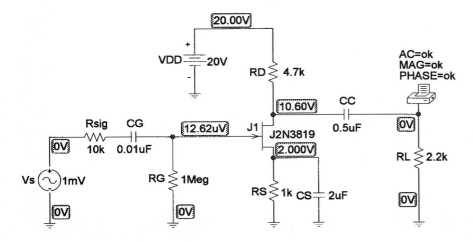

Figure 11.39   Schematic network for Example 11.10.

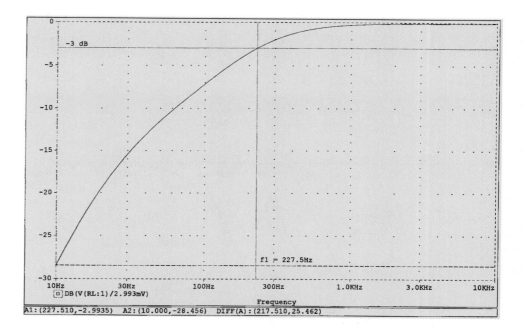

Figure 11.40 dB response for the low-frequency region in the network of Example 11.10.

Electronics Workbench can also provide a frequency plot of the gain and phase response of a BJT or a JFET network by first constructing the network or calling it from storage. Since the network of Fig. 11.39 is the same as that analyzed using EWB in Chapter 9 (Fig. 9.64), it will be retrieved again and displayed as Fig. 11.41 with its dc levels at the drain and source terminals. Next the sequence **Simulate-Analyses-AC Analysis** is applied to obtain the **Ac Analysis** dialog box. Under **Frequency Parameters,** the **Start frequency** is selected as **10Hz** and the **Stop frequency** as **10kHz** to match the plot of Fig. 11.40. Next the **Sweep type** will be left at the default

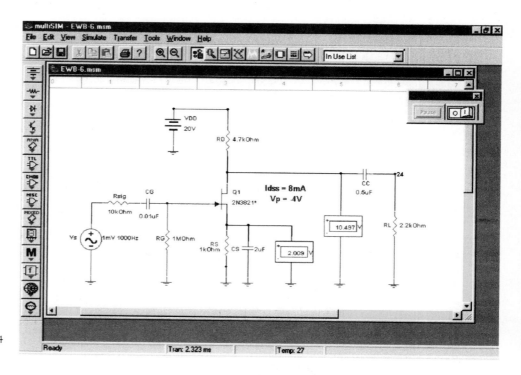

Figure 11.41 Network of Fig. 9.64 (Example 11.10) using EWB.

selection of **decade,** and the **Number of points** per decade also left at **100.** Finally, the vertical scale will be set in the linear mode since it will be the magnitude of the output voltage versus frequency rather than the dB gain as plotted in Fig. 11.40.

Next, **Output variables** will be selected in the dialog box, and node 24 will be selected from the **Variables in circuit** as **Plot during simulation.** As a result, it will appear in the **Selected variables for analysis** column. That concludes the choices that have to be made as summarized under the **Summary** heading. The only thing left to do is to select **Simulate** at the bottom of the dialog box, resulting in the plot of Fig. 11.42. At first, the plot may appear without a grid structure to help define the levels at each frequency. This is corrected by the sequence **View-Show/Hide Grid** as shown in Fig. 11.42. Always be aware that the red arrow along the left vertical column defines the plot under review. To add the grid to the phase plot, simply click on the lower graph at any point, and the red arrow will drop down. Then follow with the same sequence as above to establish the grid structure. If you want the graph to fill the entire screen, simply select the full screen option at the top right corner of the **Analysis Graphs.**

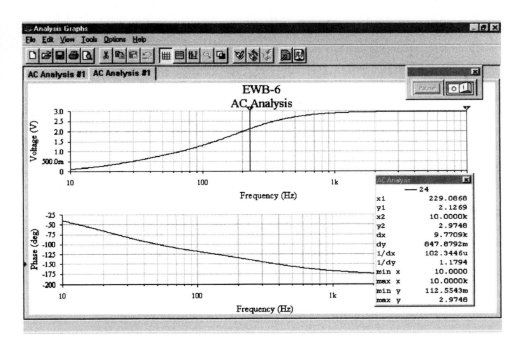

Figure 11.42 EWB plots for Example 11.10.

Finally, cursors can be added to define the level of the plotted function at any frequency. Simply select **View-Show/Hide Cursors,** and the cursors will appear on the selected graph (which is the magnitude plot in Fig. 11.42). Then click on cursor 1, and the **AC Analysis** dialog box on the screen will reveal the level of the voltage and the frequency. By clicking on cursor 1 and moving it to the right, we can try to find an **x1** value of 227.5 to match the −3-dB point of Fig. 11.40. The closest we can get for the chosen number of plot points is 229.08 Hz as shown in Fig. 11.42, but that is certainly close enough. At this frequency the output voltage (**y1**) is 2.13 V which is very close to the 0.707 level of the 2.93 gain (actually 2.07 V) obtained in Chapter 9. Cursor 2 was moved to an **x2** value of 10 kHz to obtain a voltage of 2.97 V which again is very close to the maximum gain of 2.93 in Chapter 9. Before leaving Fig. 11.42, note that the higher the frequency, the closer the phase shift is to 180° as the relatively large, low-frequency capacitors lose their effect.

11.8 MILLER EFFECT CAPACITANCE

In the high-frequency region, the capacitive elements of importance are the inter-electrode (between terminals) capacitances internal to the active device and the wiring capacitance between leads of the network. The large capacitors of the network that controlled the low-frequency response have all been replaced by their short-circuit equivalent due to their very low reactance levels.

For *inverting* amplifiers (phase shift of 180° between input and output resulting in a negative value for A_v), the input and output capacitance is increased by a capacitance level sensitive to the interelectrode capacitance between the input and output terminals of the device and the gain of the amplifier. In Fig. 11.43, this "feedback" capacitance is defined by C_f.

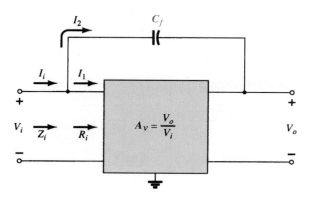

Figure 11.43 Network employed in the derivation of an equation for the Miller input capacitance.

Applying Kirchhoff's current law gives

$$I_i = I_1 + I_2$$

Using Ohm's law yields

$$I_i = \frac{V_i}{Z_i}, \quad I_1 = \frac{V_i}{R_i}$$

and

$$I_2 = \frac{V_i - V_o}{X_{C_f}} = \frac{V_i - A_v V_i}{X_{C_f}} = \frac{(1 - A_v)V_i}{X_{C_f}}$$

Substituting, we obtain

$$\frac{V_i}{Z_i} = \frac{V_i}{R_i} + \frac{(1 - A_v)V_i}{X_{C_f}}$$

and

$$\frac{1}{Z_i} = \frac{1}{R_i} + \frac{1}{X_{C_f}/(1 - A_v)}$$

but

$$\frac{X_{C_f}}{1 - A_v} = \underbrace{\frac{1}{\omega(1 - A_v)C_f}}_{C_M} = X_{C_M}$$

and

$$\frac{1}{Z_i} = \frac{1}{R_i} + \frac{1}{X_{C_M}}$$

establishing the equivalent network of Fig. 11.44. The result is an equivalent input impedance to the amplifier of Fig. 11.43 that includes the same R_i that we have dealt with in previous chapters, with the addition of a feedback capacitor magnified by the gain of the amplifier. Any interelectrode capacitance at the input terminals to the amplifier will simply be added in parallel with the elements of Fig. 11.44.

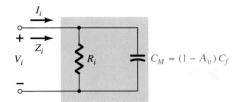

Figure 11.44 Demonstrating the impact of the Miller effect capacitance.

In general, therefore, the Miller effect input capacitance is defined by

$$C_{M_i} = (1 - A_v)C_f \tag{11.42}$$

This shows us that:

For any inverting amplifier, the input capacitance will be increased by a Miller effect capacitance sensitive to the gain of the amplifier and the inter-electrode (parasitic) capacitance between the input and output terminals of the active device.

The dilemma of an equation such as Eq. (11.42) is that at high frequencies the gain A_v will be a function of the level of C_{M_i}. However, since the maximum gain is the midband value, using the midband value will result in the highest level of C_{M_i} and the worst-case scenario. In general, therefore, the midband value is typically employed for A_v in Eq. (11.42).

The reason for the constraint that the amplifier be of the inverting variety is now more apparent when one examines Eq. (11.42). A positive value for A_v would result in a negative capacitance (for $A_v > 1$).

The Miller effect will also increase the level of output capacitance, which must also be considered when the high-frequency cutoff is determined. In Fig. 11.45, the parameters of importance to determine the output Miller effect are in place. Applying Kirchhoff's current law will result in

$$I_o = I_1 + I_2$$

with

$$I_1 = \frac{V_o}{R_o} \quad \text{and} \quad I_2 = \frac{V_o - V_i}{X_{C_f}}$$

The resistance R_o is usually sufficiently large to permit ignoring the first term of the equation compared to the second term and assuming that

$$I_o \cong \frac{V_o - V_i}{X_{C_f}}$$

Substituting $V_i = V_o/A_v$ from $A_v = V_o/V_i$ will result in

$$I_o = \frac{V_o - V_o/A_v}{X_{C_f}} = \frac{V_o(1 - 1/A_v)}{X_{C_f}}$$

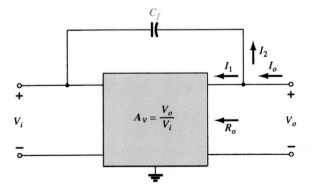

Figure 11.45 Network employed in the derivation of an equation for the Miller output capacitance.

and
$$\frac{I_o}{V_o} = \frac{1 - 1/A_v}{X_{C_f}}$$

or
$$\frac{V_o}{I_o} = \frac{X_{C_f}}{1 - 1/A_v} = \frac{1}{\omega C_f(1 - 1/A_v)} = \frac{1}{\omega C_{M_o}}$$

resulting in the following equation for the Miller output capacitance:

$$C_{M_o} = \left(1 - \frac{1}{A_v}\right)C_f \qquad (11.43)$$

For the usual situation where $A_v \gg 1$, Eq. (11.43) reduces to

$$C_{M_o} \cong C_f \Big|_{|A_v|>>1} \qquad (11.44)$$

Examples in the use of Eq. (11.43) will appear in the next two sections as we investigate the high-frequency responses of BJT and FET amplifiers.

11.9 HIGH-FREQUENCY RESPONSE— BJT AMPLIFIER

At the high-frequency end, there are two factors that will define the -3-dB point: the network capacitance (parasitic and introduced) and the frequency dependence of $h_{fe}(\beta)$.

Network Parameters

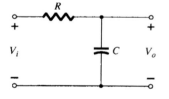

In the high-frequency region, the RC network of concern has the configuration appearing in Fig. 11.46. At increasing frequencies, the reactance X_C will decrease in magnitude, resulting in a shorting effect across the output and a decrease in gain. The derivation leading to the corner frequency for this R_C configuration follows along similar lines to that encountered for the low-frequency region. The most significant difference is in the general form of A_v appearing below:

$$A_v = \frac{1}{1 + j(f/f_2)} \qquad (11.45)$$

Figure 11.46 *R-C combination that will define a high cutoff frequency.*

which results in a magnitude plot such as shown in Fig. 11.47 that drops off at 6 dB/octave with increasing frequency. Note that f_2 is in the denominator of the frequency

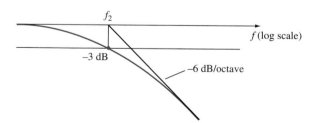

Figure 11.47 *Asymptotic plot as defined by Eq. (11.45).*

ratio rather than the numerator as occurred for f_1 in Eq. (11.21).

In Fig. 11.48, the various parasitic capacitances (C_{be}, C_{bc}, C_{ce}) of the transistor have been included with the wiring capacitances (C_{W_i}, C_{W_o}) introduced during construction. The high-frequency equivalent model for the network of Fig. 11.48 appears

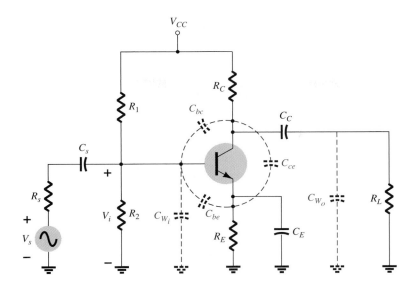

Figure 11.48 Network of Fig. 11.18 with the capacitors that affect the high-frequency response.

$$C_i = C_{W_i} + C_{be} + C_{M_i} \qquad\qquad C_o = C_{W_o} + C_{ce} + C_{M_o}$$

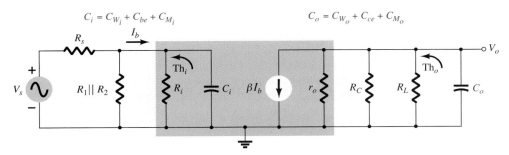

Figure 11.49 High-frequency ac equivalent model for the network of Fig. 11.48.

in Fig. 11.49. Note the absence of the capacitors C_s, C_C, and C_E, which are all assumed to be in the short-circuit state at these frequencies. The capacitance C_i includes the input wiring capacitance C_{W_i}, the transition capacitance C_{be}, and the Miller capacitance C_{M_i}. The capacitance C_o includes the output wiring capacitance C_{W_o}, the parasitic capacitance C_{ce}, and the output Miller capacitance C_{M_o}. In general, the capacitance C_{be} is the largest of the parasitic capacitances, with C_{ce} the smallest. In fact, most specification sheets simply provide the levels of C_{be} and C_{bc} and do not include C_{ce} unless it will affect the response of a particular type of transistor in a specific area of application.

Determining the Thévenin equivalent circuit for the input and output networks of Fig. 11.49 will result in the configurations of Fig. 11.50. For the input network, the -3-dB frequency is defined by

$$f_{H_i} = \frac{1}{2\pi R_{\text{Th}_i} C_i} \qquad\qquad (11.46)$$

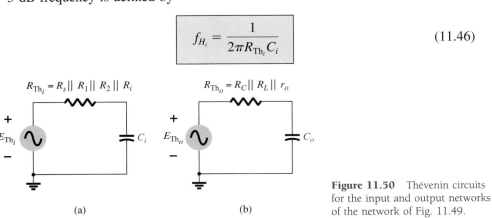

Figure 11.50 Thévenin circuits for the input and output networks of the network of Fig. 11.49.

(a)　　　　　　　(b)

with

$$R_{Th_i} = R_s \| R_1 \| R_2 \| R_i \qquad (11.47)$$

and

$$C_i = C_{W_i} + C_{be} + C_{M_i} = C_{W_i} + C_{be} + (1 - A_v)C_{bc} \qquad (11.48)$$

At very high frequencies, the effect of C_i is to reduce the total impedance of the parallel combination of R_1, R_2, R_i, and C_i in Fig. 11.49. The result is a reduced level of voltage across C_i, a reduction in I_b, and a gain for the system.

For the output network,

$$f_{H_o} = \frac{1}{2\pi R_{Th_o} C_o} \qquad (11.49)$$

with

$$R_{Th_o} = R_C \| R_L \| r_o \qquad (11.50)$$

and

$$C_o = C_{W_o} + C_{ce} + C_{M_o} \qquad (11.51)$$

At very high frequencies, the capacitive reactance of C_o will decrease and consequently reduce the total impedance of the output parallel branches of Fig. 11.49. The net result is that V_o will also decline toward zero as the reactance X_C becomes smaller. The frequencies f_{H_i} and f_{H_o} will each define a -6-dB/octave asymptote such as depicted in Fig. 11.47. If the parasitic capacitors were the only elements to determine the high cutoff frequency, the lowest frequency would be the determining factor. However, the decrease in h_{fe} (or β) with frequency must also be considered as to whether its break frequency is lower than f_{H_i} or f_{H_o}.

h_{fe} (or β) Variation

The variation of h_{fe} (or β) with frequency will approach, with some degree of accuracy, the following relationship:

$$h_{fe} = \frac{h_{fe_{mid}}}{1 + j(f/f_\beta)} \qquad (11.52)$$

The use of h_{fe} rather than β in some of this descriptive material is due primarily to the fact that manufacturers typically use the hybrid parameters when covering this issue in their specification sheets, and so on.

The only undefined quantity, f_β, is determined by a set of parameters employed in the *hybrid π* or *Giacoletto* model frequently applied to best represent the transistor in the high-frequency region. It appears in Fig. 11.51. The various parameters warrant a moment of explanation. The resistance $r_{bb'}$ includes the base contact, base bulk, and base spreading resistance. The first is due to the actual connection to the base. The second includes the resistance from the external terminal to the active region of

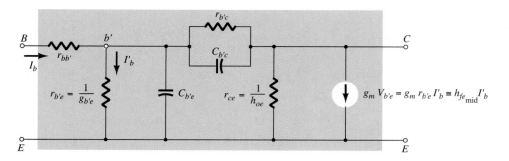

Figure 11.51 Giacoletto (or hybrid π) high-frequency transistor small-signal ac equivalent circuit.

the transistors, while the last is the actual resistance within the active base region. The resistances $r_{b'e}$, r_{ce}, and $r_{b'c}$ are the resistances between the indicated terminals when the device is in the active region. The same is true for the capacitances $C_{b'c}$ and $C_{b'e}$, although the former is a transition capacitance while the latter is a diffusion capacitance. A more detailed explanation of the frequency dependence of each can be found in a number of readily available texts.

In terms of these parameters,

$$f_\beta \text{(sometimes appearing as } f_{h_{fe}}) = \frac{g_{b'e}}{2\pi(C_{b'e} + C_{b'c})} \qquad (11.53)$$

or since the hybrid parameter h_{fe} is related to $g_{b'e}$ through $g_m = h_{fe_{mid}} g_{b'e}$,

$$f_\beta = \frac{1}{h_{fe_{mid}}} \frac{g_m}{2\pi(C_{b'e} + C_{b'c})} \qquad (11.54)$$

Taking it a step further,

$$g_m = h_{fe_{mid}} g_{b'e} = h_{fe_{mid}} \frac{1}{r_{b'e}} \cong \frac{h_{fe_{mid}}}{h_{ie}} = \frac{\beta_{mid}}{\beta_{mid}r_e} = \frac{1}{r_e}$$

and using the approximations

$$C_{b'e} \cong C_{be} \qquad \text{and} \qquad C_{b'c} \cong C_{bc}$$

will result in the following form for Eq. (11.52):

$$f_\beta \cong \frac{1}{2\pi\beta_{mid}r_e(C_{be} + C_{bc})} \qquad (11.55)$$

Equation (11.55) clearly reveals that since r_e is a function of the network design:

f_β is a function of the bias conditions.

The basic format of Eq. (11.52) is exactly the same as Eq. (11.45) if we extract the multiplying factor $h_{fe_{mid}}$, revealing that h_{fe} will drop off from its midband value with a 6-dB/octave slope as shown in Fig. 11.52. The same figure has a plot of h_{fb} (or α) versus frequency. Note the small change in h_{fb} for the chosen frequency range, revealing that the common-base configuration displays improved high-frequency characteristics over the common-emitter configuration. Recall also the absence of the Miller effect capacitance due to the noninverting characteristics of the common-base configuration. For this very reason, common-base high-frequency parameters rather than common-emitter parameters are often specified for a transistor—especially those designed specifically to operate in the high-frequency regions.

The following equation permits a direct conversion for determining f_β if f_α and α are specified.

$$f_\beta = f_\alpha(1 - \alpha) \qquad (11.56)$$

A quantity called the *gain–bandwidth product* is defined for the transistor by the condition

$$\left| \frac{h_{fe_{mid}}}{1 + j(f/f_\beta)} \right| = 1$$

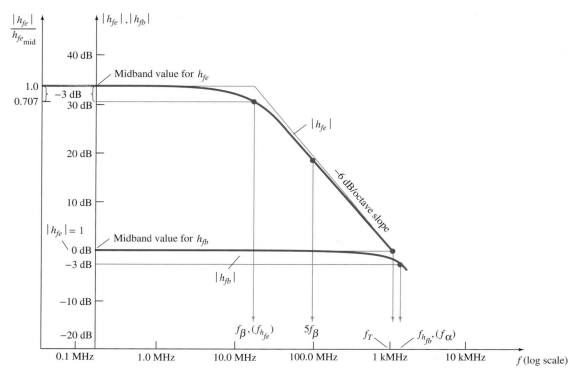

Figure 11.52 h_{fe} and h_{fb} versus frequency in the high-frequency region.

so that

$$|h_{fe}|_{dB} = 20 \log_{10} \left| \frac{h_{fe_{mid}}}{1 + j(f/f_\beta)} \right| = 20 \log_{10} 1 = 0 \text{ dB}$$

The frequency at which $|h_{fe}|_{dB} = 0$ dB is clearly indicated by f_T in Fig. 11.52. The magnitude of h_{fe} at the defined condition point $(f_T \gg f_\beta)$ is given by

$$\frac{h_{fe_{mid}}}{\sqrt{1 + (f_T/f_\beta)^2}} \cong \frac{h_{fe_{mid}}}{f_T/f_\beta} = 1$$

so that

$$f_T \cong h_{fe_{mid}} \cdot f_\beta \quad \overbrace{\phantom{f_T \cong h_{fe_{mid}} \cdot f_\beta}}^{(\cong \text{BW})} \quad \text{(gain–bandwidth product)} \tag{11.57}$$

or

$$f_T \cong \beta_{mid} f_\beta \tag{11.58}$$

with

$$f_\beta = \frac{f_T}{\beta_{mid}} \tag{11.59}$$

Substituting Eq. (11.55) for f_β in Eq. (11.57) gives

$$f_T \cong \beta_{mid} \frac{1}{2\pi \beta_{mid} r_e (C_{be} + C_{bc})}$$

and

$$f_T \cong \frac{1}{2\pi r_e (C_{be} + C_{bc})} \tag{11.60}$$

For the network of Fig. 11.48 with the same parameters as in Example 11.9, that is,

$R_s = 1$ kΩ, $R_1 = 40$ kΩ, $R_2 = 10$ kΩ, $R_E = 2$ kΩ, $R_C = 4$ kΩ, $R_L = 2.2$ kΩ

$C_s = 10$ μF, $C_C = 1$ μF, $C_E = 20$ μF

$\beta = 100$, $r_o = \infty$ Ω, $V_{CC} = 20$ V

with the addition of

$$C_{be} = 36 \text{ pF}, C_{bc} = 4 \text{ pF}, C_{ce} = 1 \text{ pF}, C_{W_i} = 6 \text{ pF}, C_{W_o} = 8 \text{ pF}$$

(a) Determine f_{H_i} and f_{H_o}.
(b) Find f_β and f_T.
(c) Sketch the frequency response for the low- and high-frequency regions using the results of Example 11.9 and the results of parts (a) and (b).
(d) Obtain a **PROBE** response for the full frequency spectrum and compare with the results of part (c).

Solution

(a) From Example 11.9:

$$R_i = 1.32 \text{ kΩ}, \qquad A_{v_{mid}}(\text{amplifier}) = -90$$

and $\quad R_{Th_i} = R_s \| R_1 \| R_2 \| R_i = 1$ kΩ $\| 40$ kΩ $\| 10$ kΩ $\| 1.32$ kΩ

$$\cong 0.531 \text{ kΩ}$$

with $\quad C_i = C_{W_i} + C_{be} + (1 - A_v)C_{be}$

$$= 6 \text{ pF} + 36 \text{ pF} + [1 - (-90)]4 \text{ pF}$$

$$= 406 \text{ pF}$$

$$f_{H_i} = \frac{1}{2\pi R_{Th_i} C_i} = \frac{1}{2\pi(0.531 \text{ kΩ})(406 \text{ pF})}$$

$$= \textbf{738.24 kHz}$$

$$R_{Th_o} = R_C \| R_L = 4 \text{ kΩ} \| 2.2 \text{ kΩ} = 1.419 \text{ kΩ}$$

$$C_o = C_{W_o} + C_{ce} + C_{M_o} = 8 \text{ pF} + 1 \text{ pF} + \left(1 - \frac{1}{-90}\right)4 \text{ pF}$$

$$= 13.04 \text{ pF}$$

$$f_{H_o} = \frac{1}{2\pi R_{Th_o} C_o} = \frac{1}{2\pi(1.419 \text{ kΩ})(13.04 \text{ pF})}$$

$$= \textbf{8.6 MHz}$$

(b) Applying Eq. (11.55) gives

$$f_\beta = \frac{1}{2\pi\beta_{mid}r_e(C_{be} + C_{bc})}$$

$$= \frac{1}{2\pi(100)(15.76 \text{ Ω})(36 \text{ pF} + 4 \text{ pF})} = \frac{1}{2\pi(100)(15.76 \text{ Ω})(40 \text{ pF})}$$

$$= \textbf{2.52 MHz}$$

$$f_T = \beta_{mid}f_\beta = (100)(2.52 \text{ MHz})$$

$$= \textbf{252 MHz}$$

(c) See Fig. 11.53. Both f_β and f_{H_o} will lower the upper cutoff frequency below the level determined by f_{H_i}. f_β is closer to f_{H_i} and therefore will have a greater impact than f_{H_o}. In any event, the bandwidth will be less than that defined solely by f_{H_i}. In fact, for the parameters of this network the upper cutoff frequency will be relatively close to 600 kHz.

In general, therefore, the lowest of the upper-cutoff frequencies defines a maximum possible bandwidth for a system.

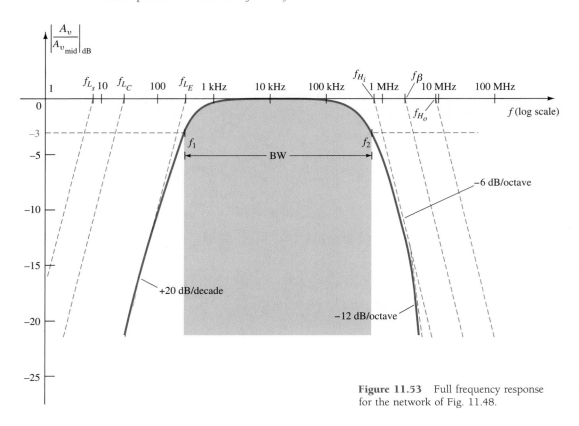

Figure 11.53 Full frequency response for the network of Fig. 11.48.

(d) In order to obtain a PSpice analysis for the full frequency range, the parasitic capacitances have to be added to the network as shown in Fig. 11.54.

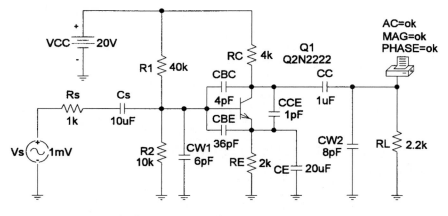

Figure 11.54 Network of Figure 11.27 with parasitic capacitances in place.

An **Analysis** will result in the plot of Fig. 11.55 using the **Trace Expression** appearing at the bottom of the plot. The vertical scale was changed from -60 to 0 dB to -30 to 0 dB to highlight the area of interest using the **Y-Axis Settings.** The low cutoff frequency of 324 Hz is as determined primarily by f_{L_E}, and the high cutoff frequency is near 667 kHz. Even though f_{H_o} is more than a decade higher than f_{H_i}, it will have an impact on the high cutoff frequency. In total, however, the PSpice analysis has been a welcome verification of the hand-written approach.

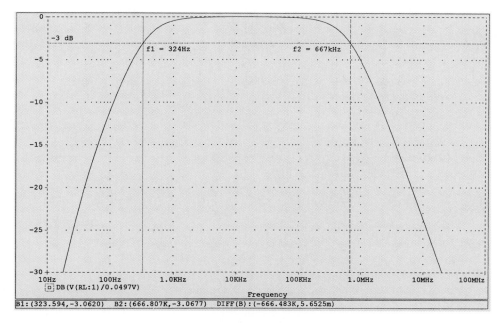

Figure 11.55 Full frequency response for the network of Fig. 11.54.

11.10 HIGH-FREQUENCY RESPONSE— FET AMPLIFIER

The analysis of the high-frequency response of the FET amplifier will proceed in a very similar manner to that encountered for the BJT amplifier. As shown in Fig. 11.56, there are interelectrode and wiring capacitances that will determine the high-frequency characteristics of the amplifier. The capacitors C_{gs} and C_{gd} typically vary from 1 to 10 pF, while the capacitance C_{ds} is usually quite a bit smaller, ranging from 0.1 to 1 pF.

Since the network of Fig. 11.56 is an inverting amplifier, a Miller effect capacitance will appear in the high-frequency ac equivalent network appearing in Fig. 11.57. At high frequencies, C_i will approach a short-circuit equivalent and V_{gs} will drop in value and reduce the overall gain. At frequencies where C_o approaches its short-circuit equivalent, the parallel output voltage V_o will drop in magnitude.

The cutoff frequencies defined by the input and output circuits can be obtained by first finding the Thévenin equivalent circuits for each section as shown in Fig. 11.58. For the input circuit,

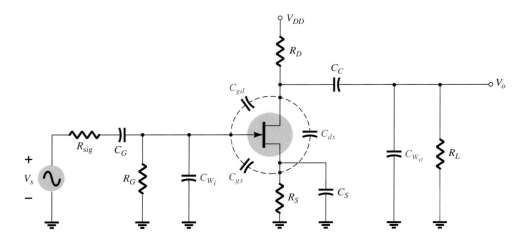

Figure 11.56 Capacitive elements that affect the high-frequency response of a JFET amplifier.

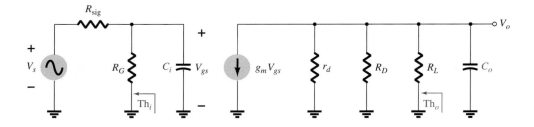

Figure 11.57 High-frequency ac equivalent circuit for Fig. 11.56.

$$f_{H_i} = \frac{1}{2\pi R_{Th_i} C_i} \tag{11.61}$$

and

$$R_{Th_i} = R_{sig} \| R_G \tag{11.62}$$

with

$$C_i = C_{W_i} + C_{gs} + C_{M_i} \tag{11.63}$$

and

$$C_{M_i} = (1 - A_v) C_{gd} \tag{11.64}$$

for the output circuit,

$$f_{H_o} = \frac{1}{2\pi R_{Th_o} C_o} \tag{11.65}$$

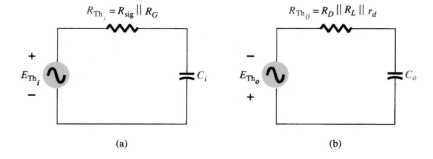

Figure 11.58 The Thévenin equivalent circuits for the (a) input circuit and (b) output circuit.

with
$$R_{Th_o} = R_D \| R_L \| r_d \qquad (11.66)$$

and
$$C_o = C_{W_o} + C_{ds} + C_{M_o} \qquad (11.67)$$

and
$$C_{M_o} = \left(1 - \frac{1}{A_V}\right)C_{gd} \qquad (11.68)$$

EXAMPLE 11.12

(a) Determine the high cutoff frequencies for the network of Fig. 11.56 using the same parameters as Example 11.10:

$$C_G = 0.01 \ \mu F, \qquad C_C = 0.5 \ \mu F, \qquad C_S = 2 \ \mu F$$

$$R_{sig} = 10 \ k\Omega, \quad R_G = 1 \ M\Omega, \quad R_D = 4.7 \ k\Omega, \quad R_S = 1 \ k\Omega, \quad R_L = 2.2 \ k\Omega$$

$$I_{DSS} = 8 \ mA, \qquad V_P = -4 \ V, \qquad r_d = \infty \ \Omega, \qquad V_{DD} = 20 \ V$$

with the addition of

$$C_{gd} = 2 \ pF, \quad C_{gs} = 4 \ pF, \quad C_{ds} = 0.5 \ pF, \quad C_{W_i} = 5 \ pF, \quad C_{W_o} = 6 \ pF$$

(b) Review a **PROBE** response for the full frequency range and note whether it supports the conclusions of Example 11.10 and the calculations above.

Solution

(a) $R_{Th_i} = R_{sig} \| R_G = 10 \ k\Omega \| 1 \ M\Omega = 9.9 \ k\Omega$

From Example 11.10, $A_v = -3$.

$$C_i = C_{W_i} + C_{gs} + (1 - A_v)C_{gd}$$

$$= 5 \ pF + 4 \ pF + (1 + 3)2 \ pF$$

$$= 9 \ pF + 8 \ pF$$

$$= 17 \ pF$$

$$f_{H_1} = \frac{1}{2\pi R_{Th_i} C_i}$$

$$= \frac{1}{2\pi(9.9 \ k\Omega)(17 \ pF)} = \mathbf{945.67 \ kHz}$$

$$R_{Th_o} = R_D \| R_L$$

$$= 4.7 \ k\Omega \| 2.2 \ k\Omega$$

$$\cong 1.5 \ k\Omega$$

$$C_o = C_{W_o} + C_{ds} + C_{M_o} = 6 \ pF + 0.5 \ pF + \left(1 - \frac{1}{-3}\right)2 \ pF = 9.17 \ pF$$

$$f_{H_o} = \frac{1}{2\pi(1.5 \ k\Omega)(9.17 \ pF)} = \mathbf{11.57 \ MHz}$$

The results above clearly indicate that the input capacitance with its Miller effect capacitance will determine the upper cutoff frequency. This is typically the case due to the smaller value of C_{ds} and the resistance levels encountered in the output circuit.

(b) Using PSpice Windows, the schematic for the network will appear as shown in Fig. 11.59.

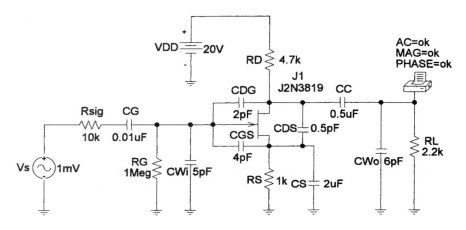

Figure 11.59 Network of Figure 11.56 with assigned values.

Under **Analysis,** the **AC Sweep** is set to **Decade** with **Pts/Decade** at 1000, **Start Freq.:** at 10 Hz, and **End Freq.:** at 10 MHz. Under the **Add Traces** dialog box, the **Trace Expression** is entered as **DB(V(RL:1)/2.993mV),** and the plot of Fig. 11.60 is obtained. Just for a moment, consider how much time it must have taken to obtain a plot such as in Fig. 11.60 without computer methods for a network as complicated as Fig. 11.59. Often, we forget how computer systems have helped us through some painstaking, lengthy, and boring series of calculations.

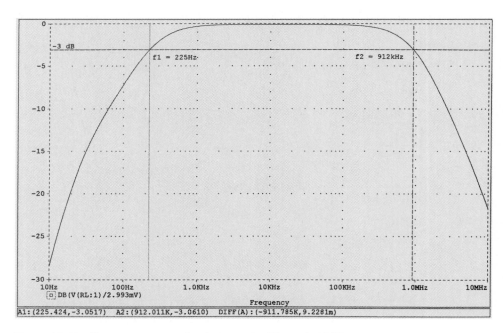

Figure 11.60 Frequency response for the network of Example 11.12.

Using the cursor, we find the lower and upper cutoff frequencies to be 225 Hz and 921 kHz, respectively, providing a nice match with the calculated values.

Even though the analysis of the past few sections has been limited to two configurations, the exposure to the general procedure for determining the cutoff frequencies should support the analysis of any other transistor configuration. Keep in mind that the Miller capacitance is limited to inverting amplifiers and that f_α is significantly greater than f_β if the common-base configuration is encountered. There is a great deal more literature on the analysis of single-stage amplifiers that goes beyond the coverage of this chapter. However, the content of this chapter should provide a firm foundation for any future analysis of frequency effects.

11.11 MULTISTAGE FREQUENCY EFFECTS

For a second transistor stage connected directly to the output of a first stage, there will be a significant change in the overall frequency response. In the high-frequency region, the output capacitance C_o must now include the wiring capacitance (C_{W_1}), parasitic capacitance (C_{be}), and Miller capacitance (C_{M_i}) of the following stage. Further, there will be additional low-frequency cutoff levels due to the second stage that will further reduce the overall gain of the system in this region. For each additional stage, the upper cutoff frequency will be determined primarily by that stage having the lowest cutoff frequency. The low-frequency cutoff is primarily determined by that stage having the highest low-frequency cutoff frequency. Obviously, therefore, one poorly designed stage can offset an otherwise well-designed cascaded system.

The effect of increasing the number of *identical* stages can be clearly demonstrated by considering the situations indicated in Fig. 11.61. In each case, the upper and lower cutoff frequencies of each of the cascaded stages are identical. For a single stage, the cutoff frequencies are f_1 and f_2 as indicated. For two identical stages in cascade, the drop-off rate in the high- and low-frequency regions has increased to -12 dB/octave or -40 dB/decade. At f_1 and f_2, therefore, the decibel drop is now -6 dB rather than the defined band frequency gain level of -3 dB. The -3-dB point has shifted to f_1' and f_2' as indicated, with a resulting drop in the bandwidth. A -18-dB/octave or -60-dB/decade slope will result for a three-stage system of identical stages with the indicated reduction in bandwidth (f_1'' and f_2'').

Assuming identical stages, an equation for each band frequency as a function of the number of stages (n) can be determined in the following manner: For the low-frequency region,

$$A_{v_{\text{low, (overall)}}} = A_{v_{1_{\text{low}}}} A_{v_{2_{\text{low}}}} A_{v_{3_{\text{low}}}} \cdots A_{v_{n_{\text{low}}}}$$

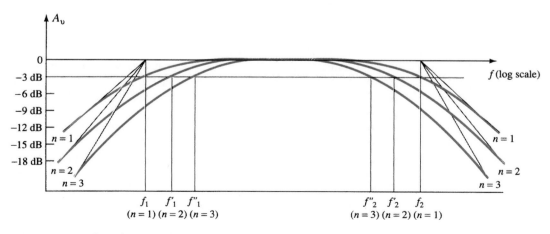

Figure 11.61 Effect of an increased number of stages on the cutoff frequencies and the bandwidth.

but since each stage is identical, $A_{v_{1_{low}}} = A_{v_{2_{low}}} = $ etc. and

$$A_{v_{low, (overall)}} = (A_{v_{1_{low}}})^n$$

or

$$\frac{A_{v_{low}}}{A_{v_{mid}}}(overall) = \left(\frac{A_{v_{low}}}{A_{v_{mid}}}\right)^n = \frac{1}{(1 - jf_1/f)^n}$$

Setting the magnitude of this result equal to $1/\sqrt{2}(-3 \text{ dB level})$ results in

$$\frac{1}{\sqrt{[1 + (f_1/f_1')^2]^n}} = \frac{1}{\sqrt{2}}$$

or

$$\left\{\left[1 + \left(\frac{f_1}{f_1'}\right)^2\right]^{1/2}\right\}^n = \left\{\left[1 + \left(\frac{f_1}{f_1'}\right)^2\right]^n\right\}^{1/2} = (2)^{1/2}$$

so that

$$\left[1 + \left(\frac{f_1}{f_1'}\right)^2\right]^n = 2$$

and

$$1 + \left(\frac{f_1}{f_1'}\right)^2 = 2^{1/n}$$

with the result that

$$\boxed{f_1' = \frac{f_1}{\sqrt{2^{1/n} - 1}}} \tag{11.69}$$

In a similar manner, it can be shown that for the high-frequency region,

$$\boxed{f_2' = (\sqrt{2^{1/n} - 1})f_2} \tag{11.70}$$

Note the presence of the same factor $\sqrt{2^{1/n} - 1}$ in each equation. The magnitude of this factor for various values of n is listed below.

n	$\sqrt{2^{1/n} - 1}$
2	0.64
3	0.51
4	0.43
5	0.39

For $n = 2$, consider that the upper cutoff frequency $f_2' = 0.64f_2$ or 64% of the value obtained for a single stage, while $f_1' = (1/0.64)f_1 = 1.56f_1$. For $n = 3, f_2' = 0.51f_2$ or approximately $\frac{1}{2}$ the value of a single stage with $f_1' = (1/0.51) f_1 = 1.96f_1$ or approximately *twice* the single-stage value.

For the *RC*-coupled transistor amplifier, if $f_2 = f_\beta$, or if they are close enough in magnitude for both to affect the upper 3-dB frequency, the number of stages must be increased by a factor of 2 when determining f_2' due to the increased number of factors $1/(1 + jf/f_x)$.

A decrease in bandwidth is not always associated with an increase in the number of stages if the midband gain can remain fixed and independent of the number of stages. For instance, if a single-stage amplifier produces a gain of 100 with a bandwidth of 10,000 Hz, the resulting gain–bandwidth product is $10^2 \times 10^4 = 10^6$. For a two-stage system the same gain can be obtained by having two stages with a gain of 10 since ($10 \times 10 = 100$). The bandwidth of each stage would then increase by a factor of 10 to 100,000 due to the lower gain requirement and fixed gain–bandwidth product of 10^6. Of course, the design must be such as to permit the increased bandwidth and establish the lower gain level.

11.12 SQUARE-WAVE TESTING

A sense for the frequency response of an amplifier can be determined experimentally by applying a square-wave signal to the amplifier and noting the output response. The shape of the output waveform will reveal whether the high or low frequencies are being properly amplified. The use of *square-wave testing* is significantly less time-consuming than applying a series of sinusoidal signals at different frequencies and magnitudes to test the frequency response of the amplifier.

The reason for choosing a square-wave signal for the testing process is best described by examining the *Fourier series* expansion of a square wave composed of a series of sinusoidal components of different magnitudes and frequencies. The summation of the terms of the series will result in the original waveform. In other words, even though a waveform may not be sinusoidal, it can be reproduced by a series of sinusoidal terms of different frequencies and magnitudes.

The Fourier series expansion for the square wave of Fig. 11.62 is

$$v = \frac{4}{\pi} V_m \left(\sin 2\pi f_s t + \frac{1}{3} \sin 2\pi (3f_s)t + \frac{1}{5} \sin 2\pi (5f_s)t + \frac{1}{7} \sin 2\pi (7f_s)t \right.$$
$$\left. + \frac{1}{9} \sin 2\pi (9f_s)t + \cdots + \frac{1}{n} \sin 2\pi (nf_s)t \right) \quad (11.71)$$

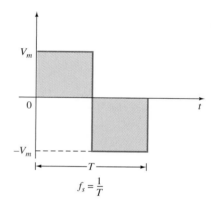

Figure 11.62 Square wave.

The first term of the series is called the *fundamental* term and in this case has the same frequency, f_s, as the square wave. The next term has a frequency equal to three times the fundamental and is referred to as the *third harmonic*. Its magnitude is one-third the magnitude of the fundamental term. The frequencies of the succeeding terms are odd multiples of the fundamental term, and the magnitude decreases with each higher harmonic. Figure 11.63 demonstrates how the summation of terms of a Fourier series can result in a nonsinusoidal waveform. The generation of the square wave of Fig. 11.62 would require an infinite number of terms. However, the summation of just the fundamental term and the third harmonic in Fig. 11.63a clearly results in a waveform that is beginning to take on the appearance of a square wave. Including the fifth and seventh harmonics as in Fig. 11.63b takes us a step closer to the waveform of Fig. 11.62.

Since the ninth harmonic has a magnitude greater than 10% of the fundamental term $\left[\frac{1}{9} (100\%) = 11.1\% \right]$, the fundamental term through the ninth harmonic are the major contributors to the Fourier series expansion of the square-wave function. It is therefore reasonable to assume that if the application of a square wave of a particular frequency results in a nice clean square wave at the output, then the fundamental

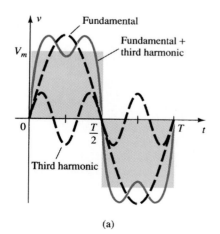

(a)

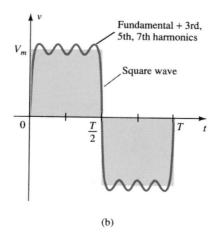

(b)

Figure 11.63 Harmonic content of a square wave.

through the ninth harmonic are being amplified without visual distortion by the amplifier. For instance, if an audio amplifier with a bandwidth of 20 kHz (audio range is from 20 Hz to 20 kHz) is to be tested, the frequency of the applied signal should be at least 20 kHz/9 = 2.22 kHz.

If the response of an amplifier to an applied square wave is an undistorted replica of the input, the frequency response (or BW) of the amplifier is obviously sufficient for the applied frequency. If the response is as shown in Fig. 11.64a and b, the low frequencies are not being amplified properly and the low cutoff frequency has to be investigated. If the waveform has the appearance of Fig. 11.64c, the high-frequency components are not receiving sufficient amplification and the high cutoff frequency (or BW) has to be reviewed.

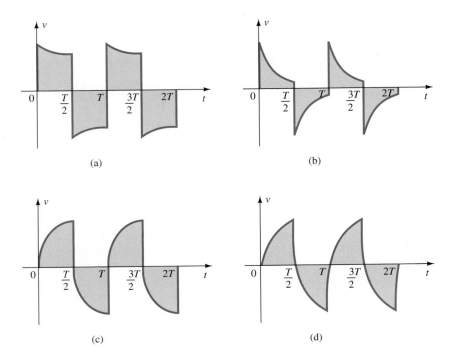

Figure 11.64 (a) Poor low frequency response; (b) very poor low-frequency response; (c) poor high-frequency response; (d) very poor high-frequency response.

The actual high cutoff frequency (or BW) can be determined from the output waveform by carefully measuring the rise time defined between 10% and 90% of the peak value, as shown in Fig. 11.65. Substituting into the following equation will provide the upper cutoff frequency, and since BW = $f_{H_i} - f_{L_o} \cong f_{H_i}$, the equation also provides an indication of the BW of the amplifier.

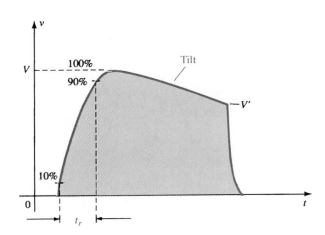

Figure 11.65 Defining the rise time and tilt of a square wave response.

$$\boxed{\text{BW} \cong f_{H_i} = \frac{0.35}{t_r}} \qquad (11.72)$$

The low cutoff frequency can be determined from the output response by carefully measuring the tilt of Fig. 11.65 and substituting into one of the following equations:

$$\boxed{\% \text{ tilt} = P\% = \frac{V - V'}{V} \times 100\%} \qquad (11.73)$$

$$\boxed{\text{tilt} = P = \frac{V - V'}{V}} \qquad \text{(decimal form)} \qquad (11.74)$$

The low cutoff frequency is then determined from

$$\boxed{f_{L_o} = \frac{P}{\pi} f_s} \qquad (11.75)$$

The application of a 1-mV, 5-kHz square wave to an amplifier resulted in the output waveform of Fig. 11.66.
(a) Write the Fourier series expansion for the square wave through the ninth harmonic.
(b) Determine the bandwidth of the amplifier.
(c) Calculate the low cutoff frequency.

EXAMPLE 11.13

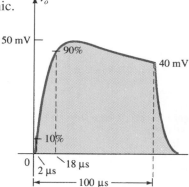

Figure 11.66 Example 11.13

Solution

(a) $v_i = \dfrac{4 \text{ mV}}{\pi} \bigg(\sin 2\pi (5 \times 10^3)t + \dfrac{1}{3} \sin 2\pi(15 \times 10^3)t + \dfrac{1}{5} \sin 2\pi(25 \times 10^3)t$

$\qquad + \dfrac{1}{7} \sin 2\pi(35 \times 10^3)t + \dfrac{1}{9} \sin 2\pi(45 \times 10^3)t \bigg)$

(b) $t_r = 18 \ \mu\text{s} - 2 \ \mu\text{s} = 16 \ \mu\text{s}$

$\text{BW} = \dfrac{0.35}{t_r} = \dfrac{0.35}{16 \ \mu\text{s}} = \textbf{21,875 Hz} \cong 4.4 f_s$

(c) $P = \dfrac{V - V'}{V} = \dfrac{50 \text{ mV} - 40 \text{ mV}}{50 \text{ mV}} = 0.2$

$f_{L_o} = \dfrac{P}{\pi} f_s = \left(\dfrac{0.2}{\pi} \right)(5 \text{ kHz}) = \textbf{318.31 Hz}$

11.13 SUMMARY

Important Conclusions and Concepts

1. The logarithm of a number will give you the **power to which the base must be brought to obtain the same number.** If the base is 10, it is referred to as the **common logarithm;** if the base is $e = 2.71828\ldots$, it is called the **natural logarithm.**

2. Since the decibel rating of any piece of equipment is a **comparison between levels**, a reference level must be selected for each area of application. For audio systems the reference level is generally accepted as **1 mW**. When using voltage levels to determine the gain in dB between two points, any difference in resistance level is generally ignored.

3. The dB gain of cascaded systems is simply the **sum** of the dB gains of each stage.

4. It is the **capacitive elements** of a network that determine the **bandwidth** of a system. The **larger** capacitive elements of the basic design will determine the **low** cutoff frequency, whereas the **smaller** parasitic capacitors will determine the **high** cutoff frequencies.

5. The frequencies at which the gain drops to 70.7% of the mid-band value are called the **cutoff, corner, band, break,** or **half-power** frequencies.

6. The **narrower** the bandwidth, the **smaller** the range of frequencies that will permit a transfer of power to the load that is at least 50% of the mid-band level.

7. A change in frequency by a factor of **2**, equivalent to **1 octave**, results in a **6-dB change in gain**. For a **10:1** change in frequency, equivalent to **1 decade**, there is a **20-dB change in gain**.

8. For any **inverting** amplifier, the input capacitance will be increased by a **Miller effect** capacitance determined by the **gain** of the amplifier and the **interelectrode** (parasitic) capacitance between the input and output terminals of the active device.

9. A **3-dB drop in beta** (h_{fe}) will occur at a frequency defined by f_β that is sensitive to the **dc operating conditions** of the transistor. This variation in beta can define the upper cutoff frequency of the design.

10. The **high and low cutoff frequencies** of an amplifier can be determined by the response of the system to a **square-wave input.** The general appearance will immediately reveal whether the low- or high-frequency response of the system is too limited for the applied frequency, whereas a more detailed examination of the response will reveal the actual bandwidth of the amplifier.

Equations

Logarithms:

$$a = b^x, \qquad x = \log_b a$$

$$\log_{10} \frac{a}{b} = \log_{10} a - \log_{10} b$$

$$\log_{10} ab = \log_{10} a + \log_{10} b$$

$$G_{dB} = 10 \log_{10} \frac{P_2}{P_1} = 20 \log_{10} \frac{V_2}{V_1}$$

$$G_{dB_T} = G_{dB_1} + G_{dB_2} + G_{dB_3} + \cdots + G_{dB_n}$$

Low-frequency response:

$$A_v = \frac{1}{1 - j(f_1/f)}$$

$$f_1 = \frac{1}{2\pi RC}$$

BJT low-frequency response:

$$f_{L_s} = \frac{1}{2\pi(R_s + R_i)C_s}$$

$$R_i = R_1\|R_2\|\beta r_e$$

$$f_{L_C} = \frac{1}{2\pi(R_o + R_L)C_C}$$

$$R_o = R_C\|r_o$$

$$f_{L_E} = \frac{1}{2\pi R_e C_E}$$

$$R_e = R_E\|\left(\frac{R_s'}{\beta} + r_e\right)$$

$$R_s' = R_s\|R_1\|R_2$$

FET low-frequency response:

$$f_{L_G} = \frac{1}{2\pi(R_{\text{sig}} + R_i)C_G}$$

$$R_i = R_G$$

$$f_{L_C} = \frac{1}{2\pi(R_o + R_L)C_C}$$

$$R_o = R_D\|r_d$$

$$f_{L_S} = \frac{1}{2\pi R_{eq}C_S}$$

$$R_{eq} = \frac{R_S}{1 + R_S(1 + g_m r_d)/(r_d + R_D\|R_L)} \cong R_S\|\frac{1}{g_m}\bigg|_{r_d \cong \infty\,\Omega}$$

Miller effect capacitance:

$$C_{M_i} = (1 - A_v)C_f$$

$$C_{M_o} = \left(1 - \frac{1}{A_v}\right)C_f$$

BJT high-frequency response:

$$A_v = \frac{1}{1 + j(f/f_2)}$$

$$f_{H_i} = \frac{1}{2\pi R_{\text{Th}_i}C_i}$$

$$R_{\text{Th}_i} = R_s\|R_1\|R_2\|R_i$$

$$C_i = C_{W_i} + C_{be} + C_{M_i}$$

$$f_{H_o} = \frac{1}{2\pi R_{\text{Th}_o}C_o}$$

$$R_{\text{Th}_o} = R_C\|R_L\|r_o$$

$$C_o = C_{W_o} + C_{ce} + C_{M_o}$$

$$h_{fe} = \frac{h_{fe_{mid}}}{1 + j(f/f_\beta)}$$

$$f_\beta \cong \frac{1}{2\pi\beta_{mid}r_e(C_{be} + C_{bc})}$$

$$f_T \cong h_{fe_{mid}} \cdot f_\beta$$

FET high-frequency response:

$$f_{H_i} = \frac{1}{2\pi R_{Th_i}C_i}$$

$$R_{Th_i} = R_{sig}\|R_G$$

$$C_i = C_{W_i} + C_{gs} + C_{M_i}$$

$$C_{M_i} = (1 - A_v)C_{gd}$$

$$f_{H_o} = \frac{1}{2\pi R_{Th_o}C_o}$$

$$R_{Th_o} = R_D\|R_L\|r_d$$

$$C_o = C_{W_o} + C_{ds} + C_{M_o}$$

$$C_{M_o} = \left(1 - \frac{1}{A_v}\right)C_{gd}$$

Multistage effects:

$$f_1' = \frac{f_1}{\sqrt{2^{1/n} - 1}}$$

$$f_2' = (\sqrt{2^{1/n} - 1})f_2$$

Square-wave testing:

$$BW \cong f_{H_i} = \frac{0.35}{t_r}$$

$$f_{L_o} = \frac{P}{\pi}f_s$$

$$P = \frac{V - V'}{V}$$

11.14   COMPUTER ANALYSIS

The computer analysis of this chapter was integrated into the chapter for emphasis and a clear demonstration of the power of the PSpice software package. The complete frequency response of a single-stage or multistage system can be determined in a relatively short period of time to verify theoretical calculations or provide an immediate indication of the low and high cutoff frequencies of the system. The exercises in the chapter will provide an opportunity to apply the PSpice software package to a variety of networks.

§ 11.2 Logarithms

1. (a) Determine the common logarithm of the following numbers: 10^3, 50, and 0.707.
   (b) Determine the natural logarithm of the same numbers appearing in part (a).
   (c) Compare the solutions of parts (a) and (b).

2. (a) Determine the common logarithm of the number 2.2×10^3.
   (b) Determine the natural logarithm of the number of part (a) using Eq. (11.4).
   (c) Determine the natural logarithm of the number of part (a) using natural logarithms and compare with the solution of part (b).

3. Determine:
   (a) $20 \log_{10} \frac{40}{8}$ using Eq. (11.6) and compare with $20 \log_{10} 5$.
   (b) $10 \log_{10} \frac{1}{20}$ using Eq. (11.7) and compare with $10 \log_{10} 0.05$.
   (c) $\log_{10}(40)(0.125)$ using Eq. (11.8) and compare with $\log_{10} 5$.

4. Calculate the power gain in decibels for each of the following cases.
   (a) $P_o = 100$ W, $P_i = 5$ W.
   (b) $P_o = 100$ mW, $P_i = 5$ mW.
   (c) $P_o = 100$ mW, $P_i = 20$ μW.

5. Determine G_{dBm} for an output power level of 25 W.

6. Two voltage measurements made across the same resistance are $V_1 = 25$ V and $V_2 = 100$ V. Calculate the power gain in decibels of the second reading over the first reading.

7. Input and output voltage measurements of $V_i = 10$ mV and $V_o = 25$ V are made. What is the voltage gain in decibels?

*8. (a) The total decibel gain of a three-stage system is 120 dB. Determine the decibel gain of each stage if the second stage has twice the decibel gain of the first and the third has 2.7 times the decibel gain of the first.
   (b) Determine the voltage gain of each stage.

*9. If the applied ac power to a system is 5 μW at 100 mV and the output power is 48 W, determine:
   (a) The power gain in decibels.
   (b) The voltage gain in decibels if the output impedance is 40 kΩ.
   (c) The input impedance.
   (d) The output voltage.

§ 11.4 General Frequency Considerations

10. Given the characteristics of Fig. 11.67, sketch:
    (a) The normalized gain.
    (b) The normalized dB gain (and determine the bandwidth and cutoff frequencies).

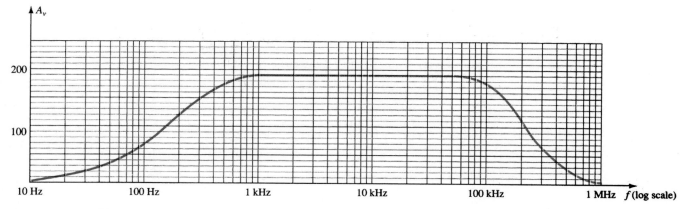

Figure 11.67 Problem 10

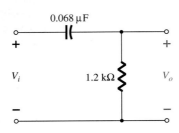

Figure 11.68 Problems 11, 12, and 32

§ 11.5 Low-Frequency Analysis—Bode Plot

11. For the network of Fig. 11.68:
    (a) Determine the mathematical expression for the magnitude of the ratio V_o/V_i.
    (b) Using the results of part (a), determine V_o/V_i at 100 Hz, 1 kHz, 2 kHz, 5 kHz, and 10 kHz, and plot the resulting curve for the frequency range of 100 Hz to 10 kHz. Use a log scale.
    (c) Determine the break frequency.
    (d) Sketch the asymptotes and locate the −3-dB point.
    (e) Sketch the frequency response for V_o/V_i and compare to the results of part (b).

12. For the network of Fig. 11.68:
    (a) Determine the mathematical expression for the angle by which V_o leads V_i.
    (b) Determine the phase angle at $f =$ 100 Hz, 1 kHz, 2 kHz, 5 kHz, and 10 kHz, and plot the resulting curve for the frequency range of 100 Hz to 10 kHz.
    (c) Determine the break frequency.
    (d) Sketch the frequency response of θ for the same frequency spectrum of part (b) and compare results.

13. (a) What frequency is 1 octave above 5 kHz?
    (b) What frequency is 1 decade below 10 kHz?
    (c) What frequency is 2 octaves below 20 kHz?
    (d) What frequency is 2 decades above 1 kHz?

§ 11.6 Low-Frequency Response—BJT Amplifier

14. Repeat the analysis of Example 11.9 with $r_o =$ 40 kΩ. What is the effect on $A_{v_{mid}}, f_{L_S}, f_{L_C}, f_{L_E},$ and the resulting cutoff frequency?

15. For the network of Fig. 11.69:
    (a) Determine r_e.
    (b) Find $A_{v_{mid}} = V_o/V_i$.
    (c) Calculate Z_i.
    (d) Find $A_{v_{Smid}} = V_o/V_s$.
    (e) Determine $f_{L_S}, f_{L_C},$ and f_{L_E}.
    (f) Determine the low cutoff frequency.
    (g) Sketch the asymptotes of the Bode plot defined by the cutoff frequencies of part (e).
    (h) Sketch the low-frequency response for the amplifier using the results of part (f).

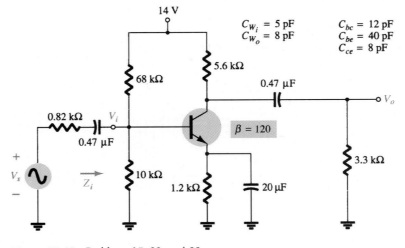

Figure 11.69 Problems 15, 22, and 33

* 16. Repeat Problem 15 for the emitter-stabilized network of Fig. 11.70.

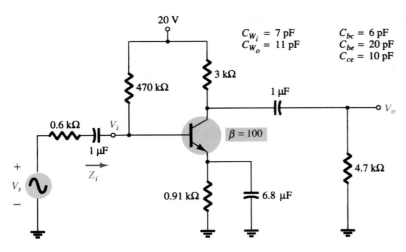

Figure 11.70 Problems 16 and 23

* 17. Repeat Problem 15 for the emitter-follower network of Fig. 11.71.

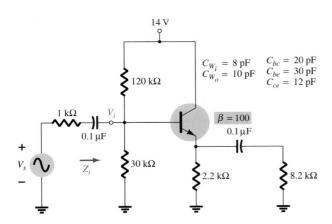

Figure 11.71 Problems 17 and 24

* 18. Repeat Problem 15 for the common-base configuration of Fig. 11.72. Keep in mind that the common-base configuration is a noninverting network when you consider the Miller effect.

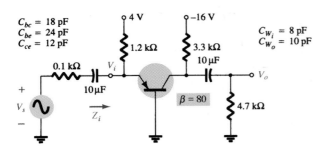

Figure 11.72 Problems 18, 25, and 34

19. For the network of Fig. 11.73:
(a) Determine V_{GS_Q} and I_{D_Q}.
(b) Find g_{m0} and g_m.
(c) Calculate the midband gain of $A_v = V_o/V_i$.
(d) Determine Z_i.
(e) Calculate $A_{v_s} = V_o/V_s$.
(f) Determine f_{L_G}, f_{L_C}, and f_{L_S}.
(g) Determine the low cutoff frequency.
(h) Sketch the asymptotes of the Bode plot defined by part (f).
(i) Sketch the low-frequency response for the amplifier using the results of part (f).

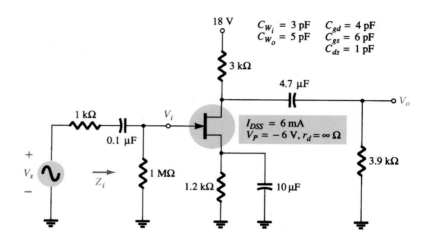

Figure 11.73 Problems 19, 20, 26, and 35

* **20.** Repeat the analysis of Problem 19 with $r_d = 100 \text{ k}\Omega$. Does it have an impact of any consequence on the results? If so, which elements?

* **21.** Repeat the analysis of Problem 19 for the network of Fig. 11.74. What effect did the voltage-divider configuration have on the input impedance and the gain A_{v_s} compared to the biasing arrangement of Fig. 11.73?

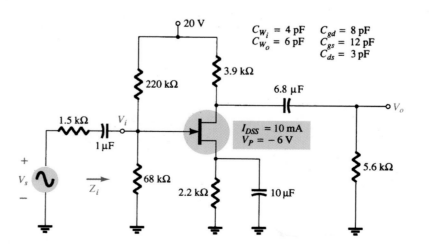

Figure 11.74 Problems 21 and 27

§ 11.9 High-Frequency Response—BJT Amplifier

22. For the network of Fig. 11.69:
    (a) Determine f_{H_i} and f_{H_o}.
    (b) Assuming that $C_{b'e} = C_{be}$ and $C_{b'c} = C_{bc}$, find f_β and f_T.
    (c) Sketch the frequency response for the high-frequency region using a Bode plot and determine the cutoff frequency.

* 23. Repeat the analysis of Problem 22 for the network of Fig. 11.70.

* 24. Repeat the analysis of Problem 22 for the network of Fig. 11.71.

* 25. Repeat the analysis of Problem 22 for the network of Fig. 11.72.

§ 11.10 High-Frequency Response—FET Amplifier

26. For the network of Fig. 11.73:
    (a) Determine g_{m_0} and g_m.
    (b) Find A_v and A_{v_s} in the mid-frequency range.
    (c) Determine f_{H_i} and f_{H_o}.
    (d) Sketch the frequency response for the high-frequency region using a Bode plot and determine the cutoff frequency.

* 27. Repeat the analysis of Problem 26 for the network of Fig. 11.74.

§ 11.11 Multistage Frequency Effects

28. Calculate the overall voltage gain of four identical stages of an amplifier, each having a gain of 20.

29. Calculate the overall upper 3-dB frequency for a four-stage amplifier having an individual stage value of $f_2 = 2.5$ MHz.

30. A four-stage amplifier has a lower 3-dB frequency for an individual stage of $f_1 = 40$ Hz. What is the value of f_1 for this full amplifier?

§ 11.12 Square-Wave Testing

* 31. The application of a 10-mV, 100-kHz square wave to an amplifier resulted in the output waveform of Fig. 11.75.

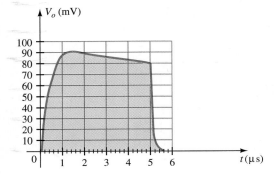

Figure 11.75   Problem 31

f

(a) Write the Fourier series expansion for the square wave through the ninth harmonic.
(b) Determine the bandwidth of the amplifier to the accuracy available by the waveform of Fig. 11.75.
(c) Calculate the low cutoff frequency.

§ 11.14 Computer Analysis

32. Using PSpice Windows, determine the frequency response of V_o/V_i for the high-pass filter of Fig. 11.68.

33. Using PSpice Windows, determine the frequency response of V_o/V_s for the BJT amplifier of Fig. 11.69.

34. Repeat Problem 33 for the network of Fig. 11.72 using Electronics Workbench.

35. Repeat Problem 33 for the JFET configuration of Fig. 11.73 using Electronics Workbench.

*Please Note: Asterisks indicate more difficult problems.

Compound Configurations

12.1 INTRODUCTION

In the present chapter, we introduce a number of circuit connections that, although not standard common-emitter, common-collector, or common-base, are still quite important, being widely used in either discrete or in integrated circuits. The cascade connection provides stages in series, while the cascode connection places one transistor on top of another. Both these connection forms are found in practical circuits. The Darlington connection and the feedback pair connection provide multiple transistors connected for operation as a single transistor for improved performance, usually with much larger current gain.

The CMOS connection, using both p-type enhancement and n-type enhancement MOSFET transistors in a very low power operating circuit, is introduced in this chapter. Much of the newest digital circuitry uses CMOS circuits to either permit portable operation at very low battery power or to allow very high packing density in integrated circuits with lowest power dissipation in the small space used by an IC chip.

Both discrete circuits and integrated circuits use the current source connection. The current mirror connection provides constant current to various other circuits and is especially important in linear integrated circuits.

The differential amplifier is the basic part of operational amplifier circuits (to be covered fully in Chapter 13). The basic differential circuit connection and its operation is introduced in this chapter. Although placed at the end of the chapter, it is nevertheless a most important circuit connection. A bipolar-JFET circuit used in ICs is the BiFET connection, while the bipolar-MOSFET connection is called a BiMOS connection. Both of these are used in linear integrated circuits.

12.2 CASCADE CONNECTION

A popular connection of amplifier stages is the cascade connection. Basically, a cascade connection is a series connection with the output of one stage then applied as input to the second stage. Figure 12.1 shows a cascade connection of two FET amplifier stages. The cascade connection provides a multiplication of the gain of each stage for a larger overall gain.

The gain of the overall cascade amplifier is the product of stage gains A_{v_1} and A_{v_2},

$$A_v = A_{v_1}A_{v_2} = (-g_{m_1}R_{D_1})(-g_{m_2}R_{D_2}) \tag{12.1}$$

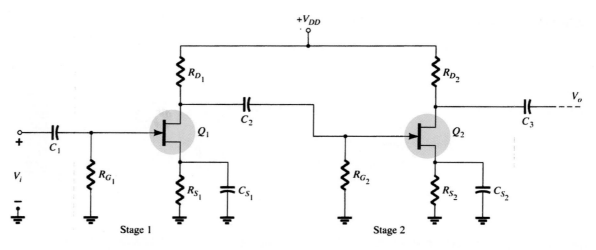

Figure 12.1   Cascaded FET amplifier.

The input impedance of the cascade amplifier is that of stage 1,

$$Z_i = R_{G_1} \tag{12.2}$$

while the output impedance is that of stage 2,

$$Z_o = R_{D_2} \tag{12.3}$$

The main function of cascading stages is the larger overall gain achieved. Since dc bias and ac calculations for a cascade amplifier follow those derived for the individual stages, an example will demonstrate the various calculations to determine dc bias and ac operation.

EXAMPLE 12.1

Calculate the dc bias, voltage gain, input impedance, output impedance, and the resulting output voltage for the cascade amplifier shown in Fig. 12.2. Calculate the load voltage if a 10-kΩ load is connected across the output.

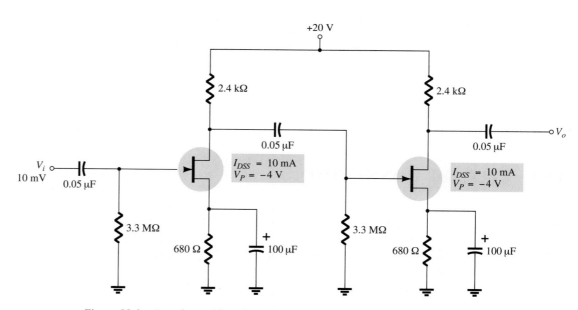

Figure 12.2   Cascade amplifier circuit for Example 12.1.

Solution

Both amplifier stages have the same dc bias. Using dc bias techniques from Chapter 6 results in

$$V_{GS_Q} = -1.9 \text{ V}, \qquad I_{D_Q} = 2.8 \text{ mA}$$

Both transistors have

$$g_{m0} = \frac{2I_{DSS}}{|V_P|} = \frac{2(10 \text{ mA})}{|-4 \text{ V}|} = 5 \text{ mS}$$

and at the dc bias point,

$$g_m = g_{m0}\left(1 - \frac{V_{GS_Q}}{V_P}\right) = (5 \text{ mS})\left(1 - \frac{-1.9 \text{ V}}{-4 \text{ V}}\right) = \textbf{2.6 mS}$$

The voltage gain of each stage is

$$A_{v_1} = A_{v_2} = -g_m R_D = -(2.6 \text{ mS})(2.4 \text{ k}\Omega) = \textbf{-6.2}$$

The cascade amplifier voltage gain is

$$\text{Eq. (12.1):} \quad A_v = A_{v_1}A_{v_2} = (-6.2)(-6.2) = 38.4$$

The output voltage is then

$$V_o = A_v V_i = (38.4)(10 \text{ mV}) = 384 \text{ mV}$$

The cascade amplifier input impedance is

$$Z_i = R_G = \textbf{3.3 M}\Omega$$

The cascade amplifier output impedance (assuming that $r_d = \infty$) is

$$Z_o = R_D = \textbf{2.4 k}\Omega$$

The output voltage across a 10-kΩ load would then be

$$V_L = \frac{R_L}{Z_o + R_L} V_o = \frac{10 \text{ k}\Omega}{2.4 \text{ k}\Omega + 10 \text{ k}\Omega}(384 \text{ mV}) = \textbf{310 mV}$$

BJT Cascade Amplifier

An *RC*-coupled cascade amplifier built using BJTs is shown in Fig. 12.3. As before, the advantage of cascading stages is the large overall voltage gain. DC bias is ob-

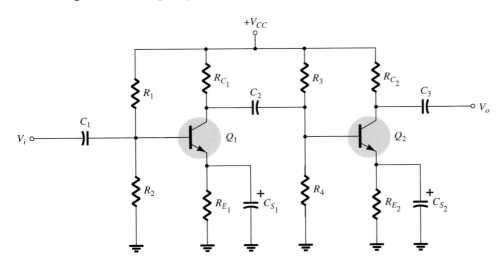

Figure 12.3 Cascaded BJT amplifier (*RC* coupled).

tained using the procedures of Chapter 4. The voltage gain of each stage is

$$A_v = \frac{-R_C \| R_L}{r_e} \tag{12.4}$$

The amplifier input impedance is that of stage 1,

$$Z_i = R_1 \| R_2 \| \beta r_e \tag{12.5}$$

and the output impedance of the amplifier is that of stage 2,

$$Z_o = R_C \| r_o \tag{12.6}$$

The next example demonstrates the analysis of a cascade BJT amplifier showing the large voltage gain achieved.

EXAMPLE 12.2

Calculate the voltage gain, output voltage, input impedance, and output impedance for the cascade BJT amplifier of Fig. 12.4. Calculate the output voltage resulting if a 10-kΩ load is connected to the output.

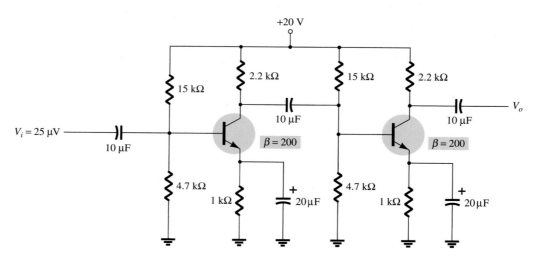

Figure 12.4 *RC*-coupled BJT amplifier for Example 12.2.

Solution

DC bias analysis results in

$$V_B = 4.7 \text{ V}, \qquad V_E = 4.0 \text{ V}, \qquad V_C = 11 \text{ V}, \qquad I_E = 4.0 \text{ mA}$$

At the bias point,

$$r_e = \frac{26}{I_E} = \frac{26}{4.0} = 6.5 \ \Omega$$

The voltage gain of stage 1 is then

$$A_{v_1} = -\frac{R_C \| (R_1 \| R_2 \| \beta r_e)}{r_e}$$

$$= -\frac{(2.2 \text{ k}\Omega) \| [15 \text{ k}\Omega \| 4.7 \text{ k}\Omega \| (200)(6.5 \ \Omega)]}{6.5 \ \Omega}$$

$$= -\frac{665.2 \ \Omega}{6.5 \ \Omega} = -102.3$$

while the voltage gain of stage 2 is

$$A_{v_2} = -\frac{R_C}{r_e} = -\frac{2.2 \text{ k}\Omega}{6.5 \text{ }\Omega} = -338.46$$

for an overall voltage gain of

$$A_v = A_{v_1}A_{v_2} = (-102.3)(-338.46) = \textbf{34,624}$$

The output voltage is then

$$V_o = A_vV_i = (34,624)(25 \text{ }\mu\text{V}) = \textbf{0.866 V}$$

The amplifier input impedance is

$$Z_i = R_1\|R_2\|\beta r_e = 4.7 \text{ k}\Omega\|15 \text{ k}\Omega\|(200)(6.5 \text{ }\Omega)$$
$$= \textbf{953.6 }\Omega$$

while the amplifier output impedance is

$$Z_o = R_C = \textbf{2.2 k}\Omega$$

If a 10-kΩ load is connected to the amplifier output, the resulting voltage across the load is

$$V_L = \frac{R_L}{Z_o + R_L}V_o = \frac{10 \text{ k}\Omega}{2.2 \text{ k}\Omega + 10 \text{ k}\Omega}(0.866 \text{ V}) = \textbf{0.71 V}$$

A combination of FET and BJT stages can also be used to provide high voltage gain and high input impedance, as demonstrated by the next example.

EXAMPLE 12.3

For the cascade amplifier of Fig. 12.5, use the dc bias calculated in Examples 12.1 and 12.2 to calculate input impedance, output impedance, voltage gain, and the resulting output voltage.

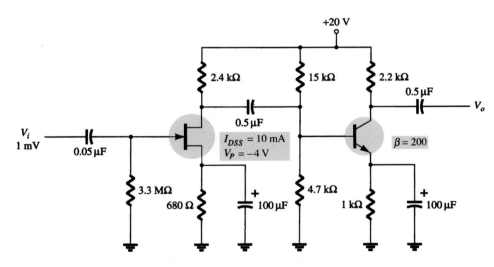

Figure 12.5 Cascaded JFET-BJT amplifier for Example 12.3.

Solution

Since R_i (stage 2) = 15 kΩ‖4.7 kΩ‖200(6.5 Ω) = 953.6 Ω, the gain of stage 1 (when loaded by stage 2) is

$$A_{v_1} = -g_m[R_D\|R_i \text{ (stage 2)}]$$
$$= -2.6 \text{ mS}(2.4 \text{ k}\Omega\|953.6 \text{ }\Omega) = -1.77$$

From Example 12.2, the voltage gain of stage 2 is $A_{v_2} = -338.46$. The overall voltage gain is then

$$A_v = A_{v_1}A_{v_2} = (-1.77)(-338.46) = \mathbf{599.1}$$

The output voltage is then

$$V_o = A_vV_i = (599.1)(1 \text{ mV}) \approx \mathbf{0.6 \text{ V}}$$

The input impedance of the amplifier is that of stage 1,

$$Z_i = \mathbf{3.3 \text{ M}\Omega}$$

while the output impedance is that of stage 2,

$$Z_o = R_D = \mathbf{2.2 \text{ k}\Omega}$$

12.3   CASCODE CONNECTION

A cascode connection has one transistor on top of (in series with) another. Figure 12.6 shows a cascode configuration with a common-emitter (CE) stage feeding a common-base (CB) stage. This arrangement is designed to provide a high input impedance with low voltage gain to ensure that the input Miller capacitance (see Chapter 11) is at a minimum with the CB stage providing good high-frequency operation. A practical BJT version of a cascode amplifier is provided in Fig. 12.7.

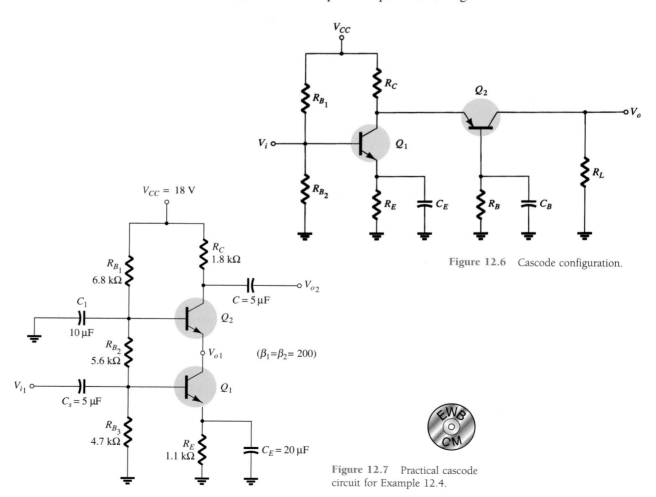

Figure 12.6   Cascode configuration.

Figure 12.7   Practical cascode circuit for Example 12.4.

EXAMPLE 12.4

Calculate the voltage gain for the cascode amplifier of Fig. 12.7.

Solution

DC bias analysis using procedures of Chapter 4 result in

$$V_{B_1} = 4.9 \text{ V}, \qquad V_{B_2} = 10.8 \text{ V}, \qquad I_{C_1} \approx I_{C_2} = 3.8 \text{ mA}$$

The dynamic resistance of each transistor is then

$$r_e = \frac{26}{I_E} = \frac{26}{3.8} = 6.8 \text{ }\Omega$$

The voltage gain of stage 1 (common-emitter) is approximately

$$A_{v_1} = -\frac{R_C}{r_e} = -\frac{r_e}{r_e} = -1$$

The voltage gain of stage 2 (common-base) is

$$A_{v_2} = \frac{R_C}{r_e} = \frac{1.8 \text{ k}\Omega}{6.8 \text{ }\Omega} = 265$$

resulting in an overall cascode amplifier gain of

$$A_v = A_{v_1}A_{v_2} = (-1)(265) = \mathbf{-265}$$

As expected, the CE stage with a gain of -1 provides the higher input impedance of a CE stage (over that of a CB stage). With a voltage gain of only -1, the Miller input capacitance is kept quite small. A large voltage gain is then provided by the CB stage, resulting in a large overall gain ($A_v = -265$).

12.4 DARLINGTON CONNECTION

A very popular connection of two bipolar junction transistors for operation as one "superbeta" transistor is the Darlington connection shown in Fig. 12.8. The main feature of the Darlington connection is that the composite transistor acts as a single unit with a current gain that is the product of the current gains of the individual transistors. If the connection is made using two separate transistors having current gains of β_1 and β_2, the Darlington connection provides a current gain of

$$\beta_D = \beta_1\beta_2 \tag{12.7}$$

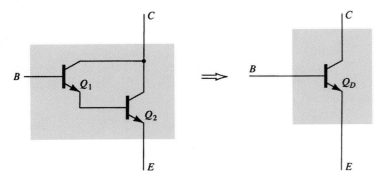

Figure 12.8 Makeup of Darlington transistor.

If the two transistors are matched so that $\beta_1 = \beta_2 = \beta$, the Darlington connection provides a current gain of

$$\beta_D = \beta^2 \tag{12.8}$$

A Darlington transistor connection provides a transistor having a very large current gain, typically a few thousand.

EXAMPLE 12.5

What current gain is provided by a Darlington connection of two identical transistors each having a current gain of $\beta = 200$?

Solution

$$\text{Eq. (12.8):} \quad \beta_D = \beta^2 = (200)^2 = \mathbf{40{,}000}$$

Packaged Darlington Transistor

Since the Darlington connection is popular, one can obtain a single package containing two BJTs internally connected as a Darlington transistor. Figure 12.9 provides some specification sheet data on a typical Darlington pair. The current gain listed is that of the overall Darlington-connected transistor, the device providing externally only three terminals (base, emitter, and collector). One may consider the unit a single Darlington transistor having very high current gain when compared to other typical single transistors.

Type 2N999
N-P-N Darlington-Connected
Silicon Transistor Package

Parameter	Test Conditions	Min.	Max.
V_{BE}	$I_C = 100\,\text{mA}$		1.8 V
$h_{FE}\ (\beta_D)$	$I_C = 10\,\text{mA}$	4000	
	$I_C = 100\,\text{mA}$	7000	70,000

Figure 12.9 Specification information on Darlington transistor package (2N999).

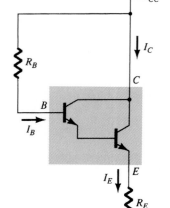

Figure 12.10 Basic Darlington bias circuit.

DC Bias of Darlington Circuit

A basic Darlington circuit is shown in Fig. 12.10. A Darlington transistor having very high current gain, β_D, is used. The base current may be calculated from

$$I_B = \frac{V_{CC} - V_{BE}}{R_B + \beta_D R_E} \tag{12.9}$$

While this equation is the same as for a regular transistor, the value of β_D is much greater and the value of V_{BE} is larger, as indicated by the data in the spec sheet of Fig. 12.9. The emitter current is then

$$I_E = (\beta_D + 1)I_B \approx \beta_D I_B \tag{12.10}$$

Dc voltages are

$$V_E = I_E R_E \tag{12.11}$$

$$V_B = V_E + V_{BE} \tag{12.12}$$

Calculate the dc bias voltages and currents in the circuit of Fig. 12.11.

EXAMPLE 12.6

Solution

The base current is

$$\text{Eq. (12.9):} \quad I_B = \frac{18\text{ V} - 1.6\text{ V}}{3.3\text{ M}\Omega + 8000(390\ \Omega)} \approx \mathbf{2.56\ \mu A}$$

The emitter current is then

$$\text{Eq. (12.10):} \quad I_E \approx 8000(2.56\ \mu A) = \mathbf{20.48\ mA} \approx I_C$$

The emitter dc voltage is

$$\text{Eq. (12.11):} \quad V_E = 20.48\text{ mA}(390\ \Omega) \approx \mathbf{8\ V}$$

and the base voltage is

$$\text{Eq. (12.12):} \quad V_B = 8\text{ V} + 1.6\text{ V} = \mathbf{9.6\ V}$$

The collector voltage is the supply value of

$$V_C = \mathbf{18\ V}$$

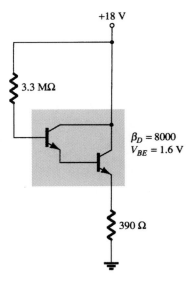

Figure 12.11 Circuit for Example 12.6.

AC Equivalent Circuit

A Darlington emitter-follower circuit is shown in Fig. 12.12. The ac input signal is applied to the base of the Darlington transistor through capacitor C_1, with the ac output, V_o, obtained from the emitter through capacitor C_2. An ac equivalent circuit is drawn in Fig. 12.13. The Darlington transistor is replaced by an ac equivalent circuit comprised of an input resistance, r_i, and an output current source, $\beta_D I_b$.

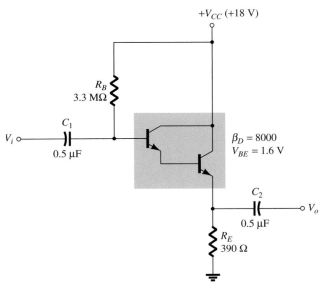

Figure 12.12 Darlington emitter-follower circuit.

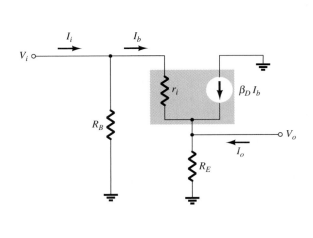

Figure 12.13 AC equivalent circuit of Darlington emitter-follower.

AC INPUT IMPEDANCE

The ac base current through r_i is

$$I_b = \frac{V_i - V_o}{r_i} \qquad (12.13)$$

Since

$$V_o = (I_b + \beta_D I_b)R_E \qquad (12.14)$$

we can use Eq. (12.13) in Eq. (12.14) to obtain

$$I_b r_i = V_i - V_o = V_i - I_b(1 + \beta_D)R_E$$

Solving for V_i,

$$V_i = I_b[r_i + (1 + \beta_D)R_E] \approx I_b(r_i + \beta_D R_E)$$

The ac input impedance looking into the transistor base is then

$$\frac{V_i}{I_b} = r_i + \beta_D R_E$$

and that looking into the circuit is

$$Z_i = R_B \| (r_i + \beta_D R_E) \qquad (12.15)$$

EXAMPLE 12.7

Calculate the input impedance of the circuit of Fig. 12.12 if $r_i = 5\ \text{k}\Omega$.

Solution

Eq. (12.15): $Z_i = 3.3\ \text{M}\Omega \| [5\ \text{k}\Omega + (8000)(390\ \Omega)] = \mathbf{1.6\ M\Omega}$

AC CURRENT GAIN

The ac output current through R_E is (see Fig. 12.13)

$$I_o = I_b + \beta_D I_b = (\beta_D + 1)I_b \approx \beta_D I_b$$

The transistor current gain is then

$$\frac{I_o}{I_b} = \beta_D$$

The ac current gain of the circuit is

$$A_i = \frac{I_o}{I_i} = \frac{I_o}{I_b}\frac{I_b}{I_i}$$

We can use the current-divider rule to express I_b/I_i:

$$I_b = \frac{R_B}{(r_i + \beta_D R_E) + R_B} I_i \approx \frac{R_B}{R_B + \beta_D R_E} I_i$$

so that the ac circuit current gain is

$$A_i = \beta_D \frac{R_B}{R_B + \beta_D R_E} = \frac{\beta_D R_B}{R_B + \beta_D R_E} \qquad (12.16)$$

EXAMPLE 12.8

Calculate the ac current gain of the circuit in Fig. 12.12.

Solution

Eq. (12.16): $A_i = \dfrac{\beta_D R_B}{R_B + \beta_D R_E} = \dfrac{(8000)(3.3 \text{ M}\Omega)}{3.3 \text{ M}\Omega + (8000)(390 \ \Omega)} = \mathbf{4112}$

AC OUTPUT IMPEDANCE

The ac output impedance can be determined for the ac circuit shown in Fig. 12.14a. The output impedance seen by load R_L is determined by applying a voltage V_o and measuring the current I_o (with input V_s set to zero). Figure 12.14b shows this situation. Solving for I_o yields

$$I_o = \frac{V_o}{R_E} + \frac{V_o}{r_i} - \beta_D I_b = \frac{V_o}{R_E} + \frac{V_o}{r_i} - \beta_D\left(\frac{V_o}{r_i}\right)$$

$$= \left(\frac{1}{R_E} + \frac{1}{r_i} + \frac{\beta_D}{r_i}\right)V_o$$

Solving for Z_o gives

$$Z_o = \frac{V_o}{I_o} = \frac{1}{1/R_E + 1/r_i + \beta_D/r_i}$$

$$= R_E \| r_i \| \frac{r_i}{\beta_D} \approx \frac{r_i}{\beta_D} \qquad (12.17)$$

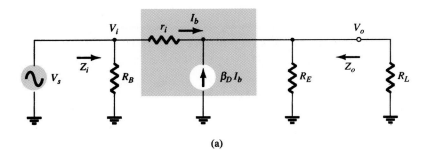

(a)

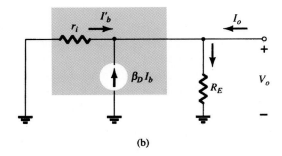

(b)

Figure 12.14   AC equivalent circuit to determine Z_o.

EXAMPLE 12.9

Calculate the output impedance of the circuit in Fig. 12.12.

Solution

$$\text{Eq. (12.17):} \quad Z_o = 390 \ \Omega \| 5 \ \text{k}\Omega \| \frac{5 \ \text{k}\Omega}{8000} \approx \frac{5 \ \text{k}\Omega}{8000} = \mathbf{0.625 \ \Omega}$$

AC VOLTAGE GAIN

The ac voltage gain for the circuit of Fig. 12.12 can be determined using the ac equivalent circuit of Fig. 12.15. Since

$$V_o = (I_b + \beta_D I_b)R_E = I_b(R_E + \beta_D R_E)$$

and

$$V_i = I_b r_i + (I_b + \beta_D I_b)R_E$$

from which we obtain

$$V_i = I_b(r_i + R_E + \beta_D R_E)$$

so that

$$V_o = \frac{V_i}{r_i + (R_E + \beta_D R_E)}(R_E + \beta_D R_E)$$

$$A_v = \frac{V_o}{V_i} = \frac{R_E + \beta_D R_E}{r_i + (R_E + \beta_D R_E)} \approx 1 \qquad (12.18)$$

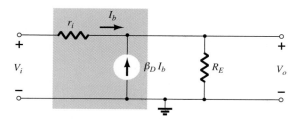

Figure 12.15 AC equivalent circuit to determine A_v.

EXAMPLE 12.10

Calculate the ac voltage gain A_v for the circuit of Fig. 12.12.

Solution

$$A_v = \frac{390 \ \Omega + (8000)(390 \ \Omega)}{5 \ \text{k}\Omega + [390 \ \Omega + (8000)(390 \ \Omega)]} = \mathbf{0.998}$$

12.5 FEEDBACK PAIR

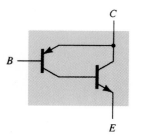

Figure 12.16 Feedback pair connection.

The feedback pair connection (see Fig. 12.16) is a two-transistor circuit that operates like the Darlington circuit. Notice that the feedback pair uses a *pnp* transistor driving an *npn* transistor, the two devices acting effectively much like one *pnp* transistor. As with a Darlington connection, the feedback pair provides very high current gain (the product of the transistor current gains). A typical application (see Chapter 15) uses a Darlington connection and a feedback pair connection to provide complementary transistor operation. A practical circuit using a feedback pair is provided in Fig. 12.17. Some consideration of the dc bias and ac operation will provide better understanding of how the connection works.

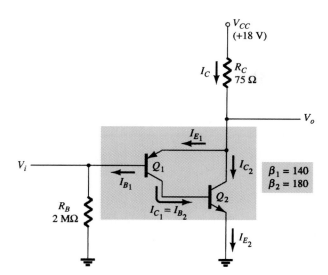

Figure 12.17 Operation of feedback pair.

DC Bias

The dc bias calculations that follow use practical simplifications wherever possible to provide simpler results. From the Q_1 base-emitter loop, one obtains

$$V_{CC} - I_C R_C - V_{EB_1} - I_{B_1} R_B = 0$$

$$V_{CC} - \beta_1 \beta_2 I_{B_1} R_C - V_{EB_1} - I_{B_1} R_B = 0$$

The base current is then

$$I_{B_1} = \frac{V_{CC} - V_{EB_1}}{R_B + \beta_1 \beta_2 R_C} \tag{12.19}$$

The collector current of Q_1 is

$$I_{C_1} = \beta_1 I_{B_1} = I_{B_2}$$

which is also the base Q_2 current. The transistor Q_2 collector current is

$$I_{C_2} = \beta_2 I_{B_2} \approx I_{E_2}$$

so that the current through R_C is

$$I_C = I_{E_1} + I_{C_2} \approx I_{C_1} + I_{C_2} \approx I_{C_2} \tag{12.20}$$

Calculate the dc bias currents and voltages for the circuit of Fig. 12.17 to provide V_o at one-half the supply voltage ($I_C R_C = 9$ V).

EXAMPLE 12.11

Solution

$$I_{B_1} = \frac{18\text{ V} - 0.7\text{ V}}{2\text{ M}\Omega + (140)(180)(75\ \Omega)} = \frac{17.3\text{ V}}{3.89 \times 10^6} = 4.45\ \mu\text{A}$$

The base Q_2 current is then

$$I_{B_2} = I_{C_1} = \beta_1 I_{B_1} = 140(4.45\ \mu\text{A}) = 0.623\text{ mA}$$

resulting in a Q_2 collector current of

$$I_{C_2} = \beta_2 I_{B_2} = 180(0.623\text{ mA}) = 112.1\text{ mA}$$

and the current through R_C is then

Eq. (12.20): $I_C = I_{E_1} + I_{C_2} = 0.623 \text{ mA} + 112.1 \text{ mA} \approx I_{C_2} = 112.1 \text{ mA}$

The dc voltage at the output is thus

$$V_o(\text{dc}) = V_{CC} - I_C R_C = 18 \text{ V} - 112.1 \text{ mA}(75 \text{ }\Omega) = 9.6 \text{ V}$$

and

$$V_i(\text{dc}) = V_o(\text{dc}) - V_{BE} = 9.6 \text{ V} - 0.7 \text{ V} = 8.9 \text{ V}$$

AC Operation

The ac equivalent circuit for that of Fig. 12.17 is drawn in Fig. 12.18. The circuit is first drawn in Fig. 12.18a to show clearly each transistor and the base and collector resistor placement. The ac equivalent circuit is then redrawn in Fig. 12.18b to permit analysis.

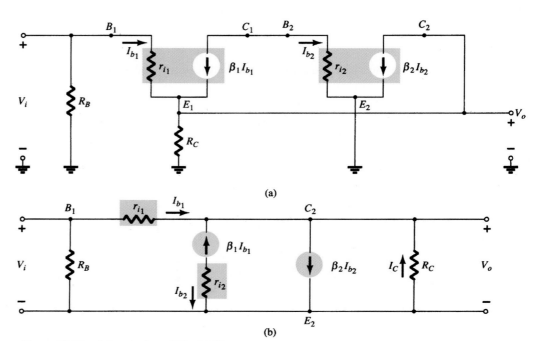

Figure 12.18   AC equivalent of Fig. 12.17.

AC INPUT IMPEDANCE, Z_i

The ac input impedance seen looking into the base of transistor Q_1 is determined (refer to Fig. 12.18b) as follows:

$$I_{b_1} = \frac{V_i - V_o}{r_{i_1}}$$

where

$$V_o = -I_C R_C \approx (-\beta_1 I_{b_1} + \beta_2 I_{b_2})R_C \approx (\beta_2 I_{b_2})R_C$$

so that

$$I_{b_1} r_{i_1} = V_i - V_o \approx V_i - \beta_2 I_{b_2} R_C$$

$$I_{b_1} r_{i_1} + \beta_2(\beta_1 I_{b_1})R_C = V_i \quad (\text{since } I_{b_2} = I_{C_1} = \beta_1 I_{b_1})$$

$$\frac{V_i}{I_{b_1}} = r_{i_1} + \beta_1 \beta_2 R_C$$

Including the base-bias resistance,

$$Z_i \approx R_B \| (r_{i_1} + \beta_1 \beta_2 R_C) \qquad (12.21)$$

AC CURRENT GAIN, A_i

The ac current gain can be determined as follows:

$$I_o = \beta_2 I_{b_2} - \beta_1 I_{b_1} - I_{b_1}$$
$$= \beta_2(\beta_1 I_{b_1}) - (1 + \beta_1)I_{b_1} \approx \beta_1 \beta_2 I_{b_1}$$

$$\frac{I_o}{I_{b_1}} = \beta_1 \beta_2$$

Including R_B, the current gain is

$$A_i = \frac{I_o}{I_i} = \frac{I_o}{I_{b_1}} \frac{I_{b_1}}{I_i} = \beta_1 \beta_2 \frac{R_B}{R_B + Z_i} \qquad (12.22)$$

AC OUTPUT IMPEDANCE, Z_o

Z_o can be obtained by applying a voltage, V_o, with V_i set to 0. The resulting analysis provides that

$$Z_o = \frac{V_o}{I_o} = R_C \| r_{i_1} \| \frac{r_{i_1}}{\beta_1} \| \frac{r_{i_1}}{\beta_1 \beta_2} \cong \frac{r_{i_1}}{\beta_1 \beta_2} \qquad (12.23)$$

which results in a low output impedance.

AC VOLTAGE GAIN, A_v

The output voltage V_o is

$$V_o = -I_C R_C \approx \beta_1 \beta_2 I_{b_1} R_C$$

Since

$$I_{b_1} = \frac{V_i - V_o}{r_{i_1}}$$

$$V_o = V_i - I_{b_1} r_{i_1} = V_i - \frac{V_o}{\beta_1 \beta_2 R_C} r_{i_1}$$

$$A_v = \frac{V_o}{V_i} = \frac{1}{1 + r_{i_1}/(\beta_1 \beta_2 R_C)} = \frac{\beta_1 \beta_2 R_C}{\beta_1 \beta_2 R_C + r_{i_1}} \qquad (12.24)$$

Calculate the ac circuit values of Z_i, Z_o, A_i, and A_v for the circuit of Fig. 12.17. Assume that $r_{i_1} = 3 \text{ k}\Omega$.

EXAMPLE 12.12

Solution

$$Z_i \approx R_B \| (r_{i_1} + \beta_1 \beta_2 R_C) = 2 \text{ M}\Omega \| [3 \text{ k}\Omega + (140)(180)(75 \text{ }\Omega)]$$
$$\approx \mathbf{974 \text{ k}\Omega}$$

$$A_i = \beta_1 \beta_2 \frac{R_B}{R_B + Z_i} = (140)(180)\left(\frac{2 \text{ M}\Omega}{2 \text{ M}\Omega + 974 \text{ k}\Omega}\right)$$
$$= \mathbf{16.95 \times 10^3}$$

$$Z_o \approx \frac{r_{i_1}}{\beta_1\beta_2} = \frac{3 \times 10^3}{(140)(180)} = \mathbf{0.12\ \Omega}$$

and
$$A_v = \frac{\beta_1\beta_2 R_C}{\beta_1\beta_2 R_C + r_{i_1}} = \frac{(140)(180)(75\ \Omega)}{(140)(180)(75\ \Omega) + 3000\ \Omega}$$
$$= \mathbf{0.9984} \approx \mathbf{1}$$

Example 12.12 shows that the feedback pair connection provides operation with voltage gain very near 1 (just as with a Darlington emitter follower), a very high current gain, a very low output impedance, and a high input impedance.

12.6 CMOS CIRCUIT

A form of circuit popular in digital circuitry uses both *n*-channel and *p*-channel enhancement MOSFET transistors (see Fig. 12.19). This complementary MOSFET or CMOS circuit uses these opposite (or complementary)-type transistors. The input, V_i, is applied to both gates with the output taken from the connected drains. Before going into the operation of the CMOS circuit, let's review the operation of the enhancement MOSFET transistors.

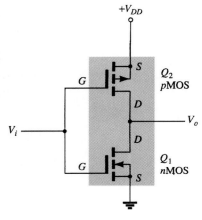

Figure 12.19 CMOS inverter circuit.

*n*MOS On/Off Operation

The drain characteristic of an *n*-channel enhancement MOSFET or *n*MOS transistor is shown in Fig. 12.20a. With 0 V applied to the gate–source, there is no drain current. Not until V_{GS} is raised past the device threshold level, V_T, does any current result. With an input of, say, $+5$ V, the *n*MOS device is fully on with current I_D present. In summary:

An input of 0 V leaves the nMOS off, while an input of $+5$ V turns the nMOS on.

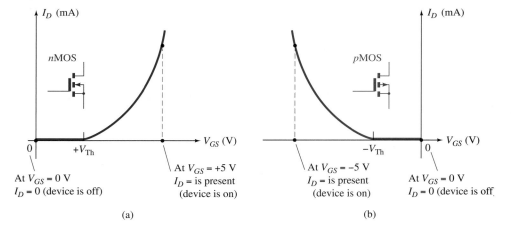

Figure 12.20 Enhancement MOSFET characteristic showing off and on conditions: (a) *n*MOS; (b) *p*MOS.

*p*MOS On/Off Operation

The drain characteristic for a *p*-channel MOSFET or *p*MOS transistor is shown in Fig. 12.20b. When 0 V is applied, the device is off (no drain current present), while for an input of -5 V (greater than the threshold voltage), the device is on with drain current present. In summary:

$V_{GS} = 0$ V leaves pMOS off; $V_{GS} = -5$ V turns pMOS on.

Operation of CMOS Circuit

Consider next how the actual CMOS circuit of Fig. 12.21 operates for input of 0 or +5 V.

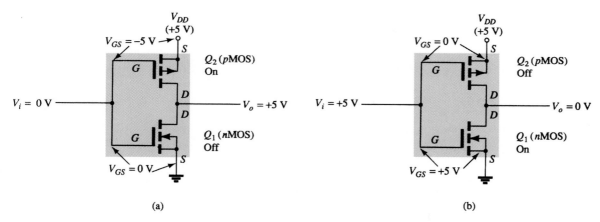

Figure 12.21   Operation of CMOS circuit: (a) output +5 V; (b) output 0 V.

0-V INPUT

When 0 V is applied as input to the CMOS circuit, it provides 0 V to both nMOS and pMOS gates. Figure 12.21a shows that

For nMOS (Q_1): $V_{GS} = V_i - 0\text{ V} = 0\text{ V} - 0\text{ V} = 0\text{ V}$

For pMOS (Q_2): $V_{GS} = V_i - (+5\text{ V}) = 0\text{ V} - 5\text{ V} = -5\text{ V}$

Input of 0 V to an nMOS transistor Q_1 leaves that device off. The same 0-V input, however, results in the gate–source voltage of pMOS transistor Q_2 being -5 V (gate at 0 V is 5 V less than source at $+5$ V), resulting in that device turning on. The output, V_o, is then $+5$ V.

+5-V INPUT

When $V_i = +5$ V, it provides $+5$ V to both gates. Figure 12.21b shows that

For nMOS (Q_1): $V_{GS} = V_i - 0\text{ V} = +5\text{ V} - 0\text{ V} = +5\text{ V}$

For pMOS (Q_2): $V_{GS} = V_i - (+5\text{ V}) = +5\text{ V} - 5\text{ V} = 0\text{ V}$

This input results in transistor Q_1 being turned on and transistor Q_2 remaining off, the output then near 0 V, through conducting transistor Q_2. The CMOS connection of Fig. 12.19 provides operation as a logic inverter with V_o the opposite of V_i, as shown in Table 12.1.

TABLE 12.1	Operation of CMOS Circuit		
$V_i(V)$	Q_1	Q_2	$V_o(V)$
0	Off	On	+5
+5	On	Off	0

12.7 CURRENT SOURCE CIRCUITS

The concept of a power supply provides a start in our consideration of current source circuits. A practical voltage source (see Fig. 12.22a) is a voltage supply in series with a resistance. An ideal voltage source has $R = 0$, while a practical source includes some small resistance. A practical current source (see Fig. 12.22b) is a current supply in parallel with a resistance. An ideal current source has $R = \infty$, while a practical current source includes some very large resistance.

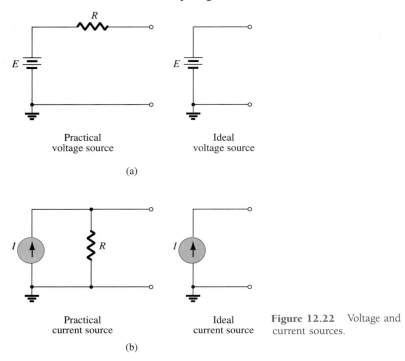

Practical
voltage source

Ideal
voltage source

(a)

Practical
current source

Ideal
current source

(b)

Figure 12.22 Voltage and current sources.

An ideal current source provides a constant current regardless of the load connected to it. There are many uses in electronics for a circuit providing a constant current at a very high impedance. Constant-current circuits can be built using FET devices, bipolar devices, and a combination of these components. There are circuits used in discrete form and others more suitable for operation in integrated circuits. We consider some forms of both types in this section and in Section 12.8.

JFET Current Source

A simple JFET current source is that of Fig. 12.23. With V_{GS} set to 0 V, the drain current is fixed at

$$I_D = I_{DSS} = 10 \text{ mA}$$

The device therefore operates like a current source of value 10 mA. While the actual JFET does have an output resistance, the ideal current source would be a 10-mA supply, as shown in Fig. 12.23.

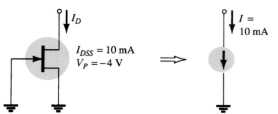

Figure 12.23 JFET constant-current source.

Determine the load current I_D and output voltage V_o for the circuit of Fig. 12.24 for:
(a) $R_D = 1.2$ kΩ.
(b) $R_D = 3.3$ kΩ.

EXAMPLE 12.13

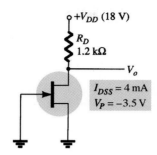

Figure 12.24 JFET current source for Example 12.13.

Solution

Since $V_{GS} = 0$ V, $I_D = I_{DSS} =$ **4 mA.**

(a) $V_o = V_{DD} - I_D R_D = 18$ V $- (4$ mA$)(1.2$ kΩ$) =$ **13.2 V**

(b) $V_o = V_{DD} - I_D R_D = 18$ V $- (4$ mA$)(3.3$ kΩ$) =$ **4.8 V**

Notice that the output voltage changes with R_D, but the current through R_D remains 4 mA since the JFET operates as a constant-current source.

Bipolar Transistor Constant-Current Source

Bipolar transistors can be connected in a circuit which acts as a constant-current source in a number of ways. Figure 12.25 shows a circuit using a few resistors and an *npn* transistor for operation as a constant-current circuit. The current through I_E can be determined as follows. Assuming that the base input impedance is much larger than R_1 or R_2,

$$V_B = \frac{R_1}{R_1 + R_2}(-V_{EE})$$

and

$$V_E = V_B - 0.7 \text{ V}$$

with

$$I_E = \frac{V_E - (-V_{EE})}{R_E} \approx I_C \qquad (12.25)$$

where I_C is the constant current provided by the circuit of Fig. 12.25.

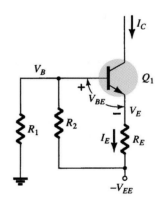

Figure 12.25 Discrete constant-current source.

Calculate the constant current I in the circuit of Fig. 12.26.

EXAMPLE 12.14

Solution

$$V_B = \frac{R_1}{R_1 + R_2}(-V_{EE}) = \frac{5.1 \text{ k}\Omega}{5.1 \text{ k}\Omega + 5.1 \text{ k}\Omega}(-20 \text{ V}) = -10 \text{ V}$$

$$V_E = V_B - 0.7 \text{ V} = -10 \text{ V} - 0.7 \text{ V} = -10.7 \text{ V}$$

$$I = I_E = \frac{V_E - (-V_{EE})}{R_E} = \frac{-10.7 \text{ V} - (-20 \text{ V})}{2 \text{ k}\Omega}$$

$$= \frac{9.3 \text{ V}}{2 \text{ k}\Omega} = \textbf{4.65 mA}$$

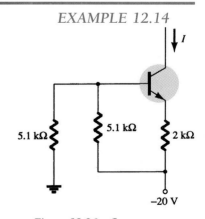

Figure 12.26 Constant-current source for Example 12.14.

Transistor/Zener Constant-Current Source

Replacing resistor R_2 with a Zener diode, as shown in Fig. 12.27, provides an improved constant-current source over that of Fig. 12.25. The Zener diode results in a constant current calculated using the base-emitter KVL (Kirchoff Voltage Loop) equation. The value of I can be calculated using

$$I \approx I_E = \frac{V_Z - V_{BE}}{R_E} \tag{12.26}$$

A major point to consider is that the constant current depends on the Zener diode voltage, which remains quite constant and the emitter resistor R_E. The voltage supply V_{EE} has no effect on the value of I.

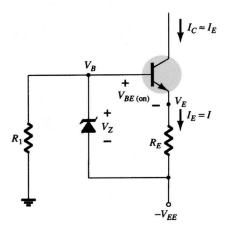

Figure 12.27 Constant-current circuit using Zener diode.

EXAMPLE 12.15

Calculate the constant current I in the circuit of Fig. 12.28.

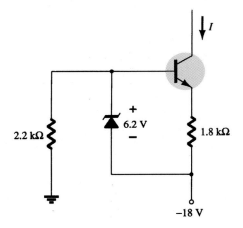

Figure 12.28 Constant-current circuit for Example 12.15.

Solution

$$\text{Eq. (12.26):} \quad I = \frac{V_Z - V_{BE}}{R_E} = \frac{6.2\text{ V} - 0.7\text{ V}}{1.8\text{ k}\Omega} = 3.06\text{ mA} \approx \mathbf{3\text{ mA}}$$

12.8 CURRENT MIRROR CIRCUITS

A current mirror circuit (see Fig. 12.29) provides a constant current and is used primarily in integrated circuits. The constant current is obtained from an output current, which is the reflection or mirror of a constant current developed on one side of the

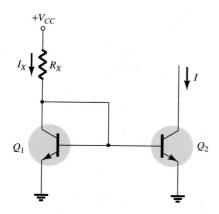

Figure 12.29 Current mirror circuit.

circuit. The circuit is particularly suited to IC manufacture since the circuit requires that the transistors used have identical base-emitter voltage drops and identical values of beta—results best achieved when transistors are formed at the same time in IC manufacture. In Fig. 12.29, the current I_X, set by transistor Q_1 and resistor R_X, is mirrored in the current I through transistor Q_2.

The currents I_X and I can be obtained using the circuit currents listed in Fig. 12.30. We assume that the emitter current (I_E) for both transistors is the same (Q_1 and Q_2 being fabricated near each other on the same chip). The two transistor base currents are then approximately

$$I_B = \frac{I_E}{\beta + 1} \approx \frac{I_E}{\beta}$$

The collector current of each transistor is then

$$I_C \approx I_E$$

Finally, the current through resistor R_X, I_X, is

$$I_X = I_E + \frac{2I_E}{\beta} = \frac{\beta I_E}{\beta} + \frac{2I_E}{\beta} = \frac{\beta + 2}{\beta} I_E \approx I_E$$

In summary, the constant current provided at the collector of Q_2 mirrors that of Q_1. Since

$$I_X = \frac{V_{CC} - V_{BE}}{R_X} \qquad (12.27)$$

the current I_X set by V_{CC} and R_X is mirrored in the current into the collector of Q_2.

Transistor Q_1 is referred to as a diode-connected transistor because the base and collector are shorted together.

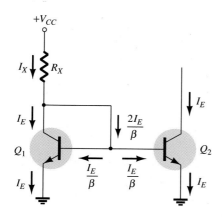

Figure 12.30 Circuit currents for current-mirror circuit.

12.8 Current Mirror Circuits

EXAMPLE 12.16   Calculate the mirrored current, I, in the circuit of Fig. 12.31.

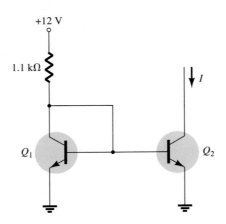

Figure 12.31   Current mirror circuit for Example 12.16.

Solution

Eq. (12.27): $I = I_X = \dfrac{V_{CC} - V_{BE}}{R_X} = \dfrac{12\text{ V} - 0.7\text{ V}}{1.1\text{ k}\Omega} = \mathbf{10.27\text{ mA}}$

EXAMPLE 12.17   Calculate the current, I, through each of the transistors Q_2 and Q_3 in the circuit of Fig. 12.32.

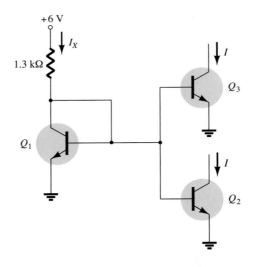

Figure 12.32   Current mirror circuit for Example 12.17.

Solution

The current I_X is

$$I_X = I_E + \frac{3I_E}{\beta} = \frac{\beta + 3}{\beta} I_E \approx I_E$$

Therefore,

$$I \approx I_X = \frac{V_{CC} - V_{BE}}{R_X} = \frac{6\text{ V} - 0.7\text{ V}}{1.3\text{ k}\Omega} = \mathbf{4.08\text{ mA}}$$

Figure 12.33 shows another form of current mirror to provide higher output impedance than that of Fig. 12.29. The current through R_X is

$$I_X = \frac{V_{CC} - 2V_{BE}}{R_X} \approx I_E + \frac{I_E}{\beta} = \frac{\beta + 1}{\beta} I_E \approx I_E$$

Assuming that Q_1 and Q_2 are well matched, the output current, I, is held constant at

$$I \approx I_E = I_X$$

Again we see that the output current I is a mirrored value of the current set by the fixed current through R_X.

Figure 12.34 shows still another form of current mirror. The JFET provides a constant current set at the value of I_{DSS}. This current is mirrored, resulting in a current through Q_2 of the same value:

$$I = I_{DSS}$$

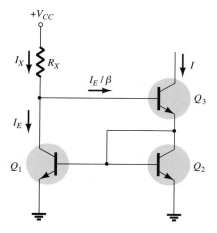

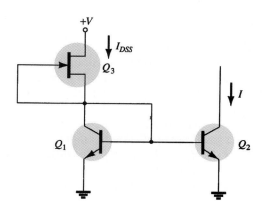

Figure 12.33 Current mirror circuit with higher output impedance.

Figure 12.34 Current mirror connection.

12.9 DIFFERENTIAL AMPLIFIER CIRCUIT

The differential amplifier circuit is an extremely popular connection used in IC units. This connection can be described by considering the basic differential amplifier shown in Fig. 12.35. Notice that the circuit has two separate inputs, two separate outputs, and that the emitters are connected together. While most differential amplifier circuits use two separate voltage supplies, the circuit can also operate using a single supply.

A number of input signal combinations are possible:

If an input signal is applied to either input with the other input connected to ground, the operation is referred to as "single-ended."

If two opposite polarity input signals are applied, the operation is referred to as "double-ended."

If the same input is applied to both inputs, the operation is called "common-mode."

In single-ended operation, a single input signal is applied. However, due to the common-emitter connection, the input signal operates both transistors, resulting in output from *both* collectors.

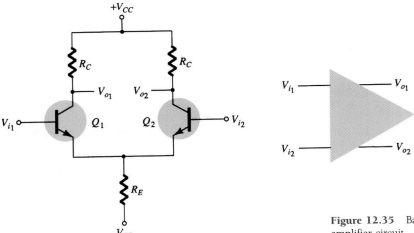

Figure 12.35 Basic differential amplifier circuit.

In double-ended operation, two input signals are applied, the difference of the inputs resulting in outputs from both collectors due to the difference of the signals applied to both inputs.

In common-mode operation, the common input signal results in opposite signals at each collector, these signals canceling so that the resulting output signal is zero. As a practical matter, the opposite signals do not completely cancel and a small signal results.

The main feature of the differential amplifier is the very large gain when opposite signals are applied to the inputs as compared to the very small gain resulting from common inputs. The ratio of this difference gain to the common gain is called *common-mode rejection*. These concepts are discussed fully in Chapter 13. For the present, the operation of the differential amplifier circuit will be fully covered.

DC Bias

Let's first consider the dc bias operation of the circuit of Fig. 12.35. With ac inputs obtained from voltage sources, the dc voltage at each input is essentially connected to 0 V, as shown in Fig. 12.36. With each base voltage at 0 V, the common-emitter dc bias voltage is

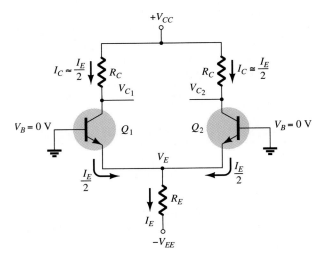

Figure 12.36 DC bias of differential amplifier circuit.

$$V_E = 0 \text{ V} - V_{BE} = -0.7 \text{ V}$$

The emitter dc bias current is then

$$I_E = \frac{V_E - (-V_{EE})}{R_E} \approx \frac{V_{EE} - 0.7 \text{ V}}{R_E} \tag{12.28}$$

Assuming that the transistors are well matched (as would occur in an IC unit),

$$I_{C_1} = I_{C_2} = \frac{I_E}{2} \tag{12.29}$$

resulting in a collector voltage of

$$V_{C_1} = V_{C_2} = V_{CC} - I_C R_C = V_{CC} - \frac{I_E}{2} R_C \tag{12.30}$$

Calculate the dc voltages and currents in the circuit of Fig. 12.37.

EXAMPLE 12.18

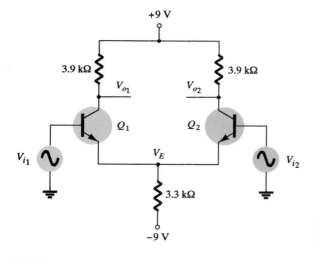

+9 V

3.9 kΩ V_{o_1} V_{o_2} 3.9 kΩ

Q_1 Q_2

V_{i_1} V_E V_{i_2}

3.3 kΩ

−9 V

Figure 12.37 Differential amplifier circuit for Example 12.18.

Solution

$$\text{Eq. (12.28):} \quad I_E = \frac{V_{EE} - 0.7 \text{ V}}{R_E} = \frac{9 \text{ V} - 0.7 \text{ V}}{3.3 \text{ k}\Omega} \approx \textbf{2.5 mA}$$

The collector current is then

$$\text{Eq. (12.29):} \quad I_C = \frac{I_E}{2} = \frac{2.5 \text{ mA}}{2} = \textbf{1.25 mA}$$

resulting in a collector voltage of

$$\text{Eq. (12.30):} \quad V_C = V_{CC} - I_C R_C = 9 \text{ V} - (1.25 \text{ mA})(3.9 \text{ k}\Omega) \approx \textbf{4.1 V}$$

The common-emitter voltage is thus −0.7 V, while the collector bias voltage is near 4.1 V for both outputs.

AC Operation of Circuit

An ac connection of a differential amplifier is shown in Fig. 12.38. Separate input signals are applied as V_{i_1} and V_{i_2}, with separate outputs resulting as V_{o_1} and V_{o_2}. To carry out ac analysis, the circuit is redrawn in Fig. 12.39. Each transistor is replaced by its ac equivalent.

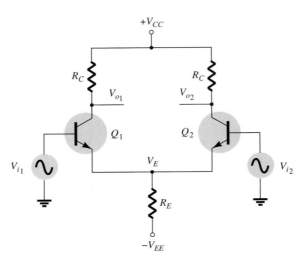

Figure 12.38 AC connection of differential amplifier.

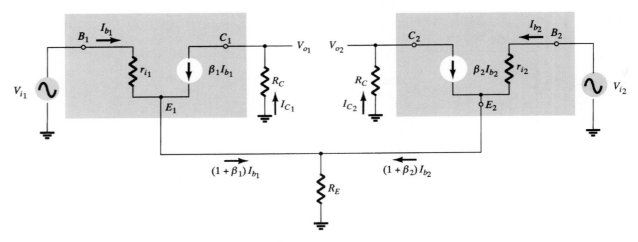

Figure 12.39 AC equivalent of differential amplifier circuit.

SINGLE-ENDED AC VOLTAGE GAIN

To calculate the single-ended ac voltage gain, V_o/V_i, apply signal to one input with the other connected to ground, as shown in Fig. 12.40. The ac equivalent of this connection is drawn in Fig. 12.41. The ac base current can be calculated using the base

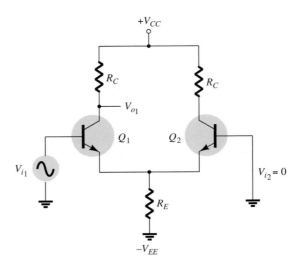

Figure 12.40 Connection to calculate $A_{v_1} = V_{o_1}/V_{i_1}$.

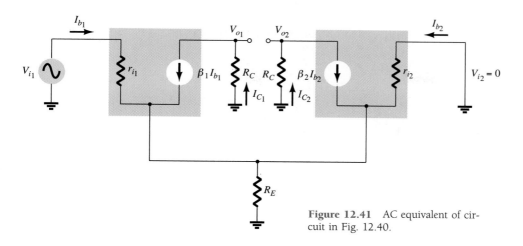

Figure 12.41 AC equivalent of circuit in Fig. 12.40.

1 input KVL equation. If one assumes that the two transistors are well matched, then

$$I_{b_1} = I_{b_2} = I_b$$

$$r_{i_1} = r_{i_2} = r_i$$

With R_E very large (ideally infinite), the circuit for obtaining the KVL equation simplifies to that of Fig. 12.42, from which we can write

$$V_{i_1} - I_b r_i - I_b r_i = 0$$

so that

$$I_b = \frac{V_{i_1}}{2r_i}$$

If we also assume that

$$\beta_1 = \beta_2 = \beta$$

then

$$I_C = \beta I_b = \beta \frac{\beta V_{i_1}}{2r_i}$$

and the output voltage magnitude at either collector is

$$V_o = I_C R_C = \beta \frac{V_{i_1}}{2r_1} R_C = \frac{\beta R_C}{2\beta r_e} V_i$$

for which the single-ended voltage gain magnitude at either collector is

$$A_v = \frac{V_o}{V_{i_1}} = \frac{R_C}{2r_e} \tag{12.31}$$

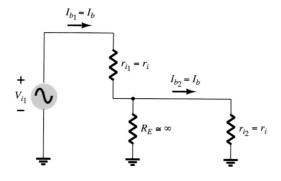

Figure 12.42 Partial circuit to calculate I_b.

12.9 Differential Amplifier Circuit

EXAMPLE 12.19

Calculate the single-ended output voltage, V_{o_1}, for the circuit of Fig. 12.43.

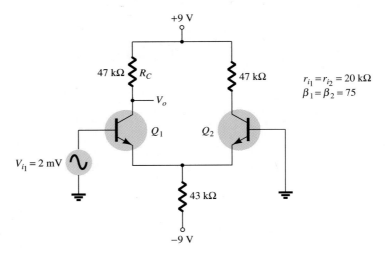

Figure 12.43 Circuit for Examples 12.19 and 12.20.

Solution

The dc bias calculations provide

$$I_E = \frac{V_{EE} - 0.7 \text{ V}}{R_E} = \frac{9 \text{ V} - 0.7 \text{ V}}{43 \text{ k}\Omega} = 193 \ \mu\text{A}$$

The collector dc current is then

$$I_C = \frac{I_E}{2} = 96.5 \ \mu\text{A}$$

so that
$$V_C = V_{CC} - I_C R_C = 9 \text{ V} - (96.5 \ \mu\text{A})(47 \text{ k}\Omega) = 4.5 \text{ V}$$

The value of r_e is

$$r_e = \frac{26}{0.0965} \cong 269 \ \Omega$$

The ac voltage gain magnitude can be calculated using Eq. (12.31):

$$A_v = \frac{R_C}{2r_e} = \frac{(47 \text{ k}\Omega)}{2(269 \ \Omega)} = 87.4$$

providing an output ac voltage of magnitude
$$V_o = A_v V_i = (87.4)(2 \text{ mV}) = 174.8 \text{ mV} = \mathbf{0.175 \ V}$$

DOUBLE-ENDED AC VOLTAGE GAIN

A similar analysis could also be used to show that for the condition of signals applied to both inputs, the differential voltage gain magnitude would be

$$A_d = \frac{V_o}{V_d} = \frac{\beta R_C}{2r_i} \tag{12.32}$$

where $V_d = V_{i_1} - V_{i_2}$.

Common-Mode Operation of Circuit

While a differential amplifier provides large amplification of the difference signal applied to both inputs, it should also provide as small an amplification of the signal common to both inputs. An ac connection showing common input to both transistors is shown in Fig. 12.44. The ac equivalent circuit is then drawn in Fig. 12.45, from which we can write

$$I_b = \frac{V_i - 2(\beta + 1)I_b R_E}{r_i}$$

which can be rewritten as

$$I_b = \frac{V_i}{r_i + 2(\beta + 1)R_E}$$

The output voltage magnitude is then

$$V_o = I_C R_C = \beta I_b R_C = \frac{\beta V_i R_C}{r_i + 2(\beta + 1)R_E}$$

providing a voltage gain magnitude of

$$A_c = \frac{V_o}{V_i} = \frac{\beta R_C}{r_i + 2(\beta + 1)R_E} \tag{12.33}$$

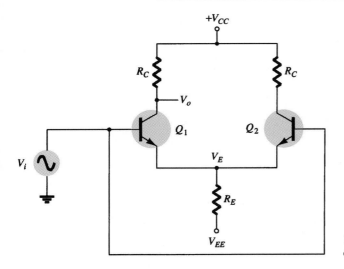

Figure 12.44   Common-mode connection.

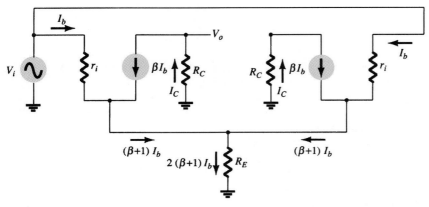

Figure 12.45   AC circuit in common-mode connection.

EXAMPLE 12.20

Calculate the common-mode gain for the amplifier circuit of Fig. 12.43.

Solution

$$\text{Eq. (12.33):} \quad A_c = \frac{V_o}{V_i} = \frac{\beta R_C}{r_i + 2(\beta + 1)R_E} = \frac{75(47 \text{ k}\Omega)}{20 \text{ k}\Omega + 2(76)(43 \text{ k}\Omega)} = \mathbf{0.54}$$

Use of Constant-Current Source

A good differential amplifier has a very large difference gain, A_d, which is much larger than the common-mode gain, A_c. The common-mode rejection ability of the circuit can be considerably improved by making the common-mode gain as small as possible (ideally, 0). From Eq. (12.33), we see that the larger R_E, the smaller A_c. One popular method for increasing the ac value of R_E is using a constant-current source circuit. Figure 12.46 shows a differential amplifier with constant-current source to provide a large value of resistance from common emitter to ac ground. The major improvement of this circuit over that in Fig. 12.35 is the much larger ac impedance for R_E obtained using the constant-current source. Figure 12.47 shows the ac equivalent circuit for the circuit of Fig. 12.46. A practical constant-current source is shown as a high impedance, in parallel with the constant current.

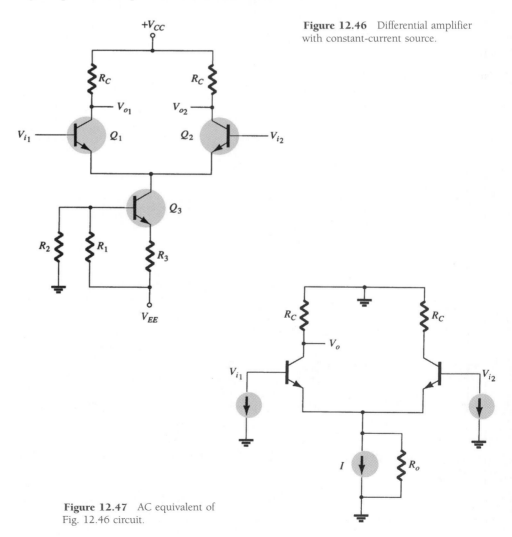

Figure 12.46 Differential amplifier with constant-current source.

Figure 12.47 AC equivalent of Fig. 12.46 circuit.

Calculate the common-mode gain for the differential amplifier of Fig. 12.48.

EXAMPLE 12.21

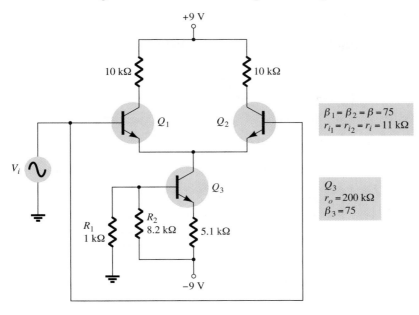

Figure 12.48 Circuit for Example 12.21.

Solution

Using $R_E = r_o = 200$ kΩ gives

$$A_c = \frac{\beta R_C}{r_i + 2(\beta + 1)R_E} = \frac{75(10 \text{ k}\Omega)}{11 \text{ k}\Omega + 2(76)(200 \text{ k}\Omega)} = \mathbf{24.7 \times 10^{-3}}$$

12.10 BIFET, BIMOS, AND CMOS DIFFERENTIAL AMPLIFIER CIRCUITS

While the preceding section provided an introduction to the differential amplifier using bipolar devices, units commercially available also use JFET and MOSFET transistors to build these types of circuits. An IC unit containing a differential amplifier built using both bipolar (Bi) and junction field-effect (FET) transistors is referred to as a *BiFET circuit*. An IC unit made using both bipolar (Bi) and MOSFET (MOS) transistors is called a *BiMOS circuit*. Finally, a circuit built using opposite type MOSFET transistors is a *CMOS circuit*.

The circuits used below to show the various multidevice circuits are mostly symbolic, since the actual circuits used in ICs are much more complex. Figure 12.49 shows a BiFET circuit with JFET transistors at the inputs and bipolar transistors to provide the current source (using a current mirror circuit). The current mirror ensures that each JFET is operated at the same bias current. For ac operation, the JFET provides a high input impedance (much higher than that provided using only bipolar transistors).

Figure 12.50 shows a circuit using MOSFET input transistors and bipolar transistors for the current sources, the BiMOS unit providing even higher input impedance than the BiFET due to the use of MOSFET transistors.

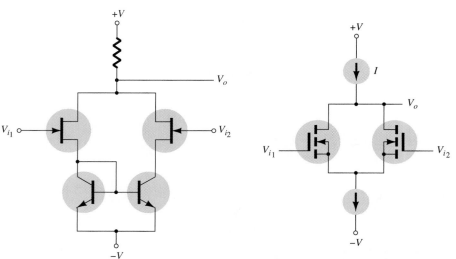

Figure 12.49 BiFET differential amplifier circuit.

Figure 12.50 BiMOS differential amplifier circuit.

Finally, a differential amplifier circuit can be built using complementary MOS-FET transistors as shown in Fig. 12.51. The pMOS transistors provide the opposite inputs, while the nMOS transistors operate as the constant-current source. A single output is taken from the common point between nMOS and pMOS transistors on one side of the circuit. This type of CMOS differential amplifier is particularly well suited for battery operation due to the low power dissipation of a CMOS circuit.

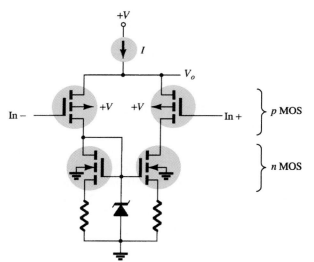

Figure 12.51 CMOS differential amplifier.

12.11 SUMMARY

Important Conclusions and Concepts

1. A cascade connection is a series connection.
2. For a cascade connection, amplification is the product of the stage gains.
3. A cascode connection provides a high input impedance and a low output impedance.
4. A Darlington connection provides two transistors connected as one "super" transistor.

5. A feedback pair, using an *npn* and a *pnp* transistor, provides an equivalent "super" transistor.

6. A CMOS circuit provides a very low power inverter or other digital logic circuit.

7. A current source provides a circuit element with a constant current (not dependent on the voltage across the element).

8. A current mirror circuit provides a constant-current source in an IC circuit.

9. A differential amplifier circuit provides a flexible, high-gain circuit, which is a basic stage of an op-amp circuit.

10. Common-mode operation provides the cancellation of noise in a differential amplifier circuit.

11. Differential-mode operation provides the high gain of a differential stage connection.

12. A BiFET circuit is constructed using **Bi**polar and **FET** transistors.

13. A BiMOS circuit is constructed using **Bi**polar and **MOSFET** transistors.

Equations

Cascade connection:

$$A_v = A_{v_1} A_{v_2} = (-g_{m_1} R_{D_1})(-g_{m_2} R_{D_2})$$

$$A_v = \frac{-R_C \| R_L}{r_e}$$

$$Z_i = R_1 \| R_2 \| \beta r_e$$

Darlington connection:

$$\beta_D = \beta_1 \beta_2$$

$$\beta_D = \beta^2$$

$$Z_i = R_B \| (r_i + \beta_D R_E)$$

$$A_i = \beta_D \frac{R_B}{(R_B + \beta_D R_E)}$$

$$Z_o = R_E \| r_i \| \frac{r_i}{\beta_D} \approx \frac{r_i}{\beta_D}$$

$$A_v = \frac{V_o}{V_i} = \frac{R_E + \beta_D R_E}{r_i + (R_E + \beta_D R_E)} \approx 1$$

Feedback pair:

$$Z_i = R_B \| (r_i + \beta_1 \beta_2 R_C)$$

$$A_i = \beta_1 \beta_2 \frac{R_B}{R_B + Z_i}$$

$$Z_o \approx \frac{r_{i_1}}{\beta_1 \beta_2}$$

$$A_v = \frac{\beta_1 \beta_2 R_C}{(\beta_1 \beta_2 R_C + r_{i_1})}$$

Differential amplifier circuit:

$$A_v = \frac{V_o}{V_i} = \frac{R_C}{2r_e}$$

$$A_d = \frac{V_o}{V_i} = \frac{\beta R_C}{2r_i}$$

$$A_c = \frac{\beta R_C}{r_i + 2(\beta + 1)R_E}$$

12.12 COMPUTER ANALYSIS

Computer analysis of various compound circuits can be obtained using PSpice, Electronics Workbench, or Mathcad.

PSpice Windows

Using PSpice, one must first use the schematic capture program to draw the circuit. Then, the device values and parameters must be set, the analysis run, and the results observed by examining the schematic, output file, or running **PROBE** to view various waveforms.

While the BJT transistor models have default values of **Beta** and the JFET models have default values of **VTO** and **KP,** these values can be changed to those of the particular circuit being analyzed. Also, numerous other device values are defined for the BJT and JFET device models provided. These can be changed or, to more closely match the calculated circuit values, all but the main device values can be deleted.

Program 12.1—Cascade JFET Amplifier

The JFET multistage circuit of Fig. 12.2 can be drawn using **Schematic Capture** as shown in Fig. 12.52. Draw the circuit by placing each of the components, devices, and signal sources as shown. Select the component values and set them the same as those of Fig. 12.2.

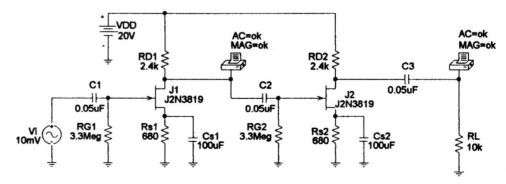

Figure 12.52 Design Center circuit to analyze cascaded JFET amplifier.

Some of the steps in drawing the circuit are as follows:

Select **VDC** and set **NAME** to V_{DD} and **VALUE** to 20 V.

Select **VSIN** and set **NAME** to V_i and **VALUE** to 10 mV.

Place **VPRINT1** symbols at the drain output of stage 1 and at the load output and set these to display the ac magnitude (set **AC** = ok and **MAG** = ok).

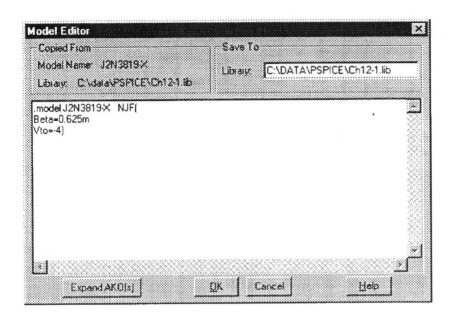

Figure 12.53 Display resulting JFET model definition.

Select the JFET part J2N3819 and set the JFET values to that used in Fig. 12.2 (**Beta** = 0.625E-3 and **VP** = −4V). This is done by selecting the JFET device, using **Edit-Model-Edit-Instance Model (Text),** and setting the two device parameters (Beta and $V_{to}[=V_P]$). To get the same results as that for Fig. 12.2, delete all other JFET device parameters as shown in Fig. 12.53.

Set all resistor and capacitor values as shown in Fig. 12.52.

Use the **Analysis Setup** to do a Linear ac sweep of V_i for 1 point at $f = 10$ kHz. Save the schematic.

Run the Circuit Analysis by pressing the **Simulate** button.

Examine the **Analysis Output** to see the results. An edited output file is shown in Fig. 12.54. Notice that the JFET devices have **VTO** = −4 and **Beta** = 625E-6 (as those in Fig. 12.2). The dc bias value shows that $I_{DQ} = 2.78$ mA (that in Fig. 12.2 being 2.8 mA). The ac analysis shows that the ac voltage magnitude at the output is 3.226E-01 = 322.6 mV (that for the rounded results of Fig. 12.2 being 310 mV).

The dc values in various circuit locations determined by the analysis can be displayed by pressing the **Enable Bias Voltage Display** button. The resulting display is shown in Fig. 12.55. Repeated pressing of this button will turn the display of dc voltages on the schematic on or off.

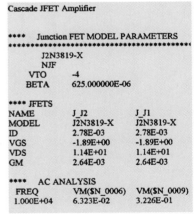

Cascade JFET Amplifier

**** Junction FET MODEL PARAMETERS
```
        J2N3819-X
        NJF
  VTO      -4
  BETA     625.000000E-06
```

**** JFETS
NAME	J_J2	J_J1
MODEL	J2N3819-X	J2N3819-X
ID	2.78E-03	2.78E-03
VGS	-1.89E+00	-1.89E+00
VDS	1.14E+01	1.14E+01
GM	2.64E-03	2.64E-03

**** AC ANALYSIS
FREQ	VM($N_0006)	VM($N_0009)
1.000E+04	6.323E-02	3.226E-01

Figure 12.54 PSpice output for circuit of Figure 12.52 (edited).

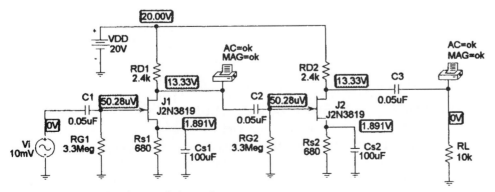

Figure 12.55 Display showing dc bias values.

A summary of the results obtained from the output results is as follows:

Dc bias results (for each transistor):

$$V_G = 0 \text{ V}, \qquad V_D = 13.33 \text{ V} \qquad V_S = 1.89 \text{ V}$$

JFET parameters (for each transistor):

$$I_{DQ} = 2.78 \text{ mA}, \qquad V_{GSQ} = -1.89 \text{ V}$$

$$g_m = 2.64 \text{ mS} \ (g_m = 2.6 \text{ mS in Example 12.1})$$

Ac results:

$$A_{v_1} = V_{o_1}/V_i = 6.323 \times 10^{-2}/1 \times 10^{-2} = 6.3 \ (= 6.2 \text{ in Example 12.1})$$

$$A_{v_2} = V_o/V_{o_1} = 3.226 \times 10^{-1}/6.323 = 10^{-2} = 5.1$$

$$V_o = 3.226 \times 10^{-1} = 322.6 \text{ mV} \qquad (V_L = 310 \text{ mV in Example 12.1})$$

The ac voltage gain and output ac voltage obtained in Example 12.1 and determined using PSpice compare quite well.

Program 12.2—Cascade BJT Amplifier

The cascade BJT amplifier of Fig. 12.4 is drawn in the Design Center schematic of Fig. 12.56. The schematic indicates displaying ac magnitudes for the collector output of the first stage and the output across the load, R_L. The BJT models have been edited so that only the default value of **IS** = 6.734E-15 and the desired value of **BF** = 200 are used. The ac sweep is set for one frequency point at $f = 10$ kHz.

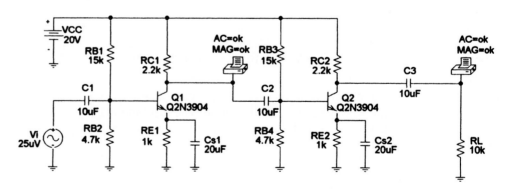

Figure 12.56 Design Center circuit to analyze two-stage BJT amplifier.

Some of the steps in drawing the circuit are as follows:

Select **VDC** and set **NAME** to V_{CC} and **VALUE** to 20 V.
Select **VSIN** and set **NAME** to V_i and **VALUE** to 25 μV.
Place **VPRINT1** symbols at the drain output of stage 1 and at the load output and set these to display the ac magnitude (set **AC** = ok and **MAG** = ok).
Select the BJT part Q2N3904 and set the BJT values to those used in Fig. 12.4

(**BF**-200) This is done by selecting the BJT device, using **Edit-Model-Edit Instance Model (Text),** and setting the device parameter (**BF**[=Beta]). To get the same results as that for Fig. 12.4, delete all other BJT device parameters.

Set all resistor and capacitor values as shown in Fig. 12.56.

Use the **Analysis Setup** to do a Linear ac sweep of V_i for 1 point at $f = 10$ kHz. Save the schematic.

Now run the Circuit Analysis by pressing the **Simulate** button.

Examine the Analysis Output to see the results. An edited output file is shown in Fig. 12.57. Notice that the BJT devices have **BF** = 200 (as those in Fig. 12.4). The dc bias value shows that $V_B = 4.7004$ (= 4.7 V for Fig. 12.4), $V_E = 3.993$ V (= 4 V for Fig. 12.4), $V_C = 11.245$ V (= 11 V for Fig. 12.4), and $I_{CQ} = 3.98$ mA (that in Fig. 12.4 being 4 mA). The ac analysis shows that the ac voltage magnitude at the output is 7.031E-01 = 703.1 mV (while that for the rounded results of Fig. 12.4 was 710 mV).

Program 12.3—Darlington Circuit

The Darlington circuit of Fig. 12.11 is drawn in the Design Center schematic of Fig. 12.58a. The BJT model has been edited for device model values of **IS** = 100E-18 and **BF** = 89.4. These correspond to the circuit of Fig. 12.11. The Analysis Setup provides for a single ac point at $f = 10$ kHz. The resulting Output Listing is shown in Fig. 12.58b.

Some of the steps in drawing the circuit are as follows:

Select **VDC** and set **NAME** to V_{CC} and **VALUE** to 18 V.

Select **VSIN** and set **NAME** to V_i and **VALUE** to 100 mV.

Place **VPRINT1** symbols at the load output and set it to display the ac magnitude (set **AC** = ok and **MAG** = ok).

Cascade BJT Amplifier

```
****    BJT MODEL PARAMETERS
        Q2N3904-X
        NPN
     IS  6.734000E-15
     BF  200

****    SMALL SIGNAL BIAS SOLUTION
NODE   VOLTAGE          NODE   VOLTAGE
($N_0001)  20.0000    ($N_0002)  4.7004
($N_0003)  11.2450    ($N_0004)  4.7004
($N_0005)   3.9993    ($N_0006)  3.9993
($N_0007)  11.2450    ($N_0008)  0.0000
($N_0009)   0.0000

**** BIPOLAR JUNCTION TRANSISTORS
NAME          Q_Q2          Q_Q1
MODEL      Q2N3904-X     Q2N3904-X
IB          1.99E-05      1.99E-05
IC          3.98E-03      3.98E-03
VBE         7.01E-01      7.01E-01
VBC        -6.54E+00     -6.54E+00
VCE         7.25E+00      7.25E+00
BETADC      2.00E+02      2.00E+02
GM          1.54E-01      1.54E-01
RPI         1.30E+03      1.30E+03
RO          1.00E+12      1.00E+12
BETAAC      2.00E+02      2.00E+02
FT          2.45E+18      2.45E+18

****    AC ANALYSIS
FREQ         VM($N_0003)   VM($N_0009)
1.000E+04    2.554E-03     7.031E-01
```

Figure 12.57 Output listing for the circuit of Figure 12.56 (edited).

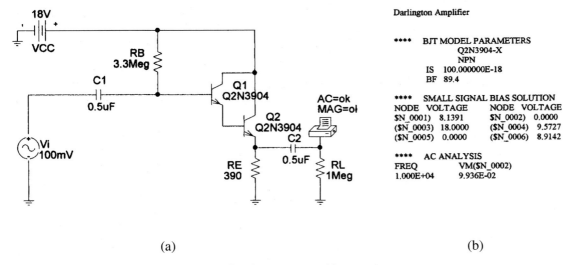

(a)

Darlington Amplifier

```
****    BJT MODEL PARAMETERS
        Q2N3904-X
        NPN
     IS  100.000000E-18
     BF  89.4

****    SMALL SIGNAL BIAS SOLUTION
NODE   VOLTAGE          NODE   VOLTAGE
$N_0001)   8.1391    $N_0002)   0.0000
($N_0003)  18.0000    ($N_0004)   9.5727
($N_0005)   0.0000    ($N_0006)   8.9142

****    AC ANALYSIS
FREQ         VM($N_0002)
1.000E+04    9.936E-02
```

(b)

Figure 12.58 (a) Design Center schematic of Darlington circuit; (b) output listing for circuit of part (a) (edited).

Select the BJT part Q2N3904 and set the BJT values to that used in Fig. 12.11 (**BF** = 89.4 and **IS** = 100E-18). This is done by selecting the BJT device, using **Edit-Model-Edit-Instance Model (Text),** and setting the device parameters (**BF** [=Beta]and **IS**). To get the same results as that for Fig. 12.11, delete all other BJT device parameters.

Set all resistor and capacitor values as shown in Fig. 12.58a.

Use the **Analysis Setup** to do a Linear ac sweep of V_i for 1 point at f = 10 kHz. Save the schematic.

Run the Circuit Analysis by pressing the **Simulate** button.

Examine the Analysis Output to see the results. An edited output file is shown in Fig. 12.58b. Dc bias shows V_E = 8.1391 V (= 8 V in the circuit of Fig. 12.11) and V_B = 9.5727 V (= 9.6 V in the circuit of Fig. 12.11). For an input of 100 mV at frequency of 10 kHz, the resulting output is 99.36 mV.

A summary of the results obtained from the output results is as follows:
Dc bias:

$$V_{B_1} = 9.65 \text{ V} \qquad V_E = 8.06 \text{ V}$$

Providing VBE (Darlington) = 1.59 V.

Transistor parameters:

$$I_{B_1} = 2.53 \ \mu\text{A}, I_{C_1} = 0.23 \text{ mA} \ (\beta_1 = 0.23 \text{ mA}/2.53 \ \mu\text{A} = 90.9)$$

$$I_{B_2} = 229 \ \mu\text{A}, I_{C_2} = 20.4 \text{ mA} \ (\beta_2 = 20.4 \text{ mA}/229 \ \mu\text{A} = 89.1)$$

For a Darlington beta of

$$\beta_D = \beta_1\beta_2 = (90.9)(89.1) = 8100$$

It is difficult to force the PSpice transistor model to exactly match the ideal transistor model used in Fig. 12.12. Notice that PSpice results provide

$$V_{BE_1} = 0.736 \text{ V}, \qquad V_{BE_2} = 0.852 \text{ V}$$

while the model used in Fig. 12.12 specifies $V_{BE}(D)$ = 1.6 V (about the same as 0.736 V + 0.852 V).

Ac operation: For an input of V_i = 100 mV, the output in the PSpice listing is

$$V_o = 9.93\text{E-2} = 99.36 \text{ mV}$$

Providing an amplifier gain of

$$A_v = V_o/V_i = 9.936 \times 10^{-2}/1 \times 10^{-1} = 0.9936$$

while the results in Example 12.10 provide A_v = 0.998, quite close.

Electronics Workbench

This same circuit can be analyzed using Electronics Workbench. Figure 12.59 shows a schematic created using EWB. For comparison with the PSpice circuit of Fig. 12.58a, the EWB circuit has a dc voltmeter placed at the base of the input transistor **(Q1),** and another dc voltmeter placed at the emitter of the output transistor **(Q2).** The comparable dc voltages are listed next:

PSpice	EWB
$V_{B_1} = 9.5727$ V	$V_{B_1} = 9.617$ V
$V_{E_1} = 8.1391$ V	$V_{E_1} = 7.793$ V

The slight difference in answers is understandable, considering that EWB uses some different transistor parameters than PSpice does.

For the input of $V_i = 100$ mV at 10 kHz, the output ac voltage is seen to be V_o (ac) = 99.36 mV, in both the EWB and PSpice solutions.

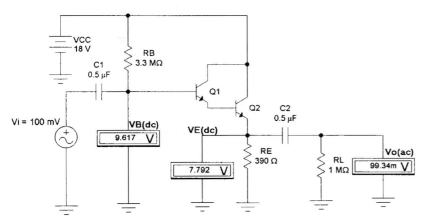

Figure 12.59 Circuit of Fig. 12.58a analyzed using EWB.

Program 12.4 — CMOS Inverter Circuit

A CMOS Inverter circuit is drawn in the Design Center schematic of Fig. 12.60. The input is provided by a pulse voltage source, V_i, which varies between 0 and 5 V.

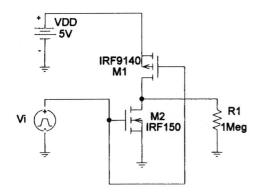

Figure 12.60 CMOS inverter circuit.

Some of the steps in drawing the circuit are as follows:

Select **VDC** and set **NAME** to V_{DD} and **VALUE** to 5 V.

Select pulse voltage source, **VPULSE,** and set **NAME** to V_i, **V1** = 0 V, **V2** = 5 V, **PW** = 1 ms, and **PER** = 2 ms. See Fig. 12.61.

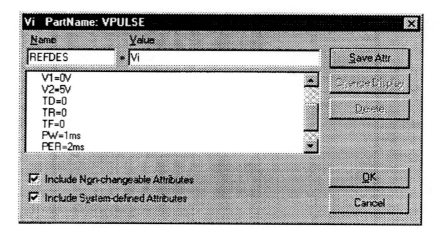

Figure 12.61 Setting values in VPULSE source.

Select *p*MOS device IRF9140 and set $V_{to} = -2$ and **KP** = 20E-6.

Select *n*MOS device IRF150 and set $V_{to} = 2$ and **KP** = 20E-6.

Place resistor **R1** of 1Meg as output load.

Save the schematic.

Run the Circuit Analysis by pressing the **Simulate** button.

Examine the Analysis Output to see the results. An edited output file is shown in Fig. 12.62.

CMOS Inverter Circuit

****	MOSFET MODEL PARAMETERS	
	IRF9140-X	IRF150-X
	PMOS	NMOS
VTO	-2	2
KP	20.000000E-06	20.000000E-06

**** MOSFETS		
NAME	M_M1	M_M2
MODEL	IRF9140-X	IRF150-X
ID	-4.92E-06	4.93E-12
VGS	-5.00E+00	0.00E+00
VDS	-8.31E-02	4.92E+00
VTH	-2.00E+00	2.00E+00
VDSAT	-3.00E+00	0.00E+00
GM	1.66E-06	0.00E+00
GDS	5.83E-05	0.00E+00

Figure 12.62 Output listing for CMOS inverter (edited).

Run **PROBE.** Select and display V_i. Add a second plot, and select and display the voltage across R_1. The resulting waveforms are shown in Fig. 12.63.

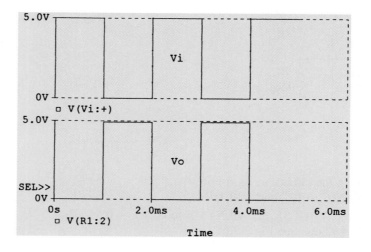

Figure 12.63 Input and output waveforms.

Mathcad

The dc bias solution of Example 12.11 and the ac calculations for the same circuit in Example 12.12 are solved in the following Mathcad example. Mathcad provides the user the ability to enter circuit values and solve the many circuit equations, providing output of all the desired circuit currents, voltages, and other impedance and gains desired.

dc bias calculations:

$Vcc := 18$ $RB := 2 \cdot 10^6$ $beta1 := 140$ $beta2 := 180$ $RC := 75$

$IB1 := \dfrac{(Vcc - 0.7)}{(RB + (beta1 \cdot beta2 \cdot RC))}$ $IB1 = 4.447 \cdot 10^{-6}$

$IB2 := beta1 \cdot IB1$ $IB2 = 6.226 \cdot 10^{-4}$

$IC1 := IB2$

$IE1 := IB1 + IC1$ $IE1 = 6.271 \cdot 10^{-4}$

$IC2 := beta2 \cdot IB2$ $IC2 = 0.112$

$IC := IE1 + IC2$ $IC = 0.113$

$Vodc := Vcc - IC \cdot RC$ $Vodc = 9.548$

$Vidc := Vodc - 0.7$ $Vidc = 8.848$

ac calculations:

$ri1 := 3000$

$RBe := ri1 + beta1 \cdot beta2 \cdot RC$

$Zi := \dfrac{(RB \cdot RBe)}{(RB + RBe)}$ $Zi = 9.725 \cdot 10^5$

$Ai := (beta1 \cdot beta2) \cdot \left[\dfrac{RB}{(RB + Zi)} \right]$ $Ai = 1.696 \cdot 10^4$

$Zo := \dfrac{ri1}{(beta1 \cdot beta2)}$ $Zo = 0.119$

$Av := \dfrac{(beta1 \cdot beta2 \cdot RC)}{(beta1 \cdot beta2 \cdot RC + ri1)}$ $Av = 0.998$

---

§ 12.2 Cascade Connection

1. For the JFET cascade amplifier in Fig. 12.64, calculate the dc bias conditions for the two identical stages, using JFETs with $I_{DSS} = 8$ mA and $V_P = -4.5$ V.

2. For the JFET cascade amplifier of Fig. 12.64, using identical JFETs with $I_{DSS} = 8$ mA and $V_P = -4.5$ V, calculate the voltage gain of each stage, the overall gain of the amplifier, and the output voltage, V_o.

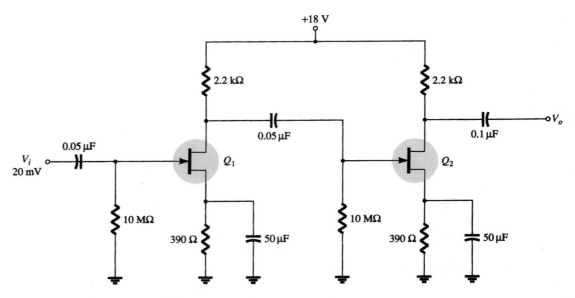

Figure 12.64 Problems 1–5, 30 and 31

3. If both JFETs in the cascade amplifier of Fig. 12.64 are changed to those having specifications $I_{DSS} = 12$ mA and $V_P = -3$ V, calculate the resulting dc bias of each stage.

4. If both JFETs in the cascade amplifier of Fig. 12.64 are changed to those having the specifications $I_{DSS} = 12$ mA, $V_P = -3$ V, and $y_{os} = 25$ μS, calculate the resulting voltage gain for each stage, the overall voltage gain, and the output voltage, V_o.

5. For the cascade amplifier of Fig. 12.64, using JFETs with specifications $I_{DSS} = 12$ mA, $V_P = -3$ V, and $y_{os} = 25$ μS, calculate the circuit input impedance (Z_i) and output impedance (Z_o).

6. For the BJT cascade amplifier of Fig. 12.65, calculate the dc bias voltages and collector current for each stage.

7. Calculate the voltage gain of each stage and the overall ac voltage gain for the BJT cascade amplifier circuit of Fig. 12.65.

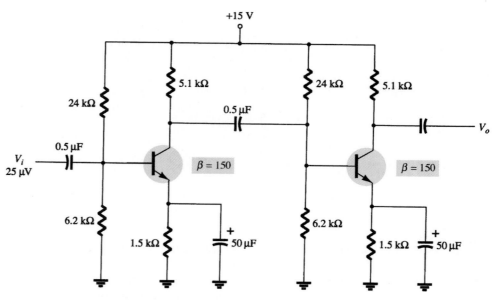

Figure 12.65 Problems 6–8 and 32

Chapter 12 Compound Configurations

8. For the circuit of Fig. 12.65, calculate the input impedance (Z_i) and output impedance (Z_o).

9. For the cascade amplifier of Fig. 12.66, calculate the dc bias voltages and collector current of each stage.

10. For the amplifier circuit of Fig. 12.66, calculate the voltage gain of each stage and the overall amplifier voltage gain.

11. Calculate the input impedance (Z_i) and output impedance (Z_o) for the amplifier circuit of Fig. 12.66.

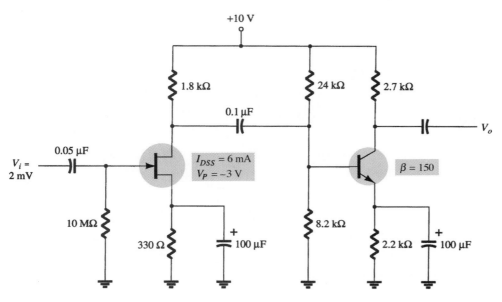

Figure 12.66 Problems 9–11

§ **12.3 Cascode Connection**

12. In the cascode amplifier circuit of Fig. 12.67, calculate the dc bias voltages V_{B_1}, V_{B_2}, and V_{C_2}.

*13. For the cascode amplifier circuit of Fig. 12.67, calculate the voltage gain, A_v, and output voltage, V_o.

14. Calculate the ac voltage across a 10-kΩ load connected at the output of the circuit in Fig. 12.67.

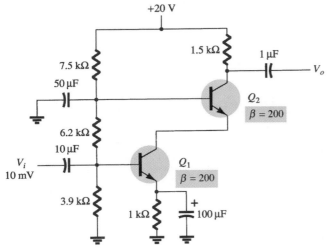

Figure 12.67 Problems 12–14

§ 12.4 Darlington Connection

15. For the circuit of Fig. 12.68, calculate the dc bias voltage, V_{E_2}, and emitter current, I_{E_2}.

* **16.** For the circuit of Fig. 12.68, calculate the amplifier voltage gain.

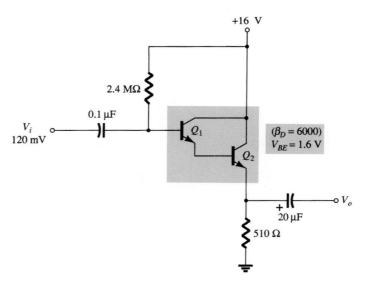

Figure 12.68 Problems 15, 16 and 33

§ 12.5 Feedback Pair

17. For the feedback pair circuit of Fig. 12.69, calculate the dc bias values of V_{B_1}, V_{C_2}, and I_C.

* **18.** Calculate the output ac voltage for the circuit of Fig. 12.69.

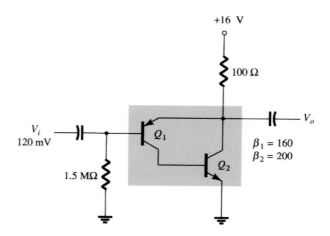

Figure 12.69 Problems 17 and 18

Chapter 12 Compound Configurations

§ 12.6 CMOS Circuit

19. Determine which transistors are off and which are on on the circuit of Fig. 12.70 for an input of:

(a) $V_1 = 0$ V and $V_2 = 0$ V.

(b) $V_1 = +5$ V and $V_2 = +5$ V.

(c) $V_1 = 0$ V and $V_2 = +5$ V.

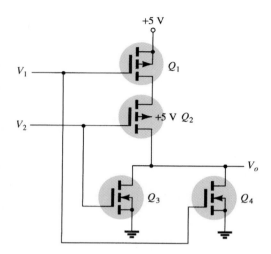

Figure 12.70 Problems 19, 20, and 34

20. For the circuit of Fig. 12.70, complete the voltage table below.

V_1	V_2	V_o
0 V	0 V	
0 V	+5 V	
+5 V	0 V	
+5 V	+5 V	

§ 12.7 Current Source Circuits

21. Calculate the current through the 2-kΩ load in the circuit of Fig. 12.71.

22. For the circuit of Fig. 12.72, calculate the current I.

***23.** Calculate the current I in the circuit of Fig. 12.73.

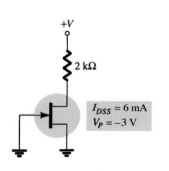

Figure 12.71   Problem 21

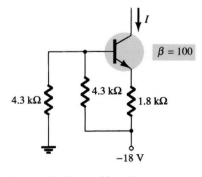

Figure 12.72   Problem 22

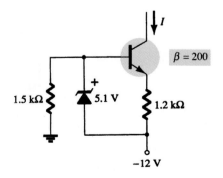

Figure 12.73   Problem 23

§ 12.8 Current Mirror Circuits

24. Calculate the mirrored current I in the circuit of Fig. 12.74.

∗**25.** Calculate collector currents for Q_1 and Q_2 in Fig. 12.75.

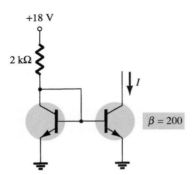

Figure 12.74   Problem 24

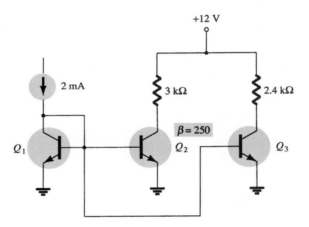

Figure 12.75   Problem 25

§ 12.9 Differential Amplifier Circuit

26. Calculate dc bias values of I_C and V_C for the matched transistors of Fig. 12.76.

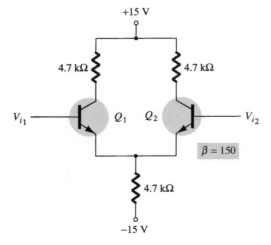

Figure 12.76   Problem 26

27. Calculate the dc bias values of I_C and V_C for the matched transistors of Fig. 12.77.

*__28.__ Calculate V_o in the circuit in Fig. 12.78.

*__29.__ Calculate V_o in the circuit of Fig. 12.79.

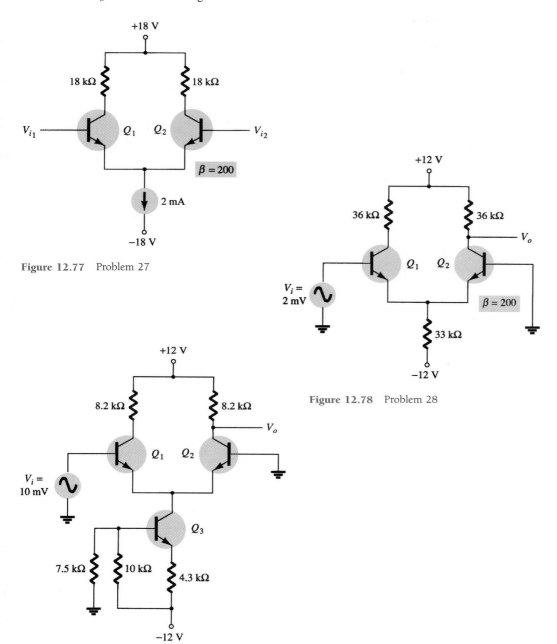

Figure 12.77 Problem 27

Figure 12.78 Problem 28

Figure 12.79 Problem 29

§ **12.12 Computer Analysis**

*__30.__ Use the Design Center to draw a schematic circuit of the cascade JFET amplifier as in Fig. 12.64. Set the JFET parameters for I_{DSS} 12 mA and $V_P = 3$ V, and have the analysis determine the dc bias.

*__31.__ Use the Design Center to draw a schematic circuit for a cascade JFET amplifier as shown in Fig. 12.64. Set the analysis to calculate the ac output voltage, V_o, for $I_{DSS} = 12$ mA and $V_P = -3$ V.

*32. Use the Design Center to draw a schematic circuit for the cascade BJT amplifier of Fig. 12.65. Have the analysis calculate the output ac voltage at each stage.

*33. Use the Design Center to draw a schematic circuit for the Darlington amplifier of Fig. 12.68. Have the analysis determine the dc bias and output ac voltage.

*34. Use the Design Center to draw a schematic for the CMOS circuit of Fig. 12.70. Have the analysis determine the output dc voltages for the following sets of input voltages:
(a) $V_1 = 0$ V and $V_2 = 0$ V.
(b) $V_1 = 0$ V and $V_2 = 5$ V.
(c) $V_1 = 5$ V and $V_2 = 5$ V.

---

*Please Note: Asterisks indicate more difficult problems.

Operational Amplifiers

13.1 INTRODUCTION

An operational amplifier, or op-amp, is a very high gain differential amplifier with high input impedance and low output impedance. Typical uses of the operational amplifier are to provide voltage amplitude changes (amplitude and polarity), oscillators, filter circuits, and many types of instrumentation circuits. An op-amp contains a number of differential amplifier stages to achieve a very high voltage gain.

Figure 13.1 shows a basic op-amp with two inputs and one output as would result using a differential amplifier input stage. Recall from Chapter 12 that each input results in either the same or an opposite polarity (or phase) output, depending on whether the signal is applied to the plus ($+$) or the minus ($-$) input.

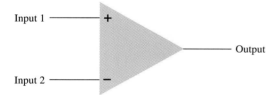

Figure 13.1 Basic op-amp.

Single-Ended Input

Single-ended input operation results when the input signal is connected to one input with the other input connected to ground. Figure 13.2 shows the signals connected

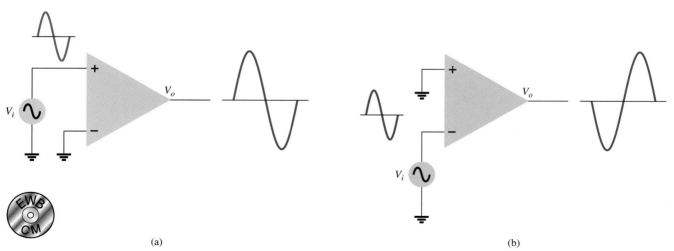

(a)

(b)

Figure 13.2 Single-ended operation.

for this operation. In Fig. 13.2a, the input is applied to the plus input (with minus input at ground), which results in an output having the same polarity as the applied input signal. Figure 13.2b shows an input signal applied to the minus input, the output then being opposite in phase to the applied signal.

Double-Ended (Differential) Input

In addition to using only one input, it is possible to apply signals at each input—this being a double-ended operation. Figure 13.3a shows an input, V_d, applied between the two input terminals (recall that neither input is at ground), with the resulting amplified output in phase with that applied between the plus and minus inputs. Figure 13.3b shows the same action resulting when two separate signals are applied to the inputs, the difference signal being $V_{i_1} - V_{i_2}$.

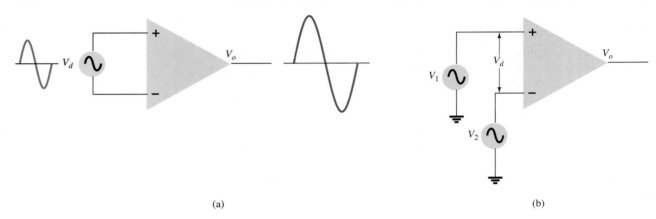

(a)   (b)

Figure 13.3   Double-ended (differential) operation.

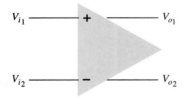

Figure 13.4   Double-ended output.

Double-Ended Output

While the operation discussed so far had a single output, the op-amp can also be operated with opposite outputs, as shown in Fig. 13.4. An input applied to either input will result in outputs from both output terminals, these outputs always being opposite in polarity. Figure 13.5 shows a single-ended input with a double-ended output. As shown, the signal applied to the plus input results in two amplified outputs of opposite polarity. Figure 13.6 shows the same operation with a single output measured

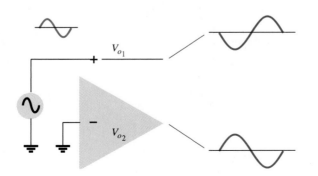

Figure 13.5   Double-ended output with single-ended input.

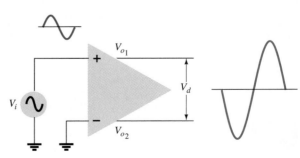

Figure 13.6   Double-ended output.

between output terminals (not with respect to ground). This difference output signal is $V_{o_1} - V_{o_2}$. The difference output is also referred to as a *floating signal* since neither output terminal is the ground (reference) terminal. Notice that the difference output is twice as large as either V_{o_1} or V_{o_2} since they are of opposite polarity and subtracting them results in twice their amplitude [i.e., $10\,V - (-10\,V) = 20\,V$]. Figure 13.7 shows a differential input, differential output operation. The input is applied between the two input terminals and the output taken from between the two output terminals. This is fully differential operation.

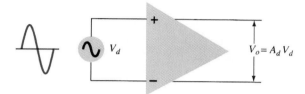

$$V_o = A_d V_d$$

Figure 13.7 Differential-input, differential-output operation.

Common-Mode Operation

When the same input signals are applied to both inputs, common-mode operation results, as shown in Fig. 13.8. Ideally, the two inputs are equally amplified, and since they result in opposite polarity signals at the output, these signals cancel, resulting in 0-V output. Practically, a small output signal will result.

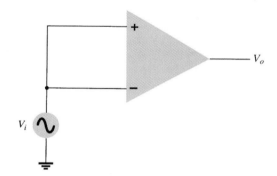

$$V_o$$

$$V_i$$

Figure 13.8 Common-mode operation.

Common-Mode Rejection

A significant feature of a differential connection is that the signals which are opposite at the inputs are highly amplified, while those which are common to the two inputs are only slightly amplified—the overall operation being to amplify the difference signal while rejecting the common signal at the two inputs. Since noise (any unwanted input signal) is generally common to both inputs, the differential connection tends to provide attenuation of this unwanted input while providing an amplified output of the difference signal applied to the inputs. This operating feature, referred to as common-mode rejection, is discussed more fully in the next section.

13.2 DIFFERENTIAL AND COMMON-MODE OPERATION

One of the more important features of a differential circuit connection, as provided in an op-amp, is the circuit's ability to greatly amplify signals that are opposite at the two inputs, while only slightly amplifying signals that are common to both inputs. An

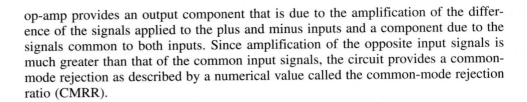

op-amp provides an output component that is due to the amplification of the difference of the signals applied to the plus and minus inputs and a component due to the signals common to both inputs. Since amplification of the opposite input signals is much greater than that of the common input signals, the circuit provides a common-mode rejection as described by a numerical value called the common-mode rejection ratio (CMRR).

Differential Inputs

When separate inputs are applied to the op-amp, the resulting difference signal is the difference between the two inputs.

$$\boxed{V_d = V_{i_1} - V_{i_2}} \qquad (13.1)$$

Common Inputs

When both input signals are the same, a common signal element due to the two inputs can be defined as the average of the sum of the two signals.

$$\boxed{V_c = \tfrac{1}{2}(V_{i_1} + V_{i_2})} \qquad (13.2)$$

Output Voltage

Since any signals applied to an op-amp in general have both in-phase and out-of-phase components, the resulting output can be expressed as

$$\boxed{V_o = A_d V_d + A_c V_c} \qquad (13.3)$$

where V_d = difference voltage given by Eq. (13.1)
$\quad\ V_c$ = common voltage given by Eq. (13.2)
$\quad\ A_d$ = differential gain of the amplifier
$\quad\ A_c$ = common-mode gain of the amplifier

Opposite Polarity Inputs

If opposite polarity inputs applied to an op-amp are ideally opposite signals. $V_{i_1} = -V_{i_2} = V_s$, the resulting difference voltage is

$$\text{Eq. (13.1):} \quad V_d = V_{i_1} - V_{i_2} = V_s - (-V_s) = 2V_s$$

while the resulting common voltage is

$$\text{Eq. (13.2):} \quad V_c = \tfrac{1}{2}(V_{i_1} + V_{i_2}) = \tfrac{1}{2}[V_s + (-V_s)] = 0$$

so that the resulting output voltage is

$$\text{Eq. (13.3):} \quad V_o = A_d V_d + A_c V_c = A_d(2V_s) + 0 = 2A_d V_s$$

This shows that when the inputs are an ideal opposite signal (no common element), the output is the differential gain times twice the input signal applied to one of the inputs.

Same Polarity Inputs

If the same polarity inputs are applied to an op-amp, $V_{i_1} = V_{i_2} = V_s$, the resulting difference voltage is

$$\text{Eq. (13.1):} \quad V_d = V_{i_1} - V_{i_2} = V_s - V_s = 0$$

while the resulting common voltage is

$$\text{Eq. (13.2):} \quad V_c = \tfrac{1}{2}(V_{i_1} + V_{i_2}) = \tfrac{1}{2}(V_s + V_s) = V_s$$

so that the resulting output voltage is

$$\text{Eq. (13.3):} \quad V_o = A_d V_d + A_c V_c = A_d(0) + A_c V_s = A_c V_s$$

This shows that when the inputs are ideal in-phase signals (no difference signal), the output is the common-mode gain times the input signal, V_s, which shows that only common-mode operation occurs.

Common-Mode Rejection

The solutions above provide the relationships that can be used to measure A_d and A_c in op-amp circuits.

1. *To measure A_d:* Set $V_{i_1} = -V_{i_2} = V_s = 0.5$ V, so that

$$\text{Eq. (13.1):} \quad V_d = (V_{i_1} - V_{i_2}) = (0.5 \text{ V} - (-0.5 \text{ V}) = 1 \text{ V}$$

and $\quad$ Eq. (13.2): $\quad V_c = \tfrac{1}{2}(V_{i_1} + V_{i_2}) = \tfrac{1}{2}[0.5 \text{ V} + (-0.5 \text{ V})] = 0 \text{ V}$

Under these conditions the output voltage is

$$\text{Eq. (13.3):} \quad V_o = A_d V_d + A_c V_c = A_d(1 \text{ V}) + A_c(0) = A_d$$

Thus, setting the input voltages $V_{i_1} = -V_{i_2} = 0.5$ V results in an output voltage numerically equal to the value of A_d.

2. *To measure A_c:* Set $V_{i_1} = V_{i_2} = V_s = 1$ V, so that

$$\text{Eq. (13.1):} \quad V_d = (V_{i_1} - V_{i_2}) = (1 \text{ V} - 1 \text{ V}) = 0 \text{ V}$$

and $\quad$ Eq. (13.2): $\quad V_c = \tfrac{1}{2}(V_{i_1} + V_{i_2}) = \tfrac{1}{2}(1 \text{ V} + 1 \text{ V}) = 1 \text{ V}$

Under these conditions the output voltage is

$$\text{Eq. (13.3):} \quad V_o = A_d V_d + A_c V_c = A_d(0 \text{ V}) + A_c(1 \text{ V}) = A_c$$

Thus, setting the input voltages $V_{i_1} = V_{i_2} = 1$ V results in an output voltage numerically equal to the value of A_c.

Common-Mode Rejection Ratio

Having obtained A_d and A_c (as in the measurement procedure discussed above), we can now calculate a value for the common-mode rejection ratio (CMRR), which is defined by the following equation:

$$\boxed{\text{CMRR} = \frac{A_d}{A_c}} \tag{13.4}$$

The value of CMRR can also be expressed in logarithmic terms as

$$\boxed{\text{CMRR (log)} = 20 \log_{10} \frac{A_d}{A_c}} \quad \text{(dB)} \tag{13.5}$$

EXAMPLE 13.1

Calculate the CMRR for the circuit measurements shown in Fig. 13.9.

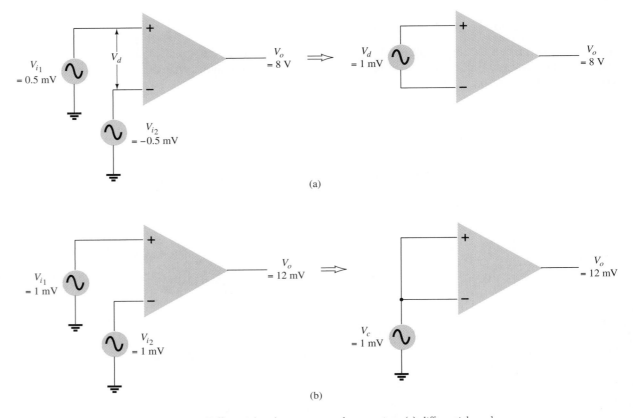

(a)

(b)

Figure 13.9 Differential and common-mode operation: (a) differential-mode; (b) common-mode.

Solution

From the measurement shown in Fig. 13.9a, using the procedure in step 1 above, we obtain

$$A_d = \frac{V_o}{V_d} = \frac{8\text{ V}}{1\text{ mV}} = 8000$$

The measurement shown in Fig. 13.9b, using the procedure in step 2 above, gives us

$$A_c = \frac{V_o}{V_c} = \frac{12\text{ mV}}{1\text{ mV}} = 12$$

Using Eq. (13.4), the value of CMRR is

$$\text{CMRR} = \frac{A_d}{A_c} = \frac{8000}{12} = \textbf{666.7}$$

which can also be expressed as

$$\text{CMRR} = 20\log_{10}\frac{A_d}{A_c} = 20\log_{10}666.7 = \textbf{56.48 dB}$$

It should be clear that the desired operation will have A_d very large with A_c very small. That is, the signal components of opposite polarity will appear greatly amplified at the output, whereas the signal components that are in phase will mostly cancel out so that the common-mode gain, A_c, is very small. Ideally, the value of the CMRR is infinite. Practically, the larger the value of CMRR, the better the circuit operation.

We can express the output voltage in terms of the value of CMRR as follows:

Eq. (13.3): $V_o = A_d V_d + A_c V_c = A_d V_d \left(1 + \dfrac{A_c V_c}{A_d V_d}\right)$

Using Eq. (13.4), we can write the above as

$$V_o = A_d V_d \left(1 + \dfrac{1}{\text{CMRR}} \dfrac{V_c}{V_d}\right) \qquad (13.6)$$

Even when both V_d and V_c components of signal are present, Eq. (13.6) shows that for large values of CMRR, the output voltage will be due mostly to the difference signal, with the common-mode component greatly reduced or rejected. Some practical examples should help clarify this idea.

EXAMPLE 13.2

Determine the output voltage of an op-amp for input voltages of $V_{i_1} = 150\ \mu V$, $V_{i_2} = 140\ \mu V$. The amplifier has a differential gain of $A_d = 4000$ and the value of CMRR is:
(a) 100.
(b) 10^5.

Solution

Eq. (13.1): $V_d = V_{i_1} - V_{i_2} = (150 - 140)\ \mu V = 10\ \mu V$

Eq. (13.2): $V_c = \dfrac{1}{2}(V_{i_1} + V_{i_2}) = \dfrac{150\ \mu V + 140\ \mu V}{2} = 145\ \mu V$

(a) Eq. (13.6): $V_o = A_d V_d \left(1 + \dfrac{1}{\text{CMRR}} \dfrac{V_c}{V_d}\right)$

$= (4000)(10\ \mu V)\left(1 + \dfrac{1}{100}\dfrac{145\ \mu V}{10\ \mu V}\right)$

$= 40\ \text{mV}(1.145) = \mathbf{45.8\ mV}$

(b) $V_o = (4000)(10\ \mu V)\left(1 + \dfrac{1}{10^5}\dfrac{145\ \mu V}{10\ \mu V}\right) = 40\ \text{mV}(1.000145) = \mathbf{40.006\ mV}$

Example 13.2 shows that the larger the value of CMRR, the closer the output voltage is to the difference input times the difference gain with the common-mode signal being rejected.

13.3   OP-AMP BASICS

An operational amplifier is a very high gain amplifier having very high input impedance (typically a few megohms) and low output impedance (less than 100 Ω). The basic circuit is made using a difference amplifier having two inputs (plus and minus) and at least one output. Figure 13.10 shows a basic op-amp unit. As discussed ear-

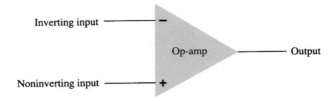

Figure 13.10 Basic op-amp.

lier, the plus $(+)$ input produces an output that is in phase with the signal applied, while an input to the minus $(-)$ input results in an opposite polarity output. The ac equivalent circuit of the op-amp is shown in Fig. 13.11a. As shown, the input signal applied between input terminals sees an input impedance, R_i, typically very high. The output voltage is shown to be the amplifier gain times the input signal taken through an output impedance, R_o, which is typically very low. An ideal op-amp circuit, as shown in Fig. 13.11b, would have infinite input impedance, zero output impedance, and an infinite voltage gain.

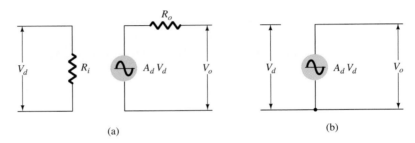

Figure 13.11 Ac equivalent of op-amp circuit: (a) practical; (b) ideal.

Basic Op-Amp

The basic circuit connection using an op-amp is shown in Fig. 13.12. The circuit shown provides operation as a constant-gain multiplier. An input signal, V_1, is applied through resistor R_1 to the minus input. The output is then connected back to the same minus input through resistor R_f. The plus input is connected to ground. Since the signal V_1 is essentially applied to the minus input, the resulting output is opposite in phase to the input signal. Figure 13.13a shows the op-amp replaced by its ac equivalent circuit. If we use the ideal op-amp equivalent circuit, replacing R_i by an infinite resistance and R_o by zero resistance, the ac equivalent circuit is that shown in Fig. 13.13b. The circuit is then redrawn, as shown in Fig. 13.13c, from which circuit analysis is carried out.

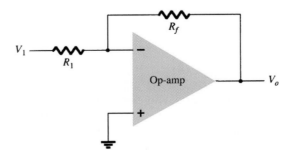

Figure 13.12 Basic op-amp connection.

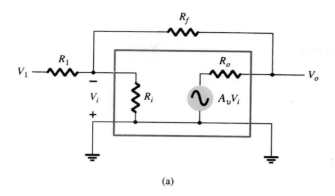

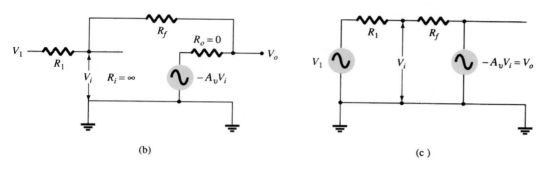

Figure 13.13 Operation of op-amp as constant-gain multiplier: (a) op-amp ac equivalent circuit; (b) ideal op-amp equivalent circuit; (c) redrawn equivalent circuit.

Using superposition, we can solve for the voltage V_1 in terms of the components due to each of the sources. For source V_1 only ($-A_vV_i$ set to zero),

$$V_{i_1} = \frac{R_f}{R_1 + R_f} V_1$$

For source $-A_vV_i$ only (V_1 set to zero),

$$V_{i_2} = \frac{R_1}{R_1 + R_f}(-A_vV_i)$$

The total voltage V_i is then

$$V_i = V_{i_1} + V_{i_2} = \frac{R_f}{R_1 + R_f}V_1 + \frac{R_1}{R_1 + R_f}(-A_vV_i)$$

which can be solved for V_i as

$$V_i = \frac{R_f}{R_f + (1 + A_v)R_1}V_1 \qquad (13.7)$$

If $A_v \gg 1$ and $A_v R_1 \gg R_f$, as is usually true, then

$$V_i = \frac{R_f}{A_v R_1}V_1$$

Solving for V_o/V_i, we get

$$\frac{V_o}{V_i} = \frac{-A_vV_i}{V_i} = \frac{-A_v R_fV_1}{V_i A_v R_1} = -\frac{R_f}{R_1}\frac{V_1}{V_i}$$

so that

$$\boxed{\frac{V_o}{V_1} = -\frac{R_f}{R_1}} \qquad (13.8)$$

The result, in Eq. (13.8), shows that the ratio of overall output to input voltage is dependent only on the values of resistors R_1 and R_f—provided that A_v is very large.

Unity Gain

If $R_f = R_1$, the gain is

$$\text{voltage gain} = -\frac{R_f}{R_1} = -1$$

so that the circuit provides a unity voltage gain with 180° phase inversion. If R_f is exactly R_1, the voltage gain is exactly 1.

Constant Magnitude Gain

If R_f is some multiple of R_1, the overall amplifier gain is a constant. For example, if $R_f = 10R_1$, then

$$\text{voltage gain} = -\frac{R_f}{R_1} = -10$$

and the circuit provides a voltage gain of exactly 10 along with an 180° phase inversion from the input signal. If we select precise resistor values for R_f and R_1, we can obtain a wide range of gains, the gain being as accurate as the resistors used and is only slightly affected by temperature and other circuit factors.

Virtual Ground

The output voltage is limited by the supply voltage of, typically, a few volts. As stated before, voltage gains are very high. If, for example, $V_o = -10$ V and $A_v = 20{,}000$, the input voltage would then be

$$V_i = \frac{-V_o}{A_v} = \frac{10 \text{ V}}{20{,}000} = 0.5 \text{ mV}$$

If the circuit has an overall gain (V_o/V_1) of, say, 1, the value of V_1 would then be 10 V. Compared to all other input and output voltages, the value of V_i is then small and may be considered 0 V.

Note that although $V_i \approx 0$ V, it is not exactly 0 V. (The output voltage is a few volts due to the very small input V_i times a very large gain A_v.) The fact that $V_i \approx 0$ V leads to the concept that at the amplifier input there exists a virtual short circuit or virtual ground.

The concept of a virtual short implies that although the voltage is nearly 0 V, there is no current through the amplifier input to ground. Figure 13.14 depicts the virtual ground concept. The heavy line is used to indicate that we may consider that a short

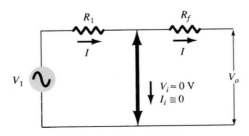

Figure 13.14 Virtual ground in an op-amp.

exists with $V_i \approx 0$ V but that this is a virtual short so that no current goes through the short to ground. Current goes only through resistors R_1 and R_f as shown.

Using the virtual ground concept, we can write equations for the current I as follows:

$$I = \frac{V_1}{R_1} = -\frac{V_o}{R_f}$$

which can be solved for V_o/V_1:

$$\frac{V_o}{V_1} = -\frac{R_f}{R_1}$$

The virtual ground concept, which depends on A_v being very large, allowed a simple solution to determine the overall voltage gain. It should be understood that although the circuit of Fig. 13.14 is not physically correct, it does allow an easy means for determining the overall voltage gain.

13.4 PRACTICAL OP-AMP CIRCUITS

The op-amp can be connected in a large number of circuits to provide various operating characteristics. In this section, we cover a few of the most common of these circuit connections.

Inverting Amplifier

The most widely used constant-gain amplifier circuit is the inverting amplifier, as shown in Fig. 13.15. The output is obtained by multiplying the input by a fixed or constant gain, set by the input resistor (R_1) and feedback resistor (R_f)—this output also being inverted from the input. Using Eq. (13.8) we can write

$$V_o = -\frac{R_f}{R_1} V_1$$

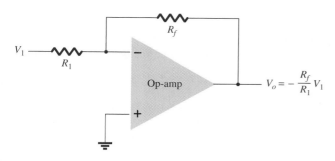

Figure 13.15 Inverting constant-gain multiplier.

If the circuit of Fig. 13.15 has $R_1 = 100$ kΩ and $R_f = 500$ kΩ, what output voltage results for an input of $V_1 = 2$ V?

EXAMPLE 13.3

Solution

$$\text{Eq. (13.8):} \quad V_o = -\frac{R_f}{R_1} V_1 = -\frac{500 \text{ k}\Omega}{100 \text{ k}\Omega} (2 \text{ V}) = \mathbf{-10 \ V}$$

Noninverting Amplifier

The connection of Fig. 13.16a shows an op-amp circuit that works as a noninverting amplifier or constant-gain multiplier. It should be noted that the inverting amplifier connection is more widely used because it has better frequency stability (discussed later). To determine the voltage gain of the circuit, we can use the equivalent representation shown in Fig. 13.16b. Note that the voltage across R_1 is V_1 since $V_i \approx 0$ V. This must be equal to the output voltage, through a voltage divider of R_1 and R_f, so that

$$V_1 = \frac{R_1}{R_1 + R_f} V_o$$

which results in

$$\frac{V_o}{V_1} = \frac{R_1 + R_f}{R_1} = 1 + \frac{R_f}{R_1} \tag{13.9}$$

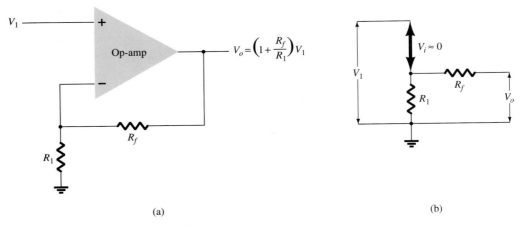

(a) (b)

Figure 13.16 Noninverting constant-gain multiplier.

EXAMPLE 13.4

Calculate the output voltage of a noninverting amplifier (as in Fig. 13.16) for values of $V_1 = 2$ V, $R_f = 500$ kΩ, and $R_1 = 100$ kΩ.

Solution

$$\text{Eq. (13.9):} \quad V_o = \left(1 + \frac{R_f}{R_1}\right) V_1 = \left(1 + \frac{500 \text{ kΩ}}{100 \text{ kΩ}}\right)(2 \text{ V}) = 6(2 \text{ V}) = \textbf{+12 V}$$

Unity Follower

The unity-follower circuit, as shown in Fig. 13.17a, provides a gain of unity (1) with no polarity or phase reversal. From the equivalent circuit (see Fig. 13.17b) it is clear that

$$V_o = V_1 \tag{13.10}$$

and that the output is the same polarity and magnitude as the input. The circuit operates like an emitter- or source-follower circuit except that the gain is exactly unity.

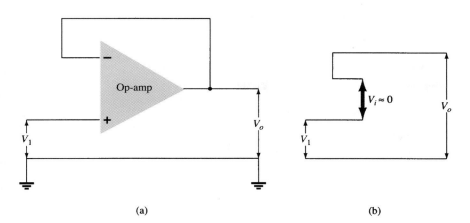

Figure 13.17 (a) Unity follower; (b) virtual-ground equivalent circuit.

Summing Amplifier

Probably the most used of the op-amp circuits is the summing amplifier circuit shown in Fig. 13.18a. The circuit shows a three-input summing amplifier circuit, which provides a means of algebraically summing (adding) three voltages, each multiplied by a constant-gain factor. Using the equivalent representation shown in Fig. 13.18b, the output voltage can be expressed in terms of the inputs as

$$V_o = -\left(\frac{R_f}{R_1} V_1 + \frac{R_f}{R_2} V_2 + \frac{R_f}{R_3} V_3 \right) \qquad (13.11)$$

In other words, each input adds a voltage to the output multiplied by its separate constant-gain multiplier. If more inputs are used, they each add an additional component to the output.

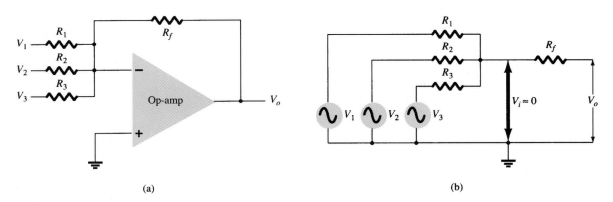

Figure 13.18 (a) Summing amplifier; (b) virtual-ground equivalent circuit.

Calculate the output voltage of an op-amp summing amplifier for the following sets of voltages and resistors. Use $R_f = 1$ MΩ in all cases.
(a) $V_1 = +1$ V, $V_2 = +2$ V, $V_3 = +3$ V, $R_1 = 500$ kΩ, $R_2 = 1$ MΩ, $R_3 = 1$ MΩ.
(b) $V_1 = -2$ V, $V_2 = +3$ V, $V_3 = +1$ V, $R_1 = 200$ kΩ, $R_2 = 500$ kΩ, $R_3 = 1$ MΩ.

EXAMPLE 13.5

Solution

Using Eq. (13.11):

(a) $V_o = -\left[\dfrac{1000 \text{ k}\Omega}{500 \text{ k}\Omega}(+1 \text{ V}) + \dfrac{1000 \text{ k}\Omega}{1000 \text{ k}\Omega}(+2 \text{ V}) + \dfrac{1000 \text{ k}\Omega}{1000 \text{ k}\Omega}(+3 \text{ V})\right]$

$= -[2(1 \text{ V}) + 1(2 \text{ V}) + 1(3 \text{ V})] = \mathbf{-7\ V}$

(b) $V_o = -\left[\dfrac{1000 \text{ k}\Omega}{200 \text{ k}\Omega}(-2 \text{ V}) + \dfrac{1000 \text{ k}\Omega}{500 \text{ k}\Omega}(+3 \text{ V}) + \dfrac{1000 \text{ k}\Omega}{1000 \text{ k}\Omega}(+1 \text{ V})\right]$

$= -[5(-2 \text{ V}) + 2(3 \text{ V}) + 1(1 \text{ V})] = \mathbf{+3\ V}$

Integrator

So far, the input and feedback components have been resistors. If the feedback component used is a capacitor, as shown in Fig. 13.19a, the resulting connection is called an *integrator*. The virtual-ground equivalent circuit (Fig. 13.19b) shows that an expression for the voltage between input and output can be derived in terms of the current I, from input to output. Recall that virtual ground means that we can consider the voltage at the junction of R and X_C to be ground (since $V_i \approx 0$ V) but that no current goes into ground at that point. The capacitive impedance can be expressed as

$$X_C = \frac{1}{j\omega C} = \frac{1}{sC}$$

where $s = j\omega$ is in the Laplace notation.* Solving for V_o/V_1 yields

$$I = \frac{V_1}{R} = -\frac{V_o}{X_C} = \frac{-V_o}{1/sC} = -sCV_o$$

$$\frac{V_o}{V_1} = \frac{-1}{sCR} \tag{13.12}$$

The expression above can be rewritten in the time domain as

$$v_o(t) = -\frac{1}{RC}\int v_1(t)\, dt \tag{13.13}$$

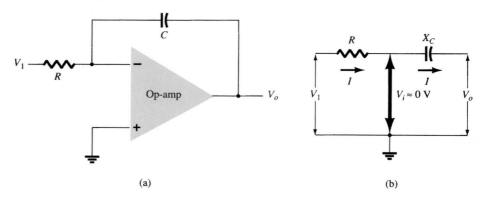

(a)　　　　　　(b)

Figure 13.19 Integrator.

---

*Laplace notation allows expressing differential or integral operations which are part of calculus in algebraic form using the operator s. Readers unfamiliar with calculus should ignore the steps leading to Eq. (13.13) and follow the physical meaning used thereafter.

Equation (13.13) shows that the output is the integral of the input, with an inversion and scale multiplier of $1/RC$. The ability to integrate a given signal provides the analog computer with the ability to solve differential equations and therefore provides the ability to electrically solve analogs of physical system operation.

The integration operation is one of summation, summing the area under a waveform or curve over a period of time. If a fixed voltage is applied as input to an integrator circuit, Eq. (13.13) shows that the output voltage grows over a period of time, providing a ramp voltage. Equation (13.13) can thus be understood to show that the output voltage ramp (for a fixed input voltage) is opposite in polarity to the input voltage and is multiplied by the factor $1/RC$. While the circuit of Fig. 13.19 can operate on many varied types of input signals, the following examples will use only a fixed input voltage, resulting in a ramp output voltage.

As an example, consider an input voltage, $V_1 = 1$ V, to the integrator circuit of Fig. 13.20a. The scale factor of $1/RC$ is

$$-\frac{1}{RC} = \frac{1}{(1\text{ M}\Omega)(1\ \mu\text{F})} = -1$$

so that the output is a negative ramp voltage as shown in Fig. 13.20b. If the scale factor is changed by making $R = 100$ kΩ, for example, then

$$-\frac{1}{RC} = \frac{1}{(100\text{ k}\Omega)(1\ \mu\text{F})} = -10$$

and the output is then a steeper ramp voltage, as shown in Fig. 13.20c.

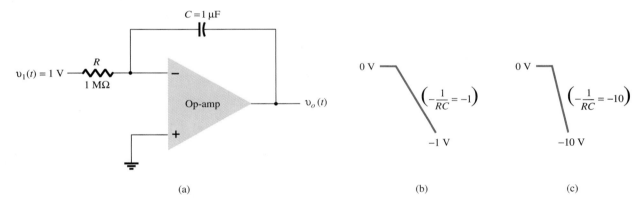

(a) (b) (c)

Figure 13.20 Operation of integrator with step input.

More than one input may be applied to an integrator, as shown in Fig. 13.21, with the resulting operation given by

$$v_o(t) = -\left[\frac{1}{R_1 C} \int v_1(t)\, dt + \frac{1}{R_2 C} \int v_2(t)\, dt + \frac{1}{R_3 C} \int v_3(t)\, dt \right] \quad (13.14)$$

An example of a summing integrator as used in an analog computer is given in Fig. 13.21. The actual circuit is shown with input resistors and feedback capacitor, whereas the analog-computer representation indicates only the scale factor for each input.

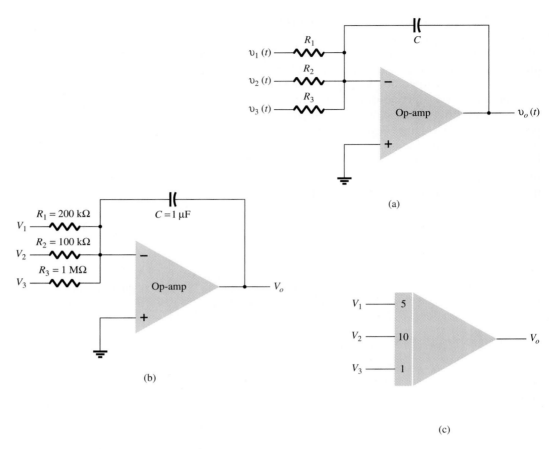

(a)

(b)

(c)

Figure 13.21 (a) Summing-integrator circuit; (b) component values; (c) analog-computer, integrator-circuit representation.

Differentiator

A differentiator circuit is shown in Fig. 13.22. While not as useful as the circuit forms covered above, the differentiator does provide a useful operation, the resulting relation for the circuit being

$$v_o(t) = -RC\frac{dv_1(t)}{dt} \tag{13.15}$$

where the scale factor is $-RC$.

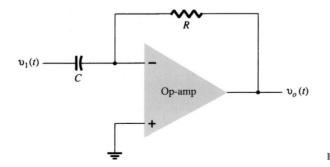

Figure 13.22 Differentiator circuit.

13.5 OP-AMP SPECIFICATIONS—DC OFFSET PARAMETERS

Before going into various practical applications using op-amps, we should become familiar with some of the parameters used to define the operation of the unit. These specifications include both dc and transient or frequency operating features, as covered next.

Offset Currents and Voltages

While the op-amp output should be 0 V when the input is 0 V, in actual operation there is some offset voltage at the output. For example, if one connected 0 V to both op-amp inputs and then measured 26 mV(dc) at the output, this would represent 26 mV of unwanted voltage generated by the circuit and not by the input signal. Since the user may connect the amplifier circuit for various gain and polarity operations, however, the manufacturer specifies an input offset voltage for the op-amp. The output offset voltage is then determined by the input offset voltage and the gain of the amplifier, as connected by the user.

The output offset voltage can be shown to be affected by two separate circuit conditions. These are: (1) an input offset voltage, V_{IO}, and (2) an offset current due to the difference in currents resulting at the plus ($+$) and minus ($-$) inputs.

INPUT OFFSET VOLTAGE, V_{IO}

The manufacturer's specification sheet provides a value of V_{IO} for the op-amp. To determine the effect of this input voltage on the output, consider the connection shown in Fig. 13.23. Using $V_o = AV_i$, we can write

$$V_o = AV_i = A\left(V_{IO} - V_o\frac{R_1}{R_1 + R_f}\right)$$

Solving for V_o, we get

$$V_o = V_{IO}\frac{A}{1 + A[R_1/(R_1 + R_f)]} \approx V_{IO}\frac{A}{A[R_1/(R_1 + R_f)]}$$

from which we can write

$$\boxed{V_o(\text{offset}) = V_{IO}\frac{R_1 + R_f}{R_1}} \tag{13.16}$$

Equation (13.16) shows how the output offset voltage results from a specified input offset voltage for a typical amplifier connection of the op-amp.

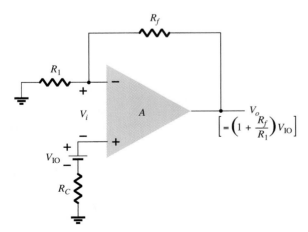

$$\left[= \left(1 + \frac{R_f}{R_1}\right)V_{IO}\right]$$

Figure 13.23 Operation showing effect of input offset voltage, V_{IO}.

EXAMPLE 13.6   Calculate the output offset voltage of the circuit in Fig. 13.24. The op-amp spec lists $V_{IO} = 1.2$ mV.

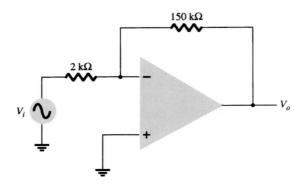

Figure 13.24 Op-amp connection for Examples 13.6 and 13.7.

Solution

Eq. (13.16): $V_o(\text{offset}) = V_{IO}\dfrac{R_1 + R_f}{R_1} = (1.2 \text{ mV})\left(\dfrac{2 \text{ k}\Omega + 150 \text{ k}\Omega}{2 \text{ k}\Omega}\right) = \mathbf{91.2 \text{ mV}}$

OUTPUT OFFSET VOLTAGE DUE TO INPUT OFFSET CURRENT, I_{IO}

An output offset voltage will also result due to any difference in dc bias currents at both inputs. Since the two input transistors are never exactly matched, each will operate at a slightly different current. For a typical op-amp connection, such as that shown in Fig. 13.25, an output offset voltage can be determined as follows. Replacing the bias currents through the input resistors by the voltage drop that each develops, as shown in Fig. 13.26, we can determine the expression for the resulting output voltage. Using superposition, the output voltage due to input bias current I_{IB}^{+}, denoted by V_o^{+}, is

$$V_o^{+} = I_{IB}^{+} R_C\left(1 + \frac{R_f}{R_1}\right)$$

while the output voltage due to only I_{IB}^{-}, denoted by V_o^{-}, is

$$V_o^{-} = I_{IB}^{-} R_1\left(-\frac{R_f}{R_1}\right)$$

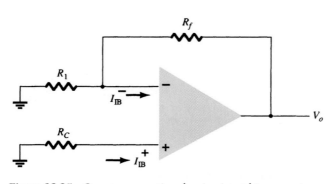

Figure 13.25 Op-amp connection showing input bias currents.

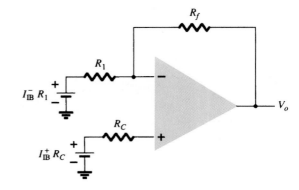

Figure 13.26 Redrawn circuit of Fig. 13.25.

for a total output offset voltage of

$$V_o(\text{offset due to } I_{\text{IB}}^+ \text{ and } I_{\text{IB}}^-) = I_{\text{IB}}^+ R_C\left(1 + \frac{R_f}{R_1}\right) - I_{\text{IB}}^- R_1 \frac{R_f}{R_1} \qquad (13.17)$$

Since the main consideration is the difference between the input bias currents rather than each value, we define the offset current I_{IO} by

$$I_{\text{IO}} = I_{\text{IB}}^+ - I_{\text{IB}}^-$$

Since the compensating resistance R_C is usually approximately equal to the value of R_1, using $R_C = R_1$ in Eq. (13.17) we can write

$$V_o(\text{offset}) = I_{\text{IB}}^+(R_1 + R_f) - I_{\text{IB}}^- R_f$$
$$= I_{\text{IB}}^+ R_f - I_{\text{IB}}^- R_f = R_f(I_{\text{IB}}^+ - I_{\text{IB}}^-)$$

resulting in

$$\boxed{V_o(\text{offset due to } I_{\text{IO}}) = I_{\text{IO}} R_f} \qquad (13.18)$$

EXAMPLE 13.7

Calculate the offset voltage for the circuit of Fig. 13.24 for op-amp specification listing $I_{\text{IO}} = 100$ nA.

Solution

Eq. (13.18): $V_o = I_{\text{IO}} R_f = (100 \text{ nA})(150 \text{ k}\Omega) = \mathbf{15\ mV}$

TOTAL OFFSET DUE TO V_{IO} AND I_{IO}

Since the op-amp output may have an output offset voltage due to both factors covered above, the total output offset voltage can be expressed as

$$|V_o(\text{offset})| = |V_o(\text{offset due to } V_{\text{IO}})| + |V_o(\text{offset due to } I_{\text{IO}})| \qquad (13.19)$$

The absolute magnitude is used to accommodate the fact that the offset polarity may be either positive or negative.

EXAMPLE 13.8

Calculate the total offset voltage for the circuit of Fig. 13.27 for an op-amp with specified values of input offset voltage, $V_{\text{IO}} = 4$ mV and input offset current $I_{\text{IO}} = 150$ nA.

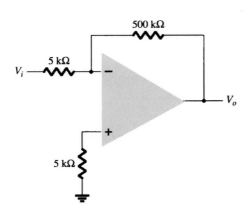

Figure 13.27 Op-amp circuit for Example 13.8.

Solution

The offset due to V_{IO} is

$$\text{Eq. (13.16):} \quad V_o(\text{offset due to } V_{IO}) = V_{IO}\frac{R_1 + R_f}{R_1} = (4 \text{ mV})\left(\frac{5 \text{ k}\Omega + 500 \text{ k}\Omega}{5 \text{ k}\Omega}\right)$$

$$= 404 \text{ mV}$$

$$\text{Eq. (13.18):} \quad V_o(\text{offset due to } I_{IO}) = I_{IO}R_f = (150 \text{ nA})(500 \text{ k}\Omega) = 75 \text{ mV}$$

resulting in a total offset

$$\text{Eq. (13.19):} \quad V_o(\text{total offset}) = V_o(\text{offset due to } V_{IO}) + V_o(\text{offset due to } I_{IO})$$

$$= 404 \text{ mV} + 75 \text{ mV} = \textbf{479 mV}$$

INPUT BIAS CURRENT, I_{IB}

A parameter related to I_{IO} and the separate input bias currents I_{IB}^+ and I_{IB}^- is the average bias current defined as

$$I_{IB} = \frac{I_{IB}^+ + I_{IB}^-}{2} \tag{13.20}$$

One could determine the separate input bias currents using the specified values I_{IO} and I_{IB}. It can be shown that for $I_{IB}^+ > I_{IB}^-$

$$I_{IB}^+ = I_{IB} + \frac{I_{IO}}{2} \tag{13.21}$$

$$I_{IB}^- = I_{IB} - \frac{I_{IO}}{2} \tag{13.21}$$

EXAMPLE 13.9

Calculate the input bias currents at each input of an op-amp having specified values of $I_{IO} = 5$ nA and $I_{IB} = 30$ nA.

Solution

Using Eq. (13.21):

$$I_{IB}^+ = I_{IB} + \frac{I_{IO}}{2} = 30 \text{ nA} + \frac{5 \text{ nA}}{2} = \textbf{32.5 nA}$$

$$I_{IB}^- = I_{IB} - \frac{I_{IO}}{2} = 30 \text{ nA} - \frac{5 \text{ nA}}{2} = \textbf{27.5 nA}$$

13.6 OP-AMP SPECIFICATIONS— FREQUENCY PARAMETERS

An op-amp is designed to be a high-gain, wide-bandwidth amplifier. This operation tends to be unstable (oscillate) due to positive feedback (see Chapter 17). To ensure stable operation, op-amps are built with internal compensation circuitry, which also causes the very high open-loop gain to diminish with increasing frequency. This gain reduction is referred to as *roll-off*. In most op-amps, roll-off occurs at a rate of 20 dB per decade (-20 dB/decade) or 6 dB per octave (-6 dB/octave). (Refer to Chapter 11 for introductory coverage of dB and frequency response.)

Note that while op-amp specifications list an open-loop voltage gain (A_{VD}), the user typically connects the op-amp using feedback resistors to reduce the circuit voltage gain to a much smaller value (closed-loop voltage gain, A_{CL}). A number of circuit improvements result from this gain reduction. First, the amplifier voltage gain is a more stable, precise value set by the external resistors; second, the input impedance of the circuit is increased over that of the op-amp alone; third, the circuit output impedance is reduced from that of the op-amp alone; and finally, the frequency response of the circuit is increased over that of the op-amp alone.

Gain–Bandwidth

Because of the internal compensation circuitry included in an op-amp, the voltage gain drops off as frequency increases. Op-amp specifications provide a description of the gain versus bandwidth. Figure 13.28 provides a plot of gain versus frequency for a typical op-amp. At low frequency down to dc operation the gain is that value listed by the manufacturer's specification A_{VD} (voltage differential gain) and is typically a very large value. As the frequency of the input signal increases the open-loop gain drops off until it finally reaches the value of 1 (unity). The frequency at this gain value is specified by the manufacturer as the unity-gain bandwidth, B_1. While this value is a frequency (see Fig. 13.28) at which the gain becomes 1, it can be considered a bandwidth, since the frequency band from 0 Hz to the unity-gain frequency is also a bandwidth. One could therefore refer to the point at which the gain reduces to 1 as the unity-gain frequency (f_1) or unity-gain bandwidth (B_1).

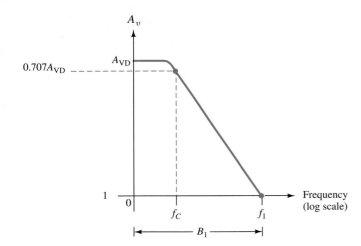

Figure 13.28 Gain versus frequency plot.

Another frequency of interest is that shown in Fig. 13.28, at which the gain drops by 3 dB (or to 0.707 the dc gain, A_{VD}), this being the cutoff frequency of the op-amp, f_C. In fact, the unity-gain frequency and cutoff frequency are related by

$$f_1 = A_{VD} f_C \tag{13.22}$$

Equation (13.22) shows that the unity-gain frequency may also be called the gain–bandwidth product of the op-amp.

EXAMPLE 13.10

Determine the cutoff frequency of an op-amp having specified values $B_1 = \mathbf{1\ MHz}$ and $A_{VD} = 200$ V/mV.

Solution

Since $f_1 = B_1 = 1$ MHz, we can use Eq. (13.22) to calculate

$$f_C = \frac{f_1}{A_{VD}} = \frac{1\ \text{MHz}}{200\ \text{V/mV}} = \frac{1 \times 10^6}{200 \times 10^3} = \mathbf{5\ Hz}$$

Slew Rate, SR

Another parameter reflecting the op-amp's ability to handling varying signals is slew rate, defined as

slew rate = maximum rate at which amplifier output can change in volts per microsecond (V/μs)

$$\boxed{\text{SR} = \frac{\Delta V_o}{\Delta t} \quad \text{V}/\mu\text{s}} \quad \text{with } t \text{ in } \mu\text{s} \qquad (13.23)$$

The slew rate provides a parameter specifying the maximum rate of change of the output voltage when driven by a large step-input signal.[*] If one tried to drive the output at a rate of voltage change greater than the slew rate, the output would not be able to change fast enough and would not vary over the full range expected, resulting in signal clipping or distortion. In any case, the output would not be an amplified duplicate of the input signal if the op-amp slew rate is exceeded.

EXAMPLE 13.11

For an op-amp having a slew rate of SR = 2 V/μs, what is the maximum closed-loop voltage gain that can be used when the input signal varies by 0.5 V in 10 μs?

Solution

Since $V_o = A_{CL}V_i$, we can use

$$\frac{\Delta V_o}{\Delta t} = A_{CL}\frac{\Delta V_i}{\Delta t}$$

from which we get

$$A_{CL} = \frac{\Delta V_o/\Delta t}{\Delta V_i/\Delta t} = \frac{\text{SR}}{\Delta V_i/\Delta t} = \frac{2\ \text{V}/\mu\text{s}}{0.5\ \text{V}/10\ \mu\text{s}} = \mathbf{40}$$

Any closed-loop voltage gain of magnitude greater than 40 would drive the output at a rate greater than the slew rate allows, so the maximum closed-loop gain is 40.

---

[*]The closed-loop gain is that obtained with the output connected back to the input in some way.

Maximum Signal Frequency

The maximum frequency that an op-amp may operate at depends on both the bandwidth (BW) and slew rate (SR) parameters of the op-amp. For a sinusoidal signal of general form

$$v_o = K \sin(2\pi f t)$$

the maximum voltage rate of change can be shown to be

$$\text{signal maximum rate of change} = 2\pi f K \quad \text{V/s}$$

To prevent distortion at the output, the rate of change must also be less than the slew rate, that is,

$$2\pi f K \leq \text{SR}$$

$$\omega K \leq \text{SR}$$

so that

$$\boxed{\begin{aligned} f &\leq \frac{\text{SR}}{2\pi K} \quad \text{Hz} \\ \omega &\leq \frac{\text{SR}}{K} \quad \text{rad/s} \end{aligned}} \tag{13.24}$$

Additionally, the maximum frequency, f, in Eq. (13.24), is also limited by the unity-gain bandwidth.

For the signal and circuit of Fig. 13.29, determine the maximum frequency that may be used. Op-amp slew rate is SR = 0.5 V/μs.

EXAMPLE 13.12

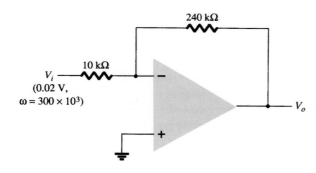

V_i
(0.02 V,
$\omega = 300 \times 10^3$)

240 kΩ

10 kΩ

V_o

Figure 13.29 Op-amp circuit for Example 13.12.

Solution

For a gain of magnitude

$$A_{\text{CL}} = \left| \frac{R_f}{R_1} \right| = \frac{240 \text{ k}\Omega}{10 \text{ k}\Omega} = 24$$

the output voltage provides

$$K = A_{\text{CL}} V_i = 24(0.02 \text{ V}) = 0.48 \text{ V}$$

Eq. (13.24): $\quad \omega \leq \dfrac{\text{SR}}{K} = \dfrac{0.5 \text{ V}/\mu\text{s}}{0.48 \text{ V}} = \mathbf{1.1 \times 10^6 \text{ rad/s}}$

Since the signal's frequency, $\omega = 300 \times 10^3$ rad/s, is less than the maximum value determined above, no output distortion will result.

13.7 OP-AMP UNIT SPECIFICATIONS

In this section, we discuss how the manufacturer's specifications are read for a typi-cal op-amp unit. A popular bipolar op-amp IC is the 741 described by the informa-tion provided in Fig. 13.30. The op-amp is available in a number of packages, an 8-pin DIP and a 10-pin flatpack being among the more usual forms.

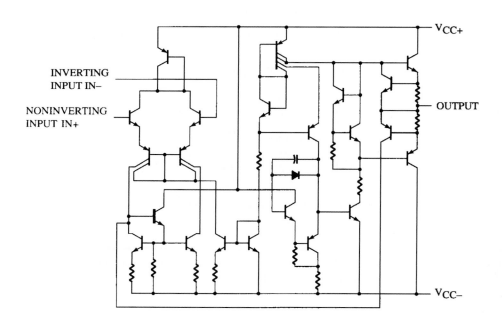

absolute maximum ratings over operating free-air temperature range (unless otherwise noted)

		uA741M	uA741C	UNIT
Supply voltage V_{CC+} (see Note 1)		22	18	V
Supply voltage V_{CC-} (see Note 1)		−22	−18	V
Differential input voltage (see Note 2)		±30	±30	V
Input voltage any input (see Notes 1 and 3)		±15	±15	V
Voltage between either offset null terminal (N1/N2) and V_{CC-}		±0.5	±0.5	V
Duration of output short-circuit (see Note 4)		unlimited	unlimited	
Continuous total power dissipation at (or below) 25°C free-air temperature (see Note 5)		500	500	mW
Operating free-air temperature range		−55 to 125	0 to 70	°C
Storage temperature range		−65 to 150	−65 to 150	°C
Lead temperature 1,6 mm (1/16 inch) from case for 60 seconds	FH, FK, J, JG, or U package	300	300	°C
Lead temperature 1,6 mm (1/16 inch) from case for 10 seconds	D, N, or P package		260	°C

NOTES: 1. All voltage values, unless otherwise noted, are with respect to the midpoint between V_{CC+} and V_{CC-}.
   2. Differential voltages are at the noninverting input terminal with respect to the inverting input terminal.
   3. The magnitude of the input voltage must never exceed the magnitude of the supply voltage or 15 volts, whichever is less.
   4. The output may be shorted to ground or either power supply. For the uA741M only, the unlimited duration of the short-circuit applies at (or below) 125°C case temperature or 75°C free-air temperature.
   5. For operation above 25°C free-air temperature, refer to Dissipation Derating Curves, Section 2. In the J and JG packages, uA741M chips are alloy mounted; uA741C chips are glass mounted.

Figure 13.30   741 op-amp specifications.

electrical characteristics at specified free-air temperature, $V_{CC+} = 15$ V, $V_{CC-} = -15$ V

PARAMETER		TEST CONDITIONS †		uA741M			uA741C			UNIT
				MIN	TYP	MAX	MIN	TYP	MAX	
V_{IO}	Input offset voltage	$V_O = 0$	25°C		1	5		1	6	mV
			Full range			6			7.5	
$\Delta V_{IO(adj)}$	Offset voltage adjust range	$V_O = 0$	25°C		± 15			± 15		mV
I_{IO}	Input offset current	$V_O = 0$	25°C		20	200		20	200	nA
			Full range			500			300	
I_{IB}	Input bias current	$V_O = 0$	25°C		80	500		80	500	nA
			Full range			1500			800	
V_{ICR}	Common-mode input voltage range		25°C	± 12	± 13		± 12	± 13		V
			Full range	± 12			± 12			
V_{OM}	Maximum peak output voltage swing	$R_L = 10$ kΩ	25°C	± 12	± 14		± 12	± 14		V
		$R_L \geq 10$ kΩ	Full range	± 12			± 12			
		$R_L = 2$ kΩ	25°C	± 10	± 13		± 10	± 13		
		$R_L \geq 2$ kΩ	Full range	± 10			± 10			
A_{VD}	Large-signal differential voltage amplification	$R_L \geq 2$ kΩ	25°C	50	200		20	200		V/mV
		$V_O = \pm 10$ V	Full range	25			15			
r_i	Input resistance		25°C	0.3	2		0.3	2		MΩ
r_o	Output resistance	$V_O = 0$ See note 6	25°C		75			75		Ω
C_i	Input capacitance		25°C		1.4			1.4		pF
CMRR	Common-mode rejection ratio	$V_{IC} = V_{ICR}$ min	25°C	70	90		70	90		dB
			Full range	70			70			
k_{SVS}	Supply voltage sensitivity $\Delta V_{IO}/\Delta V_{CC})$	$V_{CC} = \pm 9$ V to ± 15 V	25°C		30	150		30	150	μV/V
			Full range			150			150	
I_{OS}	Short-circuit output current		25°C		± 25	± 40		± 25	± 40	mA
I_{CC}	Supply current	No load, $V_O = 0$	25°C		1.7	2.8		1.7	2.8	mA
			Full range			3.3			3.3	
P_D	Total power dissipation	No load, $V_O = 0$	25°C		50	85		50	85	mW
			Full range			100			100	

operating characteristics, $V_{CC+} = 15$ V, $V_{CC-} = -15$ V, $T_A = 25$°C

PARAMETER		TEST CONDITIONS	uA741M			uA741C			UNIT
			MIN	TYP	MAX	MIN	TYP	MAX	
t_r	Rise time	$V_I = 20$ mV, $R_L = 2$ kΩ,		0.3			0.3		μs
	Overshoot factor	$C_L = 100$ pF, See Figure 1		5%			5%		
SR	Slew rate at unity gain	$V_I = 10$ V, $R_L = 2$ kΩ, $C_L = 100$ pF, See Figure 1		0.5			0.5		V/μs

Figure 13.30 Continued.

Absolute Maximum Ratings

The absolute maximum ratings provide information on what largest voltage supplies may be used, how large the input signal swing may be, and at how much power the device is capable of operating. Depending on the particular version of 741 used, the largest supply voltage is a dual supply of ±18 V or ±22 V. In addition, the IC can internally dissipate from 310 to 570 mW, depending on the IC package used. Table 13.1 summarizes some typical values to use in examples and problems.

TABLE 13.1 Absolute Maximum Ratings	
Supply voltage	±22 V
Internal power dissipation	500 mW
Differential input voltage	±30 V
Input voltage	±15 V

Determine the current draw from a dual power supply of ±12 V if the IC dissipates 500 mW.

EXAMPLE 13.13

Solution

If we assume that each supply provides half the total power to the IC, then

$$P = VI$$

$$250 \text{ mW} = 12 \text{ V}(I)$$

so that each supply must provide a current of

$$I = \frac{250 \text{ mW}}{12 \text{ V}} = \mathbf{20.83 \text{ mA}}$$

Electrical Characteristics

Electrical characteristics include many of the parameters covered earlier in this chapter. The manufacturer provides some combination of typical, minimum, or maximum values for various parameters as deemed most useful to the user. A summary is provided in Table 13.2.

TABLE 13.2 μA741 Electrical Characteristics: $V_{CC} = \pm 15$ V, $T_A = 25°C$

Characteristic	MIN	TYP	MAX	Unit
V_{IO} Input offset voltage		1	6	mV
I_{IO} Input offset current		20	200	nA
I_{IB} Input bias current		80	500	nA
V_{ICR} Common-mode input voltage range	± 12	± 13		V
V_{OM} Maximum peak output voltage swing	± 12	± 14		V
A_{VD} Large-signal differential voltage amplification	20	200		V/mV
r_i Input resistance	0.3	2		MΩ
r_o Output resistance		75		Ω
C_i Input capacitance		1.4		pF
CMRR Common-mode rejection ratio	70	90		dB
I_{CC} Supply current		1.7	2.8	mA
P_D Total power dissipation		50	85	mW

V_{IO} **Input offset voltage:** The input offset voltage is seen to be typically 1 mV, but can go as high as 6 mV. The output offset voltage is then computed based on the circuit used. If the worst condition possible is of interest, the maximum value should be used. Typical values are those more commonly expected when using the op-amp.

I_{IO} **Input offset current:** The input offset current is listed to be typically 20 nA, while the largest value expected is 200 nA.

I_{IB} **Input bias current:** The input bias current is typically 80 nA and may be as large as 500 nA.

V_{ICR} **Common-mode input voltage range:** This parameter lists the range that the input voltage may vary over (using a supply of ± 15 V), about ± 12 to ± 13 V. Inputs larger in amplitude than this value will probably result in output distortion and should be avoided.

V_{OM} **Maximum peak output voltage swing:** This parameter lists the largest value the output may vary (using a ± 15-V supply). Depending on the circuit closed-loop gain, the input signal should be limited to keep the output from varying by an amount no larger than ± 12 V, in the worst case, or by ± 14 V, typically.

A_{VD} **Large-signal differential voltage amplification:** This is the open-loop voltage gain of the op-amp. While a minimum value of 20 V/mV or 20,000 V/V is listed, the manufacturer also lists a typical value of 200 V/mV or 200,000 V/V.

r_i **Input resistance:** The input resistance of the op-amp when measured under open-loop is typically 2 MΩ but could be as little as 0.3 MΩ or 300 kΩ. In a closed-loop circuit, this input impedance can be much larger, as discussed previously.

r_o **Output resistance:** The op-amp output resistance is listed as typically 75 Ω. No minimum or maximum value is given by the manufacturer for this op-amp. Again, in a closed-loop circuit, the output impedance can be lower, depending on the circuit gain.

C_i **Input capacitance:** For high-frequency considerations, it is helpful to know that the input to the op-amp has typically 1.4 pF of capacitance, a generally small value compared even to stray wiring.

CMRR Common-mode rejection ratio: The op-amp parameter is seen to be typically 90 dB but could go as low as 70 dB. Since 90 dB is equivalent to 31622.78, the op-amp amplifies noise (common inputs) by over 30,000 times less than difference inputs.

I_{CC} **Supply current:** The op-amp draws a total of 2.8 mA, typically from the dual voltage supply, but the current drawn could be as little as 1.7 mA. This parameter helps the user determine the size of the voltage supply to use. It also can be used to calculate the power dissipated by the IC ($P_D = 2V_{CC}I_{CC}$).

P_D **Total power dissipation:** The total power dissipated by the op-amp is typically 50 mW but could go as high as 85 mW. Referring to the previous parameter, the op-amp will dissipate about 50 mW when drawing about 1.7 mA using a dual 15-V supply. At smaller supply voltages, the current drawn will be less and the total power dissipated will also be less.

Using the specifications listed in Table 13.2, calculate the typical output offset voltage for the circuit connection of Fig. 13.31.

EXAMPLE 13.14

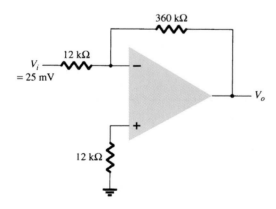

Figure 13.31 Op-amp circuit for Examples 13.14, 13.15, and 13.17.

Solution

The output offset due to V_{IO} is calculated to be

Eq. (13.16): $V_o(\text{offset}) = V_{IO}\dfrac{R_1 + R_f}{R_1} = (1 \text{ mV})\left(\dfrac{12 \text{ k}\Omega + 360 \text{ k}\Omega}{12 \text{ k}\Omega}\right) = 31 \text{ mV}$

The output voltage due to I_{IO} is calculated to be

Eq. (13.18): $V_o(\text{offset}) = I_{IO}R_f = 20 \text{ nA}(360 \text{ k}\Omega) = 7.2 \text{ mV}$

Assuming that these two offsets are the same polarity at the output, the total output offset voltage is then

$$V_o(\text{offset}) = 31 \text{ mV} + 7.2 \text{ mV} = \mathbf{38.2 \text{ mV}}$$

EXAMPLE 13.15

For the typical characteristics of the 741 op-amp ($r_o = 75 \text{ }\Omega, A = 200 \text{ k}\Omega$), calculate the following values for the circuit of Fig. 13.31.
(a) A_{CL}.
(b) Z_i.
(c) Z_o.

Solution

(a) Eq. (13.8): $\dfrac{V_o}{V_i} = -\dfrac{R_f}{R_1} = -\dfrac{360 \text{ k}\Omega}{12 \text{ k}\Omega} = \mathbf{-30} \cong \dfrac{1}{\beta}$

(b) $Z_i = R_1 = \mathbf{12 \text{ k}\Omega}$

(c) $Z_o = \dfrac{r_o}{(1 + \beta A)} = \dfrac{75 \text{ }\Omega}{1 + \left(\dfrac{1}{30}\right)(200 \text{ k}\Omega)} = \mathbf{0.011 \text{ }\Omega}$

Operating Characteristics

Another group of values used to describe the operation of the op-amp over varying signals is provided in Table 13.3.

TABLE 13.3 Operating Characteristics: $V_{CC} = \pm 15 \text{ V}, T_A = 25°C$

Parameter	MIN	TYP	MAX	Unit
B_1 Unity gain bandwidth		1		MHz
t_r Rise time		0.3		μs

EXAMPLE 13.16

Calculate the cutoff frequency of an op-amp having characteristics given in Tables 13.2 and 13.3.

Solution

$$\text{Eq. (13.22):} \quad f_C = \frac{f_1}{A_{VD}} = \frac{B_1}{A_{VD}} = \frac{1 \text{ MHz}}{20,000} = \mathbf{50 \text{ Hz}}$$

EXAMPLE 13.17

Calculate the maximum frequency of the input signal for the circuit in Fig. 13.31, with an input of $V_i = 25 \text{ mV}$.

Solution

For a closed-loop gain of $A_{CL} = 30$ and an input of $V_i = 25 \text{ mV}$, the output gain factor is calculated to be

$$K = A_{CL}V_i = 30(25 \text{ mV}) = 750 \text{ mV} = 0.750 \text{ V}$$

Using Eq. (13.24), the maximum signal frequency, f_{max}, is

$$f_{max} = \frac{SR}{2\pi K} = \frac{0.5 \text{ V}/\mu s}{2\pi(0.750 \text{ V})} = \textbf{106 kHz}$$

Op-Amp Performance

The manufacturer provides a number of graphical descriptions to describe the performance of the op-amp. Figure 13.32 includes some typical performance curves comparing various characteristics as a function of supply voltage. The open-loop voltage gain is seen to get larger with a larger supply voltage value. While the previous tabular information provided information at a particular supply voltage, the performance curve shows how the voltage gain is affected by using a range of supply voltage values.

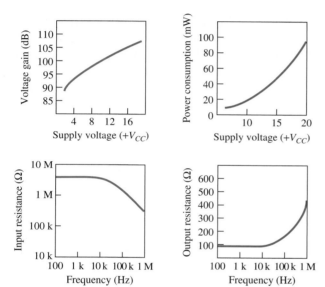

Figure 13.32 Performance curves.

Using Fig. 13.32, determine the open-loop voltage gain for a supply voltage of $V_{CC} = \pm 12$ V.

EXAMPLE 13.18

Solution

From the curve in Fig. 13.32, $A_{VD} \approx 104$ dB. This is a linear voltage gain of

$$A_{VD}\text{(dB)} = 20 \log_{10} A_{VD}$$

$$104 \text{ dB} = 20 \log A_{VD}$$

$$A_{VD} = \text{antilog} \frac{104}{20} = 158.5 \times 10^3$$

Another performance curve in Fig. 13.32 shows how power consumption varies as a function of supply voltage. As shown, the power consumption increases with larger values of supply voltage. For example, while the power dissipation is about 50 mW at $V_{CC} = \pm 15$ V, it drops to about 5 mW with $V_{CC} = \pm 5$ V. Two other curves show how the input and output resistances are affected by frequency, the input impedance dropping and the output resistance increasing at higher frequency.

13.8   SUMMARY

Important Conclusions and Concepts

1. Differential operation involves the use of opposite polarity inputs.
2. Common-mode operation involves the use of the same polarity inputs.
3. Common-mode rejection compares the gain for differential inputs to that for common inputs.
4. An op-amp is an **op**erational **amp**lifier.
5. The basic features of an op-amp are:

   Very high input impedance (typically megohms)
   Very high voltage gain (typically a few hundred thousand and greater)
   Low output impedance (typically less than 100 Ω)

6. Virtual ground is the concept based on the practical fact that the differential input voltage between plus (+) and minus (−) inputs is nearly (virtually) zero volts—when calculated as the output voltage (at most, that of the voltage supply) divided by the very high voltage gain of the op-amp.
7. Basic op-amp connections include:

   Inverting amplifier
   Noninverting amplifier
   Unity-gain amplifier
   Summing amplifier
   Integrator amplifier

8. Op-amp specs include:

   Offset voltages and currents
   Frequency parameters
   Gain–bandwidth
   Slew rate

Equations

$$\text{CMRR} = 20 \log_{10} \frac{A_d}{A_c}$$

Inverting amplifier:

$$\frac{V_o}{V_i} = -\frac{R_f}{R_1}$$

Noninverting amplifier:

$$\frac{V_o}{V_i} = 1 + \frac{R_f}{R_1}$$

Unity follower:

$$V_o = V_1$$

Summing amplifier:

$$V_o = -\left(\frac{R_f}{R_1}V_1 + \frac{R_f}{R_2}V_2 + \frac{R_f}{R_3}V_3\right)$$

Integrator amplifier:

$$v_o(t) = -\frac{1}{RC}\int v_1(t)\,dt$$

$$\text{Slew rate (SR)} = \frac{\Delta V_o}{\Delta t} \quad \text{V}/\mu s$$

13.9   COMPUTER ANALYSIS

PSpice Windows

Program 13.1—Inverting Op-Amp

An inverting op-amp, shown in Fig. 13.33, is considered first. With the dc voltage display turned on, the result after running an analysis shows that for an input of 2 V and a circuit gain of −5,

$$A_v = -R_F/R_1 = -500 \text{ k}\Omega/100 \text{ k}\Omega = -5$$

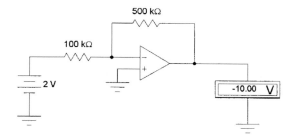

Figure 13.33   Inverting op-amp using ideal model.

The output is exactly −10 V.

$$V_o = A_v V_i = -5(2 \text{ V}) = -10 \text{ V}$$

The input to the minus terminal is −50.01 μV, which is virtually ground or 0 V.

A practical inverting op-amp circuit is drawn in Fig. 13.34. Using the same resistor values as in Fig. 13.33 with a practical op-amp unit, the μA741, the resulting output is −9.96 V, near the ideal value of −10 V. This slight difference from the ideal is due to the actual gain and input impedance of the μA741 op-amp unit.

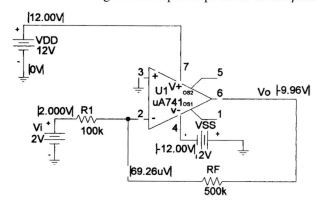

Figure 13.34   Practical inverting op-amp circuit.

Before the analysis is done, selecting **Analysis Setup, Transfer Function,** and then **Output** of **V(RF:2)** and **Input Source** of V_i will provide the small-signal characteristics in the output listing. The circuit gain is seen to be

$$V_o/V_i = -5$$

$$\text{Input resistance at } V_i = 1 \times 10^5$$

$$\text{Output resistance at } V_o = 4.95 \times 10^{-3}$$

Program 13.2—Noninverting Op-Amp

Figure 13.35 shows a noninverting op-amp circuit. The bias voltages are displayed on the figure. The theoretical gain of the amplifier circuit should be

$$A_v = (1 + RF/R1) = 1 + 500 \text{ k}\Omega/100 \text{ k}\Omega = 6$$

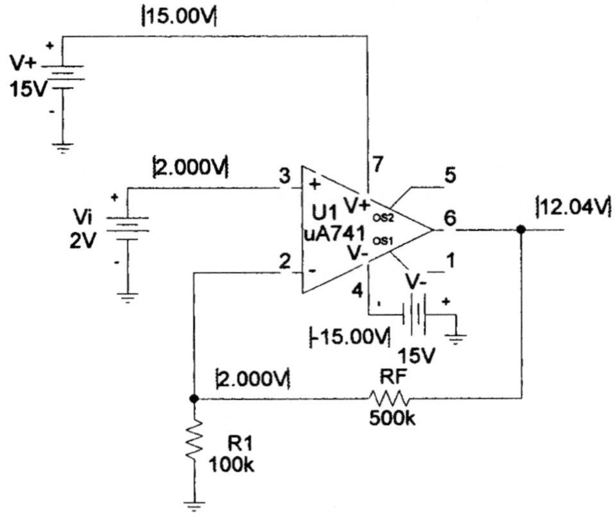

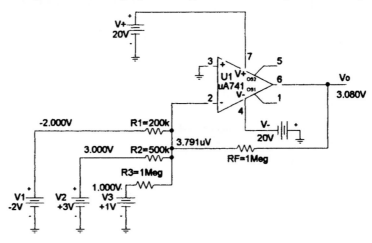

Figure 13.35 Design Center schematic for noninverting op-amp circuit.

For an input of 2 V, the resulting output will be

$$V_o = A_v V_i = 5(2 \text{ V}) = 10 \text{ V}$$

The output is noninverted from the input.

Program 13.3—Summing Op-Amp Circuit

A summing op-amp circuit such as that in Example 13.5 is shown in Fig. 13.36. Bias voltages also are displayed in Fig. 13.36, showing the resulting output at 3 V, as was

Figure 13.36 Summing amplifier for Program 13.3.

calculated in Example 13.5. Notice how well the virtual ground concept works with the minus input being only 3.791 μV.

Program 13.4—Unity-Gain Op-Amp Circuit

Figure 13.37 shows a unity-gain op-amp circuit with bias voltages displayed. For an input of +2 V, the output is exactly +2 V.

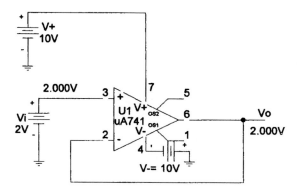

Figure 13.37 Unity-gain amplifier.

Program 13.5—Op-Amp Integrator Circuit

An op-amp integrator circuit is shown in Fig. 13.38. The input is selected as **VPULSE,** which is set to be a step input as follows:

Set **ac** = 0, **dc** = 0, **V1** = 0 V, **V2** = 2 V, **TD** = 0, **TR** = 0, **TF** = 0, **PW** = 10 ms, and **PER** = 20 ms. This provides a step from 0 to 2 V, with no time delay, rise time or fall time, having a period of 10 ms and repeating after a period of 20 ms. For this problem, the voltage rises instantly to 2 V, then stays there for a sufficiently long time for the output to drop as a ramp voltage from the maximum supply level of +20 V to the lowest level of −20 V. Theoretically, the output for the circuit of Fig. 13.38 is

$$v_o(t) = -1/RC \int v_i(t)\, dt$$

$$v_o(t) = -1/(10\ \text{k}\Omega)(0.01\ \mu\text{F}) \int 2\, dt = -10,000 \int 2\, dt = -20,000t$$

This is a negative ramp voltage dropping at a rate (slope) of −20,000 V/s. This ramp voltage will drop from +20 V to −20 V in

$$40\ \text{V}/20,000 = 2 \times 10^{-3} = 2\ \text{ms}$$

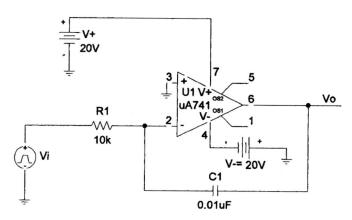

Figure 13.38 Op-amp integrator circuit.

Figure 13.39 shows the input step waveform and the resulting output ramp waveform obtained using **PROBE.**

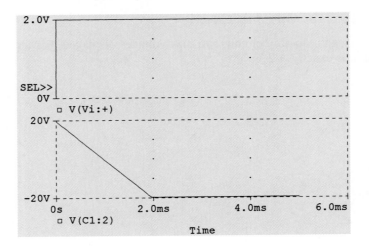

Figure 13.39 Probe waveform for integrator circuit.

Electronics Workbench

The same integrator circuit can be constructed and operated using EWB. Figure 13.40a shows the integrator circuit built using EWB, with an oscilloscope connected to the op-amp output. The oscilloscope graph obtained is shown in Fig. 13.40b, the linear output waveform going from $+20$ V down to -20 V in a period of about 2 ms.

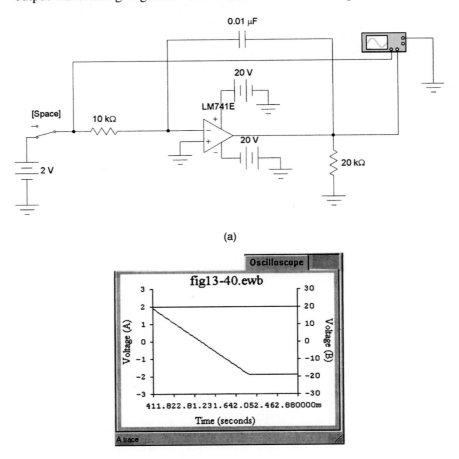

(a)

(b)

Figure 13.40 EWB integrator circuit: (a) circuit; (b) waveform.

Program 13.6—Multistage Op-Amp Circuit

A multistage op-amp circuit is shown in Fig. 13.41. The input to stage 1 of 200 mV provides an output of 200 mV to stages 2 and 3. Stage 2 is an inverting amplifier with gain $-200\,k\Omega/20\,k\Omega = -10$, with an output from stage 2 of $-10(200\,mV) = -2\,V$. State 3 is a noninverting amplifier with gain of $(1 + 200\,k\Omega/10\,k\Omega = 21)$, resulting in an output of $21(200\,mV) = 4.2\,V$.

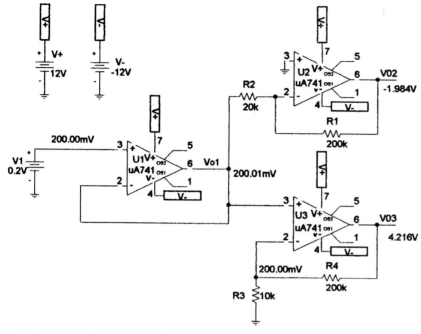

Figure 13.41   Multistage op-amp circuit.

---

§ 13.2 Differential and Common-Mode Operation

1. Calculate the CMRR (in dB) for the circuit measurements of $V_d = 1\,mV$, $V_o = 120\,mV$, and $V_C = 1\,mV$, $V_o = 20\,\mu V$.

2. Determine the output voltage of an op-amp for input voltages of $V_{i_1} = 200\,\mu V$ and $V_{i_2} = 140\,\mu V$. The amplifier has a differential gain of $A_d = 6000$ and the value of CMRR is:
   (a) 200.
   (b) 10^5.

§ 13.4 Practical Op-Amp Circuits

3. What is the output voltage in the circuit of Fig. 13.42?

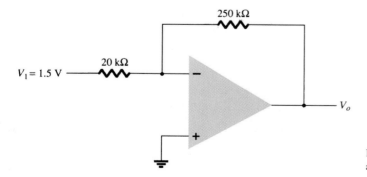

Figure 13.42   Problems 3 and 25

4. What is the range of the voltage-gain adjustment in the circuit of Fig. 13.43?

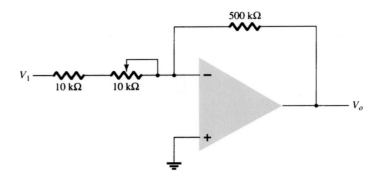

Figure 13.43 Problem 4

5. What input voltage results in an output of 2 V in the circuit of Fig. 13.44?

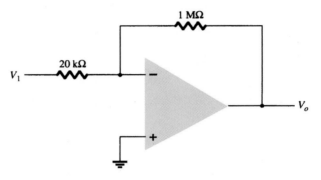

Figure 13.44 Problem 5

6. What is the range of the output voltage in the circuit of Fig. 13.45 if the input can vary from 0.1 to 0.5 V?

7. What output voltage results in the circuit of Fig. 13.46 for an input of $V_1 = -0.3$ V?

8. What input must be applied to the input of Fig. 13.46 to result in an output of 2.4 V?

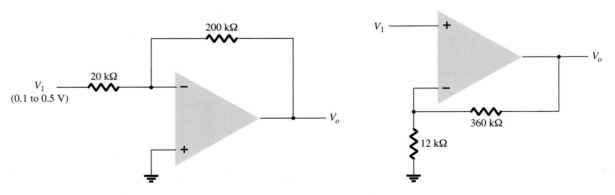

Figure 13.45 Problem 6

Figure 13.46 Problems 7, 8, and 26

Chapter 13 **Operational Amplifiers**

9. What range of output voltage is developed in the circuit of Fig. 13.47?

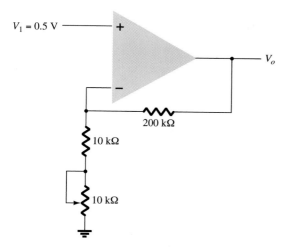

Figure 13.47 Problem 9

10. Calculate the output voltage developed by the circuit of Fig. 13.48 for $R_f = 330$ kΩ.

11. Calculate the output voltage of the circuit in Fig. 13.48 for $R_f = 68$ kΩ.

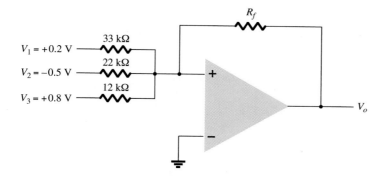

Figure 13.48 Problems 10, 11, and 27

12. Sketch the output waveform resulting in Fig. 13.49.

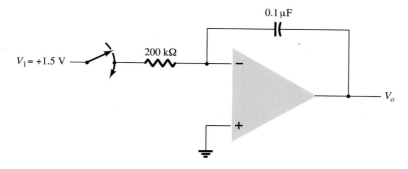

Figure 13.49 Problem 12

13. What output voltage results in the circuit of Fig. 13.50 for $V_1 = +0.5$ V?

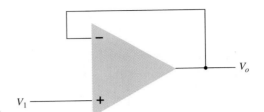

Figure 13.50 Problem 13

14. Calculate the output voltage for the circuit of Fig. 13.51.

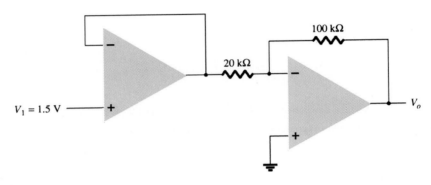

Figure 13.51 Problems 14 and 28

15. Calculate the output voltages V_2 and V_3 in the circuit of Fig. 13.52.

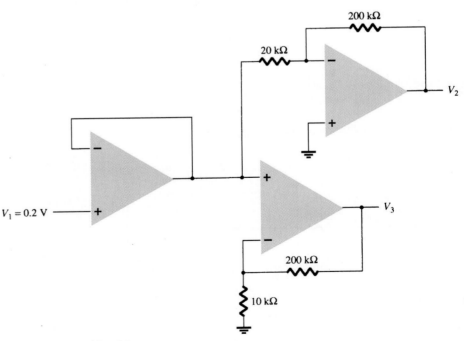

Figure 13.52 Problem 15

Chapter 13   Operational Amplifiers

16. Calculate the output voltage, V_o, in the circuit of Fig. 13.53.

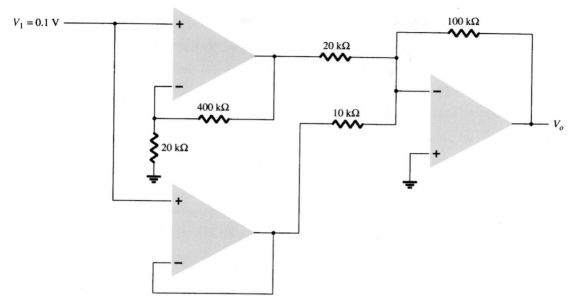

Figure 13.53   Problems 16 and 29

17. Calculate V_o in the circuit of Fig. 13.54.

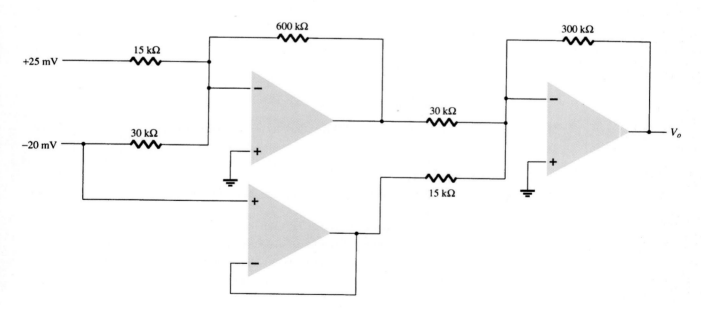

Figure 13.54   Problem 17

§ 13.5 Op-Amp Specifications—DC Offset Parameters

∗**18.** Calculate the total offset voltage for the circuit of Fig. 13.55 for an op-amp with specified values of input offset voltage $V_{IO} = 6$ mV and input offset current $I_{IO} = 120$ nA.

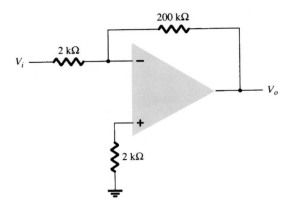

Figure 13.55 Problems 18, 22, 23, and 24

* **19.** Calculate the input bias current at each input of an op-amp having specified values of $I_{IO} = 4$ nA and $I_{IB} = 20$ nA.

§ 13.6 Op-Amp Specifications—Frequency Parameters

20. Determine the cutoff frequency of an op-amp having specified values $B_1 = 800$ kHz and $A_{VD} = 150$ V/mV.

* **21.** For an op-amp having a slew rate of SR $= 2.4$ V/μs, what is the maximum closed-loop voltage gain that can be used when the input signal varies by 0.3 V in 10 μs?

* **22.** For an input of $V_1 = 50$ mV in the circuit of Fig. 13.55, determine the maximum frequency that may be used. The op-amp slew rate SR $= 0.4$ V/μs.

* **23.** Using the specifications listed in Table 14.2, calculate the typical offset voltage for the circuit connection of Fig. 13.55.

* **24.** For the typical characteristics of the 741 op-amp, calculate the following values for the circuit of Fig. 13.55:
    (a) A_{CL}.
    (b) Z_i.
    (c) Z_o.

§ 13.9 Computer Analysis

* **25.** Use Schematic Capture or EWB to draw a circuit to determine the output voltage in the circuit of Fig. 13.42.

* **26.** Use Schematic Capture or EWB to calculate the output voltage in the circuit of Fig. 13.46 for the input of $V_i = 0.5$ V.

* **27.** Use Schematic Capture or EWB to calculate the output voltage in the circuit of Fig. 13.48 for $R_f = 68$ kΩ.

* **28.** Use Schematic Capture or EWB to calculate the output voltage in the circuit of Fig. 13.51.

* **29.** Use Schematic Capture or EWB to calculate the output voltage in the circuit of Fig. 13.53.

* **30.** Use Schematic Capture or EWB to calculate the output voltage in the circuit of Fig. 13.54.

* **31.** Use Schematic Capture or EWB to obtain the output waveform for a 2 V step input to an integrator circuit, as shown in Fig. 13.20, with values of $R = 40$ kΩ and $C = 0.003$ μF.

---

*Please Note: Asterisks indicate more difficult problems.

Chapter 13 Operational Amplifiers

Op-Amp Applications

14.1 CONSTANT-GAIN MULTIPLIER

One of the most common op-amp circuits is the inverting constant-gain multiplier, which provides a precise gain or amplification. Figure 14.1 shows a standard circuit connection with the resulting gain being given by

$$A = -\frac{R_f}{R_1} \qquad (14.1)$$

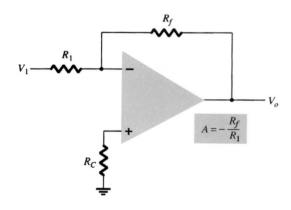

$$A = -\frac{R_f}{R_1}$$

Figure 14.1 Fixed-gain amplifier.

Determine the output voltage for the circuit of Fig. 14.2 with a sinusoidal input of 2.5 mV.

EXAMPLE 14.1

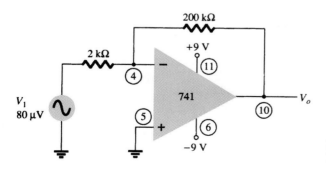

Figure 14.2 Circuit for Example 14.1.

Solution

The circuit of Fig. 14.2 uses a 741 op-amp to provide a constant or fixed gain, calculated from Eq. (14.1) to be

$$A = -\frac{R_f}{R_1} = -\frac{200 \text{ k}\Omega}{2 \text{ k}\Omega} = -100$$

The output voltage is then

$$V_o = AV_i = -100(2.5 \text{ mV}) = -250 \text{ mV} = \mathbf{-0.25 \text{ V}}$$

A noninverting constant-gain multiplier is provided by the circuit of Fig. 14.3, with the gain given by

$$\boxed{A = 1 + \frac{R_f}{R_1}}$$ (14.2)

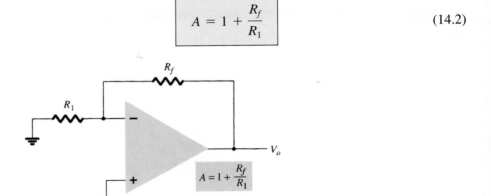

$$A = 1 + \frac{R_f}{R_1}$$

Figure 14.3 Noninverting fixed-gain amplifier.

EXAMPLE 14.2

Calculate the output voltage from the circuit of Fig. 14.4 for an input of $120 \ \mu V$.

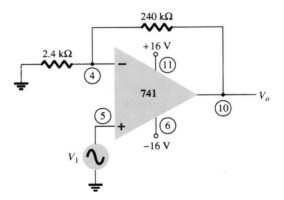

Figure 14.4 Circuit for Example 14.2.

Solution

The gain of the op-amp circuit is calculated using Eq. (14.2) to be

$$A = 1 + \frac{R_f}{R_1} = 1 + \frac{240 \text{ k}\Omega}{2.4 \text{ k}\Omega} = 1 + 100 = 101$$

The output voltage is then

$$V_o = AV_i = 101(120 \ \mu V) = \mathbf{12.12 \text{ mV}}$$

Chapter 14 Op-Amp Applications

Multiple-Stage Gains

When a number of stages are connected in series, the overall gain is the product of the individual stage gains. Figure 14.5 shows a connection of three stages. The first stage is connected to provide noninverting gain as given by Eq. (14.2). The next two stages provide an inverting gain given by Eq. (14.1). The overall circuit gain is then noninverting and calculated by

$$A = A_1 A_2 A_3$$

where $A_1 = 1 + R_f/R_1$, $A_2 = -R_f/R_2$, and $A_3 = -R_f/R_3$.

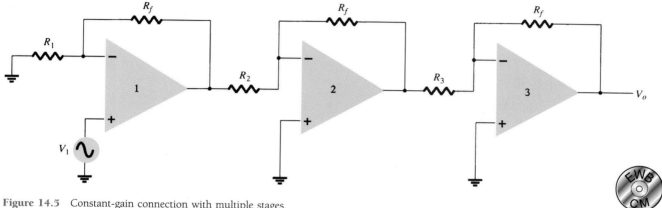

Figure 14.5 Constant-gain connection with multiple stages.

Calculate the output voltage using the circuit of Fig. 14.5 for resistor components of value $R_f = 470 \text{ k}\Omega$, $R_1 = 4.3 \text{ k}\Omega$, $R_2 = 33 \text{ k}\Omega$, and $R_3 = 33 \text{ k}\Omega$ for an input of $80 \ \mu\text{V}$.

EXAMPLE 14.3

Solution

The amplifier gain is calculated to be

$$A = A_1 A_2 A_3 = \left(1 + \frac{R_f}{R_1}\right)\left(-\frac{R_f}{R_2}\right)\left(-\frac{R_f}{R_3}\right)$$

$$= \left(1 + \frac{470 \text{ k}\Omega}{4.3 \text{ k}\Omega}\right)\left(-\frac{470 \text{ k}\Omega}{33 \text{ k}\Omega}\right)\left(-\frac{470 \text{ k}\Omega}{33 \text{ k}\Omega}\right)$$

$$= (110.3)(-14.2)(-14.2) = 22.2 \times 10^3$$

so that

$$V_o = A V_i = 22.2 \times 10^3 (80 \ \mu\text{V}) = \mathbf{1.78 \ V}$$

Show the connection of an LM124 quad op-amp as a three-stage amplifier with gains of $+10$, -18, and -27. Use a 270-kΩ feedback resistor for all three circuits. What output voltage will result for an input of $150 \ \mu\text{V}$?

EXAMPLE 14.4

Solution

For the gain of $+10$:

$$A_1 = 1 + \frac{R_f}{R_1} = +10$$

$$\frac{R_f}{R_1} = 10 - 1 = 9$$

$$R_1 = \frac{R_f}{9} = \frac{270 \text{ k}\Omega}{9} = 30 \text{ k}\Omega$$

For the gain of -18:

$$A_2 = -\frac{R_f}{R_2} = -18$$

$$R_2 = \frac{R_f}{18} = \frac{270 \text{ k}\Omega}{18} = 15 \text{ k}\Omega$$

For the gain of -27:

$$A_3 = -\frac{R_f}{R_3} = -27$$

$$R_3 = \frac{R_f}{27} = \frac{270 \text{ k}\Omega}{27} = 10 \text{ k}\Omega$$

The circuit showing the pin connections and all components used is in Fig. 14.6. For an input of $V_1 = 150 \ \mu\text{V}$, the output voltage will be

$$V_o = A_1 A_2 A_3 V_1 = (10)(-18)(-27)(150 \ \mu\text{V}) = 4860(150 \ \mu\text{V})$$
$$= \mathbf{0.729 \ V}$$

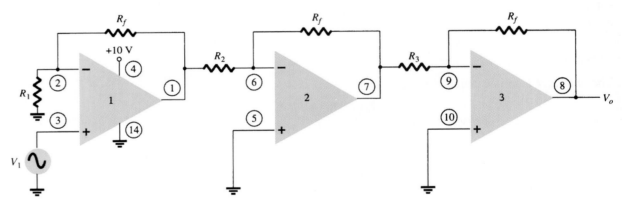

Figure 14.6   Circuit for Example 14.4 (using LM124).

A number of op-amp stages could also be used to provide separate gains, as demonstrated in the next example.

EXAMPLE 14.5

Show the connection of three op-amp stages using an LM348 IC to provide outputs that are 10, 20, and 50 times larger than the input. Use a feedback resistor of $R_f = 500 \text{ k}\Omega$ in all stages.

Solution

The resistor component for each stage is calculated to be

$$R_1 = -\frac{R_f}{A_1} = -\frac{500 \text{ k}\Omega}{-10} = 50 \text{ k}\Omega$$

$$R_2 = -\frac{R_f}{A_2} = -\frac{500 \text{ k}\Omega}{-20} = 25 \text{ k}\Omega$$

$$R_3 = -\frac{R_f}{A_3} = -\frac{500 \text{ k}\Omega}{-50} = 10 \text{ k}\Omega$$

The resulting circuit is drawn in Fig. 14.7.

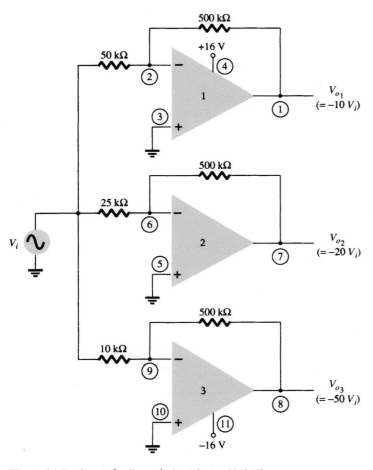

Figure 14.7 Circuit for Example 14.5 (using LM348).

14.2 VOLTAGE SUMMING

Another popular use of an op-amp is as a summing amplifier. Figure 14.8 shows the connection with the output being the sum of the three inputs, each multiplied by a different gain. The output voltage is

$$V_o = -\left(\frac{R_f}{R_1}V_1 + \frac{R_f}{R_2}V_2 + \frac{R_f}{R_3}V_3\right) \qquad (14.3)$$

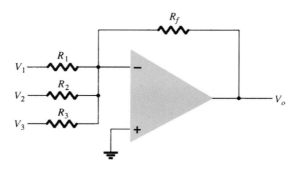

Figure 14.8 Summing amplifier.

EXAMPLE 14.6

Calculate the output voltage for the circuit of Fig. 14.9. The inputs are $V_1 = 50$ mV sin$(1000t)$ and $V_2 = 10$ mV sin$(3000t)$.

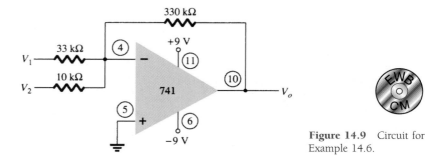

Figure 14.9 Circuit for Example 14.6.

Solution

The output voltage is

$$V_o = -\left(\frac{330 \text{ k}\Omega}{33 \text{ k}\Omega} V_1 + \frac{330 \text{ k}\Omega}{10 \text{ k}\Omega} V_2\right) = -(10V_1 + 33V_2)$$

$$= -[10(50 \text{ mV}) \sin(1000t) + 33(10 \text{ mV}) \sin(3000t)]$$

$$= -[\mathbf{0.5 \sin(1000t) + 0.33 \sin(3000t)}]$$

Voltage Subtraction

Two signals can be subtracted, one from the other, in a number of ways. Figure 14.10 shows two op-amp stages used to provide subtraction of input signals. The resulting output is given by

Figure 14.10 Circuit to subtract two signals.

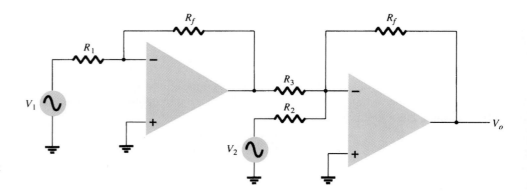

$$V_o = -\left[\frac{R_f}{R_3}\left(-\frac{R_f}{R_1}V_1\right) + \frac{R_f}{R_2}V_2\right]$$

$$V_o = -\left(\frac{R_f}{R_2}V_2 - \frac{R_f}{R_3}\frac{R_f}{R_1}V_1\right) \qquad (14.4)$$

Determine the output for the circuit of Fig. 14.10 with components $R_f = 1\ M\Omega$, $R_1 = 100\ k\Omega$, $R_2 = 50\ k\Omega$, and $R_3 = 500\ k\Omega$.

EXAMPLE 14.7

Solution

The output voltage is calculated to be

$$V_o = -\left(\frac{1\ M\Omega}{50\ k\Omega}V_2 - \frac{1\ M\Omega}{500\ k\Omega}\frac{1\ M\Omega}{100\ k\Omega}V_1\right) = -(20V_2 - 20V_1) = \mathbf{-20(V_2 - V_1)}$$

The output is seen to be the difference of V_2 and V_1 multiplied by a gain factor of -20.

Another connection to provide subtraction of two signals is shown in Fig. 14.11. This connection uses only one op-amp stage to provide subtracting two input signals. Using superposition the output can be shown to be

$$V_o = \frac{R_3}{R_1 + R_3}\frac{R_2 + R_4}{R_2}V_1 - \frac{R_4}{R_2}V_2 \qquad (14.5)$$

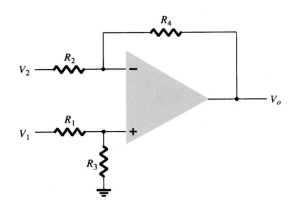

Figure 14.11 Subtraction circuit.

Determine the output voltage for the circuit of Fig. 14.12.

EXAMPLE 14.8

Figure 14.12 Circuit for Example 14.8.

Solution

The resulting output voltage can be expressed as

$$V_o = \left(\frac{20\text{ k}\Omega}{20\text{ k}\Omega + 20\text{ k}\Omega}\right)\left(\frac{100\text{ k}\Omega + 100\text{ k}\Omega}{100\text{ k}\Omega}\right)V_1 - \frac{100\text{ k}\Omega}{100\text{ k}\Omega}V_2$$

$$= V_1 - V_2$$

The resulting output voltage is seen to be the difference of the two input voltages.

14.3   VOLTAGE BUFFER

A voltage buffer circuit provides a means of isolating an input signal from a load by using a stage having unity voltage gain, with no phase or polarity inversion, and acting as an ideal circuit with very high input impedance and low output impedance. Figure 14.13 shows an op-amp connected to provide this buffer amplifier operation. The output voltage is determined by

$$\boxed{V_o = V_1} \tag{14.6}$$

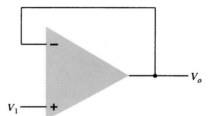

Figure 14.13   Unity-gain (buffer) amplifier.

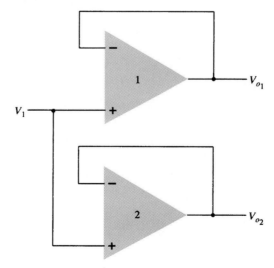

Figure 14.14   Use of buffer amplifier to provide output signals.

Figure 14.14 shows how an input signal can be provided to two separate outputs. The advantage of this connection is that the load connected across one output has no (or little) effect on the other output. In effect, the outputs are buffered or isolated from each other.

EXAMPLE 14.9

Show the connection of a 741 as a unity-gain circuit.

Solution

The connection is shown in Fig. 14.15.

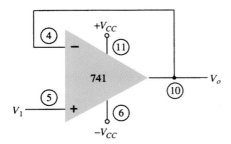

Figure 14.15 Connection for Example 14.9.

14.4 CONTROLLED SOURCES

Operational amplifiers can be used to form various types of controlled sources. An input voltage can be used to control an output voltage or current, or an input current can be used to control an output voltage or current. These types of connections are suitable for use in various instrumentation circuits. A form of each type of controlled source is provided next.

Voltage-Controlled Voltage Source

An ideal form of a voltage source whose output V_o is controlled by an input voltage V_1 is shown in Fig. 14.16. The output voltage is seen to be dependent on the input voltage (times a scale factor k). This type of circuit can be built using an op-amp as shown in Fig. 14.17. Two versions of the circuit are shown, one using the inverting input, the other the noninverting input. For the connection of Fig. 14.17a, the output voltage is

$$V_o = -\frac{R_f}{R_1}V_1 = kV_1 \qquad (14.7)$$

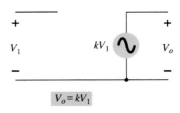

Figure 14.16 Ideal voltage-controlled voltage source.

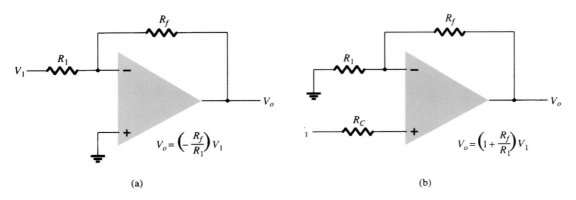

Figure 14.17 Practical voltage-controlled voltage source circuits.

while that of Fig. 14.17b results in

$$V_o = \left(1 + \frac{R_f}{R_1}\right)V_1 = kV_1 \qquad (14.8)$$

Voltage-Controlled Current Source

An ideal form of circuit providing an output current controlled by an input voltage is that of Fig. 14.18. The output current is dependent on the input voltage. A practical circuit can be built, as in Fig. 14.19, with the output current through load resistor R_L controlled by the input voltage V_1. The current through load resistor R_L can be seen to be

$$I_o = \frac{V_1}{R_1} = kV_1 \qquad (14.9)$$

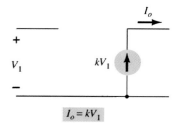

Figure 14.18 Ideal voltage-controlled current source.

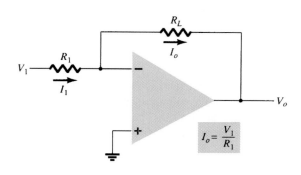

Figure 14.19 Practical voltage-controlled current source.

Current-Controlled Voltage Source

An ideal form of a voltage source controlled by an input current is shown in Fig. 14.20. The output voltage is dependent on the input current. A practical form of the circuit is built using an op-amp as shown in Fig. 14.21. The output voltage is seen to be

$$V_o = -I_1 R_L = kI_1 \qquad (14.10)$$

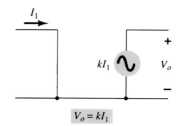

Figure 14.20 Ideal current-controlled voltage source.

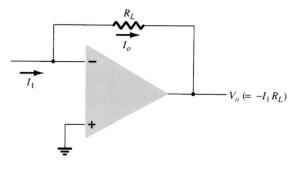

Figure 14.21 Practical form of current-controlled voltage source.

Current-Controlled Current Source

An ideal form of a circuit providing an output current dependent on an input current is shown in Fig. 14.22. In this type of circuit, an output current is provided dependent on the input current. A practical form of the circuit is shown in Fig. 14.23. The input current I_1 can be shown to result in the output current I_o so that

$$I_o = I_1 + I_2 = I_1 + \frac{I_1 R_1}{R_2} = \left(1 + \frac{R_1}{R_2}\right) I_1 = k I_1 \qquad (14.11)$$

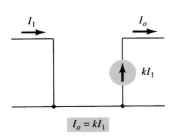

Figure 14.22 Ideal current-controlled current source.

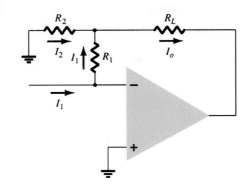

Figure 14.23 Practical form of current-controlled current source.

EXAMPLE 14.10

(a) For the circuit of Fig. 14.24a, calculate I_L.
(b) For the circuit of Fig. 14.24b, calculate V_o.

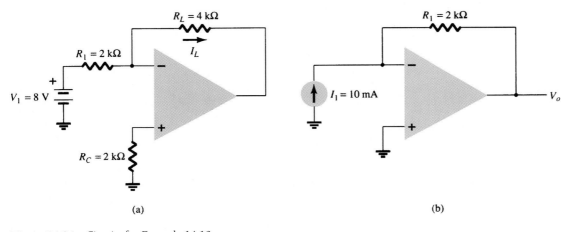

(a)   (b)

Figure 14.24 Circuits for Example 14.10.

Solution

(a) For the circuit of Fig. 14.24a,

$$I_L = \frac{V_1}{R_1} = \frac{8\ \text{V}}{2\ \text{k}\Omega} = \mathbf{4\ mA}$$

(b) For the circuit of Fig. 14.24b,

$$V_o = -I_1 R_1 = -(10\ \text{mA})(2\ \text{k}\Omega) = \mathbf{-20\ V}$$

14.5 INSTRUMENTATION CIRCUITS

A popular area of op-amp application is in instrumentation circuits such as dc or ac voltmeters. A few typical circuits will demonstrate how op-amps can be used.

DC Millivoltmeter

Figure 14.25 shows a 741 op-amp used as the basic amplifier in a dc millivoltmeter. The amplifier provides a meter with high input impedance and scale factors dependent only on resistor value and accuracy. Notice that the meter reading represents millivolts of signal at the circuit input. An analysis of the op-amp circuit provides the circuit transfer function

$$\left|\frac{I_o}{V_1}\right| = \frac{R_f}{R_1}\left(\frac{1}{R_S}\right) = \left(\frac{100 \text{ k}\Omega}{100 \text{ k}\Omega}\right)\left(\frac{1}{10 \text{ }\Omega}\right) = \frac{1 \text{ mA}}{10 \text{ mV}}$$

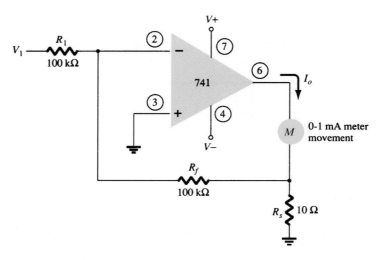

Figure 14.25 Op-amp dc millivoltmeter.

Thus, an input of 10 mV will result in a current through the meter of 1 mA. If the input is 5 mV, the current through the meter will be 0.5 mA, which is half-scale deflection. Changing R_f to 200 kΩ, for example, would result in a circuit scale factor of

$$\left|\frac{I_o}{V_1}\right| = \left(\frac{200 \text{ k}\Omega}{100 \text{ k}\Omega}\right)\left(\frac{1}{10 \text{ }\Omega}\right) = \frac{1 \text{ mA}}{5 \text{ mV}}$$

showing that the meter now reads 5 mV, full scale. It should be kept in mind that building such a millivoltmeter requires purchasing an op-amp, a few resistors, diodes, capacitors, and a meter movement.

AC Millivoltmeter

Another example of an instrumentation circuit is the ac millivoltmeter shown in Fig. 14.26. The circuit transfer function is

$$\left|\frac{I_o}{V_1}\right| = \frac{R_f}{R_1}\left(\frac{1}{R_S}\right) = \left(\frac{100 \text{ k}\Omega}{100 \text{ k}\Omega}\right)\left(\frac{1}{10 \text{ }\Omega}\right) = \frac{1 \text{ mA}}{10 \text{ mV}}$$

which appears the same as the dc millivoltmeter, except that in this case the signal handled is an ac signal. The meter indication provides a full-scale deflection for an ac input voltage of 10 mV, while an ac input of 5 mV will result in half-scale deflection with the meter reading interpreted in millivolt units.

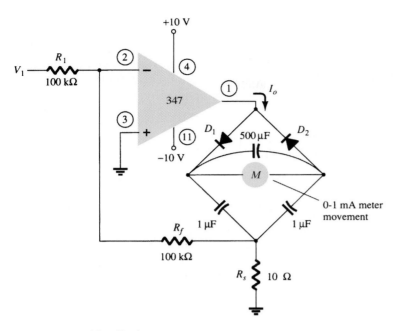

Figure 14.26 AC millivoltmeter using op-amp.

Display Driver

Figure 14.27 shows op-amp circuits that can be used to drive a lamp display or LED display. When the noninverting input to the circuit in Fig. 14.27a goes above the inverting input, the output at terminal 1 goes to the positive saturation level (near +5 V in this example) and the lamp is driven on when transistor Q_1 conducts. As shown in the circuit, the output of the op-amp provides 30 mA of current to the base of transistor Q_1, which then drives 600 mA through a suitably selected transistor (with $\beta > 20$) capable of handling that amount of current. Figure 14.27b shows an op-amp circuit that can supply 20 mA to drive an LED display when the noninverting input goes positive compared to the inverting input.

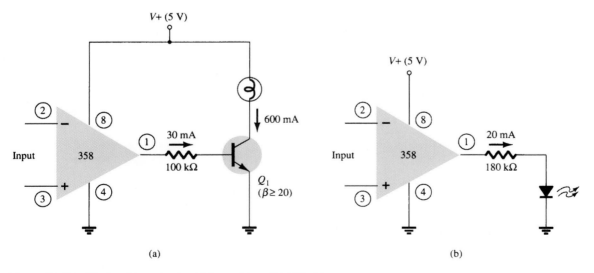

Figure 14.27 Display driver circuits: (a) lamp driver; (b) LED driver.

Instrumentation Amplifier

A circuit providing an output based on the difference between two inputs (times a scale factor) is shown in Fig. 14.28. A potentiometer is provided to permit adjusting the scale factor of the circuit. While three op-amps are used, a single-quad op-amp IC is all that is necessary (other than the resistor components). The output voltage can be shown to be

$$\frac{V_o}{V_1 - V_2} = 1 + \frac{2R}{R_P}$$

so that the output can be obtained from

$$V_o = \left(1 + \frac{2R}{R_P}\right)(V_1 - V_2) = k(V_1 - V_2) \tag{14.12}$$

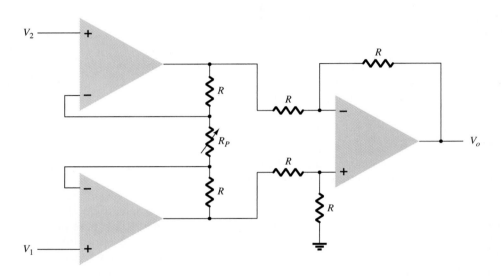

Figure 14.28 Instrumentation amplifier.

EXAMPLE 14.11

Calculate the output voltage expression for the circuit of Fig. 14.29.

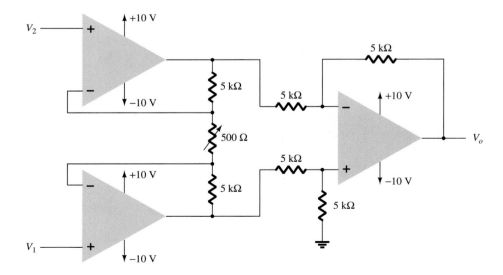

Figure 14.29 Circuit for Example 14.11.

Chapter 14   Op-Amp Applications

Solution

The output voltage can then be expressed using Eq. (14.12) as

$$V_o = \left(1 + \frac{2R}{R_P}\right)(V_1 - V_2) = \left[1 + \frac{2(5000)}{500}\right](V_1 - V_2)$$

$$= 21(V_1 - V_2)$$

14.6 ACTIVE FILTERS

A popular application uses op-amps to build active filter circuits. A filter circuit can be constructed using passive components: resistors and capacitors. An active filter additionally uses an amplifier to provide voltage amplification and signal isolation or buffering.

A filter that provides a constant output from dc up to a cutoff frequency f_{OH} and then passes no signal above that frequency is called an ideal low-pass filter. The ideal response of a low-pass filter is shown in Fig. 14.30a. A filter that provides or passes signals above a cutoff frequency f_{OL} is a high-pass filter, as idealized in Fig. 14.30b. When the filter circuit passes signals that are above one ideal cutoff frequency and below a second cutoff frequency, it is called a bandpass filter, as idealized in Fig. 14.30c.

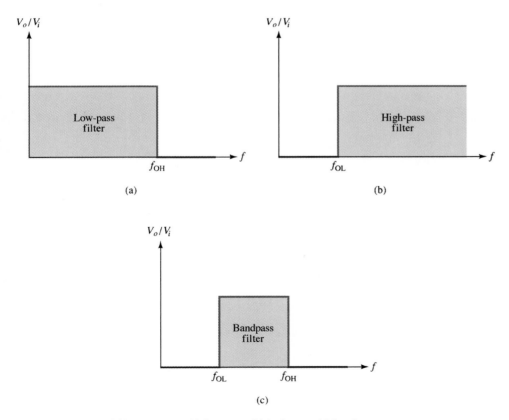

Figure 14.30 Ideal filter response: (a) low-pass; (b) high-pass; (c) bandpass.

Low-Pass Filter

A first-order, low-pass filter using a single resistor and capacitor as in Fig. 14.31a has a practical slope of −20 dB per decade, as shown in Fig. 14.31b (rather than the ideal response of Fig. 14.30a). The voltage gain below the cutoff frequency is constant at

$$A_v = 1 + \frac{R_F}{R_G} \tag{14.13}$$

at a cutoff frequency of

$$f_{OH} = \frac{1}{2\pi R_1 C_1} \tag{14.14}$$

Connecting two sections of filter as in Fig. 14.32 results in a second-order low-pass filter with cutoff at −40 dB per decade—closer to the ideal characteristic of Fig. 14.30a.

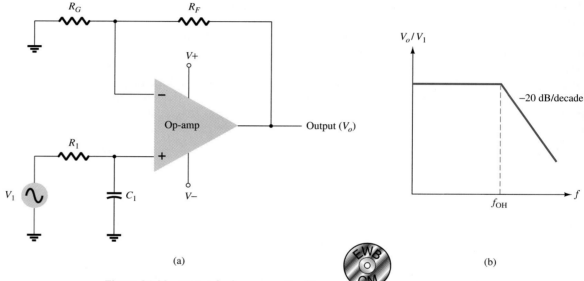

Figure 14.31 First-order low-pass active filter.

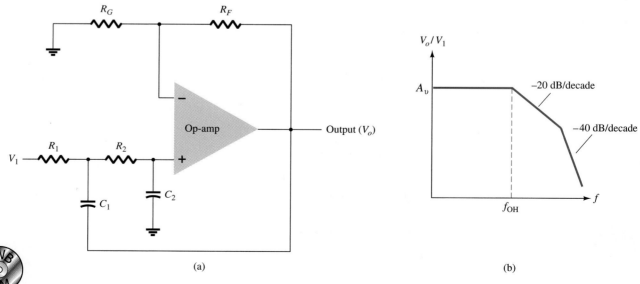

Figure 14.32 Second-order low-pass active filter.

The circuit voltage gain and cutoff frequency are the same for the second-order circuit as for the first-order filter circuit, except that the filter response drops at a faster rate for a second-order filter circuit.

EXAMPLE 14.12

Calculate the cutoff frequency of a first-order low-pass filter for $R_1 = 1.2$ kΩ and $C_1 = 0.02$ μF.

Solution

$$f_{OH} = \frac{1}{2\pi R_1 C_1} = \frac{1}{2\pi(1.2 \times 10^3)(0.02 \times 10^{-6})} = \textbf{6.63 kHz}$$

High-Pass Active Filter

First- and second-order high-pass active filters can be built as shown in Fig. 14.33. The amplifier gain is calculated using Eq. (14.13). The amplifier cutoff frequency is

$$f_{OL} = \frac{1}{2\pi R_1 C_1} \tag{14.15}$$

with a second-order filter $R_1 = R_2$, and $C_1 = C_2$ results in the same cutoff frequency as in Eq. (14.15).

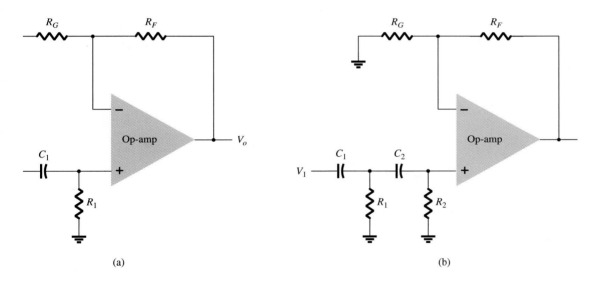

Figure 14.33   High-pass filter: (a) first order; (b) second order; (c) response plot.

EXAMPLE 14.13 Calculate the cutoff frequency of a second-order high-pass filter as in Fig. 14.33b for $R_1 = R_2 = 2.1$ kΩ, $C_1 = C_2 = 0.05$ μF, and $R_G = 10$ kΩ, $R_F = 50$ kΩ.

Solution

$$\text{Eq. (14.13):}\quad A_v = 1 + \frac{R_F}{R_G} = 1 + \frac{50 \text{ k}\Omega}{10 \text{ k}\Omega} = 6$$

The cutoff frequency is then

$$\text{Eq. (14.15):}\quad f_{OL} = \frac{1}{2\pi R_1 C_1} = \frac{1}{2\pi(2.1 \times 10^3)(0.05 \times 10^{-6})} \approx \textbf{1.5 kHz}$$

Bandpass Filter

Figure 14.34 shows a bandpass filter using two stages, the first a high-pass filter and the second a low-pass filter, the combined operation being the desired bandpass response.

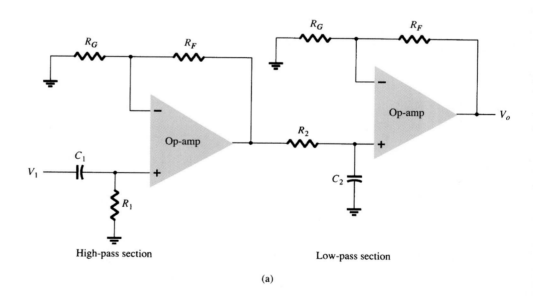

(a)

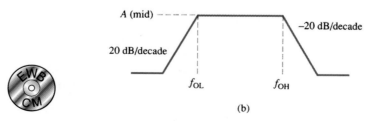

(b)

Figure 14.34 Bandpass active filter.

Calculate the cutoff frequencies of the bandpass filter circuit of Fig. 14.34 with $R_1 = R_2 = 10$ kΩ, $C_1 = 0.1$ μF, and $C_2 = 0.002$ μF.

EXAMPLE 14.14

Solution

$$f_{\text{OL}} = \frac{1}{2\pi R_1 C_1} = \frac{1}{2\pi(10 \times 10^3)(0.1 \times 10^{-6})} = \textbf{159.15 Hz}$$

$$f_{\text{OH}} = \frac{1}{2\pi R_2 C_2} = \frac{1}{2\pi(10 \times 10^3)(0.002 \times 10^{-6})} = \textbf{7.96 kHz}$$

14.7 SUMMARY

Equations

Constant-gain multiplier:

$$A = -\frac{R_f}{R_1}$$

Noninverting constant-gain multiplier:

$$A = 1 + \frac{R_f}{R_1}$$

Voltage summing amplifier:

$$A = -\left[\frac{R_f}{R_1} V_1 + \frac{R_f}{R_2} V_2 + \frac{R_f}{R_3} V_3 \right]$$

Voltage buffer:

$$V_o = V_1$$

Low-pass active filter cutoff frequency:

$$f_{\text{OH}} = \frac{1}{2\pi R_1 C_1}$$

High-pass active filter cutoff frequency:

$$f_{\text{OL}} = \frac{1}{2\pi R_1 C_1}$$

14.8 COMPUTER ANALYSIS

Many of the practical op-amp applications covered in this chapter can be analyzed using PSpice. Analysis of various problems will be used to display the resulting dc bias or, using **PROBE,** to display resulting waveforms. As always, first use **Schematic** drawing to draw the circuit diagram and set the desired analysis, then use **Simulation** to analyze the circuit. Finally, examine the resulting **Output** or use **PROBE** to view various waveforms.

Program 14.1—Summing Op-Amp

A summing op-amp using a 741 IC is shown in Fig. 14.35. Three dc voltage inputs are summed, with a resulting output dc voltage determined as follows:

$$V_o = -[(100 \text{ k}\Omega/20 \text{ k}\Omega)(+2 \text{ V}) + (100 \text{ k}\Omega/50 \text{ k}\Omega)(-3 \text{ V})$$
$$+ (100 \text{ k}\Omega/10 \text{ k}\Omega)(+1 \text{ V})]$$

$$= -[(10 \text{ V}) + (-6 \text{ V}) + (10 \text{ V})] = -[20 \text{ V} - 6 \text{ V}] = -14 \text{ V}$$

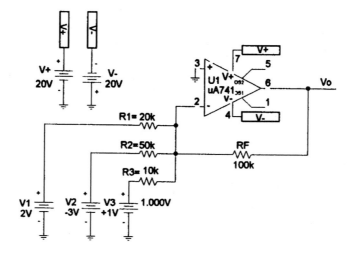

Figure 14.35   Summing amplifier using μA741 op-amp.

The steps in drawing the circuit and doing the analysis are as follows. Using **Get New Part:**

> Select **uA741.**
>
> Select **R** and repeatedly place three input resistors and feedback resistor; set resistor values and change resistor names, if desired.
>
> Select **VDC** and place three input voltages and two supply voltages; set voltage values and change voltage names, if desired.
>
> Select **GLOBAL** (global connector) and use to identify supply voltages and make connection to op-amp power input terminals (4 and 7)

Now that the circuit is drawn and all part names and values set as in Fig. 14.35, press the **Simulation** button to have PSpice analyze the circuit. Since no specific analysis has been chosen, only the dc bias will be carried out.

Press the **Enable Bias Voltage Display** button to see the dc voltages at various points in the circuit. The bias voltages displayed in Fig. 14.35 shows the output to be -13.99 V (compared to the calculated value of -14 V above).

Program 14.2—Op-Amp DC Voltmeter

A dc voltmeter built using a μA741 op-amp is provided by the schematic of Fig. 14.36. From the material presented in Section 14.5, the transfer function of the circuit is

$$I_o/V_1 = (R_F/R_1)(1/R_S) = (1 \text{ M}\Omega/1 \text{ M}\Omega)(1/10 \text{ k}\Omega)$$

The full-scale setting of this voltmeter (for I_o full scale at 1 mA) is then

$$V_1(\text{full scale}) = (10 \text{ k}\Omega)(1 \text{ mA}) = 10 \text{ V}$$

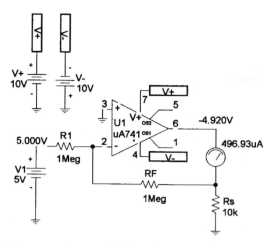

Figure 14.36 Op-amp dc voltmeter.

Thus, an input of 10 V will result in a meter current of 1 mA—the full-scale deflection of the meter. Any input less then 10 V will result in a proportionately smaller meter deflection.

The steps in drawing the circuit and doing the analysis are as follows. Using **Get New Part:**

Select **μA741.**

Select **R** and repeatedly place input resistor, feedback resistor; and meter setting resistor; set resistor values and change resistor names, if desired.

Select **VDC** and place input voltage and two supply voltages; set voltage values and change voltage names, if desired.

Select **GLOBAL** (global connector) and use to identify supply voltages and make connection to op-amp power input terminals (4 and 7)

Select **IPROBE** and use as meter movement.

Now that the circuit is drawn and all part names and values set as in Fig. 14.36, press the **Simulation** button to have PSpice analyze the circuit. Since no specific analysis has been chosen, only the dc bias will be carried out.

Figure 14.36 shows that an input of 5 V will result in a current of 0.5 mA, with the meter reading of 0.5 being read as 5 V (since 1 mA full scale will occur for 10 V input).

Program 14.3—Low-Pass Active Filter

Figure 14.37 shows the schematic of a low-pass active filter using EWB. This first-order filter circuit passes frequencies from dc up to the cutoff frequency determined

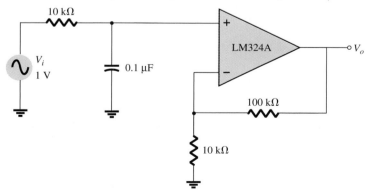

Figure 14.37 Low-pass active filter.

by resistor R_1 and capacitor C_1 using

$$f_{OH} = 1/(2\pi R_1 C_1)$$

For the circuit of Fig. 14.37, this is

$$f_{OH} = 1/(2\pi R_1 C_1) = 1/(2\pi \cdot 10 \text{ k}\Omega \cdot 0.1 \text{ }\mu\text{F}) = 159 \text{ Hz}$$

Figure 14.38 is the result using the **Analysis Setup-AC frequency** and then choosing an ac sweep of 100 points per decade from 1 Hz to 10 kHz. After running the analysis, the **Analysis Graph** is created as shown in Fig. 14.38. The cutoff frequency obtained is seen to be 158.5 Hz, very close to that calculated above.

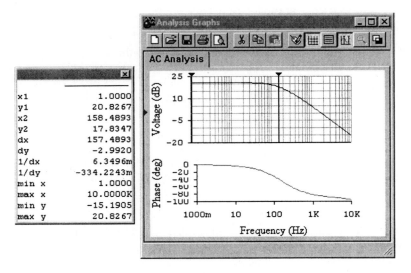

x1	1.0000
y1	20.8267
x2	158.4893
y2	17.8347
dx	157.4893
dy	−2.9920
1/dx	6.3496m
1/dy	−334.2243m
min x	1.0000
max x	10.0000K
min y	−15.1905
max y	20.8267

Figure 14.38 AC analysis of low-pass filter.

Program 14.4—High-Pass Active Filter

Figure 14.39 shows the schematic of a high-pass active filter using EWB. This first-order filter circuit passes frequencies above a cutoff frequency determined by resistor R_1 and capacitor C_1 using

$$f_{OL} = 1/(2\pi R_1 C_1)$$

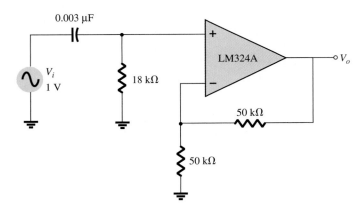

Figure 14.39 High-pass active filter.

Chapter 14   Op-Amp Applications

For the circuit of Fig. 14.39, this is

$$f_{OH} = 1/(2\pi R_1 C_1) = 1/(2\pi \cdot 18 \text{ k}\Omega \cdot 0.003 \ \mu\text{F}) = 2.95 \text{ kHz}$$

The **Analysis** is set for an ac sweep of 100 points per decade from 10 Hz to 100 kHz. After running the analysis, the output showing the output voltage in dB units is that shown in Fig. 14.40. The cutoff frequency obtained is seen to be 2.9 kHz, very close to that calculated above.

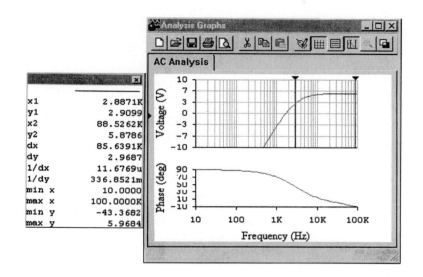

Figure 14.40 dB output plot for the active high-pass filter circuit of Fig. 14.39.

Program 14.5—Second-Order High-Pass Active Filter

Figure 14.41 shows the schematic of a second-order high-pass active filter. This second-order filter circuit passes frequencies above a cutoff frequency determined by resistor R_1 and capacitor C_1 using

$$f_{OL} = 1/(2\pi R_1 C_1)$$

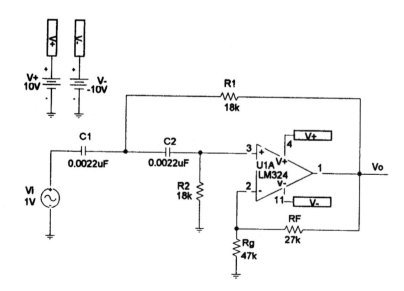

Figure 14.41 Second-order high-pass active filter.

For the circuit of Fig. 14.41 this is

$$f_{OL} = 1/(2\pi R_1 C_1) = 1/(2\pi \cdot 18 \text{ k}\Omega \cdot 0.0022 \ \mu\text{F}) = 4 \text{ kHz}$$

The **Analysis Setup** is set for an ac sweep of 20 points per decade from 100 Hz to 100 kHz, as shown in Fig. 14.42. After running the analysis, a **PROBE** output showing the output voltage (V_o) is shown in Fig. 14.43. The cutoff frequency obtained using **PROBE** is seen to be **fL** = 4 kHz, the same as that calculated above.

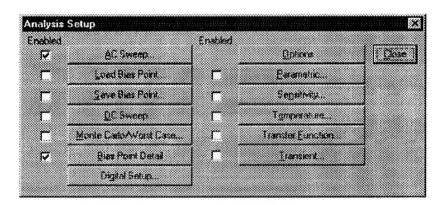

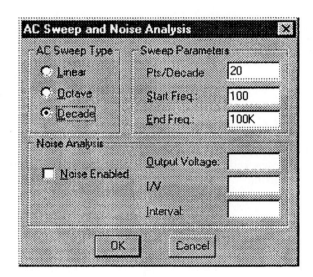

Figure 14.42   Analysis Setup for Fig. 14.41.

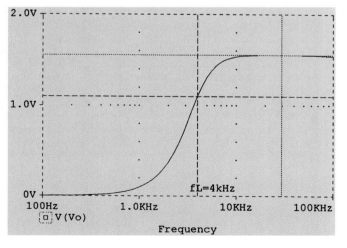

Figure 14.43   Probe plot of V_o for second-order high-pass active filter.

Figure 14.44 shows the **PROBE** plot of the dB gain versus frequency, showing that over a decade (from about 200 Hz to about 2 kHz) the gain changes by about 40 dB—as expected for a second-order filter.

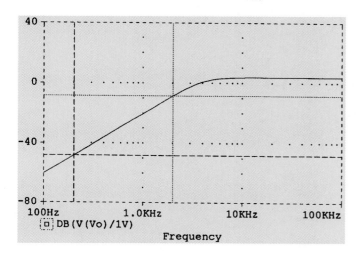

Figure 14.44 Probe plot of dB (V_o/V_i) for second-order high-pass active filter.

Program 14.6—Bandpass Active Filter

Figure 14.45 shows a bandpass active filter circuit. Using the values of Example 14.14, the bandpass frequencies are

$$f_{OL} = 1/(2\pi R_1 C_1) = 1/(2\pi \cdot 10 \text{ k}\Omega \cdot 0.1 \text{ }\mu\text{F}) = 159 \text{ Hz}$$

$$f_{OH} = 1/(2\pi R_2 C_2) = 1/(2\pi \cdot 10 \text{ k}\Omega \cdot 0.002 \text{ }\mu\text{F}) = 7.96 \text{ kHz}$$

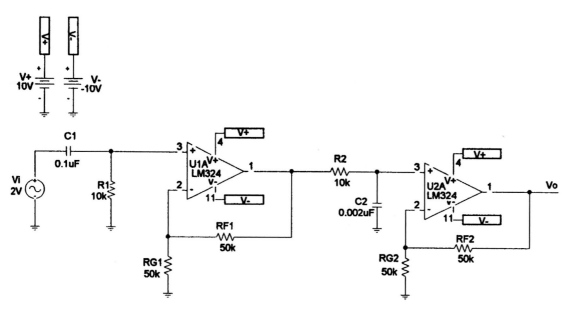

Figure 14.45 Bandpass active filter.

The sweep is set at 10 points per decade from 10 Hz to 1 MHz. The **PROBE** plot of V_o in Fig. 14.46 shows the low cutoff frequency at about 153 Hz and the upper cutoff frequency at about 8.2 kHz, these values matching those calculated above quite well.

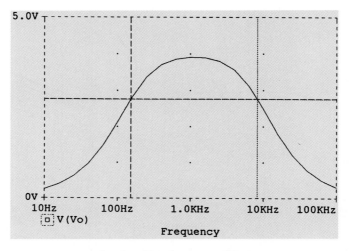

Figure 14.46   Probe plot of bandpass active filter.

PROBLEMS

§ 14.1 Constant-Gain Multiplier

1. Calculate the output voltage for the circuit of Fig. 14.47 for an input of $V_i = 3.5$ mV rms.

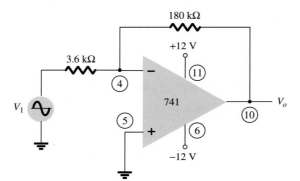

Figure 14.47   Problem 1

2. Calculate the output voltage of the circuit of Fig. 14.48 for an input of 150 mV rms.

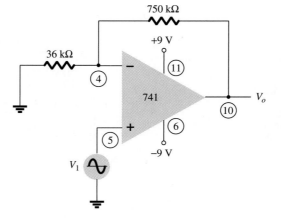

Figure 14.48   Problem 2

* **3.** Calculate the output voltage in the circuit of Fig. 14.49.

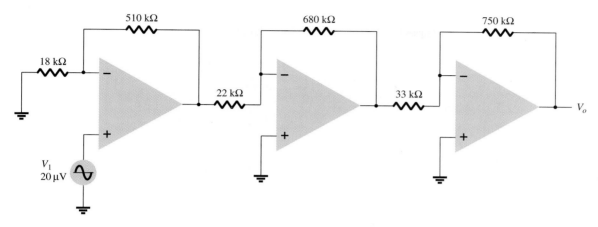

Figure 14.49 Problem 3

* **4.** Show the connection of an LM124 quad op-amp as a three-stage amplifier with gains of +15, −22, and −30. Use a 420-kΩ feedback resistor for all stages. What output voltage results for an input of $V_1 = 80\ \mu V$?

5. Show the connection of two op-amp stages using an LM358 IC to provide outputs that are 15 and −30 times larger than the input. Use a feedback resistor, $R_F = 150\ k\Omega$, in all stages.

§ **14.2 Voltage Summing**

6. Calculate the output voltage for the circuit of Fig. 14.50 with inputs of $V_1 = 40\ mV$ rms and $V_2 = 20\ mV$ rms.

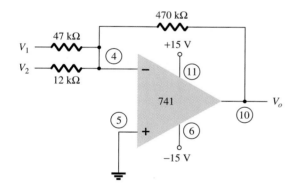

Figure 14.50 Problem 6

7. Determine the output voltage for the circuit of Fig. 14.51.

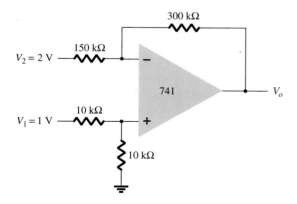

Figure 14.51 Problem 7

8. Determine the output voltage for the circuit of Fig. 14.52.

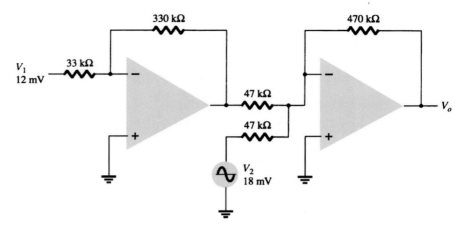

Figure 14.52 Problem 8

§ 14.3 Voltage Buffer

9. Show the connection (including pin information) of an LM124 IC stage connected as a unity-gain amplifier.

10. Show the connection (including pin information) of two LM358 stages connected as unity-gain amplifiers to provide the same output.

§ 14.4 Controlled Sources

11. For the circuit of Fig. 14.53, calculate I_L.

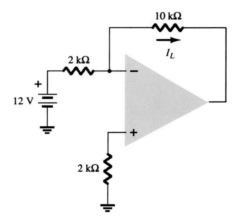

Figure 14.53 Problem 11

12. Calculate V_o for the circuit of Fig. 14.54.

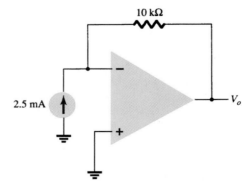

Figure 14.54 Problem 12

§ 14.5 Instrumentation Circuits

13. Calculate the output current I_o in the circuit of Fig. 14.55.

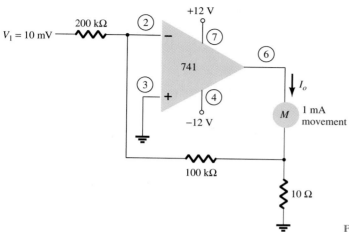

Figure 14.55 Problem 13

∗**14.** Calculate V_o in the circuit of Fig. 14.56.

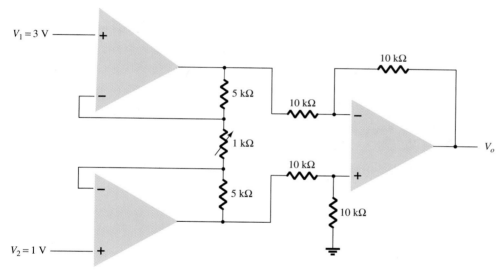

Figure 14.56 Problem 14

§ 14.6 Active Filters

15. Calculate the cutoff frequency of a first-order low-pass filter in the circuit of Fig. 14.57.

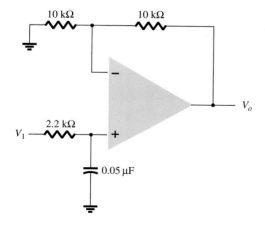

Figure 14.57 Problem 15

16. Calculate the cutoff frequency of the high-pass filter circuit in Fig. 14.58.

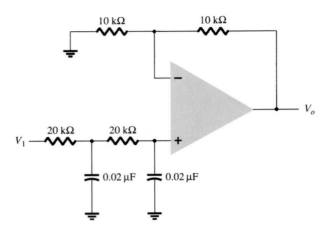

Figure 14.58 Problem 16

17. Calculate the lower and upper cutoff frequencies of the bandpass filter circuit in Fig. 14.59.

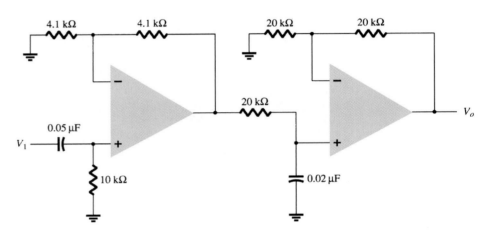

Figure 14.59 Problem 17

§ **14.8 Computer Analysis**

∗ **18.** Use Design Center to draw the schematic of Fig. 14.60 and determine V_o.

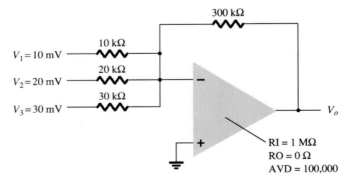

Figure 14.60 Problem 18

*19. Use Design Center to calculate I(VSENSE) in the circuit of Fig. 14.61.

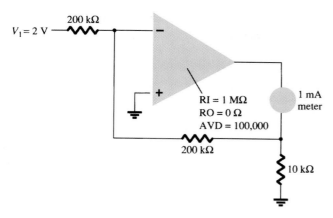

Figure 14.61 Problem 19

*20. Use EWB to plot the response of the low-pass filter circuit in Fig. 14.62.

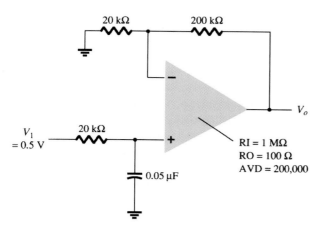

Figure 14.62 Problem 20

*21. Use EWB to plot the response of the high-pass filter circuit in Fig. 14.63.

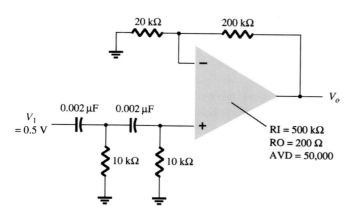

Figure 14.63 Problem 21

* **22.** Use Design Center to plot the response of the bandpass filter circuit in Fig. 14.64.

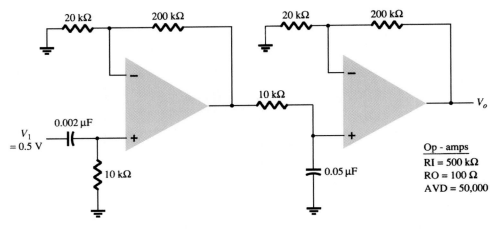

Figure 14.64 Problem 22

*Please Note: Asterisks indicate more difficult problems.

Power Amplifiers

CHAPTER 15

P_L

15.1 INTRODUCTION—DEFINITIONS AND AMPLIFIER TYPES

An amplifier receives a signal from some pickup transducer or other input source and provides a larger version of the signal to some output device or to another amplifier stage. An input transducer signal is generally small (a few millivolts from a cassette or CD input, or a few microvolts from an antenna) and needs to be amplified sufficiently to operate an output device (speaker or other power-handling device). In small-signal amplifiers, the main factors are usually amplification linearity and magnitude of gain. Since signal voltage and current are small in a small-signal amplifier, the amount of power-handling capacity and power efficiency are of little concern. A voltage amplifier provides voltage amplification primarily to increase the voltage of the input signal. Large-signal or power amplifiers, on the other hand, primarily provide sufficient power to an output load to drive a speaker or other power device, typically a few watts to tens of watts. In the present chapter, we concentrate on those amplifier circuits used to handle large-voltage signals at moderate to high current levels. The main features of a large-signal amplifier are the circuit's power efficiency, the maximum amount of power that the circuit is capable of handling, and the impedance matching to the output device.

One method used to categorize amplifiers is by class. Basically, amplifier classes represent the amount the output signal varies over one cycle of operation for a full cycle of input signal. A brief description of amplifier classes is provided next.

Class A: The output signal varies for a full 360° of the cycle. Figure 15.1a shows

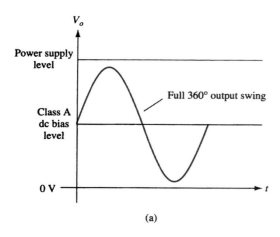

(a)

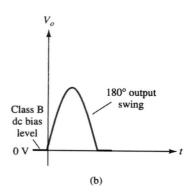

(b)

Figure 15.1 Amplifier operating classes.

747

that this requires the Q-point to be biased at a level so that at least half the signal swing of the output may vary up and down without going to a high-enough voltage to be limited by the supply voltage level or too low to approach the lower supply level, or 0 V in this description.

Class B: A class B circuit provides an output signal varying over one-half the input signal cycle, or for 180° of signal, as shown in Fig. 15.1b. The dc bias point for class B is therefore at 0 V, with the output then varying from this bias point for a half-cycle. Obviously, the output is not a faithful reproduction of the input if only one half-cycle is present. Two class B operations—one to provide output on the positive-output half-cycle and another to provide operation on the negative-output half-cycle are necessary. The combined half-cycles then provide an output for a full 360° of operation. This type of connection is referred to as push-pull operation, which is discussed later in this chapter. Note that class B operation by itself creates a very distorted output signal since reproduction of the input takes place for only 180° of the output signal swing.

Class AB: An amplifier may be biased at a dc level above the zero base current level of class B and above one-half the supply voltage level of class A; this bias condition is class AB. Class AB operation still requires a push-pull connection to achieve a full output cycle, but the dc bias level is usually closer to the zero base current level for better power efficiency, as described shortly. For class AB operation, the output signal swing occurs between 180° and 360° and is neither class A nor class B operation.

Class C: The output of a class C amplifier is biased for operation at less than 180° of the cycle and will operate only with a tuned (resonant) circuit, which provides a full cycle of operation for the tuned or resonant frequency. This operating class is therefore used in special areas of tuned circuits, such as radio or communications.

Class D: This operating class is a form of amplifier operation using pulse (digital) signals, which are on for a short interval and off for a longer interval. Using digital techniques makes it possible to obtain a signal that varies over the full cycle (using sample-and-hold circuitry) to recreate the output from many pieces of input signal. The major advantage of class D operation is that the amplifier is on (using power) only for short intervals and the overall efficiency can practically be very high, as described next.

Amplifier Efficiency

The power efficiency of an amplifier, defined as the ratio of power output to power input, improves (gets higher) going from class A to class D. In general terms, we see that a class A amplifier, with dc bias at one-half the supply voltage level, uses a good amount of power to maintain bias, even with no input signal applied. This results in very poor efficiency, especially with small input signals, when very little ac power is delivered to the load. In fact, the maximum efficiency of a class A circuit, occurring for the largest output voltage and current swing, is only 25% with a direct or series-fed load connection and 50% with a transformer connection to the load. Class B operation, with no dc bias power for no input signal, can be shown to provide a maximum efficiency that reaches 78.5%. Class D operation can achieve power efficiency over 90% and provides the most efficient operation of all the operating classes. Since class AB falls between class A and class B in bias, it also falls between their efficiency ratings—between 25% (or 50%) and 78.5%. Table 15.1 summarizes the operation of the various amplifier classes. This table provides a relative comparison of the output cycle operation and power efficiency for the various class types. In class B operation, a push-pull connection is obtained using either a transformer coupling or by using complementary (or quasi-complementary) operation with *npn* and *pnp* transistors to provide operation on opposite polarity cycles. While transformer oper-

TABLE 15.1 Comparison of Amplifier Classes

	A	AB	Class B	C*	D
Operating cycle	360°	180° to 360°	180°	Less than 180°	Pulse operation
Power efficiency	25% to 50%	Between 25% (50%) and 78.5%	78.5%		Typically over 90%

Class C is usually not used for delivering large amounts of power, thus the efficiency is not given here.

ation can provide opposite cycle signals, the transformer itself is quite large in many applications. A transformerless circuit using complementary transistors provides the same operation in a much smaller package. Circuits and examples are provided later in this chapter.

15.2 SERIES-FED CLASS A AMPLIFIER

The simple fixed-bias circuit connection shown in Fig. 15.2 can be used to discuss the main features of a class A series-fed amplifier. The only differences between this circuit and the small-signal version considered previously is that the signals handled by the large-signal circuit are in the range of volts and the transistor used is a power transistor that is capable of operating in the range of a few to tens of watts. As will be shown in this section, this circuit is not the best to use as a large-signal amplifier because of its poor power efficiency. The beta of a power transistor is generally less than 100, the overall amplifier circuit using power transistors that are capable of handling large power or current while not providing much voltage gain.

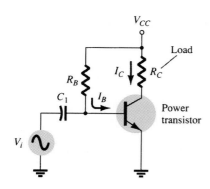

Figure 15.2 Series-fed class A large-signal amplifier.

DC Bias Operation

The dc bias set by V_{CC} and R_B fixes the dc base-bias current at

$$I_B = \frac{V_{CC} - 0.7 \text{ V}}{R_B} \tag{15.1}$$

with the collector current then being

$$I_C = \beta I_B \tag{15.2}$$

with the collector–emitter voltage then

$$V_{CE} = V_{CC} - I_C R_C \tag{15.3}$$

To appreciate the importance of the dc bias on the operation of the power amplifier, consider the collector characteristic shown in Fig. 15.3. An ac load line is drawn using the values of V_{CC} and R_C. The intersection of the dc bias value of I_B with the dc load line then determines the operating point (Q-point) for the circuit. The quiescent-point values are those calculated using Eqs. (15.1) through (15.3). If the dc bias collector current is set at one-half the possible signal swing (between 0 and V_{CC}/R_C), the largest collector current swing will be possible. Additionally, if the quiescent collector–emitter voltage is set at one-half the supply voltage, the largest voltage swing will be possible. With the Q-point set at this optimum bias point, the power considerations for the circuit of Fig. 15.2 are determined as described below.

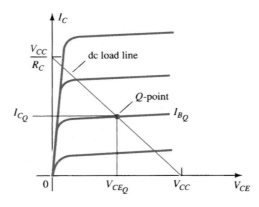

Figure 15.3 Transistor characteristic showing load line and Q-point.

AC Operation

When an input ac signal is applied to the amplifier of Fig. 15.2, the output will vary from its dc bias operating voltage and current. A small input signal, as shown in Fig. 15.4, will cause the base current to vary above and below the dc bias point, which will then cause the collector current (output) to vary from the dc bias point set as well

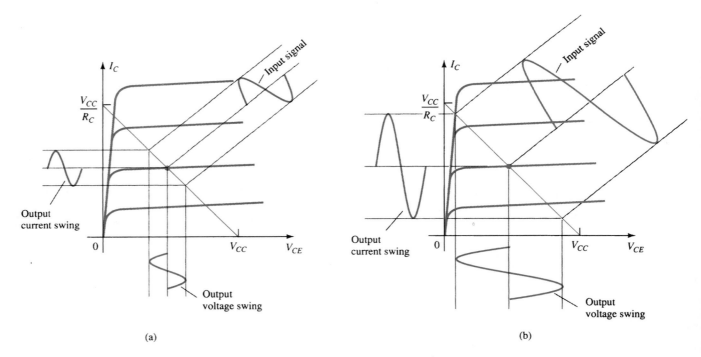

(a)   (b)

Figure 15.4 Amplifier input and output signal variation.

as the collector–emitter voltage to vary around its dc bias value. As the input signal is made larger, the output will vary further around the established dc bias point until either the current or the voltage reaches a limiting condition. For the current this limiting condition is either zero current at the low end or V_{CC}/R_C at the high end of its swing. For the collector–emitter voltage, the limit is either 0 V or the supply voltage, V_{CC}.

Power Considerations

The power into an amplifier is provided by the supply. With no input signal, the dc current drawn is the collector bias current, I_{C_Q}. The power then drawn from the supply is

$$P_i(\text{dc}) = V_{CC}I_{C_Q} \qquad (15.4)$$

Even with an ac signal applied, the average current drawn from the supply remains the same, so that Eq. (15.4) represents the input power supplied to the class A series-fed amplifier.

OUTPUT POWER

The output voltage and current varying around the bias point provide ac power to the load. This ac power is delivered to the load, R_C, in the circuit of Fig. 15.2. The ac signal, V_i, causes the base current to vary around the dc bias current and the collector current around its quiescent level, I_{C_Q}. As shown in Fig. 15.4, the ac input signal results in ac current and ac voltage signals. The larger the input signal, the larger the output swing, up to the maximum set by the circuit. The ac power delivered to the load (R_C) can be expressed in a number of ways.

Using rms signals: The ac power delivered to the load (R_C) may be expressed using

$$P_o(\text{ac}) = V_{CE}(\text{rms})I_C(\text{rms}) \qquad (15.5\text{a})$$

$$P_o(\text{ac}) = I_C^2(\text{rms})R_C \qquad (15.5\text{b})$$

$$P_o(\text{ac}) = \frac{V_C^2(\text{rms})}{R_C} \qquad (15.5\text{c})$$

Using peak signals: The ac power delivered to the load may be expressed using

$$P_o(\text{ac}) = \frac{V_{CE}(\text{p})I_C(\text{p})}{2} \qquad (15.6\text{a})$$

$$P_o(\text{ac}) = \frac{I_C^2(\text{p})}{2R_C} \qquad (15.6\text{b})$$

$$P_o(\text{ac}) = \frac{V_{CE}^2(\text{p})}{2R_C} \qquad (15.6\text{c})$$

Using peak-to-peak signals: The ac power delivered to the load may be expressed using

$$P_o(\text{ac}) = \frac{V_{CE}(\text{p-p})I_C(\text{p-p})}{8} \qquad (15.7\text{a})$$

$$P_o(\text{ac}) = \frac{I_C^2(\text{p-p})}{8} R_C \qquad (15.7b)$$

$$P_o(\text{ac}) = \frac{V_{CE}^2(\text{p-p})}{8R_C} \qquad (15.7c)$$

Efficiency

The efficiency of an amplifier represents the amount of ac power delivered (transferred) from the dc source. The efficiency of the amplifier is calculated using

$$\% \; \eta = \frac{P_o(\text{ac})}{P_i(\text{dc})} \times 100\% \qquad (15.8)$$

MAXIMUM EFFICIENCY

For the class A series-fed amplifier, the maximum efficiency can be determined using the maximum voltage and current swings. For the voltage swing it is

$$\text{maximum } V_{CE}(\text{p-p}) = V_{CC}$$

For the current swing it is

$$\text{maximum } I_C(\text{p-p}) = \frac{V_{CC}}{R_C}$$

Using the maximum voltage swing in Eq. (15.7a) yields

$$\text{maximum } P_o(\text{ac}) = \frac{V_{CC}(V_{CC}/R_C)}{8}$$

$$= \frac{V_{CC}^2}{8R_C}$$

The maximum power input can be calculated using the dc bias current set to one-half the maximum value:

$$\text{maximum } P_i(\text{dc}) = V_{CC}(\text{maximum } I_C) = V_{CC}\frac{V_{CC}/R_C}{2}$$

$$= \frac{V_{CC}^2}{2R_C}$$

We can then use Eq. (15.8) to calculate the maximum efficiency:

$$\text{maximum } \% \; \eta = \frac{\text{maximum } P_o(\text{ac})}{\text{maximum } P_i(\text{dc})} \times 100\%$$

$$= \frac{V_{CC}^2/8R_C}{V_{CC}^2/2R_C} \times 100\%$$

$$= 25\%$$

The maximum efficiency of a class A series-fed amplifier is thus seen to be 25%. Since this maximum efficiency will occur only for ideal conditions of both voltage swing and current swing, most series-fed circuits will provide efficiencies of much less than 25%.

Calculate the input power, output power, and efficiency of the amplifier circuit in Fig. 15.5 for an input voltage that results in a base current of 10 mA peak.

EXAMPLE 15.1

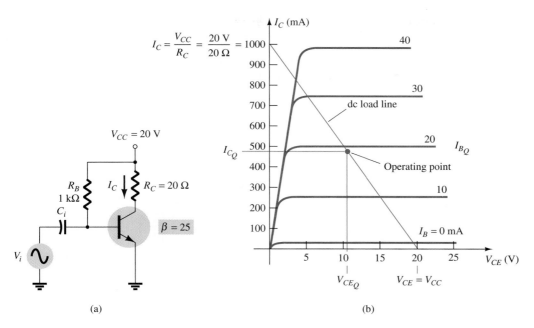

$$I_C = \frac{V_{CC}}{R_C} = \frac{20 \text{ V}}{20 \text{ }\Omega} = 1000$$

Figure 15.5   Operation of a series-fed circuit for Example 15.1.

Solution

Using Eqs. (15.1) through (15.3), the *Q*-point can be determined to be

$$I_{B_Q} = \frac{V_{CC} - 0.7 \text{ V}}{R_B} = \frac{20 \text{ V} - 0.7 \text{ V}}{1 \text{ k}\Omega} = 19.3 \text{ mA}$$

$$I_{C_Q} = \beta I_B = 25(19.3 \text{ mA}) = 482.5 \text{ mA} \cong 0.48 \text{ A}$$

$$V_{CE_Q} = V_{CC} - I_C R_C = 20 \text{ V} - (0.48 \text{ }\Omega)(20 \text{ }\Omega) = 10.4 \text{ V}$$

This bias point is marked on the transistor collector characteristic of Fig. 15.5b. The ac variation of the output signal can be obtained graphically using the dc load line drawn on Fig. 15.5b by connecting $V_{CE} = V_{CC} = 20$ V with $I_C = V_{CC}/R_C = 1000$ mA = 1 A, as shown. When the input ac base current increases from its dc bias level, the collector current rises by

$$I_C(\text{p}) = \beta I_B(\text{p}) = 25(10 \text{ mA peak}) = 250 \text{ mA peak}$$

Using Eq. (15.6b) yields

$$P_o(\text{ac}) = \frac{I_C^2(\text{p})}{2}R_C = \frac{(250 \times 10^{-3} \text{ A})^2}{2}(20 \text{ }\Omega) = \textbf{0.625 W}$$

Using Eq. (15.4) results in

$$P_i(\text{dc}) = V_{CC}I_{C_Q} = (20 \text{ V})(0.48 \text{ A}) = \textbf{9.6 W}$$

The amplifier's power efficiency can then be calculated using Eq. (15.8):

$$\% \eta = \frac{P_o(\text{ac})}{P_i(\text{dc})} \times 100\% = \frac{0.625 \text{ W}}{9.6 \text{ W}} \times 100\% = \textbf{6.5\%}$$

15.3 TRANSFORMER-COUPLED CLASS A AMPLIFIER

A form of class A amplifier having maximum efficiency of 50% uses a transformer to couple the output signal to the load as shown in Fig. 15.6. This is a simple circuit form to use in presenting a few basic concepts. More practical circuit versions are covered later. Since the circuit uses a transformer to step voltage or current, a review of voltage and current step-up and step-down is presented next.

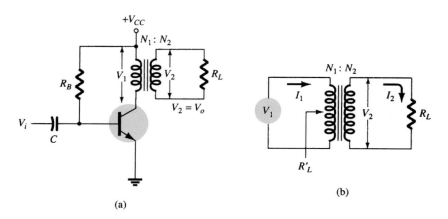

(a)

(b)

Figure 15.6 Transformer-coupled audio power amplifier.

Transformer Action

A transformer can increase or decrease voltage or current levels according to the turns ratio, as explained below. In addition, the impedance connected to one side of a transformer can be made to appear either larger or smaller (step up or step down) at the other side of the transformer, depending on the square of the transformer winding turns ratio. The following discussion assumes ideal (100%) power transfer from primary to secondary, that is, no power losses are considered.

VOLTAGE TRANSFORMATION

As shown in Fig. 15.7a, the transformer can step up or step down a voltage applied to one side directly as the ratio of the turns (or number of windings) on each side. The voltage transformation is given by

$$\frac{V_2}{V_1} = \frac{N_2}{N_1} \tag{15.9}$$

Equation (15.9) shows that if the number of turns of wire on the secondary side is larger than on the primary, the voltage at the secondary side is larger than the voltage at the primary side.

CURRENT TRANSFORMATION

The current in the secondary winding is inversely proportional to the number of turns in the windings. The current transformation is given by

$$\frac{I_2}{I_1} = \frac{N_1}{N_2} \tag{15.10}$$

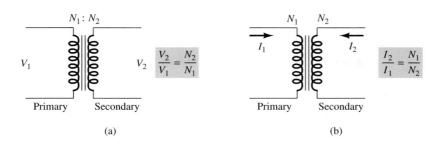

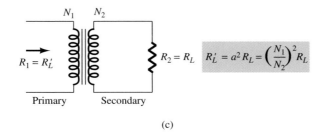

This relationship is shown in Fig. 15.7b. If the number of turns of wire on the secondary is greater than that on the primary, the secondary current will be less than the current in the primary.

IMPEDANCE TRANSFORMATION

Since the voltage and current can be changed by a transformer, an impedance "seen" from either side (primary or secondary) can also be changed. As shown in Fig. 15.7c, an impedance R_L is connected across the transformer secondary. This impedance is changed by the transformer when viewed at the primary side (R_L'). This can be shown as follows:

$$\frac{R_L}{R_L'} = \frac{R_2}{R_1} = \frac{V_2/I_2}{V_1/I_1} = \frac{V_2}{I_2}\frac{I_1}{V_1} = \frac{V_2}{V_1}\frac{I_1}{I_2} = \frac{N_2}{N_1}\frac{N_2}{N_1} = \left(\frac{N_2}{N_1}\right)^2$$

If we define $a = N_1/N_2$, where a is the turns ratio of the transformer, the above equation becomes

$$\frac{R_L'}{R_L} = \frac{R_1}{R_2} = \left(\frac{N_1}{N_2}\right)^2 = a^2 \qquad (15.11)$$

We can express the load resistance reflected to the primary side as:

$$R_1 = a^2 R_2 \qquad \text{or} \qquad R_L' = a^2 R_L \qquad (15.12)$$

where R_L' is the reflected impedance. As shown in Eq. (15.12), the reflected impedance is related directly to the square of the turns ratio. If the number of turns of the secondary is smaller than that of the primary, the impedance seen looking into the primary is larger than that of the secondary by the square of the turns ratio.

Calculate the effective resistance seen looking into the primary of a 15:1 transformer connected to an 8-Ω load.

EXAMPLE 15.2

Solution

Eq. (15.12): $R_L' = a^2 R_L = (15)^2 (8\ \Omega) = 1800\ \Omega = \mathbf{1.8\ k\Omega}$

EXAMPLE 15.3   What transformer turns ratio is required to match a 16-Ω speaker load so that the effective load resistance seen at the primary is 10 kΩ?

Solution

$$\text{Eq. (15.11):} \quad \left(\frac{N_1}{N_2}\right)^2 = \frac{R_L'}{R_L} = \frac{10\ k\Omega}{16\ \Omega} = 625$$

$$\frac{N_1}{N_2} = \sqrt{625} = \mathbf{25:1}$$

Operation of Amplifier Stage

DC LOAD LINE

The transformer (dc) winding resistance determines the dc load line for the circuit of Fig. 15.6. Typically, this dc resistance is small (ideally 0 Ω) and, as shown in Fig. 15.8, a 0-Ω dc load line is a straight vertical line. A practical transformer winding resistance would be a few ohms, but only the ideal case will be considered in this discussion. There is no dc voltage drop across the 0-Ω dc load resistance, and the load line is drawn straight vertically from the voltage point, $V_{CE_Q} = V_{CC}$.

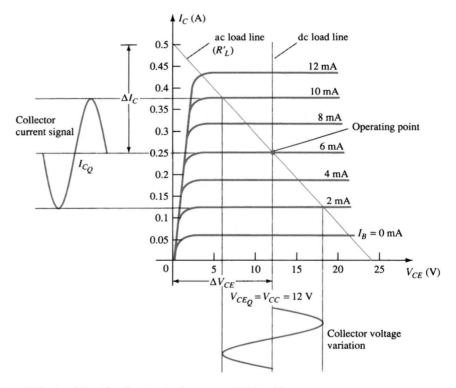

Figure 15.8   Load lines for class A transformer-coupled amplifier.

QUIESCENT OPERATING POINT

The operating point in the characteristic curve of Fig. 15.8 can be obtained graphically at the point of intersection of the dc load line and the base current set by the circuit. The collector quiescent current can then be obtained from the operating point. In class A operation, keep in mind that the dc bias point sets the conditions for the maximum undistorted signal swing for both collector current and collector–emitter voltage. If the input signal produces a voltage swing less than the maximum possible, the efficiency of the circuit at that time will be less than 25%. The dc bias point is therefore important in setting the operation of a class A series-fed amplifier.

AC LOAD LINE

To carry out ac analysis, it is necessary to calculate the ac load resistance "seen" looking into the primary side of the transformer, then draw the ac load line on the collector characteristic. The reflected load resistance (R_L') is calculated using Eq. (15.12) using the value of the load connected across the secondary (R_L) and the turns ratio of the transformer. The graphical analysis technique then proceeds as follows. Draw the ac load line so that it passes through the operating point and has a slope equal to $-1/R_L'$ (the reflected load resistance), the load line slope being the negative reciprocal of the ac load resistance. Notice that the ac load line shows that the output signal swing can exceed the value of V_{CC}. In fact, the voltage developed across the transformer primary can be quite large. It is therefore necessary after obtaining the ac load line to check that the possible voltage swing does not exceed transistor maximum ratings.

SIGNAL SWING AND OUTPUT AC POWER

Figure 15.9 shows the voltage and current signal swings from the circuit of Fig. 15.6. From the signal variations shown in Fig. 15.9, the values of the peak-to-peak signal swings are

$$V_{CE}(\text{p-p}) = V_{CE_{max}} - V_{CE_{min}}$$

$$I_C(\text{p-p}) = I_{C_{max}} - I_{C_{min}}$$

The ac power developed across the transformer primary can then be calculated using

$$P_o(\text{ac}) = \frac{(V_{CE_{max}} - V_{CE_{min}})(I_{C_{max}} - I_{C_{min}})}{8} \tag{15.13}$$

The ac power calculated is that developed across the primary of the transformer. Assuming an ideal transformer (a highly efficient transformer has an efficiency of well over 90%), the power delivered by the secondary to the load is approximately that calculated using Eq. (15.13). The output ac power can also be determined using the voltage delivered to the load.

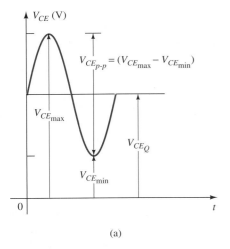
(a)

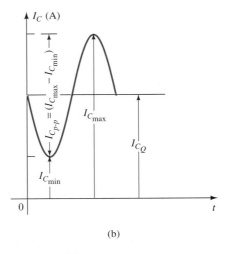
(b)

Figure 15.9 Graphical operation of transformer-coupled class A amplifier.

For the ideal transformer, the voltage delivered to the load can be calculated using Eq. (15.9):

$$V_L = V_2 = \frac{N_2}{N_1} V_1$$

The power across the load can then be expressed as

$$P_L = \frac{V_L^2(\text{rms})}{R_L}$$

and equals the power calculated using Eq. (15.5c).

Using Eq. (15.10) to calculate the load current yields

$$I_L = I_2 = \frac{N_1}{N_2} I_C$$

with the output ac power then calculated using

$$P_L = I_L^2(\text{rms})R_L$$

EXAMPLE 15.4

Calculate the ac power delivered to the 8-Ω speaker for the circuit of Fig. 15.10. The circuit component values result in a dc base current of 6 mA, and the input signal (V_i) results in a peak base current swing of 4 mA.

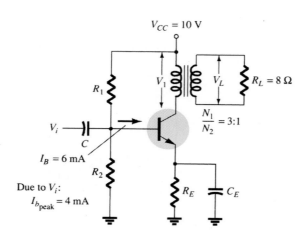

$V_{CC} = 10$ V

R_1

V_i

C

$I_B = 6$ mA

R_2

Due to V_i:

$I_{b_{peak}} = 4$ mA

V_1 V_L $R_L = 8\ \Omega$

$\dfrac{N_1}{N_2} = 3:1$

R_E C_E

Figure 15.10 Transformer-coupled class A amplifier for Example 15.4.

Solution

The dc load line is drawn vertically (see Fig. 15.11) from the voltage point:

$$V_{CE_Q} = V_{CC} = 10\text{ V}$$

For $I_B = 6$ mA, the operating point on Fig. 15.11 is

$$V_{CE_Q} = 10\text{ V} \qquad \text{and} \qquad I_{C_Q} = 140\text{ mA}$$

The effective ac resistance seen at the primary is

$$R_L' = \left(\frac{N_1}{N_2}\right)^2 R_L = (3)^2(8) = 72\ \Omega$$

The ac load line can then be drawn of slope $-1/72$ going through the indicated operating point. To help draw the load line, consider the following procedure. For a current swing of

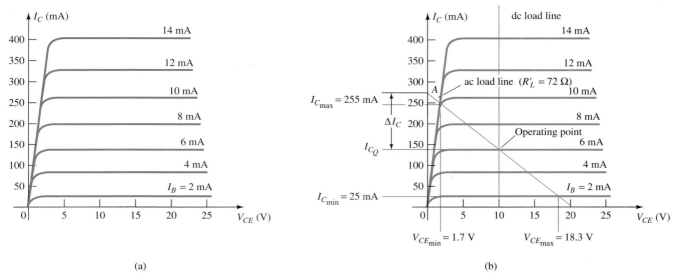

Figure 15.11 Transformer-coupled class A transistor characteristic for Examples 15.4 and 15.5: (a) device characteristic; (b) dc and ac load lines.

$$I_C = \frac{V_{CE}}{R'_L} = \frac{10\ \text{V}}{72\ \Omega} = 139\ \text{mA}$$

mark a point (A):

$$I_{CE_Q} + I_C = 140\ \text{mA} + 139\ \text{mA} = 279\ \text{mA along the } y\text{-axis}$$

Connect point A through the Q-point to obtain the ac load line. For the given base current swing of 4 mA peak, the maximum and minimum collector current and collector–emitter voltage obtained from Fig. 15.11 are

$$V_{CE_{min}} = 1.7\ \text{V} \qquad I_{C_{min}} = 25\ \text{mA}$$

$$V_{CE_{max}} = 18.3\ \text{V} \qquad I_{C_{max}} = 255\ \text{mA}$$

The ac power delivered to the load can then be calculated using Eq. (15.13):

$$P_o(\text{ac}) = \frac{(V_{CE_{max}} - V_{CE_{min}})(I_{C_{max}} - I_{C_{min}})}{8}$$

$$= \frac{(18.3\ \text{V} - 1.7\ \text{V})(255\ \text{mA} - 25\ \text{mA})}{8} = \textbf{0.477 W}$$

Efficiency

So far we have considered calculating the ac power delivered to the load. We next consider the input power from the battery, power losses in the amplifier, and the overall power efficiency of the transformer-coupled class A amplifier.

The input (dc) power obtained from the supply is calculated from the supply dc voltage and the average power drawn from the supply:

$$\boxed{P_i(\text{dc}) = V_{CC}I_{C_Q}} \tag{15.14}$$

For the transformer-coupled amplifier, the power dissipated by the transformer is small (due to the small dc resistance of a coil) and will be ignored in the present calcula-

P_L

tions. Thus the only power loss considered here is that dissipated by the power transistor and calculated using

$$P_Q = P_i(\text{dc}) - P_o(\text{ac}) \qquad (15.15)$$

where P_Q is the power dissipated as heat. While the equation is simple, it is nevertheless significant when operating a class A amplifier. The amount of power dissipated by the transistor is the difference between that drawn from the dc supply (set by the bias point) and the amount delivered to the ac load. When the input signal is very small, with very little ac power delivered to the load, the maximum power is dissipated by the transistor. When the input signal is larger and power delivered to the load is larger, less power is dissipated by the transistor. In other words, the transistor of a class A amplifier has to work hardest (dissipate the most power) when the load is disconnected from the amplifier, and the transistor dissipates the least power when the load is drawing maximum power from the circuit.

EXAMPLE 15.5

For the circuit of Fig. 15.10 and results of Example 15.4, calculate the dc input power, power dissipated by the transistor, and efficiency of the circuit for the input signal of Example 15.4.

Solution

$$\text{Eq. (15.14):} \quad P_i(\text{dc}) = V_{CC}I_{C_Q} = (10 \text{ V})(140 \text{ mA}) = \textbf{1.4 W}$$

$$\text{Eq. (15.15):} \quad P_Q = P_i(\text{dc}) - P_o(\text{ac}) = 1.4 \text{ W} - 0.477 \text{ W} = \textbf{0.92 W}$$

The efficiency of the amplifier is then

$$\% \, \eta = \frac{P_o(\text{ac})}{P_i(\text{dc})} \times 100\% = \frac{0.477 \text{ W}}{1.4 \text{ W}} \times 100\% = \textbf{34.1\%}$$

MAXIMUM THEORETICAL EFFICIENCY

For a class A transformer-coupled amplifier, the maximum theoretical efficiency goes up to 50%. Based on the signals obtained using the amplifier, the efficiency can be expressed as

$$\% \, \eta = 50\left(\frac{V_{CE_{max}} - V_{CE_{min}}}{V_{CE_{max}} + V_{CE_{min}}}\right)^2 \% \qquad (15.16)$$

The larger the value of $V_{CE_{max}}$ and the smaller the value of $V_{CE_{min}}$, the closer the efficiency approaches the theoretical limit of 50%.

EXAMPLE 15.6

Calculate the efficiency of a transformer-coupled class A amplifier for a supply of 12 V and outputs of:
(a) $V(\text{p}) = 12$ V.
(b) $V(\text{p}) = 6$ V.
(c) $V(\text{p}) = 2$ V.

Solution

Since $V_{CE_Q} = V_{CC} = 12$ V, the maximum and minimum of the voltage swing are

(a) $V_{CE_{max}} = V_{CE_Q} + V(\text{p}) = 12 \text{ V} + 12 \text{ V} = 24 \text{ V}$

$\quad V_{CE_{min}} = V_{CE_Q} - V(\text{p}) = 12 \text{ V} - 12 \text{ V} = 0 \text{ V}$

resulting in

$$\% \eta = 50\left(\frac{24\text{ V} - 0\text{ V}}{24\text{ V} + 0\text{ V}}\right)^2 \% = \mathbf{50\%}$$

(b) $V_{CE_{\text{max}}} = V_{CE_Q} + V(\text{p}) = 12\text{ V} + 6\text{ V} = 18\text{ V}$

$\quad V_{CE_{\text{min}}} = V_{CE_Q} - V(\text{p}) = 12\text{ V} - 6\text{ V} = 6\text{ V}$

resulting in

$$\% \eta = 50\left(\frac{18\text{ V} - 6\text{ V}}{18\text{ V} + 6\text{ V}}\right)^2 \% = \mathbf{12.5\%}$$

(c) $V_{CE_{\text{max}}} = V_{CE_Q} + V(\text{p}) = 12\text{ V} + 2\text{ V} = 14\text{ V}$

$\quad V_{CE_{\text{min}}} = V_{CE_Q} - V(\text{p}) = 12\text{ V} - 2\text{ V} = 10\text{ V}$

resulting in

$$\% \eta = 50\left(\frac{14\text{ V} - 10\text{ V}}{14\text{ V} + 10\text{ V}}\right)^2 \% = \mathbf{1.39\%}$$

Notice how dramatically the amplifier efficiency drops from a maximum of 50% for $V(\text{p}) = V_{CC}$ to slightly over 1% for $V(\text{p}) = 2$ V.

15.4   CLASS B AMPLIFIER OPERATION

Class B operation is provided when the dc bias leaves the transistor biased just off, the transistor turning on when the ac signal is applied. This is essentially no bias, and the transistor conducts current for only one-half of the signal cycle. To obtain output for the full cycle of signal, it is necessary to use two transistors and have each conduct on opposite half-cycles, the combined operation providing a full cycle of output signal. Since one part of the circuit pushes the signal high during one half-cycle and the other part pulls the signal low during the other half-cycle, the circuit is referred to as a *push-pull circuit*. Figure 15.12 shows a diagram for push-pull operation. An ac input signal is applied to the push-pull circuit, with each half operating on alternate half-cycles, the load then receiving a signal for the full ac cycle. The power transistors used in the push-pull circuit are capable of delivering the desired power to the load, and the class B operation of these transistors provides greater efficiency than was possible using a single transistor in class A operation.

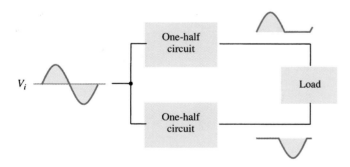

Figure 15.12   Block representation of push-pull operation.

Input (DC) Power

The power supplied to the load by an amplifier is drawn from the power supply (or power supplies; see Fig. 15.13) that provides the input or dc power. The amount of this input power can be calculated using

$$P_i(\text{dc}) = V_{CC}I_{\text{dc}} \tag{15.17}$$

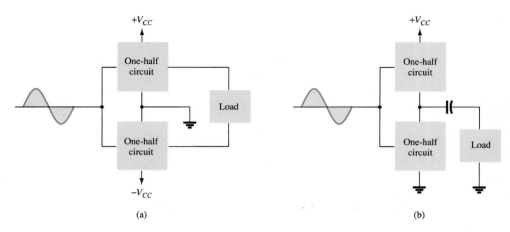

Figure 15.13 Connection of push-pull amplifier to load: (a) using two voltage supplies; (b) using one voltage supply.

where I_{dc} is the average or dc current drawn from the power supplies. In class B operation, the current drawn from a single power supply has the form of a full-wave rectified signal, while that drawn from two power supplies has the form of a half-wave rectified signal from each supply. In either case, the value of the average current drawn can be expressed as

$$I_{dc} = \frac{2}{\pi}I(p)$$ (15.18)

where $I(p)$ is the peak value of the output current waveform. Using Eq. (15.18) in the power input equation (Eq. 15.17) results in

$$P_i(dc) = V_{CC}\left(\frac{2}{\pi}I(p)\right)$$ (15.19)

Output (AC) Power

The power delivered to the load (usually referred to as a resistance, R_L) can be calculated using any one of a number of equations. If one is using an rms meter to measure the voltage across the load, the output power can be calculated as

$$P_o(ac) = \frac{V_L^2(rms)}{R_L}$$ (15.20)

If one is using an oscilloscope, the peak, or peak-to-peak, output voltage measured can be used:

$$P_o(ac) = \frac{V_L^2(p\text{-}p)}{8R_L} = \frac{V_L^2(p)}{2R_L}$$ (15.21)

The larger the rms or peak output voltage, the larger the power delivered to the load.

Efficiency

The efficiency of the class B amplifier can be calculated using the basic equation:

$$\% \, \eta = \frac{P_o(ac)}{P_i(dc)} \times 100\%$$

Using Eqs. (15.19) and (15.21) in the efficiency equation above results in

$$\% \ \eta = \frac{P_o(\text{ac})}{P_i(\text{dc})} \times 100\% = \frac{V_L^2(\text{p})/2R_L}{V_{CC}[(2/\pi)I(\text{p})]} \times 100\% = \frac{\pi}{4} \frac{V_L(\text{p})}{V_{CC}} \times 100\% \quad (15.22)$$

(using $I(\text{p}) = V_L(\text{p})/R_L$). Equation (15.22) shows that the larger the peak voltage, the higher the circuit efficiency, up to a maximum value when $V_L(\text{p}) = V_{CC}$, this maximum efficiency then being

$$\text{maximum efficiency} = \frac{\pi}{4} \times 100\% = 78.5\%$$

Power Dissipated by Output Transistors

The power dissipated (as heat) by the output power transistors is the difference between the input power delivered by the supplies and the output power delivered to the load.

$$P_{2Q} = P_i(\text{dc}) - P_o(\text{ac}) \quad (15.23)$$

where P_{2Q} is the power dissipated by the two output power transistors. The dissipated power handled by each transistor is then

$$P_Q = \frac{P_{2Q}}{2} \quad (15.24)$$

For a class B amplifier providing a 20-V peak signal to a 16-Ω load (speaker) and a power supply of $V_{CC} = 30$ V, determine the input power, output power, and circuit efficiency.

EXAMPLE 15.7

Solution

A 20-V peak signal across a 16-Ω load provides a peak load current of

$$I_L(\text{p}) = \frac{V_L(\text{p})}{R_L} = \frac{20 \text{ V}}{16 \ \Omega} = 1.25 \text{ A}$$

The dc value of the current drawn from the power supply is then

$$I_{\text{dc}} = \frac{2}{\pi} I_L(\text{p}) = \frac{2}{\pi}(1.25 \text{ A}) = 0.796 \text{ A}$$

and the input power delivered by the supply voltage is

$$P_i(\text{dc}) = V_{CC}I_{\text{dc}} = (30 \text{ V})(0.796 \text{ A}) = \textbf{23.9 W}$$

The output power delivered to the load is

$$P_o(\text{ac}) = \frac{V_L^2(\text{p})}{2R_L} = \frac{(20 \text{ V})^2}{2(16 \ \Omega)} = \textbf{12.5 W}$$

for a resulting efficiency of

$$\% \ \eta = \frac{P_o(\text{ac})}{P_i(\text{dc})} \times 100\% = \frac{12.5 \text{ W}}{23.9 \text{ W}} \times 100\% = \textbf{52.3\%}$$

P_L

Maximum Power Considerations

For class B operation, the maximum output power is delivered to the load when $V_L(\text{p}) = V_{CC}$:

$$\boxed{\text{maximum } P_o(\text{ac}) = \frac{V_{CC}^2}{2R_L}} \qquad (15.25)$$

The corresponding peak ac current $I(\text{p})$ is then

$$I(\text{p}) = \frac{V_{CC}}{R_L}$$

so that the maximum value of average current from the power supply is

$$\text{maximum } I_{\text{dc}} = \frac{2}{\pi} I(\text{p}) = \frac{2V_{CC}}{\pi R_L}$$

Using this current to calculate the maximum value of input power results in

$$\boxed{\text{maximum } P_i(\text{dc}) = V_{CC}(\text{maximum } I_{\text{dc}}) = V_{CC}\left(\frac{2V_{CC}}{\pi R_L}\right) = \frac{2V_{CC}^2}{\pi R_L}} \qquad (15.26)$$

The maximum circuit efficiency for class B operation is then

$$\text{maximum } \% \, \eta = \frac{P_o(\text{ac})}{P_i(\text{dc})} \times 100\% = \frac{V_{CC}^2/2R_L}{V_{CC}[(2/\pi)(V_{CC}/R_L)]} \times 100\%$$

$$= \frac{\pi}{4} \times 100\% = \mathbf{78.54\%} \qquad (15.27)$$

When the input signal results in less than the maximum output signal swing, the circuit efficiency is less than 78.5%.

For class B operation, the maximum power dissipated by the output transistors does not occur at the maximum power input or output condition. The maximum power dissipated by the two output transistors occurs when the output voltage across the load is

$$V_L(\text{p}) = 0.636V_{CC} \qquad \left(= \frac{2}{\pi} V_{CC}\right)$$

for a maximum transistor power dissipation of

$$\boxed{\text{maximum } P_{2Q} = \frac{2V_{CC}^2}{\pi^2 R_L}} \qquad (15.28)$$

EXAMPLE 15.8

For a class B amplifier using a supply of $V_{CC} = 30$ V and driving a load of 16 Ω, determine the maximum input power, output power, and transistor dissipation.

Solution

The maximum output power is

$$\text{maximum } P_o(\text{ac}) = \frac{V_{CC}^2}{2R_L} = \frac{(30 \text{ V})^2}{2(16 \text{ }\Omega)} = \mathbf{28.125 \text{ W}}$$

The maximum input power drawn from the voltage supply is

$$\text{maximum } P_i(\text{dc}) = \frac{2V_{CC}^2}{\pi RL} = \frac{2(30 \text{ V})^2}{\pi(16 \text{ }\Omega)} = \textbf{35.81 W}$$

The circuit efficiency is then

$$\text{maximum } \% \text{ } \eta = \frac{P_o(\text{ac})}{P_i(\text{dc})} \times 100\% = \frac{28.125 \text{ W}}{35.81 \text{ W}} \times 100\% = 78.54\%$$

as expected. The maximum power dissipated by each transistor is

$$\text{maximum } P_Q = \frac{\text{maximum } P_{2Q}}{2} = 0.5\left(\frac{2V_{CC}^2}{\pi^2 R_L}\right) = 0.5\left[\frac{2(30 \text{ V})^2}{\pi^2 16 \text{ }\Omega}\right] = \textbf{5.7 W}$$

Under maximum conditions a pair of transistors, each handling 5.7 W at most, can deliver 28.125 W to a 16-Ω load while drawing 35.81 W from the supply.

The maximum efficiency of a class B amplifier can also be expressed as follows:

$$P_o(\text{ac}) = \frac{V_L^2(\text{p})}{2R_L}$$

$$P_i(\text{dc}) = V_{CC}I_{\text{dc}} = V_{CC}\left[\frac{2V_L(\text{p})}{\pi R_L}\right]$$

so that

$$\% \text{ } \eta = \frac{P_o(\text{ac})}{P_i(\text{dc})} \times 100\% = \frac{V_L^2(\text{p})/2R_L}{V_{CC}[(2/\pi)(V_L(\text{p})/R_L)]} \times 100\%$$

$$\% \text{ } \eta = 78.54\frac{V_L(\text{p})}{V_{CC}}\% \tag{15.29}$$

Calculate the efficiency of a class B amplifier for a supply voltage of $V_{CC} = 24$ V with peak output voltages of:
(a) $V_L(\text{p}) = 22$ V.
(b) $V_L(\text{p}) = 6$ V.

EXAMPLE 15.9

Solution

Using Eq. (15.29) gives

(a) $\% \text{ } \eta = 78.54\frac{V_L(\text{p})}{V_{CC}}\% = 78.54\left(\frac{22 \text{ V}}{24 \text{ V}}\right) = \textbf{72\%}$

(b) $\% \text{ } \eta = 78.54\left(\frac{6 \text{ V}}{24 \text{ V}}\right)\% = \textbf{19.6\%}$

Notice that a voltage near the maximum [22 V in part (a)] results in an efficiency near the maximum, while a small voltage swing [6 V in part (b)] still provides an efficiency near 20%. Similar power supply and signal swings would have resulted in much poorer efficiency in a class A amplifier.

15.5   CLASS B AMPLIFIER CIRCUITS

A number of circuit arrangements for obtaining class B operation are possible. We will consider the advantages and disadvantages of a number of the more popular circuits in this section. The input signals to the amplifier could be a single signal, the circuit then providing two different output stages, each operating for one-half the cycle. If the input is in the form of two opposite polarity signals, two similar stages could be used, each operating on the alternate cycle because of the input signal. One means of obtaining polarity or phase inversion is using a transformer, the transformer-coupled amplifier having been very popular for a long time. Opposite polarity inputs can easily be obtained using an op-amp having two opposite outputs or using a few op-amp stages to obtain two opposite polarity signals. An opposite polarity operation can also be achieved using a single input and complementary transistors (*npn* and *pnp,* or *n*MOS and *p*MOS).

Figure 15.14 shows different ways to obtain phase-inverted signals from a single input signal. Figure 15.14a shows a center-tapped transformer to provide opposite

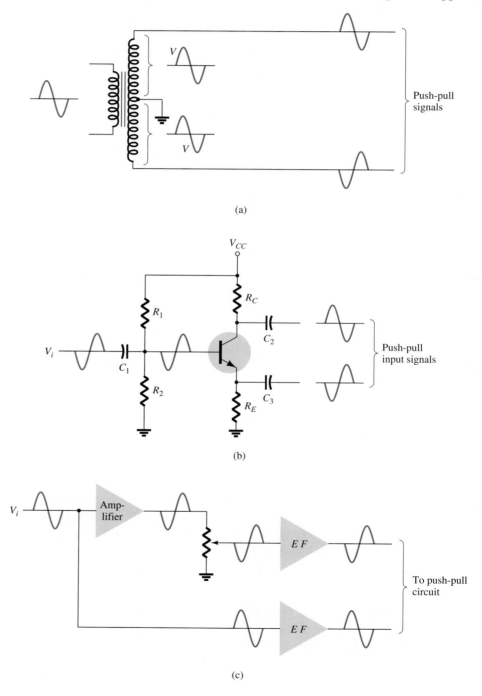

Figure 15.14   Phase-splitter circuits.

Chapter 15   Power Amplifiers

phase signals. If the transformer is exactly center-tapped, the two signals are exactly opposite in phase and of the same magnitude. The circuit of Fig. 15.14b uses a BJT stage with in-phase output from the emitter and opposite phase output from the collector. If the gain is made nearly 1 for each output, the same magnitude results. Probably most common would be using op-amp stages, one to provide an inverting gain of unity and the other a noninverting gain of unity, to provide two outputs of the same magnitude but of opposite phase.

Transformer-Coupled Push–Pull Circuits

The circuit of Fig. 15.15 uses a center-tapped input transformer to produce opposite polarity signals to the two transistor inputs and an output transformer to drive the load in a push-pull mode of operation described next.

During the first half-cycle of operation, transistor Q_1 is driven into conduction whereas transistor Q_2 is driven off. The current I_1 through the transformer results in the first half-cycle of signal to the load. During the second half-cycle of the input signal, Q_2 conducts whereas Q_1 stays off, the current I_2 through the transformer resulting in the second half-cycle to the load. The overall signal developed across the load then varies over the full cycle of signal operation.

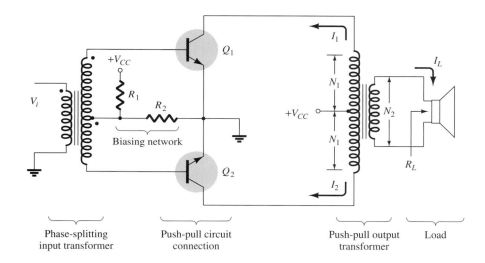

Figure 15.15 Push-pull circuit.

Complementary-Symmetry Circuits

Using complementary transistors (*npn* and *pnp*) it is possible to obtain a full cycle output across a load using half-cycles of operation from each transistor, as shown in Fig. 15.16a. While a single input signal is applied to the base of both transistors, the transistors, being of opposite type, will conduct on opposite half-cycles of the input. The *npn* transistor will be biased into conduction by the positive half-cycle of signal, with a resulting half-cycle of signal across the load as shown in Fig. 15.16b. During the negative half-cycle of signal, the *pnp* transistor is biased into conduction when the input goes negative, as shown in Fig. 15.16c.

During a complete cycle of the input, a complete cycle of output signal is developed across the load. One disadvantage of the circuit is the need for two separate voltage supplies. Another, less obvious disadvantage with the complementary circuit is shown in the resulting crossover distortion in the output signal (see Fig. 15.16d). *Crossover distortion* refers to the fact that during the signal crossover from positive to negative (or vice versa) there is some nonlinearity in the output signal. This results from the fact that the circuit does not provide exact switching of one transistor off and the other on at the zero-voltage condition. Both transistors may be partially off

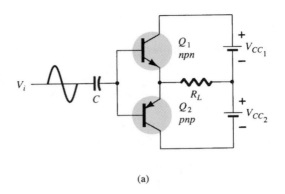

(a)

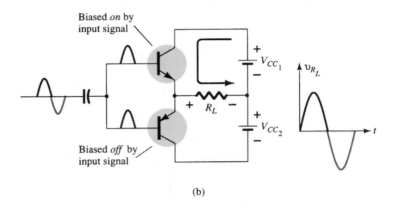

(b)

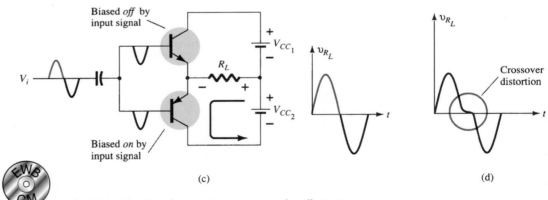

(c)   (d)

Figure 15.16 Complementary-symmetry push-pull circuit.

so that the output voltage does not follow the input around the zero-voltage condition. Biasing the transistors in class AB improves this operation by biasing both transistors to be on for more than half a cycle.

A more practical version of a push-pull circuit using complementary transistors is shown in Fig. 15.17. Note that the load is driven as the output of an emitter-follower so that the load resistance of the load is matched by the low output resistance of the driving source. The circuit uses complementary Darlington-connected transistors to provide higher output current and lower output resistance.

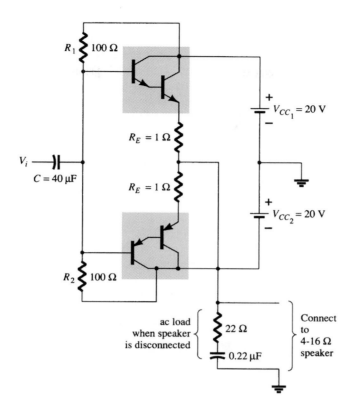

Figure 15.17 Complementary-symmetry push-pull circuit using Darlington transistors.

Quasi-Complementary Push–Pull Amplifier

In practical power amplifier circuits, it is preferable to use *npn* transistors for both high-current-output devices. Since the push-pull connection requires complementary devices, a *pnp* high-power transistor must be used. A practical means of obtaining complementary operation while using the same, matched *npn* transistors for the output is provided by a quasi-complementary circuit, as shown in Fig. 15.18. The push-pull op-

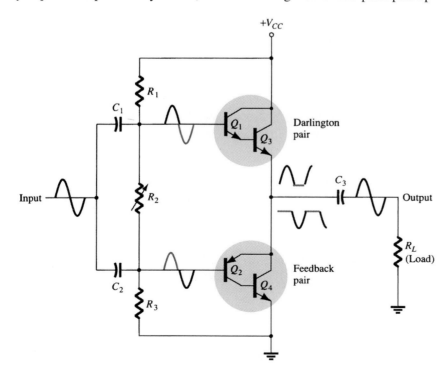

Figure 15.18 Quasi-complementary push-pull transformerless power amplifier.

15.5 **Class B Amplifier Circuits** 769

eration is achieved by using complementary transistors (Q_1 and Q_2) before the matched *npn* output transistors (Q_3 and Q_4). Notice that transistors Q_1 and Q_3 form a Darlington connection that provides output from a low-impedance emitter-follower. The connection of transistors Q_2 and Q_4 forms a feedback pair, which similarly provides a low-impedance drive to the load. Resistor R_2 can be adjusted to minimize crossover distortion by adjusting the dc bias condition. The single input signal applied to the push-pull stage then results in a full cycle output to the load. The quasi-complementary push-pull amplifier is presently the most popular form of power amplifier.

EXAMPLE 15.10

For the circuit of Fig. 15.19, calculate the input power, output power, and power handled by each output transistor and the circuit efficiency for an input of 12 V rms.

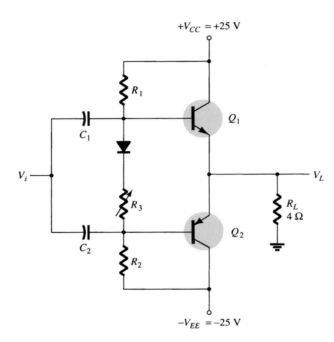

Figure 15.19 Class B power amplifier for Examples 15.10–15.12.

Solution

The peak input voltage is

$$V_i(\text{p}) = \sqrt{2}\, V_i\,(\text{rms}) = \sqrt{2}\,(12\text{ V}) = 16.97\text{ V} \approx 17\text{ V}$$

Since the resulting voltage across the load is ideally the same as the input signal (the amplifier has, ideally, a voltage gain of unity),

$$V_L(\text{p}) = 17\text{ V}$$

and the output power developed across the load is

$$P_o(\text{ac}) = \frac{V_L^2(\text{p})}{2R_L} = \frac{(17\text{ V})^2}{2(4\ \Omega)} = \mathbf{36.125\ W}$$

The peak load current is

$$I_L(\text{p}) = \frac{V_L(\text{p})}{R_L} = \frac{17\text{ V}}{4\ \Omega} = 4.25\text{ A}$$

from which the dc current from the supplies is calculated to be

$$I_{\text{dc}} = \frac{2}{\pi}I_L(\text{p}) = \frac{2(4.25\text{ A})}{\pi} = 2.71\text{ A}$$

so that the power supplied to the circuit is

$$P_i(\text{dc}) = V_{CC}I_{\text{dc}} = (25 \text{ V})(2.71 \text{ A}) = \textbf{67.75 W}$$

The power dissipated by each output transistor is

$$P_Q = \frac{P_{2Q}}{2} = \frac{P_i - P_o}{2} = \frac{67.75 \text{ W} - 36.125 \text{ W}}{2} = \textbf{15.8 W}$$

The circuit efficiency (for the input of 12 V, rms) is then

$$\% \, \eta = \frac{P_o}{P_i} \times 100\% = \frac{36.125 \text{ W}}{67.75 \text{ W}} \times 100\% = \textbf{53.3\%}$$

For the circuit of Fig. 15.19, calculate the maximum input power, maximum output power, input voltage for maximum power operation, and the power dissipated by the output transistors at this voltage.

EXAMPLE 15.11

Solution

The maximum input power is

$$\text{maximum } P_i(\text{dc}) = \frac{2V_{CC}^2}{\pi R_L} = \frac{2(25 \text{ V})^2}{\pi 4 \, \Omega} = \textbf{99.47 W}$$

The maximum output power is

$$\text{maximum } P_o(\text{ac}) = \frac{V_{CC}^2}{2R_L} = \frac{(25 \text{ V})^2}{2(4 \, \Omega)} = \textbf{78.125 W}$$

[Note that the maximum efficiency is achieved:]

$$\% \, \eta = \frac{P_o}{P_i} \times 100\% = \frac{78.125 \text{ W}}{99.47 \text{ W}} \, 100\% = 78.54\%$$

To achieve maximum power operation the output voltage must be

$$V_L(\text{p}) = V_{CC} = 25 \text{ V}$$

and the power dissipated by the output transistors is then

$$P_{2Q} = P_i - P_o = 99.47 \text{ W} - 78.125 \text{ W} = \textbf{21.3 W}$$

For the circuit of Fig. 15.19, determine the maximum power dissipated by the output transistors and the input voltage at which this occurs.

EXAMPLE 15.12

Solution

The maximum power dissipated by both output transistors is

$$\text{maximum } P_{2Q} = \frac{2V_{CC}^2}{\pi^2 R_L} = \frac{2(25 \text{ V})^2}{\pi^2 4 \, \Omega} = \textbf{31.66 W}$$

This maximum dissipation occurs at

$$V_L = 0.636 V_L(\text{p}) = 0.636(25 \text{ V}) = \textbf{15.9 V}$$

(Notice that at $V_L = 15.9$ V the circuit required the output transistors to dissipate 31.66 W, while at $V_L = 25$ V they only had to dissipate 21.3 W.)

15.6 AMPLIFIER DISTORTION

A pure sinusoidal signal has a single frequency at which the voltage varies positive and negative by equal amounts. Any signal varying over less than the full 360° cycle is considered to have distortion. An ideal amplifier is capable of amplifying a pure sinusoidal signal to provide a larger version, the resulting waveform being a pure single-frequency sinusoidal signal. When distortion occurs the output will not be an exact duplicate (except for magnitude) of the input signal.

Distortion can occur because the device characteristic is not linear, in which case nonlinear or amplitude distortion occurs. This can occur with all classes of amplifier operation. Distortion can also occur because the circuit elements and devices respond to the input signal differently at various frequencies, this being frequency distortion.

One technique for describing distorted but period waveforms uses Fourier analysis, a method that describes any periodic waveform in terms of its fundamental frequency component and frequency components at integer multiples—these components are called *harmonic components* or *harmonics*. For example, a signal that is originally 1000 Hz could result, after distortion, in a frequency component at 1000 Hz (1 kHz) and harmonic components at 2 kHz(2×1 kHz), 3 kHz(3×1 kHz), 4 kHz(4×1 kHz), and so on. The original frequency of 1 kHz is called the *fundamental frequency*; those at integer multiples are the harmonics. The 2-kHz component is therefore called a *second harmonic*, that at 3 kHz is the *third harmonic*, and so on. The fundamental frequency is not considered a harmonic. Fourier analysis does not allow for fractional harmonic frequencies—only integer multiples of the fundamental.

Harmonic Distortion

A signal is considered to have harmonic distortion when there are harmonic frequency components (not just the fundamental component). If the fundamental frequency has an amplitude, A_1, and the nth frequency component has an amplitude, A_n, a harmonic distortion can be defined as

$$\% \ n\text{th harmonic distortion} = \% \ D_n = \frac{|A_n|}{|A_1|} \times 100\% \tag{15.30}$$

The fundamental component is typically larger than any harmonic component.

EXAMPLE 15.13
Calculate the harmonic distortion components for an output signal having fundamental amplitude of 2.5 V, second harmonic amplitude of 0.25 V, third harmonic amplitude of 0.1 V, and fourth harmonic amplitude of 0.05 V.

Solution

Using Eq. (15.30) yields

$$\% \ D_2 = \frac{|A_2|}{|A_1|} \times 100\% = \frac{0.25 \text{ V}}{2.5 \text{ V}} \times 100\% = \mathbf{10\%}$$

$$\% \ D_3 = \frac{|A_3|}{|A_1|} \times 100\% = \frac{0.1 \text{ V}}{2.5 \text{ V}} \times 100\% = \mathbf{4\%}$$

$$\% \ D_4 = \frac{|A_4|}{|A_1|} \times 100\% = \frac{0.05 \text{ V}}{2.5 \text{ V}} \times 100\% = \mathbf{2\%}$$

TOTAL HARMONIC DISTORTION

When an output signal has a number of individual harmonic distortion components, the signal can be seen to have a total harmonic distortion based on the individual elements as combined by the relationship of the following equation:

$$\% \text{ THD} = \sqrt{D_2^2 + D_3^2 + D_4^2 + \cdots} \times 100\% \qquad (15.31)$$

where THD is total harmonic distortion.

Calculate the total harmonic distortion for the amplitude components given in Example 15.13.

EXAMPLE 15.14

Solution

Using the computed values of $D_2 = 0.10$, $D_3 = 0.04$, and $D_4 = 0.02$ in Eq. (15.31),

$$\% \text{ THD} = \sqrt{D_2^2 + D_3^2 + D_4^2} \times 100\%$$
$$= \sqrt{(0.10)^2 + (0.04)^2 + (0.02)^2} \times 100\% = 0.1095 \times 100\%$$
$$= \mathbf{10.95\%}$$

An instrument such as a spectrum analyzer would allow measurement of the harmonics present in the signal by providing a display of the fundamental component of a signal and a number of its harmonics on a display screen. Similarly, a wave analyzer instrument allows more precise measurement of the harmonic components of a distorted signal by filtering out each of these components and providing a reading of these components. In any case, the technique of considering any distorted signal as containing a fundamental component and harmonic components is practical and useful. For a signal occurring in class AB or class B, the distortion may be mainly even harmonics, of which the second harmonic component is the largest. Thus, although the distorted signal theoretically contains all harmonic components from the second harmonic up, the most important in terms of the amount of distortion in the classes presented above is the second harmonic.

SECOND HARMONIC DISTORTION

Figure 15.20 shows a waveform to use for obtaining second harmonic distortion. A collector current waveform is shown with the quiescent, minimum, and maximum signal levels, and the time at which they occur is marked on the waveform. The sig-

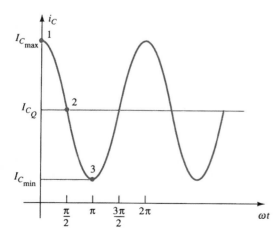

Figure 15.20 Waveform for obtaining second harmonic distortion.

nal shown indicates that some distortion is present. An equation that approximately describes the distorted signal waveform is

$$i_C \approx I_{C_Q} + I_0 + I_1 \cos \omega t + I_2 \cos \omega t \tag{15.32}$$

The current waveform contains the original quiescent current I_{C_Q}, which occurs with zero input signal; an additional dc current I_0, due to the nonzero average of the distorted signal; the fundamental component of the distorted ac signal, I_1; and a second harmonic component I_2, at twice the fundamental frequency. Although other harmonics are also present, only the second is considered here. Equating the resulting current from Eq. (15.32) at a few points in the cycle to that shown on the current waveform provides the following three relations:

At point 1 ($\omega t = 0$):

$$i_C = I_{C_{\max}} = I_{C_Q} + I_0 + I_1 \cos 0 + I_2 \cos 0$$

$$I_{C_{\max}} = I_{C_Q} + I_0 + I_1 + I_2$$

At point 2 ($\omega t = \pi/2$):

$$i_C = I_{C_Q} = I_{C_Q} + I_0 + I_1 \cos \frac{\pi}{2} + I_2 \cos \frac{2\pi}{2}$$

$$I_{C_Q} = I_{C_Q} + I_0 - I_2$$

At point 3 ($\omega t = \pi$):

$$i_C = I_{C_{\min}} = I_{C_Q} + I_0 + I_1 \cos \pi + I_2 \cos 2\pi$$

$$I_{C_{\min}} = I_{C_Q} + I_0 - I_1 + I_2$$

Solving the preceding three equations simultaneously gives the following results:

$$I_0 = I_2 = \frac{I_{C_{\max}} + I_{C_{\min}} - 2I_{C_Q}}{4}, \qquad I_1 = \frac{I_{C_{\max}} - I_{C_{\min}}}{2}$$

Referring to Eq. (15.30), the definition of second harmonic distortion may be expressed as

$$D_2 = \left| \frac{I_2}{I_1} \right| \times 100\%$$

Inserting the values of I_1 and I_2 determined above gives

$$D_2 = \left| \frac{\frac{1}{2}(I_{C_{\max}} + I_{C_{\min}}) - I_{C_Q}}{I_{C_{\max}} - I_{C_{\min}}} \right| \times 100\% \tag{15.33}$$

In a similar manner, the second harmonic distortion can be expressed in terms of measured collector–emitter voltages:

$$D_2 = \left| \frac{\frac{1}{2}(V_{CE_{\max}} + V_{CE_{\min}}) - V_{CE_Q}}{V_{CE_{\max}} - V_{CE_{\min}}} \right| \times 100\% \tag{15.34}$$

EXAMPLE 15.15

Calculate the second harmonic distortion, if an output waveform displayed on an oscilloscope provides the following measurements:
(a) $V_{CE_{\min}} = 1$ V, $V_{CE_{\max}} = 22$ V, $V_{CE_Q} = 12$ V.
(b) $V_{CE_{\min}} = 4$ V, $V_{CE_{\max}} = 20$ V, $V_{CE_Q} = 12$ V.

Solution

Using Eq. (15.34), we get

(a) $D_2 = \left| \dfrac{\frac{1}{2}(22\text{ V} + 1\text{ V}) - 12\text{ V}}{22\text{ V} - 1\text{ V}} \right| \times 100\% = \mathbf{2.38\%}$

(b) $D_2 = \left| \dfrac{\frac{1}{2}(20\text{ V} + 4\text{ V}) - 12\text{ V}}{20\text{ V} - 4\text{ V}} \right| \times 100\% = \mathbf{0\%}$ \qquad (no distortion)

Power of Signal Having Distortion

When distortion does occur, the output power calculated for the undistorted signal is no longer correct. When distortion is present, the output power delivered to the load resistor R_C due to the fundamental component of the distorted signal is

$$P_1 = \frac{I_1^2 R_C}{2} \tag{15.35}$$

The total power due to all the harmonic components of the distorted signal can then be calculated using

$$P = (I_1^2 + I_2^2 + I_3^2 + \cdots)\frac{R_C}{2} \tag{15.36}$$

The total power can also be expressed in terms of the total harmonic distortion,

$$P = (1 + D_2^2 + D_3^2 + \cdots)I_1^2\frac{R_C}{2} = (1 + \text{THD}^2)P_1 \tag{15.37}$$

For harmonic distortion reading of $D_2 = 0.1$, $D_3 = 0.02$, and $D_4 = 0.01$, with $I_1 = 4$ A and $R_C = 8\ \Omega$, calculate the total harmonic distortion, fundamental power component, and total power.

EXAMPLE 15.16

Solution

The total harmonic distortion is

$$\text{THD} = \sqrt{D_2^2 + D_3^2 + D_4^2} = \sqrt{(0.1)^2 + (0.02)^2 + (0.01)^2} \approx \mathbf{0.1}$$

The fundamental power, using Eq. (15.35), is

$$P_1 = \frac{I_1^2 R_C}{2} = \frac{(4\text{ A})^2(8\ \Omega)}{2} = \mathbf{64\ W}$$

The total power calculated using Eq. (15.37) is then

$$P = (1 + \text{THD}^2)P_1 = [1 + (0.1)^2]64 = (1.01)64 = \mathbf{64.64\ W}$$

(Note that the total power is due mainly to the fundamental component even with 10% second harmonic distortion.)

Graphical Description of Harmonic Components of Distorted Signal

A distorted waveform such as that which occurs in class B operation can be represented using Fourier analysis as a fundamental with harmonic components. Figure 15.21a shows a positive half-cycle such as the type that would result in one side of a class B

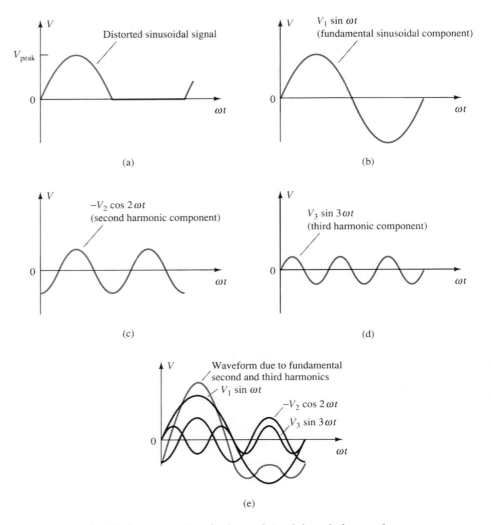

Figure 15.21 Graphical representation of a distorted signal through the use of harmonic components.

amplifier. Using Fourier analysis techniques, the fundamental component of the distorted signal can be obtained, as shown in Fig. 15.21b. Similarly, the second and third harmonic components can be obtained and are shown in Fig. 15.21c and d, respectively. Using the Fourier technique, the distorted waveform can be made by adding the fundamental and harmonic components, as shown in Fig. 15.21e. In general, any periodic distorted waveform can be represented by adding a fundamental component and all harmonic components, each of varying amplitude and at various phase angles.

15.7 POWER TRANSISTOR HEAT SINKING

While integrated circuits are used for small-signal and low-power applications, most high-power applications still require individual power transistors. Improvements in production techniques have provided higher power ratings in small-sized packaging cases, have increased the maximum transistor breakdown voltage, and have provided faster-switching power transistors.

The maximum power handled by a particular device and the temperature of the transistor junctions are related since the power dissipated by the device causes an increase in temperature at the junction of the device. Obviously, a 100-W transistor will provide more power capability than a 10-W transistor. On the other hand, proper heat-sinking techniques will allow operation of a device at about one-half its maximum power rating.

We should note that of the two types of bipolar transistors—germanium and silicon—silicon transistors provide greater maximum temperature ratings. Typically, the maximum junction temperature of these types of power transistors is

Silicon: 150–200°C

Germanium: 100–110°C

For many applications the average power dissipated may be approximated by

$$P_D = V_{CE}I_C \tag{15.38}$$

This power dissipation, however, is allowed only up to a maximum temperature. Above this temperature, the device power dissipation capacity must be reduced (or derated) so that at higher case temperatures the power-handling capacity is reduced, down to 0 W at the device maximum case temperature.

The greater the power handled by the transistor, the higher the case temperature. Actually, the limiting factor in power handling by a particular transistor is the temperature of the device's collector junction. Power transistors are mounted in large metal cases to provide a large area from which the heat generated by the device may radiate (be transferred). Even so, operating a transistor directly into air (mounting it on a plastic board, for example) severely limits the device power rating. If, instead (as is usual practice), the device is mounted on some form of heat sink, its power-handling capacity can approach the rated maximum value more closely. A few heat sinks are shown in Fig. 15.22. When the heat sink is used, the heat produced by the transistor dissipating power has a larger area from which to radiate (transfer) the heat into the air, thereby holding the case temperature to a much lower value than would result without the heat sink. Even with an infinite heat sink (which, of course, is not available), for which the case temperature is held at the ambient (air) temperature, the junction will be heated above the case temperature and a maximum power rating must be considered.

Since even a good heat sink cannot hold the transistor case temperature at ambient (which, by the way, could be more than 25°C if the transistor circuit is in a confined area where other devices are also radiating a good amount of heat), it is necessary to derate the amount of maximum power allowed for a particular transistor as a function of increased case temperature.

Figure 15.22 Typical power heat sinks.

Figure 15.23 shows a typical power derating curve for a silicon transistor. The curve shows that the manufacturer will specify an upper temperature point (not necessarily 25°C), after which a linear derating takes place. For silicon, the maximum power that should be handled by the device does not reduce to 0 W until the case temperature is 200°C.

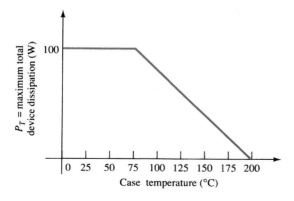

Figure 15.23 Typical power derating curve for silicon transistors.

P_L

It is not necessary to provide a derating curve since the same information could be given simply as a listed derating factor on the device specification sheet. Stated mathematically, we have

$$P_D(\text{temp}_1) = P_D(\text{temp}_0) - (\text{Temp}_1 - \text{Temp}_0)(\text{derating factor}) \qquad (15.39)$$

where the value of Temp_0 is the temperature at which derating should begin, the value of Temp_1 is the particular temperature of interest (above the value Temp_0), $P_D(\text{temp}_0)$ and $P_D(\text{temp}_1)$ are the maximum power dissipations at the temperatures specified, and the derating factor is the value given by the manufacturer in units of watts (or milliwatts) per degree of temperature.

EXAMPLE 15.17

Determine what maximum dissipation will be allowed for an 80-W silicon transistor (rated at 25°C) if derating is required above 25°C by a derating factor of 0.5 W/°C at a case temperature of 125°C.

Solution

$$P_D(125°C) = P_D(25°C) - (125°C - 25°C)(0.5 \text{ W/}°C)$$

$$= 80 \text{ W} - 100°C(0.5 \text{ W/}°C) = \mathbf{30 \text{ W}}$$

It is interesting to note what power rating results from using a power transistor without a heat sink. For example, a silicon transistor rated at 100 W at (or below) 100°C is rated only 4 W at (or below) 25°C, the free-air temperature. Thus, operated without a heat sink, the device can handle a maximum of only 4 W at the room temperature of 25°C. Using a heat sink large enough to hold the case temperature to 100°C at 100 W allows operating at the maximum power rating.

Thermal Analogy of Power Transistor

Selection of a suitable heat sink requires a considerable amount of detail that is not appropriate to our present basic considerations of the power transistor. However, more detail about the thermal characteristics of the transistor and its relation to the power dissipation of the transistor may help provide a clearer understanding of power as limited by temperature. The following discussion should prove useful.

A picture of how the junction temperature (T_J), case temperature (T_C), and ambient (air) temperature (T_A) are related by the device heat-handling capacity—a temperature coefficient usually called thermal resistance—is presented in the thermal-electric analogy shown in Fig. 15.24.

In providing a thermal-electrical analogy, the term *thermal resistance* is used to describe heat effects by an electrical term. The terms in Fig. 15.24 are defined as follows:

$$\theta_{JA} = \text{total thermal resistance (junction to ambient)}$$

$$\theta_{JC} = \text{transistor thermal resistance (junction to case)}$$

$$\theta_{CS} = \text{insulator thermal resistance (case to heat sink)}$$

$$\theta_{SA} = \text{heat-sink thermal resistance (heat sink to ambient)}$$

Using the electrical analogy for thermal resistances, we can write

$$\boxed{\theta_{JA} = \theta_{JC} + \theta_{CS} + \theta_{SA}} \qquad (15.40)$$

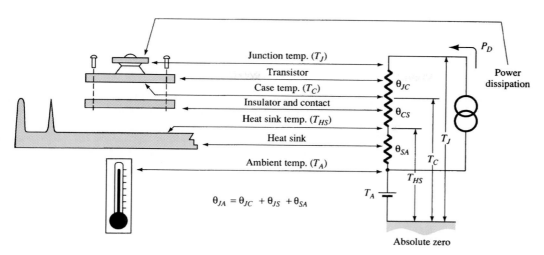

Figure 15.24 Thermal-to-electrical analogy.

The analogy can also be used in applying Kirchhoff's law to obtain

$$T_J = P_D\theta_{JA} + T_A \qquad (15.41)$$

The last relation shows that the junction temperature "floats" on the ambient temperature and that the higher the ambient temperature, the lower the allowed value of device power dissipation.

The thermal factor θ provides information about how much temperature drop (or rise) results for a given amount of power dissipation. For example, the value of θ_{JC} is usually about 0.5°C/W. This means that for a power dissipation of 50 W, the difference in temperature between case temperature (as measured by a thermocouple) and the inside junction temperature is only

$$T_J - T_C = \theta_{JC}P_D = (0.5°\text{C/W})(50\text{ W}) = 25°\text{C}$$

Thus, if the heat sink can hold the case at, say, 50°C, the junction is then only at 75°C. This is a relatively small temperature difference, especially at lower power-dissipation levels.

The value of thermal resistance from junction to free air (using no heat sink) is, typically,

$$\theta_{JA} = 40°\text{C/W} \qquad \text{(into free air)}$$

For this thermal resistance, only 1 W of power dissipation results in a junction temperature 40°C greater than the ambient.

A heat sink can now be seen to provide a low thermal resistance between case and air—much less than the 40°C/W value of the transistor case alone. Using a heat sink having

$$\theta_{SA} = 2°\text{C/W}$$

and with an insulating thermal resistance (from case to heat sink) of

$$\theta_{CS} = 0.8°\text{C/W}$$

and finally, for the transistor,

$$\theta_{CJ} = 0.5°\text{C/W}$$

we can obtain

$$\theta_{JA} = \theta_{SA} + \theta_{CS} + \theta_{CJ}$$

$$= 2.0°\text{C/W} + 0.8°\text{C/W} + 0.5°\text{C/W} = 3.3°\text{C/W}$$

So with a heat sink, the thermal resistance between air and the junction is only 3.3°C/W, compared to 40°C/W for the transistor operating directly into free air. Using the value of θ_{JA} above for a transistor operated at, say, 2 W, we calculate

$$T_J - T_A = \theta_{JA}P_D = (3.3°C/W)(2\ W) = 6.6°C$$

In other words, the use of a heat sink in this example provides only a 6.6°C increase in junction temperature as compared to an 80°C rise without a heat sink.

EXAMPLE 15.18

A silicon power transistor is operated with a heat sink ($\theta_{SA} = 1.5°C/W$). The transistor, rated at 150 W (25°C), has $\theta_{JC} = 0.5°C/W$, and the mounting insulation has $\theta_{CS} = 0.6°C/W$. What maximum power can be dissipated if the ambient temperature is 40°C and $T_{J_{max}} = 200°C$?

Solution

$$P_D = \frac{T_J - T_A}{\theta_{JC} + \theta_{CS} + \theta_{SA}} = \frac{200°C - 40°C}{0.5°C/W + 0.6°C/W + 1.5°C/W} \approx \mathbf{61.5\ W}$$

15.8 CLASS C AND CLASS D AMPLIFIERS

Although class A, class AB, and class B amplifiers are most used as power amplifiers, class D amplifiers are popular because of their very high efficiency. Class C amplifiers, while not used as audio amplifiers, do find use in tuned circuits as used in communications.

Class C Amplifier

A class C amplifier, as that shown in Fig. 15.25, is biased to operate for less than 180° of the input signal cycle. The tuned circuit in the output, however, will provide a full cycle of output signal for the fundamental or resonant frequency of the tuned circuit (L and C tank circuit) of the output. This type of operation is therefore limited to use at one fixed frequency, as occurs in a communications circuit, for example. Operation of a class C circuit is not intended primarily for large-signal or power amplifiers.

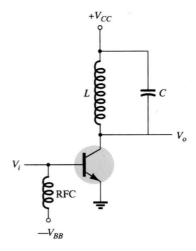

Figure 15.25 Class C amplifier circuit.

Class D Amplifier

A class D amplifier is designed to operate with digital or pulse-type signals. An efficiency of over 90% is achieved using this type of circuit, making it quite desirable in power amplifiers. It is necessary, however, to convert any input signal into a pulse-type waveform before using it to drive a large power load and to convert the signal back to a sinusoidal-type signal to recover the original signal. Figure 15.26 shows how a sinusoidal signal may be converted into a pulse-type signal using some form of sawtooth or chopping waveform to be applied with the input into a comparator-type op-amp circuit so that a representative pulse-type signal is produced. While the letter D is used to describe the next type of bias operation after class C, the D could also be considered to stand for "Digital," since that is the nature of the signals provided to the class D amplifier.

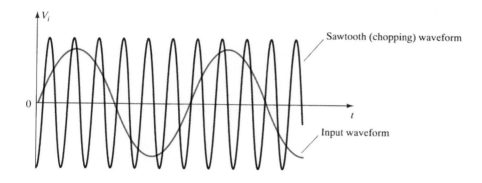

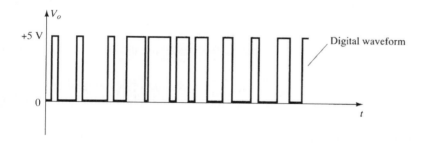

Figure 15.26 Chopping of sinusoidal waveform to produce digital waveform.

Figure 15.27 shows a block diagram of the unit needed to amplify the class D signal and then convert back to the sinusoidal-type signal using a low-pass filter. Since the amplifier's transistor devices used to provide the output are basically either off or

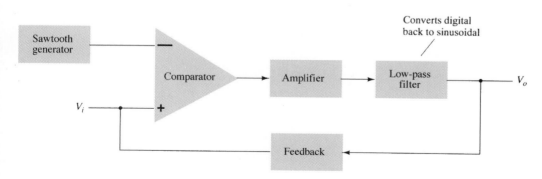

Figure 15.27 Block diagram of class D amplifier.

on, they provide current only when they are turned on, with little power loss due to their low on-voltage. Since most of the power applied to the amplifier is transferred to the load, the efficiency of the circuit is typically very high. Power MOSFET devices have been quite popular as the driver devices for the class D amplifier.

15.9 SUMMARY

Important Conclusions and Concepts

1. Amplifier classes:

   Class A—the output stage conducts for a full 360° (a full waveform cycle).

   Class B—the output stages each conduct for 180° (together providing a full cycle).

   Class AB—the output stages each conduct between 180° and 360° (providing a full cycle at less efficiency).

   Class C—the output stage conducts for less than 180° (used in tuned circuits).

   Class D—has operation using digital or pulsed signals.

2. Amplifier efficiency:

   Class A—maximum efficiency of 25% (without transformer) and 50% (with transformer).

   Class B—maximum efficiency of 78.5%.

3. Power considerations:
   a. Input power is provided by the dc power supply.
   b. Output power is that delivered to the load.
   c. Power dissipated by active devices is essentially the difference between the input and output powers.
4. Push-pull (complementary) operation is typically the opposite of that of devices with one on at a time—one "pushing" for half the cycle and the other "pulling" for half the cycle.
5. **Harmonic distortion** refers to the nonsinusoidal nature of a periodic waveform—the distortion being defined as that at the periodic frequency and multiples of that frequency.
6. **Heat sink** is the use of metal cases or frames and fans to remove the heat generated in a circuit element.

Equations

$$P_i(\text{dc}) = V_{CC}I_{CQ}$$

$$P_o(\text{ac}) = V_{CE}(\text{rms})I_C(\text{rms})$$
$$= I_C^2(\text{rms})R_C$$
$$= \frac{V_C^2(\text{rms})}{R_C}$$

$$P_o(\text{ac}) = \frac{V_{CE}(\text{p})I_C(\text{p})}{2}$$

$$= \frac{I_C^2(\text{p})}{2R_C}$$

$$= \frac{V_{CE}^2(\text{p})}{2R_C}$$

$$P_o(\text{ac}) = \frac{V_{CE}(\text{p-p})I_C(\text{p-p})}{8}$$

$$= \frac{I_C^2(\text{p-p})}{8}R_C$$

$$= \frac{V_{CE}^2(\text{p-p})}{8R_C}$$

$$\% \, \eta = \frac{P_o(\text{ac})}{P_i(\text{dc})} \times 100\%$$

Transformer action:

$$\frac{V_2}{V_1} = \frac{N_2}{N_1}$$

$$\frac{I_2}{I_1} = \frac{N_1}{N_2}$$

Class B operation:

$$I_{\text{dc}} = \frac{2}{\pi}I(\text{p})$$

$$P_i(\text{dc}) = V_{CC}\left(\frac{2}{\pi}I(\text{p})\right)$$

$$P_o(\text{ac}) = \frac{V_L^2(\text{rms})}{R_L}$$

$$\text{maximum } P_o(\text{ac}) = \frac{V_{CC}^2}{2R_L}$$

$$\text{maximum } P_i(\text{dc}) = V_{CC}(\text{maximum } I_{\text{dc}}) = V_{CC}\left(\frac{2V_{CC}}{\pi R_L}\right) = \frac{2V_{CC}^2}{\pi R_L}$$

$$\text{maximum } P_{2Q} = \frac{2V_{CC}^2}{\pi^2 R_L}$$

Harmonic distortion:

$$\% \, n\text{th harmonic distortion} = \% \, D_n = \frac{|A_n|}{|A_1|} \times 100\%$$

Heat sink:

$$\theta_{JA} = \theta_{JC} + \theta_{CS} + \theta_{SA}$$

15.10 COMPUTER ANALYSIS

Program 15.1—Series-Fed Class A Amplifier

Using Design Center, the circuit of a series-fed class A amplifier is drawn as shown in Fig. 15.28. Figure 15.29 shows some of the analysis output. Edit the transistor model for values of only **BF** = 90 and **IS** = 2E-15. This keeps the transistor model more ideal so that PSpice calculations better match those below.

The dc bias of the collector voltage is shown to be

$$V_c(\text{dc}) = 12.47 \text{ V}$$

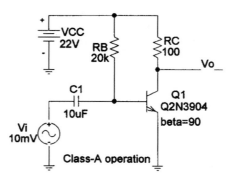

Figure 15.28 Series-fed class A amplifier.

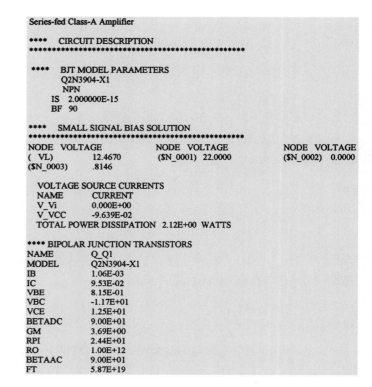

Series-fed Class-A Amplifier

****   CIRCUIT DESCRIPTION
****

  ****   BJT MODEL PARAMETERS
         Q2N3904-X1
         NPN
      IS   2.000000E-15
      BF   90

****   SMALL SIGNAL BIAS SOLUTION
****

NODE VOLTAGE		NODE VOLTAGE		NODE VOLTAGE
(VL)	12.4670	($N_0001)	22.0000	($N_0002) 0.0000
($N_0003)	.8146			

      VOLTAGE SOURCE CURRENTS
      NAME   CURRENT
      V_Vi   0.000E+00
      V_VCC   -9.639E-02
      TOTAL POWER DISSIPATION   2.12E+00 WATTS

**** BIPOLAR JUNCTION TRANSISTORS

NAME	Q_Q1
MODEL	Q2N3904-X1
IB	1.06E-03
IC	9.53E-02
VBE	8.15E-01
VBC	-1.17E+01
VCE	1.25E+01
BETADC	9.00E+01
GM	3.69E+00
RPI	2.44E+01
RO	1.00E+12
BETAAC	9.00E+01
FT	5.87E+19

Figure 15.29 Analysis output for the circuit of Fig. 15.28.

With transistor beta set to 90, the ac gain is calculated as follows:

$$I_E = I_c = 95 \text{ mA (from analysis output of PSpice)}$$

$$r_e = 26 \text{ mV}/95 \text{ mA} = 0.27 \text{ }\Omega$$

For a gain of

$$A_v = -R_c/r_e = -100/0.27 = -370$$

The output voltage is then

$$V_o = A_v V_i = (-370) \cdot 10 \text{ mV} = -3.7 \text{ V(peak)}$$

The output waveform obtained using **probe** is shown in Fig. 15.30. For a peak-to-peak output of

$$V_o(\text{p-p}) = 15.6 \text{ V} - 8.75 \text{ V} = 6.85 \text{ V}$$

the peak output is

$$V_o(\text{p}) = 6.85 \text{ V}/2 = 3.4 \text{ V}$$

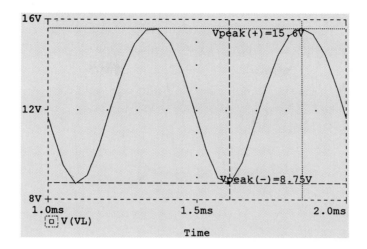

Figure 15.30 Probe output for the circuit of Fig. 15.28.

which compares well with that calculated above.

From the circuit output analysis, the input power is

$$P_i = V_{CC}I_C = (22 \text{ V}) \cdot (95 \text{ mA}) = 2.09 \text{ W}$$

From the probe ac data, the output power is

$$P_o(\text{ac}) = V_o(\text{p-p})^2/[8 \cdot R_L] = (6.85)^2/[8 \cdot 100] = 58 \text{ mW}$$

The efficiency is then

$$\%\eta = P_o/P_i \cdot 100\% = (58 \text{ mW}/2.09 \text{ W}) \cdot 100\% = 2.8\%$$

A larger input signal would increase the ac power delivered to the load and increase the efficiency (the maximum being 25%).

Program 15.2—Quasi-Complementary Push-Pull Amplifier

Figure 15.31 shows a quasi-complementary push-pull class B power amplifier. For the input of $V_i = 20 \text{ V(p)}$, the output waveform obtained using **probe** is shown in Fig. 15.32.

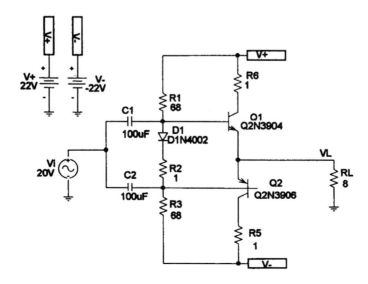

Figure 15.31 Quasi-complementary class B power amplifier.

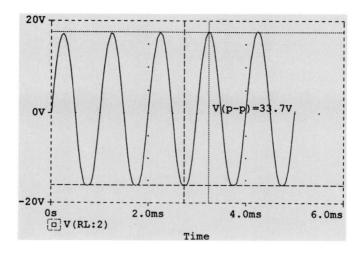

Figure 15.32 Probe output of the circuit in Fig. 15.31.

The resulting ac output voltage is seen to be

$$V_o(\text{p-p}) = 33.7 \text{ V}$$

so that

$$P_o = V_o^2(\text{p-p})/(8 \cdot R_L) = (33.7 \text{ V})^2/(8 \cdot 8 \text{ }\Omega) = 17.7 \text{ W}$$

The input power for that amplitude signal is

$$P_i = V_{CC}I_{\text{dc}} = V_{CC}[(2/\pi)(V_o(\text{p-p})/2)/R_L]$$

$$= (22 \text{ V}) \cdot [(2/\pi)(33.7 \text{ V}/2)/8] = 29.5 \text{ W}$$

The circuit efficiency is then

$$\%\eta = P_o/P_i \cdot 100\% = (17.7 \text{ W}/29.5 \text{ W}) \cdot 100\% = 60\%$$

Program 15.3—Op-Amp Push-Pull Amplifier

Figure 15.33 shows an op-amp push-pull amplifier providing ac output to an 8-Ω load. As shown, the op-amp provides a gain of

$$A_v = -R_F/R_1 = -47 \text{ k}\Omega/18 \text{ k}\Omega = -2.6$$

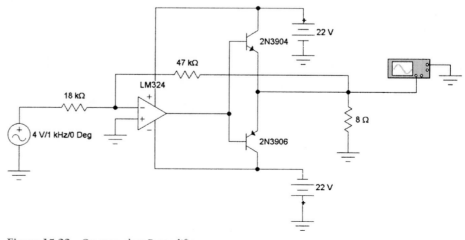

Figure 15.33 Op-amp class B amplifier.

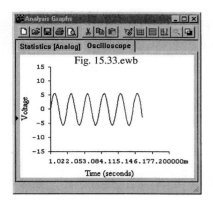

Figure 15.34 Probe output for the circuit of Fig. 15.33.

For the input, $V_i = 1$ V, the output is

$$V_o(\text{p}) = A_v V_i = -2.6 \cdot (1 \text{ V}) = -2.6 \text{ V}$$

Figure 15.34 shows the oscilloscope display of the output voltage.
The output power, input power, and circuit efficiency are then calculated to be

$$P_o = V_o^2(\text{p-p})/(8 \cdot R_L) = (20.4 \text{ V})^2/(8 \cdot 8 \text{ }\Omega) = 6.5 \text{ W}$$

The input power for that amplitude signal is

$$P_i = V_{CC}I_{\text{dc}} = V_{CC}[(2/\pi)(V_o(\text{p-p})/2)/R_L]$$
$$= (12 \text{ V}) \cdot [(2/\pi) \cdot (20.4 \text{ V}/2)/8] = 9.7 \text{ W}$$

The circuit efficiency is then

$$\%\eta = P_o/P_i \cdot 100\% = (6.5 \text{ W}/9.7 \text{ W}) \cdot 100\% = 67\%$$

MATHCAD

The calculations for the class B power amplifier of Example 15.7 and for the class B power amplifier of Example 15.12 are shown below. Using Mathcad, one can enter any desired value of **VCC, RL,** and **VL peak,** with all the calculations immediately providing the new results.

Class-B Power Amplifier (Example 15.7)

VCC := 30   RL := 16   VLpeak := 20

$\text{ILpeak} := \dfrac{\text{VLpeak}}{\text{RL}}$ ILpeak = 1.25

$\text{Idc} := 2 \cdot \dfrac{\text{ILpeak}}{3.14159}$ Idc = 0.796

Pidc := VCC·Idc   Pidc = 23.873

$\text{Poac} := \dfrac{\text{VLpeak}^2}{(2 \cdot \text{RL})}$ Poac = 12.5

$n := \left(\dfrac{\text{Poac}}{\text{Pidc}}\right) \cdot 100$ n = 52.36

Class-B Power Amplifier (Example 15.12)

VCC := 25   RL := 4

$\text{maxPidc} := \dfrac{(2 \cdot \text{VCC}^2)}{(3.14159) \cdot \text{RL}}$ maxPidc = 99.472

$\text{maxPoac} := \dfrac{(\text{VCC}^2)}{(2 \cdot \text{RL})}$ maxPoac = 78.125

$n := \left(\dfrac{\text{maxPoac}}{\text{maxPidc}}\right) \cdot 100$ n = 78.54

P2Q := maxPidc − maxPoac   P2Q = 21.347

The harmonic distortion calculations of Examples 15.13 and 15.14 are shown for a select set of values for **A1** through **A4.**

Examples 15.13 and 15.14
Harmonic Distortion Calculations

A1 := 2.5   A2 := 0.25   A3 := 0.1   A4 := 0.05

$$D2 := \left(\frac{|A2|}{|A1|} \right) \cdot 100 \qquad D2 = 10 \ \%$$

$$D3 := \left(\frac{|A3|}{|A1|} \right) \cdot 100 \qquad D3 = 4 \ \%$$

$$D4 := \left(\frac{|A4|}{|A1|} \right) \cdot 100 \qquad D4 = 2 \ \%$$

$$THD := \sqrt{D2^2 + D3^2 + D4^2} \qquad THD = 10.954 \ \%$$

PROBLEMS

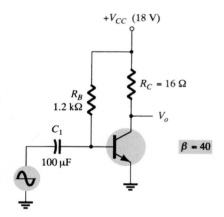

Figure 15.35 Problems 1–4, 26

§ 15.2 Series-Fed Class A Amplifier

1. Calculate the input and output power for the circuit of Fig. 15.35. The input signal results in a base current of 5 mA rms.

2. Calculate the input power dissipated by the circuit of Fig. 15.35 if R_B is changed to 1.5 kΩ.

3. What maximum output power can be delivered by the circuit of Fig. 15.35 if R_B is changed to 1.5 kΩ?

4. If the circuit of Fig. 15.35 is biased at its center voltage and center collector operating point, what is the input power for a maximum output power of 1.5 W?

§ 15.3 Transformer-Coupled Class A Amplifier

5. A class A transformer-coupled amplifier uses a 25:1 transformer to drive a 4-Ω load. Calculate the effective ac load (seen by the transistor connected to the larger turns side of the transformer).

6. What turns ratio transformer is needed to couple to an 8-Ω load so that it appears as an 8-kΩ effective load?

7. Calculate the transformer turns ratio required to connect four parallel 16-Ω speakers so that they appear as an 8-kΩ effective load.

*8. A transformer-coupled class A amplifier drives a 16-Ω speaker through a 3.87:1 transformer. Using a power supply of $V_{CC} = 36$ V, the circuit delivers 2 W to the load. Calculate:
(a) $P(ac)$ across transformer primary.
(b) $V_L(ac)$.
(c) $V(ac)$ at transformer primary.
(d) The rms values of load and primary current.

9. Calculate the efficiency of the circuit of Problem 8 if the bias current is $I_{C_Q} = 150$ mA.

10. Draw the circuit diagram of a class A transformer-coupled amplifier using an *npn* transistor.

§ 15.4 Class B Amplifier Operation

11. Draw the circuit diagram of a class B *npn* push-pull power amplifier using transformer-coupled input.

12. For a class B amplifier providing a 22-V peak signal to an 8-Ω load and a power supply of $V_{CC} = 25$ V, determine:
(a) Input power.
(b) Output power.
(c) Circuit efficiency.

13. For a class B amplifier with $V_{CC} = 25$ V driving an 8-Ω load, determine:
    (a) Maximum input power.
    (b) Maximum output power.
    (c) Maximum circuit efficiency.

* 14. Calculate the efficiency of a class B amplifier for a supply voltage of $V_{CC} = 22$ V driving a 4-Ω load with peak output voltages of:
    (a) $V_L(p) = 20$ V.
    (b) $V_L(p) = 4$ V.

§ 15.5 Class B Amplifier Circuits

15. Sketch the circuit diagram of a quasi-complementary amplifier, showing voltage waveforms in the circuit.

16. For the class B power amplifier of Fig. 15.36, calculate:
    (a) Maximum $P_o(ac)$.
    (b) Maximum $P_i(dc)$.
    (c) Maximum $\%\eta$.
    (d) Maximum power dissipated by both transistors.

* 17. If the input voltage to the power amplifier of Fig. 15.36 is 8-V rms, calculate:
    (a) $P_i(dc)$.
    (b) $P_o(ac)$.
    (c) $\%\eta$.
    (d) Power dissipated by both power output transistors.

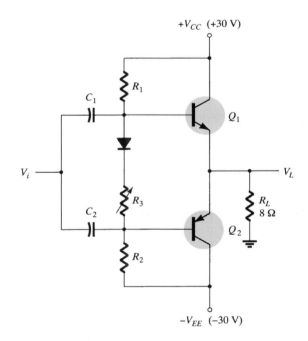

Figure 15.36   Problems 16 and 17, 27

* 18. For the power amplifier of Fig. 15.37, calculate:
    (a) $P_o(ac)$.
    (b) $P_i(dc)$.
    (c) $\%\eta$.
    (d) Power dissipated by both output transistors.

§ 15.6 Amplifier Distortion

19. Calculate the harmonic distortion components for an output signal having fundamental amplitude of 2.1 V, second harmonic amplitude of 0.3 V, third harmonic component of 0.1 V, and fourth harmonic component of 0.05 V.

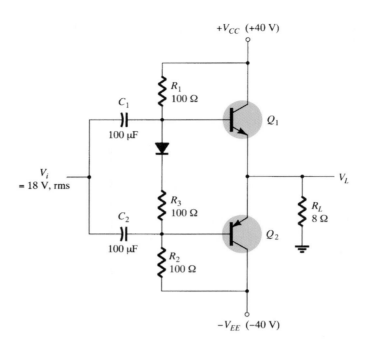

Figure 15.37 Problem 18

20. Calculate the total harmonic distortion for the amplitude components of Problem 19.

21. Calculate the second harmonic distortion for an output waveform having measured values of $V_{CE_{min}} = 2.4$ V, $V_{CE_Q} = 10$ V, and $V_{CE_{max}} = 20$ V.

22. For distortion readings of $D_2 = 0.15$, $D_3 = 0.01$, and $D_4 = 0.05$, with $I_1 = 3.3$ A and $R_C = 4\ \Omega$, calculate the total harmonic distortion fundamental power component and total power.

§ 15.7 Power Transistor Heat Sinking

23. Determine the maximum dissipation allowed for a 100-W silicon transistor (rated at 25°C) for a derating factor of 0.6 W/°C at a case temperature of 150°C.

∗24. A 160-W silicon power transistor operated with a heat sink ($\theta_{SA} = 1.5$°C/W) has θ_{JC} 0.5°C/W and a mounting insulation of $\theta_{CS} = 0.8$°C/W. What maximum power can be handled by the transistor at an ambient temperature of 80°C? (The junction temperature should not exceed 200°C.)

25. What maximum power can a silicon transistor ($T_{J_{max}} = 200$°C) dissipate into free air at an ambient temperature of 80°C?

§ 15.9 Computer Applications

∗26. Use Design Center to draw the schematic of Fig. 15.35 with $V_i = 9.1$ mV.

∗27. Use Design Center to draw the schematic of Fig. 15.36 with $V_i = 25$ V(p). Determine the circuit efficiency.

∗28. Use EWB to draw the schematic of an op-amp class B amplifier as in Fig. 15.33. Use $R_1 = 10$ kΩ, $R_F = 50$ kΩ, and $V_i = 2.5$ V(p). Determine the circuit efficiency.

*Please Note: Asterisks indicate more difficult problems.

Linear-Digital ICs

16.1   INTRODUCTION

While there are many ICs containing only digital circuits and many that contain only linear circuits, there are a number of units that contain both linear and digital circuits. Among the linear/digital ICs are comparator circuits, digital/analog converters, interface circuits, timer circuits, voltage-controlled oscillator (VCO) circuits, and phase-locked loops (PLLs).

The comparator circuit is one to which a linear input voltage is compared to another reference voltage, the output being a digital condition representing whether the input voltage exceeded the reference voltage.

Circuits that convert digital signals into an analog or linear voltage, and those that convert a linear voltage into a digital value, are popular in aerospace equipment, automotive equipment, and compact disk (CD) players, among many others.

Interface circuits are used to enable connecting signals of different digital voltage levels, from different types of output devices, or from different impedances so that both the driver stage and the receiver stage operate properly.

Timer ICs provide linear and digital circuits to use in various timing operations, as in a car alarm, a home timer to turn lights on or off, and a circuit in electromechanical equipment to provide proper timing to match the intended unit operation. The 555 timer has long been a popular IC unit. A voltage-controlled oscillator provides an output clock signal whose frequency can be varied or adjusted by an input voltage. One popular application of a VCO is in a phase-locked loop unit, as used in various communication transmitters and receivers.

16.2   COMPARATOR UNIT OPERATION

A comparator circuit accepts input of linear voltages and provides a digital output that indicates when one input is less than or greater than the second. A basic comparator circuit can be represented as in Fig. 16.1a. The output is a digital signal that stays at a high voltage level when the noninverting ($+$) input is greater than the voltage at the inverting ($-$) input and switches to a lower voltage level when the noninverting input voltage goes below the inverting input voltage.

Figure 16.1b shows a typical connection with one input (the inverting input in this example) connected to a reference voltage, the other connected to the input signal voltage. As long as V_{in} is less than the reference voltage level of $+2$ V, the output remains at a low voltage level (near -10 V). When the input rises just above $+2$ V,

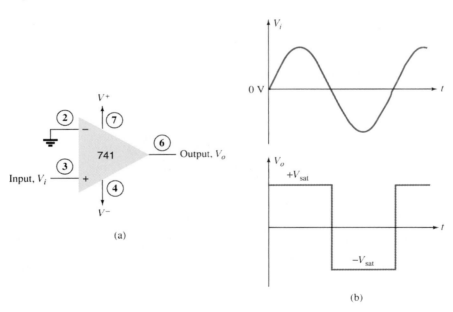

Figure 16.1 Comparator unit: (a) basic unit; (b) typical application.

the output quickly switches to a high-voltage level (near +10 V). Thus the high output indicates that the input signal is greater than +2 V.

Since the internal circuit used to build a comparator contains essentially an op-amp circuit with very high voltage gain, we can examine the operation of a comparator using a 741 op-amp, as shown in Fig. 16.2. With reference input (at pin 2) set to 0 V, a sinusoidal signal applied to the noninverting input (pin 3) will cause the output to switch between its two output states, as shown in Fig. 16.2b. The input V_i going even a fraction of a millivolt above the 0-V reference level will be amplified by the very high voltage gain (typically over 100,000) so that the output rises to its positive output saturation level and remains there while the input stays above $V_{ref} = 0$ V. When the input drops just below the 0-V reference level, the output is driven to its lower saturation level and stays there while the input remains below $V_{ref} = 0$ V. Figure 16.2b clearly shows that the input signal is linear while the output is digital.

Figure 16.2 Operation of 741 op-amp as comparator.

In general use, the reference level need not be 0 V but can be any desired positive or negative voltage. Also, the reference voltage may be connected to either plus or minus input and the input signal then applied to the other input.

Use of Op-Amp as Comparator

Figure 16.3a shows a circuit operating with a positive reference voltage connected to the minus input and the output connected to an indicator LED. The reference voltage level is set at

$$V_{ref} = \frac{10 \text{ k}\Omega}{10 \text{ k}\Omega + 10 \text{ k}\Omega}(+12 \text{ V}) = +6 \text{ V}$$

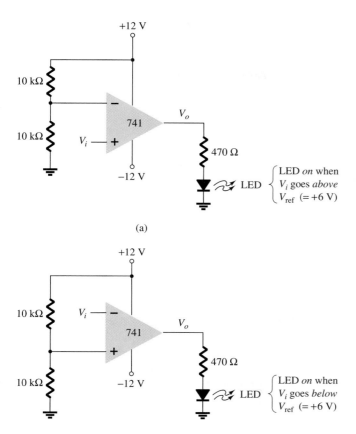

Figure 16.3 A 741 op-amp used as a comparator.

Since the reference voltage is connected to the inverting input, the output will switch to its positive saturation level when the input, V_i, goes more positive than the $+6$-V reference voltage level. The output, V_o, then drives the LED on as an indication that the input is more positive than the reference level.

As an alternative connection, the reference voltage could be connected to the non-inverting input as shown in Fig. 16.3b. With this connection, the input signal going below the reference level would cause the output to drive the LED on. The LED can thus be made to go on when the input signal goes above or below the reference level, depending on which input is connected as signal input and which as reference input.

Using Comparator IC Units

While op-amps can be used as comparator circuits, separate IC comparator units are more suitable. Some of the improvements built into a comparator IC are faster switching between the two output levels, built-in noise immunity to prevent the output from oscillating when the input passes by the reference level, and outputs capable of directly driving a variety of loads. A few popular IC comparators are covered next, describing their pin connections and how they may be used.

311 COMPARATOR

The 311 voltage comparator shown in Fig. 16.4 contains a comparator circuit that can operate as well from dual power supplies of ± 15 V as from a single $+5$-V supply (as used in digital logic circuits). The output can provide a voltage at one of two distinct levels or can be used to drive a lamp or a relay. Notice that the output is taken

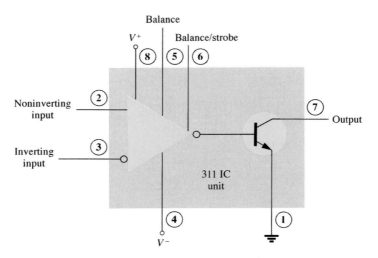

Figure 16.4 A 311 comparator (eight-pin DIP unit).

from a bipolar transistor to allow driving a variety of loads. The unit also has balance and strobe inputs, the strobe input allowing gating of the output. A few examples will show how this comparator unit can be used in some common applications.

A zero-crossing detector that senses (detects) the input voltage crossing through 0 V is shown using the 311 IC in Fig. 16.5. The inverting input is connected to ground (the reference voltage). The input signal going positive drives the output transistor on, with the output then going low (-10 V in this case). The input signal going negative (below 0 V) will drive the output transistor off, the output then going high (to $+10$ V). The output is thus an indication of whether the input is above or below 0 V. When the input is any positive voltage, the output is low, while any negative voltage will result in the output going to a high voltage level.

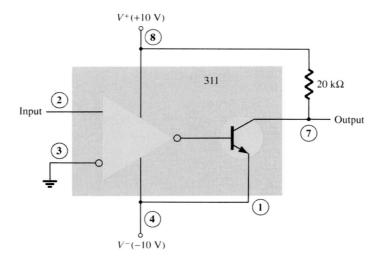

Figure 16.5 Zero-crossing detector using a 311 IC.

Figure 16.6 shows how a 311 comparator can be used with strobing. In this example, the output will go high when the input goes above the reference level—but only if the TTL strobe input is off (or 0 V). If the TTL strobe input goes high, it drives the 311 strobe input at pin 6 low, causing the output to remain in the off state (with output high) regardless of the input signal. In effect, the output remains high

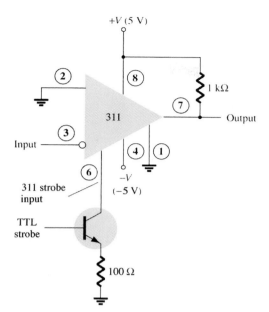

Figure 16.6 Operation of a 311 comparator with strobe input.

unless strobed. If strobed, the output then acts normally, switching from high to low depending on the input signal level. In operation, the comparator output will respond to the input signal only during the time the strobe signal allows such operation.

Figure 16.7 shows the comparator output driving a relay. When the input goes below 0 V, driving the output low, the relay is activated, closing the normally open (N.O.) contacts at that time. These contacts can then be connected to operate a large variety of devices. For example, a buzzer or bell wired to the contacts can be driven on whenever the input voltage drops below 0 V. As long as the voltage is present at the input terminal, the buzzer will remain off.

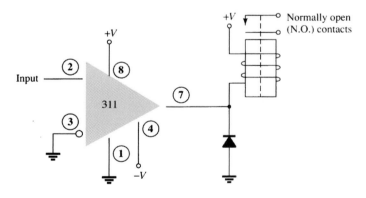

Figure 16.7 Operation of a 311 comparator with relay output.

339 COMPARATOR

The 339 IC is a quad comparator containing four independent voltage comparator circuits connected to external pins as shown in Fig. 16.8. Each comparator has inverting and noninverting inputs and a single output. The supply voltage applied to a pair of pins powers all four comparators. Even if one wishes to use one comparator, all four will be drawing power.

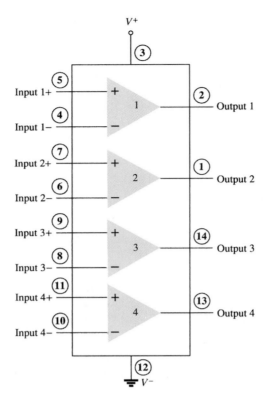

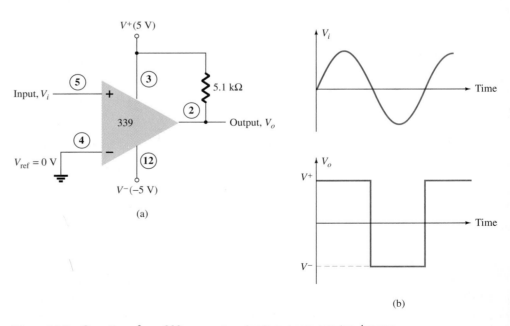

Figure 16.8 Quad comparator IC (339).

To see how these comparator circuits can be used, Fig. 16.9 shows one of the 339 comparator circuits connected as a zero-crossing detector. Whenever the input signal goes above 0 V, the output switches to V^+. The input switches to V^- only when the input goes below 0 V.

A reference level other than 0 V can also be used, and either input terminal could be used as the reference, the other terminal then being connected to the input signal. The operation of one of the comparator circuits is described next.

Figure 16.9 Operation of one 339 comparator circuit as a zero-crossing detector.

The differential input voltage (difference voltage across input terminals) going positive drives the output transistor off (open circuit), while a negative differential input voltage drives the output transistor on—the output then at the supply low level.

If the negative input is set at a reference level V_{ref}, the positive input goes above V_{ref} and results in a positive differential input with output driven to the open-circuit state. When the noninverting input goes below V_{ref}, resulting in a negative differential input, the output will be driven to V^-.

If the positive input is set at the reference level, the inverting input going below V_{ref} results in the output open circuit while the inverting input going above V_{ref} results in the output at V^-. This operation is summarized in Fig. 16.10.

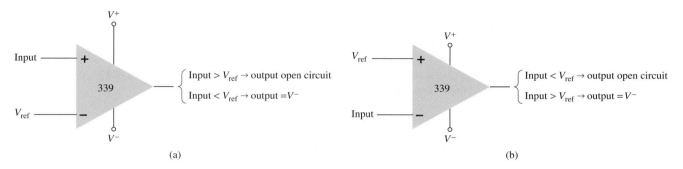

(a)   (b)

Figure 16.10   Operation of a 339 comparator circuit with reference input: (a) minus input; (b) plus input.

Since the output of one of these comparator circuits is from an open-circuit collector, applications in which the outputs from more than one circuit can be wire-ORed are possible. Figure 16.11 shows two comparator circuits connected with common output and also with common input. Comparator 1 has a +5-V reference voltage

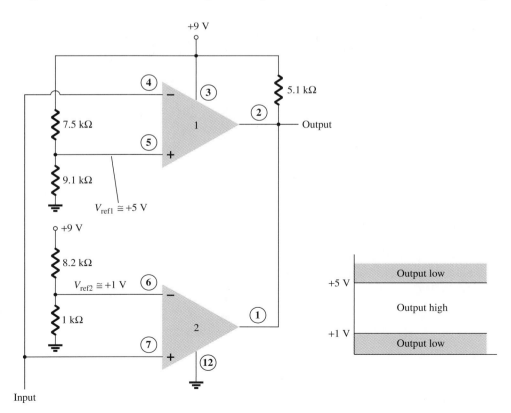

Figure 16.11   Operation of two 339 comparator circuits as a window detector.

input connected to the noninverting input. The output will be driven low by comparator 1 when the input signal goes above $+5$ V. Comparator 2 has a reference voltage of $+1$ V connected to the inverting input. The output of comparator 2 will be driven low when the input signal goes below $+1$ V. In total, the output will go low whenever the input is below $+1$ V or above $+5$ V, as shown in Fig. 16.11, the overall operation being that of a voltage window detector. The high output indicates that the input is within a voltage window of $+1$ to $+5$ V (these values being set by the reference voltage levels used).

16.3 DIGITAL–ANALOG CONVERTERS

Many voltages and currents in electronics vary continuously over some range of values. In digital circuitry the signals are at either one of two levels, representing the binary values of 1 or zero. An analog–digital converter (ADC) obtains a digital value representing an input analog voltage, while a digital–analog converter (DAC) changes a digital value back into an analog voltage.

Digital-to-Analog Conversion

LADDER NETWORK CONVERSION

Digital-to-analog conversion can be achieved using a number of different methods. One popular scheme uses a network of resistors, called a *ladder network*. A ladder network accepts inputs of binary values at, typically, 0 V or V_{ref} and provides an output voltage proportional to the binary input value. Figure 16.12a shows a ladder network with four input voltages, representing 4 bits of digital data and a dc voltage output. The output voltage is proportional to the digital input value as given by the relation

$$V_o = \frac{D_0 \times 2^0 + D_1 \times 2^1 \times D_2 \times 2^2 + D_3 \times 2^3}{2^4} V_{\text{ref}} \qquad (16.1)$$

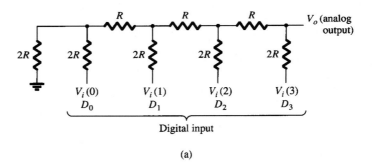

(a)

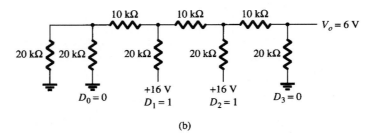

(b)

Figure 16.12 Four-stage ladder network used as a DAC: (a) basic circuit; (b) circuit example with 0110 input.

In the example shown in Fig. 16.12b, the output voltage resulting should be

$$V_o = \frac{0 \times 1 + 1 \times 2 + 1 \times 4 + 0 \times 8}{16}(16 \text{ V}) = 6 \text{ V}$$

Therefore, 0110_2, digital, converts to 6 V, analog.

The function of the ladder network is to convert the 16 possible binary values from 0000 to 1111 into one of 16 voltage levels in steps of $V_{ref}/16$. Using more sections of ladder allows having more binary inputs and greater quantization for each step. For example, a 10-stage ladder network could extend the number of voltage steps or the voltage resolution to $V_{ref}/2^{10}$ or $V_{ref}/1024$. A reference voltage of $V_{ref} = 10$ V would then provide output voltage steps of 10 V/1024 or approximately 10 mV. More ladder stages provide greater voltage resolution. In general, the voltage resolution for n ladder stages is

$$\boxed{\frac{V_{ref}}{2^n}} \tag{16.2}$$

Figure 16.13 shows a block diagram of a typical DAC using a ladder network. The ladder network, referred in the diagram as an *R-2R ladder*, is sandwiched between the reference current supply and current switches connected to each binary input, the resulting output current proportional to the input binary value. The binary input turns on selected legs of the ladder, the output current being a weighted summing of the reference current. Connecting the output current through a resistor will produce an analog voltage, if desired.

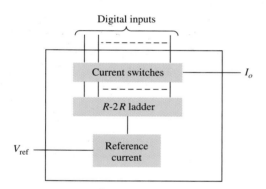

Figure 16.13 DAC IC using R-2R ladder network.

Analog-to-Digital Conversion

DUAL-SLOPE CONVERSION

A popular method for converting an analog voltage into a digital value is the dual-slope method. Figure 16.14a shows a block diagram of the basic dual-slope converter. The analog voltage to be converted is applied through an electronic switch to an integrator or ramp-generator circuit (essentially a constant current charging a capacitor to produce a linear ramp voltage). The digital output is obtained from a counter operated during both positive and negative slope intervals of the integrator.

The method of conversion proceeds as follows. For a fixed time interval (usually the full count range of the counter), the analog voltage connected to the integrator raises the voltage at the comparator input to some positive level. Figure 16.14b shows that at the end of the fixed time interval the voltage from the integrator is greater for the larger input voltage. At the end of the fixed count interval, the count is set to zero and the electronic switch connects the integrator to a reference or fixed input voltage.

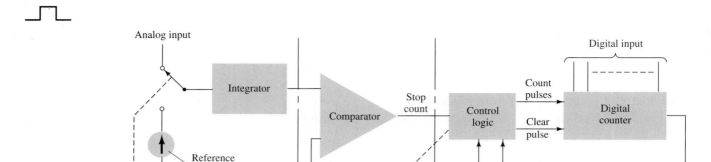

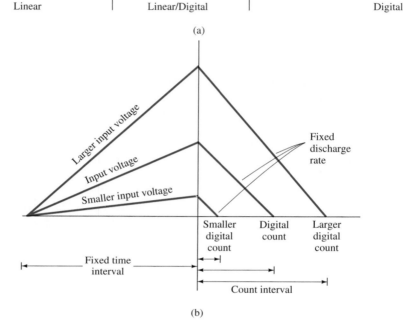

Figure 16.14 Analog-to-digital conversion using dual-slope method: (a) logic diagram; (b) waveform.

The integrator output (or capacitor input) then decreases at a fixed rate. The counter advances during this time, while the integrator's output decreases at a fixed rate until it drops below the comparator reference voltage, at which time the control logic receives a signal (the comparator output) to stop the count. The digital value stored in the counter is then the digital output of the converter.

Using the same clock and integrator to perform the conversion during positive and negative slope intervals tends to compensate for clock frequency drift and integrator accuracy limitations. Setting the reference input value and clock rate can scale the counter output as desired. The counter can be a binary, BCD, or other form of digital counter, if desired.

LADDER-NETWORK CONVERSION

Another popular method of analog-to-digital conversion uses a ladder network along with counter and comparator circuits (see Fig. 16.15). A digital counter advances from a zero count while a ladder network driven by the counter outputs a staircase voltage, as shown in Fig. 16.15b, which increases one voltage increment for each count step. A comparator circuit, receiving both staircase voltage and analog input

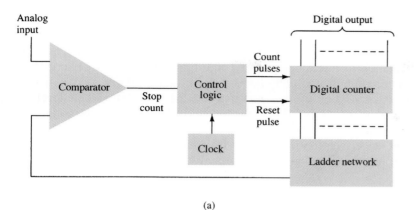

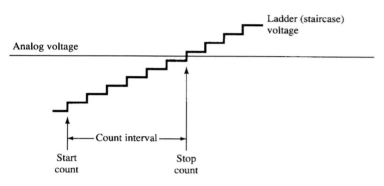

Figure 16.15 Analog-to-digital conversion using ladder network: (a) logic diagram; (b) waveform.

voltage, provides a signal to stop the count when the staircase voltage rises above the input voltage. The counter value at that time is the digital output.

The amount of voltage change stepped by the staircase signal depends on the number of count bits used. A 12-stage counter operating a 12-stage ladder network using a reference voltage of 10 V would step each count by a voltage of

$$\frac{V_{\text{ref}}}{2^{12}} = \frac{10 \text{ V}}{4096} = 2.4 \text{ mV}$$

This would result in a conversion resolution of 2.4 mV. The clock rate of the counter would affect the time required to carry out a conversion. A clock rate of 1 MHz operating a 12-stage counter would need a maximum conversion time of

$$4096 \times 1 \text{ } \mu s = 4096 \text{ } \mu s \approx 4.1 \text{ ms}$$

The minimum number of conversions that could be carried out each second would then be

$$\text{number of conversions} = 1/4.1 \text{ ms} \approx 244 \text{ conversions/second}$$

Since on the average, with some conversions requiring little count time and others near maximum count time, a conversion time of 4.1 ms/2 = 2.05 ms would be needed, and the average number of conversions would be 2 × 244 = 488 conversions/second. A slower clock rate would result in fewer conversions per second. A converter using fewer count stages (and less conversion resolution) would carry out more conversions per second. The conversion accuracy depends on the accuracy of the comparator.

16.4 TIMER IC UNIT OPERATION

Another popular analog–digital integrated circuit is the versatile 555 timer. The IC is made of a combination of linear comparators and digital flip-flops as described in Fig. 16.16. The entire circuit is usually housed in an 8-pin package as specified in Fig. 16.16. A series connection of three resistors sets the reference voltage levels to the two comparators at $2V_{CC}/3$ and $V_{CC}/3$, the output of these comparators setting or resetting the flip-flop unit. The output of the flip-flop circuit is then brought out through an output amplifier stage. The flip-flop circuit also operates a transistor inside the IC, the transistor collector usually being driven low to discharge a timing capacitor.

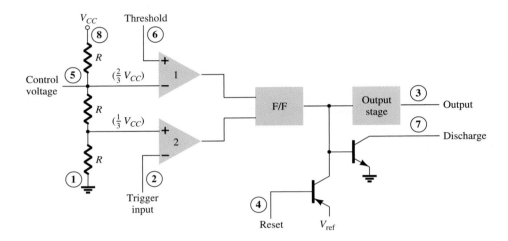

Figure 16.16 Details of 555 timer IC.

Astable Operation

One popular application of the 555 timer IC is as an astable multivibrator or clock circuit. The following analysis of the operation of the 555 as an astable circuit includes details of the different parts of the unit and how the various inputs and outputs are utilized. Figure 16.17 shows an astable circuit built using an external resistor and capacitor to set the timing interval of the output signal.

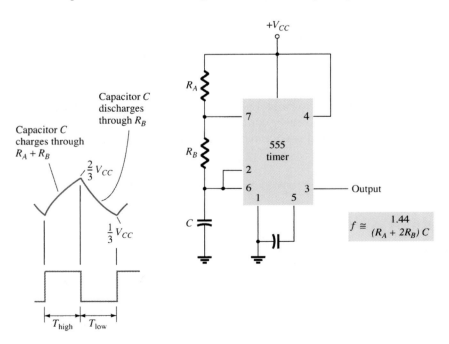

$$f \cong \frac{1.44}{(R_A + 2R_B)\,C}$$

Figure 16.17 Astable multivibrator using 555 IC.

Capacitor C charges toward V_{CC} through external resistors R_A and R_B. Referring to Fig. 16.17, the capacitor voltage rises until it goes above $2V_{CC}/3$. This voltage is the threshold voltage at pin 6, which drives comparator 1 to trigger the flip-flop so that the output at pin 3 goes low. In addition, the discharge transistor is driven on, causing the output at pin 7 to discharge the capacitor through resistor R_B. The capacitor voltage then decreases until it drops below the trigger level ($V_{CC}/3$). The flip-flop is triggered so that the output goes back high and the discharge transistor is turned off, so that the capacitor can again charge through resistors R_A and R_B toward V_{CC}.

Figure 16.18a shows the capacitor and output waveforms resulting from the astable circuit. Calculation of the time intervals during which the output is high and low can be made using the relations

$$T_{\text{high}} \approx 0.7(R_A + R_B)C \qquad (16.3)$$

$$T_{\text{low}} \approx 0.7R_B C \qquad (16.4)$$

The total period is

$$T = \text{period} = T_{\text{high}} + T_{\text{low}} \qquad (16.5)$$

The frequency of the astable circuit is then calculated using*

$$f = \frac{1}{T} \approx \frac{1.44}{(R_A + 2R_B)C} \qquad (16.6)$$

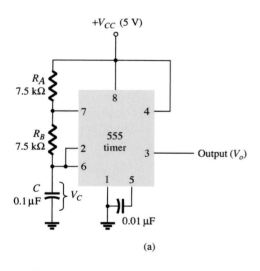

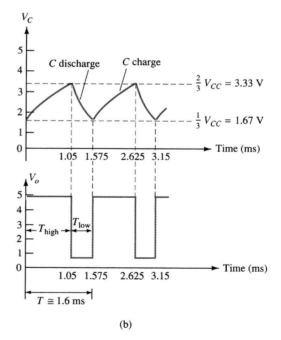

(a)

(b)

Figure 16.18 Astable multivibrator for Example 16.1: (a) circuit; (b) waveforms.

---

*The period can be directly calculated from

$$T = 0.693(R_A + 2R_B)C \approx 0.7(R_A + 2R_B)C$$

and the frequency from

$$f \approx \frac{1.44}{(R_A + 2R_B)C}$$

EXAMPLE 16.1   Determine the frequency and draw the output waveform for the circuit of Fig. 16.18a.

Solution

Using Eqs. (16.3) through (16.6) yields

$$T_{\text{high}} = 0.7(R_A + R_B)C = 0.7(7.5 \times 10^3 + 7.5 \times 10^3)(0.1 \times 10^{-6})$$

$$= 1.05 \text{ ms}$$

$$T_{\text{low}} = 0.7R_B C = 0.7(7.5 \times 10^3)(0.1 \times 10^{-6}) = 0.525 \text{ ms}$$

$$T = T_{\text{high}} + T_{\text{low}} = 1.05 \text{ ms} + 0.525 \text{ ms} = 1.575 \text{ ms}$$

$$f = \frac{1}{T} = \frac{1}{1.575 \times 10^{-3}} \approx \textbf{635 Hz}$$

The waveforms are drawn in Fig. 16.18b.

Monostable Operation

The 555 timer can also be used as a one-shot or monostable multivibrator circuit, as shown in Fig. 16.19. When the trigger input signal goes negative, it triggers the one-shot, with output at pin 3 then going high for a time period.

$$T_{\text{high}} = 1.1R_A C \tag{16.7}$$

Referring back to Fig. 16.16, the negative edge of the trigger input causes comparator 2 to trigger the flip-flop, with the output at pin 3 going high. Capacitor C charges toward V_{CC} through resistor R_A. During the charge interval, the output remains high. When the voltage across the capacitor reaches the threshold level of $2V_{CC}/3$, comparator 1 triggers the flip-flop, with output going low. The discharge transistor also goes low, causing the capacitor to remain at near 0 V until triggered again.

Figure 16.19b shows the input trigger signal and the resulting output waveform for the 555 timer operated as a one-shot. Time periods for this circuit can range from microseconds to many seconds, making this IC useful for a range of applications.

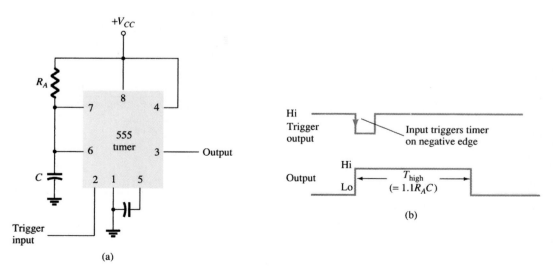

Figure 16.19   Operation of 555 timer as one-shot: (a) circuit; (b) waveforms.

Chapter 16   Linear-Digital ICs

EXAMPLE 16.2

Determine the period of the output waveform for the circuit of Fig. 16.20 when triggered by a negative pulse.

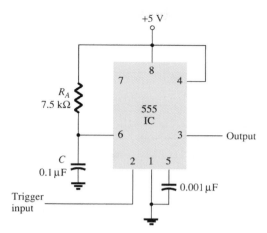

Figure 16.20 Monostable circuit for Example 16.2.

Solution

Using Eq. (16.7), we obtain

$$T_{\text{high}} = 1.1R_A C = 1.1(7.5 \times 10^3)(0.1 \times 10^{-6}) = \textbf{0.825 ms}$$

16.5 VOLTAGE-CONTROLLED OSCILLATOR

A voltage-controlled oscillator (VCO) is a circuit that provides a varying output signal (typically of square-wave or triangular-wave form) whose frequency can be adjusted over a range controlled by a dc voltage. An example of a VCO is the 566 IC unit, which contains circuitry to generate both square-wave and triangular-wave signals whose frequency is set by an external resistor and capacitor and then varied by an applied dc voltage. Figure 16.21a shows that the 566 contains current sources to charge and discharge an external capacitor C_1 at a rate set by external resistor R_1 and the modulating dc input voltage. A Schmitt trigger circuit is used to switch the current sources between charging and discharging the capacitor, and the triangular voltage developed across the capacitor and square wave from the Schmitt trigger are provided as outputs through buffer amplifiers.

Figure 16.21b shows the pin connection of the 566 unit and a summary of formula and value limitations. The oscillator can be programmed over a 10-to-1 frequency range by proper selection of an external resistor and capacitor, and then modulated over a 10-to-1 frequency range by a control voltage, V_C.

A free-running or center-operating frequency, f_o, can be calculated from

$$f_o = \frac{2}{R_1 C_1}\left(\frac{V^+ - V_C}{V^+}\right) \tag{16.8}$$

with the following practical circuit value restrictions:

1. R_1 should be within the range $2\ k\Omega \leq R_1 \leq 20\ k\Omega$.
2. V_C should be within range $\frac{3}{4}V^+ \leq V_C \leq V^+$.

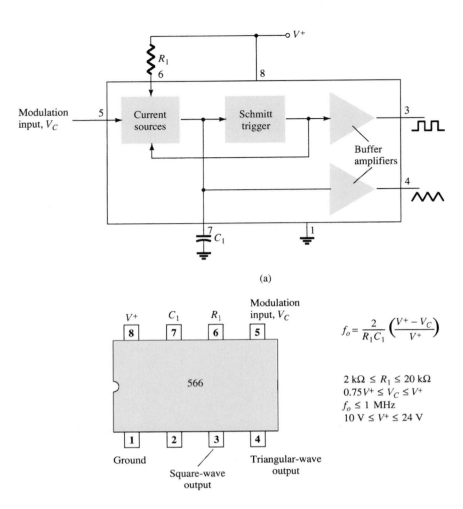

(a)

$$f_o = \frac{2}{R_1 C_1}\left(\frac{V^+ - V_C}{V^+}\right)$$

$2\ k\Omega \leq R_1 \leq 20\ k\Omega$
$0.75 V^+ \leq V_C \leq V^+$
$f_o \leq 1\ MHz$
$10\ V \leq V^+ \leq 24\ V$

Figure 16.21 A 566 function generator: (a) block diagram; (b) pin configuration and summary of operating data.

3. f_o should be below 1 MHz.
4. V^+ should range between 10 V and 24 V.

Figure 16.22 shows an example in which the 566 function generator is used to provide both square-wave and triangular-wave signals at a fixed frequency set by R_1, C_1, and V_C. A resistor divider R_2 and R_3 sets the dc modulating voltage at a fixed value

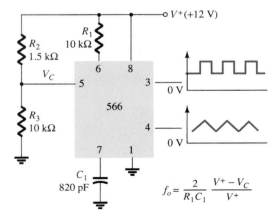

$$f_o = \frac{2}{R_1 C_1}\frac{V^+ - V_C}{V^+}$$

Figure 16.22 Connection of 566 VCO unit.

$$V_C = \frac{R_3}{R_2 + R_3}V^+ = \frac{10 \text{ k}\Omega}{1.5 \text{ k}\Omega + 10 \text{ k}\Omega}(12 \text{ V}) = 10.4 \text{ V}$$

(which falls properly in the voltage range $0.75V^+ = 9$ V and $V^+ = 12$ V). Using Eq. (16.8) yields

$$f_o = \frac{2}{(10 \times 10^3)(820 \times 10^{-12})}\left(\frac{12 - 10.4}{12}\right) \approx 32.5 \text{ kHz}$$

The circuit of Fig. 16.23 shows how the output square-wave frequency can be adjusted using the input voltage, V_C, to vary the signal frequency. Potentiometer R_3 allows varying V_C from about 9 V to near 12 V, over the full 10-to-1 frequency range. With the potentiometer wiper set at the top, the control voltage is

$$V_C = \frac{R_3 + R_4}{R_2 + R_3 + R_4}(V^+) = \frac{5 \text{ k}\Omega + 18 \text{ k}\Omega}{510 \text{ }\Omega + 5 \text{ k}\Omega + 18 \text{ k}\Omega}(+12 \text{ V}) = 11.74 \text{ V}$$

resulting in a lower output frequency of

$$f_o = \frac{2}{(10 \times 10^3)(220 \times 10^{-12})}\left(\frac{12 - 11.74}{12}\right) \approx 19.7 \text{ kHz}$$

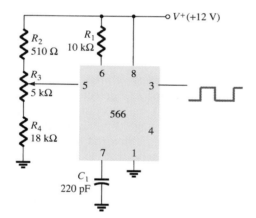

Figure 16.23 Connection of 566 as a VCO unit.

With the wiper arm of R_3 set at the bottom, the control voltage is

$$V_C = \frac{R_4}{R_2 + R_3 + R_4}(V^+) = \frac{18 \text{ k}\Omega}{510 \text{ }\Omega + 5 \text{ k}\Omega + 18 \text{ k}\Omega}(+12 \text{ V}) = 9.19 \text{ V}$$

resulting in an upper frequency of

$$f_o = \frac{2}{(10 \times 10^3)(220 \times 10^{-12})}\left(\frac{12 - 9.19}{12}\right) \approx 212.9 \text{ kHz}$$

The frequency of the output square wave can then be varied using potentiometer R_3 over a frequency range of at least 10 to 1.

Rather than varying a potentiometer setting to change the value of V_C, an input modulating voltage, V_{in}, can be applied as shown in Fig. 16.24. The voltage divider sets V_C at about 10.4 V. An input ac voltage of about 1.4 V peak can drive V_C around the bias point between voltages of 9 and 11.8 V, causing the output frequency to vary over about a 10-to-1 range. The input signal V_{in} thus frequency-modulates the output voltage around the center frequency set by the bias value of $V_C = 10.4$ V ($f_o = 121.2$ kHz).

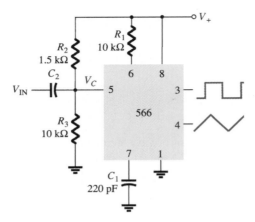

Figure 16.24 Operation of VCO with frequency-modulating input.

16.6 PHASE-LOCKED LOOP

A phase-locked loop (PLL) is an electronic circuit that consists of a phase detector, a low-pass filter, and a voltage-controlled oscillator connected as shown in Fig. 16.25. Common applications of a PLL include: (1) frequency synthesizers that provide multiples of a reference signal frequency [e.g., the carrier frequency for the multiple channels of a citizens' band (CB) unit or marine-radio-band unit can be generated using a single-crystal-controlled frequency and its multiples generated using a PLL]; (2) FM demodulation networks for FM operation with excellent linearity between the input signal frequency and the PLL output voltage; (3) demodulation of the two data transmission or carrier frequencies in digital-data transmission used in frequency-shift keying (FSK) operation; and (4) a wide variety of areas including modems, telemetry receivers and transmitters, tone decoders, AM detectors, and tracking filters.

An input signal, V_i, and that from a VCO, V_o, are compared by a phase comparator (refer to Fig. 16.25) providing an output voltage, V_e, that represents the phase difference between the two signals. This voltage is then fed to a low-pass filter that provides an output voltage (amplified if necessary) that can be taken as the output voltage from the PLL and is used internally as the voltage to modulate the VCO's frequency. The closed-loop operation of the circuit is to maintain the VCO frequency locked to that of the input signal frequency.

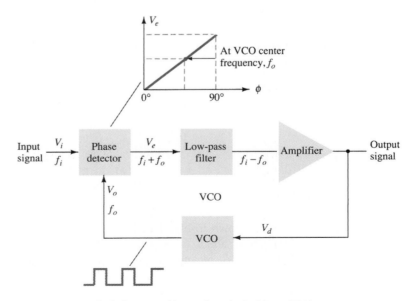

Figure 16.25 Block diagram of basic phase-locked loop (PLL).

Basic PLL Operation

The basic operation of a PLL circuit can be explained using the circuit of Fig. 16.25 as reference. We will first consider the operation of the various circuits in the phase-locked loop when the loop is operating in lock (the input signal frequency and the VCO frequency are the same). When the input signal frequency is the same as that from the VCO to the comparator, the voltage, V_d, taken as output is the value needed to hold the VCO in lock with the input signal. The VCO then provides output of a fixed-amplitude square-wave signal at the frequency of the input. Best operation is obtained if the VCO center frequency, f_o, is set with the dc bias voltage midway in its linear operating range. The amplifier allows this adjustment in dc voltage from that obtained as output of the filter circuit. When the loop is in lock, the two signals to the comparator are of the same frequency, although not necessarily in phase. A fixed phase difference between the two signals to the comparator results in a fixed dc voltage to the VCO. Changes in the input signal frequency then result in change in the dc voltage to the VCO. Within a capture-and-lock frequency range, the dc voltage will drive the VCO frequency to match that of the input.

While the loop is trying to achieve lock, the output of the phase comparator contains frequency components at the sum and difference of the signals compared. A low-pass filter passes only the lower-frequency component of the signal so that the loop can obtain lock between input and VCO signals.

Owing to the limited operating range of the VCO and the feedback connection of the PLL circuit, there are two important frequency bands specified for a PLL. The capture range of a PLL is the frequency range centered about the VCO free-running frequency, f_o, over which the loop can acquire lock with the input signal. Once the PLL has achieved capture, it can maintain lock with the input signal over a somewhat wider frequency range called the *lock range*.

Applications

The PLL can be used in a wide variety of applications, including (1) frequency demodulation, (2) frequency synthesis, and (3) FSK decoders. Examples of each of these follow.

FREQUENCY DEMODULATION

FM demodulation or detection can be directly achieved using the PLL circuit. If the PLL center frequency is selected or designed at the FM carrier frequency, the filtered or output voltage of the circuit of Fig. 16.25 is the desired demodulated voltage, varying in value proportional to the variation of the signal frequency. The PLL circuit thus operates as a complete intermediate-frequency (IF) strip, limiter, and demodulator as used in FM receivers.

One popular PLL unit is the 565, shown in Fig. 16.26a. The 565 contains a phase detector, amplifier, and voltage-controlled oscillator, which are only partially connected internally. An external resistor and capacitor, R_1 and C_1, are used to set the free-running or center frequency of the VCO. Another external capacitor, C_2, is used to set the low-pass filter passband, and the VCO output must be connected back as input to the phase detector to close the PLL loop. The 565 typically uses two power supplies, V^+ and V^-.

Figure 16.26b shows the PLL connected to work as an FM demodulator. Resistor R_1 and capacitor C_1 set the free-running frequency, f_o.

$$\boxed{f_o = \frac{0.3}{R_1 C_1}} \tag{16.9}$$

$$= \frac{0.3}{(10 \times 10^3)(220 \times 10^{-12})} = 136.36 \text{ kHz}$$

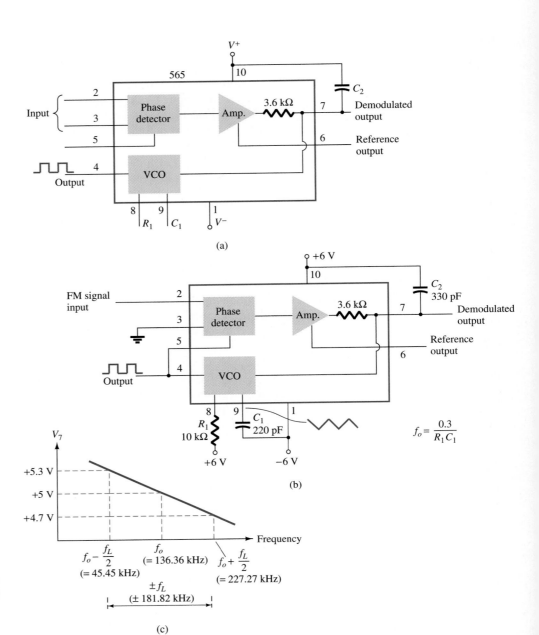

Figure 16.26 Phase-locked loop (PLL): (a) basic block diagram: (b) PLL connected as a frequency demodulator: (c) output voltage vs. frequency plot.

with limitation $2 \text{ k}\Omega \leq R_1 \leq 20 \text{ k}\Omega$. The lock range is

$$f_L = \pm \frac{8 f_o}{V}$$

$$= \pm \frac{8(136.36 \times 10^3)}{6} = \pm 181.8 \text{ kHz}$$

for supply voltages $V = \pm 6$ V. The capture range is

$$f_C = \pm \frac{1}{2\pi} \sqrt{\frac{2\pi f_L}{R_2 C_2}}$$

$$= \pm \frac{1}{2\pi} \sqrt{\frac{2\pi(181.8 \times 10^3)}{(3.6 \times 10^3)(330 \times 10^{-12})}} = 156.1 \text{ kHz}$$

The signal at pin 4 is a 136.36-kHz square wave. An input within the lock range of 181.8 kHz will result in the output at pin 7 varying around its dc voltage level set with input signal at f_o. Figure 16.26c shows the output at pin 7 as a function of the input signal frequency. The dc voltage at pin 7 is linearly related to the input signal frequency within the frequency range $f_L = 181.8$ kHz around the center frequency 136.36 kHz. The output voltage is the demodulated signal that varies with frequency within the operating range specified.

FREQUENCY SYNTHESIS

A frequency synthesizer can be built around a PLL as shown in Fig. 16.27. A frequency divider is inserted between the VCO output and the phase comparator so that the loop signal to the comparator is at frequency f_o while the VCO output is Nf_o. This output is a multiple of the input frequency as long as the loop is in lock. The input signal can be stabilized at f_1 with the resulting VCO output at Nf_1 if the loop is set up to lock at the fundamental frequency (when $f_o = f_1$). Figure 16.27b shows an example using a 565 PLL as frequency multiplier and a 7490 as divider. The input V_i at frequency f_1 is compared to the input (frequency f_o) at pin 5. An output at Nf_o ($4f_o$ in the present example) is connected through an inverter circuit to provide an input at pin 14 of the 7490, which varies between 0 and +5 V. Using the output at pin 9, which is divided by 4 from that at the input to the 7490, the signal at pin 4 of the PLL is four times the input frequency as long as the loop remains in lock. Since the VCO can vary over only a limited range from its center frequency, it may be nec-

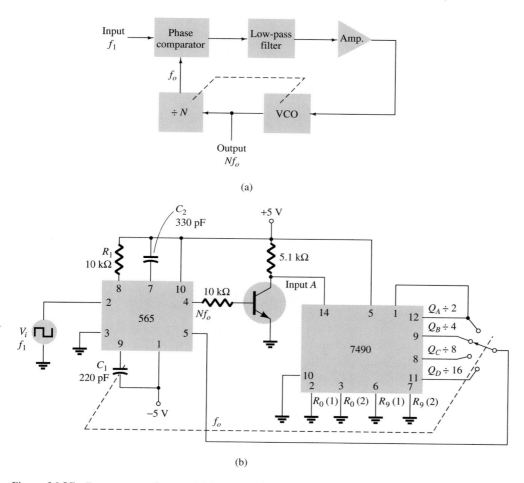

Figure 16.27 Frequency synthesizer: (a) block diagram; (b) implementation using 565 PLL unit.

essary to change the VCO frequency whenever the divider value is changed. As long as the PLL circuit is in lock, the VCO output frequency will be exactly N times the input frequency. It is only necessary to readjust f_o to be within the capture-and-lock range, the closed loop then resulting in the VCO output becoming exactly Nf_1 at lock.

FSK DECODERS

An FSK (frequency-shift keyed) signal decoder can be built as shown in Fig. 16.28. The decoder receives a signal at one of two distinct carrier frequencies, 1270 Hz or 1070 Hz, representing the RS-232C logic levels or mark (-5 V) or space ($+14$ V), respectively. As the signal appears at the input, the loop locks to the input frequency and tracks it between two possible frequencies with a corresponding dc shift at the output.

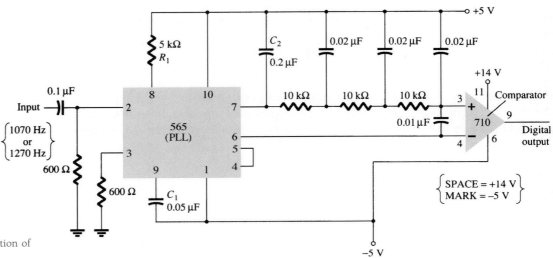

Figure 16.28 Connection of 565 as FSK decoder.

The RC ladder filter (three sections of $C = 0.02$ μF and $R = 10$ kΩ) is used to remove the sum frequency component. The free-running frequency is adjusted with R_1 so that the dc voltage level at the output (pin 7) is the same as that at pin 6. Then an input at frequency 1070 Hz will drive the decoder output voltage to a more positive voltage level, driving the digital output to the high level (space or $+14$ V). An input at 1270 Hz will correspondingly drive the 565 dc output less positive with the digital output, which then drops to the low level (mark or -5 V).

16.7   INTERFACING CIRCUITRY

Connecting different types of circuits, either in digital or analog circuits, may require some sort of interfacing circuit. An interface circuit may be used to drive a load or to obtain a signal as a receiver circuit. A driver circuit provides the output signal at a voltage or current level suitable to operate a number of loads, or to operate such devices as relays, displays, or power units. A receiver circuit essentially accepts an input signal, providing high input impedance to minimize loading of the input signal. Furthermore, the interface circuits may include strobing, which provides connecting the interface signals during specific time intervals established by the strobe.

Figure 16.29a shows a dual-line driver, each driver accepting input of TTL signals, providing output capable of driving TTL or MOS device circuits. This type of interface circuit comes in various forms, some as inverting and others as noninverting units. The circuit of Fig. 16.29b shows a dual-line receiver having both inverting and noninverting inputs so that either operating condition can be selected. As an example, connection of an input signal to the inverting input would result in an inverted output from the receiver unit. Connecting the input to the noninverting input would provide the same interfacing except that the output obtained would have the same polarity as the received signal. The driver-receiver unit of Fig. 16.29 provides an output when the strobe signal is present (high in this case).

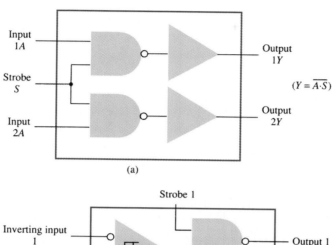

$$(Y = \overline{A \cdot S})$$

(a)

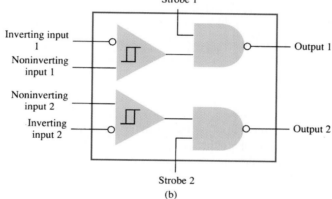

(b)

Figure 16.29 Interface units: (a) dual-line drivers (SN75150); (b) dual-line receivers (SN75152).

Another type of interface circuit is that used to connect various digital input and output units, signals with devices such as keyboards, video terminals, and printers. One of the EIA electronic industry standards is referred to as RS-232C. This standard states that a digital signal represents a mark (logic-1) and a space (logic-0). The definitions of mark and space vary with the type of circuit used (although a full reading of the standard will spell out the acceptable limits of mark and space signals).

RS-232C-to-TTL Converter

For TTL circuits, $+5$ V is a mark and 0 V is a space. For RS-232C, a mark could be -12 V and a space $+12$ V. Figure 16.30a provides a tabulation of some mark and space definitions. For a unit having outputs defined by RS-232C that is to operate into another unit operating with a TTL signal level, an interface circuit as shown in Fig. 16.30b could be used. A mark output from the driver (at -12 V) would be clipped by the diode so that the input to the inverter circuit is near 0 V, resulting in an output of $+5$ V (TTL mark). A space output at $+12$ V would drive the inverter output low for a 0-V output (a space).

Another example of an interface circuit converts the signals from a TTY current loop into TTL levels as shown in Fig. 16.30c. An input mark results when 20 mA of current is drawn from the source through the output line of the teletype (TTY). This current then goes through the diode element of an opto-isolator, driving the output transistor on. The input to the inverter going low results in a $+5$-V signal from the

	Current Loop	RS-232-C	TTL
MARK	20 mA	−12 V	+5 V
SPACE	0 mA	+12 V	0 V

(a)

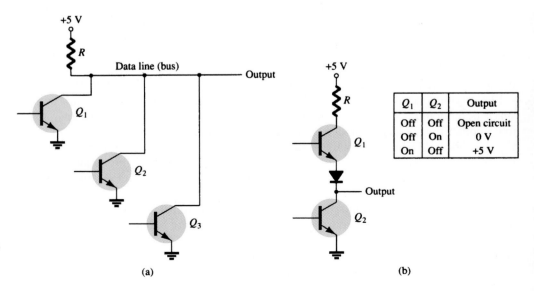

(b)

Figure 16.30 Interfacing signal standards and converter circuits.

(c)

7407 inverter output so that a mark from the teletype results in a mark to the TTL input. A space from the teletype current loop provides no current, with the opto-isolator transistor remaining off and the inverter output then 0 V, which is a TTL space signal.

Another means of interfacing digital signals is made using open-collector output or tri-state output. When a signal is output from a transistor collector (see Fig. 16.31) that is not connected to any other electronic component, the output is open-collector. This permits connecting a number of signals to the same wire or bus. Any transistor going on then provides a low output voltage, while all transistors remaining off provide a high output voltage.

Q_1	Q_2	Output
Off	Off	Open circuit
Off	On	0 V
On	Off	+5 V

Figure 16.31 Connections to data lines: (a) open-collector output; (b) tri-state output.

(a)

(b)

16.8 SUMMARY

Important Conclusions and Concepts

1. A comparator provides an output of either maximum high or maximum low when one input goes above or below the other.
2. A DAC is a digital-to-analog converter.
3. An ADC is an analog-to-digital converter.
4. Timer IC:
   a. An astable circuit acts as a clock.
   b. A monostable circuit acts as a one-shot or timer.
5. A phase-locked loop (PLL) circuit contains a phase detector, a low-pass filter, and a voltage-controlled oscillator (VCO).
6. There are two standard types of interfacing circuits: the RS-232-C and the TTL.

16.9 COMPUTER ANALYSIS

PSpice Windows

Many of the practical op-amp applications covered in this chapter can be analyzed using PSpice. Analysis of various problems can display the resulting dc bias, or one can use **PROBE** to display resulting waveforms.

Program 16.1—Comparator Circuit Used to Drive an LED

Using Design Center, draw the circuit of a comparator circuit with output driving an LED indicator as shown in Fig. 16.32. To be able to view the magnitude of the dc output voltage, place a **VPRINT1** component at V_o with **DC** and **MAG** selected. To view the dc current through the LED, place an **IPRINT** component in series with the LED current meter as shown in Fig. 16.32. The **Analysis Setup** provides for a dc sweep as shown in Fig. 16.33. The **DC Sweep** is set, as shown, for V_i from 4 to 8 V in 1-V steps. After running the simulation, some of the resulting analysis output obtained is shown in Fig. 16.34.

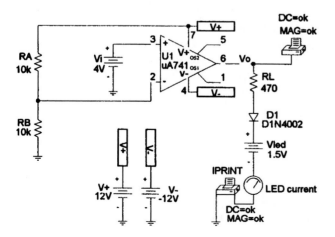

Figure 16.32 Comparator circuit used to drive an LED.

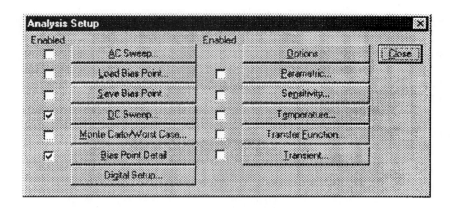

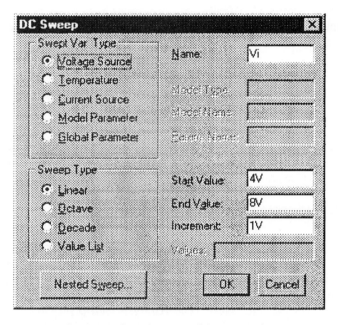

Figure 16.33 Analysis Setup for a dc sweep of the circuit of Fig. 16.32.

The circuit of Fig. 16.32 shows a voltage divider which provides 6 V to the minus input so that any input (V_i) below 6 V will result in the output at the minus saturation voltage (near -10 V). Any input above $+6$ V results in the output going to the positive saturation level (near $+10$ V). The LED will therefore be driven *on* by any input above the reference level of $+6$ V and left *off* by any input below $+6$ V. The listing of Fig. 16.34 shows a table of the output voltage and a table of the LED current for inputs from 4 to 8 V. The table shows that the LED current is nearly 0 for inputs up to $+6$ V and that a current of about 20 mA lights the LED for inputs at $+6$ V or above.

Program 16.2—Comparator Operation

The operation of a comparator IC can be demonstrated using a 741 op-amp as shown in Fig. 16.35. The input is a 5 V, peak sinusoidal signal. The **Analysis Setup** provides for **Transient** analysis with **Print Step** of **20 ns** and **Final Time** of **3 ms.** Since the input signal is applied to the noninverting input, the output is in-phase with the input. When the input goes above 0 V, the output goes to the positive saturation level,

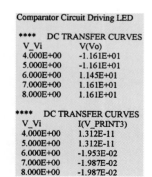

Comparator Circuit Driving LED

**** DC TRANSFER CURVES
V_Vi	V(Vo)
4.000E+00	-1.161E+01
5.000E+00	-1.161E+01
6.000E+00	1.145E+01
7.000E+00	1.161E+01
8.000E+00	1.161E+01

**** DC TRANSFER CURVES
V_Vi	I(V_PRINT3)
4.000E+00	1.312E-11
5.000E+00	1.312E-11
6.000E+00	-1.953E-02
7.000E+00	-1.987E-02
8.000E+00	-1.987E-02

Figure 16.34 Analysis output (edited) for circuit of Fig. 16.32.

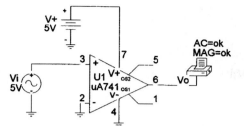

Figure 16.35 Schematic for a comparator.

near +5 V. When the input goes below 0 V, the output goes to the negative saturation level—this being 0 V since the minus voltage input is set to that value. Figure 16.36 shows a **PROBE** output of input and output voltages.

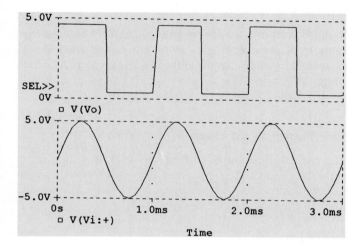

Figure 16.36 Probe output for the comparator of Fig. 16.35.

Program 16.3—Operation of 555 Timer as Oscillator

Figure 16.37 shows a 555 timer connected as an oscillator. Equations (16.3) and (16.4) can be used to calculate the charge and discharge times as follows:

$$T_{high} = 0.7(R_A + R_B)C = 0.7(7.5 \text{ k}\Omega + 7.15 \text{ k}\Omega)(0.1 \text{ }\mu\text{F}) = 1.05 \text{ ms}$$

$$T_{low} = 0.7R_B C = 0.7(7.5 \text{ k}\Omega)(0.1 \text{ }\mu\text{F}) = 0.525 \text{ ms}$$

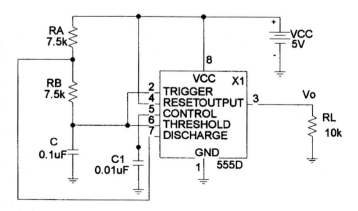

Figure 16.37 Schematic of a 555 timer oscillator.

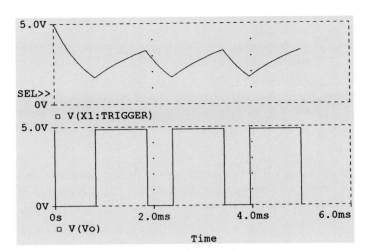

Figure 16.38 Probe output for the 555 oscillator of Fig. 16.37.

The resulting trigger and output waveforms are shown in Fig. 16.38. When the trigger charges to the upper trigger level, the output goes to the low output level of 0 V. The output stays low until the trigger input discharges to the low trigger level, at which time the output goes to the high level of $+5$ V.

Electronics Workbench

Program 16.4—The 555 Timer as an Oscillator

Figure 16.39 shows the same oscillator circuit as in Program 16.3, this time using Electronics Workbench to build the circuit and to show resulting waveforms on an oscilloscope. Using the oscilloscope instrument, the waveform across the capacitor and that from the output are shown in Fig. 16.39.

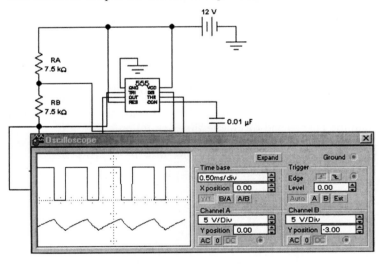

Figure 16.39 Timer oscillator using EWB.

PROBLEMS

§ 16.2 Comparator Unit Operation

1. Draw the diagram of a 741 op-amp operated from ± 15-V supplies with $V_i(-) = 0$ V and $V_i(+) = +5$ V. Include terminal pin connections.

2. Sketch the output waveform for the circuit of Fig. 16.40.

3. Draw a circuit diagram of a 311 op-amp showing an input of 10 V rms applied to the inverting input and the plus input to ground. Identify all pin numbers.

4. Draw the resulting output waveform for the circuit of Fig. 16.41.

5. Draw the circuit diagram of a zero-crossing detector using a 339 comparator stage with ± 12-V supplies.

Chapter 16   Linear-Digital ICs

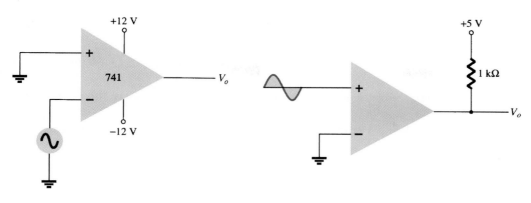

Figure 16.40   Problem 2

Figure 16.41   Problem 4

6. Sketch the output waveform for the circuit of Fig. 16.42.

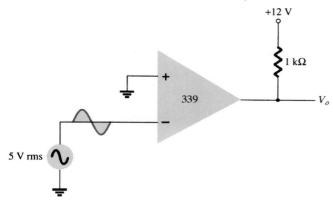

Figure 16.42   Problem 6

* **7.** Describe the operation of the circuit in Fig. 16.43.

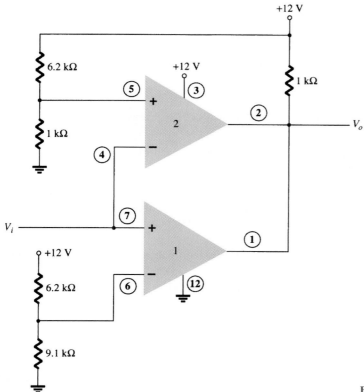

Figure 16.43   Problem 7

Problems

8. Sketch a five-stage ladder network using 15-kΩ and 30-kΩ resistors.

9. For a reference voltage of 16 V, calculate the output voltage for an input of 11010 to the circuit of Problem 8.

10. What voltage resolution is possible using a 12-stage ladder network with a 10-V reference voltage?

11. For a dual-slope converter, describe what occurs during the fixed time interval and the count interval.

12. How many count steps occur using a 12-stage digital counter at the output of an ADC?

13. What is the maximum count interval using a 12-stage counter operated at a clock rate of 20 MHz?

§ 16.4 Timer IC Unit Operation

14. Sketch the circuit of a 555 timer connected as an astable multivibrator for operation at 350 kHz. Determine the value of capacitor, C, needed using $R_A = R_B = 7.5$ kΩ.

15. Draw the circuit of a one-shot using a 555 timer to provide one time period of 20 μs. If $R_A = 7.5$ kΩ, what value of C is needed?

16. Sketch the input and output waveforms for a one-shot using a 555 timer triggered by a 10-kHz clock for $R_A = 5.1$ kΩ and $C = 5$ nF.

§ 16.5 Voltage-Controlled Oscillator

17. Calculate the center frequency of a VCO using a 566 IC as in Fig. 16.22 for $R_1 = 4.7$ kΩ, $R_2 = 1.8$ kΩ, $R_3 = 11$ kΩ, and $C_1 = 0.001$ μF.

*__18.__ What frequency range results in the circuit of Fig. 16.23 for $C_1 = 0.001$ μF?

19. Determine the capacitor needed in the circuit of Fig. 16.22 to obtain a 200-kHz output.

§ 16.6 Phase-Locked Loop

20. Calculate the VCO free-running frequency for the circuit of Fig. 16.26b with $R_1 = 4.7$ kΩ and $C_1 = 0.001$ μF.

21. What value of capacitor, C_1, is required in the circuit of Fig. 16.26b to obtain a center frequency of 100 kHz?

22. What is the lock range of the PLL circuit in Fig. 16.26b for $R_1 = 4.7$ kΩ and $C_1 = 0.001$ μF?

§ 16.7 Interfacing Circuitry

23. Describe the signal conditions for current-loop and RS-232C interfaces.

24. What is a data bus?

25. What is the difference between open-collector and tri-state output?

§ 16.9 Computer Analysis

*__26.__ Use Design Center to draw a schematic circuit as in Fig. 16.32, using an LM111 with $V_i = 5$ V rms applied to minus ($-$) input and $+5$ V rms applied to plus ($+$) input. Use Probe to view the output waveform.

*__27.__ Use Design Center to draw a schematic circuit as in Fig. 16.35. Examine the output listing for the results.

*__28.__ Use EWB to draw a 555 oscillator with resulting output with $t_{low} = 2$ ms, $t_{high} = 5$ ms.

*Please note: Asterisks indicate more difficult problems.

Feedback and Oscillator Circuits

17.1 FEEDBACK CONCEPTS

Feedback has been mentioned previously. In particular, feedback was used in op-amp circuits as described in Chapters 13 and 14. Depending on the relative polarity of the signal being fed back into a circuit, one may have negative or positive feedback. Negative feedback results in decreased voltage gain, for which a number of circuit features are improved as summarized below. Positive feedback drives a circuit into oscillation as in various types of oscillator circuits.

A typical feedback connection is shown in Fig. 17.1. The input signal, V_s, is applied to a mixer network, where it is combined with a feedback signal, V_f. The difference of these signals, V_i, is then the input voltage to the amplifier. A portion of the amplifier output, V_o, is connected to the feedback network (β), which provides a reduced portion of the output as feedback signal to the input mixer network.

If the feedback signal is of opposite polarity to the input signal, as shown in Fig. 17.1, negative feedback results. While negative feedback results in reduced overall voltage gain, a number of improvements are obtained, among them being:

1. Higher input impedance.
2. Better stabilized voltage gain.
3. Improved frequency response.
4. Lower output impedance.
5. Reduced noise.
6. More linear operation.

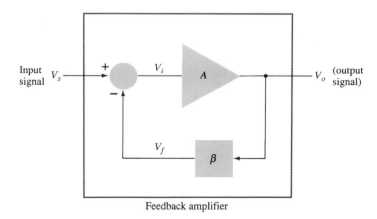

Feedback amplifier

Figure 17.1 Simple block diagram of feedback amplifier.

17.2 FEEDBACK CONNECTION TYPES

There are four basic ways of connecting the feedback signal. Both *voltage* and *current* can be fed back to the input either in *series* or *parallel*. Specifically, there can be:

1. Voltage-series feedback (Fig. 17.2a).
2. Voltage-shunt feedback (Fig. 17.2b).
3. Current-series feedback (Fig. 17.2c).
4. Current-shunt feedback (Fig. 17.2d).

In the list above, *voltage* refers to connecting the output voltage as input to the feedback network; *current* refers to tapping off some output current through the feedback network. *Series* refers to connecting the feedback signal in series with the input signal voltage; *shunt* refers to connecting the feedback signal in shunt (parallel) with an input current source.

Series feedback connections tend to *increase* the input resistance, while shunt feedback connections tend to *decrease* the input resistance. Voltage feedback tends to *decrease* the output impedance, while current feedback tends to *increase* the output impedance. Typically, higher input and lower output impedances are desired for most cascade amplifiers. Both of these are provided using the voltage-series feedback connection. We shall therefore concentrate first on this amplifier connection.

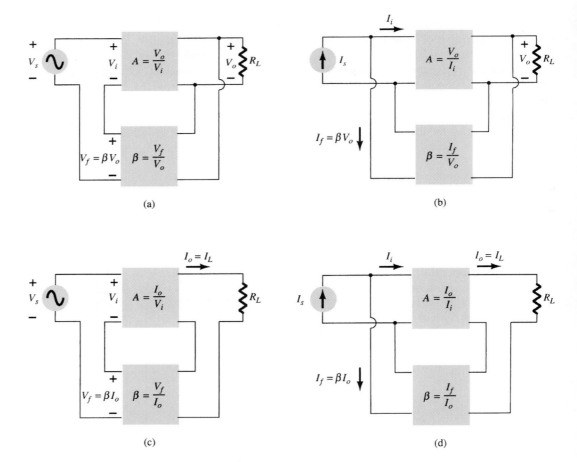

Figure 17.2 Feedback amplifier types: (a) voltage-series feedback, $A_f = V_o/V_s$; (b) voltage-shunt feedback, $A_f = V_o/I_s$; (c) current-series feedback, $A_f = I_o/V_s$; (d) current-shunt feedback, $A_f = I_o/I_s$.

Gain with Feedback

In this section we examine the gain of each of the feedback circuit connections of Fig. 17.2. The gain without feedback, A, is that of the amplifier stage. With feedback, β, the overall gain of the circuit is reduced by a factor $(1 + \beta A)$, as detailed below. A summary of the gain, feedback factor, and gain with feedback of Fig. 17.2 is provided for reference in Table 17.1.

TABLE 17.1 Summary of Gain, Feedback, and Gain with Feedback from Fig. 17.2

		Voltage-Series	Voltage-Shunt	Current-Series	Current-Shunt
Gain without feedback	A	$\dfrac{V_o}{V_i}$	$\dfrac{V_o}{I_i}$	$\dfrac{I_o}{V_i}$	$\dfrac{I_o}{I_i}$
Feedback	β	$\dfrac{V_f}{V_o}$	$\dfrac{I_f}{V_o}$	$\dfrac{V_f}{I_o}$	$\dfrac{I_f}{I_o}$
Gain with feedback	A_f	$\dfrac{V_o}{V_s}$	$\dfrac{V_o}{I_s}$	$\dfrac{I_o}{V_s}$	$\dfrac{I_o}{I_s}$

VOLTAGE-SERIES FEEDBACK

Figure 17.2a shows the voltage-series feedback connection with a part of the output voltage fed back in series with the input signal, resulting in an overall gain reduction. If there is no feedback ($V_f = 0$), the voltage gain of the amplifier stage is

$$A = \frac{V_o}{V_s} = \frac{V_o}{V_i} \tag{17.1}$$

If a feedback signal, V_f, is connected in series with the input, then

$$V_i = V_s - V_f$$

Since

$$V_o = AV_i = A(V_s - V_f) = AV_s - AV_f = AV_s - A(\beta V_o)$$

then

$$(1 + \beta A)V_o = AV_s$$

so that the overall voltage gain *with* feedback is

$$\boxed{A_f = \frac{V_o}{V_s} = \frac{A}{1 + \beta A}} \tag{17.2}$$

Equation (17.2) shows that the gain *with* feedback is the amplifier gain reduced by the factor $(1 + \beta A)$. This factor will be seen also to affect input and output impedance among other circuit features.

VOLTAGE-SHUNT FEEDBACK

The gain with feedback for the network of Fig. 17.2b is

$$A_f = \frac{V_o}{I_s} = \frac{A I_i}{I_i + I_f} = \frac{A I_i}{I_i + \beta V_o} = \frac{A I_i}{I_i + \beta A I_i}$$

$$\boxed{A_f = \frac{A}{1 + \beta A}} \tag{17.3}$$

Input Impedance with Feedback

VOLTAGE-SERIES FEEDBACK

A more detailed voltage-series feedback connection is shown in Fig. 17.3. The input impedance can be determined as follows:

$$I_i = \frac{V_i}{Z_i} = \frac{V_s - V_f}{Z_i} = \frac{V_s - \beta V_o}{Z_i} = \frac{V_s - \beta A V_i}{Z_i}$$

$$I_i Z_i = V_s - \beta A V_i$$

$$V_s = I_i Z_i + \beta A V_i = I_i Z_i + \beta A I_i Z_i$$

$$Z_{if} = \frac{V_s}{I_i} = Z_i + (\beta A)Z_i = Z_i(1 + \beta A) \tag{17.4}$$

The input impedance with series feedback is seen to be the value of the input impedance without feedback multiplied by the factor $(1 + \beta A)$ and applies to both voltage-series (Fig. 17.2a) and current-series (Fig. 17.2c) configurations.

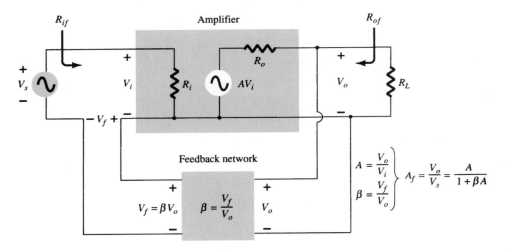

Figure 17.3   Voltage-series feedback connection.

VOLTAGE-SHUNT FEEDBACK

A more detailed voltage-shunt feedback connection is shown in Fig. 17.4. The input impedance can be determined to be

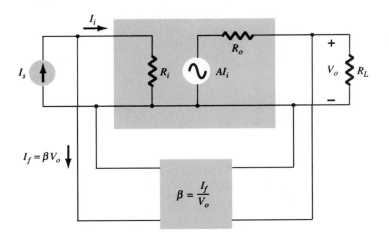

Figure 17.4   Voltage-shunt feedback connection.

$$Z_{if} = \frac{V_i}{I_s} = \frac{V_i}{I_i + I_f} = \frac{V_i}{I_i + \beta V_o}$$

$$= \frac{V_i/I_i}{I_i/I_i + \beta V_o/I_i}$$

$$\boxed{Z_{if} = \frac{Z_i}{1 + \beta A}} \qquad (17.5)$$

This reduced input impedance applies to the voltage-series connection of Fig. 17.2a and the voltage-shunt connection of Fig. 17.2b.

Output Impedance with Feedback

The output impedance for the connections of Fig. 17.2 are dependent on whether voltage or current feedback is used. For voltage feedback, the output impedance is decreased, while current feedback increases the output impedance.

VOLTAGE-SERIES FEEDBACK

The voltage-series feedback circuit of Fig. 17.3 provides sufficient circuit detail to determine the output impedance with feedback. The output impedance is determined by applying a voltage, V, resulting in a current, I, with V_s shorted out ($V_s = 0$). The voltage V is then

$$V = IZ_o + AV_i$$

For $V_s = 0$, $\qquad V_i = -V_f$

so that $\qquad V = IZ_o - AV_f = IZ_o - A(\beta V)$

Rewriting the equation as

$$V + \beta A V = IZ_o$$

allows solving for the output resistance with feedback:

$$\boxed{Z_{of} = \frac{V}{I} = \frac{Z_o}{1 + \beta A}} \qquad (17.6)$$

Equation (17.6) shows that with voltage-series feedback the output impedance is reduced from that without feedback by the factor $(1 + \beta A)$.

CURRENT-SERIES FEEDBACK

The output impedance with current-series feedback can be determined by applying a signal V to the output with V_s shorted out, resulting in a current I, the ratio of V to I being the output impedance. Figure 17.5 shows a more detailed connection with current-series feedback. For the output part of a current-series feedback connection shown in Fig. 17.5, the resulting output impedance is determined as follows. With $V_s = 0$,

$$V_i = V_f$$

$$I = \frac{V}{Z_o} - AV_i = \frac{V}{Z_o} - AV_f = \frac{V}{Z_o} - A\beta I$$

$$Z_o(1 + \beta A)I = V$$

$$\boxed{Z_{of} = \frac{V}{I} = Z_o(1 + \beta A)} \qquad (17.7)$$

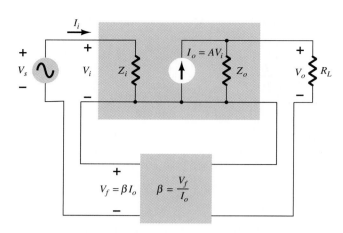

Figure 17.5 Current-series feedback connection.

A summary of the effect of feedback on input and output impedance is provided in Table 17.2.

TABLE 17.2 Effect of Feedback Connection on Input and Output Impedance

Voltage-Series	Current-Series	Voltage-Shunt	Current-Shunt
Z_{if} $Z_i(1 + \beta A)$	$Z_i(1 + \beta A)$	$\dfrac{Z_i}{1 + \beta A}$	$\dfrac{Z_i}{1 + \beta A}$
(increased)	(increased)	(decreased)	(decreased)
Z_{of} $\dfrac{Z_o}{1 + \beta A}$	$Z_o(1 + \beta A)$	$\dfrac{Z_o}{1 + \beta A}$	$Z_o(1 + \beta A)$
(decreased)	(increased)	(decreased)	(increased)

EXAMPLE 17.1

Determine the voltage gain, input, and output impedance with feedback for voltage series feedback having $A = -100, R_i = 10 \text{ k}\Omega, R_o = 20 \text{ k}\Omega$ for feedback of (a) $\beta = -0.1$ and (b) $\beta = -0.5$.

Solution

Using Eqs. (17.2), (17.4), and (17.6), we obtain

(a) $A_f = \dfrac{A}{1 + \beta A} = \dfrac{-100}{1 + (-0.1)(-100)} = \dfrac{-100}{11} = \mathbf{-9.09}$

$Z_{if} = Z_i (1 + \beta A) = 10 \text{ k}\Omega (11) = \mathbf{110 \text{ k}\Omega}$

$Z_{of} = \dfrac{Z_o}{1 + \beta A} = \dfrac{20 \times 10^3}{11} = \mathbf{1.82 \text{ k}\Omega}$

(b) $A_f = \dfrac{A}{1 + \beta A} = \dfrac{-100}{1 + (-0.5)(-100)} = \dfrac{-100}{51} = \mathbf{-1.96}$

$Z_{if} = Z_i (1 + \beta A) = 10 \text{ k}\Omega (51) = \mathbf{510 \text{ k}\Omega}$

$Z_{of} = \dfrac{Z_o}{1 + \beta A} = \dfrac{20 \times 10^3}{51} = \mathbf{392.16 \text{ }\Omega}$

Example 17.1 demonstrates the trade-off of gain for improved input and output resistance. Reducing the gain by a factor of 11 (from 100 to 9.09) is complemented by a reduced output resistance and increased input resistance by the same factor of 11.

Reducing the gain by a factor of 51 provides a gain of only 2 but with input resistance increased by the factor of 51 (to over 500 kΩ) and output resistance reduced from 20 kΩ to under 400 Ω. Feedback offers the designer the choice of trading away some of the available amplifier gain for other improved circuit features.

Reduction in Frequency Distortion

For a negative-feedback amplifier having $\beta A \gg 1$, the gain with feedback is $A_f \cong 1/\beta$. It follows from this that if the feedback network is purely resistive, the gain with feedback is not dependent on frequency even though the basic amplifier gain is frequency dependent. Practically, the frequency distortion arising because of varying amplifier gain with frequency is considerably reduced in a negative-voltage feedback amplifier circuit.

Reduction in Noise and Nonlinear Distortion

Signal feedback tends to hold down the amount of noise signal (such as power-supply hum) and nonlinear distortion. The factor $(1 + \beta A)$ reduces both input noise and resulting nonlinear distortion for considerable improvement. However, it should be noted that there is a reduction in overall gain (the price required for the improvement in circuit performance). If additional stages are used to bring the overall gain up to the level without feedback, it should be noted that the extra stage(s) might introduce as much noise back into the system as that reduced by the feedback amplifier. This problem can be somewhat alleviated by readjusting the gain of the feedback-amplifier circuit to obtain higher gain while also providing reduced noise signal.

Effect of Negative Feedback on Gain and Bandwidth

In Eq. (17.2), the overall gain with negative feedback is shown to be

$$A_f = \frac{A}{1 + \beta A} \cong \frac{A}{\beta A} = \frac{1}{\beta} \qquad \text{for } \beta A \gg 1$$

As long as $\beta A \gg 1$, the overall gain is approximately $1/\beta$. We should realize that for a practical amplifier (for single low- and high-frequency breakpoints) the open-loop gain drops off at high frequencies due to the active device and circuit capacitances. Gain may also drop off at low frequencies for capacitively coupled amplifier stages. Once the open-loop gain A drops low enough and the factor βA is no longer much larger than 1, the conclusion of Eq. (17.2) that $A_f \cong 1/\beta$ no longer holds true.

Figure 17.6 shows that the amplifier with negative feedback has more bandwidth (B_f) than the amplifier without feedback (B). The feedback amplifier has a higher upper 3-dB frequency and smaller lower 3-dB frequency.

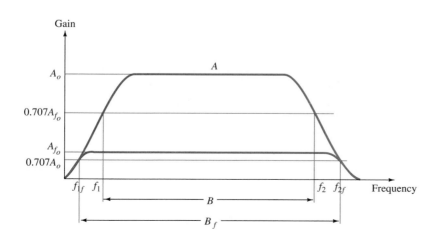

Figure 17.6 Effect of negative feedback on gain and bandwidth.

A_f

It is interesting to note that the use of feedback, while resulting in a lowering of voltage gain, has provided an increase in B and in the upper 3-dB frequency particularly. In fact, the product of gain and frequency remains the same so that the gain–bandwidth product of the basic amplifier is the same value for the feedback amplifier. However, since the feedback amplifier has lower gain, the net operation was to *trade* gain for bandwidth (we use bandwidth for the upper 3-dB frequency since typically $f_2 \gg f_1$).

Gain Stability with Feedback

In addition to the β factor setting a precise gain value, we are also interested in how stable the feedback amplifier is compared to an amplifier without feedback. Differentiating Eq. (17.2) leads to

$$\left|\frac{dA_f}{A_f}\right| = \frac{1}{|1 + \beta A|}\left|\frac{dA}{A}\right| \tag{17.8}$$

$$\left|\frac{dA_f}{A_f}\right| \cong \left|\frac{1}{\beta A}\right|\left|\frac{dA}{A}\right| \qquad \text{for } \beta A \gg 1 \tag{17.9}$$

This shows that magnitude of the relative change in gain $\left|\dfrac{dA_f}{A_f}\right|$ is reduced by the factor $|\beta A|$ compared to that without feedback $\left(\left|\dfrac{dA}{A}\right|\right)$.

EXAMPLE 17.2

If an amplifier with gain of -1000 and feedback of $\beta = -0.1$ has a gain change of 20% due to temperature, calculate the change in gain of the feedback amplifier.

Solution

Using Eq. (17.9), we get

$$\left|\frac{dA_f}{A_f}\right| \cong \left|\frac{1}{\beta A}\right|\left|\frac{dA}{A}\right| = \left|\frac{1}{-0.1(-1000)}(20\%)\right| = \mathbf{0.2\%}$$

The improvement is 100 times. Thus, while the amplifier gain changes from $|A| = 1000$ by 20%, the gain with feedback changes from $|A_f| = 100$ by only 0.2%.

17.3 PRACTICAL FEEDBACK CIRCUITS

Examples of practical feedback circuits will provide a means of demonstrating the effect feedback has on the various connection types. This section provides only a basic introduction to this topic.

Voltage-Series Feedback

Figure 17.7 shows an FET amplifier stage with voltage-series feedback. A part of the output signal (V_o) is obtained using a feedback network of resistors R_1 and R_2. The feedback voltage V_f is connected in series with the source signal V_s, their difference being the input signal V_i.

Without feedback the amplifier gain is

$$A = \frac{V_o}{V_i} = -g_m R_L \tag{17.10}$$

where R_L is the parallel combination of resistors:

$$R_L = R_D R_o (R_1 + R_2) \tag{17.11}$$

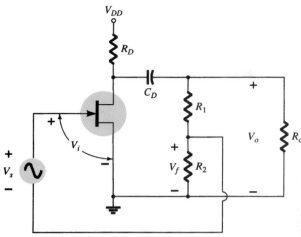

Figure 17.7 FET amplifier stage with voltage-series feedback.

The feedback network provides a feedback factor of

$$\beta = \frac{V_f}{V_o} = \frac{-R_2}{R_1 + R_2} \qquad (17.12)$$

Using the values of A and β above in Eq. (17.2), we find the gain with negative feedback to be

$$A_f = \frac{A}{1 + \beta A} = \frac{-g_m R_L}{1 + [R_2 R_L/(R_1 + R_2)]g_m} \qquad (17.13)$$

If $\beta A \gg 1$, we have

$$A_f \cong \frac{1}{\beta} = -\frac{R_1 + R_2}{R_2} \qquad (17.14)$$

Calculate the gain without and with feedback for the FET amplifier circuit of Fig. 17.7 and the following circuit values: $R_1 = 80\ \text{k}\Omega$, $R_2 = 20\ \text{k}\Omega$, $R_o = 10\ \text{k}\Omega$, $R_D = 10\ \text{k}\Omega$, and $g_m = 4000\ \mu\text{S}$.

EXAMPLE 17.3

Solution

$$R_L \cong \frac{R_o R_D}{R_o + R_D} = \frac{10\ \text{k}\Omega\ (10\ \text{k}\Omega)}{10\ \text{k}\Omega + 10\ \text{k}\Omega} = 5\ \text{k}\Omega$$

Neglecting 100 kΩ resistance of R_1 and R_2 in series

$$A = -g_m R_L = -(4000 \times 10^{-6}\ \mu\text{S})(5\ \text{k}\Omega) = \mathbf{-20}$$

The feedback factor is

$$\beta = \frac{-R_2}{R_1 + R_2} = \frac{-20\ \text{k}\Omega}{80\ \text{k}\Omega + 20\ \text{k}\Omega} = -0.2$$

The gain with feedback is

$$A_f = \frac{A}{1 + \beta A} = \frac{-20}{1 + (-0.2)(-20)} = \frac{-20}{5} = \mathbf{-4}$$

17.3 Practical Feedback Circuits

829

Figure 17.8 shows a voltage-series feedback connection using an op-amp. The gain of the op-amp, A, without feedback, is reduced by the feedback factor

$$\beta = \frac{R_2}{R_1 + R_2} \qquad (17.15)$$

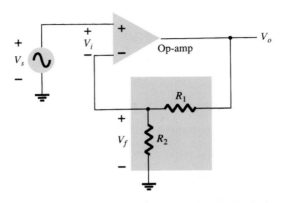

Figure 17.8 Voltage-series feedback in an op-amp connection.

EXAMPLE 17.4

Calculate the amplifier gain of the circuit of Fig. 17.8 for op-amp gain $A = 100,000$ and resistances $R_1 = 1.8 \text{ k}\Omega$ and $R_2 = 200 \ \Omega$.

Solution

$$\beta = \frac{R_2}{R_1 + R_2} = \frac{200 \ \Omega}{200 \ \Omega + 1.8 \text{ k}\Omega} = 0.1$$

$$A_f = \frac{A}{1 + \beta A} = \frac{100,000}{1 + (0.1)(100,000)}$$

$$= \frac{100,000}{10,001} = 9.999$$

Note that since $\beta A \gg 1$,

$$A_f \cong \frac{1}{\beta} = \frac{1}{0.1} = \mathbf{10}$$

The emitter-follower circuit of Fig. 17.9 provides voltage-series feedback. The signal voltage, V_s, is the input voltage, V_i. The output voltage, V_o, is also the feedback

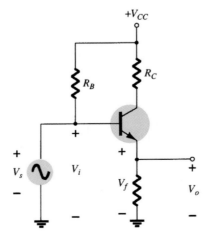

Figure 17.9 Voltage-series feedback circuit (emitter-follower).

voltage in series with the input voltage. The amplifier, as shown in Fig. 17.9, provides the operation *with* feedback. The operation of the circuit without feedback provides $V_f = 0$, so that

$$A = \frac{V_o}{V_s} = \frac{h_{fe}I_b R_E}{V_s} = \frac{h_{fe}R_E(V_s/h_{ie})}{V_s} = \frac{h_{fe}R_E}{h_{ie}}$$

and

$$\beta = \frac{V_f}{V_o} = 1$$

The operation with feedback then provides that

$$A_f = \frac{V_o}{V_s} = \frac{A}{1 + \beta A} = \frac{h_{fe}R_E/h_{ie}}{1 + (1)(h_{fe}R_E/h_{ie})}$$

$$= \frac{h_{fe}R_E}{h_{ie} + h_{fe}R_E}$$

For $h_{fe}R_E \gg h_{ie}$,

$$A_f \cong 1$$

Current-Series Feedback

Another feedback technique is to sample the output current (I_o) and return a proportional voltage in series with the input. While stabilizing the amplifier gain, the current-series feedback connection increases input resistance.

Figure 17.10 shows a single transistor amplifier stage. Since the emitter of this stage has an unbypassed emitter, it effectively has current-series feedback. The current through resistor R_E results in a feedback voltage that opposes the source signal applied so that the output voltage V_o is reduced. To remove the current-series feedback, the emitter resistor must be either removed or bypassed by a capacitor (as is usually done).

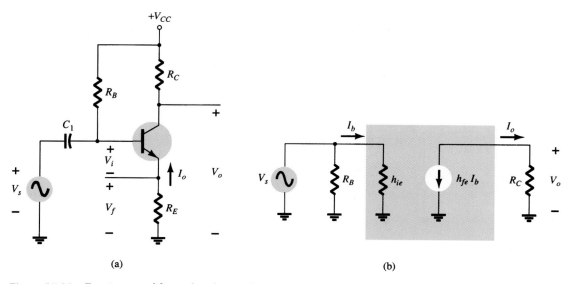

(a)   **(b)**

Figure 17.10   Transistor amplifier with unbypassed emitter resistor (R_E) for current-series feedback: (a) amplifier circuit; (b) ac equivalent circuit without feedback.

17.3   Practical Feedback Circuits

A_f

WITHOUT FEEDBACK

Referring to the basic format of Fig. 17.2a and summarized in Table 17.1, we have

$$A = \frac{I_o}{V_i} = \frac{-I_b h_{fe}}{I_b h_{ie} + R_E} = \frac{-h_{fe}}{h_{ie} + R_E} \tag{17.16}$$

$$\beta = \frac{V_f}{I_o} = \frac{-I_o R_E}{I_o} = -R_E \tag{17.17}$$

The input and output impedances are

$$Z_i = R_B \| (h_{ie} + R_E) \cong h_{ie} + R_E \tag{17.18}$$

$$Z_o = R_C \tag{17.19}$$

WITH FEEDBACK

$$A_f = \frac{I_o}{V_s} = \frac{A}{1 + \beta A} = \frac{-h_{fe}/h_{ie}}{1 + (-R_E)\left(\dfrac{-h_{fe}}{h_{ie} + R_E}\right)} \cong \frac{-h_{fe}}{h_{ie} + h_{fe}R_E} \tag{17.20}$$

The input and output impedance is calculated as specified in Table 17.2.

$$Z_{if} = Z_i(1 + \beta A) \cong h_{ie}\left(1 + \frac{h_{fe}R_E}{h_{ie}}\right) = h_{ie} + h_{fe}R_E \tag{17.21}$$

$$Z_{of} = Z_o(1 + \beta A) = R_C\left(1 + \frac{h_{fe}R_E}{h_{ie}}\right) \tag{17.22}$$

The voltage gain (A) with feedback is

$$A_{vf} = \frac{V_o}{V_s} = \frac{I_o R_C}{V_s} = \left(\frac{I_o}{V_s}\right)R_C = A_f R_C \cong \frac{-h_{fe}R_C}{h_{ie} + h_{fe}R_E} \tag{17.23}$$

EXAMPLE 17.5   Calculate the voltage gain of the circuit of Fig. 17.11.

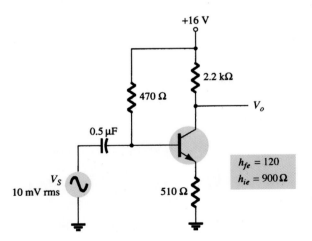

$h_{fe} = 120$
$h_{ie} = 900 \, \Omega$

Figure 17.11 BJT amplifier with current-series feedback for Example 17.5.

Solution

Without feedback,

$$A = \frac{I_o}{V_i} = \frac{-h_{fe}}{h_{ie} + R_E} = \frac{-120}{900 + 510} = -0.085$$

$$\beta = \frac{V_f}{I_o} = -R_E = -510$$

The factor $(1 + \beta A)$ is then

$$1 + \beta A = 1 + (-0.085)(-510) = 44.35$$

The gain with feedback is then

$$A_f = \frac{I_o}{V_s} = \frac{A}{1 + \beta A} = \frac{-0.085}{44.35} = -1.92 \times 10^{-3}$$

and the voltage gain with feedback A_{vf} is

$$A_{vf} = \frac{V_o}{V_s} = A_f R_C = (-1.92 \times 10^{-3})(2.2 \times 10^3) = \mathbf{-4.2}$$

Without feedback $(R_E = 0)$, the voltage gain is

$$A_v = \frac{-R_C}{r_e} = \frac{-2.2 \times 10^3}{7.5} = \mathbf{-293.3}$$

Voltage-Shunt Feedback

The constant-gain op-amp circuit of Fig. 17.12a provides voltage-shunt feedback. Referring to Fig. 17.2b and Table 17.1 and the op-amp ideal characteristics $I_i = 0$, $V_i = 0$, and voltage gain of infinity, we have

$$A = \frac{V_o}{I_i} = \infty \tag{17.24}$$

$$\beta = \frac{I_f}{V_o} = \frac{-1}{R_o} \tag{17.25}$$

The gain with feedback is then

$$A_f = \frac{V_o}{I_s} = \frac{V_o}{I_i} = \frac{A}{1 + \beta A} = \frac{1}{\beta} = -R_o \tag{17.26}$$

This is a transfer resistance gain. The more usual gain is the voltage gain with feedback,

$$A_{vf} = \frac{V_o}{I_s} \frac{I_s}{V_1} = (-R_o)\frac{1}{R_1} = \frac{-R_o}{R_1} \tag{17.27}$$

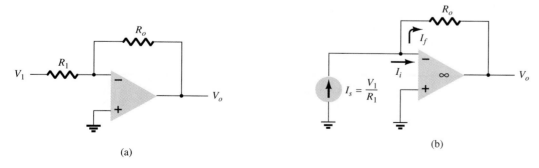

(a)

(b)

Figure 17.12 Voltage-shunt negative feedback amplifier: (a) constant-gain circuit; (b) equivalent circuit.

17.3 Practical Feedback Circuits

The circuit of Fig. 17.13 is a voltage-shunt feedback amplifier using an FET with no feedback, $V_f = 0$.

$$A = \frac{V_o}{I_i} \cong -g_m R_D R_S \qquad (17.28)$$

The feedback is

$$\beta = \frac{I_f}{V_o} = \frac{-1}{R_F} \qquad (17.29)$$

With feedback, the gain of the circuit is

$$A_f = \frac{V_o}{I_s} = \frac{A}{1 + \beta A} = \frac{-g_m R_D R_S}{1 + (-1/R_F)(-g_m R_D R_S)}$$

$$= \frac{-g_m R_D R_S R_F}{R_F + g_m R_D R_S} \qquad (17.30)$$

The voltage gain of the circuit with feedback is then

$$A_{vf} = \frac{V_o}{I_s} \frac{I_s}{V_s} = \frac{-g_m R_D R_S R_F}{R_F + g_m R_D R_S} \left(\frac{1}{R_S} \right)$$

$$= \frac{-g_m R_D R_F}{R_F + g_m R_D R_S} = (-g_m R_D) \frac{R_F}{R_F + g_m R_D R_S} \qquad (17.31)$$

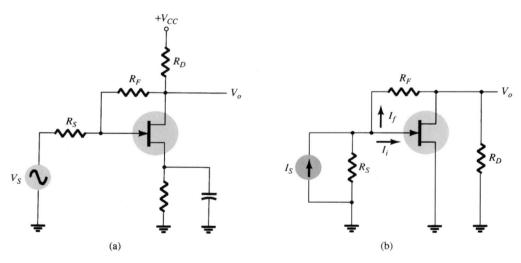

Figure 17.13 Voltage-shunt feedback amplifier using an FET: (a) circuit; (b) equivalent circuit.

EXAMPLE 17.6

Calculate the voltage gain with and without feedback for the circuit of Fig. 17.13a with values of $g_m = 5$ mS, $R_D = 5.1$ kΩ, $R_S = 1$ kΩ, and $R_F = 20$ kΩ.

Solution

Without feedback, the voltage gain is

$$A_v = -g_m R_D = -(5 \times 10^{-3})(5.1 \times 10^3) = \mathbf{-25.5}$$

With feedback the gain is reduced to

$$A_{vf} = (-g_m R_D) \frac{R_F}{R_F + g_m R_D R_S}$$

$$= (-25.5) \frac{20 \times 10^3}{(20 \times 10^3) + (5 \times 10^{-3})(5.1 \times 10^3)(1 \times 10^3)}$$

$$= -25.5(0.44) = \mathbf{-11.2}$$

17.4 FEEDBACK AMPLIFIER—PHASE AND FREQUENCY CONSIDERATIONS

So far we have considered the operation of a feedback amplifier in which the feedback signal was *opposite* to the input signal—negative feedback. In any practical circuit this condition occurs only for some mid-frequency range of operation. We know that an amplifier gain will change with frequency, dropping off at high frequencies from the mid-frequency value. In addition, the phase shift of an amplifier will also change with frequency.

If, as the frequency increases, the phase shift changes then some of the feedback signal will *add* to the input signal. It is then possible for the amplifier to break into oscillations due to positive feedback. If the amplifier oscillates at some low or high frequency, it is no longer useful as an amplifier. Proper feedback-amplifier design requires that the circuit be stable at *all* frequencies, not merely those in the range of interest. Otherwise, a transient disturbance could cause a seemingly stable amplifier to suddenly start oscillating.

Nyquist Criterion

In judging the stability of a feedback amplifier, as a function of frequency, the βA product and the phase shift between input and output are the determining factors. One of the most popular techniques used to investigate stability is the Nyquist method. A Nyquist diagram is used to plot gain and phase shift as a function of frequency on a complex plane. The Nyquist plot, in effect, combines the two Bode plots of gain versus frequency and phase shift versus frequency on a single plot. A Nyquist plot is used to quickly show whether an amplifier is stable for all frequencies and how stable the amplifier is relative to some gain or phase-shift criteria.

As a start, consider the *complex plane* shown in Fig. 17.14. A few points of various gain (βA) values are shown at a few different phase-shift angles. By using the

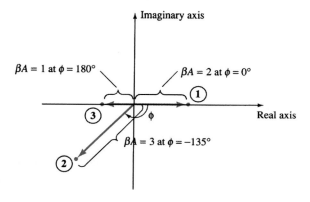

Figure 17.14 Complex plane showing typical gain-phase points.

positive real axis as reference (0°), a magnitude of $\beta A = 2$ is shown at a phase shift of 0° at point 1. Additionally, a magnitude of $\beta A = 3$ at a phase shift of $-135°$ is shown at point 2 and a magnitude/phase of $\beta A = 1$ at 180° is shown at point 3. Thus points on this plot can represent *both* gain magnitude of βA and phase shift. If the points representing gain and phase shift for an amplifier circuit are plotted at increasing frequency, then a Nyquist plot is obtained as shown by the plot in Fig. 17.15. At the origin, the gain is 0 at a frequency of 0 (for *RC*-type coupling). At increasing frequency, points $f_1, f_2,$ and f_3 and the phase shift increased, as did the magnitude of βA. At a representative frequency f_4, the value of A is the vector length from the origin to point f_4 and the phase shift is the angle ϕ. At a frequency f_5, the phase shift is 180°. At higher frequencies, the gain is shown to decrease back to 0.

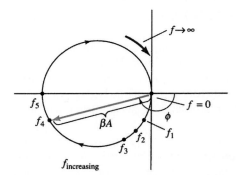

Figure 17.15 Nyquist plot.

The Nyquist criterion for stability can be stated as follows:

The amplifier is unstable if the Nyquist curve plotted encloses (encircles) the − 1 point, and it is stable otherwise.

An example of the Nyquist criterion is demonstrated by the curves in Fig. 17.16. The Nyquist plot in Fig. 17.16a is stable since it does not encircle the −1 point, whereas that shown in Fig. 17.16b is unstable since the curve does encircle the −1 point. Keep in mind that encircling the −1 point means that at a phase shift of 180° the loop gain (βA) is greater than 1; therefore, the feedback signal is in phase with the input and large enough to result in a larger input signal than that applied, with the result that oscillation occurs.

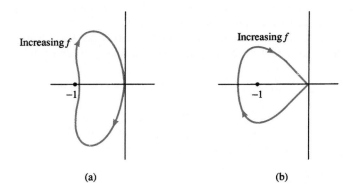

Figure 17.16 Nyquist plots showing stability conditions; (a) stable; (b) unstable.

Gain and Phase Margins

From the Nyquist criterion, we know that a feedback amplifier is stable if the loop gain (βA) is less than unity (0 dB) when its phase angle is 180°. We can additionally determine some margins of stability to indicate how close to instability the amplifier is. That is, if the gain (βA) is less than unity but, say, 0.95 in value, this would not be as relatively stable as another amplifier having, say, $(\beta A) = 0.7$ (both measured at 180°). Of course, amplifiers with loop gains 0.95 and 0.7 are both stable, but one is closer to instability, if the loop gain increases, than the other. We can define the following terms:

Gain margin (GM) is defined as the negative of the value of $|\beta A|$ in decibels at the frequency at which the phase angle is 180°. Thus, 0 dB, equal to a value of $\beta A = 1$, is on the border of stability and any negative decibel value is stable. The GM may be evaluated in decibels from the curve of Fig. 17.17.

Phase margin (PM) is defined as the angle of 180° minus the magnitude of the angle at which the value $|\beta A|$ is unity (0 dB). The PM may also be evaluated directly from the curve of Fig. 17.17.

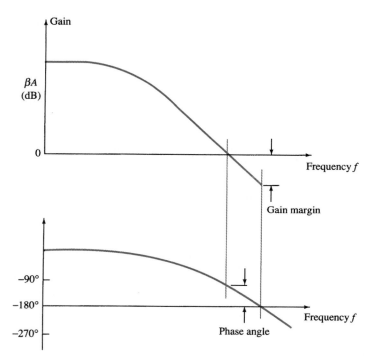

Figure 17.17 Bode plots showing gain and phase margins.

17.5 OSCILLATOR OPERATION

The use of positive feedback that results in a feedback amplifier having closed-loop gain $|A_f|$ greater than 1 and satisfies the phase conditions will result in operation as an oscillator circuit. An oscillator circuit then provides a varying output signal. If the output signal varies sinusoidally, the circuit is referred to as a *sinusoidal oscillator.* If the output voltage rises quickly to one voltage level and later drops quickly to another voltage level, the circuit is generally referred to as a *pulse* or *square-wave oscillator.*

To understand how a feedback circuit performs as an oscillator, consider the feedback circuit of Fig. 17.18. When the switch at the amplifier input is open, no oscillation occurs. Consider that we have a *fictitious* voltage at the amplifier input (V_i). This results in an output voltage $V_o = AV_i$ after the amplifier stage and in a voltage $V_f = \beta(AV_i)$ after the feedback stage. Thus, we have a feedback voltage $V_f = \beta AV_i$, where βA is referred to as the *loop gain.* If the circuits of the base amplifier and feedback network provide βA of a correct magnitude and phase, V_f can be made equal to

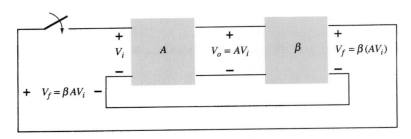

Figure 17.18 Feedback circuit used as an oscillator.

V_i. Then, when the switch is closed and fictitious voltage V_i is removed, the circuit will continue operating since the feedback voltage is sufficient to drive the amplifier and feedback circuits resulting in a proper input voltage to sustain the loop operation. The output waveform will still exist after the switch is closed if the condition

$$\beta A = 1 \qquad (17.32)$$

is met. This is known as the *Barkhausen criterion* for oscillation.

In reality, no input signal is needed to start the oscillator going. Only the condition $\beta A = 1$ must be satisfied for self-sustained oscillations to result. In practice, βA is made greater than 1 and the system is started oscillating by amplifying noise voltage, which is always present. Saturation factors in the practical circuit provide an "average" value of βA of 1. The resulting waveforms are never exactly sinusoidal. However, the closer the value βA is to exactly 1, the more nearly sinusoidal is the waveform. Figure 17.19 shows how the noise signal results in a buildup of a steady-state oscillation condition.

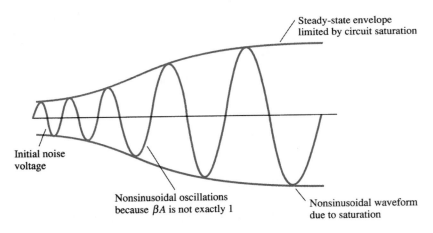

Figure 17.19 Buildup of steady-state oscillations.

Another way of seeing how the feedback circuit provides operation as an oscillator is obtained by noting the denominator in the basic feedback equation (17.2), $A_f = A/(1 + \beta A)$. When $\beta A = -1$ or magnitude 1 at a phase angle of 180°, the denominator becomes 0 and the gain with feedback, A_f, becomes infinite. Thus, an infinitesimal signal (noise voltage) can provide a measurable output voltage, and the circuit acts as an oscillator even without an input signal.

The remainder of this chapter is devoted to various oscillator circuits that use a variety of components. Practical considerations are included so that workable circuits in each of the various cases are discussed.

17.6 PHASE-SHIFT OSCILLATOR

An example of an oscillator circuit that follows the basic development of a feedback circuit is the *phase-shift oscillator*. An idealized version of this circuit is shown in Fig. 17.20. Recall that the requirements for oscillation are that the loop gain, βA, is greater than unity *and* that the phase shift around the feedback network is 180° (providing positive feedback). In the present idealization, we are considering the feedback network to be driven by a perfect source (zero source impedance) and the output of the feedback network to be connected into a perfect load (infinite load impedance). The idealized case will allow development of the theory behind the operation of the phase-shift oscillator. Practical circuit versions will then be considered.

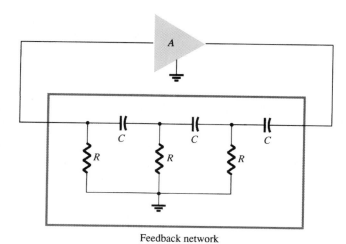

Feedback network

Figure 17.20 Idealized phase-shift oscillator.

Concentrating our attention on the phase-shift network, we are interested in the attenuation of the network at the frequency at which the phase shift is exactly 180°. Using classical network analysis, we find that

$$f = \frac{1}{2\pi RC\sqrt{6}} \qquad (17.33)$$

$$\beta = \frac{1}{29} \qquad (17.34)$$

and the phase shift is 180°.

For the loop gain βA to be greater than unity, the gain of the amplifier stage must be greater than $1/\beta$ or 29:

$$A > 29 \qquad (17.35)$$

When considering the operation of the feedback network, one might naively select the values of R and C to provide (at a specific frequency) 60°-phase shift per section for three sections, resulting in a 180° phase shift, as desired. This, however, is not the case, since each section of the RC in the feedback network loads down the previous one. The net result that the *total* phase shift be 180° is all that is important. The frequency given by Eq. (17.33) is that at which the *total* phase shift is 180°. If one measured the phase shift per RC section, each section would not provide the same phase shift (although the overall phase shift is 180°). If it were desired to obtain exactly a 60° phase shift for each of three stages, then emitter-follower stages would be needed for each RC section to prevent each from being loaded from the following circuit.

FET Phase-Shift Oscillator

A practical version of a phase-shift oscillator circuit is shown in Fig. 17.21a. The circuit is drawn to show clearly the amplifier and feedback network. The amplifier stage is self-biased with a capacitor bypassed source resistor R_S and a drain bias resistor R_D. The FET device parameters of interest are g_m and r_d. From FET amplifier theory, the amplifier gain magnitude is calculated from

$$|A| = g_m R_L \tag{17.36}$$

where R_L in this case is the parallel resistance of R_D and r_d

$$R_L = \frac{R_D r_d}{R_D + r_d} \tag{17.37}$$

We shall assume as a very good approximation that the input impedance of the FET amplifier stage is infinite. This assumption is valid as long as the oscillator operating frequency is low enough so that FET capacitive impedances can be neglected. The output impedance of the amplifier stage given by R_L should also be small compared to the impedance seen looking into the feedback network so that no attenuation due to loading occurs. In practice, these considerations are not always negligible, and the amplifier stage gain is then selected somewhat larger than the needed factor of 29 to assure oscillator action.

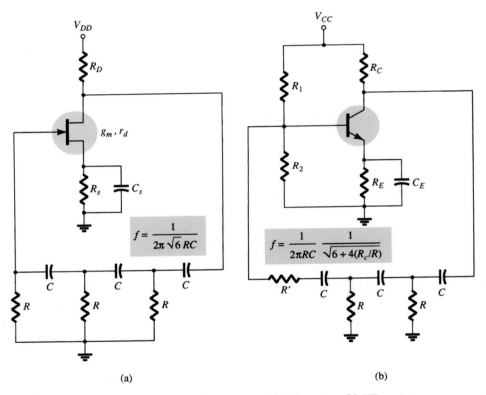

Figure 17.21   Practical phase-shift oscillator circuits: (a) FET version; (b) BJT version.

EXAMPLE 17.7

It is desired to design a phase-shift oscillator (as in Fig. 17.21a) using an FET having $g_m = 5000 \ \mu S$, $r_d = 40 \ k\Omega$, and feedback circuit value of $R = 10 \ k\Omega$. Select the value of C for oscillator operation at 1 kHz and R_D for $A > 29$ to ensure oscillator action.

Solution

Equation (17.33) is used to solve for the capacitor value. Since $f = 1/2\pi RC\sqrt{6}$, we can solve for C:

$$C = \frac{1}{2\pi Rf\sqrt{6}} = \frac{1}{(6.28)(10 \times 10^3)(1 \times 10^3)(2.45)} = \mathbf{6.5\ nF}$$

Using Eq. (17.36), we solve for R_L to provide a gain of, say, $A = 40$ (this allows for some loading between R_L and the feedback network input impedance):

$$|A| = g_m R_L$$

$$R_L = \frac{|A|}{g_m} = \frac{40}{5000 \times 10^{-6}} = 8\ k\Omega$$

Using Eq. (17.37), we solve for $R_D = \mathbf{10\ k\Omega}.$

Transistor Phase-Shift Oscillator

If a transistor is used as the active element of the amplifier stage, the output of the feedback network is loaded appreciably by the relatively low input resistance (h_{ie}) of the transistor. Of course, an emitter-follower input stage followed by a common-emitter amplifier stage could be used. If a single transistor stage is desired, however, the use of voltage-shunt feedback (as shown in Fig. 17.21b) is more suitable. In this connection, the feedback signal is coupled through the feedback resistor R' in *series* with the amplifier stage input resistance (R_i).

Analysis of the ac circuit provides the following equation for the resulting oscillator frequency:

$$f = \frac{1}{2\pi RC} \frac{1}{\sqrt{6 + 4(R_C/R)}} \tag{17.38}$$

For the loop gain to be greater than unity, the requirement on the current gain of the transistor is found to be

$$h_{fe} > 23 + 29\frac{R}{R_C} + 4\frac{R_C}{R} \tag{17.39}$$

IC Phase-Shift Oscillator

As IC circuits have become more popular, they have been adapted to operate in oscillator circuits. One need buy only an op-amp to obtain an amplifier circuit of stabilized gain setting and incorporate some means of signal feedback to produce an oscillator circuit. For example, a phase-shift oscillator is shown in Fig. 17.22. The output

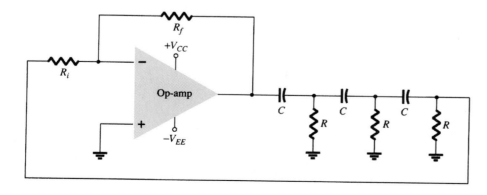

Figure 17.22 Phase-shift oscillator using op-amp.

of the op-amp is fed to a three-stage RC network, which provides the needed 180° of phase shift (at an attenuation factor of 1/29). If the op-amp provides gain (set by resistors R_i and R_f) of greater than 29, a loop gain greater than unity results and the circuit acts as an oscillator [oscillator frequency is given by Eq. (17.33)].

17.7 WIEN BRIDGE OSCILLATOR

A practical oscillator circuit uses an op-amp and RC bridge circuit, with the oscillator frequency set by the R and C components. Figure 17.23 shows a basic version of a Wien bridge oscillator circuit. Note the basic bridge connection. Resistors R_1 and R_2 and capacitors C_1 and C_2 form the frequency-adjustment elements, while resistors R_3 and R_4 form part of the feedback path. The op-amp output is connected as the bridge input at points a and c. The bridge circuit output at points b and d is the input to the op-amp.

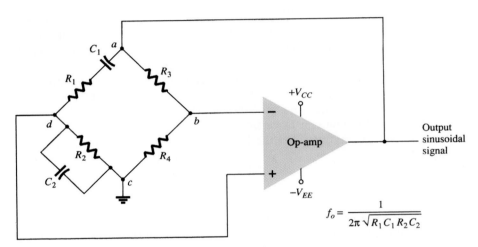

Figure 17.23 Wien bridge oscillator circuit using op-amp amplifier.

Neglecting loading effects of the op-amp input and output impedances, the analysis of the bridge circuit results in

$$\frac{R_3}{R_4} = \frac{R_1}{R_2} + \frac{C_2}{C_1} \tag{17.40}$$

and

$$f_o = \frac{1}{2\pi\sqrt{R_1 C_1 R_2 C_2}} \tag{17.41}$$

If, in particular, the values are $R_1 = R_2 = R$ and $C_1 = C_2 = C$, the resulting oscillator frequency is

$$f_o = \frac{1}{2\pi RC} \tag{17.42}$$

and

$$\frac{R_3}{R_4} = 2 \tag{17.43}$$

Thus a ratio of R_3 to R_4 greater than 2 will provide sufficient loop gain for the circuit to oscillate at the frequency calculated using Eq. (17.42).

Calculate the resonant frequency of the Wien bridge oscillator of Fig. 17.24.

EXAMPLE 17.8

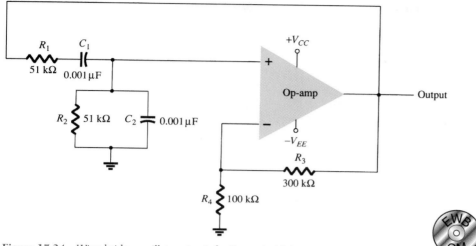

Figure 17.24   Wien bridge oscillator circuit for Example 17.8.

Solution

Using Eq. (17.42) yields

$$f_o = \frac{1}{2\pi RC} = \frac{1}{2\pi(51 \times 10^3)(0.001 \times 10^{-6})} = \mathbf{3120.7 \ Hz}$$

Design the *RC* elements of a Wien bridge oscillator as in Fig. 17.24 for operation at $f_o = 10$ kHz.

EXAMPLE 17.9

Solution

Using equal values of *R* and *C* we can select $R = 100$ kΩ and calculate the required value of *C* using Eq. (17.42):

$$C = \frac{1}{2\pi f_o R} = \frac{1}{6.28(10 \times 10^3)(100 \times 10^3)} = \frac{10^{-9}}{6.28} = \mathbf{159 \ pF}$$

We can use $R_3 = 300$ kΩ and $R_4 = 100$ kΩ to provide a ratio R_3/R_4 greater than 2 for oscillation to take place.

17.8   TUNED OSCILLATOR CIRCUIT

Tuned-Input, Tuned-Output Oscillator Circuits

A variety of circuits can be built using that shown in Fig. 17.25 by providing tuning in both the input and output sections of the circuit. Analysis of the circuit of Fig. 17.25 reveals that the following types of oscillators are obtained when the reactance elements are as designated:

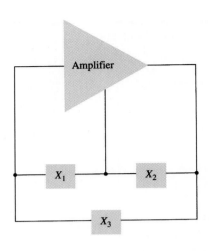

Figure 17.25 Basic configuration of resonant circuit oscillator.

| | Reactance Element | | |
Oscillator Type	X_1	X_2	X_3
Colpitts oscillator	C	C	L
Hartley oscillator	L	L	C
Tuned input, tuned output	LC	LC	—

Colpitts Oscillator

FET COLPITTS OSCILLATOR

A practical version of an FET Colpitts oscillator is shown in Fig. 17.26. The circuit is basically the same form as shown in Fig. 17.25 with the addition of the components needed for dc bias of the FET amplifier. The oscillator frequency can be found to be

$$f_o = \frac{1}{2\pi\sqrt{LC_{eq}}} \qquad (17.44)$$

where

$$C_{eq} = \frac{C_1 C_2}{C_1 + C_2} \qquad (17.45)$$

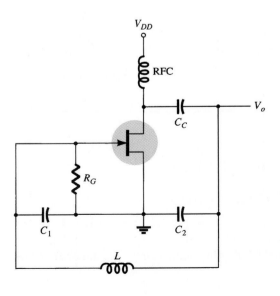

Figure 17.26 FET Colpitts oscillator.

TRANSISTOR COLPITTS OSCILLATOR

A transistor Colpitts oscillator circuit can be made as shown in Fig. 17.27. The circuit frequency of oscillation is given by Eq. (17.44).

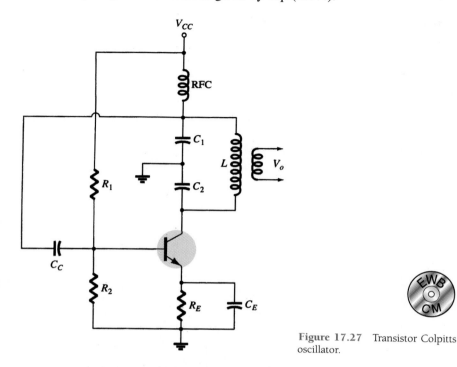

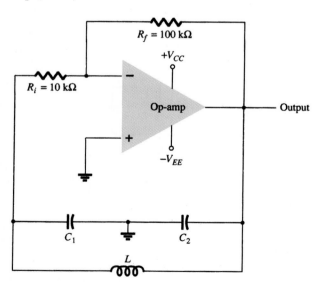

Figure 17.27 Transistor Colpitts oscillator.

IC COLPITTS OSCILLATOR

An op-amp Colpitts oscillator circuit is shown in Fig. 17.28. Again, the op-amp provides the basic amplification needed while the oscillator frequency is set by an LC feedback network of a Colpitts configuration. The oscillator frequency is given by Eq. (17.44).

Figure 17.28 Op-amp Colpitts oscillator.

Hartley Oscillator

If the elements in the basic resonant circuit of Fig. 17.25 are X_1 and X_2 (inductors) and X_3 (capacitor), the circuit is a Hartley oscillator.

FET HARTLEY OSCILLATOR

An FET Hartley oscillator circuit is shown in Fig. 17.29. The circuit is drawn so that the feedback network conforms to the form shown in the basic resonant circuit (Fig. 17.25). Note, however, that inductors L_1 and L_2 have a mutual coupling, M, which must be taken into account in determining the equivalent inductance for the resonant tank circuit. The circuit frequency of oscillation is then given approximately by

$$f_o = \frac{1}{2\pi\sqrt{L_{eq}C}} \qquad (17.46)$$

with
$$L_{eq} = L_1 + L_2 + 2M \qquad (17.47)$$

TRANSISTOR HARTLEY OSCILLATOR

Figure 17.30 shows a transistor Hartley oscillator circuit. The circuit operates at a frequency given by Eq. (17.46).

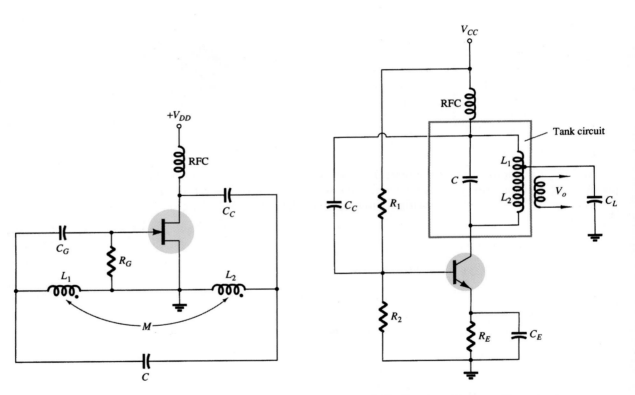

Figure 17.29 FET Hartley oscillator.

Figure 17.30 Transistor Hartley oscillator circuit.

17.9   CRYSTAL OSCILLATOR

A crystal oscillator is basically a tuned-circuit oscillator using a piezoelectric crystal as a resonant tank circuit. The crystal (usually quartz) has a greater stability in holding constant at whatever frequency the crystal is originally cut to operate. Crystal oscillators are used whenever great stability is required, such as in communication transmitters and receivers.

Characteristics of a Quartz Crystal

A quartz crystal (one of a number of crystal types) exhibits the property that when mechanical stress is applied across the faces of the crystal, a difference of potential develops across opposite faces of the crystal. This property of a crystal is called the *piezoelectric effect.* Similarly, a voltage applied across one set of faces of the crystal causes mechanical distortion in the crystal shape.

When alternating voltage is applied to a crystal, mechanical vibrations are set up—these vibrations having a natural resonant frequency dependent on the crystal. Although the crystal has electromechanical resonance, we can represent the crystal action by an equivalent electrical resonant circuit as shown in Fig. 17.31. The inductor L and capacitor C represent electrical equivalents of crystal mass and compliance, while resistance R is an electrical equivalent of the crystal structure's internal friction. The shunt capacitance C_M represents the capacitance due to mechanical mounting of the crystal. Because the crystal losses, represented by R, are small, the equivalent crystal Q (quality factor) is high—typically 20,000. Values of Q up to almost 10^6 can be achieved by using crystals.

The crystal as represented by the equivalent electrical circuit of Fig. 17.31 can have two resonant frequencies. One resonant condition occurs when the reactances of the series RLC leg are equal (and opposite). For this condition, the *series-resonant* impedance is very low (equal to R). The other resonant condition occurs at a higher frequency when the reactance of the series-resonant leg equals the reactance of capacitor C_M. This is a parallel resonance or antiresonance condition of the crystal. At this frequency, the crystal offers a very high impedance to the external circuit. The impedance versus frequency of the crystal is shown in Fig. 17.32. In order to use the crystal properly, it must be connected in a circuit so that its low impedance in the series-resonant operating mode or high impedance in the antiresonant operating mode is selected.

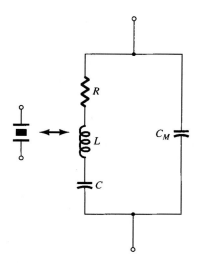

Figure 17.31 Electrical equivalent circuit of a crystal.

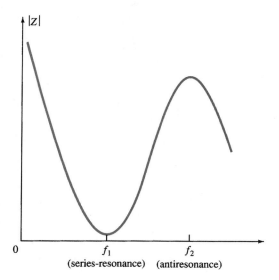

Figure 17.32 Crystal impedance versus frequency.

Series-Resonant Circuits

To excite a crystal for operation in the series-resonant mode, it may be connected as a series element in a feedback path. At the series-resonant frequency of the crystal, its impedance is smallest and the amount of (positive) feedback is largest. A typical transistor circuit is shown in Fig. 17.33. Resistors R_1, R_2, and R_E provide a voltage-divider stabilized dc bias circuit. Capacitor C_E provides ac bypass of the emitter

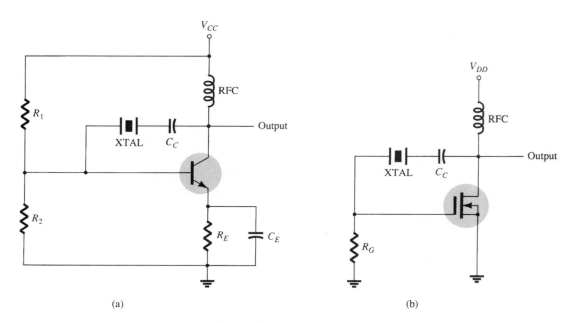

Figure 17.33 Crystal-controlled oscillator using crystal in series-feedback path: (a) BJT circuit; (b) FET circuit.

resistor, and the RFC coil provides for dc bias while decoupling any ac signal on the power lines from affecting the output signal. The voltage feedback from collector to base is a maximum when the crystal impedance is minimum (in series-resonant mode). The coupling capacitor C_C has negligible impedance at the circuit operating frequency but blocks any dc between collector and base.

The resulting circuit frequency of oscillation is set, then, by the series-resonant frequency of the crystal. Changes in supply voltage, transistor device parameters, and so on have no effect on the circuit operating frequency, which is held stabilized by the crystal. The circuit frequency stability is set by the crystal frequency stability, which is good.

Parallel-Resonant Circuits

Since the parallel-resonant impedance of a crystal is a maximum value, it is connected in shunt. At the parallel-resonant operating frequency, a crystal appears as an inductive reactance of largest value. Figure 17.34 shows a crystal connected as the induc-

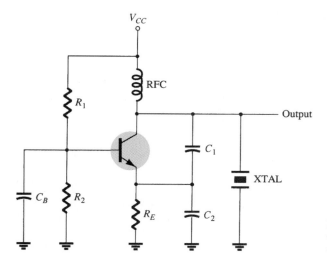

Figure 17.34 Crystal-controlled oscillator operating in parallel-resonant mode.

tor element in a modified Colpitts circuit. The basic dc bias circuit should be evident. Maximum voltage is developed across the crystal at its parallel-resonant frequency. The voltage is coupled to the emitter by a capacitor voltage divider—capacitors C_1 and C_2.

A *Miller crystal-controlled oscillator* circuit is shown in Fig. 17.35. A tuned *LC* circuit in the drain section is adjusted near the crystal parallel-resonant frequency. The maximum gate–source signal occurs at the crystal antiresonant frequency controlling the circuit operating frequency.

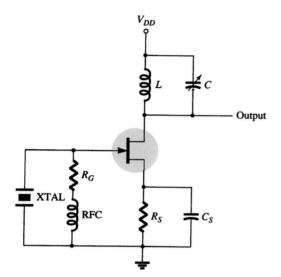

Figure 17.35 Miller crystal-controlled oscillator.

Crystal Oscillator

An op-amp can be used in a crystal oscillator as shown in Fig. 17.36. The crystal is connected in the series-resonant path and operates at the crystal series-resonant frequency. The present circuit has a high gain so that an output square-wave signal results as shown in the figure. A pair of Zener diodes is shown at the output to provide output amplitude at exactly the Zener voltage (V_Z).

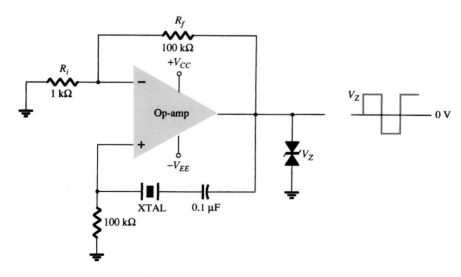

Figure 17.36 Crystal oscillator using op-amp.

17.10 UNIJUNCTION OSCILLATOR

A particular device, the unijunction transistor can be used in a single-stage oscillator circuit to provide a pulse signal suitable for digital-circuit applications. The unijunction transistor can be used in what is called a *relaxation oscillator* as shown by the basic circuit of Fig. 17.37. Resistor R_T and capacitor C_T are the timing components that set the circuit oscillating rate. The oscillating frequency may be calculated using Eq. (17.48), which includes the unijunction transistor *intrinsic stand-off ratio η* as a factor (in addition to R_T and C_T) in the oscillator operating frequency.

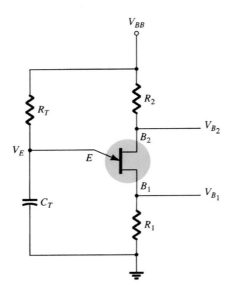

Figure 17.37 Basic unijunction oscillator circuit.

$$f_o \cong \frac{1}{R_T C_T \ln[1/(1 - \eta)]}$$

(17.48)

Typically, a unijunction transistor has a stand-off ratio from 0.4 to 0.6. Using a value of $\eta = 0.5$, we get

$$f_o \cong \frac{1}{R_T C_T \ln[1/(1 - 0.5)]} = \frac{1.44}{R_T C_T \ln 2} = \frac{1.44}{R_T C_T}$$

$$\cong \frac{1.5}{R_T C_T}$$

(17.49)

Capacitor C_T is charged through resistor R_T toward supply voltage V_{BB}. As long as the capacitor voltage V_E is below a stand-off voltage (V_P) set by the voltage across $B_1 - B_2$ and the transistor stand-off ratio η

$$V_P = \eta V_{B_1} V_{B_2} - V_D$$

(17.50)

the unijunction emitter lead appears as an open circuit. When the emitter voltage across capacitor C_T exceeds this value (V_P), the unijunction circuit fires, discharging the capacitor, after which a new charge cycle begins. When the unijunction fires, a voltage rise is developed across R_1 and a voltage drop is developed across R_2 as shown in Fig. 17.38. The signal at the emitter is a sawtooth voltage waveform that at base 1 is a positive-going pulse and at base 2 is a negative-going pulse. A few circuit variations of the unijunction oscillator are provided in Fig. 17.39.

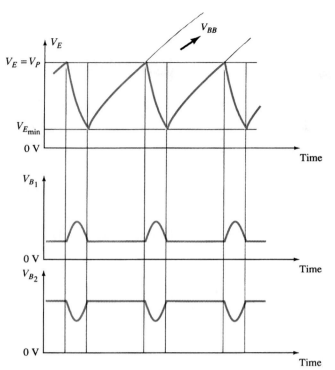

Figure 17.38   Unijunction oscillator waveforms.

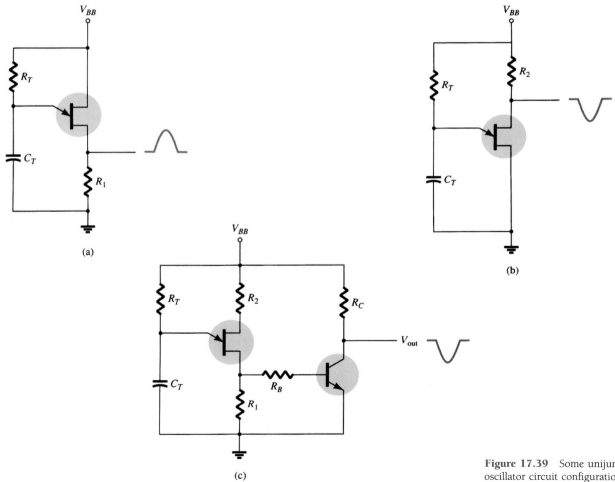

Figure 17.39   Some unijunction oscillator circuit configurations.

17.11   SUMMARY

Equations

Voltage-series feedback:

$$A_f = \frac{V_o}{V_s} = \frac{A}{1 + \beta A}$$

$$Z_{if} = \frac{V_s}{I_i} = Z_i + (\beta A)Z_i = Z_i(1 + \beta A)$$

$$Z_{of} = \frac{V}{I} = \frac{Z_o}{(1 + \beta A)}$$

Voltage-shunt feedback:

$$A_f = \frac{A}{1 + \beta A}$$

$$Z_{if} = \frac{Z_i}{(1 + \beta A)}$$

Current-series feedback:

$$Z_{if} = \frac{V}{I} = Z_i(1 + \beta A)$$

$$Z_{of} = \frac{V}{I} = Z_o(1 + \beta A)$$

Current shunt feedback:

$$Z_{if} = \frac{Z_i}{(1 + \beta A)}$$

$$Z_{of} = \frac{V}{I} = Z_o(1 + \beta A)$$

Phase-shift oscillator:

$$f = \frac{1}{2\pi RC\sqrt{6}}$$

$$\beta = \frac{1}{29}$$

Wien bridge oscillator:

$$f_o = \frac{1}{2\pi\sqrt{R_1 C_1 R_2 C_2}}$$

Colpitts oscillator:

$$f_o = \frac{1}{2\pi\sqrt{L C_{\text{eq}}}}$$

where

$$C_{\text{eq}} = \frac{C_1 C_2}{C_1 + C_2}$$

Hartley oscillator:

$$f_o = \frac{1}{2\pi\sqrt{L_{\text{eq}} C}}$$

where

$$L_{\text{eq}} = L_1 + L_2 + 2M$$

Unijunction oscillator:

$$f_o \cong \frac{1}{R_T C_T \ln[1/(1-\eta)]}$$

17.12 COMPUTER ANALYSIS

Electronics Workbench

Example 17.10—IC Phase-Shift Oscillator

Using EWB, a phase-shift oscillator is drawn as shown in Fig. 17.40. The diode network helps the circuit go into self-oscillation, with the output frequency calculated using

$$f_o = 1/(2\pi\sqrt{6}RC)$$
$$= 1/[2\pi\sqrt{6}(20 \times 10^3)(0.001 \times 10^{-6})] = 3{,}248.7 \text{ Hz}$$

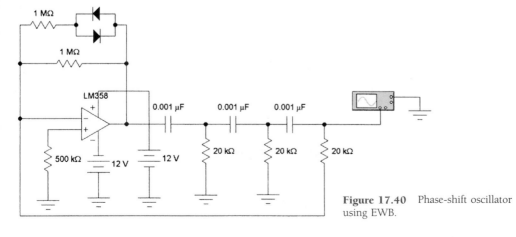

Figure 17.40 Phase-shift oscillator using EWB.

The oscilloscope waveform in Fig. 17.41 shows a cycle in about 3 divisions. The measured frequency for the scope set at 0.1 ms/div is

$$f_{\text{measured}} = 1/(3 \text{ div} \times 0.1 \text{ ms/div}) = 3{,}333 \text{ Hz}$$

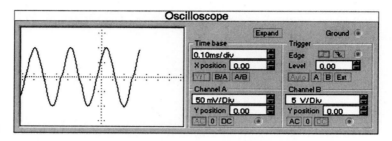

Figure 17.41 Oscilloscope waveform.

Example 17.11—IC Wien Bridge Oscillator

Using EWB, an IC Wien bridge oscillator is constructed as shown in Fig. 17.42. The oscillator frequency is calculated using

$$f_o = 1/(2\pi\sqrt{R_1 C_1 R_2 C_2})$$

which, for $R_1 = R_2 = R$ and $C_1 = C_2 = C$, is

$$f_o = 1/(2\pi\sqrt{RC})$$

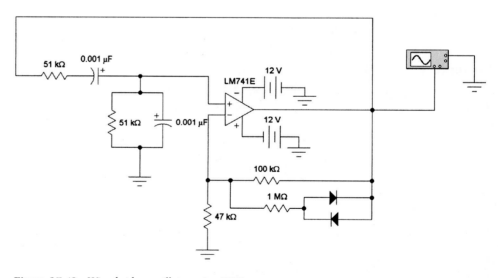

Figure 17.42 Wien bridge oscillator using EWB.

Example 17.12—IC Colpitts Oscillator

Using EWB, a Colpitts oscillator is constructed as shown in Fig. 17.43. The oscillator frequency for this circuit is then

$$f_o = 1/(2\pi\sqrt{LC_{\text{eq}}})$$

Chapter 17 Feedback and Oscillator Circuits

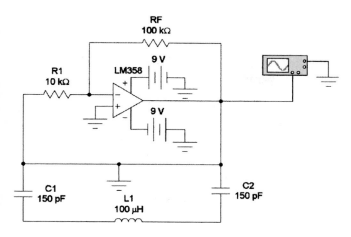

Figure 17.43 IC Colpitts oscillator using EWB.

Example 17.13—Crystal Oscillator

Using EWB, a crystal oscillator circuit is drawn as shown in Fig. 17.44a. The oscillator frequency is set by that of the crystal. The waveform in Fig. 17.44b shows the period to be about

$$1.8 \text{ div} \times 5 \ \mu s/\text{div} = 9 \ \mu s$$

The frequency is then

$$f = 1/T = 1/9 \ \mu s = 111.1 \text{ kHz}$$

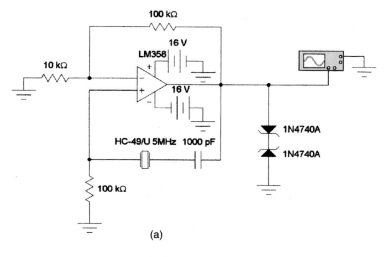

(a)

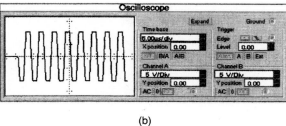

(b)

Figure 17.44 Crystal oscillator using EWB.

A_f

MATHCAD

Example 17.1(a) shows how the calculations of Example 17.1(a) are obtained using Mathcad. The calculations are done for a beta of -0.1.

Example 17.1(a)–Voltage Series Feedback

$$A := -100 \qquad Ri := 10 \cdot 10^3 \qquad Ro := 20 \cdot 10^3 \qquad beta := -0.1$$

$$Af := \frac{A}{(1 + beta \cdot A)} \qquad Af = -9.091$$

$$Zi := Ri$$
$$Zif := Zi \cdot (1 + beta \cdot A) \qquad Zif = 1.1 \bullet 10^5$$

$$Zo := Ro$$
$$Zof := \frac{Zo}{(1 + beta \cdot A)} \qquad Zof = 1.818 \bullet 10^3$$

Example 17.1(b) shows the same calculations for beta $= -0.5$ as in Example 17.1(b).

Example 17.1(b)–Voltage Series Feedback

$$A := -100 \qquad Ri := 10 \cdot 10^3 \qquad Ro := 20 \cdot 10^3 \qquad beta := -0.5$$

$$Af := \frac{A}{(1 + beta \cdot A)} \qquad Af = -1.961$$

$$Zi := Ri$$
$$Zif := Zi \cdot (1 + beta \cdot A) \qquad Zif = 5.1 \bullet 10^5$$

$$Zo := Ro$$
$$Zof := \frac{Zo}{(1 + beta \cdot A)} \qquad Zof = 392.157$$

Example 17.3 provides the same calculations as in Example 17.3.

Example 17.3 — Voltage Series Feedback

$$R1 := 80 \cdot 10^3 \qquad R2 := 20 \cdot 10^3 \qquad Ro := 10 \cdot 10^3 \qquad RD := 10 \cdot 10^3 \qquad gm := 4000 \cdot 10^{-6}$$

$$RL := Ro \frac{RD}{(Ro + RD)} \qquad RL = 5 \cdot 10^3$$

$$A := -gm \cdot RL \qquad A = -20$$

$$beta := \frac{-R2}{(R1 + R2)} \qquad beta = -0.2$$

$$Af := \frac{A}{(1 + beta \cdot A)} \qquad Af = -4$$

Chapter 17 Feedback and Oscillator Circuits

Example 17.5 provides the same calculations as in Example 17.5.

Example 17.5 – Current Series Feedback.

$RC := 2.2 \cdot 10^3$ $\qquad$ $RE := 510$ $\qquad$ $hfe := 120$ $\qquad$ $hie := 900$

$A := \dfrac{-hfe}{(hie + RE)}$ $\qquad$ $A = -0.085$

$beta := -RE$ $\qquad$ $beta = -510$

$1 + beta \cdot A = 44.404$

$Af := \dfrac{A}{(1 + beta \cdot A)}$ $\qquad$ $Af = -1.917 \cdot 10^{-3}$

$Avf := Af \cdot RC$ $\qquad$ $Avf = -4.217$

$re := \dfrac{hie}{hfe}$ $\qquad$ $re = 7.5$

$Av := \dfrac{-RC}{re}$ $\qquad$ $Av = -293.333$

§ 17.2 Feedback Connection Types

1. Calculate the gain of a negative-feedback amplifier having $A = -2000$ and $\beta = -1/10$.

2. If the gain of an amplifier changes from a value of -1000 by 10%, calculate the gain change if the amplifier is used in a feedback circuit having $\beta = -1/20$.

3. Calculate the gain, input, and output impedances of a voltage-series feedback amplifier having $A = -300$, $R_i = 1.5$ kΩ, $R_o = 50$ kΩ, and $\beta = -1/15$.

§ 17.3 Practical Feedback Circuits

*4. Calculate the gain with and without feedback for an FET amplifier as in Fig. 17.7 for circuit values $R_1 = 800$ kΩ, $R_2 = 200$ Ω, $R_o = 40$ kΩ, $R_D = 8$ kΩ, and $g_m = 5000$ μS.

5. For a circuit as in Fig. 17.11 and the following circuit values, calculate the circuit gain and the input and output impedances with and without feedback: $R_B = 600$ kΩ, $R_E = 1.2$ kΩ, $R_C = 4.7$ kΩ, and $\beta = 75$. Use $V_{CC} = 16$ V.

§ 17.6 Phase-Shift Oscillator

6. An FET phase-shift oscillator having $g_m = 6000$ μS, $r_d = 36$ kΩ, and feedback resistor $R = 12$ kΩ is to operate at 2.5 kHz. Select C for specified oscillator operation.

7. Calculate the operating frequency of a BJT phase-shift oscillator as in Fig. 17.21b for $R = 6$ kΩ, $C = 1500$ pF, and $R_C = 18$ kΩ.

§ 17.7 Wien Bridge Oscillator

8. Calculate the frequency of a Wien bridge oscillator circuit (as in Fig. 17.23) when $R = 10$ kΩ and $C = 2400$ pF.

PROBLEMS

§ 17.8 Tuned Oscillator Circuit

9. For an FET Colpitts oscillator as in Fig. 17.26 and the following circuit values determine the circuit oscillation frequency: $C_1 = 750$ pF, $C_2 = 2500$ pF, and $L = 40$ μH.

10. For the transistor Colpitts oscillator of Fig. 17.27 and the following circuit values, calculate the oscillation frequency: $L = 100$ μH, $L_{RFC} = 0.5$ mH, $C_1 = 0.005$ μF, $C_2 = 0.01$ μF, and $C_C = 10$ μF.

11. Calculate the oscillator frequency for an FET Hartley oscillator as in Fig. 17.29 for the following circuit values: $C = 250$ pF, $L_1 = 1.5$ mH, $L_2 = 1.5$ mH, and $M = 0.5$ mH.

12. Calculate the oscillation frequency for the transistor Hartley circuit of Fig. 17.30 and the following circuit values: $L_{RFC} = 0.5$ mH, $L_1 = 750$ μH, $L_2 = 750$ μH, $M = 150$ μH, and $C = 150$ pF.

§ 17.9 Crystal Oscillator

13. Draw circuit diagrams of (a) a series-operated crystal oscillator and (b) a shunt-excited crystal oscillator.

§ 17.10 Unijunction Oscillator

14. Design a unijunction oscillator circuit for operation at (a) 1 kHz and (b) 150 kHz.

*Please Note: Asterisks indicate more difficult problems.

Power Supplies (Voltage Regulators)

18

18.1 INTRODUCTION

The present chapter introduces the operation of power supply circuits built using filters, rectifiers, and then voltage regulators. (Refer to Chapter 2 for the initial description of diode rectifier circuits.) Starting with an ac voltage, a steady dc voltage is obtained by rectifying the ac voltage, then filtering to a dc level and, finally, regulating to obtain a desired fixed dc voltage. The regulation is usually obtained from an IC voltage regulator unit, which takes a dc voltage and provides a somewhat lower dc voltage, which remains the same even if the input dc voltage varies or the output load connected to the dc voltage changes.

A block diagram containing the parts of a typical power supply and the voltage at various points in the unit is shown in Fig. 18.1. The ac voltage, typically 120 V rms, is connected to a transformer, which steps that ac voltage down to the level for the desired dc output. A diode rectifier then provides a full-wave rectified voltage that is initially filtered by a simple capacitor filter to produce a dc voltage. This resulting dc voltage usually has some ripple or ac voltage variation. A regulator circuit can use this dc input to provide a dc voltage that not only has much less ripple voltage but also remains the same dc value even if the input dc voltage varies somewhat or the load connected to the output dc voltage changes. This voltage regulation is usually obtained using one of a number of popular voltage regulator IC units.

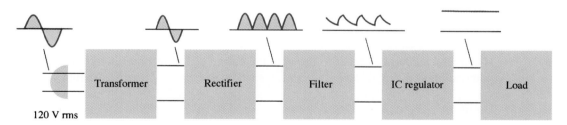

Figure 18.1 Block diagram showing parts of a power supply.

18.2 GENERAL FILTER CONSIDERATIONS

A rectifier circuit is necessary to convert a signal having zero average value into one that has a nonzero average. The output resulting from a rectifier is a pulsating dc voltage and not yet suitable as a battery replacement. Such a voltage could be used in, say, a battery charger, where the average dc voltage is large enough to provide a charging current for the battery. For dc supply voltages, as those used in a radio, stereo system, computer, and so on, the pulsating dc voltage from a rectifier is not good enough. A filter circuit is necessary to provide a steadier dc voltage.

Filter Voltage Regulation and Ripple Voltage

Before going into the details of a filter circuit, it would be appropriate to consider the usual methods of rating filter circuits so that we can compare a circuit's effectiveness as a filter. Figure 18.2 shows a typical filter output voltage, which will be used to define some of the signal factors. The filtered output of Fig. 18.2 has a dc value and some ac variation (ripple). Although a battery has essentially a constant or dc output voltage, the dc voltage derived from an ac source signal by rectifying and filtering will have some ac variation (ripple). The smaller the ac variation with respect to the dc level, the better the filter circuit's operation.

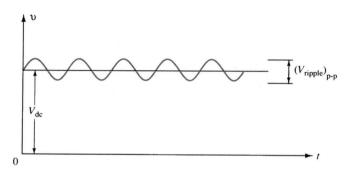

Figure 18.2 Filter voltage waveform showing dc and ripple voltages.

Consider measuring the output voltage of a filter circuit using a dc voltmeter and an ac (rms) voltmeter. The dc voltmeter will read only the average or dc level of the output voltage. The ac (rms) meter will read only the rms value of the ac component of the output voltage (assuming the ac signal is coupled through a capacitor to block out the dc level).

Definition: Ripple

$$r = \frac{\text{ripple voltage (rms)}}{\text{dc voltage}} = \frac{V_r(\text{rms})}{V_{dc}} \times 100\% \qquad (18.1)$$

EXAMPLE 18.1

Using a dc and ac voltmeter to measure the output signal from a filter circuit, we obtain readings of 25 V dc and 1.5 V rms. Calculate the ripple of the filter output voltage.

Solution

$$r = \frac{V_r(\text{rms})}{V_{dc}} \times 100\% = \frac{1.5\ \text{V}}{25\ \text{V}} \times 100\% = \textbf{6\%}$$

VOLTAGE REGULATION

Another factor of importance in a power supply is the amount the dc output voltage changes over a range of circuit operation. The voltage provided at the output under no-load condition (no current drawn from the supply) is reduced when load current is drawn from the supply (under load). The amount the dc voltage changes between the no-load and load conditions is described by a factor called voltage regulation.

Definition: Voltage regulation

$$\text{Voltage regulation} = \frac{\text{no-load voltage} - \text{full-load voltage}}{\text{full-load voltage}}$$

$$\boxed{\%\,\text{V.R.} = \frac{V_{NL} - V_{FL}}{V_{FL}} \times 100\%} \tag{18.2}$$

A dc voltage supply provides 60 V when the output is unloaded. When connected to a load, the output drops to 56 V. Calculate the value of voltage regulation.

EXAMPLE 18.2

Solution

Eq. (18.2): $\%\,\text{V.R.} = \dfrac{V_{NL} - V_{FL}}{V_{FL}} \times 100\% = \dfrac{60\ \text{V} - 56\ \text{V}}{56\ \text{V}} \times 100\% = \mathbf{7.1\%}$

If the value of full-load voltage is the same as the no-load voltage, the voltage regulation calculated is 0%, which is the best expected. This means that the supply is a perfect voltage source for which the output voltage is independent of the current drawn from the supply. The smaller the voltage regulation, the better the operation of the voltage supply circuit.

RIPPLE FACTOR OF RECTIFIED SIGNAL

Although the rectified voltage is not a filtered voltage, it nevertheless contains a dc component and a ripple component. We will see that the full-wave rectified signal has a larger dc component and less ripple than the half-wave rectified voltage.

Half-wave: For a half-wave rectified signal, the output dc voltage is

$$V_{dc} = 0.318 V_m \tag{18.3}$$

The rms value of the ac component of the output signal can be calculated (see Appendix C) to be

$$V_r(\text{rms}) = 0.385 V_m \tag{18.4}$$

The percent ripple of a half-wave rectified signal can then be calculated as

$$r = \frac{V_r(\text{rms})}{V_{dc}} \times 100\% = \frac{0.385 V_m}{0.318 V_m} \times 100\% = 121\% \tag{18.5}$$

Full-wave: For a full-wave rectified voltage the dc value is

$$V_{dc} = 0.636 V_m \tag{18.6}$$

The rms value of the ac component of the output signal can be calculated (see Appendix C) to be

$$V_r(\text{rms}) = 0.308 V_m \tag{18.7}$$

The percent ripple of a full-wave rectified signal can then be calculated as

$$r = \frac{V_r(\text{rms})}{V_{dc}} \times 100\% = \frac{0.308 V_m}{0.636 V_m} \times 100\% = 48\% \tag{18.8}$$

In summary, a full-wave rectified signal has less ripple than a half-wave rectified signal and is thus better to apply to a filter.

18.3 CAPACITOR FILTER

A very popular filter circuit is the capacitor-filter circuit shown in Fig. 18.3. A capacitor is connected at the rectifier output, and a dc voltage is obtained across the capacitor. Figure 18.4a shows the output voltage of a full-wave rectifier before the signal is filtered, while Fig. 18.4b shows the resulting waveform after the filter capacitor is connected at the rectifier output. Notice that the filtered waveform is essentially a dc voltage with some ripple (or ac variation).

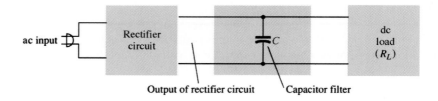

Figure 18.3 Simple capacitor filter.

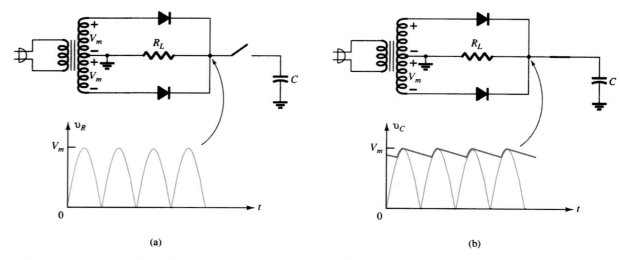

(a)   (b)

Figure 18.4 Capacitor filter operation: (a) full-wave rectifier voltage; (b) filtered output voltage.

Figure 18.5a shows a full-wave bridge rectifier and the output waveform obtained from the circuit when connected to a load (R_L). If no load were connected across the capacitor, the output waveform would ideally be a constant dc level equal in value to the peak voltage (V_m) from the rectifier circuit. However, the purpose of obtaining a

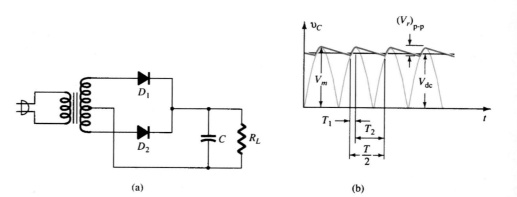

Figure 18.5 Capacitor filter: (a) capacitor filter circuit; (b) output voltage waveform.

(a)   (b) dc voltage is to provide this voltage for use by various electronic circuits, which then constitute a load on the voltage supply. Since there will always be a load on the filter output, we must consider this practical case in our discussion.

Output Waveform Times

Figure 18.5b shows the waveform across a capacitor filter. Time T_1 is the time during which diodes of the full-wave rectifier conduct, charging the capacitor up to the peak rectifier voltage, V_m. Time T_2 is the time interval during which the rectifier voltage drops below the peak voltage, and the capacitor discharges through the load. Since the charge–discharge cycle occurs for each half-cycle for a full-wave rectifier, the period of the rectified waveform is $T/2$, one-half the input signal frequency. The filtered voltage, as shown in Fig. 18.6, shows the output waveform to have a dc level V_{dc} and a ripple voltage V_r (rms) as the capacitor charges and discharges. Some details of these waveforms and the circuit elements are considered next.

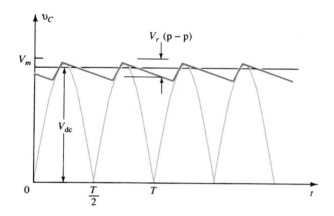

Figure 18.6 Approximate output voltage of capacitor filter circuit.

RIPPLE VOLTAGE, V_r (RMS)

Appendix C provides the details for determining the value of the ripple voltage in terms of the other circuit parameters. The ripple voltage can be calculated from

$$V_r(\text{rms}) = \frac{I_{dc}}{4\sqrt{3}\,fC} = \frac{2.4I_{dc}}{C} = \frac{2.4V_{dc}}{R_L C} \qquad (18.9)$$

where I_{dc} is in milliamperes, C is in microfarads, and R_L is in kilohms.

Calculate the ripple voltage of a full-wave rectifier with a 100-μF filter capacitor connected to a load drawing 50 mA.

EXAMPLE 18.3

Solution

$$\text{Eq. (18.9):} \quad V_r(\text{rms}) = \frac{2.4(50)}{100} = \textbf{1.2 V}$$

DC VOLTAGE, V_{dc}

From Appendix C, we can express the dc value of the waveform across the filter capacitor as

$$V_{dc} = V_m - \frac{I_{dc}}{4fC} = V_m - \frac{4.17I_{dc}}{C} \qquad (18.10)$$

where V_m is the peak rectifier voltage, I_{dc} is the load current in milliamperes, and C is the filter capacitor in microfarads.

EXAMPLE 18.4

If the peak rectified voltage for the filter circuit of Example 18.3 is 30 V, calculate the filter dc voltage.

Solution

Eq. (18.10): $V_{dc} = V_m - \dfrac{4.17I_{dc}}{C} = 30 - \dfrac{4.17(50)}{100} = \mathbf{27.9\ V}$

Filter Capacitor Ripple

Using the definition of ripple [Eq. (18.1)], Eq. (18.9), and Eq. (18.10), with $V_{dc} \approx V_m$, we can obtain the expression for the output waveform ripple of a full-wave rectifier and filter-capacitor circuit.

$$r = \frac{V_r(\text{rms})}{V_{dc}} \times 100\% = \frac{2.4\,I_{dc}}{CV_{dc}} \times 100\% = \frac{2.4}{R_L C} \times 100\% \qquad (18.11)$$

where I_{dc} is in milliamperes, C is in microfarads, V_{dc} is in volts, and R_L is in kilohms.

EXAMPLE 18.5

Calculate the ripple of a capacitor filter for a peak rectified voltage of 30 V, capacitor $C = 50\ \mu\text{F}$, and a load current of 50 mA.

Solution

Eq. (18.11): $r = \dfrac{2.4\,I_{dc}}{CV_{dc}} \times 100\% = \dfrac{2.4(50)}{100(27.9)} \times 100\% = \mathbf{4.3\%}$

We could also calculate the ripple using the basic definition

$$r = \frac{V_r(\text{rms})}{V_{dc}} \times 100\% = \frac{1.2\ \text{V}}{27.9\ \text{V}} \times 100\% = \mathbf{4.3\%}$$

Diode Conduction Period and Peak Diode Current

From the previous discussion, it should be clear that larger values of capacitance provide less ripple and higher average voltage, thereby providing better filter action. From this one might conclude that to improve the performance of a capacitor filter it is only necessary to increase the size of the filter capacitor. The capacitor, however, also affects the peak current drawn through the rectifying diodes, and as will be shown next, the larger the value of the capacitor, the larger the peak current drawn through the rectifying diodes.

Recall that the diodes conduct during period T_1 (see Fig. 18.5), during which time the diode must provide the necessary average current to charge the capacitor. The shorter this time interval, the larger the amount of the charging current. Figure 18.7 shows this relation for a half-wave rectified signal (it would be the same basic oper-

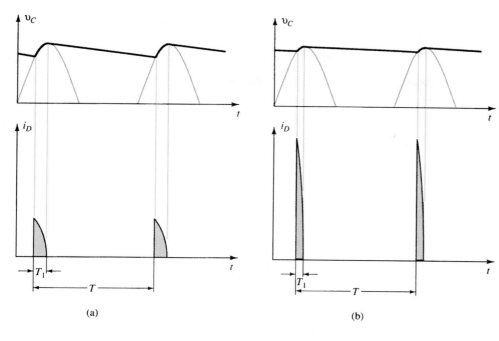

Figure 18.7 Output voltage and diode current waveforms: (a) small C; (b) large C.

ation for full-wave). Notice that for smaller values of capacitor, with T_1 larger, the peak diode current is less than for larger values of filter capacitor.

Since the average current drawn from the supply must equal the average diode current during the charging period, the following relation can be used (assuming constant diode current during charge time):

$$I_{dc} = \frac{T_1}{T} I_{peak}$$

from which we obtain

$$I_{peak} = \frac{T}{T_1} I_{dc} \qquad (18.12)$$

where T_1 = diode conduction time
$T = 1/f$ ($f = 2 \times 60$ for full-wave)
I_{dc} = average current drawn from filter
I_{peak} = peak current through conducting diodes

18.4   RC FILTER

It is possible to further reduce the amount of ripple across a filter capacitor by using an additional RC filter section as shown in Fig. 18.8. The purpose of the added RC section is to pass most of the dc component while attenuating (reducing) as much of

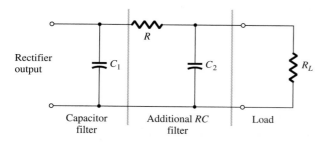

Capacitor
filter

Additional RC
filter

Load

Figure 18.8 RC filter stage.

the ac component as possible. Figure 18.9 shows a full-wave rectifier with capacitor filter followed by an *RC* filter section. The operation of the filter circuit can be analyzed using superposition for the dc and ac components of signal.

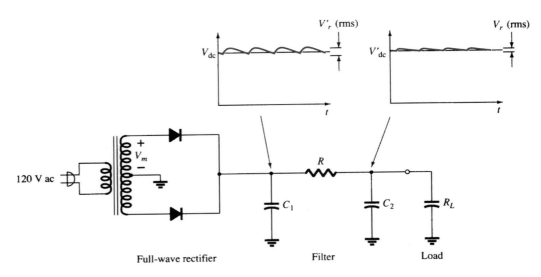

Figure 18.9 Full-wave rectifier and *RC* filter circuit.

DC Operation of *RC* Filter Section

Figure 18.10a shows the dc equivalent circuit to use in analyzing the *RC* filter circuit of Fig. 18.9. Since both capacitors are open-circuit for dc operation, the resulting output dc voltage is

$$V'_{dc} = \frac{R_L}{R + R_L} V_{dc}$$

(18.13)

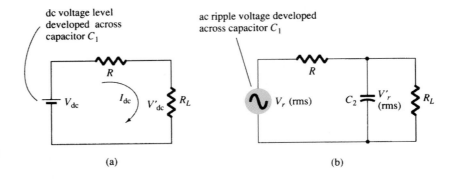

Figure 18.10 (a) DC and (b) ac equivalent circuits of *RC* filter.

(a) (b)

EXAMPLE 18.6

Calculate the dc voltage across a 1-kΩ load for an *RC* filter section ($R = 120\ \Omega$, $C = 10\ \mu F$). The dc voltage across the initial filter capacitor is $V_{dc} = 60$ V.

Solution

Eq. (18.13): $V'_{dc} = \dfrac{R_L}{R + R_L} V_{dc} = \dfrac{1000}{120 + 1000} (60\ \text{V}) = \mathbf{53.6\ V}$

Chapter 18 **Power Supplies (Voltage Regulators)**

AC Operation of RC Filter Section

Figure 18.10b shows the ac equivalent circuit of the RC filter section. Due to the voltage-divider action of the capacitor ac impedance and the load resistor, the ac component of voltage resulting across the load is

$$V'_r(\text{rms}) \approx \frac{X_C}{R}V_r(\text{rms}) \qquad (18.14)$$

For a full-wave rectifier with ac ripple at 120 Hz, the impedance of a capacitor can be calculated using

$$X_C = \frac{1.3}{C} \qquad (18.15)$$

where C is in microfarads and X_C is in kilohms.

Calculate the dc and ac components of the output signal across load R_L in the circuit of Fig. 18.11. Calculate the ripple of the output waveform.

EXAMPLE 18.7

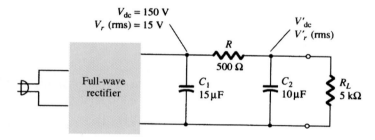

Figure 18.11 RC filter circuit for Example 18.7.

Solution

DC Calculation:

$$\text{Eq. (18.13):}\quad V'_{dc} = \frac{R_L}{R + R_L}V_{dc} = \frac{5\text{ k}\Omega}{500 + 5\text{ k}\Omega}(150\text{ V}) = \mathbf{136.4\ V}$$

AC Calculation:
The RC section capacitive impedance is

$$\text{Eq. (18.15):}\quad X_C = \frac{1.3}{C} = \frac{1.3}{10} = 0.13\text{ k}\Omega = 130\ \Omega$$

The ac component of the output voltage, calculated using Eq. (18.14), is

$$V'_r(\text{rms}) = \frac{X_C}{R}V_r(\text{rms}) = \frac{130}{500}(15\text{ V}) = \mathbf{3.9\ V}$$

The ripple of the output waveform is then

$$r = \frac{V'_r(\text{rms})}{V'_{dc}} \times 100\% = \frac{3.9\text{ V}}{136.4\text{ V}} \times 100\% = \mathbf{2.86\%}$$

18.5 DISCRETE TRANSISTOR VOLTAGE REGULATION

Two types of transistor voltage regulators are the series voltage regulator and the shunt voltage regulator. Each type of circuit can provide an output dc voltage that is regulated or maintained at a set value even if the input voltage varies or if the load connected to the output changes.

Series Voltage Regulation

The basic connection of a series regulator circuit is shown in the block diagram of Fig. 18.12. The series element controls the amount of the input voltage that gets to the output. The output voltage is sampled by a circuit that provides a feedback voltage to be compared to a reference voltage.

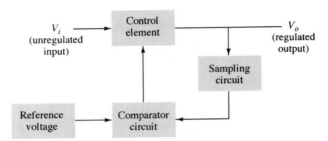

Figure 18.12 Series regulator block diagram.

1. If the output voltage increases, the comparator circuit provides a control signal to cause the series control element to decrease the amount of the output voltage—thereby maintaining the output voltage.

2. If the output voltage decreases, the comparator circuit provides a control signal to cause the series control element to increase the amount of the output voltage.

SERIES REGULATOR CIRCUIT

A simple series regulator circuit is shown in Fig. 18.13. Transistor Q_1 is the series control element, and Zener diode D_Z provides the reference voltage. The regulating operation can be described as follows:

1. If the output voltage decreases, the increased base-emitter voltage causes transistor Q_1 to conduct more, thereby raising the output voltage—maintaining the output constant.

2. If the output voltage increases, the decreased base-emitter voltage causes transistor Q_1 to conduct less, thereby reducing the output voltage—maintaining the output constant.

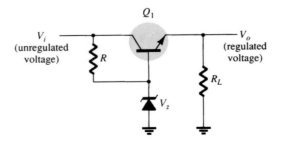

Figure 18.13 Series regulator circuit.

Calculate the output voltage and Zener current in the regulator circuit of Fig. 18.14 for $R_L = 1\ k\Omega$.

EXAMPLE 18.8

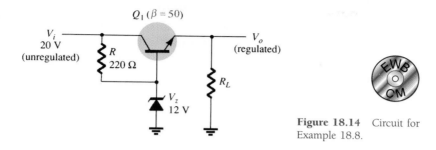

Figure 18.14 Circuit for Example 18.8.

Solution

$$V_o = V_Z - V_{BE} = 12\ V - 0.7\ V = \textbf{11.3 V}$$

$$V_{CE} = V_i - V_o = 20\ V - 11.3\ V = 8.7\ V$$

$$I_R = \frac{20\ V - 12\ V}{220\ \Omega} = \frac{8\ V}{220\ \Omega} = 36.4\ mA$$

For $R_L = 1\ k\Omega$,

$$I_L = \frac{V_o}{R_L} = \frac{11.3\ V}{1\ k\Omega} = 11.3\ mA$$

$$I_B = \frac{I_C}{\beta} = \frac{11.3\ mA}{50} = 226\ \mu A$$

$$I_Z = I_R - I_B = 36.4\ mA - 226\ \mu A \approx \textbf{36 mA}$$

IMPROVED SERIES REGULATOR

An improved series regulator circuit is that of Fig. 18.15. Resistors R_1 and R_2 act as a sampling circuit, Zener diode D_Z providing a reference voltage, and transistor Q_2 then controls the base current to transistor Q_1 to vary the current passed by transistor Q_1 to maintain the output voltage constant.

If the output voltage tries to increase, the increased voltage sampled by R_1 and R_2, increased voltage V_2, causes the base-emitter voltage of transistor Q_2 to go up

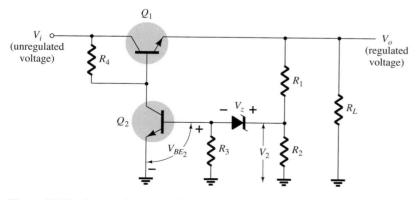

Figure 18.15 Improved series regulator circuit.

18.5 **Discrete Transistor Voltage Regulation**

869

(since V_Z remains fixed). If Q_2 conducts more current, less goes to the base of transistor Q_1, which then passes less current to the load, reducing the output voltage—thereby maintaining the output voltage constant. The opposite takes place if the output voltage tries to decrease, causing less current to be supplied to the load, to keep the voltage from decreasing.

The voltage V_2 provided by sensing resistors R_1 and R_2 must equal the sum of the base-emitter voltage of Q_2 and the Zener diode, that is,

$$V_{BE_2} + V_Z = V_2 = \frac{R_2}{R_1 + R_2} V_o \qquad (18.16)$$

Solving Eq. (18.16) for the regulated output voltage, V_o,

$$\boxed{V_o = \frac{R_1 + R_2}{R_2}(V_Z + V_{BE_2})} \qquad (18.17)$$

EXAMPLE 18.9

What regulated output voltage is provided by the circuit of Fig. 18.15 for the following circuit elements: $R_1 = 20$ kΩ, $R_2 = 30$ kΩ, and $V_Z = 8.3$ V?

Solution

From Eq. (18.17), the regulated output voltage will be

$$V_o = \frac{20 \text{ k}\Omega + 30 \text{ k}\Omega}{30 \text{ k}\Omega}(8.3 \text{ V} + 0.7 \text{ V}) = \textbf{15 V}$$

OP-AMP SERIES REGULATOR

Another version of series regulator is that shown in Fig. 18.16. The op-amp compares the Zener diode reference voltage with the feedback voltage from sensing resistors R_1 and R_2. If the output voltage varies, the conduction of transistor Q_1 is controlled to maintain the output voltage constant. The output voltage will be maintained at a value of

$$\boxed{V_o = \left(1 + \frac{R_1}{R_2}\right)V_Z} \qquad (18.18)$$

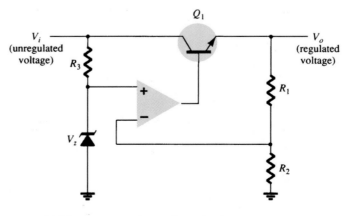

Figure 18.16 Op-amp series regulator circuit.

Chapter 18 Power Supplies (Voltage Regulators)

Calculate the regulated output voltage in the circuit of Fig. 18.17.

EXAMPLE 18.10

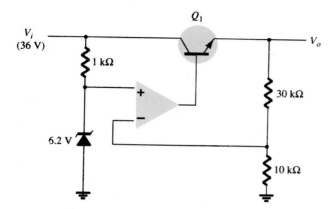

Figure 18.17 Circuit for Example 18.10.

Solution

$$\text{Eq. (18.18):} \quad V_o = \left(1 + \frac{30 \text{ k}\Omega}{10 \text{ k}\Omega}\right) 6.2 \text{ V} = \mathbf{24.8 \text{ V}}$$

CURRENT-LIMITING CIRCUIT

One form of short-circuit or overload protection is current limiting, as shown in Fig. 18.18. As load current I_L increases, the voltage drop across the short-circuit sensing resistor R_{SC} increases. When the voltage drop across R_{SC} becomes large enough, it will drive Q_2 on, diverting current from the base of transistor Q_1, thereby reducing the load current through transistor Q_1, preventing any additional current to load R_L. The action of components R_{SC} and Q_2 provides limiting of the maximum load current.

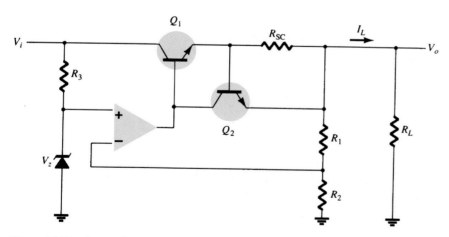

Figure 18.18 Current-limiting voltage regulator.

FOLDBACK LIMITING

Current limiting reduces the load voltage when the current becomes larger than the limiting value. The circuit of Fig. 18.19 provides foldback limiting, which reduces both the output voltage and output current protecting the load from overcurrent, as well as protecting the regulator.

18.5 Discrete Transistor Voltage Regulation

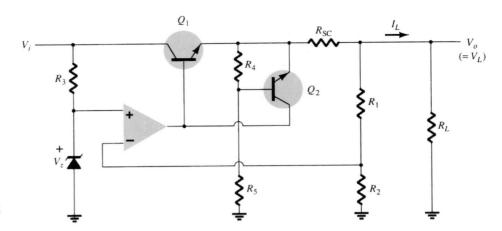

Figure 18.19 Foldback-limiting series regulator circuit.

Foldback limiting is provided by the additional voltage-divider network of R_4 and R_5 in the circuit of Fig. 18.19 (over that of Fig. 18.17). The divider circuit senses the voltage at the output (emitter) of Q_1. When I_L increases to its maximum value, the voltage across R_{SC} becomes large enough to drive Q_2 on, thereby providing current limiting. If the load resistance is made smaller, the voltage driving Q_2 on becomes less, so that I_L drops when V_L also drops in value—this action being foldback limiting. When the load resistance is returned to its rated value, the circuit resumes its voltage regulation action.

Shunt Voltage Regulation

A shunt voltage regulator provides regulation by shunting current away from the load to regulate the output voltage. Figure 18.20 shows the block diagram of such a voltage regulator. The input unregulated voltage provides current to the load. Some of the current is pulled away by the control element to maintain the regulated output voltage across the load. If the load voltage tries to change due to a change in the load, the sampling circuit provides a feedback signal to a comparator, which then provides a control signal to vary the amount of the current shunted away from the load. As the output voltage tries to get larger, for example, the sampling circuit provides a feedback signal to the comparator circuit, which then provides a control signal to draw increased shunt current, providing less load current, thereby keeping the regulated voltage from rising.

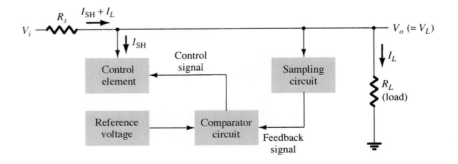

Figure 18.20 Block diagram of shunt voltage regulator.

BASIC TRANSISTOR SHUNT REGULATOR

A simple shunt regulator circuit is shown in Fig. 18.21. Resistor R_S drops the unregulated voltage by an amount that depends on the current supplied to the load, R_L. The voltage across the load is set by the Zener diode and transistor base-emitter volt-

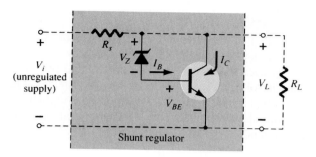

Figure 18.21 Transistor shunt voltage regulator.

age. If the load resistance decreases, a reduced drive current to the base of Q_1 results, shunting less collector current. The load current is thus larger, thereby maintaining the regulated voltage across the load. The output voltage to the load is

$$V_L = V_Z + V_{BE} \qquad (18.19)$$

Determine the regulated voltage and circuit currents for the shunt regulator of Fig. 18.22.

EXAMPLE 18.11

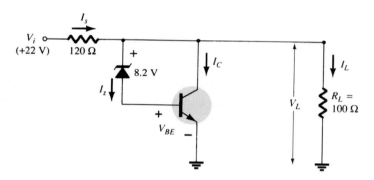

Figure 18.22 Circuit for Example 18.11.

Solution

The load voltage is

Eq. (18.19): $\quad V_L = 8.2 \text{ V} + 0.7 \text{ V} = \textbf{8.9 V}$

For the given load,

$$I_L = \frac{V_L}{R_L} = \frac{8.9 \text{ V}}{100 \text{ }\Omega} = \textbf{89 mA}$$

With the unregulated input voltage at 22 V, the current through R_S is

$$I_S = \frac{V_i - V_L}{R_S} = \frac{22 \text{ V} - 8.9 \text{ V}}{120} = \textbf{109 mA}$$

so that the collector current is

$$I_C = I_S - I_L = 109 \text{ mA} - 89 \text{ mA} = \textbf{20 mA}$$

(The current through the Zener and transistor base–emitter is smaller than I_C by the transistor beta.)

IMPROVED SHUNT REGULATOR

The circuit of Fig. 18.23 shows an improved shunt voltage regulator circuit. The Zener diode provides a reference voltage so that the voltage across R_1 senses the output voltage. As the output voltage tries to change, the current shunted by transistor Q_1 is varied to maintain the output voltage constant. Transistor Q_2 provides a larger base current to transistor Q_1 than the circuit of Fig. 18.21, so that the regulator handles a larger load current. The output voltage is set by the Zener voltage and that across the two transistor base–emitters,

$$V_o = V_L = V_Z + V_{BE_2} + V_{BE_1} \tag{18.20}$$

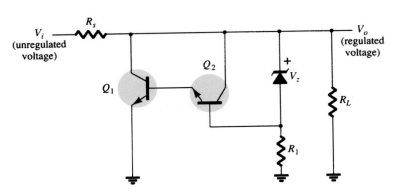

Figure 18.23 Improved shunt voltage regulator circuit.

SHUNT VOLTAGE REGULATOR USING OP-AMP

Figure 18.24 shows another version of a shunt voltage regulator using an op-amp as voltage comparator. The Zener voltage is compared to the feedback voltage obtained from voltage divider R_1 and R_2 to provide the control drive current to shunt element Q_1. The current through resistor R_S is thus controlled to drop a voltage across R_S so that the output voltage is maintained.

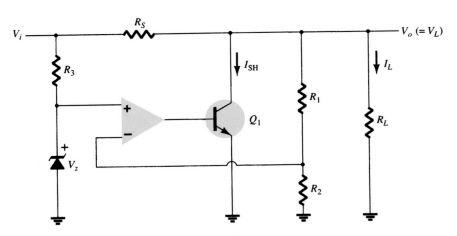

Figure 18.24 Shunt voltage regulator using op-amp.

Switching Regulation

A type of regulator circuit that is quite popular for its efficient transfer of power to the load is the switching regulator. Basically, a switching regulator passes voltage to the load in pulses, which are then filtered to provide a smooth dc voltage. Figure 18.25

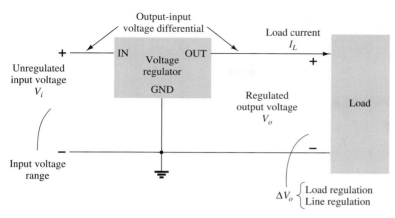

Figure 18.25 Block representation of three-terminal voltage regulator.

shows the basic components of such a voltage regulator. The added circuit complexity is well worth the improved operating efficiency obtained.

18.6 IC VOLTAGE REGULATORS

Voltage regulators comprise a class of widely used ICs. Regulator IC units contain the circuitry for reference source, comparator amplifier, control device, and overload protection all in a single IC. Although the internal construction of the IC is somewhat different from that described for discrete voltage regulator circuits, the external operation is much the same. IC units provide regulation of either a fixed positive voltage, a fixed negative voltage, or an adjustably set voltage.

A power supply can be built using a transformer connected to the ac supply line to step the ac voltage to a desired amplitude, then rectifying that ac voltage, filtering with a capacitor and RC filter, if desired, and finally regulating the dc voltage using an IC regulator. The regulators can be selected for operation with load currents from hundreds of milliamperes to tens of amperes, corresponding to power ratings from milliwatts to tens of watts.

Three-Terminal Voltage Regulators

Figure 18.25 shows the basic connection of a three-terminal voltage regulator IC to a load. The fixed voltage regulator has an unregulated dc input voltage, V_i, applied to one input terminal, a regulated output dc voltage, V_o, from a second terminal, with the third terminal connected to ground. For a selected regulator, IC device specifications list a voltage range over which the input voltage can vary to maintain a regulated output voltage over a range of load current. The specifications also list the amount of output voltage change resulting from a change in load current (load regulation) or in input voltage (line regulation).

Fixed Positive Voltage Regulators

The series 78 regulators provide fixed regulated voltages from 5 to 24 V. Figure 18.26 shows how one such IC, a 7812, is connected to provide voltage regulation with output from this unit of +12 V dc. An unregulated input voltage V_i is filtered by capacitor C_1 and connected to the IC's IN terminal. The IC's OUT terminal provides a regulated +12 V, which is filtered by capacitor C_2 (mostly for any high-frequency noise). The third IC terminal is connected to ground (GND). While the input voltage may vary over some permissible voltage range and the output load may vary over some

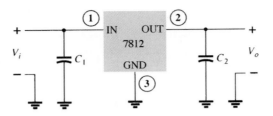

Figure 18.26 Connection of 7812 voltage regulator.

acceptable range, the output voltage remains constant within specified voltage variation limits. These limitations are spelled out in the manufacturer's specification sheets. A table of positive voltage regulator ICs is provided in Table 18.1.

TABLE 18.1 Positive Voltage Regulators in 7800 Series		
IC Part	Output Voltage (V)	Minimum V_i (V)
7805	+5	7.3
7806	+6	8.3
7808	+8	10.5
7810	+10	12.5
7812	+12	14.6
7815	+15	17.7
7818	+18	21.0
7824	+24	27.1

The connection of a 7812 in a complete voltage supply is shown in the connection of Fig. 18.27. The ac line voltage (120 V rms) is stepped down to 18 V rms across each half of the center-tapped transformer. A full-wave rectifier and capacitor filter then provides an unregulated dc voltage, shown as a dc voltage of about 22 V, with ac ripple of a few volts as input to the voltage regulator. The 7812 IC then provides an output that is a regulated +12 V dc.

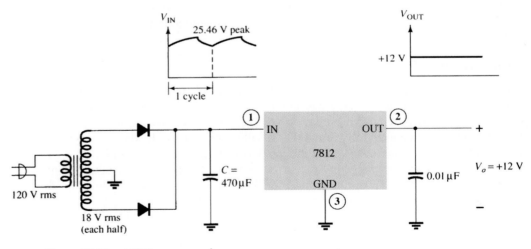

Figure 18.27 +12 V power supply.

POSITIVE VOLTAGE REGULATOR SPECIFICATIONS

The specifications sheet of voltage regulators is typified by that shown in Fig. 18.28 for the group of series 7800 positive voltage regulators. Some consideration of a few of the more important parameters should be made.

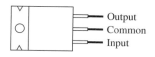
Output
Common
Input

Nominal output voltage	Regulator
5 V	7805
6 V	7806
8 V	7808
10 V	7810
12 V	7812
15 V	7815
18 V	7818
24 V	7824

Absolute maximum ratings:

Input voltage   40 V
Continuous total dissipation 2 W
Operating free-air
   temperature range −65 to 150°C

μA 7812C electrical characteristics:

Parameter	Min.	Typ.	Max.	Units
Output voltage	11.5	12	12.5	V
Input regulation		3	120	mV
Ripple rejection	55	71		dB
Output regulation		4	100	mV
Output resistance		0.018		Ω
Dropout voltage		2.0		V
Short-circuit output current		350		mA
Peak output current		2.2		A

Figure 18.28 Specification sheet data for voltage regulator ICs.

Output voltage: The specification for the 7812 shows that the output voltage is typically +12 V but could be as low as 11.5 V or as high as 12.5 V.

Output regulation: The output voltage regulation is seen to be typically 4 mV, to a maximum of 100 mV (at output currents from 0.25 to 0.75 A). This information specifies that the output voltage can typically vary only 4 mV from the rated 12 V dc.

Short-circuit output current: The amount of current is limited to typically 0.35 A if the output were to be short-circuited (presumably by accident or by another faulty component).

Peak output current: While the rated maximum current is 1.5 A for this series of IC, the typical peak output current that might be drawn by a load is 2.2 A. This shows that although the manufacturer rates the IC as capable of providing 1.5 A, one could draw somewhat more current (possibly for a short period of time).

Dropout voltage: The dropout voltage, typically 2 V, is the minimum amount of voltage across the input–output terminals that must be maintained if the IC is to operate as a regulator. If the input voltage drops too low or the output rises so that at least 2 V is not maintained across the IC input–output, the IC will no longer provide voltage regulation. One therefore maintains an input voltage large enough to assure that the dropout voltage is provided.

Fixed Negative Voltage Regulators

The series 7900 ICs provide negative voltage regulators, similar to those providing positive voltages. A list of negative voltage regulator ICs is provided in Table 18.2. As shown, IC regulators are available for a range of fixed negative voltages, the selected IC providing the rated output voltage as long as the input voltage is maintained greater than the minimum input value. For example, the 7912 provides an output of −12 V as long as the input to the regulator IC is more negative than −14.6 V.

TABLE 18.2 Negative Voltage Regulators in 7900 Series		
IC Part	Output Voltage (V)	Minimum V_i (V)
7905	−5	−7.3
7906	−6	−8.4
7908	−8	−10.5
7909	−9	−11.5
7912	−12	−14.6
7915	−15	−17.7
7918	−18	−20.8
7924	−24	−27.1

EXAMPLE 18.12

Draw a voltage supply using a full-wave bridge rectifier, capacitor filter, and IC regulator to provide an output of +5 V.

Solution

The resulting circuit is shown in Fig. 18.29.

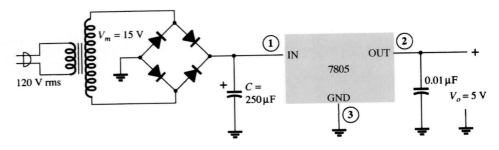

Figure 18.29 +5-V power supply.

EXAMPLE 18.13

For a transformer output of 15 V and a filter capacitor of 250 μF, calculate the minimum input voltage when connected to a load drawing 400 mA.

Solution

The voltages across the filter capacitor are

$$V_r(\text{peak}) = \sqrt{3}\, V_r(\text{rms}) = \sqrt{3}\,\frac{2.4 I_{dc}}{C} = \sqrt{3}\,\frac{2.4(400)}{250} = 6.65 \text{ V}$$

$$V_{dc} = V_m - V_r(\text{peak}) = 15 \text{ V} - 6.65 \text{ V} = 8.35 \text{ V}$$

Since the input swings around this dc level, the minimum input voltage can drop to as low as

$$V_i(\text{low}) = V_{dc} - V_r(\text{peak}) = 15 \text{ V} - 6.65 \text{ V} = \textbf{8.35 V}$$

Since this voltage is greater than the minimum required for the IC regulator (from Table 18.1, $V_i = 7.3$ V), the IC can provide a regulated voltage to the given load.

EXAMPLE 18.14

Determine the maximum value of load current at which regulation is maintained for the circuit of Fig. 18.29.

Chapter 18   Power Supplies (Voltage Regulators)

Solution

To maintain $V_i(\text{min}) \geq 7.3$ V,

$$V_r(\text{peak}) \leq V_m - V_i(\text{min}) = 15 \text{ V} - 7.3 \text{ V} = 7.7 \text{ V}$$

so that

$$V_r(\text{rms}) = \frac{V_r(\text{peak})}{\sqrt{3}} = \frac{7.7 \text{ V}}{1.73} = 4.4 \text{ V}$$

The value of load current is then

$$I_{dc} = \frac{V_r(\text{rms})C}{2.4} = \frac{(4.4 \text{ V})(250)}{2.4} = \textbf{458 mA}$$

Any current above this value is too large for the circuit to maintain the regulator output at +5 V.

Adjustable Voltage Regulators

Voltage regulators are also available in circuit configurations that allow the user to set the output voltage to a desired regulated value. The LM317, for example, can be operated with the output voltage regulated at any setting over the range of voltage from 1.2 to 37 V. Figure 18.30 shows how the regulated output voltage of an LM317 can be set.

Resistors R_1 and R_2 set the output to any desired voltage over the adjustment range (1.2 to 37 V). The output voltage desired can be calculated using

$$V_o = V_{\text{ref}}\left(1 + \frac{R_2}{R_1}\right) + I_{\text{adj}}R_2 \qquad (18.21)$$

with typical IC values of

$$V_{\text{ref}} = 1.25 \text{ V} \qquad \text{and} \qquad I_{\text{adj}} = 100 \ \mu\text{A}$$

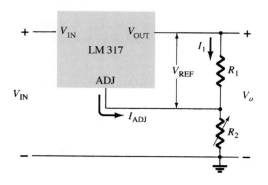

Figure 18.30 Connection of LM317 adjustable-voltage regulator.

Determine the regulated voltage in the circuit of Fig. 18.30 with $R_1 = 240 \ \Omega$ and $R_2 = 2.4$ kΩ. *EXAMPLE 18.15*

Solution

$$\text{Eq. (18.21):} \quad V_o = 1.25 \text{ V}\left(1 + \frac{2.4 \text{ k}\Omega}{240 \ \Omega}\right) + (100 \ \mu\text{A})(2.4 \text{ k}\Omega)$$

$$= 13.75 \text{ V} + 0.24 \text{ V} = \textbf{13.99 V}$$

EXAMPLE 18.16   Determine the regulated output voltage of the circuit in Fig. 18.31.

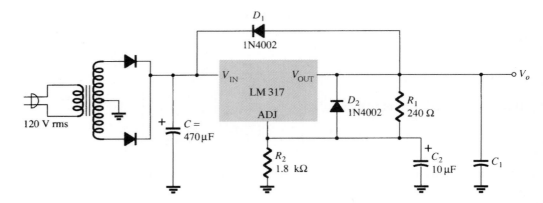

Figure 18.31   Positive adjustable-voltage regulator for Example 18.16.

Solution

The output voltage calculated using Eq. (18.21) is

$$V_o = 1.25 \text{ V}\left(1 + \frac{1.8 \text{ k}\Omega}{240 \text{ }\Omega}\right) + (100 \text{ }\mu\text{A})(1.8 \text{ k}\Omega) \approx \mathbf{10.8 \text{ V}}$$

A check of the filter capacitor voltage shows that an input–output difference of 2 V can be maintained up to at least 200 mA load current.

18.7   PRACTICAL APPLICATIONS

Power Supplies

Power supplies are a part of every electronic device, so a wide variety of circuits are used to accommodate such factors as power rating, size of circuit, cost, desired regulation, etc. This section will outline a number of practical supplies and chargers.

SIMPLE DC SUPPLY

A simple way to drop the ac voltage, without a bulky and expensive transformer, is to use a capacitor in series with the line voltage. This type of supply, shown in Fig. 18.32, uses few parts and is thus very simple. A half-wave rectifier (or bridge rectifier) with a filter circuit is used to get a voltage with a dc component. This circuit has a number of negatives: There is no isolation from the ac line; a minimal current must always be drawn; and the load current cannot be excessive. So the simple

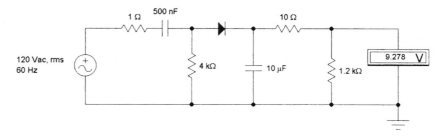

Figure 18.32   Simple dc supply.

dc supply can be used to provide a poorly regulated dc voltage when light current draw is desired in an inexpensive device.

DC SUPPLY WITH TRANSFORMER INPUT

The next type of power supply uses a transformer to step down the ac line voltage. The transformer can be either a wall mount (external) or a chassis mount (internal). A rectifier is used after the transformer, followed by a capacitor filter and possibly a regulator. The regulator becomes a problem as the power requirements increase. Heat sink size, cooling, and power requirements become a major obstacle to these types of supplies.

Figure 18.33 shows a simple half-wave rectified supply with an isolating step-down transformer. This relatively simple circuit provides no regulation.

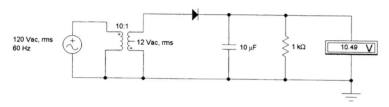

Figure 18.33   DC supply with transformer input.

Figure 18.34 shows probably the best standard power supply—with transformer isolation and voltage step-down; a bridge rectifier; a dual filter with choke; and a regulator circuit made of a Zener reference, a parallel regulation transistor, and an op-amp with feedback to aid the regulation. This circuit obviously provides excellent voltage regulation.

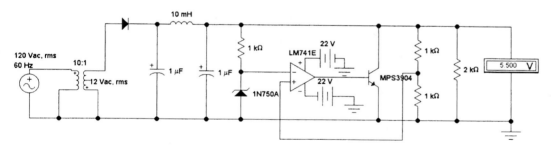

Figure 18.34   Series-regulated supply with transformer input and IC regulation.

CHOPPER SUPPLY

Today's modern power supplies convert ac to dc using a chopper circuit such as that shown in Fig. 18.35. The ac input is connected to the circuit through various line

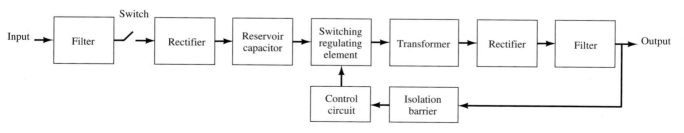

Figure 18.35   Block diagram of chopper power supply.

conditioners and filters. This removes any electrical noise. The input is then rectified and lightly filtered. The high dc voltage is chopped at a rate of approximately 100 kHz. The rate and the duration of the chopping are controlled by a special-function integrated circuit. A torroidal isolation transformer couples the chopped dc to a filtering and rectifying circuit. The output of the power supply is fed back to the control integrated circuit. By monitoring the output, the IC can regulate the output voltage. Although this type of power supply is more complicated, it has many advantages over traditional supplies. Here are a few: It operates over a very large range of input ac voltages; it operates independently of the input frequency; it can be made very small in size; and it operates over a large range of current demands and low heat dissipation.

SPECIAL TV HORIZONTAL HIGH-VOLTAGE SUPPLY

Television sets require a very high dc voltage to operate the picture tube (CRT). In early TV sets this voltage was supplied by a high-voltage transformer with very high voltage rated capacitors. The circuit was very bulky, heavy, and dangerous. TV sets utilize two basic frequencies to scan the screen: 60 Hz (vertical oscillator) and 15 kHz (horizontal oscillator). Using the horizontal oscillator, a high-voltage dc supply can be built. The circuit is known as a *flyback power supply* (see Fig. 18.36). The low dc voltage is pulsed into a small flyback transformer. The flyback transformer is a step-up autotransformer. The output is rectified and filtered with a small-value capacitor. The flyback transformer can be small, and the filter capacitor can be a small, low-value unit, because the frequency is very high. This type of circuit is lightweight and very reliable.

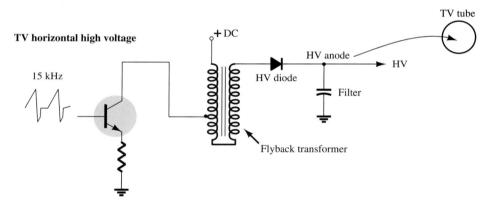

Figure 18.36 TV horizontal high-voltage supply.

BATTERY CHARGER CIRCUITS

Battery charger circuits employ variations of the power supply circuits mentioned above. Figure 18.37a shows the basics of a simple charging circuit using a transformer setting with a selector switch to determine the charge rate current provided. For NiCad batteries the voltage that supplies the battery must be greater than the battery being charged. The current must also be controlled and limited. Figure 18.37b shows a typical NiCad charging circuit. For a lead-acid battery, the voltage must be controlled so as not to exceed the battery's rated voltage. The charge current is determined by the power supply's capability, the power rating of the battery, and the amount of charge required. Figure 18.37c shows a simple lead-acid charging circuit.

Batteries can be charged using traditional dc supplies or from more elaborate chopper supplies. The major problem with charging batteries is determining when the battery is completely charged. Many exotic circuits exist to check the battery status.

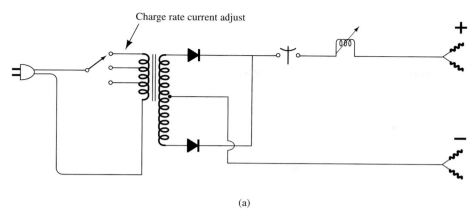

(a)

Simple lead acid battery charger

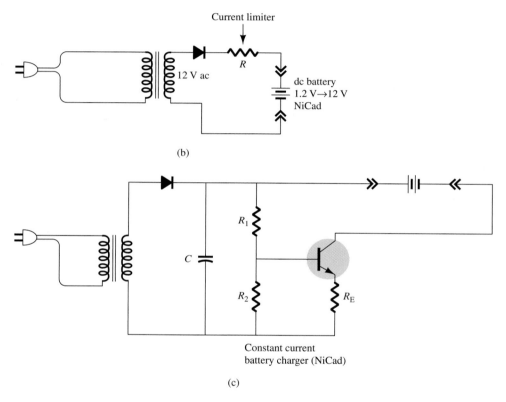

(b)

(c)

Figure 18.37 Battery charger circuits.

18.8   SUMMARY

Equations

Ripple:

$$r = \frac{\text{ripple voltage (rms)}}{\text{dc voltage}} = \frac{V_r(\text{rms})}{V_{dc}} \times 100\%$$

Voltage regulation:

$$\% \text{V.R.} = \frac{V_{NL} - V_{FL}}{V_{FL}} \times 100\%$$

Half-wave rectifier:

$$V_{dc} = 0.318V_m$$

$$V_r(\text{rms}) = 0.385V_m$$

$$r = \frac{0.385V_m}{0.318V_m} \times 100\% = 121\%$$

Full-wave rectifier:

$$V_{dc} = 0.636V_m$$

$$V_r(\text{rms}) = 0.308V_m$$

$$r = \frac{0.308V_m}{0.636V_m} \times 100\% = 48\%$$

Simple capacitor filter:

$$V_r(\text{rms}) = \frac{I_{dc}}{4\sqrt{3}\,fC} = \frac{2.4I_{dc}}{C} = \frac{2.4V_{dc}}{R_LC}$$

$$V_{dc} = V_m - \frac{I_{dc}}{4fC} = \frac{4.17I_{dc}}{C}$$

$$r = \frac{V_r(\text{rms})}{V_{dc}} \times 100\% = \frac{2.4I_{dc}}{CV_{dc}} \times 100\% = \frac{2.4}{R_LC} \times 100\%$$

RC filter:

$$V'_{dc} = \frac{R_L}{R + R_L}V_{dc}$$

$$X_C = \frac{1.3}{C}$$

$$V'_r(\text{rms}) = \frac{X_C}{R}V_r(\text{rms})$$

Op-amp series regulator:

$$V_o = \left(1 + \frac{R_1}{R_2}\right)V_Z$$

18.9   COMPUTER ANALYSIS

Program 18.1—Op-Amp Series Regulator

The op-amp series regulator circuit of Fig. 18.16 can be analyzed using PSpice Windows Design Center, with the resulting schematic drawn as shown in Fig. 18.38. The **Analysis Setup** was used to provide a dc voltage sweep from 8 to 15 V in 0.5-V increments. Diode D_1 provides a Zener voltage of 4.7 V ($V_Z = 4.7$), and transistor Q_1 is set to beta = 100. Using Eq. (18.18),

$$V_o = \left(1 + \frac{R_1}{R_2}\right)V_Z = \left(1 + \frac{1\text{ k}\Omega}{1\text{ k}\Omega}\right)4.7\text{ V} = 9.4\text{ V}$$

Notice in Fig. 18.38 that the regulated output voltage is 9.25 V when the input is 10 V. Figure 18.39 shows the **PROBE** output for the dc voltage sweep. Notice also that after the input goes above about 9 V, the output is held regulated at about 9.3 V.

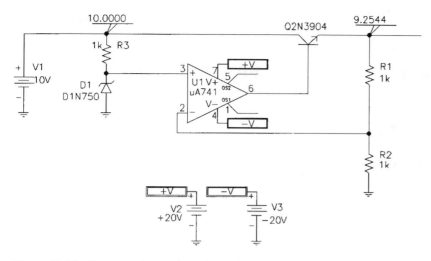

Figure 18.38 Op-amp series regulator drawn using PSpice Design Center.

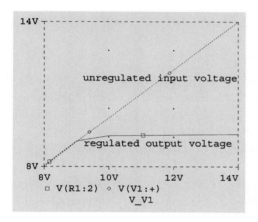

Figure 18.39 Probe output showing the voltage regulation of Fig. 18.38.

Program 18.2—Shunt Voltage Regulator Using Op-Amp

The shunt voltage regulator circuit of Fig. 18.40 was drawn using PSpice Windows Design Center. With the Zener voltage set at 4.7 V and transistor beta set at 100, the output is 9.255 V when the input is 10 V. A dc sweep from 8 V to 15 V is shown in the

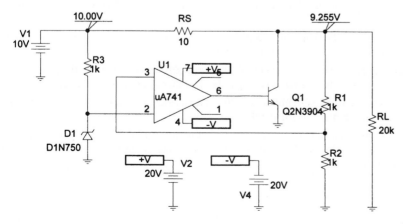

Figure 18.40 Shunt voltage regulator using op-amp.

PROBE output in Fig. 18.41. The circuit provides good voltage regulation for inputs from about 9.5 to over 14 V, the output being held at the regulated value of about 9.3 V.

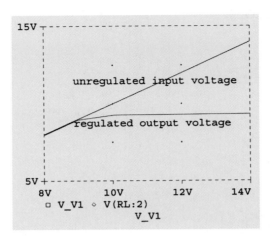

Figure 18.41 Probe output for the dc voltage sweep of Fig. 18.40.

PROBLEMS

§ 18.2 General Filter Considerations

1. What is the ripple factor of a sinusoidal signal having peak ripple of 2 V on an average of 50 V?

2. A filter circuit provides an output of 28 V unloaded and 25 V under full-load operation. Calculate the percent voltage regulation.

3. A half-wave rectifier develops 20 V dc. What is the value of the ripple voltage?

4. What is the rms ripple voltage of a full-wave rectifier with output voltage 8 V dc?

§ 18.3 Capacitor Filter

5. A simple capacitor filter fed by a full-wave rectifier develops 14.5 V dc at 8.5% ripple factor. What is the output ripple voltage (rms)?

6. A full-wave rectified signal of 18 V peak is fed into a capacitor filter. What is the voltage regulation of the filter if the output is 17 V dc at full load?

7. A full-wave rectified voltage of 18 V peak is connected to a 400-μF filter capacitor. What are the ripple and dc voltages across the capacitor at a load of 100 mA?

8. A full-wave rectifier operating from the 60-Hz ac supply produces a 20-V peak rectified voltage. If a 200-μF capacitor is used, calculate the ripple at a load of 120 mA.

9. A full-wave rectifier (operating from a 60-Hz supply) drives a capacitor-filter circuit ($C = 100 \ \mu$F), which develops 12 V dc when connected to a 2.5-kΩ load. Calculate the output voltage ripple.

10. Calculate the size of the filter capacitor needed to obtain a filtered voltage having 15% ripple at a load of 150 mA. The full-wave rectified voltage is 24 V dc, and the supply is 60 Hz.

*11. A 500-μF capacitor provides a load current of 200 mA at 8% ripple. Calculate the peak rectified voltage obtained from the 60-Hz supply and the dc voltage across the filter capacitor.

12. Calculate the size of the filter capacitor needed to obtain a filtered voltage with 7% ripple at a load of 200 mA. The full-wave rectified voltage is 30 V dc, and the supply is 60 Hz.

13. Calculate the percent ripple for the voltage developed across a 120-μF filter capacitor when providing a load current of 80 mA. The full-wave rectifier operating from the 60-Hz supply develops a peak rectified voltage of 25 V.

§ 18.4 RC Filter

14. An *RC* filter stage is added after a capacitor filter to reduce the percent of ripple to 2%. Calculate the ripple voltage at the output of the *RC* filter stage providing 80 V dc.

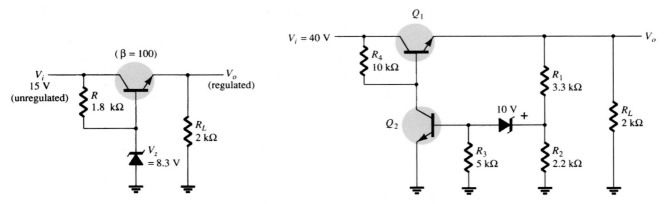
* **15.** An *RC* filter stage ($R = 33 \, \Omega$, $C = 120 \, \mu F$) is used to filter a signal of 24 V dc with 2 V rms operating from a full-wave rectifier. Calculate the percent ripple at the output of the *RC* section for a 100-mA load. Also, calculate the ripple of the filtered signal applied to the *RC* stage.

* **16.** A simple capacitor filter has an input of 40 V dc. If this voltage is fed through an *RC* filter section ($R = 50 \, \Omega$, $C = 40 \, \mu F$), what is the load current for a load resistance of 500 Ω?

17. Calculate the rms ripple voltage at the output of an *RC* filter section that feeds a 1-kΩ load when the filter input is 50 V dc with 2.5-V rms ripple from a full-wave rectifier and capacitor filter. The *RC* filter section components are $R = 100 \, \Omega$ and $C = 100 \, \mu F$.

18. If the no-load output voltage for Problem 17 is 50 V, calculate the percent voltage regulation with a 1-kΩ load.

§ 18.5 Discrete Transistor Voltage Regulation

* **19.** Calculate the output voltage and Zener diode current in the regulator circuit of Fig. 18.42.

20. What regulated output voltage results in the circuit of Fig. 18.43?

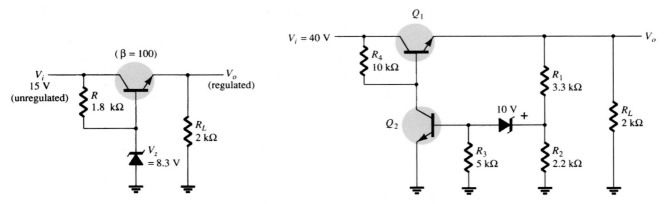

Figure 18.42 Problem 19   **Figure 18.43** Problem 20

21. Calculate the regulated output voltage in the circuit of Fig. 18.44.

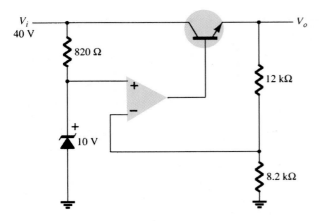

Figure 18.44 Problem 21

22. Determine the regulated voltage and circuit currents for the shunt regulator of Fig. 18.45.

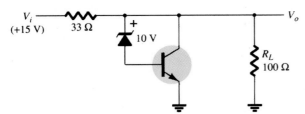

Figure 18.45 Problem 22

Problems   **887**

§ 18.6 IC Voltage Regulators

23. Draw the circuit of a voltage supply comprised of a full-wave bridge rectifier, capacitor filter, and IC regulator to provide an output of +12 V.

*** 24.** Calculate the minimum input voltage of the full-wave rectifier and filter capacitor network in Fig. 18.46 when connected to a load drawing 250 mA.

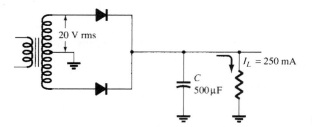

Figure 18.46 Problem 24

*** 25.** Determine the maximum value of load current at which regulation is maintained for the circuit of Fig. 18.47.

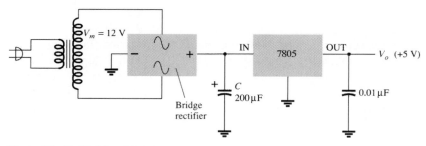

Figure 18.47 Problem 25

26. Determine the regulated voltage in the circuit of Fig. 18.30 with $R_1 = 240\ \Omega$ and $R_2 = 1.8\ \text{k}\Omega$.

27. Determine the regulated output voltage from the circuit of Fig. 18.48.

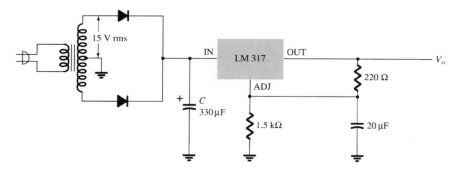

Figure 18.48 Problem 27

§ 18.9 Computer Analysis

*** 28.** Modify the circuit of Fig. 18.38 to include a load resistor, R_L. Keeping the input voltage fixed at 10 V, do a sweep of the load resistor from 100 Ω to 20 kΩ, showing the output voltage using Probe.

*** 29.** For the circuit of Fig. 18.40, do a sweep showing the output voltage for R_L varied from 5 kΩ to 20 kΩ.

*** 30.** Run a PSpice analysis of the circuit of Fig. 18.19 for $V_Z = 4.7$ V, beta (Q_1) = beta (Q_2) = 100, and vary V_i from 5 V to 20 V.

*Please Note: Asterisks indicate more difficult problems.

Other
Two-Terminal
Devices

19.1 INTRODUCTION

There are a number of two-terminal devices having a single *p-n* junction like the semiconductor or Zener diode but with different modes of operation, terminal characteristics, and areas of application. A number, including the Schottky, tunnel, varactor, photodiode, and solar cell, will be introduced in this chapter. In addition, two-terminal devices of a different construction, such as the photoconductive cell, LCD (liquid-crystal display), and thermistor, will be examined.

19.2 SCHOTTKY BARRIER (HOT-CARRIER) DIODES

In recent years, there has been increasing interest in a two-terminal device referred to as a *Schottky-barrier, surface-barrier,* or *hot-carrier* diode. Its areas of application were first limited to the very high frequency range due to its quick response time (especially important at high frequencies) and a lower noise figure (a quantity of real importance in high-frequency applications). In recent years, however, it is appearing more and more in low-voltage/high-current power supplies and ac-to-dc converters. Other areas of application of the device include radar systems, Schottky TTL logic for computers, mixers and detectors in communication equipment, instrumentation, and analog-to-digital converters.

Its construction is quite different from the conventional *p-n* junction in that a metal-semiconductor junction is created such as shown in Fig. 19.1. The semiconductor is

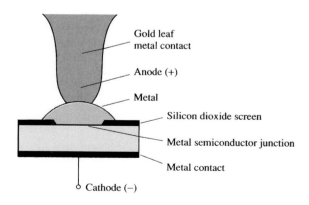

Gold leaf
metal contact

Anode (+)

Metal

Silicon dioxide screen

Metal semiconductor junction

Metal contact

Cathode (−)

Figure 19.1 Passivated hot-carrier diode.

normally n-type silicon (although p-type silicon is sometimes used), while a host of different metals, such as molybdenum, platinum, chrome, or tungsten, are used. Different construction techniques will result in a different set of characteristics for the device, such as increased frequency range, lower forward bias, and so on. Priorities do not permit an examination of each technique here, but information will usually be provided by the manufacturer. In general, however, Schottky diode construction results in a more uniform junction region and a high level of ruggedness.

In both materials, the electron is the majority carrier. In the metal, the level of minority carriers (holes) is insignificant. When the materials are joined, the electrons in the n-type silicon semiconductor material immediately flow into the adjoining metal, establishing a heavy flow of majority carriers. Since the injected carriers have a very high kinetic energy level compared to the electrons of the metal, they are commonly called "hot carriers." In the conventional p-n junction, there was the injection of minority carriers into the adjoining region. Here the electrons are injected into a region of the same electron plurality. Schottky diodes are therefore unique in that conduction is entirely by majority carriers. The heavy flow of electrons into the metal creates a region near the junction surface depleted of carriers in the silicon material— much like the depletion region in the p-n junction diode. The additional carriers in the metal establish a "negative wall" in the metal at the boundary between the two materials. The net result is a "surface barrier" between the two materials, preventing any further current. That is, any electrons (negatively charged) in the silicon material face a carrier-free region and a "negative wall" at the surface of the metal.

The application of a forward bias as shown in the first quadrant of Fig. 19.2 will reduce the strength of the negative barrier through the attraction of the applied positive potential for electrons from this region. The result is a return to the heavy flow of electrons across the boundary, the magnitude of which is controlled by the level of the applied bias potential. The barrier at the junction for a Schottky diode is less than that of the p-n junction device in both the forward- and reverse-bias regions. The result is therefore a higher current at the same applied bias in the forward- and reverse-bias regions. This is a desirable effect in the forward-bias region but highly undesirable in the reverse-bias region.

The exponential rise in current with forward bias is described by Eq. (1.4) but with η dependent on the construction technique (1.05 for the metal whisker type of construction, which is somewhat similar to the germanium diode). In the reverse-bias region, the current I_s is due primarily to those electrons in the metal passing into the

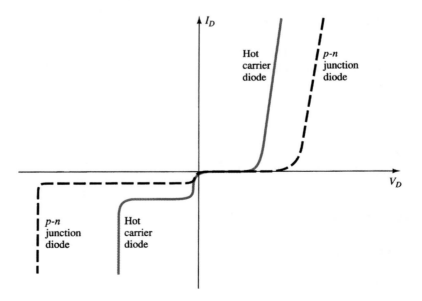

Figure 19.2 Comparison of characteristics of hot-carrier and p-n junction diodes.

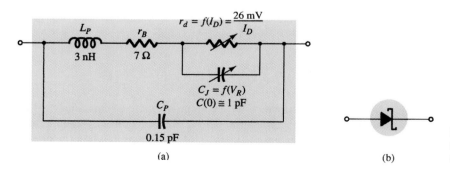

semiconductor material. One of the areas of continuing research on the Schottky diode centers on reducing the high leakage currents that result with temperatures over 100°C. Through design, improvement units are now becoming available that have a temperature range from −65 to +150°C. At room temperature, I_s is typically in the microampere range for low-power units and milliampere range for high-power devices, although it is typically larger than that encountered using conventional p-n junction devices with the same current limits. In addition, the PIV of Schottky diodes is usually significantly less than that of a comparable p-n junction unit. Typically, for a 50-A unit, the PIV of the Schottky diode would be about 50 V as compared to 150 V for the p-n junction variety. Recent advances, however, have resulted in Schottky diodes with PIVs greater than 100 V at this current level. It is obvious from the characteristics of Fig. 19.2 that the Schottky diode is closer to the ideal set of characteristics than the point contact and has levels of V_T less than the typical silicon semiconductor p-n junction. The level of V_T for the "hot-carrier" diode is controlled to a large measure by the metal employed. There exists a required trade-off between temperature range and level of V_T. An increase in one appears to correspond to a resulting increase in the other. In addition, the lower the range of allowable current levels, the lower the value of V_T. For some low-level units, the value of V_T can be assumed to be essentially zero on an approximate basis. For the middle and high range, however, a value of 0.2 V would appear to be a good representative value.

The maximum current rating of the device is presently limited to about 75 A, although 100-A units appear to be on the horizon. One of the primary areas of application of this diode is in *switching power supplies* that operate in the frequency range of 20 kHz or more. A typical unit at 25°C may be rated at 50 A at a forward voltage of 0.6 V with a recovery time of 10 ns for use in one of these supplies. A p-n junction device with the same current limit of 50 A may have a forward voltage drop of 1.1 V and a recovery time of 30 to 50 ns. The difference in forward voltage may not appear significant, but consider the power dissipation difference: $P_{\text{hot carrier}} =$ (0.6 V)(50 A) = 30 W compared to $P_{p-n} = $ (1.1 V)(50 A) = 55 W, which is a measurable difference when efficiency criteria must be met. There will, of course, be a higher dissipation in the reverse-bias region for the Schottky diode due to the higher leakage current, but the total power loss in the forward- and reverse-bias regions is still significantly improved as compared to the p-n junction device.

Recall from our discussion of reverse recovery time for the semiconductor diode that the injected minority carriers accounted for the high level of t_{rr} (the reverse recovery time). The absence of minority carriers at any appreciable level in the Schottky diode results in a reverse recovery time of significantly lower levels, as indicated above. This is the primary reason Schottky diodes are so effective at frequencies approaching 20 GHz, where the device must switch states at a very high rate. For higher frequencies the point-contact diode, with its very small junction area, is still employed.

The equivalent circuit for the device (with typical values) and a commonly used symbol appear in Fig. 19.3. A number of manufacturers prefer to use the standard diode symbol for the device since its function is essentially the same. The inductance

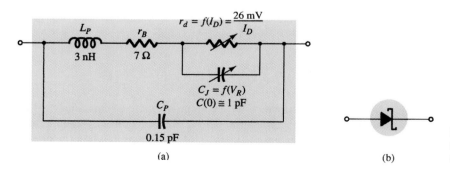

(a)

(b)

Figure 19.3 Schottky (hot-carrier) diode: (a) equivalent circuit; (b) symbol.

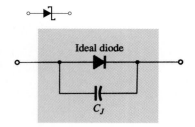

Figure 19.4 Approximate equivalent circuit for the Schottky diode.

L_P and capacitance C_P are package values, and r_B is the series resistance, which includes the contact and bulk resistance. The resistance r_d and capacitance C_J are values defined by equations introduced in earlier sections. For many applications, an excellent approximate equivalent circuit simply includes an ideal diode in parallel with the junction capacitance as shown in Fig. 19.4.

A number of hot-carrier rectifiers manufactured by Motorola Semiconductor Products, Inc., appear in Fig. 19.5 with their specifications and terminal identification. Note that the maximum forward voltage drop V_F does not exceed 0.65 V for any of the devices, while this was essentially V_T for a silicon diode.

		I_o, Average rectified forward current (amperes)													
		0.5 A	1.0 A		3.0 A		3.0 A	5.0 A	15 A		25 A		40 A		
	Case	51-02 (DO-7) Glass	59-04 Plastic		267 Plastic		60 Metal		257 (DO-4) Metal				257 (DO-5) Metal	430-2 (DO-21) Metal	
V_{RRM} (Volts)	Anode Cathode:				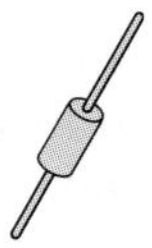		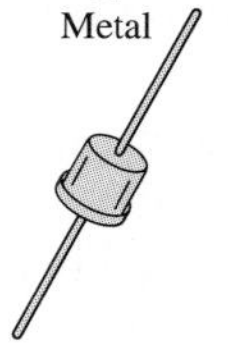								
20		MBR020	IN5817	MBR120P	IN5820	MBR320P	MBR320M	IN5823	IN5826	MBR1520	IN5829	MBR2520	IN5832	MBR4020	MBR4020PF
30		MBR030	IN5818	MBR130P	IN5821	MBR330P	MBR330M	IN5824	IN5827	MBR1530	IN5830	MBR2530	IN5833	MBR4030	MBR4030PF
35				MBR135P		MBR335P	MBR335M			MBR1535		MBR2535		MBR4035	MBR4035PF
40			IN5819	MBR140P	IN5822	MBR340P	MBR340M	IN5825	IN5828	MBR1540	IN5831	MBR2540	IN5834	MBR4040	
I_{FSM} (Amps)		5.0	100	50	250	200	500	500	500	500	800	800	800	800	800
T_C @ Rated I_o (°C)									85	80	85	80	75	70	50
T_J Max		125°C	125°C	125°C	125°C	125°C	125°C	125°C	125°C	125°C	125°C	125°C	125°C	125°C	125°C
Max V_F @ $I_{FM} = I_o$		0.50 V	*0.60 V	0.65 V	*0.525 V	0.60 V	0.45 V@5A	*0.38 V	*0.50 V	0.55 V	*0.48 V	0.55 V	*0.59 V	0.63 V	0.63 V

. . . Schottky barrier devices, ideal for use in low-voltage, high-frequency power supplies and as free-wheeling diodes. These units feature very low forward voltages and switching times estimated at less than 10 ns. They are offered in current ranges of 0.5 to 5.0 amperes and in voltages to 40.

V_{RRM} —respective peak reverse voltage
I_{FSM} —forward current, surge peak
I_{FM} —forward current, maximum

Figure 19.5 Motorola Schottky barrier devices. (Courtesy Motorola Semiconductor Products, Incorporated.)

Three sets of curves for the Hewlett-Packard 5082-2300 series of general-purpose Schottky barrier diodes are provided in Fig. 19.6. Note at $T = 100$°C in Fig. 19.6a that V_F is only 0.1 V at a current of 0.01 mA. Note also that the reverse current has been limited to nanoamperes in Fig. 19.6b and the capacitance to 1 pF in Fig. 19.6c to ensure a high switching rate.

19.3 VARACTOR (VARICAP) DIODES

Varactor [also called varicap, VVC (voltage-variable capacitance), or tuning] diodes are semiconductor, voltage-dependent, variable capacitors. Their mode of operation depends on the capacitance that exists at the *p-n* junction when the element is reverse-biased. Under reverse-bias conditions, it was established that there is a region of un-

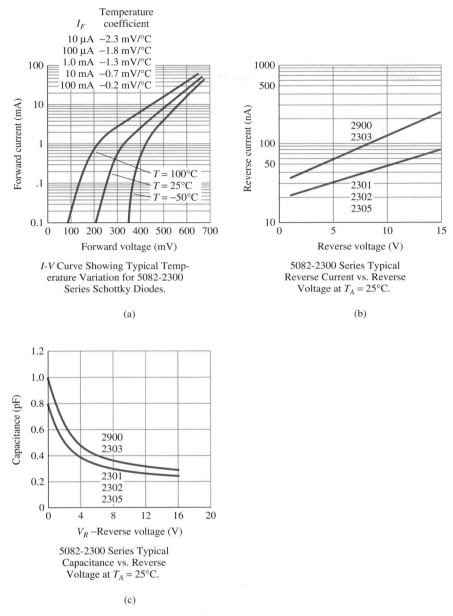

Temperature
I_F coefficient

10 μA −2.3 mV/°C
100 μA −1.8 mV/°C
1.0 mA −1.3 mV/°C
10 mA −0.7 mV/°C
100 mA −0.2 mV/°C

$T = 100°C$
$T = 25°C$
$T = -50°C$

I-V Curve Showing Typical Temperature Variation for 5082-2300
Series Schottky Diodes.

(a)

2900
2303
2301
2302
2305

5082-2300 Series Typical
Reverse Current vs. Reverse
Voltage at $T_A = 25°C$.

(b)

2900
2303
2301
2302
2305

V_R −Reverse voltage (V)

5082-2300 Series Typical
Capacitance vs. Reverse
Voltage at $T_A = 25°C$.

(c)

Figure 19.6 Characteristic curves for Hewlett-Packard 5082-2300 series of
general-purpose Schottky barrier diodes. (Courtesy Hewlett-Packard Corporation.)

covered charge on either side of the junction that together the regions make up the
depletion region and define the depletion width W_d. The transition capacitance (C_T)
established by the isolated uncovered charges is determined by

$$C_T = \epsilon \frac{A}{W_d}$$

(19.1)

where ϵ is the permittivity of the semiconductor materials, A the *p-n* junction area,
and W_d the depletion width.

As the reverse-bias potential increases, the width of the depletion region increases,
which in turn reduces the transition capacitance. The characteristics of a typical com-

mercially available varicap diode appear in Fig. 19.7. Note the initial sharp decline in C_T with increase in reverse bias. The normal range of V_R for VVC diodes is limited to about 20 V. In terms of the applied reverse bias, the transition capacitance is given approximately by

$$C_T = \frac{K}{(V_T + V_R)^n} \qquad (19.2)$$

where K = constant determined by the semiconductor material and construction technique

V_T = knee potential as defined in Section 1.6

V_R = magnitude of the applied reverse-bias potential

$n = \frac{1}{2}$ for alloy junctions and $\frac{1}{3}$ for diffused junctions

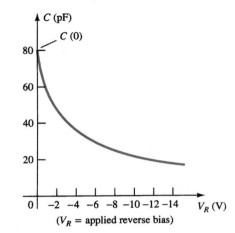

Figure 19.7 Varicap characteristics: C (pF) versus V_R.

In terms of the capacitance at the zero-bias condition $C(0)$, the capacitance as a function of V_R is given by

$$C_T(V_R) = \frac{C(0)}{(1 + |V_R/V_T|)^n} \qquad (19.3)$$

The symbols most commonly used for the varicap diode and a first approximation for its equivalent circuit in the reverse-bias region are shown in Fig. 19.8. Since we are in the reverse-bias region, the resistance in the equivalent circuit is very large in magnitude—typically 1 MΩ or larger—while R_S, the geometric resistance of the diode, is, as indicated in Fig. 19.8, very small. The magnitude of C_T will vary from about 2 to 100 pF depending on the varicap considered. To ensure that R_R is as large (for minimum leakage current) as possible, silicon is normally used in varicap diodes. The fact that the device will be employed at very high frequencies requires that we include the inductance L_S even though it is measured in nanohenries. Recall that

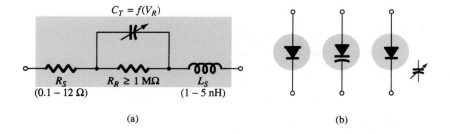

Figure 19.8 Varicap diode: (a) equivalent circuit in the reverse-bias region; (b) symbols.

(a)

(b)

$X_L = 2\pi fL$ and a frequency of 10 GHz with $L_S = 1$ nH will result in an $X_{L_S} = 2\pi fL = (6.28)(10^{10} \text{ Hz})(10^{-9} \text{ F}) = 62.8\ \Omega$. There is obviously, therefore, a frequency limit associated with the use of each varicap diode.

Assuming the proper frequency range and a low value of R_S and X_{L_S} compared to the other series elements, then the equivalent circuit for the varicap of Fig. 19.8a can be replaced by the variable capacitor alone. The complete data sheet and its characteristic curves appear in Figs. 19.9 and 19.10, respectively. The C_3/C_{25} ratio in Fig. 19.9 is the ratio of capacitance levels at reverse-bias potentials of 3 and 25 V. It provides a quick estimate of how much the capacitance will change with reverse-bias potential. The figure of merit is a quantity of consideration in the application of the device and is a measure of the ratio of energy stored by the capacitive device per cycle to the energy dissipated (or lost) per cycle. Since energy loss is seldom considered a positive attribute, the higher its relative value the better. The resonant frequency of the device is determined by $f_o = 1/2\pi\sqrt{LC}$ and affects the range of application of the device.

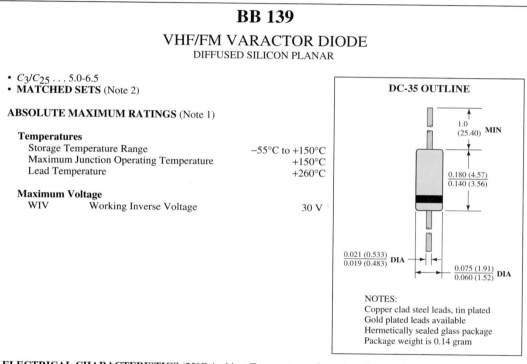

BB 139

VHF/FM VARACTOR DIODE
DIFFUSED SILICON PLANAR

- C_3/C_{25} . . . 5.0-6.5
- **MATCHED SETS** (Note 2)

DC-35 OUTLINE

ABSOLUTE MAXIMUM RATINGS (Note 1)

Temperatures

Storage Temperature Range	$-55°C$ to $+150°C$
Maximum Junction Operating Temperature	$+150°C$
Lead Temperature	$+260°C$

Maximum Voltage

WIV	Working Inverse Voltage	30 V

NOTES:
Copper clad steel leads, tin plated
Gold plated leads available
Hermetically sealed glass package
Package weight is 0.14 gram

ELECTRICAL CHARACTERISTICS (25°C Ambient Temperature unless otherwise noted)

SYMBOL	CHARACTERISTIC	MIN	TYP	MAX	UNITS	TEST CONDITIONS
BV	Breakdown Voltage	30			V	$I_R = 100\ \mu A$
I_R	Reverse Current		10	50	nA	$V_R = 28$ V
			0.1	0.5	μA	$V_R = 28$ V, $T_A = 60°C$
C	Capacitance		29		pF	$V_R = 3.0$ V, $f = 1$ MHz
		4.3	5.1	6.0	pF	$V_R = 25$ V, $f = 1$ MHz
C_3/C_{25}	Capacitance Ratio	5.0	5.7	6.5		$V_R = 3$ V/25 V, $f = 1$ MHz
Q	Figure of Merit		150			$V_R = 3.0$ V, $f = 100$ MHz
R_S	Series Resistance		0.35		Ω	$C = 10$ pF, $f = 600$ MHz
L_S	Series Inductance		2.5		nH	1.5 mm from case
f_o	Series Resonant Frequency		1.4		GHz	$V_R = 25$ V

NOTES;
1. These ratings are limiting values above which the serviceability of the diodes may be impaired.
2. The capacitance diffrence between any two diodes in one set is less than 3% over the reverse voltage range of 0.5 V to 28 V

Figure 19.9 Electrical characteristics for a VHF/FM Fairchild varactor diode. (Courtesy Fairchild Camera and Instrument Corporation.)

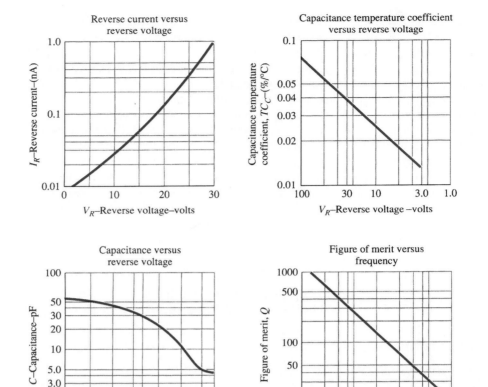

Figure 19.10 Characteristic curves for a VHF/FM Fairchild varactor diode. (Courtesy

In Fig. 19.10, most quantities are self-explanatory. However, the *capacitance temperature coefficient* is defined by

$$TC_C = \frac{\Delta C}{C_0(T_1 - T_0)} \times 100\% \qquad \%/°C \qquad (19.4)$$

where ΔC is the change in capacitance due to the temperature change $T_1 - T_0$ and C_0 is the capacitance at T_0 for a particular reverse-bias potential. For example, Fig. 19.9 indicates that $C_0 = 29$ pF with $V_R = 3$ V and $T_0 = 25°C$. A change in capacitance ΔC could then be determined using Eq. (19.4) simply by substituting the new temperature T_1 and the TC_C as determined from the graph (= 0.013). At a new V_R, the value of TC_C would change accordingly. Returning to Fig. 19.9, note that the maximum frequency appearing is 600 MHz. At this frequency,

$$X_L = 2\pi fL = (6.28)(600 \times 10^6 \text{ Hz})(2.5 \times 10^{-9} \text{ F}) = 9.42 \ \Omega$$

normally a quantity of sufficiently small magnitude to be ignored.

Some of the high-frequency (as defined by the small capacitance levels) areas of application include FM modulators, automatic-frequency-control devices, adjustable bandpass filters, and parametric amplifiers.

Application

In Fig. 19.11, the varactor diode is employed in a tuning network. That is, the resonant frequency of the parallel L-C combination is determined by $f_p = 1/2\pi \sqrt{L_2 C_T'}$ (high-Q system) with the level of $C_T' = C_T + C_C$ determined by the applied reverse-

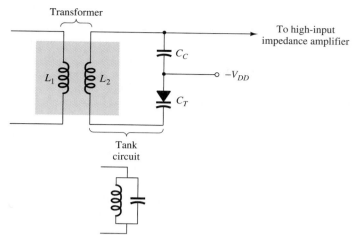

Transformer

L_1 L_2

C_C

To high-input
impedance amplifier

$-V_{DD}$

C_T

Tank
circuit

Figure 19.11 Tuning
network employing a
varactor diode.

bias potential V_{DD}. The coupling capacitor C_C is present to provide isolation between
the shorting effect of L_2 and the applied bias. The selected frequencies of the tuned
network are then passed on to the high input amplifier for further amplification.

19.4 POWER DIODES

There are a number of diodes designed specifically to handle the high-power and high-
temperature demands of some applications. The most frequent use of power diodes
occurs in the rectification process, in which ac signals (having zero average value)
are converted to ones having an average or dc level. As noted in Chapter 2, when used
in this capacity, diodes are normally referred to as *rectifiers*.

The majority of the power diodes are constructed using silicon because of its
higher current, temperature, and PIV ratings. The higher current demands require that
the junction area be larger, to ensure that there is a low forward diode resistance. If
the forward resistance were too large, the I^2R losses would be excessive. The current
capability of power diodes can be increased by placing two or more in parallel, and
the PIV rating can be increased by stacking the diodes in series.

Various types of power diodes and their current rating have been provided in
Fig. 19.12a. The high temperatures resulting from the heavy current require, in many

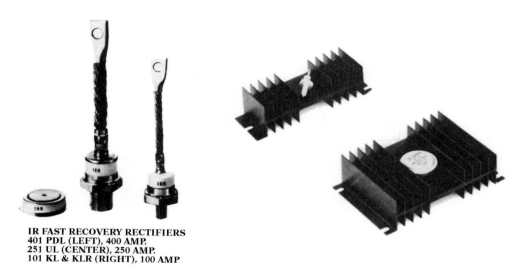

**IR FAST RECOVERY RECTIFIERS
401 PDL (LEFT), 400 AMP.
251 UL (CENTER), 250 AMP.
101 KL & KLR (RIGHT), 100 AMP**

(a)

(b)

Figure 19.12 Power diodes and
heat sinks. (Courtesy International
Rectifier Corporation.)

cases, that heat sinks be used to draw the heat away from the element. A few of the various types of heat sinks available are shown in Fig. 19.12b. If heat sinks are not employed, stud diodes are designed to be attached directly to the chassis, which in turn will act as the heat sink.

19.5 TUNNEL DIODES

The tunnel diode was first introduced by Leo Esaki in 1958. Its characteristics, shown in Fig. 19.13, are different from any diode discussed thus far in that it has a negative-resistance region. In this region, an increase in terminal voltage results in a reduction in diode current.

The tunnel diode is fabricated by doping the semiconductor materials that will form the *p-n* junction at a level one hundred to several thousand times that of a typical semiconductor diode. This will result in a greatly reduced depletion region, of the order of magnitude of 10^{-6} cm, or typically about $\frac{1}{100}$ the width of this region for a typical semiconductor diode. It is this thin depletion region that many carriers can "tunnel" through, rather than attempt to surmount, at low forward-bias potentials that accounts for the peak in the curve of Fig. 19.13. For comparison purposes, a typical semiconductor diode characteristic has been superimposed on the tunnel-diode characteristic of Fig. 19.13.

This reduced depletion region results in carriers "punching through" at velocities that far exceed those available with conventional diodes. The tunnel diode can therefore be used in high-speed applications such as in computers, where switching times in the order of nanoseconds or picoseconds are desirable.

You will recall from Section 1.15 that an increase in the doping level will drop the Zener potential. Note the effect of a very high doping level on this region in Fig. 19.13. The semiconductor materials most frequently used in the manufacture of tunnel diodes are germanium and gallium arsenide. The ratio I_P/I_V is very important for computer applications. For germanium, it is typically 10:1, while for gallium arsenide, it is closer to 20:1.

The peak current, I_P, of a tunnel diode can vary from a few microamperes to several hundred amperes. The peak voltage, however, is limited to about 600 mV. For this reason, a simple VOM with an internal dc battery potential of 1.5 V can severely damage a tunnel diode if applied improperly.

The tunnel diode equivalent circuit in the negative-resistance region is provided in Fig. 19.14, with the symbols most frequently employed for tunnel diodes. The val-

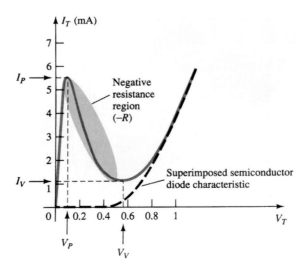

Figure 19.13 Tunnel diode characteristics.

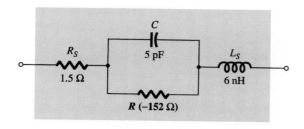

(a)

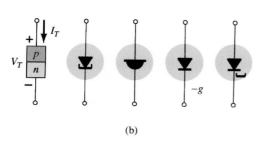

(b)

Figure 19.14 Tunnel diode: (a) equivalent circuit; (b) symbols.

TABLE 19.1 Specifications: Ge 1N2939

	Minimum	Typical	Maximum	
Absolute maximum ratings (25°C)				
Forward current (−55 to +100°C)		5 mA		
Reverse current (−55 to +100°C)		10 mA		
Electrical characteristics (25°C)($\frac{1}{8}$-in. leads)				
I_P	0.9	1.0	1.1	mA
I_V		0.1	0.14	mA
V_P	50	60	65	mV
V_V		350		mV
Reverse voltage ($I_R = 1.0$ mA)			30	mV
Forward peak point current voltage, V_{fp}	450	500	600	mV
I_P/I_V		10		
$-R$		−152		Ω
C		5	15	pF
L_S		6		nH
R_S		1.5	4.0	Ω

ues for each parameter are for the 1N2939 tunnel diode whose specifications appear in Table 19.1. The inductor L_S is due mainly to the terminal leads. The resistor R_S is due to the leads, ohmic contact at the lead–semiconductor junction, and the semiconductor materials themselves. The capacitance C is the junction diffusion capacitance, and the R is the negative resistance of the region. The negative resistance finds application in oscillators to be described later.

Note the lead length of $\frac{1}{8}$ in. included in the specifications. An increase in this length will cause L_S to increase. In fact, it was given for this device that L_S will vary 1 to 12 nH, depending on lead length. At high frequencies ($X_{L_S} = 2\pi f L_S$), this factor can take its toll.

The fact that $V_{fp} = 500$ mV (typ.) and $I_{forward}$ (max.) = 5 mA indicates that tunnel diodes are low-power devices $[P_D = (0.5$ V$)(5$ mA$) = 2.5$ mW$]$, which is also excellent for computer applications. A rendering of the device appears in Fig. 19.15.

Figure 19.15 A Ge IN2939 tunnel diode. (Courtesy Powerex, Inc.)

19.5 Tunnel Diodes

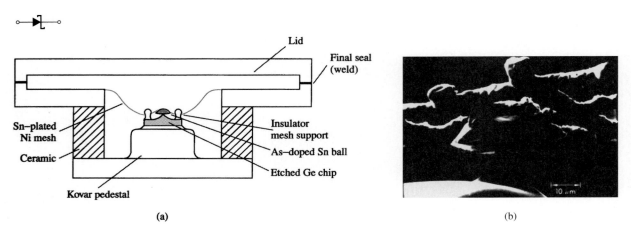

Figure 19.16 Tunnel diode: (a) construction; (b) photograph. (Courtesy *COM SAT Technical Review*, P. F. Varadi and T. D. Kirkendall.)

Although the use of tunnel diodes in present-day high-frequency systems has been dramatically stalled by manufacturing techniques that suggest alternatives to the tunnel diode, its simplicity, linearity, low power drain, and reliability ensure its continued life and application. The basic construction of an advance design tunnel diode appears in Fig. 19.16 with a photograph of the actual junction.

In Fig. 19.17, the chosen supply voltage and load resistance have defined a load line that intersects the tunnel diode characteristics at three points. Keep in mind that the load line is determined solely by the network and the characteristics by the device. The intersections at *a* and *b* are referred to as *stable* operating points, due to the positive resistance characteristic. That is, at either of these operating points, a slight disturbance in the network will not set the network into oscillations or result in a significant change in the location of the *Q*-point. For instance, if the defined operating point is at *b*, a slight increase in supply voltage *E* will move the operating point up the curve since the voltage across the diode will increase. Once the disturbance has passed, the voltage across the diode and the associated diode current will return to the levels defined by the *Q*-point at *b*. The operating point defined by *c* is an *unstable* one because a slight change in the voltage across or current through the diode will result in the *Q*-point moving to either *a* or *b*. For instance, the slightest increase in *E* will cause the voltage across the tunnel diode to increase above its level at *c*. In this region, however, an increase in V_T will cause a decrease in I_T and a further increase in V_T. This increased level in V_T will result in a continuing decrease in I_T, and so on. The result is an increase in V_T and a change in I_T until the stable operating point at *b* is established. A slight drop in supply voltage would result in a transition to stability at point *a*. In other words, point *c* can be defined as the operating point using

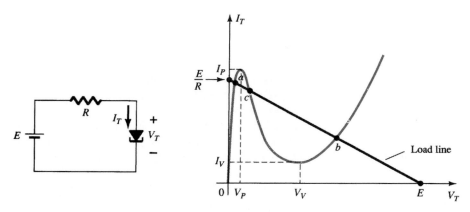

Figure 19.17 Tunnel diode and resulting load line.

the load-line technique, but once the system is energized, it will eventually stabilize at location *a* or *b*.

The availability of a negative resistance region can be put to good use in the design of oscillators, switching networks, pulse generators, and amplifiers.

Applications

In Fig. 19.18a, a *negative-resistance oscillator* was constructed using a tunnel diode. The choice of network elements is designed to establish a load line such as shown in Fig. 19.18b. Note that the only intersection with the characteristics is in the unstable negative-resistance region—a stable operating point is not defined. When the power is turned on, the terminal voltage of the supply will build up from 0 V to a final value of E volts. Initially, the current I_T will increase from 0 mA to I_P, resulting in a storage of energy in the inductor in the form of a magnetic field. However, once I_P is reached, the diode characteristics suggest that the current I_T must now decrease with increase in voltage across the diode. This is a contradiction to the fact that

$$E = I_T R + I_T (-R_T)$$

and

$$E = \underbrace{I_T}_{\text{less}} \underbrace{(R - R_T)}_{\text{less}}$$

If both elements of the equation above were to decrease, it would be impossible for the supply voltage to reach its set value. Therefore, for the current I_T to continue rising, the point of operation must shift from point 1 to point 2. However, at point 2, the voltage V_T has jumped to a value greater than the applied voltage (point 2 is to the right of any point on the network load line). To satisfy Kirchhoff's voltage law,

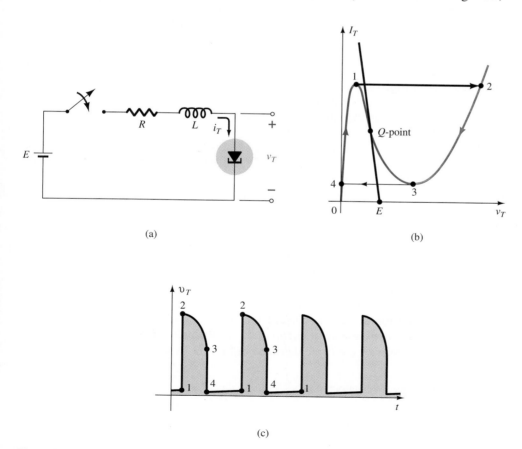

(a)

(b)

(c)

Figure 19.18 Negative-resistance oscillator.

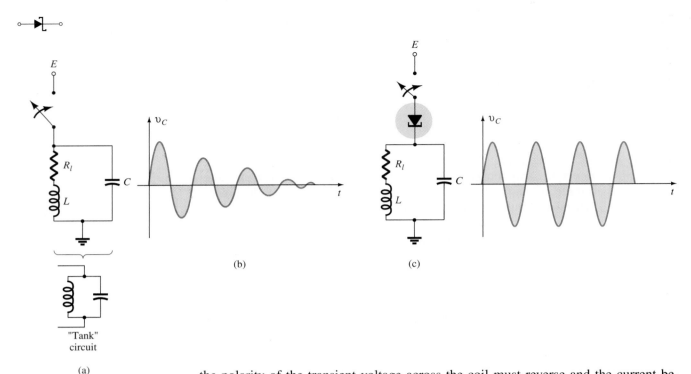

(b)

(c)

"Tank"
circuit

(a)

Figure 19.19 Sinusoidal
oscillator.

the polarity of the transient voltage across the coil must reverse and the current be-gin to decrease as shown from 2 to 3 on the characteristics. When V_T drops to V_V, the characteristics suggest that the current I_T will begin to increase again. This is unac-ceptable since V_T is still more than the applied voltage and the coil is discharging through the series circuit. The point of operation must shift to point 4 to permit a con-tinuation of the decrease in I_T. However, once at point 4, the potential levels are such that the tunnel current can again increase from 0 mA to I_P as shown on the charac-teristics. The process will repeat itself again and again, never settling in on the oper-ating point defined for the unstable region. The resulting voltage across the tunnel diode appears in Fig. 19.18c and will continue as long as the dc supply is energized. The result is an oscillatory output established by a fixed supply and a device with a negative-resistance characteristic. The waveform of Fig. 19.18c has extensive appli-cation in timing and computer logic circuitry.

A tunnel diode can also be used to generate a sinusoidal voltage using simply a dc supply and a few passive elements. In Fig. 19.19a, the closing of the switch will result in a sinusoidal voltage that will decrease in amplitude with time. Depending on the elements employed, the time period can be from one almost instantaneous to one measurable in minutes using typical parameter values. This *damping* of the oscilla-tory output with time is due to the dissipative characteristics of the resistive elements. By placing a tunnel diode in series with the tank circuit as shown in Fig. 19.19c, the negative resistance of the tunnel diode will offset the resistive characteristics of the tank circuit, resulting in the *undamped* response appearing in the same figure. The design must continue to result in a load line that will intersect the characteristics only in the negative-resistance region. In another light, the sinusoidal generator of Fig. 19.19 is simply an extension of the pulse oscillator of Fig. 19.18, with the addition of the capacitor to permit an exchange of energy between the inductor and the capacitor dur-ing the various phases of the cycle depicted in Fig. 19.18b.

19.6 PHOTODIODES

The interest in light-sensitive devices has been increasing at an almost exponential rate in recent years. The resulting field of *optoelectronics* will be receiving a great deal of research interest as efforts are made to improve efficiency levels. Through the advertising media, the layperson has become quite aware that light sources offer a unique source of energy. This energy, transmitted as discrete packages called *photons*, has a level directly related to the frequency of the traveling light wave as determined by the following equation:

$$W = hf \quad \text{joules} \qquad (19.5)$$

where h is called Planck's constant and is equal to 6.624×10^{-34} joule-second. It clearly states that since h is a constant, the energy associated with incident light waves is directly related to the frequency of the traveling wave.

The frequency is, in turn, directly related to the wavelength (distance between successive peaks) of the traveling wave by the following equation:

$$\lambda = \frac{v}{f} \qquad (19.6)$$

where λ = wavelength, meters
v = velocity of light, 3×10^8 m/s
f = frequency of the traveling wave, hertz

The wavelength is usually measured in angstrom units (Å) or micrometers (μm), where

$$1 \text{ Å} = 10^{-10} \text{ m} \quad \text{and} \quad 1 \text{ } \mu\text{m} = 10^{-6} \text{ m}$$

The wavelength is important because it will determine the material to be used in the optoelectronic device. The relative spectral response for Ge, Si, and selenium is provided in Fig. 19.20. The visible-light spectrum has also been included with an indication of the wavelength associated with the various colors.

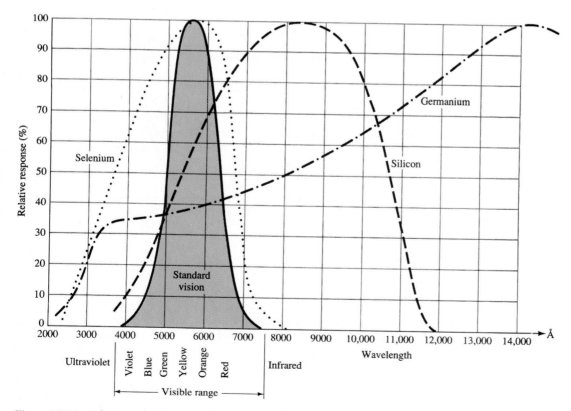

Figure 19.20 Relative spectral response for Si, Ge, and selenium as compared to the human eye.

The number of free electrons generated in each material is proportional to the *intensity* of the incident light. Light intensity is a measure of the amount of *luminous flux* falling in a particular surface area. Luminous flux is normally measured in *lumens* (lm) or watts. The two units are related by

$$1 \text{ lm} = 1.496 \times 10^{-10} \text{ W}$$

The light intensity is normally measured in lm/ft^2, footcandles (fc), or W/m^2, where

$$1 \text{ lm/ft}^2 = 1 \text{ fc} = 1.609 \times 10^{-9} \text{ W/m}^2$$

The photodiode is a semiconductor *p-n* junction device whose region of operation is limited to the reverse-bias region. The basic biasing arrangement, construction, and symbol for the device appear in Fig. 19.21.

Recall from Chapter 1 that the reverse saturation current is normally limited to a few microamperes. It is due solely to the thermally generated minority carriers in the *n*- and *p*-type materials. The application of light to the junction will result in a transfer of energy from the incident traveling light waves (in the form of photons) to the atomic structure, resulting in an increased number of minority carriers and an increased level of reverse current. This is clearly shown in Fig. 19.22 for different intensity levels. The *dark* current is that current that will exist with no applied illumination. Note that the current will only return to zero with a positive applied bias equal to V_T. In addition, Fig. 19.21 demonstrates the use of a lens to concentrate the light on the junction region. Commercially available photodiodes appear in Fig. 19.23.

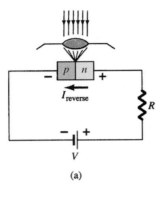

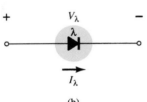

Figure 19.21 Photodiode: (a) basic biasing arrangement and construction; (b) symbol.

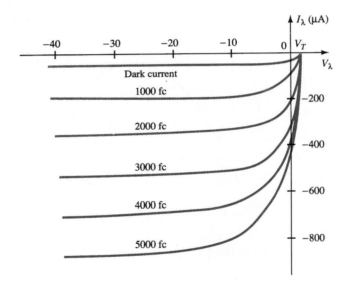

Figure 19.22 Photodiode characteristics.

The almost equal spacing between the curves for the same increment in luminous flux reveals that the reverse current and luminous flux are almost linearly related. In other words, an increase in light intensity will result in a similar increase in reverse current. A plot of the two to show this linear relationship appears in Fig. 19.24 for a fixed voltage V_λ of 20 V. On the relative basis, we can assume that the reverse current is essentially zero in the absence of incident light. Since the rise and fall times (change-of-state parameters) are very small for this device (in the nanosecond range),

Figure 19.23 Photodiodes (Courtesy EG&G VACTEC, Inc.)

the device can be used for high-speed counting or switching applications. Returning to Fig. 19.20, we note that Ge encompasses a wider spectrum of wavelengths than Si. This would make it suitable for incident light in the infrared region as provided by lasers and IR (infrared) light sources, to be described shortly. Of course, Ge has a higher dark current than silicon, but it also has a higher level of reverse current. The level of current generated by the incident light on a photodiode is not such that it could be used as a direct control, but it can be amplified for this purpose.

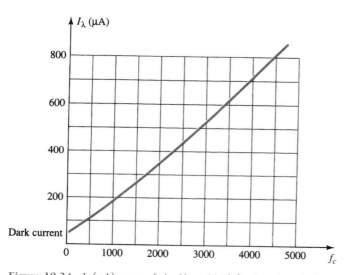

Figure 19.24 I_λ (μA) versus f_C (at $V_\lambda = 20$ V) for the photodiode of Fig. 19.22.

Applications

In Fig. 19.25, the photodiode is employed in an alarm system. The reverse current I_λ will continue to flow as long as the light beam is not broken. If interrupted, I_λ drops to the dark current level and sounds the alarm. In Fig. 19.26, a photodiode is used to count items on a conveyor belt. As each item passes the light beam is broken, I_λ drops to the dark current level and the counter is increased by one.

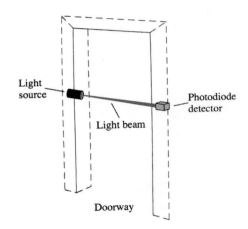

Figure 19.25 Using a photodiode in an alarm system.

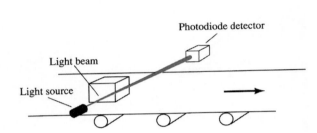

Figure 19.26 Using a photodiode in a counter operation.

19.7 PHOTOCONDUCTIVE CELLS

The photoconductive cell is a two-terminal semiconductor device whose terminal resistance will vary (linearly) with the intensity of the incident light. For obvious reasons, it is frequently called a *photoresistive device*. A typical photoconductive cell and the most widely used graphical symbol for the device appear in Fig. 19.27.

The photoconductive materials most frequently used include cadmium sulfide (CdS) and cadmium selenide (CdSe). The peak spectral response of CdS occurs at approximately 5100 Å and for CdSe at 6150 Å (note Fig. 19.20). The response time

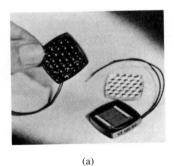

(a)

Figure 19.27 Photoconductive cell: (a) appearance; (b) symbol. [(a) Courtesy International Rectifier Corporation.]

(b)

of CdS units is about 100 ms, and 10 ms for CdSe cells. The photoconductive cell does not have a junction like the photodiode. A thin layer of the material connected between terminals is simply exposed to the incident light energy.

As the illumination on the device increases in intensity, the energy state of a larger number of electrons in the structure will also increase because of the increased availability of the photon packages of energy. The result is an increasing number of relatively "free" electrons in the structure and a decrease in the terminal resistance. The sensitivity curve for a typical photoconductive device appears in Fig. 19.28. Note the linearity (when plotted using a log-log scale) of the resulting curve and the large change in resistance ($100 \text{ k}\Omega \rightarrow 100 \ \Omega$) for the indicated change in illumination.

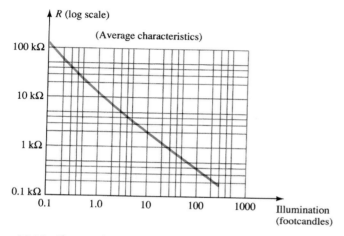

Figure 19.28 Photoconductive cell-terminal characteristics (GE type B425).

In an effort to demonstrate the wealth of material available on each device from manufacturers, consider the CdS (cadmium sulfide) photoconductive cell described in Fig. 19.29. Note again the concern with temperature and response time.

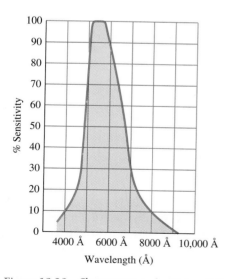

Variation of Conductance With Temperature and Light					
Footcandles	*0.01*	*0.1*	*1.0*	*10*	*100*
Temperature		*% Conductance*			
−25°C	103	104	104	102	106
0	98	102	102	100	103
25°C	100	100	100	100	100
50°C	98	102	103	104	99
75°C	90	106	108	109	104

Response Time Versus Light					
Footcandles	*0.01*	*0.1*	*1.0*	*10*	*100*
Rise (seconds)	0.5	0.095	0.022	0.005	0.002
Decay (seconds)	0.125	0.021	0.005	0.002	0.001

Figure 19.29 Characteristics of a Clairex CdS photoconductive cell. (Courtesy Clairex Electronics.)

Application

One rather simple, but interesting, application of the device appears in Fig. 19.30. The purpose of the system is to maintain V_o at a fixed level even though V_i may fluctuate from its rated value. As indicated in the figure, the photoconductive cell, bulb, and resistor all form part of this voltage-regulator system. If V_i should drop in magnitude for any number of reasons, the brightness of the bulb would also decrease. The decrease in illumination would result in an increase in the resistance (R_λ) of the photoconductive cell to maintain V_o at its rate level as determined by the voltage-divider rule, that is,

$$V_o = \frac{R_\lambda V_i}{R_\lambda + R_1} \tag{19.7}$$

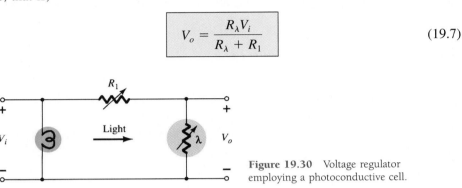

Figure 19.30 Voltage regulator employing a photoconductive cell.

19.8 IR EMITTERS

Infrared-emitting diodes are solid-state gallium arsenide devices that emit a beam of radiant flux when forward-biased. The basic construction of the device is shown in Fig. 19.31. When the junction is forward-biased, electrons from the *n*-region will recombine with excess holes of the *p*-material in a specially designed recombination region sandwiched between the *p*- and *n*-type materials. During this recombination process, energy is radiated away from the device in the form of photons. The generated photons will either be reabsorbed in the structure or leave the surface of the device as radiant energy, as shown in Fig. 19.31.

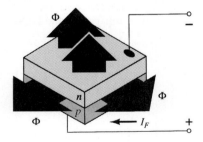

Figure 19.31 General structure of a semiconductor IR-emitting diode. (Courtesy RCA Solid State Division.)

The radiant flux in mW versus the dc forward current for a typical device appears in Fig. 19.32. Note the almost linear relationship between the two. An interesting pattern for such devices is provided in Fig. 19.33. Note the very narrow pattern for devices with an internal collimating system. One such device appears in Fig. 19.34, with its internal construction and graphical symbol. A few areas of application for such devices include card and paper-tape readers, shaft encoders, data-transmission systems, and intrusion alarms.

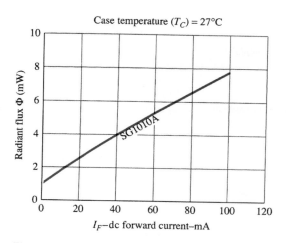

Figure 19.32 Typical radiant flux versus dc forward current for an IR-emitting diode. (Courtesy RCA Solid State Division.)

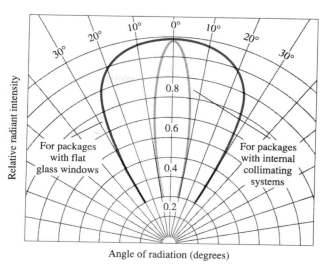

Figure 19.33 Typical radiant intensity patterns of RCA IR-emitting diodes. (Courtesy RCA Solid State Division.)

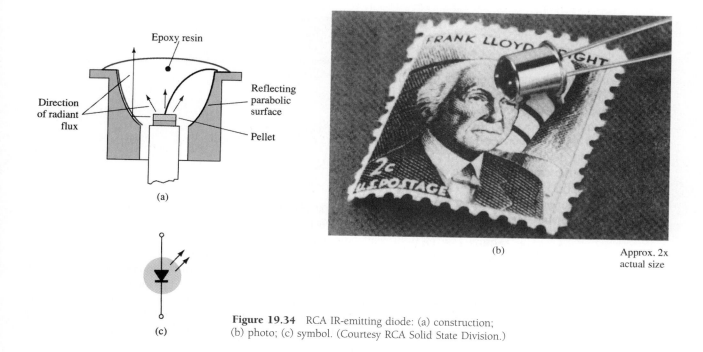

Figure 19.34 RCA IR-emitting diode: (a) construction; (b) photo; (c) symbol. (Courtesy RCA Solid State Division.)

19.9 LIQUID-CRYSTAL DISPLAYS

The liquid-crystal display (LCD) has the distinct advantage of having a lower power requirement than the LED. It is typically in the order of microwatts for the display, as compared to the same order of milliwatts for LEDs. It does, however, require an external or internal light source and is limited to a temperature range of about 0° to 60°C. Lifetime is an area of concern because LCDs can chemically degrade. The types receiving the major interest today are the field-effect and dynamic-scattering units. Each will be covered in some detail in this section.

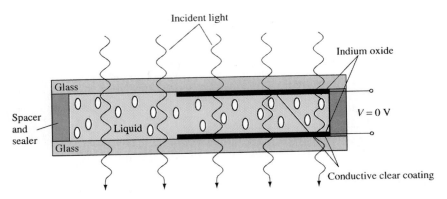

Figure 19.35   Nematic liquid crystal with no applied bias.

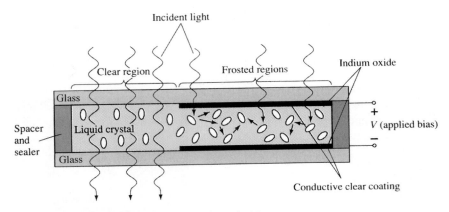

Figure 19.36   Nematic liquid crystal with applied bias.

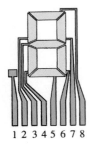

Figure 19.37   LCD eight-segment digit display.

A liquid crystal is a material (normally organic for LCDs) that will flow like a liquid but whose molecular structure has some properties normally associated with solids. For the light-scattering units, the greatest interest is in the *nematic liquid crystal,* having the crystal structure shown in Fig. 19.35. The individual molecules have a rodlike appearance as shown in the figure. The indium oxide conducting surface is transparent, and under the condition shown in the figure, the incident light will simply pass through and the liquid-crystal structure will appear clear. If a voltage (for commercial units the threshold level is usually between 6 and 20 V) is applied across the conducting surfaces, as shown in Fig. 19.36, the molecular arrangement is disturbed, with the result that regions will be established with different indices of refraction. The incident light is therefore reflected in different directions at the interface between regions of different indices of refraction (referred to as *dynamic scattering*—first studied by RCA in 1968) with the result that the scattered light has a frosted-glass appearance. Note in Fig. 19.36, however, that the frosted look occurs only where the conducting surfaces are opposite each other and the remaining areas remain translucent.

A digit on an LCD display may have the segment appearance shown in Fig. 19.37. The black area is actually a clear conducting surface connected to the terminals below for external control. Two similar masks are placed on opposite sides of a sealed thick layer of liquid-crystal material. If the number 2 were required, the terminals 8, 7, 3, 4, and 5 would be energized, and only those regions would be frosted while the other areas would remain clear.

As indicated earlier, the LCD does not generate its own light but depends on an external or internal source. Under dark conditions, it would be necessary for the unit to have its own internal light source either behind or to the side of the LCD. During the day, or in lighted areas, a reflector can be put behind the LCD to reflect the light back through the display for maximum intensity. For optimum operation, current watch manufacturers are using a combination of the transmissive (own light source) and reflective modes called *transflective*.

The *field-effect* or *twisted nematic* LCD has the same segment appearance and thin layer of encapsulated liquid crystal, but its mode of operation is very different. Similar to the dynamic-scattering LCD, the field-effect LCD can be operated in the reflective or transmissive mode with an internal source. The transmissive display appears in Fig. 19.38. The internal light source is on the right, and the viewer is on the left. This figure is most noticeably different from Fig. 19.35 in that there is an addition of a *light polarizer*. Only the vertical component of the entering light on the right can pass through the vertical-light polarizer on the right. In the field-effect LCD, either the clear conducting surface to the right is chemically etched or an organic film is applied to orient the molecules in the liquid crystal in the vertical plane, parallel to the cell wall. Note the rods to the far right in the liquid crystal. The opposite conducting surface is also treated to ensure that the molecules are 90° out of phase in the direction shown (horizontal) but still parallel to the cell wall. In between the two walls of the liquid crystal there is a general drift from one polarization to the other, as shown in the figure. The left-hand light polarizer is also such that it permits the passage of only the vertically polarized incident light. If there is no applied voltage to the conducting surfaces, the vertically polarized light enters the liquid-crystal region and follows the 90° bending of the molecular structure. Its horizontal polarization at the left-hand vertical light polarizer does not allow it to pass through, and the viewer sees a uniformly dark pattern across the entire display. When a threshold voltage is applied (for commercial units from 2 to 8 V), the rodlike molecules align themselves with the field (perpendicular to the wall) and the light passes directly through without the 90° shift. The vertically incident light can then press directly through the second vertically polarized screen, and a light area is seen by the viewer. Through proper excitation of the segments of each digit, the pattern will appear as shown in Fig. 19.39.

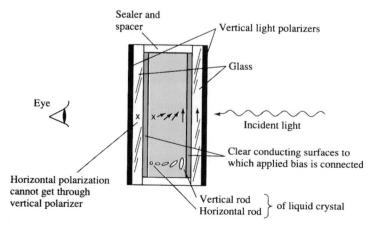

Figure 19.38 Transmissive field-effect LCD with no applied bias.

Figure 19.39 Reflective-type LCD. (Courtesy RCA Solid State Division.)

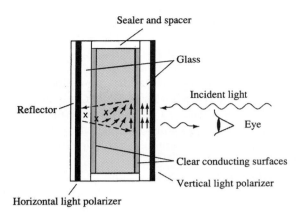

Figure 19.40 Reflective field-effect LCD with no applied bias.

Sealer and spacer

Glass

Incident light

Eye

Reflector

Clear conducting surfaces

Vertical light polarizer

Horizontal light polarizer

Figure 19.41 Transmissive-type LCD. (Courtesy RCA Solid State Division.)

The reflective-type field-effect LCD is shown in Fig. 19.40. In this case, the horizontally polarized light at the far left encounters a horizontally polarized filter and passes through to the reflector, where it is reflected back into the liquid crystal, bent back to the other vertical polarization, and returned to the observer. If there is no applied voltage, there is a uniformly lit display. The application of a voltage results in a vertically incident light encountering a horizontally polarized filter at the left, which it will not be able to pass through and will be reflected. A dark area results on the crystal, and the pattern as shown in Fig. 19.41 appears.

Field-effect LCDs are normally used when a source of energy is a prime factor (e.g., in watches, portable instrumentation, etc.) since they absorb considerably less power than the light-scattering types—the microwatt range compared to the low-milliwatt range. The cost is typically higher for field-effect units, and their height is limited to about 2 in. while light-scattering units are available up to 8 in. in height.

A further consideration in displays is turn-on and turn-off time. LCDs are characteristically much slower than LEDs. LCDs typically have response times in the range 100 to 300 ms, while LEDs are available with response times below 100 ns. However, there are numerous applications, such as in a watch, where the difference between 100 ns and 100 ms ($\frac{1}{10}$ of a second) is of little consequence. For such applications, the lower power demand of LCDs is a very attractive characteristic. The lifetime of LCD units is steadily increasing beyond the 10,000+ hours limit. Since the color generated by LCD units is dependent on the source of illumination, there is a greater range of color choice.

19.10 SOLAR CELLS

In recent years, there has been increasing interest in the solar cell as an alternative source of energy. When we consider that the power density received from the sun at sea level is about 100 mW/cm^2 (1 kW/m^2), it is certainly an energy source that requires further research and development to maximize the conversion efficiency from solar to electrical energy.

The basic construction of a silicon *p-n* junction solar cell appears in Fig. 19.42. As shown in the top view, every effort is made to ensure that the surface area perpendicular to the sun is a maximum. Also, note that the metallic conductor connected to the *p*-type material and the thickness of the *p*-type material are such that they ensure that a maximum number of photons of light energy will reach the junction. A photon of light energy in this region may collide with a valence electron and impart to it sufficient energy to leave the parent atom. The result is a generation of free electrons and holes. This phenomenon will occur on each side of the junction. In the

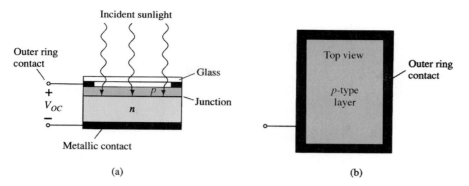

Figure 19.42 Solar cell: (a) cross section; (b) top view.

p-type material, the newly generated electrons are minority carriers and will move rather freely across the junction as explained for the basic *p-n* junction with no applied bias. A similar discussion is true for the holes generated in the *n*-type material. The result is an increase in the minority-carrier flow, which is opposite in direction to the conventional forward current of a *p-n* junction. This increase in reverse current is shown in Fig. 19.43. Since $V = 0$ anywhere on the vertical axis and represents a short-circuit condition, the current at this intersection is called the *short-circuit current* and is represented by the notation I_{SC}. Under open-circuit conditions ($i_d = 0$), the *photovoltaic* voltage V_{OC} will result. This is a logarithmic function of the illumination, as shown in Fig. 19.44. V_{OC} is the terminal voltage of a battery under no-load (open-circuit) conditions. Note, however, in the same figure that the short-circuit current is a linear function of the illumination. That is, it will double for the same increase in illumination (f_{C_1} and $2f_{C_1}$ in Fig. 19.44) while the change in V_{OC} is less for this region. The major increase in V_{OC} occurs for lower-level increases in illumination. Eventually, a further increase in illumination will have very little effect on V_{OC}, although I_{SC} will increase, causing the power capabilities to increase.

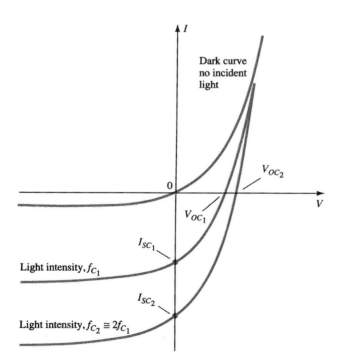

Figure 19.43 Short-circuit current and open-circuit voltage versus light intensity for a solar cell.

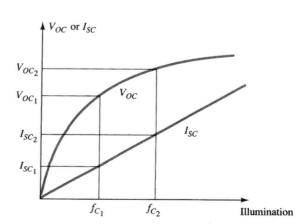

Figure 19.44 V_{OC} and I_{SC} versus illumination for a solar cell.

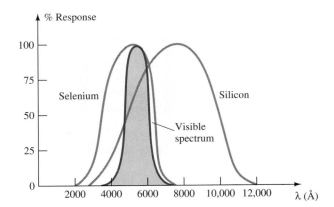

Figure 19.45 Spectral response of Se, Si, and the naked eye.

Selenium and silicon are the most widely used materials for solar cells, although gallium arsenide, indium arsenide, and cadmium sulfide, among others, are also used. The wavelength of the incident light will affect the response of the *p-n* junction to the incident photons. Note in Fig. 19.45 how closely the selenium cell response curve matches that of the eye. This fact has widespread application in photographic equipment such as exposure meters and automatic exposure diaphragms. Silicon also overlaps the visible spectrum but has its peak at the 0.8 μm (8000 Å) wavelength, which is in the infrared region. In general, silicon has a higher conversion efficiency, greater stability, and is less subject to fatigue. Both materials have excellent temperature characteristics. That is, they can withstand extreme high or low temperatures without a significant drop-off in efficiency. Typical solar cells, with their electrical characteristics, appear in Fig. 19.46.

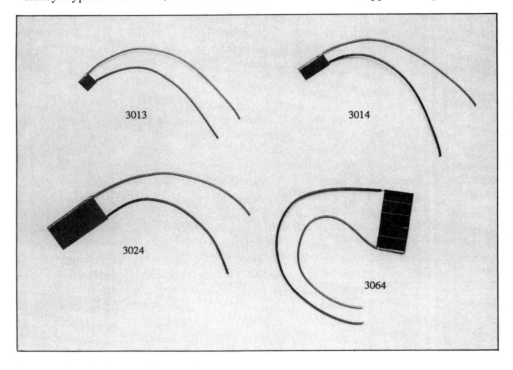

Electrical Characteristics

Part No.	Active Area	Test Voltage	Minimum current Test voltage
3013	0.032 in² (0.21 cm²)	0.4 V	4.2 mA
3014	0.065 in² (0.42 cm²)	0.4 V	8.4 mA
3024	0.29 in² (1.87 cm²)	0.4 V	38 mA
3064	0.325 in² (2.1 cm²)	2 V	8.4 mA

Figure 19.46 Typical solar cells and their electrical characteristics. (Courtesy EG&G VACTEC, Inc.)

A very recent innovation in the use of solar cells appears in Fig. 19.47. The series arrangement of solar cells permits a voltage beyond that of a single element. The performance of a typical four-cell array appears in the same figure. At a current of approximately 2.6 mA, the output voltage is about 1.6 V, resulting in an output power of 4.16 mW. The Schottky barrier diode is included to prevent battery current drain through the power converter. That is, the resistance of the Schottky diode is so high to charge flowing down through (+ to −) the power converter that it will appear as an open circuit to the rechargeable battery and not draw current from it.

It might be of interest to note that the Lockheed Missiles and Space Company is working on a grant from the National Aeronautics and Space Administration to develop a massive solar-array wing for the space shuttle. The wing will measure 13.5 ft by 105 ft when extended and will contain 41 panels, each carrying 3060 silicon solar cells. The wing can generate a total of 12.5 kW of electrical power.

The efficiency of operation of a solar cell is determined by the electrical power output divided by the power provided by the light source. That is,

$$\eta = \frac{P_{o(\text{electrical})}}{P_{i(\text{light energy})}} \times 100\% = \frac{P_{\text{max(device)}}}{(\text{area in cm}^2)(100 \text{ mW/cm}^2)} \times 100\% \qquad (19.8)$$

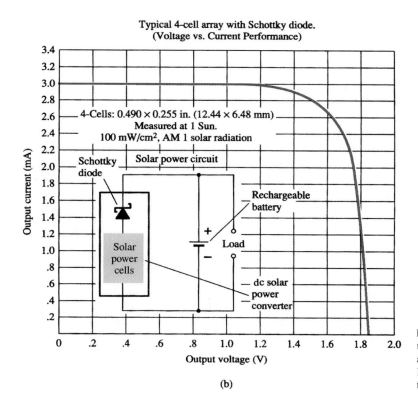

(a)

Typical 4-cell array with Schottky diode.
(Voltage vs. Current Performance)

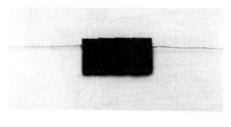

(b)

Figure 19.47 International Rectifier four-cell array: (a) appearance; (b) characteristics. (Courtesy International Rectifier Corporation.)

Typical levels of efficiency range from 10% to 40%—a level that should improve measurably if the present interest continues. A typical set of output characteristics for silicon solar cells of 10% efficiency with an active area of 1 cm² appears in Fig. 19.48. Note the optimum power locus and the almost linear increase in output current with luminous flux for a fixed voltage.

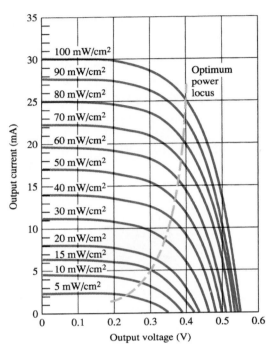

Figure 19.48 Typical output characteristics for silicon solar cells of 10% efficiency having an active area of 1 cm². Cell temperature is 30°C.

19.11 THERMISTORS

The thermistor is, as the name implies, a temperature-sensitive resistor; that is, its terminal resistance is related to its body temperature. It is not a junction device and is constructed of Ge, Si, or a mixture of oxides of cobalt, nickel, strontium, or manganese. The compound employed will determine whether the device has a positive or negative temperature coefficient.

The characteristics of a representative thermistor with a negative temperature coefficient are provided in Fig. 19.49, with the commonly used symbol for the device. Note in particular that at room temperature (20°C) the resistance of the thermistor is approximately 5000 Ω, while at 100°C (212°F) the resistance has decreased to 100 Ω. A temperature span of 80°C has therefore resulted in a 50 : 1 change in resistance. It is typically 3% to 5% per degree change in temperature. There are fundamentally two ways to change the temperature of the device: internally and externally. A simple change in current through the device will result in an internal change in temperature. A small applied voltage will result in a current too small to raise the body temperature above that of the surroundings. In this region, as shown in Fig. 19.50, the ther-

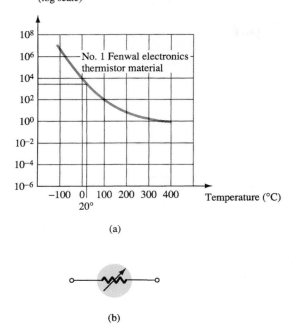

Specific resistance – (ohm–cm, the resistance between faces of 1 cm³ of the material) (log scale)

(a)

(b)

Figure 19.49 Thermistor: (a) typical set of characteristics; (b) symbol.

mistor will act like a resistor and have a positive temperature coefficient. However, as the current increases, the temperature will rise to the point where the negative temperature coefficient will appear as shown in Fig. 19.50. The fact that the rate of internal flow can have such an effect on the resistance of the device introduces a wide vista of applications in control, measuring techniques, and so on. An external change would require changing the temperature of the surrounding medium or immersing the device in a hot or cold solution.

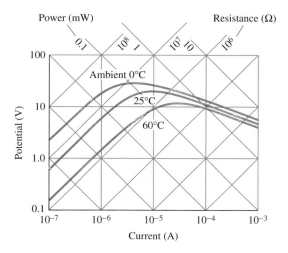

Figure 19.50 Steady-state voltage–current characteristics of Fenwal Electronics BK65VI Thermistor. (Courtesy Fenwal Electronics, Incorporated.)

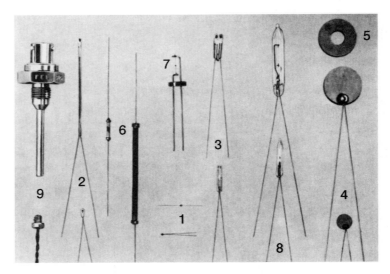

Figure 19.51 Various types of thermistors: (1) beads; (2) glass probes; (3) iso-curve interchangeable probes and beads; (4) disks; (5) washers; (6) rods; (7) specially mounted beads; (8) vacuum and gas-filled probes; (9) special probe assemblies. (Courtesy Fenwal Electronics, Incorporated.)

A photograph of a number of commercially available thermistors is provided in Fig. 19.51.

Application

A simple temperature-indicating circuit appears in Fig. 19.52. Any increase in the temperature of the surrounding medium will result in a decrease in the resistance of the thermistor and an increase in the current I_T. An increase in I_T will produce an increased movement deflection, which when properly calibrated will accurately indicate the higher temperature. The variable resistance was added for calibration purposes.

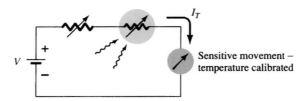

Figure 19.52 Temperature-indicating circuit.

19.12 SUMMARY

Important Conclusions and Concepts

1. The Schottky (hot-carrier) diode has a **lower threshold voltage** (about 0.2 V), a **larger reverse saturation current,** and a **smaller PIV** than the conventional *p-n* junction variety. It can also be used at higher frequencies because of the reduced reverse recovery time.

2. The varactor (varicap) diode has a **transition capacitance** sensitive to the applied reverse-bias potential that is a maximum at zero volts and that **decreases exponentially** with increasing reverse-bias potentials.

3. The **current capability** of power diodes can be increased by placing two or more in **parallel,** and the **PIV rating** can be increased by stacking the diodes in series.

4. The chassis itself can be used as a **heat sink** for power diodes.

5. **Tunnel diodes** are unique in that they have a **negative resistance region** at voltage levels less than the typical *p-n* junction threshold voltage. This characteristic is particularly useful in oscillators to establish an oscillating waveform from a switched dc power supply. Due to its reduced depletion region, it is also considered a **high-frequency device** for applications where switching times in nanoseconds or picoseconds are required.

6. The region of operation for **photodiodes** is the **reverse-bias region.** The resulting diode current increases almost **linearly** with an increase in incident light. The **wavelength** of the incident light will determine which material will result in the best response; selenium has a good match with the naked eye, and silicon is better for incident light with higher wavelengths.

7. A photoconductive cell is one whose terminal resistance will **decrease exponentially** with an **increase in incident light.**

8. An **infrared-emitting diode** will emit a beam of radiant flux when **forward-biased.** The strength of the emitted flux pattern is almost **linearly related** to the dc forward current through the device.

9. **LCDs** have a much **lower power absorption level** than LEDs, but their lifetime is much **shorter,** and they require an **internal or external light source.**

10. The **solar cell** is capable of converting light energy in the form of photons into electrical energy in the form of a difference in potential or **voltage.** The terminal voltage will **initially increase quite rapidly** with the application of light, but then the increase will occur at an increasingly **slower rate.** In other words, the terminal voltage will reach a **saturation level** at some point where any further increase in incident light will have little effect on the magnitude of the terminal voltage.

11. A **thermistor** can have regions of **positive or negative temperature coefficients** determined by the construction material or the temperature of the material. The change in temperature can be due to **internal effects** such as caused by the current through the thermistor or due to **external effects** of heating or cooling.

Equations

Varactor diode:

$$C_T(V_R) = \frac{C(0)}{(1 + |V_R/V_T|)^n}$$

where

$$n = 1/2 \text{ alloy junction}$$

$$n = 1/3 \text{ diffused junction}$$

$$TC_C = \frac{\Delta C}{C_0(T_1 - T_0)} \times 100\% \qquad \%/°C$$

Photodiodes:

$$\lambda = \frac{v}{f} = \frac{3 \times 10^8 \text{ m/s}}{f}$$

$$1 \text{ Å} = 10^{-10} \text{ m} \qquad \text{and} \qquad 1 \text{ lm} = 1.496 \times 10^{-10} \text{ W}$$

$$1 \text{ fc} = 1 \text{ lm/ft}^2 = 1.609 \times 10^{-9} \text{ W/m}^2$$

Solar cells:

$$\eta = \frac{P_{o(\text{electrical})}}{P_{i(\text{light energy})}} \times 100\%$$

$$= \frac{P_{\max(\text{device})}}{(\text{area in cm}^2)(100 \text{ mW/cm}^2)} \times 100\%$$

PROBLEMS

§ 19.2 Schottky Barrier (Hot-Carrier) Diodes

1. (a) Describe in your own words how the construction of the hot-carrier diode is significantly different from the conventional semiconductor diode.
   (b) In addition, describe its mode of operation.

2. (a) Consult Fig. 19.2. How would you compare the dynamic resistances of the diodes in the forward-bias regions?
   (b) How do the levels of I_s and V_Z compare?

3. Referring to Fig. 19.5, how does the maximum surge current I_{FSM} relate to the average rectified forward current? Is it typically greater than 20:1? Why is it possible to have such high levels of current? What noticeable difference is there in construction as the current rating increases?

4. Referring to Fig. 19.6a, at what temperature is the forward voltage drop 300 mV at a current of 1 mA? Which current levels have the highest levels of temperature coefficients? Assume a linear progression between temperature levels.

*5. For the curve of Fig. 19.6b denoted 2900/2303, determine the percent change in I_R for a change in reverse voltage from 5 to 10 V. At what reverse voltage would you expect to reach a reverse current of 1 μA? Note the log scale for I_R.

*6. Determine the percent change in capacitance between 0 and 2 V for the 2900/2303 curve of Fig. 19.6c. How does this compare to the change between 8 and 10 V?

§ 19.3 Varactor (Varicap) Diodes

7. (a) Determine the transition capacitance of a diffused junction varicap diode at a reverse potential of 4.2 V if $C(0) = 80$ pF and $V_T = 0.7$ V.
   (b) From the information of part (a), determine the constant K in Eq. (19.2).

8. (a) For a varicap diode having the characteristics of Fig. 19.7, determine the difference in capacitance between reverse-bias potentials of -3 and -12 V.
   (b) Determine the incremental rate of change $(\Delta C/\Delta V_r)$ at $V = -8$ V. How does this value compare with the incremental change determined at -2 V?

*9. (a) The resonant frequency of a series RLC network is determined by $f_0 = 1/(2\pi\sqrt{LC})$. Using the value of f_0 and L_S provided in Fig. 19.9, determine the value of C.
   (b) How does the value calculated in part (a) compare with that determined by the curve in Fig. 19.10 at $V_R = 25$ V?

10. Referring to Fig. 19.10, determine the ratio of capacitance at $V_R = 3$ V to $V_R = 25$ V and compare to the value of C_3/C_{25} given in Fig. 19.9 (maximum = 6.5).

11. Determine T_1 for a varactor diode if $C_0 = 22$ pF, $TC_C = 0.02\%/°C$, and $\Delta C = 0.11$ pF due to an increase in temperature above $T_0 = 25°C$.

12. What region of V_R would appear to have the greatest change in capacitance per change in reverse voltage for the BB139 varactor diode of Figs. 19.9 and 19.10? Be aware that the scales are nonlinear.

*13. If $Q = X_L/R = 2\pi fL/R$, determine the figure of merit (Q) at 600 MHz using the fact that $R_S = 0.35 \ \Omega$ and $L_S = 2.5$ nH. Comment on the change in Q with frequency and the support or nonsupport of the curve in Fig. 19.10.

§ 19.4 Power Diodes

14. Consult a manufacturer's data book and compare the general characteristics of a high-power device (>10 A) to a low-power unit (<100 mA). Is there a significant change in the data and characteristics provided? Why?

Chapter 19   Other Two-Terminal Devices

§ 19.5 Tunnel Diodes

15. What are the essential differences between a semiconductor junction diode and a tunnel diode?

*16. Note in the equivalent circuit of Fig. 19.14 that the capacitor appears in parallel with the negative resistance. Determine the reactance of the capacitor at 1 MHz and 100 MHz if $C = 5$ pF, and determine the total impedance of the parallel combination (with $R = -152\ \Omega$) at each frequency. Is the magnitude of the inductive reactance anything to be overly concerned about at either of these frequencies if $L_S = 6$ nH?

*17. Why do you believe the maximum reverse current rating for the tunnel diode can be greater than the forward current rating? (*Hint*: Note the characteristics and consider the power rating.)

18. Determine the negative resistance for the tunnel diode of Fig. 19.13 between $V_T = 0.1$ V and $V_T = 0.3$ V.

19. Determine the stable operating points for the network of Fig. 19.17 if $E = 2$ V, $R = 0.39$ kΩ, and the tunnel diode of Fig. 19.13 is employed. Use typical values from Table 19.1.

*20. For $E = 0.5$ V and $R = 51\ \Omega$, sketch v_T for the network of Figure 19.18 and the tunnel diode of Fig. 19.13.

21. Determine the frequency of oscillation for the network of Fig. 19.19 if $L = 5$ mH, $R_1 = 10\ \Omega$, and $C = 1\ \mu$F.

§ 19.6 Photodiodes

22. Determine the energy associated with the photons of green light if the wavelength is 5000 Å. Give your answer in joules and electron volts.

23. (a) Referring to Fig. 19.20, what would appear to be the frequencies associated with the upper and lower limits of the visible spectrum?
    (b) What is the wavelength in microns associated with the peak relative response of silicon?
    (c) If we define the bandwidth of the spectral response of each material to occur at 70% of its peak level, what is the bandwidth of silicon?

24. Referring to Fig. 19.22, determine I_λ if $V_\lambda = 30$ V and the light intensity is 4×10^{-9} W/m^2.

25. (a) Which material of Fig. 19.20 would appear to provide the best response to yellow, red, green, and infrared (less than 11,000 Å) light sources?
    (b) At a frequency of 0.5×10^{15} Hz, which color has the maximum spectral response?

*26. Determine the voltage drop across the resistor of Fig. 19.21 if the incident flux is 3000 fc, $V_\lambda = 25$ V, and $R = 100$ kΩ. Use the characteristics of Fig. 19.22.

§ 19.7 Photoconductive Cells

*27. What is the approximate rate of change of resistance with illumination for a photoconductive cell with the characteristics of Fig. 19.28 for the ranges (a) $0.1 \rightarrow 1$ kΩ, (b) $1 \rightarrow 10$ kΩ, and (c) $10 \rightarrow$ kΩ? (Note that this is a log scale.) Which region has the greatest rate of change in resistance with illumination?

28. What is the "dark current" of a photodiode?

29. If the illumination on the photoconductive diode in Fig. 19.30 is 10 fc, determine the magnitude of V_i to establish 6 V across the cell if R_1 is equal to 5 kΩ. Use the characteristics of Fig. 19.28.

*30. Using the data provided in Fig. 19.29, sketch a curve of percent conductance versus temperature for 0.01, 1.0, and 100 fc. Are there any noticeable effects?

*31. (a) Sketch a curve of rise time versus illumination using the data from Fig. 19.29.
    (b) Repeat part (a) for the decay time.
    (c) Discuss any noticeable effects of illumination in parts (a) and (b).

32. Which colors is the CdS unit of Fig. 19.29 most sensitive to?

§ 19.8 IR Emitters

33. (a) Determine the radiant flux at a dc forward current of 70 mA for the device of Fig. 19.32.
    (b) Determine the radiant flux in lumens at a dc forward current of 45 mA.

∗34. (a) Through the use of Fig. 19.33, determine the relative radiant intensity at an angle of 25°
    for a package with a flat glass window.

   (b) Plot a curve of relative radiant intensity versus degrees for the flat package.

∗35. If 60 mA of dc forward current is applied to an SG1010A IR emitter, what will be the incident
   radiant flux in lumens 5° off the center if the package has an internal collimating system? Refer to Figs. 19.32 and 19.33.

§ 19.9 Liquid-Crystal Displays

36. Referring to Fig. 19.37, which terminals must be energized to display number 7?

37. In your own words, describe the basic operation of an LCD.

38. Discuss the relative differences in mode of operation between an LED and an LCD display.

39. What are the relative advantages and disadvantages of an LCD display as compared to an LED
   display?

§ 19.10 Solar Cells

40. A 1-cm by 2-cm solar cell has a conversion efficiency of 9%. Determine the maximum power
   rating of the device.

∗41. If the power rating of a solar cell is determined on a very rough scale by the product $V_{OC} I_{SC}$,
   is the greatest rate of increase obtained at lower or higher levels of illumination? Explain your
   reasoning.

42. (a) Referring to Fig. 19.48, what power density is required to establish a current of 24 mA at
    an output voltage of 0.25 V?

   (b) Why is 100 mW/cm^2 the maximum power density in Fig. 19.48?

   (c) Determine the output current if the power is 40 mW/cm^2 and the output voltage is 0.3 V.

∗43. (a) Sketch a curve of output current versus power density at an output voltage of 0.15 V using
    the characteristics of Fig. 19.48.

   (b) Sketch a curve of output voltage versus power density at a current of 19 mA.

   (c) Is either of the curves from parts (a) and (b) linear within the limits of the maximum power
    limitation?

§ 19.11 Thermistors

∗44. For the thermistor of Fig. 19.49, determine the dynamic rate of change in specific resistance
   with temperature at $T = 20°C$. How does this compare to the value determined at $T = 300°C$?
   From the results, determine whether the greatest change in resistance per unit change in temperature occurs at lower or higher levels of temperature. Note the vertical log scale.

45. Using the information provided in Fig. 19.49, determine the total resistance of a 2-cm length
   of the material having a perpendicular surface area of 1 cm^2 at a temperature of 0°C. Note the
   vertical log scale.

46. (a) Referring to Fig. 19.50, determine the current at which a 25°C sample of the material changes
    from a positive to negative temperature coefficient. (Figure 19.50 is a log scale.)

   (b) Determine the power and resistance levels of the device (Fig. 19.50) at the peak of the 0°C
    curve.

   (c) At a temperature of 25°C, determine the power rating if the resistance level is 1 MΩ.

47. In Fig. 19.52, $V = 0.2$ V and $R_{variable} = 10$ Ω. If the current through the sensitive movement is
   2 mA and the voltage drop across the movement is 0 V, what is the resistance of the thermistor?

---

*Please Note: Asterisks indicate more difficult problems.

20.1 INTRODUCTION

In this chapter, a number of important devices not discussed in detail in earlier chapters are introduced. The two-layer semiconductor diode has led to three-, four-, and even five-layer devices. A family of four-layer *pnpn* devices will first be considered: SCR (silicon-controlled rectifier), SCS (silicon-controlled switch), GTO (gate turn-off switch), LASCR (light-activated SCR), followed by an increasingly important device—the UJT (unijunction transistor). Those four-layer devices with a control mechanism are commonly referred to as *thyristors,* although the term is most frequently applied to the SCR (silicon-controlled rectifier). The chapter closes with an introduction to the phototransistor, opto-isolators, and the PUT (programmable unijunction transistor).

pnpn DEVICES

20.2 SILICON-CONTROLLED RECTIFIER

Within the family of *pnpn* devices, the silicon-controlled rectifier (SCR) is unquestionably of the greatest interest today. It was first introduced in 1956 by Bell Telephone Laboratories. A few of the more common areas of application for SCRs include relay controls, time-delay circuits, regulated power suppliers, static switches, motor controls, choppers, inverters, cycloconverters, battery chargers, protective circuits, heater controls, and phase controls.

In recent years, SCRs have been designed to *control* powers as high as 10 MW with individual ratings as high as 2000 A at 1800 V. Its frequency range of application has also been extended to about 50 kHz, permitting some high-frequency applications such as induction heating and ultrasonic cleaning.

20.3 BASIC SILICON-CONTROLLED RECTIFIER OPERATION

As the terminology indicates, the SCR is a rectifier constructed of silicon material with a third terminal for control purposes. Silicon was chosen because of its high temperature and power capabilities. The basic operation of the SCR is different from the fundamental two-layer semiconductor diode in that a third terminal, called a *gate,* determines when the rectifier switches from the open-circuit to short-circuit state. It is

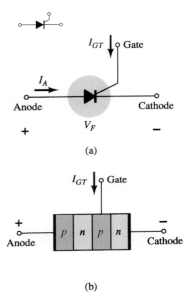

Figure 20.1 (a) SCR symbol; (b) basic construction.

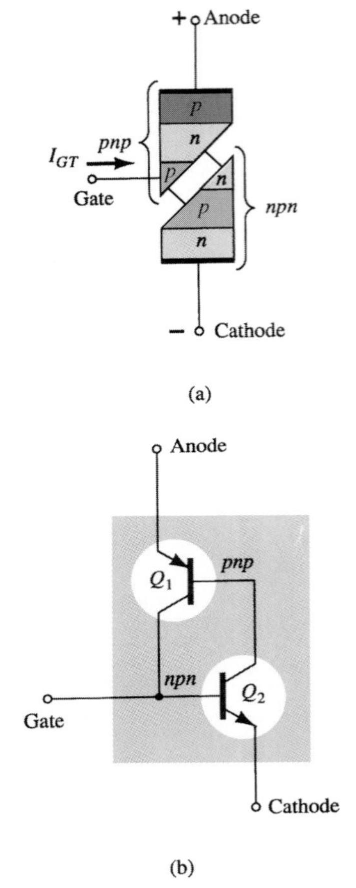

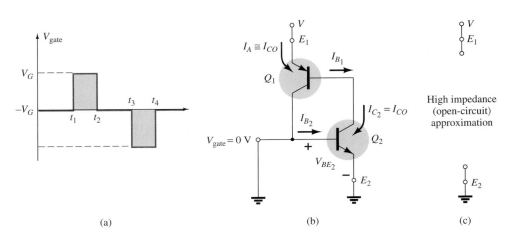

Figure 20.2 SCR two-transistor equivalent circuit.

not enough to simply forward-bias the anode-to-cathode region of the device. In the conduction region, the dynamic resistance of the SCR is typically 0.01 to 0.1 Ω. The reverse resistance is typically 100 kΩ or more.

The graphic symbol for the SCR is shown in Fig. 20.1 with the corresponding connections to the four-layer semiconductor structure. As indicated in Fig. 20.1a, if forward conduction is to be established, the anode must be positive with respect to the cathode. This is not, however, a sufficient criterion for turning the device on. A pulse of sufficient magnitude must also be applied to the gate to establish a turn-on gate current, represented symbolically by I_{GT}.

A more detailed examination of the basic operation of an SCR is best effected by splitting the four-layer *pnpn* structure of Fig. 20.1b into two three-layer transistor structures as shown in Fig. 20.2a and then considering the resultant circuit of Fig. 20.2b.

Note that one transistor for Fig. 20.2 is an *npn* device while the other is a *pnp* transistor. For discussion purposes, the signal shown in Fig. 20.3a will be applied to the gate of the circuit of Fig. 20.2b. During the interval $0 \rightarrow t_1$, $V_{\text{gate}} = 0$ V, the circuit of Fig. 20.2b will appear as shown in Fig. 20.3b ($V_{\text{gate}} = 0$ V is equivalent to the gate terminal being grounded as shown in the figure). For $V_{B_{E_2}} = V_{\text{gate}} = 0$ V, the base current $I_{B_2} = 0$ and I_{C_2} will be approximately I_{CO}. The base current of Q_1, $I_{B_1} = I_{C_2} = I_{CO}$, is too small to turn Q_1 on. Both transistors are therefore in the "off" state, resulting in a high impedance between the collector and emitter of each transistor and the open-circuit representation for the controlled rectifier as shown in Fig. 20.3c.

Figure 20.3 "Off" state of the SCR.

At $t = t_1$, a pulse of V_G volts will appear at the SCR gate. The circuit conditions established with this input are shown in Fig. 20.4a. The potential V_G was chosen sufficiently large to turn Q_2 on ($V_{BE_2} = V_G$). The collector current of Q_2 will then rise to a value sufficiently large to turn Q_1 on ($I_{B_1} = I_{C_2}$). As Q_1 turns on, I_{C_1} will increase, resulting in a corresponding increase in I_{B_2}. The increase in base current for Q_2 will result in a further increase in I_{C_2}. The net result is a regenerative increase in the collector current of each transistor. The resulting anode-to-cathode resistance ($R_{SCR} = V/I_A$) is then small because I_A is large, resulting in the short-circuit representation for the SCR as indicated in Fig. 20.4b. The regenerative action described above results in SCRs having typical turn-on times of 0.1 to 1 μs. However, high-power devices in the range 100 to 400 A may have 10- to 25-μs turn-on times.

In addition to gate triggering, SCRs can also be turned on by significantly raising the temperature of the device or raising the anode-to-cathode voltage to the breakover value shown on the characteristics of Fig. 20.7.

The next question of concern is: How long is the turn-off time and how is turn-off accomplished? An SCR *cannot* be turned off by simply removing the gate signal, and only a special few can be turned off by applying a negative pulse to the gate terminal as shown in Fig. 20.3a at $t = t_3$.

The two general methods for turning off an SCR are categorized as the anode current interruption and the forced-commutation technique.

The two possibilities for current interruption are shown in Fig. 20.5. In Fig. 20.5a, I_A is zero when the switch is opened (series interruption), while in Fig. 20.5b, the same condition is established when the switch is closed (shunt interruption).

Forced commutation is the "forcing" of current through the SCR in the direction opposite to forward conduction. There are a wide variety of circuits for performing this function, a number of which can be found in the manuals of major manufacturers in this area. One of the more basic types is shown in Fig. 20.6. As indicated in the figure, the turn-off circuit consists of an *npn* transistor, a dc battery V_B, and a pulse generator. During SCR conduction, the transistor is in the "off" state, that is, $I_B = 0$ and the collector-to-emitter impedance is very high (for all practical purposes an open circuit). This high impedance will isolate the turn-off circuitry from affecting the operation of the SCR. For turn-off conditions, a positive pulse is applied to the base of the transistor, turning it heavily on, resulting in a very low impedance from collector to emitter (short-circuit representation). The battery potential will then appear directly across the SCR as shown in Fig. 20.6b, forcing current through it in the reverse direction for turn-off. Turn-off times of SCRs are typically 5 to 30 μs.

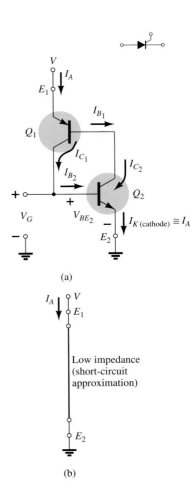

(a)

(b)

Figure 20.4 "On" state of the SCR.

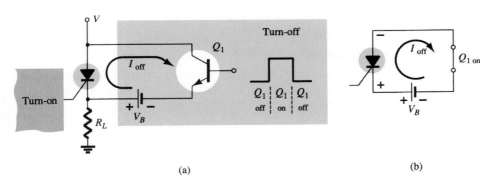

(a)

(b)

Figure 20.6 Forced-commutation technique.

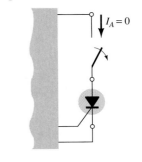

(a)

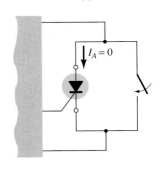

(b)

Figure 20.5 Anode current interruption.

20.4 SCR CHARACTERISTICS AND RATINGS

The characteristics of an SCR are provided in Fig. 20.7 for various values of gate current. The currents and voltages of usual interest are indicated on the characteristic. A brief description of each follows.

1. *Forward breakover voltage* $V_{(BR)F*}$ is that voltage above which the SCR enters the conduction region. The asterisk ($*$) is a letter to be added that is dependent on the condition of the gate terminal as follows:

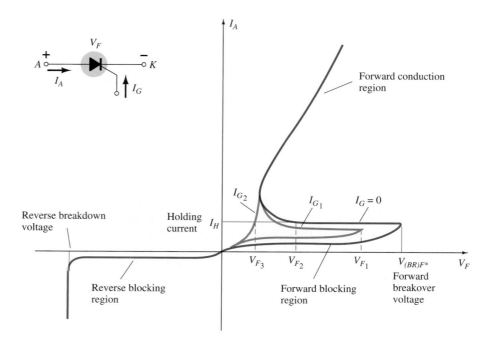

Figure 20.7 SCR characteristics.

$$O = \text{open circuit from } G \text{ to } K$$
$$S = \text{short circuit from } G \text{ to } K$$
$$R = \text{resistor from } G \text{ to } K$$
$$V = \text{fixed bias (voltage) from } G \text{ to } K$$

2. *Holding current* (I_H) is that value of current below which the SCR switches from the conduction state to the forward blocking region under stated conditions.

3. *Forward and reverse blocking regions* are the regions corresponding to the open-circuit condition for the controlled rectifier which *block* the flow of charge (current) from anode to cathode.

4. *Reverse breakdown voltage* is equivalent to the Zener or avalanche region of the fundamental two-layer semiconductor diode.

It should be immediately obvious that the SCR characteristics of Fig. 20.7 are very similar to those of the basic two-layer semiconductor diode except for the horizontal offshoot before entering the conduction region. It is this horizontal jutting region that gives the gate control over the response of the SCR. For the characteristic having the solid blue line in Fig. 20.7 ($I_G = 0$), V_F must reach the largest required breakover voltage ($V_{(BR)F*}$) before the "collapsing" effect will result and the SCR can enter the conduction region corresponding to the *on* state. If the gate current is increased to I_{G_1}, as shown in the same figure by applying a bias voltage to the gate terminal, the value of V_F required for the conduction (V_{F_1}) is considerably less. Note also that I_H drops with increase in I_G. If increased to I_{G_2}, the SCR will fire at very low values of voltage (V_{F_3}) and the characteristics begin to approach those of the basic *p-n* junction diode. Looking at the characteristics in a completely different sense, for a particular V_F voltage, say V_{F_2} (Fig. 20.7), if the gate current is increased from $I_G = 0$ to I_{G_1} or more, the SCR will fire.

The gate characteristics are provided in Fig. 20.8. The characteristics of Fig. 20.8b are an expanded version of the shaded region of Fig. 20.8a. In Fig. 20.8a, the three gate ratings of greatest interest, P_{GFM}, I_{GFM}, and V_{GFM}, are indicated. Each is included on the characteristics in the same manner employed for the transistor. Except for portions of the shaded region, any combination of gate current and voltage that falls within this region will fire any SCR in the series of components for which these characteristics are provided. Temperature will determine which sections of the shaded region must be avoided. At −65°C the minimum current that will trigger the series of SCRs is 100 mA, while at +150°C only 20 mA are required. The effect of temperature on the minimum gate voltage is usually not indicated on curves of this type since gate potentials of 3 V or more are usually obtained easily. As indicated on Fig. 20.8b, a minimum of 3 V is indicated for all units for the temperature range of interest.

Other parameters usually included on the specification sheet of an SCR are the turn-on time (t_{on}), turn-off time (t_{off}), junction temperature (T_J), and case temperature (T_C), all of which should by now be, to some extent, self-explanatory.

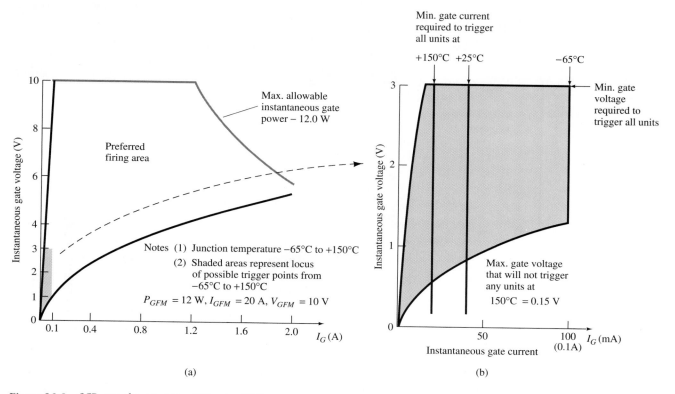

Figure 20.8 SCR gate characteristics (GE series C38).

20.5 SCR CONSTRUCTION AND TERMINAL IDENTIFICATION

The basic construction of the four-layer pellet of an SCR is shown in Fig. 20.9a. The complete construction of a thermal fatigue-free, high-current SCR is shown in Fig. 20.9b. Note the position of the gate, cathode, and anode terminals. The pedestal acts as a heat sink by transferring the heat developed to the chassis on which the SCR is mounted. The case construction and terminal identification of SCRs will vary with the application. Other case-construction techniques and the terminal identification of each are indicated in Fig. 20.10.

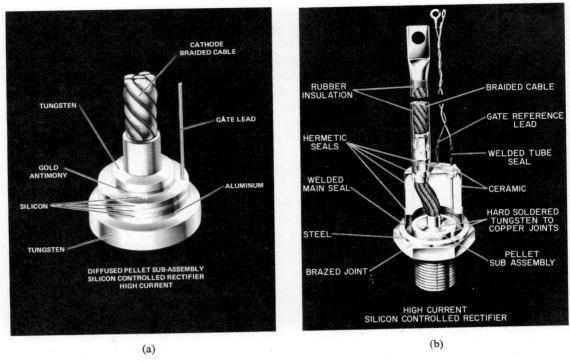

Figure 20.9 (a) Alloy-diffused SCR pellet; (b) thermal fatigue-free SCR construction. (Courtesy General Electric Company.)

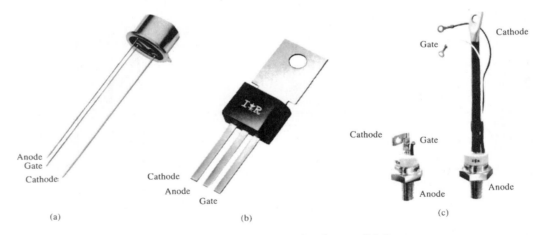

Figure 20.10 SCR case construction and terminal identification. [(a) Courtesy General Electric Company; (b) and (c) courtesy International Rectifier Corporation.]

20.6 SCR APPLICATIONS

A few of the possible applications for the SCR are listed in the introduction to the SCR (Section 20.2). In this section, we consider five: a static switch, phase-control system, battery charger, temperature controller, and single-source emergency-lighting system.

Series Static Switch

A half-wave *series static switch* is shown in Fig. 20.11a. If the switch is closed as shown in Fig. 20.11b, a gate current will flow during the positive portion of the input signal, turning the SCR on. Resistor R_1 limits the magnitude of the gate current. When the SCR turns on, the anode-to cathode voltage (V_F) will drop to the conduction value, resulting in a greatly reduced gate current and very little loss in the gate circuitry. For the negative region of the input signal, the SCR will turn off since the anode is negative with respect to the cathode. The diode D_1 is included to prevent a reversal in gate current.

The waveforms for the resulting load current and voltage are shown in Fig. 20.11b. The result is a half-wave-rectified signal through the load. If less than 180° conduction is desired, the switch can be closed at any phase displacement during the positive portion of the input signal. The switch can be electronic, electromagnetic, or mechanical, depending on the application.

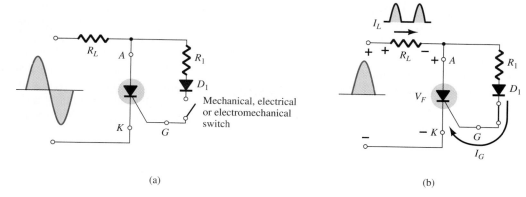

Figure 20.11 Half-wave series static switch.

Variable-Resistance Phase Control

A circuit capable of establishing a conduction angle between 90° and 180° is shown in Fig. 20.12a. The circuit is similar to that of Fig. 20.11a except for the addition of a variable resistor and the elimination of the switch. The combination of the resistors R and R_1 will limit the gate current during the positive portion of the input signal. If R_1 is set to its maximum value, the gate current may never reach turn-on magnitude. As R_1 is decreased from the maximum, the gate current will increase from the same input voltage. In this way, the required turn-on gate current can be established in any point between 0° and 90° as shown in Fig. 20.12b. If R_1 is low, the SCR will fire almost immediately, resulting in the same action as that obtained from the circuit of Fig. 20.11a (180° conduction). However, as indicated above, if R_1 is increased, a larger input voltage (positive) will be required to fire the SCR. As shown in Fig. 20.12b, the control cannot be extended past a 90° phase displacement since the input is at its maximum at this point. If it fails to fire at this and lesser values of input voltage on the

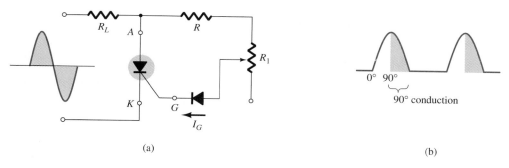

Figure 20.12 Half-wave variable-resistance phase control.

positive slope of the input, the same response must be expected from the negatively sloped portion of the signal waveform. The operation here is normally referred to in technical terms as *half-wave variable-resistance phase control*. It is an effective method of controlling the rms current and therefore power to the load.

Battery-Charging Regulator

A third popular application of the SCR is in a *battery-charging regulator*. The fundamental components of the circuit are shown in Fig. 20.13. You will note that the control circuit has been blocked off for discussion purposes.

As indicated in the figure, D_1 and D_2 establish a full-wave-rectified signal across SCR_1 and the 12-V battery to be charged. At low battery voltages, SCR_2 is in the "off" state for reasons to be explained shortly. With SCR_2 open, the SCR_1 controlling circuit is exactly the same as the series static switch control discussed earlier in this section. When the full-wave-rectified input is sufficiently large to produce the required turn-on gate current (controlled by R_1), SCR_1 will turn on and charging of the battery will commence. At the start of charging, the low battery voltage will result in a low voltage V_R as determined by the simple voltage-divider circuit. Voltage V_R is in turn too small to cause 11.0-V Zener conduction. In the off state, the Zener is effectively an open circuit, maintaining SCR_2 in the "off" state since the gate current is zero. The capacitor C_1 is included to prevent any voltage transients in the circuit from accidentally turning on SCR_2. Recall from your fundamental study of circuit analysis that the voltage cannot change instantaneously across a capacitor. In this way, C_1 prevents transient effects from affecting the SCR.

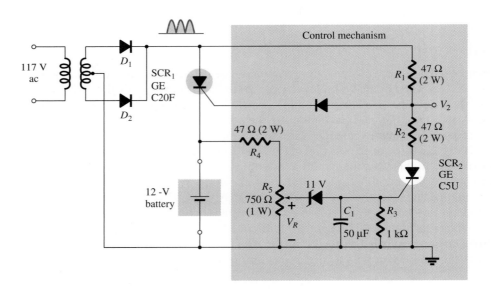

Figure 20.13 Battery-charging regulator.

As charging continues, the battery voltage rises to a point where V_R is sufficiently high to both turn on the 11.0-V Zener and fire SCR_2. Once SCR_2 has fired, the short-circuit representation for SCR_2 will result in a voltage-divider circuit determined by R_1 and R_2 that will maintain V_2 at a level too small to turn SCR_1 on. When this occurs, the battery is fully charged and the open-circuit state of SCR_1 will cut off the charging current. Thus the regulator recharges the battery whenever the voltage drops and prevents overcharging when fully charged.

Temperature Controller

The schematic diagram of a 100-W heater control using an SCR appears in Fig. 20.14. It is designed such that the 100-W heater will turn on and off as determined by thermostats. Mercury-in-glass thermostats are very sensitive to temperature change. In fact, they can sense changes as small as 0.1°C. It is limited in application, however, in that it can handle only very low levels of current—below 1 mA. In this application, the SCR serves as a current amplifier in a load-switching element. It is not an amplifier in the sense that it magnifies the current level of the thermostat. Rather it is a device whose higher current level is controlled by the behavior of the thermostat.

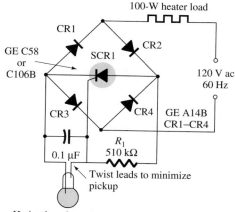

Figure 20.14 Temperature controller. (Courtesy General Electric Semiconductor Products Division.)

It should be clear that the bridge network is connected to the ac supply through the 100-W heater. This will result in a full-wave-rectified voltage across the SCR. When the thermostat is open, the voltage across the capacitor will charge to a gate-firing potential through each pulse of the rectified signal. The charging time constant is determined by the RC product. This will trigger the SCR during each half-cycle of the input signal, permitting a flow of charge (current) to the heater. As the temperature rises, the conductive thermostat will short-circuit the capacitor, eliminating the possibility of the capacitor charging to the firing potential and triggering the SCR. The 510-kΩ resistor will then contribute to maintaining a very low current (less than 250 μA) through the thermostat.

Emergency-Lighting System

The last application for the SCR to be described is shown in Fig. 20.15. It is a single-source emergency-lighting system that will maintain the charge on a 6-V battery to ensure its availability and also provide dc energy to a bulb if there is a power shortage. A full-wave-rectified signal will appear across the 6-V lamp due to diodes D_2 and D_1. The capacitor C_1 will charge to a voltage slightly less than a difference between the peak value of the full-wave-rectified signal and the dc voltage across R_2 established by the 6-V battery. In any event, the cathode of SCR$_1$ is higher than the anode and the gate-to-cathode voltage is negative, ensuring that the SCR is nonconducting. The battery is being charged through R_1 and D_1 at a rate determined by R_1. Charging will only take place when the anode of D_1 is more positive than its cathode. The dc level of the full-wave-rectified signal will ensure that the bulb is lit when the power is on. If the power should fail, the capacitor C_1 will discharge through D_1, R_1, and R_3 until the cathode of SCR$_1$ is less positive than the anode. At the same time,

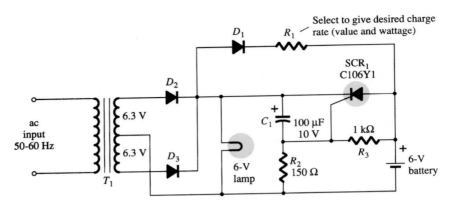

Figure 20.15 Single-source emergency-lighting system. (Courtesy General Electric Semiconductor Products Division.)

Figure 20.15 Single-source emergency-lighting system. (Courtesy General Electric Semiconductor Products Division.)

the junction of R_2 and R_3 will become positive and establish sufficient gate-to-cathode voltage to trigger the SCR. Once fired, the 6-V battery would discharge through the SCR_1 and energize the lamp and maintain its illumination. Once power is restored, the capacitor C_1 will recharge and re-establish the nonconducting state of SCR_1 as described above.

20.7 SILICON-CONTROLLED SWITCH

The silicon-controlled switch (SCS), like the silicon-controlled rectifier, is a four-layer *pnpn* device. All four semiconductor layers of the SCS are available due to the addition of an anode gate, as shown in Fig. 20.16a. The graphic symbol and transistor equivalent circuit are shown in the same figure. The characteristics of the device are essentially the same as those for the SCR. The effect of an anode gate current is very similar to that demonstrated by the gate current in Fig. 20.7. The higher the anode gate current, the lower the required anode-to-cathode voltage to turn the device on.

The anode gate connection can be used to turn the device either on or off. To turn on the device, a negative pulse must be applied to the anode gate terminal, while a positive pulse is required to turn off the device. The need for the type of pulse indicated above can be demonstrated using the circuit of Fig. 20.16c. A negative pulse at the anode gate will forward-bias the base-to-emitter junction of Q_1, turning it on. The resulting heavy collector current I_{C_1} will turn on Q_2, resulting in a regenerative action and the on state for the SCS device. A positive pulse at the anode gate will reverse-bias the base-to-emitter junction of Q_1, turning it off, resulting in the open-circuit

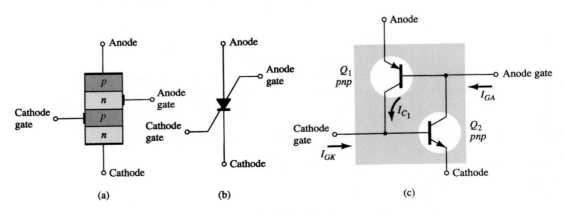

Figure 20.16 Silicon-controlled switch (SCS): (a) basic construction; (b) graphic symbol; (c) equivalent transistor circuit.

Chapter 20 *pnpn* and Other Devices

"off" state of the device. In general, the triggering (turn-on) anode gate current is larger in magnitude than the required cathode gate current. For one representative SCS device, the triggering anode gate current is 1.5 mA while the required cathode gate current is 1 μA. The required turn-on gate current at either terminal is affected by many factors. A few include the operating temperature, anode-to-cathode voltage, load placement, and type of cathode, gate-to-cathode or anode gate-to-anode connection (short-circuit, open-circuit, bias, load, etc.). Tables, graphs, and curves are normally available for each device to provide the type of information indicated above.

Three of the more fundamental types of turn-off circuits for the SCS are shown in Fig. 20.17. When a pulse is applied to the circuit of Fig. 20.17a, the transistor conducts heavily, resulting in a low-impedance ($\cong$ short-circuit) characteristic between collector and emitter. This low-impedance branch diverts anode current away from the SCS, dropping it below the holding value and consequently turning it off. Similarly, the positive pulse at the anode gate of Fig. 20.17b will turn the SCS off by the mechanism described earlier in this section. The circuit of Fig. 20.17c can be turned either off *or* on by a pulse of the proper magnitude at the cathode gate. The turn-off characteristic is possible only if the correct value of R_A is employed. It will control the amount of regenerative feedback, the magnitude of which is critical for this type of operation. Note the variety of positions in which the load resistor R_L can be placed. There are a number of other possibilities that can be found in any comprehensive semiconductor handbook or manual.

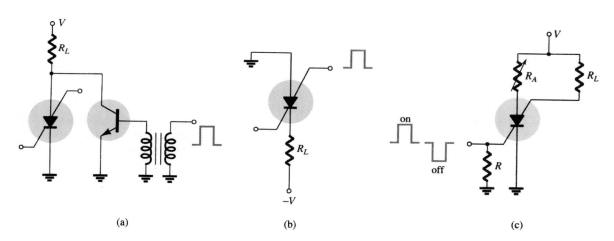

(a)   (b)   (c)

Figure 20.17 SCS turn-off techniques.

An advantage of the SCS over a corresponding SCR is the reduced turn-off time, typically within the range 1 to 10 μs for the SCS and 5 to 30 μs for the SCR. Some of the remaining advantages of the SCS over an SCR include increased control and triggering sensitivity and a more predictable firing situation. At present, however, the SCS is limited to low power, current, and voltage ratings. Typical maximum anode currents range from 100 to 300 mA with dissipation (power) ratings of 100 to 500 mW.

Voltage Sensor

Sensitivity to temperature-, light-, or radiation-sensitive resistors whose resistance increases due to the application of any of the three energy sources described above can be accommodated by simply interchanging the location of R_S and the variable resistor. The terminal identification of an SCS is shown in Fig. 20.18 with a packaged SCS.

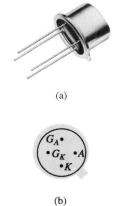

(a)

(b)

Figure 20.18 Silicon-controlled switch (SCS): (a) device; (b) terminal identification. (Courtesy General Electric Company.)

A few of the more common areas of application include a wide variety of computer circuits (counters, registers, and timing circuits), pulse generators, voltage sensors, and oscillators. One simple application for an SCS as a voltage-sensing device is shown in Fig. 20.19. It is an alarm system with *n* inputs from various stations. Any single input will turn that particular SCS on, resulting in an energized alarm relay and light in the anode gate circuit to indicate the location of the input (disturbance).

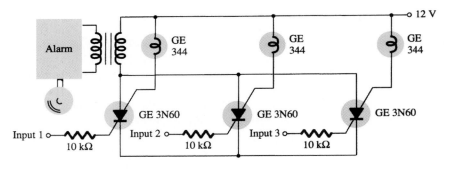

Figure 20.19 SCS alarm circuit.

Alarm Circuit

One additional application of the SCS is in the alarm circuit of Fig. 20.20. R_S represents a temperature-, light-, or radiation-sensitive resistor, that is, an element whose resistance will decrease with the application of any of the three energy sources listed above. The cathode gate potential is determined by the divider relationship established by R_S and the variable resistor. Note that the gate potential is at approximately 0 V if R_S equals the value set by the variable resistor since both resistors will have 12 V across them. However, if R_S decreases, the potential of the junction will increase until the SCS is forward-biased, causing the SCS to turn on and energize the alarm relay.

The 100-kΩ resistor is included to reduce the possibility of accidental triggering of the device through a phenomenon known as *rate effect*. It is caused by the stray capacitance levels between gates. A high-frequency transient can establish sufficient base current to turn the SCS on accidentally. The device is reset by pressing the reset button, which in turn opens the conduction path of the SCS and reduces the anode current to zero.

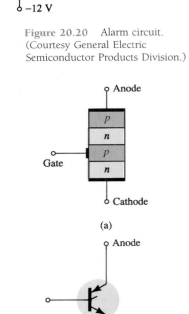

Figure 20.20 Alarm circuit. (Courtesy General Electric Semiconductor Products Division.)

Figure 20.21 Gate turn-off switch (GTO): (a) basic construction; (b) symbol.

20.8 GATE TURN-OFF SWITCH

The gate turn-off switch (GTO) is the third *pnpn* device to be introduced in this chapter. Like the SCR, however, it has only three external terminals, as indicated in Fig. 20.21a. Its graphical symbol is also shown in Fig. 20.21b. Although the graphical symbol is different from either the SCR or the SCS, the transistor equivalent is exactly the same and the characteristics are similar.

The most obvious advantage of the GTO over the SCR or SCS is the fact that it can be turned on *or* off by applying the proper pulse to the cathode gate (without the anode gate and associated circuitry required for the SCS). A consequence of this turn-off capability is an increase in the magnitude of the required gate current for triggering. For an SCR and GTO of similar maximum rms current ratings, the gate-triggering current of a particular SCR is 30 μA while the triggering current of the GTO is 20 mA. The turn-off current of a GTO is slightly larger than the required triggering current. The maximum rms current and dissipation ratings of GTOs manufactured today are limited to about 3 A and 20 W, respectively.

A second very important characteristic of the GTO is improved switching characteristics. The turn-on time is similar to the SCR (typically 1 μs), but the turn-off time of about the *same* duration (1 μs) is much smaller than the typical turn-off time of an SCR (5 to 30 μs). The fact that the turn-off time is similar to the turn-on time rather than considerably larger permits the use of this device in high-speed applications.

A typical GTO and its terminal identification are shown in Fig. 20.22. The GTO gate input characteristics and turn-off circuits can be found in a comprehensive manual or specification sheet. The majority of the SCR turn-off circuits can also be used for GTOs.

Sawtooth Generator

Some of the areas of application for the GTO include counters, pulse generators, multivibrators, and voltage regulators. Figure 20.23 is an illustration of a simple sawtooth generator employing a GTO and a Zener diode.

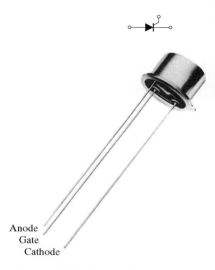

Figure 20.22 Typical GTO and its terminal identification. (Courtesy General Electric Company.)

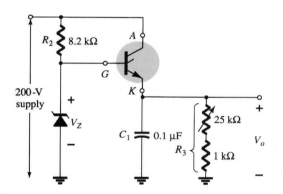

Figure 20.23 GTO sawtooth generator.

When the supply is energized, the GTO will turn on, resulting in the short-circuit equivalent from anode to cathode. The capacitor C_1 will then begin to charge toward the supply voltage as shown in Fig. 20.23. As the voltage across the capacitor C_1 charges above the Zener potential, a reversal in gate-to-cathode voltage will result, establishing a reversal in gate current. Eventually, the negative gate current will be large enough to turn the GTO off. Once the GTO turns off, resulting in the open-circuit representation, the capacitor C_1 will discharge through the resistor R_3. The discharge time will be determined by the circuit time constant $\tau = R_3 C_1$. The proper choice of R_3 and C_1 will result in the sawtooth waveform of Fig. 20.23. Once the output potential V_o drops below V_Z, the GTO will turn on and the process will repeat.

20.9 LIGHT-ACTIVATED SCR

The next in the series of *pnpn* devices is the light-activated SCR (LASCR). As indicated by the terminology, it is an SCR whose state is controlled by the light falling upon a silicon semiconductor layer of the device. The basic construction of an LASCR is shown in Fig. 20.24a. As indicated in Fig. 20.24a, a gate lead is also provided to permit triggering the device using typical SCR methods. Note also in the figure that the mounting surface for the silicon pellet is the anode connection for the device. The

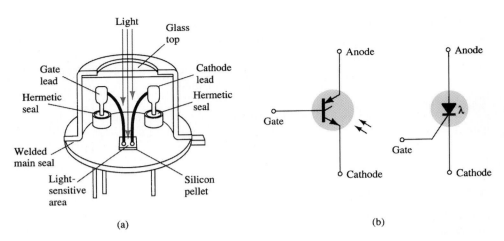

Figure 20.24 Light-activated SCR (LASCR): (a) basic construction; (b) symbols.

graphical symbols most commonly employed for the LASCR are provided in Fig. 20.24b. The terminal identification and a typical LASCR appear in Fig. 20.25a.

Some of the areas of application for the LASCR include optical light controls, relays, phase control, motor control, and a variety of computer applications. The maximum current (rms) and power (gate) ratings for LASCRs commercially available today are about 3 A and 0.1 W. The characteristics (light triggering) of a representative LASCR are provided in Fig. 20.25b. Note in this figure that an increase in junction temperature results in a reduction in light energy required to activate the device.

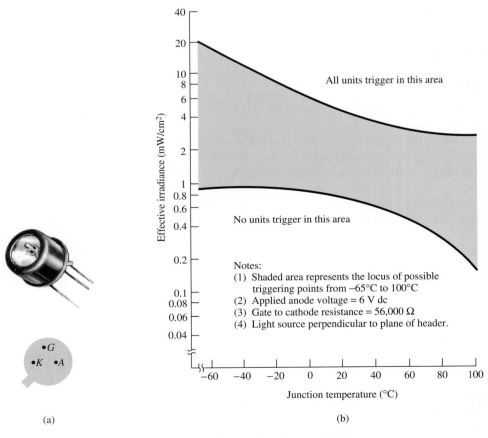

Figure 20.25 LASCR: (a) appearance and terminal identification; (b) light-triggering characteristics. (Courtesy General Electric Company.)

AND/OR Circuits

One interesting application of an LASCR is in the AND and OR circuits of Fig. 20.26. Only when light falls on LASCR$_1$ *and* LASCR$_2$ will the short-circuit representation for each be applicable and the supply voltage appear across the load. For the OR circuit, light energy applied to LASCR$_1$ *or* LASCR$_2$ will result in the supply voltage appearing across the load.

The LASCR is most sensitive to light when the gate terminal is open. Its sensitivity can be reduced and controlled somewhat by the insertion of a gate resistor, as shown in Fig. 20.26.

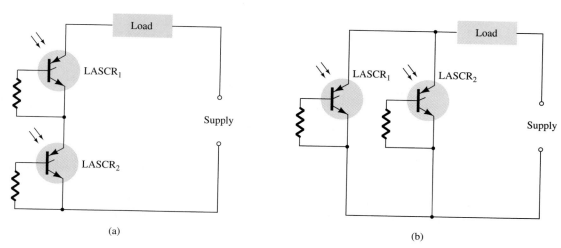

(a)   (b)

Figure 20.26 LASCR optoelectronic logic circuitry: (a) AND gate: input to LASCR$_1$ *and* LASCR$_2$ required for energization of the load; (b) OR gate: input to either LASCR$_1$ *or* LASCR$_2$ will energize the load.

Latching Relay

A second application of the LASCR appears in Fig. 20.27. It is the semiconductor analog of an electromechanical relay. Note that it offers complete isolation between the input and switching element. The energizing current can be passed through a light-emitting diode or a lamp, as shown in the figure. The incident light will cause the LASCR to turn on and permit a flow of charge (current) through the load as established by the dc supply. The LASCR can be turned off using the reset switch S_1. This system offers the additional advantages over an electromechanical switch of long life, microsecond response, small size, and the elimination of contact bounce.

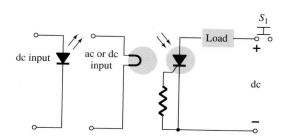

Figure 20.27 Latching relay.
(Courtesy Powerex, Inc.)

20.10 SHOCKLEY DIODE

The Shockley diode is a four-layer *pnpn* diode with only two external terminals, as shown in Fig. 20.28a with its graphical symbol. The characteristics (Fig. 20.28b) of the device are exactly the same as those encountered for the SCR with $I_G = 0$. As indicated by the characteristics, the device is in the off state (open-circuit representation) until the breakover voltage is reached, at which time avalanche conditions develop and the device turns on (short-circuit representation).

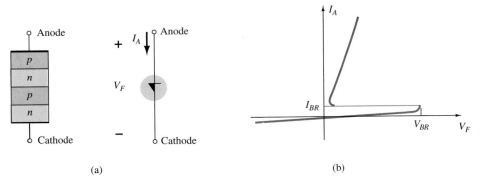

Figure 20.28 Shockley diode: (a) basic construction and symbol; (b) characteristics.

Trigger Switch

One common application of the Shockley diode is shown in Fig. 20.29, where it is employed as a trigger switch for an SCR. When the circuit is energized, the voltage across the capacitor will begin to change toward the supply voltage. Eventually, the voltage across the capacitor will be sufficiently high to first turn on the Shockley diode and then the SCR.

Figure 20.29 Shockley diode application—trigger switch for an SCR.

20.11 DIAC

The diac is basically a two-terminal parallel-inverse combination of semiconductor layers that permits triggering in either direction. The characteristics of the device, presented in Fig. 20.30a, clearly demonstrate that there is a breakover voltage in either direction. This possibility of an *on* condition in either direction can be used to its fullest advantage in ac applications.

The basic arrangement of the semiconductor layers of the diac is shown in Fig. 20.30b, along with its graphical symbol. Note that neither terminal is referred to

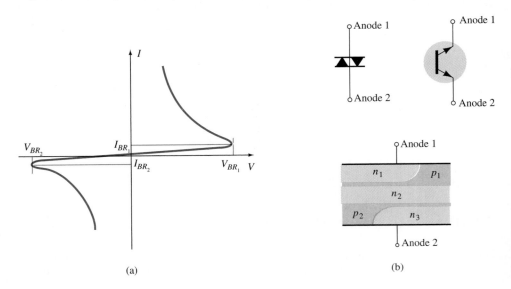

Figure 20.30 Diac: (a) characteristics; (b) symbols and basic construction. (Courtesy General Electric Company.)

Chapter 20 *pnpn* and Other Devices

as the cathode. Instead, there is an anode 1 (or electrode 1) and an anode 2 (or electrode 2). When anode 1 is positive with respect to anode 2, the semiconductor layers of particular interest are $p_1 n_2 p_2$ and n_3. For anode 2 positive with respect to anode 1, the applicable layers are $p_2 n_2 p_1$ and n_1.

For the unit appearing in Fig. 20.30, the breakdown voltages are very close in magnitude but may vary from a minimum of 28 V to a maximum of 42 V. They are related by the following equation provided in the specification sheet:

$$V_{BR_1} = V_{BR_2} \pm 0.1 V_{BR_2}$$

(20.1)

The current levels (I_{BR_1} and I_{BR_2}) are also very close in magnitude for each device. For the unit of Fig. 20.30, both current levels are about 200 μA = 0.2 mA.

Proximity Detector

The use of the diac in a proximity detector appears in Fig. 20.31. Note the use of an SCR in series with the load and the programmable unijunction transistor (to be described in Section 20.13) connected directly to the sensing electrode.

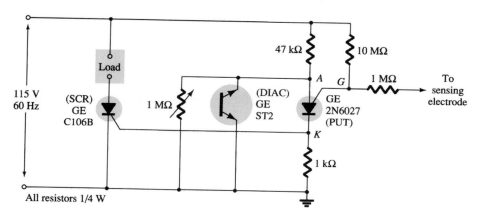

Figure 20.31   Proximity detector or touch switch. (Courtesy Powerex, Inc.)

As the human body approaches the sensing electrode, the capacitance between the electrode and ground will increase. The programmable UJT (PUT) is a device that will fire (enter the short-circuit state) when the anode voltage (V_A) is at least 0.7 V (for silicon) greater than the gate voltage (V_G). Before the programmable device turns on, the system is essentially as shown in Fig. 20.32. As the input voltage rises, the diac voltage V_G will follow as shown in the figure until the firing potential is reached. It will then turn on and the diac voltage will drop substantially, as shown. Note that

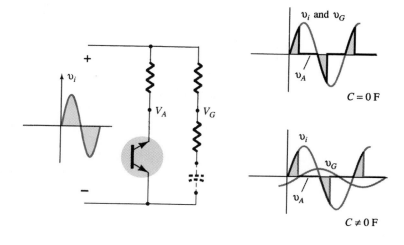

Figure 20.32   Effect of capacitive element on the behavior of the network of Fig. 20.31.

the diac is in essentially an open-circuit state until it fires. Before the capacitive element is introduced, the voltage V_G will be the same as the input. As indicated in the figure, since both V_A and V_G follow the input, V_A can never be greater than V_G by 0.7 V and turn on the device. However, as the capacitive element is introduced, the voltage V_G will begin to lag the input voltage by an increasing angle, as indicated in the figure. There is therefore a point established where V_A can exceed V_G by 0.7 V and cause the programmable device to fire. A heavy current is established through the PUT at this point, raising the voltage V_K and turning on the SCR. A heavy SCR current will then exist through the load, reacting to the presence of the approaching person.

A second application of the diac appears in the next section (Fig. 20.34) as we consider an important power-control device: the triac.

20.12   TRIAC

The triac is fundamentally a diac with a gate terminal for controlling the turn-on conditions of the bilateral device in either direction. In other words, for either direction the gate current can control the action of the device in a manner very similar to that demonstrated for an SCR. The characteristics, however, of the triac in the first and third quadrants are somewhat different from those of the diac, as shown in Fig. 20.33c. Note the holding current in each direction not present in the characteristics of the diac.

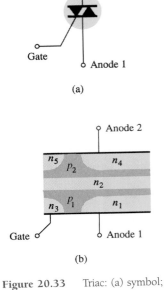

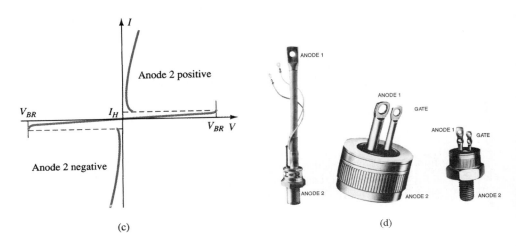

Figure 20.33   Triac: (a) symbol; (b) basic construction; (c) characteristics; (d) photographs.

The graphical symbol for the device and the distribution of the semiconductor layers are provided in Fig. 20.33 with photographs of the device. For each possible direction of conduction, there is a combination of semiconductor layers whose state will be controlled by the signal applied to the gate terminal.

Phase (Power) Control

One fundamental application of the triac is presented in Fig. 20.34. In this capacity, it is controlling the ac power to the load by switching on and off during the positive and negative regions of input sinusoidal signal. The action of this circuit during the positive portion of the input signal is very similar to that encountered for the Shockley diode in Fig. 20.29. The advantage of this configuration is that during the negative portion of the input signal, the same type of response will result since both the diac and triac can fire in the reverse direction. The resulting waveform for the current through the load is provided in Fig. 20.34. By varying the resistor R, the conduction angle can be controlled. There are units available today that can handle in excess of 10-kW loads.

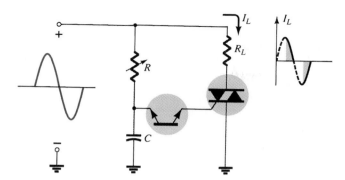

Figure 20.34 Triac application: phase (power) control.

OTHER DEVICES

20.13 UNIJUNCTION TRANSISTOR

Recent interest in the unijunction transistor (UJT) has, like that for the SCR, been increasing at an exponential rate. Although first introduced in 1948, the device did not become commercially available until 1952. The low cost per unit combined with the excellent characteristics of the device have warranted its use in a wide variety of applications. A few include oscillators, trigger circuits, sawtooth generators, phase control, timing circuits, bistable networks, and voltage- or current-regulated supplies. The fact that this device is, in general, a low-power-absorbing device under normal operating conditions is a tremendous aid in the continual effort to design relatively efficient systems.

The UJT is a three-terminal device having the basic construction of Fig. 20.35. A slab of lightly doped (increased resistance characteristic) n-type silicon material has two base contacts attached to both ends of one surface and an aluminum rod alloyed to the opposite surface. The p-n junction of the device is formed at the boundary of the aluminum rod and the n-type silicon slab. The single p-n junction accounts for the terminology *unijunction*. It was originally called a duo (double) base diode due to the presence of two base contacts. Note in Fig. 20.35 that the aluminum rod is alloyed to the silicon slab at a point closer to the base 2 contact than the base 1 contact and that the base 2 terminal is made positive with respect to the base 1 terminal by V_{BB} volts. The effect of each will become evident in the paragraphs to follow.

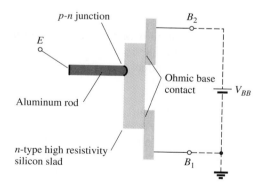

Figure 20.35 Unijunction transistor (UJT): basic construction.

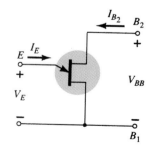

Figure 20.36 Symbol and basic biasing arrangement for the unijunction transistor.

The symbol for the unijunction transistor is provided in Fig. 20.36. Note that the emitter leg is drawn at an angle to the vertical line representing the slab of n-type material. The arrowhead is pointing in the direction of conventional current (hole) flow when the device is in the forward-biased, active, or conducting state.

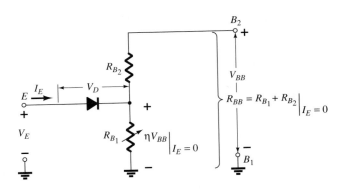

Figure 20.37 UJT equivalent circuit.

The circuit equivalent of the UJT is shown in Fig. 20.37. Note the relative simplicity of this equivalent circuit: two resistors (one fixed, one variable) and a single diode. The resistance R_{B_1} is shown as a variable resistor since its magnitude will vary with the current I_E. In fact, for a representative unijunction transistor, R_{B_1} may vary from 5 kΩ down to 50 Ω for a corresponding change of I_E from 0 to 50 μA. The interbase resistance R_{BB} is the resistance of the device between terminals B_1 and B_2 when $I_E = 0$. In equation form,

$$R_{BB} = (R_{B_1} + R_{B_2})\big|_{I_E=0} \qquad (20.2)$$

(R_{BB} is typically within the range of 4 to 10 kΩ.) The position of the aluminum rod of Fig. 20.35 will determine the relative values of R_{B_1} and R_{B_2} with $I_E = 0$. The magnitude of $V_{R_{B_1}}$ (with $I_E = 0$) is determined by the voltage-divider rule in the following manner:

$$V_{R_{B_1}} = \frac{R_{B_1}}{R_{B_1} + R_{B_2}} \cdot V_{BB} = \eta V_{BB}\bigg|_{I_E=0} \qquad (20.3)$$

The Greek letter η (eta) is called the *intrinsic stand-off* ratio of the device and is defined by

$$\eta = \frac{R_{B_1}}{R_{B_1} + R_{B_2}}\bigg|_{I_E=0} = \frac{R_{B_1}}{R_{BB}} \qquad (20.4)$$

For applied emitter potentials (V_E) greater than $V_{R_{B_1}}(= \eta V_{BB})$ by the forward voltage drop of the diode V_D (0.35 $\rightarrow$ 0.70 V), the diode will fire. Assume the short-circuit representation (on an ideal basis), and I_E will begin to flow through R_{B_1}. In equation form, the emitter firing potential is given by

$$V_P = \eta V_{BB} + V_D \qquad (20.5)$$

The characteristics of a representative unijunction transistor are shown for $V_{BB} = 10$ V in Fig. 20.38. Note that for emitter potentials to the left of the peak point, the magnitude of I_E is never greater than I_{EO} (measured in microamperes). The current I_{EO} corresponds very closely with the reverse leakage current I_{CO} of the conventional bipolar transistor. This region, as indicated in the figure, is called the cutoff region. Once conduction is established at $V_E = V_P$, the emitter potential V_E will drop with increase in I_E. This corresponds exactly with the decreasing resistance R_{B_1} for increasing current I_E, as discussed earlier. This device, therefore, has a *negative resistance* region that is stable enough to be used with a great deal of reliability in the areas of application listed earlier. Eventually, the valley point will be reached, and any further increase in I_E will place the device in the saturation region. In this region,

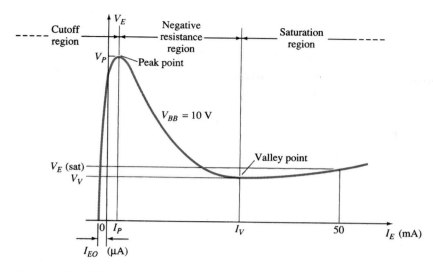

Figure 20.38 UJT static emitter-characteristic curve.

the characteristics approach that of the semiconductor diode in the equivalent circuit of Fig. 20.37.

The decrease in resistance in the active region is due to the holes injected into the *n*-type slab from the aluminum *p*-type rod when conduction is established. The increased hole content in the *n*-type material will result in an increase in the number of free electrons in the slab, producing an increase in conductivity (G) and a corresponding drop in resistance ($R\downarrow = 1/G\uparrow$). Three other important parameters for the unijunction transistor are I_P, V_V, and I_V. Each is indicated on Fig. 20.38. They are all self-explanatory.

The emitter characteristics as they normally appear are provided in Fig. 20.39. Note that I_{EO} (μA) is not in evidence since the horizontal scale is in milliamperes.

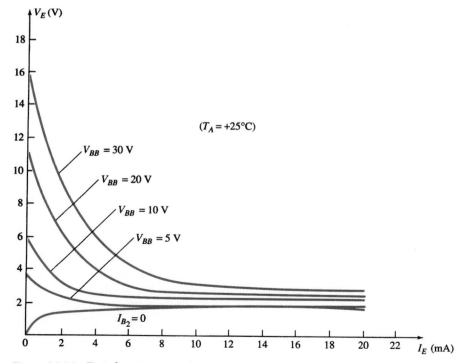

Figure 20.39 Typical static emitter-characteristic curves for a UJT.

The intersection of each curve with the vertical axis is the corresponding value of V_P. For fixed values of η and V_D, the magnitude of V_P will vary as V_{BB}, that is,

$$V_P\uparrow = \eta V_{BB}\uparrow + V_D$$
$$\underbrace{\qquad\qquad}_{\text{fixed}}$$

A typical set of specifications for the UJT is provided in Fig. 20.40b. The discussion of the last few paragraphs should make each quantity readily recognizable. The terminal identification is provided in the same figure with a photograph of a representative UJT. Note that the base terminals are opposite each other while the emitter terminal is between the two. In addition, the base terminal to be tied to the higher potential is closer to the extension on the lip of the casing.

Absolute maximum ratings (25°C):

Power dissipation	300 mW
RMS emitter current	50 mA
Peak emitter current	2 A
Emitter reverse voltage	30 V
Interbase voltage	35 V
Operating temperature range	−65°C to +125°C
Storage temperature range	−65°C to +150°C

Electrical characteristics (25°C):

		Min.	Typ.	Max.
Intrinsic standoff ratio		0.56	0.65	
$(V_{BB} = 10\text{ V})$	η	0.56	0.65	0.75
Interbase resistance (kΩ)				
$(V_{BB} = 3\text{ V}, I_E = 0)$	R_{BB}	4.7	7	9.1
Emitter saturation voltage				
$(V_{BB} = 10\text{ V}, I_E = 50\text{ mA})$	$V_{E\text{(sat)}}$		2	
Emitter reverse current				
$(V_{BB} = 3\text{ V}, I_{B1} = 0)$	I_{EO}		0.05	12
Peak point emitter current	I_P (μA)		0.04	5
$(V_{BB} = 25\text{ V})$				
Valley point current				
$(V_{BB} = 20\text{ V})$	I_V (mA)	4	6	

(a)	(b)	(c)

Figure 20.40 UJT: (a) appearance; (b) specification sheet; (c) terminal identification. (Courtesy General Electric Company.)

SCR Triggering

One rather common application of the UJT is in the triggering of other devices such as the SCR. The basic elements of such a triggering circuit are shown in Fig. 20.41. The resistor R_1 must be chosen to ensure that the load line determined by R_1 passes through the device characteristics in the negative resistance region, that is, to the right of the peak point but to the left of the valley point as shown in Fig. 20.42. If the load line fails to pass to the right of the peak point, the device cannot turn on. An equation for R_1 that will ensure a turn-on condition can be established if we consider the peak point at which $I_{R_1} = I_P$ and $V_E = V_P$. (The equality $I_{R_1} = I_P$ is valid since the charging current of the capacitor, at this instant, is zero. That is, the capacitor is at this particular instant changing from a charging to a discharging state.) Then $V - I_{R_1}R_1 = V_E$ and $R_1 = (V - V_E)/I_{R_1} = (V - V_P)/I_P$ at the peak point. To ensure firing,

Chapter 20   *pnpn* and Other Devices

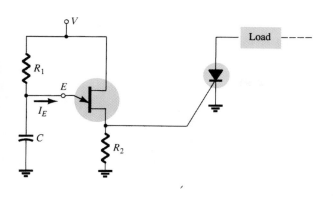

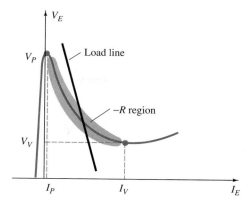

Figure 20.41 UJT triggering of an SCR.

Figure 20.42 Load line for a triggering application.

$$R_1 < \frac{V - V_P}{I_P} \qquad (20.6)$$

At the valley point $I_E = I_V$ and $V_E = V_V$, so that

$$V - I_{R_1}R_1 = V_E$$

becomes

$$V - I_V R_1 = V_V$$

and

$$R_1 = \frac{V - V_V}{I_V}$$

or to ensure turning off,

$$R_1 > \frac{V - V_V}{I_V} \qquad (20.7)$$

The range of R_1 is therefore limited by

$$\frac{V - V_V}{I_V} < R_1 < \frac{V - V_P}{I_P} \qquad (20.8)$$

The resistance R_2 must be chosen small enough to ensure that the SCR is not turned on by the voltage V_{R_2} of Fig. 20.43 when $I_E \cong 0$ A. The voltage V_{R_2} is then given by:

$$V_{R_2} \cong \frac{R_2 V}{R_2 + R_{BB}}\bigg|_{I_E = 0 \text{ A}} \qquad (20.9)$$

The capacitor C will determine, as we shall see, the time interval between triggering pulses and the time span of each pulse.

At the instant the dc supply voltage V is applied, the voltage $v_E = v_C$ will charge toward V volts from V_V as shown in Fig. 20.44 with a time constant $\tau = R_1 C$.

The general equation for the charging period is

$$v_C = V_V + (V - V_V)(1 - e^{-t/R_1 C}) \qquad (20.10)$$

As noted in Fig. 20.44, the voltage across R_2 is determined by Eq. (20.9) during this charging period. When $v_C = v_E = V_P$, the UJT will enter the conduction state and the capacitor will discharge through R_{B_1} and R_2 at a rate determined by the time constant $\tau = (R_{B_1} + R_2)C$.

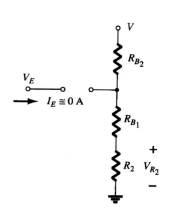

Figure 20.43 Triggering network when $I_E \cong 0$ A.

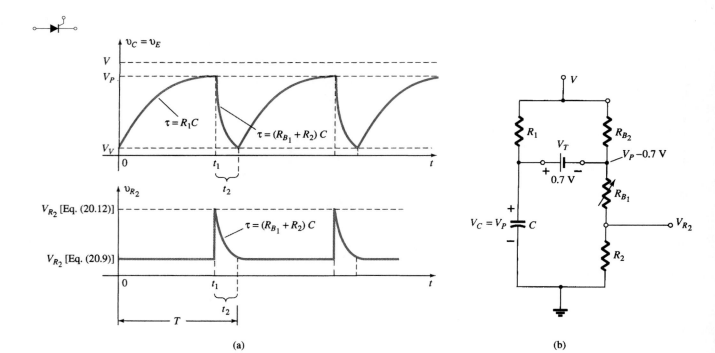

Figure 20.44 (a) Charging and discharging phases for trigger network of Fig. 20.41; (b) equivalent network when UJT turns on.

The discharge equation for the voltage $v_C = v_E$ is the following:

$$v_C \cong V_P e^{-t/(R_{B_1} + R_2)C} \qquad (20.11)$$

Equation (20.11) is complicated somewhat by the fact that R_{B_1} will decrease with increasing emitter current and the other elements of the network, such as R_1 and V, will affect the discharge rate and final level. However, the equivalent network appears as shown in Fig. 20.44 and the magnitudes of R_1 and R_{B_2} are typically such that a Thévenin network for the network surrounding the capacitor C will be only slightly affected by these two resistors. Even though V is a reasonably high voltage, the voltage-divider contribution to the Thévenin voltage can be ignored on an approximate basis.

Using the reduced equivalent of Fig. 20.45 for the discharge phase will result in the following approximation for the peak value of V_{R_2}:

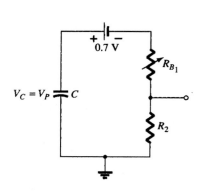

Figure 20.45 Reduced equivalent network when UJT turns on.

$$V_{R_2} \cong \frac{R_2(V_P - 0.7)}{R_2 + R_{B_1}} \qquad (20.12)$$

The period t_1 of Fig. 20.44 can be determined in the following manner:

$$v_C \text{ (charging)} = V_V + (V - V_V)(1 - e^{-t/R_1C})$$

$$= V_V + V - V_V - (V - V_V)e^{-t/R_1C}$$

$$= V - (V - V_V)e^{-t/R_1C}$$

when $v_C = V_P$, $t = t_1$, and $V_P = V - (V - V_V)e^{-t_1/R_1C}$, or

$$\frac{V_P - V}{V - V_V} = -e^{-t_1/R_1C}$$

and

$$e^{-t_1/R_1C} = \frac{V - V_P}{V - V_V}$$

Using logs, we have

$$\log_e e^{-t_1/R_1C} = \log_e \frac{V - V_P}{V - V_V}$$

and

$$\frac{-t_1}{R_1C} = \log_e \frac{V - V_P}{V - V_P}$$

with

$$\boxed{t_1 = R_1C \log_e \frac{V - V_V}{V - V_P}} \tag{20.13}$$

For the discharge period the time between t_1 and t_2 can be determined from Eq. (20.11) as follows:

$$v_C \,(\text{discharging}) = V_P \, e^{-t/(R_{B_1} + R_2)C}$$

Establishing t_1 as $t = 0$ gives us

$$v_C = V_V \text{ at } t = t_2$$

and

$$V_V = V_P e^{-t_2/(R_{B_1} + R_2)C}$$

or

$$e^{-t_2/(R_{B_1} + R_2)C} = \frac{V_V}{V_P}$$

Using logs yields

$$\frac{-t_2}{(R_{B_1} + R_2)C} = \log_e \frac{V_V}{V_P}$$

and

$$\boxed{t_2 = (R_{B_1} + R_2)C \log_e \frac{V_P}{V_V}} \tag{20.14}$$

The period of time to complete one cycle is defined by T in Fig. 20.44. That is,

$$\boxed{T = t_1 + t_2} \tag{20.15}$$

Relaxation Oscillator

If the SCR were dropped from the configuration, the network would behave as a *relaxation oscillator*, generating the waveform of Fig. 20.44. The frequency of oscillation is determined by

$$\boxed{f_{\text{osc}} = \frac{1}{T}} \tag{20.16}$$

In many systems, $t_1 \gg t_2$ and

$$T \cong t_1 = R_1C \log_e \frac{V - V_V}{V - V_P}$$

Since $V \gg V_V$ in many instances,

$$T \cong t_1 = R_1C \log_e \frac{V}{V - V_P}$$

$$= R_1C \log_e \frac{1}{1 - V_P/V}$$

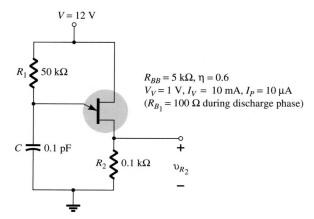

but $\eta = V_P/V$ if we ignore the effects of V_D in Eq. (20.5) and

$$T \cong R_1 C \log_e \frac{1}{1 - \eta}$$

or $$f \cong \frac{1}{R_1 C \log_e [1/(1 - \eta)]}$$ (20.17)

EXAMPLE 20.1

Given the relaxation oscillator of Fig. 20.46:
(a) Determine R_{B_1} and R_{B_2} at $I_E = 0$ A.
(b) Calculate V_P, the voltage necessary to turn on the UJT.
(c) Determine whether R_1 is within the permissible range of values as determined by Eq. (20.8) to ensure firing of the UJT.
(d) Determine the frequency of oscillation if $R_{B_1} = 100\ \Omega$ during the discharge phase.
(e) Sketch the waveform of v_C for a full cycle.
(f) Sketch the waveform of v_{R_2} for a full cycle.

$V = 12$ V

R_1 50 kΩ

$R_{BB} = 5$ kΩ, $\eta = 0.6$
$V_V = 1$ V, $I_V = 10$ mA, $I_P = 10$ μA
($R_{B_1} = 100\ \Omega$ during discharge phase)

$C = 0.1$ pF

R_2 0.1 kΩ v_{R_2}

Figure 20.46   Example 20.1.

Solution

(a) $\eta = \dfrac{R_{B_1}}{R_{B_1} + R_{B_2}}$

$0.6 = \dfrac{R_{B_1}}{R_{BB}}$

$R_{B_1} = 0.6 R_{BB} = 0.6(5\text{ k}\Omega) = \mathbf{3\ k\Omega}$

$R_{B_2} = R_{BB} - R_{B_1} = 5\text{ k}\Omega - 3\text{ k}\Omega = \mathbf{2\ k\Omega}$

(b) At the point where $v_C = V_P$, if we continue with $I_E = 0$ A, the network of Fig. 20.47 will result, where

$$V_P = 0.7\text{ V} + \underbrace{\frac{(R_{B_1} + R_2)12\text{ V}}{R_{B_1} + R_{B_2} + R_2}}_{R_{BB}}$$

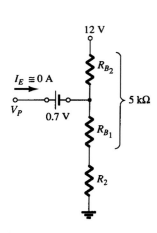

Figure 20.47   Network to determine V_P, the voltage required to turn on the UJT.

$$= 0.7 \text{ V} + \frac{(3 \text{ k}\Omega + 0.1 \text{ k}\Omega)12 \text{ V}}{5 \text{ k}\Omega + 0.1 \text{ k}\Omega} = 0.7 \text{ V} + 7.294 \text{ V}$$

$$\cong \mathbf{8\ V}$$

(c) $\dfrac{V - V_V}{I_V} < R_1 < \dfrac{V - V_P}{I_P}$

$\dfrac{12 \text{ V} - 1 \text{ V}}{10 \text{ mA}} < R_1 < \dfrac{12 \text{ V} - 8 \text{ V}}{10 \text{ } \mu\text{A}}$

$1.1 \text{ k}\Omega < R_1 < 400 \text{ k}\Omega$

The resistance $R_1 = 50 \text{ k}\Omega$ falls within this range.

(d) $t_1 = R_1 C \log_e \dfrac{V - V_V}{V - V_P}$

$= (50 \text{ k}\Omega)(0.1 \text{ pF}) \log_e \dfrac{12 \text{ V} - 1 \text{ V}}{12 \text{ V} - 8 \text{ V}}$

$= 5 \times 10^{-3} \log_e \dfrac{11}{4} = 5 \times 10^{-3}(1.01)$

$= 5.05 \text{ ms}$

$t_2 = (R_{B_1} + R_2)C \log_e \dfrac{V_P}{V_V}$

$= (0.1 \text{ k}\Omega + 0.1 \text{ k}\Omega)(0.1 \text{ pF}) \log_e \dfrac{8}{1}$

$= (0.02 \times 10^{-6})(2.08)$

$= 41.6 \text{ } \mu\text{s}$

and
$$T = t_1 + t_2 = 5.05 \text{ ms} + 0.0416 \text{ ms}$$

$$= 5.092 \text{ ms}$$

with
$$f_{\text{osc}} = \frac{1}{T} = \frac{1}{5.092 \text{ ms}} \cong \mathbf{196\ Hz}$$

Using Eq. (20.17) gives us

$$f \cong \frac{1}{R_1 C \log_e [1/(1 - \eta)]}$$

$$= \frac{1}{5 \times 10^{-3} \log_e 2.5}$$

$$= \mathbf{218\ Hz}$$

(e) See Fig. 20.48.

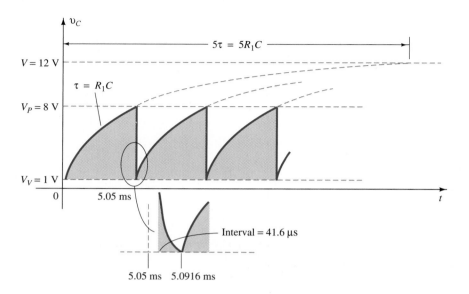

Figure 20.48 The voltage v_C for the relaxation oscillator of Fig. 20.46.

(f) During the charging phase, (Eq. 20.9)

$$V_{R_2} = \frac{R_2 V}{R_2 + R_{BB}} = \frac{0.1 \text{ k}\Omega(12 \text{ V})}{0.1 \text{ k}\Omega + 5 \text{ k}\Omega} = \mathbf{0.235 \text{ V}}$$

When $v_C = V_P$ (Eq. 20.12)

$$V_{R_2} \cong \frac{R_2(V_P - 0.7 \text{ V})}{R_2 + R_{B_1}} = \frac{0.1 \text{ k}\Omega(8 \text{ V} - 0.7 \text{ V})}{0.1 \text{ k}\Omega + 0.1 \text{ k}\Omega}$$

$$= \mathbf{3.65 \text{ V}}$$

The plot of v_{R_2} appears in Fig. 20.49.

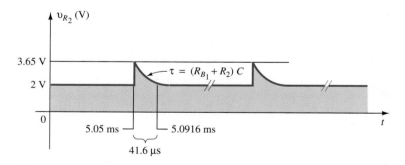

Figure 20.49 The voltage v_{R_2} for the relaxation oscillator of Fig. 20.46.

20.14 PHOTOTRANSISTORS

The fundamental behavior of photoelectric devices was introduced earlier with the description of the photodiode. This discussion will now be extended to include the phototransistor, which has a photosensitive collector–base *p-n* junction. The current induced by photoelectric effects is the base current of the transistor. If we assign the notation I_λ for

Chapter 20 *pnpn* and Other Devices

the photoinduced base current, the resulting collector current, on an approximate basis, is

$$I_C \cong h_{fe} I_\lambda \qquad (20.18)$$

A representative set of characteristics for a phototransistor is provided in Fig. 20.50 with the symbolic representation of the device. Note the similarities between these curves and those of a typical bipolar transistor. As expected, an increase in light intensity corresponds with an increase in collector current. To develop a greater degree of familiarity with the light-intensity unit of measurement, milliwatts per square centimeter, a curve of base current versus flux density appears in Fig. 20.51a. Note the exponential increase in base current with increasing flux density. In the same figure, a sketch of the phototransistor is provided with the terminal identification and the angular alignment.

Some of the areas of application for the phototransistor include punch-card readers, computer logic circuitry, lighting control (highways, etc.), level indication, relays, and counting systems.

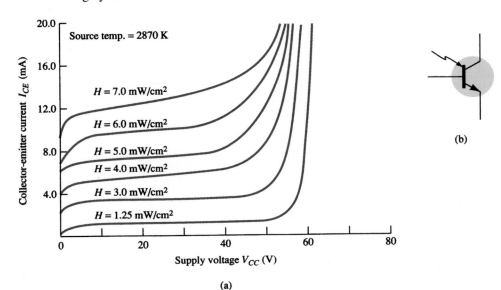

Figure 20.50 Phototransistor: (a) collector characteristics (MRD300); (b) symbol. (Courtesy Motorola, Inc.)

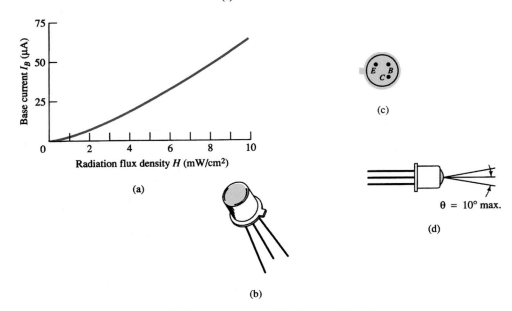

Figure 20.51 Phototransistor: (a) base current versus flux density; (b) device; (c) terminal identification; (d) angular alignment. (Courtesy Motorola, Inc.)

High-Isolation AND Gate

A high-isolation AND gate is shown in Fig. 20.52 using three phototransistors and three LEDs (light-emitting diodes). The LEDs are semiconductor devices that emit light at an intensity determined by the forward current through the device. With the aid of discussions in Chapter 1, the circuit behavior should be relatively easy to understand. The terminology *high isolation* simply refers to the lack of an electrical connection between the input and output circuits.

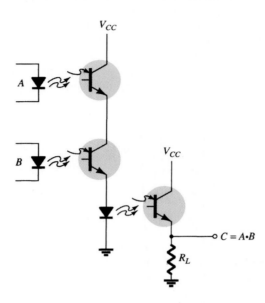

Figure 20.52 High-isolation AND gate employing phototransistors and light-emitting diodes (LEDs).

20.15 OPTO-ISOLATORS

The *opto-isolator* is a device that incorporates many of the characteristics described in the preceding section. It is simply a package that contains both an infrared LED and a photodetector such as a silicon diode, transistor Darlington pair, or SCR. The wavelength response of each device is tailored to be as identical as possible to permit the highest measure of coupling possible. In Fig. 20.53, two possible chip configurations are provided, with a photograph of each. There is a transparent insulating

Figure 20.53 Two Litronix opto-isolators. (Courtesy Siemens Components, Inc.)

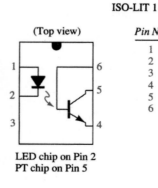

ISO-LIT 1

(Top view)

Pin No.	Function
1	anode
2	cathode
3	nc
4	emitter
5	collector
6	base

LED chip on Pin 2
PT chip on Pin 5

ISO-LIT Q1

(Top view)

Pin No.	Function
1	anode
2	cathode
3	cathode
4	anode
5	anode
6	cathode
7	cathode
8	anode
9	emitter
10	collector
11	collector
12	emitter
13	emitter
14	collector
15	collector
16	emitter

cap between each set of elements embedded in the structure (not visible) to permit the passage of light. They are designed with response times so small that they can be used to transmit data in the megahertz range.

The maximum ratings and electrical characteristics for the IL-1 model are provided in Fig. 20.54. Note that I_{CEO} is measured in nanoamperes and that the power dissipation of the LED and transistor are about the same.

(a) Maximum Ratings

Gallium arsenide LED (each channel) IL-1
Power dissipation @ 25°C	200 mW
Derate linearly from 25°C	2.6 mW/°C
Continuous forward current	150 mA

Detector silicon phototransistor (each channel) IL-1
Power dissipation @ 25°C	200 mW
Derate linearly from 25°C	2.6 mW/°C
Collector-emitter breakdown voltage	30 V
Emitter-collector breakdown voltage	7 V
Collector-base breakdown voltage	70 V

Package IL-1
Total package dissipation at 25°C ambient (LED plus detector)	250 mW
Derate linearly from 25°C	3.3 mW/°C
Storage temperature	−55°C to +150°C
Operating temperature	−55°C to +100°C

(b) Electrical Characteristics per Channel (at 25°C Ambient)

Parameter	Min.	Typ.	Max.	Unit	Test Conditions
Gallium arsenide LED					
Forward voltage		1.3	1.5	V	$I_F = 60$ mA
Reverse current		0.1	10	μA	$V_R = 3.0$ V
Capacitance		100		pF	$V_R = 0$ V
Phototransistor detector					
BV_{CEO}	30			V	$I_C = 1$ mA
I_{CEO}		5.0	50	nA	$V_{CE} = 10$ V, $I_F = 0$ A
Collector-emitter capacitance		2.0		pF	$V_{CE} = 0$ V
BV_{ECO}	7			V	$I_E = 100$ μA
Coupled characteristics					
dc current transfer ratio	0.2	0.35			$I_F = 10$ mA, $V_{CE} = 10$ V
Capacitance, input to output		0.5		pF	
Breakdown voltage	2500			V	DC
Resistance, input to output		100		GΩ	
V_{sat}			0.5	V	$I_C = 1.6$ mA, $I_F = 16$ mA
Propagation delay					
$t_{D\,on}$		6.0		μs	$R_L = 2.4$ kΩ, $V_{CE} = 5$ V
$t_{D\,off}$		25		μs	$I_F = 16$ mA

Figure 20.54 Litronix IL-1 opto-isolator.

The typical optoelectronic characteristic curves for each channel are provided in Figs. 20.55 through 20.59. Note the very pronounced effect of temperature on the output current at low temperatures but the fairly level response at or above room temperature (25°C). As mentioned earlier, the level of I_{CEO} is improving steadily with improved design and construction techniques (the lower the better). In Fig. 20.55, we do not reach 1 μA until the temperature rises above 75°C. The transfer characteristics of Fig. 20.56 compare the input LED current (which establishes the luminous flux) to the resulting collector current of the output transistor (whose base current is determined by the incident flux). In fact, Fig. 20.57 demonstrates that the V_{CE} voltage affects the resulting collector current only very slightly. It is interesting to note in Fig. 20.58 that the switching time of an opto-isolator decreases with increased current, while for many devices it is exactly the reverse. Consider that it is only 2 μs for a collector current of 6 mA and a load R_L of 100 Ω. The relative output versus temperature appears in Fig. 20.59.

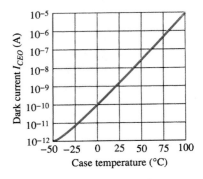

Figure 20.55 Dark current (I_{CEO}) versus temperature.

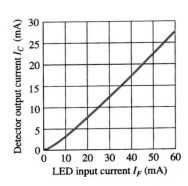

Figure 20.56 Transfer characteristics.

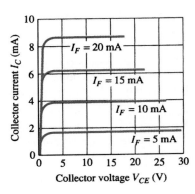

Figure 20.57 Detector output characteristics.

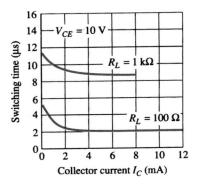

Figure 20.58 Switching time versus collector current.

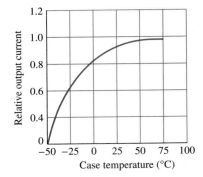

Figure 20.59 Relative output versus temperature.

The schematic representation for a transistor coupler appears in Fig. 20.53. The schematic representations for a photodiode, photo-Darlington, and photo-SCR opto-isolator appear in Fig. 20.60.

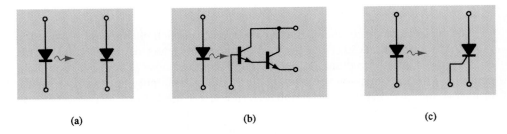

(a)　　　　　　　　　　(b)　　　　　　　　　　(c)

Figure 20.60 Opto-isolators: (a) photodiode; (b) photo-Darlington; (c) photo-SCR.

20.16 PROGRAMMABLE UNIJUNCTION TRANSISTOR

Although there is a similarity in name, the actual construction and mode of operation of the programmable unijunction transistor (PUT) is quite different from the unijunction transistor. The fact that the *I–V* characteristics and applications of each are similar prompted the choice of labels.

As indicated in Fig. 20.61, the PUT is a four-layer *pnpn* device with a gate connected directly to the sandwiched *n*-type layer. The symbol for the device and the basic biasing arrangement appears in Fig. 20.62. As the symbol suggests, it is essentially an SCR with a control mechanism that permits a duplication of the characteristics of the typical SCR. The term *programmable* is applied because R_{BB}, η, and V_P as defined for the UJT can be controlled through the resistors R_{B_1}, R_{B_2}, and the supply voltage V_{BB}. Note in Fig. 20.62 that through an application of the voltage-divider rule, when $I_G = 0$:

$$V_G = \frac{R_{B_1}}{R_{B_1} + R_{B_2}} V_{BB} = \eta V_{BB} \qquad (20.19)$$

where

$$\eta = \frac{R_{B_1}}{R_{B_1} + R_{B_2}}$$

as defined for the UJT.

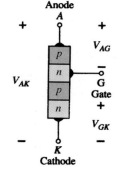

Figure 20.61 Programmable UJT (PUT)

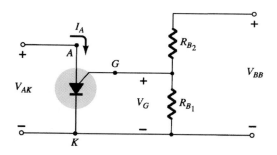

Figure 20.62 Basic biasing arrangement for the PUT.

The characteristics of the device appear in Fig. 20.63. As noted on the diagram, the "off" state (*I* low, *V* between 0 and V_P) and the "on" state ($I \geq I_V$, $V \geq V_V$) are separated by the unstable region as occurred for the UJT. That is, the device cannot stay in the unstable state—it will simply shift to either the "off" or "on" stable states.

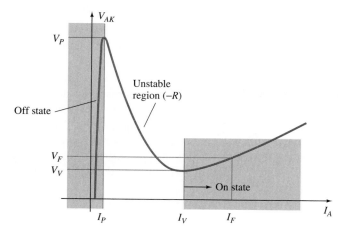

Figure 20.63 PUT characteristics.

The firing potential (V_P) or voltage necessary to "fire" the device is given by

$$V_P = \eta V_{BB} + V_D \qquad (20.20)$$

as defined for the UJT. However, V_P represents the voltage drop V_{AK} in Fig. 20.61 (the forward voltage drop across the conducting diode). For silicon, V_D is typically 0.7 V. Therefore,

$$V_{AK} = V_{AG} + V_{GK}$$

$$V_P = V_D + V_G$$

and

$$V_P = \eta V_{BB} + 0.7 \text{ V} \qquad \text{silicon} \qquad (20.21)$$

We noted above, however, that $V_G = \eta V_{BB}$ with the result that

$$V_P = V_G + 0.7 \qquad \text{silicon} \qquad (20.22)$$

Recall that for the UJT both R_{B_1} and R_{B_2} represent the bulk resistance and ohmic base contacts of the device—both inaccessible. In the development above, we note that R_{B_1} and R_{B_2} are external to the device, permitting an adjustment of η and hence V_G above. In other words, the PUT provides a measure of control on the level of V_P required to turn on the device.

Although the characteristics of the PUT and UJT are similar, the peak and valley currents of the PUT are typically lower than those of a similarly rated UJT. In addition, the minimum operating voltage is also less for a PUT.

If we take a Thévenin equivalent of the network to the right of the gate terminal in Fig. 20.62, the network of Fig. 20.64 will result. The resulting resistance R_S is important because it is often included in specification sheets since it affects the level of I_V.

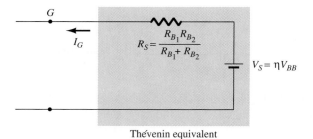

Thévenin equivalent

Figure 20.64 Thévenin equivalent for the network to the right of the gate terminal in Fig. 20.62.

The basic operation of the device can be reviewed through reference to Fig. 20.63. A device in the "off" state will not change state until the voltage V_P as defined by V_G and V_D is reached. The level of current until I_P is reached is very low, resulting in an open-circuit equivalent since $R = V\text{(high)}/I\text{ (low)}$ will result in a high resistance level. When V_P is reached, the device will switch through the unstable region to the "on" state, where the voltage is lower but the current higher, resulting in a terminal resistance $R = V\text{(low)}/I\text{(high)}$, which is quite small, representing short-circuit equivalent on an approximate basis. The device has therefore switched from essentially an open-circuit to a short-circuit state at a point determined by the choice of R_{B_1}, R_{B_2}, and V_{BB}. Once the device is in the "on" state, the removal of V_G will not turn the device off. The level of voltage V_{AK} must be dropped sufficiently to reduce the current below a holding level.

Determine R_{B_1} and V_{BB} for a silicon PUT if it is determined that $\eta = 0.8$, $V_P = 10.3$ V, and $R_{B_2} = 5$ kΩ.

EXAMPLE 20.2

Solution

$$\text{Eq. (20.4):} \quad \eta = \frac{R_{B_1}}{R_{B_1} + R_{B_2}} = 0.8$$

$$R_{B_1} = 0.8(R_{B_1} + R_{B_2})$$

$$0.2R_{B_1} = 0.8R_{B_2}$$

$$R_{B_1} = 4R_{B_2}$$

$$R_{B_1} = 4(5 \text{ k}\Omega) = \mathbf{20 \text{ k}\Omega}$$

$$\text{Eq. (20.20):} \quad V_P = \eta V_{BB} + V_D$$

$$10.3 \text{ V} = (0.8)(V_{BB}) + 0.7 \text{ V}$$

$$9.6 \text{ V} = 0.8V_{BB}$$

$$V_{BB} = \mathbf{12 \text{ V}}$$

Relaxation Oscillator

One popular application of the PUT is in the relaxation oscillator of Fig. 20.65. The instant the supply is connected, the capacitor will begin to charge toward V_{BB} volts since there is no anode current at this point. The charging curve appears in Fig. 20.66. The period T required to reach the firing potential V_P is given approximately by

$$T \cong RC \log_e \frac{V_{BB}}{V_{BB} - V_P} \tag{20.23}$$

or when $V_P \cong \eta V_{BB}$

$$T \cong RC \log_e \left(1 + \frac{R_{B_1}}{R_{B_2}}\right) \tag{20.24}$$

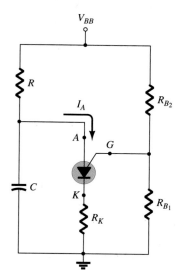

Figure 20.65 PUT relaxation oscillator.

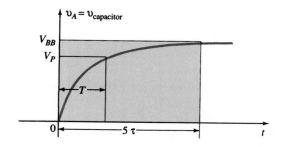

Figure 20.66 Charging wave for the capacitor C of Fig. 20.65.

The instant the voltage across the capacitor equals V_P, the device will fire and a current $I_A = I_P$ will be established through the PUT. If R is too large, the current I_P cannot be established and the device will not fire. At the point of transition,

$$I_P R = V_{BB} - V_P$$

and

$$R_{\max} = \frac{V_{BB} - V_P}{I_P} \tag{20.25}$$

The subscript is included to indicate that any R greater than R_{max} will result in a current less than I_P. The level of R must also be such to ensure it is less than I_V if oscillations are to occur. In other words, we want the device to enter the unstable region and then return to the "off" state. From reasoning similar to that above:

$$R_{\text{min}} = \frac{V_{BB} - V_V}{I_V} \qquad (20.26)$$

The discussion above requires that R be limited to the following for an oscillatory system:

$$R_{\text{min}} < R < R_{\text{max}}$$

The waveforms of v_A, v_G, and v_K appear in Fig. 20.67. Note that T determines the maximum voltage v_A can charge to. Once the device fires, the capacitor will rapidly discharge through the PUT and R_K, producing the drop shown. Of course, v_K will peak at the same time due to the brief but heavy current. The voltage v_G will rapidly drop down from V_G to a level just greater than 0 V. When the capacitor voltage drops to a low level, the PUT will once again turn off and the charging cycle will be repeated. The effect on V_G and V_K is shown in Fig. 20.67.

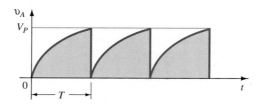

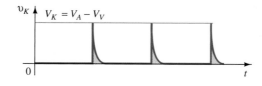

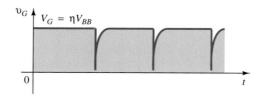

Figure 20.67 Waveforms for PUT oscillator of Fig. 20.65.

EXAMPLE 20.3

If $V_{BB} = 12$ V, $R = 20$ kΩ, $C = 1$ μF, $R_K = 100$ Ω, $R_{B_1} = 10$ kΩ, $R_{B_2} = 5$ kΩ, $I_P = 100$ μA, $V_V = 1$ V, and $I_V = 5.5$ mA, determine:
(a) V_P.
(b) R_{max} and R_{min}.
(c) T and frequency of oscillation.
(d) The waveforms of v_A, v_G, and v_K.

Solution

(a) Eq. 20.20: $V_P = \eta V_{BB} + V_D$

$$= \frac{R_{B_1}}{R_{B_1} + R_{B_2}} V_{BB} + 0.7 \text{ V}$$

$$= \frac{10\ \text{k}\Omega}{10\ \text{k}\Omega + 5\ \text{k}\Omega}(12\ \text{V}) + 0.7\ \text{V}$$

$$= (0.67)(12\ \text{V}) + 0.7\ \text{V} = \mathbf{8.7\ V}$$

(b) From Eq. (20.25): $R_{\text{max}} = \dfrac{V_{BB} - V_P}{I_P}$

$$= \frac{12\ \text{V} - 8.7\ \text{V}}{100\ \mu\text{A}} = \mathbf{33\ k\Omega}$$

From Eq. (20.26): $R_{\text{min}} = \dfrac{V_{BB} - V_V}{I_V}$

$$= \frac{12\ \text{V} - 1\ \text{V}}{5.5\ \text{mA}} = \mathbf{2\ k\Omega}$$

$$R:\quad 2\ \text{k}\Omega < 20\ \text{k}\Omega < 33\ \text{k}\Omega$$

(c) Eq. (20.23): $T = RC \log_e \dfrac{V_{BB}}{V_{BB} - V_P}$

$$= (20\ \text{k}\Omega)(1\ \mu\text{F}) \log_e \frac{12\ \text{V}}{12\ \text{V} - 8.7\ \text{V}}$$

$$= 20 \times 10^{-3} \log_e (3.64)$$

$$= 20 \times 10^{-3}(1.29)$$

$$= \mathbf{25.8\ ms}$$

$$f = \frac{1}{T} = \frac{1}{25.8\ \text{ms}} = \mathbf{38.8\ Hz}$$

(d) As indicated in Fig. 20.68.

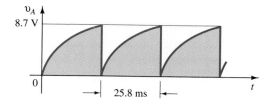

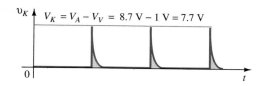

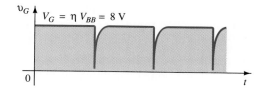

Figure 20.68 Waveforms for the oscillator of Example 20.3.

20.17 SUMMARY

Important Conclusions and Concepts

1. **The silicon-controlled rectifier** (SCR) is a rectifier whose state **is controlled by the magnitude of the gate current.** The forward-bias voltage across the device will determine the level of gate current required to "fire" (turn on) the device. The **higher** the level of biasing voltage, the **less** the required gate current.

2. In addition to gate triggering, an SCR can be **turned on with zero gate current** simply by applying **sufficient voltage** across the device. The higher the gate current, however, the less the required biasing voltage to turn the SCR on.

3. The **silicon-controlled switch** has both **an anode gate and a cathode gate** for controlling the state of the device, although the anode gate is now connected to an n-type layer with the cathode gate connected to a p-type layer. The result is that **a negative pulse at the anode gate will turn the device on, while a positive pulse will turn it off.** The reverse is true for the cathode gate.

4. A **gate turn-off switch** (GTO) looks similar in construction to the SCR with only **one gate connection,** but the GTO has the added advantage of being able to turn the device **off and on** at the gate terminal. However, this added option of being able to turn the device off at the gate has resulted in a much **higher gate current** to turn the device on.

5. The **LASCR** is a light-activated SCR whose state can be controlled by **light falling on a semiconductor layer** of the device **or by triggering the gate terminal** in a manner described for SCRs. The higher the junction temperature of the device, the less the required incident light to turn the device on.

6. The **Shockley diode** has essentially the **same characteristics of an SCR with zero gate current.** It is turned on by simply increasing the forward-bias voltage across the device beyond the breakover level.

7. The **diac** is essentially a **Shockley diode that can fire in either direction.** The application of sufficient voltage of either polarity will turn the device on.

8. The **triac** is fundamentally a **diac with a gate terminal to control the action of the device** in either direction.

9. The **unijunction transistor** is a three-terminal device with a p-n junction formed between an aluminum rod and an n-type silicon slab. Once the emitter firing potential is reached, the emitter voltage will drop with an increase in emitter current, establishing a **negative-resistance region** excellent for oscillator applications. Once the valley point is reached, the characteristics of the device **take on those of a semiconductor diode.** The higher the applied voltage across the device, the higher the emitter firing potential.

10. The **phototransistor** is a three-terminal device having characteristics **very similar to those of a BJT** with a base and collector current sensitive to the incident light intensity. The base current that results is essentially **linearly related to the applied light** with a level almost independent of the voltage across the device until breakdown results.

11. **Opto-isolators** contain an **infrared LED** and a **photodetector** to provide a linkage between systems that does not require a direct connection. The output detector current **is less than but linearly related to the applied input LED current.** Further, the collector current is essentially independent of the collector-to-emitter voltage.

12. The **PUT** (programmable unijunction transistor) is, as the name implies, a device with the **characteristics of a UJT** but with the added capability of **being able to control the firing potential.** In general, the peak, valley, and minimum operating voltages of PUTs are less than those of UJTs.

Equations

Diac:

$$V_{BR_1} = V_{BR_2} \pm 0.1V_{BR_2}$$

UJT:

$$R_{BB} = (R_{B_1} + R_{B_2})|_{I_E=0}$$

$$V_{R_{B_1}} = \frac{R_{B_1}}{R_{B_1} + R_{B_2}} \cdot V_{BB} = \eta V_{BB}\Big|_{I_E=0}$$

$$\eta = \frac{R_{B_1}}{R_{BB}}$$

$$V_P = \eta V_{BB} + V_D$$

Phototransistor:

$$I_C \cong h_{fe}I_\lambda$$

PUT:

$$V_G = \frac{R_{B_1}}{R_{B_1} + R_{B_2}} \cdot V_{BB} = \eta V_{BB}$$

$$V_P = \eta V_{BB} + V_D$$

§ 20.3 Basic Silicon-Controlled Rectifier Operation

1. Describe in your own words the basic behavior of the SCR using the two-transistor equivalent circuit.

2. Describe two techniques for turning an SCR off.

3. Consult a manufacturer's manual or specification sheet and obtain a turn-off network. If possible, describe the turn-off action of the design.

§ 20.4 SCR Characteristics and Ratings

* 4. (a) At high levels of gate current, the characteristics of an SCR approach those of what two-terminal device?
   (b) At a fixed anode-to-cathode voltage less than $V_{(BR)F*}$, what is the effect on the firing of the SCR as the gate current is reduced from its maximum value to the zero level?
   (c) At a fixed gate current greater than $I_G = 0$, what is the effect on the firing of the SCR as the gate voltage is reduced from $V_{(BR)F*}$?
   (d) For increasing levels of I_G, what is the effect on the holding current?

5. (a) Using Fig. 20.8, will a gate current of 50 mA fire the device at room temperature (25°C)?
   (b) Repeat part (a) for a gate current of 10 mA.
   (c) Will a gate voltage of 2.6 V trigger the device at room temperature?
   (d) Is $V_G = 6$ V, $I_G = 800$ mA a good choice for firing conditions? Would $V_G = 4$ V, $I_G = 1.6$ A be preferred? Explain.

§ 20.6 SCR Applications

6. In Fig. 20.11b, why is there very little loss in potential across the SCR during conduction?

7. Fully explain why reduced values of R_1 in Fig. 20.12 will result in an increased angle of conduction.

* **8.** Refer to the charging network of Fig. 20.13.
    - (a) Determine the dc level of the full-wave rectified signal if a 1 : 1 transformer were employed.
    - (b) If the battery in its uncharged state is sitting at 11 V, what is the anode-to-cathode voltage drop across SCR_1?
    - (c) What is the maximum possible value of V_R ($V_{GK} \cong 0.7$ V)?
    - (d) At the maximum value of part (c), what is the gate potential of SCR_2?
    - (e) Once SCR_2 has entered the short-circuit state, what is the level of V_2?

§ 20.7 Silicon-Controlled Switch

9. Fully describe in your own words the behavior of the networks of Fig. 20.17.

§ 20.8 Gate Turn-Off Switch

10. (a) In Fig. 20.23, if $V_Z = 50$ V, determine the maximum possible value the capacitor C_1 can charge to ($V_{GK} \cong 0.7$ V).
    - (b) Determine the approximate discharge time (5τ) for $R_3 = 20$ kΩ.
    - (c) Determine the internal resistance of the GTO if the rise time is one-half the decay period determined in part (b).

§ 20.9 Light-Activated SCR

11. (a) Using Fig. 20.25b, determine the minimum irradiance required to fire the device at room temperature (25°C).
    - (b) What percent reduction in irradiance is allowable if the junction temperature is increased from 0°C (32°F) to 100°C (212°F)?

§ 20.10 Shockley Diode

12. For the network of Fig. 20.29, if $V_{BR} = 6$ V, $V = 40$ V, $R = 10$ kΩ, $C = 0.2$ μF, and V_{GK} (firing potential) = 3 V, determine the time period between energizing the network and the turning on of the SCR.

§ 20.11 Diac

13. Using whatever reference you require, find an application of a diac and explain the network behavior.

14. If V_{BR_2} is 6.4 V, determine the range for V_{BR_1} using Eq. (20.1).

§ 20.12 Triac

15. Repeat Problem 13 for the triac.

§ 20.13 Unijunction Transistor

16. For the network of Fig. 20.41, in which $V = 40$ V, $\eta = 0.6$, $V_V = 1$ V, $I_V = 8$ mA, and $I_P = 10$ μA, determine the range of R_1 for the triggering network.

17. For a unijunction transistor with $V_{BB} = 20$ V, $\eta = 0.65$, $R_{B_1} = 2$ kΩ ($I_E = 0$), and $V_D = 0.7$ V, determine:
    - (a) R_{B_2}.
    - (b) R_{BB}.
    - (c) $V_{R_{B_1}}$.
    - (d) V_P.

* **18.** Given the relaxation oscillator of Fig. 20.69:
    - (a) Find R_{B_1} and R_{B_2} at $I_E = 0$ A.
    - (b) Determine V_P, the voltage necessary to turn on the UJT.
    - (c) Determine whether R_1 is within the permissible range of values defined by Eq. (20.8).
    - (d) Determine the frequency of oscillation if $R_{B_1} = 200$ Ω during the discharge phase.

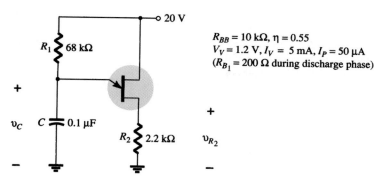

$$R_{BB} = 10 \text{ k}\Omega, \eta = 0.55$$
$$V_V = 1.2 \text{ V}, I_V = 5 \text{ mA}, I_P = 50 \text{ μA}$$
$$(R_{B_1} = 200 \text{ Ω during discharge phase})$$

Figure 20.69 Problem 18

(e) Sketch the waveform of v_C for two full cycles.

(f) Sketch the waveform of v_{R_2} for two full cycles.

(g) Determine the frequency using Eq. (20.17) and compare to the value determined in part (d). Account for any major differences.

§ 20.14 Phototransistors

19. For a phototransistor having the characteristics of Fig. 20.51, determine the photoinduced base current for a radiant flux density of 5 mW/cm². If $h_{fe} = 40$, find I_C.

* 20. Design a high-isolation OR-gate employing phototransistors and LEDs.

§ 20.15 Opto-Isolators

21. (a) Determine an average derating factor from the curve of Fig. 20.59 for the region defined by temperatures between $-25°C$ and $+50°C$.

(b) Is it fair to say that for temperatures greater than room temperature (up to 100°C), the output current is somewhat unaffected by temperature?

22. (a) Determine from Fig. 20.55 the average change in I_{CEO} per degree change in temperature for the range 25 to 50°C.

(b) Can the results of part (a) be used to determine the level of I_{CEO} at 35°C? Test your theory.

23. Determine from Fig. 20.56 the ratio of LED output current to detector input current for an output current of 20 mA. Would you consider the device to be relatively efficient in its purpose?

* 24. (a) Sketch the maximum-power curve of $P_D = 200$ mW on the graph of Fig. 20.57. List any noteworthy conclusions.

(b) Determine β_{dc} (defined by I_C/I_F) for the system at $V_{CE} = 15$ V, $I_F = 10$ mA.

(c) Compare the results of part (b) with those obtained from Fig. 20.56 at $I_F = 10$ mA. Do they compare? Should they? Why?

* 25. (a) Referring to Fig. 20.58, determine the collector current above which the switching time does not change appreciably for $R_L = 1$ kΩ and $R_L = 100$ Ω.

(b) At $I_C = 6$ mA, how does the ratio of switching times for $R_L = 1$ kΩ and $R_L = 100$ Ω compare to the ratio of resistance levels?

§ 20.16 Programmable Unijunction Transistor

26. Determine η and V_G for a PUT with $V_{BB} = 20$ V and $R_{B_1} = 3R_{B_2}$.

27. Using the data provided in Example 20.3, determine the impedance of the PUT at the firing and valley points. Are the approximate open- and short-circuit states verified?

28. Can Eq. (20.24) be derived exactly as shown from Eq. (20.23)? If not, what element is missing in Eq. (20.24)?

*   **29.** (a) Will the network of Example 20.3 oscillate if V_{BB} is changed to 10 V? What minimum value of V_{BB} is required (V_V a constant)?
    (b) Referring to the same example, what value of R would place the network in the stable "on" state and remove the oscillatory response of the system?
    (c) What value of R would make the network a 2-ms time-delay network? That is, provide a pulse v_K 2 ms after the supply is turned on and then stay in the "on" state.

---

*Please Note: Asterisks indicate more difficult problems.

CHAPTER

Oscilloscope and Other Measuring Instruments

21

21.1 INTRODUCTION

One of the basic functions of electronic circuits is the generation and manipulation of electronic waveshapes. These electronic signals may represent audio information, computer data, television signals, timing signals (as used in radar), and so on. The common meters used in electronic measurement are the multimeter—analog or digital, to enable measuring dc or ac voltages, currents, or impedances. Most meters provide ac measurements that are correct for nondistorted sinusoidal signals only. The oscilloscope, on the other hand, displays the exact waveform, and the viewer can decide what to make of the various readings observed.

The cathode ray oscilloscope (CRO) provides a visual presentation of any waveform applied to the input terminals. A cathode ray tube (CRT), much like a television tube, provides the visual display showing the form of the signal applied as a waveform on the front screen. An electron beam is deflected as it sweeps across the tube face, leaving a display of the signal applied to input terminals.

While multimeters provide numeric information about an applied signal, the oscilloscope allows the actual form of the waveform to be displayed. A wide range of oscilloscopes is available, some suited to measure signals below a specified frequency, others to provide measuring signals of the shortest time span. A CRO may be built to operate from a few hertz up to hundreds of megahertz; CROs may also be used to measure time spans from fractions of a nanosecond (10^{-9}) to many seconds.

21.2 CATHODE RAY TUBE—THEORY AND CONSTRUCTION

The CRT is the "heart" of the CRO, providing visual display of an input signal's waveform. A CRT contains four basic parts:

1. An electron gun to produce a stream of electrons.
2. Focusing and accelerating elements to produce a well-defined beam of electrons.
3. Horizontal and vertical deflecting plates to control the path of the electron beam.
4. An evacuated glass envelope with a phosphorescent screen, which glows visibly when struck by the electron beam.

Figure 21.1 shows the basic construction of a CRT. We will first consider the device's basic operation. A cathode (K) containing an oxide coating is heated indirectly

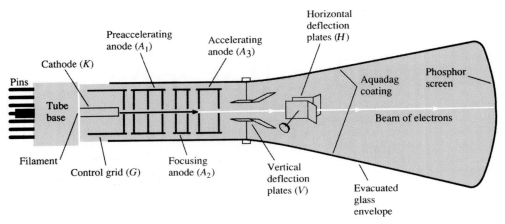

Figure 21.1 Cathode ray tube: basic construction.

by a filament, resulting in the release of electrons from the cathode surface. A control grid (G) provides for control of the number of electrons passing farther into the tube. A voltage on the control grid determines how many of the electrons freed by heating are allowed to continue moving toward the face of the tube. After the electrons pass the control grid, they are focused into a tight beam and accelerated to a higher velocity by the focusing and accelerating anodes. The parts discussed so far comprise the electron gun of the CRT.

The high-velocity, well-defined electron beam then passes through two sets of deflection plates. The first set of plates is oriented to deflect the electron beam vertically, up or down. The direction of the vertical deflection is determined by the voltage polarity applied to the deflecting plates. The amount of deflection is set by the magnitude of the applied voltage. The beam is also deflected horizontally (left or right) by a voltage applied to the horizontal deflecting plates. The deflected beam is then further accelerated by very high voltages applied to the tube, with the beam finally striking a phosphorescent material on the inside face of the tube. This phosphor glows when struck by the energetic electrons—the visible glow seen at the front of the tube by the person using the scope.

The CRT is a self-contained unit with leads brought out through a base to pins. Various types of CRTs are manufactured in a variety of sizes, with different phosphor materials and deflection electrode placement. We can now consider how the CRT is used in an oscilloscope.

21.3 CATHODE RAY OSCILLOSCOPE OPERATION

For operation as an oscilloscope, the electron beam is deflected horizontally by a sweep voltage and vertically by the voltage to be measured. While the electron beam is moved across the face of the CRT by the horizontal sweep signal, the input signal deflects the beam vertically, resulting in a display of the input signal waveform. One sweep of the beam across the face of the tube, followed by a "blank" period during which the beam is turned off while being returned to the starting point across the tube face, constitutes one sweep of the beam.

A steady display is obtained when the beam repeatedly sweeps across the tube with exactly the same image each sweep. This requires a synchronization, starting the sweep at the same point in a repetitive waveform cycle. If the signal is properly synchronized, the display will be stationary. In the absence of sync, the picture will appear to drift or move horizontally across the screen.

Basic Parts of a CRO

The basic parts of a CRO are shown in Fig. 21.2. We will first consider the CRO's operation for this simplified block diagram. To obtain a noticeable beam deflection from a centimeter to a few centimeters, the usual voltage applied to the deflection plates must be on the order of tens to hundreds of volts. Since the signals measured using a CRO are typically only a few volts, or even a few millivolts, amplifier circuits are needed to increase the input signal to the voltage levels required to operate the tube. There are amplifier sections for both the vertical and the horizontal deflection of the beam. To adjust the level of a signal, each input goes through an attenuator circuit, which can adjust the amplitude of the display.

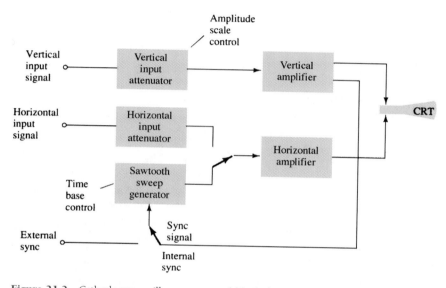

Figure 21.2 Cathode ray oscilloscope: general block diagram.

21.4   VOLTAGE SWEEP OPERATION

When the vertical input is 0 V, the electron beam may be positioned at the vertical center of the screen. If 0 V is also applied to the horizontal input, the beam is then at the center of the CRT face and remains a stationary dot. The vertical and horizontal positioning controls allow moving the dot anywhere on the tube face. Any dc voltage applied to an input will result in shifting the dot. Figure 21.3 shows a CRT face with a centered dot and with a dot moved by a positive horizontal voltage (to the right) and a negative vertical input voltage (down from center).

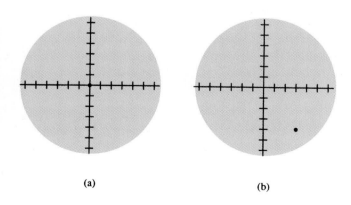

Figure 21.3 Dot on CRT screen due to stationary electron beam: (a) centered dot due to stationary electron beam; (b) off-center stationary dot.

Horizontal Sweep Signal

To view a signal on the CRT face, it is necessary to deflect the beam across the CRT with a horizontal sweep signal so that any variation of the vertical signal can be observed. Figure 21.4 shows the resulting straight-line display for a positive voltage applied to the vertical input using a linear (sawtooth) sweep signal applied to the horizontal channel. With the electron beam held at a constant vertical distance, the horizontal voltage, going from negative to zero to positive voltage, causes the beam to move from the left side of the tube, to the center, to the right side. The resulting display is a straight line above the vertical center with the dc voltage properly displayed as a straight line.

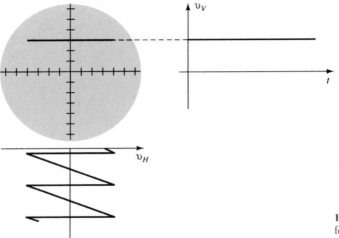

Figure 21.4 Scope display for dc vertical signal and linear horizontal sweep signal.

The sweep voltage is shown to be a continuous waveform, not just a single sweep. This is necessary if a long-term display is to be seen. A single sweep across the tube face would quickly fade out. By repeating the sweep, the display is generated over and over, and if enough sweeps are generated per second, the display appears present continuously. If the sweep rate is slowed down (as set by the time-scale controls of the scope), the actual travel of the beam across the tube face can be observed.

Applying only a sinusoidal signal to the vertical inputs (no horizontal sweep) results in a vertical straight line as shown in Fig. 21.5. If the sweep speed (frequency

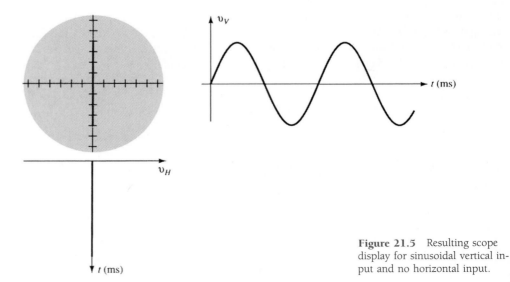

Figure 21.5 Resulting scope display for sinusoidal vertical input and no horizontal input.

Chapter 21   Oscilloscope and Other Measuring Instruments

of the sinusoidal signal) is reduced, it is possible to see the electron beam moving up and down along a straight-line path.

Use of Linear Sawtooth Sweep to Display Vertical Input

To view a sinusoidal signal, it is necessary to use a sweep signal on the horizontal channel so that the signal applied to the vertical channel can be seen on the tube face. Figure 21.6 shows the resulting CRO display from a horizontal linear sweep and a sinusoidal input to the vertical channel. For one cycle of the input signal to appear as shown in Fig. 21.6a, it is necessary that the signal and linear sweep frequencies be synchronized. If there is any difference, the display will appear to move (not be synchronized) unless the sweep frequency is some multiple of the sinusoidal frequency. Lowering the sweep frequency allows more cycles of the sinusoidal signal to be displayed, whereas increasing the sweep frequency results in less of the sinusoidal vertical input to be displayed, thereby appearing as a magnification of a part of the input signal.

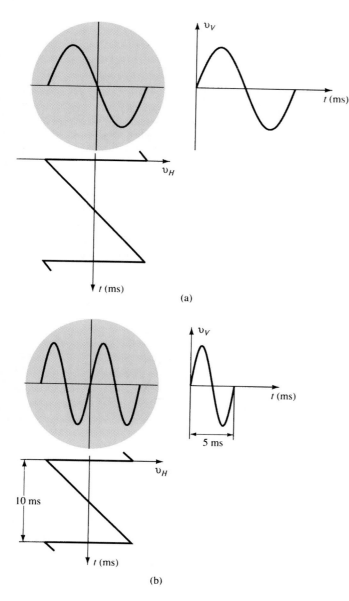

Figure 21.6 Display of sinusoidal vertical input and horizontal sweep input: (a) display of vertical input signal using linear sweep signal for horizontal deflection; (b) scope display for a sinusoidal vertical input and a horizontal sweep speed equal to one-half that of the vertical signal.

EXAMPLE 21.1

Determine how many cycles of a 2-kHz sinusoidal signal are viewed if the sweep frequency is:
(a) 2 kHz.
(b) 4 kHz.
(c) 1 kHz.

Solution

(a) When the two signals have the same frequency, a full cycle will be seen.
(b) When the sweep frequency is increased to 4 kHz, a half-cycle will be seen.
(c) When the sweep frequency is reduced to 1 kHz, two cycles will be seen.

Figure 21.7 shows a pulse-type waveform applied as vertical input with a horizontal sweep, resulting in a scope display of the pulse signal. The numbering at each waveform permits following the display for variation of input and sweep voltage during one cycle.

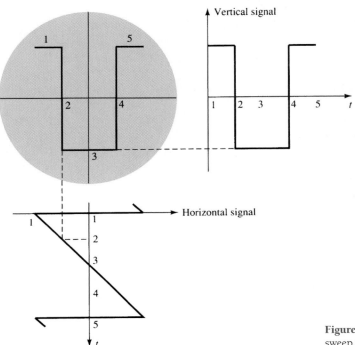

Figure 21.7 Use of the linear sweep for a pulse-type waveform.

21.5 SYNCHRONIZATION AND TRIGGERING

A CRO display can be adjusted by setting the sweep speed (frequency) to display either one cycle, a number of cycles, or part of a cycle. This is a very valuable feature of the CRO. Figure 21.8 shows the display resulting for a few cycles of the sweep signal. Each time the horizontal sawtooth sweep voltage goes through a linear sweep cycle (from maximum negative to zero to maximum positive), the electron beam is caused to move horizontally across the tube face, from left to center to right. The sawtooth voltage then drops quickly back to the negative starting voltage, with the beam back to the left side. During the time the sweep voltage goes quickly negative (re-

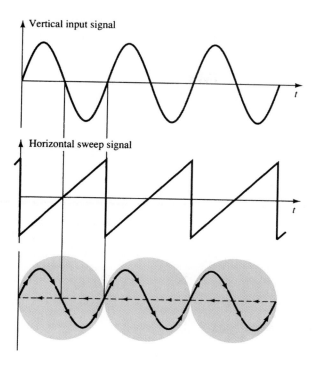

Figure 21.8   Steady scope display—input and sweep signals synchronized.

trace), the beam is blanked (the grid voltage prevents the electrons from hitting the tube face).

To see a steady display each time the beam is swept across the face of the tube, it is necessary to start the sweep at the same point in the input signal cycle. In Fig. 21.9, the sweep frequency is too low and the CRO display will have an apparent "drift" to the left. Figure 21.10 shows the result of setting the sweep frequency too high, with an apparent drift to the right.

It should be obvious that adjusting the sweep frequency to exactly the same as the signal frequency to obtain a steady sweep is impractical. A more practical proce-

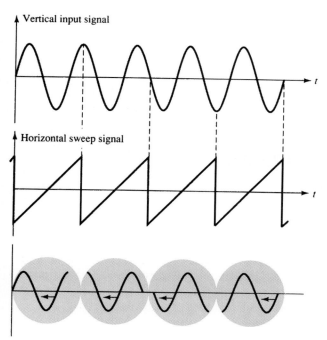

Figure 21.9   Sweep frequency too *low*—apparent drift to left.

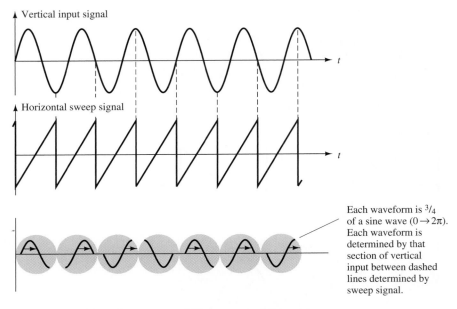

Each waveform is $^3/_4$ of a sine wave $(0 \rightarrow 2\pi)$. Each waveform is determined by that section of vertical input between dashed lines determined by sweep signal.

Figure 21.10 Sweep frequency too *high*—apparent drift to right.

dure is to wait until the signal reaches the same point in a cycle to start the trace. This triggering has a number of features, as described next.

Triggering

The usual method of synchronizing uses a portion of the input signal to trigger a sweep generator so that the sweep signal is locked or synchronized to the input signal. Using a portion of the same signal to be viewed to provide the synchronizing signal assures synchronization. Figure 21.11 shows a block diagram of how a trigger signal is derived in a single-channel display. The trigger signal source is obtained from the line frequency (60 Hz) for viewing signals related to the line voltage, from an external signal (one other than that to be viewed), or more likely, from a signal derived from that applied as vertical input. The selector switch on the scope being set

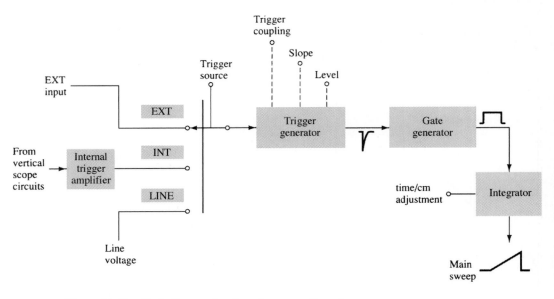

Figure 21.11 Block diagram showing trigger operation of scope.

to INTERNAL will provide a part of the input signal to the trigger generator circuit. The output of the trigger generator is a trigger signal that is used to start the main sweep of the scope, which lasts a time set by the time/cm adjustment. Figure 21.12 shows triggering being started at various points in a signal cycle.

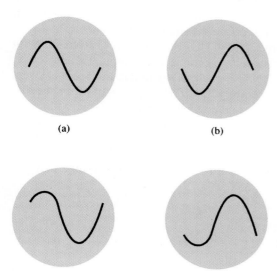

(a)　　　　(b)

(c)　　　　(d)

Figure 21.12 Triggering at various points of signal level (*Note:* sine starts at same point in cycle each sweep and is therefore synchronized): (a) positive-going zero level; (b) negative-going zero level; (c) positive-voltage trigger level; (d) negative-voltage trigger level.

The trigger sweep operation can also be seen by looking at some of the resulting waveforms. From a given input signal, a trigger waveform is obtained to provide for a sweep signal. As seen in Fig. 21.13, the sweep is started at a time in the input signal cycle and lasts a period set by the sweep length controls. Then the scope waits until the input reaches an identical point in its cycle before starting another sweep operation. The length of the sweep determines how many cycles will be viewed, while the triggering assures that synchronization takes place.

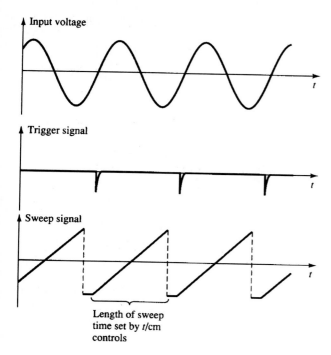

Length of sweep time set by *t*/cm controls

Figure 21.13 Triggered sweep.

21.5 Synchronization and Triggering

973

21.6 MULTITRACE OPERATION

Most modern oscilloscopes provide for viewing two or more traces on the scope face at the same time. This allows comparing amplitude, special waveform features, and other important waveform characteristics. A multiple trace can be obtained using more than one electron gun, with the separate beams creating separate displays. More often, however, a single electron beam is used to create the multiple images.

Two methods of developing two traces are CHOPPED and ALTERNATE. With two input signals applied, an electronic switch first connects one input, then the other, to the deflection circuitry. In the ALTERNATE mode of operation, the beam is swept across the tube face displaying however many cycles of one input signal are to be displayed. Then the input switches (alternates) to the second input and displays the same number of cycles of the second signal. Figure 21.14a shows the operation with alternate display. In the CHOPPED mode of operation (Fig. 21.14b), the beam repeatedly switches between the two input signals during one sweep of the beam. As long as the signal is of relatively low frequency, the action of switching is not visible and two separate displays are seen.

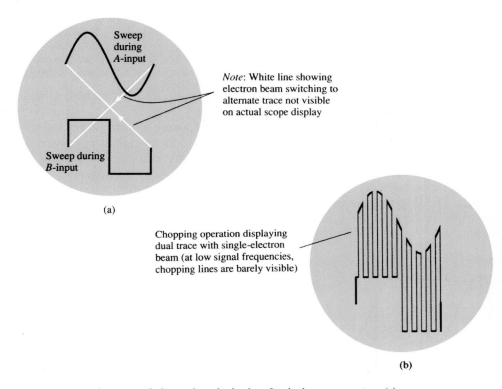

Figure 21.14 Alternate and chopped mode displays for dual-trace operation: (a) alternate mode for dual-trace using single electron beam; (b) chopped mode for dual-trace using single electron beam.

21.7 MEASUREMENT USING CALIBRATED CRO SCALES

The oscilloscope tube face has a calibrated scale to use in making amplitude or time measurements. Figure 21.15 shows a typical calibrated scale. The boxes are divided into centimeters (cm), 4 cm on each side of center. Each centimeter (box) is further divided into 0.2-cm intervals.

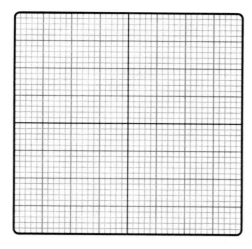

Figure 21.15 Calibrated scope face.

Amplitude Measurements

The vertical scale is calibrated in either volts per centimeter (V/cm) or millivolts per centimeter (mV/cm). Using the scale setting of the scope and the signal measured off the face of the scope, one typically can measure peak-to-peak or peak voltages for an ac signal.

Calculate the peak-to-peak amplitude of the sinusoidal signal in Fig. 21.16 if the scope scale is set to 5 mV/cm.

EXAMPLE 21.2

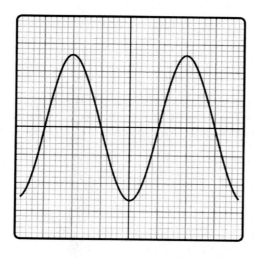

Figure 21.16 Waveform for Example 21.2.

Solution

The peak-to-peak amplitude is

$$2 \times 2.6 \text{ cm} \times 5 \text{ mV/cm} = \textbf{26 mV}$$

Note that a scope provides easy measurement of peak-to-peak values, whereas a multimeter typically provides measurement of rms (for a sinusoidal waveform).

EXAMPLE 21.3   Calculate the amplitude of the pulse signal in Fig. 21.17 (scope setting 100 mV/cm).

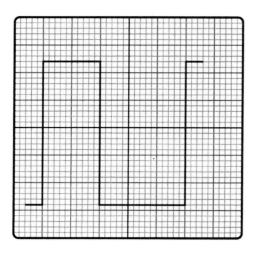

Figure 21.17 Waveform for Example 21.3.

Solution

The peak-to-peak amplitude is

$$(2.8 \text{ cm} + 2.4 \text{ cm}) \times 100 \text{ mV/cm} = \textbf{520 mV} = \textbf{0.52 V}$$

Time Measurements

PERIOD

The horizontal scale of the scope can be used to measure time, in seconds (s), milliseconds (ms), microseconds (μs), or nanoseconds (ns). The interval of a pulse from start to end is the period of the pulse. When the signal is repetitive, the period is one cycle of the waveform.

EXAMPLE 21.4   Calculate the period of the waveform shown in Fig. 21.18 (scope setting at 20 μs/cm).

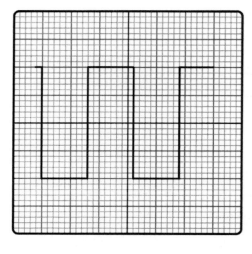

Figure 21.18 Waveform for Example 21.4.

Solution

For the waveform of Fig. 21.18,

$$\text{period} = T = 3.2 \text{ cm} \times 20 \text{ } \mu\text{s/cm} = \textbf{64 } \boldsymbol{\mu}\textbf{s}$$

FREQUENCY

The measurement of a repetitive waveform's period can be used to calculate the signal's frequency. Since frequency is the reciprocal of the period,

$$f = \frac{1}{T} \tag{21.1}$$

EXAMPLE 21.5

Determine the frequency of the waveform shown in Fig. 21.18 (scope setting at 5 μs/cm).

Solution

From the waveform

$$\text{period} = T = 3.2 \text{ cm} \times 5 \text{ } \mu\text{s/cm} = 16 \text{ } \mu\text{s}$$

$$f = \frac{1}{T} = \frac{1}{16 \text{ } \mu\text{s}}$$

$$= \textbf{62.5 kHz}$$

PULSE WIDTH

The time interval that a waveform is high (or low) is the pulse width of the signal. When the waveform edges go up and down instantly, the width is measured from start (leading edge) to end (trailing edge) (see Fig. 21.19a). For a waveform with edges that rise or fall over some time, the pulse width is measured between the 50% points as shown in Fig. 21.19b.

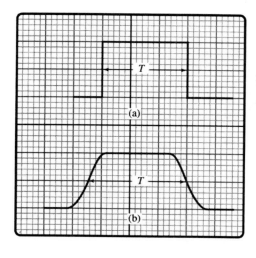

Figure 21.19 Pulse-width measurement.

EXAMPLE 21.6

Determine the pulse width of the waveform in Fig. 21.20.

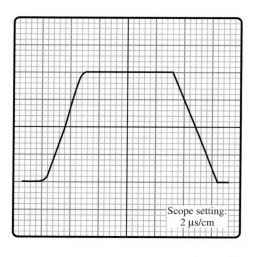

Figure 21.20 Waveform for Example 21.6.

Solution

For a reading of 4.6 cm at the midpoint of the waveform, the pulse width is

$$T_{PW} = 4.6 \text{ cm} \times 2 \text{ } \mu s/\text{cm} = \textbf{9.2 } \boldsymbol{\mu s}$$

PULSE DELAY

The time interval between pulses is called the pulse delay. For waveforms, as shown in Fig. 21.21, the pulse delay is measured between the midpoint (50% point) at the start of each pulse.

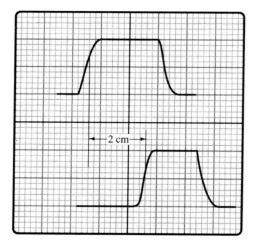

Scope setting: 50 μs/cm

Figure 21.21 Waveform for Example 21.7.

EXAMPLE 21.7

Determine the pulse delay for the waveforms of Fig. 21.21.

Solution

From the waveforms in Fig. 21.21,

$$\text{pulse delay} = T_{PD} = 2 \text{ cm} \times 50 \text{ } \mu s/\text{cm} = \textbf{100 } \boldsymbol{\mu s}$$

21.8 SPECIAL CRO FEATURES

The CRO has become more sophisticated and specialized in use. The range of amplitude measurements, the scales of time measurements, the number of traces displayed, the methods of providing sweep triggering, and the types of measurements are different depending on the area of specialized scope usage.

Delayed Sweep

A useful CRO feature uses two time bases to provide selection of a small part of the signal for viewing. One time base selects the overall signal viewed on the scope, while a second permits selecting a small part of the viewed signal to be displayed in an expanded mode. The main time base is referred to as the A time base, while the second time base, referred to as B, displays the signal after a selected delay time.

Figure 21.22 provides a block diagram showing the operation of the two time bases. With front-panel controls set to operate from the A sweep, a main sweep signal is set to view a number of cycles of the input signal. The controls then allow setting the B sweep using a variable setting dial, with the B sweep usually an intensified interval that can be moved over the face of the displayed sweep. When the desired portion of the displayed sweep is set, the controls are moved to display the delayed part of the signal, which is seen at the second time base setting as a magnified display. Figure 21.23 shows a pulse-type signal first viewed using the A sweep and then the selected portion on a magnified sweep setting.

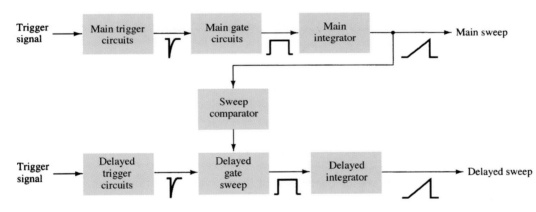

Figure 21.22 Operation of delayed sweep—block diagram.

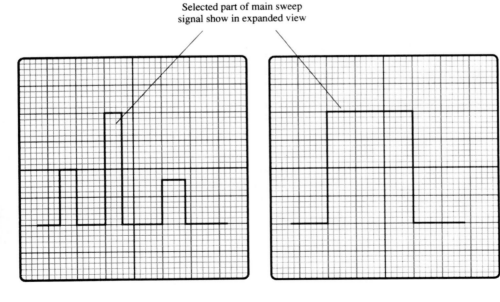

Selected part of main sweep signal show in expanded view

Figure 21.23 Main and delayed sweeps.

21.9   SIGNAL GENERATORS

A signal generator provides an ac signal of adjustable amplitude and varying frequency to use when operating an amplifier or other linear circuit. The frequency can typically be adjusted from a few hertz to a few megahertz. The signal amplitude can be adjusted from millivolts to a few volts of amplitude. While the signal is typically a sinusoidal waveform, pulse waveforms or even triangular waveforms are often available.

Waveform Generator IC (8038)

A precision waveform generator is provided by the 8038 IC unit shown in Fig. 21.24. The single 14-pin IC is capable of producing highly accurate sinusoidal, square, or triangular waveforms to use in operating or testing other equipment. Consideration of the IC's operation will help understand how any commercially available signal generator operates. This particular IC can provide output frequency that may be adjusted from less than 1 Hz up to about 300 kHz. The range of commercial units can be considerably higher. As indicated in Fig. 21.24, the IC provides three types of output waveform, and all at the same frequency, the frequency being selected by the user.

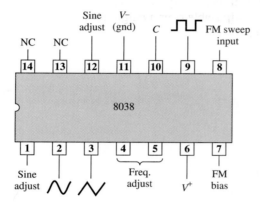

Figure 21.24   8038 waveform generator IC.

Figure 21.25 shows the connection of the IC when used to provide an adjustable frequency output. The frequency of the unit would then be

$$f = \frac{0.15}{RC}$$

(21.2)

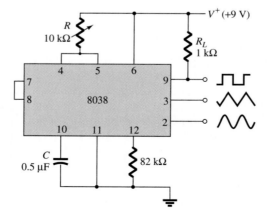

Figure 21.25   Connection of 8038 as variable frequency generator.

Referring to Fig. 21.25, determine the lowest and highest frequencies obtained when varying the 10-kΩ potentiometer from its minimum to its maximum setting.

EXAMPLE 21.8

Solution

Using Eq. (21.2), for a potentiometer set at 0, $R = 10 \ \Omega$:

$$f = \frac{0.15}{(10 \ \Omega)(0.5 \ \mu F)} = \textbf{30 kHz}$$

For a potentiometer set to its maximum,

$$f = \frac{0.15}{(10 \ k\Omega)(0.5 \ \mu F)} = \textbf{30 Hz}$$

ADJUSTABLE OUTPUT AMPLITUDE

The connection of Fig. 21.26 shows how to provide adjustment of the sinusoidal waveform amplitude with the sinusoidal output provided through a buffered driver. The 310 op-amp is a unity-gain buffer providing the sinusoidal output from a low-impedance output. [The 310 has a voltage gain near unity (1), with an output impedance of about 1 Ω.] The output frequency is adjustable over a range from about 30 Hz to 30 kHz, with an amplitude adjustable up to about 9 V peak.

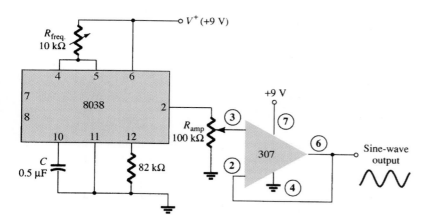

Figure 21.26 Sinusoidal waveform generator with adjustable frequency and amplitude.

5-V (TTL) PULSE GENERATOR

A circuit providing a 5-V pulse waveform for use with TTL digital circuits is shown in Fig. 21.27. The 8038 IC provides a rectangular or pulse waveform at a fixed

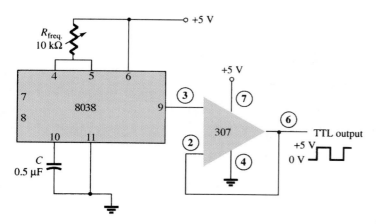

Figure 21.27 TTL signal waveform generator.

output between 0 and +5 V. The frequency of the output can be varied from about 30 Hz to 30 kHz when adjusting the value of the 10-kΩ potentiometer. A commercial signal generator would probably include switched capacitors to provide frequency over a range of values. As long as the supply uses an IC regulator to provide the +5-V supply voltage, the output will be a well-defined value, as is typically used in TTL circuits. The 310 unity follower provides the output from a low-impedance source, making it possible to connect the output to a number of loads without affecting the amplitude or frequency of the signal waveform.

21.10   COMPUTER ANALYSIS

Using the Oscilloscope in EWB

Electronics Workbench provides a number of instruments, one of which is an oscilloscope. Figure 21.28 shows the front screen of the EWB oscilloscope. The instrument is a two-channel oscilloscope with inputs for channel A, channel B, or both. There are user settings for the **Time base;** for the amplitude settings of **Channel A** or **Channel B;** and a **Trigger** section to select the source of the trigger signal, the edge to be triggered on, and even a level to trigger on. This is essentially a standard, simple, two-channel oscilloscope as found in any educational lab.

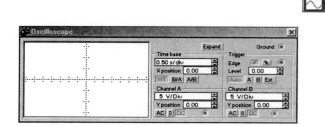

Figure 21.28   EWB two-channel oscilloscope.

The oscilloscope is used by connecting a signal—usually from a circuit—to the desired channel inputs. Figure 21.29a shows a signal generator providing a 10-kHz input which is viewed on the oscilloscope. If the **Signal source** is changed to a triangular waveform, the oscilloscope display is that of Fig. 21.29b.

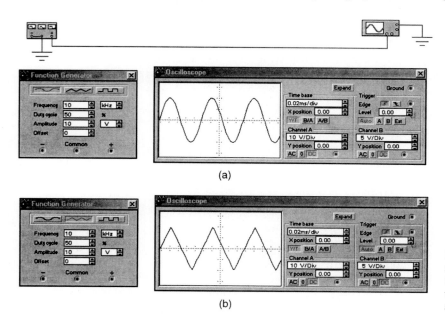

(a)

(b)

Figure 21.29   (a) EWB using signal generator with oscilloscope; (b) EWB using a signal generator with a triangular waveform and an oscilloscope.

APPENDICES

A Making the Chips that Run the World

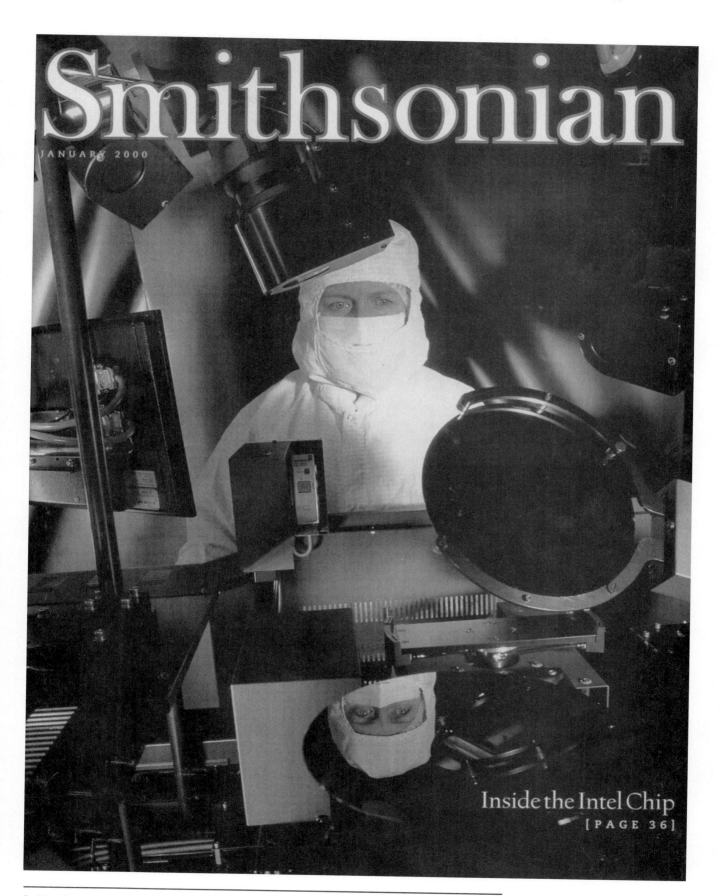

Smithsonian

JANUARY 2000

Inside the Intel Chip
[PAGE 36]

Smithsonian cover reprinted courtesy of Smithsonian Magazine. © 2000 Kay Chernush. Text courtesy
Mr. Jake Page.

These finished chips, each about a half-inch across, have not yet been cut apart. Opposite: Workers don their "bunny suits" before entering clean rooms.

Photograph © 2000 Kay Chernush

Making the Chips that Run the World

A PIECE OF CAKE: PUT 9½ MILLION TRANSISTORS IN A SPACE THE SIZE OF YOUR THUMBNAIL AND ALLOW ZERO CONTAMINATION

BY JAKE PAGE

PHOTOGRAPHS BY KAY CHERNUSH

LIKE UNCOUNTED MILLIONS OF PEOPLE in the world, I have Intel inside. It says so right in front of me on a silvery sticker glued to my laptop.

But I also have Intel *outside*.

Two miles west and a bit south of my house as the crow flies, looming on the edge of a high sandy mesa overlooking the Rio Grande in New Mexico, is a 218-acre installation that includes three Intel fabrication plants, or Fabs. Up there they make the microprocessors, the chips that run the show, that Intel has inside laptops like mine.

All I knew about any of this until recently was that microprocessors, the very core of our computers, are so complicated that they defy common understanding. I also knew from the local media that the newest Intel Fab in New Mexico was a huge room the size of about five football fields, and it cost more than $4,000 a square foot.

I got curious about such extremes of scale and, having talked my way inside, I soon had my first hands-on experience of the magic of this new engine powering the global village. I assisted in constructing a chip from variously colored bits of Play-Doh. Yes, kindergarten is in store for newbies at Intel. It is called a Functional Area Macro Overview Class or, in a place just as ebulliently awash in acronyms as the Department of Defense, a FAM.

Among my classmates was Terry McDer-

If all goes well, each silicon atom will line up in the proper position, creating a single crystal.

mott, a linebacker-size gent with twinkly eyes who was once a local TV sportscaster and, prior to that, a catcher and infielder in the Los Angeles Dodgers organization. Now one of his chores in life is to shepherd people like me through the Intel chip-making labyrinth.

From the FAM, I learned first that chips, none any larger than my thumbnail, may be made of some 20 infinitesimally thin layers, and that they are produced on wafers of silicon eight inches in diameter, with anywhere from 100 to 600 chips to a wafer. When the industry began in 1960, the wafers were about the same size as communion wafers, hence the name. Think of a wafer as a pizza, Terry said, in a more secular vein. The bigger the pizza, the more pepperoni you can get on it. And of course, the smaller the pepperoni slices, the more you can get on it, too. Smaller is better: cheaper and faster.

We were slow. It took the better part of two hours for us to make the first few layers of a chip with our Play-Doh, cutting out shapes with stencils they call masks, placing them carefully one upon the prior one, making little holes through several layers like elevators between floors so the layers could communicate with each other. During the process, we were interrupted by Intel's musical cue over a loudspeaker that meant it was time for everyone to stop whatever they were doing and begin five to ten minutes of stretching exercises. They really mean everybody, whether you're a technician in the Fab or a "carpetdweller" (office worker) in a cubicle.

Life in an Intel Fab begins and ends with these wafers, now 8 inches but before long 12 inches (more room for pepperoni). The wafers arrive at the plant in cassettes of 25 and months later go forth much altered but still in cassettes of 25. The wafers are made of almost pure silicon, the second most plentiful element in the earth's crust and what is known as a semiconductor: it can easily be persuaded to be a conductor or an insulator, which means that it can conduct electricity or

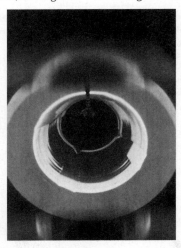

A crystal forming from molten silicon is slowly drawn from the furnace in which it was heated to 2,700 degrees F. At right, Rosemary Gerald checks the temperature of a finished crystal while Gary Burgess works the furnace controls.

not. Each transistor that will be built into the finished chip is an electrical switch that can be on or off. A positive charge fed into a transistor's gate opens it, in a sense, and a negative charge closes it. In the many layers of a Pentium III chip they can install as many as 9½ million transistors.

A bit shell-shocked by so many zeros, I naively asked where the wafers came from and soon enough found myself at the approximate beginnings of the entire chip-making flow in one of the plants that grows silicon crystals and makes wafers out of them—a place in Phoenix, Arizona, called Sumitomo Sitix.

Silicon arrives at Sitix in a form so pure that only one atom of impurity is permitted in ten billion atoms of silicon. In an atmosphere of inert argon gas, the silicon is melted at 2,700 degrees Fahrenheit in disposable $2,000 crucibles placed in a soldierly row of large blue ovens. Boron atoms are added to increase its electrical conductivity. Into this broth, a needlelike apparatus lowers a tiny silicon seed crystal, which is then pulled ever so slowly upward as it revolves. If all goes well in the next 72 hours, as the silicon freezes onto the crystal, each silicon atom will line up in the proper position, creating a huge single silicon crystal four feet long that looks for all the world like a 250-pound chromium-plated salami.

The giant cylindrical crystals are then robotically hauled off to be ground down to the proper diameter (eight inches), x-rayed, and given thermal and other stresses to simulate whatever conditions the customer may later subject them to. Then they are whisked off, again automatically, to be sawed into chunks and then wafers by what Sitix chief operating officer Robert Gill likens to a giant bread slicer. The saw consists of high-tension steel wires, each about the thickness of a human hair, wrapped around three drums, taut as the rows of wire in a piano. The "blades" move back and forth, dragging a mix of oil and Carborundum through the silicon crystal, and in seven hours the big crystal chunks have been neatly sliced into wafers, ¹⁄₃₂ of an inch thick.

A major criterion for a wafer is that it be flat, so that electrons will have uniform pathways to follow. On average, its elevation does not vary more than one micron—1/25,000 of an inch (a human hair is about 100 microns in diameter). Bob Gill explained the

Single silicon crystal

These rooms are the most fanatically clean, most thoroughly sanitized on the planet.

degree of accuracy this way: suppose instead of an eight-inch wafer you have a road grader and you are told to grade a circular area that is two miles in diameter. The terrain cannot vary in elevation from one place to another by more than two-thirds of an inch.

Back in micronland, it takes a lot of grinding, polishing, heating and cleaning to achieve this sort of tolerance. All of this is done without the touch of the human hand, the hand being far too gross an implement for this kind of work. Within glassed-in stations, little metal arms thrust forward and pull backward like the beaks of mechanical herons, feeding wafers into each successive operation while other mechanical arms seem to be saluting with Prussian precision. After each process, the wafers move along in special cassettes that are often kept submerged in water, the better to protect their surfaces from unwanted intrusions.

A big container of wafers in water is very heavy, and a robot that looks like a high-tech laundry bin bustles along a track, picking them up and depositing them at their next station. The robot (which has evidently been sexed and is referred to in the third person as "he") plays the Beatles' "Let It Be," one of 1,001 songs in his repertoire, to warn of his approach. If a visitor unwittingly stands in his track, he will stop a few inches away and politely ask the interloper to get out of the way.

Thousands of readings are taken for each wafer, to monitor thickness, bow, warp, taper and flatness. The wafers are cleaned chemically and mechanically, visually inspected under exceedingly bright lights—a particle half a micron across on the highly polished surface will light up like a beacon—and reinspected by laser for particles that are only two-tenths of a micron. The wafers are sorted into four bins: "reclean," "repolish," "scrap" or "good." Selected good ones are then rechecked by x-ray, atomic absorption spectrograph and scanning electron microscope for the likes of unwanted ions and metals. Those judged pristine go forth (in special shipping boxes, of

course) to the places where microprocessors and other chips are made.

The most crucial steps in wafer making (that is, the trickiest steps) are growing the giant crystals and polishing the wafers, but if there is a higher power in all this, cleanliness is right next to it. Much of wafer-making takes place in what are called clean rooms. Such rooms are even more important to the actual making of microprocessors on these wafers, so we will interrupt the product flow here to rehearse the precise nature of the most fanatically clean, most thoroughly sanitized places on the planet.

Technically speaking, they are called laminar flow clean rooms, and they were invented by Sandia National Laboratories in Albuquerque, New Mexico, happily at just about the time they came to be required for making microprocessors. Laminar flow means that particles of a fluid or gas all move in parallel tiers or rows, and in this case it means that the air in a clean room all moves continuously from the ceiling down to the floor, where it vanishes through grates. Then it flows through large fans back into the room via the ceiling, which is made up of a huge array of filters. In this manner, all the air in a given room can be changed at least six times a minute—even in a Fab the size of five football fields. (By comparison, the average home air-conditioning system changes the air in a house about twice a day.) In a clean room, the flowing air carries away with it all the nasty stuff that most air and most rooms are filled with, a process aided by the fact that the air pressure inside is greater than that outside, keeping most dirty air from entering in the first place.

A cubic foot of the air you are currently breathing likely contains several hundred grains of pollen and fungal spores, as well as carbon monoxide, radon gas, scent molecules, spider legs, fragments of soil, fur, a bit of carbon from a faraway fire, dust mites from your carpet, flakes of your skin, hair, lint, bacteria and viruses and 15-micron-wide droplets from when you sneezed (SMITHSONIAN, April 1995). In all, that cubic foot of air typically contains about a million specks half a micron or larger of one thing or another (ten million if there's a smoker present), any one of which could wreck a microprocessor.

Intel's Fab 11 is a Class One

Wafer slices from a cylindrical crystal wait in quartz "boats" to be examined for defects.

John Hallam places wafers in a machine that will polish them to extraordinarily close tolerances.

clean room, meaning among other things that anything more than one half-micron particle in a cubic foot of air is strictly verboten, but in practice Intel folk take beady-eyed notice if the sensors located all over the place note any particles at all.

To understand why chips require such extraordinary degrees of cleanliness, think of a chip in terms of the Manhattan metaphor. To wit: imagine a detailed map of Manhattan, with every street, every alley, showing. Then reduce that map down to a piece of paper a quarter-inch by a quarter-inch. If you stuck a pin in the map now, it would create a pothole from Times Square practically to Harlem. One obvious result is that traffic would come to an everlasting halt. Even a mere flake of Godzilla skin would short out the entire city by lying across four or five north-south avenues.

So how do clean rooms cope with us slovenly types? Clothes, they like to say, by Omar the Tentmaker.

Or in other words, the "bunny suit." Intel "employees" in multicolored bunny suits danced on the TV screen in whimsical company ads not long ago. Subsequently, many real Intel employees requested pastel bunny suits—each costs about $800—but they only have white.

It took me about 15 minutes, with a lot of instruction and patient assistance, to put on a bunny suit. (I had been told that I wasn't allowed to wear any makeup, hair spray, gel, perfume or aftershave.) The outfit includes blue booties over the shoes, like those worn by the cast of *ER*. A floppy snoodlike affair over the hair, and another one around the lower face for those with facial hair. Soft white gloves. A helmet with a hood to be rigged with two PVC-type pipes fixed inside near the opening for the face and extruding out the back of the hood. Donning of the actual suit (which opens at the top) involves inserting the feet first and then the arms. I note as I sit on a bench struggling into the suit that the regulars can do the whole procedure standing up.

Cinch a belt around the waist, and then put on the helmet. The hood needs to be stuffed under the shoulders of the suit, and the suit zipped up across the shoulders. A

Workers monitor furnaces in which a layer of silicon dioxide is added to wafers. "Guns" read bar codes on the boats.

clear plastic spit shield is affixed over the lower part of the face opening. Then the pipes from the helmet are attached to an air-filtration system in a box that hooks onto the belt, along with a battery pack. The pipes begin sucking exhalations away amid a slightly breathy hum. Then I put on more booties, which reach almost to the knees and fasten tightly around the legs with snaps. Finally, yellow latex gloves carefully tucked under the cuffs of the suit go on and a pair of clear plastic safety glasses. At this point, bundled up like a birthday present and standing in the steady room temperature of 72 degrees F, I wished I had worn something cooler, such as a pair of shorts and a T-shirt, like most clean room employees.

On first entering a Fab, a newbie will be forgiven for imagining he or she has entered a 21st-century James Bond movie. The vastly large structure is broken up into long corridors and shorter cross-corridors. It is here that newly

arrived pure silicon wafers will undergo a process not completely unlike what the FAM did with Play-Doh. I knew that much, of course, but the Fab had an eerie look. It is mostly white, though down some corridors one sees areas lit by a lemony yellow light. The floors are all made of grating for the air circulation. Here and there, rods about two feet long with little cylindrical nubbins on the end hang down from the high ceiling, looking like devices to listen in for sedition. Also overhead, plastic boxes filled with wafers zip around silently on tracks, veering off into one or another cross-corridor where people in white bunny suits, standing alone or in small groups, are desultorily watching large and mostly motionless machines behind glass. It's as if, on some command by a titanically ambitious Dr. No, the place will suddenly burst into action and the world will abruptly end. But no . . .

Ann Tiao, whose title is Inline Defect Engineer and who

acted as our guide, explained that the little bugs hanging from the ceiling are "ionizers." In the course of being worked on by all the machines, the chips-to-be (called dice, plural of "die") collect ions, and a buildup of negative or positive ions could lead to an electrostatic discharge that would in turn blow apart circuits on the dice. So the ionizers pump great quantities of both positive and negative ions into the superclean air to neutralize any ion buildup on any surface.

The bunny-suited troops, Tiao assured me, were in fact extremely busy monitoring the behavior of the "tools," the huge machines that accomplish one or another process along the flow and that, at a cost of some $5 million apiece, give new meaning to that humble term. And these multi-million-dollar tools can misbehave, it seems. A technician's job is to know the idiosyncrasies of his or her tool and make the proper adjustments when it strays. So there is craft here, too, and artisanship.

Tiao led us around to each of the tools needed to make a layer of transistors on a wafer. The entire process is, of course, based on a design drawn up long before, a complicated three-dimensional map of a microprocessor.

Part of the process is not unlike lithography, the means of printing by which this magazine is produced. In the Fab this stage is called "litho," and its numerous steps take place in those areas bathed in yellow light. On top of an insulating layer of silicon dioxide, the wafer is coated with a substance called photoresist. A stencil is placed over the photoresist layer and is exposed to ultraviolet light (UV).

The portions of photoresist that are struck by the UV become soluble and are washed away, leaving a photoresist pattern on top of the silicon dioxide. Each chip on a wafer is exposed in this manner, and from there the wafer goes to a scanning electron microscope, where a technician makes sure the patterns are the right width and that everything is properly aligned. If not, the wafer will be stripped down to the insulating layer and redone.

From here, the wafers are shuttled off to have the circuits etched so they become vertical-sided channels only a quarter-micron across. Etching can be a bit messy, however, creating particles and such, so the wafers go immediately to be cleaned in sulfuric acid, which removes the remaining photoresist and any particles. The acid is then rinsed off by water so pure it becomes an even stronger solvent than it naturally is.

After this step (and virtually all the others) the wafers are subjected to an optical inspection to detect imperfections that can be fixed—practical when a few dice are involved but not in a "white out," in which many are spoiled. It is far better to junk a spoiled wafer earlier rather than later, since each step is expensive. Even with the astounding economies of scale involved in producing millions of these little chips a week, those that are micro-

It's as if the place will suddenly burst into action and the world will abruptly end.

processors will represent more than 10 percent of the cost of a personal computer.

The litho and etch processes are repeated again and again. Thus, a chip is built up of layers of different materials, using different patterns. If a layer is too thick, or if electricity can't be sent through it, the chip is tossed. The final test is called "sort." A tool with needlelike probes taps test points on each chip's surface—10,000 taps, or tests, a second. Any chip that fails any of the tests is adorned with a drop of ink, the Black Spot, the stigma of shame, meaning, of course, doom.

The tour over, Terry McDermott and I followed Tiao down a quarter-mile-long corridor like a couple of tired ballplayers headed for the showers. We shed our bunny suits and were taken in hand by Bill Westmoreland, site environmental manager. He insisted that we go through what is called the Subfab, the place where

Individual memory chips that have been found defective are marked with a black dot and will be discarded.

Moore's Law states that the number of transistors on each chip will double every year.

80 percent of the working part of the tools reside. It is really the basement, but unlike any basement I had ever imagined.

In the Subfab of Fab 11, one could eat one's oatmeal off the floor. It is a vast, high-ceilinged place with no loose wires drooping, no dark spiderwebbed corners, no leaking pipes or dank spots where the sump pump fails to function. Here instead is the place where there are surely more miles of overhead pipes, tubes and lines than anywhere else on earth. Each of these thousands upon thousands of conduits is labeled so one can tell which chemical or gas or what voltage it is carrying to the tools upstairs or out of the Fab. Here is a place packed with giant compressors, chillers, heaters, a single pipeline that is well over six feet in diameter, and gigantic, looming scrubbers.

A plant like this has the potential for an enormous environmental impact. In an arid land, it uses four million gallons of water a day—as much as nine of Albuquerque's golf courses. But 85 percent of the water is eventually discharged to the Rio Grande. Acids and bases are neutralized. The nasty part of hydrofluoric acid, for example, is precipitated out into inert calcium fluoride "cakes."

Many people whose job it is to see that the machines do their work never see what happens next to their handiwork—the tasks of cutting the chips from each wafer and "packaging" them. In the so-called packaging process, diamond-blade saws cut out each individual chip from the wafers, and each one is glued with a thermal-cure adhesive onto a square plastic or ceramic substrate anywhere from 0.6 to 1.4 inches across. Looking at a chip now attached to a green plastic square put me in mind of looking down on a tennis stadium, with the chip perhaps as center court.

To see how it's done, I went to an Intel development lab in Chandler, Arizona, where a new chip design called for a lot of tinkering with the tools. It was a smaller, faster chip—produced, I assumed, in accordance with what the semiconductor industry calls, in tones of awe, Moore's Law. Moore's Law is named for Gordon Moore, one of the original brains behind Intel. His law, enunciated in the 1960s, states that the number of transistors on a chip will double every year.

For the new chip, some 500 gold wires—the electrical leads—had to be attached to the chip and then to the substrate in three layers, a task done by an astonishingly complex and fast-moving "sewing machine" according to a carefully worked out maplike blueprint fed into the sewing machine's brain. The process is called thermal sonic bonding. The gold thread is $1\frac{1}{5}$ mils thick (one mil is 1/1,000 inch), or one-seventh the thickness of the average human

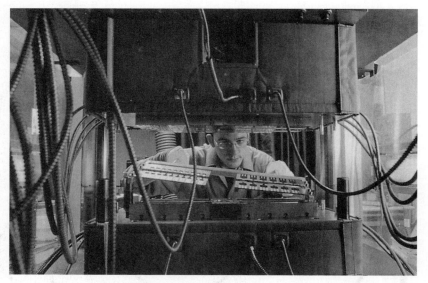

At Intel's Chandler, Arizona, facility, Jonathan McFarland operates a press that encapsulates chips with an epoxy compound to protect them from the environment.

hair. The sewing machine (technically, a wire-bonding machine) forms a tiny gold ball on the end of the thread, taps it onto the proper aluminum spot on the chip and bonds it there with a blast of ultrasound. It then draws the wire over to the plastic substrate and repeats the process. It will place from four to seven such leads per second, a frantic business indeed when peered at through a microscope.

The chip and its filigree of gold are too fragile to be left

At Intel, stretching sessions are scheduled twice during every 12-hour shift, in addition to the three breaks allowed.

unprotected. The whole thing is covered over with an epoxy seal, never to see the light again. Its destiny is to be placed on a circuit board and shipped off to somewhere in the world where it might be put inside a laptop like mine with a silvery label affixed that says "Intel Inside."

Later on, I asked Gordon Moore, Intel chairman emeritus, if there was really much need for chips with exponentially more transistors, more power, more speed. He said, "Think of a personal computer that recognized your face, your voice, and understood well enough to know if you were saying 'too,' 'to,' or 'two.' That would take a lot more power. It's maybe two generations away." In Silicon Valley-speak, two generations is about four years. There will always be a need for greater power and speed, Moore

added. The software programmers can always use it. "We have a saying: Intel giveth and Microsoft taketh away."

Asked how he had come up with Moore's Law, he said that it was very early in the semiconductor industry and many people didn't recognize the industry's possibilities. He simply wanted to startle the world into seeing its importance, so he announced the doubling law. With a beatific smile, he said modestly that he was surprised and delighted to find that it had, in fact, worked out that way. ♜

Jake Page reported on a unique ranchers' group on the Mexico-New Mexico-Arizona border in June 1997. In the past 19 years, Kay Chernush has photographed everything from gems to toxic waste.

APPENDIX B

Hybrid Parameters— Conversion Equations (Exact and Approximate)

B.1   EXACT

Common-Emitter Configuration

$$h_{ie} = \frac{h_{ib}}{(1 + h_{fb})(1 - h_{rb}) + h_{ob}h_{ib}} = h_{ic}$$

$$h_{re} = \frac{h_{ib}h_{ob} - h_{rb}(1 + h_{fb})}{(1 + h_{fb})(1 - h_{rb}) + h_{ob}h_{ib}} = 1 - h_{rc}$$

$$h_{fe} = \frac{-h_{fb}(1 - h_{rb}) - h_{ob}h_{ib}}{(1 + h_{fb})(1 - h_{rb}) + h_{ob}h_{ib}} = -(1 + h_{fc})$$

$$h_{oe} = \frac{h_{ob}}{(1 + h_{fb})(1 - h_{rb}) + h_{ob}h_{ib}} = h_{oc}$$

Common-Base Configuration

$$h_{ib} = \frac{h_{ie}}{(1 + h_{fe})(1 - h_{re}) + h_{ie}h_{oe}} = \frac{h_{ic}}{h_{ic}h_{oc} - h_{fc}h_{rc}}$$

$$h_{rb} = \frac{h_{ie}h_{oe} - h_{re}(1 + h_{fe})}{(1 + h_{fe})(1 - h_{re}) + h_{ie}h_{oe}} = \frac{h_{fc}(1 - h_{rc}) + h_{ic}h_{oc}}{h_{ic}h_{oc} - h_{fc}h_{rc}}$$

$$h_{fb} = \frac{-h_{fe}(1 - h_{re}) - h_{ie}h_{oe}}{(1 + h_{fe})(1 - h_{re}) + h_{ie}h_{oe}} = \frac{h_{rc}(1 + h_{fc}) - h_{ic}h_{oc}}{h_{ic}h_{oc} - h_{fc}h_{rc}}$$

$$h_{ob} = \frac{h_{oe}}{(1 + h_{fe})(1 - h_{re}) + h_{ie}h_{oe}} = \frac{h_{oc}}{h_{ic}h_{oc} - h_{fc}h_{rc}}$$

Common-Collector Configuration

$$h_{ic} = \frac{h_{ib}}{(1 + h_{fb})(1 - h_{rb}) + h_{ob}h_{ib}} = h_{ie}$$

$$h_{rc} = \frac{1 + h_{fb}}{(1 + h_{fb})(1 - h_{rb}) + h_{ob}h_{ib}} = 1 - h_{re}$$

$$h_{fc} = \frac{h_{rb} - 1}{(1 + h_{fb})(1 - h_{rb}) + h_{ob}h_{ib}} = -(1 + h_{fe})$$

$$h_{oc} = \frac{h_{ob}}{(1 + h_{fb})(1 - h_{rb}) + h_{ob}h_{ib}} = h_{oe}$$

B.2 APPROXIMATE

Common-Emitter Configuration

$$h_{ie} \cong \frac{h_{ib}}{1 + h_{fb}} \cong \beta r_e$$

$$h_{re} \cong \frac{h_{ib}h_{ob}}{1 + h_{fb}} - h_{rb}$$

$$h_{fe} \cong \frac{-h_{fb}}{1 + h_{fb}} \cong \beta$$

$$h_{oe} \cong \frac{h_{ob}}{1 + h_{fb}}$$

Common-Base Configuration

$$h_{ib} \cong \frac{h_{ie}}{1 + h_{fe}} \cong \frac{-h_{ic}}{h_{fc}} \cong r_e$$

$$h_{rb} \cong \frac{h_{ie}h_{oe}}{1 + h_{fe}} - h_{re} \cong h_{rc} - 1 - \frac{h_{ic}h_{oc}}{h_{fc}}$$

$$h_{fb} \cong \frac{-h_{fe}}{1 + h_{fe}} \cong -\frac{(1 + h_{fc})}{h_{fc}} \cong -\alpha$$

$$h_{ob} \cong \frac{h_{oe}}{1 + h_{fe}} \cong \frac{-h_{oc}}{h_{fc}}$$

Common-Collector Configuration

$$h_{ic} \cong \frac{h_{ib}}{1 + h_{fb}} \cong \beta r_e$$

$$h_{rc} \cong 1$$

$$h_{fc} \cong \frac{-1}{1 + h_{fb}} \cong -\beta$$

$$h_{oc} \cong \frac{h_{ob}}{1 + h_{fb}}$$

APPENDIX

C

Ripple Factor and Voltage Calculations

C.1 RIPPLE FACTOR OF RECTIFIER

The ripple factor of a voltage is defined by

$$r = \frac{\text{rms value of ac component of signal}}{\text{average value of signal}}$$

which can be expressed as

$$r = \frac{V_r(\text{rms})}{V_{dc}}$$

Since the ac voltage component of a signal containing a dc level is

$$v_{ac} = v - V_{dc}$$

the rms value of the ac component is

$$V_r(\text{rms}) = \left[\frac{1}{2\pi} \int_0^{2\pi} v_{ac}^2 \, d\theta \right]^{1/2}$$

$$= \left[\frac{1}{2\pi} \int_0^{2\pi} (v - V_{dc})^2 \, d\theta \right]^{1/2}$$

$$= \left[\frac{1}{2\pi} \int_0^{2\pi} (v^2 - 2vV_{dc} + V_{dc}^2) \, d\theta \right]^{1/2}$$

$$= [V^2(\text{rms}) - 2V_{dc}^2 + V_{dc}^2]^{1/2}$$

$$= [V^2(\text{rms}) - V_{dc}^2]^{1/2}$$

where $V(\text{rms})$ is the rms value of the total voltage. For the half-wave rectified signal,

$$V_r(\text{rms}) = [V^2(\text{rms}) - V_{dc}^2]^{1/2}$$

$$= \left[\left(\frac{V_m}{2}\right)^2 - \left(\frac{V_m}{\pi}\right)^2 \right]^{1/2}$$

$$= V_m \left[\left(\frac{1}{2}\right)^2 - \left(\frac{1}{\pi}\right)^2 \right]^{1/2}$$

$$\boxed{V_r(\text{rms}) = 0.385 V_m \qquad \text{(half-wave)}}$$

(C.1)

For the full-wave rectified signal,

$$V_r(\text{rms}) = \left[V^2(\text{rms}) - V_{\text{dc}}^2 \right]^{1/2}$$

$$= \left[\left(\frac{V_m}{\sqrt{2}} \right)^2 - \left(\frac{2V_m}{\pi} \right)^2 \right]^{1/2}$$

$$= V_m \left(\frac{1}{2} - \frac{4}{\pi^2} \right)^{1/2}$$

$$\boxed{V_r(\text{rms}) = 0.308 V_m \qquad \text{(full-wave)}} \qquad (C.2)$$

C.2 RIPPLE VOLTAGE OF CAPACITOR FILTER

Assuming a triangular ripple waveform approximation as shown in Fig. C.1, we can write (see Fig. C.2)

$$V_{\text{dc}} = V_m - \frac{V_r(\text{p-p})}{2} \qquad (C.3)$$

During capacitor-discharge, the voltage change across C is

$$V_r(\text{p-p}) = \frac{I_{\text{dc}} T_2}{C} \qquad (C.4)$$

From the triangular waveform in Fig. C.1,

$$V_r(\text{rms}) = \frac{V_r(\text{p-p})}{2\sqrt{3}} \qquad (C.5)$$

(obtained by calculations not shown).

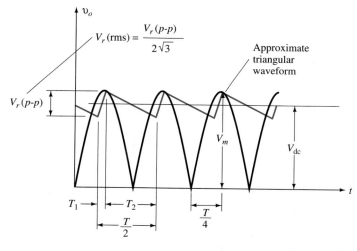

Figure C.1 Approximate triangular ripple voltage for capacitor filter.

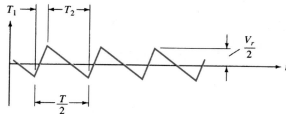

Figure C.2 Ripple voltage.

Using the waveform details of Fig. C.1 results in

$$\frac{V_r(\text{p-p})}{T_1} = \frac{V_m}{T/4}$$

$$T_1 = \frac{V_r(\text{p-p})(T/4)}{V_m}$$

Also,
$$T_2 = \frac{T}{2} - T_1 = \frac{T}{2} - \frac{V_r(\text{p-p})(T/4)}{V_m} = \frac{2TV_m - V_r(\text{p-p})T}{4V_m}$$

$$T_2 = \frac{2V_m - V_r(\text{p-p})}{V_m} \frac{T}{4} \qquad\qquad (C.6)$$

Since Eq. (C.3) can be written as

$$V_{\text{dc}} = \frac{2V_m - V_r(\text{p-p})}{2}$$

we can combine the last equation with Eq. (C.6):

$$T_2 = \frac{V_{\text{dc}}}{V_m} \frac{T}{2}$$

which, inserted into Eq. (C.4), gives

$$V_r(\text{p-p}) = \frac{I_{\text{dc}}}{C}\left(\frac{V_{\text{dc}}}{V_m}\frac{T}{2}\right)$$

$$T = \frac{1}{f}$$

$$V_r(\text{p-p}) = \frac{I_{\text{dc}}}{2fC}\frac{V_{\text{dc}}}{V_m} \qquad\qquad (C.7)$$

Combining Eqs. (C.5) and (C.7), we solve for $V_r(\text{rms})$:

$$\boxed{V_r(\text{rms}) = \frac{V_r(\text{p-p})}{2\sqrt{3}} = \frac{I_{\text{dc}}}{4\sqrt{3}fC}\frac{V_{\text{dc}}}{V_m}} \qquad\qquad (C.8)$$

C.3   RELATION OF V_{dc} AND V_m TO RIPPLE, r

The dc voltage developed across a filter capacitor from a transformer providing a peak voltage, V_m, can be related to the ripple as follows:

$$r = \frac{V_r(\text{rms})}{V_{\text{dc}}} = \frac{V_r(\text{p-p})}{2\sqrt{3}V_{\text{dc}}}$$

$$V_{\text{dc}} = \frac{V_r(\text{p-p})}{2\sqrt{3}r} = \frac{V_r(\text{p-p})/2}{\sqrt{3}r} = \frac{V_r(\text{p})}{\sqrt{3}r} = \frac{V_m - V_{\text{dc}}}{\sqrt{3}r}$$

$$V_m - V_{\text{dc}} = \sqrt{3}rV_{\text{dc}}$$

$$V_m = (1 + \sqrt{3}r)V_{\text{dc}}$$

$$\boxed{\frac{V_m}{V_{dc}} = 1 + \sqrt{3}r}$$

(C.9)

The relation of Eq. (C.9) applies to both half-wave and full-wave rectifier-capacitor filter circuits and is plotted in Fig. C.3. As an example, at a ripple of 5% the dc voltage is $V_{dc} = 0.92V_m$, or within 10% of the peak voltage, whereas at 20% ripple the dc voltage drops to only $0.74V_m$, which is more than 25% less than the peak value. Note that V_{dc} is within 10% of V_m for ripple less than 6.5%. This amount of ripple represents the borderline of the light-load condition.

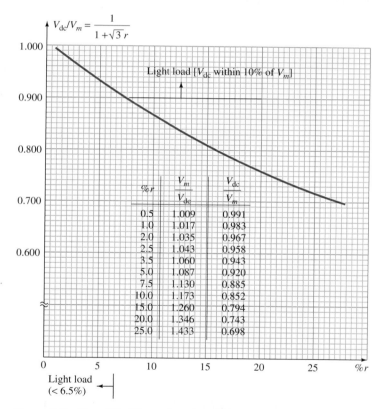

Figure C.3 Plot of V_{dc}/V_m as a function of %r.

C.4 RELATION OF V_r(RMS) AND V_m TO RIPPLE, r

We can also obtain a relation between V_r(rms), V_m, and the amount of ripple for both half-wave and full-wave rectifier-capacitor filter circuits as follows:

$$\frac{V_r(\text{p-p})}{2} = V_m - V_{dc}$$

$$\frac{V_r(\text{p-p})/2}{V_m} = \frac{V_m - V_{dc}}{V_m} = 1 - \frac{V_{dc}}{V_m}$$

$$\frac{\sqrt{3}V_r(\text{rms})}{V_m} = 1 - \frac{V_{dc}}{V_m}$$

Using Eq. (C.9), we get

$$\frac{\sqrt{3}V_r(\text{rms})}{V_m} = 1 - \frac{1}{1 + \sqrt{3}r}$$

$$\frac{V_r(\text{rms})}{V_m} = \frac{1}{\sqrt{3}}\left(1 - \frac{1}{1 + \sqrt{3}r}\right) = \frac{1}{\sqrt{3}}\left(\frac{1 + \sqrt{3}r - 1}{1 + \sqrt{3}r}\right)$$

$$\boxed{\frac{V_r(\text{rms})}{V_m} = \frac{r}{1 + \sqrt{3}r}} \qquad \text{(C.10)}$$

Equation (C.10) is plotted in Fig. C.4.

Since V_{dc} is within 10% of V_m for ripple $\leq 6.5\%$,

$$\frac{V_r(\text{rms})}{V_m} \cong \frac{V_r(\text{rms})}{V_{dc}} = r \qquad \text{(light load)}$$

and we can use $V_r(\text{rms})/V_m = r$ for ripple $\leq 6.5\%$.

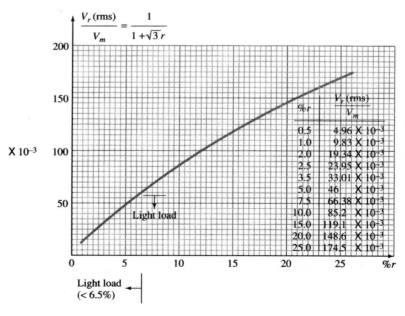

Figure C.4 Plot of $V_r(\text{rms})/V_m$ as a function of $\%r$.

Appendix C   Ripple Factor and Voltage Calculations

C.5 RELATION BETWEEN CONDUCTION ANGLE, % RIPPLE, AND I_{peak}/I_{dc} FOR RECTIFIER-CAPACITOR FILTER CIRCUITS

In Fig. C.1, we can determine the angle at which the diode starts to conduct, θ, as follows: Since

$$v = V_m \sin\theta = V_m - V_r(\text{p-p}) \qquad \text{at} \qquad \theta = \theta_1$$

$$\theta_1 = \sin^{-1}\left[1 - \frac{V_r(\text{p-p})}{V_m}\right]$$

Using Eq. (C.10) and $V_r(\text{rms}) = V_r(\text{p-p})/2\sqrt{3}$ gives

$$\frac{V_r(\text{p-p})}{V_m} = \frac{2\sqrt{3}V_r(\text{rms})}{V_m}$$

so that

$$1 - \frac{V_r(\text{p-p})}{V_m} = 1 - \frac{2\sqrt{3}V_r(\text{rms})}{V_m} = 1 - 2\sqrt{3}\left(\frac{r}{1 + \sqrt{3}r}\right)$$

$$= \frac{1 - \sqrt{3}r}{1 + \sqrt{3}r}$$

and

$$\boxed{\theta_1 = \sin^{-1}\frac{1 - \sqrt{3}r}{1 + \sqrt{3}r}} \tag{C.11}$$

where θ_1 is the angle at which conduction starts.

When the current becomes zero after charging the parallel impedances R_L and C, we can determine that

$$\theta_2 = \pi - \tan^{-1}\omega R_L C$$

An expression for $\omega R_L C$ can be obtained as follows:

$$r = \frac{V_r(\text{rms})}{V_{dc}} = \frac{(I_{dc}/4\sqrt{3}fC)(V_{dc}/V_m)}{V_{dc}} = \frac{V_{dc}/R_L}{4\sqrt{3}fC}\frac{1}{V_m}$$

$$= \frac{V_{dc}/V_m}{4\sqrt{3}fCR_L} = \frac{2\pi\left(\dfrac{1}{1 + \sqrt{3}r}\right)}{4\sqrt{3}\omega CR_L}$$

so that

$$\omega R_L C = \frac{2\pi}{4\sqrt{3}(1 + \sqrt{3}r)r} = \frac{0.907}{r(1 + \sqrt{3}r)}$$

Thus conduction stops at an angle:

$$\boxed{\theta_2 = \pi - \tan^{-1}\frac{0.907}{(1 + \sqrt{3}r)r}} \tag{C.12}$$

From Eq. (15.10b), we can write

$$\frac{I_{peak}}{I_{dc}} = \frac{I_p}{I_{dc}} = \frac{T}{T_1} = \frac{180°}{\theta} \qquad \text{(full-wave)}$$

$$= \frac{360°}{\theta} \qquad \text{(half-wave)} \tag{C.13}$$

A plot of I_p/I_{dc} as a function of ripple is provided in Fig. C.5 for both half-wave and full-wave operation.

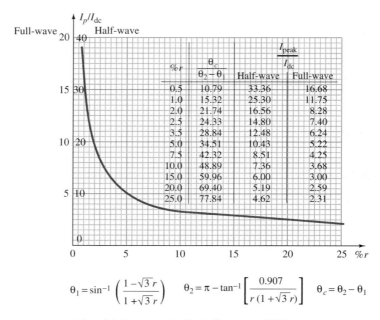

Table in figure:

%r	$\dfrac{\theta_c}{\theta_2 - \theta_1}$	$\dfrac{I_{peak}}{I_{dc}}$ Half-wave	$\dfrac{I_{peak}}{I_{dc}}$ Full-wave
0.5	10.79	33.36	16.68
1.0	15.32	25.30	11.75
2.0	21.74	16.56	8.28
2.5	24.33	14.80	7.40
3.5	28.84	12.48	6.24
5.0	34.51	10.43	5.22
7.5	42.32	8.51	4.25
10.0	48.89	7.36	3.68
15.0	59.96	6.00	3.00
20.0	69.40	5.19	2.59
25.0	77.84	4.62	2.31

$$\theta_1 = \sin^{-1}\left(\frac{1 - \sqrt{3}\,r}{1 + \sqrt{3}\,r}\right) \qquad \theta_2 = \pi - \tan^{-1}\left[\frac{0.907}{r\,(1 + \sqrt{3}\,r)}\right] \qquad \theta_c = \theta_2 - \theta_1$$

Figure C.5 Plot of I_p/I_{dc} versus %r for half-wave and full-wave operation.

Charts and Tables

APPENDIX

D

TABLE D.1 Greek Alphabet and Common Designations

Name	Capital	Lowercase	Used to Designate
alpha	A	α	Angles, area, coefficients
beta	B	β	Angles, flux density, coefficients
gamma	Γ	γ	Conductivity, specific gravity
delta	Δ	δ	Variation, density
epsilon	E	ϵ	Base of natural logarithms
zeta	Z	ζ	Impedance, coefficients, coordinates
eta	H	η	Hysteresis coefficient, efficiency
theta	Θ	θ	Temperature, phase angle
iota	I	ι	
kappa	K	κ	Dielectric constant, susceptibility
lambda	Λ	λ	Wavelength
mu	M	μ	Micro, amplification factor, permeability
nu	N	ν	Reluctivity
xi	Ξ	ξ	
omicron	O	o	
pi	Π	π	Ratio of circumference to diameter = 3.1416
rho	P	ρ	Resistivity
sigma	Σ	σ	Sign of summation
tau	T	τ	Time constant, time phase displacement
upsilon	Υ	υ	
phi	Φ	ϕ	Magnetic flux, angles
chi	X	χ	
psi	Ψ	ψ	Dielectric flux, phase difference
omega	Ω	ω	Capital: ohms; lowercase: angular velocity

TABLE D.2 Standard Values of Commercially Available Resistors

Ohms (Ω)					Kilohms (kΩ)		Megohms (MΩ)	
0.10	**1.0**	**10**	**100**	**1000**	**10**	**100**	**1.0**	**10.0**
0.11	1.1	11	110	1100	11	110	1.1	11.0
0.12	**1.2**	**12**	**120**	**1200**	**12**	**120**	**1.2**	**12.0**
0.13	1.3	13	130	1300	13	130	1.3	13.0
0.15	**1.5**	**15**	**150**	**1500**	**15**	**150**	**1.5**	**15.0**
0.16	1.6	16	160	1600	16	160	1.6	16.0
0.18	**1.8**	**18**	**180**	**1800**	**18**	**180**	**1.8**	**18.0**
0.20	2.0	20	200	2000	20	200	2.0	20.0
0.22	**2.2**	**22**	**220**	**2200**	**22**	**220**	**2.2**	**22.0**
0.24	2.4	24	240	2400	24	240	2.4	
0.27	**2.7**	**27**	**270**	**2700**	**27**	**270**	**2.7**	
0.30	3.0	30	300	3000	30	300	3.0	
0.33	**3.3**	**33**	**330**	**3300**	**33**	**330**	**3.3**	
0.36	3.6	36	360	3600	36	360	3.6	
0.39	**3.9**	**39**	**390**	**3900**	**39**	**390**	**3.9**	
0.43	4.3	43	430	4300	43	430	4.3	
0.47	**4.7**	**47**	**470**	**4700**	**47**	**470**	**4.7**	
0.51	5.1	51	510	5100	51	510	5.1	
0.56	**5.6**	**56**	**560**	**5600**	**56**	**560**	**5.6**	
0.62	6.2	62	620	6200	62	620	6.2	
0.68	**6.8**	**68**	**680**	**6800**	**68**	**680**	**6.8**	
0.75	7.5	75	750	7500	75	750	7.5	
0.82	**8.2**	**82**	**820**	**8200**	**82**	**820**	**8.2**	
0.91	9.1	91	910	9100	91	910	9.1	

TABLE D.3 Typical Capacitor Component Values

pF				μF				
10	100	1000	10000	0.10	1.0	10	100	1000
12	120	1200						
15	150	1500	15000	0.15	1.5	18	180	1800
22	220	2200	22000	0.22	2.2	22	220	2200
27	270	2700						
33	330	3300	33000	0.33	3.3	33	330	3300
39	390	3900						
47	470	4700	47000	0.47	4.7	47	470	4700
56	560	5600						
68	680	6800	68000	0.68	6.8			
82	820	8200						

CHAPTER 1

3. Conduction in only one direction

5. (a) 150 kΩ (b) 12.5 kΩ
(c) 800 kΩ (d) 3 $\mu\Omega$
$R_{Si}:R_{Cu} = 50 \times 10^9 : 1$

9. 18 J

21. 56.35 mA

23. (b) 1 (c) For $V = 0$ V, $e^0 = 1$ and $I_D = 0$ mA

27. 325 Ω

29. -10 V : 100 MΩ, -30 V : 300 MΩ

31. $R_{DC} = 76\ \Omega$
$r_d = 3\ \Omega$
$R_{DC} \gg r_d$

33. $I_D = 1$ mA, $r_d = 52\ \Omega$ vs. 55 Ω(#32)
$I_D = 15$ mA, $r_d = 1.73\ \Omega$ vs. 2 Ω(#32)

35. 22.5 Ω vs. 24.4 Ω(#34)

37. Using the best approximation to the curve beyond $V_D = 0.7$ V
$r_{av} = 4\ \Omega$

39. Decrease rapidly with increase in reverse-bias voltage

41. Log scale, $T = 25°C : I_R = 0.5$ nA
$T = 100°C : I_R = 60$ nA
Yes, at 95°C I_R increased to 64 nA.

43. $T = 25°C : P_{max} = 500$ mW, $I_{F_{max}} = 714.29$ mA
$T = 100°C : P_{max} = 260$ mW, $I_{F_{max}} = 371.43$ mA

45. (a) $V_R = -25$ V : $C_T \simeq 0.75$ pF
$V_R = -10$ V : $C_T \simeq 1.25$ pF
$\Delta C_T/\Delta V_R = 0.033$ pF/V
(b) $V_R = -10$ V : $C_T \simeq 1.25$ pF
$V_R = -1$ V : $C_T \simeq 3$ pF
$\Delta C_T/\Delta V_R = 0.194$ pF/V
(c) 0.194 pF/V : 0.033 pF/V $\simeq 5.88:1$
Increased sensitivity near $V_D = 0$ V

49. $I_F = 1$ mA, $I_R = 0.5$ mA

$t_s = 3$ ns, $t_t = 6$ ns

51. $T_1 = 129.17°$

53. 20V : $T_C \simeq 0.06\%/°C$
5V : $T_C \simeq -0.025\%/°C$

55. 0.2 mA : $\simeq 400\ \Omega$
1 mA $\simeq 95\Omega$, 10 mA $\simeq 13\Omega$
Nonlinear relationship between I_Z and dynamic impedance

57. $V_F = 2.3$ V

59. (a) $I_{peak(max)} = 37$ mA
(b) $I_{peak(max)} = 56$ mA

CHAPTER 2

1. (a) $I_{D_Q} \simeq 21.5$ mA, $V_{D_Q} \simeq 0.92$ V, $V_R = 7.08$ V
(b) $I_{D_Q} \simeq 22.2$ mA, $V_{D_Q} = 0.7$ V, $V_R = 7.3$ V
(c) $I_{D_Q} = 24.24$ mA, $V_{D_Q} = 0$ V, $V_R = 8$ V

3. $R = 0.62$ kΩ

5. (a) $I = 0$ mA (b) $I = 0.965$ A
(c) $I = 1$ A

7. (a) $V_o = 9.5$ V (b) $V_o = 7$ V

9. (a) $V_{o_1} = 11.3$ V, $V_{o_2} = 0.3$ V
(b) $V_{o_1} = -9$ V, $V_{o_2} = -6.6$ V

11. (a) $V_o = 9.7$ V, $I = 9.7$ mA
(b) $V_o = 14.6$ V, $I = 0.553$ mA

13. $V_o = 6.2$ V, $I_D = 1.55$ mA

15. $V_o = 9.3$ V

17. $V_o = 10$ V

19. $V_o = -0.7$ V

21. $V_o = 4.7$ V

23. $v_i : V_m = 6.98$ V, v_d : pos. max = 0.7 V, neg. peak = -6.98 V
i_d : pos. pulse of 2.85 mA

25. Pos. pulse, peak = 155.56 V, $V_{dc} = 49.47$ V

27. (a) $I_{D_{max}} = 20$ mA
(b) $I_{max} = 36.71$ mA

(c) $I_D = 18.36$ mA
(d) $I_D = 36.71$ mA $> I_{D_{max}} = 20$ mA

29. Full rectified waveform, peak = -100 V; PIV = 100 V

31. Full rectified waveform, peak = 56.67 V; $V_{dc} = 36.04$ V

33. (a) Pos. pulse of 3.28 V
(b) Pos. pulse of 14.3 V

35. (a) Clipped at 4.7 V
(b) Pos. clip at 0.7 V, neg. peak = -12 V

37. (a) 0 V to 40 V swing
(b) -5 V to 35 V swing

39. (a) 28 ms (b) 56 : 1
(c) -1.3 V to -21.3 V swing

41. Network of Fig. 2.179 with battery reversed

43. (a) $R_s = 20\ \Omega$, $V_Z = 12$ V
(b) $P_{Z_{max}} = 2.4$ W

45. $R_s = 0.5$ kΩ, $I_{ZM} = 40$ mA

47. $V_o = 339.36$ V

CHAPTER 3

3. Forward and reverse biased

9. $I_C = 7.921$ mA, $I_B = 79.21\ \mu A$

11. $V_{CB} = 1$ V : $V_{BE} = 800$ mV
$V_{CB} = 10$ V : $V_{BE} = 770$ mV
$V_{CB} = 20$ V : $V_{BE} = 750$ mV
Only slight

13. (a) $I_C \simeq 4.5$ mA (b) $I_C \simeq 4.5$ mA
(c) negligible (d) $I_C = I_E$

15. (a) $I_C = 3.992$ mA (b) $\alpha = 0.993$
(c) $I_E = 2$ mA

17. $A_v = 50$

21. (a) $\beta_{dc} = 117.65$ (b) $\alpha_{dc} = 0.992$
(c) $I_{CEO} = 0.3$ mA (d) $I_{CBO} = 2.4\ \mu A$

23. (a) $\beta_{dc} = 83.75$ (b) $\beta_{dc} = 170$
(c) $\beta_{dc} = 113.33$

25. $\beta_{dc} = 116$, $\alpha_{dc} = 0.991$,
$I_E = 2.93$ mA
31. $I_C = I_{C_{max}}$, $V_{CB} = 5$ V
$V_{CB} = V_{CB_{max}}$, $I_C = 2$ mA
$I_C = 4$ mA, $V_{CB} = 7.5$ V
$V_{CB} = 10$ V, $I_C = 3$ mA
33. $I_C = I_{C_{max}}$, $V_{CE} = 3.125$ V
$V_{CE} = V_{CE_{max}}$, $I_C = 20.83$ mA
$I_C = 100$ mA, $V_{CE} = 6.25$ V
$V_{CE} = 20$ V, $I_C = 31.25$ mA
35. $h_{FE}: I_C = 0.1$ mA, $h_{FE} \cong 43$
$I_C = 10$ mA, $h_{FE} \cong 98$
$h_{fe}: I_C = 0.1$ mA, $h_{fe} \cong 72$
$I_C = 10$ mA, $h_{fe} \cong 160$
37. $I_C = 1$ mA, $h_{fe} \cong 120$
$I_C = 10$ mA, $h_{fe} \cong 160$
39. $\beta_{ac} = 190$ (b) $\beta_{dc} = 201.7$
(c) $\beta_{ac} = 200$ (d) $\beta_{dc} = 230.77$ (f) Yes

CHAPTER 4

1. (a) $I_{B_Q} = 32.55$ μA
(b) $I_{C_Q} = 2.93$ mA
(c) $V_{CE_Q} = 8.09$ V
(d) $V_C = 8.09$ V
(e) $V_B = 0.7$ V (f) $V_E = 0$ V
3. (a) $I_C = 3.98$ mA (b) $V_{CC} = $
15.96 V (c) $\beta = 199$ (d) $R_B = 763$ kΩ
5. (b) $R_B = 812$ kΩ (c) $I_{C_Q} = 3.4$ mA,
$V_{CE_Q} = 10.75$ V (d) $\beta_{dc} = 136$
(e) $\alpha = 0.992$ (f) $I_{C_{sat}} = 7$ mA
(h) $P_D = 36.55$ mW (i) $P_s = 71.92$ mW
(j) $P_R = 35.37$ mW
7. (a) $R_C = 2.2$ kΩ (b) $R_E = 1.2$ kΩ
(c) $R_B = 356$ kΩ (d) $V_{CE} = 5.2$ V
(e) $V_B = 3.1$ V
9. $I_{C_{sat}} = 5.13$ mA
11. $I_C = 2.93$ mA, $V_{CE} = 8.09$ V
(b) $I_C = 4.39$ mA, $V_{CE} = 4.15$ V
(c) $\%\Delta I_C = 49.83\%$, $\%\Delta V_{CE} = 48.70\%$
(d) $I_C = 2.92$ mA, $V_{CE} = 8.61$ V
(e) $I_C = 3.93$ mA, $V_{CE} = 4.67$ V
(f) $\%\Delta I_C = 34.59\%$, $\%\Delta V_{CE} = 46.76\%$
13. (a) $I_C = 1.28$ mA (b) $V_E = 1.54$ V
(c) $V_B = 2.24$ V (d) $R_1 = 39.4$ kΩ
15. $I_{C_{sat}} = 3.49$ mA
17. (a) $I_C = 2.28$ mA (b) $V_{CE} = 8.2$ V
(c) $I_B = 19.02$ μA (d) $V_E \cong 2.28$ V
(e) $V_B = 2.98$ V Approx. approach valid
19. (a) $R_C = 2.4$ kΩ, $R_E = 0.8$ kΩ
(b) $V_E = 4$ V (c) $V_B = 4.7$ V
(d) $R_2 = 5.84$ kΩ
(e) $\beta_{dc} = 129.8$
(f) 103.84 k$\Omega = 58.4$ kΩ (checks)
21. I. (a) $I_C = 2.43$ mA, $V_{CE} = 7.55$ V
(b) $I_C = 2.33$ mA, $V_{CE} = 7.98$ V
(c) **Approx.** approach : $\%\Delta I_C = 0\%$,
$\%\Delta V_{CE} = 0\%$

Exact approach : $\%\Delta I_C = 2.19\%$,
$\%\Delta V_{CE} = 2.68\%$
(d) $\%\Delta I_C = 2.19\%$ vs. 49.83% for
Prob. 11, $\%\Delta V_{CE} = 2.68\%$ vs. 49.70%
for Prob. 11
(e) Voltage-divider configuration least
sensitive
II. $\%\Delta I_C$ and $\%\Delta V_{CE}$ are quite small
23. (a) $I_C = 2.01$ mA
(b) $V_C = 17.54$ V (c) $V_E = 3.02$ V
(d) $V_{CE} = 14.52$ V
25. V_C from 5.98 V to 8.31 V
27. (a) $I_B = 13.04$ μA
(b) $I_C = 2.56$ mA
(c) $\beta = 196.32$
(d) $V_{CE} = 8$ V
29. (a) $I_B = 13.95$ μA
(b) $I_C = 1.81$ mA
(c) $V_E = -4.42$ V
(d) $V_{CE} = 5.95$ V
31. (a) $I_E = 3.32$ mA (b) $V_C = 4.02$ V
(c) $V_{CE} = 4.72$ V
33. $R_B = 430$ kΩ, $R_C = 1.6$ kΩ,
$R_E = 390$ Ω
35. $R_E = 1.1$ kΩ, $R_C = 1.6$ kΩ,
$R_1 = 51$ kΩ, $R_2 = 15$ kΩ
37. $R_B = 43$ kΩ, $R_C = 0.62$ kΩ
39. (a) Open circuit, damaged transistor
(b) Open at collector terminal, shorted
base–emitter junction
(c) Open circuit, open transistor
41. (a) $R_B \downarrow$, $I_B \downarrow$, $I_C \downarrow$, $V_C \uparrow$
(b) $\beta \downarrow$, $I_C \downarrow$ (c) Unchanged
(d) $V_{CC} \downarrow$, $I_B \downarrow$, $I_C \downarrow$
(e) $\beta \downarrow$, $I_C \downarrow$, $V_{R_C} \downarrow$, $V_{R_E} \downarrow$, $V_{CE} \uparrow$
43. (a) R_B open, $I_B = 0$ μA,
$I_C = I_{CEO} \cong 0$ mA, $V_C \cong V_{CC} = 18$ V
(b) $\beta \uparrow$, $I_C \uparrow$, $V_{R_C} \uparrow$, $V_{R_E} \uparrow$, $V_{CE} \downarrow$
(c) $R_C \downarrow$, $I_B \uparrow$, $I_C \uparrow$, $V_E \uparrow$
(d) Drop to relatively low
voltage $\cong 0.06$ V
(e) Open base terminal
45. $V_C = -13.53$ V, $I_B = 17.5$ μA
47. (a) $S(I_{CO}) = 91$
(b) $S(V_{BE}) = -1.92 \times 10^{-4}$ S
(c) $S(\beta) = 32.56 \times 10^{-6}$ A
(d) $\Delta I_C = 1.66$ mA
49. (a) $S(I_{CO}) = 11.08$
(b) $S(V_{BE}) = -1.27 \times 10^{-3}$ S
(c) $S(\beta) = 2.41 \times 10^6$ A
(d) $\Delta I_C = 0.411$ mA
51. $S(I_{CO})$—Voltage divider less than
the other three
$S(V_{BE})$—Voltage divider more sensitive
than the other three (which have similar
levels)
$S(\beta)$—Voltage divider least sensitive
with fixed bias very sensitive
In general, Voltage divider least

sensitive and fixed bias the most
sensitive.

CHAPTER 5

3. (a) $V_{DS} \cong 1.4$ V (b) $r_d = 233.33$ Ω
(c) $V_{DS} \cong 1.6$ V (d) $r_d = 533.33$ Ω
(e) $V_{DS} \cong 1.4$ V (f) $r_d = 933.33$ Ω
(g) $r_d = 414.81$ Ω (h) $r_d = 933.2$ Ω
(i) In general, yes
11. (a) $I_D = 9$ mA (b) $I_D = 1.653$ mA
(c) $I_D = 0$ mA (d) $I_D = 0$ mA
15. $I_{DSS} = 12$ mA
17. $V_{DS} = 25$ V, $I_D = 4.8$ mA
$I_D = 10$ mA, $V_{DS} = 12$ V
$I_D = 7$ mA, $V_{DS} = 17.14$ V
19. Yes
21. $I_D = 4$ mA (exact match)
29. $I_{DSS} = 11.11$ mA
31. $V_{DS} = 25$ V
35. $V_T = 2$ V, $k = 5.31 \times 10^{-4}$
$I_D = 5.31 \times 10^{-4} (V_{GS} - 2 \text{ V})^2$
37. $V_{GS} = 27.36$ V

CHAPTER 6

1. (c) $I_{D_Q} \cong 4.7$ mA, $V_{DS_Q} = 6.36$ V
(d) $I_{D_Q} \cong 4.69$ mA, $V_{DS_Q} = 6.37$ V
3. (a) $I_D = 3.125$ mA (b) $V_{DS} = 9$ V
(c) $V_{GG} = 1.5$ V
5. $V_D = 18$ V
7. $I_{D_Q} \cong 2.6$ mA, $V_{GS} = -1.95$ V
9. (a) $I_{D_Q} = 3.33$ mA
(b) $V_{GS_Q} \cong -1.7$ V
(c) $I_{DSS} = 10.06$ mA (d) $V_D = 11.34$ V
(e) $V_{DS} = 9.64$ V
11. $V_S = 1.4$ V
13. (a) $I_{D_Q} \cong 5.8$ mA, $V_{GS_Q} \cong -0.85$ V,
$I_{D_Q} \uparrow$, $V_{GS_Q} \downarrow$ (b) 216 Ω
15. (a) $I_{D_Q} \cong 2.7$ mA, $V_{GS_Q} = -2$ V
(b) $V_{DS} = 8.12$ V, $V_S = 2$ V
17. (a) $I_{D_Q} \cong 2.9$ mA, $V_{GS_Q} = -1.2$ V
(b) $V_{DS} = 9.27$ V, $V_D = 10.52$ V
19. (a) $I_{D_Q} \cong 8.25$ mA
(b) $V_{GS_Q} = V_{DS_Q} = 7.9$ V
(c) $V_D = 12.1$ V, $V_S = 4.21$ V
(d) $V_{DS} = 7.89$ V
21. (a) $V_G \cong 3.3$ V
(b) $V_{GS_Q} = -1.25$ V, $I_{D_Q} = 3.75$ mA
(c) $I_E = 3.75$ mA (d) $I_B = 23.44$ μA
(e) $V_D = 11.56$ V (f) $V_C = 15.88$ V
23. $R_S = 0.43$ kΩ, $R_D = 1.3$ kΩ
25. $R_D = 1.42$ kΩ, $R_G = 10$ MΩ
27. D-S short-circuited; actual I_{DSS} or V_P
or combination thereof larger in
magnitude than specified.
29. (a) $I_{D_Q} = 3$ mA, $V_{GS_Q} = 1.55$ V

(b) $V_{DS} = -9.87$ V (c) $V_D = -11.4$ V
31. $I_{D_Q} = 4.68$ mA vs. 4.69 mA of #1,
$V_{DS_Q} = 6.38$ V vs. 6.37 V of #1
33. $I_{D_Q} = 3.3$ mA (same), $V_{GS_Q} = -1.47$ V vs. -1.5 V of #12

CHAPTER 7

1. (a) 0 (b) Clipping (c) 80.4%
3. 1 kHz: $X_C = 15.92$ Ω
100 kHz: $X_C = 0.1592$ Ω
Yes, better at 100 kHz
7. (a) $Z_o = 50$ kΩ
(b) $I_L = 5.747$ mA
9. (a) $I_i = 8$ μA (b) $Z_i = 500$ Ω
(c) $V_o = -720$ mV (d) $I_o = 1.41$ mA
(e) $A_i = 176.25$ (f) $A_i = 176.47$
11. (a) $r_e = 15$ Ω (b) $Z_i = 15$ Ω
(c) $I_c = 3.168$ mA (d) $V_o = 6.97$ V
(e) $A_v = 145.21$ (f) $I_b = 32$ μA
13. (a) $r_e = 8.571$ Ω (b) $I_b = 25$ μA
(c) $I_c = 3.5$ mA (d) $A_i = 132.84$
(e) $A_v = -298.89$
19. (a) $V_o = -160V_i$
(b) $I_b = 9.68 \times 10^{-4} V_i$
(c) $I_b = 1 \times 10^{-3} V_i$ (d) 3.2%
(e) Valid first approximation
21. (a) $V_o = -180V_i$
(b) $I_b = 2.32 \times 10^{-4} V_i$
(c) $I_b = 2.5 \times 10^{-4} V_i$
(d) 7.2% (e) Yes, less than 10%
23. (a) $h_{fe} = 100$ (b) $h_{ie} = 120$
25. (a) $h_{ie} = 1.5$ kΩ (b) $h_{ie} = 6.5$ kΩ
27. $h_{fe} = 100$, $h_{ie} = 2$ kΩ
29. $r_e = 15$ Ω, $\beta = 100$, $r_o = 30.3$ kΩ
31. (a) 75% (b) 70%
33. (a) $h_{oe} = 200$ μS (b) 5 kΩ, not a good approximation
35. (a) h_{fe} (b) h_{oe}
(c) Maximum : $h_{oe} \cong 30$ (normalized)
Minimum : $h_{oe} \cong 0.1$ (normalized)
At low levels of I_C
(d) Midregion

CHAPTER 8

1. (a) $Z_i = 497.47$ Ω , $Z_o = 2.2$ kΩ
(b) $A_v = -264.74$, $A_i = 60$
(c) $Z_i = 497.47$ Ω, $Z_o = 1.98$ kΩ
(d) $A_v = -238.27$, $A_i = 53.88$
3. (a) $I_B = 23.85$ μA, $I_C = 2.38$ mA,
$r_e = 10.79$ Ω
(b) $Z_i = 1.08$ kΩ, $Z_o = 4.3$ kΩ
(c) $A_v = -398.52$, $A_i = 100$
(d) $A_v = -348.47$, $A_i = 87.52$
5. 30.68 V
7. (a) 5.34 Ω (b) $Z_i = 118.37$ kΩ,

$Z_o = 2.2$ kΩ (c) $A_v = 1.81$, $A_i = 97.39$
(d) $Z_i = 105.95$ kΩ, $Z_o = 2.2$ kΩ,
$A_v = -1.81$, $A_i = 87.17$
9. (a) 5.34 Ω (b) $Z_i = 746.17$ Ω,
$Z_o = 2.2$ kΩ
(c) $A_v = -411.99$, $A_i = 139.73$
(d) $Z_i = 746.17$ Ω, $Z_o = 1.98$ kΩ,
$A_v = -370.79$, $A_i = 125.76$
11. (a) $r_e = 8.72$ Ω, $\beta r_e = 959.2$ Ω
(b) $Z_i = 142.25$ kΩ, $Z_o = 8.69$ Ω
(c) $A_v \simeq 0.997$, $A_i = 52.53$
13. (a) $I_B = 4.61$ μA, $I_C = 0.922$ mA
(b) 28.05 Ω (c) $Z_i = 7.03$ kΩ,
$Z_o = 27.66$ Ω
(d) $A_v = 0.986$, $A_i = -3.47$
15. $A_v = 163.2$, $A_i = 0.9868$
17. $R_C = 1.6$ kΩ, $R_F = 33.59$ kΩ,
$V_{CC} = 5.28$ V
19. (a) $Z_i = 0.62$ kΩ, $Z_o = 1.66$ kΩ
(b) $A_v = -209.82$, $A_i = 72.27$
21. (a) 8.31 Ω
(b) $h_{fe} = 60$, $h_{ie} = 498.6$ Ω
(c) $Z_i = 497.47$ Ω, $Z_o = 2.2$ kΩ
(d) $A_v = -264.74$, $A_i = 56.73$
(e) $Z_i = 497.47$ Ω, $Z_o = 1.98$ kΩ
(f) $A_v = -238.27$, $A_i = 53.88$
23. (a) $Z_i = 9.38$ kΩ, $Z_o \simeq 2.7$ kΩ
(b) $A_v = 283.43$, $A_i \simeq -1$
(c) $\alpha = 0.992$, $\beta = 124$, $r_e = 9.45$ Ω,
$r_o = 1$ MΩ
25. (a) $Z_i = 816.21$ Ω
(b) $A_v = -357.68$
(c) $A_i = 132.43$, (d) $Z_o = 2.14$ kΩ
27. (a) No! (b) R_2 disconnected at base

CHAPTER 9

1. 6 mS
3. 8.75 mA
5. 12.5 mA
7. 2.4 mS
9. 40 kΩ, -180
11. (a) 4 mS (b) 2.8 mS (c) 2.8 mS
(d) 2 mS (e) 2 mS
13. (a) 0.75 mS (b) 100 kΩ
15. $g_m = 5.6$ mS, $r_d = 66.7$ kΩ
17. $Z_i = 1$ MΩ, $Z_o = 1.72$ kΩ,
$A_v = -5.375$
19. $Z_i = 10$ MΩ, $Z_o = 2.83$ kΩ,
$A_v = -8.49$
21. $Z_i = 1$ MΩ, $Z_o = 730$ Ω,
$A_v = -2.19$
23. $Z_i = 9.7$ MΩ, $Z_o = 1.96$ kΩ,
$V_o = -214.4$ mV
25. $Z_i = 9.7$ MΩ, $Z_o = 1.82$ kΩ,
$V_o = 198.8$ mV
27. $Z_i = 10$ MΩ, $Z_o = 512.9$ Ω,
$A_v = 0.754$

29. $Z_i = 10$ MΩ, $Z_o \cong 1$ kΩ,
$A_v = 0.66$
31. $Z_i = 386.1$ Ω, $Z_o = 2.92$ kΩ,
$V_o = 0.636$ mV
33. 11.73 mV
35. $Z_i = 10$ MΩ, $Z_o = 1.68$ kΩ,
$A_v = -9.07$
37. $Z_i = 9$ MΩ, $Z_o = 242.1$ Ω,
$A_v = 0.816$
39. $Z_i = 1.73$ MΩ, $Z_o = 2.15$ kΩ,
$A_v = -4.77$
41. 203 mV
43. -3.51 mV
45. $R_S = 180$ Ω, $R_D = 2$ kΩ

CHAPTER 10

1. (a) $A_{v_{NL}} = -326.22$, $Z_i = 1.01$ kΩ,
$Z_o = 3.3$ kΩ
(c) $A_v = -191.65$
(d) $A_i = 41.18$
(e) The same
3. $R_L = 4.7$ kΩ: $A_v = -191.65$
$R_L = 2.2$ kΩ: $A_v = -130.49$
$R_L = 0.5$ kΩ: $A_v = -42.42$
As $R_L \downarrow$, $A_v \downarrow$
5. (a) $A_{v_{NL}} = -557.36$, $Z_i = 616.52$ Ω,
$Z_o = 4.3$ kΩ
(c) $A_v = -214.98$, $A_{v_s} = -81.91$
(d) $A_i = 49.04$
(e) $A_{v_s} = -120.12$, As $R_L \uparrow$, $A_{v_s} \uparrow$
(f) $A_{v_s} = -118.67$, As $R_s \downarrow$, $A_{v_s} \uparrow$
(b) Unaffected
7. (b) $A_v = -160$ vs. -162.4(#6)
9. (a) $A_{v_{NL}} = -3.61$, $Z_i = 81.17$ kΩ,
$Z_o = 3$ kΩ
(c) $A_v = -2.2$, $A_{v_s} = -2.18$
(d) None
(e) A_v—none, $A_{v_s} = -2.17$, As
$R_s \uparrow$, $A_{v_s} \downarrow$ (only slightly for moderate
changes in R_s since Z_i typically so large)
11. (a) $Z_i = 10.74$ Ω, $Z_o = 4.7$ kΩ,
$A_{v_{NL}} = 435.59$
(c) $A_v = 236.83$, $A_{v_s} = 22.97$
(d) The same
(e) $A_v = 138.88$, $A_{v_s} = 2.92$,
A_{v_s} very sensitive to increase in R_s due
to small Z_i, $R_L \downarrow$, $A_v \downarrow$, $A_{v_s} \downarrow$
13. (a) $A_{v_{NL}} = 0.737$, $Z_i = 2$ MΩ,
$Z_o = 0.867$ kΩ
(c) $A_{v_s} \simeq A_v = 0.529$
(d) $A_{v_s} \simeq A_v = 0.622$,
$R_L \uparrow$, $A_{v_s} \simeq A_v \uparrow$
(e) Little effect since $R_i \gg R_{sig}$
(f) No effect on Z_i or Z_o
15. (a) $A_{v_1} = -97.67$, $A_{v_2} = -189$
(b) $A_{v_T} = 18.46 \times 10^3$,

$A_{v_{s(T)}} = 11.54 \times 10^3$
(c) $A_{i_1} = 97.67, A_{i_2} = 70$
(d) $A_{i_T} = 6.84 \times 10^3$
(e) No effect
(f) No effect
(g) In phase

CHAPTER 11

1. (a) 3, 1.699, −0.151
(b) 6.908, 3.912, −0.347
(c) Results differ by magnitude of 2.3
3. (a) Same : 13.98
(b) Same : −13.01
(c) Same : 0.699
5. $G_{dBm} = 43.98$ dBm
7. $G_{dB} = 67.96$ dB
9. (a) $G_{dB} = 69.83$ dB
(b) $G_v = 82.83$ dB
(c) $R_i = 2$ kΩ
(d) $V_o = 1385.64$ V
11. (a) $|A_v| = 1/\sqrt{1 + (1950.43 \text{ Hz}/f)^2}$
(b) 100 Hz : $|A_v| = 0.051$
1 kHz : $|A_v| = 0.456$
2 kHz : $|A_v| = 0.716$
5 kHz : $|A_v| = 0.932$
10 kHz : $|A_v| = 0.982$
(c) $f_1 = 1950.43$ Hz
13. (a) 10 kHz (b) 1 kHz
(c) 5 kHz (d) 100 kHz
15. (a) $r_e = 28.48$ Ω
(b) $A_{v_{mid}} = -72.91$
(c) $Z_i = 2.455$ kΩ
(d) $A_{v_s} = -54.68$
(e) $f_{L_S} = 103.4$ Hz, $f_{L_C} = 38.05$ Hz,
$f_{L_E} = 235.79$ Hz
(f) $f_1 \simeq f_{L_E}$
17. (a) $r_e = 30.23$ Ω
(b) $A_{v_{mid}} \simeq 0.983$
(c) $Z_i = 21.13$ kΩ
(d) $A_{v_{s mid}} \simeq 0.955$
(e) $f_{L_s} = 71.92$ Hz, $f_{L_C} = 193.16$ Hz
(f) $f_1 \simeq f_{L_C}: f_1 \simeq 210$ Hz (PSpice)
19. (a) $V_{GS_Q} = -2.45$ V, $I_{D_Q} = 2.1$ mA
(b) $g_{mo} = 2$ mS, $g_m = 1.18$ mS
(c) $A_{v_{mid}} = -2$
(d) $Z_i = 1$ MΩ
(e) $A_{v_s} \simeq A_v = -2$
(f) $f_{L_G} = 1.59$ Hz, $f_{L_C} = 4.91$ Hz,
$f_{L_S} = 32.04$ Hz
(g) $f_1 \simeq 32$ Hz
21. (a) $V_{GS_Q} = -2.55$ V, $I_{D_Q} = 3.3$ mA
(b) $g_{mo} = 3.33$ mS, $g_m = 1.91$ mS
(c) $A_{v_{mid}} = -4.39$
(d) $Z_i = 51.94$ kΩ
(e) $A_{v_{s mid}} = -4.27$
(f) $f_{L_G} = 2.98$ Hz, $f_{L_C} = 2.46$ Hz,
$f_{L_S} = 41$ Hz

(g) $f_1 \simeq f_{L_S} = 41$ Hz
Z_i considerably less but still sufficiently greater than R_{sig} to result in minimum effect on A_{v_s}; reduced Z_i, however, can raise level of f_{L_G}
23. (a) $f_{H_i} \simeq 293$ kHz, $f_{H_o} = 3.22$ MHz
(b) $f_\beta = 8.03$ MHz, $f_T = 883.3$ MHz
25. (a) $f_{H_i} \simeq 584$ MHz, $f_{H_o} = 2.93$ MHz
(b) $f_\beta = 5.01$ MHz, $f_T = 400.8$ MHz
27. (a) $g_{mo} = 3.33$ mS, $g_m = 1.91$ mS
(b) $A_{v_{mid}} = -4.39, A_{v_{s mid}} = -4.27$
(c) $f_{H_i} = 1.84$ MHz, $f_{H_o} = 3.68$ MHz
29. $f_2' = 1.09$ MHz
31. (a) $v = 12.73 \times 10^{-3}$ [sin $2\pi(100 \times 10^3)t + \frac{1}{3}\sin 2\pi(300 \times 10^3)t + \frac{1}{5}\sin 2\pi(500 \times 10^3)t + \frac{1}{7}\sin 2\pi(700 \times 10^3)t + \frac{1}{9}\sin 2\pi(900 \times 10^3)t$] (b) BW = 500 kHz
(c) $f_{L_o} \simeq 3.53$ kHz

CHAPTER 12

1. $V_G = 0$ V, $V_S = 1.4$ V, $V_D = 9.86$ V
3. $V_G = 0$ V, $V_S = 1.4$ V, $V_D = 10.3$ V
5. $Z_i = 10$ MΩ, $Z_o = 2.1$ kΩ
7. $A_{V1} = -75.8, A_{V2} = -311.9,$
$A_V = 23,642$
9. $V_B = 2.55$ V, $V_E = 1.85$ V,
$V_C = 2.7$ V, $I_C = 0.84$ mA
11. $Z_i = 10$ MΩ, $Z_o = 2.7$ kΩ
13. $A_V = -214, V_o = -2.14$ V
15. $V_{E2} = 8.06$ V, $I_{E2} = 15.8$ mA
17. $V_{B1} = 4.88$ V, $V_{C2} = 5.58$ V,
$I_C = 104.2$ mA
19. (a) Q_1 on, Q_2 on, Q_3 off, Q_4 off
(b) Q_1 off, Q_2 off, Q_3 on, Q_4 on
(c) Q_1 on, Q_2 off, Q_3 on, Q_4 off
21. $I_D = 6$ mA
23. $I = 3.67$ mA
25. $I = 2$ mA
27. $I_C = 1$ mA, $V_C = 0$ V
29. $V_o = 1.89$ V

CHAPTER 13

1. CMRR = 75.56 dB
3. $V_o = -18.75$ V
5. $V_i = -40$ mV
7. $V_o = -9.3$ V
9. V_o ranges from 5.5 to 10.5 V
11. $V_o = -3.39$ V
13. $V_o = 0.5$ V
15. $V_2 = -2$ V, $V_3 = 4.2$ V
17. $V_o = 6.4$ V
19. $I_{IB}^+ = 22$ nA, $I_{IB}^- = 18$ nA
21. $A_{CL} = 80$
23. V_o(offset) = 105 mV

CHAPTER 14

1. $V_o = -175$ mV, rms
3. $V_o = 412$ mV
7. $V_o = -2.5$ V
11. $I_L = 6$ mA
13. $I_o = 0.5$ mA
15. $f_{OH} = 1.45$ kHz
17. $f_{OL} = 318.3$ Hz, $f_{OH} = 397.9$ Hz

CHAPTER 15

1. $P_i = 10.4$ W, $P_o = 640$ mW
3. $P_o = 2.1$ W
5. R(eff) = 2.5 kΩ
7. $a = 44.7$
9. $\%\eta = 37\%$
13. (a) Maximum $P_i = 49.7$ W
(b) Maximum $P_o = 39.06$ W
(c) Maximum $\%\eta = 78.5\%$
17. (a) $P_i = 27$ W (b) $P_o = 8$ W
(c) $\%\eta = 29.6\%$ (d) $P_{2Q} = 19$ W
19. $\%D_2 = 14.3\%, \%D_3 = 4.8\%,$
$\%D_4 = 2.4\%$
21. $\%D_2 = 6.8\%$
23. $P_D = 25$ W
25. $P_D = 3$ W

CHAPTER 16

9. $V_o = 13$ V
13. Period = 204.8 μs
17. $f_o = 60$ kHz
19. $C = 133$ pF
21. $C_1 = 300$ pF

CHAPTER 17

1. $A_f = -9.95$
3. $A_f = -14.3, R_{if} = 31.5$ kΩ,
$R_{of} = 2.4$ kΩ
5. Without feedback : $A_v = -303.2$,
$Z_i = 1.18$ kΩ, $Z_o = 4.7$ kΩ
With feedback : $A_{vf} = -3.82, Z_{if} = 45.8$ kΩ
7. $f_o = 4.2$ kHz
9. $f_o = 1.05$ MHz
11. $f_o = 159.2$ kHz

CHAPTER 18

1. Ripple factor = 0.028
3. Ripple voltage = 24.2 V
5. $V_r = 1.2$ V

7. $V_r = 0.6$ V rms, $V_{dc} = 17$ V

9. $V_r = 0.12$ V rms

11. $V_m = 13.7$ V

13. $\%r = 7.2\%$

15. $\%r = 8.3\%$, $\%r' = 3.1\%$

17. $V_r = 0.325$ V rms

19. $V_o = 7.6$ V, $I_Z = 3.66$ mA

21. $V_o = 24.6$ V

25. $I_{dc} = 225$ mA

27. $V_o = 9.9$ V

CHAPTER 19

3. 30 : 1 or better is typical, short period of time, casing design

5. 124% increase, $V_R \simeq 25$ V

7. (a) $C_T = 41.85$ pF
(b) $k \simeq 71 \times 10^{-12}$

9. (a) $C = 5.17$ pF
(b) Graph, $C \simeq 5$ pF

11. $T_1 = 50°C$

13. $Q = 26.93$, Q drops significantly with increase in frequency

19. $I_T = 5$ mA, $V_T = 60$ mV
$I_T = 2.8$ mA, $V_T = 900$ mV

21. $f_p \simeq 2228$ Hz

23. (a) 3750 Å → 7500 Å
(b) $\simeq 8400$ Å (c) BW = 4200 Å

25. (a) Silicon (b) Orange

27. (a) $\simeq 0.9$ Ω/fc (b) $\simeq 380$ Ω/fc
(c) $\simeq 78$ kΩ/fc Low-illumination region

29. $V_i = 21$ V

31. As fc increases, t_r and t_d decrease exponentially

33. (a) $\phi \simeq 5$ mW (b) 2.27 lm

35. $\phi = 3.44$ mW

41. Lower levels

45. $R \simeq 20$ kΩ

47. R(thermistor) = 90 Ω

CHAPTER 20

5. (a) Yes (b) No (c) No (d) Yes, no

11. (a) $\simeq 0.7$ mW/cm^2 (b) 82.35%

17. (a) $R_{B_2} = 1.08$ kΩ
(b) $R_{BB} = 3.08$ kΩ
(c) $V_{R_{B_1}} = 13$ V (d) $V_P = 13.7$ V

19. $I_B = 25$ μA, $I_C = 1$ mA

21. (a) For decreasing temperatures, 0.53%/°C (b) Yes

23. $I_C/I_F = 0.44$ Relatively efficient

25. (a) $I_C \geq 3$ mA (b) $\Delta R : \Delta t \simeq 2.3 : 1$

27. $Z_p = 87$ kΩ, $Z_V = 181.8$Ω, To a degree

29. Yes, 8.18 V (b) $R < 2$ kΩ
(c) $R = 1.82$ kΩ

Index

CRO (*cont.*)
  operation, 966–67
  pulse delay measurement, 978
  pulse width measurement, 977–78
  synchronization, 970–72
  time measurement, 976–77
  triggering, 972–73
  voltage sweep, 967–70
Crossover distortion, 767–68
Crystal, 4
Current-controlled devices, 245
Current gain:
  ac operation, 641
  Darlington circuit, 636–37
Current mirror circuits, 226–27, 646–49
Current-series feedback amplifier, 831–33
Current source circuit, 644–46
Curve tracer, 35, 36–37, 153–54, 260–62
Cutoff, 164–66, 225, 578, 586–87

D

Dacey, G.C., 246
Damping, 902
Dark current, 904, 906
Darlington connection, 633–42, 663–65
Darlington transistor, 633, 634, 769
DC bias:
  Darlington circuit, 634–35
  differential amplifier circuit, 650–51
  feedback pair, 639–40
DC load line, 756
DC power supply, 880–83
DC resistance, 20–21, 26
Decade, 582
Decibels, 573–76
De Forest, Lee, 131
Depletion region, 10–13, 266
Depletion-type MOSFET, 263–68, 487–89
Design, 195–201, 322–24
Diac, 938–40
Dielectric insulator, 263–64
Differential amplifier, 649–58
Differential inputs, 678
Differential-mode operation, 677–81
Differentiator, 690
Diffusion:
  capacitance, 33, 605–6
Digital-analog converters, 798–801
Diodes:
  ac resistance, 21–26
  AND/OR gates, 72–73
  applications, 55–130
  approximate equivalent model, 59–60, 62–63
  array, 45
  average resistance, 25–26
  center-tapped transformer, 79–80
  clampers, 88–92
  clippers, 81–87
  computer analysis, 47–51, 113–20
  conduction period, 864–65
  construction, 10–17
  covalent bonding, 4
  dc resistance, 20–21, 26
  depletion region, 10–13

  diffusion capacitance, 33
  equivalent circuits, 26–29, 891
  extrinsic materials, 7–10
  forward bias, 12–15
  full-wave rectification, 77–80
  gallium arsenide, 898
  germanium, 15–17, 62, 63, 898
  half-wave rectification, 74–77
  heat sinks, 897–98
  hot-carrier, 889–92
  ideal, 1–3, 60–61, 62–63
  intrinsic carriers, 5
  light-emitting, 40–44, 912
  load-line analysis, 56–61
  majority carriers, 10
  minority carriers, 10
  multipliers, 98–100
  n- and p-type materials, 10–13
  no-bias, 10–12
  n-type materials, 7–8
  parallel (and series-parallel) configurations, 69–72
  peak current, 864–65
  photodiodes, 902–6
  PIV, 77, 80
  p-n junction, 898–902
  power, 897–98
  power dissipation, 30
  protective configuration, 103
  PRV, 77
  PSpice, 48–49, 113–18
  p-type materials, 9–13
  resistance, 20–26
  reverse bias, 10, 12
  reverse recovery time, 34
  reverse saturation current, 12, 13, 16
  Schottky barrier, 889–92
  semiconductor, 54
  series configuration, 64–69
  Shockley, 937–38
  silicon, 4–10, 15–17
  sinusoidal, 902
  specification sheet, 29–33, 892, 893, 895, 899
  surface-barrier, 889–92
  symbols, 34–35
  temperature effects, 16–17
  testing, 35–37
  threshold voltage, 16
  transition capacitance, 33
  tunnel, 898–902
  varactor, 892–97
  varicap, 892–97
  Zener, 14–15, 37–40
Direct-coupled amplifier, 577–78
Distortion, 772–76
DMM, 35–36, 154
Donor atoms, 8
Doping, 4–9
Double-ended output, 676–77
Double-ended input, 676
Double-ended voltage gain, 649, 650
Doubler, 98–100
Drain current, 271
Drain-feedback configuration, 490–93, 496
Dual-slope conversion, 799–800
Dynamic resistance, 21–25
Dynamic scattering, 910